2nd International Conference on Sustainable Development in Civil, Urban and Transportation Engineering (CUTE 2018)

IOP Conference Series: Earth and Environmental Science
Volume 143

Ho Chi Minh City, Vietnam
17 - 19 April 2018

Editors:

Quoc-Bao Bui
Radim Cajka
Minh-Tung Tran
Tu-Anh Trinh
Ansar-Ul-Haque Yasar
Geert Wets
Monika Woloszyn

ISBN: 978-1-5108-6155-8
ISSN: 1755-1307

Printed from e-media with permission by:

Curran Associates, Inc.
57 Morehouse Lane
Red Hook, NY 12571

Some format issues inherent in the e-media version may also appear in this print version.

Printed by Curran Associates, Inc. (2018)

For permission requests, please contact the Institute of Physics
at the address below.

Institute of Physics
Dirac House, Temple Back
Bristol BS1 6BE UK

Phone: 44 1 17 929 7481
Fax: 44 1 17 920 0979

techtracking@iop.org

Additional copies of this publication are available from:

Curran Associates, Inc.
57 Morehouse Lane
Red Hook, NY 12571 USA
Phone: 845-758-0400
Fax: 845-758-2633
Email: curran@proceedings.com
Web: www.proceedings.com

Table of contents

Volume 143

2nd International Conference on Sustainable Development in Civil, Urban and Transportation Engineering (CUTE 2018)

17–19 April 2018, Ho Chi Minh City, Vietnam

Accepted papers received: 18 March 2018
Published online: 12 April 2018

Preface

Papers

Preface

Sustainable development is actually an important topic which needs the attention of the society. The civil engineering is figured as one of the most impacting domains on the environment because of numerous problems such as the energy consumption, the CO2 emission and the natural resource depletion. Alternative solutions have been searched during the last decades to reduce environmental impacts of the civil engineering sector. Different approaches have been proposed: reducing the energy consumption by using low-embodied energy materials ("eco-materials"); developing the energy-efficient buildings with high thermal performance, with renewable energy integrations (solar, wind, geothermal, etc.), or by using the eco-architecture principles to optimize the energy consumption of the buildings. Today, sustainable development in civil engineering is not still restrained at a building scale but the city scale should be considered: the transport of the occupants between their residence and their work, the energy loss in the energy distribution network, etc.

In this context, the 2nd International Conference on Sustainable Development in Civil, Urban and Transportation Engineering (CUTE 2018) is organized at Ton Duc Thang University, Ho Chi Minh City, from 17 to 19 April 2018. This conference is organized by Ton Duc Thang University (Vietnam) in cooperation with Université Savoie Mont Blanc (France), Hasselt University (Belgium), Technical University of Ostrava (Czech), Wrocław University of Technology (Poland) and University of Trieste (Italy). This is the second edition after the first conference which was organized in 2016. The main scope of CUTE 2018 conference is to propose an exchange platform for different approaches which can contribute to the sustainability of the civil engineering sector: investigations on sustainable materials and structures; intelligent principles in architecture design; sustainable solutions in transportation and urban planning.

The host organization Ton Duc Thang University is gratefully acknowledged for the administrative and logistical support. The co-organizer universities are thanked for their cooperation and support.

The sponsors of the conference are gratefully acknowledged for their financial supports.

The organizers would like also to thank Honorary Chairs, Conference Chairs, International Scientific Committee, and Local Organizing Committee.

Editors

Quoc-Bao BUI, Radim CAJKA, Minh-Tung TRAN, Tu-Anh TRINH
Ansar-Ul-Haque YASAR, Geert WETS, Monika WOLOSZYN

IOP Conf. Series: Earth and Environmental Science **143** (2018) 011002 doi:10.1088/1755-1315/143/1/011002

Peer review statement

All papers published in this volume of *IOP Conference Series: Earth and Environmental Science* have been peer reviewed through processes administered by the proceedings Editors. Reviews were conducted by expert referees to the professional and scientific standards expected of a proceedings journal published by IOP Publishing.

A new concept of precast concrete retaining wall: from laboratory model to the in-situ tests

T T Bui[1*], H V Tran[1], A Limam[2], M Bost[3], Q B Bui[4], P Robit[5]

[1]University of Lyon, INSA Lyon, SMS-ID, France
[2]University of Lyon, France
[3]GERS-RRO, IFSTTAR, Bron, France
[4]Sustainable Developments in Civil Engineering Research Group, Faculty of Civil Engineering, Ton Duc Thang University, Ho Chi Minh City, Vietnam
[5]GTS Group NGE Lyon, France

[*]Corresponding author, Email: tan-trung.bui@insa-lyon.fr

Abstract. A new concept for the soil nail walls is here proposed and validated through experimental and numerical approaches. This process, based on the use of precast elements that are easier to install, is cheaper and more aesthetic than the classical methods, but the main advantage is reducing the cement consumption which conducts to divided carbon footprint by three. In order to characterize the structural capacity of this new process, this article present an investigation on two *in-situ* representative walls, one in shotcrete which is the old way of construction, and the other, consisting the precast reinforced concrete slabs, which is the new process. We thus have a demonstrator on a real scale, and perfectly representative, since the constructive modes, as well as the mechanical, thermal, and hydric loadings are the real ones associated with the environment in situ. Substantial instrumentation has been realized over a long period (nearly 2 years), enabling to follow the evolution of the displacements of each wall and the efforts in the anchor nails. To determine the bearing capacity of the constituent element of the precast nail wall, an experimental study coupled with a numerical simulation has been conducted in the laboratory on a single precast slab. This study allows the evaluation of the load associated to crack initiation and the bearing capacity associated to the ultimate state, at the scale of the constituent elements. Finally, in order to evaluate the behaviour of the two concepts of nail walls in the case of extreme solicitation, a dynamic loading induced by an explosion has been conducted on the site.

1. Introduction

Soil nailing is a soil retention technique using grouted tension-resisting steel elements that can be used to retain excavations and stabilize steep cut slopes under static and seismic conditions [1]. The case histories and post-earthquake investigations show that soil nail walls have performed well during strong ground motions in contrast with the generally poor performance of gravity retaining structures [2-3]. Knowing that, for the fabrication of a soil nail wall in shotcrete, the quantity of concrete lost by projection rebound is significant, therefore this constructive approach is expensive and damaging from the environmental point of view analysis. An alternative process to shotcrete nail walls was proposed. This process is based on the use of precast elements that are easier to install, cheaper, good quality in bearing capacity as well as aesthetics are better than the shotcrete. To gauge all the parameters, a real scale test was realized and instrumented. To study this new technique, and to compare it with the conventional technique (shotcrete), two representative walls are made: one corresponds to the

IOP Conf. Series: Earth and Environmental Science **143** (2018) 012001 doi:10.1088/1755-1315/143/1/012001

conventional shotcrete nail wall (wall 7.50 m high and 10 m wide) and the other to the precast slabs nail wall of same size (Figure 1).

The first phase of this study is to determine the bearing capacity and the associated failure mode of each precast slab which is the constituent element of the precast nail wall in order to verify its design. Then, an experimental study coupled with numerical simulation has been conducted in the laboratory [4], on a precast slab submitted to representative mechanical solicitation. This study allows validating the load associated to crack initiation, cracking behaviour (opening and propagation), and finally, the ultimate state or bearing capacity as well as the associated failure mode.

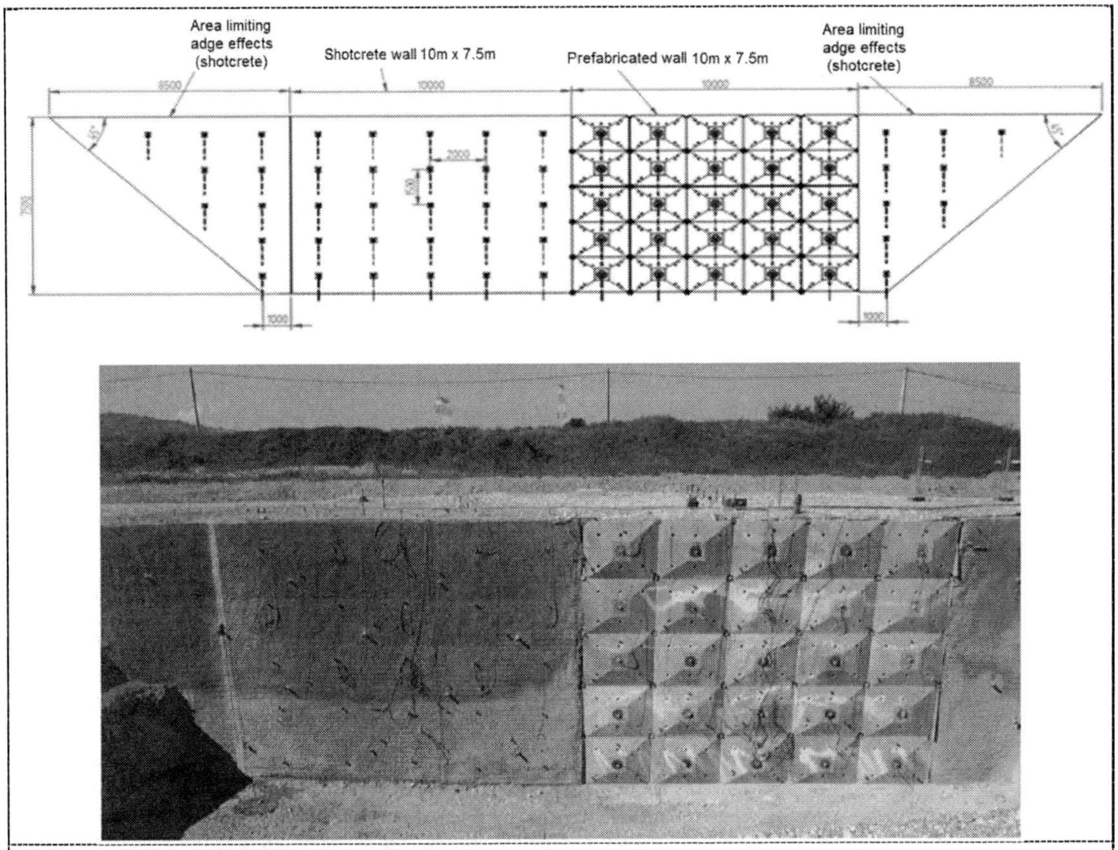

Figure 1. In situ full-scale wall (Rives, Rhône-Alpes, France): shotcrete nail wall - left side; precast nail wall – right side. Unit in [mm]

The second phase is the validation of the behaviour of the full-scale nail wall [5], constructed from the precast slabs which have been tested in the laboratory at the first phase. To assess the efficiency of this new design, a second wall using the shotcrete technique has been constructed next by the precast soil nail wall (Figure 1). The behaviour of the two structures was then analysed thanks to an in situ two years continuous monitoring, where horizontal displacements efforts on the wall and on the anchor nails have been carried out and analysed. Thus, the qualification of the behaviour of the nail wall consisting of precast slabs, compared to the shotcrete wall, could be evaluated, not only on the long term mechanical behaviour, but also on the whole construction phase (economy of material, construction time, ...).

In order to evaluate the behaviour of the two concepts in the case of extreme stress, it is proposed here in the final phase to apply dynamic loading induced by an explosion. Different intensities of the explosion are conducted to evaluate for each level, the behaviour of each configuration, the efforts taken by the nails and the damage of the walls.

2. Laboratory tests

In order to evaluate the behavior of the precast slab under representative mechanical solicitation, two tests are carried out on identical specimens (Figure 2). The objective is to quantify the onset of cracking as well as to evaluate the bearing capacity, and the failure mode. The principle is applying the specimen by a quasi-static force on the central zone on the downstream (top) side. This zone, in fact corresponds to the place where the nail applies on the slab through a hole.

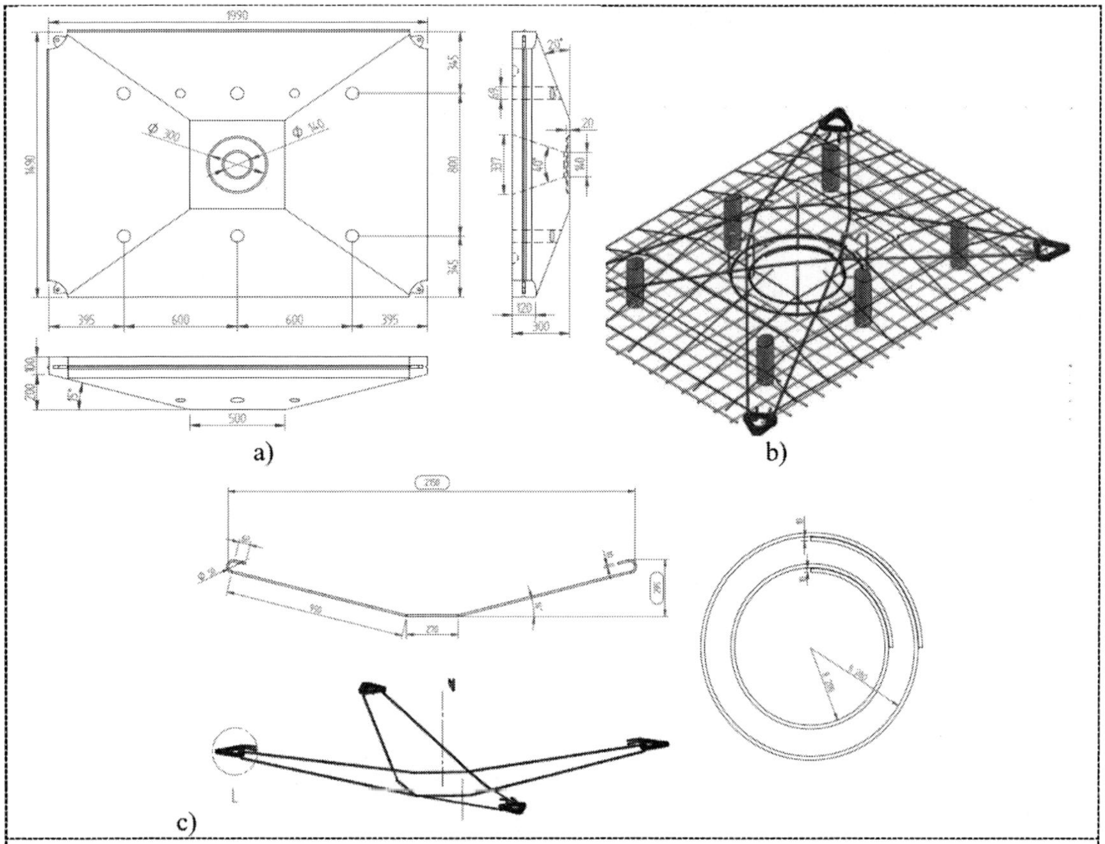

Figure 2. Dimension of the precast concrete slab

2.1. Specimens and test setup

The specimen is supported on four supports (Figure 3). A quasi-static loading was applied (imposed force) on the central zone by a hydraulic jack with a load speed of 1kN/second (60kN/minute). The specimen is instrumented by five displacement sensors LVDT (C1- C5). The specimens were cast using ready-mixed concrete, and a normal strength concrete of class C40/50 was used. The compressive strength of the concrete mixture used for the slabs was determined from compression tests on 11×22-cm cylindrical specimens, performed on the day of slab testing with an average strength of 56.3MPa.

The reinforcement had the yield strength f_y=500 MPa and the ultimate strength f_u=540 MPa. The concrete cover was 30 mm. The amount, the spacing, and the diameter of longitudinal and transverse reinforcement are given in Figure 2. For the lower reinforcement layout, the longitudinal and transversal reinforcement consisted of Ø8 mm placed at an identical distance of 100 mm. For the upper reinforcement layout, the longitudinal and transversal reinforcement consisted of Ø6 mm placed at an identical distance of 200 mm. Two rebar circles placed at the centre of the slab with the diameters of Ø10 mm are used to reinforce the concrete under concentrated load. The cross longitudinal bars used to connecter the slabs through four corners had the diameter of Ø10 mm (Figure 2b).

IOP Conf. Series: Earth and Environmental Science 143 (2018) 012001 doi:10.1088/1755-1315/143/1/012001

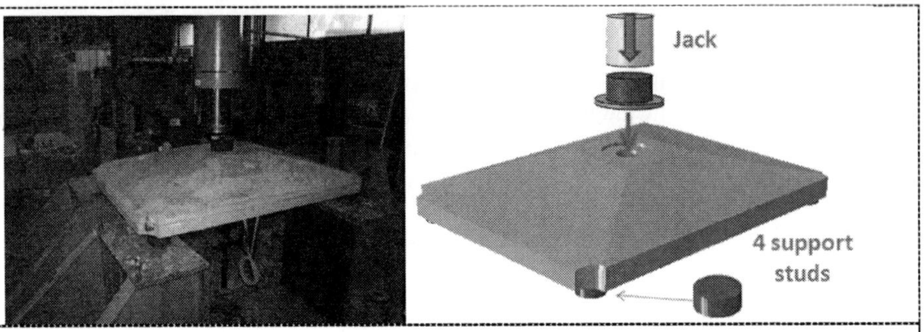

Figure 3. Test setup of the precast concrete slab

2.2. Experimental results
The load/deflection curves of the two tests conducted are shown in Figure 4. The limit of elasticity obtained on the load /deflection curves is about 140kN. The ultimate load is 407kN (specimen 1) and 413kN (specimen 2). Both tests give very similar results, for the bearing capacity the difference is less than 1.5%, and an excellent reproducibility of all the phases characterising the behaviour, from the elastic part at the beginning of loading, to the ultimate state.

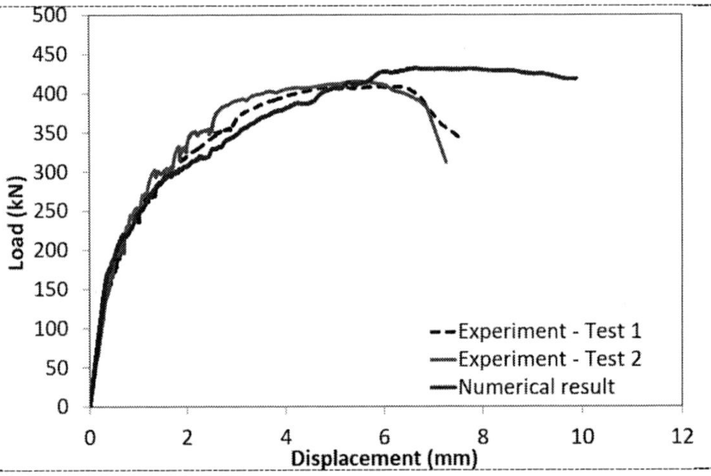

Figure 4. Load/displacement curves: experimental vs numerical results

The failure modes observed for specimen 1 are specified in Figure 5 below. On the bottom face, the rupture due to bending in two ways has been found with the cracks corresponding to the flexural rebar's positions. On the top face, at the four corner of the specimen shear cracks were observed.

Figure 5. Failure modes of specimen 1: bottom face and top face

2.3. Numerical results

The numerical investigations were conducted with Abaqus software. The failure of RC slabs is sudden, in particular for failure in shear. This could cause numerical instability and convergence problems in the traditional static analysis (implicit analysis). Therefore, the explicit method was used with very small load increments considering a so-called smooth-step function to develop quasi-static analysis. All calculations were made force controlled with an explicit solution technique. In addition, the computing time corresponds to the critical increment size and for FEM with a great number of degrees of freedom; it is considerably shorter in comparison to the computing time under application of implicit solvers. The concrete damaged plasticity model (CDP) in Abaqus was used for concrete material and the elastic-plastic model was applied to reinforcement steel.

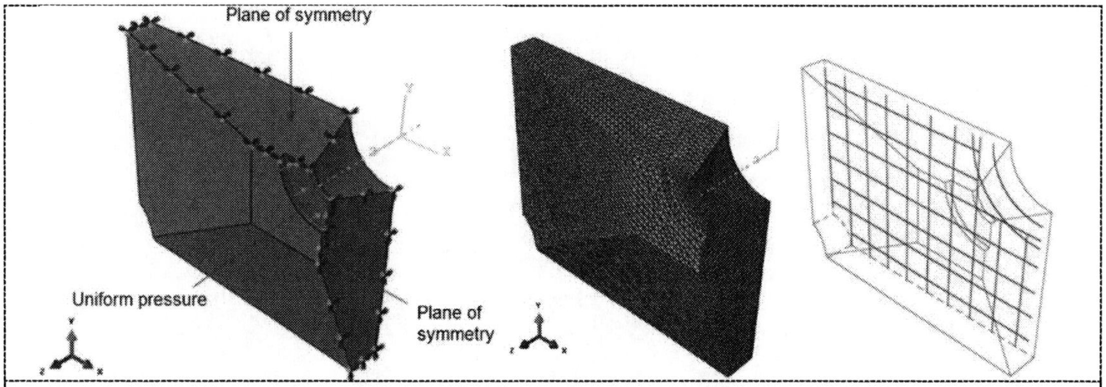

Figure 6. Mesh of the precast slab

The CDP model in Abaqus is based on the models proposed by Lubliner et al. (1989) [3] and (Lee & Fenves, 1998) [7]. The model proposed by Lubliner et al. (1989) [6] for monotonic loading and has been developed later by Lee and Fenves (1998) [7] to consider the dynamic and cyclic loadings. The constitutive theory in this section aims to capture the effects of irreversible damage associated with the failure mechanisms that occur in concrete and other quasi-brittle materials under fairly low confining pressures (less than four or five times the ultimate compressive stress in uniaxial compression loading). The nonlinear behaviour of concrete is attributed to the damage and plasticity processes. The plasticity behaviour can be characterized by several phenomena such as strain softening, progressive deterioration, etc. The damage process can be attributed to micro-cracking. Damage is associated with the concrete's failure mechanisms and therefore results in a reduction in the elastic stiffness. The model based on the concept of isotropic damaged elasticity in combination with isotropic tensile and compressive plasticity to represent the inelastic behaviour of concrete i.e. tensile cracking and compressive crushing.

With the complexity of the selected geometric shape, a 3D finite element model developed on the Abaqus code has been chosen. The modeled geometry is identical to the real geometry and taken into account the presence of the reinforcement (steel) in the concrete slab. Only one quarter of the slab is modeled, taking into account symmetry conditions. The slab rests on 4 supports (Uz blocked unilaterally). Uniform pressure is applied to the central disk of the top surface (Figure 6) in accordance with the experimental test. A perfect adhesion between steel rebars and concrete is assumed. This modeling is fairly faithful to the test.

Table 1. Properties of concrete used in numerical model

Characterictics	Value
Compressive strength f_{cm} (MPa)	56.3
Tensile strength f_{ctm} (MPa)	3.52
Young's modulus (MPa)	35500

Table 2. Properties of steel used in numerical model

Characterictics	Valeurs
Young's modulus (GPa)	210
Yield strength f_{yk} (MPa)	500
Rupture strength (MPa)	540

The characteristics of C40/50 concrete taken from the experimental tests are given in Table 1. For steel reinforcement type Fe500, an elastoplastic model with isotropic hardening is used. The characteristics of the reinforcements are specified in Table 2.

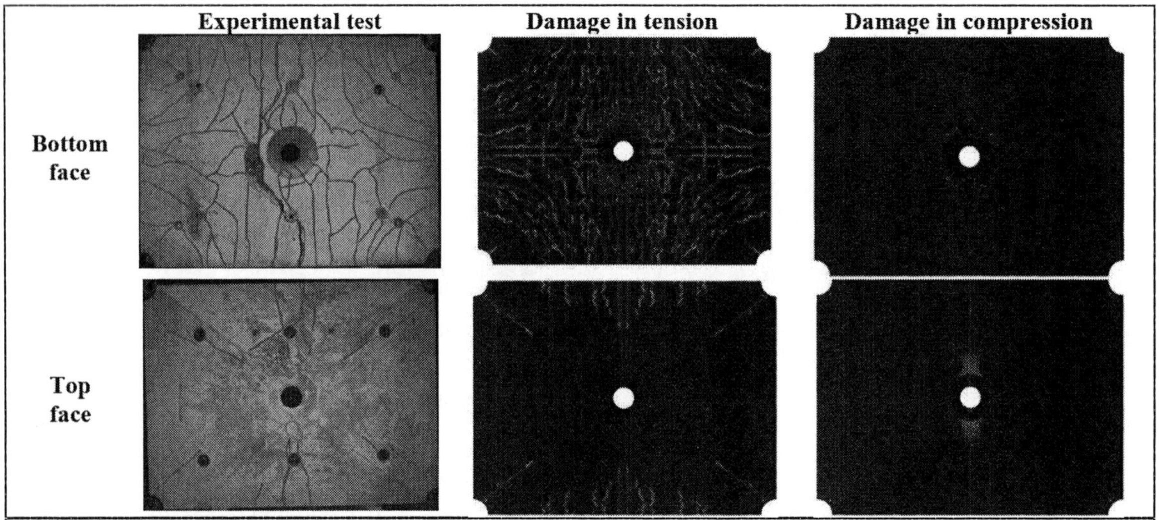

Figure 7. Failure modes: experimental vs numerical results (Specimen 2)

The load/deflection curve obtained numerically at the central point (point instrumented by the sensor C3) is compared with those of the tests (Figure 4). The ultimate load obtained by the numerical computation is of 452 kN, a difference of 11% compared to the experimental result of the test 1 ($P_{exp-test1}$ = 408kN) and of 9% compared to the result of experimental test 2 ($P_{exp-test2}$ = 415kN). The elastic limit (beginning of cracking) is also quite well reproduced by calculation ($P_{elas-FEM}$ = 160kN, $P_{elas-exp}$ = 140kN) a difference of 14% is noticed. The rigidity obtained numerically is similar to that found experimentally, this for the linear phase, but also in the non-linearity phase. The curves in this case are in good adequacy, thus showing the relevance of modeling.

Numerical computation leads to a final failure mode characteristic of bending with essentially two cracks crossing the half-length of the slab. It is confirmed by observing the plastic deformations of longitudinal reinforcement when the ultimate load is reached. The cracking facies are similar between simulation and experiment (Figure7). Using the variables of damage we can visualize the cracking. On the bottom face (tensile face), most cracks are due to traction. On the top face, there is a zone in the center around the hole, where the rupture is obtained by damage in compression of the concrete. This damage is perfectly similar (location) to that observed experimentally (crushing or bursting of the concrete around the hole). Tensile fractures on the top face have developed and prolong the cracks observed initially in the bottom face. They appear at the four corners of the scale, and also on the long span, no crack appears on the smallest span (width) of the scale. This reflects a bending failure but only in one direction, which consolidates a beam-like behavior (one way direction of solicitation). The first cracking observed is the longitudinal crack due to bending. We recall that this test is very conservative in view of the loading that the slab will undergo in situ.

3. Static measurements in situ

On the site located on the city of Rives in the Rhône-Alpes region, France, GTS has built two nail walls. The first comes from a classic construction process, namely a monolithic shotcrete wall which is anchored to the ground with anchor nails. The second concept is an innovation proposed by GTS. The wall here consists of reinforced concrete slabs, each manufactured in prefabrication factory, which are nailed to the ground using an anchor nails.

For the experimental site, five excavations of 1.5m in deep are made all along the wall reaching a theoretical excavated volume of 2200 m3 for a depth of 7.5m. Each excavation phase includes the placement of sealed nails, one nail every two meters horizontally constituting a rectangular mesh with an anchor for each slab for precast elements. This wall consists of five slabs for each row (Figure 8).

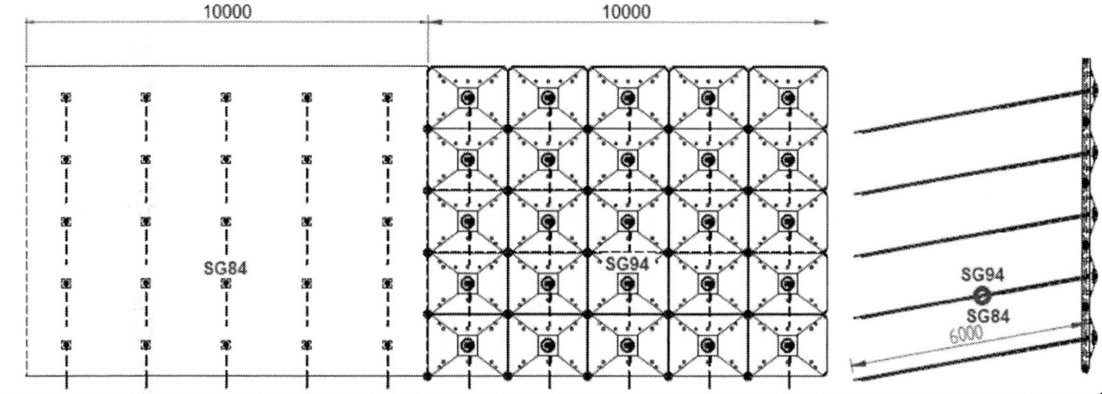

Figure 8. Instrumentation of strain gauges (SG84 and SG94) on the anchor nails

The two walls are instrumented with various sensors, and similarly for the wall of shotcrete and that consisting of precast slabs. The measurement channels quantify either the forces in the nails, using strain gauges, or the pressure on the wall using pressure sensors. We also have a force sensor positioned on a nail. This measure is used to serve as verification. The instrumented points are symmetrically so that the behavior of the shotcrete wall and that of precast slabs can be compared. The study of the quasi-static behavior, which spread over a period of 2 years, is to compare between the two wall behaviors.

Each bar (anchor nail) is equipped with two strain gauges positioned symmetrically on the flats of the bar. This is to take into account any bending of the bar. Figure 9 presents the measures of the two strain gauges in the anchor nail of the shotcrete wall (SG84_01 & SG84_02) and the ones of the precast wall (SG94_01 & SG94_02). The positions of the strain gauges SG84 and SG94 are described in Figure 8. The frequency chosen for our tests is an acquisition every 20 minutes.

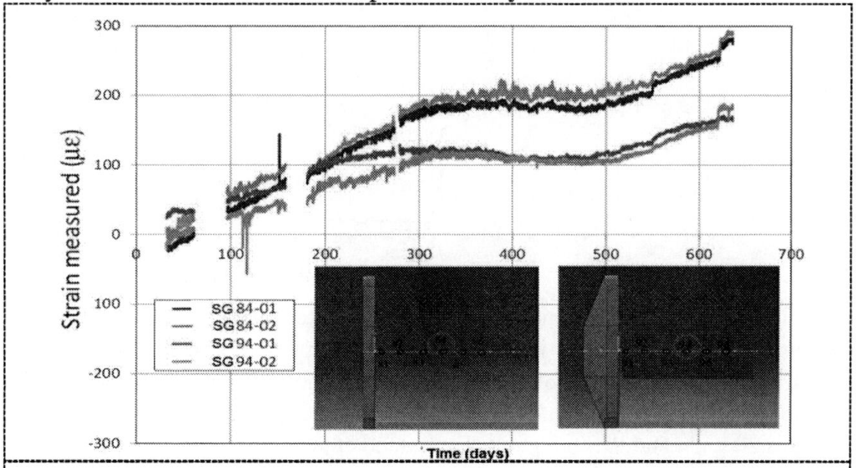

Figure 9. Strain measured on the anchor nails of two walls: SG84 vs SG94

It is noted that the effort is incremented over time, it remains however limited since less than 30 kN (100 με correspond to about 10 kN). It is found that all the sensors give substantially the same orders of magnitude. For this case, perfect equality between the two strain gauges (SG84_01 compared to SG84_02, and SG94_01 compared to SG94_02). We conclude that the nail undergoes a uniform traction. The value of the effort in the precast wall, about 20 kN, is less than the value in the shotcrete wall, about 30 kN (Figure 9). For duration of over 300 days, a plateau equivalent to a load level is reached for two both systems of shotcrete wall and precast wall. Before 500 days we see that the effort varies again and increases (Figure 9).

In this paper, only the measures of two sensors SG84 and SG94 are presented. If the analyses are based on the results of all the sensors, we found that on the majority of the sensors efforts remain low (less than 20 kN with some maximum to 60 kN). Moreover, in many cases there is flexion in the head nail especially for the slabs of precast wall. Apart from a few points where the effort is incremented (from 20 to 40 kN in 2 years), the pushing efforts remain limited. They do not increase in a monotonous way. This is certainly due to a significant "cohesion" of the soil upstream of the nail walls.

4. Explosion tests in situ

In order to evaluate the behavior of the two concepts of nail walls in the case of an extreme load, it is here proposed to apply a dynamic loading induced by an explosion. Three tests are carried out with different intensities of the explosion to evaluate for each level, the behavior of each configuration, the efforts taken by the nails and see the damage of the walls.

Two first tests are conducted with an explosion of a low intensity. For the first shot the amount of explosive that remains low is positioned away from the wall, while for the second shot, the same amount of explosive is positioned close to the wall. Dynamic measurements are carried out during these two tests, as well as measurements in statics, after the two shots, the object being to compare the state of the structure before and after application of the dynamic loading. The third test is an extreme load, with a large amount of explosive (163kg), located near the walls. The objectives of this last test are to characterize the peaks of effort before collapse as well as the type of collapse for each configuration. In this paper, only the results of the third test are presented. Position of the explosives is given in Figure 10.

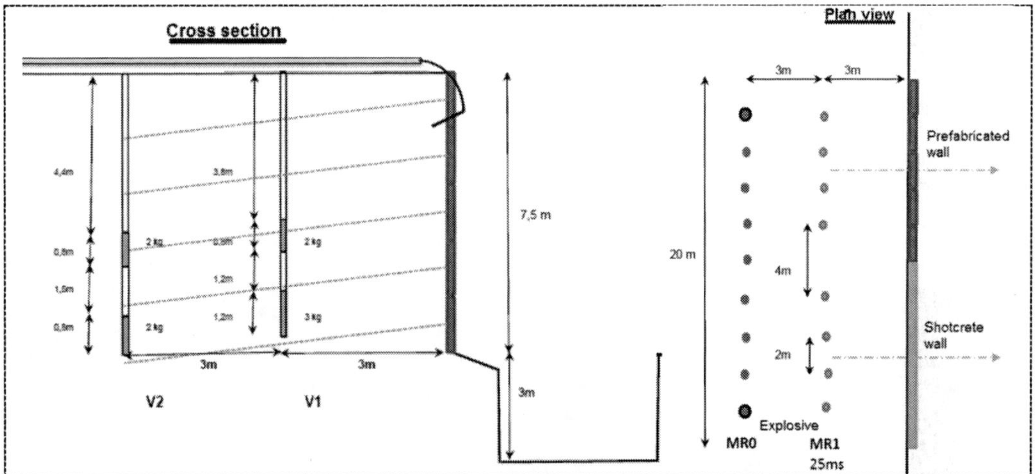

Figure 10. Position of the explosives

For the dynamic measurement associated with "ultimate" test (165kg of explosive), the initial zero of the various sensors was recalibrated, in order to clearly identify the maximum value of deformation induced in the nails (and therefore the force) induced by the shot of explosive undergone. It also allows us to easily compare behavior between different measurement points. We recall that the limit of the force sensor used is 20 tons, and that the gauges are glued at several points along the anchoring nails, in order to measure the distribution of deformation along the nail. From the deformation, we go back to the equivalent force. On each measuring point there are two gauges glued in opposition side. If the

measured strain is positive, the cable is in traction. If the measured strain is negative, the cable is compressive or flexural (thus with a gauge in compression and one in traction). In a measurement position, if a gauge gives a positive deformation, while the one opposite gives a negative deformation, then the nail works well in flexion, this is particularly the case at the neighborhood of the slab.
Note that the elastic limit of the nails (Ø25mm) of the wall is 670MPa, which corresponds to a maximum elastic deformation of 3190µm/m and a maximum equivalent force of 330kN. So, for a deformation higher than 3190µm/m, we can conclude to the plastic of the nail, and we do not have directly the value of the force. The acquisition frequency of the dynamic measurement is 10kHz (ten thousand points per second). Note that for dynamic testing, we found that some sensors did not work.

Important damage is observed at the base (foot) of the wall. (Figure 11) For the shotcrete wall, the mode of ruin is due to bending. This wall behaves like a continuous structure. Its behavior is similar to reinforced concrete voile subjected to dynamic out-of-plane loading (pressure). For the precast wall, the mode of ruin is also observed at the foot of the structure (two lines of low slabs). In particular, the failure mainly focuses on the 3 slabs in the center of each line, there are 5 slabs in total on each line, so 2 slabs at the edges ensure boundary conditions like two lateral supports. An important point to note while observing the collapse of the wall in slabs: the ruin is due to the rupture at the level of the joints (point of connection of the slabs) we do not see damage in the body of the slabs. These last ones undergo essentially rotations.

Figure 11. Collapse mode of two nail walls under explosion

Finally, it appears clearly on the film associated with the test, that the collapse first appears on the shotcrete wall, the collapse of the latter causes that of the precast wall. The latter because of the possible rotation of the slabs is more resistant and shows more resilience compared to the monolithic wall, in that it can develop a local "collapse" associated with large rotations on one or two slabs without the wall collapses.

For the precast wall, we note just before t = 0.24s the presence of peaks and oscillation that reflect the occurrence of the explosion (Figure 12). Beyond 5/100th of a second translate the behavior to collapse. There is indeed a flexion of the slab: the nail is subjected the compression (for a strain gauge) and for the other of the traction. The measured deformations remain however limited, although the rotation at the nail head of the slab is consequent.

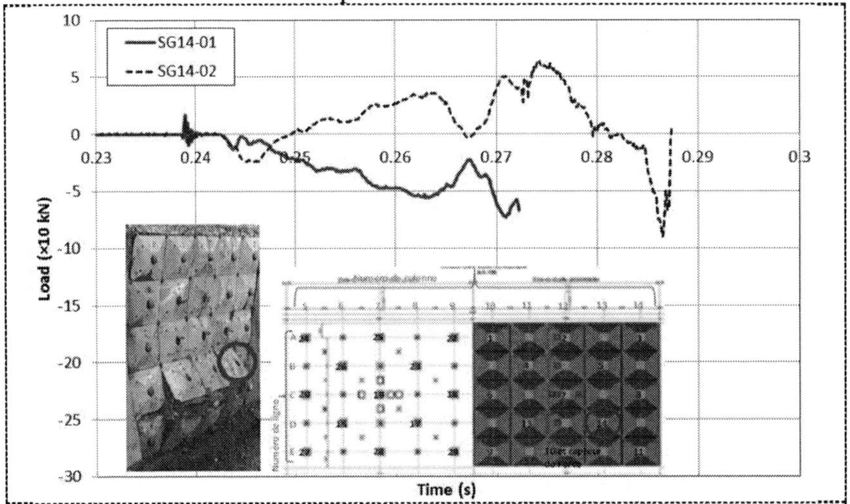

Figure 12. Load measured on the anchor nails of precast wall: SG14

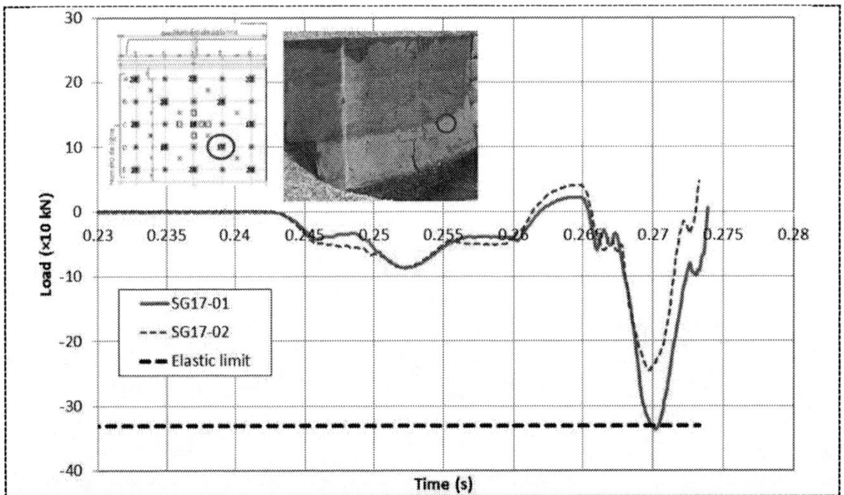

Figure 13. Load measured on the anchor nails of shotcrete wall: SG17

For the shotcrete wall (Figure 13), traction is obvious and this is the main mode of deformation (little bending), which is explained by the continuous and monolithic character of the wall. The latter undergoes essentially thrust and thus an "extension" out-of-plane, the local rotations are obtained ultimately to rupture and essentially to the right of the lines of rupture that do not pass through anchors. There are effort thresholds leading to the plastic of the anchor nail. The first compressive phase translates the propagation of the wave in the nail. It is transformed into traction once arrived at height of the wall.

5. Conclusion

This study presents the behaviour of a new concept of the soil nail wall using precast slabs. To gauge accurately the differences with classical shotcrete soil nail wall, a comparative in situ study was conducted including all the phases from the construction to the in-situ static measurements over two years and the in-situ dynamic tests under explosion demonstrated the robustness of the precast soil nail wall in comparing with the shotcrete wall. The laboratory testing on the precast slabs confirm the performance in mechanical properties of the constituent elements of the precast soil nail walls. The numerical model successfully reproduced the nonlinear behaviour of the experimental tests. Therefore this numerical model can be used in perspective to model the full wall constituted from the precast slabs elements. Furthermore, the use of precast slabs for the construction of nail walls provides solutions to many existing problems currently compared with shotcrete wall including:

- The delay: no phase reinforcement / formwork / drying
- Hygiene: no dust and noise when making the wall
- Safety: no risk related to concrete projection by air pressure
- Local environmental impact: no loss of concrete left in place
- The carbon footprint is divided by three by reducing the consumption of cement and use of eco-materials
- Geometric tolerances: the thicknesses of concrete are respected
- Weather dependence: no problem with this technique

References

[1] Chavan D, Mondal G and Prashant A 2017 Seismic analysis of nail soil slope considering interface effects *Soil Dyn Earthq Eng* vol 100 pp 480–490.

[2] Colin JG, Chouery-Curtis VE and Berg RR 1992 Field observations of reinforced soil structures under seismic loading *Proceedings of the international symposium on earth reinforcement* Fukuoka Japan vol 1 pp 223–228.

[3] Tatsuoka F, Tateyama M, Koseki J and Uchimura T 1995 Geotextile-reinforced soil retaining wall and their seismic behaviour *Proceedings of 10th Asian regional conferonce on soil mechanics and foundation engineering* vol 2 pp 26–49.

[4] Bui T T and Limam A 2015 Paroi clouée préfabriquée: Evaluation de la capacité portante d'une écaille constitutive: approche expérimentale et numérique *Internal report.*

[5] Bui T T and Limam A 2015 Parois clouées: Suivi du comportement sur chantier *Internal report.*

[6] Lee J and Fenves G 1998 Plastic-damage model for cyclic loading of concrete structures *Journal of Engineering Mechanics* vol 124 pp 892–900.

[7] Lubliner J, Oliver J, Oller S and Oñate E 1989 A plastic-damage model for concrete *International Journal of Solids and Structures* vol 25 pp 299–326.

Acknowledgments

This work was funded by contract from GTS Group NGE Lyon in France.

IOP Conf. Series: Earth and Environmental Science **143** (2018) 012002 doi:10.1088/1755-1315/143/1/012002

Prefabricated Vertical Drain (PVD) and Deep Cement Mixing (DCM)/Stiffened DCM (SDCM) techniques for soft ground improvement

D T Bergado[1,*], P V Long[2], S Chaiyaput[3] and A S Balasubramaniam[4,5]

[1]Asian Institute of Technology and Geotechnical Specialist, Index International Group, Bangkok, Thailand
[2]Vina Mekong Consultants (VMEC), Ho Chi Minh City, Vietnam
[3]Faculty of Engineering, King Mongkut's Institute of Technology Ladkrabang, Bangkok, Thailand
[4]Asian Institute of Technology, Bangkok, Thailand
[5]Griffith University, Gold Coast Campus, Brisbane, Australia

*Corresponding author: dbergado@gmail.com

Abstract. Soft ground improvement techniques have become most practical and popular methods to increase soil strength, soil stiffness and reduce soil compressibility including the soft Bangkok clay. This paper focuses on comparative performances of prefabricated vertical drain (PVD) using surcharge, vacuum and heat preloading as well as the cement-admixed clay of Deep Cement Mixing (DCM) and Stiffened DCM (SDCM) methods. The Vacuum-PVD can increase the horizontal coefficient of consolidation, C_h, resulting in faster rate of settlement at the same magnitudes of settlement compared to Conventional PVD. Several field methods of applying vacuum preloading are also compared. Moreover, the Thermal PVD and Thermal Vacuum PVD can increase further the coefficient of horizontal consolidation, C_h, with the associated reduction of k_h/k_s values by reducing the drainage retardation effects in the smear zone around the PVD which resulted in faster rates of consolidation and higher magnitudes of settlements. Furthermore, the equivalent smear effect due to non-uniform consolidation is also discussed in addition to the smear due to the mechanical installation of PVDs. In addition, a new kind of reinforced deep mixing method, namely Stiffened Deep Cement Mixing (SDCM) pile is introduced to improve the flexural resistance, improve the field quality control, and prevent unexpected failures of the Deep Cement Mixing (DCM) pile. The SDCM pile consists of DCM pile reinforced with the insertion of precast reinforced concrete (RC) core. The full scale test embankment on soft clay improved by SDCM and DCM piles was also analysed. Numerical simulations using the 3D PLAXIS Foundation finite element software have been done to understand the behavior of SDCM and DCM piles. The simulation results indicated that the surface settlements decreased with increasing lengths of the RC cores, and, at lesser extent, increasing sectional areas of the RC cores in the SDCM piles. In addition, the lateral movements decreased by increasing the lengths (longer than 4 m) and, the sectional areas of the RC cores in the SDCM piles. The results of the numerical simulations closely agreed with the observed data and successfully verified the parameters affecting the performances and behavior of both SDCM and DCM piles.

1. Introduction

Prefabricated vertical drains (PVDs) is the cheapest method of soft ground improvement, this is the reason why the PVD method has widely adopted and applied for construction on soft ground [1], [2], [3], [4]. PVDs with embankment preloading method functions by draining the pore water mostly in horizontal direction from the surrounding soft soil faster during the consolidation process by shortening the drainage path and by taking advantage of the higher horizontal hydraulic conductivity in the soft ground deposit. Then, the pore water can flow freely along the drain vertically towards the permeable drainage layers. However, the PVD installation using a mandrel causes disturbances in the clay surrounding the PVD resulting in lower horizontal hydraulic conductivity in the smear zone [3-6]. PVDs with embankment preloading combined with vacuum pressure (called Vacuum PVD) has been utilized to minimize the instability problem and to accelerate the rate of consolidation. Kjellman [7] first proposed the vacuum consolidation in early 1950s. Subsequently, the studies of vacuum consolidation continued up to the present [8-14]. Vacuum consolidation can reduce the pore pressure and maintain constant total stress instead of increasing the total stress. The effective stress is increased due to the reduced (less than atmospheric) pressure in the soil mass. In addition, with heat application, the horizontal hydraulic conductivity in smear zone, k_s, around the PVD increased which was first proposed by Abuel-Naga et al. [15]. This technique is called Thermal PVD by using the thermal treatment up to 90°C combining with PVD. The PVD combined with heat works by the reduction of the smear effect due to the increased clay permeability at elevated temperature. Consequently, faster rate of consolidation was achieved but with larger magnitude of settlement [16]. The effect of temperature on the soil permeability was also studied previously and reported that the hydraulic conductivity of soil increased with increasing the temperature [15], [17]. The effect of heat on the volume change and compressibility of clays were also studied [18].

Moreover, the one kind of ground improvement, which is continuously developed, is Deep Cement Mixing (DCM) pile. Although DCM pile has many advantages with various applications, failure caused by pile failure can occur especially when subjected to the lateral loads. Moreover, the unexpected lower strength than the design commonly occurs due to lack of quality control during construction. Thus, DCM pile still fails by pile failure mode which is lower than the soil failure mode particularly at the top of DCM pile due to low strength and stiffness [19]. Consequently, the bearing capacity of DCM pile can be lower than the design load of 10 tons due to pile failure.

2. Prefabricated Vertical Drain (PVD)

2.1. Laboratory tests using PVDs, Vacuum PVDs and Thermal PVDs

2.1.1. Large consolidometer

The large scale consolidometer (Figure 1) consisted of a cylinder cell of 0.45m in inner diameter and 0.95m in height made of polyvinyl chloride (PVC) with a thickness of 10 mm rested on steel base which can resist pressure less than 500 kPa. The air pressure was applied through the upper pedestal to the top of the piston and the vacuum pressure is applied through the shaft of the piston to the bottom of piston and the PVD cap, respectively. Air pressure was transformed to vacuum pressure by a vacuum generator which was connected directly with an air pump during consolidation test. A natural rubber membrane with a thickness of 3mm was installed in the chamber above the piston to prevent the leakage of the air pressure and/or vacuum pressure through the piston. The natural rubber membrane was folded in vertical direction initially to allow the vertical displacement of the piston during consolidation. Geotextiles were placed on top and bottom of the soil specimens to prevent clogging of the loading piston. Dial gauges were placed on top of the shaft for settlement measurements. A pore pressure transducer was installed and connected with data logger to monitor pore pressure in the specimen during consolidation.

Flexible heater wire with a capacity of 120°C and power of 6W per meter was attached to the PVD for the reconstituted specimen improved with Thermal PVD and Thermal-Vacuum PVD.

Thermocouples were installed at radial distances of 25, 50, 100 and 200 mm and were connected to the digital data logger to monitor the temperature and heat transfer during the consolidation test (Figure 2). Before and after the laboratory tests, the water contents and shear strengths were also measured at the aforementioned locations, the temperature sensor boxes were used to maintain the heat temperature at heat source and shut down automatically upon reaching the controlled heat of 90°C, at the heat source.

Customized vane shear equipment was used to measure the undrained shear strength before and after consolidometer tests. The vane blade made of stainless steel, is 20mm in diameter and 40mm in height. It is attached to an adjustable stainlesssteel rod 5mm in diameter, capable of measuring the shear strengths at different locations and depths. The rod was held in place by a removable rectangular steel adapter above the circular plate rotator. The adapter was connected to the force sensor, which was attached on top of the circular rotator when the vane is turned. The force sensor was connected to the transducer to determine the readings. The maximum torque of each radial distance was calculated by multiplication of force and radial arm distance. The vane shear tests were done at radial distances of 25mm, 50mm, 100mm and 200mm, respectively, at two different depths.

The vacuum generator was utilized to transform the air pressure to vacuum pressure. This apparatus was connected directly with air pump during consolidation test. The water generated from suction by vacuum pressure was stored in a closed container.

A KPD-200 kPa excess pore pressure transducer was connected with data logger to monitor the pore pressure in the specimen during consolidation using a probe inserted into the wall of the consolidometer cell at the desired level. The excess pore pressures were monitored in the consolidomer test by using the data logger.

Figure 1. Large scale consolidometers for Thermal PVD and Thermal Vacuum PVD.

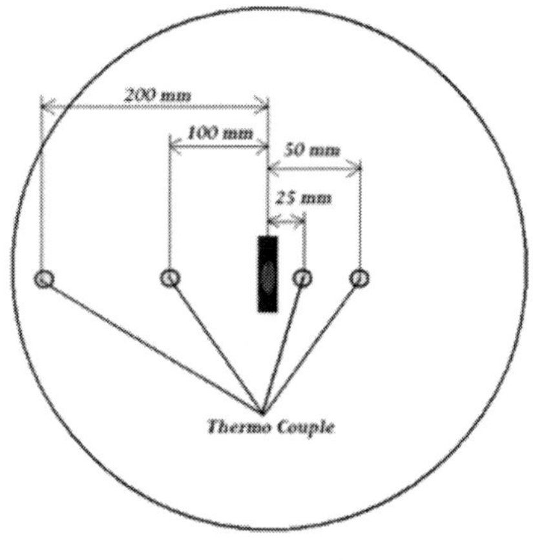

Figure 2. Configurations of PVD and thermo couples.

Table 1. Physical properties of soft Bangkok clay in Suvarnabhumi Airport.

Physical properties	
Liquid limit (%)	102.24
Plastic limit (%)	39.55
Water content (%)	112.69
Plasticity index	62.69

Total unit weight (kN/m^3)	14.70
Specific gravity	2.66

2.1.2. Test procedures
The soil samples used in this study were obtained from a site which is located at the area of Second Bangkok International Airport (SBIA), Thailand. The soft clay samples were collected from 3.0 to 4.0m depths. Disturbed samples were obtained by digging the soils up to required depth with a backhoe. Then, the disturbed soil samples were placed in covered plastic containers for storage. Undisturbed samples were collected by 10 inches piston samples and immediately covered with wax after sampling to prevent loss of moisture. The physical properties of the soft Bangkok clay are shown in Table 1. The PVD material used was CeTeau drain (CT-D911). The disturbed samples were mixed by using a mixer. Water was added until the water content was slightly greater than the liquid limit. The mixed soil was placed into the large scale consolidometer cell layer by layer until to the desired height. The appropriate loads were applied for the reconstitution process to obtain the desired water content and void ratio equal to the initial conditions.

2.1.3. Consolidometer test program
For the reconstitution using the new large consolidometer, a 50 kPa pressure was applied. Drainage was allowed to flow to the top and bottom of the apparatus. Silicone grease was applied to the insides of the large consolidometer to reduce the friction. Throughout the whole process, the settlements were monitored. After reconstitution, the water content, shear strength, and specimen height were determined. The vertical pressure of 50 kPa was increased to 100 kPa after reconstitution under the applied vertical pressure in the specimen improved with PVD. For the specimen improved with PVD and vacuum pressure (Vacuum PVD), a vertical pressure of 50 kPa and vacuum pressure of -50 kPa were applied. For the Thermal PVD, the specimen was improved with PVD combined with vertical pressure of 100 kPa together with application of heat up to 90 °C. For the Thermal Vacuum PVD, the specimen was improved with PVD combined with heat up to 90 °C together with vertical pressure of 50 kPa and vacuum pressure of -50 kPa. The temperatures were controlled by an electronic thermal control unit that received the signal from thermocouples. Drainage was allowed to flow only one way. Settlement was monitored during the test until the soil specimen reached to 90% consolidation. The method of Asaoka [20](1978) was used to estimate the degree of consolidation and the magnitude of final settlement. The large scale consolidometer tests of specimens improved with Conventional PVD, Vacuum PVD, Thermal PVD and Thermal Vacuum PVD are shown in Figure 1.

2.1.4. Effect on the coefficient of consolidation, Ch, and permeability ratio, k_h/k_s
The values of Ch for all tests were back-calculated using the equations from Hansbo [1] for radial consolidation with PVD is given as follows:

$$U_h = 1 - \exp\left[\frac{-8T_h}{F}\right] \tag{1}$$

where U_h is the degree of consolidation for horizontal drainage; T_h is the time factor for horizontal drainage; F is the factor which expresses the additive effect due to the spacing of the drains, F(n), smear effect, F_s, and well-resistance, F_r. The values of F(n), Fs and Fr are given by the following equations:

$$F(n) = \ln\left[\frac{D_e}{d_w}\right] - \frac{3}{4} \tag{2}$$

$$F_s = \left[\frac{k_h}{k_s} - 1\right]\ln\left[\frac{d_s}{d_w}\right] \tag{3}$$

$$F_r = \pi z(L - z)\frac{k_h}{q_w} \tag{4}$$

where D_e is the diameter of the equivalent soil cylinder, d_w is the equivalent diameter of the drain, k_h is the coefficient of horizontal permeability, k_s is the horizontal permeability of the smear zone, d_s is the diameter of the smear zone, z is the distance from the drainage end of the drain, L is the length of the drain for double drainage and twice the length of the drain for single drainage, q_w is the discharge capacity of the drain at hydraulic gradient of 1 (one). The time factor, T_h, for horizontal drainage can be calculated using:

$$T_h = \frac{C_h t}{D_e^2} \tag{5}$$

where C_h is the coefficient of horizontal consolidation and t is the time elapsed after the application of the load.

2.1.5. Consolidation behavior of tested samples

The comparison of settlements of specimens improved with and without vacuum and heat in the large scale consolidometers test are shown in Figure 3. The specimen improved with Vacuum PVD has higher rate of settlement than the specimen improved with Conventional PVD but the final settlement magnitudes are the same. In contrast, much faster and higher settlements were achieved from Thermal PVD and Thermal Vacuum PVD. However, the performance of Thermal PVD is slightly lower than Thermal Vacuum PVD. Among the tests, the Thermal Vacuum PVD combination achieved the fastest rate of settlement and the largest final settlement. There is 25% of increase in final settlement in Thermal PVD with almost the same settlement magnitude with the Thermal Vacuum PVD. Similar results to Pothiraksanon et al. [21], the use of the Thermal PVD result in increased settlements of about 30%. In addition, Saowapakpiboon et al. [11] showed the differences in the rate of settlement of Thermal PVD was highest due to volume contraction and rate of consolidation due to thermal effects. These results show that the higher the temperature, the higher the volume contraction and the higher the rate of consolidation resulting from the increased soil permeability due to reduced pore water viscosity [17]. The compressibility increases due thermally induced irreversible contraction of saturated normally consolidated soft Bangkok clay. Furthermore, Abuel-Naga et al. [15] indicated substantial increase in the clay permeability up to more than 3 times when the temperature was raised up to 90 degrees centigrade under fully drained constant stress conditions oedomoter tests in the laboratory. Moreover, the compression line of soft Bangkok clay shifts to the left with approximately similar slope upon heating when the thermal loading is applied at normally consolidated state [18]. Thus, heating the clay specimen combined with both heat and vacuum pressure (Thermal Vacuum PVD) has achieved the highest settlement and also faster rate of consolidation than Thermal PVD, Vacuum PVD and Conventional PVD, respectively.

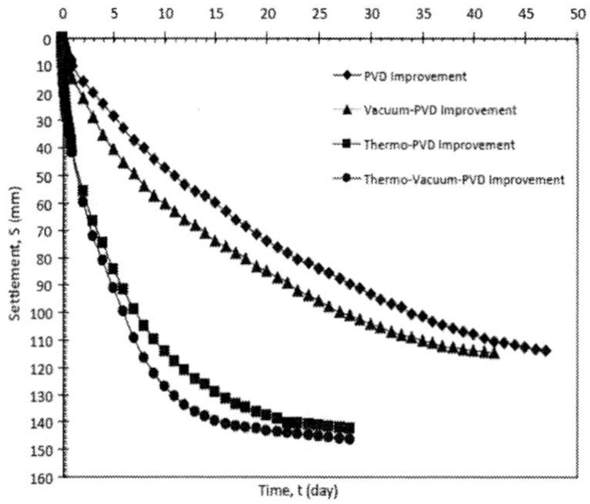

Figure 3. Comparison of settlement behavior using large consolidometer with reconstituted specimen.

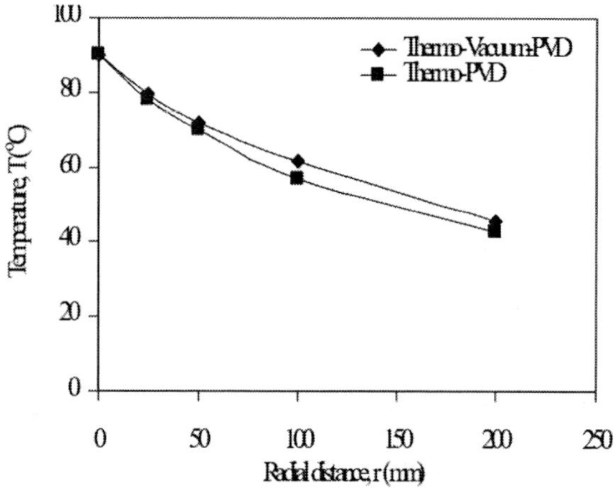

Figure 4. Comparison of temperatures with radial distances in Thermal PVD and Thermal Vacuum PVD.

2.1.6. Heat transfer

The heat transfers in the soil specimens at specific temperatures with respect to the specific distances from PVD were observed. Flexible wire heaters attacked to the core of PVD were used as heat source to increase the temperature up to 90°C. Temperatures of the heat source were measured by thermocouples which were embedded at 150mm depth and at radial distances of 25mm, 50mm 100mm and 200mm from the center of the specimen . The temperature decreased with distance from the heat source. The Thermal Vacuum PVD has similar heat transfer characteristics with the Thermal PVD. The radius, r, of 25 mm corresponds to the location of the smear zone with a mandrel dimension of 18.2 mm x 81.90 mm, and the diameter of the disturbed zone can be calculated as 87mm. Figure 4 shows the temperature of 80°C in the smear zone and decreased with distance from the heat source. It took 40 hours for the temperature to reach equilibrium in the large consolidometer. The result is consistent with the previous work of Abuel-Naga et al. [15]. The temperature change around the PVD decreased as the radial distance increased and become constant thereafter.

2.1.7. Vacuum and thermal effect on water content

The reduction in water content was measured before and after consolidation tests at the radial distances of 25mm, 50mm, 100mm and 200mm from the center of the specimens at depths 10 mm and 40 mm. Lower water contents were observed nearer to the heat source with significant reduction in water contents. The levels of water content reduction decreased with the distance from the heat source. The percent decrease of water contents after consolidation tests at the smear zone of Vacuum PVD was slightly higher than that Conventional PVD. The Thermal PVD had much higher percent decrease of water contents than the Vacuum PVD while the Thermal Vacuum PVD had the highest percent decrease of water contents but nearly similar to that of Thermal PVD.

2.1.8. Effect of vacuum and heat on shear strength of clay

A comparison of the increase in the shear strength in all the four tests after improvement in the reconstituted specimens is plotted in Figure 5. For the smear zone, the Conventional PVD and Vacuum PVD were not much different in the increase of shear strengths. For Thermal PVD, the shear strength had increased but slightly less than the Thermal Vacuum PVD around the smear zone. The results demonstrated that much higher shear strength developed in the smear zone using Thermal PVD and Thermal Vacuum PVD. The increase in shear strength is due to the reduction of water contents and because the clay particle changed to flocculated structure with volume contraction as the soil temperature increased [18].

2.1.9. Vacuum and thermal effects on excess pore pressure

The excess pore pressures were measured at the large scale consolidometer tests during consolidation by using excess pore pressure transducers which were inserted through the wall of the cylinder cell into the smeared and undisturbed zone and monitored by digital data logger. For specimen with Conventional PVD, the excess pore pressure had increased to maximum of 80 kPa and decreased to about 10 kPa after 47 days in the undisturbed zone. For Thermal PVD, the specimen at the undisturbed zone had demonstrated higher excess pore water pressures to 80 kPa and then rapidly reduced to 1 kPa after 28 days due to the effects of thermally induced pore water pressures and volume changes. This behavior demonstrates that the permeability in the smear zone was increased at higher temperature. Consequently, the excess pore water pressures dissipated more rapidly [21]. The thermally induced excess pore pressures were larger because the thermal expansion coefficient of the pore water is approximately 15 times larger than the thermal expansion of the clay solid skeleton [17]. In the case of, Vacuum PVD specimen, the maximum excess pore pressure in the smear zone had increased about 40 kPa and dissipated rapidly after 8 days and decreased to about -40 kPa after 40 days. The excess pore pressures in the undisturbed zone increased to maximum of 40 kPa and transformed to minus excess pore pressure after 13 days and decreased to about -40 kPa after 40 days. The excess pore pressure had reduced faster at the smear zone than in the undisturbed zone due to the proximity to vacuum source. The excess pore pressure of Thermal Vacuum PVD specimen in the undisturbed zone increased to 40 kPa and reduced very fast to -9. 1 kPa after only 6 days and then decreased to about- 43 kPa after 28 days. The vacuum preloading generates negative (suction) excess pore pressure equivalent to the applied vacuum pressure. This behavior is similar to the results of Indraratna et al. [22] which indicated that the assumption of constant vacuum pressure distribution over the soil surface and linearly decreasing vacuum pressure along the drain length. Moreover, Chai et al. [9] showed the measured and simulated excess pore pressure and vacuum pressure variation with time. The simulated values and dissipation rates were higher than the measured values.

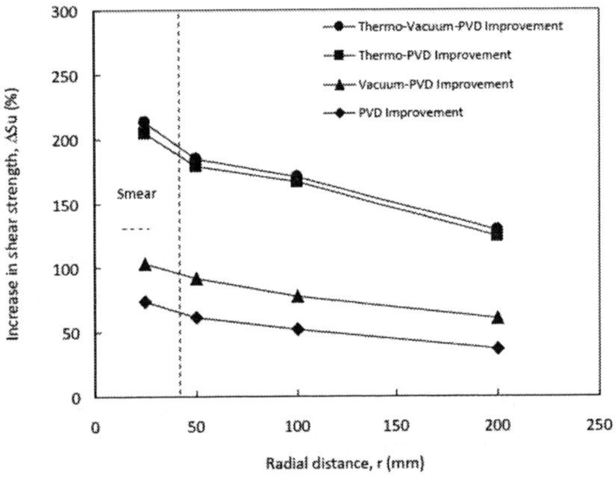

Figure 5. Comparison of shear strengths with radial distances after consolidation test

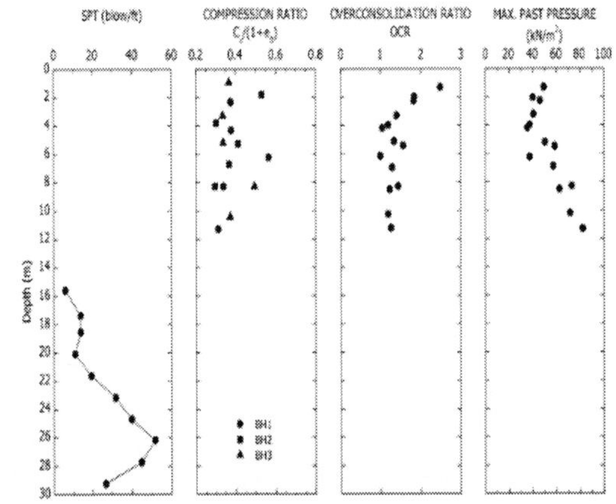

Figure 6. Soil parameters of SBIA project [3]

2.1.10. C_h and k_h/k_s values

The test results from reconstituted specimens in the large scale consolidometer were back-calculated to determine the values of C_h and k_h/k_s by using the method by Hansbo [2]. The back-calculated values of reconstituted specimen improved with Conventional PVD obtained Ch value of 1.93 m^2/yr with k_h/k_s of 3.0. For reconstituted specimen improved with Vacuum-PVD, the back-calculated Ch value consisted of 2.23 m^2/yr with k_h/k_s of 2.7. Similar results were observed by Mohamedelhassan and Shang [23] wherein the vacuum and surcharge preloading of the same magnitudes produced similar amounts of compression. Recently, Lam et al. [12] presented numerical and analytical analyses of field data from Conventional PVD and Vacuum PVD while Long et al. [13], [14] presented field results of Vacuum PVD with and without airtight membranes. Meanwhile, Chai and Rondonuwu [24] has investigated the effects of surcharge loading rate for minimizing the lateral displacement of Vacuum PVD improved soft clay.

Similarly, for the reconstituted sample in large consolidometer with heat (Thermo-PVD), the corresponding back-calculated C_h and k_h/k_s are 4.17 m^2/yr and 1.4, respectively . These values show that the thermal effects increased the permeability of the smear zone resulted in the decrease k_h/k_s values and increase in the Ch values. Due to the elevated temperature, the viscosity of the pore water decreased, and the soil permeability increased in the smear zone [6]. Moreover, for the reconstituted specimen improved with Thermal Vacuum PVD, the C_h and k_h/k_s are 4.38 m^2/yr and 1.1, respectively. The predicted curve agreed well with the observed values of reconstituted specimens in large scale consolidometer tests. In summary, the values of the C_h and k_h/k_s for all the specimens are also tabulated in Table 2. The percent increases in C_h of the reconstituted specimen improved with Vacuum PVD, Thermal PVD and Thermal Vacuum PVD consisted of 15.5%, 116.1% and 126.4% respectively. The reductions in k_h/k_s were 10.0%, 53.3% and 63.3%, respectively. Thus, the vacuum pressure can increase the horizontal coefficient of consolidation, C_h because the vacuum pressure generated negative pore water pressure along the drain, so the effective stress of soil increased which resulted in faster rate of settlement with the same magnitudes of settlement compared to Conventional PVD (See Figure 3). The high temperature can further increase the coefficient of horizontal consolidation, C_h with the reduction of k_h/k_s because the soil permeability increased and consequently reduced the drainage retardation effects in the smear zone around the PVD. Moreover, the soil elements also changed the volume of particle arrangement from the thermal effect with much faster rates of consolidation and higher magnitudes of settlement [17] (Abuel-Naga, et al., 2007).

Table 2. Summary of changes in flow parameters for reconstituted specimens using large consolidometer.

Flow parameters	PVD	Vacuum-PVD	Thermo-PVD	Thermo-Vacuum-PVD
C_h (m^2/yr)	1.93	2.23	4.17	4.38
k_h/k_s	3.00	2.70	1.40	1.10

2.2. Field test using PVDs with and without vacuum pressure

2.2.1. Site investigations and field construction

Both vacuum-PVD and conventional PVD system at Suvarnabhumi Airport, Thailand were reported by COFRA [25]. The soil profile at the site can be divided into 8 sublayers as shown in Table 3 and it consists of a 2.0 m thick weathered clay layer overlying a very soft layer which extends from 2.0 m to 10.0 m depth. Underneath the soft clay layer of 3.0 m thickness, a 3.0 m thick medium clay layer can be found. The light-brown stiff clay layer can be encountered at 15.0 m to 30.0 m depth. The groundwater level was found at about 0.50 m depth. In Table 3, the maximum past pressure (σ'_{vm}) is derived from the given OCR value taken as an average value of each layer. The soil profiles within the site are relatively uniform with some small variations in the soil thickness. The typical soil properties is also given in Figure 6 taken from Bergado et al. [3].

In the Conventional PVD method, the PVD was installed to 10 m depth with a spacing of 0.85 m and arranged in a triangular pattern. This method had the embankment height in 4.2 m with loading in 2 stages.

Typically, a 2:1 side slope was used for the low embankment with height less than 2.5 m. However, a 4:1 side slope was adopted for the high embankment to reduce the effect from erosion due to rainfall. The high embankment is usually constructed along with counterweight berms for stability purpose. The cross-section of the test embankment using conventional PVD method is shown in Figure 7. Several types of monitoring instruments were used, including settlement plates, settlement benchmark, deep settlement gauges, piezometers, inclinometers and observation wells. It is worth to mention that Abuel-Naga et al. [15] has devised a design chart for Conventional PVD incorporating varied soil permeabilities in the smear zone.

Table 3. Compressibility parameters at Suvarnabhumi Airport.

Type	Unit weight [kN/m^3]	Compressibility			POP (kPa)	C_v theory [m^2/yr]
		RR	CR	C_a		
weathered clay	18.50	0.035	0.350	0.014	45	-
very soft clay1	13.80	0.050	0.500	0.020	37	0.79
very soft clay2	14.00	0.042	0.420	0.017	59	0.79
soft clay	15.00	0.040	0.400	0.016	100	0.79
soft to medium clay	15.70	0.030	0.300	0.012	110	0.79
stiff clay1	18.50	0.008	0.080	0.003	300	-
stiff clay2	19.00	0.008	0.080	0.003	500	-
stiff clay3	20.40	0.000	0.000	0.000	500	-

VCM- DT method - Vacuum PVD method using direct connection method without airtight membranes, the PVD was installed into 10 m depth with a spacing of 0.85 m and arranged in a triangular pattern. The instrumentation equipments were installed to monitor the field behavior. For the vacuum-PVD, similar instrumentation equipments were installed to monitor the field behavior. The locations of the inclinometers, piezometers and the settlement plate are shown in Figures. 8a, b. The following boundary conditions were used in the design of vacuum-PVD: installation time of drains of 2 months, maximum pumping time of 8 months, vacuum pressure of -60 kPa at 5 m depth, depth or length of PVD of 10 m below ground surface and 60 % consolidation requirement. The embankment was 2.8 m high with unit weight of 18 kN/m^3. The embankment was constructed in two phases, namely: Phase 1 (1.5 m height, day 0) and Phase 2 (1.3 m height, day 14). The load-time relations of PVD improvement with and without vacuum preloading are shown in Figures 9a, b.

Herein, the Vacuum PVD system used the direct connection method without sealing geomembranes at the ground surface. Long et al. [13], [14] has presented full scale case histories of Vacuum PVD with and without airtight membranes. Chai and Rodonuwu [24] (2015) devised a surcharge loading rate to minimize the lateral displacement of vacuum PVD scheme in soft ground improvement. Moreover, Lam et al. [12] presents further analyses of field data from Vacuum PVD using numerical and analytical techniques. Figure 10 shows the data of vacuum pressures measured using the VCM-DT method.

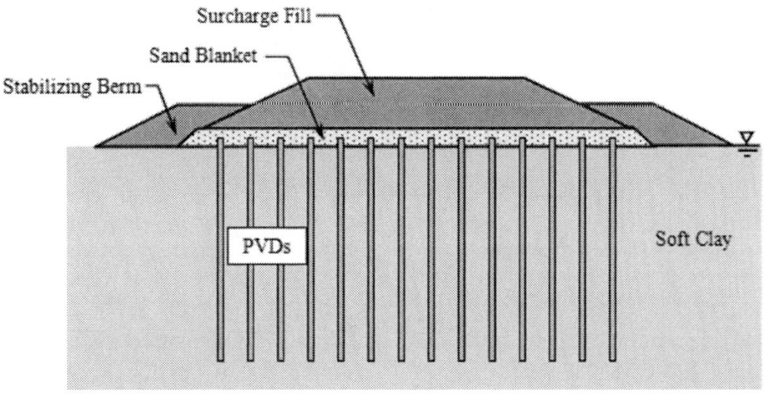

Figure 7. Cross-section of test embankment using conventional PVD method.

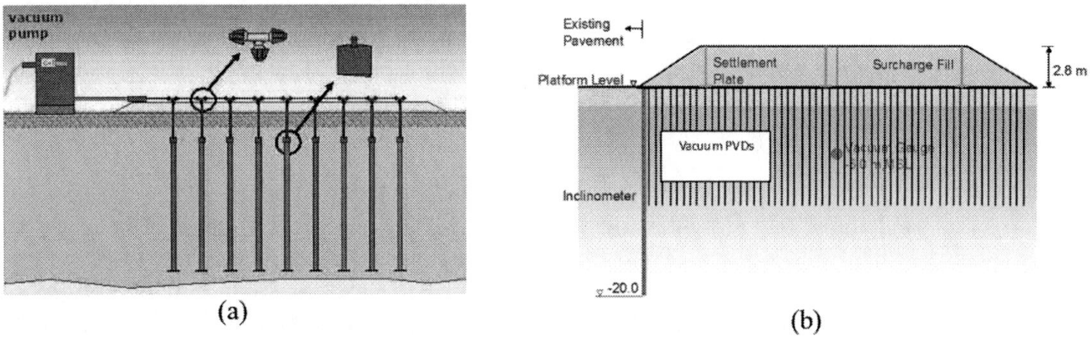

Figure 8. Vacuum PVD by direct connection method: (a) Cross-section and (b) Monitoring instruments.

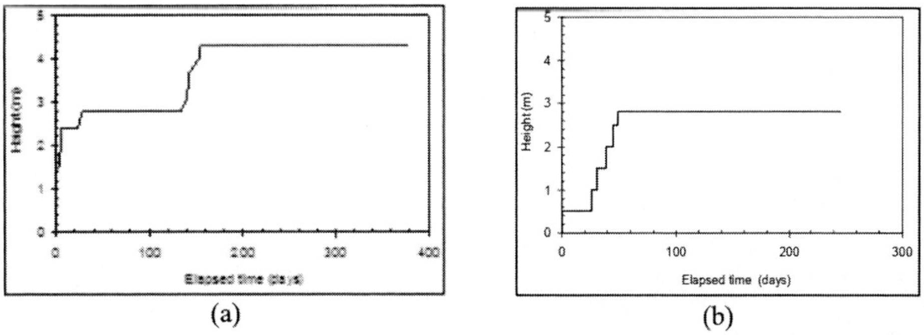

Figures 9. a,b The load-time relationships: (a) Conventional PVD method and (b) Vacuum PVD method.

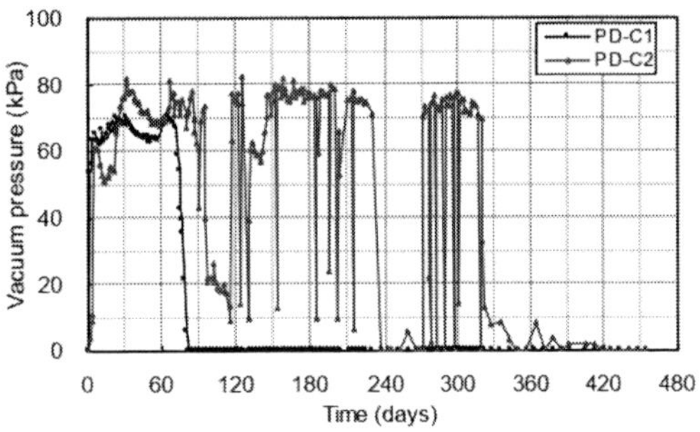

Figure 10. Measured vacuum pressure in PVDs at Sections C1 and C2 (VCM-DT method).

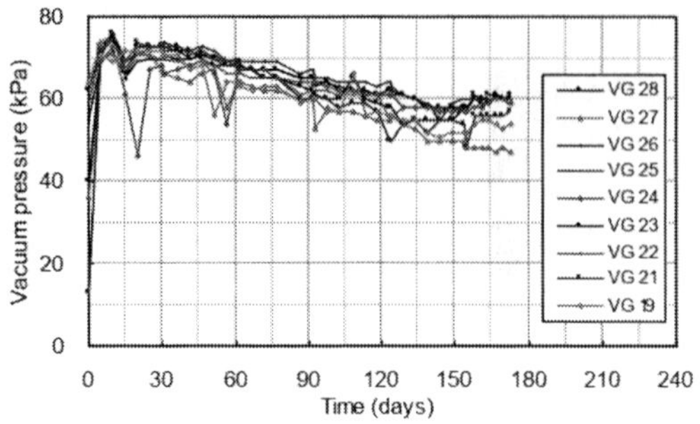

Figure 11. Measured vacuum pressure in sand blanket with airtight membrane (VCM-MS method).

2.2.2 Other methods of applying vacuum PVD preloading

VCM-MS method - The sand blanket is covered with an airtight membrane to enable the transfer of vacuum pressure from the sand blanket to the underlying soft soil through the PVDs. Vacuum pressures can be applied in the sand blanket through a sub-drain system using flexible perforated pipes placed at spacing of several rows of PVDs. Measured data of vacuum pressures in the sand blanket as given in Figure 11 indicated that the vacuum pressure at early stage was higher than 70 kPa and then gradually decreased to about 50 to 55 kPa at the end of preloading.

VCM-MB method - Vacuum pressure can be applied to PVDs horizontal drains using perforated pipe and horizontal band drains with airtight membrane (VCM-MB) as shown in Figure 12 . The measured vacuum pressures at the mid-depth of PVDs for two trial sections (D1 and D2) are presented in Figure 13 .It can be seen that the effective vacuum pressure of about 70 to 80 kPa was achieved for both sections in early stages. However, after about 120 days, the vacuum pressures in Section D2 was gradually decreased to about 20 kPa at the end of preloading (EOP).

VCM-MT method. For this method, horizontal drainage system under airtight membrane consisted of horizontal drainage using flexible, corrugated and perforated tubes connected to longitudinal main drains consisting of PVC or HDPE pipes . The PVDs were connected to horizontal pipes by winding with string tie (Figure 14a) and using 4-way or T-shaped connector for connection between horizontal drainage tube and longitudinal main pipe (Figure 14b). The measured vacuum pressures at the mid-depth of PVDs in various construction sections are presented in Figure 15. From this figure, it can be seen that the vacuum pressure in PVD was not dependent on the PVD lengths.

Figure 12. VCM-MB using airtight membrane and horizontal band drains.

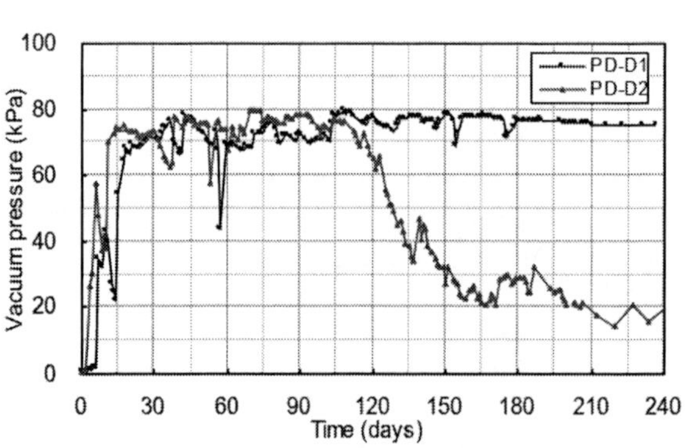

Figure 13. Measured vacuum pressures in Sections D1 and D2 (VCM-MB Method)

(a)

(b)

Figure 14. VCM-MT using airtight membrane and horizontal perforated tubes.

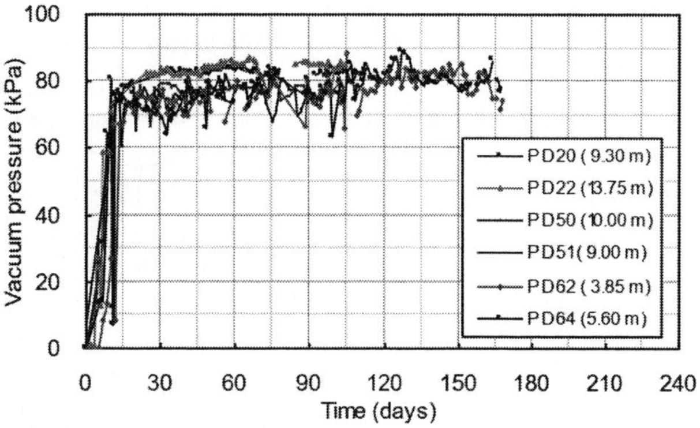

Figure 15. Measured vacuum pressures in PVDs in sections using VCM-MT method.

2.2.3. Settlement predictions of field test using PVDs with and without vacuum

The final settlement was calculated using the Asaoka [20] graphical method. This method is based on the field monitored data. The horizontal coefficient of consolidation, C_h, was also back-calculated at different periods depending on the time of PVD installation. Before the PVD installation, the vertical drainage was mainly assumed in the calculation of the degree of consolidation. After the PVDs installation, the horizontal drainage mainly governed in the calculation of the degree of consolidation. For the settlement analysis, the subsoil was divided into 5 layers (Table 3) and the whole surcharge fill was divided into different loading stages. The design parameters of PVDs with and without vacuum for settlement analysis are tabulated in Tables 4 and 5, respectively.

Table 4. Parameters for settlement analyses related to Vacuum PVD.

Item	Unit	Values
Drain Type; CeTeau drain type CT-D911		
Equivalent diameter of the drain, d_w =)b)t/(2	mm	51.75
Diameter of the equivalent soil cylinder, $D_e = 1.05S$	m	0.8925
Smear zone diameter, $d_s = 2d_m$	mm	191.49
Hydraulic conductivity ratio, k_h/k_s		2-10
Discharge capacity, q_w	m^3/yr	100

Table 5. Parameters for settlement analyses related to conventional PVD.

Item	Unit	Values

Drain Type; CeTeau drain type CT-D911		
Equivalent diameter of the drain, $d_w = (b(t)/2$	mm	51.50
Diameter of the equivalent soil cylinder, $D_e = 1.05S$	m	0.8925
Smear zone diameter, $d_s = 2d_m$	mm	186.00
Hydraulic conductivity ratio, k_h/k_s		2-10
Discharge capacity, q_w	m^3/yr	1000

2.2.4 Field test results using PVDs with and without vacuum pressure

The method of Asaoka [20] for prediction of settlement magnitudes and Hansbo [1] based on Equation (1) to (5) for prediction of settlement rates were combined together to analyze the field observation data of two stations of PVD improved soft ground by surcharge load and another two stations of PVD improved soft ground with surcharge load combined with vacuum preloading. The measured settlements of those stations were then compared with the prediction. The comparison of settlement behavior using PVD without and with vacuum are plotted with time in Figures 16a, b. The PVD with vacuum clearly indicates faster rate of settlements. The reduction of time to reach the 90 %degree of consolidation using surcharge load combined with vacuum pressure was higher than PVDs without vacuum pressure by about 1.4 to 1.5 times due to the higher values of horizontal coefficient of consolidation. Figure 17a shows the measured and predicted settlements at station X =14012, Y =11567-12633 by using PVD with surcharge load. The back-calculated C_h value was 2.03 m^2/yr with k_h/k_s values of 7.0. The final settlement predicted from Asaoka method was 1485.89 mm. Figure 17b show the measured and predicted settlements at station X =13560, Y =11567-12600. The C_h value was 2.31 m^2/yr with k_h/k_s values of 7.4. The final settlement from Asaoka [20] (1978) method was 1773.41 mm.

Similar procedures were used in the back-calculation of compressibility parameters from measured and predicted settlements at station X =12566.400, Y =12570.000 by using PVD with surcharge and vacuum preloading. The C_h value was 3.21 m^2/yr with k_h/k_s values of 6.2. The final settlement predicted by Asaoka [20] method was 1356.62 mm. The corresponding back-calculated values were also obtained from the measured and predicted settlements at another station X =12566.400, Y =11683.50. The back-calculated C_h value was 3.80 m^2/yr with k_h/k_s values of 7.1. The final settlement predicted by Asaoka [20] method of the station was 1614.94 mm.

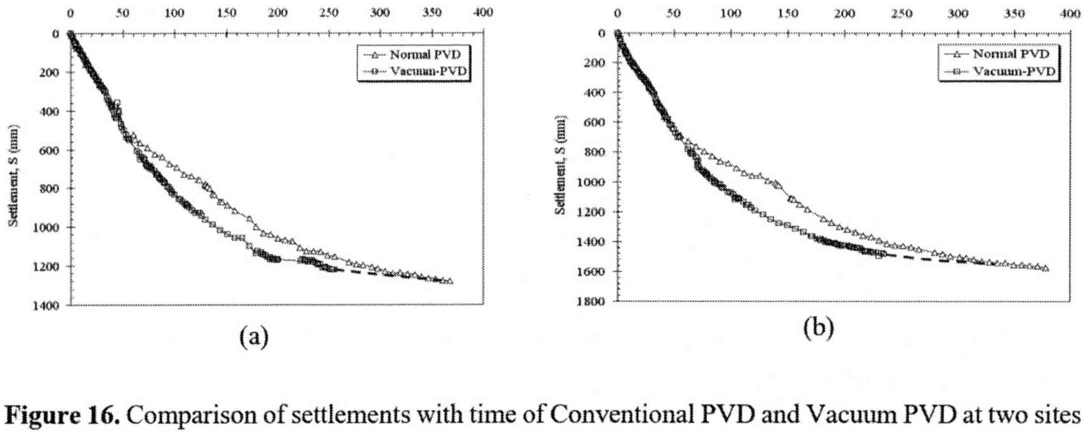

Figure 16. Comparison of settlements with time of Conventional PVD and Vacuum PVD at two sites in SBIA.

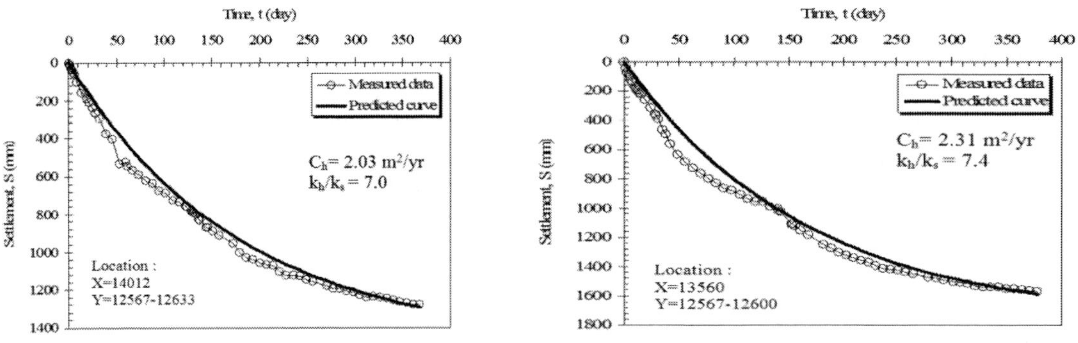

(a) (b)

Figure 17. Back-calculation of compressibility parameters from field settlement observations of conventional PVD.

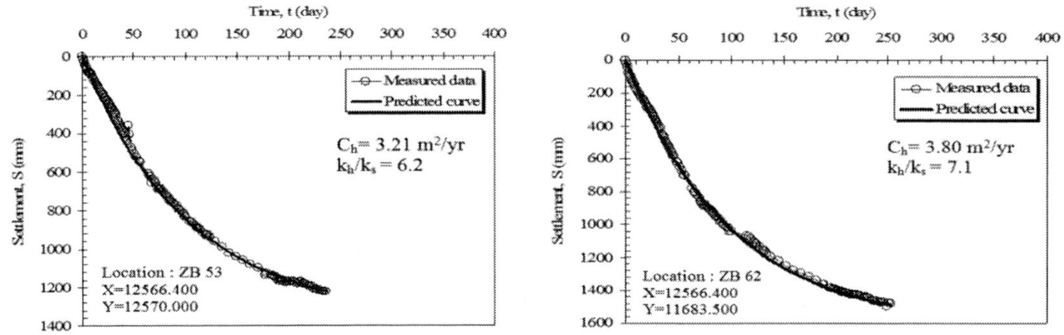

(a)

(b)

Figure 18. a, b Back-calculation of compressibility parameters from field settlement observations of Vacuum PVD.

Figure 18a shows the measured and predicted settlements at station X = 12566.400, Y = 12570.000 by using PVD with surcharge and vacuum preloading. The C_h value was 3.21 m^2/yr with k_h/k_s values of 6.2. The final settlement predicted by Asaoka [20]method was 1356.62 mm. Figure 18b shows the measured and predicted settlements at another station X = 12566.400, Y = 11683.50. The back-calculated C_h value was 3.80 m^2/yr with k_h/k_s values of 7.1. The final settlement predicted by Asaoka [20] method of the station was 1614.94 mm.

2.2.5 PVDCON method

In PVDCON, finite element formulation considers the effects of PVDs by modifying 1D continuity equation of consolidation as follows [26].

$$\frac{k_v}{\gamma_w}\frac{\partial^2(u-p_{vac})}{\partial z^2}-\frac{8k_h(u-p_{vac})}{\gamma_w D_e^2\mu}+\frac{\partial\varepsilon_v}{\partial t}=0 \tag{6}$$

$$\mu=\ln\frac{n}{s}+\frac{k_h}{k_s}\ln s-\frac{3}{4}+\pi\frac{2l^2k_h}{3q_w} \tag{7}$$

where γ_w is the unit weight of water, z is depth, t is time ε_v is volumetric strain, u is excess pore water pressure, p_{vac} is the final vacuum pressures, k_v is hydraulic conductivity in the vertical direction, k_h is hydraulic conductivity in the horizontal direction, l is drainage length, D is the diameter of unit cell, q_w

is discharge capacity of PVD, n=D/d_w (d_w is the equivalent diameter of PVD) and s=d_s/d_w, k_h, k_s and d_s were defined previously.

For Conventional PVD, at OCR of 1.2, the k_h/k_s of 4 to 6 and C_h=2C_v were obtained where C_v = 1.58 m²/year as shown in Figure 19. For Vacuum PVD with vacuum pressure at -60 kPa, and OCR of 1.2, the k_h/k_s of 4 to 8 and C_h=3C_v were obtained where C_v = 1.58 m²/year as shown in Figure 20. In summary, the addition of vacuum pressure to PVD increased the coefficient of horizontal consolidation, C_h as expected. The other parameters may not be affected that much, particularly, the permeability ratio, k_h/k_s, because the vacuum preloading seems to increase both the permeabilities of the smeared and the undisturbed zones.

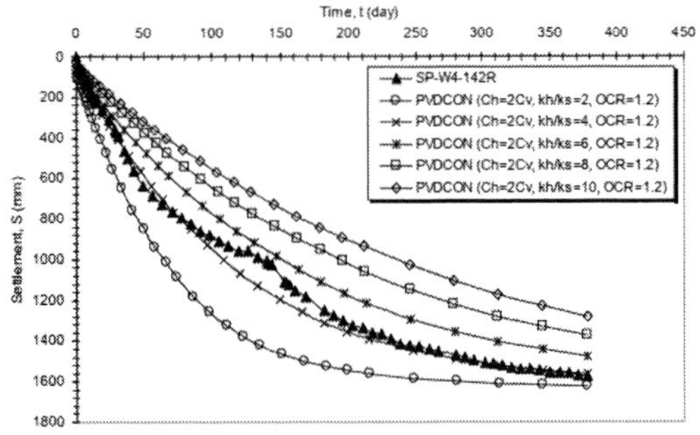

Figure 19. Observed and predicted settlements using PVDCON of Conventional PVD.

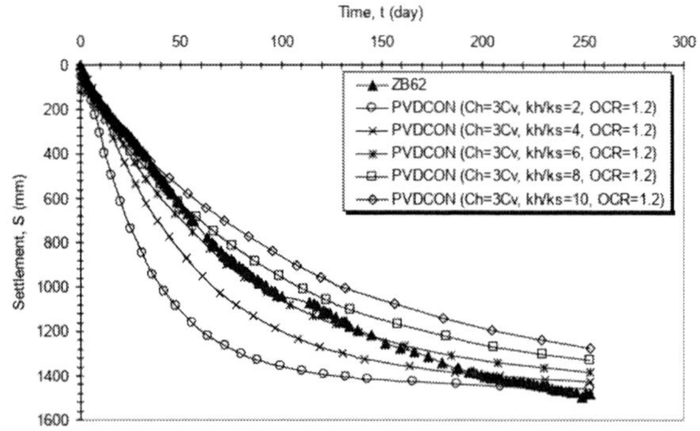

Figure 20. Observed and predicted settlements using PVDCON of Vacuum PVD.

2.3. Field test using PVDs with and without heat

2.3.1. Site investigations and field construction

The embankment tests site was located inside the campus of the Asian Institute of Technology (AIT) which is 42 km north of Bangkok, Thailand. This AIT campus is located within the Central Plain of Thailand which is a flat with deltaic-marine deposits widely known as "soft Bangkok clay ." The soil profile at the site can be divided into 3 sublayers as shown in Figure 21. It consists of a 2.0 m thick clay backfill layer overlying soft clay layer which extends from 2.0 m to 9.0 m depth. Underneath the soft clay layer, a 2.0 m thick medium clay layer can be found. The groundwater level was found at about 2.0 m depth [27].

Embankments improved without and with heat using solar power were constructed in the dimensions of 11 m x 11 m at the bottom and 3 m x 3 m at the top with 6 m height as shown in Figure 22. The PVD/PVTD (Thermal PVD) was installed into 8 m depth with a spacing of 1.0 m and arranged in a square pattern. The embankment fill material (Ayuttaya sand) was properly compacted. Geogrid, gabion cages and geotextile were used for the slope protection. The instrumentations were installed to monitor the performances of embankments improved without and with heat using solar power. The following instrumentations were installed, namely: 2 piezometers (below 3.0 m and 6.0 m from ground surface level), 5 settlement plates (below 0.0, 3.0, and 6.0 m from ground surface level) and 4 thermo-couples (below 3.0 m and 6.0 m from ground surface level for PVTD embankment). The PVTD (Thermal PVD) unit consists of a U-tube made of cross-linked polyethylene plastic (PEX) attached to both sides of conventional PVD (Figure 23). Preheated water at about 70 to 90°C was stored in the tank and then, was circulated through the attached U-tube by water pump to raise the soil temperature underneath the embankment. A solar panel system was used to heat the circulated water from ambient temperature (25°C) to 72°C. Then an electrical heater was utilized to raise its temperature from 72°C to 90°C, especially during cloudy day and at night.

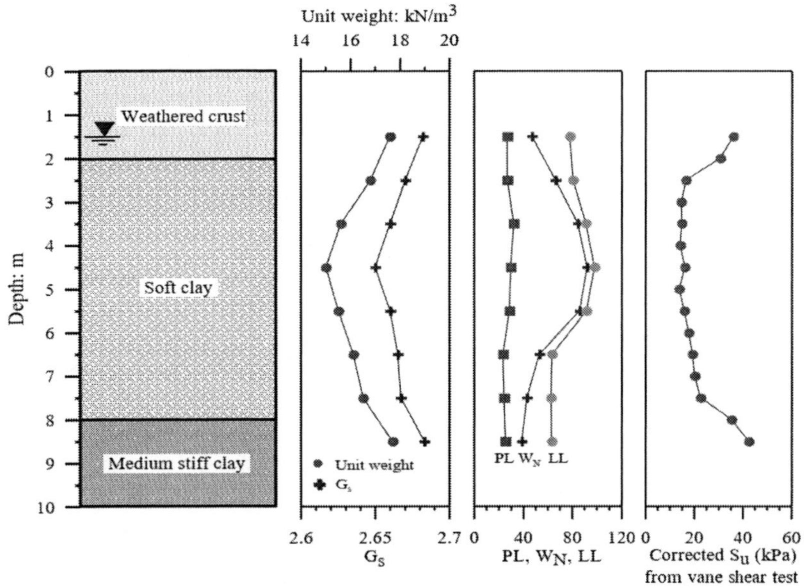

Figure 21. Soil properties at the site (AIT Campus, Thailand).

2.3.2 Field test results using PVDs with and without heat

The settlement results of surface and deep settlements of two embankments with Conventional and Thermal PVD are shown in Figure 24a. Subsequently, the comparison of surface settlements from the full scale embankment improved without and with Thermal PVD is shown in Figure 24b. For the Conventional PVD embankment, the final settlement was 262.20 mm. On the other hand, the thermal PVD embankment indicated larger final settlement of 407.62 mm which had an increase of about 55%. These results demonstrate that the higher the temperature, the higher in volume contraction and the higher rate of consolidation. Thus, Thermal PVD has achieved higher settlement and also faster rate of consolidation than the Conventional PVD [15]. The compressibility increases due thermally induced irreversible contraction of underlying saturated normally consolidated (NC) soft clay. Furthermore, Abuel-Naga et al. [15] indicated that the Thermal PVD resulted in substantial amount of settlement as much as 5 times using undisturbed specimen of soft Bangkok clay in the laboratory model test. Moreover, the compression line of soft Bangkok clay shifts to the left with approximately similar slope upon heating when thermal loading is applied at normally consolidated state. Furthermore, the normally consolidated specimens contracted irreversibly and non-linearly under drained heating condition [17].

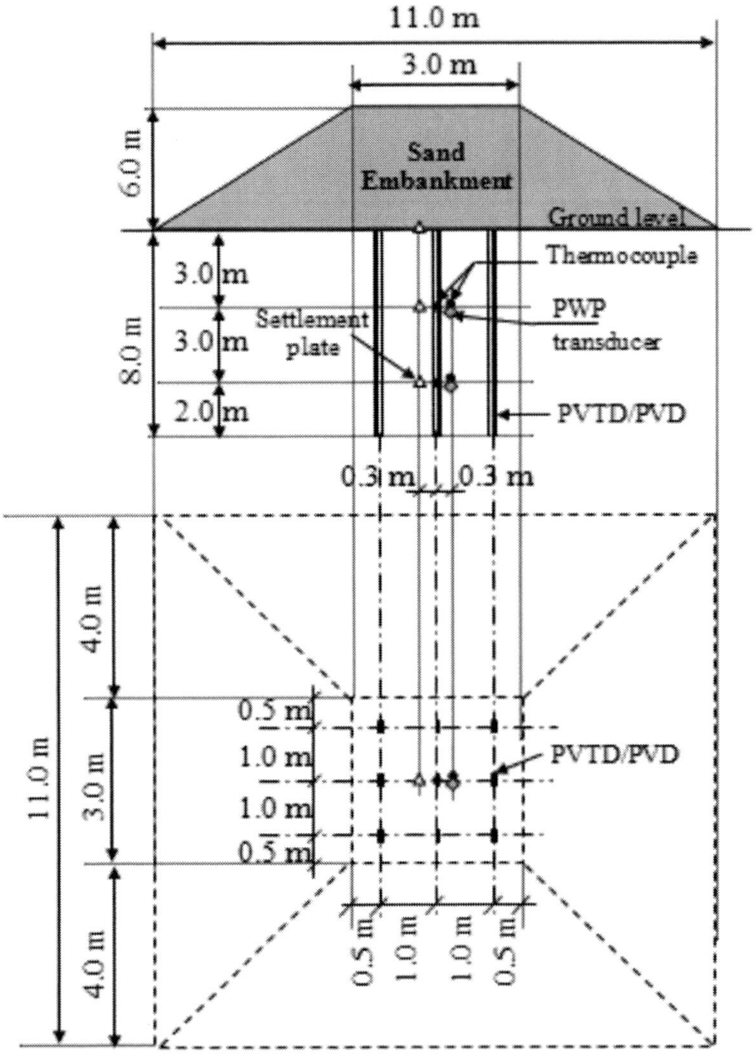

Figure 22. Layout of test embankment with thermal PVD.

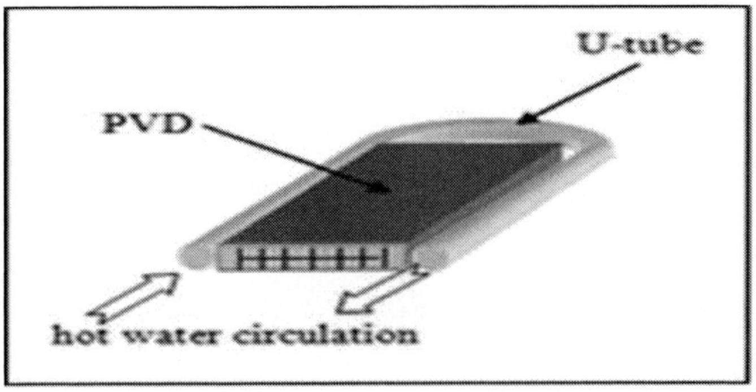

Figure 23. Configuration of Thermal PVD.

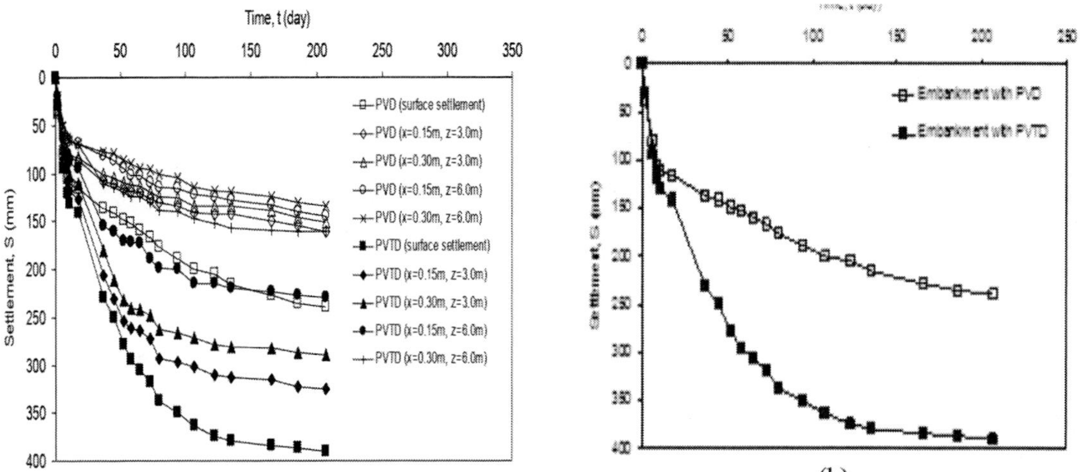

(a) (b)

Figure 24. Comparisons of Conventional and Thermal PVDs surface settlements.

The excess pore pressures were measured at the embankments improved with PVD and PVTD as shown in Figure 25. For PVD embankment, the excess pore pressure had increased to maximum of 30 kPa but after 200 days decreased to about 7 kPa. For PVTD embankment had demonstrated excess pore water pressures to 30 kPa then after 200 days it reduced to 2 kPa due to the effects of thermally induced pore water pressures and volume changes. This behavior can be explained that the increase in hydraulic conductivity at elevated temperature as the result of the decrease in water viscosity. The temperature-radial distances relationship at different time of the PVTD embankment at depth of 3 m was monitored. The initial temperature was measured which showed constant temperature of 29°C at time of 0 day with T/T0 value of 1.16 based on the standard temperature (T_0) is 25°C which is the reference temperature. The steady state condition of temperature was obtained after about 60 days and the soil temperature becomes almost constant (T/T_0=3.0) at the radial distance r/r_w of 6, where r_w is the equivalent drain radius of the Thermal PVD (PVTD) (r_w =26 mm).

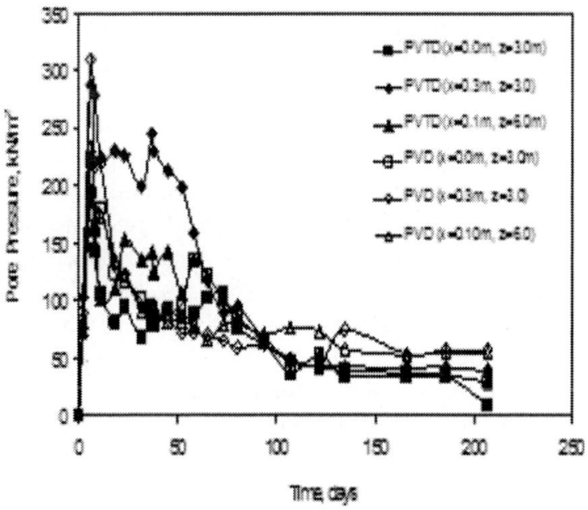

Figure 25. Excess pore pressures for Conventional and Thermal PVDs.

The measured and predicted time-settlement curves for the embankment with conventional PVD and thermal PVD are shown in Figure 26 a,b. The values of Ch and kh/ks for the embankment with PVD only were obtained as 6.7 m^2/yr and 6.2, respectively. For the embankment with PVTD, the corresponding values were 8.5 m^2/yr and 4.1, respectively. Consequently, the use of PVTD increased the permeability of the smear zone resulting in the decrease in k_h/k_s of about 33.9% and an increase in C_h by 27.2%. Therefore, the increase in the average horizontal hydraulic conductivity beyond the smear zone is 8.5/6.7 =1.27 times. The elevated temperature increased the horizontal permeability in the smear zone to 1.9 times as much as the horizontal permeability in the smear zone without heat.

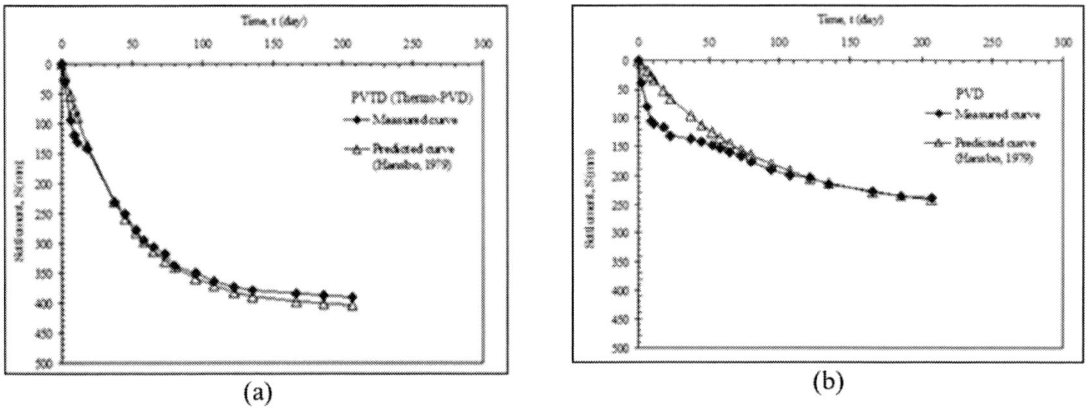

Figure 26. a, b Back-calculation of C_h values from observed settlements (a) Thermal PVD and (b) Conventional PVD.

Thus, the temperature effect on hydraulic conductivity of smear zone is higher due to the gradient behavior of heat transfer around linear heat source which generate higher temperature that coincides approximately with the smear zone [15]. Therefore, it can be concluded that Thermal PVD (PVTD) system offers a solution to partially remove the smear effect and, consequently, accelerate the consolidation process .

2.3.3 Smear effect and back-calculated C_h value

The rate of consolidation of PVD improved ground depends on PVD spacing, well resistance (F_r), and smear resistance (Fs). Well resistance is a function of PVD length and the ratio of horizontal permeability to PVD discharge capacity, kh/qw. The value of kh/qw is often smaller than 0.0001 for most PVDs in practical cases. Thus, the value of well resistance Fr becomes negligible in comparison with the values of Fn and Fs. Balasubramaniam et al. [28], Bergado et al. [3] and Long et al. [14] also indicated that the well resistance has very little effect when the in-situ discharge capacity of PVD greater than 50 m^3/year as seen in Figure 27. Therefore, with a known value of PVD spacing, the back-calculated C_h value mainly depends on the assumed values of R_s= k_h/k_s, and d_s/d_m. Moreover, with an assumed value of smear zone, ds/dm, the back-calculated C_h value is directly proportional to the value of R_s as seen in Figure 28 [29] (Long et al., 2013). From this figure, it can be seen that the back-calculated value of C_h at SBIA can be varied from 1.6 m2/year to higher than 5m2/year depending on the assumed values of kh/ks and ds/dm. This can be explained for the different values of Ch for Nong Ngu Hao soft clay as back-calculated by various researchers [28], [3], [29].

Regarding to the smear effect under vacuum consolidation, back calculated results conducted by Long et al. [14] illustrated that the conductivity ratio, $R_s = k_h/k_s$, in vacuum consolidation is about 1.33 times smaller than that in conventional surcharge preloading. This is reasonable because under vacuum consolidation with high hydraulic gradient at the soil/PVD interface, the clogging and blinding in the PVD filter can be improved and soil particles in smear zone around PVD might be re-arranged for higher conductivity in smear zone, k_s.

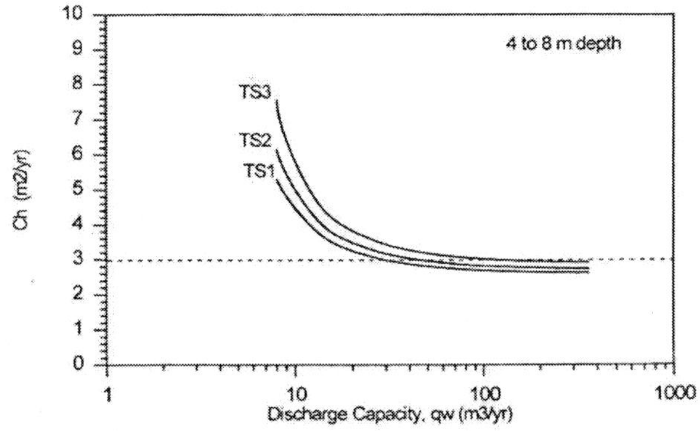

Figure 27. Back-calculated C_h value of test embankments at SBIA site using $d_s/d_m = 2$ and $k_h/k_s = 5$. [3] (Bergado et al., 2002)

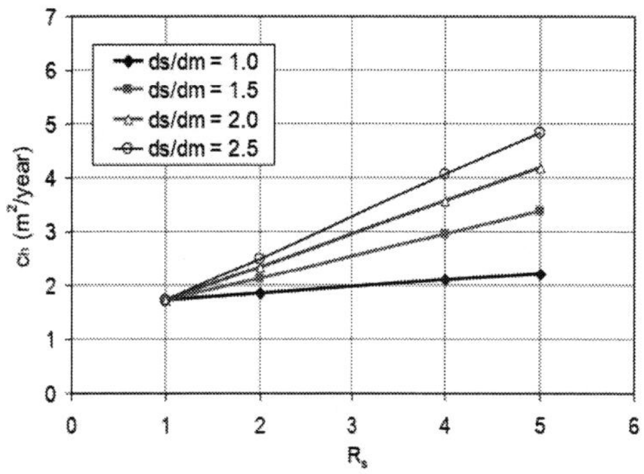

Figure 28. Back-calculated c_h values at TS3 – SBIA. [29] (Long et al., 2013)

2.3.4 Equivalent smear effect due to non-uniform consolidation

Zhou and Chai [30] conducted finite element analyses (FEA) on the behavior of PVD unit cell considering only horizontal drainage and considering local variation of permeability within the unit cell. Both (a) equal stress (free strain) and (b) equal strain in the top boundary were simulated. Taylor (1948) equation was adopted for variation of permeability (k) with void ratio (e) as follows:

$$k = k_o \cdot 10^{-(e_0 - e)/C_k} \tag{8}$$

From void ratio and initial value of k, the distributions of k at different degree of consolidation (DOC) can be calculated using Equa. (8). As illustrated in Figure 29, designate the permeability (k) at the periphery of the unit cell as k_m and at a radial distance, r, as k_r and the permeability ratio, $n=k_m/k_r$. The variation of n with radial distance is shown in Figure 30. The value of n dramatically varies within r = 100 mm and its value can be high at the early stage of consolidation. With the increase in DOC, the n value gradually decreases.

The average degrees of consolidation (DOC) are shown in Figure 31 from the results of FEA for both free and equal strain conditions where not much difference between them. The results from Hansbo (1981) without smear is also plotted for comparison. Considering a smear zone diameter, d_s, to be 4 times the equivalent diameter, d_w, of a PVD with d_s=0.21, the back-fitted $(k_h/k_s)_e$ of 2.4 is obtained. If the mechanical smear effect is denoted as $(k_h/ks)_m$ and the equivalent smear as $(k_h/k_s)_e$, then the overall smear effect, k_h/k_s, can be expressed as follows:

$$k_h/k_s = (k_h/k_s)_m \times (k_h/k_s)_e \tag{9}$$

Most back-fitted field values of k_h/k_s were much larger than the laboratory measured values. By separating the smear effects into mechanical component and non-uniform consolidation component, the mechanical component can be evaluated reliably. For example, the back-fitted field value of smear effect in Saga Airport, Japan was found to be k_h/k_s = 10 by Chai and Miura [26]. However, the corresponding laboratory value obtained in the laboratory by Madhav et al. (1993) was only half at $(k_h/k_s)_m$= 5. Assuming the above-mentioned $(k_h/k_s)_e$= 2.4 can be used, then the overall k_h/k_s = 12 is obtained which is close to the back-fitted value of 10. Similar values were obtained in similar situation in PVD improvement of soft Bangkok clay.

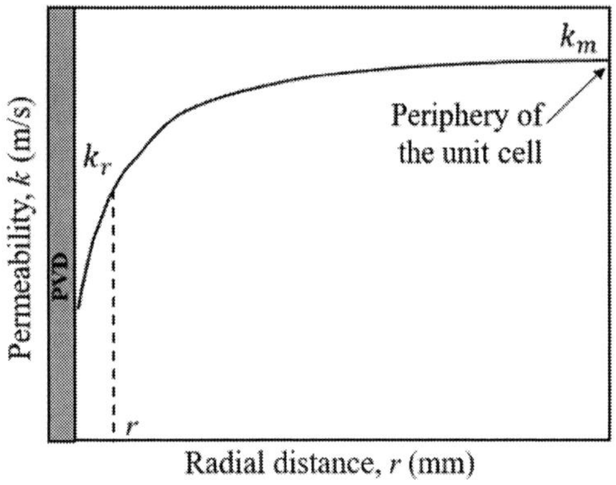

Figure 29. Illustrations of locations of k_r and k_m [30]

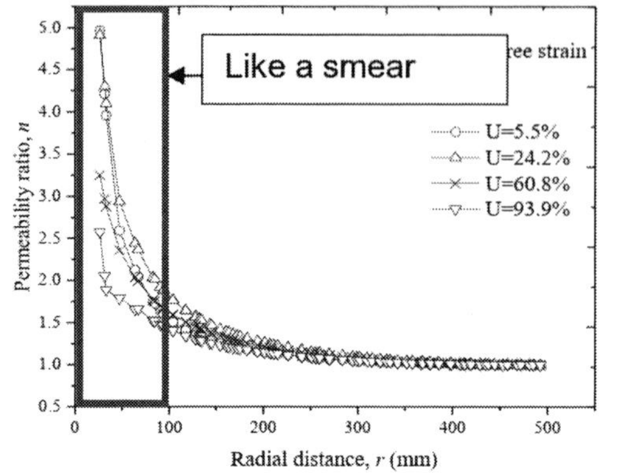

Figure 30. Permeability ratio variation with radial drainage [30]

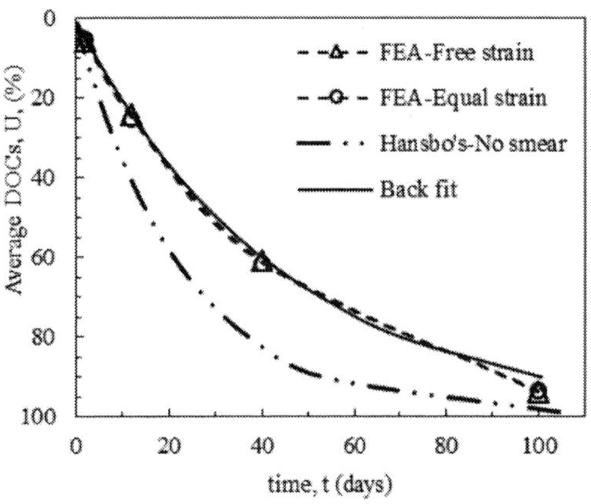

Figure 31. Comparison of average degree of consolidation (DOC) [30]

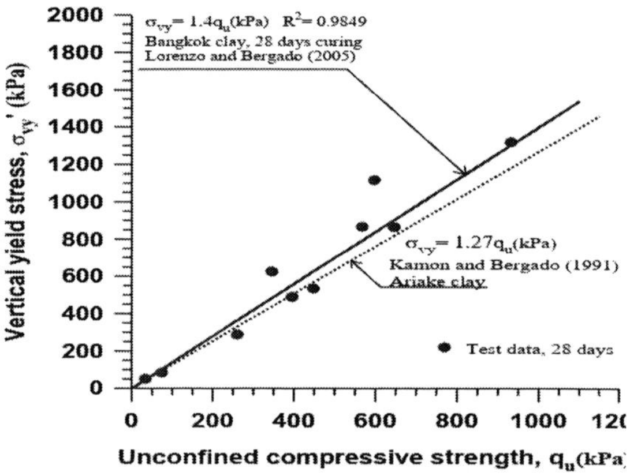

Figure 32. Relation between unconfined compression strength and vertical yield stress.

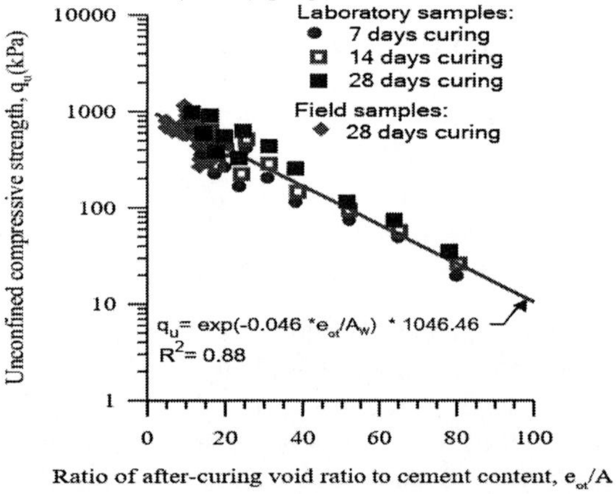

Figure 33. Unconfined compression strength versus ratio of after curing void ratio to cement content.

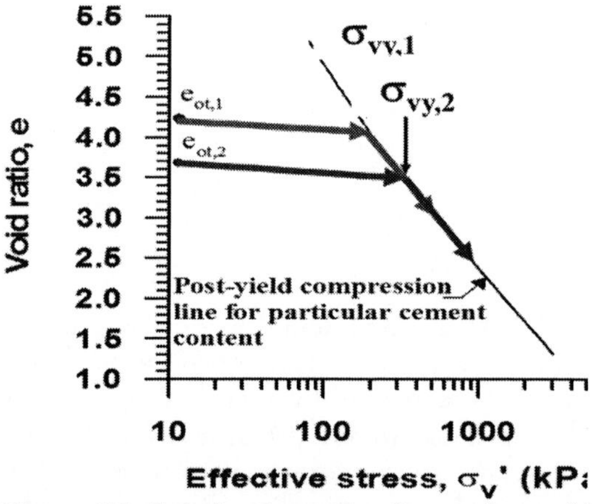

Figure 34. Relation between after curing void ratio and effective stress.

3. Deep cement mixing (DCM) and stiffened DCM

3.1 Prediction of shear strength and compressibility for DCM

For cement admixed clay, the shear strength and compressibility can be predicted in the laboratory tests. Figure 32 shows the relationship between the unconfined compression strength and the vertical yield stress. For the prediction of shear strength, the ratio of after curing void ratio and cement content can be correlated with the unconfined compression strength as indicated in Figure 33. Afterwards, the data on the vertical yield stress and after curing void ratio can be combined to predict the compressibility once the post-yield compression line is established for a particular cement content (Figure 34).

3.2 Effect of curing stress on the strength of DCM

Jongpradist [31] et al (2011) investigated the influence of curing stress on one-dimensional yielding of cement-admixed clay. The curing stress compressed the treated clay and, consequently, increased the vertical yield stress. However, the curing stress had no significant influence in the location on the post-yield compression line (Figure 34). The curing stress reduced the after curing void ratio, and subsequently, increased the shear strength of the cement-admixed clay (Figure 33).

3.3 Stiffened Deep Cement Mixing (SDCM)

Although DCM pile has many advantages with various applications, failure caused by pile failure can occur especially when subjected to the lateral loads. Moreover, the unexpected lower strength than the design commonly occurs due to lack of quality control during construction. DCM pile can fail by pile failure mode which is lower than the soil failure mode particularly at the top of DCM pile due to low strength and stiffness as shown in Figure 35a. In addition, as shown in Figure 35b, from the test results by Petchgate et al. [19], about half of DCM piles failed by pile failure. Consequently, the bearing capacity of DCM pile can be lower than the design load of 10 tons. Both pile and soil failures are illustrated in Figure 35a.

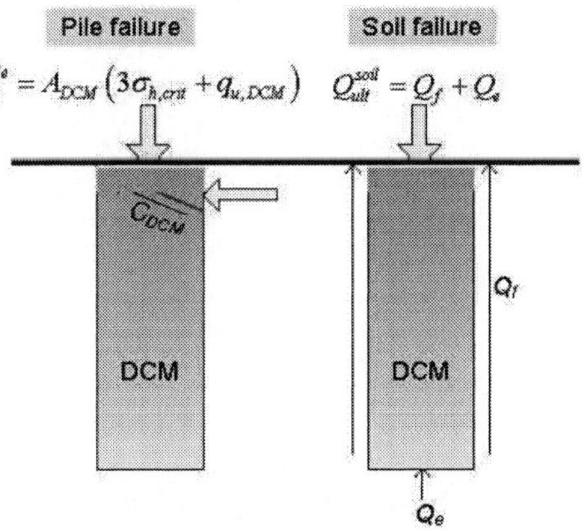

(a)

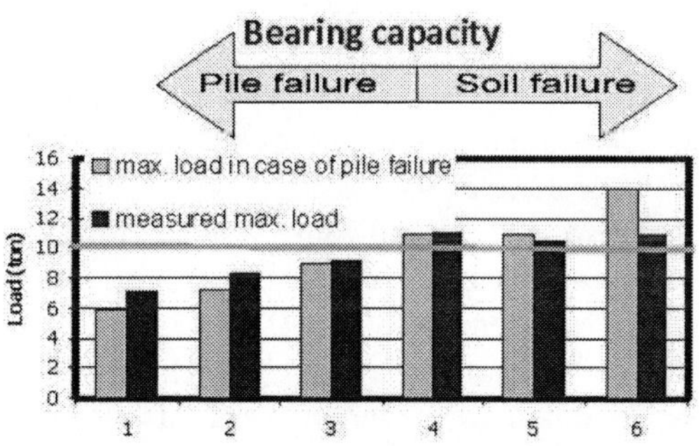

(b)

Figure 35. Bearing capacity of DCM piles on soft Bangkok clay (Petchgate et al., 2003).

Stiffened Deep Cement Mixed (SDCM) piles are a composite structure of concrete piles and deep cement mixed piles combining the advantages of both components as shown in Figure 36a,b. A pre-stressed reinforced concrete (RC) core is inserted into the center of a DCM pile immediately after the construction by wet mixing method. The two parts of the composite piles work together by supporting and transferring the vertical load effectively to the DCM pile and to the surrounding soil. In the SDCM pile, the DCM pile forms the surrounding outer layer supporting the RC core pile increasing its stiffness and resisting compressive stress along the pile.

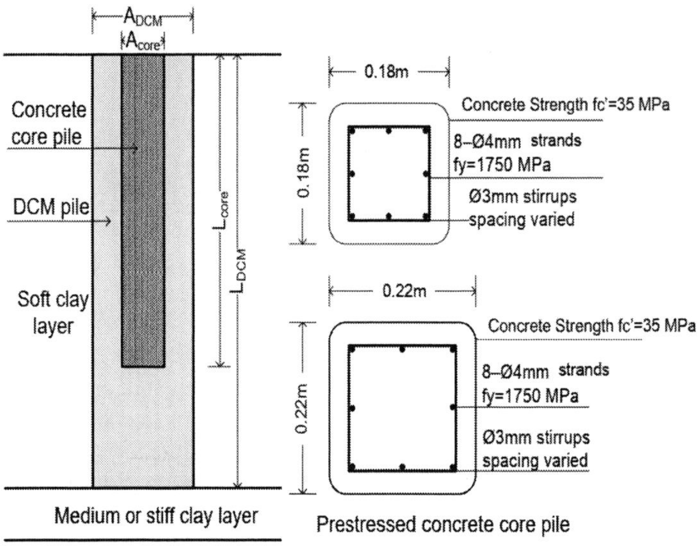

Figure 36. (a) Schematic of SDCM pile, (b) Details of prestressed concrete core piles.

3.4 Full scale field load and embankment tests

The full scale axial and lateral pile load tests were performed by Shinwuttiwong [32] and Jamsawang [33] and the full scale embankment load test was conducted by Jamsawang [33] within the campus of Asian Institute of Technology (AIT). The site is situated in the Central Plains of Thailand famous for its thick layer deposit of soft Bangkok clay. The foundation soils and their properties at the site are shown earlier in Figure 21. Other parameters are shown in Table 6. The strength of the concrete piles was found to be 35 MPa. Two lengths of concrete core piles were used in the field test, namely: 4.0 m and 6.0 m. However, for the numerical simulation the length of the concrete pile was varied from 1.00 m to 7.00 m with 1.0 m increase to evaluate the effect of the lengths of the core pile on the capacity of the SDCM pile. The Mohr-Coulomb model was recommended to simulate for mass concrete core pile instead of linear elastic model because its stiffness can be overestimated if the tensile strain is large enough to crack the concrete [34].

Jamsawang [33] and Jamsawang et al. [35], [36] constructed the full scale test embankment on improved soft Bangkok clay using two different method namely: stiffened deep cement mixing (SDCM) pile and deep cement mixing (DCM) pile. The DCM pile consisted of 7m long and 0.6 m in diameter. The objectives of this research work were to investigate ground improvement performances under embankment loading and to verify the related design parameters. Surface settlements and lateral movements were monitored during and after the embankment construction for two years. Figure 37 shows the plan layout of the embankment together with the DCM and SDCM piles. Figure 38 indicates the section view of the test embankment.

3.5 Numerical simulations using PLAXIS 3D Foundation

The embankment is supported by two types of piles consisting of the 16-SDCM piles and 16-DCM piles. For the purpose of simulation, the length of concrete core piles in SDCM piles were varied from 3.00 to 7.00 m with varied sectional dimensions from 0.22x0.22 to 0.30x0.30 m. The embankment discretization model followed the standard procedures of PLAXIS Foundation 3D software (Brinkgreve and Broere, 2006). The soil parameters and models used in the numerical simulations are tabulated in Table 6. The soft soil model *(SSM)* was used for the soft clay layer and the Mohr-Coulomb model *(MCM)* was used for other elements including DCM and SDCM piles. The basic soil elements were represented by 15 node wedge elements except the plate and interface elements. The DCM pile was modeled by volume elements that can simulate deformation stresses. The prestressed concrete core pile was modeled as "massive pile" composed of volume elements. The interface elements were modeled as pairs of corresponding nodes with zero distance between each pair. Interface elements required strength reduction factor, R_{inter}, for soil strengths mobilized at the interface. R_{inter} varies from 0.35 to 0.43 with average of 0.40 [36], [37], [38]. The first phase was the initial stage that was set-up as the in-situ state (k_0 procedure) to generate the initial in-situ stresses. In the second phase, the DCM pile and concrete core pile were installed. Next step was the excavation stage of the uppermost 1.00 m of soil. The subsequent steps consisted of filling with compacted silty sand at the first phase at the base and, subsequently, filling by compacted weathered clay. Afterwards, the surface settlement at the top of SDCM, DCM and surrounding soil were checked after 60, 90, 120, 150, 180, 240, 300, 360, 420, 510 and 600 days, respectively.

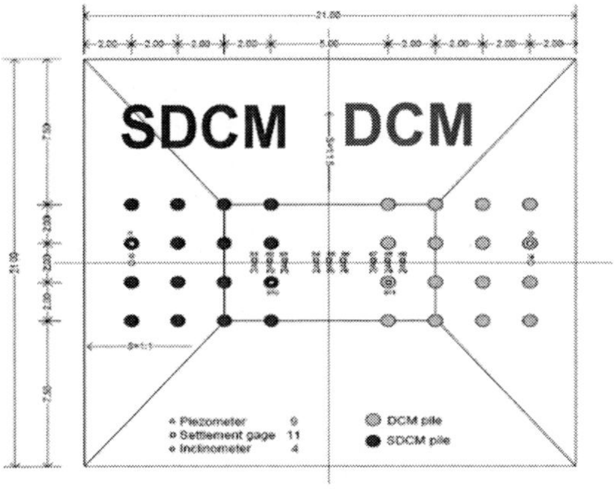

Figure 37. Plan view of the full scale test embankment test.

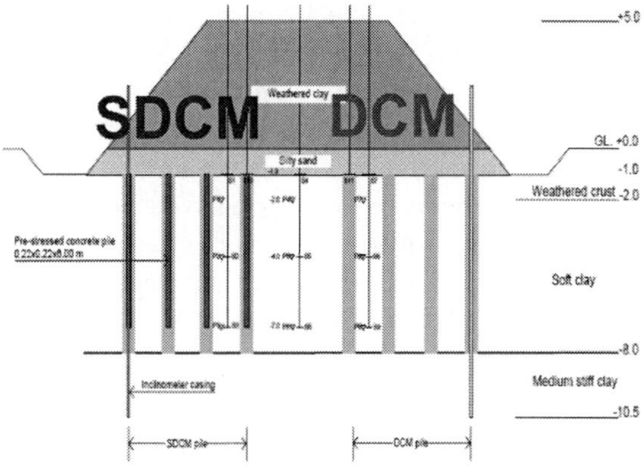

Figure 38. Section view and location of instruments of the SDCM embankment.

Table 6. Parameters for numerical simulation.

Materials		Model	γ (kN/m³)	Material Behavior	E'_{ref} (kPa)	ν	λ^*	κ^*	c' (kPa)	ϕ' (deg)	k_x (m/day)	OCR	Tensile Strength (kPa)
Subsoil	Depth (m)												
Weathered crust	0-2.0	MCM	17	Undrained	2500	0.25			10	23	1x10^{-3}		
Soft clay	2.0-8.0	SSM	15	Undrained			0.10	0.02	2	23	4x10^{-4}	1.5	
Medium stiff clay	8.0-10.0	MCM	18	Undrained	5000	0.25			10	25	2x10^{-4}		
Stiff Clay	10.0-30.0	MCM	19	Undrained	9000	0.25			30	26	4x10^{-4}		
Foundation													
Concrete core pile		MCM	24	Drained	2.8x10^{7}	0.15			8000	40			5000
DCM pile (with interface elements)		MCM	15	Undrained	30000-60000	0.33			100-300	30	0.012	0-100	
Steel plate		LEM	-	Non-porous	2.1x10^{8}	0.15							

3.6 Results of full scale field tests, analyses and discussions

The surface settlements were measured at the top of DCM, SDCM piles and the unimproved ground in the middle of the embankment)untreated clay .(The observed settlements are plotted in Figure 39 together with the simulated values .Both the magnitude and rate of settlements from simulations agreed well with the observed data from field tests as illustrated in Figure 39 . Consequently, the parameters involved were derived and verified .The parametric study was further conducted by varying the sectional areas of the concrete core pile to 0.22 x 0.22m and 0.30 x 0.30m as well as varying the lengths of concrete core piles to 4, 5, 6 and 7m to study their effects on the embankment settlements . The effects of lengths and sectional areas of the concrete core piles of SDCM piles on the ultimate settlement of embankment simulation are illustrated in Figure 40 . It can be summarized that the ultimate settlement at 600 days after consolidation proportionally decreased with increasing lengths of concrete core piles from 4 to 6m and only slightly decreased from lengths of 6 to 7m. Moreover, the ultimate settlement only slightly decreased when increasing the sectional areas of the concrete core piles from 0.22 to 0.30m .

Figure 41 shows the summary of the effect of core pile length on the settlement at 600 days after consolidation in surrounding clay of SDCM pile .The settlement of surrounding clay of SDCM at surface and 4m depth decreased with increasing the lengths of concrete core pile and only slightly decreased with increasing the sectional areas .Therefore, it can be concluded that the ultimate settlements proportionally reduced with increasing lengths of concrete core pile .In addition, both the sectional area and length of concrete core pile have no effect in the subsurface settlement at 7m depth.

Differential settlements occur in the subsurface at various depths because the stresses proportionally decreased from the surface to the depths of 4 and 7m, respectively .Moreover, the stresses in the surrounding clay of SDCM and DCM piles as well as the unimproved zone are plotted together in Figure 42 .The stresses of surrounding clay of SDCM is the lowest meaning that lowest settlement occur at the surface and 4 m depth .For the 7 m depth, the stresses are only slightly different and the settlements were similar.

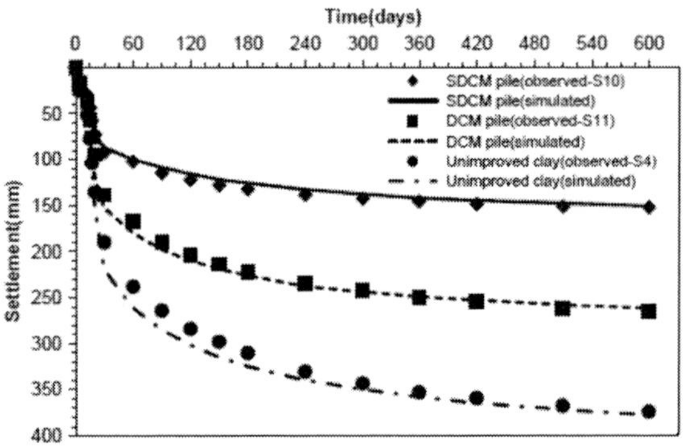

Figure 39. Comparison of observed and simulated surface settlements.

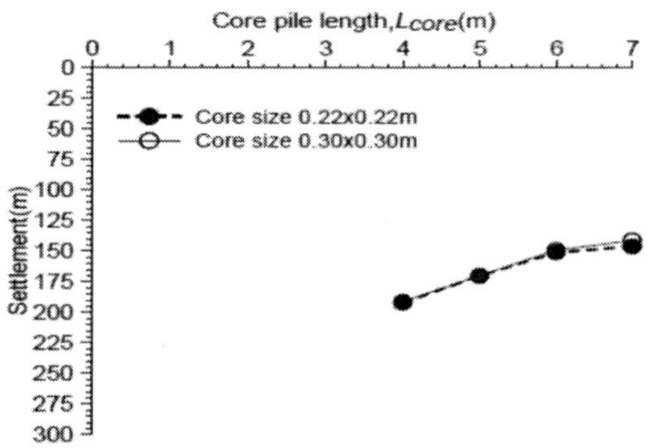

Figure 40. Effect of lengths and sectional areas of concrete core pile on ultimate settlements of SDCM pile.

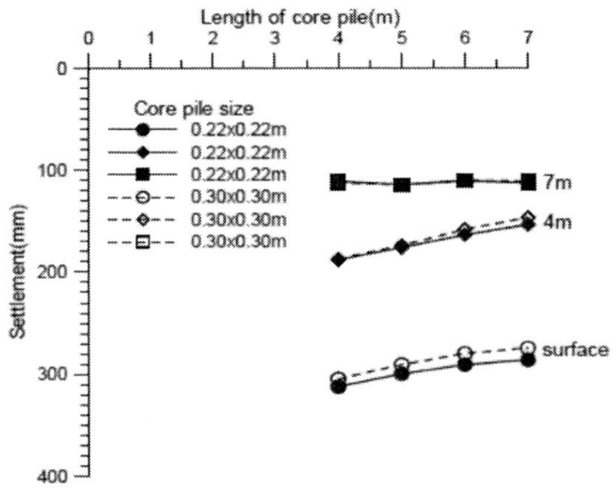

Figure 41. Effects of core pile lengths on ultimate surface and subsurface settlements in surrounding clay of SDCM pile.

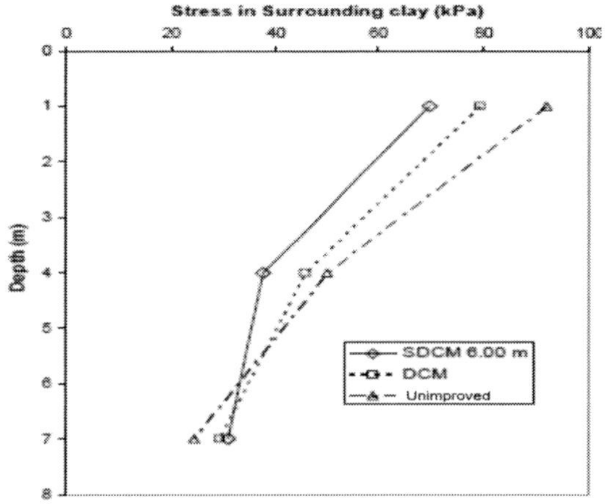

Figure 42. Stresses in surrounding clay of unimproved zone, SDCM and DCM piles piles at depth 1, 4 and 7 m.

The effect of length of concrete core pile on the lateral movements are also studied through the simulations by varying the lengths of the concrete core pile from 4 to 7m as well as their sectional areas consisting of 0.22 x 0.22 m and 0.30 x 0.30 m. The simulated and observed results of the lateral movements for both DCM and SDCM piles at different periods after construction are illustrated in Figure 43. It can be clearly observed that the lateral movements obtained from inclinometers for SDCM piles were much lower than that of DCM piles. The measured and simulated lateral movement data agreed well. The effects of concrete core pile lengths longer than 4m on lateral movements of SDCM piles, at 570 days after construction, are illustrated in Figure 44. The lateral movement reduced with increasing lengths of concrete core piles longer than 4m. From subsequent simulations as shown in Figure 45, increasing the lengths and, to a lesser degree, the sectional areas of concrete core piles, reduced the lateral movements of the SDCM piles for concrete core piles longer than 4m. The lateral movement significantly reduced with increasing lengths as well as with increasing the sectional areas of concrete core piles. It can be summarized that increasing both the lengths and sectional areas of core piles reduced the lateral movement.

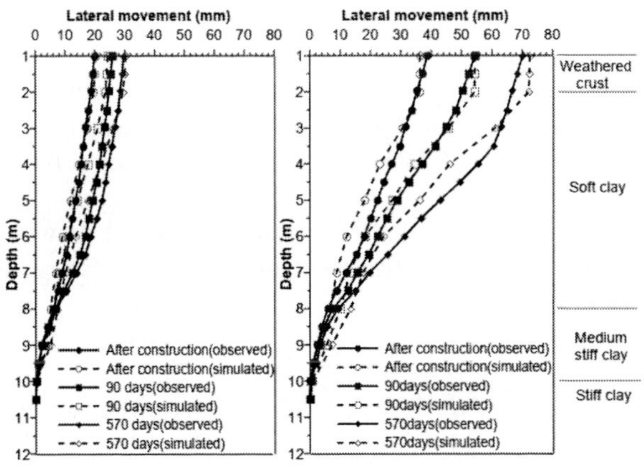

Figure 43. Simulated and observed lateral movements of SDCM and DCM piles at different periods after construction.

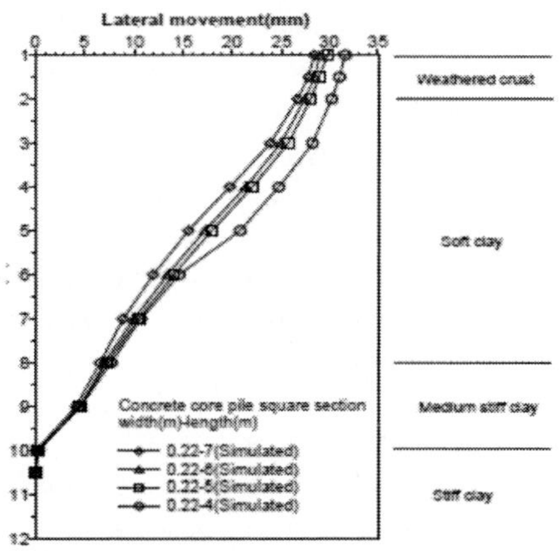

Figure 44. Effects of concrete core pile lengths on lateral movement profiles of SDCM pile with 0.22x0.22m core pile from simulations.

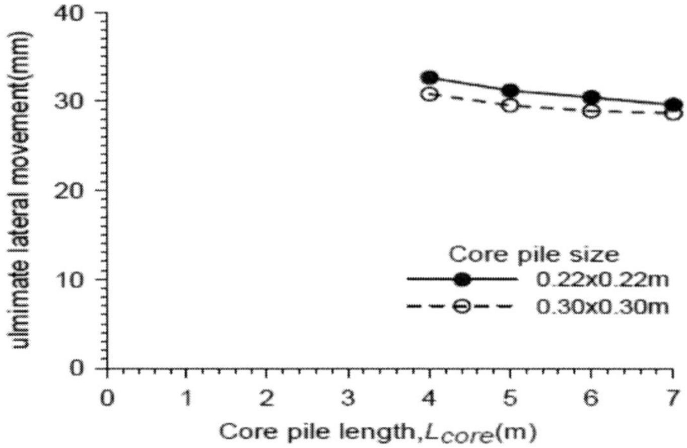

Figure 45. Effects of sectional areas and lengths of concrete core piles on the maximum lateral movement of SDCM pile.

4. Conclusions

4.1 PVD method

- Due to the increase in hydraulic conductivity of the smear zone caused by the vacuum effects, the Vacuum PVD yielded slightly lower k_h/k_s values while the coefficient of horizontal consolidation, C_h, increased. From back-calculated results of the field test data, the average C_h values were 2.17 m^2/yr. and 3.51 m^2/yr from the Conventional PVD and Vacuum PVD, respectively, with corresponding decrease in k_h/k_s values from 7.2 to 6.6.
- The Vacuum PVD improved soft ground reduced the time to 90% degree of consolidation by one-third due to the consequent higher values of the coefficient of horizontal consolidation and subsequent increase in the rate of settlements.
- For the Thermal PVD (PVTD) improvement, the consolidation rate was faster and the settlements was larger compared to the Conventional PVD. The permeability of the smear zone increased due to the increase in temperature which caused the reduction of viscosity in the pore water. Consequently, the k_h/k_s values slightly decreased while the coefficient of horizontal consolidation, C_h, increased.
- The back-calculated C_h values for embankment improved with Conventional PVD and Thermal PVD (PVTD) in the field were 6.69 and 8.51 m^2/yr, respectively, and the corresponding k_h/k_s values were 6.2 and 4.1. The increased permeability in the smear zone of PVTD resulted in the decrease in k_h/k_s of about 33.9% and increase in C_h by 27.2% as well as increase in horizontal permeability in the smear zone by 1.92 times.
- The utilization of solar heating technique in the Thermal PVD improvement of soft ground is sustainable and cost effective method to achieve the required settlement at much faster and shorter time period.

4.2 DCM/SDCM method

- The full scale embankment loading test supported by SDCM and DCM piles was constructed, monitored and, consequently, simulated by using PLAXIS Foundation 3D finite element software in order to study and verify the design parameters.
- The results indicated that the longer reinforced concrete (RC) core pile can reduce the vertical displacements of SDCM pile as well as the subsurface portions of the surrounding soil. The settlements reduced with increasing lengths of RC core piles from 4 to 6m but slightly reduced

from 6 to 7m core pile length. Moreover, the length of RC core pile affected both the surface and subsurface settlements at 4m depth but did not affect the subsurface settlements at 7m depth.

- In the case of lateral deformations, the lengths and sectional areas of RC core pile reduced the lateral movement of the embankment. The longer the lengths, the lower the lateral movements. Furthermore, larger sectional areas of RC core pile also reduced the lateral movements although with smaller effects. It was also found that the RC core piles need to be longer than 4 m in order to effectively reduce the lateral movements.

References

[1] Hansbo S 1979 Consolidation of clay by band-shaped prefabricated drains *Ground Engineering 12* No.5 pp. 16-25.

[2] Hansbo S 1981 Consolidation of Fine-Grained Soils by Prefabricated Drains *Proc 10th International Conference on Soil Mechanics and Foundation Engineering* Stockholm **1** pp. 677-682.

[3] Bergado D T, Balasubramanian A S, Fannin R J and Holtz R D 2002 Prefabricated Vertical Drain (PVD) in Soft Bangkok Clay: A case of NBIA Project *Canadian Geotechnical Journal* **39** No.2 pp. 304-315.

[4] Abuel-Naga H, Bergado D T and Gniel J 2015 Design Chart for PVD Improved Ground *Geotextiles and Geomembrane* **43** No 6.

[5] Bergado D T, Asakami H, Alfaro M and Balasuramaniam A S 1991 Smear Effects of Vertical Drains on Soft Bangkok Clay *Journal of Geotechnical Engineering ASCE* **117** No.10 pp. 1509-1529.

[6] Abuel-Naga H M and Bouazza A 2009 Equivalent Diameter of a Prefabricated Vertical Drain *Geotextiles and Geomembranes* **27** pp. 227-231.

[7] Kjellmann W 1952 Consolidation of Clay Soil by Means of Atmospheric Pressure. *Proc on Soil Stabilization Conference.* Boston USA **1** pp. 258-263.

[8] Chai J C, Carter J P and Hayashi S 2006 Vacuum Consolidation and Its Combination with Embankment Loading *Canadian Geotechnical Journal* **43** pp. 985-996.

[9] Chai J C, Miura N and Nomura T 2007 Experimental Investigation on Optimum Installation Depth of PVD under Vacuum Consolidation *Proceedings 3rd China-Japan Joint Geotecnical Symposium* China.

[10] Saowapakpiboon J, Bergado D T, Thann Y M, Voottipruex P 2009 Assessing the Performance of PVD with Vacuum and Heat Preloading *Geosynthetics International* **16** No.5 pp. 384-392.

[11] Saowapakpiboon J, Bergado D T, Youwai S, Chai J C, Wanthong P and Voottipruex P 2010 Measured and Predicted Performance of Prefabricated Vertical Drains (PVDs) with and without Vacuum Preloading *Geotextiles and Geomembranes* **28** pp. 1-11.

[12] Lam L G, Bergado D T and Hino T 2015 PVD Improvement of Soft Bangkok Clay with and without Vacuum Preloading using Analytical and Numerical Analyses *Geotextiles and Geomembranes* **43** No.6 pp. 547–557.

[13] Long, P.V., Nguyen, L.V., Bergado, D.T. and Balasubramaniam, A.S. 2015. Performance of PVD Improved Soft Ground using Vacuum Consolidation Method with and without Airtight Membranes. Goetextiles and Geomembranes **43** No. 6 pp. 473–483.

[14] Long P V, Nguyen L V, Tri T D and Balasubramaniam A S 2016 Performance and Analyses of Thick Soft Deposit Improved by PVD with Surcharge Preloading and Vacuum Consolidation *– A Case Study at CMIT, Proceedings 11th SEAGS-2nd AGGSEA Conference* Kuala Lumpur Malaysia.

[15] Abuel-Naga H M, Bergado D T and Chaiprakaikeow S 2006 Innovative Thermal Technique for Enhancing the Performance of Prefabricated Vertical Drain during the Preloading Process *Geotextiles and Geomembranes* **24** pp. 359-370.

[16] Pothiraksanon C, Saowapakpiboon J, Bergado D T and Thann N M 2008 Reduction of Smear Effects around PVD using Thermo-PVD *Ground Improvement* **161** (G14) pp. 179-187.

[17] Abuel-Naga H M, Bergado D T, Bouazza A and Ramana G V 2007 Volume Change Behavior of Saturated Clays under Drained Heating Conditions: Experimental Results and Constitutive Modelling *Canadian Geotechnical Journal* **44** pp. 942-956.

[18] Abuel-Naga H M, Bergado D T and Bouazza A 2008 Thermal Conductivity Evolution of Saturated Clay under Consolidation Process *International Journal of Geomechanics* ASCE **8** No.2 pp. 114-122.

[19] Petchgate K, Jongpradist P and Panmanajareonphol S 2003 Field Pile Load Test of Soil-Cement Column in Soft Clay *Proceedings of the International Symposium 2003 on Soil/Ground Improvement and Geosynthetics in Waste Containment and Erosion Control Applications* 2-3 December 2003 Asian Institute of Technology, Thailand pp. 175-184.

[20] Asaoka A 1978 Observational Procedure of Settlement Prediction *Soils and Foundations* **18** No.4 pp. 87-101

[21] Pothiraksanon C, Saowapakpiboon J, Bergado D T, Voottipruex P and Abuel-Naga H 2010 Soft Ground Improvement with PVD and Thermo PVD with Solar Power *Ground Improvement Journal* **163** Issue G11 pp. 1-8.

[22] Indraratna B, Rujikatkamjorn C and Sathananthan L 2005 Analysis and Numerical Solution for a Single Vertical Drain Including Effects of Vacuum Preloading *Canadian Geotechnical Journal* **42** pp. 994-1014.

[23] Mohamedelhassan E and Shang J Q 2012 Vacuum and Surcharge Combined One-Dimensional Consolidation of Clay Soils *Canadian Geotechnical Journal* **39** pp. 1126-1138.

[24] Chai J C and Rondonuwu S G 2015 Surcharge Loading Rate for Minimizing Lateral Displacement of PVD Improved Deposit with Vacuum Pressure *Geotextiles and Geomembranes* **43** No. 6.

[25] COFRA (Thailand) 1996 *Design and Proposal for the Execution of a Test Field*: Vacuum Consolidation for the Landslide Road System for Second Bangkok International Airport, Bangkok, Thailand.

[26] Chai J C and Miura N 1999 Investigations of Factors Affecting Vertical Drain Behaviour. *ASCE J. of Geotech and Geoenvi Engg* **125** No. 3 pp. 216-226.

[27] Bergado D T, Ahmed S, Sampaco C L and Balasubramaniam A S 1990 Settlement of Bangna-Bangpakong Highway on Soft Bangkok Clay *ASCE Journal of Geotechnical Engineering* **116** No.1 pp. 136-154.

[28] Balasubramaniam A S, Bergado D T and Phienwej N 1995 The Full Scale Field of Prefabricated Vertical Drains for the Second Bangkok Airport *Final Report Division of Geotechnical and Transportation Engineerinf.* AIT Bangkok Thailand 259 pages.

[29] Long, P.V., Bergado, D.T., Nguyen, L.V. and Balasubramaniam, A.S. 2013 Design and Performance of Soft Ground Improvement using Prefabricated Vertical Drains with and without Vacuum Consolidation *Geotechnical Engineering Journal of SEAGS & AGSSEA* **44** No. 4 pp. 37-52.

[30] Chai J C and Zhou Y 2016 Equivalent Smear Effect due to Non-Uniform Consolidation around a PVD *Goetechnique* pp. 1-10 [http://dx.doi.org/10.1680/jgeot.16.P.087].

[31] Jongpradist P, Youwai S, Manorat S, Kongkitkul W and Chucheepkul S 2011 Influence of Curing Stress on One-Dimensional Yielding of Cement-Admixed Bangkok Clay at High Water Content *Soil and Foundations* **51** No. 2 pp. 351-357.

[32] Shinwuttiwong W 2007 Full Scale Behavior of SDCM Piles under Axial and Lateral Loading with Simulations *M. Eng. Thesis No GE-07-04* Asian Institute of Technology, Bangkok, Thailand.

[33] Jamsawang P 2008 Full Scale Tests on Stiffened Deep Cement Mixing (SDCM) Pile Including 3d Finite Element Simulation *D. Eng. Diss. No. GE-08-01, Asian Institute of Technology.* Bangkok Thailand.

[34] Tand K E and Vipulanandan C 2008 Comparison of Computed vs. Measured Lateral Load/Deflection Response of ACIP piles *PLAXIS Bulletin* Issue **23** pp. 10-13.

[35] Jamsawang P, Bergado DT, Bandari A and Voottipruex P 2008 Investigation and Simulation of

Behavior of Stiffened Deep Cement Mixing (SDCM) Piles *International Journal of Geotechnical Engineering* **2** No.3 pp.229-246.
[36] Jamsawang P, Bergado D T, Bhandari A and Voottipruex P 2009 Behavior of Stiffened Deep Cement Mixing (SDCM) Pile in Laboratory *Lowland Technology International, June.*
[37] Bhandari A 2006 Laboratory Investigation of Stiffened Deep Cement Mixed (SDCM) Pile. M.Eng *Thesis No. GT-78-6, Asian Institute of Technology* AIT Bangkok Thailand.
[38] Wu M, Zhao X and Dou Y M 2005 Application of Stiffened Deep Cement Mixed Column in Ground Improvement *Proceedings Intl. Conf. on Deep Mixing Best Practices and Recent Advances*. Stockholm Sweden.

Numerical Analyses of Subsoil-structure Interaction in Original Non-commercial Software based on FEM

R Cajka[1], J Vaskova[2*] J Vasek[3]

[1] Department of Structures, Faculty of Civil Engineering, VŠB – Technical University of Ostrava, Ludvika Podeste 1875/17, 708 33 Ostrava-Poruba, tel. 00420 597 321 900, Czech Republic
[2] Department of Structures, Faculty of Civil Engineering, VŠB – Technical University of Ostrava, Ludvika Podeste 1875/17, 708 33 Ostrava-Poruba, tel. 00420 597 321 987, Czech Republic,
[3] Department of Structures, Faculty of Civil Engineering, VŠB – Technical University of Ostrava, Ludvika Podeste 1875/17, 708 33 Ostrava-Poruba, Czech Republic

*Corresponding author: jana.vaskova@vsb.cz

Abstract. For decades attention has been paid to interaction of foundation structures and subsoil and development of interaction models. Given that analytical solutions of subsoil-structure interaction could be deduced only for some simple shapes of load, analytical solutions are increasingly being replaced by numerical solutions (eg. FEM – Finite element method). Numerical analyses provides greater possibilities for taking into account the real factors involved in the subsoil-structure interaction and was also used in this article. This makes it possible to design the foundation structures more efficiently and still reliably and securely. Currently there are several software that, can deal with the interaction of foundations and subsoil. It has been demonstrated that non-commercial software called MKPINTER (created by Cajka) provides appropriately results close to actual measured values. In MKPINTER software stress-strain analysis of elastic half-space by means of Gauss numerical integration and Jacobean of transformation is done. Input data for numerical analysis were observed by experimental loading test of concrete slab. The loading was performed using unique experimental equipment which was constructed in the area Faculty of Civil Engineering, VŠB-TU Ostrava. The purpose of this paper is to compare resulting deformation of the slab with values observed during experimental loading test.

1. Introduction

With the beginning and subsequent development of computer technology, numerical methods have also been used to solve the interactions between base and subsoil. Currently (and also in the past) many experts have been dealt with numerical model based on Finite Element Method (FEM) – eg. Zienkiewicz a Taylor [1], Kralik, Jendzelovsky [2], Kolar, Nemec [3]. The results of the Finite Element analysis of the loaded half-space are also published by Cajka and Labudkova in [4, 5, 6] and by Cajka, Burkovic and Fojtik in [7]. Numerical analyses of stress in a half-space are dealt with in articles [8, 9], in article [10] published by Han, Bao and Wang and in articles [11, 12, 13] published by Cajka. Some of widely used software, which solve interaction tasks, are for example Scia Engineer, Ansys, Trimas, MKPINTER, RFEM a RF-Soilin or Plaxis. Despite many commercial software and non-commercial software, a computational model was not still found to accurately capture the behaviour in interaction of the foundation structure and subsoil. The difficulty of designing an accurate static design of foundation structures lies in several aspects, for example the influence of physical-nonlinear behaviour of the structure or the uncertainty associated with the description of the properties and behaviour of the foundation soil, because it is natural material and its properties cannot be determined unambiguously. Despite these uncertainties in input parameters, Finite Element Method (FEM) is an excellent computing method for a whole range of tasks, including interaction tasks. The Finite Element Method (FEM) allows solution of a so-called complete interaction system including "subsoil - foundation structure - upper structure". Complete interaction system is more demanding to compute, but provides more accurate results than a simplified interaction system "subsoil - foundation structure". However, the question is, how accurate are input data and parameters entering the calculations. Both, 2D and 3D finite elements, can be using to solve of numerical analyses of the complete interaction systems and also simplified interaction systems. Results from 3D numerical models have already been described by the authors of this article, for example, in [14, 15, 16]. In the 3D numerical models, in which the subsoil was modelled as a linear elastic homogeneous isotropic continuum, the results are greatly dependent on the chosen parameters entered into the calculation. These parameters are e.g. the size of the modelled area representing the subsoil, the choice of boundary conditions or the mesh size. A parametric study [14, 15, 16] of the dependence of the vertical deformations on the variable depth of the subsoil model has clearly demonstrated that with the increasing depth of the subsoil model there also increases the difference between the calculated deformations for different boundary conditions. It is problematic resp. ambiguous to determine the size of an area representing the subsoil in that it does not apply to solving subsoil-structure interaction using 2D interaction models (called contact subsoil models). All characteristics of the subsoil are concentrated into contact joints between the structure and subsoil in such interaction models. Such 2D surface models can be in principle considered as a type of meshless method. 2D surface models utilize far fewer elements, forming a mesh greater than 3D models. The subject of this article is a numerical analysis in the original non-commercial software (created by Cajka) created to solve the subsoil-structure interaction. Surface (2D) subsoil models in MKPINTER is based on the theory of elastic half-space using the modified structural strength of the soil. Given that this is a surface (2D) model, the results are not dependent on the selected depth of the area representing the subsoil, as is the case of 3D models. The paper investigates the stress-strain deformation in the interaction between the foundation structure and subsoil. The influence of the properties of the foundation structure on the subsoil and, in turn, those of the subsoil on the foundation structure were observed in a specific task – experimental loading test of concrete slab on ground.

2. Experimental loading test of concrete slab on ground

Experimental loading test of concrete slab was realized using the unique experimental facility built in campus of Faculty of Civil Engineering, VŠB - Technical University of Ostrava in the Czech Republic. For this experimental loading test, numerical analyses have been done in the software MKPINTER based on the finite element method (FEM). Concrete slab was chosen for simplicity in the implementation of an experiment aimed at verifying the test methods and equipment (Figure 1).

Figure 1. Experimental loading test of concrete slab on ground

Concrete slab corresponded to class of concrete C45/55. The dimensions of the slab were 500 x 500 x 48 mm. The upper layer of subsoil consists of loess loam with F4 consistency. The following input parameters were used to create the numerical model. Plain concrete slab was specified by volume weight $\gamma = 25$ kN.m^{-3}, Poisson coefficient $\mu = 0.2$ and modulus of elasticity of concrete E = 36,3 GPa. Subsoil was specified by volume weight $\gamma = 18.5$ kN.m^{-3}, Poisson coefficient $\mu = 0.35$, and modulus of deformability $E_{def} = 2.65$ MPa. During the loading test, the concrete slab was loaded in the centre by the pressure applied by a hydraulic press. Dimensions of the area under load were 100 x 100 mm. The load at the moment of failure was 18.640 kN. Measurements regarding to the interaction of foundation–subsoil is dealt with in such [17, 18, 19].

3. MKPINTER - original non-commercial software based on FEM

The calculation of the deformations and internal forces of the slab is done by the finite element method using isoparametric plate elements with shear effect. The course of contact stress affects the deformation of the slab and subsoil. In order to cause the contact stress of the same deformation of the slab and subsoil in this nonlinear interaction, the solution is performed by an iterative method. The calculation of tension, subsidence and contact stress is solved by the universal method of calculation using Jacobian transformation. In MKPINTER, interactions are solved by numerical integration of stress computations and settlement of the elastic half-space modified by structural strength. Accuracy of results affects the number of integration points. More detailed descriptions and other results from software MKPINTER are given in [11, 12]. In numerical model, the concrete slab model is loaded in the center on an area of 100 x 100 mm and its finite element mesh is evident from Figure 2. A cross section marked A - A', which contains nodes 37, 38, 39, 40, 41, 42, 43, 44, 45, is guided in the center.

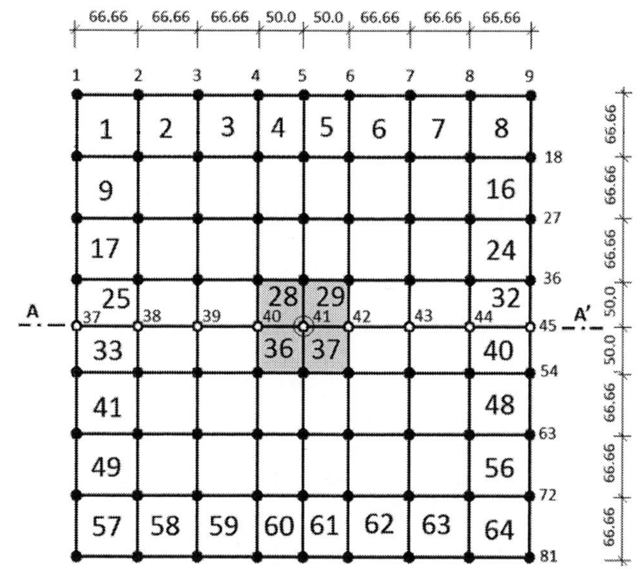

Figure 2. Finite element mesh of slab model

At nodes 37-45 in cut A - A' are presented the results of subsidence of the modified half-space, slab deformations, contact stress and the course of the specific bending moment. Figure 3 and Figure 4 show the final deformation of the slab, specific bending moments as well as the contact stress obtained after the 9th iteration for 12 integration points. Contact stress under a stiff foundation has theoretically infinite values on the edges and corners of the slab. Such stress values in a real soil environment cannot occur because there is a plasticization of the soil. These stress peaks can be reduced in the software MKPINTER. It enables the reallocation (increasing) of the stress values also in the central part of the course of contact stress.

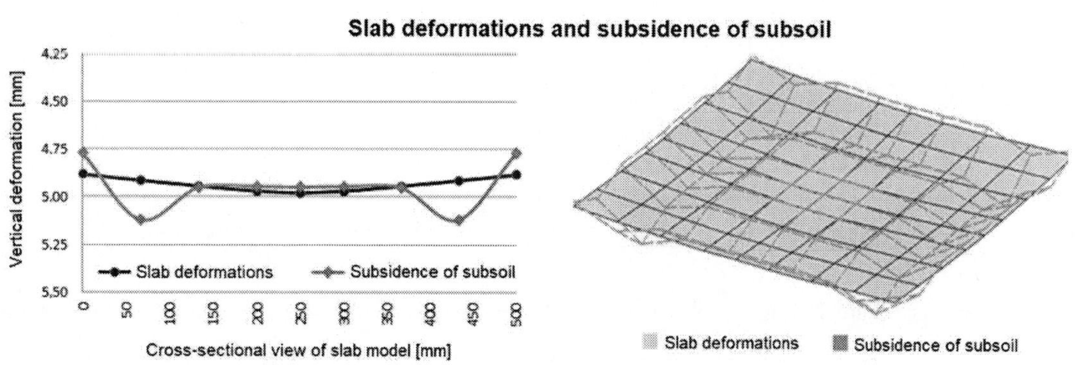

Figure 3. Final slab deformations and subsidence of subsoil

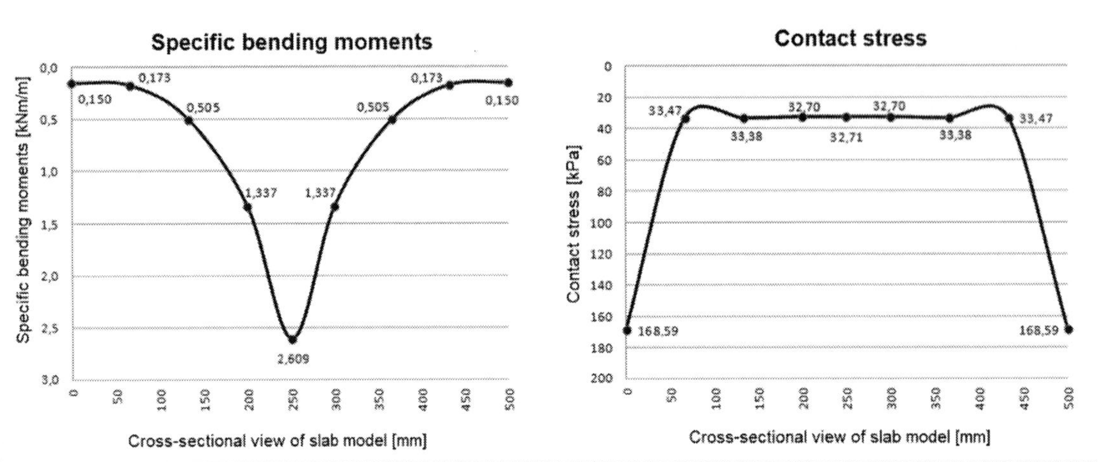

Figure 4. Final specific bending moments and the contact stress

This interaction model provides results for individual iterations. For comparison, 0th iteration, 4th iteration, and last (9th) iteration are selected. In Figure 5, it is possible to see how the course of settlement after these iterations progressively approaches the deformation of the slab. Approximately halfway through the computation process (after 4th iteration), there is still a noticeable difference between subsoiling and deformation of the slab even though it is already similar to the resulting deformation obtained after 9th iteration.

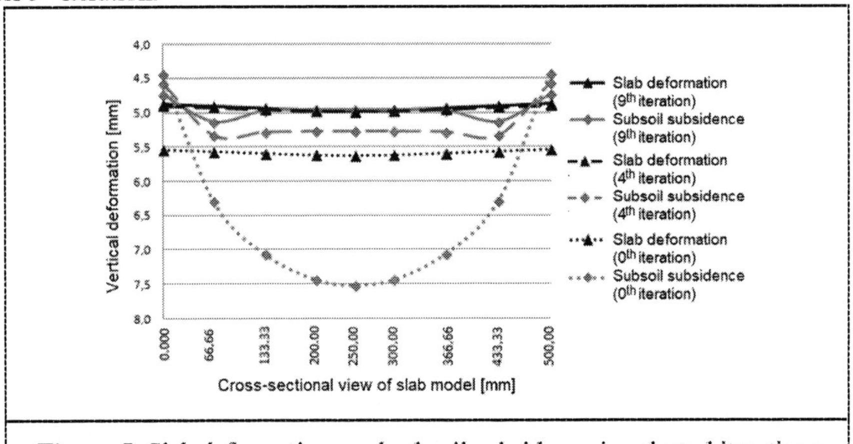

Figure 5. Slab deformations and subsoil subsidence in selected iterations

Figure 6 and Figure 7 show the deformation of the foundation slab and the subsoil subsidence after each iteration.

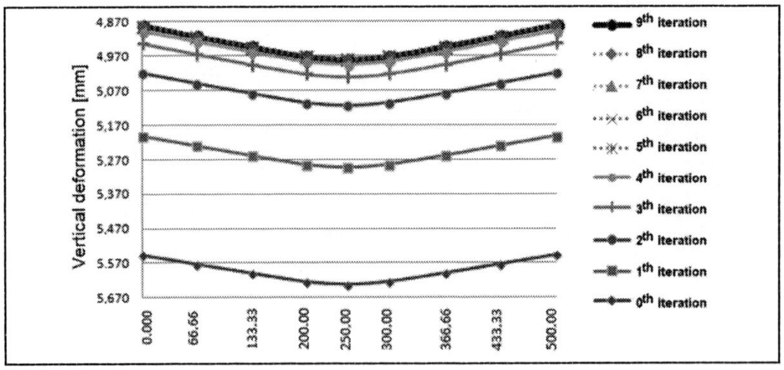

Figure 6. Slab deformations after each iteration

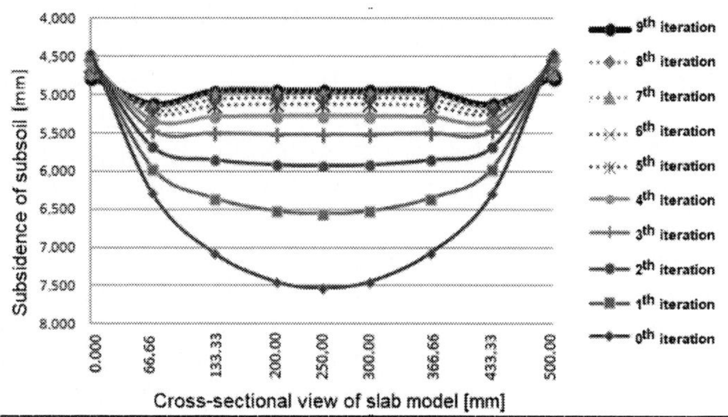

Figure 7. Subsidence of subsoil after each iteration

Figure 8 shows the distribution of the contact stress after each iteration during the calculation process.

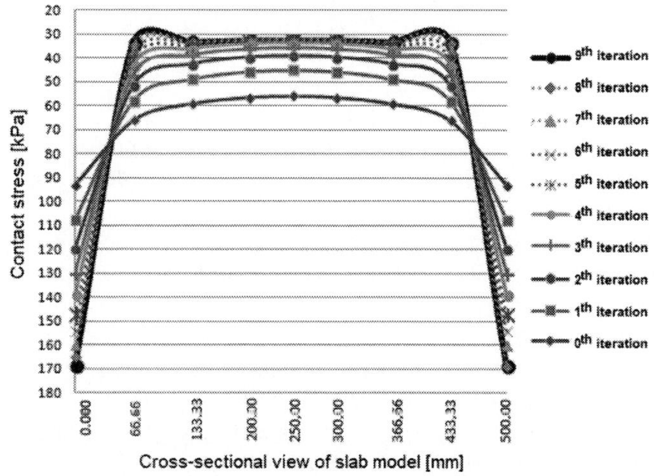

Figure 8. Contact stress after each iteration

The course of the specific bending moments during the computational process is not mentioned because the differences between the results of the individual iterations were not very noticeable in the graphical presentation.

4. Summary

During experimental loading test the break out of the slab model occurred when the load reached the value 18.6 kN. The measured subsidence in the middle of the slab is 3.7 mm. This is by 27% less than subsidence calculated in original non-commercial software MKPINTER. In MKPINTER the calculation of the deformations and internal forces of the slab is done by the finite element method using isoparametric plate elements with shear effect. The numerical calculation is solved by the universal method of calculation using Jacobian transformation. In MKPINTER, interactions are solved by numerical integration of stress computations and settlement of the elastic half-space modified by structural strength. MKPINTER calculates subsoil-structure interaction using 2D interaction model. An important parameter for numerical analyses of subsoil-structure interaction is the degree of discretising which influences both the results and the number of degrees of freedoms, being thus important for the time needed for calculation and for the quantity of the processed data. In comparison the 2D model of the subsoil, which is the issue of this article, and 3D models [14, 15, 16] show that calculation time plays the main role. With increasing size of the numerical model of the subsoil an increase in the number of finite element and computational time is required to solve the task. With the use of 2D models there can be reduced the calculation time, which could take about a few hours for using 3D models (of course according to the size of the mesh). 2D surface models can be in principle considered as a type of meshless method. It is not desirable to extend the calculation time during structure analysis. It especially is true when the results of 2D and 3D models are not much different. The results of the 3D model of the subsoil depend on the geometry, the size of subsoil model, and boundary conditions [14, 15, 16].

Poor experience with this addition can lead to a numerical model which has results out of the limit state. Caution should be exercised. To obtain reliable results from the subsoil-structure interaction there is an optimal combination of experimental tests, laboratory and field tests and numerical analyses. All these approaches are also combined at the Faculty of Civil Engineering, VSB - Technical University of Ostrava. A unique experimental facility [20] for monitoring the interaction of foundations and subsoil was built on its facilities. The results of the numerical analyses of 2D models and 3D models are (and will be) compared with the actual measurements obtained during the loading test of foundation slabs-on-ground (reinforced concrete slabs, steel fibre-reinforced concrete slabs and prestressed concrete slabs).

1. References

[1] Zienkiewicz O C, Taylor R L 2000 The finite element method *Solid Mechanics* 479

[2] Kralik J, Jendzelovsky, 1994 N Contact problem of reinforced concrete girder and non-linear Winkler foundations *Geomechanics 93* 233-6

[3] Kolar V, Nemec I 1989 Modelling of soil structure interactions (Elsevier Science)

[4] Cajka R, Labudkova J 2013 Influence of parameters of a 3D numerical model on deformation arising in interaction of a foundation structure and subsoil *Recent Advances in Urban Planning and Construction* 119-124

[5] Cajka R, Labudkova J 2015Comparison of results of analyses the foundation slab calculated by two fem programs *Advanced Materials Research* 1065-1069

[6] Cajka R, Labudkova J 2015 Fibre concrete foundation slab experiment and FEM analysis *Key Engineering Materials* **627** 441-444

[7] Cajka R, Burkovic K, Fojtik R 2013 Experimental soil – concrete plate interaction test and numerical models *Key Engineering Materials* **33** 577-578

[8] Frydrysek K, Janco R, Gondek H 2013 Solutions of beams, frames and 3d structures on elastic foundation using FEM *International Journal of Mechanics* **7** 362-9

[9] Kralik J, Simonovic M 1994 Elasto-plastic analysis of deformation soil body with 3D-finite and infinite elements of the International Conference Geomechanics **93** 229-32

[10] Han H, Bao W, Wang T 1997 Numerical simulation for the problem of infinite elastic foundation *Computer Methods in Applied Mechanics and Engineering* **147** 369-385

[11] Cajka R 2013 Accuracy of stress analysis using numerical integration of elastic half-space *Applied Mechanics and Materials* 1127-1135

[12] Cajka R 2013 Analysis of stress in half-space using Jacobian of transformation and gauss numerical integration *Advanced Materials Research* 178-186

[13] Cajka R 2014Numerical solution of temperature field for stress analysis of plate structures *Applied mechanics and materials* **470** 177-87

[14] Labudkova J, Cajka R 2017Numerical analyses of interaction of steel-fibre reinforced concrete slab model with subsoil *Frattura ed Integrità Strutturale* **39** 47-55

[15] Labudkova J, Cajka R 2016 Numerical modeling of the subsoil-structure interaction *Key Engineering Materials* **691** 333-43

[16] Labudkova J, Cajka R 2015Application of an inhomogeneous half-space for the solution of foundation-subsoil interaction based on the finite element method *Proceedings of the Fifteenth International Conference on Civil, Structural and Environmental Engineering Computing* 9

[17] Barros J A O, Figuieiras J A 1998 Experimental behaviour of fibre concrete slabs on soil *Mechanics of Cohesive-Frictional Materials* **3** 277-90

[18] Sorelli L G, Meda A, Plizzari G A 2006 Steel fiber concrete slabs on ground: A structural matter *ACI Structural Journal* **103** 551-8
[19] Alani A, Aboutalebi M 2012 Analysis of the subgrade stiffness effect on the behaviour of ground-supported concrete slabs *Structural Concrete* **2** 102-8
[20] Cajka R, Krivy V, Sekanina D 2011 Design and development of a testing device for experimental measurements of foundation slabs on the subsoil *Transactions of the VŠB - Technical University of Ostrava - Civil Engineering Series* **11 (1)** 1-6

Acknowledgments
This outcome has been achieved with the financial support of the project GACR No. 16-08937S „State of stress and strain of fiber reinforced composites in interaction with the soil environment“.

Static behavior of the weld in the joint of the steel support element using experiment and numerical modeling

M Krejsa[1], J Brozovsky[1], D Mikolasek[1], P Parenica[1] and L Koubova[1]

[1] Department of Structural Mechanics, Faculty of Civil Engineering, VSB-Technical University of Ostrava, Ludvika Podeste 1875/17, 708 33 Ostrava – Poruba, Czech Republic

E-mail: martin.krejsa@vsb.cz

Abstract. The paper is focused on the numerical modeling of welded steel bearing elements using commercial software system ANSYS, which is based on the finite element method - FEM. It is important to check and compare the results of FEM analysis with the results of physical verification test, in which the real behavior of the bearing element can be observed. The results of the comparison can be used for calibration of the computational model.

The article deals with the physical test of steel supporting elements, whose main purpose is obtaining of material, geometry and strength characteristics of the fillet and butt welds including heat affected zone in the basic material of welded steel bearing element. The pressure test was performed during the experiment, wherein the total load value and the corresponding deformation of the specimens under the load was monitored. Obtained data were used for the calibration of numerical models of test samples and they are necessary for further stress and strain analysis of steel supporting elements.

1. Introduction

Welded details are indispensable in modern steel structures. With their use it can be achieved the homogeneous nature of the joint, while not weakening the original support element as is the case with mechanical fasteners. Welds and their use, of course, also have their limitations and disadvantages. Among the disadvantages can be stress in the vicinity of the weld, which results in residual stress. The material around the weld during cooling takes shape differently than the rest of thermally unaffected parts of the welded construction, which leads to deformations of the support element and the said residues. Welds and joints of welded parts also cause local geometric, stiffness and material non-linearity, which in real constructions creates local stress peaks.

Problems associated with numerical modeling of building structures and supporting elements have been published in the past in a number of valuable scientific papers (e.g. [2, 4, 6, 7, 14, 20]). For most of these purposes, the Finite Element Method – FEM is used (e.g. [19]), which is very suitable for these numerical studies. This method is applied in a number of commercial software that are more or less widely used in engineering practice [13, 16]. The ANSYS software system is probably the most widely used in academic and research sphere. ANSYS allows modeling of the entire construction or its supporting elements in detail (e.g. [8, 18]) and for the purposes of the research work of this article is very appropriate.

It is necessary to mention several publications that were inspiring for our research work. Based on ANSYS simulation software, the model of resistance welding drying plate - thin-wall pressure vessel with multiple welds, large size and large deformation, was established to study the deformation and stress distribution under different temperatures and pressures in [11]. The interest on calculating welding residual stresses from the welding process, through finite element (FE) simulation, has increased in the last period (e.g. [5]). The effects of weld geometry size on bearing capacity of submerged arc welded pipe with double-side longitudinal seam under bearing internal pressure conditions were simulated and analyzed by finite element simulation based on the ANSYS software in [17]. The finite element software ANSYS was also used to set up the full-scale sectional model of the welded connection of the U-rib and the deck plate and the refined model for the local part of the welded connection was established with the focus on the stress in the weld toes and weld roots of the welded connections in [15]. Design optimization of spot welded structures to attain maximum strength by using the ANSYS, general purpose finite element analysis software was presented in [3]. A comparative study was conducted in [1] to determine the structural hot-spot stress at the weld toe of the welded joint in the orthotropic steel bridge deck by use the finite element models of the welded joint and the finite element software ANSYS with the solid and the shell elements. The software ANSYS was also used to establish the three-dimensional finite element model for the steel and concrete joint section of railway hybrid box girder cable-stayed bridge to investigate the force characteristics and force transfer mechanism in [21]. The welding process and in accordance with the occurrence mechanism and characteristics of the welding residual stress in the constructional details of the complex structure of the steel girder bridge were proposed in [22]. Finite element models built by ANSYS were used to simulate girth weld of marine steel tubular piles with focus on pipe splices in [12].

Basic problems of numerical modeling of fillet and butt welds in steel structural elements were published by authors in [9, 10]. The paper [9] presented basic research of modeling of fillet welds without taking account the influence of heat-affected zones. There were also published description of interesting methods for statical analysis of welded joint behavior, such as 3D strain measurement using laser sensor Dantecdynamics Q100 or infrared thermographic non-destructive testing. The short international conference paper [10] dealt with the brief description of physical tests of steel supporting elements, whose main purpose is obtaining the material, geometry and strength characteristics of the fillet and butt welds. Both publications show that this is a topical theme and research in this topic is useful in engineering practice, which is complemented in this paper by research of investigating the chosen working diagrams for steel components, and influence of the heat-affected zones on resistances and stress strains in the analyzed types of joints. Furthermore, the effects of changes in stiffness of different types of materials, sharp edges and thickness of the bonded sheets were analyzed on the overall strain and strain of details. Numerical models were validated by physical models that were measured using multiple types of measuring instruments (extensometers, strain gauges and optical laser topography and temperature analysis in the infrared spectrum of light waves).

2. Description of numerical models

The accuracy of static calculations achieved by numerical methods depends on several basic factors. The first is the geometric description of the solved problem, which is well fulfilled with the current state of computation. The detail in the numerical model can be described on a very fine scale now. The second factor influencing the accuracy of the solution is the choice of the FEM element. The mathematical approximation of the modelled object is always a simplification of reality. With current numerical modelling options, the selection of suitable finite elements is sufficient to correctly describe nodal deformations. With proper FEM mesh selection and a correct definition of boundary conditions (load and external connections, together with appropriate load increments), solution of the numerical modelling converges to satisfactory results. The third factor is the choice of the stress-strain diagram of the behavior of the individual materials, which depends on the results under analysis and demands for the accuracy of the solution. Together with these requirements, computing time and hardware demands also come into play. With more complex material models, however, higher demands are placed on users, both on the correct interpretation of the results and on their evaluation.

Seven types of welded samples with different load effects on the weld were created and tested in the framework of numerical modeling of welding in the joint of the steel support element. The first three (#1, #2 – see Figs. 1 & 2, and #3 – see Figs. 3 & 4) were pressure tested, focusing on fillet welds behavior (details in [9]). The remaining four samples (#4 – see Figs. 5, 6 & 12, #5 – see Figs. 7, 8 & 14, #6 and #7) were stressed with a focus on the behavior in the fillet and butt welds (Figs. 9 & 10). The models were calculated for different FEM mesh densities and for different types of material stress-strain diagrams. The following is a more detailed description of these parameters.

Figure 1. Test specimen #2, details of fillet weld damage.

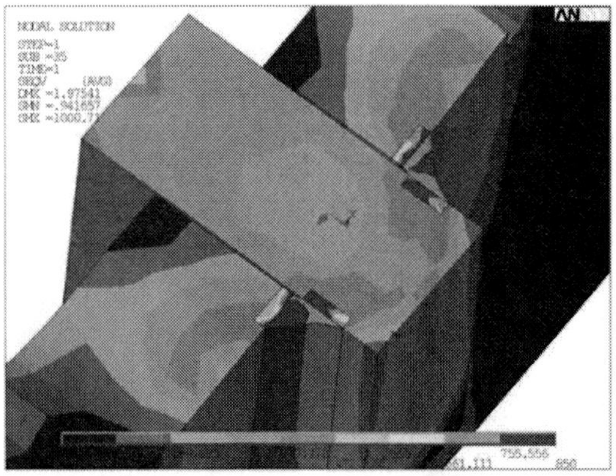

Figure 2. Test specimen #2, distribution of von Misses stresses at 100% limit load.

Figure 3. Test specimen #3, details of fillet weld damage.

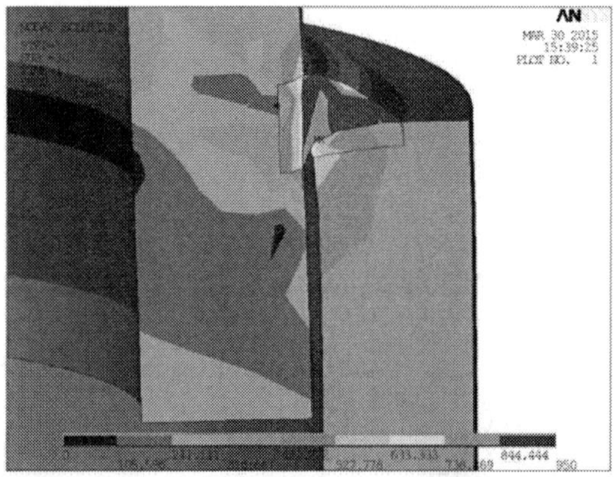

Figure 4. Test specimen #3, distribution of von Misses stresses at 100% limit load.

2.1. Modeling of welds in ANSYS

Welds were modeled in the ANSYS using SOLID186 - homogeneous structural solid element. The SOLID186 is a higher order 3-D 20-node solid element that exhibits quadratic displacement behavior. The element is defined by 20 nodes having three degrees of freedom per node: translations in the *x*, *y*, and *z* directions. The element supports plasticity, hyperelasticity, creep, stress stiffening, large deflection, and large strain capabilities. It also has mixed formulation capability for simulating deformations of nearly incompressible elastoplastic materials, and fully incompressible hyperelastic materials. This finite element is suitable for numerical models with large stress differences. Numerical models also respect structural non-linearity through contact elements in areas that were not welded and bonded with weld metal.

2.2. FEM mesh density

Numerical models have been assembled for the different density of the FEM mesh. The FEM mesh density has been optimized from the welds to the external supports throughout the whole model volume. The aim of this parametric study was to determine the effect of the FEM mesh density on the resulting deformations and stresses in the area of welded joint. The FEM mesh density affects not only the accuracy of the obtained results but also the creation of areas with local numerical peaks of stress. Apart from these, however, the FEM mesh density is also tied to the computational demands of numerical models - with the speed of calculation, memory requirements and effectiveness of the results analysis. A detailed study of the effect of FEM mesh density on the resulting behavior of the numerical models can not be mentioned here due to the limited scope, and it is necessary to refer the readers to one of the other articles the authors prepare to publish, but the results show that the effect of the FEM density is not critical.

Figure 5. Test specimen #4 consists of two flat steel elements and fillet weld, during the tensile test.

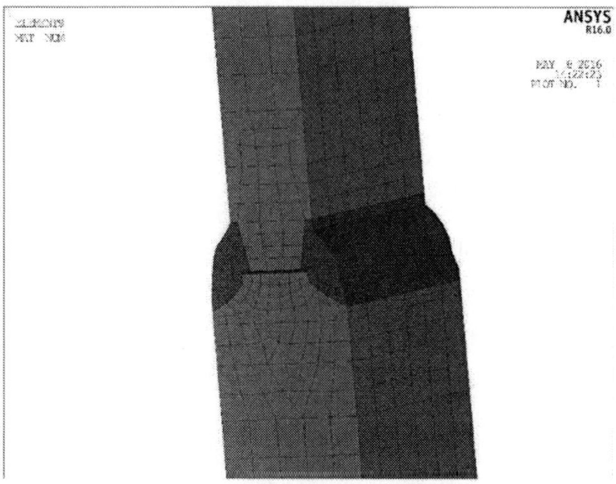

Figure 6. Test specimen #4, detail of FEM in numerical model.

Figure 7. Test specimen #5 consists of two flat steel elements and butt weld, detail of damage.

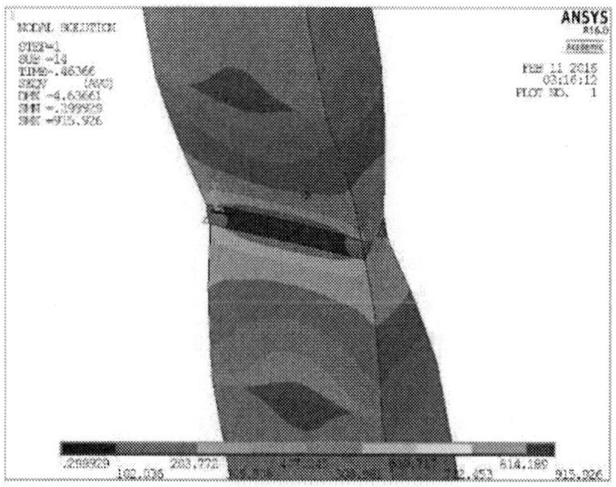

Figure 8. Numerical model of test specimen #5, distribution of von Mises stress.

IOP Conf. Series: Earth and Environmental Science **143** (2018) 012004 doi:10.1088/1755-1315/143/1/012004

2.3. Heat-affected weld zone

The heat-affected zone (HAZ) can be determined on the basis of a standard EN ISO 9015-1:2011 (Destructive tests on welds in metallic materials - Hardness testing) using Vickers hardness test. This makes it possible to obtain hardness distributions around the fillet and butt welds (see Fig. 11).

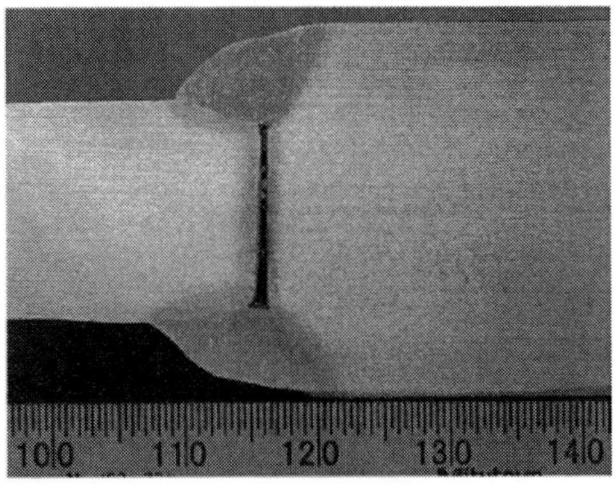

Figure 9. Fillet weld of test specimen #4 in detail during the Vickers hardness test.

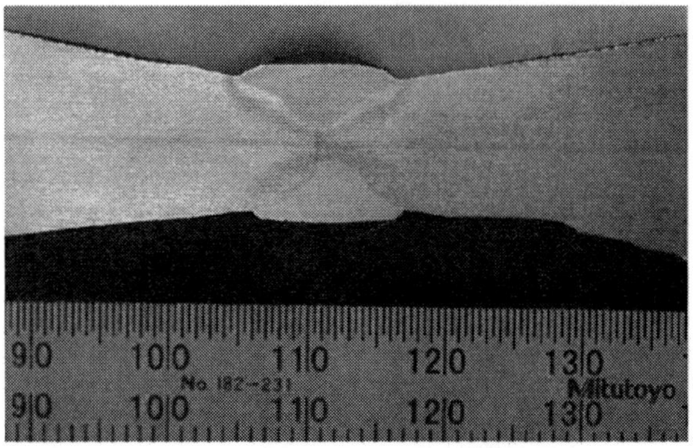

Figure 10. Butt weld of test specimen #5 in detail during the Vickers hardness test.

The heat-affected zone was included in the computational models of the tensioned specimens. The heat-affected zone was divided into several areas, which are different in the stress-strain diagrams (see Figs. 13 & 15). These zones have different strength, yield stress, and strain behavior during loading. Compared to the base material, the heat-affected zone has higher stiffness and strength but is more brittle. Local extremal stress and strain accumulate here, and there is an initiation of a material failure. It has been shown that the use of a heat-affected zone in numerical modeling of welds is necessary.

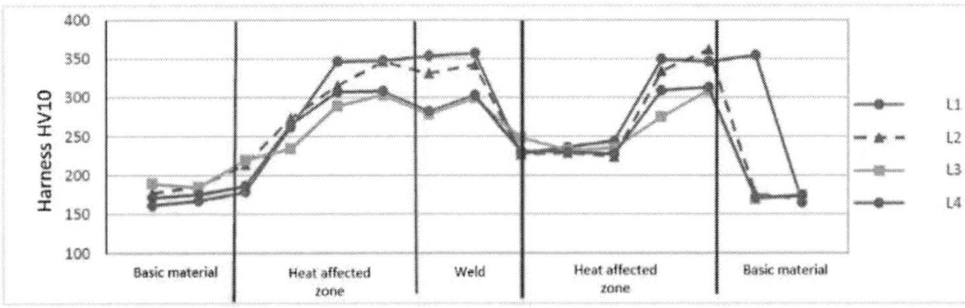

Figure 11. Graphical dependence of hardness distribution in butt weld and its heat-affected zone for four measurements using Vickers hardness test.

2.4. Stress-strain diagrams in numerical models

Stress-strain diagrams entering mathematical models play a significant role in achieving the correct results compared to physical tests. In case of requirements for the analysis of heavily plasticized materials, it is necessary to use more precise working diagrams. In principle, three basic material models are used for the calculations described here.

The first one is a linear material model that respects Hooke's law and has an informative value only to identify details where stress is to reaching the yield strength. This material model can advantageously be used for rapid numerical analysis before proceeding to use more advanced material models.

The second material model is an elastic-plastic material with bilinear hardening, which is very practical, relatively quick to apply and easily accessible. The advantage of this model is also its transparency and it can also be assessed and evaluated with regard to the basic knowledge of the materials used. This material model should be used in cases where knowledge of the stiffness of the behavior of the welded joint is not important. Its use can achieve the correct global deformations and also the maximum bearing capacity of the supporting element. However, the redistribution of the strain in the details of joints and welds is not correct because the stress does not correspond to the real values measured on the physically tested samples.

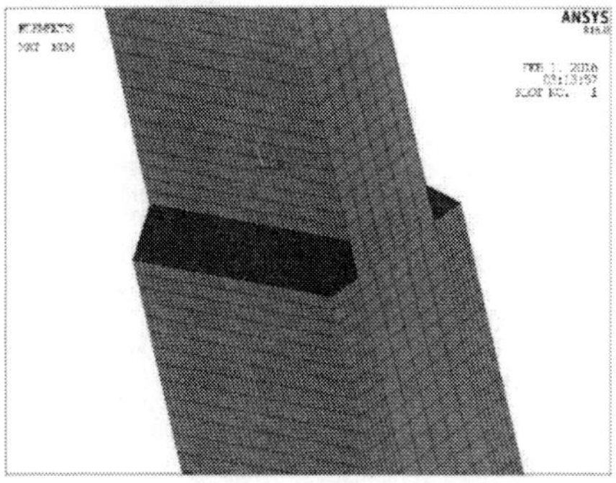

Figure 12. Numerical model of test specimen #4.

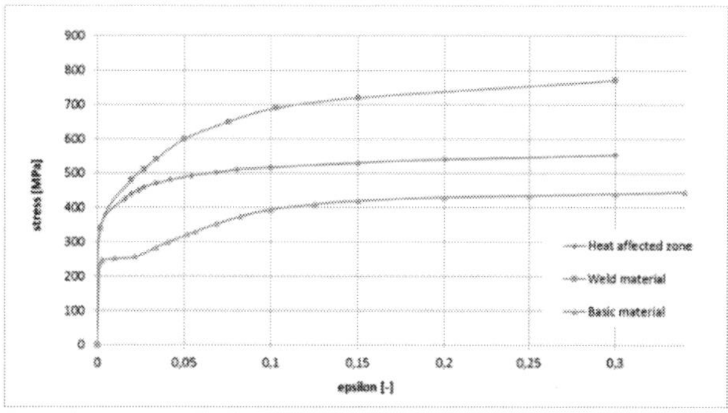

Figure 13. Used stress-strain diagrams for numerical modelling of test specimen #4.

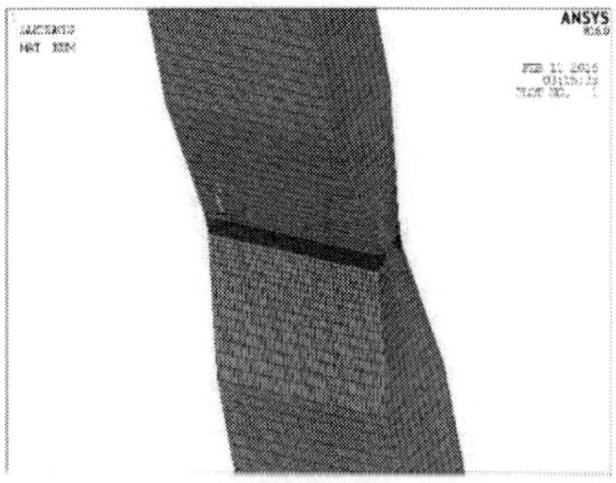

Figure 14. Numerical model of test specimen #5.

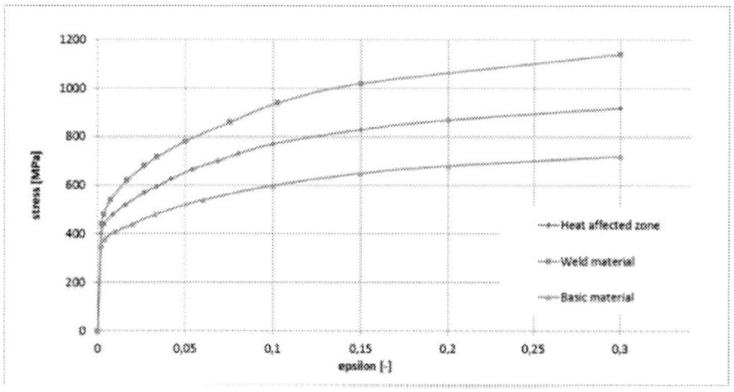

Figure 15. Used stress-strain diagrams for numerical modelling of test specimen #5.

The third material model has a general curve for hardening that characterizes the actual behavior of the material under load. This curve can have a continuous character, and can be approximated by a polyline. This material model most closely matches the actual behavior of the welded joint under analysis or detail under loading. This claim is based on the fact that the resulting stress-strain diagrams of numerical models were subjected to detailed analysis and compared with the experimental results. Overall, the comparison between the results of numerical modeling using the third material model and experimental data actually showed the best agreement – see Figs. 16 and 17. However, obtaining this curve is usually complex and time-consuming and costly.

3. Analysis of numerical modeling results

Analysis of numerical models was done in two ways. The first and basic is the physical test. Real measurement of deformations and strains together with input load is the most important benchmark for numerical modeling. A comparison showed very good agreement (see Figs. 16 & 17). The second way of evaluating is the experience with the material and the loadings, the standard assessments and the provisions that have an implemented estimate of the behavior of the phenomena studied. The experience and the feelings of the model maker that the numerical model under analysis creates and harmonizes is also essential. Even the exact numerical model may exhibit deviations from the physically measured values.

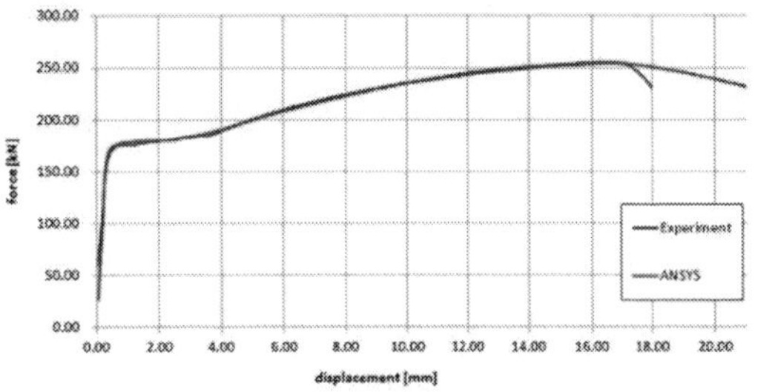

Figure 16. Comparison of stress-strain diagrams - numerical model vs. experiment, specimen #4.

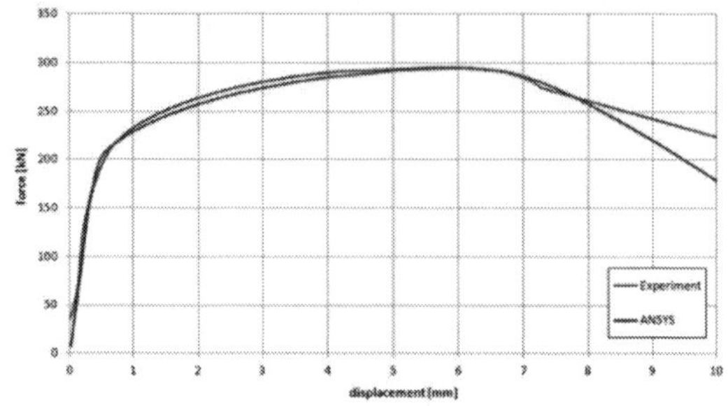

Figure 17. Comparison of stress-strain diagrams - numerical model vs. experiment, specimen #5.

In evaluating the numerical tests carried out, the following facts were found:

- It has been found that the analysed numerical models are not too sensitive to the increased density of the FEM network.
- Numerical models had to be modelled with taking account to the heat-affected zone for reasons of consistency with physical tests.
- The actual geometry of the entire welded joint and the heat-affected zone was modelled for a more precise match of the stress-strain curve behaviour of the numerical model with the physical tests.
- Numerical models included advanced material models for compliance with physical tests until sample failure.

4. Conclusion

In conclusion, we can summarize the newly acquired experience useful in the numerical modeling of the welds of supporting elements in steel structures. When creating numerical models in common cases, it is sufficient to use a FEM mesh with standard density, define description of geometry of the structure without considering initial imperfections and use the bilinear stress-strain diagram for the description of the used base material and the heat-affected area behavior. The reason is that the use of building structures in terms of stress occurs at the level of stress on the yield point and the effects of the load are below the level of plastic utilization of the supporting elements. The stress peaks are redistributed in the design using a bilinear work diagram, and the static behavior of the structure and its supporting elements can be evaluated at the strain according to standards.

If a more accurate analysis is needed, it is appropriate to use the numerical modeling procedure described in the article. This analysis is suitable for complex details especially for welded structures where there is multi-axis strain, notches and heat-affected zones with modified material properties. The residual stress produced by steel structures also plays its role here. In this regard, numerical models give valuable insight into the behavior of the structure and redistribution of stresses in it. Numerical models, however, need to be supplemented by physical tests.

References

[1] Di S-K, Wen C and Ye X-W 2015 Finite element analysis of structural hot-spot stress for orthotropic steel bridge deck *Journal of Zhejiang University (Engineering Science)* **49**(2) 225-231

[2] Drienovska J and Tvrda K 2017 Deflection of a Beam Considering the Creep *Procedia Engineering* **190** 459-463

[3] Ertas, A H 2015 Design optimization of spot welded structures to attain maximum strength *Steel and Composite Structures* **19**(4) 995-1009
[4] Kala Z 2015 Reliability analysis of the lateral torsional buckling resistance and the ultimate limit state of steel beams with random imperfections *Journal of Civil Engineering and Management* **21**(7) 902-911
[5] Knoedel P, Gkatzogiannis S and Ummenhofer T 2017 Practical aspects of welding residual stress simulation *Journal of Constructional Steel Research* **132** 83-96
[6] Kormanikova E, Riecky D and Zmindak M 2011 Strength of composites with fibres *Computational Methods in Applied Sciences* **24** 167-183
[7] Kotrasova K and Grajciar I 2014 Dynamic analysis of liquid storage cylindrical tanks due to earthquake *Advanced Materials Research* **969** 119-124
[8] Kralik J 2017 Actual problems of the safety and reliability of the NPP Structures in Slovakia *Key Engineering Materials* **738** 261-272
[9] Krejsa M, Brozovsky J, Mikolasek D, Parenica P, Flodr J, Materna A, Halama R and Kozak J 2017 Numerical modeling of steel fillet welded joint *Advances in Engineering Software* (in press) DOI: 10.1016/j.advengsoft.2017.03.013
[10] Krejsa M, Brozovsky J, Mikolasek D, Parenica P, Koubova L and Materna A 2017 Numerical Modeling of Fillet and Butt Welds in Steel Structural Elements with Verification Using Experiment *Procedia Engineering* **190** 318-325 DOI: 10.1016/j.proeng.2017.05.344
[11] Li H, Duan D, Huang C and Qi H 2017 Finite Element Analysis and Design of the Resistance Welding Structure of the Drying Plate *Journal of Tianjin University Science and Technology* **50**(10) 1093-1098
[12] Li Y, Zhou X-P, Qi Z-M and Zhang Y-B 2014 Numerical study on girth weld of marine steel tubular piles *Applied Ocean Research* **44** 112-118
[13] Major M and Major I 2017 Modelling of wave phenomena in the Zahorski material based on modified library for ADINA software *Applied Mathematical Modelling* **46** 727-735
[14] Malikova L and Seitl S 2017 Application of the Williams expansion near a bi-material interface *Key Engineering Materials* **754** 206-209
[15] Mei D-P and Liao G-X 2017 Notch stress analysis of welded connection of new type of edge upset U-rib and deck plate *Bridge Construction* **47**(1) 65-70
[16] Nemec I, Stekbauer H, Vaneckova A and Vlk Z 2017 Explicit and implicit method in nonlinear seismic analysis *MATEC Web of Conferences* **107** 00066
[17] Qiao G, Liu Y, Han X, Wang X and Xiao F 2017 Simulation study on effects of geometry size of weld joint on bearing capacity of steel pipe *Transactions of the China Welding Institution* **38**(3) 33-36
[18] Salajka V, Hradil P and Kala J 2013 Assess of the nuclear power plant structures residual life and earthquake resistance *Applied Mechanics and Materials* **284-287** 1247-1250
[19] Vales J, Kala Z, Martinasek J and Omishore A 2016 FE nonlinear analysis of lateral-torsional buckling resistance *International Journal of Mechanics* **10** 235-241
[20] Vican J, Gocal J, Melis B, Kotes P and Kotula P 2008 Real behaviour and remaining lifetime of bridge structures *Komunikacie* **10**(2) 30-37
[21] Yao Y-D, Yang Y-Q, Liu Z-B, Pu Q-H and Shi Z 2015 Finite element analysis of steel and concrete joint section of railway hybrid box girder cable-stayed bridge *Bridge Construction* **45**(1) 45-50
[22] Wei J, Wang D-M, Liu X-C, Wu Z-Q and Gao Z-Y 2015 Computer simulation of welding residual stress in steel girder bridge *Bridge Construction* **45**(4) 94-99

Acknowledgments
This contribution has been completed thanks to the financial support provided to VSB-Technical University of Ostrava by the Czech Ministry of Education, Youth and Sports from the budget for conceptual development of science, research and innovations for the 2018 year.

IOP Conf. Series: Earth and Environmental Science **143** (2018) 012005 doi:10.1088/1755-1315/143/1/012005

Experimental sensitivity analysis of subsoil-slab behaviour regarding degree of fibre-concrete slab reinforcement

E Hrubesova[1], H Lahuta[1], M Mohyla[1], T B Quang[2] and N D Phi[2]

[1] VSB - Technical University of Ostrava, Faculty of Civil Engineering, Czech Republic
[2] Ton Duc Thang University, Faculty of Civil Engineering, Ho Chi Minh City, Viet Nam

Corresponding email: eva.hrubesova@vsb.cz

Abstract. The paper is focused on the sensitivity analysis of behaviour of the subsoil – foundation system as regards the variant properties of fibre-concrete slab resulting into different relative stiffness of the whole cooperating system. The character of slab and its properties are very important for the character of external load transfer, but the character of subsoil cannot be neglected either because it determines the stress-strain behaviour of the all system and consequently the bearing capacity of structure. The sensitivity analysis was carried out based on experimental results, which include both the stress values in soil below the foundation structure and settlements of structure, characterized by different quantity of fibres in it. Flat dynamometers GEOKON were used for the stress measurements below the observed slab, the strains inside slab were registered by tensometers, the settlements were monitored geodetically. The paper is focused on the comparison of soil stresses below the slab for different quantity of fibres in structure. The results obtained from the experimental stand can contribute to more objective knowledge of soil – slab interaction, to the evaluation of real carrying capacity of the slab, to the calibration of corresponding numerical models, to the optimization of quantity of fibres in the slab, and finally, to higher safety and more economical design of slab.

1. Character of subsoil – structure interaction and its determinant factors

Interaction of the building structure with the rock environment determines the basic behaviour of the foundation structure and the whole upper building structure. The building structure, due to its load, causes both stress and deformation changes in the subsoil and as a result of these changes, additional forces and deformations can be brought back into the upper structure. The qualitative and quantitative nature of this interaction depends on many determinant factors. One of them is the relative stiffness of the entire cooperative system, which is determined by both the properties of the structure itself and the subsoil.

Soil materials below the slab are generally very heterogeneous and their behaviour under external loading depends on the following fundamental factors:

- the strength and deformation characteristics of the soil, in the case of cohesive soils, especially with respect to their humidity;
- consolidation processes in the soil (especially in the case of soft, low-permeable clay soils);
- changes of soil properties in time (soil rheology);
- sizes and load characteristics;
- history of loading;
- loading speed;

Relative stiffness of the collaborative “structure-soil“system is often defined by the Schultz relationship:

$$k = \frac{E_f}{E_s}\left(\frac{t}{l}\right)^3 \qquad (1)$$

where E_f = Young's modulus of the slab material, E_s = Young's modulus of the soil (weighted average module of layers), t = thickness of the slab, l = length of the slab. For relative stiffness k<1 is the foundation classified as flexible, for k>1, it is considered as a rigid foundation.

The behaviour of rigid and flexible foundations is significantly different in qualitative and quantitative terms, which is reflected in qualitative and quantitative differences of internal forces in the structure. It is generally known that, in the case of a flexible foundation (k <1) a significant deflection of the foundation structure occurs, and the contact stress is uniform [1]. For a rigid foundation (relative stiffness greater than 1), the contact stress concentrations are manifested below the edges and corners of the foundation and consequently, depending on the shear strength and corresponding development of the plastic strains, they are subsequently redistributed [2] - Fig. 1. At a small load, the plastic manifestation of soil occurs in a very small area below the foundation edge (or corner). In this case the distribution of the contact stresses is very close to the theoretical result with infinitely large contact stresses below the edge (corner) of the rigid foundation [3, 4, 5, 6]. When the load increases, the plastic zone expands, and the contact stress corresponds to a saddle-shape, adequate settlements are approximately uniform [7]. Further increase of the load causes redistribution of the contact stresses towards the axis of the foundation and the development of the contact stresses corresponds to the parabolic shape. If the shear strength of the foundation soil is exceeded, the stress concentration increases below the foundation axis and the contact stresses correspond to a bell-shaped pattern. The nature of the contact stresses therefore depend very much on the strength characteristics of the soils [8, 9]. In the case of the cohesive subsoil, the contact stress in the rigid foundations is predominantly saddle-shaped, in the case of non-cohesive subsoil, the contact stress often has a parabolic shape.

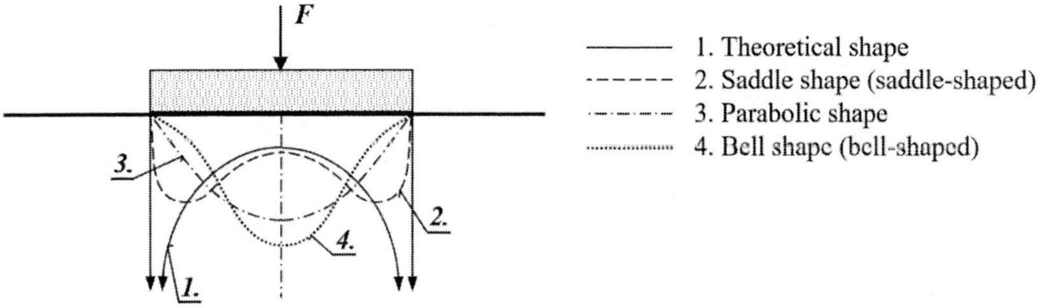

Figure 1. Various shapes of contact stresses under the rigid foundation.

To ensure reliable design of the foundation from the point of view of the interaction with the subsoil, the change of stiffness of the entire cooperating system must be considered during the loading. This change in stiffness results from the changes in both the soil material and the structural material. When a certain load value is exceeded, the softening of some parts of the structure may occur due to microcracks. On the other hand, the development of plastic strains, seasonal precipitation fluctuations and the consolidation of the soil environment can also induce a change in the stiffness and strength of the subsoil during loading. These aspects can then change the relative stiffness of the foundation and consequently change the entire investigated interaction.

2. Characterization of experimental works

Our experimental research was focused on the behaviour of the foundation fibre-concrete slab of square shape supported directly by the cohesive soil. Dimension of the square concrete slab was 2 x 2 m, thickness of 0.15 m. The basic matrix of slab corresponds to the concrete C25/30, three different quantities of fibres DRAMIX 3D 65/60 BG were added to the concrete [10].

The geological profile below the slab corresponds to loam F4 (minimal thickness of 5 meters), character of deeper soil is not known in detail, but a soft soil is assumed in deeper horizons too. The original soil was rather inhomogeneous due to the building intervention into the ground during the construction process of the stand. Therefore the subsoil was homogenized before the start of the experiment. Homogenization included the digging of the original soil to the depth of approximately 1.2 m (depth corresponding approximately to the depth of the location of the stand foundation), mixing of excavated soil so as to achieve quasi-homogenized material, back-refilling the dig with the homogenized soil and compaction in partial layers [11, 12]. Water level was not reached until bottom of the dig. Homogenized subsoil was tested using both laboratory and in-situ methods. Static load plate test was used not only for the evaluation of the deformational modulus of the subsoil, but also for the verification of quality of homogenization. Material parameters of quasi-homogenized subsoil obtained from laboratory and field testing (carried out in July 2017) are shown in Table 1.

Table 1. Material parameters of quasi-homogenized subsoil under tested foundation.

	Unit weight	Deformational modulus	Poisson's number	Cohesion	Friction angle
	kN/m^3	MPa		kPa	degree
Homogenized subsoil	19	12.9	0.35	9.3	19.3

The deformation and strength characteristics of the tested fibre-concrete materials were determined on cylindrical (diameter of 150 mm, height of 300 mm) and cubic samples (150 x 150 x 150 mm) in the Laboratory of Building Materials at the Faculty of Civil Engineering of VSB-Technical university Ostrava and their average values are presented in the following table No. 2. This table contains also the calculated relative stiffness according Schultz formula (corresponding to the Young modulus of soil 12.9 MPa) and information related to the precipitation in the previous 14 days before testing. Three slabs differing in the content of fibres were tested, the slabs labelled G05, G06 and G07 corresponded to the addition of fibres of 25 kg/m^3, 50 kg/m^3 and 75 kg/m^3 respectively.

From the calculated relative stiffness of the tested slabs (at a constant value of the Young´s modulus 12.9 MPa) follows that the slabs can be considered as flexible foundations. The mentioned precipitation before the tests of individual plates shows that the real relative stiffness of the foundation could be affected not only by the change in the contents of the fibres in the slabs but also by the potential degradation of the deformation properties in their subsoil, especially in the near-surface layers.

The slabs were loaded in the central part by gradual variable surcharge applied by a hydraulic press (Figure 2). Loading was performed using a special strutted steel structure (experimental stand), applied step by step at uniform velocity until the slab failed (partial loading step corresponds to 75 kN) [13]. The experimental measurements of slab settlement and vertical stresses in the subsoil below the slab were carried out during the loading process.

Results of monitoring related to settlements and strain inside the slab are not the focus of this paper, they were presented in papers [14, 15, 16, 17, 18]. This paper is focused on monitoring of vertical stresses in subsoil below the slab, which was performed using 9 vibrations wire flat cells GEOKON [19] located at three depth levels (marked X), as is shown in Figure 3 (localization on contact surface (X=1), 0.35 m below the contact surface (X=2), 0.8 m below the contact surface (X=3)). In each of the three depth levels three flat cells are located, one cell is placed in the centre of the slab (Y=1), the second one at the centre of the edge of the slab (Y=2) and the third cell is located at the corner (Y=3). Thus, flat cells located at the contact surface are marked 1.1., 1.2, 1.3, cells at 0.35 m (resp. 0.8 m) below the surface are marked 2.1, 2.2, 2.3 (resp. 3.1, 3.2, 3.3).

Table 2. Characteristics of tested slabs labelled G05, G06 and G07.

Slab label	Precipitation (mm)	Date of the test	Content of fibres kg/m^3	Compressive strength (MPa) (cylindrical samples – an average of 3 tests)	Compressive strength (MPa) (cubic samples – an average of 6 tests)	Young's module of elasticity (GPa) (cylindrical samples – an average of 3 tests)	Relative stiffness (Schultz)
G05	23	21.6.2017	25	24.86	30.9	19.7	0.644
G06	44	18.9.2017	50	22.1	24.4	18.5	0.605
G07	17	26.10.2017	75	23.6	28.2	18.3	0.598

Figure 2. Scheme of experimental measurements on the stand

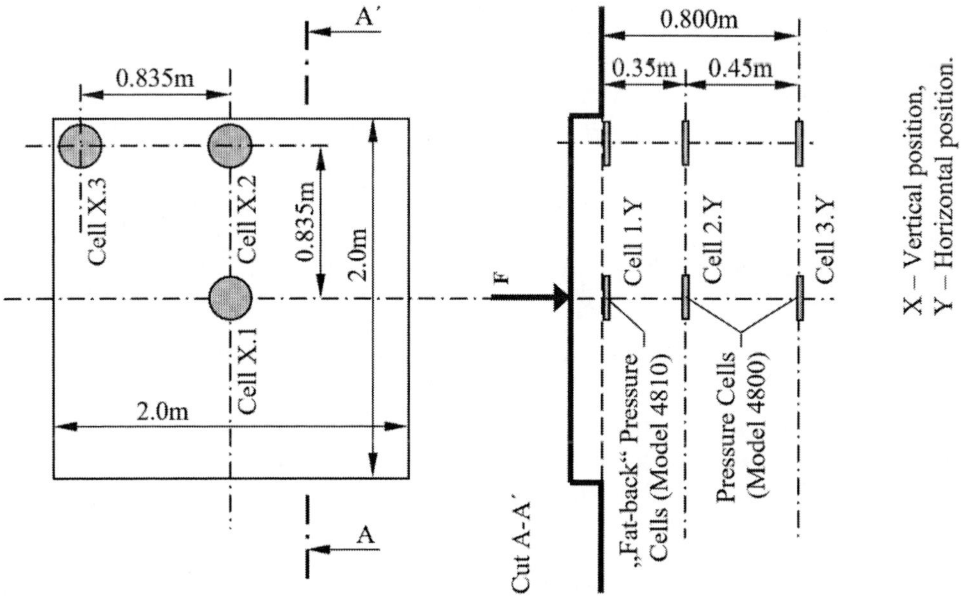

Figure 3. Localization of flat cells in experiment (in plan and cross section)

3. Results of experimental works

In the following Figure 4 are presented the total vertical stresses below the three variant slabs marked G05, G06 and G07. Evaluation is carried out under the assumption that the self-weight of the slab and original stress (soil weight above the measuring cell) is removed.

The first row in the Figure 4 corresponds to stresses in two various depths below the centre point of the slab (one sensor stopped working at the depth of 0.35 m, so only two depths were investigated and evaluated) for all tested slabs G05, G06, G07. The second row represents the stresses in three various depths below the centre of the slab edge, the 3rd row illustrates the situation below the corner of slab.

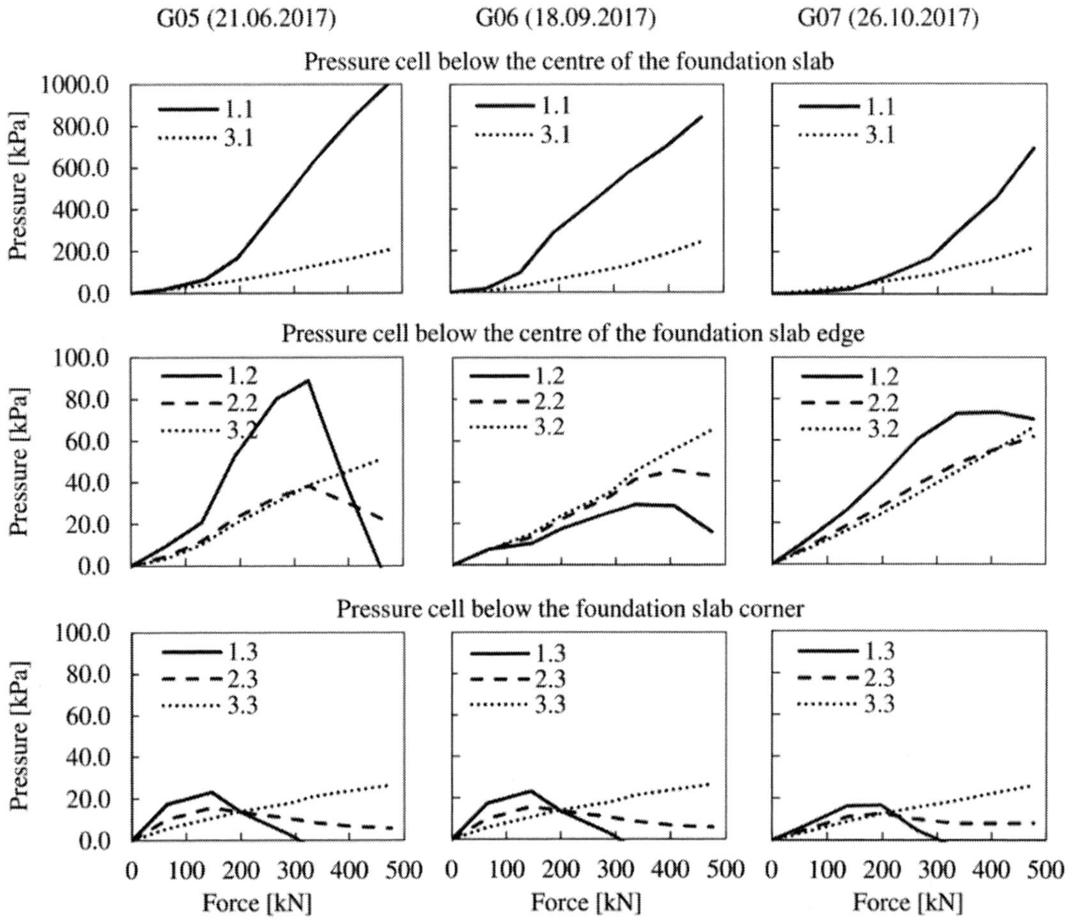

Figure 4. Assessment of monitored stresses below the slabs G05, G06, G07

Results obtained from the cell 1.1 (contact cell located below the centre point of slab) are presented in Fig. 5. Contact stresses below the centre of slab G05 (the slab with the high relative stiffness) increase significantly until the loading 180 kN is achieved, at higher load we registered a lower gradient of contact stresses at this monitoring point.

In the case of slab G06, the contact stress increases with the lower gradient at lower load level, when this load of 300 kN is exceeded, the gradient of the contact stresses continuously increases. Due to the softening of the subsoil due to the increased moisture in the subsoil layer below this slab G06 (higher moisture is induced by stronger rainfall before testing) the relative stiffness increases. Compared with other slabs, it is very important to note that relative stiffness of foundation increases due to the expected (but not quantified) softening of the subsoil below this slab G06 due to the significant precipitation prior to its testing. This impact is registered and documented (in accordance with the theory) by decreasing of contact measured stresses in the central zone below the slab G06 and by increasing of contact stresses below the corner.

In the case of the G07 slab test, the increased gradient of the contact stress is evident mainly in the middle of the loading interval (200-300 kN). Outside of this loading interval the gradient of contact stresses is lower.

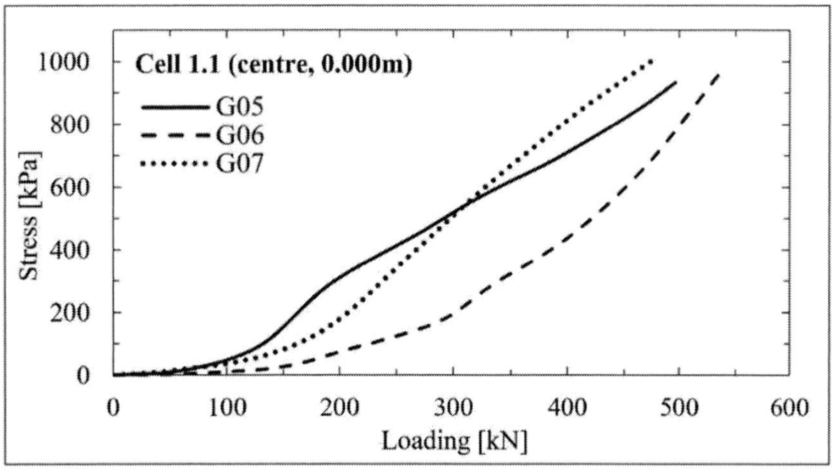

Figure 5. Load - contact stress curve below the centre of slab

Fig. 6 shows a load - contact stress curve below the centre of the slab edge. The results of the measurements document the reduction in the stresses at a specific load level resulting from a significant deflection of the slab that was observed during the test. This deflection leads to an uplift of the edges and corners of slab and a gap between the soil and slab opens in some part of contact surface (contact surface is reduced).

A more significant impact of the uplift of the slab corners is documented in Figure 7. The results presented in Fig. 6 and Fig. 7 show that a loss of contact between the slab and the subsoil at the slab corners occurs at lower load compared to the load required to create a gap on the edge of the slab.

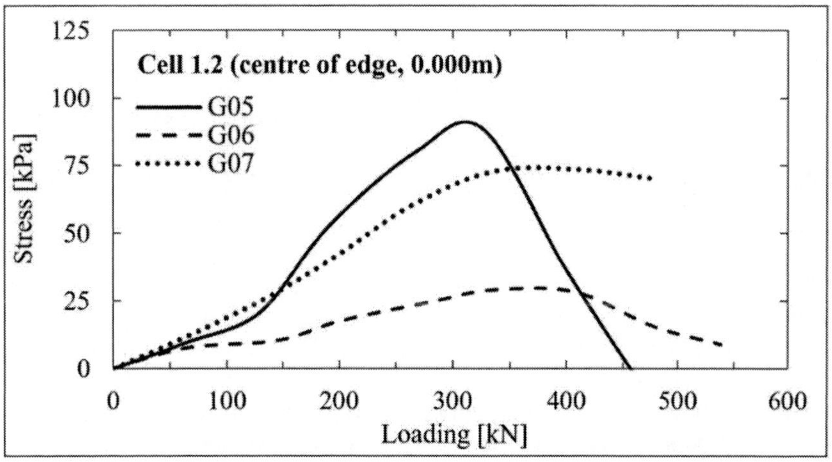

Figure 6. Load - contact stress curve below the centre of slab edge

The test results did not show too great differences in the final load value, which corresponds to the extrusion of the slab (the maximum extruding force is approximately equal to 500 kN).
The pressure cell located at the maximum investigated depth below the slab (depth of 0.8 m) does not record any significant differences in the values of the vertical total stresses under the slabs (Fig. 8). Some minor differences are visible for G05 slab, but there are observed for the load higher than 350 kN only.

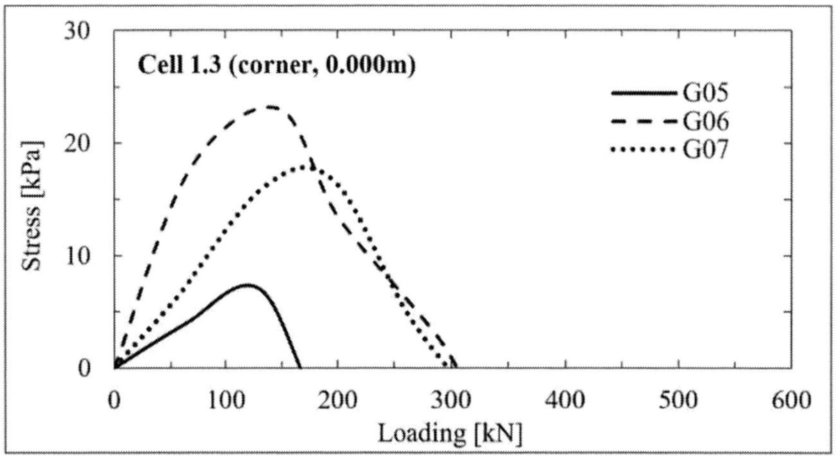

Figure 7. Load - contact stress curve below the corner of slab

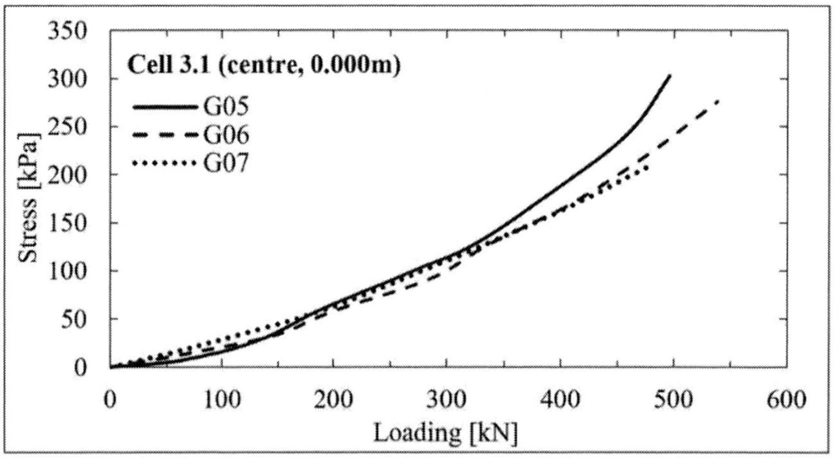

Figure 8. Load - stress curve below the centre of slab (depth 0.8 m)

4. Conclusions

- Slab-subsoil interaction is a complex problem depending on many factors, including the properties of the structure and the subsoil. This interaction has a significant rheological character confirmed by the results of measurements. The relative stiffness of the foundation may not be identical throughout the whole load cycle. Corresponding changes in stiffness can be caused, among other factors, by changes in the subsoil below slab (influence of soil moisture variation, etc.).
- Experimental stress measurements indicated some decrease in the effective contact surface between the slab and the subsoil during the loading due to the uplift of the corners and edges of slab.

- The vertical contact stresses induced by the vertical axial load of the slab reach the maximum values below its centre, regardless of the slab variant. Lower contact stresses are reached below the edge of the slab, the lowest stresses are monitored below the corner of the slab. In this context, however, it is important to note, that the uplift of the edges and corners of the slab is manifested during test. This phenomenon results into the reduction of vertical stresses at these measuring points during the loading.
- Different contact stresses are evident for the individual tested slabs, but it is not possible to clearly distinguish the effect of the different degree of reinforcement of slab (among of fibres) on the one side and change in the properties of the subsoil on the other side.
- Some differences in vertical total stress are identified for specific slabs at a depth of 0.35 m below the contact surface. Because the monitoring pressure cell located below the centre of the slab showed a defect at the beginning of measurement already, the values of vertical total stresses below the centre could not be evaluated.
  The performed measurements show, that up to the load approx. of 300-400 kN the significant differences of the vertical stresses below the centre of the edge of the slabs G05 and G06 are not indicated. If this load limit is exceeded, the vertical stresses below the slabs G05 and G06 gradually drop probably due to the uplift of edges and corners of the slabs. A different situation occurs at this depth under the slab G07. Below this slab the stresses below the centre of the edge increase with the increasing loading up to the maximal applied load (unlike the situation below the slabs G05 a G06 the monitoring did not detect a significant stress drop).
- The influence of corner uplift corresponding to the specific loading is manifested at the depth of 0.35 m under the contact surface too. At this depth level the vertical stress under the corner of slab G05 drops to a force of 150 kN, under the slab G06 to 180 kN and under the slab G07 to 200 kN. The maximal value of the induced vertical stress at this depth achieves approx. of 20 kPa independently of the type of slab.
- At a depth of 0.8 m below the contact surface (maximum investigated depth), the effect of the different contents of the fibres in the slab as well as the potential degradation of soil characteristic is eliminated already (registered stresses are practically identical).

References

[1] Boswell L F and Scott C R 1975 A flexible circular plate on a heterogeneous elastic „half-space" influence coefficients for contact stress and settlement *Geotechnique* 25 **3** pp 604-610

[2] Myslivec A, Jesenak, J and Eichler J 1970 *Mechanika zemin* Praha SNTL

[3] Davis E H and Poulos H G 1968 The use of elastic theory for settlement prediction under three-dimensional conditions *Geotechnique* 18 **1** London pp 67-91

[4] Balakrishna C K, Murthy B R S and Nagaraj T S 1992 Stress distribution beneath rigid circular foundations on sands *International Journal for Numerical and Analytical Methods in Geomechanics* 16 **1** pp 65-72

[5] Delpak R, Rowlands G O and Siraty A 1992 Modelling of strip foundations for serviceability assessment *Structural engineer London* 70 **23-24** pp 412-420

[6] Bose S K and Das S C 1997 Nonlinear finite element analysis of stresses and deformations beneath rigid footings *Computers and Structures* 62 **3** pp 487-492

[7] Abdullah W S 2008 New elastoplastic method for calculating the contact pressure distribution under rigid foundations *Jordan Journal of Civil Engineering* 2 **1** pp 71-89

[8] Yamin M M, Ashteyat A M, Al-Mohd I and Mahmoud E 2017 Numerical study of contact stresses under foundations resting on cohesionless soil: Effects of foundation rigidity and applied stress level *KSCE Journal of Civil Engineering* 21 **4** pp 1107-1114

[9] Brown P T 1969 Numerical analysis of uniformly loaded circular rafts on elastic layers of finite depth *Geotechnique* 19 **2** pp 301-306

[10] Sucharda O, Bilek V, Smirakova M, Kubosek J and Cajka R 2017 Comparative evaluation of mechanical properties of fibre-reinforced concrete and approach to modelling of bearing capacity ground slab *Periodica Polytechnica Civil Engineering* 61 **4** pp 972-986

[11] Stolarik M, Hrubesova E and Pinka M 2015 Dynamic response of the experimental foundation slab - seismic measurement and analysis *International Multidisciplinary Scientific GeoConference Surveying Geology and Mining Ecology Management SGEM* 2 **1** pp 261-268

[12] Mohyla M, Vojtasik K, Stolarik M, Pinka M and Lahuta H 2017 Experimentally measurement and analysis of stress under foundation slab *International Journal of GEOMATE* 13 **35** pp 128-135

[13] Lahuta H, Hrubesova E, Duris L and Petrasova, T 2015 Behaviour subsoil of slab foundation under loading International Multidisciplinary Scientific GeoConference Surveying Geology and Mining Ecology Management SGEM 2 1 pp 119-126

[14] Cajka R, Burkovic K and Buchta V 2014 Foundation slab in interaction with subsoil *Advanced Materials Research* Vol 838-841 pp 375-380

[15] Hrubesova E, Lahuta H, Duris L and Jaafar M 2015 Mathematical modeling of foundation-subsoil interaction *International Multidisciplinary Scientific GeoConference Surveying Geology and Mining Ecology Management, SGEM* 2 **1** pp 437-444

[16] Cajka R, Labudkova J and Mynarcik P 2016 Numerical solution of soil - foundation interaction and comparison of results with experimetal measurements *International Journal of GEOMATE* 11 **1** pp 2116-2122

[17] Mynarcik, P, Labudkova J and Koktan J 2016 Experimental and numerical analysis of interaction between subsoil and post-ensioned slab-on-ground *Jurnal Teknologi* 78 **5-4** pp 23-27

[18] Mynarcik P and Cajka R 2016 Experimental testing of post-tensioned concrete industrial floor model - Subsidence analysis *International Journal of Mechanics* 10 pp 33-38

[19] Mohyla M., Vojtasik K, Hrubesova E, Stolarik M and Pinka M 2016 The optimalization of pressure cell installation via numerical simulation *International Multidisciplinary Scientific GeoConference Surveying Geology and Mining Ecology Management SGEM* 1 pp 1029-1036

Acknowledgments
This paper was elaborated with the financial support of the project GACR No. 16-08937S "State of stress and strain of fibre reinforced composites in interaction with the soil environment ".

IOP Conf. Series: Earth and Environmental Science **143** (2018) 012006 doi:10.1088/1755-1315/143/1/012006

Design of foundations with sliding joint at areas affected with underground mining

P Matečková[1], M Šmiřáková[1] and P Maňásek[2]

[1] VSB – Technical University of Ostrava, Faculty of Civil Engineering, L. Podeste 1875, 708 33 Ostrava - Poruba, Czech Republic

[2] IP system, Inc., U panelarny 573/3, 772 00 Olomouc, Czech Republic

*Corresponding author: pavlina.mateckova@vsb.cz

Abstract. Underground mining always influences also landscape on surface. If there are buildings on the surface they are affected with terrain deformation which comprises terrain inclination, curvature, shift and horizontal deformation. Ostrava – Karvina region is specific with underground mining very close to densely inhabited area. About 25 years ago there were mines even in the city of Ostrava. Recommendations and rules for design of building structures at areas affected with underground mining have been therefore analysed in long term. This paper is focused on deformation action caused by terrain horizontal deformation - expansion or compression. Through the friction between foundation structure and subsoil in footing bottom the foundation structure has to resist significant normal forces. The idea of sliding joint which eliminates the friction and decreases internal forces comes from the last century. Sliding joint made of asphalt belt has been analysed at Faculty of Civil Engineering, VSB – Technical University of Ostrava in long term. The influence of vertical and horizontal load and the effect of temperature in temperature controlled room have been examined. Testing, design and utilization of sliding joint is presented.

1. Introduction

Building structures in areas affected by underground mining demand specific treatment due to expected terrain deformation. Terrain deformation comprises subsidence, declination, curvature, and horizontal deformation. The most demanding, and also most expensive, are requirements for terrain horizontal deformation. Through the friction between subsoil and foundations, the foundation structure is loaded with significant normal forces. The idea of elimination of this friction with layer of sliding joint and thus decrease the normal forces comes from the 1970's, figure 1. Several materials for sliding joint between subsoil and foundation were considered in the beginning. The bitumen asphalt belt, given its rheological properties, has been chosen as an effective material for sliding joints.

Primary experiments of asphalt belt sliding joints were realized in the 1980's and authors of this experiment were Balcarek and Bradac [1]. The scheme of the testing is shown in figure 2. A concrete block weighing 2208 kg was placed on the asphalt belt on an inclined plane, and the displacement was measured for different tilt angles, and consequently different horizontal and vertical force. This primary experiments were carried out for asphalt belts common in time of testing and proved rheological properties of asphalt sliding joint. Though many significant aspects were not taken into account (e.g. influence of a different vertical load) experiment results were the basis for sliding joint design principles mentioned in still valid national code [2] and lead to a linear dependence of shear stress and deformation rate. In simplified way it is also possible to involve the influence of temperature.

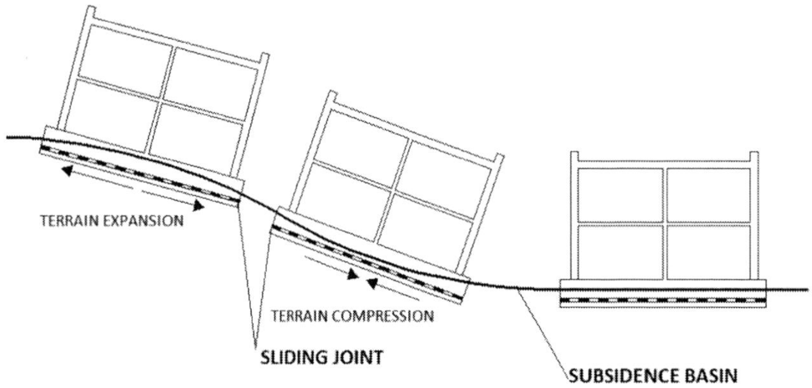

Figure 1. Scheme of sliding joint application

Motivation for new testing was mainly wide range of new asphalt belt types in the beginning of new century. With new testing there was effort to take into account the influence of vertical and horizontal load and later also influence of temperature.

The utilization of sliding joints is not limited to areas affected with underground mining, it is valuable instrument in case of horizontal deformation load due to pre – stressing, in the paper [3] it is presented utilization of sliding joint made of plastic foil for decrease of horizontal deformation load in case of pre-steressing of oil tanks. Experiments have been carried out also by Hornikova et al [4] concerning assessment of slab-on-grade friction coefficient for elimination of horizontal deformation load due to shrinkage and pre-stressing of concrete floors.

Though the bitumen sliding joint was successfully applied in a few buildings [5], sliding joints have not been widely used yet. Experiments presented in this paper should contribute to a wider utilization of sliding joint and thus enable design of more durable and sustainable building structures.

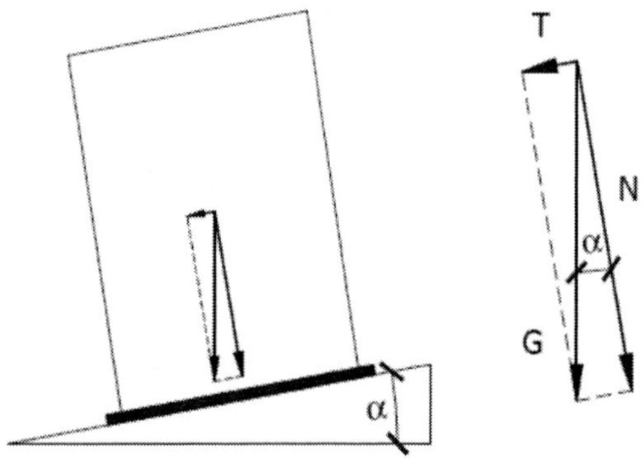

Figure 2. Sliding joint pilot testing

2. Sliding joint testing

2.1. Testing principle

Shear resistance is the main material characteristic when using asphalt belt as sliding joint. It was found out that the shear resistance of slide joints is primarily dependant on the deformation velocity. When the deformation velocity is slow the shear resistance of the bitumen sliding joint is low. The terrain deformation velocity could be estimated on the basis of exploitation plan.

However, testing of the shear resistance for particular deformation velocity is problematic. It was decided to appoint experimentally the deformation velocity for different shear stresses. Using linear regression, it is possible to appoint the shear resistance of a slide joint as a function of deformation rate.

2.2. Testing equipment

At VSB – Technical University of Ostrava unique equipment was designed for shear resistance measurement, figure 3. Experiments started in 2008 and have proceeded continuously. Asphalt belt specimens are placed in between concrete blocks with a dimension of 300 x 300 x 100 mm. Specimens are exposed to a vertical load. A horizontal load is applied after a one-day delay. Displacement of the middle concrete block is measured for 6 days, and sometimes also for more days. Selected materials have been tested also for arbitrary temperature when the testing equipment was placed in a temperature controlled room.

Figure 3. Equipment for sliding joint testing

2.3. Testing results

Primitive asphalt is refined with oxidization or modified with an admixture of polymers. Depending on the type of admixture polymer there are asphalt belts modified with rubber, usually styrene-butadiene-styrene (SBS asphalt) and thermoplastics, mostly amorphous polypropylene (APP asphalt). Oxidized and modified asphalts possess different temperature sensitivity, elasticity and plasticity or adhesiveness also in correlation with the amount of admixture.

A few types of different trademarks bitumen asphalt belts have been tested since the year 2008. The thickness of asphalt belts is between 3-5 mm, and they are predominantly covered with mineral gritting. Four types were oxidized, nine types SBS modified and 1 type APP modified. Six types of asphalt belt were tested with dependence of temperature.

The specimens were exposed to vertical load 100 kPa and 500 kPa. Applied horizontal load is 1.0 kN and 2.0 kN, such that the deformation velocity corresponds with expected terrain deformation velocity.

In the figure 4 there are measured displacements for one type of oxidized asphalt belt and in the figure 5 there are measured displacements for one type of SBS modified asphalt belt, both for the temperature 20 °C.

The testing temperature was 20 °C. Shear resistance of some specimens was tested also for temperature 10 °C, which corresponds better with temperature expected in footing bottom, appropriate test results are published in papers [6], appointing of expected temperature in footing bottom is discussed in the paper [7].

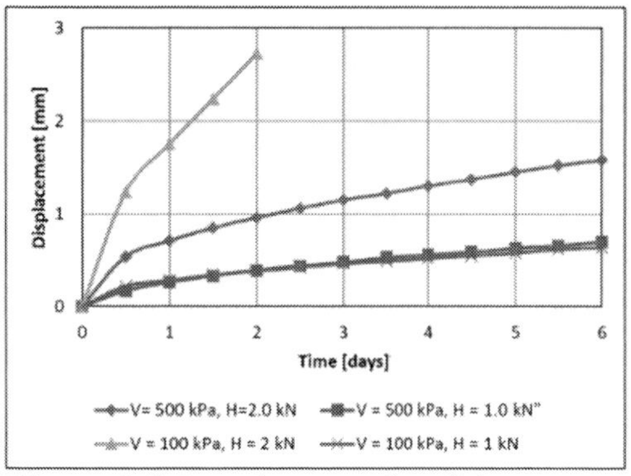

Figure 4. Selected test results of shear resistance, oxidized asphalt belt, 20°C

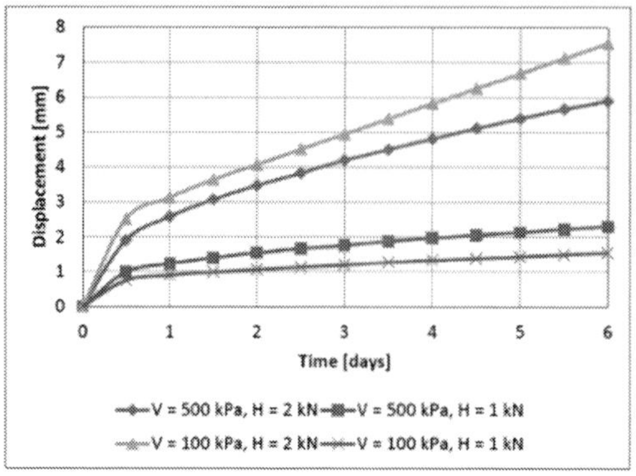

Figure 5. Selected test results of shear resistance, SBS modified asphalt belt, 20°C

3. Design of sliding joint

3.1. Shear stress in footing bottom

A simple method for appointing the shear stress between a foundation and subsoil without a sliding joint brings forth Czech code [2]. Shear stress between a foundation and subsoil are settled as a function of terrain horizontal deformation ε, dimensions of the foundation structure and oedometric modulus of subsoil E_{oed}.

This code includes also simple formulas for appointing the shear stress with a sliding joint. Shear stress between a foundation and subsoil is settled as a function of terrain horizontal deformation rate, temperature is taken into consideration in a simplified way, figure 6. Type of the asphalt belt and vertical load are not taken into consideration.

The advantage of calculating shear stress based on test results is possibility to take into account many parameters, e.g. type of asphalt belt, expected vertical load, temperature in footing bottom and time of deformation duration. Sher stress functions based on experiments are in the figure 6. In the charts it is visible that the formula according to current code for the temperature 10°C, leads to similar shear resistance as tested SBS modified asphalt belt.

Results of asphalt belt testing can be used also for numerical analysis of soil-structure interaction. Shear resistance is defined with parameters C_{1x}, analogically to parameter of vertical resistance defined in Winkler one parametrical model, more details are in the paper [8].

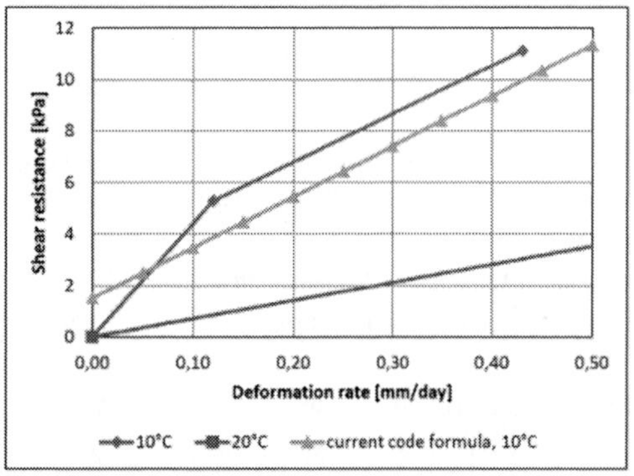

Figure 6. Shear resistance of sliding joint, SBS modified asphalt belt, vertical load 500 kPa

3.2. Pilot example

Stress function is used in model example of strip foundation situated in an area attached to underground mining. The length is 20 m, and width 1.0 m. It is supposed that the strip foundation is exposed to a vertical load 0.5 MPa, laid on gravel. According to the data obtained from mining company it is expected that relative horizontal terrain deformation is 5 mm/month, at first the subsoil is compressed and then expanded, each phase will be lasting 16 months. Comparison of shear stress between subsoil and strip foundation with and without sliding joint is in the figure 7. In the figure 8 significant decrease of normal force acting on foundation structure when using sliding joint is presented. The decrease is remarkable both for sliding joint made of oxidised asphalt and sliding joint made of SBS asphalt belt.

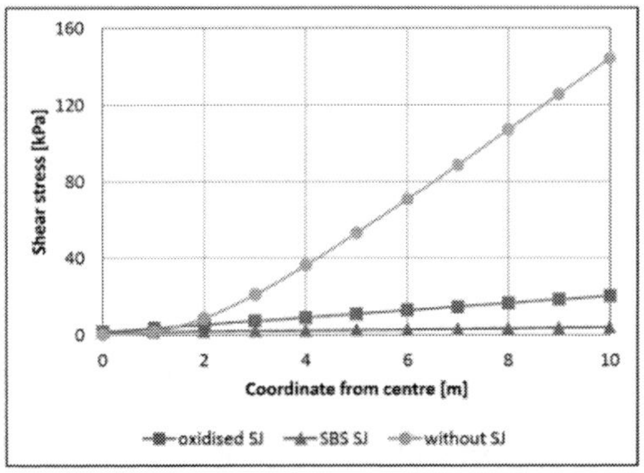

Figure 7. Shear stress with and without sliding joint in pilot example, vertical load 500 kPa, temperature 10°C, SBS modified/oxidised asphalt belt

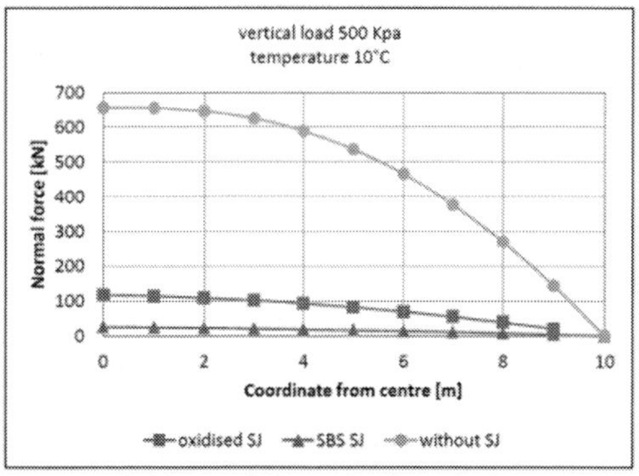

Figure 8. Normal force in foundation structure with and without sliding joint in pilot example, vertical load 500 kPa, temperature 10°C, SBS modified/oxidised asphalt belt

4. Planned testing

Research at technical university in Ostrava is focused on soil structure interaction, both in the experimental field and in the consecutive mathematical modelling [9]. On the yard of the technical university testing devise was built up for testing of simple model foundation structures, figure 9. It is possible to analyse experimentally foundations up to dimensions 2.0 m x 2.0 m. Vertical mechanical load is applied using a hydraulic press which is placed between the tested structure and steel frame. Limit value of mechanical load is 1000 kN. Since 2012 research team has tested a few slabs made of plain concrete, reinforced concrete, pre-stressed concrete and recently also fiber concrete. In 2018 it is planned to concrete also foundation slab with sliding joint and apply horizontal load together with vertical load. Experiments should complement the small-size laboratory testing of sliding joints.

Figure 9. Soil and foundation interaction testing

5. Conclusion

In the paper shear resistance testing of different types of asphalt belt was presented. Obtained data are utilized for design of sliding joint. When the friction in footing bottom is decreased normal forces in foundation structure could be significantly decreased.

Laboratory testing of sliding joints comprises testing of different types of sliding joints exposed to various vertical and horizontal load and influence of temperature. Testing of big-size sliding joint specimen with dimensions 2 x 2 m is planned in 2018.

Sliding joint has been already applied in a few buildings. Experiments which have been carried out at VSB – Technical University of Ostrava should contribute to a wider utilization of sliding joint and thus enable design of more durable and sustainable building structures.

References

[1] Balcarek V and Bradac J 1982 Utilization of bitumen insulating stripes as sliding joints for buildings on undermined area, *Journal Civil Engineering*, 2/1982 (In Czech)

[2] CSN 730039 20015 *Design of constructions on the mining subsidence areas* Czech code, UNMZ Prague (In Czech)

[3] Vitek J L, Strasky J, Zich M and Kasal P 2010 Prestressed concrete oil tanks *Proceedings 3rd International Fib Congress and Exhibition, Incorporating the PCI Annual Convention and Bridge Conference: Think Globally, Build Locally* Washington, USA

[4] Hornikova K, Foglar M, Kolisko J and Kolar J 2016 Experimental and numerical assessment of slab-on grade friction coefficient. *Journal Beton TKS* 6/2016, 42-46 (In Czech)

[5] Cajka R, Labudek P, Burkovic K and Cajka M 2014 Construction of a green golf club buildings on undermined area. *4th International Conference on Green Building, Materials and Civil Engineering,* 861-867.

[6] Cajka R, Janulikova M, Mateckova P and Stara M 2011 Modelling of foundation structures with slide joints of temperature dependent characteristics. *Proceedings of the Thirteenth International Conference on Civil, Structural and Environmental Engineering Computing*, Crete, Greece. DOI:10.4203/ccp.96.208

[7] Cajka R, Mateckova P 2011 Temperature distribution of slide joint in reinforced concrete foundation structures. *17th International Conference on Engineering Mechanics*. Svratka, Czech Republic. WOS: 000313492700017

[8] Cajka R 2013 Analytical derivation of friction parameters for FEM calculation of the state of stress in foundation structures on undermined territories. *Acta Montanistica Slovaca* 18.4, 254-261. WOS:000343184100006.

[9] Labudkova J and Cajka R 2017 3D numerical model in nonlinear analysis of the interaction between subsoil and SFCR slab. *Int J Geomate* 13(35):120-127.

Acknowledgments

This outcome has been achieved with the financial support of the project GACR No. 16-08937S "State of stress and strain of fiber reinforced composites in interaction with the soil environment".

Thermal changes of the environment and their influence on reinforced concrete structures

R Fojtik[1*], R Cajka[1]

[1]Faculty of Civil Engineering, Technical University of Ostrava, Ludvika Podeste 1875/17, 708 00, Ostrava-Poruba, Czech Republic

*Corresponding author: roman.fojtik@vsb.cz

Abstract. The thermal expansion of concrete elements concerns both monolithic and prefabricated structures. Inappropriate design of dilation segments may cause minor but even larger failures. Critical environment factors are temperature-changing operations, such as unheated underground garages, where temperature fluctuations may occur depending on the exterior conditions. This paper numerically and experimentally analyses the thermal deformation of selected girders in the underground garages and the consequent structure failures, their causes, possible prevention and appropriate remediation.

1. Introduction

Every structure bears a load that is designed for. Temperature load is one of the loads that act more or less on every structure. This load results from temperature expansion [1]. The phenomenon of thermal expansion is essential when designing huge or long structures as bridges, halls, frames, pipelines etc. [2], [3], [4], [5]. Thermal expansion is a material property and it is unique to each material [6]. It is common to combine materials with similar thermal expansion because of minimization of unfavorable stress states caused by temperature changes.

Similar thermal expansion coefficients α [1/°K] of steel (12 * 10-6 [1/°K]) and concrete (10 * 10-6 [1/°K]) allow using these two materials as a composite material called reinforced concrete. The reinforced concrete is one of the most frequently used materials for building structures. When designing these structures, it is necessary to consider thermal expansion which can be prevented by dilatation or reinforcement of the structure [7], [8], [9]. Otherwise, it may cause minor but even larger failures.

Temperature fluctuation is very often present open and unheated garage areas, for example shopping mall underground garages or garage of IT4Innovations national supercomputing center of Ostrava [10], [11].

Failures from the temperature change are almost perpendicular to the direction of failure source (thermal expansion). It is advisable to control the size of the cracks changing with temperature in order to determine the cracks caused by thermal expansion. This paper analyzes one yearlong monitoring of the creation and development of the cracks on shopping mall structure.

2. Structure and diagnostic methods

Cracks caused by thermal expansion were long-term monitored on reinforced concrete frame building. The main supporting structure is built out of prefabricated reinforced concrete members. The building

IOP Conf. Series: Earth and Environmental Science **143** (2018) 012007 doi:10.1088/1755-1315/143/1/012007

is supported by reinforcement concrete piles. Prefabricated RC girders transfer the load from the floor to the prefabricated RC columns underneath it.

It is the two-story multifunctional centre of polygonal ground plan located in the Czech Republic. The total area is approximately 160 x 145 m, the span between RC columns ; 8 x 8 m in 1st Basement and 16 x 16 m in Ground floor. The girders are mounted on the column heads every 8 m. There are prestressed reinforced concrete panels lied down on the girders. The structure was designed in 2013 and put into operation in 2014.

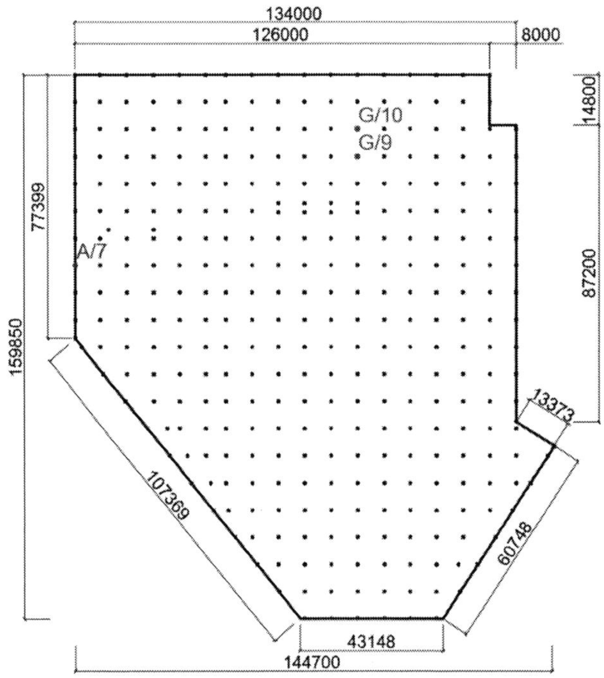

Figure 1. Ground plan of 1st basement with sensor location marked

String tension sensors TSR/5.5 were attached to the columns holding up thermally stressed girders. Monitored columns can be seen in Figure 1. The sensors were recording cracks movements in the columns. The recording was carried out regularly by reading measures from the data logger. Temperature influences the girders volume changes, therefore the temperature was recorded as well.

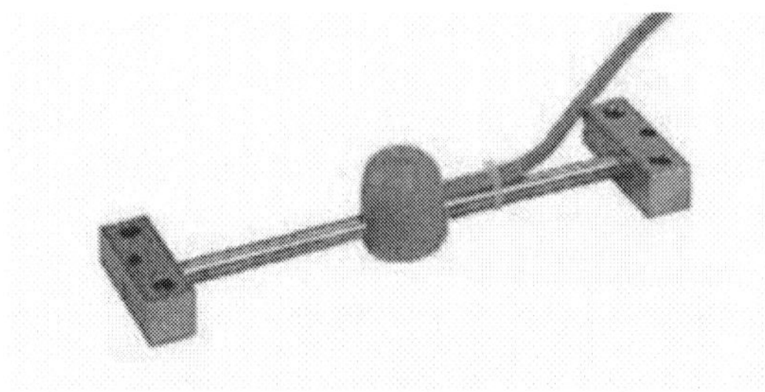

Figure 2. String tension sensor TSR/5,5

String tension sensor TSR/5.5 contains a steel string tensioned between 2 fastening blocks on each end. The string is protected by steel tube which can move freely in fastening blocks. The waterproof electromagnetic coil is fixed in the middle of the length of the string tension sensor. The active length of the string is 139 mm. String frequency is approximately 1000 Hz when unstressed. String stress and frequency change due to deformation of the structure.

3. Thermal expansion of reinforced concrete

Thermal expansion is one of the basics mechanical properties of all materials including material as reinforced concrete for construction of structures. Thermal expansion is a habit of material to change length (volume) of the material as a response to temperature change. Almost every substance expands during heating that means molecules moving faster and its equilibrium positions are further apart.

3.1. Theoretical solution

Application of thermal expansion to the particular reinforced concrete frame formed by girders (simple beam) and columns give us a local requirement for dilatation of each girder [12], [13], [14], [15]. For the application of theoretical principles for a change of length depending on temperature by using the formula

$$\Delta l = \alpha * l[\Delta \mathrm{T}] \tag{1}$$

where Δl is a change of the bar length, α is a coefficient of thermal expansion, l is an initial length and ΔT is a change of temperature, we can get an extension of the element (length is 8 [m] and temperature change is 1 [°C]) about 8*10-5 [m]. We can get roughly a change of tension/compression stress about 0.341 [MPa] on the C35/45 concrete member fixed at both ends and girder temperature change equals to 1 [°C] by the formula

$$\sigma = E * \frac{\Delta l}{l}[\mathrm{MPa}] \tag{2}$$

By theoretical principle the strength of the concrete element equals to 1.47 MPa, the stress will reach the strength of unreinforced concrete when temperate change is 4.31 °C.

Boundary conditions are crucial in this case. According to static scheme and considerations of civil engineers, the girder is simply supported beam with sliding supports secured by steel stud connector.

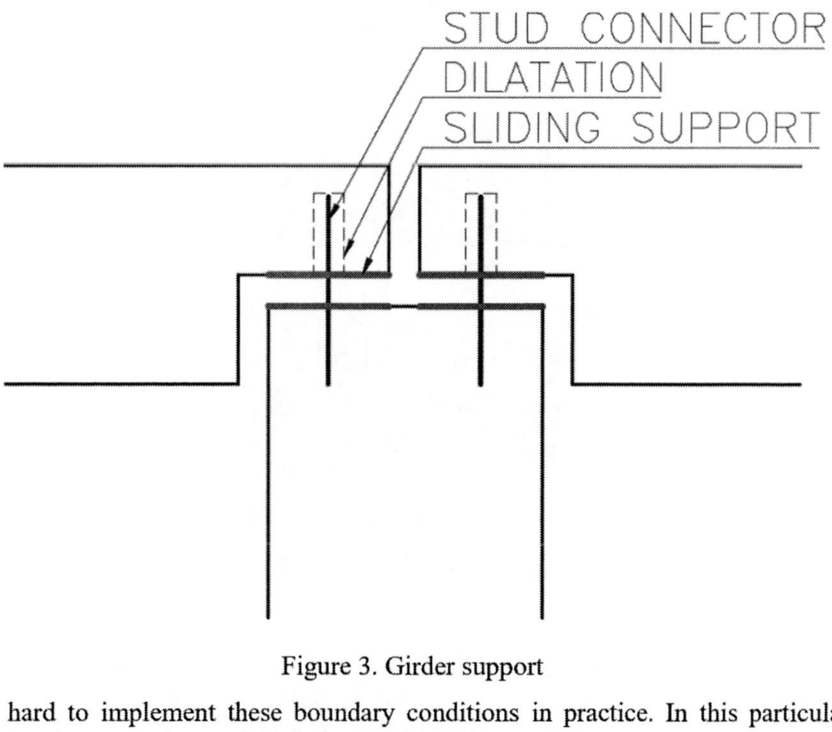

Figure 3. Girder support

It is very hard to implement these boundary conditions in practice. In this particular case, the simplified static scheme can be used as shown in Figure 4.

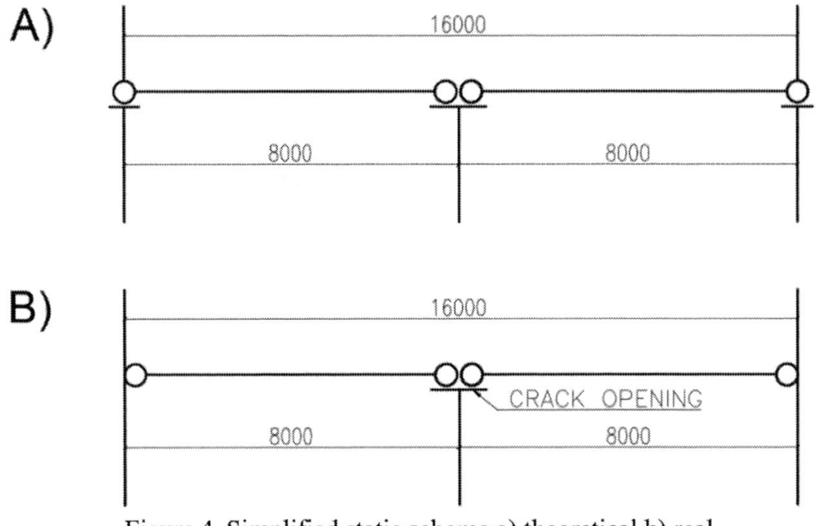

Figure 4. Simplified static scheme a) theoretical b) real

Figure 4a shows assumed static scheme and Figure 4b) real behavior of the structure. Movements of the outer ends of the girders are restrained due to heavy load by upper structures that´s why dilatation movements are realized only above the center column where the normal force is much smaller. The dilatation is limited by stud connectors (see Figure 3), if the maximum displacement of dilatation is reached, the static scheme will change to pinned girder and unfavorable tensile stress will occur in concrete due to thermal extension (change of temperature).

Figure 5. Monitored column failures of the shopping mall caused by thermal expansion of girders

The static scheme has been changed and the tensile strength of column was reached, therefore cracks occurred in concrete. Diagnostics of the structure showed that the crack occurred at the location of the steel stud connectors which reached their maximum displacement dilatation (see Figure 5). Diagnostics showed also that the number of column head stirrups is insufficient for transporting forces from girder thermal changes [16].

3.2. *Experimental solution*

The crack size change which responded mainly to the changes of the thermal longitudinal expansion of the concrete element was monitored during the diagnostics. The structure had been shrunk and crept during diagnostic, only deformations from the thermal expansion were monitored. Monitoring of the structure lasted one year so every season of the year could be monitored together with temperature changes. The temperature was one of the main monitored parameters of diagnostic and it fluctuated between 3 °C and 23 °C.

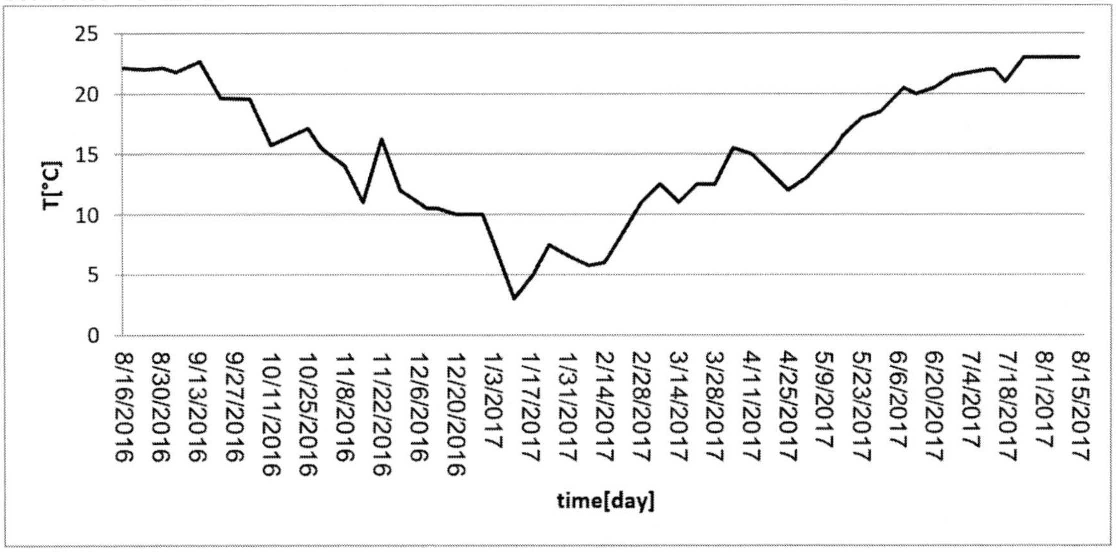

Figure 6. Record of temperatures

Record of temperatures during a year can be seen in Figure 6. This is the temperature of the interior of the unheated garage area. The temperature was always measured on the structure close to the sensor location. Open space area caused delaying of temperature changes including reduction of outside environment temperature. The temperature has not reached a negative or extremely high value in this area. ΔT was approximately around 20 °C.

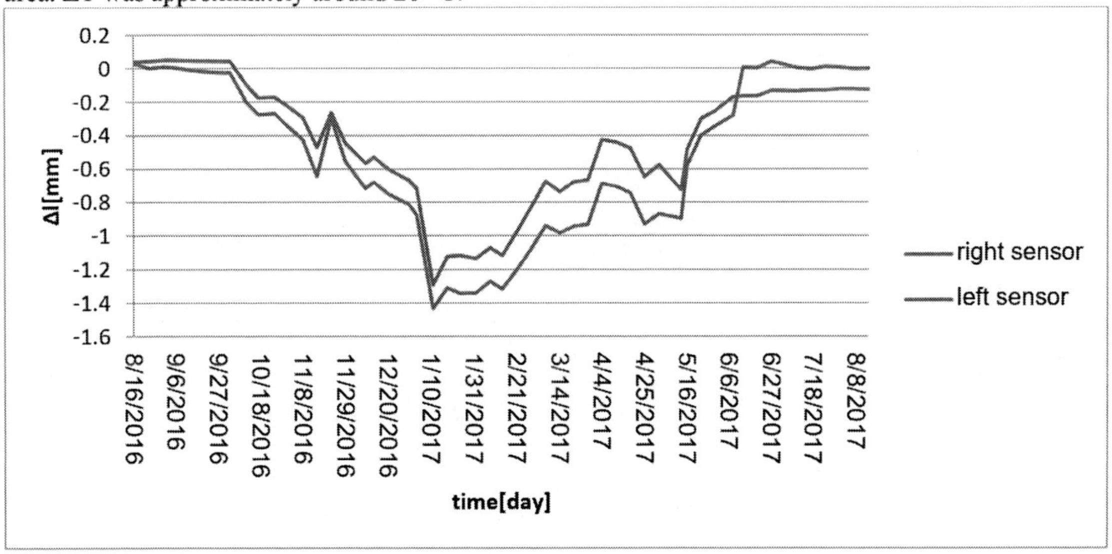

Figure 7. Record of measured crack deformations

Change of the crack size was monitored by TSR/5.5 string tension sensor always on the left and the right side, so it was possible to see any asymmetry which would mean that the rotation in horizontal plane occurred. Figure 7 shows changing of the crack size which is slightly parallel to the temperatures shown in Figure 6.

3.3. Comparison

Based on measured temperatures and known parameters of girder it is possible to obtain a graphic representation of the girder theoretical behaviour according to the application of thermal expansion shown in chapter 3.1. The result is shown in Figure 8. This figure shows a curve variation of girder extension according to the temperature [17], [18], [19], [20].

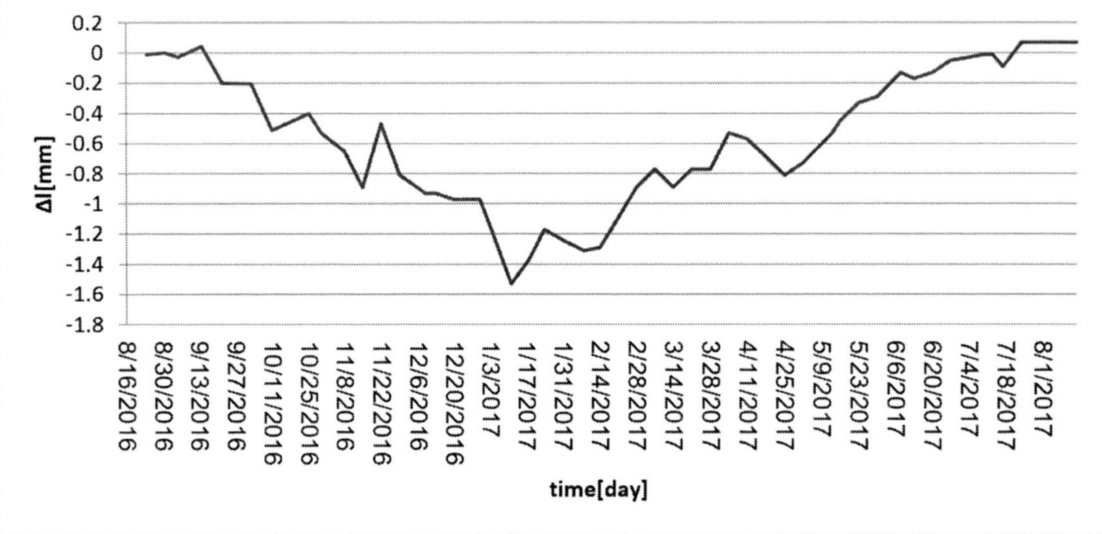

Figure 8. Graphical record of the theoretical crack deformation

Comparison between theoretical and measured values is shown in Figure 9. It is obvious that the values are similar for both approaches.

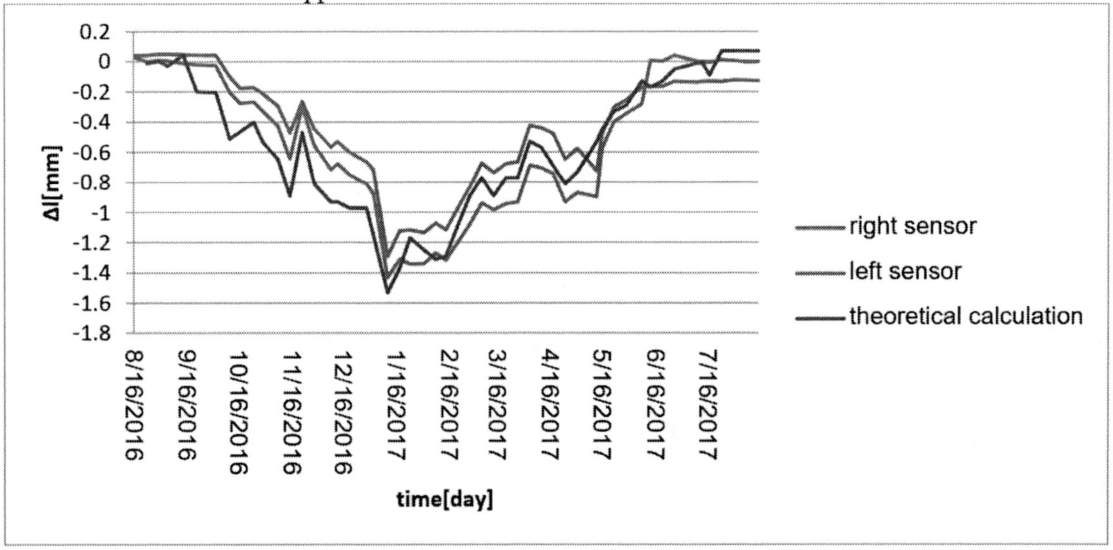

Figure 9. Graphical record of the experimental and theoretical crack deformation

Difference between theoretically and experimentally obtained values is primarily caused by boundary conditions of girder supports [21], [22].

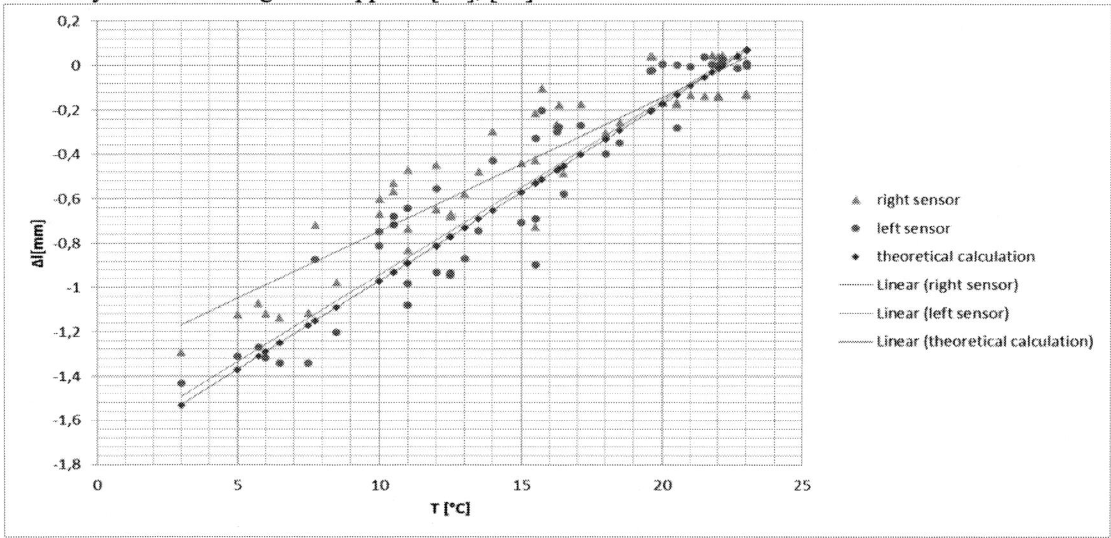

Figure 10. Dependence of crack deformation on temperature from experimental and theoretical values

Figure 10 shows that the values obtained from the theoretical calculation are comparable with the values obtained from the left sensor. Values from the right sensor deviate from the theoretical calculation, especially at low temperatures. That means there is horizontal asymmetry caused by boundary conditions at girder supports.

According to the crack size changes which corresponds to displacements of a beam fixed at one end and pinned at another end, the assumption of the static scheme from the Figure 4b) is confirmed. Due to change of static scheme, unequal distribution of tension stress caused by girder thermal shrinkage occurs, therefore the phenomena of thermal shrinkage works on the side of lower stiffness only.

4. Conclusion

It is necessary to consider the physical properties of materials during design. Specifically is thermal expansion associated with material and especially with temperature change, therefore it can cause large failures of the structures. Thermal expansion in structures can be prevented by dilatation or reinforcement of the structure.

Sometimes only dilatation itself does not have to be sufficient when boundary conditions change as described in this article. Restoration of this particular case is possible for example with classics steel bandage replacing the absence of reinforcement in the tensile direction [23], [24]. It is appropriate to apply this steel bandage during the highest shrinkage of the girder. Otherwise, it could cause failures of others structure elements as a result of redistribution of the girder thermal expansion effects.

For a designer, it is necessary to consider the real structure behavior including thermal expansion of materials and its possible consequences. It is fundamental to prevent potential failures in order to avoid damage or collapse of the structure.

References

[1] Cajka R, Burkovic K, Manasek P and Labudkova J 2015 *Changes in the State of Stress in Enclosing Panels after Additional Thermal Insulation* Advanced Materials and Structural Engineering. Proceedings of the International Conference on Advanced Materials and Engineering Structural Technology (ICAMEST 2015) April 25-26 Qingdao China. CRC Press Taylor & Francis Group ISBN 9781138027862

[2] Cajka R and Mateckova P 2010 *Different types of pre-stressed hollow core panels and their fire*

resistance according to eurocodes. Journal of Structural Fire Engineering. ISSN 20402317, DOI: 10.1260/2040-2317.1.4.243, pp. 243-248

[3] Cajka R and Fojtik R 2013 *Development of Temperature and Stress during Foundation Slab Concreting of National Supercomputer Centre IT4.* Procedia Engineering. vol 65 ISSN 1877-7058 doi: 10.1016/j.proeng.2013.09.035 pp 230-235

[4] Cajka R and Mateckova P 2013 *Fire Resistance of Ceiling Slab Concreted in Trapezoidal Sheet* Procedia Engineering. vol 65 ISSN 1877-7058 doi: 10.1016/j.proeng.2013.09.061 pp 393-396

[5] Abdel-Fattah M T, Abdel-Fattah, T T and Hemada, A A 2018 *Nonlinear Finite-Element Analysis of Integral Abutment Bridges due to Cyclic Thermal Changes* Journal of Bridge Engineering vol 22 DOI: 10.1061/(ASCE)BE.1943-5592.0001183

[6] Chen H R and Choi J H 2011 *Analysis of shrinkage and thermal stresses in concrete slabs reinforced with GFRP rebars* Journal of Materials in Civil Engineering DOI: 10.1061/(ASCE)MT.1943-5533.0000216 pp 612-627

[7] Janulikova M and Stara M 2013 *Reducing the Shear Stress in the Footing Bottom of Concrete and Masonry Structures* Procedia Engineering vol 65 ISSN 1877-7058, doi: 10.1016/j.proeng.2013.09.044 pp 284-289

[8] Janulikova M 2014 *Comparison of the shear resistance in the sliding joint between asphalt belts and modern PVC foils* Applied Mechanics and Materials vol 501-504 ISSN: 16627482, ISBN: 978-303835005-7 DOI: 10.4028/www.scientific.net/AMM.501-504.945 pp 945-948

[9] Bilcik J, Sonnenschein R and Gazovicova N 2017 *Causes of Early-Age Thermal Cracking of Concrete Foundation Slabs and their Reinforcement to Control the Cracking* Slovak Journal of Civil Engineering doi:10.1515/sjce-2017-0013 pp. 8-14

[10] Cajka R 2013 *Numerical Solution of Temperature Field for Stress Analysis of Plate Structures* Applied Mechanics and Materials vol. 470 Trans Tech Publications Switzerland, ISSN: 16609336 ISBN: 978-3-03785-961-2 doi:10.4028/www.scientific.net/AMM.470.177 pp 177-187

[11] Iqbal M 2007 *Thermal movements in parking structures* ACI Structural Journal vol. 104 (5) pp 542-548

[12] de Borst R and Peeters P P J M 1989 *Analysis of concrete structures under thermal loading* Computer Methods in Applied Mechanics and Engineering vol. 77 (3) DOI: 10.1016/0045-7825(89)90079-0 pp 293-310

[13] Bernardi P, Cerioni R, Michelini E and Sirico A 2016 *Numerical modeling of the cracking behavior of RC and SFRC shear-critical beams* Engineering Fracture Mechanics vol. 167 DOI: 10.1016/j.engfracmech.2016.04.008 pp 151-166

[14] Bažant Z P 1972 *Thermodynamics of interacting continua with surfaces and creep analysis of concrete structures* Nuclear Engineering and Design vol 20 (2) DOI: 10.1016/0029-5493(72)90124-0 pp 477-505

[15] Wu Y and Luna R 2001 *Numerical implementation of temperature and creep in mass concrete* Finite elements in analysis and design vol. 37 (2) DOI: 10.1016/S0168-874X(00)00022-6 pp. 97-106

[16] Klemczak B, Flaga K and Knoppik-Wrobel A 2017 *Analytical model for evaluation of thermal-shrinkage strains and stresses in RC wall-on-slab structures* Archives of Civil and Mechanical Engineering vol. 7 Issue 1 DOI: 10.1016/j.acme.2016.08.006 pp. 75-95

[17] Ge Y, Elshafie MZEB, Dirar S and Middleton CR 2014 *The response of embedded strain sensors in concrete beams subjected to thermal loading* Construction and Building Materials vol 70 DOI: 10.1016/j.conbuildmat.2014.07.102 pp. 279-290

[18] Choi S and Won MC 2010 *Thermal Strain and Drying Shrinkage of Concrete Structures in the Field* ACI Materials Journal vol. 107 Issue 5 pp. 498-507

[19] Rostasy FS, Krauss M, Laube M, Rusack T and Budelmann H 2007 *Online-monitoring and calculation of the concrete stresses caused by thermal restraint for a trough structure at the main station of Berlin* Bautechnik vol. 84 Issue 4 DOI: 10.1002/bate.200710020 pp. 235-242

[20] Tennyson RC, Banthia N, Rivera E, Huffman S and Sturrock I 2007 *Monitoring structures using long gauge length fibre optic sensors* Canadian Journal of Civil Engineering vol. 34 Issue 3 DOI: 10.1139/L06-155 pp. 422-429

[21] Kromanis R and Kripakaran P 2014 *Predicting thermal response of bridges using regression models derived from measurement histories* Computers & Structures vol. 136 DOI: 10.1016/j.compstruc.2014.01.026 pp.64-77

[22] Hedegaard BD and French CEW and Shield CK 2016 *Effects of Cyclic Temperature on the Time-Dependent Behavior of Posttensioned Concrete Bridges* Journal of Structural Engineering vol. 142 Issue 10 DOI: 10.1061/(ASCE)ST.1943-541X.0001538

[23] Shin H C and Lange D A 2004 *Effects of shrinkage and temperature in bonded concrete overlays* ACI Materials Journal vol. 101 (5) pp. 358-364

[24] Shin A H C and Lange D A 2012 *Effects of overlay thickness on surface cracking and debonding in bonded concrete overlays* Canadian Journal of Civil Engineering vol 39 (3) DOI: 10.1139/L2012-007 pp. 304-312

IOP Conf. Series: Earth and Environmental Science **143** (2018) 012008 doi:10.1088/1755-1315/143/1/012008

Influence of microclimate on the sustainability and reliability of weathering steel bridge

M Kubzova[1, 3], V Krivy[1] and K Kreislova[2]

[1] Department of Building Structures, Faculty of Civil Engineering, VSB - Technical University of Ostrava, L. Podeste 1875, 708 00 Ostrava, Czech Republic
[2] SVUOM Ltd., U Mestanskeho Pivovaru 934/4, 170 00 Prague 7, Czech Republic
[3] Correspondence: monika.kubzova@vsb.cz; Phone: +420-734-206-860

Abstract. Reliability and sustainability of bridge structures designed from weathering steel are influenced by the development of a sufficiently protective layer of corrosion products on its surface. The development of this protective layer is affected by several parameters such as air pollution around the bridge structure, the microclimate under the bridge, the location of surface within the bridge structure and the time of wetness. Design of structural details also significantly influences the development of the protective corrosion layer.
The article deals with the results of the experimental tests carried out on the road bridge located in the city of Ostrava in the Czech Republic. The development of the protective corrosion layer on the surface of the bridge is significantly influenced by the intensive traffic under the bridge construction and the design solution of the bridge itself. Attention is focused mainly on the influence of chloride deposition on the protective function of the corrosion layer. Corrosion samples were placed on the bridge to evaluate the influence of the above-mentioned parameters. The deposition rate of chlorides spreading from the road to surfaces of the steel structure is also measured.

1. Introduction

Steels with increased resistance to atmospheric corrosion, weathering steels, belong to a group of low-alloy steels. The content of alloying elements in such steels does not exceed 2 wt.%. The elementary alloying elements that affect the corrosion resistance of steel are Cu, Cr, P and Ni [1, 2]. Weathering steel is mainly used for structures in outdoor, not only because of its increased corrosion resistance in the air but also with respect to its aesthetic properties. Weathering steel is mainly used in the design of bridge structures, transmission towers, facade of building, but also in the statues and monuments. Increased corrosion resistance of steel is due to the development of a sufficiently protective corrosion layer on the steel surface after 3 to 7 years of direct exposure.

The development of corrosion layers significantly depends on environmental conditions of the structure's location [3]. One of the basic environmental parameters that affect the development of protective patina is the time of wetness of the metal surface. For the favourable development of the protective patina, it is important to alternately wet and dry the surface. In cases where an increased moisture build-up occurs on the surface of the structure, non-protective, large corrosion products may develop, the presence of which is associated with increased corrosion weakening of the structure bearing elements.

It is therefore obvious that the design of the bridge structure has, in addition to the basic parameters of the environment, a significant influence on the development of corrosion layers [4, 5]. Other environmental factors that significantly influence the development of the protective corrosion

layer are the concentration of corrosion stimulators in the air, especially concentration of SO_2 and chlorides. The concentration of SO_2 in the air reached maximum values in Europe in the 1970s and 1980s; after this period there was a significant decrease in SO_2 concentration [6, 7], see Figure 1.
In 1970 in industrial areas of the Czech Republic (Prague, Ostrava, Northern Bohemia), the experimentally measured SO_2 concentration in the atmosphere was almost 200 $\mu g.m^{-3}$. Currently, the SO_2 concentration in the air is approximately 20 times lower and ranges around 5 - 15 $\mu g.m^{-3}$. On the figure 1 there is illustrated decrease of concentration of SO_2 experimentally measured in industrial location Kopisty (Czech Republic).

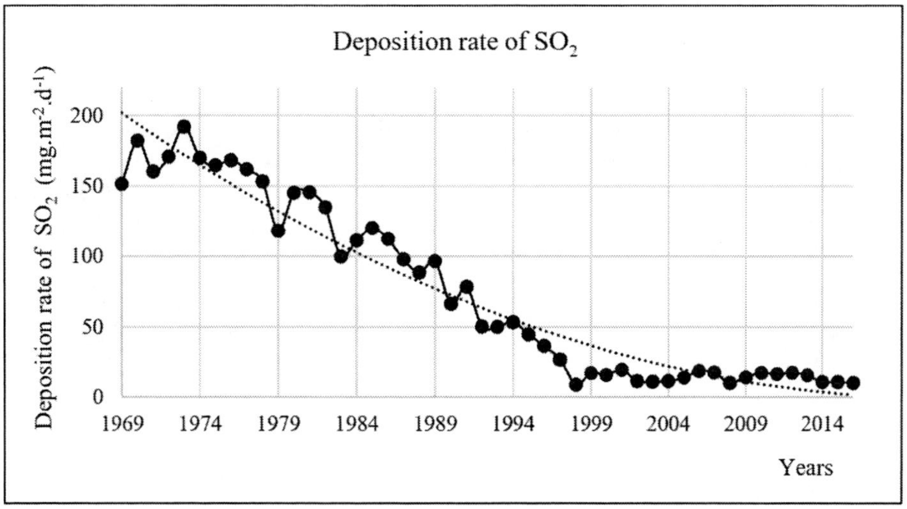

Figure 1. Changes of concentration of SO_2 during 1969-2015 in industrial location Kopisty (Czech Republic)

The decrease in concentration of SO_2 is significant and the reason for it is mainly the introduction of desulphurization units into industrial production and the reduction of fossil fuel consumption. In the past, the main corrosive stimulator in the Czech Republic was SO_2. At present, due to its low concentration in the atmosphere, the influence of chlorides is increased as a significant corrosion stimulator. In seaside areas, chlorides spread along with the seawater aerosol. In inland areas, sources of chlorides are primarily de-icing salts used in winter maintenance of roads. Therefore, the structures located near motorways with high road traffic are the ones that are most affected by chloride deposition. The negative effect of chlorides not only shows on roads and adjacent structures, but also, for example, on vegetation in a relatively wide area around the road [8].

Increased attention needs to be paid to the development of corrosion processes on bridge structures leading above busy roads. The local microclimate with increased atmospheric corrosivity can be expected mainly in design solutions, creating *tunnel-like conditions* [9] (see Figure 2). The aerosol and dust raised by the traffic cannot be dispersed in partially tunnel conditions under bridge into the environment, leading to increased settling of dusts on the supporting elements of the bridge structure. These surfaces are not sufficiently ventilated and washed by rain due to closed conditions. The unfavourable development of corrosion products can be expected mainly on horizontal surfaces that are more affected by deposition and retention of dusts with corrosive stimulators than vertical surfaces. In steel structures, these are mainly the upper surfaces of the lower flanges of the main girders of bridge [10].

For design practice, it is very important to predict the development of corrosion processes with sufficient accuracy, both in designing new structures and in assessing the reliability and lifetime of bridges already in use. Workplaces all over the world have been carrying out long-term research projects in this field, where experimental verification of atmospheric corrosivity of the environment and evaluation of the influence of various corrosive agents on the behaviour of corrosion processes are carried out [1, 3]. In the submitted article, there are presented and evaluated selected results of experimental corrosion tests carried out on selected bridge structures. In particular, the results of the tests carried out on a bridge in Ostrava, which with its design solution creates the conditions partially corresponding to the conditions in the tunnel, the so-called *tunnel-like conditions*. The values of corrosion losses after one and three years of exposure of corrosive samples located on typical surfaces of the bridge structure are evaluated and discussed. In addition to the corrosion samples, sampling devices for direct measurement of chloride deposition were installed on the bridge structure. The article presents the measured values of chloride deposition in relation to a particular season of the year. Simultaneously, a comparison of measured values of chloride deposition with a neighbouring bridge located on the same road is performed, which has not been affected by the intensive road traffic under the bridge structure.

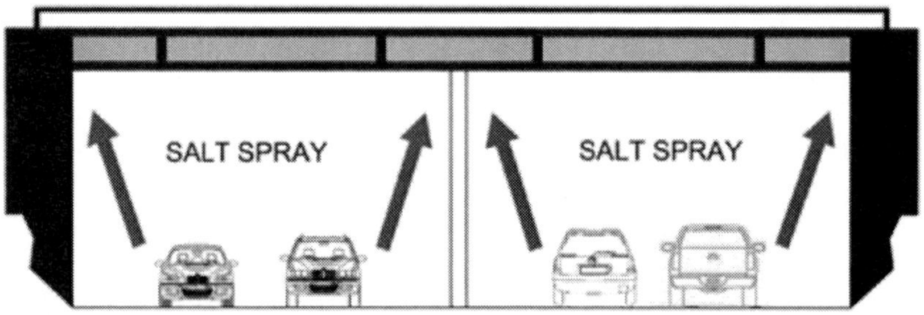

Figure 2. Schematic drawing of *tunnel-like conditions* under the bridge

2. Methods for experimental testing

The program of experimental corrosion tests includes in total 10 weathering steel bridges [4]. The road bridge above the Dl motorway in Ostrava is the only one bridge from the program that is influenced by tunnel-like conditions. The locality can be characterized in accordance with [11] as an environment with a degree of atmospheric corrosivity C2 to C3. The bridge structure was put into operation in 2001. The steel bridge moves road and tram traffic above the busy D1 motorway. The structure is designed using main girders that are coupled with a reinforced concrete slab. The supporting steel structure is made of weathering steel Atmofix B (S355J2W). The construction design of the bridge structure is shown in Figure 3.

Figure 3. Views of a tested bridge structure

Corrosion test samples and sampling devices for the measurement of chloride deposition were placed on selected surfaces of the bridge structure. Corrosion samples 150x100x1.5 mm (S355J2WP steel), designed in accordance with [12], are intended to monitor the development of corrosion layers on the steel surface over time. The side of the samples adjacent to the steel structure is protected by a special tape in order to develop corrosion products only on the front exposed surface of the sample. Corrosion samples are attached to the steel structure by means of stainless steel clips in such a way that they fit the evaluated surface of the bridge. A total of three corrosion samples were placed on each selected surface of the bridge, which are supposed to be sequentially removed after 1, 3 and 10 years of exposure. The thickness of the corrosion layer on the samples using a magnetic-induction method is measured continuously for all evaluated surfaces of the bridge, as well as the corrosion loss values. Elemental and diffraction analyses to determine the representation of individual phases in the corrosion layer were also performed on selected samples. Based on the ratio of individual phases in the patina layer, the values of the *PA* indexes for the evaluation of protective properties of the formed corrosion layer [13, 14] are determined. Selected surfaces of the bridge structure on which the corrosion samples were placed, are shown in Figure 5. Surfaces S8 to S11 are oriented southward (in the driving direction on the motorway under the bridge), the other samples are oriented northward (opposite-direction on the motorway under the bridge).

Sampling devices for chloride deposition measurements were designed in accordance with [15]. The devices for measuring the deposition of chlorides using the wet candle method and the dry plate method have been installed on selected surfaces of the structure. The wet candle method is based on the installation of a bottle made of inert material with prescribed solution. Surgical gauze for the collection of chlorides is wound onto a wick placed in the bottle, which are consequently transferred to the solution in the bottle via the gauze. The dry plate method consists of installing a frame of prescribed dimensions, onto which a special industrial textile for the deposition of chloride-containing dust deposit is stretched out. A small perforation of the textile is important to prevent the textile from being blown through, resulting in collected dust being blown away. As part of experimental measurements, dry plates are oriented both horizontally and vertically. Exposure of the sampling device is one calendar month. Subsequently, the sampling equipment is replaced with a new one and the collected samples are subjected to a chemical analysis to determine the amount of deposited chlorides.

The chloride deposition sampling device was installed on four selected positions on the evaluated bridge (see Figure 4):

- position P1 – outer girder No. I – north (corresponding with surfaces S3 and S4)
- position P2 – inner girder No. III – north (corresponding with surface S12)
- position P3 – inner girder No. III – south (corresponding with surface S9)
- position P4 – outer girder No. XI – south (corresponding with surface S11)

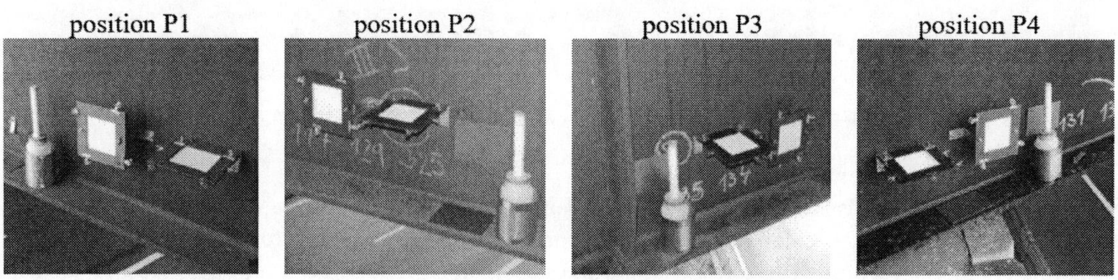

Figure 4. Tested surfaces with devices for measuring of deposition rate of chlorides

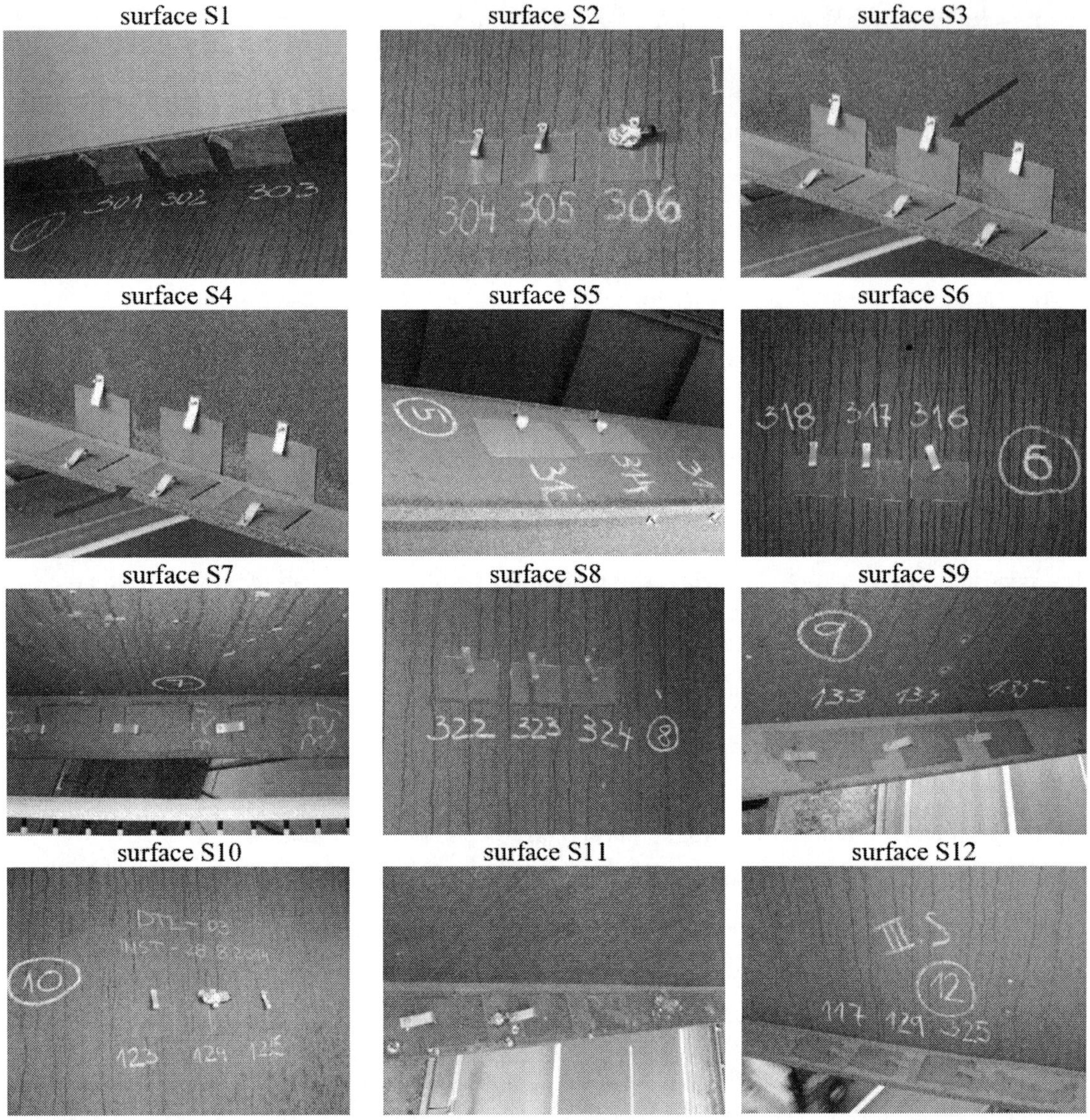

Figure 5. Tested surfaces with corrosion samples after one year of exposure

3. Results of experimental tests

In order to evaluate the influence of chlorides (or the influence of the specific environment under the bridge structure) on the development of corrosion layers, several fundamental experimental parameters can be used, such as average thickness of corrosion products layer, corrosion losses, chlorine weight content in corrosion products, *PA* index values, and actual chloride deposition in the air. Due to the extensive character of the experimentally determined data for the purposes of this article, two sets of data have been selected for the microclimate impact assessment under the bridge structure, corrosion losses and chloride deposition.

3.1. Corrosion losses after one and three years of exposure

Corrosion samples were installed on the evaluated bridge structure in 2014. In 2015 the samples were removed after 1 year of exposure and in 2017 after 3 years of exposure. The corrosion loss value was determined using a laboratory analysis in collected samples. Determined corrosion loss values corresponding to selected typical surfaces of a bridge structure are shown in Table 1.

Table 1. Corrosion losses on tested surfaces of bridge after 1 year and 3 years of exposure.

Tested surface	Corrosion losses [μm]	
	1 year	3 years
S1 - bottom surface of upper flange – external girder No. I	8.54	12.51
S2 - external wall – external girder No. I	14.61	19.46
S3 - external wall above bottom flange – external girder No. I	19.17	32.14
S4 - upper surface of bottom flange – external girder No. I	35.23	42.33
S5 - bottom surface of bottom flange – external girder No. I	10.81	18.85
S6 - internal wall – internal girder No. V	7.26	10.14
S7 - upper surface of bottom flange – internal girder No. V	27.15	38.81
S8 - internal wall – internal girder No. III	11.96	19.62
S9 - upper surface of bottom flange – internal girder No. III	27.06	41.81
S10 - external wall – external girder No. XII	10.50	17.40
S11 - upper surface of bottom flange – external girder No. XII	20.22	26.30
S12 - upper surface of bottom flange – internal girder No. III	24.10	43.04

From the results shown in Table 1, several interesting observations can be formulated. It is obvious that the course of corrosion processes is strongly influenced by the position and orientation of the surface on the structure. The highest values of corrosion losses were found on the upper horizontal surfaces of the lower flanges of the main girders (surfaces S4, S7, S9, and S12). Significantly lower values of corrosion losses were identified on the horizontal surfaces when viewing them from below (surfaces S1 and S5) and on the walls of the main girders (positions S2, S3, S6, S8, and S10).

The observed corrosion loss values after 1 and 3 years of exposure of corrosion samples have not shown any significant differences between the outer and inner surfaces of the bridge. However, the visual development of patina on corrosion samples and mostly on the surfaces of bridge structure is very different. While compact adhesive corrosion layers have developed on external surfaces, the formation of non-adhesive corrosion layers with visible surface contamination is evident on internal horizontal surfaces, see Figures 6 and 7. The results of chloride deposition measurements described in a later section of the article have shown that there are no significant differences in the deposition of chlorides at the outer and inner surfaces of the bridge. The unfavourable development of the patina near the inner horizontal surfaces is thus caused by a combination of two negative phenomena, namely: (a) high deposition of dust deposits and chlorides from traffic below the bridge structure; (b) limitation of the natural cleaning of the structure's inner surfaces by rainfall and wind.

IOP Conf. Series: Earth and Environmental Science **143** (2018) 012008 doi:10.1088/1755-1315/143/1/012008

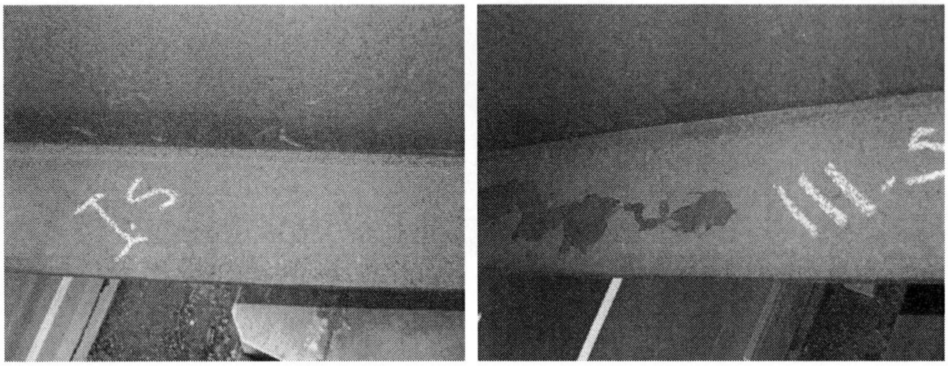

Figure 6. Upper surface of a bottom flange of the external girder No. I (left) and of the internal girder No. III (right)

Figure 7. Layer of corrosion products on corrosion samples after 3 years of exposure on the surface S4 (left) and surface S9 (right)

3.2. Deposition rate of chlorides

The sampling devices for chloride deposition measurement were installed on the assessed bridge structure in December 2016. At the time of compilation of the article, results from 10 months of exposure were available including the entire winter season 2016/2017, when de-icing salt for road maintenance was used in the assessed area. The chloride content was analysed in accordance with [15] using the spectrophotometric method. The graphs in Figures 8 to 10 show the deposition rates of chlorides from 12/2016 to 9/2017 for individual measurement methods.

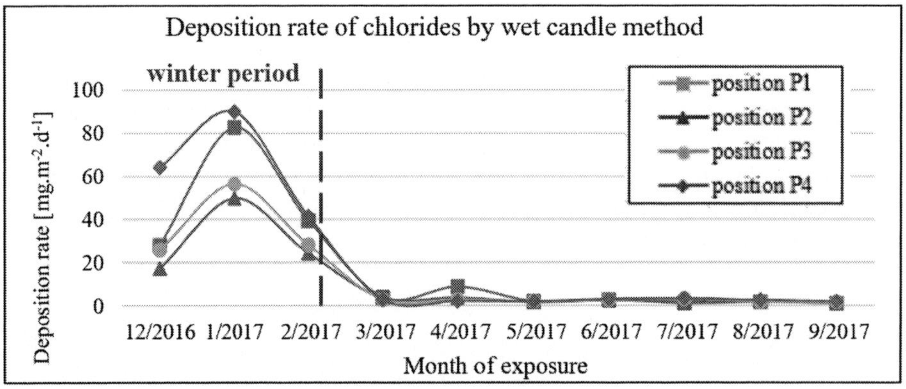

Figure 8. Deposition rate of chlorides measurement by wet candle method

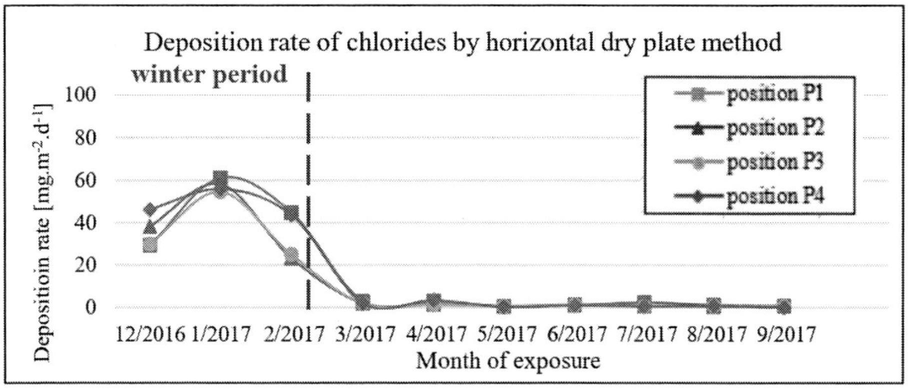

Figure 9. Deposition rate of chlorides measurement by dry plate method (horizontal)

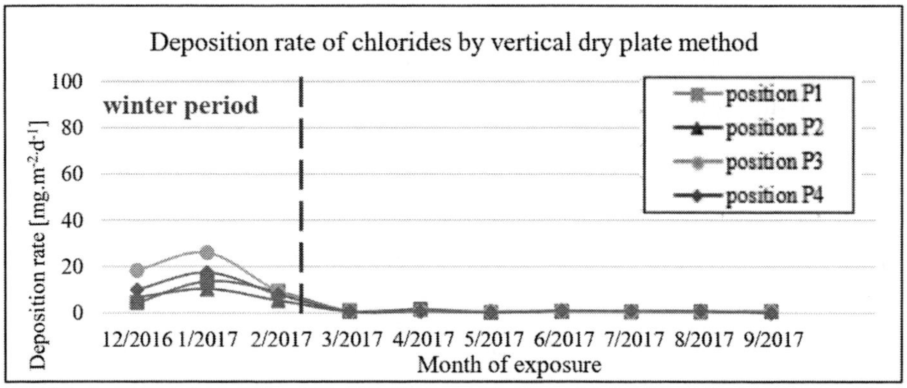

Figure 10. Deposition rate of chlorides measurement by dry plate method (vertical)

The results of the measurements show a significant influence of the season of the year on the amount of deposited chlorides. High deposition rates were only measured in those months when de-icing salt was used for road maintenance in the assessed locality (December 2016 to February 2017). For the rest of the year (i.e. during a period with no winter road maintenance) the deposition rates of chlorides are low and vary, with a maximum of 3 $mg \cdot m^{-2} \cdot d^{-1}$. Experimentally collected data did not confirm the findings from other research projects [16, 17] when increased amounts of chloride were identified even two months after the end of winter road maintenance. However, for a broader generalization of results, it will be necessary to analyse data from a longer time series of measurements.

The findings show that the external surfaces (P1 and P4) show a slightly higher deposition of chlorides compared to the internal surfaces (P2 and P3). This statement, however, does not apply generally to all measurements made. The observed phenomenon is probably related to the fact that the external surfaces of the bridge are directly exposed to the raised dust deposits and aerosols, the source of which is the traffic under the bridge structure.

Inner surfaces are partially protected from these negative phenomena. However, these surfaces are also protected against natural rainfall and wind cleaning. Despite the lower values of deposition rates, corrosion products that do not meet the requirements for sufficient patina protection have developed on unventilated and unclean horizontal internal surfaces. There are significant differences between the findings concerning deposition rates corresponding to the three measurement methods used. For most measurements, the highest deposition rates of chlorides were obtained by the wet candle method (the maximum measured deposition rate was 90 $mg \cdot m^{-2} \cdot d^{-1}$). In some isolated cases, however, the highest deposition rate of chlorides was measured using a dry plate placed in a horizontal position. The vertical positioning of the dry plate in all cases led to significantly lower values compared to other measurement methods (approximately 3 to 5 times the decrease in the values of the dry plate in the horizontal position).

3.3. Comparison results of two bridges with different microclimate under the bridge

To evaluate the impact of road transport under the bridge structure, it is possible to compare the experimentally detected data with similar results obtained on the neighbouring bridge structure. The reference bridge is located on the same road as the assessed bridge structure (the distance between the two bridges is approximately 200 m). The design of both bridges is similar – it is a steel bridge with main girders coupled with upper bridge deck in both cases. The second bridge moves road traffic above the railway line and thus cannot be affected by the deposition of dust impurities and chlorides from the transport under the bridge.

Table 2 shows the comparable values of chloride deposition rates obtained using the wet candle method placed on the inner surfaces of both compared bridges. Significantly higher deposition rates were found on a bridge structure constructed above the motorway (in January 2017 as big as a 9 times difference in values was detected). Comparing the results from the sampling devices located on the two reference bridges thus shows the significant influence of the specific microclimate under the bridge structure. It is evident that road transport under the bridge structure is the main source of chloride deposition on bridge bearing elements. The results from the bridge located above the railway show that with a suitable bridge design (a beam bridge with an upper bridge deck and sufficient overhang of the bridge deck above the main girders) the amount of deposited chlorides from the road transport on the bridge is significantly reduced.

Table 2. Deposition rate of chlorides on both bridges measured by wet candle method

Tested bridge	Deposition rate of chlorides [$mg \cdot m^{-2} \cdot d^{-1}$]			
	12/2016	1/2017	2/2017	3/2017
Tested bridge above the motorway D1	25.8	56.4	28.1	2.8
Tested bridge above the railway	9.4	6.3	5.9	1.5

The different microclimate under both compared bridges was also reflected in the results of corrosion losses after one and three years of exposure of corrosion samples. The comparable values for inner surfaces of the walls of the main girders in both bridges are listed in Table 3. In the bridge structure above the motorway, the corrosion losses are more than doubled compared to the structure not affected by road traffic under the bridge.

Table 3. Corrosion losses on tested surfaces of the bridge after 1 year and 3 years of exposure.

Tested bridge	Corrosion losses [μm]	
	1 year exposure	3 year exposure
Tested bridge above the motorway D1	11.96	19.62
Tested bridge above the railway	5.26	8.98

4. Discussion of results

Experimental values of corrosion losses and chloride deposition provide basic information on the influence of a specific environment under the bridge structure on the development of corrosion products on the surface of the structure. From the measured data, several interesting findings can be derived.

4.1. Influence of location and orientation of the surface

The results of the atmospheric tests show that the course of corrosion processes is significantly influenced by the orientation and the placement of the surface on the bridge structure [4, 18]. The development of corrosion products on typical surfaces of the bridge structure can be very different. The greatest differences were observed in corrosion losses measured after one year of corrosion surface exposure (in the tested bridge structure there was nearly 5 times the difference between S4 and S6 surfaces).

The differences after three years of exposure slightly reduced (compared to one year exposure), in that a maximum of 4 times the difference between S4 and S12 surfaces was detected in the tested bridge structure. It is reasonable to expect that during the next exposure, the differences in the corrosion losses of individual typical surfaces of the structure will continue to decrease, yet they will remain significant. In prediction of corrosion losses, it is therefore necessary to consider the influence of the positioning and orientation of the surface within the supporting structure [19], in addition to the general environmental characteristics (expressed by atmospheric corrosivity category of the environment) and microclimatic factors (for example, the influence of traffic under the bridge).

4.2. The bridge design and character of an obstacle to overcome

When designing the load-bearing structure of road bridges, it is necessary to pay increased attention to the choice of design solution with respect to minimization of the effect of adverse phenomena resulting from the use of de-icing salts in winter road maintenance. A suitable construction system is a girder bridge with an upper deck, which has a sufficient overhang over the bridge girders. Measurements of chloride deposition on the bridge over the railway have shown that the upper bridge deck sufficiently protects the steel supporting structure from depositing impurities and chlorides, the source of which is road traffic on the bridge.

Intensive road traffic under the bridge structure can, on the contrary, act as a significant factor contributing to degradation associated with inadequate protection of corrosion products. Raised dust and aerosols containing substances from traffic under the bridge may settle on the elements of the supporting bridge structure. Increased deposition is particularly likely in cases where the design of the bridge does not allow the dispersion of impurities into the area outside the bridge structure. Inappropriate design solutions primarily include the bridges where the supporting system creates conditions similar to those in the tunnels. The combination of intensive road traffic under the bridge together with the environment that does not allow dispersion of contaminated impurities and aerosols is the cause of unfavourable development of patina on the inner surfaces of the lower flanges of the main bridge girders above the D1 motorway in Ostrava. The findings from the locally adverse development of patina on a particular bridge structure can be generalized: the most endangered surfaces are the horizontal areas, the positioning of which within the supporting structure does not allow for regular cleaning by wind and rainfall.

4.3. Influence of the season

Chloride deposition is expected to reach the highest values in the winter. In March 2017 (the climatic conditions in this month did not require the use of de-icing salts in the assessed locality), there was a sharp decrease in a deposition rate regardless of the measurement method. Background values for the rest of the year when winter maintenance is not performed in the locality of the assessed bridge are very low and range up to 3 $mg{\cdot}m^{-2}{\cdot}d^{-1}$. It is therefore very important that the bridge maintenance crew perform regular cleaning of the structure with jet water after the winter season ends. With this simple basic maintenance recommended for weathering steel bridges, the adverse effects caused by increased chloride deposition during the winter season can be significantly eliminated.

5. Conclusions

The article presents and evaluates the results of experimental measurements carried out on the bridge structure above the D1 motorway in Ostrava. The research activities carried out on this bridge are part of extensive experimental corrosion testing project carried out on a total of 10 bridge structures designed from weathering steels in the Czech Republic. Details of the whole project can be found, for example, in [4].

The existing results of the project clearly show that the course of corrosion processes is, in addition to the general characteristics of the environment expressed as a degree of atmospheric corrosivity, significantly influenced by the positioning and orientation of the area within the supporting structure as well as by the specifics of the local microclimate. From the large number of experimentally detected data, more general conclusions can be drawn up about the course of corrosion processes. Significant differences in corrosion processes can be observed between the horizontal and vertical surfaces of the supporting bridge structure; the orientation of the surface outside or inside the bridge is also important.

One of the decisive factors affecting the local microclimate in the vicinity of the bridge is the influence of chlorides, of which deposition on the elements of the bridge structure is mainly due to the traffic on adjacent roads. The source of chlorides is the de-icing salts used in winter road maintenance. The previous results of chloride deposition measurements show some important findings. The measurements show that in the bridges with an appropriate construction design and arrangement, the amount of deposited chlorides, the source of which is the traffic across the bridge structure can be significantly eliminated. A higher chloride deposition may occur due to intensive road traffic under the bridge structure. Currently, there is not enough experimentally validated data available to define specific dimensional and structural requirements that when being complied with, the chloride deposition from the transport under the bridge decreases to a level that doesn't limit a favourable development of patina on bridge structure elements. Therefore, it is important to adhere to at least the basic principles of designing structures – that is to design the structures allowing the dispersion of raised impurities and aerosol into the surrounding environment, and in particular to ensure the basic maintenance involving regular cleaning of the steel structure after the end of the winter period.

6. References

[1] Morcillo M, Díaz I, Chico B, Cano H and de la Fuente D 2014 *Corros. Sci.* Weathering steels: From empirical development to scientific design. A review. 83 pp 6–31.

[2] Krivy V and Konecny P 2013 *Procedia Eng.* Real material properties of weathering steels used in bridge structures. 57 pp 624–633.

[3] Morcillo M, Chico B, Díaz I, Cano H and de la Fuente D 2013 *Corros. Sci.* Atmospheric corrosion data of weathering steels. A review. 77 pp 6–24.

[4] Krivy V, Urban V and Kreislova K 2016 *Eng. Fail. Anal.* Development and failures of corrosion layers on typical surfaces of weathering steel bridges. 69 pp 147–160.

[5] Krivy V, Urban V and Kubzova M 2016 *Procedia Eng.* Thickness of corrosion layers on typical surfaces of weathering steel bridges. 96 pp 56–62.

[6] Kreislova K, Geiplova H, Bartak, Z and Majtas, D 2017 *Koroze Ochr. Mater.* Atmospheric corrosion models. 61 pp 59–66. (in Czech)

[7] Odnevall Wallinder I, Bahar B, Leygraf C and Tidblad J 2007 *J Environ Monit* Modelling and mapping of copper runoff for Europe. 9 pp 66–73.

[8] Dolling C and Hudson R 2003 *Rev. Métall.* Weathering steel bridges in the U.K. 100 pp 1125–1133.

[9] Kreislova K and Knotkova D 2014 SVÚOM: Praha, Czechia, *Atmosférická Koroze*, ISBN 978-80-87444-11-5. (In Czech)

[10] Alcantara J, Fuente D, Chico B, Simancas J, Diaz I and Morcillo M 2017 *Mater* Marine Atmospheric Corrosion of Carbon Steel: A Review. 10-406.

[11] International Organization for Standardization. Corrosion of Metals and Alloys - Corrosivity of Atmospheres - Classification, Determination and Estimation; ISO 9223; International Organization for Standardization: Geneva, Switzerland, 2012.

[12] International Organization for Standardization. Corrosion of Metals and Alloys - Removal of corrosion products from corrosion test specimens; ISO 8407; International Organization for Standardization: Geneva, Switzerland, 2009.

[13] Kamimura T, Hara S, Miuyki H, Yamashita M and Uchida M 2006 *Corros. Sci.* Composition and protective ability of rust layer formed on weathering steel exposed to various environments. 48 pp 2799–2812.

[14] Hara S, Kamimura T, Miyuki H and Yamashita M 2007 *Corros. Sci.* Taxonomy for protective ability of rust layer using its composition formed on weathering steel bridge. 49 pp 1131–1142.
[15] International Organization for Standardization. Corrosion of Metals and Alloys - Corrosivity of Atmospheres - Measurement of Environmental Parameters Affecting Corrosivity of Atmospheres; ISO 9225; International Organization for Standardization: Geneva, Switzerland, 2012.
[16] Iwasaki E 2016 Scattering of Deicing Salt and Corrosion of Steel Bridges. In Proceedings of the Challenges in Design and Construction of an Innovative and Sustainable Built Environment, 19th IABSE Congress, Stockholm, Sweden, pp 1436–1441.
[17] Yamaguchi, E. Maintenance of weathering steel bridges. Steel Constr. Today Tomorrow 2015, 45, 12–15.
[18] Krivy V, Urban V and Kubzova M 2016 *Procedia Eng.* Thickness of corrosion layers on typical surfaces of weathering steel bridges. International Conference on Sustainable Development of Civil, Urban and Transportation Engineering. Ho Chi Minh City, VIETNAM, 142 pp 56-62. ISSN 1877-7058
[19] Krivy V, Marek P, Kreislova, K and Knotkova D. 2013 *Stahlbau. Berlin: Ernst & Sohn,* Bestimmung der Dickenzuschläge für wetterfesten Stahl im Brückenbau, 82(8) pp 583-588. ISSN 0038-9145

Acknowledgments
This paper has been achieved with the financial support of the Ministry of Education, specifically by the Student Research Grant Competition of the Technical University of Ostrava under identification number SP2018/103 and part of this research has been achieved with support of project under identification number MSMT 8X17039.

IOP Conf. Series: Earth and Environmental Science 143 (2018) 012009 doi:10.1088/1755-1315/143/1/012009

Influence of foundation mass and surface roughness on dynamic response of beam on dynamic foundation subjected to the moving load

Tinh Tran Quoc [1], Toan Khong Trong[1,*] and Hai Luong Van[2]

[1] Faculty of Civil Engineering, HUTECH University, Ho Chi Minh City, Vietnam
[2] Faculty of Civil Engineering, Ho Chi Minh City University of Technology (HCMUT), Vietnam National University HCMC, Viet nam

*Corresponding author: kt.toan@hutech.edu.vn

Abstract. In this paper, Improved Moving Element Method (IMEM) is used to analyze the dynamic response of Euler-Bernoulli beam structures on the dynamic foundation model subjected to the moving load. The effects of characteristic foundation model parameters such as Winkler stiffness, shear layer based on the Pasternak model, viscoelastic dashpot and characteristic parameter of mass on foundation. Beams are modeled by moving elements while the load is fixed. Based on the principle of the publicly virtual balancing and the theory of moving element method, the motion differential equation of the system is established and solved by means of the numerical integration based on the Newmark algorithm. The influence of mass on foundation and the roughness of the beam surface on the dynamic response of beam are examined in details.

Keywords: Improved Moving Element Method, dynamic foundation model, mass on foundation, roughness amplitude, roughness wavelength, moving load.

1. Introduction

The beam type structures on foundation is one of the most widely used in many fields of engineering like civil, industry, traffic infrastructures,... Especially, the structural elements which support the moving load of transportation means such as the foundation under moving load of vehicles of transport, railway...etc. Most published research papers, as consideration of analytical solutions for beam resting on foundation, the researchers have mainly employed the Winkler-Type Foundation [1]. This is the most classic model known as the one-parameter model assuming the soil below foundation is replaced by non-mass springs, linear elastic springs and springs are considered independent of each other. One of the short comings of the Winkler hypothesis is that assumes the foundation to closely spaced independent linear springs which are not affected beyond loaded region, thus the deformation of foundation is just limited load without taking into account for the affect of adjacent regions.

Consequently, it creates the interruption between loading and unloading foundation, but in reality the surface of soil foundation is not indicated any interruptions. From that point, these foundations have yet reflected the real characteristic response of soil foundation subjected to moving load 'Figure 1'.

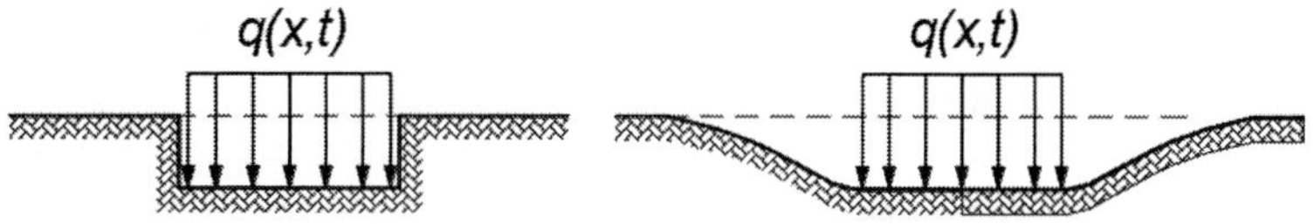

(a) Winkler foundation (b) Practical soil foundation

Figure 1. Displacement of elastic foundation under uniform pressure.

One of the ways to overcome the drawbacks in the Winkler model is to find a way describing the continuous interaction among the springs by adding to the above surface of non-mass spring layer such as bending beams, tensile stresses, shear layer or in consideration of the possible sliding of soil foundation, these layers have unchanged parameters and the characteristics of the continuous interaction of the springs in the Winkler model, the parameter is so called the second parameter. Igniting up this idea proposed by Filonenko-Borodich [2] described the continuous interaction of springs by introducing a thin elastic tensile membrane under a constant tensile force T. Following this proposal, Pasternak [3] hypothesized that the top surface of the springs are fully connected to a beam whose beam is only subjected to sloping deformation with shear modulus G. The common characteristics of those foundation models which are regardless to the affects of soil foundation mass rest on response of the upper structure. In recent years, Q Do Kien and T Khong Trong [4] used the experimental results show that the foundation mass involved in oscillation has a significant effect on the dynamic behavior of the plate. Then T Pham Dinh et al. [5] also employed the experimental results to determine the effect parameters of the foundationameters of foundation massd Mr. Hoang Phuong Hoa (2016)[5] has also employed the experimental results to define the affe mass on the kinetic behavior of a free-step system.The results also show that the influence of the characteristic parameters on the effect of the foundation mass αF on oscillation has a significant effect.

P Nguyen Trong et al. [6] systematized the foundation models and proposed a new foundation model used in the behavioral analysis of interaction structure on foundation. The author proposed a new foundation model incorporating the complete parameters such as elastic stiffness and stiffness of shear layer based on the Pasternak-Type Foundation, the coefficient of viscoelastic foundation and in particular, considering the effect of foundation mass on the behavior of the beam structure called a "dynamic foundation model". Following up the afore-mentioned studies, P NguyenTrong et al. [7] analyzed the affect of foundation mass parameter in the dynamic foundation model placed on the beam's separate oscillation. The results show that the foundation mass parameters have a significant effect on the dynamic response of the beam, which increases the overall vibration mass of the beams, thereby reducing the oscillation frequency of the system.

Recently, many models of structures resting on viscoelastic and Pasternak foundation have been developed. H Luong Van et al. [8] and P Phung Van et al. [9] analyzed dynamics response of composite plates resting on viscoelastic foundation. P PhungVan et al. [10] analyzed dynamics response of Mindlin plates on viscoelastic foundation subjected to a moving sprung vehicle. T Nguyen Thoi et al. [11] anylazed the dynamics response of composite plates on the Pasternak foundation subjected to moving mass. P Lou and F T K Au [12] have studied the response of Euler-Bernoulli beam under moving mass vehicles by employing an Finite Element Method (FEM). FEM has been used widely to solve many complicated problems, but encountered issues when the mass moves to the margin of the elements and also from one element to another while vector of moving mass must be updated at every time step. So as to make good those shortcomings.

C G Koh et al. [13] has proposed to put a moving coordinate to solve the proposed a moving mass of railway track. This method is called Moving Element Method (MEM). In this method, the railway is considered as an infinite Euler-Bernoulli resting on beam on Winkler foundation and the train is simplified by a "mass-spring-dashpot" system. T Tran Minh et al. [14] has employed MEM to study the dynamic response of express railway under inconstant speed of moving mass. K K Ang et al. [15] has studied a calculation to employ MEM to examine the dynamic response of the rail on viscoelastic foundation with moving mass. K K Ang and J Dai [16] analyzed the reaction of the high-speed railway on foundation which has inconstant stiffness, the author employed the Moving Element Method to have analytical solutions for the response of the train. K K Ang et al. [17] has used MEM to research the dynamic response of the railway system. The railway model is as a mass spring system which includes train body, cross section and wheels. Recently, T Tran Minh et al. [18] also utilized the Moving Element Method to analyze the dynamics of the express railway. In which the railway track is modelled based on Euler-Bernoulli beam on the elastic two-parameter, the impacts of reducing velocity process and the roughness levels of railway track are also investigated. MEM has a lot of advantages such as the load would never approach the margin because the limited elements system always moves, and the moving load would not have to move from this element to another, so it avoids updating the mass vector. This methods enable the limited elements with different lengths and each interaction distance can be divided more effective. However, the weak point of MEM is that must be re-updated the the stiffness matrix and dashpot matrix at every time step. It resulted in increasing the volume of calculation, prolonging the time of analysis and wasting the resources.

T Nguyen Van [19] has recently analyzed the dynamics of structural beam on Pasternak foundation utilized the Improved Moving Element Method (IMEM). The author represented a new method based on the MEM, an aim to provide the method of solving the main differential equations in a faster manner and saving resources.

Within the context of this paper, the author will use the IMEM has been proposed to investigate the two-parameter viscoelasticity under moving load in consideration of the simultaneous effect of characteristic parameter for mass foundation and the influence of the roughness of beam surface. The mass matrices, stiffness matrices and damping matrices for moving elements are illustrated in details afterward. The obtained results are helpful documents for studying and designing the structural beam under practical moving load.

2. Theoretical basis

The two parameter viscoelastic foundation includes the effect of mass foundation referred to "the dynamics foundation". The dynamics foundation model has investigated the complete mentioned-above parameters such as elastic stiffness $\bar{k}_w$, cross section, dashpots $\bar{c}$, foundation mass m.

The dynamic response of a finite length Euler-Bernoulli beam with Young's elastic modulus E, moment of inertia I and mass per unit length of the rail beam $\bar{m}$ as shown in 'Figure 2'.

The continuity in the dynamics foundation model is specified by the parameter of shear layer k_s based on the cut layer of Pasternak foundation model. According to P Nguyen Trong et al. [6], The equation of motion expressing the relation between force and displacement at each position foundation under the effect of the load $q(x, t)$ which can be expressed as follows:

$$q(x,t)=\bar{k}_w w(x,t)-k_s\frac{d^2w(x,t)}{dx^2}+\bar{c}\frac{dw(x,t)}{dt}+(\bar{m}+m)\frac{d^2w(x,t)}{dt^2} \quad (1)$$

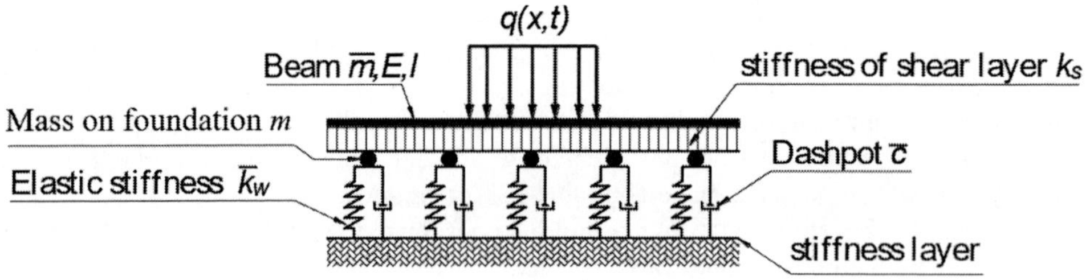

Figure 2. Beam model on dynamic foundation.

IOP Conf. Series: Earth and Environmental Science **143** (2018) 012009 doi:10.1088/1755-1315/143/1/012009

The beam model on dynamic response under moving load pursuant to P. Nguyen Trong et al. [6] combined with the overhanging mass system proposed by C G Koh et al. [13] paper model is shown in the following:

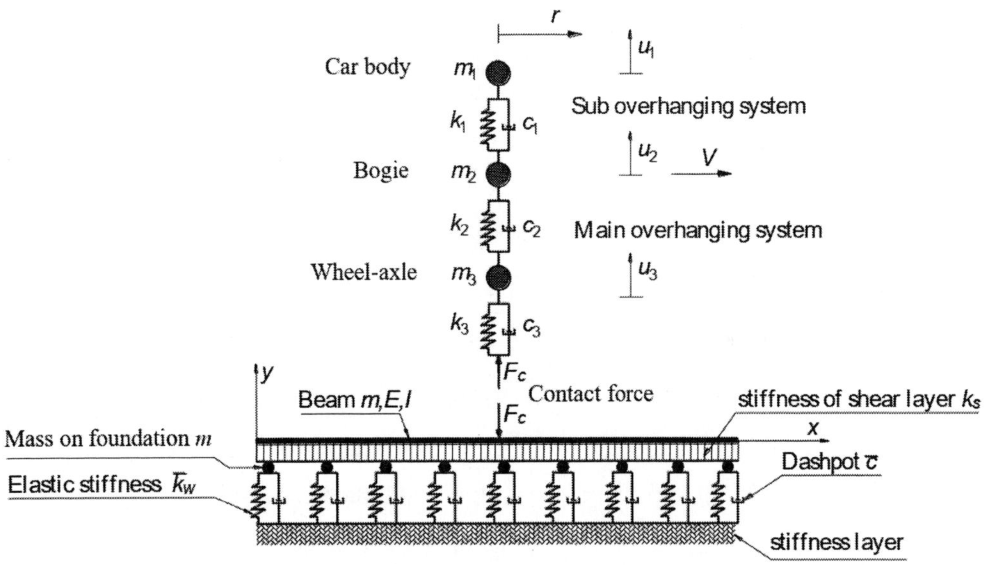

Figure 3. The Beam resting on the dynamic foundation model subjected to a moving load.

The concentrated mass *m*, dynamic response of mass on oscillating foundation model can be written as follows:

$$m = \alpha_F \rho_F H_F \tag{2}$$

In which

α_F	experimental parameters characterize the effect of the mass foundation parameter;
ρ_F	density mass foundation;
H_F	depth of foundation;
$\beta_F = \alpha_F H_F$	characteristic parameter affected by depth of foundation H_F and experimental parameters α_F describe the continuous interaction of elastic springs Winkler model. The Eq. (2) can be re-written as follows:

$$m = \beta_F H_F \tag{3}$$

When applied to a system of force F such that the displacement of a section Δl=1 unit, can be expressed as follow:

$$F = \Delta l.k = k \tag{4}$$

The moving vehicles have three mass: m1 m2 m3 and denote one direction so it has three numeral degrees of freedom.

IOP Conf. Series: Earth and Environmental Science **143** (2018) 012009 doi:10.1088/1755-1315/143/1/012009

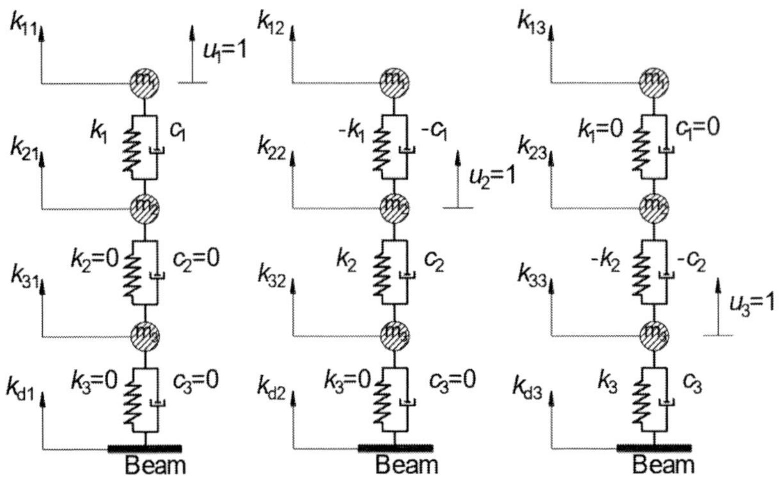

(a) Diagram 1 (b) Diagram 2 (c) Diagram 3

Figure 4. The diagrams define the mass matrices, stiffness, dashpot of vehicle motion.

Table 1. Determining the stiffness and dashpot of vehicle.

Diagram 1		Diagram 2		Diagram 3	
Stiffnes	Dashpot	Stiffnes	Dashpot	Stiffnes	Dashpot
$k_{11}=k_1$	$c_{11}=c_1$	$k_{12}=-k_1$	$c_{12}=-c_1$	$k_{13}=0$	$c_{13}=0$
$k_{21}=-k_1$	$c_{21}=-c_1$	$k_{22}=k_1+k_2$	$c_{22}=c_1+c_2$	$k_{23}=-k_2$	$c_{23}=-c_2$
$k_{31}=0$	$c_{31}=0$	$k_{32}=-k_2$	$c_{32}=-c_2$	$k_{33}=k_2+k_3$	$c_{33}=c_2+c_3$
$k_{d1}=0$	$c_{d1}=0$	$k_{d2}=0$	$c_{d2}=0$	$k_{d3}=-k_3$	$c_{d3}=-c_3$

Mass matrices can be expressed in the following:

$$\mathbf{M}_{vehicle} = \begin{bmatrix} m_1 & 0 & 0 \\ 0 & m_2 & 0 \\ 0 & 0 & m_3 \end{bmatrix} \tag{5}$$

$$\mathbf{C}_{vehicle} = \begin{bmatrix} c_1 & -c_1 & 0 \\ -c_1 & c_1+c_2 & -c_2 \\ 0 & -c_2 & c_2+c_3 \end{bmatrix} \tag{6}$$

$$\mathbf{K}_{vehicle} = \begin{bmatrix} k_1 & -k_1 & 0 \\ -k_1 & k_1+k_2 & -k_2 \\ 0 & -k_2 & k_2+k_3 \end{bmatrix} \tag{7}$$

According to the coordinates in the Figure 3, the general equation of the car model Q Do Kien and H Luong Van [21] can be expressed mathematically as follows

$$m_1\ddot{u}_1 + c_1(\dot{u}_1 - \dot{u}_2) + k_1(u_1 - u_2) = -m_1 g \tag{8}$$

$$m_2\ddot{u}_2 + c_2(\dot{u}_2 - \dot{u}_3) - c_1(\dot{u}_1 - \dot{u}_2) + k_2(u_2 - u_3) - k_1(u_1 - u_2) = -m_2 g \tag{9}$$

$$m_3\ddot{u}_3 - k_2(u_2 - u_3) - c_2(\dot{u}_2 - \dot{u}_3) = -m_3 g + F_c \tag{10}$$

in which

m_1, m_2, m_3; c_1, c_2, c_3; k_1, k_2, k_3	in turn are mass, dashpots of the vehicle, vertical springs and wheels;
$u_1, \dot{u}_1, \ddot{u}_1; u_2, \dot{u}_2, \ddot{u}_2; u_3, \dot{u}_3, \ddot{u}_3$	in turnvertical displacements, velocity, car body acceleration, and wheel-axle;
g	gravitational acceleration;
F_c	the contact force between wheels and beam, caused by the non-flat of the beam or the roughness of the beam.

The contact force Fc (with the roughness at the contact point between the moving load and the beam) is defined according to C G Koh et al. [13]as follows:

$$F_c = c_3\left(\dot{u}_d - \dot{u}_3\right) + k_3\left(u_d - u_3\right) + F_t \tag{11}$$

where: $F_t = c_3\dot{y}_t + k_3 y_t$ the track force, produced by the roughness of the beam;

u_d	denotes the vertical displacement at the contact point of the beam;
u_3	denotes the vertical displacement of the wheel-axle;
y_t	denotes the magnitude of the track irregularity at the contact point, according to C G Koh et al. [13], the track irregularity profile can be written in terms of a sinusoidal function as follows:

$$y_t = a_t\, sin\frac{2\pi S}{\lambda_t} \tag{12}$$

where a_t, λ_t denotes the amplitude and the roughness wavelength on beam, respectively;
S denotes the road section that the object is moving

In the moving element method, C G Koh et al. [13] which uses x-y coordinates. Where x axis is the beam course. The moving r-y coordinates whose origin is attached to the contact force as in Figure 5. Therefore, this coordinates moves along with the velocity V as a moving load.

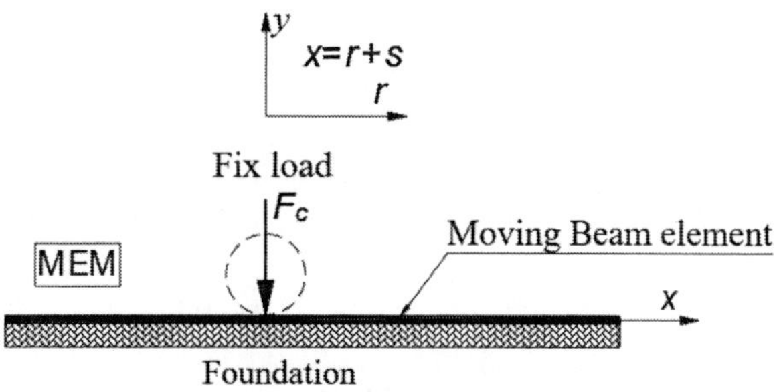

Figure 5. The coordinates of Moving Element Method (MEM).

The relationship two axes of coordinates are demonstrated as follows: $\begin{cases} x = r + s \\ y = y \end{cases}$ (13)

where: x = fixed axis; r = movable axis; s = displacement ; $V(a,t)$ = velocity function; t = moving time; a = acceleration.

The connection between the derivative operators of the coordinates when the load moves with various velocities as follows:

$$\frac{\partial^4 w(x,t)}{\partial x^4} = \frac{\partial^4 w^*(r,t)}{\partial r^4} \tag{14}$$

$$\frac{\partial^2 w(x,t)}{\partial x^2} = \frac{\partial^2 w^*(r,t)}{\partial r^2} \tag{15}$$

IOP Conf. Series: Earth and Environmental Science **143** (2018) 012009 doi:10.1088/1755-1315/143/1/012009

$$\frac{\partial w(x,t)}{\partial t}=\frac{\partial w^*(r,t)}{\partial t}\frac{\partial t}{\partial t}+\frac{\partial w^*(r,t)}{\partial r}\frac{\partial r}{\partial t}=\frac{\partial w^*(r,t)}{\partial t}-V\frac{\partial w^*(r,t)}{\partial r} \quad (16)$$

$$\frac{\partial^2 w(x,t)}{\partial t^2}=\frac{\partial^2 w^*(r,t)}{\partial t^2}-a\frac{\partial w^*(r,t)}{\partial r}\frac{\partial r}{\partial t}+V^2\frac{\partial^2 w^*(r,t)}{\partial r^2}-2V\frac{\partial^2 w^*(r,t)}{\partial r.\partial t} \quad (17)$$

where $w(x,t)$ = transverse deflection of the beam in the x-y axial coordinates; $w^*(r,t)$ = deflection of the beam in r-y coordinates.

By applying principle of virtual work and using displacement functions N, we can write $\mathbf{M}^e$, $\mathbf{C}^e$, $\mathbf{K}^e$ as generalized mass, dashpots and stiffness matrices of the beam as follows:

$$\mathbf{M}^e=(\bar{m}+m)\int_0^{l^e}\mathbf{N}^T\mathbf{N}dr \quad (18)$$

$$\mathbf{C}^e=\bar{c}\int_0^{l^e}\mathbf{N}^T\mathbf{N}dr \quad (19)$$

$$\mathbf{K}^e=EI\int_0^{l^e}(\mathbf{N}_{,rr})^T\mathbf{N}_{,rr}dr+\bar{k}_w\int_0^{l^e}\mathbf{N}^T\mathbf{N}dr-k_s\int_0^{l^e}\mathbf{N}^T\mathbf{N}_{,rr}dr \quad (20)$$

$$\mathbf{F}_1^e=2(\bar{m}+m)V\int_0^{l^e}\mathbf{N}^T\mathbf{N}_{,r}dr \quad (21)$$

$$\mathbf{F}_2^e=\left[(\bar{m}+m)a+\bar{c}V\right]\int_0^{l^e}\mathbf{N}^T\mathbf{N}_{,r}dr-(\bar{m}+m)V^2\int_0^{l^e}\mathbf{N}^T\mathbf{N}_{,rr}dr \quad (22)$$

$$\mathbf{P}=\int_0^{l^e}F_c\mathbf{N}^T dr \quad (23)$$

with $(.)_r$ and $(.)_{rr}$ in turn are first derivative and second derivative of r.

To elements of the beam, the Hermition interpolation N is written as follows:

$$N_1^e=\frac{1}{(l^e)^3}\left[2r^3-3r^2l^e+(l^e)^3\right] \quad (24)$$

$$N_2^e=\frac{1}{(l^e)^3}\left[r^3l^e-2r^2(l^e)^2+r(l^e)^3\right] \quad (25)$$

$$N_3^e=\frac{1}{(l^e)^3}\left[-2r^3+3r^2l^e\right] \quad (26)$$

$$N_4^e=\frac{1}{(l^e)^3}\left[r^3l^e-r^2(l^e)^2\right] \quad (27)$$

Based on finite element method and using numeral degree of freedom technique respectively to matrices of the general coordinates elements, the moving equation of the whole beam model on the foundation is written as follows:

$$\mathbf{M}\ddot{\mathbf{z}}+(\mathbf{C}-\mathbf{F}_1)\dot{\mathbf{z}}+(\mathbf{K}-\mathbf{F}_2)\mathbf{z}=\mathbf{P} \quad (28)$$

IOP Conf. Series: Earth and Environmental Science **143** (2018) 012009 doi:10.1088/1755-1315/143/1/012009

where: **M, C, K, P** respectively are global mass, damping and stiffness matrices and the global load vector;

$\mathbf{F}_1, \mathbf{F}_2$ denote those elements which depends on time, $\mathbf{F}_1$ and $\mathbf{F}_2$ are not forces but have the force unit so they can be considered to be Pseudo-force.

The Eq. (28) is the main differential equation of the traditional MEM, in the Eq. (28), we can see the left side is comprised of elements which change over time, those elements are the Pseudo-force $\mathbf{F}_1$ and $\mathbf{F}_2$ matrices. Therefore, when solving the problem we need to update the global mass, damping and stiffness matrices and this prolongs the processing time.

To fix this limitation of the traditional MEM, we like to move the Pseudo-forces from the left side of the Eq. (28) to the right side. This idea is called Improved Moving Element Method. After the moving, the Eq. (28) is written as follows:

$$\mathbf{M\ddot{z}} + \mathbf{C\dot{z}} + \mathbf{Kz} = \mathbf{P} + \mathbf{F}_1\dot{\mathbf{z}} + \mathbf{F}_2\mathbf{z} \quad (29)$$

Solving the differential motion Eq. (29) is put to act upon the help of computer which is based on Newmark algorithm. This algorithm is a calculation program written by Matlab language and the reliability as well as the calculation method of the program are compared to the results of other authors which are available in the reference.

3. Survey sample

3.1. Verifying the calculation program

In this part, the article examines some numerical examples to verify the correctness and the reliability of the Matlab program. The results are compared to those of other authors.

Here is the verification of the high speed train moving on beam with hanging mass which is used by C G Koh et al. [13] 'Figure 3'. The parameters of the train, the beam and the foundation are demonstrated in Table 2 and Table 3.

Table 2. Vehicle parameters.

Car Body		Bogie		Wheel-axle	
m_1	3500 kg	m_2	250 kg	m_3	350 kg
k_1	1.41×10^5 N/m	k_2	1.26×10^6 N/m	k_3	8×10^9 N/m
c_1	8.87×10^3 Ns/m	c_2	7.1×10^3 Ns/m	c_3	6.7×10^5 Ns/m

Table 3. Beam and Foundation parameters.

Beam		Foundation	
$\bar{m}$	60 kg/m	$\bar{k}_w$	$1\text{x}10^7$ N/m
E	$2\text{x}10^{11}$ N/m^2	c	4900 Ns/m
I	$3.06\text{x}10^{-5}$ m^4		
L	50m		

In the first example, the beam is displaced while the train is moving on the beam with constant velocity, without consideration of the second foundation parameter affection. (velocity V=20m/s, roughness amplitude margin a_r=0.5mm and roughness wavelength λ_r=0.5m).

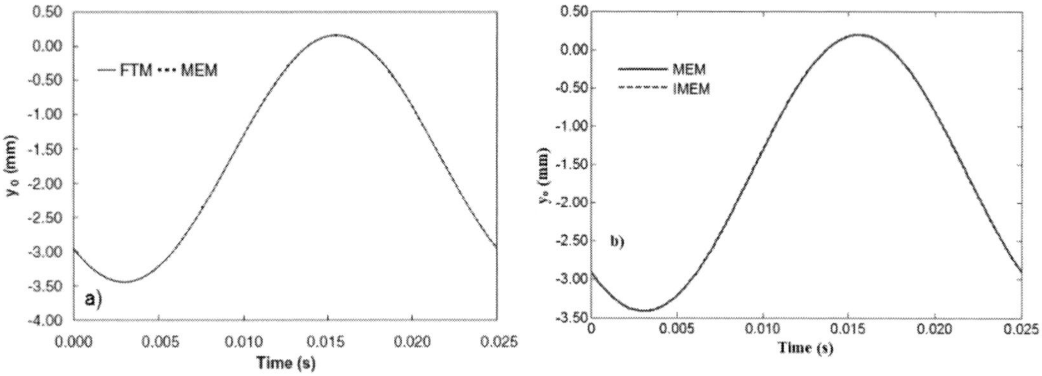

Figure 6. Beam displacement at the interaction point: (a) C G Koh et al. [13]and (b) Article

In the next verification, the beam is displaced when the train moves on the beam with changeable velocity, without consideration of the second foundation parameter affection (first velocity V= 0m/s, then moving with constant acceleration a_{max}=10m/s^2, after 2 seconds it reaches the velocity V_{max} =20m/s, then it moves with constant deceleration a_{min}= -10m/s^2, and it stops after 2 seconds. The total analyzing time is t = 6s, without consideration of the beam roughness).

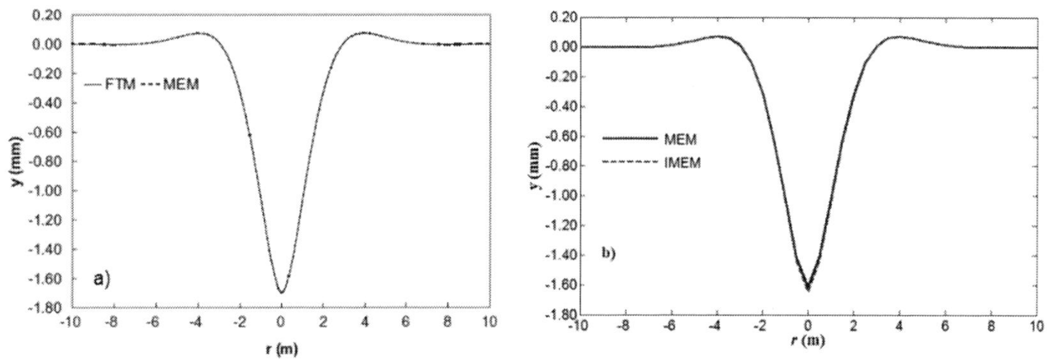

Figure 7. Beam displacement at the interaction point: (a) C G Koh et al. [13] and (b) Article.

From these surveyed results, compared to those of other authors and the results of the paper indicates that is well-matched with others as stated in the references. It proves the calculation program is reliability. Thence, we have the groundwork to continue to analyze the affection of foundation parameters, mass model, the roughness of the beam surface meeting the dynamic response of beam.

3.2. Numerical survey result

In the paper, the overhanging mass parameters and beam parameters are shown in Table 2 and Table 3. They will be used to investigate the paper's problems.

❖ **Case 1:** Survey influence of mass foundation on dynamic response of beam when the vehicle is moving with constant velocity, regardless to the affect of the beam roughness.

In the first survey, the paper researches the moving vehicles with velocity V=20m/s. The characteristic parameters of dynamics model proposed by T Pham Dinh et al. [22] written as: $\bar{k}_w$=1×10^{7}N/m, k_s=66.7×10^{5}N, c=1.5×10^{5}Ns/m and ρ_F=1800kg/m^3. The characteristic parameters β_F show the simultaneous influence of vertical depth foundation H_F and experimental parameters α_F, they are in turn to investigate such as β_F=0; β_F=0.5; β_F=1; β_F=1.5; β_F=2; β_F=2.5 and β_F=3.

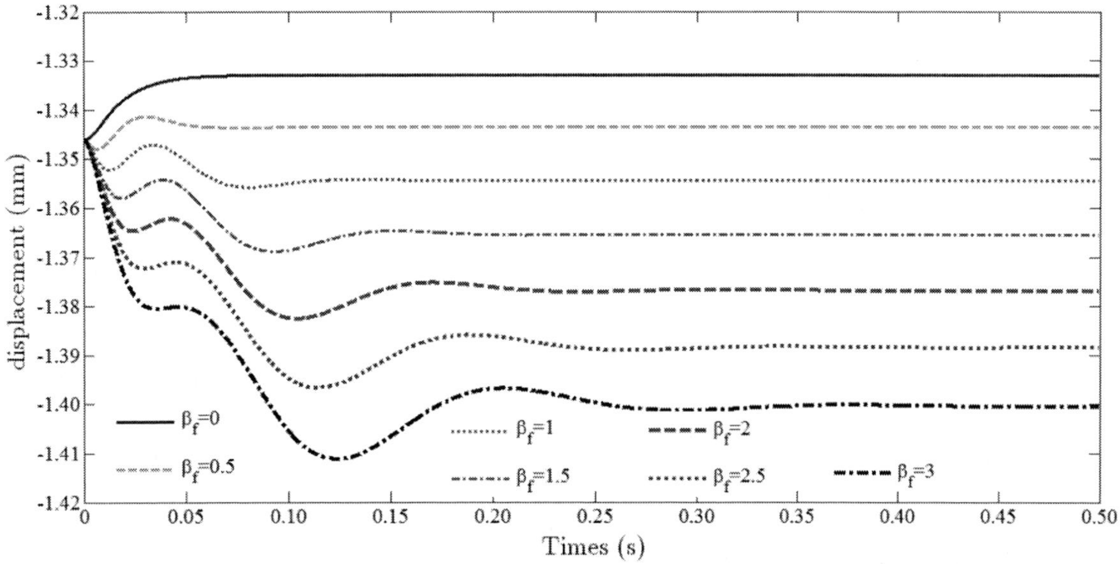

Figure 8. The displacement of beam at interaction point.

From the results as shown Figure 8 which displays the characteristic parameters for mass foundation β_F has significant influence on oscillation of beam, with the increase in β_F mass which will increase in mass foundation m participating the oscillation that is equal to the jump of participating the overall oscillation of beam, which will weaken the beam or in other way, it will soften the beam. Thus the displacement of the beam will also jump up and reach out the absolute value.

❖ **Case 2:** Conducting survey of simultaneous effect of mass foundation and load velocity of dynamic behaviour of beam.

In the second survey, the paper researches the moving vehicles on beam with constant velocity, in turn: V=10m/s; 20m/s; 30m/s; 40m/s; 50m/s; 60m/s; 70m/s; 80m/s, 90m/s; 100m/s; 110m/s; 120m/s; 130m/s; 140m/s; 150m/s; 160m/s; 170m/s; 180m/s; 190m/s; 200m/s. All characteristic parameters of dynamic foundation model is derived the same as Case 1, the characteristic parameters alone β_F for the affect of mass foundation will take turn to be conducted survey as β_F=0; β_F=0.5; β_F=1; β_F=1.5 and β_F=2.

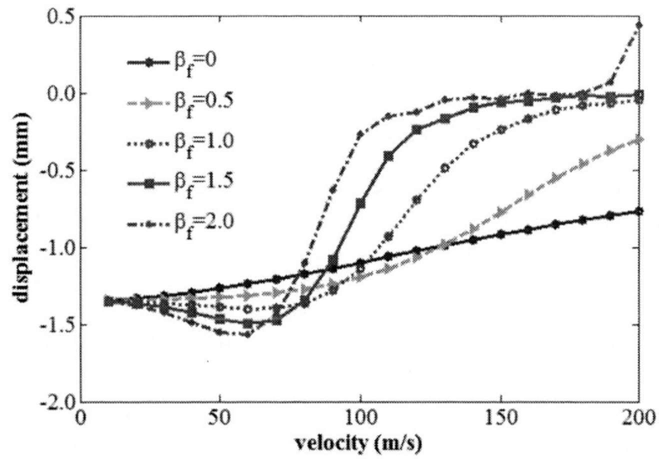

Figure 9. The max displacement of beam at interaction point.

In this problem, the effect of characteristic parameters of foundation model and dynamic response of beam subjected to moving load as given the different values of velocity parameters to be considered. The results of analysis as shown 'Figure 9'. The moving velocity is recognized V=10m/s up to V=70m/s. The larger mass on oscillating foundation is, The higher displacement of beam is, on the contrary, when velocity above V=80m/s up, if mass on oscillating foundation on beam system is high then the displacement of beam is reduced and proximity to a certain value.

The results as shown 'Figure 9' indicate the mass parameters on foundation has considerable influence on the dynamic characteristics of the system and from the point, it increases dynamic response of beam which is equivalent to the increase in characteristics parameters affected by mass on foundation. Simultaneously, the results obtained also to indicate that the characteristic parameters of moving vehicle as the velocity of movement which has a significant effect on the dynamic response of the beam structure.

❖ **Case 3:** Conducting survey of simultaneous effect of the mass on foundation and the roughness amplitude on beam of dynamic behaviour of beam.

In the following survey, the paper surveys the moving vehicle on beam with the velocity V=60m/s, the parameters on dynamic foundation model is derived from the Case 1, the specific parameters βF offer the effect of mass on foundation which will take turn to be surveyed as βF =0; βF =0.5; βF =1; βF =1.5 and βF=2. In addition, the roughness on the beam is represented as a function of time regulation on the surface of beams yt=atsin(2πS/λt) with at ,λt in terms of amplitude and wavelength of the roughness of the beam surface. In this problem keeps the whole roughness wavelength on beam λt=0.5m, the change in the roughness amplitude in turn at=0.5mm; 1.5mm; 2.0mm; 2.5mm; 3.0mm; 3.5mm and 4.0mm.

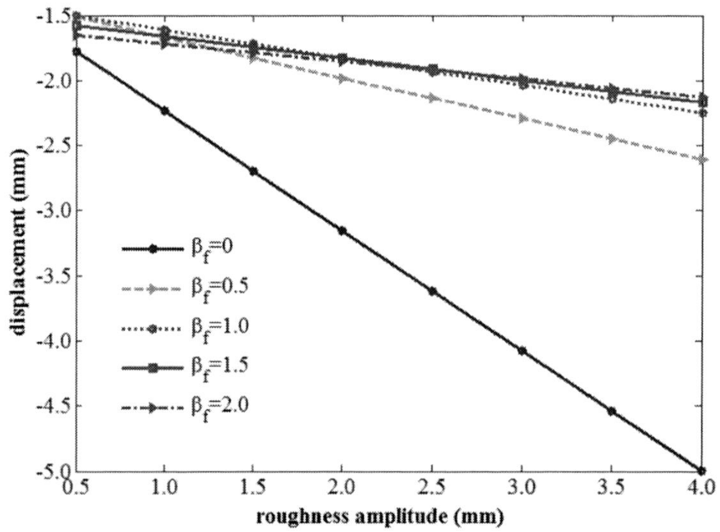

Figure 10. Maximum displacement of beam keeps the whole roughness wavelength on beam λ_t =0.5m changing the specific parameter on foundation mass β_F and the roughness amplitude.

The effect of the roughness amplitude on beam: Based on the results obtained in 'Figure 10' when keeping the whole roughness wavelength on beam λ_t=0,5m and the characteristic parameters of the foundation mass β_F only changes in the amplitude of the roughness. The results show that as increasing the value of roughness amplitude on beam the value of displacement of beam increase. The bigger value a_t is, the bigger value of beam displacement is. This proves that the displacement of beam depends very large on the amplitude of roughness on beam, when the roughness amplitude inches up, then the value of displacement on beam also inches up at the almost linear rate.

The effect of the mass characteristic parameter of foundation β_F considering the roughness on the surface of beam: The analyzed results as shown 'Figure 10' indicates that if considering the effect of roughness amplitude on beam then dynamic response of beam is specified as follows: the roughness amplitude within the approximately from a_t=0,5÷1,5mm when increasing the value of characteristic

IOP Conf. Series: Earth and Environmental Science **143** (2018) 012009 doi:10.1088/1755-1315/143/1/012009

parameters β_F of foundation mass, the displacement value also ascents. On the contrary, when the roughness amplitude from 1,5mm up above. As the increase in value of characteristic parameters β_F of mass foundation, then the displacement value also decreases, the higher value of β_F is, the lower value of displacement of beam is. But when the mass has reached the certain value then the displacement increases insignificantly. Thus, the mass parameters of foundation is fairly important, it will increase the dynamic response of system as the beam surface is flat or the roughness of amplitude is small. On the contrary, decreasing the dynamic response of system when the beam surface has greater roughness of amplitude.

❖ **Case 4:** Conducting survey of simultaneous effect of the mass on foundation and the roughness amplitude on beam of dynamic behaviour of beam.

In the survey, the parameters of the problem is derived from the Case 3, the roughness alone of the problem still keeps the whole roughness amplitude on beam at=0.5mm, the alternation of the roughness wavelength λ_t=0.5m; 1.0m; 1.5m; 2.0m; 2.5m; 3.0m; 3.5m; 4.0m and 4.5m.

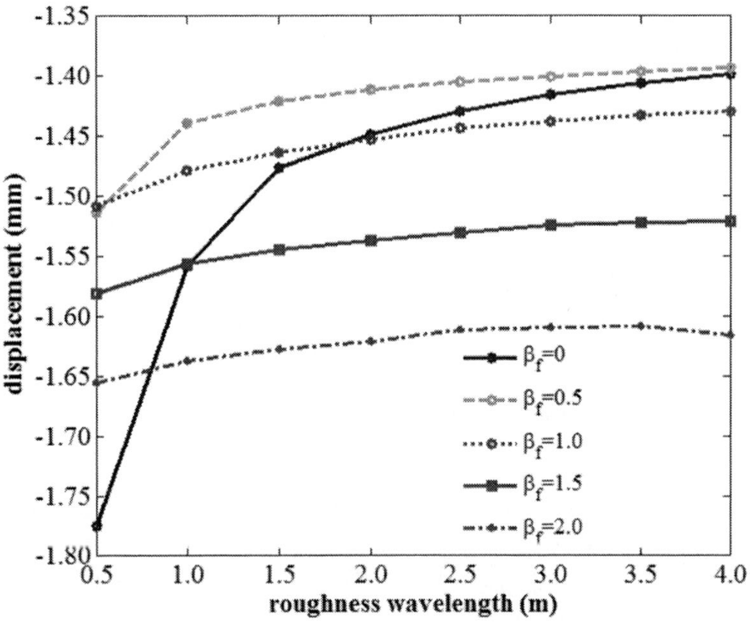

Figure 11. The maximum displacement of beam as the roughness amplitude on beam is kept as a whole by a_t=0.5mm, the change in the characteristic parameters of mass on foundation β_F and the roughness wavelength.

Based on the results of Figure 11 with the value of roughness wavelength on beam is big then the displacement of beam is reducing, on the other hand, when the wavelength reaches a certain value, the displacement diagram tends to move sideways to a certain value, specially as βF=0,5 then the displacement value will be ud=-1.5140mm; -1.4394mm; -1.4207mm; -1.4116mm; -1.4053mm; -1.4004mm; -1.3965mm and -1.3932mm corresponding to the wavelength value which in turn λt=0.5m; 1.0m; 1.5m; 2.0m; 2.5m; 3.0m; 3.5m and 4.0m

However, the value of roughness wavelength on beam reaches at a certain fixed value, if the value of characteristic parameters of mass on foundation then the displacement value of beam will increase, the bigger value βF is, the increasing of displacement value of beam is, it is specified as the roughness of wavelength λt=1.5m, the value of displacement: ud= -1.4767mm; -1.4207mm; -1.4637mm; -1.5446mm and -1.6273mm respectively to the characteristic parameters of mass on foundation in turn as written βF =0; βF =0.5; βF =1.0; βF =1.5 and βF =2.0.

IOP Conf. Series: Earth and Environmental Science **143** (2018) 012009 doi:10.1088/1755-1315/143/1/012009

4. Conclusions

The paper has presented the generalized results of dynamic beam motion analysis on moving load considering the concurrent influence of the characteristic parameters for the effect of foundation mass and the roughness of beam surfaces as well as the velocity of the moving load. The motion equation of the structural system is used based on Improved Moving Element Method. The results are obtained to display that the mass on oscillating foundation has the influence on the oscillation of beam. With the increase in the characteristic parameters of foundation mass βF will increase the mass on oscillation foundation m, it is equivalent to the increase in the general mass on oscillation foundation of system, from that point, the system will be weakened, consequently the displacement amplitude of beam will be also raised up.

On the other hand, if considering the effect of the roughness amplitude on beam then the dynamic response of beam will be greatly effected. Specially the roughness amplitude increases from the zero (flat) to at=1.5mm. The higher parameter value βF of the mass on foundation is, the bigger displacement value of beam is. In reverse, when the roughness amplitude is 1.5mm bigger as increasing the characteristic parameter value βF of the mass on foundation the displacement value of beam is decreased but when the mass reaches to a certain value, the displacement is increased insignificantly. Therefore, the parameter mass on foundation is extremely important, it creates the increase in dynamic response of system when the surface of beam is almost smooth (flat) or the roughness amplitude is small. In reserve, dynamic response of system is decreased when the beam surface has a bigger amplitude of roughness.

In addition, the velocity of moving load also has influence considerately on dynamic behaviour of beam, the study results obtained to show that the higher velocity of motion is, the lower displacement of beam is.

The value of roughness wavelength on beam is rising then the displacement of beam is going down and when the wavelength reaches up to a certain value, the displacement diagram tends to move horizontally asymptotically to a certain value.

The analytical solution results of the paper presents the study of determining accurately the parameter values of mass on oscillating foundation which is meaningful and proper with the practical application, particularly its application in analytical problem meeting the requirements on the dynamic response of the structural system interacting with the foundation models under moving loads. Besides, the unevenness of the structure surface and the element motion of velocity of load which is absolutely necessary to be considered and correct to its best feature of the structural beam.

5. References

[1] Winkler E. 1867 *Die lehre von der elastizitat und festigketit* (Prague: Dominicus)
[2] Flamenco-Borodich 1940 Some Approximate theories of elastic foundation (Moscow: Uchenyie Zapiski Moskovskogo Gosudarstvennogo Universiteta Mekhanica in Russian) pp 3–18
[3] Pasternak P L 1954 *On a new method of analysis of an elastic foundation by means of two constants* (Moscow: Gosudarstvennoe lzdatelstvo literaturi po Stroitelstvui Arkhitekture in Russian)
[4] Q Do Kien and T Khong Trong 2005 Experimental studies determine the effect of mass on foundation with separate oscillating frequency of plate on the elastic foundation *Construction journal* pp 32-35
[5] T Pham Dinh, H Hoang Phuong and P Nguyen Trong 2016 Experimental studies define the parameters influencing on dynamic response of mass on foundation of free one-parameter system *Construction journal* pp 88-91
[6] Nguyen T P, H Hoang Phuong and T Pham Dinh 2015 Systematize the foundation models and propose new foundation models used in the analytical problem of response of beam interaction with the foundation *Construction journal* pp 47-50

IOP Conf. Series: Earth and Environmental Science **143** (2018) 012009 doi:10.1088/1755-1315/143/1/012009

[7] P Nguyen Trong, H Hoang Phuong, T Pham Dinh and Q Do Kien 2015 The effects of mass parameters on dynamic foundation model of separate oscillating of beam *Construction journal* pp 95-99
[8] H Luong Van, T Nguyen Thoi, G R Liu and P Phung Van 2014 A cell-based smoothed finite element method using three-node shear-locking free Mindlin plate element (CS-FEM-MIN3) for dynamic response of laminated composite plates on viscoelastic foundation *Engineering Analysic with Boundary Element* **42** pp 8-19
[9] P Phung Van, T Nguyen Thoi, H Luong Van, C Thai Hoang and H Nguyen Xuan 2014 A cell-based smoothed discrete shear gap method (CS-FEM-DSG3) using layerwise deformation theory for dynamic response of composite plates resting on viscoelastic foundation *Computer Methods in Applied Mechanics and Engineering* **272** pp138-159
[10] P Phung Van, H Luong Van, T Nguyen Thoi and H Nguyen Xuan 2014 A cell-based smoothed discrete shear gap method (CS-FEM-DSG3) based on the C0-type higher-order shear deformation theory for dynamic responses of Mindlin plates on viscoelastic foundations subjected to a moving sprung vehicle. *International Journal for Numerical Methods in Engineering* **98** pp 988-1014
[11] T Nguyen Thoi, H Luong Van, P Phung Van, T Rabczuk and D Tran Trung 2013 Dynamic responses of composite plates on the Pasternak foundation subjected to a moving mass by a cell-based smoothed discrete shear gap (CS-FEM-DSG3) method *International Journal of Composite Materials* **3** pp 19-27
[12] P Lou and F T K Au 2013 Finite element formulae for internal forces of Bernoulli–Euler beams under moving vehicles *Journal of Sound and Vibration* **332** pp 1533-1552
[13] C G Koh, J S Y Ong, D K H Chua and J Feng 2003 Moving element for train-track dynamics *International Journal for Numerical Methods in Engineering* **56** pp 1549-1567
[14] T Tran Minh, K K Ang and H Luong Van 2014 Vertical dynamic response of non-uniform motion of high-speed rails *Journal of Sound and Vibration* (http://dx.doi.org/10.2016/j.jsv.2014.05.053)
[15] K K Ang, T Tran Minh and H Luong Van 2013 *Track vibrations during accelerating and decelerating phases of high-speed rails* (Sapporo Japan: Thirteenth East Asia-Pacific Conference on Structural Engineering and Construction EASEC 13)
[16] K K Ang and J Dai 2013 Response analysis of high speed rail system accounting for abrupt change of foundation stiffness *Journal of Sound and Vibration* **332** pp 2954-2970
[17] K K Ang, J Dai, T Tran Minh and H Luong Van 2014 Analysis of high-speed rail accounting for jumping wheel phenomenon *International Journal of Computational Methods 11* (http://dx.doi.org/10.1142/S021987621343007X)
[18] T Tran Minh, K K Ang and H Luong Van 2016 Vertical dynamic response of high-speed rails during sudden deceleration *International Journal of Computational Methods*
[19] T Nguyen Van 2016 *Analytical solutions for dynamics of inhomogeneous beam on Pasternak foundation under moving load employing the Improved Moving Element Method* (Viet Nam: Master thesis Ho Chi Minh City University)
[20] Z Feng and R Cook 1983 Beam elements on two-parameter elastic foundations *Journal of Engineering Mechanics* **109** pp1390–1402
[21] Q Do Kien and H Luong van 2010 *Structural dynamics* (Viet Nam: Ho Chi Minh City National University Publishing Company)
[22] T Pham Dinh, D Nguyen Thanh and P Nguyen Trong 2014 The effects of mass on foundation on dynamic response of beam under moving load *Construction journal* pp 83-86

IOP Conf. Series: Earth and Environmental Science **143** (2018) 012010 doi:10.1088/1755-1315/143/1/012010

Hydric characterisation of rammed earth samples for different lime concentrations

Lucile Soudani[1,2], Antonin Fabbri[1], Monika Woloszyn[2,*] , Anne-Cécile Grillet[2], Jean-Claude Morel[3]

[1] LGCB-LTDS, UMR 5513 CNRS, ENTPE, Université de Lyon, 69100 Vaulx-en-Velin, France
[2] LOCIE, Univ. Grenoble Alpes, Univ. Savoie Mont Blanc, CNRS, 73000 Chambéry, France
[3] School of Energy, Construction and Environment, Faculty of Engineering, Environment & Computing, Coventry University, UK

*corresponding author: monika.woloszyn@univ-smb.fr

Abstract. The rehabilitation of ancient rammed earth houses, as well as the use of earthen materials in modern constructions, are a growing matter of concern, especially in area such as Rhône-Alpes, France, where 40% of old constructions are in rammed earth. A current pathology observed for this type of construction is related to the rising damps, for which the water from the ground is absorbed by the wall. This situation leads to a very saturated state. As it has been proven that the compressive strength is altered by the presence of water in the pores, a better understanding on high relative humidity range is necessary to be able to predict the mechanical behavior of buildings and thus ensure a better risk assessment. The present study describes experimental results of the water uptake experiments and moisture storage at high relative humidities.

1. Introduction

Available everywhere in the world and used as a construction material for thousands years, raw earthen material can provide an answer to the world population construction needs, at an economic level as well as at an environmental one. However, it suffers from a poor scientific understanding as its knowledge is essentially practical, even though its behavior differs from other usual construction materials. As a consequence, it requires a wide range of measurements to describe its particular thermal, hydric and hygrothermal behavior. This goes from the material characterization to the full-size measurements, including wallet scale. In particular, earthen materials are able to hold a large amount of liquid water, compared to the other construction materials. A current pathology observed for this type of construction is related to the rising damps, for which the water from the ground, or massed at its surface, is absorbed by the wall. But, it has been proven that the moister the material is, the lower is its compressive strength [1], [2]. When those rising damps are too important, it thus can alter the whole building stability. Concerned about being able to assess how vulnerable a construction can be regarding to this type of pathology, studies are being carried out to characterize the capacity of this material to retain liquid water.

The water fills the pores of the material through its different states: liquid and vapor. The overall mass conservation of water (in both liquid and vapor) within a representative elementary volume of the material, and considering an isothermal case, can be written in the following form (see for example [3] for detailed derivation):

$$\frac{\partial w_L}{\partial \varphi}\frac{\partial \varphi}{\partial t} = \nabla \cdot \left(\delta_p \nabla p_v + D_L \nabla p_L\right) \quad (1)$$

Where w_L is the liquid water content, ∇ is the nabla symbol, p_v and p_L are respectively the partial pressure of vapor and the pressure of the liquid water, D_L is the liquid permeability, δ_p is the vapor transport coefficient, φ is the relative humidity. The right hand side of this equation refers therefore to transfer phenomena, while the left hand side describes water sorption capacity. The latter can be described by the sorption isotherm (water content in function of relative humidity) and then the retention curve (water content as a function of suction, for high values of relative humidity).

The relation (1) underlines that two kinds of water transport occur within the material. The first one, detected by the term $\nabla \cdot (\delta_p \nabla p_v)$ corresponds to the flow of water vapor, while the second one, $\nabla \cdot (D_L \nabla p_L)$ is the transport of liquid water. When a large amount of liquid water is flowing through the material, the liquid transport is prevailing upon the water vapor diffusion.

Knowing the behavior of the material close to saturation is a matter of concern in two main cases. First, the manufacture requires a large amount of water to be added; the material is thus saturated when it has to bear all the dead loads[4]. Second, pathologies such as rising damps can drastically increase the water content which reaches values close to saturation. The stability of the whole structure can be at stake, hence the importance of a good characterization for a better predictability of collapses.

Specifically, the retention curve is needed in order to assess water storage capacity of material. It is based on the suction (or capillary pressure), which is the difference between the atmospheric and the liquid pressure, denoted by S. The relation between the capillary pressure and the water content can be measured through different methods, more or less accurate for different types of soils and pressure ranges. The document [5] provides a good overview of the common methods:

- The most spread technic, the tensiometric measurement consists in measuring the negative pressure of liquid water using a thin saturated porous stone. The range of practical application is for pressures between 0 and 80kPa.
- In psychrometric measurements, the relative humidity is measured close to the interstitial water of the non-saturated soil, from which is deduced the suction thanks to the Kelvin law. The relative humidity is calculated through the dew point temperature of the considered atmosphere. It is a quite sophisticated method and remains unusual. The pressures thus determined don't exceed 7 MPa.
- Filter paper method is a simple and convenient method to measure suction as it only requires a 1/10 000e weighting scale. It consists in using a filter paper whose retention curve was determined before, putting it in a sample and wait for equilibrium between the filter paper and the soil. After extraction, knowing the water content of the paper, and its retention curve, it leads to the suction of the soil. Depending on which type of filter paper is used, several calibration curve can be found in the literature [16]–[19].

The link between the suction curves and the sorption curves is eventually made through the Kelvin's law (chemical equilibrium between liquid water and its vapor):

$$S = p_0 - p_L = -\frac{\rho_L RT}{M_w} \ln \varphi \tag{2}$$

with p_0 the atmospheric pressure, R the perfect gas constant, T the absolut temperature in K, M_w the molar mass of water molecules and φ the relative humidity.

Concerning transport properties, several authors have pointed out the practical troubles to measure the liquid water permeability of unsaturated soils, even if some experimental set up have been designed to overcome these difficulties [5]–[9]. Still, they remain quite sophisticated and not widespread.

Concerning earth-based materials, the use of the absorption coefficient, commonly called the A-Value, and defined by the total amount of water absorbed (in kg) per surface unit in contact with water (in meter square) and per square root of the immersion time (in seconds), is the most widespread method. The protocol of measurement is provided by the European standard EN 1015-18. Its calculation is illustrated Figure 1. The standard deals with masonry mortar but the same procedure can be applied to other porous materials. The A-value is measured on prismatic samples of 0.80x0.40x0.40 m3, initially dry, and submerged in 5 to 10 mm of water for 90 minutes. Samples are weighted after 10 and 90 minutes and the absorption speed is deduced. The water level has to be constant during the measure and the sample shouldn't stand directly on the bottom of the tank but on wedges so that the whole lower surface can be in contact with water.

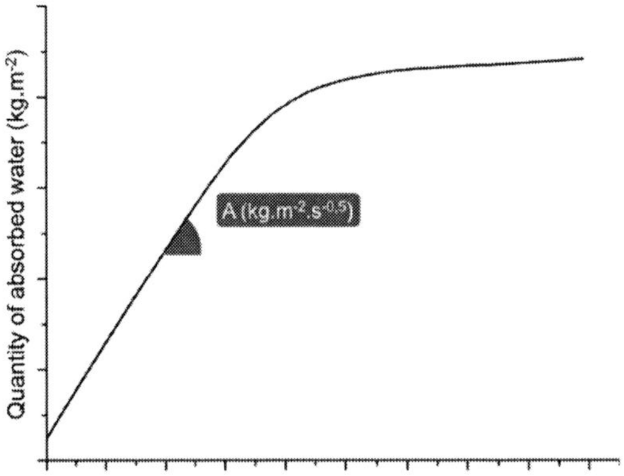

Figure 1: Calculation of the liquid absorption coefficient A

The British standard BS 3921 deals with measuring the capillary suction of water in building materials, and explains the specifications of the Initial Rate of Suction (IRS) test. The test is to be done on dry samples and weightings every minute during 5 min. However, some instructions aren't suitable for earthen materials. Hall [10] suggested another method, based on this standard. The measurement is made on cube samples of 10 cm. The samples are weighted at 1, 2, 3, 4, 5, 10, 20, 30, 40 and 60 min.

The study described in [11] consists in measuring the A-value for 15 clay brick samples. The samples are cylindrical with a diameter of 20 cm and between 1.5 and 2 cm thick. Each sample laid on a 1 cm thick paper saturated with distilled water at 20°C, put in a closed vessel to reach saturation conditions. Samples were weighted after 10, 20, 30 minutes and 1, 4, 6, 24, 48 and 72 hours.

In [12], the A-value, was determined according to the method described by Krus & Holm [13] and Hall [10] with the adaptation of the BS-3921 [14] standard to rammed earth. Indeed, some precautions need to be taken regarding the sensitivity of earth to water. In his PhD, Chabriac measured the A-value on compacted earth blocks of 0.295x0.14x0.10 m^3. The samples were weighted every 5 minutes for about one hour.

A more recent publication [15] studies the impact of stabilizers (lime, concrete) on the water transport properties. The liquid water absorption coefficient was measured on 3 types of rammed earth materials with different earth based mortars. The samples were cubic specimens of 50x50x50 mm^3 for mortars and blocs (30x20x28cm^3) and tested during around two days.

In this paper, the behavior of the material in high relative humidity range is investigated: first through suction measurements aiming at characterizing the sorption isotherm above 97% RH. Secondly, the liquid water absorption coefficient is derived from the water uptake experiment. The latter are performed for different densities and lime concentrations.

2. Material and methods

2.1. Raw earth description

The soil was extracted from a village located at less than 6km away from Saint-Antoine-L'Abbaye in Isère, France. At material scale, various measurements were made, among which the particle size distribution, given in Figure 2.

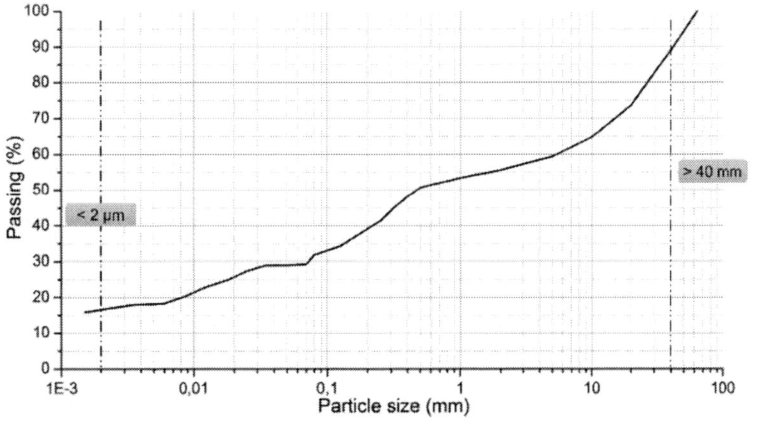

Figure 2 : Particle size distribution

It can be noticed that it includes around 10% of granulate with a diameter above 40 mm, typical of the houses of the area. The points below 2µm give the clay quantity, which reaches 16% for this material.

2.2. Definition of the lime content tested

Reference rammed earth constructions can be stabilized adding lime to the soil before compaction. The lime works on the clay minerals and can improve its strength against liquid water, particularly important for water uptake tests. The dosage is usually between 6% and 12% per dry weight and increases as the clay content increases [22].

In order to provide a better control of liquid water in the soil during construction, the natural soil was mixed with 2.5% of NHL5 lime, which represents a rather low quantity. In those conditions, the manufacturing moisture content was about 0.183 $kg.kg^{-1}$.

Samples with different lime percentages (0%, 2.5%, and 4%) were made and no curing period was considered. Indeed, lime chemically affects the soil only after a long period of time, which hadn't been waited for. In practice, the problem of liquid water migration raises starting from the construction, as a large amount of water is added. The early ages of the material are usually more affected and thus the object of following study.

2.3. Estimation of the accurate range of density variation

The dry density on site of the studied soil was measured on samples taken from a test wall realized before building the house, and was equal to 1,7. This density can be considered as low, in comparison to other modern rammed earth constructions (broad range of 1.7-2 [23]). An explanation can be the high built-in moisture content that prevents, up to a certain point, the compacting - the water being incompressible.

However, after the destruction of an experimental wallet, made by the same mason and with the same material, occurring about one and a half year after its construction, the density was measured at different spots. No regular distribution was found, nevertheless about 90% of the 23 studied samples had a density ranging from about 1.5 to 1.8. It has thus been decided to work on samples with different densities in that range.

2.4. Sample preparation and conditioning
The shape of the sample was chosen with a penetration surface wide enough in order to enable the rising damp phenomenon. The Proctor type samples were prepared: cylinders with a diameter of 10 cm and a height of approximately 6 cm. In these conditions, the penetration surface (bottom of the cylinder) was 7.85 cm^2. The soil with lime to be added was prepared about one hour before compaction, as advised in the standard EN 1015-18. In order to obtain the sample as homogenous as possible, the soil was compacted in one layer, and the on-site density of 1.7 was sought. The manufacturing of the samples is illustrated in Figure 3.

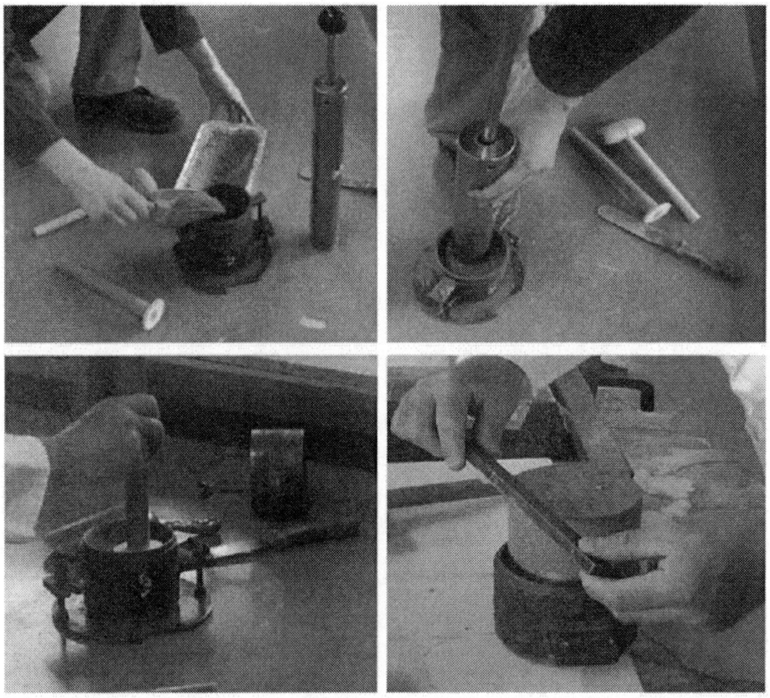

Figure 3 : Manufacturing the samples

3. Results and discussion

3.1. Sorption isotherms and retention curves

3.1.1. Sorption-desorption isotherms
Sorption/desorption were determined in a previous study [12] using saturated salt solutions and performed at 20°C. The results are given in Figure 4. As for all earthen materials, the slope of the isotherms in the high relative humidity is very important, which shows the high ability of this type of material to adsorb and hold water.

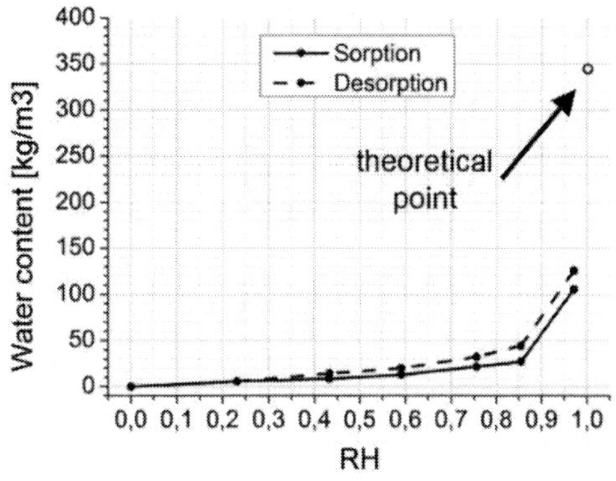

Figure 4 : Moisture storage curves in sorption and desorption

The measurement of sorption-desorption curves through salt solutions or gravimetric dynamic sorption analyzers at relative humidity higher than 95% is quite inaccurate. In addition, these methods cannot reach the saturated state, and the point RH=100% must be obtained theoretically (from the porosity). Consequently, the use of the sole sorption-desorption curves leads to high uncertainties in the range between 95% and 100% of relative humidity.

Therefore, for the suction curve, the filter paper method was chosen and the practical protocol is described in the standard [18].

3.1.2. Evolution of water content in the capillary domain
The experimental investigation was divided into two parts: the calibration of the filter paper and the suction measurement in the studied soil. Since many different calibration curves for the paper Whatman n°42 can be found, depending on the environment and other parameters, the calibration of the filter paper used was realized in the laboratory, but not detailed in the following.

The suction curves were measured on cylindrical samples of 15.2cm of diameter, and two layers of 1cm thick. They were manufactured with the studied soils at different mass water content (4%, 8%, 12%, 15% and 18%) with four samples each. Three filter papers, dried before, were put in-between the two rammed earth layers during manufacture, the one in the center having a diameter about 5mm below the two others to prevent it from soil contamination. The sample was then sealed with a plastic film and stored in a room with controlled indoor temperature and relative humidity for 7 days. The samples were weighted before and after the equilibrium period to ensure the correct sealing. After that period, the samples were destroyed with care to extract the central filter paper and retrieve its mass. The dry mass being measured previously, for each water content of the soil the corresponding mass water content of the filter paper was obtained.

Figure 5 : Suction measurement on earthen cylindrical samples

Thanks to the calibration curve of the filter paper, the retention and then the sorption isotherm could be deduced. They are both gathered in Figure 6 .

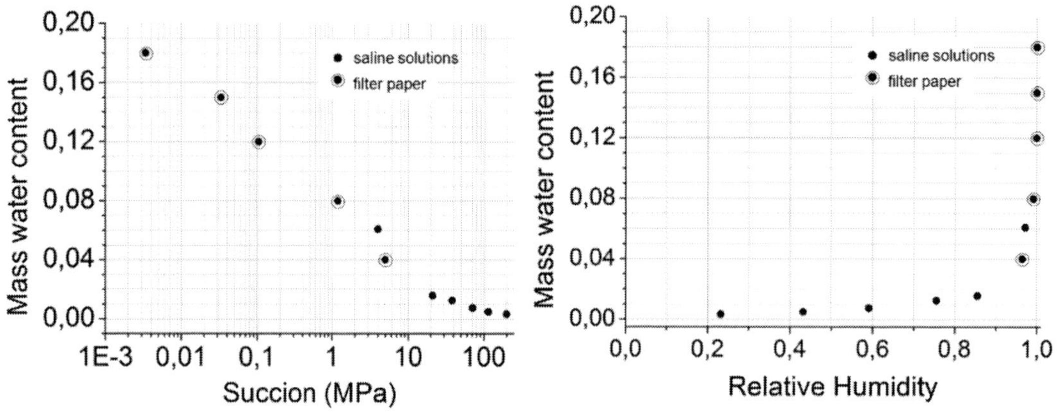

Figure 6: Final retention curve and sorption isotherm at 20°C

A good correlation can be observed between the "saline solution" data and the one measured with the filter paper. It then appears that these two methods complement one another to describe the behavior of the soil in both hygroscopic and capillary domains.

3.2. Liquid water uptake test

3.2.1. Test principle

The aim of this test is to measure the rising of water by capillarity. To ensure a one dimensional transport of liquid water, the lateral surfaces of the samples were sealed with paraffin and foil, as can be seen in Figure 7.

Figure 7 : Sealing the lateral faces

The device used for the measurement of liquid absorption coefficient was composed by a tank filled with water and a grid on wedges. Tea filter papers were added between the sample and the grid so that the tiny particles of earth didn't go in the water as the sample was becoming wet, as described in [12]. The samples were put on the grid and the grid on the water. After a certain time (shorter periods of time in the beginning and increasing), the grid was taken off and weighted. Care was taken not to keep water on the grid that wasn't absorbed by the sample, and the grid was wiped out carefully when necessary. After the weighting (happening within 30 seconds), the grid was put back on the edges and the capillary absorption could go on.

Questions have been raised concerning the time to be considered. Several references ([20], [21]) highlight the fact that the weighting operation has to be done as quick as possible. However, the time considered was strictly the time during which the sample was absorbing water, i.e. every weighting times were deleted from the whole time measurement.

The tests were realized in a room where the relative humidity nor the temperate were controlled. Therefore, to ensure a better understanding of the results, the temperature and relative humidity were measured with a USB sensor (EL-USB-2, Lascar electronics, Salisbury, UK).

3.2.2. Results

Twelve samples were tested for different densities and lime contents. The characteristics of the 12 samples are reported in Table 1.

Table 1: Densities and lime content of tested samples together with ambient temperature and relative humidity during test.

Lime	**0%**				**2.5%**				**4%**			
Name	0A1	0A2	0B1	0B2	25A1	25A2	25B1	25B2	4A1	4A2	4B1	4B2
Density	1.59	1.58	1.62	1.67	1.44	1.47	1.71	1.67	1.57	1.53	1.65	1.68
T [°C]	25.3	25.0	25.8	24.0	23.3	24.1	29.5	29.5	23.5	24.6	24.2	24.5
RH [%]	39	39	38	53	47	44	44	44	42	40	54	55

The experiment enabled us to measure the total amount of water absorbed in the sample divided by the surface area, against the square root of time. The cinematic of the rising water was measured and is synthesized in Figure 8, respectively for 0%, 2.5% and 4% of lime. Each of these graphs compare the results for samples of different densities for the same percentage of lime. They also include the linear regression for each case: the A-value (slope) and the R^2.

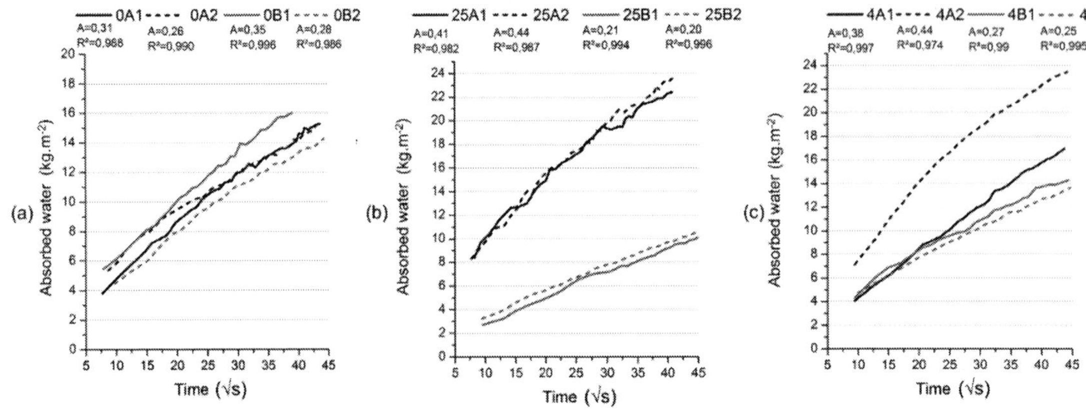

Figure 8 : Comparison of samples for three lime contents and different densities (a) 0% of lime; (b) 2.5% of lime; 4% of lime

The
Table *2* provides the A value for all samples, depending on the presence of lime or not.

Table 2: A value for each sample

Lime	0%				2.5%				4%			
Name	0A1	0A2	0B1	0B2	25A1	25A2	25B1	25B2	4A1	4A2	4B1	4B2
Density	1.59	1.58	1.62	1.67	1.44	1.47	1.71	1.67	1.57	1.53	1.65	1.68
A	0.31	0.26	0.35	0.28	0.41	0.44	0.21	0.20	0.38	0.44	0.27	0.25

For all cases, the linearity of the relation between the two parameters (absorbed water and square root of time) can be noticed. Observing a straight line (R^2 always above 97%), the liquid water absorption coefficient, which is the slope of this line, could be deduced without any trouble.

In the case of 0% lime, the curve of samples (0A1) and (0A2) are very similar, which provides a good repeatability for the measures. On the other hand, the behavior of samples (0B1) and (0B2) is slightly different: the differences in the relative humidity of the environment during the measure can be an explanation.

In the case of 2.5% lime, both the good repeatability and the difference between densities can be easily observed.

At last, the results when there is 4% lime somehow different from the previous ones: for the samples (4B1) and (4B2), the repeatability is good but, not for the samples (4A1) and (4A2) even if the slope of the curve and thus the coefficient needed is close. However, the gap between the two last samples seems to be created during a transient regime occurring in the first moment when the material is dry and isn't representative of its behavior when saturated with water.

Link with density: The first observation of this experimental campaign is the relation between the density and the A-value. All the measurements are gathered in Figure 9.

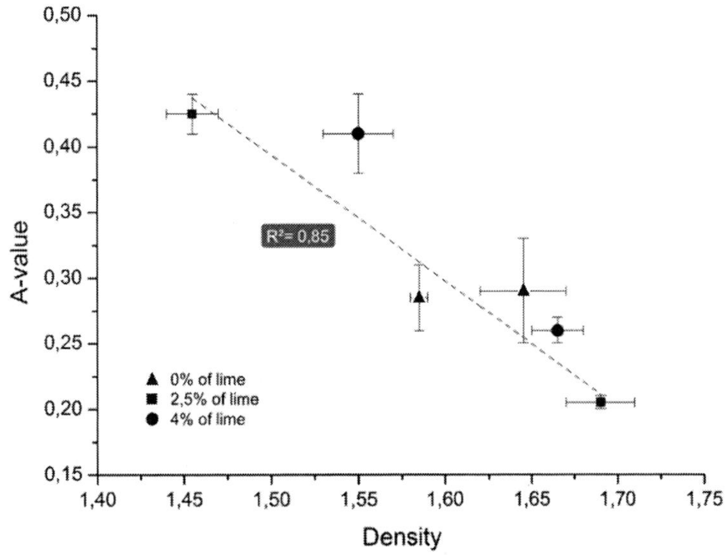

Figure 9: Comparison of A-value for different densities and lime concentrations, with their respective standard deviation

This figure shows that the A-value decreases as the density increases, which is quite intuitive. What's more, there seems to be a linear correlation between the density and the A-value, with a correlation coefficient equal to 0.85.

Link with lime concentration: The last conclusion is the fact that no significant differences are observed in terms of lime concentration variation. It seems that the quantity added doesn't impact the transport of liquid water. However, as mentioned before, the tests were done at early age of samples, therefore the chemical processes due to lime did not occur before testing. After several months, this chemical reaction might have a larger effect, as shown in [15].

4. Conclusion

To conclude, the measurement of the liquid absorption coefficient is a relatively easy-to-perform experiment, as far as a study of the possible influences (surrounding atmosphere, sealing, etc.) on the results is carried out and that the testing is conducted with care. Using filter paper to assess suction curve requires preliminary calibration of the paper, and then careful manipulations. Both tests provide interesting data to assess rammed earth behavior at high relative humidities. In addition, at early age of the samples the addition of lime was found to have no significant impact on liquid uptake.

As the next steps of this work, numerical behavior of samples will be modelled. Moreover, additional tests will be conducted in order to assess the impact of air pressure on capillary transport.

References

[1] Champiré F, Fabbri A, Morel J, Wong H and Mcgregor F 2016 "Impact of relative humidity on the mechanical behavior of compacted earth as a building material" *Constr. Build. Mater.* vol. 110 pp. 70–78.

[2] Bui Q B, Morel J C, Hans S and Meunier N 2009 "Compression behaviour of non-industrial materials in civil engineering by three scale experiments: the case of rammed earth" *Mater. Struct.* vol. 42, no. 8, pp. 1101–1116.

[3] Soudani L, Fabbri A, Morel J C, Woloszyn M, Chabriac P A and Grillet A C 2016 "Assessment of the validity of some common assumptions in hygrothermal modeling of earth based materials," *Energy Build.*, no. 116, pp. 498–511.
[4] Chabriac P A, Fabbri A, Morel J C, Laurent J P and Blanc-Gonnet J 2014 "A procedure to measure the in-situ Hygrothermal Behavior of Earth Walls," *Materials (Basel).*, vol. 7, pp. 3002–3020.
[5] Cui Y J and Delage P 2012 "L'eau dans les sols non saturés," vol. 33, no. 0, pp. 0–20.
[6] Gardner W 1958 "Some steady state solutions of the unsaturated moisture flow equation with application to evaporation from a water table," *Soil Sci.*, vol. 85, pp. 228–232.
[7] Corey A 1954 "The interrelation between gas and oil relative permeabilities," *Prod. Mon. 19*, vol. 19, pp. 38–41.
[8] Zillig W, Janssen H, Carmeliet J and Derome D 2006 "Liquid water transport in wood : Towards a mesoscopic approach," in *Research in Building Physics Engineering*. pp. 107–114.
[9] Osselin F, Fabbri A, Fen-chong T, Pereira J, Lassin A and Dangla P 2015 "Experimental investigation of the influence of supercritical state on the relative permeability of Vosges sandstone," *Comptes Rendus Mec.*, vol. 343, no. 9, pp. 495–502.
[10] Hall M and Djerbib Y May 2004 "Moisture ingress in rammed earth: Part 1—the effect of soil particle-size distribution on the rate of capillary suction," *Constr. Build. Mater.*, vol. 18, no. 4, pp. 269–280.
[11] Raimondo M, Dondi M, Gardini D, Guarini G and Mazzanti F Jul 2009 "Predicting the initial rate of water absorption in clay bricks," *Constr. Build. Mater.*, vol. 23, no. 7, pp. 2623–2630.
[12] Chabriac P A 2014 "Mesure du comportement hygrothermique du pisé," ENTPE, Université de Lyon, France.
[13] Krus M and Holm A 1999 "Simple method to approximate the liquid Transport coefficient describing the absorption and drying," in *5th symposium 'Building Physics in the Nordic Countries - Göteborg, August 24-261*. pp. 241–248.
[14] BSI, "BS-3921:1985 - Specification for clay bricks." p. 30.
[15] Gomes I, Gonçalves T D and Faria P 2016 "Hygric Behavior of Earth Materials and the Effects of their Stabilization with Cement or Lime: Study on Repair Mortars for Historical Rammed Earth Structures," *J. Mater. Civ. Eng.*
[16] Bicalho K V, Correia A G, Ferreira S, Fleureau J and M. Marinho F A 2006 "Filter paper method of soil suction measurement,".
[17] Bicalho K V, Cupertino K F and Bertolde A I 2006 "Evaluation of suction-water content calibrations of filter paper,".
[18] ASTM 1995 "Standard Test Method for Measurement of Soil Potential (Suction) Using Filter Paper," vol. 11 pp. 1–6.
[19] Nowamooz H and Chazallon C 2011 "Finite element modelling of a rammed earth wall," *Constr. Build. Mater.*, vol. 25, no. 4, pp. 2112–2121, Apr.
[20] AFNOR, "NF EN ISO 12572.pdf."
[21] Gummerson R, Hall C and Hoff W 1980 "Water Movement in Porous Building Materia . ls--II . Hydraulic Suction and Sorptlvlty of Brick and Other Masonry Materials," vol. 15, no. 3, pp. 101–108.
[22] Maniatidis V and Walker P May 2003 *A Review of Rammed Earth Construction*, no.
[23] Adam E A and Jones P J 1995 "Thermophysical Properties of Stabilised Soil Building Blocks," vol. 30, no.2.
[24] Petit B 2014 "Etude expérimentale de l'absorption d'eau du pisé," ENTPE, Université de Lyon.

Acknowledgments
We would like to thank Barthélémy Petit who carried out a large part of the experiments during his master [24] and Stéphane Cointet for his technical support.

IOP Conf. Series: Earth and Environmental Science **143** (2018) 012010 doi:10.1088/1755-1315/143/1/012010

The present work has been supported by the French Research National Agency (ANR) through the "Villes et Bâtiments Durables" program (Project Primaterre no. ANR-12-VBDU-0001) and by ADEME through CAPVENT project.

EKC Analysis and Decomposition of Influencing Factors in Building Energy Consumption of Three Municipalities in China

X Y Zhong, J X Gao*, H Ren, W G Cai

Chongqing University, Chongqing 400045

*Corresponding author's e-mail: gjxin25@163.com

Abstract. The acceleration of the urbanization process has brought new opportunities for China's development. With the rapid economic development and people's living standards improving, building energy consumption also showed a rigid growth trend. With the continuous improvement of the level of industrialization, industrial energy-saving potential declines. The construction industry to bear the task of energy-saving emission reduction will face more severe challenges. As the three municipalities of China, Beijing, Shanghai and Chongqing have significant radiation effects in the economy, urbanization level and construction industry development of the region. Therefore, it is of great significance to study the building energy consumption in the three regions with the change of urbanization level and the key factors. Based on the data of Beijing, Shanghai and Chongqing from 2001 to 2015, this paper attempts to find out whether the EKC curve of building energy consumption exists. At the same time, based on the results of the model, the data of the three regions are divided into three periods. The exponential decomposition method (LMDI) is used to find out the factors that have the greatest impact on the energy consumption of buildings in different stages. Moreover, analyzes the policy background of each stage and puts forward some policy suggestions on this basis.

1. Introduction

Urbanization is the inevitable trend of the development of human society and a sign of the progress of the human society. With the development of economy, the pace of urbanization is accelerating. The essence of urbanization is that the process of the development of social productive forces promotes the population, capital and other factors of production to centralize to the city. Some scholars point that the process of urbanization will cause a series of problems, such as traffic congestion, market competition, environmental pollution, the rapid increase of the industrial production and population etc. [1] This also leads the city to become concentrated area of carbon emissions. In fact, the relationship between urbanization and carbon emissions has been extensively studied. PARIKH et al [2] studies the cross-sectional data of 83 developing countries and concludes that urbanization has a significant positive correlation with carbon emissions. YORK et al [3] employ the data of 137 countries in 1995 to study the relationship between urbanization and carbon emissions and incorporates the quadratic items of urbanization variables in the model. The result verifies that there is a significant nonlinear relationship between urbanization and carbon emissions. Based on the panel data of 86 countries, a research by COLE [4] shows that urbanization has significant positive correlation with carbon emissions. FAN et al [5] argue that income levels have an impact on the relationship between urbanization and carbon emissions. They also divide sample data from 198 countries into groups in accordance with the World Bank's income standards. The conclusion is that the impact of urbanization on carbon emissions in low-income groups is positive, and such impact of other income groups is negative. Moreover, only the negative correlation in the high-income group countries is significant. The samples of 88 countries are divided into high, medium and low income groups by Maruotti, A [6], who argues that the urbanization of the high-income groups has a significant inhibitory effect on carbon emissions, while urbanization has a promoting effect on carbon emissions in the low-income countries.

At present, the urbanization process is developing rapidly in China. Joseph E. Stiglitz, the winner of the 2001 Nobel laureate in economics, notes that China's urbanization and the new technological revolution in the United States are the two major events affecting the development of mankind in the 21st century. This reflects the significant impact of China's urbanization on the economy and society. Some scholars have discussed the relationship between urbanization and carbon emissions in China. ZHANG et al [7] find that the effect of urbanization on carbon emissions in central China is greater than that in the eastern region. YANG et al [8] argue that the relationship between urbanization and carbon emissions in different regions is different, and the effect of urbanization on carbon emissions in developed areas is not obvious. LU et al [9] argue that urbanization contributes to Chinese carbon reduction and the contribution is more obvious in the western region. ZHAO et al [10] conclude that urbanization has a significant inhibitory effect on carbon emissions. Chen et al [11] find that urbanization can effectively reduce the level of carbon emissions by conducting a comparison analysis between the United States and China. However, some scholars have drawn different conclusions. ZHU et al [12] find that the investment and exports in the process of urbanization can lead to significant growth in carbon emissions. ZHOU et al [13] find a significant positive correlation between urbanization and carbon emissions. LIU et al [14] point out that urbanization is the main driving force of carbon emissions in China. In addition, some scholars believe that there is no linear relationship between urbanization and carbon emissions. WANG et al [15] find that urbanization has a significant positive effect on carbon emissions in the early period of urbanization and further development of urbanization will inhibit the increase in carbon emissions through studying the panel data of 9 countries. By employing the STIRPAT, HU et al [16] verify that the relationship between the urbanization and carbon emissions in China meets the EKC curve.

From the above research, it is found that most of the existing research focus on the relationship between urbanization and carbon emission while research about the relationship between urbanization and building energy consumption are obviously insufficient. Although

CHU and HU[17, 18] take account of the impact of urbanization on building energy consumption in the study of driving factors behind building energy consumption, they just regard the urbanization rate as one of the factors affecting the building energy consumption. What is the main factor of building energy consumption in China? Is the impact of urbanization on building energy consumption in different regions of China consistent? Are the effects of the different urbanization stages on building energy consumption the same? In order to solve these problems, this paper selects Beijing, Shanghai and Chongqing, which are the major municipalities of China, as the research area. Considering that these three regions have their own unique characteristics and close economic ties, this paper applies the SUR model to investigate the correlation between urbanization and the building energy consumption in these regions. On the basis, this paper expounds the energy consumption of buildings in three regions during different periods in order to find the influencing factors of building energy consumption in different urbanization stages of different regions.

2. EKC analysis of building energy consumption based on the SUR

2.1. *Variables and regression methods*

This paper investigates the existence of environmental Kuznets curve (EKC) in the building energy consumption of three cities by establishing the SUR model of building energy consumption and urbanization rate in Beijing, Shanghai and Chongqing. The energy consumption per people (E) in three cities is the dependent variable and the urbanization rate (U) is the independent variable. Considering that Beijing, Shanghai, and Chongqing are major municipalities in China and their economic relationship is very close even if the urbanization processes of these three cities are at different stages, this paper also introduces per capita GDP (G) and foreign trade dependence (T) as independent variables. Taking the three cities as the research objects, the influencing factors of building energy consumption are both relevant and different, so the unrelated regression (SUR) model has unique advantages. SUR is a system of equations consisting of a series of equations that can be estimated based on the information about the equations of the error vector, which is more efficient than a single equation.

2.2. *Model settings*

In this paper, taking the urbanization rate as the main feature, and per capita GDP, foreign trade dependence as independent variables, per capita building energy consumption as dependent variable, the following four models are established:

$$\log(E)_{it} = \alpha_i + \beta_{1i}\log(U_{it}) + \beta_{2i}\left[\log(U_{it})\right]^2 + \varepsilon_{it} \quad (1)$$

$$\log(E)_{it} = \alpha_i + \beta_{1i}\log(U_{it}) + \beta_{2i}\left[\log(U_{it})\right]^2 + \beta_{3i}\log(G_{it}) + \varepsilon_{it} \quad (2)$$

$$\log(E)_{it} = \alpha_i + \beta_{1i}\log(U_{it}) + \beta_{2i}\left[\log(U_{it})\right]^2 + \beta_{4i}\log(T_{it}) + \varepsilon_{it} \quad (3)$$

$$\log(E)_{it} = \alpha_i + \beta_{1i}\log(U_{it}) + \beta_{2i}\left[\log(U_{it})\right]^2 + \beta_{3i}\log(G_{it}) + \beta_{4i}\log(T_{it}) + \varepsilon_{it} \quad (4)$$

i = Beijing, Shanghai, Chongqing; t = 2001, ..., 2015.

2.3. *Data sources*

The urbanization rate, per capita GDP and dependence on foreign trade of three regions are from the "China Statistical Yearbook". The urbanization rate is the ratio of urban population to the total population; the per capita GDP is the ratio of regional GDP to the total population; the dependence of foreign trade is the ratio of regional total import and export to regional GDP.

Access to energy consumption per capita in the region: This method is based on the energy balance sheet. There are significant differences on energy balance information among statistical yearbook of different regions. For consistence, the China Energy Statistical Yearbook is employed as a data source for building energy consumption. The calculation formula of building energy consumption is as follows:

Building energy consumption = building energy consumption - transport energy consumption + other building energy consumption correction

Table 1. The variable statistics

Variable	statistics	Beijing	Shanghai	Chongqing
Building energy consumption	average	1781.532	914.563	331.548
	maximum	2244.870	1289.991	579.369
	Minimum	1229.138	566.484	168.240
Per capita GDP	average	64295.600	66405.600	23923.200
	maximum	106497.000	103796.000	52321.000
	Minimum	26998.000	32201.000	6219.000
Urbanization rate	average	0.837	0.853	0.499
	maximum	0.865	0.896	0.609
	Minimum	0.781	0.753	0.374
Foreign trade dependence	average	0.796	0.526	0.372
	maximum	0.854	0.569	0.458
	Minimum	0.746	0.475	0.259

2.4. Result

We first take the logarithm of the variables: per capita building energy consumption, per capita GDP, urbanization rate, dependence on foreign trade. In order to avoid the "pseudo-regression" of the time series data, the stationary sequence of the time series is tested, which is the basis for ensuring that the variables satisfy the cointegration test. In this paper, we use ADF test to test the unit root test. Through Eviews 9.0 software to get the following conclusions: The $\log(U_{it}), [\log(U_{it})]^2$ and $\log(G_{it})$ is smooth. Although the ADF statistic of $\log(T_{it}), \log(E_{it})$ is significant at 1% level, the ADF statistic is passed at a 1% level after a first order difference. Their horizontal sequences are not stable, but the first order difference is stationary, i.e. these variables are first order single I (1).

By the result of the unit root test, the above variables satisfy the cointegration test condition. In this paper, we use the Johansen method to test the covariance of variables. The statistical results show that there is at least one cointegration vector at the 10% significance level, indicating that there are cointegration relations. This paper uses Stata12.0 to test the environmental Kuznets curve of building energy consumption. The test results are shown in Table 2.

Table 2. EKC Analysis of Building Energy Consumption Based on SUR Model

		(1) sur1	(2) sur2	(3) sur3	(4) sur4
Beijing	Per capita building energy consumption U1				
	Urbanization rate	-12.48*** (10.55)	-20.97*** (14.48)	-14.02** (17.50)	-16.88 (18.63)
	The square of Urbanization rate	-42.25*** (27.21)	-60.08*** (34.12)	-45.97 (43.86)	-47.35* (45.85)
	Per capita GDP		0.131*** (0.192)		0.282*** (0.230)
	Foreign trade dependence			0.00646*** (0.872)	-0.927*** (1.145)
	Constant term	6.645*** (0.985)	4.276*** (3.338)	6.495*** (1.525)	2.715 (3.497)
	EKC inflection point	0.8624	0.83687	0.8492	0.83694
	Arrival year	2012	2005	2008	
Shanghai	Per capita building energyconsumption U1	-0.0635***	5.515***	4.805***	-4.635***
	Urbanization rate	-0.0635*** (3.509)	-5.515** (5.936)	-4.805 (3.326)	-4.635*** (6.515)
	The square of Urbanization rate	-19.661*** (9.199)	-1.236*** (12.78)	-4.202*** (8.825)	-3.829* (11.06)
	Per capita GDP		-0.249*** (0.292)		0.984*** (0.498)
	Foreign trade			-1.147*** (0.427)	-3.134*** (0.986)
	Constant term	7.062*** (0.300)	10.39*** (3.730)	6.699*** (0.365)	-6.723 (6.780)
	EKC inflection	0.89472 2009	0.8107 2004	0.56451	0.54607
Chongqing	Per capita building energy consumptionU	6.500***	3.596	6.707**	2.667*
	Urbanization	6.500***	3.596	6.707***	2.667

rate	(1.142)	(2.630)	(1.862)	(3.179)
The square of	20.572***	1.624	2.675	4.376*
	(0.789)	(1.116)	(1.205)	(1.459)
Per capita GDP		0.319***		0.445**
		(0.263)		(0.272)
Foreign trade			0.0819***	0.0488*
			(0.206)	(0.201)
Constant term	8.971***	4.273	9.146***	2.550
	(0.401)	3.878)	(0.858)	(4.190)
EKC inflection	0.854704	0.33055	0.28548	0.7374
	not reached			not reached
N	15	15	15	15
R-sq	0.716	0.687	0.651	0.648

Standard error parentheses
* p<0.1, ** p<0.05, *** p<0.01

From the results of model 1 in Table 2, we can see that there are environmental Kuznets curves in Beijing, Shanghai and Chongqing. Beijing reached the "inflection point" in 2008 and 2012 and Shanghai reached the "inflection point" in 2009. Building energy consumption in both regions are reduced with the increase of urbanization rates now. Chongqing has not yet reached the "inflection point" by now. Actually the inflection point of building energy consumption in Chongqing will appear when the urbanization rate reaches 0.74 and 0.85. Per capita GDP is added in Model 2 added , which studies the impact of economic development on the energy consumption in these cities. Beijing, Shanghai and Chongqing are in different stages of urbanization process. The economic development speed and process are also in different status. From the results, there are EKC curves in Beijing and Shanghai and the two regions reached the curve "inflection point" at year 2005 and 2004 respectively. The peaks of the two regions were earlier than the results of Model 1. Beijing's per capita GDP has a positive impact on building energy consumption, which means the building energy consumption will increase with the economic growth. In contrast to Beijing, per capita GDP in Shanghai has a negative impact on building energy consumption, indicating that with the improvement of urban economic development level, building energy consumption will reduced. Model 3 incorporates the variable of dependence on foreign trade, investigating the impact of foreign trade on building energy consumption in three regions from the perspective of external factors. The results show that foreign trade dependence in Beijing and Chongqing have a positive impact on building energy consumption, while the impact on Shanghai's building energy consumption is negative. Only the results of the model in Shanghai are meaningful in model 3. Model 4 also incorporates two variables: per capita GDP and foreign trade dependency from external and internal perspective. At this time Shanghai reaches the peak year in advance while the peak of Chongqing has not yet reached. Per capita GDP has a negative effect on the building energy consumption in Beijing and Shanghai, while foreign trade has a positive effect in Beijing and Shanghai.

IOP Conf. Series: Earth and Environmental Science 143 (2018) 012011 doi:10.1088/1755-1315/143/1/012011

From the results of model 1, 2, 3 and 4, the significance index of model 1 is obviously better than other models and its R^2 is the largest. So the fitting effect is the best. EKC curve of Beijing, Shanghai is inverted "U" shape, and the EKC curve of Chongqing is "U" shape. After adding variables such as GDP per capita and foreign trade, the results of the study were affected, and the significance of some results was reduced or even untrustworthy. These phenomena may be related to the government to adjust the economic development of the policy measures taken. Beijing and Chongqing will be will lead to building energy consumption with the economic development, only Shanghai's economic factors on the building energy consumption has the effect of emission reduction. This shows that even in developed areas in China is also difficult to achieve the effect of reducing building energy consumption to improve the environment. Therefore, the government needs to increase the intensity of building energy management. That is to say, the EKC curve of building energy consumption is not set up automatically, but requires the government and the people to raise awareness of building energy consumption and take effective policy measures and more advanced energy-saving technologies.

3. Beijing, Shanghai and Chongqing building energy consumption factor decomposition

3.1. Research methods

The second section of this paper concludes that Beijing, Shanghai have reached the EKC curve inflection point in 2012 and 2009 respectively. This section divides construction energy consumption into three stages: 2001 to 2009, 2009 to 2012 and 2012 to 2015. This paper analyzes the influencing factors of building energy consumption in three regions, trying to find out the factor with greatest impact on building energy consumption in three regions.

At present, the index decomposition method applied in the field of energy is most popular. The basic idea of the index decomposition is to decompose the change value of the target variable into a combination of several factors, and then to identify the size of the contribution of each factor to the target variable change value, to find the maximum-contribution value. Considering the practicality, operability and other factors, this paper finally employs LMDI index decomposition to analyze the driving factors behind the building energy consumption in concerned regions.

3.2. Model establishment

According to the different types of buildings, the building energy consumption is divided into rural building energy consumption, urban residential building energy consumption, and urban public building energy consumption. At the same time, according to the results and analysis of Section 2, the diving factors of building energy consumption include population, economy, urbanization rate, construction area, human behavior and technology and policy. The equation built according to the "I = PAT" model is as follows:

$$E_1 = P \times \frac{P_1}{P} \times \frac{C_1}{P_1} \times \frac{E_1}{S_1 \times B \times C_1} \times B \times S_1 \tag{5}$$

$$E_2 = P \times \frac{P_2}{P} \times \frac{C_2}{P_2} \times \frac{E_2}{S_2 \times B \times C_2} \times B \times S_2 \tag{6}$$

$$E_3 = P \times \frac{P_3}{P} \times \frac{C_3}{P_3} \times \frac{E_3}{S_3 \times B \times C_3} \times B \times S_3 \tag{7}$$

E_1, E_2, E_3 represent the energy consumption of rural buildings, the energy consumption of urban residents, and the energy consumption of urban public buildings. P is the total population of the region, representing the population factor. $\frac{P_1}{P}, \frac{P_2}{P}, \frac{P_3}{P}$ are the proportion of rural population, the proportion of urban population, the proportion of urban population, representing urbanization factors and denoted by U_1, U_2, U_3 respectively. $\frac{C_1}{P_1}, \frac{C_2}{P_2}, \frac{C_3}{P_3}$ are the per capita disposable income in rural areas, the per capita disposable income of urban residents, the per capita disposable income of urban residents, representing the economic factors and denoted by TC_1, TC_2, TC_3. $\frac{E_1}{S_1 \times B \times C_1}, \frac{E_2}{S_2 \times B \times C_2}, \frac{E_3}{S_3 \times B \times C_3}$ are the energy efficiency of rural buildings, the comparable energy efficiency of urban residential buildings, the comparable energy efficiency of urban public buildings, denoted by A_1、A_2、A_3. Building energy efficiency indicators can well reflect the characteristics of economic growth, technological progress, energy conservation policies and human behavior factors behind carbon emissions. Regional building energy efficiency represents the regional policy measures and technical level about building energy consumption in the exclusion of economic factors and human behavior factors. B is the human behavior factor. S_1、S_2、S_3 are the rural construction area, urban residential construction area, urban public construction area.

Energy consumption model is as follow:

$$E = \sum_i E_i = \sum_i PU_i TC_i A_i BS_i \quad (i = 1,2,3) \tag{8}$$

The decomposition of equation (5) is carried out based on LMDI decomposition method. In the time [0, T], the building energy consumption changes from E^0 to E^T, the change amount and the change proportion are ΔE_{tot}、D_{tot}. The decomposition forms are:

$$\Delta E_{\text{tot}} = E^T - E^0 = \Delta E_P + \Delta E_u + \Delta E_{TC} + \Delta E_A + \Delta E_B + \Delta E_S + \Delta E_\varepsilon \tag{9}$$

$$D_{\text{tot}} = E^T / E^0 = D_P D_u D_{TC} D_A D_B D_S D_\varepsilon \tag{10}$$

$E_P(D_P), \Delta E_u(D_U), \Delta E_{TC}(D_{TC}), \Delta E_A(D_A), \Delta E_B(D_B), \Delta E_S(D_S)$ represent the total population, urbanization, per capita disposable income, comparable energy efficiency, behavioral factors and the contribution of building area to changes in building energy consumption. ΔE_ε、D_ε is the decomposition residual value. Applying the LMDI decomposition method to analyse the contribution rates, the results are as follows:

Addition:

IOP Conf. Series: Earth and Environmental Science **143** (2018) 012011 doi:10.1088/1755-1315/143/1/012011

$$\Delta E_P = \sum_i \frac{E_i^T - E_i^0}{\ln E_i^T - \ln E_i^0} \ln\left(P^T / P^0\right) \tag{11}$$

$$\Delta E_U = \sum_i \frac{E_i^T - E_i^0}{\ln E_i^T - \ln E_i^0} \ln\left(U^T / U^0\right) \tag{12}$$

$$\Delta E_{TC} = \sum_i \frac{E_i^T - E_i^0}{\ln E_i^T - \ln E_i^0} \ln\left(TC^T / TC^0\right) \tag{13}$$

$$\Delta E_A = \sum_i \frac{E_i^T - E_i^0}{\ln E_i^T - \ln E_i^0} \ln\left(A^T / A^0\right) \tag{14}$$

$$\Delta E_B = \sum_i \frac{E_i^T - E_i^0}{\ln E_i^T - \ln E_i^0} \ln\left(B^T / B^0\right) \tag{15}$$

$$\Delta E_S = \sum_i \frac{E_i^T - E_i^0}{\ln E_i^T - \ln E_i^0} \ln\left(S^T / S^0\right) \tag{16}$$

Multiplication:

$$D_P = exp\left\{\frac{\ln\left(E^T / E^0\right)}{E^T - E^0} \Delta E_P\right\} \tag{17}$$

$$D_U = exp\left\{\frac{\ln\left(E^T / E^0\right)}{E^T - E^0} \Delta E_U\right\} \tag{18}$$

$$D_A = exp\left\{\frac{\ln\left(E^T / E^0\right)}{E^T - E^0} \Delta E_A\right\} \tag{19}$$

$$D_{TC} = exp\left\{\frac{\ln\left(E^T / E^0\right)}{E^T - E^0} \Delta E_{TC}\right\} \tag{20}$$

$$D_B = exp\left\{\frac{\ln\left(E^T / E^0\right)}{E^T - E^0} \Delta E_B\right\} \tag{21}$$

$$D_S = exp\left\{\frac{\ln\left(E^T / E^0\right)}{E^T - E^0} \Delta E_S\right\} \tag{22}$$

From the current research results it can be seen that human behavior can directly affect the per capita building energy consumption. It is assumed that per capita building energy consumption is proportional to human behavior, namely:

$$B = KE \tag{23}$$

The result are as follows:

$$\Delta E_A = \sum_i \frac{E_i^T\text{-E}_\text{i}^0}{\ln E_i^T - \ln E_i^0} \ln\left(A^T / A^0\right) = \sum_i \frac{E_i^T\text{-E}_\text{i}^0}{\ln E_i^T - \ln E_i^0} \ln\left(S^0 C^0 / S^T C^T\right) \tag{24}$$

$$\Delta E_B = \sum_i \frac{E_i^T\text{-E}_\text{i}^0}{\ln E_i^T - \ln E_i^0} \ln\left(B^T / B^0\right) = \sum_i \frac{E_i^T\text{-E}_\text{i}^0}{\ln E_i^T - \ln E_i^0} \ln\left(E^T / E^0\right) \tag{25}$$

The relevant data are collected from the "China Statistical Yearbook", "China Energy Statistical Yearbook" and the calculation results of Section 2.3.

3.3. Model results

By calculation, the results are shown in Table 3

Table 3. Regional Building Energy Decomposition Results

	area	years	Building energy consumption changes	population	Urbanization rate	Per capita disposable income	Building energy efficiency	Behavioral factors	construction area
Addition form	Beijing	2001~2009	1916.22	639.278	82.74	1804.2	-3192.4	1916.221	671.21
		2009~2012	-281.41	255.53	13.73	654.15	-1225.1	-281.411	301.66
		2012~2015	291.382	226.72	2.456	1441.4	-2294.9	291.382	624.53
	Shanghai	2001~2009	1912.208	485.16	2158.8	1383.25	-4818.4	1912.20	791.20
		2009~2012	-481.042	193.01	10.47	867.93	-824.98	-481.04	-246.4
		2012~2015	134.063	35.611	-30.68	672.138	-24.72	134.06	-652.33
	Chongqing	2001~2009	539.537	7.47	96.05	582.44	-966.56	539.537	274.54
		2009~2012	328.6943	35.19	59.67	481.02	-715.79	328.69	149.75

		2012~2015	309.3939	36.35	39.71	324.47	-290.88	309.39	-118.09
	area	years	Building energy consumption changes	population	Urbanization rate	Per capita disposable income	Building energy efficiency	Behavioral factors	construction area
Multiplication form	Beijing	2001~2009	1.418	1.343	1.039	2.296	0.230	2.418	1.362
		2009~2012	-0.086	1.085	1.004	1.233	0.676	0.914	1.101
		2012~2015	0.098	0.778	0.997	0.203	2.691	0.724	0.501
	Shanghai	2001~2009	2.029	1.325	3.494	2.229	0.061	3.028	1.582
		2009~2012	-0.168	1.077	1.004	1.395	0.729	0.832	0.910
		2012~2015	0.056	1.015	0.988	1.317	0.990	1.056	0.765
	Chongqing	2001~2009	1.095	1.010	1.141	2.222	0.266	2.095	1.457
		2009~2012	0.318	1.030	1.051	1.499	0.548	1.318	1.134
		2012~2015	0.227	1.024	1.027	1.240	0.825	1.227	0.925

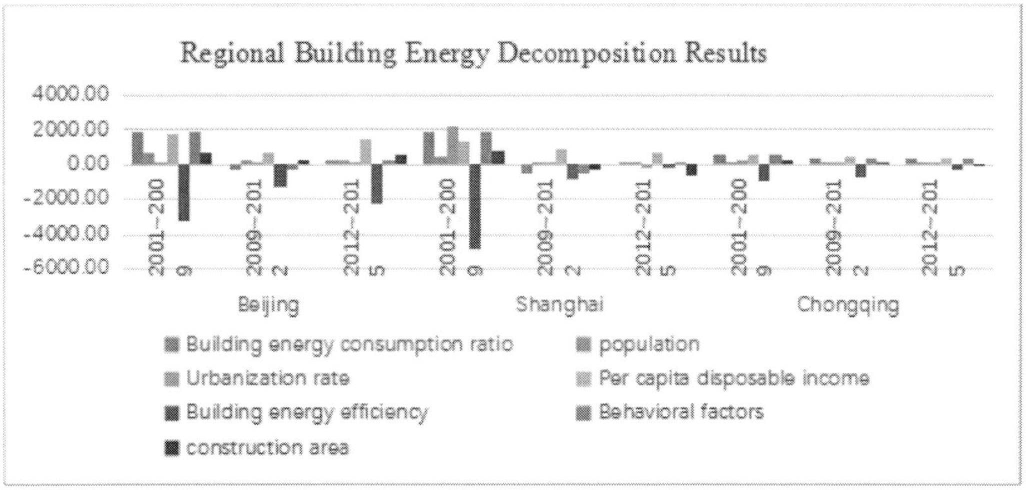

Figure 1. regional building energy consumption changes decomposition

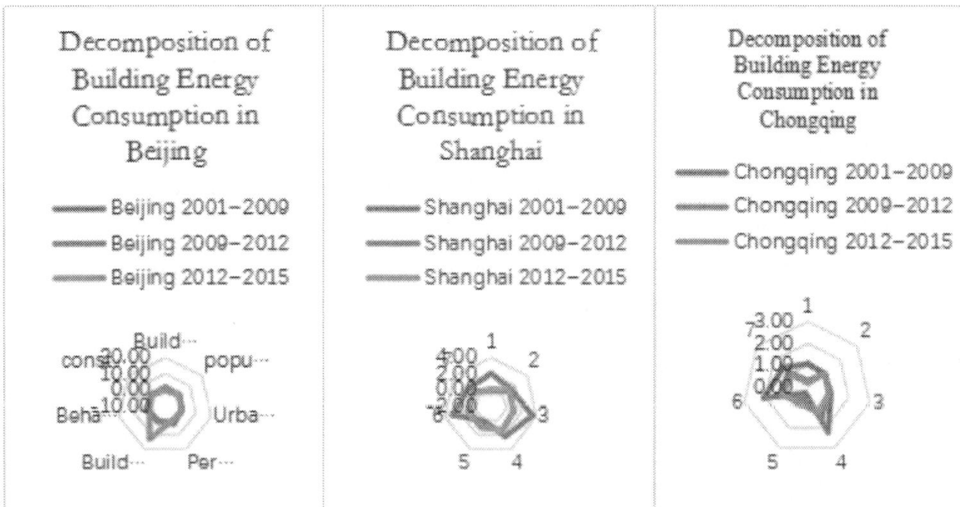

Figure 2 regional building energy consumption changes decomposition coefficient

From the results in Table 3, Fig. 1 and Fig. 2, it can be seen that the effects of different variables on building energy consumption in different regions are different in different periods.

4. Analysis on the Trend of Building Energy Consumption in Beijing, Shanghai and Chongqing

From the results of Beijing, Shanghai and Chongqing, it can be found that the impact of building energy efficiency on building energy consumption are all negative, indicating that energy conservation policies and technologies are the main drivers of building energy consumption reduction. At the same time, changes in building energy consumption are also consistent with China's various industrial policies and environmental policies. The main influencing factors include the rise of heavy industry after 2000, the implementation of energy saving and emission reduction policies, the holding of the 2008 Olympic Games and the economic stimulus plan after the global financial crisis. The year of 2006 - 2008 is Chinese "Eleventh Five - Year Plan" period and also the preparation and holding period of the Beijing Olympic Games. The "Eleventh Five-Year Plan" and the successful hosting of the Green Olympics proposed a resource-saving and environment-friendly society. The "Eleventh Five-Year" period also proposed and developed detailed energy-saving emission reduction measures. Under the influence of these policies, Chinese economic development began to focus on green development and low-carbon development. A series of economic stimulus plans were implemented in China due to 2008 global financial crisis, so some pollution-intensive industries obtained development opportunities. However, the effect of a series of stringent environmental regulation policies taken after 2005 became evident after 2009. The incentive effect of the economic stimulus policies taken after 2008 gradually appeared after 2012.These indicates that the implementation of relevant policies on the impact of building energy consumption has a certain time lag.

The per capita disposable income is the main driving factor of increasing building energy consumption in these cities. However, its effect on building energy consumption is weakening with time. Prior to 2009, the impact coefficient of per capita disposable income on building energy consumption reached a couple of points and declined after 2009. With the largest decline, the impact coefficient of the per capita disposable income on building energy consumption in beijing is only 0.203 between 2012 and 2015. The impact coefficient in Shanghai is 1.317, which is the highest among these cities. It can be seen that with the regional economic growth and social progress, the impact of economic factors on the building energy consumption decreased year by year but stay positive. This also confirms that building energy consumption can not be effectively reduced only by economic development. Contrary to the effect of per capita disposable income, the impact of building energy efficiency on building energy consumption has been negative all the time and the coefficient of influence has increased over time. It can be seen that the government's energy-saving emission reduction policies and building energy-saving technologies are key factors in reducing building energy consumption. Building energy consumption reduction is a kind of public product. A single factor does not have enough power to reduce emissions, so reducing the building energy consumption must be involved in government departments. The impact of building energy efficiency and per capita disposable income on building energy consumption is shown in Figure 3.

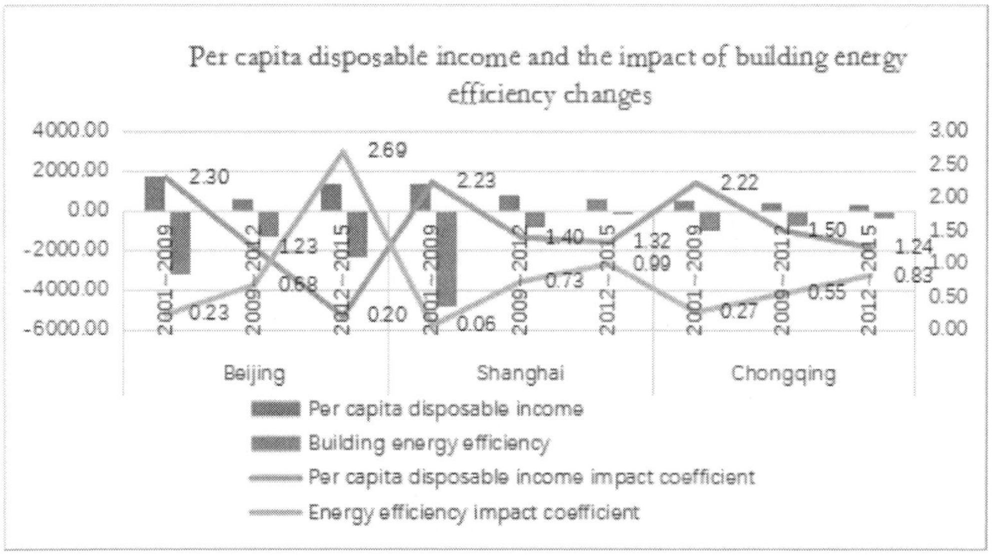

Figure 3. Building energy efficiency and per capita disposable income on the impact of building energy trends

The impact of population factors in Beijing and Shanghai on building energy consumption declined year by year, while the impact of the population factors of Chongqing on building energy consumption increased year by year before 2012 and then remained stable. It can be seen that building energy consumption doesn't grow with the growth of population in Beijing and Shanghai but the population effect of Chongqing is still relatively obvious. The impact of the proportion of urban population on building energy consumption also shows a decreasing trend year by year. In Shanghai, a highly developed city in China, the building energy consumption even decreased with the increase of urbanization rate after 2012. In Beijing, the impact of urbanization on building energy consumption is also close to zero. This also confirms the conclusion that the urbanization rate of Section 2 has an EKC curve for building energy consumption and the time of inflection point. However, it can be seen from the results, the impact of urbanization on the effect of building energy consumption and intensity in the different level of economic development regions is different. Through the trend of Figure 4, we can see that with the development of society, the standard of living and the improvement of citizens' cultural quality, both the government and the citizens' energy-saving awareness are constantly improving in the context of the government's strong advocacy of low-carbon lifestyles and the introduction of energy-saving buildings. The impact of improving building energy consumption of total population or urban population are declining, and even the impact of urbanization on the building energy consumption is negative effects, which is the future trend of social development.

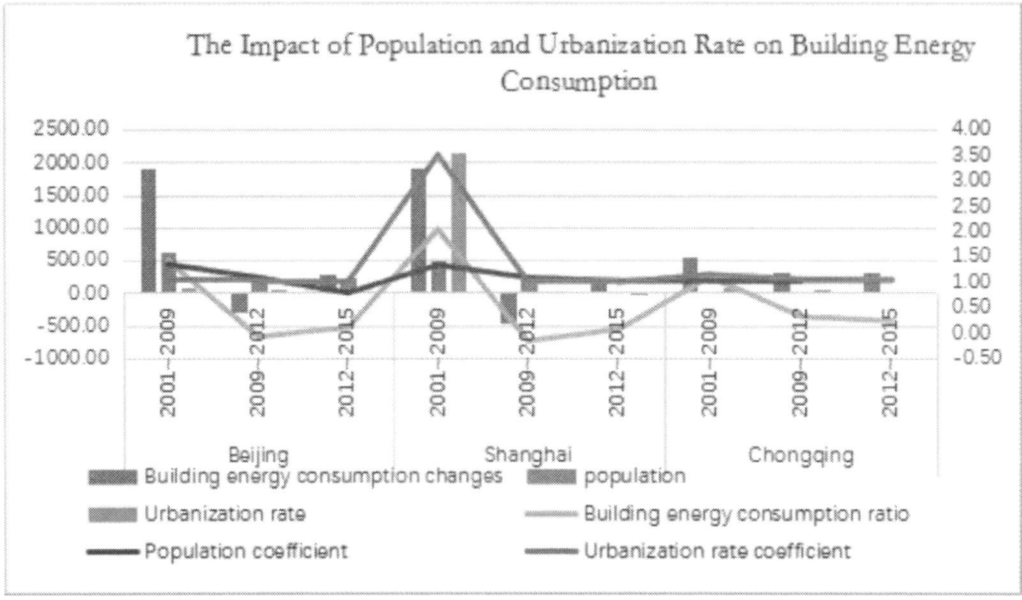

Figure 4 The Impact of Population and Urbanization Rate on Building Energy Consumption

The impact of behavioral factors on building energy consumption is positive before 2009 and the impact is negative in 2009 ~ 2012. This change is related to the country's strong advocacy of energy-saving low-carbon lifestyle. Before 2008, the Olympic year, in order to run the Olympic Games, the government has made great efforts to control the environment. The most important thing is that people also formed an environmentally friendly low-carbon lifestyle. Moreover, after 2012, the role of behavioral factors on building energy consumption changed into a positive value. This shows that the government and relevant departments should continue to vigorously promote energy conservation policies awareness.

5. Conclusion

(1) Based on the data from 2001 to 2015 in Beijing, Shanghai and Chongqing, this paper attempts to verify the existence of the environmental Kuznets curve and the inflection point of the three places on building energy consumption. Four models were constructed by using the urbanization rate, per capita GDP and foreign trade dependence as independent variables. Finally, the results show that there are environment Kuznets curve of building energy consumption in Beijing, Shanghai and Chongqing. And the inflection point have been reached in Beijing and Shanghai while the EKC curve inflection point of Chongqing has not yet reached.

(2) According to the results of the SUR model, the data of Beijing, Shanghai and Chongqing are divided into three periods: 2001 ~ 2009, 2009 ~ 2012 and 2012 ~ 2015. The LMDI decomposition method is used to find out the factor which has the largest influence on building energy consumption. The results show that the per capita disposable income is the largest promoter of increasing building energy consumption and building energy's efficiency factors are the largest contributors to reduce building energy consumption, which thus validates that the government's energy-saving environmental protection policy, and building energy-saving technology are the main driving forces for the preservation of building energy.

(3) At the end of this paper, the trend and reason of building energy consumption are analyzed. It is finally found that the economy, behavior, population, construction area or urbanization factors are unable to spontaneously reduce the building energy consumption. This shows that the government's low-carbon environmental protection policies, low-carbon energy-saving lifestyle and awareness of the popularity must be implemented in order to form a role in reducing building energy consumption.

References

[1] Krugman P 1991 Increasing Returns and Economic Geography[J] *Journal of Political Economy*, **99** 483-499.

[2] Parikh J, Shukla V 1995 Urbanization, energy use and greenhouse effects in economic development ☆: Results from a cross-national study of developing countries[J] *Angewandte Chemie* **54** 3932-3936.

[3] York R, Rosa E A, Dietz T 2003 STIRPAT, IPAT and ImPACT: analytic tools for unpacking the driving forces of environmental impacts[J] *Ecological Economics*, ,**46**:351-365.

[4] Cole M A, Neumayer E 2004 Examining the Impact of Demographic Factors on Air Pollution[J] Population & Environment **26** 5-21.

[5] Fan Y, Liu L C, Wu G 2006 Analyzing impact factors of CO emissions using the STIRPAT model[J] Environmental Impact Assessment Review **26** 377-395.

[6] Maruotti A 2011 The impact of urbanization on CO2 emissions: Evidence from developing countries[J] *Ecological Economics* **70** 1344-1353.

[7] Zhang C, Zhang Chuanguo, Lin Y 2012 Panel estimation for urbanization, energy consumption and CO2 emissions: A regional analysis in China[D] *ELSEVIER SCI LTD*

[8] Yang W 2012 The effects of population growth and urbanization on CO_2 emissions [J] *China's population, resources and environment* **22** 284-288.

[9] Lu L 2011 Study on the impact of urbanization on carbon emissions in China [J] *China Science and Technology Forum* **7** 134-140.

[10] Zhao H, Chen Y 2013 Study on the relationship between urbanization process and carbon emission reduction in China [J]. *Chinese Soft Science* **3**:184-192.

[11] Chen X, Wu B 2014 Empirical research on the relationship between economic growth, urbanization and carbon emissions -- Based on the experience of China and the United States, [J] *economic issues*, **7** 112-117.

[12] Zhu Q, Wei T 2013 The impact of population urbanization on carbon emissions from the perspective of household consumption [J] *China's population, resources and environment* **23** 21-29.

[13] Friday S, Nie M 2012 Dynamic econometric test of the factors affecting China's carbon emission intensity [J] *Management Science* **25** 99-107.

[14] Liu X, Wang Y, Song Q Y 2015 Carbon emissions from urbanization process [J] *China's population, resources and environment* **25** 61-66.

[15] Wang F, Zhou X 2012 Population structure, urbanization and carbon emission. An empirical study based on Transnational panel data [J] *Population Science in China* **2** 47-56.

[16] Hu J, Jiang Q 2015 Study on the effect of urbanization on carbon emissions from the perspective of urban agglomeration [J] *Journal of China University of Geosciences (SOCIAL SCIENCE EDITION)* **6** 11-21.

[17] Hu H 2015 Research on macro factors affecting energy consumption of civil buildings based on STIRPAT model [D] *Anhui University of Architecture.*

[18] Zhu Z, Yang Y, Wang X 2015 Study on the influence factors of building energy consumption growth based on STIRPAT model [J] *power and energy* **36** 251-255.

IOP Conf. Series: Earth and Environmental Science **143** (2018) 012012 doi:10.1088/1755-1315/143/1/012012

Bamboo fiberboards and attapulgite : does it lead to an improvement of humidity control in buildings?

D M Nguyen[1], A C Grillet[1,3], T. Goldin[1], T M Hanh Diep[2] and M Woloszyn[1]
[1] Univ. Grenoble Alpes, Univ. Savoie Mont Blanc, CNRS, LOCIE, 73000 Chambéry, France
[2] Research Center Scientific for Conservation of Natural Resources, Vietnam National University Hochiminh city (VNU) - University of Sciences, Vietnam
[3] Corresponding author: LOCIE, UMR5271 CNRS-Université de Savoie Mont Blanc, F-73000 Chambéry, France, e-mail: anne-cecile.grillet@univ-savoie.fr .

Abstract. In order to save energy used to heat or cool buildings and to improve the inhabitants comfort, control of humidity inside buildings must be improved. This can be done by using buffering materials able to absorb and release moisture when necessary. Natural fibers and mineral absorbent are good candidates to manufacture such materials. The aim of this research is to mix bamboo fibers with attapulgite to evaluate the influence of this mineral absorbent on the hygric behavior of the fiberboards. The hygric properties are slightly improved by the attapulgite and thus bamboo fiberboards can be used as building insulation materials able to participate to the indoor moisture control.

1. Introduction

Hygrothermal performance of buildings materials relates to the heat/moisture absorption, storage and release when surrounding environment changes. The hygrothermal performance of a whole material comes from the performance of each components performance. For example, the hygrothermal behavior of a building wall depends of the behavior of its elements such timber, plasterboard, insulation materials, masonry, concrete, brick, lime plaster, etc.

The hygrothermal properties include hygric behavior and thermal insulation. They contribute to the heat control in building in winter and summer through reducing heat flow through the walls as well as the need of heating/cooling, and therefore save energy consumption. Effectiveness of reducing energy demand depends on the choice of the materials, with for example good heat insulation properties through low thermal conductivity and high water vapor sorption capacity such as as natural fibers.

The hygrothermal behavior and performance of materials or building can be assessed through experimental testing, physical and numerical modeling. However, it is required to know the detailed characterization properties of such materials.

The wood and lightweight based absorbents bring an excellent hygrothermal performance through the best common buffer. The mechanism of humidity controlling materials based on absorbents was investigated. They can be classified into four groups such: silica gel, inorganic salts, inorganic natural mineral and polymer [1, 2]. The absorbents are well known to have a high equilibrium moisture content at low relative humidity level and fast sorption velocity at the beginning of absorption [3]. They are suitable for building envelopes due to their high water absorption properties. For example, the sepiolite absorbent is one kind of clay minerals; its high specific surface enhances the interfacial interactions between sepiolite and host polymer. It provides good moisture absorption and desorption capacity [4-6]. Moreover, the hygroscopic properties of porous materials are dependent on their microstructure [7, 8]. It means that the sorption ability of materials is directly proportional to the

specific surface area and total pore volume. Thus, the absorbents have good absorption capacity due to their microstructure, which improves the moisture uptake capacity for host materials.
The aim of this study is to study the hygrothermal performance of hygroscopic porous fiberboards from bamboo fibers combined with Attapulgite absorbent at different contents. The average pore diameter, pore surface area and pore size distribution of these fiberboards were investigated and classified according to pore size distribution.

2. Materials and Experiments

2.1. Materials

Bamboo fibers (bambusa stenostachya) were collected at scientific research center for conservation of natural resources, Vietnam. The bamboo fibers preparation, chemical components and length, diameter fibers distribution were exhibited in our previous papers [9]. Attapulgite is a natural porous absorbent with a high specific area, fast sorption/desorption and environmentally friendly. The composition of attapulgite is described in table 1. Its density is about 0.57 g/cm^3, and the particles main diameter distributes between 0.30 and 0.85 mm. It was purchased from Eurosorb Ltd, France.

Table 1. The components of attapulgite

Component	SiO_2	Al_2O_3	Fe_2O_3	Na_2O	K_2O	CaO	MgO	MnO	TiO_2
Amount (%)	55.6-60.5	9.0-10.1	5.7-6.7	0.03-0.11	0.96-1.30	0.42-1.95	10.7-11.4	0.61	0.32-0.6

2.2. Preparation of bamboo fiberboards

The different mixtures between bamboo fibers and attapulgite were prepared with the ratios mentioned in Table 2. Then, they were placed in an aluminum mold (150×150×6 mm) which was covered by an adhesive anti-stick coat. The bamboo fiberboards were then manufactured using an AEM3 automatic hydraulic press. The press conditions were set at 150kgf/cm^2; 160°C for 15 min. The fiberboards were removed from the mold after cooling to room temperature for at least 24 h to prevent warping.

Table 2. The constituents of materials

Materials	Bamboo fibers (wt.%)	Attapulgite (wt.%)
F100	100	0
F100-5APP	95	5
F100-10APP	90	10
F100-20APP	80	20

2.3. Pre-conditioning of samples before testing

In this study, for all the tests, the specimens were pre-conditioned at a relative humidity level of 57%RH and a temperature of 25°C before being tested. This pre-conditioning step is chosen to test the material with conditions close to those in a building.

2.4. Density fiberboards measurement

The density of fiberboards was determined according to EN 323 standard. The specimens were cut and pre-conditioned at 57%RH and 25°C before being tested. At least five specimens of each material were tested. The density value (kg/m^3) was calculated with the formula below:

$$\rho = W / (a \times b \times c)$$

where: a, b, c (m) and W (kg) are the length, width, thickness and weight of specimen respectively.

2.5. Hygric characterization

2.5.1. Test facility

In this study, all hygrothermal properties of insulation materials (except for the sorption and desorption isotherms) were measured using closed climatic chamber (called "RH-Box") (Fig. 1) which is made by two chambers connected by an airlock. Temperature and relative humidity in the RH-Box are controlled by heat exchangers and saturated salt solutions. Two fans inside the RH-Box homogenize the atmosphere when the humidity level is changed, but they are stopped when the experiments are in progress. The mass change of specimens during testing can be measured directly in the chamber through a holder connected with a microbalance located on the top of the chamber. The temperature and the relative humidity inside the RH-box are recorded with hygrothermal sensors with an accuracy of RH± 2% and T± 0.5°C.

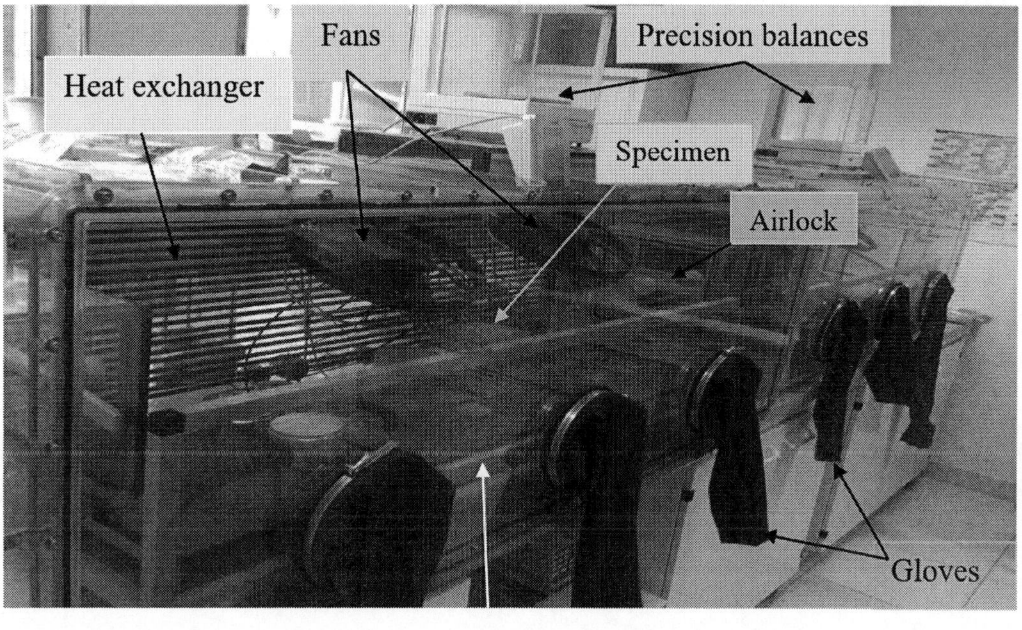

Figure 1: climate chamber RHBox

2.5.2. Kinetics water vapor sorption isotherm

In this study, the kinetics of water vapor sorption at equilibrium state was performed by using RH-Box (Figure 1). It was examined at three successive relative humidity steps: 33% (generated by $MgCl_2$ solution), 75% (NaCl solution) and 33% again. In addition, before being tested, all the specimens were pre-conditioned at 9% RH with a KOH solution at 25°C. During the period test, the specimen weight evolution is recorded, until reaching equilibrium state, by a precision balance (±0.2 mg), connected to an automatic recording system. The temperature during this experiment was kept constant at 25°C. The water vapor sorption kinetics at equilibrium curves of the specimens were obtained by plotting weight gain/loss percentages function of time.

2.5.3. Moisture buffer value

The moisture buffer values (MBVs) of the materials represent the amount of moisture passing through the open surface of specimens during cycling between 75% RH and 33% RH at 25°C. Before being tested, specimens were preconditioned at 57%RH until reaching a constant mass and five out of six faces were sealed with aluminum tape. Then the specimens were exposed in the RH Box to 75%RH for 8h and moved to 33%RH for 16h. The mass change of specimens was regularly measured during the test period for each relative humidity level. In the Nordtest project [10], the MBV is calculated from three consecutive quasi-steady state cycles (when the amplitude of mass change differ by less than 5%). At least three specimens of each sample were tested. The average values of MBV were calculated using the following equation:

$$MBV = \frac{\Delta m}{S \times (RH_{high} - RH_{low})}$$

Where: MBV is the practical moisture buffer value in g/(m^2.%RH), $RH_{high/low}$ (%) is the relative humidity at 75% and 33% respectively, Δm (g) is the mass variation during moisture adsorption and desorption at RH_{high}, RH_{low}, and S (m^2) is the open surface area of specimen.
Based on the NORDTEST Project, the MBV of materials are then classified into five categories depending on their moisture buffer efficiency: negligible (MBV: 0.0–0.2), limited (MBV: 0.2–0.5), moderate (MBV: 0.5–1.0), good (MBV: 1.0–2.0), and excellent (MBV: 2.0–upwards) [10].

2.5.4. Water vapor sorption isotherm
The capacity of water vapor storage at different relative humidity levels was measured with conditions inspired by the ISO 12571 standard [11]. At least three specimens (40 mm × 40 mm × 6 mm) of each fiberboard were cut and pre-conditioned at 57% RH and 25°C before being tested. When the tested specimens reached a constant mass at 57% RH (weight variation after two consecutive periods of 24h less than 0.1%), they were put in the humidity level of 9%RH at T=30°C. The tested specimens were regularly weighed until reaching a constant mass. After reaching equilibrium, specimens were moved to a higher relative humidity. This operation was carried out up to the maximum relative humidity of 97%RH. The reverse procedure was applied for desorption. Saturated salt solutions were used to control moisture level. The specimens were tested at five different relative humidity levels corresponding to five salt saturated solutions: 9% RH (KOH), 33%RH ($MgCl_2$), 57%RH (NaBr), 75%RH (NaCl), and 97%RH (K_2SO_4) at 30°C. The moisture content variation (ΔMC) of the specimens was calculated using the equation below:

$$\Delta MC = \frac{m_s - m_0}{m_0} * 100$$

where: m_s is the weight at different relative humidity levels and m_o is the constant weight after initial pre-conditioning at 57% RH.

3. Results and discussion

3.1. Characteristics of bamboo fibers
The various components and distribution of the length and diameter of bamboo fibers in this study are shown in figure 2. The results indicate that the length distribution ranges from 0 to 7.0 cm and is mainly concentrated around approximately 1.0–2.0 cm, whereas the diameter ranges from 0.16 to 0.52 mm and is mainly concentrated between 0.24 and 0.40 mm. It can be observed that the bamboo fibers have a relatively narrow distribution of diameter and length. Indeed, the preparation of bamboo fibers using rollers is a suitable method to extract a large amount of fibers of quite homogeneous length and small diameter. In addition, this kind of bamboo contains the same amount of components as other different bamboo fibers, but they differ from the components of other natural fibers [12].

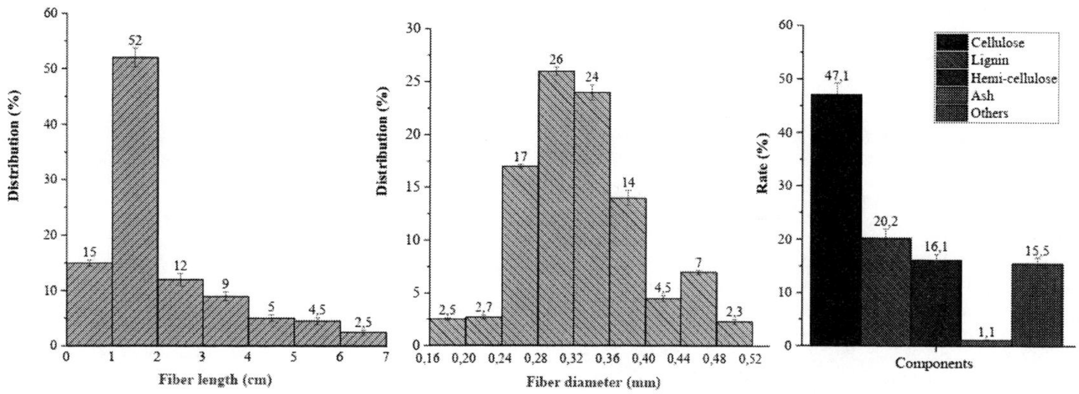

Figure 2: Length, diameter distribution and components of bamboo fibers [9]

3.2. Hygrothermal properties

3.2.1. Kinetics of vapor sorption at equilibrium state

The vapor sorption kinetics of all fiberboards were investigated as protocol in figure 3a. All specimens were pre-conditioned at 9%RH and 25°C before being tested. The moisture absorption capacity, moisture residuals and sorption behavior of these fiberboards at three equilibriums states (33-75-33% RH) can be observed in figure 3b.

In particular, there is a rapid initial uptake and release in three equilibriums states at different relative humidity levels (33% and 75%).

At the first reached equilibrium (33%RH), the F100 absorbed an amount of moisture about 3.2% after two days. During this equilibrium state (RH < 35%) the water molecules are only absorbed as monolayer on the pores surfaces of cell walls [13], leading to the small amount of moisture uptake capacity. Consequently, equilibrium is reached in a short time (about 2 days). At this humidity level, all the boards containing attapulgite showed a lightly higher moisture absorption capacity in comparison to F100 board. They presented a moisture uptake capacity range from 3.9% to 4.7% (figure 3b). Attapulgite improved the moisture absorption capacity for all the boards at this equilibrium (33%RH). This result may be due to the higher moisture absorption capacity of attapulgite at 33% than F100 board (around 4,5% against only 3,3%).

The second equilibrium state was reached at 75% RH after around 4 days of exposure. (figure 3b). The F100 exhibited the lowest moisture uptake capacity about 9%. As at 33% RH, the addition of attapulgite absorbent into F100 leads to increase its moisture uptake capacities at equilibrium state (75% RH). But this improvement is lower than at 33%RH, may be due to the lowest difference between attapulgite alone (around 9,5%) and the bamboo fibers in the board F100 [9]. Indeed, attapulgite absorbent is well known to have a good absorption capacity due to its microstructure (high specific surface area and total pore volume), leading to improve vapor absorption capacity of materials [14]. The attapulgite improves the moisture absorption ability of bamboo fiberboards because it acts as hygroscopic materials, the water molecules may be absorbed into magnesium ions and silanol groups of absorbents structure [15].

In general, in this equilibrium state (33%<RH<75%) the water molecules are absorbed as multilayer and begin to condense in porous capillaries [13].

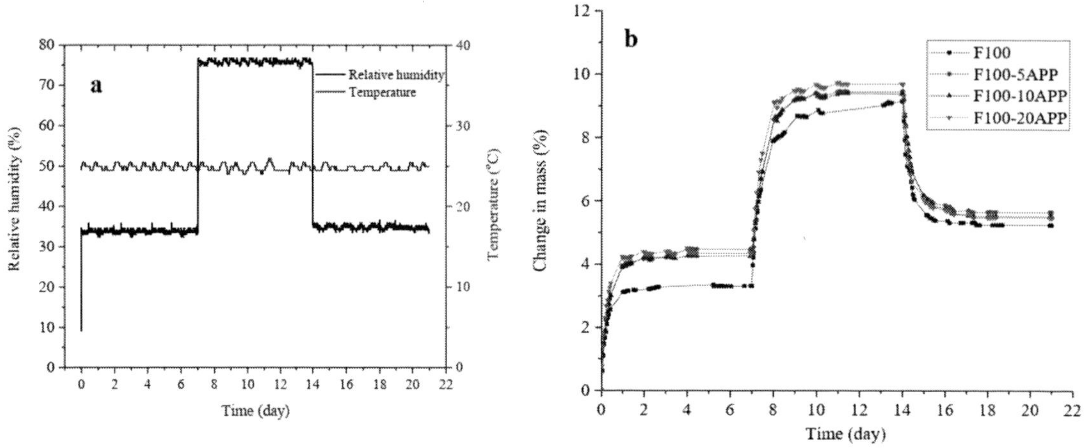

Figure 3: kinetics of water vapor sorption at equilibrium state: (a) humidity and temperature protocol (9-33-75-33% RH) at 25°C, (b): kinetics curves and moisture sorption capacity of fiberboards.

The final equilibrium state was reached at 33%RH, there was a rapid initial release in desorption process, and equilibrium state was reached after 2 or 3 days. So, these boards can quickly adapt when environmental surrounding conditions changes. But, the amount of moisture uptaken in absorption process (from 33% to 75%RH) was not totally discharged in desorption process (75%RH to 33%RH) for all the boards. The amount of residual moisture remained in boards can be seen in figure 3b. The phenomenon is explained via the capillary condensation hysteresis due to moisture trapped inside large pores with small necks (ink-bottle effect) [16].

3.2.2 MBV results

The MBV results of different fiberboards were calculated from three consecutive steady state cycles of moisture uptake and release per unit of exposed area (figure 4).

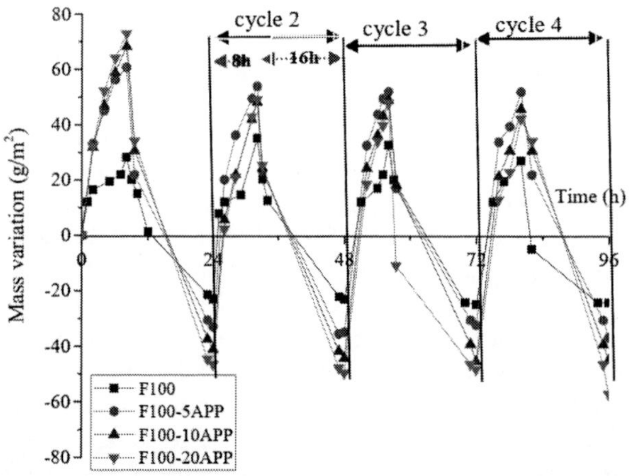

Figure 4: The three consecutive quasi-steady state cycles of moisture uptake and release of all specimens, used for MBVs calculations

Figure 5 shows the average MBV results of all fiberboards. The F100 possesses an average MBV about 1.7 g/(m^2.%RH) corresponding to a "good" moisture buffering capacity. The addition of different attapulgite absorbent contents into F100 board leads to a significant increase of average MBVs from "good" to "excellent" moisture buffer categories as in Nordtest Project. The MBVs range from 2.1 to 2.3 g/(m^2.%RH) (figure 5). As for the kinetics, it results from the higher moisture buffering performance of attapulgite absorbent than bamboo fiberboard F100. The behavior is the same as for kinetics : attapulgite increases slightly the ability of the boards to sorb moisture but the influence of the amount of attapulgite is not significant.

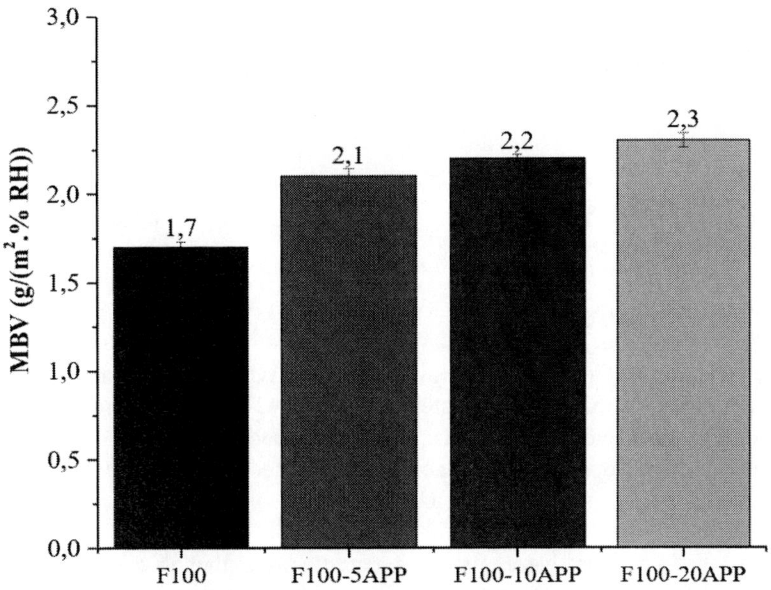

Figure 5: Average MBV results of all particleboards

3.2.3 Water vapor sorption isotherms

Figure 6 presents the vapour sorption isotherms capacity of all fiberboards at different relative humidity levels. The different components of the fiberboards are associated with the different amount of moisture storage in materials. F100 exhibited an amount of moisture absorption about 10.5% at 84% RH, which was lower than for boards containing different attapulgite contents at all relative humidity levels. In particular, while the maximum moisture absorption of F100 was 10.5% at 84% RH, the value of F100-5APP slightly increased to about 11%. The two boards F100-100APP and F100-20APP exhibited the same amount of moisture uptake about 13% at 84%RH. The results are in accordance with the previous characterizations (kinetics, MBV): adding attapulgite increases slightly the moisture absorbing capacity of the boards. But, here, the amount of attapulgite has a small influence (10% of absorbent is required to significantly increase the moisture absorption capacity for humidity level above 20%)

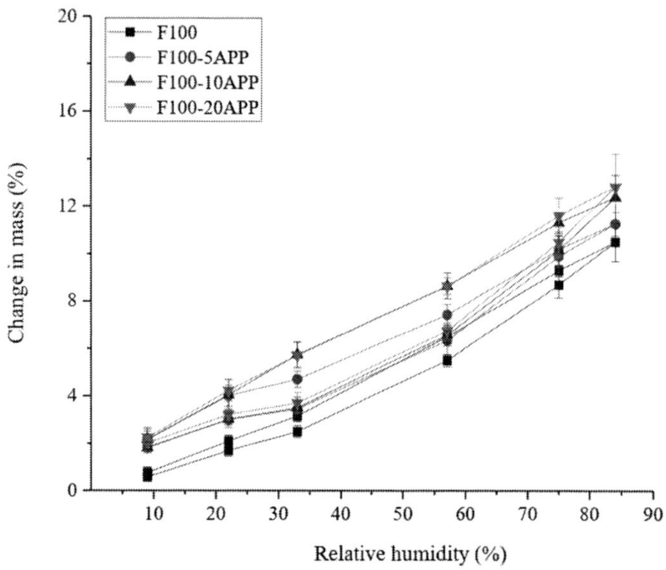

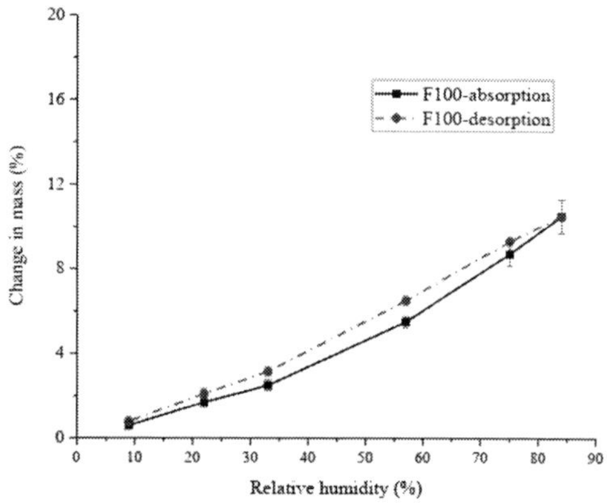

Figure 6: Vapor sorption isotherms of different fiberboards at 30°C

On the other hand, the absorption isotherm of attapulgite is close to type IV isotherm given by IUPAC classification with an interconnected mesoporous system (2-50 nm) [14, 17, 18]. The vapor absorption isotherms curves of bamboo fiberboards are similar with S-shaped accordance to type II, which characterizes the absorption of mesoporous and macroporous materials given by IUPAC [19, 20]. After adding attapulgite at different contents to the boards, the absorption isotherms curves are still kept S-shaped Type II.

The curves of water vapor desorption were always located above the absorption curves of all fiberboards. The hysteresis between absorption and desorption of these fiberboards was observed in figure 6. The hysteresis behavior is generally associated with the pores interconnection, porosity characterization and the capillary condensation of fiberboards. Moreover, it could also be attributed to the ink bottle effect, related to the moisture trapped inside the pores with small entry [16, 21]. Additionally, the high affinity of attapulgite for water may prevent discharge of water during desorption process leading to high hysteresis for fiberboards with attapulgite. Indeed, the water molecules were kept in magnesium ions and silanol groups and into the interlayers of Attapulgite structure. Thus, the water molecules which have formed hydrogen bonds during absorption procedure are not able to be released easily and then the moisture content uptaken is not entirely discharged during desorption procedure. It can explain why the hysteresis is higher for the boards with 10 and 20% of attapulgite.

4. Conclusion

Different types of bio-insulation bamboo fiberboards have been made from bamboo fibers and different attapulgite contents by thermo-pressing. The hygrothermal properties have been investigated. Addition of attapulgite into a bamboo board leads to the increase of the moisture uptake capacity at different relative humidity levels. It also increases the moisture buffer value and then the buffer ability.

The shapes of absorption curves are in good agreement with the S-shaped curves of the literature and they can be classified as type II according to IUPAC classification. To conclude, depending on the hygrothermal performance of different bamboo fiberboards, they can be used as building insulation materials which can participate to the indoor moisture control. They can be also used for the thermal insulation of ceilings or in attic spaces.

Acknowledgements
This work has been supported by the Région Rhône-Alpes (COOPERA and CMIRA programs), ADEME and Agence Nationale de la Recherche (ANR), France. It was also supported by grant from Vietnam National University Ho Chi Minh City (VNU-HCM) under project number B2018-48-01 and GEN2016-48-01/HĐ-KHCN.

References
[1] Ran MY (2002) *Mat. Rev* **16**(11) 42-44
[2] Jiang ZW (2006) *Mat. Rev* **20**(10) 8-11
[3] Wu Y, Gong G, Yu CWF and Fang P 2014 *Ind. and Built Env.* **23** 299–312
[4] Ekici S, Isikver Y and Saraydam D 2006 *Pol. Bull.* **57** 231-41
[5] Santiagao F, Mucientes AE, Osorio M and. Poblete FJ 2006)Pol. Int. **55** 843-48
[6] Yang Hl., Peng ZQ, Zhou Y, Zhao F, Zhang J, Cao XY and Hu. ZW 2011 *En. and Build.* **43** 386-92
[7] Fang P, Wu Y and Gong G 2009 *Mat. Rev.* **23**(5) 112–14
[8] Fang P, Wu Y and Gong G 2009 *Mat. Rev.* **23**(14) 475–77
[9] Nguyen DM, Grillet AC, Diep TMH, Ha-Thuc CN and Woloszyn M 2017 *Constr. and Build. Mat.* **155** 852-66.
[10] Rode C, Peuhkuri R, Hansen KK, Time B, Svennberg K, Arfvidsson J and T. Ojanen 2005. *AIVC Conference Energy Performance Regulation*, Brussels
[11] ISO 12571, Hygrothermal performance of building materials and products. Determination of hygroscopic sorption properties 2013 BSI.
[12] Hill CAS, Norton A and Newman G 2009 J.A.P.S **112** 1524–37.
[13] Latif. E, Tucker. S, Ciupala MA, Wijeyesekera DC and Newport D 2014 *Constr. and Build. Mat.* **66** 702–11.
[14] Wu Y, Gong G, Yu CWF and Fang P 2014 *Ind. and Built. Env.***23**(2) 299–312.
[15] Alvarez BA, Camara GMA and. Duch M.I 2000 *Patent USA, EP 0717928 B1.*
[16] Collet F, Bart M, Serres L and Miriel J, *Constr. and Build. Mat.* 2008 **22** 1271–80.
[17] Chen LF, Liang HW, Lu Y, Cui CH and Yu SH 2011 Langmuir **27** 8998–9004.
[18] Mohanan JL, Arachchige IU and Brock SL 2005 Science **307**(5708) 397-400.
[19] Latif E, Tucker S, Ciupala MA, Wijeyeseker DC and New D 2014 *Constr. and Build. Mat.* **66** 702–711.
[20] Rahim M, Douzane O, Tran Le A.D, Promis G and Langlet T 2016 *Constr. and Build. Mat.* **102** 679–687.
[21] Derluyn H, Derome D, J. Carmeliet, Stora E and Barbarulo R 2012 Cem. and Concr. Res. **42** 1379–88.

Time dependent variation of carrying capacity of prestressed precast beam

Tuan D Le[1,2,*], **Petr Konečný**[2,*] **and Pavlína Matečková**[2,*]
[1] Faculty of Civil Engineering, Saigon Technology University, 180 Cao Lo Str., Ward 4, Dist. 8, HCMC 700000, Vietnam
[2]VSB – Technical University of Ostrava, Faculty of Civil Engineering, L. Podeste 1875, 708 33 Ostrava-Poruba, Czech Republic

*Corresponding authors: tuan.leduc@stu.edu.vn, petr.konecny@vsb.cz, pavlina.mateckova@vsb.cz

Abstract. The article deals with the evaluation of the precast concrete element time dependent carrying capacity. The variation of the resistance is inherited property of laboratory as well as in-situ members. Thus the specification of highest, yet possible, laboratory sample resistance is important with respect to evaluation of laboratory experiments based on the test machine loading capabilities. The ultimate capacity is evaluated through the bending moment resistance of a simply supported prestressed concrete beam. The probabilistic assessment is applied. Scatter of random variables of compressive strength of concrete and effective height of the cross section is considered. Monte Carlo simulation technique is used to investigate the performance of the cross section of the beam with changes of tendons' positions and compressive strength of concrete.

1. Introduction

Prestressed concrete has been a widely used material in construction due to its conveniences in comparison with reinforced concrete such as less deep and larger in span if exposed in similar conditions, thus reducing the carbon dioxide footprint. Moreover, progressive high performance materials are being developed in order to further improve concrete performance and reduce the cement usage [1], [2], [3], [4], [5]. Industrial by-products such as slag and fly ash are used. However, it needs to be noted that prestressed concrete structures require higher quality materials in prestressing and more complicated formwork [6]. Consequently, attention should be paid to the design of these structures to get better understanding of their behavior and performance [7].

When new material and structural precast elements are being developed, experimental investigations help to verify the numerical modelling. The numerical models applicable for the design purposes should be calibrated with experiments.

When the test of the samples is prepared, the question might arise how large and how reinforced might the samples be. One wants to prepare samples large enough while address the limit of testing machines with respect to dimensions and maximum loading capacity. So, when designing the experiments, the ultimate capacity of the sample is of great importance.

There is also factor of concrete maturing because the hydration process is time dependent and concrete strength is being developed over the time. Therefore, it is necessary to address the increase of concrete resistance over the time. The EN 1992-1-1 Eurocode 2 [8] gives estimated compressive strength of concrete at various ages:

$$f_{\mathrm{cm}}(t) = \beta_{\mathrm{cc}}(t) \cdot f_{\mathrm{cm}} \tag{1}$$

where:

$f_{cm}(t)$: average strength (MPa) of concrete in compression at age t (days);

$\beta_{cc}(t)$: coefficient depends on the age of the concrete.

If measured data are available and approximation given in equation (1) does not fit measured data, other approximations are available. See e.g. [9] and [10] for applications in design of underground shotcreting works such as [11].

Design methods for reinforced and prestressed concrete structures have been developed during past decades [6] and are still under development with respect to the evaluation of advanced non-linear modelling capabilities [12], [13]. Current codes [8], [14] are based on both deterministic and probabilistic theories. However, actual application is rather deterministic from the engineer standpoint. Since the input parameters governing the resistances of reinforced as well as prestressed concrete structures show high variability, probability-based approaches [15], [16], [17], employed e.g. in [18], [19] in case of reinforced concrete, are robust tools in order to address the upper percentile of the prestressed concrete beams carrying capacity distribution. Probabilistic simulation allows addressing the question of ultimate force of foreseen experimental bending test.

In case of prestressed concrete, it is important to take of influence of rheology into account. This includes creep and shrinkage of concrete as well as relaxation loss of tendons. However, time dependent change of the prestressing force does not affect the level of ultimate bending resistance [20] in contrast to the change of concrete strength.

In this paper, the preparation of simple model for the probability-based bending resistance computation of a simply supported prestressed concrete beam is shown. The goal of the development is to prepare the procedure for the estimation of time dependent experimental simply supported beam ultimate resistance in three-point bending test (figure 1) and the newly prepared concrete mixtures. The availability of time dependent behavior of basic properties is considered. The bending moment resistance of the beam with variations of cylinder compressive strength of concrete and position of prestressing tendons in the section is evaluated. The Monte Carlo simulation based approach will be adopted in order to include scatter of input parameters. Random variations of compressive strength of concrete and effective height of the cross section are built up based on normal distributions. Procedures are composed using Matlab compatible environment [21], [22] to facilitate the simulations.

(a) (b)

Figure 1. Preparation of laboratory evaluation of carrying capacity: (a) Samples of prestressed concrete beams and (b) Test machine pressing head.

2. Material properties

In order to address necessary time dependent material behavior especially in case of precast prestressed girders, the time dependent concrete strength and modulus of elasticity is crucial of interest.

The approximation of sample concrete cylindrical strength may be based on codes [8]. If the laboratory data of strength are obtained during the concrete aging, the data may be approximated by suitable curve. Such sample case is given bellow in figure 2 on which the strength is approximated in a time by regression function, equation (2). The regression function is obtained using the spreadsheet calculation. It has good coefficient of determination R^2=0.9536. Thus, the approximation in equation (2) has a good fit.

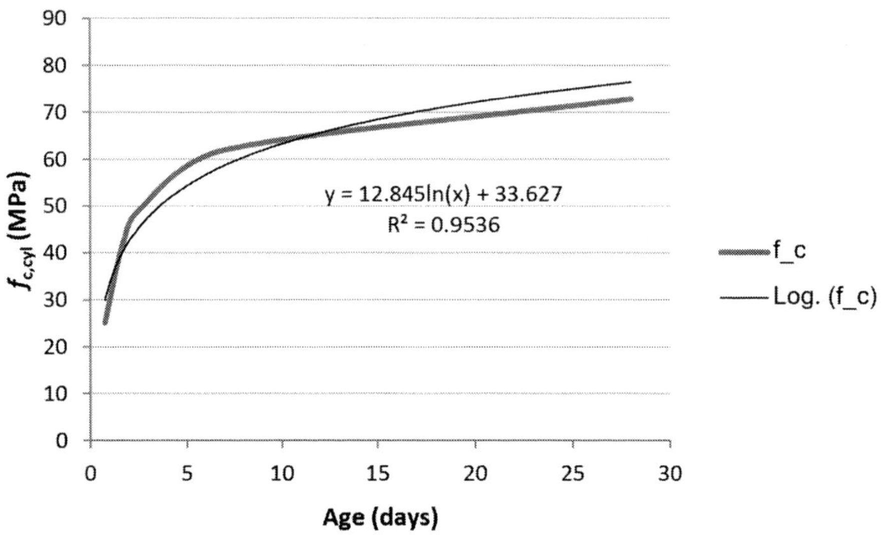

Figure 2. Approximation of considered cylinder strength time dependent behavior.

$$f_{c,cyl}(t) = 12.845\ln(t) + 33.627 \tag{2}$$

where:

$f_{c,cyl}(t)$: cylindrical strength (MPa) of a reinforced concrete tested sample at age t (days).

If the variation of concrete strength is available as given in table 1 on which available data are based on the evaluation of cubic strength, the standard deviation can be recomputed to dimensionless coefficient of variation. Thus, assumption is made here that the relative variation of cubic strength and cylinder-based strength is the same. With such assumption, the confidence bounds for the simulated cylinder-based concrete strength may be evaluated with respect to given selected percentiles (5.th and 95.th) related to its lower and upper bounds.

Table 1. Statistical parameters variation of concrete strength for 28 days.

t (days)	No. of samples	$f_{c,cube}$ (MPa)				
		$f_{c,mean}$	$f_{c,std}$	$f_{c,cov}$	$f_{c,05}$	$f_{c,95}$
28	13	87.8	3.4	0.0388	82.2	93.4

Since the probabilistic modelling is of question, the statistic characteristic of considered probability density function shall be considered as indicated on the equation (3).

$$f_{c,cyl}(t) = \mu\left(f_{c,cyl}(t)\right) + cov\left(f_{c,cube}(28)\right).\mu\left(f_{c,cyl}(t)\right) \tag{3}$$

where:

$f_{c,cyl}(t)$: cylindrical strength (MPa) of sample at age t (days);

$\mu\left(f_{c,cyl}(t)\right)$: mean value of cylindrical strength of sample at age t;

$cov\left(f_{c,cube}(28)\right)$: coefficient of variation of cubic strength of sample, t = 28 days.

3. Probabilistic modelling of beam resistance

The ultimate resistance model is specified first while the description of sample input data follows.

3.1. Computational model

Considering three-point bending test loading scheme, the ultimate force F_r for bending is:

$$F_r = \frac{4M_r}{l} \tag{4}$$

where:

M_r: ultimate bending moment resistance (kNm) given in equation (5);

l: loading span (m) of the beam.

Ultimate bending moment resistance of a critical rectangular cross section of an underreinforced beam (effect of reinforcement is ignored) is depicted as:

$$M_r = F_c(d - 0.4x) \tag{5}$$

where:

d: effective height (m) of considering cross section of the beam, the distance from extreme compression fiber to centroid of prestressed steel;

x: height of compression zone (m), calculated as:

$$x = \frac{A_p\, f_{p01}}{0.8\, f_{c,cyl}\, b} \tag{6}$$

F_c: compressive force (kN) in concrete and is determined as:

$$F_c = 0.8\, f_{c,\mathrm{cyl}}\, b\, x \tag{7}$$

A_p: cross-sectional area (m^2) of prestressing tendons;

f_{p01}: 0.1% proof-stress (kPa) of prestressing steel;

$f_{c,cyl}$: cylinder compressive strength (kPa) of concrete;

b: width (m) of cross section of the considered beam.

3.2. Monte Carlo simulation technique

Monte Carlo method is one of the widely used simulation techniques based on a process of repeated random sampling to obtain numerical results. It was first systematically used very early and can be considered as the art of approximating an expectation by the sample mean of a function of simulated random variables [23]. Steps of the approach include: creating the sample, running the model and analyzing the data.

3.3. Normal distribution generation of random variables

According to [24], a desired distribution can be modeled using the following transformation formula:

$$N(\mu,\sigma) = \mu + \sigma \times N(0,1) \tag{8}$$

where:

N(0,1): represents random numbers from the normalized normal distribution;

μ: specified mean value;

σ: specified standard deviation;

$N(\mu,\sigma)$: represents random numbers from the generated normal distribution.

If the parameter of interest is time dependent, then:

$$N(\mu,\sigma,t) = \mu(t) + \sigma(t) \times N(0,1) \tag{9}$$

where:

$\mu(t)$: specified time dependent mean value at age t (days);

$\sigma(t)$: specified time dependent standard deviation at age t;

$N(\mu,\sigma,t)$: represents time dependent normal distribution from the normalized one at age t.

4. Sample numerical example

A simply supported prestressed concrete beam with dimensions of cross section of 0.9 m wide and 0.56 m high as depicted in figure 1 is considered. The length of the beam is 7 m. There are 2 layers of bottom tendons with 8 wires in lower layer and 4 wires (their redundant segments were amputated at the face of the beam) in the upper one. The top layer of tendons was not considered in the calculation because it locates close to the neutral axis of the beam. Area of one-wire prestressing tendon is A_p = 150x10^{-6} m^2. Distance between the two layers of tendons is 0.05 m. A 0.08 m concrete cover is assumed and effect of reinforcement is ignored. The cross-sectional bending resistance at midspan of the beam will be realized through Monte Carlo simulation technique.

4.1. Input parameters

The variables include the cylinder compressive strength of concrete depicted in table 1 and the distance from extreme compression fiber to centroid of prestressed steel given in table 2.

Table 2. Information on input parameters employed in the study.

Parameter	Notation	Mean	Coefficient of variation	Transformation
Concrete strength (kPa)	$f_{c,cyl}(t)$	Equation (1)	0.0388	$\mu\left(f_{c,cyl}(t)\right) + 0.0388\times N(0,1)$
Distance from extreme compression fiber to centroid of prestressed steel (m)	d	0.52	0.0096	$d = 0.52 + 0.005\times N(0,1)$
Loading span (m)	l	6.85	-	-
Width of cross section of the considered beam (m)	b	0.9	-	-
Height of cross section of the beam (m)	h	0.56	-	-
Cross-sectional area of prestressing tendons (m^2)	A_p	150x10^{-6}	-	-
0.1% proof-stress of prestressing steel (kPa)	f_{p01}	1687x10^3	-	-
Thickness of concrete covered layer (m)	c	0.08	-	-

4.2. Distributions of random variables

Histograms of cylinder compressive strength of concrete and effective height of considering cross section of the beam are built up through using equation (8).

4.3. Results

Equation (5) is used to compute ultimate bending moment resistance of the critical cross section. It needs to be noted that the resistance is provided by concrete and prestressed tendons. The conventional reinforcing steel is neglected herein. Distribution of bending moment resistance of the cross section of the beams at different ages of concrete are displayed in figure 3.

It is likely that bending moment resistance of the cross section at different ages of concrete have normal distributions and these distributions are different with the progress of maturing of concrete.

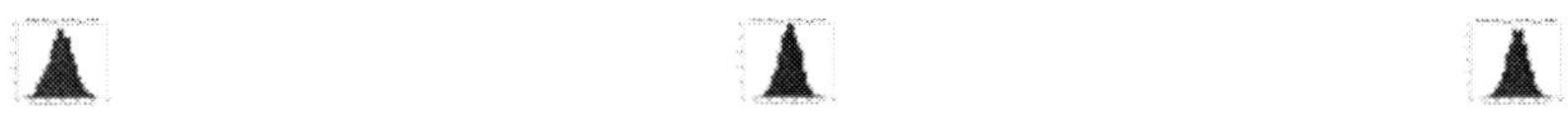

(a) (b) (c)

Figure 3. Distribution of bending moment resistance of the cross section of the beam with: (a) t = 2 days, (b) t = 28 days and (c) t = 365 days.

Table 3. Simulation results on bending moment resistance and ultimate force of cross section of the beam and their confidence bounds.

Age of concrete	Bending moment resistance of the cross section of the beam (kNm)			Ultimate force for bending of the cross section of the beam (kN)		
t (days)	M_{r05} (5%)	M_{r50} (50%)	M_{r95} (95%)	F_{r05} (5%)	F_{r50} (50%)	F_{r95} (95%)
2	1432.901	1458.424	1483.947	836.731	851.634	866.538
28	1486.796	1511.803	1536.810	868.202	882.805	897.408
365	1506.599	1532.259	1557.918	879.766	894.750	909.733

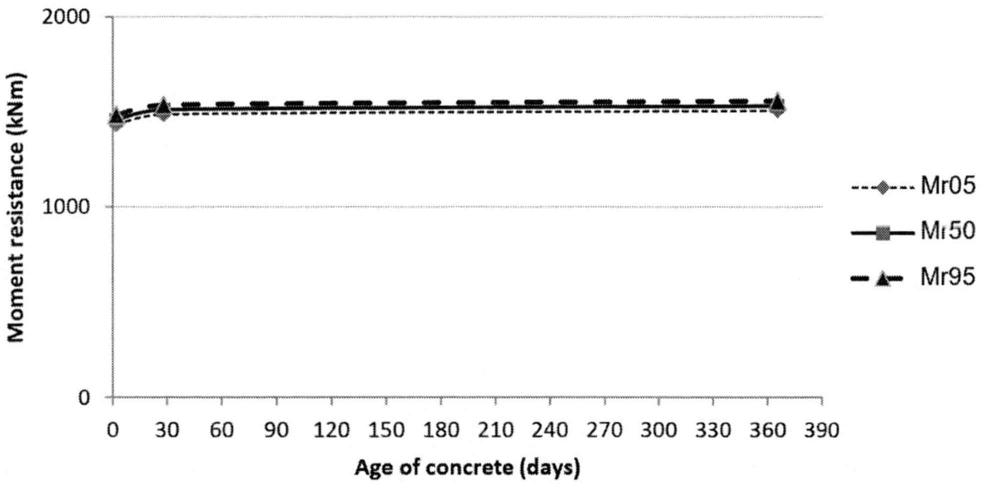

Figure 4. Bending moment resistance of the cross section of the beam and its confidence bounds.

It can be seen from figure 4 that moment resistance of the cross section increases over the time. The rise took place at concrete age of several days to one week. This trend almost slowed down after the 1st week of age and there was only a gradually light growth from 28 days onwards. The reason is the rectangular concrete beam cross section that cannot fully utilize the potential of high quality concrete. It can be also seen that the variation is almost negligible. It is within the range of ± 2 percent. If the more sophisticated shape of the beam, such as T-section or I-section, would be employed, the increase of bending resistance and its variation might be more significant.

According to three-point bending moment formula equation (4), the loading force F_r is computed in order to find the most likely ultimate value for the laboratory experiment considering the scatter of input parameters. The resulting loading forces are given in table 3.

The resulting necessary loading forces given in table 3 are below 1000 kN. If, for example, the limit for the loading unit would be 1000 kN, the samples would break under bending considering assumed simplifications.

5. Conclusions

The presented numerical procedure allows for the preliminary evaluation of the bending resistance of prestressed concrete beams. The bending resistance allows for the calculation of ultimate force necessary to break the laboratory sample valuable when considering limit of the loading unit.

The variation and time-dependency of concrete strength is considered. The variation of concrete cover is also considered. Both random parameters were modeled as normal distributions.

The results showed that bending moment resistance variation and its increase with concrete maturing was not significant. The carrying capacity growth was limited by the rectangular shape of the beam.

In addition, the random variation of bending resistance and ultimate loading force was not significant herein. Thus for further experimental preparations with rectangular shape prestressed beams, the average values for the problem specification are enough.

The current model deals only with rather simple case of carrying capacity of prestressing tendons from bottom part because tendons at top part were very near to the beam's neutral axis. In addition, the longitudinal reinforcement was not considered due to their small influence to carrying capacity of the beam. Factor of shear resistance was not also been studied yet.

In future development of this research, therefore, computation using method of limit strain on which the carrying capacity of both parts of prestressing tendons included should be exploited. Furthermore, influence of the longitudinal reinforcement to the carrying capacity should be taken into account. Expansion of this topic with inclusion of shear resistance would be a next step. Deflection, cracking, stiffness change and shape optimization of the beam are other directions of development.

References

[1] Aitcin P C 1998 High performance concrete (Taylor & Francis) ISBN-10: 0419192700, ISBN-13: 978-0419192701

[2] Konecny P, Lehner P, Ghosh P and Tran Q 2017 Variation of diffusion coefficient for selected binary and ternary concrete mixtures considering concrete aging effect *Proc. 6th Int. Conf. Non-traditional Cement & Concrete* June 19-22-2017 Brno (Preprint gr-qc/0303038)

[3] Ghosh P, Konecny P, Lehner P and Tikalsky P J 2017 Probabilistic time-dependent sensitivity analysis of HPC bridge deck exposed to chlorides *J. Computers and Concrete* vol **19** No. 3 pp 305-313 DOI: https://doi.org/10.12989/cac.2017.19.3.305

[4] Cajka R, Vaskova J and Smirakova M Stress and strain of fiber reinforced concrete composites *Proc. 6th Int. Conf. Non-traditional Cement & Concrete* June 19-22-2017 Brno (Preprint gr-qc/0303038)

[5] Keulenl A, Kakebeeke P J T, Grunewald S and Brouwers H J H Durability evaluation of aam using a plasticizing additive, enhancing fresh concrete rheology *Proc. 6th Int. Conf. Non-traditional Cement & Concrete* June 19-22-2017 Brno (Preprint gr-qc/0303038)

[6] Nawy E G 2009 *Prestressed concrete - a fundamental approach* 5th edition update (NJ Upper Saddle River: Prentice Hall)

[7] Biondini F, Frangopol D M, Bontempi F and Malerba P G 2004 Reliability of material and geometrically non-linear reinforced and prestressed concrete structures *J. Computer & Structures*

[8] EN 1992-1-1 2004 Eurocode 2: Design of concrete structures - Part 1-1: General rules and rules for buildings *The European Union Per Regulation* 305/2011

[9] Oreste P P 2003 A procedure for determining the reaction curve of shotcrete lining considering transient conditions *Rock Mechanics and Rock Engineering* vol **36 (3)** pp 209-236 DOI: 10.1007/s00603-002-0043-z
[10] Klepsatel F, Kusý P and Mařík L 2003 Construction of tunnels in rock massive (Výstavba tunelů ve skalních horninách) (Bratislava: Jaga group) ISBN 80-88905-43-5 (in Czech)
[11] Vojtasik K, Mohyla M and Hrubesova E 2017 Static design of steel-concrete lining for traffic tunnels *IOP Conference Series: Materials Science and Engineering*, vol **236 (1)** Art. no. 012083
[12] Králik J and Klabník M 2016 Nonlinear analysis of the failure of nuclear hermetic reinforced concrete structure due to extreme pressure and temperature, *Transactions of the VŠB – Technical University of Ostrava Civil Engineering Series* No. 2 vol **16** paper#17
[13] Sucharda O, Bílek V, Šmiřáková, Martina K J and Čajka R 2017 *Periodica Polytechnica Civil Engineering* vol **61** No. 4 pp 972-986
[14] Matthews S L, Bigaj-Van Vliet A J and Dieteren G G A 2016 Fib model code 2020 – A new development in structural codes: Towards a general code for both new and existing concrete structures ed Beushausen H *Proc. fib Symp. 2016 Performance-based approaches for concrete structures* (Cape Town) pp 22-31
[15] Marek P et al 2003 *Probabilistic Assessment of Structures using Monte Carlo Simulation - Background, Exercises and Software* TeReCo 2nd edition (Praha: Institute of Theoretical and Applied Mechanics, Academy of Sciences of Czech Republic)
[16] Melchers R 1999 Structural reliability analysis and prediction (Civil Engineering) (West Sussex: Wiley)
[17] Stewart M G and Rosowsky D V 1998 Time-dependent reliability of deteriorating reinforced concrete bridge decks *Structural Safety* vol **20 (1)** pp 91-109
[18] Janas P, Krejsa M, Šejnoha J and Krejsa V 2017 DOProC-based reliability analysis of structures *Structural Engineering and Mechanics* vol **64** No. 4 pp 413-426
[19] Teplý B and Vorechovská D 2012 Reinforcement corrosion: Limit states, reliability and modelling *J. Advanced Concrete Technology* vol **10** pp 353-362
[20] Navrátil J 2014 Prestressed concrete structures 2nd edition (Ostrava: VSB – Technical University of Ostrava, Faculty of Civil Engineering) ISBN: 978-80-248-3625-6
[21] MatLab: The language of technical computing *MathWorks*- on-line: www.mathworks.com
[22] GNU Octave Scientific programming language on-line: www.octave.org
[23] Anderson E C 1999 *Monte Carlo methods and importance sampling* - Lecture notes for Stat 578C, Statistical Genetics
[24] Fegan G and Gustar M 2003 *Chapter 2 – Monte Carlo Simulation* in: Marek P et al *Probabilistic Assessment of Structures using Monte Carlo Simulation - Background, Exercises and Software* TeReCo 2nd edition (Praha: Institute of Theoretical and Applied Mechanics, Academy of Sciences of Czech Republic)

Acknowledgments
The article preparation was supported from sources for conceptual development of research, development and innovations at VŠB-Technical University of Ostrava for year 2017, which were granted by the Ministry of Education, Youth and Sports of the Czech Republic.

IOP Conf. Series: Earth and Environmental Science **143** (2018) 012014 doi:10.1088/1755-1315/143/1/012014

Immediate and long term effects of compaction on the stress-strain behaviour of soil

Sarah T Noor[1], Prantick Chowdhury[2] and Tasnim Chowdhury[2]

[1] Department of Civil Engineering, University of Asia Pacific, Dhaka, Bangladesh
[2] Department of Civil Engineering, Military Institute of Science and Technology, Dhaka, Bangladesh

Email: sarah@uap-bd.edu

Abstract. This paper explores whether delay in construction after compaction can benefit from the gain in soil's strength and stability point of view. An experimental investigation has been carried out to examine the gradual development of soil's shear strength by ageing of mechanically compacted soil at three relative densities. In order to separate the gain in strength due to ageing from that occurring from the reduction in soil moisture, the soil samples prepared in moulds were kept in desiccators for different periods of time (1, 9 and 17 days) before testing unconfined compressive strength test. The soil in densely compacted state is found to gain in strength due to ageing faster than that in medium compacted state. Only due to ageing of 9 days or more, unconfined compressive strength of compacted soil is found about 1.7 to 2.4 times of that attained in day 1 after compaction.

1. Introduction

Compacted fills are often studied to investigate its vulnerability from liquefaction and moisture induced collapse point of view [1-5]. Shearing of soil can cause different types of changes in soil properties; strength, stiffness and stability, soil water potential, etc. The effect of ageing on the behaviour of remoulded soil had been studied from strength and stability point of view by several researchers mostly for agricultural applications [6- 8]. It can be noted that agriculturists concentrate mainly on the behaviour of top soil only. However, progressive increases in soil parameters (including strength, soil water potential) were also observed upon ageing since disturbance [8, 9]. During field work, greater resistance was experienced while penetrating probes in soil after some degree of ageing after remoulding [10]. This also indirectly indicates that the resistance of remoulded soil increases gradually during ageing. However, such increases in soil's stiffness and strength in nature are resulted from both desaturation and ageing. The influence of ageing need to be investigated in the laboratory only.

However, in different civil engineering projects, soil is mechanically compacted and thus becomes remoulded. During compaction, soil gets remoulded and attains greater strength and stiffness. Moreover, the compacted soils can gain additional strength due to ageing [6-11]. To date, very few investigations on soils those are not classified as sensitive were carried out, though several attempts studying the effect of ageing on regaining strength, stiffness and compressibility of sensitive clay are found [9, 12].

Previous studies investigated the effect of ageing by keeping the specimens under consolidation pressure for different periods of time [12, 13]. Increase in shear strength and decreases in pore water pressure, failure strain and angle of internal friction were noted during the experiments on specimens kept under consolidation pressure for longer periods. The objective of the present study is to extend the knowledge regarding the influence of ageing on the stress-strain relation and shear strength of low plastic silt due to compaction.

2. Experimental Program

An experimental investigation has been carried out to examine the gradual development of soil's shear strength by ageing of mechanically compacted soil at three relative densities.

2.1. Methodology

The gradual development of shear strength of low plastic silt due to mechanical compaction was studied. The soil was compacted in mould (38 mm diameter and 76 mm height) in three layers and at optimum moisture content. Figure 1 shows the preparation of compacted specimens in rubber membranes. Three different compaction efforts (such as 8 blows, 16 blows and 32 blows per layer) were applied for preparing compacted soil specimens at three relative densities: loose, medium and dense-to-very dense, respectively. The compacted specimens were kept in desiccators (water-cool and close-fitting lids) for different periods of time (1, 9 and 17 days) before conducting tests. The unconfined compressive strengths of all three types of compacted specimens were determined at the age of 1, 9 and 17 days. Figure 2 presents the photographs of unconfined compressive strength test (in-progress) and a failed specimen. The water content of the failed specimen was also determined to check that the moisture content did not decrease more than 0.5% during ageing.

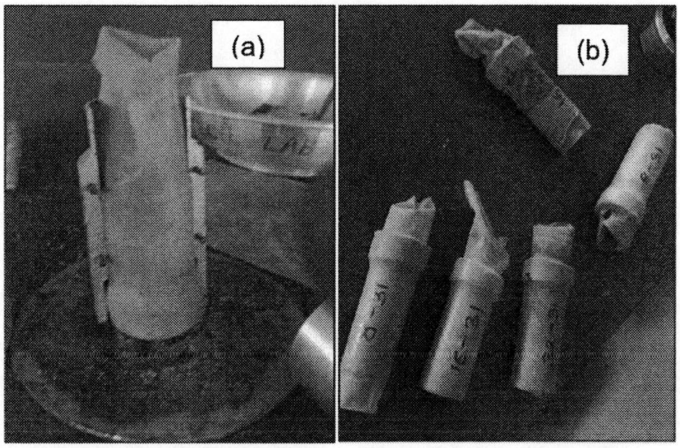

Figure 1. (a) Preparation of compacted specimen, (b) Marked specimens.

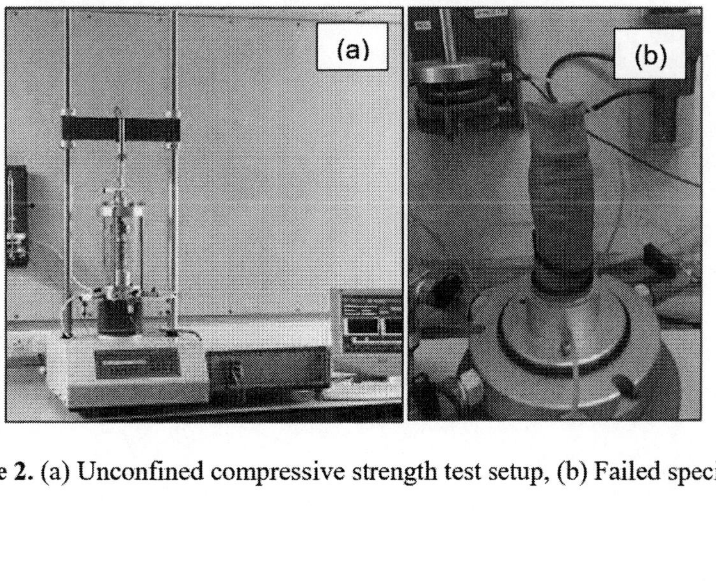

Figure 2. (a) Unconfined compressive strength test setup, (b) Failed specimen.

2.2. Soil

The soil used in this study was collected from Mirpur Cantonment, Dhaka. Different soil properties were determined by conducting different tests: Cassagrande's liquid limit, plastic limit (by thread-roll), specific gravity and standard Proctor compaction test. The soil parameters determined are given in Table 1. According to unified soil classification system, the soil was classified low plastic silt (ML). Figure 3 presents the compaction curve (standard Proctor test) of the silt soil used in this study. It was determined that the samples prepared for unconfined compressive strength tests were loose, medium and dense-to-very densely compacted when the blows per layer were 8, 16 and 32, respectively. The relative densities of loose, medium and dense-to-very densely compacted soils were determined 35%, 70% and 85%, respectively.

Table 1. Soil used in this study.

Soil Parameter	Value
Liquid limit (%)	33
Plastic limit (%)	27.4
Plasticity index (%)	5.6
Optimum Moisture Content (%)	12.6
Maximum dry density (g/cm^3)	1.575
Specific Gravity	2.61
Soil (USCS)	ML

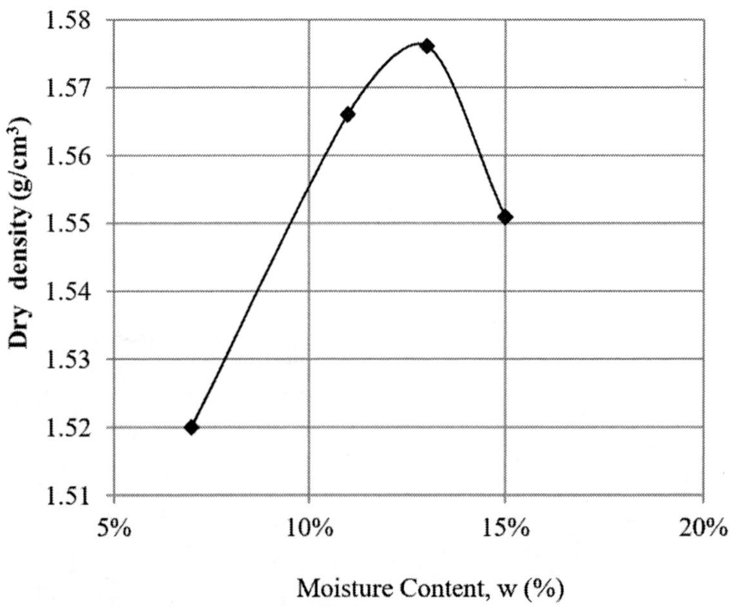

Figure 3. Compaction curve from standard Proctor test.

3. Results and Discussion

The stress-strain behaviour of the soil in three different degrees of compaction are investigated at the age of 1 day, 9 days and 17 days, as presented in figures 4 – 6, respectively. The stress-strain curve of densely compacted soil is found distinctly different from the specimens compacted at loose and medium relative densities. For loosely compacted condition, the stress-strain curves showed no peak shear stress even after ageing for 17 days. On the other hand, the peak shear stress was always observed for densely compacted soil at any age considered in this study. Further, peak shear stress in the stress-strain curve was observed at the age of 9 days or more, for medium compacted soil.

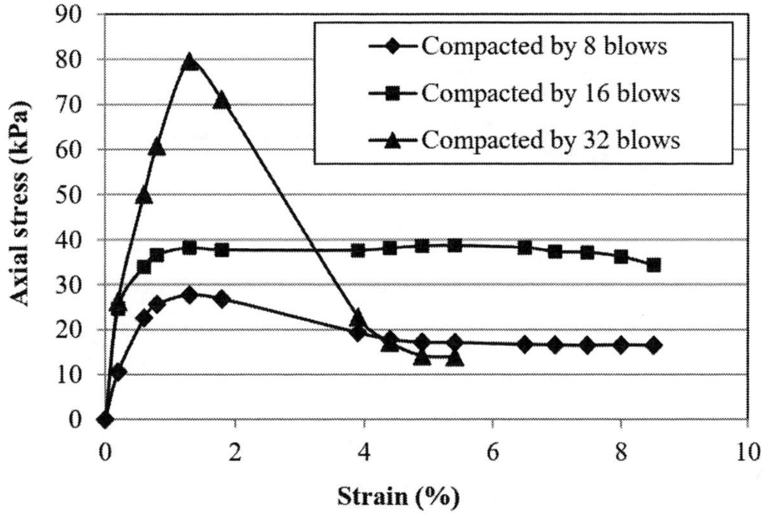

Figure 4. Unconfined compressive strength test results: Immediate effect of compaction.

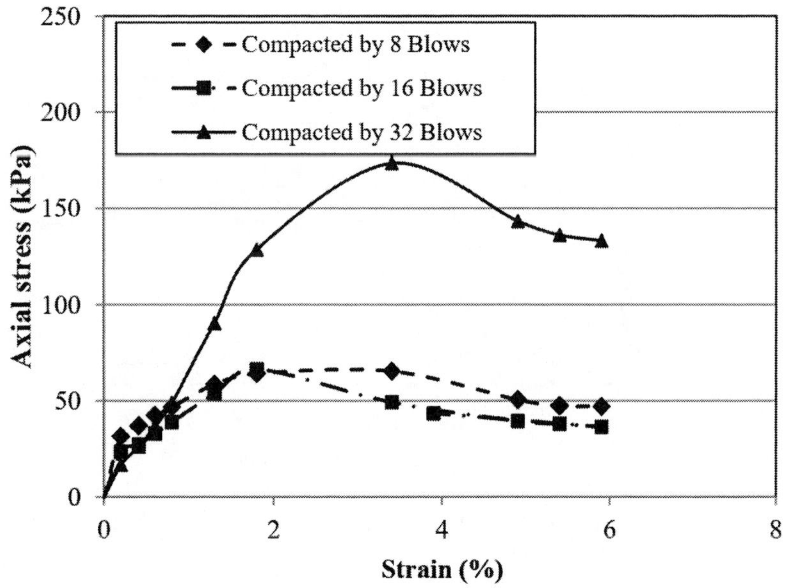

Figure 5. Unconfined compressive strength test results: 9th day after compaction.

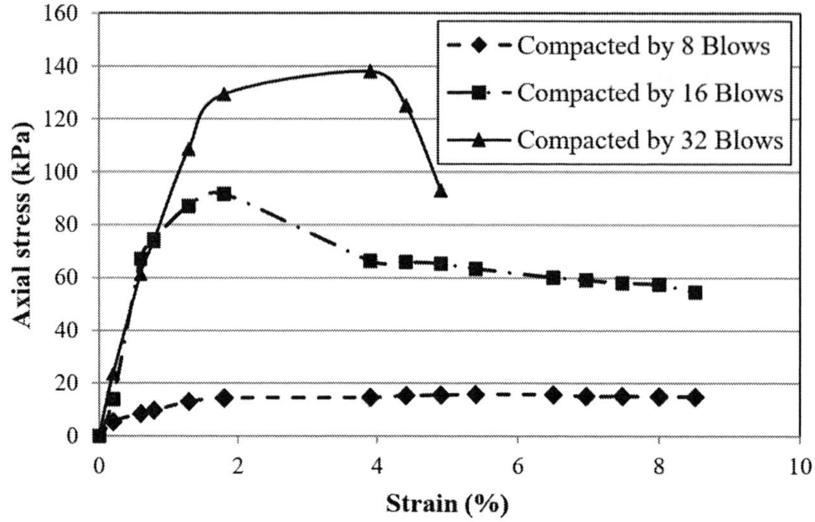

Figure 6. Unconfined compressive strength test results: 17th day after compaction.

Moreover, the effects of ageing on the strength gain of medium and densely compacted soil are studied in figures 7 and 8, respectively. The medium compacted soil gained strength gradually with time elapsed after compaction, as shown in figure 5. Moreover, it can be noted that the densely compacted soil gained sharply within the first week after compaction, while the rate of gain in strength diminishes slowly (figure 8).

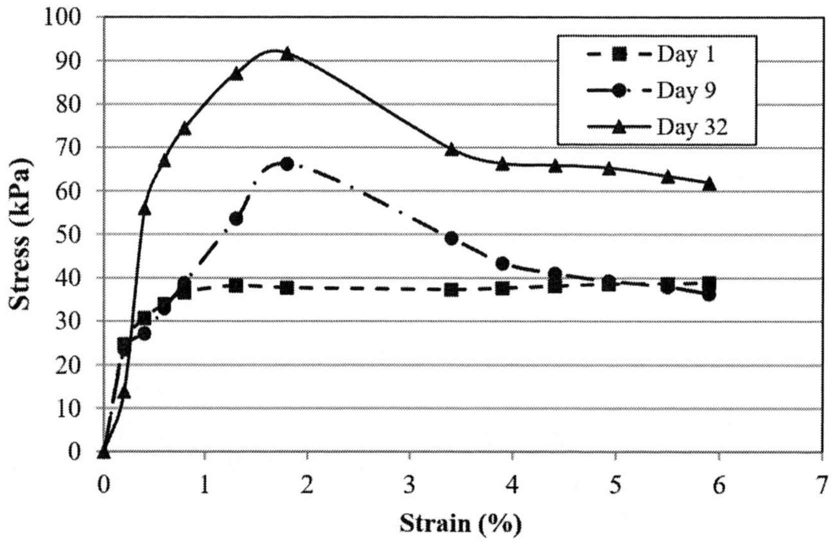

Figure 7. Unconfined compressive strength test results: Medium compacted soil.

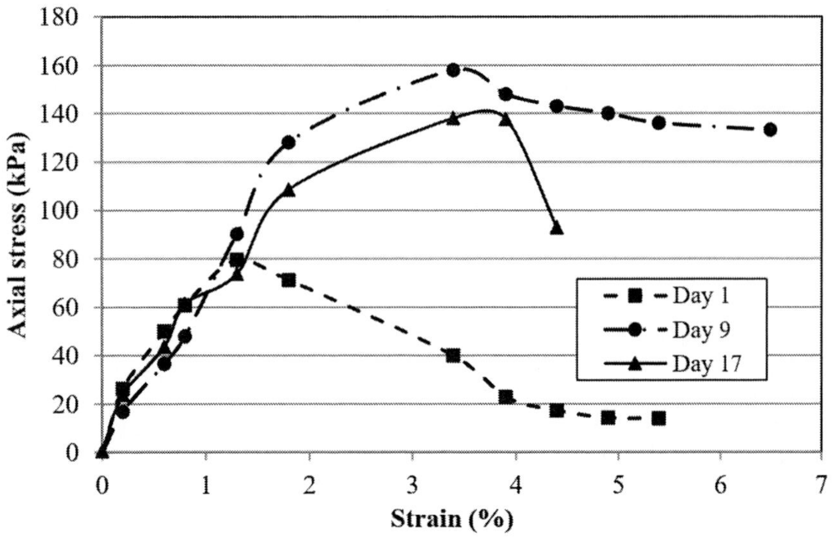

Figure 8. Unconfined compressive strength test results: Densely compacted soil.

The gain in strength of aged compacted soil is studied with respect to the strength attained immediately after compaction in terms of a strength magnification factor 'm'. The value of 'm' is obtained as the ratio of unconfined compressive strength obtained at aged sample to that obtained at 1 day aged sample after compaction.

Table 2. Ageing and gain in strength.

Relative Density (%)	m	
	Day 9	Day 17
70	1.7	2.4
80-90	2.2	1.8

It can be noted that unconfined compressive strength of compacted soil becomes 1.7 to 2.4 times of that attained in day 1 after compaction.

4. Conclusions

The present study explores whether delay in construction after compaction can benefit from soil's strength and stability point of view. Based on the experimental results, the following conclusions can be drawn:

- The soil in densely compacted state showed strength gain due to ageing faster than that in medium compacted state.
- Once the stress-strain curve shows peak shear stress, the compacted soil will not continue to gain in strength for longer period of time. It has been observed for both medium and densely compacted soil.
- Only due to ageing of 9 days or more, unconfined compressive strength of compacted soil is found about 1.7 to 2.4 times of that attained in day 1 after compaction.

References

[1] Noor, S T and Uddin, R 2017 Effect of lime stabilization on the alteration of engineering properties of cohesive soil *GCEC 2017, Proc. of the 1st Global Civil Engg. Conf.* Kuala Lumpur, Malaysia, Biswajeet Pradhan (ed.), Springer

[2] Noor S T 2017 Discussion of "The Effects of Fines on the Liquefaction Behavior in Well Graded Material" Canadian Geotechnical Journal Vol. **54** 1802-1803
[3] Noor S T 2017 Discussion of "Drag load on End-bearing Piles in Collapsible Soil due to Inundation" *Canadian Geotechnical Journal* Vol **54** 742-743
[4] Noor S T 2017 Numerical and Analytical Modelling for Predicting Drag Load Induced on Pile in Collapsible Soil because of Inundation *The Open Civil Engineering Journal* Vol **11**(1) 664-675
[5] Noor S T, Hanna A M and Mashour I 2013 Numerical Modeling of Pile in Collapsible Soil subjected to Inundation *International Journal of Geomechanics* Vol **13**(4) 514-526
[6] Utomo W H and Dexter A R 1980 Effect of ageing on compression resistance and water stability of soil aggregates disturbed by tillage. *Soil and Tillage Research* 1 127- 137
[7] Dexter A R and Tanner DW 1974 Time dependence of compressibility for remoulded and undisturbed soils *J. Soil Sci.* **25** 153--164
[8] Croney D and Coleman J D 1954 Soil structure in relation to soil suction (pF). *J. Soil Sci.* 5 75
[9] Mitchell J K 1960 Fundamental aspects of thixotropy in soils *J. Soil Mech. and Found. Div. ASCE* **26** 19--52
[10] Braunack M V and Dexter A R 1978 Compaction of aggregate beds. In: W.W. Emerson, R.D. Bond and A.R. Dexter (eds) *Modification of Soil Structure. John Wiley & Sons, Chichester, England* 119--126
[11] Shen C K, Arulanandan K and Smith W S 1973 Secondary consolidation and strength of a clay ASCE *Journal of the Geotechnical Engineering Division* **99**(1) 95-110
[12] Bjerrum L and Lo K Y 1963 Effect of aging on the shear strength properties of a normally consolidated clay *Geotechnique* **13**(1) 147-157
[13] Allam M M and Sridharan A 1979 The influence of aging on the shear strength behaviour of two fine-grained soils *Canadian Geotechnical Journal* ***16***(2) 391-397

Acknowledgments
The authors acknowledge Department of Civil Engineering, Military Institute of Science and Technology for the facility and support.

Incorporation mode effect of Nano-silica on the rheological and mechanical properties of cementitious pastes and cement mortars

B Safi[1], H Aknouche[1], H Mechakra[2], D Aboutaleb[1] and K Bouali[1]

[1] Research Unit: Materials, Processes and Environment (UR/MPE), University M'hamed Bougara of Boumerdes 35000, Algeria
[2] Dynamic Motors and Vibroacoustics Laboratory, University M'hamed Bougara of Boumerdes 35000, Algeria

Corresponding author, Email : safi_b73@univ-boumerdes.dz

Abstract. Previous research indicates that the inclusion of nanosilica (NS) modifies the properties of the fresh and hardened state, compared to the traditional mineral additions. NS decreases the setting times of the cement mortar compared to silica fume (SF) and reduce of required water while improving the cohesion of the mixtures in the fresh state. Some authors estimate that the appropriate percentage of Nano-silica should be small (1 to 5% by weight) because of difficulties caused by agglomeration to particles during mixing, while others indicate that 10% by weight, if adjustments are made to the formulation to avoid an excess of self-drying and micro cracks that could impede strength. For this purpose, the present work aim to see the effect of the introduction mode of the nanosilica on the rheological and physic mechanical properties of cement mortars. In this study, NS was used either powdered with cement or in solution with the superplasticizer (Superplasticizer doped in nanosilica). Results show that the use of nanosilica powder (replacing cement on the one hand) has a negative influence on the rheological parameters and the rheological behavior of cementitious pastes. However, the introduction of nanosilica in solution in the superplasticizer (SP) was significantly improved the rheological parameters and the rheological behavior of cementitious pastes. Indeed, more the dosage of NS-doped SP increases more the shear stress and viscosities of the cementitious pastes become more fluid and manageable. A significant reduction of shear stress and plastic viscosity were observed that due to the increase in superplasticizer. A dosage of 1.5% NS-doped SP gave adequate fluidity and the shear rate was lower.

1. Introduction

The nanotechnology of nano-structured materials develops with an astronomical speed and gains from days to day new fields of application [1-5]. It is the potential to be the key to a whole new world of today's industry, especially the production of building materials. Among the most important in the global economy, the latter is one of the major contributors to global warming by greenhouse gas (GHG) emissions and, on the other hand, is energy intensive and far from being an example of sustainable development because of the continued depletion of natural resources. The increase in energy prices and environmental requirements require a different understanding of the composition of concretes, while maintaining the same performances [5-7]. One of the possible solutions to be envisaged to mitigate this environmental and economic problem is the sustainable development of the mortar which would save not only the natural resources and the energy but also the protected environment by reduction of the wastes.

In this regard, nanotechnology helps to make concrete an intelligent material designed by the incorporation of Nano-sized substitute such as nanoparticles (NP), which is a superior and economical experience of the hour. The mastery of the science of the infinitely small makes it possible to better understand and control, at the atomic scale, the Nano-modification of cement-based materials as well as their influence on the behaviour and properties of mortar and concrete. The fresh state, the rheological functionality and the hydration process of cement paste which play an important role in the determination of the concrete structure and also on the mechanical properties and the durability [6-11]. Nanosilica is a very effective pozzolanic material. They are very fine glassy particles, about 1000 times smaller than the average cement particles [2-3]. It forms an excellent mixture with cement to improve the overall quality of the cementitious matrix to a great extent. It behaves on a scale not only as a filler to improve the microstructure, but also as an activator to promote the pozolanic reaction. Rapid and accelerated hydration is recorded in NS-containing cement pastes, but it has not yet been established whether this is due to chemical reactivity (pozzolanic activity) or high surface area. The pozzolanic reaction results in a reduction in the amount of calcium hydroxide in the concrete and reduces the price time, porosity, water absorption and permeability. It accelerates the dissolution of C-S and the formation of C-S-H with inverted activity proportional to the size and destination of nucleation sites for C-S-H. NS by their size can fill the spaces between the C-S-H gel particles, acting as a Nano-filler material [8-12].

Previous research indicates that the inclusion of nanosilica (NS) modifies the properties of the fresh and hardened state, compared to the traditional mineral additions. NS decreases the setting times of the cement mortar compared to silica fume (SF) and reduce of required water while improving the cohesion of the mixtures in the fresh state. Some authors estimate that the appropriate percentage of Nano-silica should be small (1 to 5% by weight) because of difficulties caused by agglomeration to particles during mixing, while others indicate that 10% by weight, if adjustments are made to the formulation to avoid an excess of self-drying and micro cracks that could impede strength. For this purpose, the present work aim to see the effect of the introduction mode of the nanosilica on the rheological and physic mechanical properties of cement mortars.

2. Experimental Study

2.1. Materials used

The ordinary Portland cement CEM II (42.5N type) was used in this work and it is conformed to ENV norm. The Nano-silica (NS) used is manufactured by EVONIK INDUSTRIE named AEROSIL 200 (Figure 1). It is a white amorphous powder having a silica content of 99.8% and a specific surface area of 200 m2/g with particle size about 12 nm (table 1). The chemical and physical characteristics of NS are given in Table 1. A polycarboxylate type (SP) is used as superplasticizer to ensure some workability. For mortar mixes, natural sand (S) with a maximum size of 5 mm is used in this study.

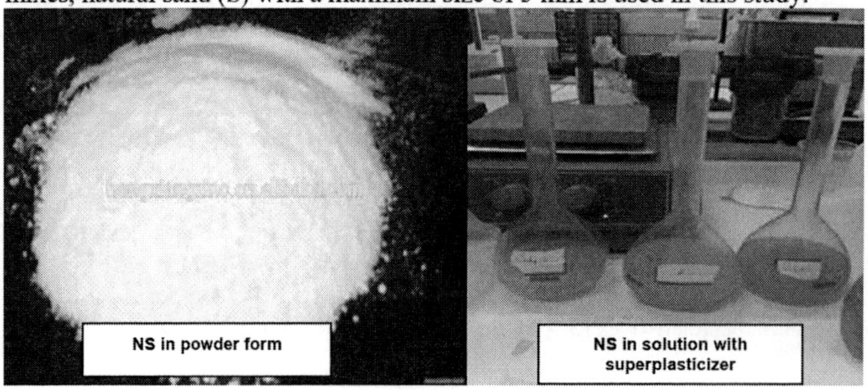

Figure 1 Nano-silica (NS) used in this work; Image obtained by SEM for Nano-silica (NS).

Table 1 Chemical composition and physical properties of cementitious materials.

Items	Chemical composition (%)	
	Ordinary Portland Cement OPC	Nano-silica NS
SiO_2	17.3	99.8
Al_2O_3	5.54	0.05
Fe_2O_3	2.80	0.005
CaO	59.34	–
MgO	1.84	–
SO_3	2.51	–
K_2O	0.54	–
Na_2O	0.25	–
P_2O_5	0.09	–
TiO_2	0.30	0.003
L.O.I	2.29	1.00
Specific gravity	3.15	-
Avg. particle size SSA (cm^2/g)	3700	200.10^4

2.2. Details of cement paste and mortar mixtures

The study was divided into two parts. Firstly, the work involves studying the effect of nanosilica in powder (2, 4, 6 and 8%.wt of cement) and solution (0.7, 1.3 and 2.8%.wt of SP) on the rheological behaviour of cementitious pastes. The saturation point is the dosage above which the superplasticizer has no effect on the rheological properties of cement paste was evaluated. This study part was carried by AR 2000 rheometer (Table2). Second work part was realized on the cement mortars with same content of NS in two cases (in powder and in SP-solution). The Control Mortar (CM) was prepared according to the European standard EN 196-1 (Table3). After, NS was used in powder form by cement substitution (2, 4, 6 and 8%.wt of cement) and used in solution with superplasticizer (0.7, 1.3 and 2.8%.wt of SP).

Table 2: Mixtures details of studied cement pastes

	NS in Solution [%]	NS in powder [%]	PC [%]	Water [g]	SP [%]	E/C
PC2NS	-	2	98	40	0,70	0,4
PC4NS	-	4	96	40	1,30	0,4
PC6NS	-	6	94	40	3,00	0,4
PC8NS	-	8	92	40	3,50	0,4
PC-NS0,7	0,7	-	100	30	0,50	0,3
PC-NS1,3	1,3	-	100	30	1,00	0,3
PC-NS2,8	2,8	-	100	30	1,50	0,3
PC-NS3,2	3,2	-	100	30	2,00	0,3

The cement mortars were mixed in a rotary kneader. Nanoparticles are not easily dispersed evenly because of their high surface energy. As a result, the mixing was carried out as follows; the nano-silica particles were stirred with the mixing water at high speed for 1 min. After, the cement was added to the mixer and mixed at medium speed for another 1 min. Then, the superplasticizer has been added and stirred at high speed for 30 s. Finaly, the sand was added gradually, then mixed for 2 min 30 s at high speed.

Table 3: Mixtures details of cement mortars

	NS in Solution [%]	NS in powder [%]	PC [g]	Water [g]	Sand [g]	SP [%]	E/C
M0	0	0	450	225	1350	0	0,5
MNS2	-	2	441	225	1350	0,70	0,5
MNS4	-	4	432	225	1350	1,30	0,5
MNS6	-	6	423	225	1350	3,00	0,5
M-NS0,7	0,7	-	450	180	1350	0,50	0,4
M-NS1,3	1,3	-	450	180	1350	1,00	0,4
M-NS2,8	2,8	-	450	180	1350	1,50	0,4
M-NS3,2	3,2	-	450	180	1350	2,00	0,4

2.3. Test Methods:

Rheological study: the rheological tests are performed on several series of cement paste samples for two studies cases of Nanosilica adding. The superplasticizer dosage is determined by testing the saturation point. Rheology tests were performed at 20 °C using a rheometer AR2000 (Figure2) equipped with a valve rotor speed imposed according to the following: presheared to 50 s-1 for 60 s followed by a rest period of 30s, then a linear ramp increasing rate of 0-435 s-1 is applied through the rheometer (Figure3). The flow curves were analyzed and modeled by the software Rheology Advantage Data Analysis (Version-V4.021/TA-Instrument).

Figure 2 : Rheometer AR2000 used for rheological study.

Figure 3 : Mexing of cement pastes with NS.

Flexural and Compressive Strength: After mixing, for each mixture, specimens of 4×4×16 cm were prepared for in accordance with ASTM specification. The test specimens were demolded after 24 h and then cured in water. The compressive strengths of cement mortar specimens were tested at 2 and 28 days of curing times by using a uniaxial hydraulic compression machine under a load control rate of 0.20 MPa/s. Three-point bending tests were carried out on prismatic samples according to ASTM C348 [13]. Half samples were subjected to compressive stress by using a hydraulic press with a capacity of 3000 KN according to ASTM C349 [14].

3. Results and discussions

3.1. Effect of Nanosilica in powder form:

Rheological study: Figure 4a and 4b show the evolution of main rheological parameters (plastic viscosity and shear stress) of nanosilica-based cementitious pastes at various superplasticizer dosages as a function of shear rate. According to the results obtained, it is clearly observed that the nanosilica (as powder adding) is more viscous which requires an extra dose of superplasticizer. However, at a higher rate of fluidity and the rate of high nanosilica causes an excess of superplasticizer.

The rheogram of figure 4b shows also the shear stress evolution of cementitious pastes based on different dosing of nanosilica and different dosage of superplasticizer. It is clear that according to these results, the nanosilica (NS) has increased shear stress and the superplasticizer has significantly reduced this stress. However, it is noted that the higher the NS dosage increases the higher the shear stress increases the pasta requires more superplasticizer to become fluid and flow at low shear stresses. We also note above 4% NS, the superplasticizer dosage has doubled [12, 15-16].

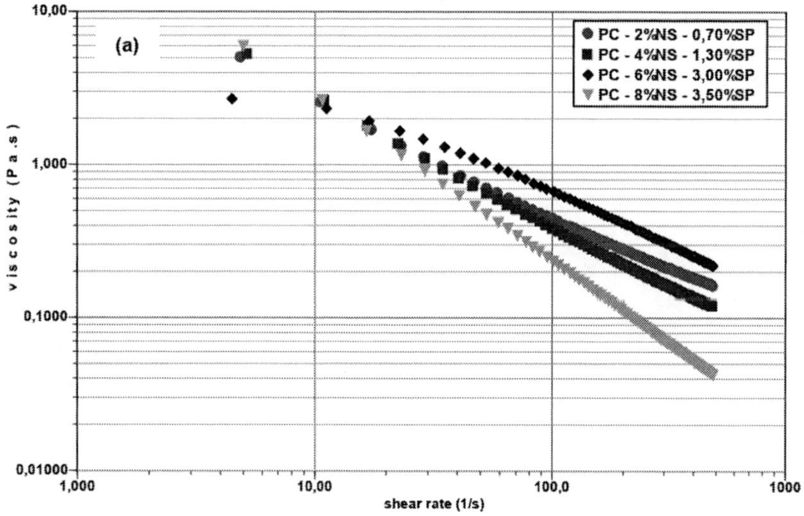

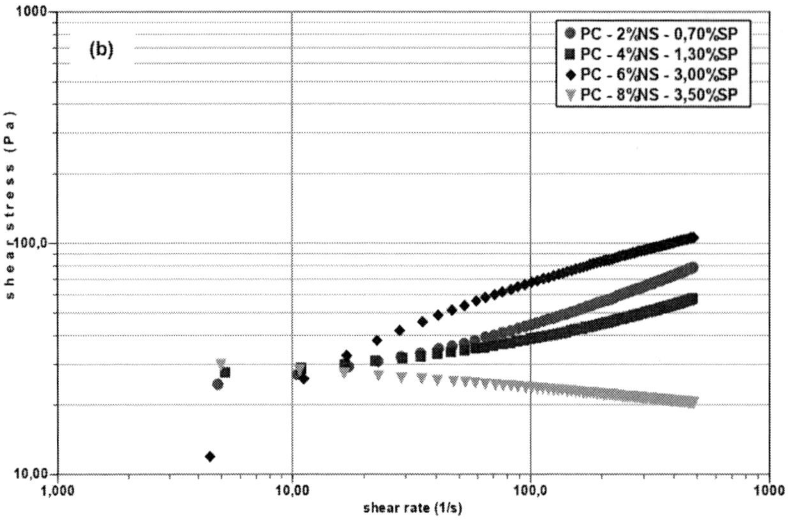

Figure 4: Evolution of main rheological parameters (plastic viscosity (a) and shear stress (b)) of nanosilica-based cementitious pastes.

Rheological behavior: The flow curves of cementitious pastes are shown in **Figure 5**. From the rheograms, all the cementitious pastes show a non-Newtonian flow. Indeed, it is clear that these pastes represent a flow threshold that means that they have a non-Newtonian behavior. According to several researchers found that the behavior of these types of pasta follows the pattern of Herschel Bulkely that depicted in the figure. It is noted that beyond 4% of NS, the behavior of the pasta changes and causes a rheological instability.

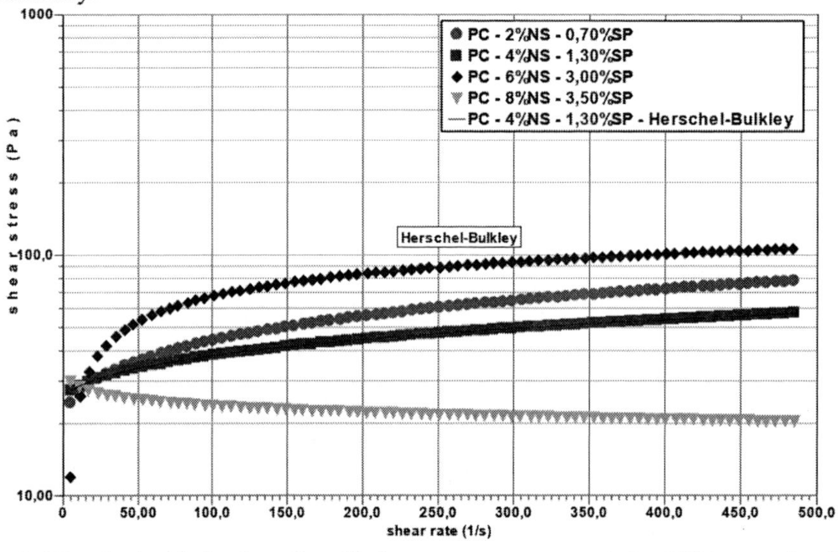

Figure 5: Rheological behavior of studied cement pastes containing NS in powder form.

3.2. Effect of Nanosilica in superplasticizer solution:

Rheological study: Figure 6a and 6b shows the variation in plastic viscosity of nanosilica (NS) -based superplasticizer (NS) -based cementitious pastes as a function of shear rate. According to the results obtained, it is clearly seen that SP doped with NS makes cementitious pastes more fluid. Indeed, the higher the SP dosage increases the viscosity of the cement paste decreases and beyond the high shear rates the viscosity of the cement pastes become constant. However, cementitious pulps with 1.5% NS-doped SP are more fluid and the shear rate is lower [12].

Figure 6b shows the variation of the shear stress as a function of NS-doped SP dosing of cementitious pastes. For all cementitious pastes studied, a significant reduction in shear stress was observed as a function of the increase in SP. This is reflected by the deflocculating and dispersive effect of SP. Indeed, in the presence of water, the SP adsorbs on the surface of the cement particles, causing electrostatic and steric inter-particle repellent effects. This results in the dispersion of the grains and the water retained in the flocculates is released, thus increasing the fluidity. It should also be noted that the saturation dosage obtained for these pastes is of the order of 1.5% of SP doped with NS.

Rheological behaviour: Figure 7 shows the flow curves of the cementitious pastes based on different dosages of the superplasticizer doped with nanosilica. According to the rheograms, these pastas also show a non-Newtonian flow, characterized by a flow threshold and which follows the model of Herschel Bulkely. It is observed that the higher the dosage of the doped superplasticizer, the more the pasta becomes more fluid. Above 1.5% of the SP doped with NS, the effect is not important, which corresponds to the saturation point.

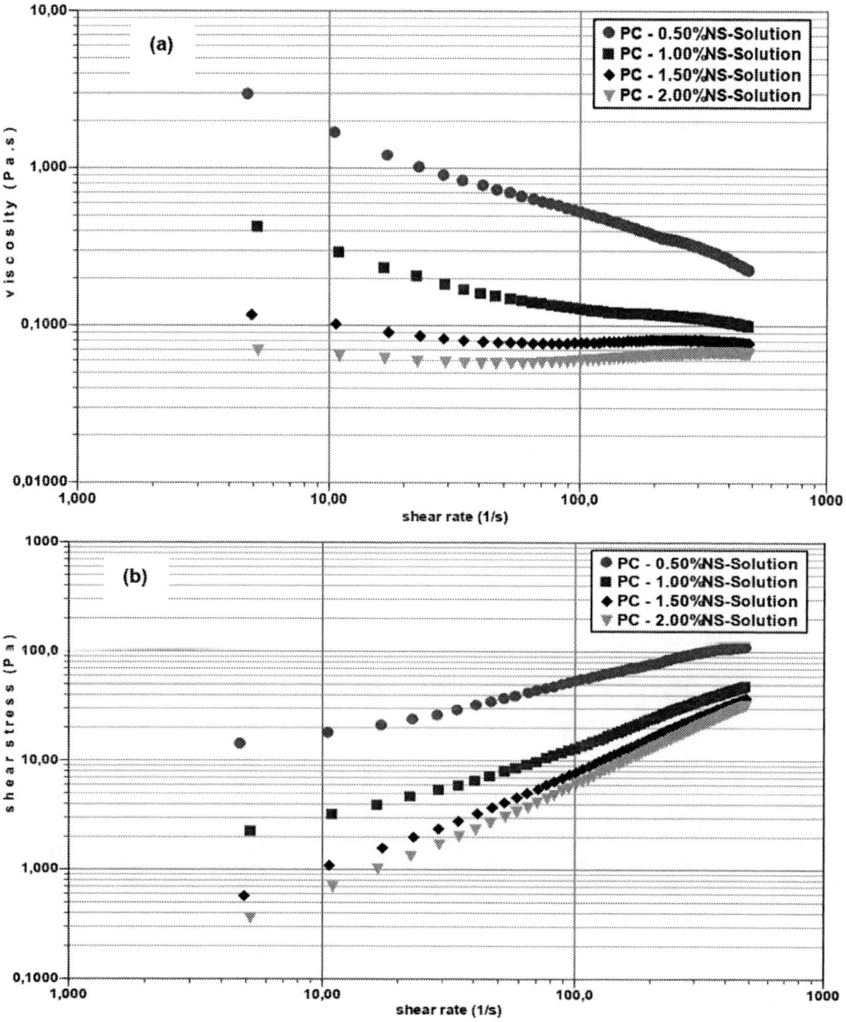

Figure 6: Evolution of main rheological parameters (plastic viscosity (a) and shear stress (b)) of nanosilica-based cementitious pastes containing NS in superplasticizer solution.

It was noted that a significant reduction in shear stress and plastic viscosity, due to the increase of the superplasticizer. This is reflected by the deflocculating effect of SP. Indeed, in the presence of water, the SP adsorbs on the surface of the cement particles, causing electrostatic and steric inter-particle repellent effects. This results in the dispersion of the grains and the water retained in the flocculates is released, thus increasing the fluidity [15-17].

3.3. Mechanical strength

The evolution of the compressive and flexural strength of nanosilicate (NS) -based mortars is given in the figure 7. According to these results, it was noted that the strength development of all mortars according to curing age is an increasing function, because the resistance (compression or flexural) increases according to age. This is explained by the effect of hydration of clinker minerals contained in the cement used. The values of resistance recorded at 28-days are in order of 57 MPa and 9 MPa respectively for compressive and flexural strength. Regarding the effect of cement substitution by NS, a slight decrease in the compressive strength of the mortars was observed and that the introduction of NS powder, causing the reduction of resistance.

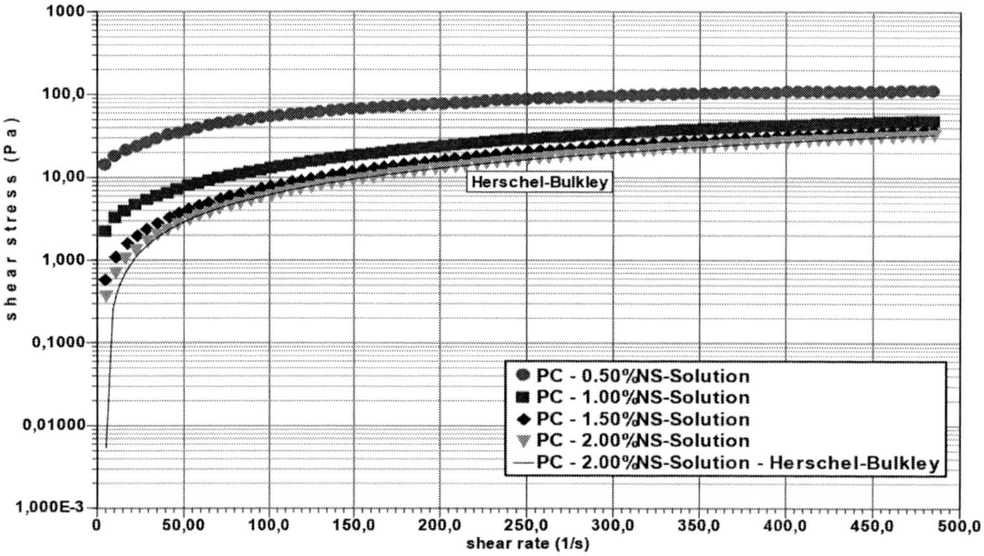

Figure 7: Rheological behavior of studied cement pastes containing NS in superplasticizer solution.

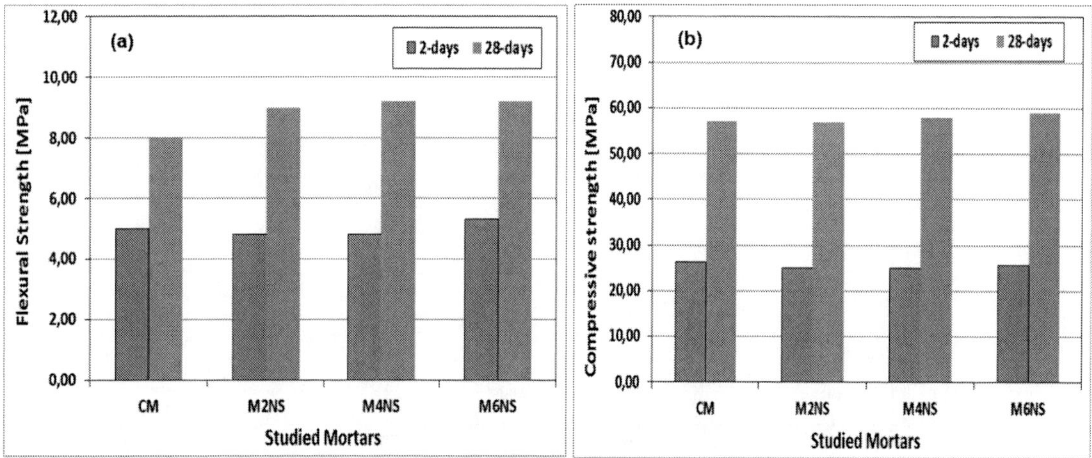

Figure 8: Mechanical Strength of studied mortars based on NS in powder form
(a) Flexural strength (b) Compression Strength.

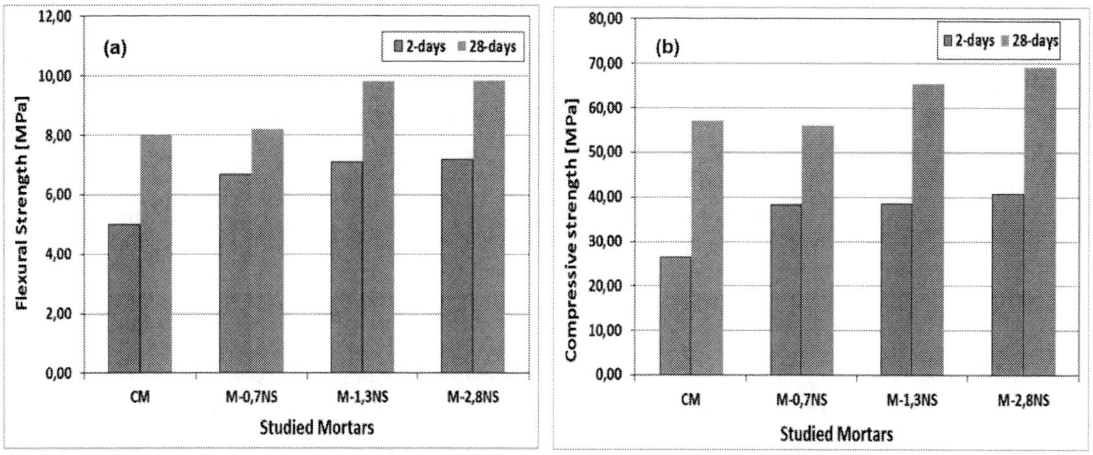

Figure 9: Mechanical Strength of studied mortars based on NS in superplasticizer solution (a) Flexural strength (b) Compression Strength.

It should be noted that during the implementation of the mortars was difficult what influenced on the resistance of the mortars [18-21]. Use up to 6% NS powder, which may have a mortar with a compressive strength of 55MPa.

The results of the mechanical tests (compressive and flexural strength) of NS-doped SP-adjuvanted mortars are presented in Figure 8. We observe that the evolution of mechanical resistance as a function of curing age is an increasing function. Indeed, the resistance (compression or flexion) increases with age. This is reflected by the effect of hydration of clinker minerals contained in the cement used. Strength values recorded at 28days are of the order of 70 MPa and 10 MPa respectively for compressive strength and flexural strength. Regarding the effect of NS-doped superplasticizer, a considerable improvement in the compressive strength of the mortars was observed and the adjuvanted mortars obtained having higher strengths than the control mortar. However, a slight decrease in compressive strength was recorded for the SP1.29% NS mortar, followed by an increase in the higher strength. This is shown the beneficial effect of a NS-doped superplasticizer [19-23].

4. Conclusion

The aim of study was to explore of the effect of introduction mode of the nanosilica on the rheological and physic mechanical properties of cement mortars. It can be concluded that:

Using of NS in powder form: As a result, the use of nanosilica powder (replacing cement on the one hand) has had a negative influence on the rheological parameters and the rheological behavior of cementitious pastes. Indeed, the higher the cement substitution rate by NS increases the shear stress and the viscosity increases, the more cementitious pastes become more viscous and the handling difficult. From the point of view of mechanical properties, it should be pointed out that the more the substitution of cement by NS in powder increases the more the implementation of the mortars become more difficult which has influenced the resistance values.

Using of NS in superplasticizer solution: The introduction of nanosilica in solution in the superplasticizer significantly improved the rheological parameters and the rheological behavior of cementitious pastes. Indeed, the higher the dosage of NS-doped SP increases the lower the shear stress and the viscosity, the more the cementing pastes become more fluid and manageable. It was noted that a significant reduction in shear stress and plastic viscosity, due to the increase in superplasticizer. A dosage of 1.5% NS-doped SP gave adequate fluidity and the shear rate was lower. This assay can be considered as saturation point assay. The effect of SP doped with NS seems very beneficial on the mechanical strengths of the mortars. As a result, the use of the nanosilica in solution in the superplasticizer has considerably improved the resistance of the mortars compared to the control mortar. The values of resistance recorded at 28days are of the order of 70 MPa and 10 MPa respectively for compressive strength and flexural strength. This is shown the beneficial effect of a NS-doped superplasticizer.

References

[1] Hanus M J and Harris A T 2013 Nanotechnology innovations for the construction industry *Progress in Materials Science* 58:1056–1102.

[2] Singh L P , Karade S R , Bhattacharyya S K, Yousuf M M and Ahalawat S 2013 Beneficial role of nanosilica in cement based materials - A review" *Construction and Building Materials* 47:1069–1077.

[3] Zhu W, Bartos P J M and Porro A 2004 Application of nanotechnology in construction *Materials and Structures* 37: 649–658.

[4] Zhang W, Li L, Yao S and Zheng G 2007 Corrosion Protection Properties of Lacquer Coatings on Steel Modified by Carbon Black Nanoparticles in NaCl Solution *Corrosion Science* 49(2): 654-661.

[5] Li H, Xiao H, Yuan J and Ou J 2004 Microstructure of Cement Mortar with Nano-particles Composites *Part B: Engineering* 35(2):185-189.

[6] Givi A N, Rashid S A, Aziz F N A, Salleh M A M 2010 Experimental investigation of the size effects of Silica nano-particles on the mechanical properties of binary blended concrete *Composites: Part B 41: Elsevier* Ltd: 673–677.

[7] Ji T 2005 Preliminary study on the water permeability and microstructure of concrete incorporating nano-Silica *Cement and Concrete Research 35 Elsevier* Ltd: 1943 – 1947.

[8] Kawashima S, Hou P, Corr D J, Shah S P 2013 Modification of cement-based materials with nanoparticles' *Cement & Concrete Composites Elsevier* Ltd:8-15.

[9] Jo B, Kim C, Tae G and Park J 2007 Characteristics of Cement Mortar with Nano-SiO2 Particles *Construction and Building Materials* 21(6):1351-1355.

[10] Sobolev K 2009 Engineering of SiO2 nano-particles for optimal performance in nano cement-based materials. In: Bittnar Z, Bartos PJM, Nemecek J, Smilauer Zeman J editors *Nanotechnol constr proc NICOM3* Prague:139–48.

[11] S Abd El Aleem, M Heikal, W M Morsi 2014 Hydration characteristic, thermal expansion and microstructure of cement containing nano-silica *Construction and Building Materials* 59: 151-160.

[12] Senff L, Hotza D, Repette W L, Ferreira V M and Labrincha JA 2010 Rheological characterisation of cement pastes with nanosilica, silica fume and superplasticiser additions" *Advances in Applied Ceramics: Structural Functional & Bioceramics* 109:213–218.

[13] ASTM C348 – 08, 2008 *Standard Test Method for Flexural Strength of Hydraulic-Cement Mortars* which appears in the Annual Book of ASTM Standards Vol. 04.01.

[14] ASTM C349 –08, 2008 *Standard Test Method for Compressive Strength of Hydraulic-Cement Mortars (Using Portions of Prisms Broken in Flexure)* which appears in the Annual Book of ASTM Standards Vol 04.01.

[15] Khomich V A, Emralieva S A, Tsyguleva M V 2016 Nanosilica Modifiers for Cement Mortars *Procedia Engineering* 152:601-607.

[16] Paratibha A, Rahul Pratap S & Yogesh A 2015 Use of nano-silica in cement based materials—A review *Cogent Engineering* 2(1):1-11.

[17] Berra M, Carassiti F, Mangialardi T, Paolini A E, & Sebastiani M 2012 Effects of nanosilica addition on workability and compressive strength of Portland cement pastes *Construction and Building Materials* 35: 666–675.

[18] Aly M, Hashmi M S J, Olabi A G, Messeiry M, Abadir E F & Hussain A I 2012 Effect of colloidal nano-silica on the mechanical and physical behaviour of waste-glass cement mortar *Materials and Design* 33: 127–135.

[19] Kawashima S, Hou P, Corr D J & Shah S P 2012 Modification of cement-based materials with nanoparticles *Cement and Concrete Composites* 36, 8–15.

[20] Kong D, Du X, Wei S, Zhang H, Yang Y, Shah S P 2012 Influence of nano-silica agglomeration on microstructure and properties of the hardened cement-based materials *Construction and Building Materials* 37: 707–715.

[21] Stefanidou M 2012 Influence of nano-SiO2 on the Portland cements pastes *Composites Part B: Engineering* 43 2706–2710.

[22] Qing Y, Zenan Z, Deyu K & Rongshen C 2007 Influence of nano-SiO2 addition on properties of hardened cement paste as compared with silica fume *Construction and Building Materials* 21 539–545.

[23] Berra M, Carassiti F, Mangialardi T, Paolini A E & Sebastiani M. 2012 Effects of nanosilica addition on workability and compressive strength of Portland cement pastes *Construction and Building Materials* 35 666–675.

IOP Conf. Series: Earth and Environmental Science **143** (2018) 012016 doi:10.1088/1755-1315/143/1/012016

A study on super-sulfated cement using Dinh Vu phosphogypsum

Nguyen Ngoc Lam

National University of Civil Engineering, Hanoi, Vietnam

Email: lamnn@nuce.edu.vn

Abstract. Super-sulfated cement (SSC) is a newly developed unburnt cementitious material. It is a kind of environmental-friendly cementitious material due to its energy-saving, carbon emission reducing, and waste-utilization. It mainly composes of phosphogysum (PG) and ground granulated blast furnace slag (GFS), with a small amount of cement. In Vietnam, the Diammonium Phosphate DAP – Dinh Vu fertilizer plant in Dinh Vu industrial zone in the northern port city of Hai Phong – has discharged millions of tons of solid waste containing gypsum after 9 years of operation. The waste has changed the color of the water, eroded metal and destroyed fauna and floral systems in the surrounding area. Notably, according to the environmental impact assessment, the gypsum landfill area is supposed to be 13 hectares and the storage time reaches up to five years. This paper presents the experimental results on SSC using a high amount of Dinh Vu phosphogypsum and GFS in comparison with those of ordinary Portland cement (PC). The results show that the setting time of SSC is much longer than that of Portland cement but the compressive strength of SSC can be obtained 45-50 MPa at the age of 28 days, similar to that of the control sample using 100% PC40, and 69MPa at the age of 90 days. This value even exceeds the compressive strength of the PC40 cement.

1. Introduction

In Vietnam today, the demand of construction materials due to the population growth and to the periodic need of infrastructure renovation is increasing. Considering Portland cement (PC) as the most widely used binder in the construction industry, the PC production, as a result, will be required intensively. This leads to a high CO_2 emission and energy demand, thus, in this aspect, searching for alternative cementitious systems is the most important issue to adapt to these demands in terms of sustainable development and lower energy consumption.

One of the most promising alternative cementing systems is super-sulfated slag cement (SSC), which is mainly composed of ground granulated blast furnace slag (GFS) and sulfate activator (such as gypsum), with a little amount of alkali activator (such as clinker cement or cement) [1, 2]. This cement can consume a large amount of gypsum in general, and an phosphogypsum (PG) in particular. Compared with ordinary Portland cement, super-sulfated cement has many advantages such as low hydration heat evolution, excellent sulfate resistance, using less clinker and more gypsum, and thus more solid waste can be consumed [3-6]. For example, Weiguo et al. [7] studied a type of lime–fly ash–phosphogypsum binder to improve the performances of lime–fly ash binder that was a typical semi-rigid road base material binder. The optimum composition of this binder was 8–12% modified lime, 18–23% phosphogypsum, and 65–74% fly ash (by weight). Moreover, Ding Sha et al. [3] investigated the concrete prepared with super-sulfated phosphogypsum-slag cement, which is made by 45% phosphogypsum, 48% GFS, 2% steel slag, and 5% clinker. The results show that the compressive strength of super-sulfated slag cement concrete can be obtained 38.6 MPa at the age of 28 days, similar to that of Ordinary Portland cement concrete.

The use of PG in producing cements has been being received much attention, because it is obtained in large amounts and is a readily available source of gypsum. For example, PG was studied to regulate the setting of ordinary Portland cement. It was shown that the impurities present in PG retarded the hydration of OPC to a large extent [8-10]. Only 15% of total PG is used for cement industry as a setting moderator, and for making gypsum plaster... The remaining 85% of PG is not used, causing an environmental trouble and requiring a large area for disposal. Therefore, attempts were made to use PG in applications such as the filler material of road and rail works, the stabilization material in base course, and building constructions [11].

In Vietnam, the Diammonium phosphate (DAP) - Dinh Vu fertilizer plant in Dinh Vu industrial zone in the northern port city of Hai Phong - has discharged about 10 millions of tons of waste after 9 years of operation [12]. According to the complex DAP fertilizer production process, gypsum (CaSO4.2H2O) is a solid waste discharged from nutrient extraction (phosphorus) found in apatite ores. This is a common solid waste, free of hazardous substances. The waste has changed the color of the water, eroded metal and destroyed fauna and floral systems in the surrounding area. Notably, according to the environmental impact assessment, the PG landfill area is supposed to be 10 hectares and the licensed storage time is three years only. However, in reality, the landfill has an area of 13 hectares and the storage time reaches up to five years [12]. Therefore, the large-scale utilization of PG is an urgent environmental problem to be solved in Vietnam.

The aim of this paper is to present some experimental results of SSC using a high volume Dinh Vu phosphogypsum, slag, Portland cement and quicklime. These results are compared with those of ordinary Portland cement to evaluate the feasibility of SSC application in construction in Vietnam.

2. Materials and Experimental methods

2.1 Materials

Ordinary Portland Cement PC40 But Son, Phosphogypsum, Ground-granulated blast-furnace slag (GFS), and quicklime (QL) were used in making super-sulfated cements. The chemical composition of these materials is represented in Table 1.

Table 1. Chemical composition of raw materials

Raw material	Chemical composition, % by weight									
	SiO_2	Fe_2O_3	Al_2O_3	CaO	MgO	Na_2O	K_2O	SO_3	TiO_2	L.O.I
PC	20.3	5.05	3.51	62.81	3.02	-	-	2.0	-	1.83
GFS	34.52	0.66	12.38	41.54	7.25	0.43	0.24	-	-	0.96
PG	8.07	0.87	2.99	30.52	0.69	0.03	0.56	38.5	0.56	21.92
QL	2.21	1.12	0.32	85.04	1.89	-	-	-	0.87	5.19

2.1.1 Phosphogypsum (PG)

In this study, PG gathered from DAP Dinh Vu fertilizer plant, has particle size finer than 11μm. The specific gravity of PG is 2.34 g/cm^3.

2.1.2 Ground-granulated blast-furnace slag (GFS)

The samples of blast-furnace slag were ground for 90 minutes in a laboratory ball mill. Physical properties and chemical composition of finely GFS used in this study are shown in Table 2.

Table 2. Properties of ground-granulated blast-furnace slag used in this study

N°	Physical properties	Unit	Value
1	Specific gravity	g/cm^3	2.94

2	Strength activity index	%	104.3
3	The mean particle size	μm	10.2
4	L.O.I	%	0.96

2.1.3 Portland cement (PC)

Ordinary Portland cement PC40 But Son was used in this study with the physical properties shown in the Table 3.

Table 3. Properties of the PC40 cement used in this study

Blaine fineness cm^2/g	Mean particle size, μm	Consistency, %	Specific gravity, g/cm^3	Compressive strength at 3 days, MPa	Compressive strength at 28 days, MPa
3890	11.0	29.5	3.15	29.8	52.2

2.1.4 Quicklime

The quicklime sieved through 0.14mm sieve was added into the mixture of binder containing GFS to combine with SiO_2 in the slag for the formation of the C-S-H product.

2.1.5 Standard sand

An ISO standard sand comply with Vietnamese standard TCVN 6227:1996 [13] was used for determining compressive strength of binder.

2.2 Experimental methods

All mixtures of the super-sulfated cement used in this study are given in Table 4.

Table 4. SSC mixtures used in the study

N°	Mixture notation	Proportion of raw materials in SSC, % by weight			
		PG	GFS	PC	QL
1	CP1	60	30	10	0
2	CP2	50	40	10	0
3	CP3	40	50	10	0
4	CP4	30	60	10	0
5	CP5	60	30	0	10
6	CP6	50	40	0	10
7	CP7	40	50	0	10
8	CP8	30	60	0	10
9	CP9	60	30	5	5
10	CP10	50	40	5	5
11	CP11	40	50	5	5
12	CP12	30	60	5	5
13	CP13	0	0	100	0

The water demand (or standard consistency), setting time, and compressive strength of SSC were determined according to Vietnamese standard TCVN 6017:2015 [14] and TCVN 6016:2011 [15].

The three SSC pastes (CP4, CP8, CP12) after mixing will be allowed to hydrate in a sealed plastic, then the hydrations were stopped at the age of 3 days by a consecutive soaking in acetone during 48 h. Afterwards, the sample were dried and powdered by using agate mortar for FTIR study. The Fourier transform infrared measurements were recorded with a IR Affinity-1S - SHIMADZU model FTIR spectrometer using the KBr pellets technique.

To determine compressive strength, tested samples were made with a binder combination of (PG, GFS, PC, QL) to standard sand of 1:3 by weight, and the water to binder ratio was fixed at 0.5, by weight. The mortar was cast into 40 mm × 40 mm × 160 mm mold by the vibration method as per TCVN 6016:2011. All samples were demolded after 48 h instead of 24 h specified in the standards due to the long setting time of SSC mortars. Afterwards, these samples were divided into 2 groups:

- Group 1: samples were cured in a sealed plastic bag until testing ages of 3, 7, 28, 90 days
- Group 2: samples were immersed in water until testing ages of 3, 7, 28, 90 days

3. Results and discussion

In this section, the setting time and the compressive strength of SSC using a high volume of PG were studied over through 12 mixtures then compared with the control sample containing 100% Portland cement to demonstrate the effectiveness and feasibility of the SSC application in construction.

3.1 Consistency and setting time

The results of standard consistency tests for SSC containing PG-GFS-PC-QL at different proportions are shown in Figure 1 and Table 5.

Table 5. Standard consistency and setting time of SSC mortars

N°	Mix notation	Standard Consistency, %	Setting time	
			Initial, minute	Final, minute
1	CP1	30.0	2340	2532
2	CP2	30.0	2346	2568
3	CP3	30.5	2352	2598
4	CP4	31.0	2340	2640
5	CP5	30.0	912	1152
6	CP6	30.5	918	1158
7	CP7	30.5	930	1170
8	CP8	31.0	906	1182
9	CP9	30.5	1818	2166
10	CP10	30.5	1824	2190
11	CP11	31.0	1806	2178
12	CP12	31.0	1800	2172
13	CP13	29.5	150	250

From the above results, the overall standard consistency for super-sulfated slag cement pastes are higher than that of the control paste containing 100% cement. This can be caused by the fact that the fineness of binder powder used in 12 mixtures was higher than that of cement in the control sample (CP13).

It is also observed that at the same amount of Portland cement PC or QL, the water requirement does not change significantly among these 12 SSC mixtures from CP1 – CP12 due to the surface area of GFS, which is almost similar to that of PG.

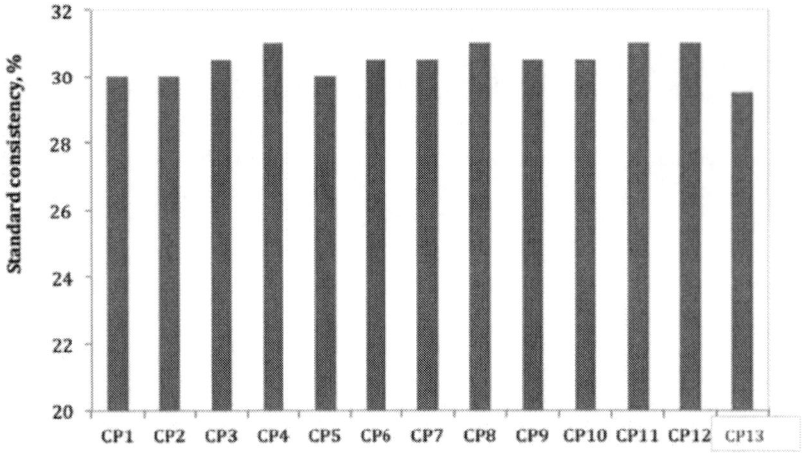

Figure 1 Standard consistency of the SSC pastes using PG

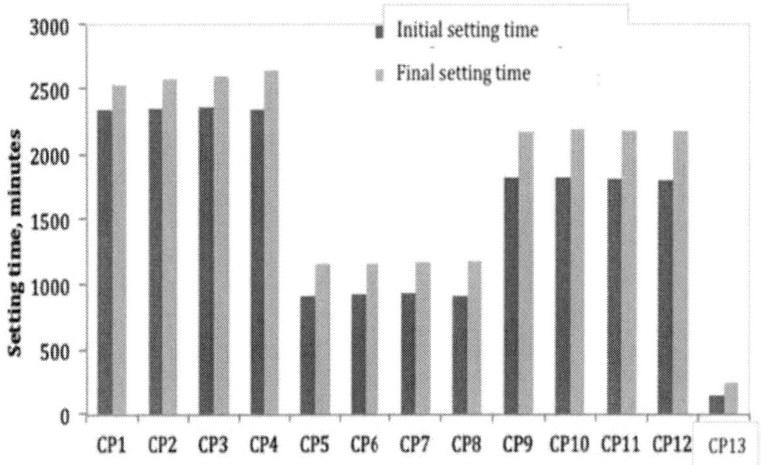

Figure 2. Setting time of the SSC pastes and the control mixture

The results in the Table 5 and Figure 2 show that the setting time of the SSC pastes is much longer than that of the control mixture (CP13 containing 100% cement). The initial and final setting time of the control mixture is 150 minutes and about 250 minutes, respectively, but the setting time of the SSC mixtures lasted longer, even up to 1800-2600 minutes. The observed retardation in setting times can be mainly due to the combined effect of a lower cement content. The results also show that the setting time of mixtures seem not to be influenced by the addition of PG or GFS.

When incorporating quicklime, the setting time of the SSC mixture is significantly decreased. The setting time of SSC mixtures using 10% quicklime is shorter than that of mixtures using 5%QL, 5% PC or 0% QL, 10% PC. This can be explained that the addition of quicklime enhances the hydration process and shortens the setting time. It is possible that the higher amount of $Ca(OH)_2$, C–S–H and ettringite formed in the mixture containing 10% quicklime, promoted a denser matrix and shortened the setting time.

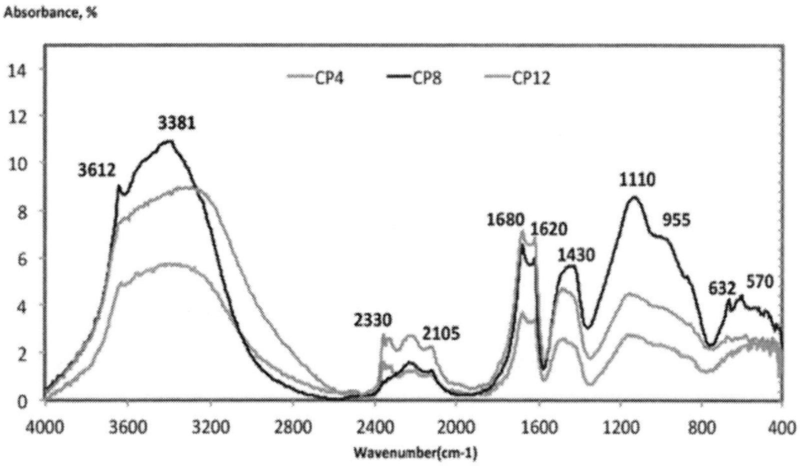

Figure 3. FTIR spectra of SSC pastes CP4, CP8 and CP12 at 3 days after mixing

The FTIR spectrum of the 3 preventative SSC samples on the Figure 3 shows a sharp band at 3630 cm^{-1} associated to O-H stretching vibrations of portlandite ($Ca(OH)_2$) and the area from 2800 to 4000 cm cm^{-1} is correspond to the bond water in the 3 compositions. The peak at 1110 cm^{-1} is due to ν3 modes of SO_4^{2-} and the week bands at 632 cm^{-1} and is due to ν4 modes of SO_4^{2-}. The strong asymmetric stretching Si-O band (ν3) is shifted to high frequencies centered at 955 cm^{-1} indicates that the formation of C-S-H [16]. Up to 3 days of hydration, the intense sulfate band (SO_4^{2-}) in ettringite at 1100-1160 cm^{-1} are observed in the 3 samples but more distinct for the CP8 sample when compared to the CP4 and CP12 samples at early age and it implies that quicklime enhance the AFt formation than Portland cement. This indicates that the initial reactions are faster due to the higher amount of ettringite present in the CP8 sample.

It is also observed that the variation in intensities of the band at 955 cm^{-1} and 570 cm^{-1} of are highest for the CP8 sample and occur from early age onwards; But in the other hand, the intensity of the band observed at 955 cm^{-1}, 1100 cm^{-1} of CP12 are less when compared to the CP4 sample. The above results explain the fact that the sample containing more QL will set quicker and improve the strength at early stage due to the activation of lime to the SSC hydration acceleration.

3.2 Compressive strength of SSC

The results of the compressive strength of 12 SSC samples and the control sample using 100% cement are shown in Table 6.

Table 6. Compressive strength of SSC samples at 3, 7, 28, 90 days

N°	Mix notation	Compressive strength of SSC samples in different curing conditions with time, **MPa**							
		At 3 days		At 7 days		At 28 days		At 90 days	
		Endogen	Water	Endogen	Water	Endogen	Water	Endogen	Water
1	CP1	2	2	22	18	32	33	43	38
2	CP2	5	5	30	29	46	38	45	39
3	CP3	5	5	31	32	45	40	60	55
4	CP4	8	8	35	32	51	50	69	55
5	CP5	3	3	14	11	21	21	24	22
6	CP6	5	5	18	15	25	26	26	24

7	CP7	6	6	20	20	29	31	35	32
8	CP8	10	10	24	25	39	32	45	39
9	CP9	4	4	14	14	28	23	35	29
10	CP10	7	7	18	16	33	28	50	43
11	CP11	9	9	19	19	38	34	53	47
12	CP12	8	8	27	28	43	41	51	49
13	CP13	28	28	34	41	50	48	55	53

Based on the results in the Table 6, the influence of raw materials content on the compressive strength of SSC samples will be analyzed in detail.

3.2.1 Effect of phosphogypsum content on compressive strength of the SSC

Based on the compressive strength results from 3 to 90 days under 2 curing conditions, the influence of phosphogypsum waste content on compressive strength of SSC samples is shown in Figure 4 and Figure 5.

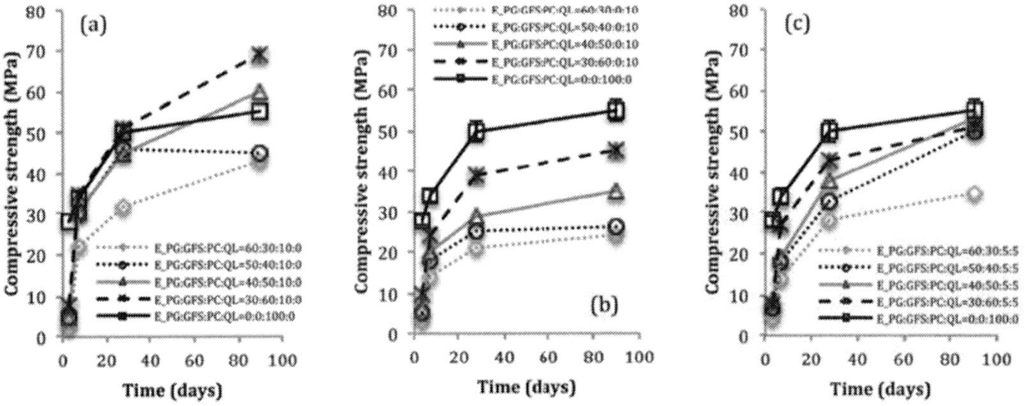

Figure 4. Compressive strength of SSC samples using (a) 10% cement, (b) 10% QL, (c) 5% QL and 5% cement under endogenous condition with time

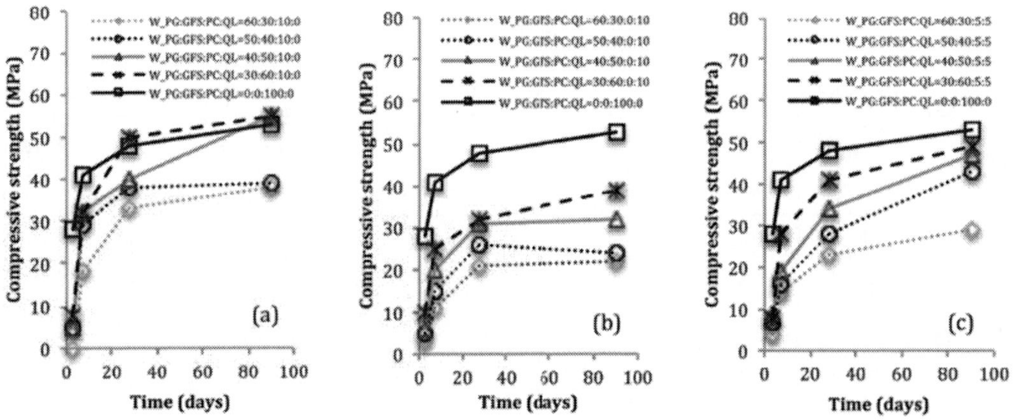

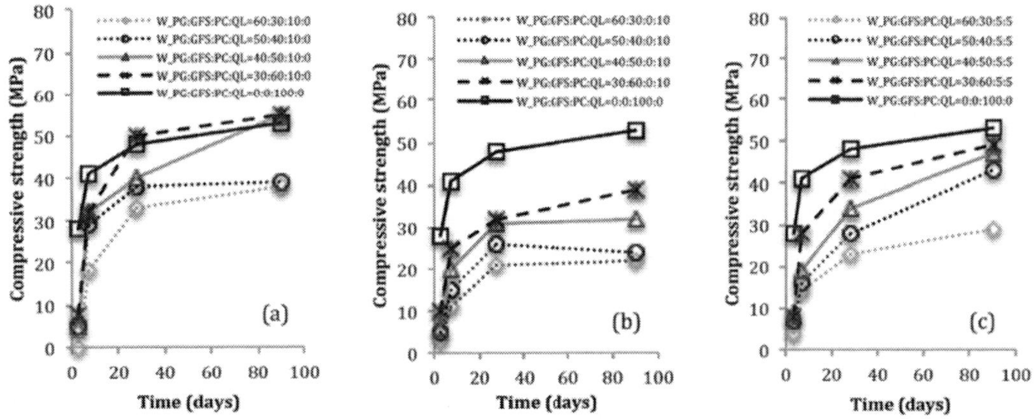

Figure 5. Compressive strength of SSC samples using (a) 10% cement, (b) 10% QL, (c) 5% QL and 5% cement under endogenous condition with time

The results in Figure 4 and Figure 5 show that the compressive strength development of all SSC samples is very low at early ages. However, the long-term compressive strength increases significantly due to the accelerated hydration process. In fact, the strength development is related to the formation of C-S-H caused by the possibility to activating GFS of cement, lime and phosphogypsum. The compressive strength of SSC samples increase rapidly from 3 to 28 days, but becomes slow down after that, i.e. from 28 to 90 days.

In addition, the results also indicate that when the amount of cement or lime in binder is kept constant, the compressive strength of SSC samples increases with the decreasing the ratio of PG or with the increasing amount of GFS in binders. This can be explained by the fact that F possesses hydraulic properties, thus, this can be activated in the presence of activators such as cement or lime in binder, whereas PG basically acts as an inert filler, only a small portion of sulfate ions are involved in the slag activation process. However, it can be seen that the compressive strength of the CP4 sample (PG: GFS: PG : PC: QL = 30: 60 : 10: 0) attains the highest value, even greater than that of the control sample containing 100% cement. The CP6 sample (PG: GFS: PC: QL = 60: 30: 0: 10) has the lowest compressive strength value under both endogenous condition and water immersion condition.

3.2.2 Compressive strength of SSC samples under different curing conditions

Due to the high amount of PG in the SSC, it is necessary to evaluate the water resistance of this type of binder to ensure its practical application. As mentioned in the research methodology, the mortar samples after demolding will be cured in a sealed plastic bag or immersed in water until the testing age for the compression test. The results are shown in Figure 6.

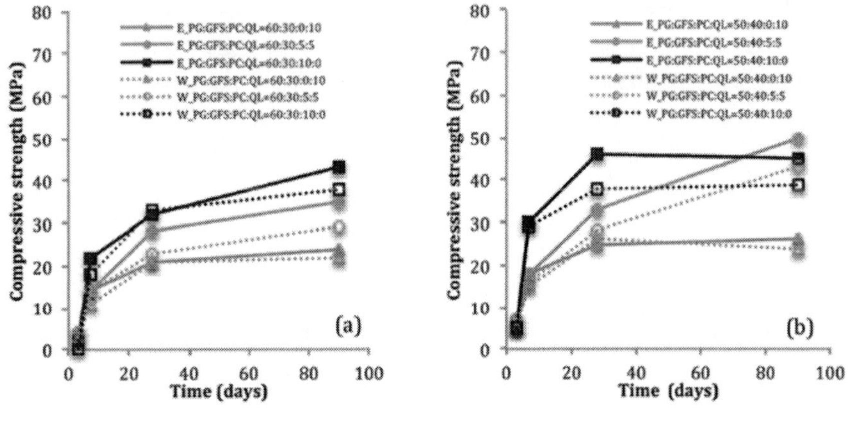

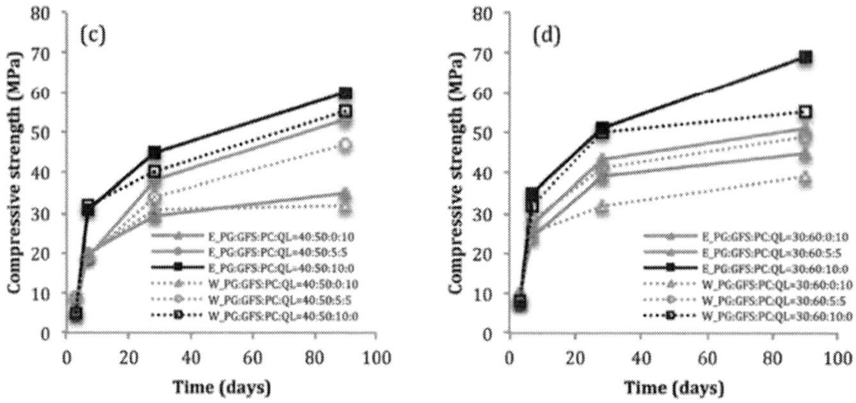

Figure 6. Compressive strength of SSC samples using different ratios of PC and QL; PG&GFS were fixed at (a) 60&30%, (b) 50&40%, (c) 40&50%, (d) 30&60% by weight of binder, respectively, under different curing conditions

The results show that the compressive strength of samples cured in the water is slightly lower than that of samples cured under endogenous condition. This difference may be caused by the fact that the binder samples cured in water are always under saturated condition, resulting in reducing the compressive strength of the sample in comparable with the samples under the endogenous condition. It also can be seen that there is no decline of the compressive strength of samples cured in the water over time. This is very important to prove the water resistance of this binder.

3.2.3 Influence of the cement and lime content on compressive strength of the SSC

Compressive strength of SSC samples from 3 to 90 days when using the same amount of PG and GFS but different amounts of PC and QL is demonstrated in Figure 7 and Figure 8.

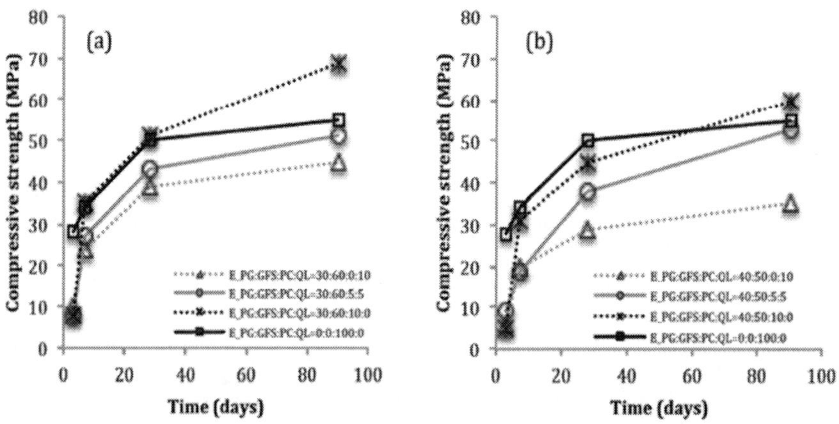

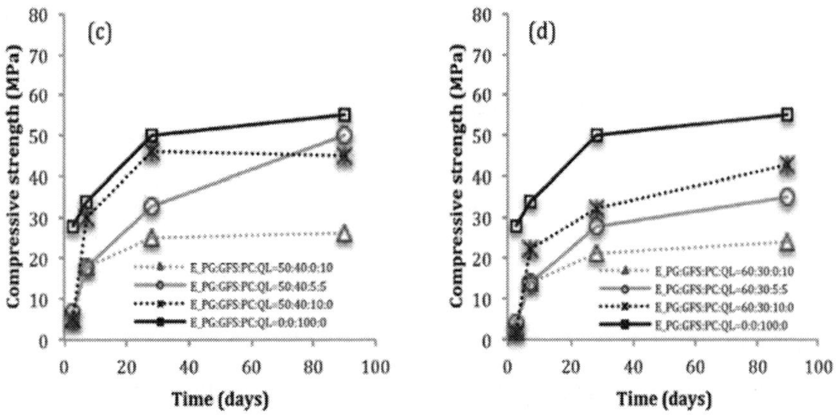

Figure 7. Compressive strength of SSC samples using diferent ratios of PC and QL with time under endogenous condition, PG&GFS were fixed at (a) 30&60%, (b) 40&50%, (c) 50&40%, (d) 60&30% by weight of binder, respectively

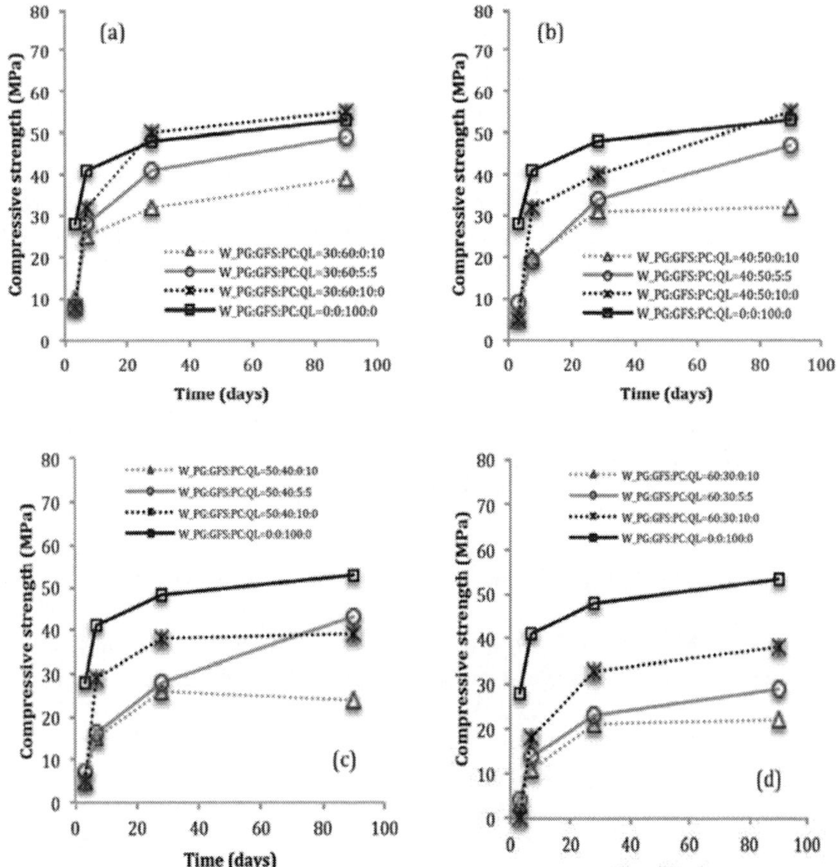

Figure 8. Compressive strength of SSC samples using diferent ratios of PC and QL with time under immersion condition, PG&GFS were fixed at (a) 30&60%, (b) 40&50%, (c) 50&40%, (d) 60&30% by weight of binder, respectively

The effect of the (PC and QL) content on compressive strength of SSC binders shows that the development of compressive strength from 3 days to 28 days is also very significant; and then becomes slow down. The compressive strength of sample using 10% PC is higher than that of mortar sample using 5% PC, 5% or 10% QL. This can be explained by the fact that calcium silicate hydrate (C-S-H) is the main product at early ages of the hydration of Portland cement and is primarily responsible for the strength of binder, while PC containing $Ca(OH)_2$ also can react with GFS to produce C-S-H but with a smaller amount of product, and this reaction occurs slowly at early ages. In addition, the $Ca(OH)_2$ formed by the hydration of cement also can react with GFS to generate more C-S-H, which refines the pore structure of SSC. Therefore, the compressive strength of SSC increases with increasing of the PC content used in binder.

4. Conclusion

In this study, the setting time and development of compressive strength of SSC using a high volume phosphogypsum have been studied. These experimental results prove a high feasibility of research into PG applications in the manufacture of a new binder as well as other non-fired building materials, especially in the context that our country is looking for developing sustainable construction. Some conclusions can be drawn as follows:

- The setting time of the SSC paste is much longer than that of the control mixture containing only Portland cement; this disadvantage can be compensated by using quicklime due to the higher amount of $Ca(OH)_2$, C–S–H and ettringite formed and promoted a denser matrix and shortened the setting time.
- Compressive strength of SSC is influenced by the amount of PG used in binder; the strength increases with decreasing of the PG content or with increasing of the GFS content in binder.
- All SSC samples are stable in water and the compressive strength of SSC using only 10% cement is higher than that of the control sample using only cement. The compressive strength of the SSC samples of 45-50 MPa, and 69 MPa can be obtained at 28 days and at 90 days, respectively. This value even exceeds the compressive strength of the PC40 cement.

References

[1] Matschei T, Bellmann F and Stark J 2005 Hydration Behaviour of Sulphate - Activated Slag Cements *Advances in Cement Research* 17(4):167-178.
[2] Kuryatnyk T, Angulski da Luz C, Ambroise J and Pera J 2008 Valorization of Phosphogypsum as Hydraulic Binder *J.Haz.Mat.* 160(2): 681–687.
[3] Ding S, Shui Z, Chen W, Lu Ji and Tian S 2014 *Properties of Super-sulfated Phosphogysum - slag Cement (SSC)* Concrete J.Wuh.Univ.Tech-Mat.Sci.Ed 29(1):109–113.
[4] Mun K J, Hyoung W K and Lee C W 2007 Basic Properties of Non- Sintering Cement Using Phosphogypsum and Waste Lime as Activator *Construction and Building Materials* 21(6):1342- 1350.
[5] Taneja C A, Singh M, Tehri S P and Raj T 1977 *12th Int Conf on silicate industry and silicate science Super sulphated cement from waste phosphogypsum* (Budapest - Hungary) pp.821-827.
[6] Mulla M S, Karjinni V V, Gundakalle V D and Thakare R B 2003 Int. Workshop and Conference on Construction and Materials *Phosphogypsum applications in cement concrete* (India) pp 677-687.
[7] Weiguo S, Mingkai Z, Qinglin Z 2007 Study on Lime–Fly Ash–Phosphogypsum Binder *Construction and Building Materials* 21(7): pp 1480-1485.
[8] Erdem E and Ölmez H 1993 The Mechanical Properties of Super- sulphated Cement Containing Phosphogypsum *Cement and Concrete Research* 23(1): pp 115-121.
[9] Stark J, Frohburg U and Mielke I 2001 Supersulfated cement with and without cement clinker *Proc of the Conf of Non-Traditional Cement and Concrete* (Brno - Czech Republic).
[10] Kumar S A 2002 A perspective study on fly ash-lime-gypsum bricks and hollow blocks for low cost housing development *Construction and Building Materials* 16(8): pp 519-525.
[11] Mohammad M S, Rami H H and Ahmad M A 1999 Potential use of phosphogypsum in concrete

Cement and Concrete Research 9(29): p.1419–1425.
[12] Vietnam Institute for Building Materials Report 2017 *Evaluation of the use of phosphogypsum as building materials*.
[13] Vietnamese standard TCVN 6227:1996 ISO *standard sand for determination of cement strength*.
[14] Vietnamese standard TCVN 6017:2015 *Cements - Test methods - Determination of setting time and soundness*.
[15] Vietnamese standard TCVN 6016:2011 *Cements - Test methods - Determination of strength*.
[16] Rikard Y, Jaglid U, Britt Marie S and Itai P 2009 Early hydration and setting of Portland cement monitored by IR, SEM and Vicat techniques *Cement and Concrete Research* 39(5):pp 433-439.

A new design concept of fully grouted rock bolts in underground construction

Quang Phich Nguyen[1,2,*], Van Manh Nguyen[3] and Ke Tuong Nguyen[4]

[1] Binh Duong University, Binh Duong, Vietnam
[2] Vietnamese Tunneling Association, Hanoi, Vietnam
[3] Hanoi University of Mining and Geology, Hanoi, Vietnam
[4] Faculty of Civil Engineering, Ton Duc Thang University, Ho Chi Minh City, Vietnam

[*] Corresponding author, Email: nqphichhumg@gmail.com

Abstract. The main problem after excavating an underground excavation is to maintain the stability of the excavation for a certain period of time. Failure in meeting this demand is a threat to safety of men and equipment. Support and reinforcement are different instruments with different mechanisms. Among the common support systems in tunnelling and mining, rock bolts have been widely used to reinforce rock mass and also to reduce geological hazards. Furthermore rock bolts can be applied under varying different geological conditions with cost-effectiveness. Although different methods are developed for grouted rock bolts design until now, the interaction mechanism of the rock bolts and rock mass is still very complicated issue. The paper addresses an analytical model for the analysis and design of fully grouted rock bolts based on the reinforcement principle. According to this concept the jointed rock mass reinforced by grouted rock bolts is considered as composite material which includes rock mass, the grout material and the bolt shank. The mechanical properties of this composite material depend on the ratio of the components. The closed-form solution was developed based on the assumption that the rock mass arround a circular tunnel remained elastic after installing fully grouted rock bolts. The main parameters of the rock-bolt system (the diameter and length of bolt shank, the space between the bolts) are then easily estimated from the obtained solution.

1. Introduction

Rock bolts are nowaday an effective and widely used support measure and an element of reinforcement in civil engineering and mining. Furthermore, rock bolts can be applied under very different geological and technical conditions to stabilize the rock mass. Rock bolts are cost effective because of their simple composition and their relatively low labour and energy consumption.

Although a great number of empirical, experimental, analytical and numerical methods have been developed to simulate bolting effects and at the ende to design rock bolts as a support means in underground excavations, informed and disussed in many publicated article, the interaction mechanisms of the rock bolts and the rock mass stay still a very complicated issue and are not well understood. There isn't a concrete theoretical concept for determining all the parameters of a rock bolts system. Design of systematic rockbolt reinforcement in tunnelling is normally done by application of empirical and rational approaches.

Several analytical models have been developed to study the interaction behavior of bolts and rock mass, and to evaluate the effect of the bolts on stress and strain behavior of the rock around the underground opening [1-17]. Samit and Anand [18] proposed an analytical model using beam-column theory to analyze the reinforcement by the rockbolts. They found that the critical buckling load of the rock beam is influenced by the rock modulus. Another analytical model has been suggested by [5] in oder to describe the reinforcement effect of rockbolts. Indraratna and Kaiser [3] have also been developed an analytical model using convergence - confinement method to analyze the effect of the bolt pattern on the extent of the yield zone and tunnel deformation. Carranza-Torres [15] has been proposed an analytical model of rockbolt reinforcement around tunnels and found that reinforcement can have a significant mechanical effect: increasing the confinement and decreasing the convergences. A simple analytical method was proposed by [12] for the analysis of the load on a single rockbolt. Those methods are based on rock-support interaction theory or convergence-confinement approach or based on composite material concept.

The analytical solutions for the composite element of the rock bolts and rock mass have been found in elsewhere [3], [9], [12], [19-20]. However, such analytical models did not consider the material components of rock bolt include the rock mass, steel bar and grout in detail.

The paper presents at first a simple analytical method for design of rock bolts based on the reinforcement principle and proposes a procedure for design of rock bolts system.

2. Analytical solution

2.1. Simplify model and concept

The problem to be solved is a circular tunnel of radius R excavated at a depth H in a homogeneous, isotropic and initially elastic rock mass. The tunnel is subjected to a hydrostatic stress field $\sigma_0 = \gamma H$, where γ is the ground unit weight. The rock mass in primary state is elastic and the mechanical properties are described by the Young modulus E_{rm}, Poisson's ratio ν_{rm} and axial compressive strength σ_{crm}. After excavation a plastic zone would be built around the tunnel perimeter, at a certain distance behind the face if no support is installed. So a grouted rock bolts system is to be installed just in time in oder to keep the rock mass elastic based on the assumption that the rock bolts system could improve the strength and decrease the deformability of the supported rock mass as a composite material. There are two zones that exist around the tunnel now, a zone of elastic rock composite and a zone of elastic rock mass (Figure 1). The rock composite has then the Young modulus $E_{rm\text{-}wb}$, the Poisson'ratio $\nu_{rm\text{-}wb}$ and axial compressive strength $\sigma_{c\text{-}wb}$. For simplification it's asumed that $\nu_{rm} = \nu_{rm\text{-}wb} = 0.5$.

To evaluate the stability of the rock mass before excavation a stability factor (n_o) is used in simple way based on the elastic solution for stress and displacement distribution. The stability factor (n_o) is defined as the ratio between axial compressive strength of rock mass σ_{crm} and the maximal tangetial stress on the perimeter $\sigma_\theta = 2\sigma_0$ from the well-known solution as following:

$$n_0 = \frac{\sigma_{crm}}{2\sigma_0} \tag{1}$$

The rock mass is stable, if $n_0 \geq 1$ and is not stable if $n_0 < 1$ and needs to support, in these case for example with grouted rock bolts.

The grouted bolts with the length of L are placed along the radial direction around the tunnel. The mechanical properties of the rock mass with bolts are characterized by elastic modulus $E_{r\text{-}wb}$, Poisson's ratio ν and axial compressive strength $\sigma_{c\text{-}wb}$ and are calculated from the mechanical parameters of the rock mass, the grouted material, the bolt rank and the geometrical parameters of the borehole, the steel bar and the opening based on the homogenization theory at any point as a funtion of r as follows [6]:

$$\frac{E_{rm-wb}}{E_{rm}} = \frac{\sigma_{c-wb}}{\sigma_c} = \alpha = 1 + \frac{\pi}{4} \frac{d_{bh}^2(E_{grt} - E_{rm}) + d_{sb}^2(E_{sb} - E_{grt})}{a^2 E_{rm}} \frac{R}{r} = 1 + A_0 \frac{R}{r} \tag{2}$$

with
$$A_0 = \frac{\pi}{4}\frac{d_{bh}^2(E_{grt} - E_{rm}) + d_{sb}^2(E_{sb} - E_{grt})}{a^2 E_{rm}} \quad (3)$$

Where: d_{bh}, d_{sb}, E_{grt}, E_{sb}, a and α are the diameter of the borehole, the diameter of the steel bar, the elastic modulus of the grout, the elastic modulus of steel bar, the distance between the rock-bolts and the reinforcement factor respectively.

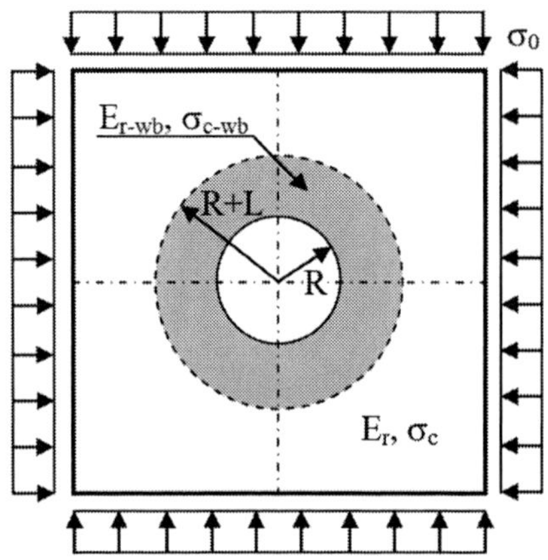

Figure 1. The axisymmetric tunnel problem

At this point it is to be mentioned that the concept of strengthening the rock masses by rock bolts as well as the use of homogenised model is not new. There are already some suggestions based on theoretical reasoning [3], [12], [18], [20] or experiment results [21-23], however there is no relationship between the reinforcement factor with the rock bolts parameter as in the formula (2).

The tunnel has then a support with uniform bolts on the perimeter of the tunnel, and it is subjected to far-field stresses $\sigma_0 = \gamma H$. The rock bolts system is to design so that the supported rock mass will be stable and remain elastic.

Solving this boundary problem, we obtaine the stress field distribution in the rock mass zone with bolts surrounding the tunnel in polar coordinates as follows:

$$\sigma_{r-wb} = 2\sigma_0 \frac{1}{B}\left[-\frac{1}{2r^2} - \frac{A}{3r^3} + \frac{1}{2R^2} + \frac{A}{3R^3}\right] \quad (4)$$

$$\sigma_{\theta-wb} = 2\sigma_0 \frac{1}{B}\left[\frac{1}{2r^2} + \frac{2A}{3r^3} + \frac{1}{2R^2} + \frac{A}{3R^3}\right] \quad (5)$$

$$\sigma_{r-w0b} = \sigma_0 - \left[\sigma_0 - \sigma_0\left(1 - \frac{1}{(R+L)^2}\frac{1}{B}\right)\right]\left(\frac{R+L}{r}\right)^2 \quad (6)$$

$$\sigma_{\theta-w0b} = \sigma_0 + \left[\sigma_0 - \sigma_0\left(1 - \frac{1}{(R+L)^2}\frac{1}{B}\right)\right]\left(\frac{R+L}{r}\right)^2 \quad (7)$$

Where: σ_{r-wb} is the radial stresses in rock mass with bolt-grout support; $\sigma_{\theta-wb}$ is the tangential stresses in rock mass with bolt-grout support; σ_{r-w0b} is the radial stresses in rock mass without bolt-grout support; $\sigma_{\theta-w0b}$ is the tangential stresses in rock mass without bolt-grout support.

$$A = \frac{\pi}{4}\frac{d_{bh}^2(E_{grt} - E_{rm}) + d_{sb}^2(E_{sb} - E_{grt})}{a^2 E_{rm}} R = A_0 R \quad (8)$$

$$B=\frac{1}{R^2}+\frac{2}{3}A\frac{(R+L)^3-R^3}{R^3(R+L)^3} \tag{9}$$

The tangential stress at the perimeter of the tunnel is determined by substitution r = R into equation (5):

$$\sigma_{\theta-wb-R}=2\sigma_0\frac{1+A_0}{1+\frac{2}{3}A_0\frac{(R+L)^3-R^3}{(R+L)^3}}=2\sigma_0\frac{(1+A_0)(R+L)^3}{\left(1+\frac{2}{3}A_0\right)(R+L)^3-\frac{2}{3}A_0R^3} \tag{10}$$

The normal displacement at the perimeter of the tunnel can be estimated by the following equations:

$$u_{n-wb-R}=\frac{3\sigma_0}{2E_{rm}}\frac{R}{1+\frac{2}{3}A_0\left(1-\frac{R^3}{(R+L)^3}\right)}=u_0\frac{1}{1+\frac{2}{3}A_0\left(1-\frac{R^3}{(R+L)^3}\right)} \tag{11}$$

Where: u_0 is the elastic radial displacement of the tunnel perimeter after excavation

The compressive strength of the rock mass after bolting is determined from equation (2) as follows:

$$\sigma_{cm-wb-R}=(1+A_0)\sigma_{cm} \tag{12}$$

The tangential stress and normal displacements at the perimeter of the tunnel before bolting can be expressed by using elastic solution as follows:

$$\sigma_{\theta 0}=2\sigma_0 \tag{13}$$

$$u_0=\frac{3\sigma_0}{2E_{rm}}R \tag{14}$$

Comparison of the stress and displacement in the rock mass with bolts and without bolts, it is shown that:

$$\sigma_{\theta-wb-R}>\sigma_{\theta 0}\,;\ u_{n-wb-R}<u_0 \tag{15}$$

This demonstrates that the displacement of the rock mass surrounding tunnel after bolting is smaller than rock mass before bolting. But the stress is greater. Therefore, the stability of the rock mass after bolting needs to be estimated carefully by calculating and adjusting the bolt parameters.

Assuming that n_0 and n are the safety factor in terms of the rock mass before and after bolting on the perimeter of the tunnel respectively, defined as follows:

$$n_0=\frac{\sigma_{cm}}{\sigma_{\theta 0}}=\frac{\sigma_{cm}}{2\sigma_0} \tag{16}$$

$$n=\frac{\sigma_{cm-wb-R}}{\sigma_{\theta-wb-R}} \tag{17}$$

If $n_0 < 1$, the rock mass surrounding tunnel is considered instable and needs to be bolted to maintain stable with $n > 1$. Hence, if the safety factor n for the rock mass is given, it is clearly that the bolt parameters (length, density, diameter of bolt, diameter of borehole, etc.) can be easily determined in relation of the mechanical properties of the rock mass and bolts.

After substituting the expressions (10) and (12) into equation (17), the safety factor of rock mass on the perimeter of the tunnel after bolting can be presented as follows:

$$n=\frac{\sigma_{cm-wb-R}}{\sigma_{\theta-wb-R}}=\frac{(1+A_0)\sigma_{cm}}{\sigma_{\theta-wb-R}}=(1+A_0)\frac{\sigma_{cm}}{2\sigma_0}\frac{\left(1+\frac{2}{3}A_0\right)(R+L)^3-\frac{2}{3}A_0R^3}{(1+A_0)(R+L)^3}=\left(1+\frac{2}{3}A_0\right)n_0-\frac{2}{3}A_0n_0\frac{R^3}{(R+L)^3} \tag{18}$$

IOP Conf. Series: Earth and Environmental Science **143** (2018) 012017 doi:10.1088/1755-1315/143/1/012017

$$\frac{(R+L)^3}{R^3} = \frac{\frac{2}{3}A_0 n_0}{\left(1+\frac{2}{3}A_0\right)n_0 - n}$$

$\Rightarrow$ (19)

The length of bolt can be expressed from equation (19) as follows:

$$L = R\left(\sqrt[3]{\frac{\frac{2}{3}A_0 n_0}{\left(1+\frac{2}{3}A_0\right)n_0 - n}} - 1\right) \tag{20}$$

Because the length of the bolt must be greater than 0 (L > 0), so that means:

$$\left(1+\frac{2}{3}A_0\right)n_0 - n > 0 \tag{21}$$

$$A_0 > \frac{3}{2}\left(\frac{n}{n_0} - 1\right)$$

$\Rightarrow$ (22)

Combination of equation (3) and (21) leads to determine the distance between the bolts at the perimeter of tunnel based on the diameter of bolt, diameter of borehole, grout properties as follows:

$$a < \sqrt{\frac{\pi}{4}\frac{d_{bh}^2\left(E_{grt} - E_{rm}\right) + d_{sb}^2\left(E_{sb} - E_{grt}\right)}{\frac{3}{2}\left(\frac{n}{n_0} - 1\right)E_{rm}}} \tag{23}$$

However, the distance between the bolts can be estimated also based on experience. It can be seen from (21) that the safety factor n of the rock mass surrounding the tunnel after bolting can be reached maximum value as following expression:

$$n_{max} = \left(1+\frac{2}{3}A_0\right)n_0 \tag{24}$$

That means the safety factor n of the rock mass after bolting is varied in the range of

$$1 < n < n_{max} \tag{25}$$

2.2. Calculation example

The case analyzed corresponds to a deep circular tunnel with R = 4.0 m, excavated at a depth H=300m in a homogeneous rock mass. The rock mas has following elastic properties: axial compressive strength, σ_{cr} = 60 MPa; unit weight, γ = 27 kN/m^3; rock mass rating index, RMR = 72; elastic modulus, E_r = 0.5GPa. The support of the tunnel is provided by rock bolts uniformly distributed along the perimeter of the tunnel. The properties of the rock bolts are: elastic modulus of steel bar, E_{sb} = 210 GPa; elastic modulus of grout, E_{grt} = 30 GPa; diameter of steel bar, d_{sb} = 25 mm; diameter of borehole, d_{bh} = 42 mm. The strength of the rock mass can be estimated according to Hoek-Brown criterion [1] as follows:

$$\sigma_{cm} = \sigma_{cr}\sqrt{s} \tag{26}$$

Where $s = \exp\left(\frac{RMR - 100}{9}\right) = \exp\left(\frac{72 - 100}{9}\right) = 0.0466$

$$\Rightarrow \sigma_{cm} = 60\sqrt{0.0466} = 12.96MPa$$

Assuming that the tunnel subjected to a hydrostatic stress field, the normal stress at the perimeter of excavated tunnel is calculated as follows:

$$s = \exp\left(\frac{RMR - 100}{9}\right) = \exp\left(\frac{72 - 100}{9}\right) = 0.0466$$

The safety factor of the rock mass before bolting is given as follows:

$$n_0 = \frac{\sigma_{cm}}{\sigma_{\theta 0}} = \frac{12.96}{16.2} = 0.8$$

Substitute the given parameters into equation (23), it is obtained the distance between bolts at the perimeter of the tunnel: $a < 0.829$ m

Using the equations (3) and (24) to determine the values of A_0 and n_{max} with the distance between bolts are 0.8 m, 0.7 m and 0.6 m. The safety factor of the rock mass requirement after bolting is n = 1.0. The length of rock bolts can be estimated then by equation (20). Table 1 lists the calculated results of rock bolts length with safety factor and distance between rock bolts are given.

Table 1. The results for the length of rock bolts

a (m)	A_0	n_{max}	n	L (m)
0.6	0.717	1.18	1	0.5
0.7	0.527	1.08	1	1.1
0.8	0.403	1.01	1	2.9

The calculated rock bolts parameters indicated that it is quite consistent with the reality conditions and other methods.

3. Conclusion

The rock bolts are widely used as supporting system in mining and tunnelling. Rockbolt systems are designed until now by application of various empirical and rational approaches. An analytical model and its theoretical solutions are presented in this paper, and a procedure for design of fully grouted rockbolt system based on the reinforcement principle is developed.

A significant difference to other methods is that the calculation proceeds with a desired or predetermined safety factor and the strengthed rock mass remain elastic after installation of the rock bolts. And that is also the limit of the method in application.

The calculated rock bolts parameters based on this method indicated that it is quite consistent with the reality conditions and other methods. Therefore, it is completely possible to apply the proposed model for designing rock bolts system.

4. References

[1] Hoek E and Brown E T 1980 *Undreground Excavation in Rock* (Institution of Mining and Metallurgy, London).
[2] Bieniawski Z T 1989 *Engineering rock mass classifications* (Wiley, New York).
[3] Indraratna B and Kaiser P K 1990 Analytical model for the design of grouted rock bolts *Int. J. Numer. Anal. Meth. Geomech.* **14** 227.
[4] Indraratna B and Kaiser P K 1990 Design for Grouted Rock Bolts Based on the Convergence Control Method *Int. J. Rock Mech. Min. Sci. & Geomech.* **27** 269.
[5] Li B and Stillborg B 1999 Analytical models for rock bolts *Int. J. Rock Mech. Min. Sci.* **36** 1013.
[6] Nguyen Q P 2001 *Rock reinforcement* (Hanoi University of Mining and Geology, in Vietnamese).
[7] Nguyen Q P 2011. *Investigation on development and completion of models for calculation and design of fully grouted rock bolts* (Research report Code number: B2009-02-76TĐ in Vietnamese).
[8] Oreste P P 2003 Analysis of structural interaction in tunnels using the convergence-confinement approach *Tunnelling and Underground Space Technology* **18** 347.

[9] Cai Y, Esaki T and Jiang Y 2004a An analytical model to predict axial load in grouted rock bolts for soft rock tunnelling *Tunnelling and Underground Space Technology* **19** 607.

[10] Cai Y, Esaki T and Jiang Y 2004b A rock bolt and rock mass interaction model *Int. J. Rock Mech. Min. Sci.* **41** 1055.

[11] Fahimifar A and Soroush H 2005 A theoretical approach for analysis of the interaction between grouted rockbolts and rock masses *Tunnelling and Underground Space Technology* **20** 333.

[12] Bobet A 2006 A simple method for analysis of point anchored rockbolts in circular tunnels in elastic ground *Rock mechanics and Rock engineering* **39** 315.

[13] Osgoui R R and Ünal E 2009 An empirical method for design of grouted bolts in rock tunnels based on the Geological Strength Index (GSI) *Engineering Geology* **107** 154.

[14] Osgoui R R and Oreste P 2010 Elasto-plastic analytical model for the design of grouted bolts in a Hoek-Brown medium *Int. J. Rock Mech. Min. Sci. & Geomech.* **34** 1651.

[15] Carranza-Torres C 2009 Analytical and numerical study of the mechanics of rockbolt reinforcement around tunnels in rock masses *Rock mechanics and Rock engineering* **42** 175.

[16] Martín LB, Tijani M, Hadj-Hassen F and Noiret A 2013 Assessment of the bolt-grout interface behaviour of fully grouted rockbolts from laboratory experiments under axial loads *Int. J. Rock Mech. Min. Sci.* **63** 50.

[17] Cao C, Nemcik J, Aziz N and Ren T 2013 Analytical study of steel bolt profile and its influence on bolt load transfer *Int. J. Rock Mech. Min. Sci.* **60** 188.

[18] Samit R and Anand B R 1996 Analysis of rockbolt reinforcement using beam-column theory *International Journal for Numerical and Analytical Methods in Geomechanics*" **34** 241.

[19] Bernaud D, Buhan P D and Maghous S 1995 Numerical simulation of the convergence of a bolt-supported tunnel through a homogenization method *Int. J. Numer. Anal. Meth. Geomech.* **19** 267.

[20] Buhan P, Bourgeois E and Hassen G 2008 Numerical simulation of bolt-supported tunnel by means of a multiphase model conceived as an improved homogenization procedure *Int. J. Numer. Anal. Meth. Geomech.* **32** 1597.

[21] Wullschläger D and Natau O 1987 The bolted rock mass as an anisotropic continuum. Material behaviour and design suggestions for rock cavities *Proceedings of the 6th International Congress on Rock Mechanics* (Montreal) p 1321.

[22] Wullschläger D 1988 Ein Verbundwerkstoffmodell für die Systemankerung im Tunnelbau *Dissertation* (Universität Karlsruhe).

[23] Sakurai S 2010 Modeling strategy for jointed rock masses reinforced by rock bolts in tunneling practice *Acta Geotechnica* **5** 121.

IOP Conf. Series: Earth and Environmental Science **143** (2018) 012018 doi:10.1088/1755-1315/143/1/012018

A study on a instability slope in Taiwan subjected to rainfalls

D H Hsiao[1*], C S Hsieh[1], L C Yeh[1], D Y Lin[1] and V T-A Phan[2]

[1]Department of Civil Engineering, National Kaohsiung University of Applied Sciences, 415 Chien Kung Road, Kaohsiung 80778, Taiwan R.O.C.
[2]Faculty of Civil Engineering, Ton Duc Thang University, Ho Chi Minh City, Vietnam

Corresponding author, E-mail address: hsiaodh@kuas.edu.tw

Abstract. After the long-term monitoring on the Chaishan area in Taiwan from 2005 to 2012 by Kaohsiung City Government, the obtained results showed that annual lateral displacements in the region are about 7-8cm to the Taiwan Strait. The geological surface profiles of Chaishan area are in sequence weathered limestone, clay layer, limestone and mudstone layer, respectively. Thus the frictional resistance between weathered soils and rock layer could decrease after infiltration of rainwater due to impervious to water of the lowest mudstone layer. Typhoon invades often Taiwan each year, resulting in rainfall infiltration and rising groundwater level, as well as increased pore water pressure within the soil mass, causing the earth movements in some parts of Chaishan, especially in the Temple A (Shan Hai Temple) accompanied with cracking phenomenon. In this paper, limit equilibrium (LE) and finite element method (FEM) are used for slope analysis, in which the slope is considered as unsaturated soil. Results showed groundwater amounts are easy to accumulate and increasing pore water pressure give resulting in decreased safety factor. Both of groundwater level and rain durations were also considered in this study.

1. Introductions

Kaohsiung Chaishan is one of the most important attractive spots in southern Taiwan due to the topography of coral reef limestone. However, due to limestone on the surface and mudstone beneath it, the rainwater can reach to the interface, so that its strength is weakened, and forms a weak surface at the interface. Hsiao et al. [1] proposed that parts of the limestone could roll down so that there is some limestones at the foot of the mountain of the Chaishan area, and its geological profiles from the ground surface was weathered mudstone layer, weathered limestone, mudstone and limestone in sequence. While continuing the research on the Chaishan area, we can find that the natural geological conditions of Chaishan are fragile and there is sustained displacement phenomenon. Jen et al. [2] identified three types of slope instability near the shore at Chaishan, including: (1) on the lower section of the slope, wave erosion exposes the mudstone and destabilizes the limestone fragments being sealed above it; (2) slumps on the re-cemented limestone fragments; (3) rock fall from cliffs near the ridge. Yang et al. [3] mentioned hydro-mechanical analysis of two unstable slopes by reducing matric suction and thus pore water pressure variations due to rain infiltration. As the effective shear strength of the sliding surface becomes smaller, the displacement of the slope gradually occurs and the slope is finally destroyed. Oh and Lu [4] analyzed two slopes in South Korea using limit equilibrium and finite element methods, of which the unsaturated part was analyzed using a suction stress characteristic curve (SSCC) and a soil water retention curve (SWRC) slope, hydromechanical properties, shear strength and rain history. Li et al. [5] discussed the role of saturated soil during water permeability and rainfall in homogeneous slope surface damage.

Here, we first detected the monitoring data of Chaishan area from 2005 to 2012, and the apparent movements in the area are known from the data, which is more obvious during the rainfall period. In this paper, both methods of finite equilibrium (LE) and finite element method (FEM) are then used to study the changes of slope movement and safety factor according to the different rainfall intensity and groundwater level variation. Based on the changes of rainfall intensity and groundwater level, Regional slope stability and the relationship between the destruction of the interface and further explore the main cause in the Chaishan area is related to rainfall. Since super Typhoon Morakot destroyed many monitoring equipment, we selected 2010 Typhoon Fanapi as a case for rainfall analysis. The objective of the study tried to understand the failure mechanism of unsaturated slope on Chaishan after the typhoon rainfalls, and expect the results could also provide the local government as a reference for subsequent disaster prevention.

2. Study area

Figure 1 is a map of Chaishan where the study area located in. It is located in the urban area of Kaohsiung City on the southwestern side of Taiwan near the Taiwan Strait. On the left side of this area, the right side of Chaishan is called Shoushan. There is a small fishery port in the area. There are still hundreds of dwellings in Chaishan, including two colleges, Sun Yat-Sen University. The scenery is very beautiful, but there seems a potential landslide on the slopes from the past literatures [1, 2]. The geology map shown in figure 2 displays most lands of Chaishan are covered with limestones and mudstone, the latter of which is more popular seen in the western areas.

Figure 2 also shows the geological profile of Shoushan, and they are comparatively simpler geology on Chaishan. The main geology in the area is composed of a few limestone mixed with mudstone, while the pervious limestone below the surface limestone and weathered collapsible soil on the slope. Regarding the Chaishan's topography, it belongs to the dip slope. As for the thickness and distribution of weathered soil layers and the limestones, we can understand them from the boring data mentioned in the later.

Figure 1. Study area and Chaishan location map

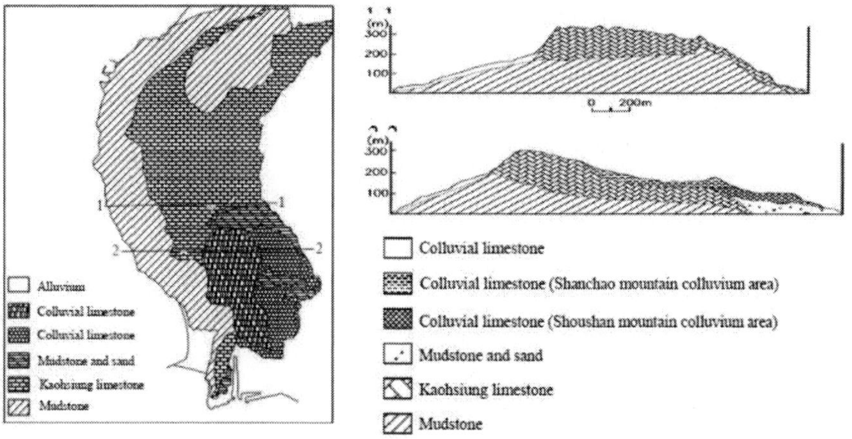

Figure 2. Shoushan geological map and geological section of the Shoushan area

Figure 3 shows the topographic map of Chaishan in Kaohsiung. The overall topography shows an east-west slope with a slope of about 20-35%. However the west side of the road seems slightly steep, possibly 40-50%, there are two temples in this area, Temple A (Shan Hai Temple) is a center for local people's beliefs, and many houses gather around these two temples. Figure 3 also shows the relevant on-site buried monitoring instruments, including drilling holes, inclinometers, piezometer, GPS stations, automatic rainfall equipment and the others, but only the inclinometers and piezometer positions are shown in the figure. A total of 21 monitoring systems, including 15 inclinometers, 6 piezometers, 15 GPS measuring stations and one automatic rainfall gauge, were set up in the zone between 2005 and 2012. At the beginning there were 18 drilling holes drilled in June 9th, after that onwards to begin monitoring. Figure 3 shows also a field investigation in site survey in the area, showing the elevation of the two sections and the Chaishan Road. The monitoring results over the long-term period (2005-2012) show that the stratigraphic slip in the area is mostly concentrated on the west side of Chaishan Road, and the displacement during the rainfall shows a significant increase. The on-site survey shows that there are obvious displacement for roads and houses in this area with obvious clearly cracks. It can be seen from the figure that there are obvious cracks and differences in height between the roads outside the houses.

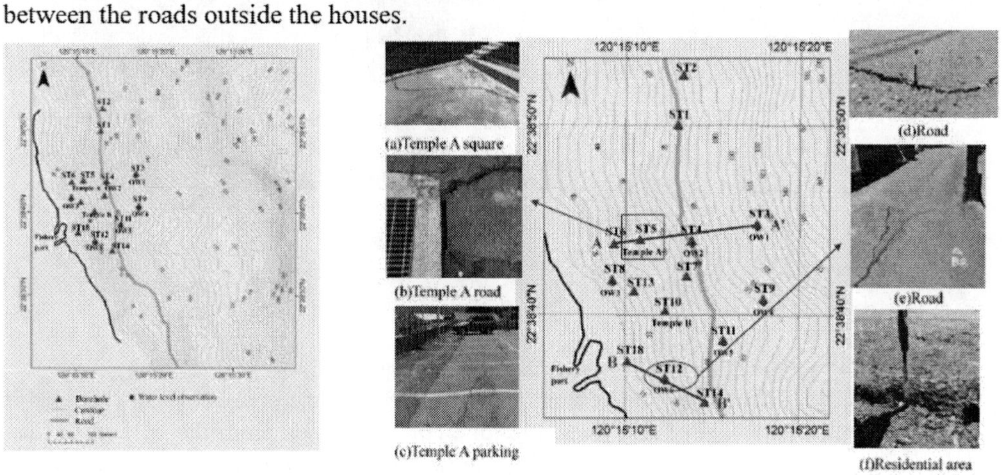

Figure 3. Topographic map (left) and field investigation (right) in site survey

Figure 4 shows the GPS and inclinometer surface displacement map between 2005 and 2012, although there are differences between the two plans but the overall road to the west side having a larger amount of displacement, as the GPS measurement is larger because there is no inclinometer embedding resistance. The annual displacement to the Taiwan Strait in Chaishan ranges from 7 to 8 cm. In the monitoring results, great displacement was found around Temple A. Based on the monitoring data, the inclinometer number near Chaishan Fishing Port was identified as ST18, in which the thick mudstone layer is underlain, there seems to be a little bulging due to the swelling behavior of mudstone. In comparison with elevation contour diagram in figure 5, we may see clearly the actual conditions on the scene from the figure.

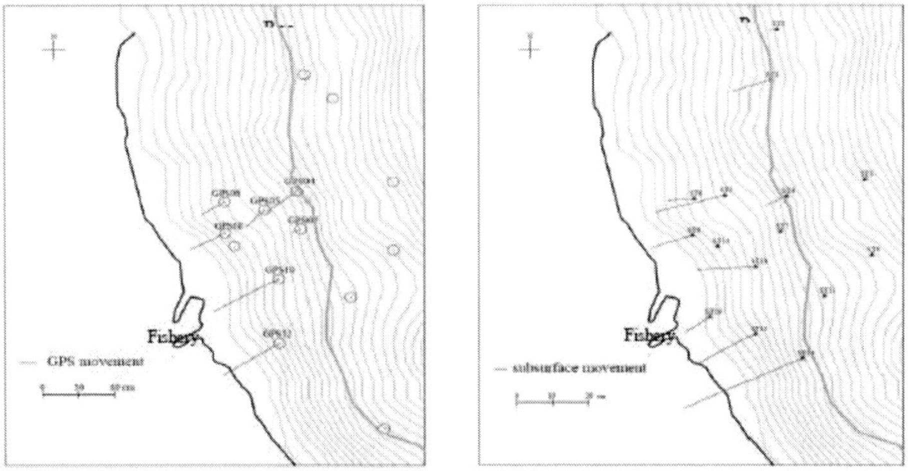

Figure 4. GPS and inclinometer displacement diagram

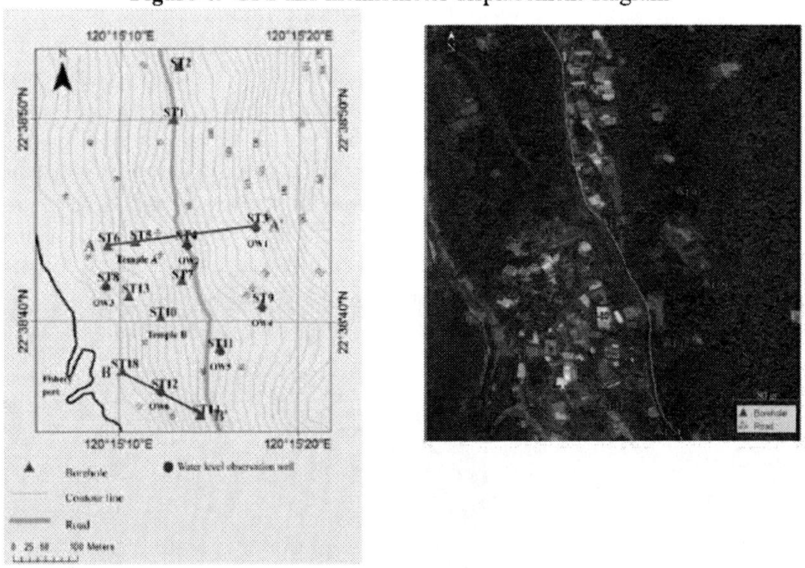

Figure 5. Elevation contour diagram and satellite picture in study area

According to the drilling data, the geological composition of the area is broadly divided into four types: mudstone, limestone, silty clay and limestone collapsible soil from the bottom to the surface. First, the drilling data are set by using the drilling holes and inclinometers in Section A-A', which extends to the east of Chaishan Road from the plaza in front of Temple A near the coast and is drawn to A-A' by ST5, ST6, BW19, BS7 and ST3 as shown below in figure 6. The figure shows also that the section is geologically impermeable to mudstone, while strata of ST5 and ST6 (around Temple A) are covered with a highly permeable limestone layer that continues over silty clay and limestone collapsible layers. Groundwater table uses water level observation wells OW1 and OW2 adjacent to this section as a basis for analyzing the effect of groundwater level. Figure 6 shows also that the lower slope of this section is near the fishing port of Chaishan in Section B-B '.

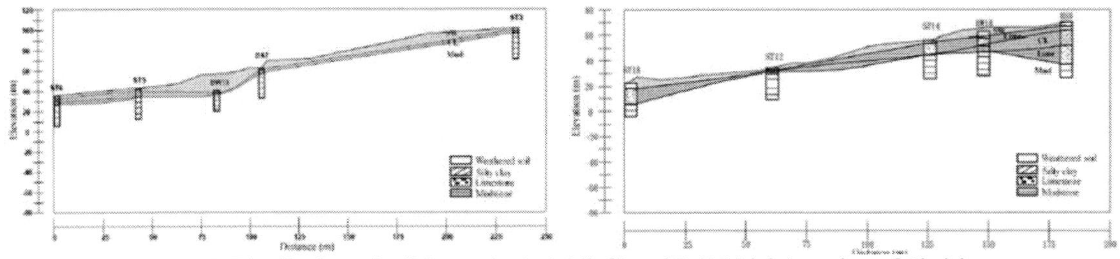

Figure 6. Distribution of soil layers in A-A ' (left) and B-B ' (right) section of Chaishan

Table 1 shows the accumulated rainfall during typhoon over the past years. Among them, the three typhoons during 2008-2010 caused major disasters in Taiwan and Kaohsiung. At that time, the accumulated rainfalls of Kalmaegi, Morakot and Fanapi in Chai Shan were 523.5, 726 and 554.5 mm respectively. As the local automatic rain gauge in 2009 was damaged during super Morakot disaster, the rainfall readings measured by the Fanapi typhoon in the second year after being repaired were used as the rainfall data for follow-up research. Figure 7 summarizes the changes in the groundwater levels of the monthly rainfall and piezometer OW1, OW2 and OW6 from 2005 to 2012. The groundwater level at section A-A 'and section B-B' is located in the rainy season showed a clear upward trend in groundwater levels. The monitoring results show that the water table is about 0.5 to 1.5 meters below ground surface, whereas groundwater table in the dry season is about 2 to 5 meters below ground surface.

Table 1. Typhoon cumulative rainfalls over the calendar year

Years	Typhoon	Duration	Accumulated rainfalls (mm)
2005	Haitang	7/16~7/20	412
2006	Bilis	7/12~7/15	303
2007	Pabuk	8/7~8/9	30
2008	Kalmaegi	7/15~7/20	523.5
2009	Morakot	8/6~8/10	726
2010	Fanapi	9/17~9/20	554.5
2011	Nanmadol	8/27~8/31	299
2012	Talim	6/16~6/21	222.5
	Tembin	8/21~8/28	174

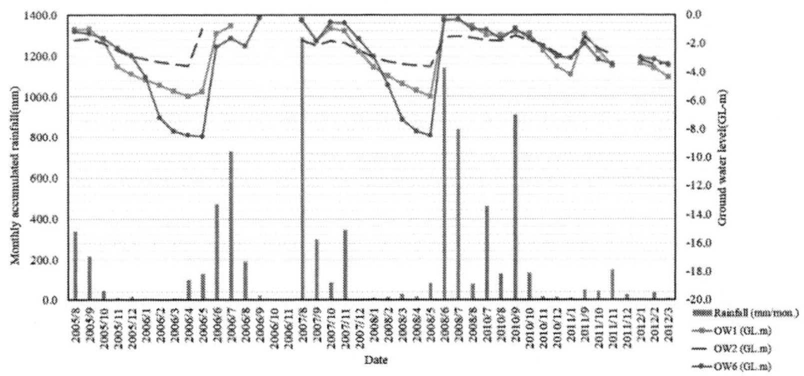

Figure 7. Monthly rainfalls and groundwater level variations from 2005 to 2012

3. Numerical simulations

In order to simulate the slope stability of the Chaishan slope during typhoon and rainfall, this paper uses the limit equilibrium (LE) and the finite element method (FEM) for slope analysis. Geo-Studio (Geoslope co.) [6] is used for the LE method as well as PLAXIS-2D (Plaxis co.) [7] is used for FEM analysis. At each stage SIMGA/W analysis is used to calculate slope stress and strain, SEEP/W analysis for seepage flow, and then the results could be put into the SLOPE/W to evaluate slope safety factor. In recent years, numerical analysis of unsaturated soils has been used extensively. The soil moisture characteristics deduced from Fredlund et al. [8], van Genuchten [9], Lu and Likos [10] and Lu et al. [11]. The empirical formulas of SWCC curve and hydraulic conductivity coefficient as shown in equations (1) and (2) were used to calculate the related soil parameters under unsaturated state of the effective stress and suction.

$$S_{eff} = \frac{\theta-\theta_r}{\theta_s-\theta_r} = \left[\frac{1}{1+\{a(u_a-u_w)\}^n}\right]^{1-1/n} \tag{1}$$

$$k_w = k_{sat}\left[\frac{\left[1-\{a(u_a-u_w)\}^{n-1}[1+\{a(u_a-u_w)\}^n]^{\frac{1}{n-1}}\right]^2}{[1+\{a(u_a-u_w)\}^n]^{\frac{1}{2}-\frac{n}{2}}}\right] \tag{2}$$

However, the soil shear strength in the unsaturated state could be written as Equation (3). The method defined according to the Mohr-Coulomb failure criterion was used to describe or predict the nonlinear relationship between soil shear strength and suction (Qi and Vanapalli [12]). Based on the Mohr-Coulomb failure criterion and the ultimate stress method, the slope safety factor is calculated as Equation (4). This study uses the SLOPE/W and PLAXIS software to calculate the safety factor of slope. The former obtained by the theory of limit equilibrium method, as shown in equation (4); but the latter is different from the limit equilibrium method. Thus performing calculation of finite element method, the re-calculation of the safety factor results will be chosen. In order to consider the convenience of numerical operations, soil strength parameters tanθ and c were usually used and gradually decrease until the slope occurs damage for shear strength reduction method, the value obtained is the slope of the reduction value used to indicate the safety factor as shown in equation (5).

$$\tau_f = c' + (\sigma - U_a) * \tan\phi' + \{S_{eff}(U_a - U_w)\} * \tan\phi' \tag{3}$$

$$\mathrm{FS} = \frac{c'+\sigma' tan\phi'}{W*sin\alpha*cos\alpha} \tag{4}$$

$$M_{sf} = \frac{tan\varphi_{input}}{tan\varphi_{reduced}} = \frac{c_{input}}{c_{reduced}} \tag{5}$$

The depth of soil layer in each section is plotted by drilling boring data. The section model built by PLAXIS-2D is shown in figure 8. The degree of mesh densification is very tight, however, due to the boundary effect, the extension of the two cross sections will be evaluated separately. The section A-A' has a total length of 235 meters, an upper slope of 101 meters and a lower slope of 35 meters, with a grid number of 15789; a section B-B' of 185 meters in length and an upper slope of 70.2 meters, The lower slope is 22.9 meters high with 4801 grids, while the two sections extend horizontally outward to 500 meters due to the boundary effect.

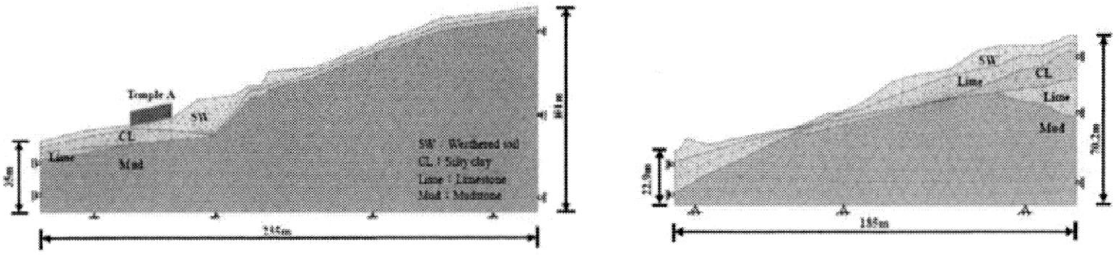

Figure 8. PLAXIS section model in A-A ' and B-B '

In this case, we use three sets of modules to analyze slope stability, SIMGA/W, SEEP/W and SLOPE/W. Since the software model can be applied to other modules only once, The parameters of the module input materials can be analyzed, as the software used to specify the destruction of the surface out of the entrance range, so do not need to consider the boundary effect of the problem, figure 10, respectively, are A-A' section and B-B' sectional model.

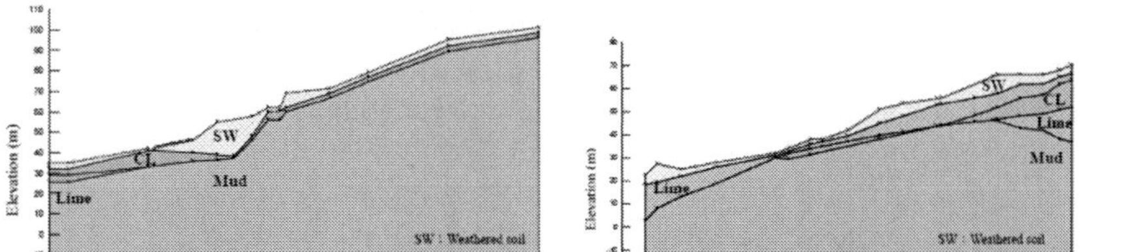

Figure 9. SLOPE/W section cross section model in A-A ' and B-B '

According to the field survey data, the water table readings of six buried holes were conducted in the Chaishan area in 2005. The water table of OW1 and OW2 adjacent to the section A-A' and the OW6 groundwater table of section B-B' measurement data is used as an assessment of the groundwater level assumptions. According to the monitoring data, the groundwater level at the two cross sections was at an average water table of about 1.5 to 2.0 meters below the surface during September 2010. Therefore, a groundwater table setting of 2.0 meters below ground surface will be used as a design groundwater table as the basis. Table 2 and 3 are the material parameters of the soil layers in sections A-A' and B-B', respectively. Due to a lot of parameters, the sensitivity analysis of the parameters will be performed later in this study, including the value of shear strength c, internal friction resistance ϕ angle and permeability coefficient k value of weathered soil and understanding they affect the slope safety factor and displacement.

Table 2. A-A' section soil parameters

Layer	SOIL/ROCK	γ_m kN/m^3	γ_{sat} kN/m^3	E' kPa	v	c kPa	Su kPa	ϕ' (°)	ψ (°)	k_x m/day	k_y m/day
I	Weathered soils	21	22	34000	0.35	20	-	20	0	1.063	1.063
II	Clay	20	21	13545	0.35	-	30	15	0	4.75×10^{-3}	4.75×10^{-3}
III	Limestone	20	21	350000	0.18	500	-	22	0	1.063	1.063
IV	Mudstone	21	22	52000	0.40	-	50	30	0	8.64×10^{-4}	8.64×10^{-4}

Table 3. B-B' section soil parameters

Layer	SOIL/ROCK	γ_m kN/m^3	γ_{sat} kN/m^3	E' kPa	v	c kPa	Su kPa	ϕ' (°)	ψ (°)	k_x m/day	k_y m/day
I	Weathered soils	21	22	34000	0.35	20	-	20	0	0.2497	0.2497
II	Limestone	20	21	350000	0.18	500	-	22	0	0.2497	0.2497
III	Clay	20	21	13545	0.35	-	30	15	0	4.75×10^{-3}	4.75×10^{-3}
IV	Mudstone	21	22	10000	0.40	-	50	30	0	8.64×10^{-4}	8.64×10^{-4}

4. Results and discussions

According to the analysis by Hsiao et al. [1] on the slopes in Chaishan region in 2015, the geological conditions in the Chaishan region are fragile and displaced during rainy season. Therefore, in the contour map of Chaishan region, the sections A-A 'and B -B 'section, the comparison between the change of slope safety coefficient and displacement under rainfall simulation and the monitoring displacement of section A-A' during Fanapi Typhoon. The section A-A 'is selected because it is close to Temple A, which has a history of three hundred years of cultural history. Local residents live near the area for many times and there are obvious cracks in the roads near the area after the site survey. The B-B 'section was selected due to the relatively low elevation of the area and its proximity is to Chaishan Fishing Port. The boring data show that the geology of the area consists mainly of limestone and impermeable mudstone with higher water permeability. Therefore, rainwater and groundwater level permeability and accumulation of the rainfalls on the surface of mudstone may cause the phenomenon of softening and swelling of mudstone.

4.1. Effect of shear strength parameters c and ϕ on slope safety factor

According to the cross section A-A', it is found that the section is thicker and forms like a bowl-shaped structure in the rear of Temple A, and is connected with silty layers and silty clay layers. The sliding surface of the area can be mainly located at the interface between the collapsible soil and the silty clay layer. Therefore, using the SLOPE/W analysis, the c values of the collapsed soil layer A-A' are defined as 20-35 kN/m^2 and then ϕ angles are 10°-45°, the variation of the safety factor is calculated. Figure 10 shows the safety factor for slope with different values of c and ϕ for the clayey soil. The table shows that when c = 20 kN/m^2 and ϕ = 10°, the FS = 0.93 and the ϕ angle continuously increases FS = 1.32 at ϕ = 30°. Although the soil cohesion increases but the angle φ remains constant, the slope is still unable to resist It can be clearly seen from the figure that when c and ϕ increase, the safety factor of the slope tends to increase. However, when c = 30 kN/m^2 and ϕ = 30°, there is a decrease of FS = 1.4 of the slope after calculation due to the phenomenon that the safety factor is decreased due to the transformation of the failure surface of the slope, which can be seen from the figure 11, the failure surface of the slope changes from arc-shaped dotted line to solid line. The reason for the decline is that the failure surface is transferred from the soil layer to the silty clay layer, as shown in figure, however, ϕ angle of the silty clay layer is smaller, so the worse after sectioning.

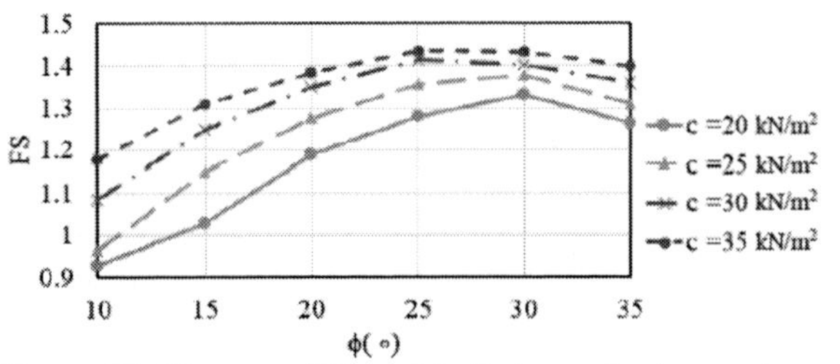

Figure 10. Changes of the safety factor owing to c and ϕ

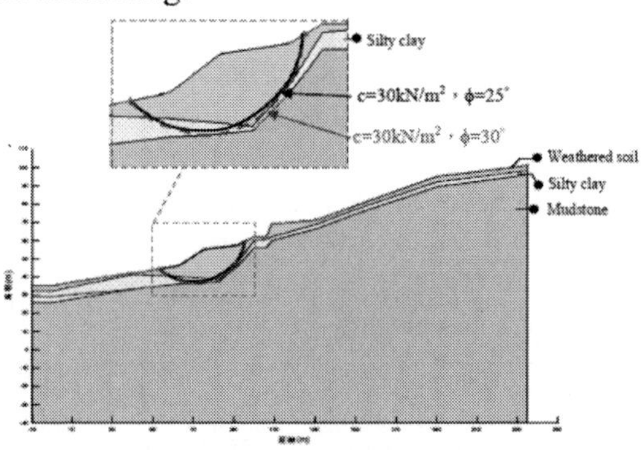

Figure 11. Change of the sliding surface for A-A' section

4.2. Effect of permeability on slope safety factor

Li et al. [5] pointed out that rainfalls and saturated permeability may be related to the rise of groundwater table or the increase of pore water pressure in the soil due to the failure mechanism of the slope, so the first two factors of rainfall duration and soil permeability will affect slope damage surface location and type of results. Although the overburden of mudstone above the B-B' section is only 10-m-thick weathered soil layers and limestone, due to the large number of limestone holes and better water permeability in this area, by replacing the collapsed soil layer with limestone. The permeability coefficient of the soil layer is discussed to affect the change of the slope safety factor when the rainfall comes. The permeability coefficient is first set from 0.8m/day to 1.5m/day. Figure 12 shows that the coefficient of permeability of collapsible soil layer is k = 0.8m/day, the FS = 1.78, however, when the rainfall persists and the rainfall continues for 4 hours, FS of the section will decrease to 0.89 and then breaks down. It can be clearly seen from the figure that on the contratory when the k value of the collapsed soil layer is larger, the rainfall has little effect on the safety factor of the slope because the surface soil rainwater can be quickly removed and the dangerous hazard to the slope is relatively low.

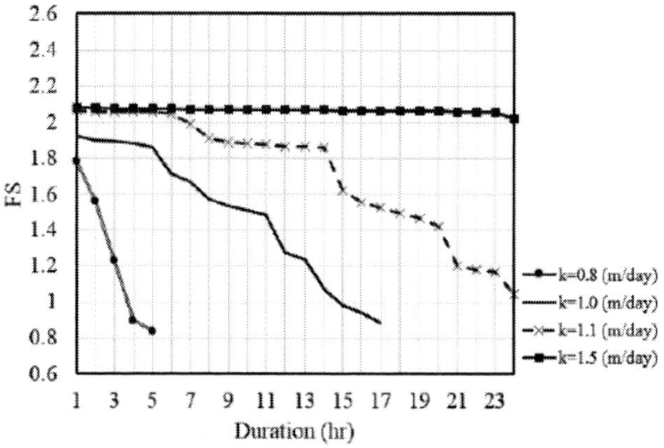

Figure 12. Different values of k impact on the safety factor for B-B 'cross-section

4.3. *Effect of slope damage surface location on safety factor*

Firstly, this study sorted out the monitoring data of each inclinometer over the years and extracted some data to provide the range of failure surface to explore the influence on the safety factor. Figure 13 is a displacement map using the inclinometer ST6, ST5 and ST3 for the section A-A'. Monitoring data show the location of the destruction of the surface, roughly draw a group of slope slip surface, and then use SLOPE/ W analysis of each group of sections of the safety factor.

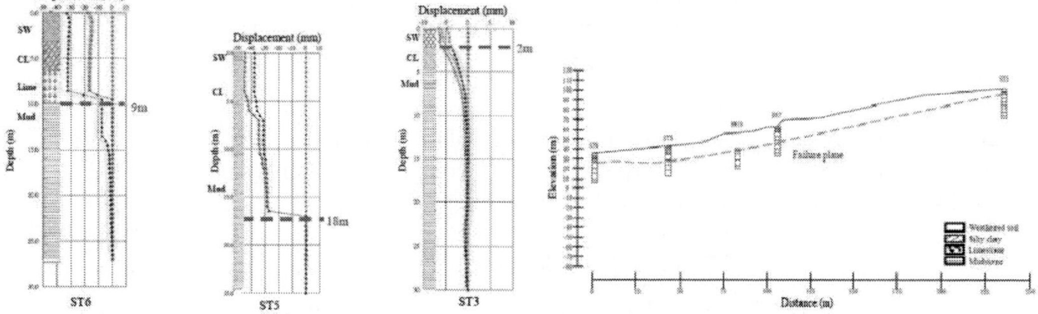

Figure 13. the inclined tube displacement and sliding surface position in A-A 'cross-sectional

The results of the analysis are shown in figure 14, in which figure 14(a) shows the safety factor for the fully specified failure surface of section A-A', where the green border is the slope sliding surface area. However, the factor of safety will overestimate the overall stability of the slope due to the fact that the slope may first be destroyed elsewhere. Therefore, in figure 14(b), the setting of the entry and exit of the failure surface, to find out the safety factor of the slope in the critical state. In the figure, the green arc-shaped checkered area is the area of failure surface of the section. Both of the above two methods belong to the boundary of the failure surface, Therefore, in figure 14(c), PLAXIS software is used to evaluate the overall slope reduction factor, and the safety factors of the three conditions are summarized in Table 4. The results show that using different methods to calculate the safety factor of slope varies greatly, among which, the failure surface FS = 1.94 is completely specified. Since this failure surface is a sliding surface drawn according to the failure section of each inclinometer, when SLOPE/W is calculating the safety factor sliding will only be made using this sliding surface, so the result may be significantly higher than the other two. FS = 1.07 shows the process of calculating the damage surface. The software calculates the node on the entrance and exit repeatedly and calculates the critical damage of the slope, however PLAXIS considers the overall displacement of section A-A' and reduces the angle ϕ until the failure of the slope, which results in a slope FS of 1.06.

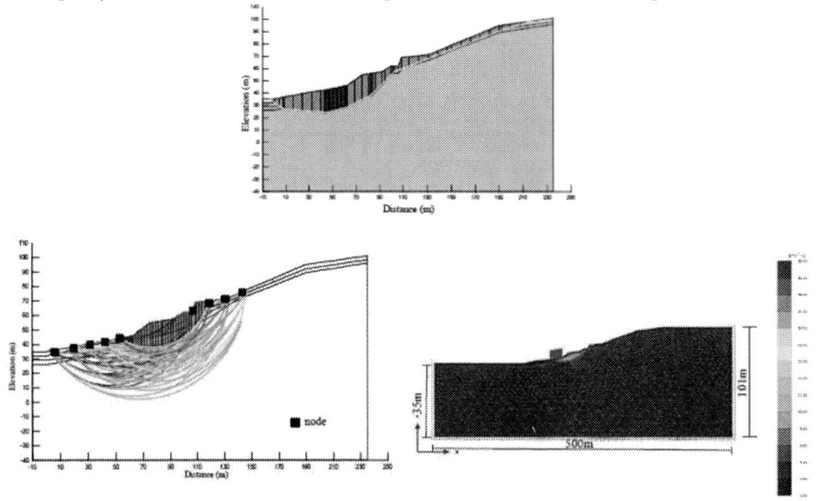

Figure 14. (a) completely designates, (b) designated A-A' section out of the entrance damage, and (c) PLAXIS for the failure plane A-A'

Table 4. Safety factor for three events in A-A' section

Completely designates the failure plane (SLOPE/W)	Designated A-A' section out of the entrance damage (SLOPE/W)	PLAXIS
1.94	1.07	1.06

As mentioned above, the section B-B' extends from the inclinometer at Chaishan fishing port at ST18 to the station ST12 at the residential area of Chai Shan tribe and finally to Chaishan Road at the inclinometer no. ST14. Table 5 using different calculation methods is quite different. In addition, the safety factor obtained from the above section B-B' compared with that from section A-A', the results show that section A-A' in Chaishan is more dangerous.

Table 5. Safety factor for three events in B-B' section

Completely designates the failure plane (SLOPE/W)	Designated A-A' section out of the entrance damage (SLOPE/W)	PLAXIS
2.68	2.07	1.32

4.4. Effect of various groundwater tables on the slope failure

Due to the special geological conditions of Chaishan, this paper simulates the failure of slopes A-A' under different groundwater levels and rainfall respectively. However, in order to find out the main failure area of slope, two rainfall intensity data for analysis rainfall intensity 5mm/hr and continuous rain for 24 hours. Since the monitoring results are in the range of 1-2m stated before over the years, the groundwater table is set as 2 meters below the ground surface. In addition, the depth of the mudstone in the section discussed is about 10 meters below the surface. Therefore, the groundwater is also located at a depth of 10 meters below the earth's surface. The result of figure 14 shows that the main displacement area is around Temple A. The simulation results show that the groundwater is located 2 meters below the surface. This may be due to the fact that the influence area reaches to 15-meter-thick silty clay layer, probably because its own drainage is not easy and existence of poor permeability of mudstone underlain below, resulting in rainwater infiltration in this area. As for 10 meters below the ground surface of figure 15, we can observe that the displacement of this area is small. The overall finding is that displacement at -2m is much larger than those at -10m.

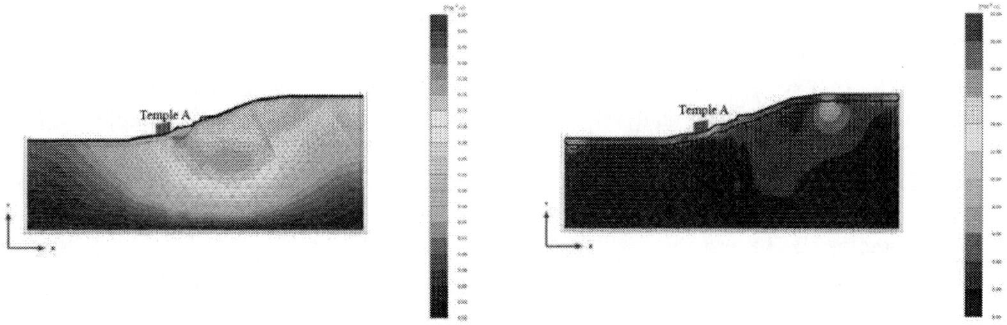

Figure 15. The displacement chart of the slope when groundwater level -2m and -10m

4.5. *Typhoon rainfalls affect the slope safety*

In order to investigate the relationship between the main causes of slope-slip and the rainfall in Chaishan, this study uses the rainfall readings measured during the Fanabi typhoon in 2010. Through the analysis of the stability of slope A-A' in Chaishan by PLAXIS and SLOPE/W software, and plot the relationship between time duration and the rainfall during the typhoon period in figure 16 and the groundwater level variation by the piezometer OW1 and OW2 adjacent to the A-A' section. The figure shows OW1 groundwater level gauge change trend is obvious, because the region geological collapse of the soil layer is relatively thin, and then take about 1.5-3m thick silty clay, so the rain can't be quickly excluded, and in this formation of temporary habitat groundwater; OW2 underground water level meter in the rear of the Temple A, the collapse of soil in this area about 19m thick, so the upward trend of groundwater level is less obvious. Further use of SLOPE/W analysis, where the A-A' section safety factor changes during typhoon, the results are shown in figure 17. In this figure, it is seen that the safety factor gradually decreases along the A-A' section during the typhoon Fanapi than during the typhoon. However, at 4:00 pm on September 19, local time, Than the highest rainfall intensity in the typhoon, the rainfall intensity as high as 86.5 mm/hr, it is clear from the figure that the section of the safety factor at this moment also significantly decreased phenomenon. Yang et al. [3] mentioned that the increase of cumulative rainfall led to a decrease of the safety factor, which may be related to that the p'-q' graph is close to K_f-line.

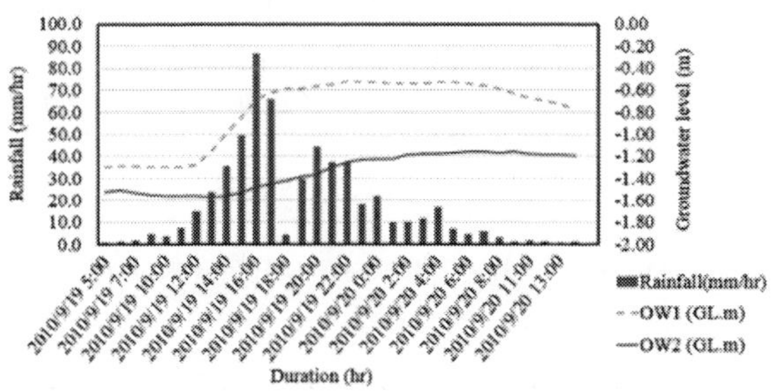

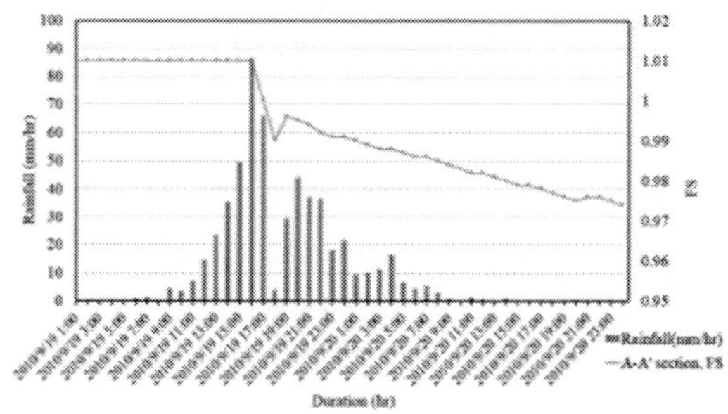

Figure 16. Variation of groundwater level during Vanaipi typhoon

Figure 17. Change of safety factor for A-A' section during the Typhoon

5. Conclusions

- The geology of limestone in Chaishan, Kaohsiung is not entirely composed of water-permeable limestone and impervious mudstone. This study indicates that it is possible that limestone from the top hills had collapsed and rolled down the Chaishan area under the influence of topography or natural forces. However, the collapsed limestone may mix with the fragments and inclusions of weathered sands and clayey soils.
- After finite equilibrium (LE) and finite element method (FEM) are used for slope analysis, the results of two-section simulation analysis are satisfactory with unsaturated soil considered. Parameters sensitivity analysis such as shear strength c value and ϕ angle, collapsible soil water permeability coefficient k value were performed. It is found that the location of failure plane will affect the safety of the slope very significantly. Through the simulation for A-A' and B-B' section, the results showed that the A-A' of the Chaishan area is more dangerous than the section B-B'.
- The result of inclinometer shows that the main sliding surface of the slope in Chaishan area is mostly located at the interface of weathered soils, limestone and mudstone. Using different rainfall intensities and the groundwater level of -2 m and -10 m, the results show that the section A-A' is at the inclinometer ST5 near Temple A. Due to the thick silty clay and poor drainage, water level is accumulated.

References

[1] Hsiao DH, Phan TAV, Kung TC, and Liao CJ 2015 Investigation and assessment of a landslide case study in southwestern Taiwan. *Disaster Advances.* **8(4)** 39-52.
[2] Jen CH, Chen JH, Lei HF, Ho LD, Zhou HW, Lin TY, Lu CH, Hsueh MY, Chyi SJ 2018 Recent movement of the Late Pleistocene Shoushan slump, southwestern Taiwan, based on landform surveys. *Quaternary International.* (In press)
[3] Yang KH, Uzuokab R, Thuoa JN, Lina GL, Nakaib Y 2017 Coupled hydro-mechanical analysis of two unstable unsaturated slopes subject to rainfall infiltration. *Engineering Geology.* **216** 13-30.
[4] Oh S, Lub N 2015 Slope stability analysis under unsaturated conditions: Case studies of rainfall-induced failure of cut slopes. *Engineering Geology.* **184** 96-103.
[5] Li WI, Lee LM, Cai H, Li HJ, Dai FC, Wang ML 2013 Combined roles of saturated permeability and rainfall characteristics on surficial failure of homogeneous soil slope. *Engineering Geology.* **153** 105-113.
[6] Geo-slope International Ltd 2015 *Geo-Studio.* Canada.
[7] Plaxis B.V. Co 2016 *PLAXIS.* Netherlands.
[8] Fredlund DG, Morgenstern NR, and Widger RA 1978 The shear strength of unsaturated soil. *Canadian Geotechnical Journal.* **15(3)** 313-321.
[9] Van Genuchten MT 1980 A closed-form equation for predicting hydraulic conductivity of unsaturated soils. *Soil Sci Soc Am J.* **44(5)** 892-898.
[10] Lu N, and Likos WJ 2006 Suction stress characteristic curve for unsaturated soil. *Journal of Geotechnical and Geoenvironmental Engineering.* **132(2)** 131-142.
[11] Lu N, Godt.,JW, and Wu DT 2010 A closed-form equation for effective stress in unsaturated soil, *Water Resour. Res.* **46(5)** 1-14
[12] Qi S, Vanapalli S K 2015 Hydro-mechanical coupling effect on surficial layer stability of unsaturated expansive soil slopes. *Computer and Geotechnics.* **70** 68-82.

Self-compacting concrete with sugarcane bagasse ash – ground blast furnace slag blended cement: fresh properties

Duc-Hien Le[1]*, Yeong-Nain Sheen[2], and My Ngoc-Tra Lam[3]

[1]Sustainable Developments in Civil Engineering Research Group, Faculty of Civil Engineering, Ton Duc Thang University, Ho Chi Minh City 70000, Vietnam
[2]Department of Civil Engineering, National Kaohsiung University of Applied Sciences, Kaohsiung 80778, Taiwan, ROC
[3]Department of Civil Engineering, Faculty of Engineering, Prince of Songkla University, Hat Yai, 10 Songkhla, Thailan

leduchien@tdt.edu.vn (D-H. Le)

Abstract. In this investigation, major properties in fresh state of self-compacting concrete (SCC) developed from sugarcane bagasse ash and granulated blast furnace slag as supplementary cementitious materials were examined through an experimental work. There were four mix groups (S0, BA10, BA20, and BA30) containing different cement replacing levels; and totally, 12 SCC mixtures and one control mixture were provided for the test. Fresh properties of the proposed SCC were evaluated through measurement of the density, slump, slump-flow, V-funnel test, T_{500} slump, Box-test, and setting time. The testing results indicated that replacing either SBA and/or BFS to OPC in SCC mixtures led to lower density, lesser flowability, and longer hardening times.

1. Introduction

When sugarcane bagasse after extraction of economical sugar was burnt for co-generation of heat, about 8-10% ash waste, which known as sugar cane bagasse ash (SBA) was produced [1]. This ash ground in to very fine particles performs potential of pozzolanic characteristics, and beneficially being used to improve concrete properties [2, 3]. In general, as presence of pozzolanic materials such as slag, fly ash, silica fume in cement mixture, the silica (SiO_2) commonly being available in these material reacts with calcium hydroxide ($Ca(OH)_2$) released during hydration of Portland cement to additionally form calcium silicate hydrate (C-H-S), leading to improve performance of concrete.

In this investigation, SBA and ground blast furnace slag (BFS) were employed as partial cement substitution materials in developing self-compacting concrete (SCC). The experimental work examined on fresh properties of the SCC such as fresh density, slump, slump flow, T_{500} and V-funnel flow time, Box-test and the setting time.

2. Methods

2.1 Materials used

- *Cement:* The type I Ordinary Portland cement conformed to the ASTM C150 [4] was used.
- *Sugarcane bagasse ash (SBA)*: Raw bagasse ash collected from landfills of a local cogeneration. It was processed with burning at 600-800 ^0C within one hour before cooling and

grinding to fine particles. The SBA has the Blaine's fineness of 4010 *g/cm*^{2} and specific gravity of 2.02. Chemically, this material was high silica content with 53.2% SiO_2, relative high of loss of ignition (LOI, 22.9%), followed by K_2O (7.3%) and MgO (2.8%). The sum of main oxides (SiO_2+ Al_2O_3+ CaO) was 63%.

- *Ground blast burnace slag (BFS)*: the BFS obtained from a steel factory was used as cement substitution like SBA.
- *Fly ash (FA)*: Type F fly ash conforming ASTM C618 [5] was used as filler material which beneficially used to replace VMA in SCC mixtures for enhancement of viscosity [6].

Figure 1 Photos of OPC, SBA, and BFS samples used in this study.

Fig. 1 and Table 1 shows the chemical compositions and physical properties of the OPC, SBA, BFS used in this study.

- *Aggregates*: coarse and fine aggregates used in this experiment are crushed limestone and river sand, respectively. These aggregates fully conforms to ASTM C33 [7] for concrete making. The fineness moduli of the coarse and fine aggregates were 6.22 and 2.87, respectively; water absorption were 1.66% for sand and 2.45% for crushed limestone.
- *Super-plasticizer (SP):* A polycarboxylate-based super-plasticizer was used in SCC mixtures.

Table 1. Chemical compositions for OPC, SBA, BFS, and FA used in this study

	OPC	SBA	BFS
Chemical composition (% by mass)			
Silicon dioxide (SiO_2)	20.8	53.2	36.61
Aluminum oxide (Al_2O_3)	4.7	6.89	13.67
Ferric oxide (Fe_2O_3)	3.13	3	0.37
Calcium oxide (CaO)	63.2	3.451	42.1
Magnesium oxide (MgO)	3.33	2.821	6.6
Sodium oxide (SO_3)	2.41	-	0.51
Potassium oxide (K_2O)	0.69	7.321	-
Sodium oxide (Na_2O)	0.21	0.677	-
Loss on ignition (LOI)	2.11	22.9	0.89
Physical properties			
Specific gravity	3.15	2.02	2.8
Fineness (*g/cm*^{2})	3530	4010	4550

2.2 Mix-proportions

SCC mixtures in this work were designed according to the densified mixture design algorithm (DMDA) [8]. This design method is appropriate for high performance concrete in which low paste volume is basically required. Cement, slag, fly ash, sugarcane bagasse ash and aggregate were included in the SCC mixtures. Table 2 demonstrated the mix-proportions proposed for this investigation. Totally, there were 13 mixtures belonging four mix groups (i.e., S0, BA10, BA20, and BA30) and one control mixture, containing OPC only.

Table 2. Mix-proportions

No.	Group	Mix type	Mix ID.	% SBA	%BFS	% OPC
1	Control	Single	BA00-G00	0	0	0
2	S0- Group (0% BFS)	Binary	BA10-S00	10	0	90
3		-	BA20-S00	20	0	80
4		-	BA 30-S00	30	0	70
5	BA10- Group (10% SBA)	Ternary	BA10-S10	10	10	80
6		-	BA10-S20	10	20	70
7		-	BA10-S30	10	30	60
8	BA20- Group (20% SBA)	Ternary	BA20-S10	20	10	70
9		-	BA20-S20	20	20	60
10		-	BA20-S30	20	30	50
11	BA30- Group (30% SBA)	Ternary	BA30-S10	30	10	60
12		-	BA30-S20	30	20	50
13		-	BA30-S30	30	30	40

Notes: Coarse and fine aggregate content were 811 kg/m^3 and 828 kg/m^3, respectively; binder content (OPC+BA+BFS) was fixed at 434 kg/m^3; water/powder of 0.45; SP/binder of 1.4%.

2.3 Testing items for fresh SCC

The fresh SCC was examined through a series of testing procedures following JSCE [9], including fresh density, slump, slump flow, T_{500} slump time (T_{500}), V-funnel time (*VF*) and Box-test. In addition, the initial and final setting time of SCC mixtures was carried out in according to ASTM C403 [10].

3. Primary results and discussion

3.1 Fresh density test

The fresh density ranges from 2070 -2289 kg/m^3, as shown in Fig. 2. It was found that density of SCC with blended cement was 1.4%–10% lesser than that of control mix (2299 kg/m^3, with OPC only). The mix group BA20 incorporated with BFS had density range to be smaller than those of BA10 and BA30; and the lowest density (2070 kg/m^3) was measured on the mixture BA20-S10. It is noticed that low density of the SBA (2.02) and BFS (2.8) in comparing with OPC (3.15) made total mixture volume to increase 1.71%, meanwhile the density decreased about 10% (calculated on mixture BA20-S10). This result would imply concrete made with blended cement contains more porosity.

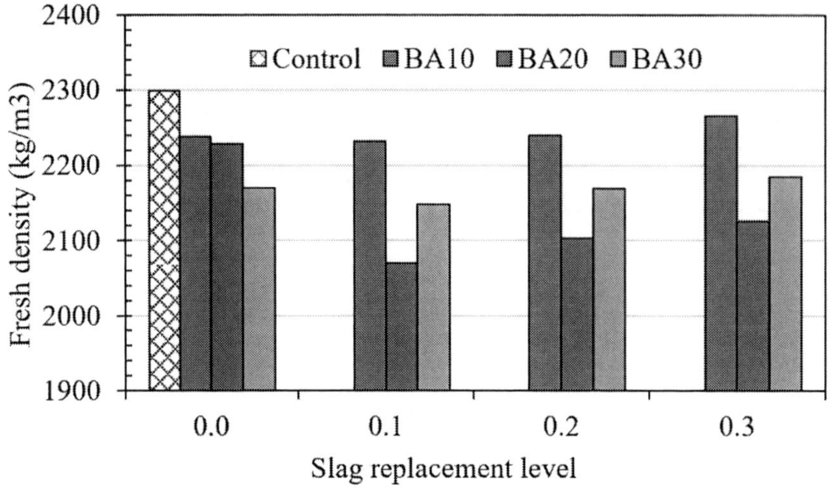

Figure 2 Measurement of fresh density.

3.2 Slump and slump flow tests

Fig. 3 show the slump results measured on all SCC mixtures, fully falling in the ranges of 200-300 *mm* for slump and 550-700 *mm* for slump flow. According to JSCE [9], these ranges were normally acceptable for many applications. Table 3 presents the limit values for SCC in fresh state, suggested by JSCE. The control mix (BA00-S00) had greatest slump and slump flow. From Fig. 3, presence of either SBA and/or BFS in mixtures led to lesser flowability. The higher amount of cement-replacing materials in binary/ternary system the lower slump/slump flow was. For example, fresh SCC mixture containing 10%, 20%, and 30% SBA had the slump flow decreased by 7%, 9%, and 13% comparing to the control mix, respectively. This behaviour would result from angularity and irregular shapes of cementitious materials, basically enhancing the friction between fine particles making fresh mixture to be harsh. In addition, more natural porosities between SBA particles tending to absorbed some of the mixing water.

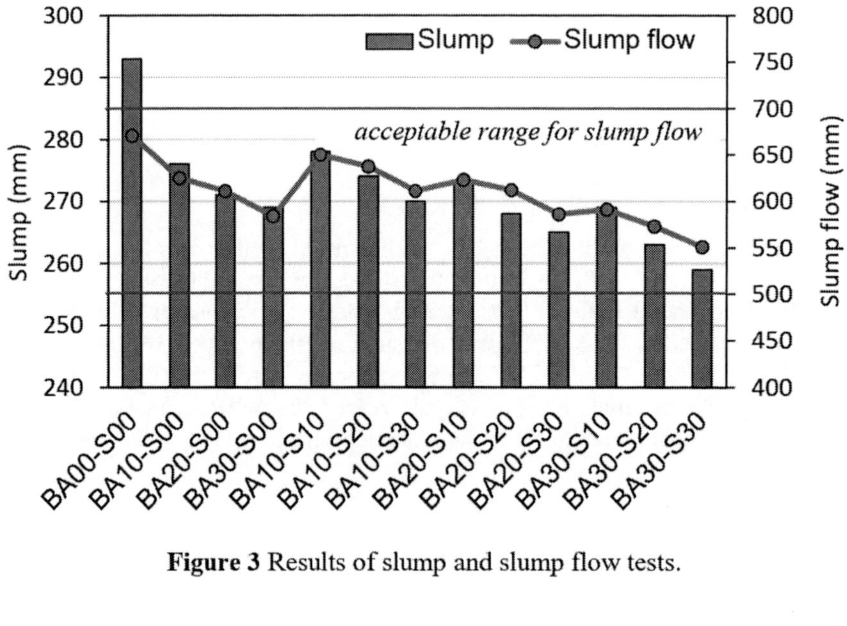

Figure 3 Results of slump and slump flow tests.

Table 3. Limit values for concrete in fresh state proposed by JSCE

Class of filling ability of concrete		**1**	**2**	**3**
Construction condition	Minimum gap between reinforcement (*mm*)	35–60	60–200	≥ 200
	Amount of reinforcement (kg/m^3)	≥ 350	100-350	≤ 100
Filling high of box-test (*mm*)		≥ 300	≥ 300	≥ 300
Flowability	Slump flow (*mm*)	650–750	600–700	500–650
Segregation resistance ability	Time required to flow through V-funnel (*sec*)	10–20	7–20	7–20
	Time required to reach 500 mm of slump flow (*sec*)	5–25	3–15	3–15

3.3 T_{500} and V-funnel time tests

The T_{500} is defined as the time required for fresh slump flow spreading up to 500 *mm* after removing the standard slump cone; where V-funnel time (in seconds) performs the flowability and stability of the SCC mixtures, defining as the elapsed time between the moment opening of the bottom outlet and the time when the light becomes visible from the bottom, observed from the top [11]. As seen in Figs. 4-5, the T_{500} and V-funnel time were in duration of 4.31-8.93 *sec* and 9.03-15.77 *sec*, respectively. These periods satisfied well for conventional applications, provided by JSCE (see Table 3). Substitution of SBA and/or BFS to OPC made the flow time of fresh concrete to be stably extended. The flow time measured on the largest amount of OPC replacement (BA30-S30) was longest, which was 112% comparing to that of control mix (BA00-S00). Especially, in each SCC mix-group associated with each level of SBA (10%, 20%, and 30%), the T_{500} was shortest at 10% BFS replacement; and it gradually increased as the level of slag increased. Similarly, increase in SBA or BFS caused the significant extension of V-funnel time.

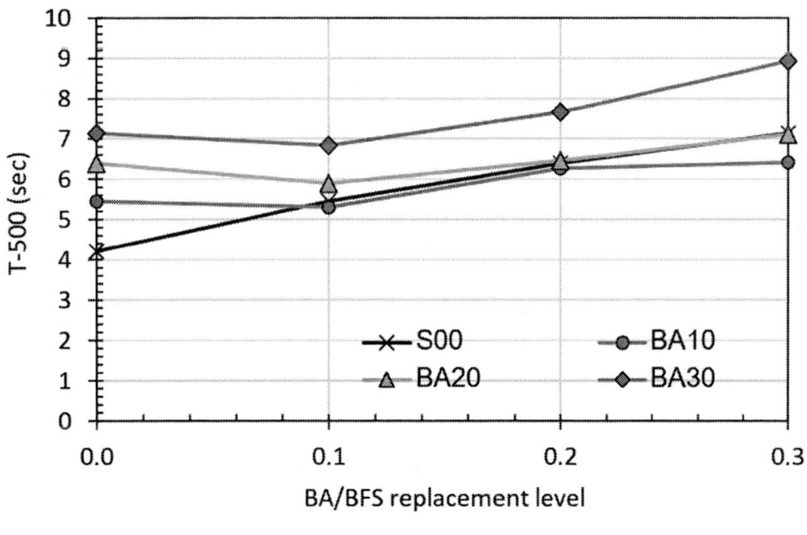

Figure 4 Effect of BA/ BFS replacement level on T_{500}.

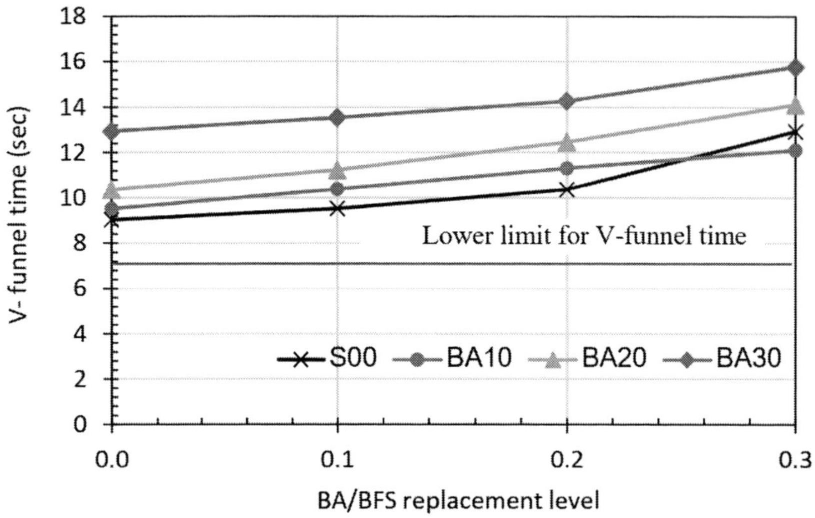

Figure 5 Variations of V-funnel time with various BA/BFS replacement level.

3.4 Box-test

The Box-test performs the degree of compactability and passing ability through the heavily congested reinforcement under concrete own-weight. Fig. 6 shows the filling height measured from the box-test, ranging from 244–330 *mm*. JSCE recommends that the height should be 300 *mm* or greater e Table 3). Hence, only BA10 group (except BA10-S30) and control mix (BA00-S00) achieved the requirement of passing ability. It was obvious that incorporation of both SBA and BFS in SCC mixture led to reduce the filling height.

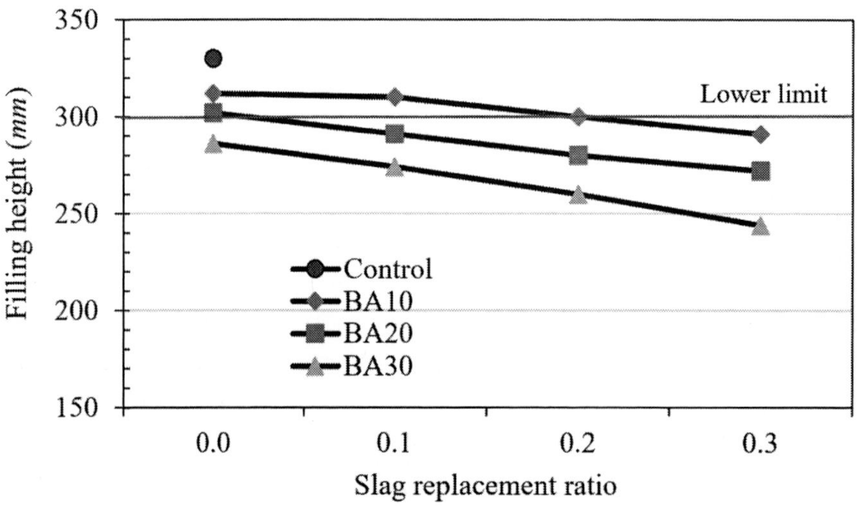

Figure 6 Variation of filling height with various slag replacement levels.

3.5 Setting time test

Fig. 7 demonstrated the initial and final setting time, varied 550 – 1131 *min.* and 750 –1467 *min.*, respectively. The longest hardening times (initial and final) were measured on mixture made of highest replacement level of minerals (BA30-S30). Incorporation of SBA and BFS caused the retardation of hardening process. For binary mixtures (OPC+SBA), when SBA substituting level varied 10%, 20%, and 30% the setting time increased 35%, 42%, 48% (initial) and 22%, 53%, and 53% (final) in comparing that of control mixture, respectively. Meanwhile, for each mix group in the ternary mixtures (OPC+SBA+BFS), when the slag increased from 10%, 20%, and 30%, the initial setting time were continuously increased 10%, 23%, and 35% (BA10 group), 8%, 24%, and 30% (BA20 group), and 10%, 24%, and 39% (BA30 group), respectively as compared with the corresponding SBA-blended binder mixture (without BFS). Less C_3A component originally existing in OPC due to cement replacement and low rate hydration of supplementary cementitious materials owing to contain high calcium silicates could mainly result in the above-mentioned time extension.

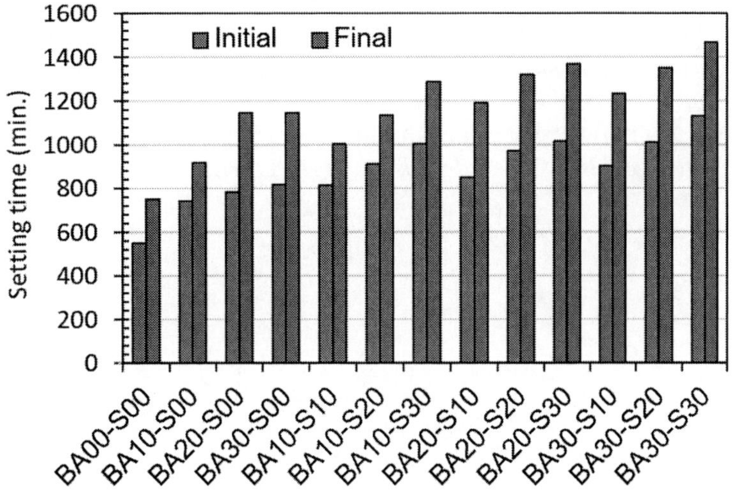

Figure 7 Results of measurement setting time.

4. Highlighted Remarks

Based on the experimental work, some of highlighted remarks could be drawn:

(1) Fresh density of SCC mixtures ranges from 2070 -2289 kg/m^3, which is similar to normal concrete. Density of SCC with blended cement was 1.4%–10% lesser than that of control mix.

(2) In comparison of control mix, adding SBA and/or BFS in mixtures was apparently resulted in lesser flowability. Angularity and irregular shapes of cementitious materials enhanced the friction between fine particles making fresh mixture to be harsh. Relationship between slump and slump flow can be expressed by a linearly best-fit.

(3) Substitution of OPC by SBA and/or BFS made the flow times (T_{500} and V-funnel) of fresh concrete to be stably extended.

(4) Incorporation of SBA and BFS caused the retardation of hardening process. The longest hardening times (initial and final) were measured on mixture made of highest replacement level of minerals (BA30-S30).

References

[1] Modani P O, Vyawahare M R 2013 Utilization of Bagasse Ash as a Partial Replacement of Fine Aggregate in Concrete *Procedia Engineering 51(Supplement C)* 25-29.

[2] Ganesan K, Rajagopal K, Thangavel K 2007 Evaluation of bagasse ash as supplementary cementitious material *Cement and Concrete Composites* 29(6) 515-524.

[3] Akkarapongtrakul A, Julphunthong P, Nochaiya T 2017 Setting time and microstructure of Portland cement-bottom ash–sugarcane bagasse ash pastes *Monatshefte für Chemie - Chemical Monthly* 148(7) 1355-1362.
[4] ASTM:C150 2002 *Standard Specification for Portland Cement.*
[5] ASTM:C618 2008 *Standard Specification for Coal Fly Ash and Raw or Calcined Natural Pozzolan for Use in Concrete.*
[6] Corinaldesi V, Moriconi G 2011 The role of industrial by-products in self-compacting concrete *Construction and Building Materials* 25(8) 3181-3186.
[7] ASTM:C33 2003 *Standard Specification for Concrete Aggregates.*
[8] Chang P K 2004 An approach to optimizing mix design for properties of high-performance concrete, *Cement and Concrete Research* 34(4) 623-629.
[9] JSCE 1998 *Guide to Construction of High Flowing Concrete.*, Gihoudou Pub., Tokyo,
[10] ASTM:C403 1999, *Standard Test Method for Time of Setting of Concrete Mixtures by Penetration Resistance.*
[11] Uysal M, Sumer M 2011 Performance of self-compacting concrete containing different mineral admixtures *Construction and Building Materials 25*(11) 4112-4120.

Undrained behavior and shear strength of clean sand containing low-plastic fines

Vu To-Anh Phan[1*] Darn-Horng Hsiao[2]

[1]Faculty of Civil Engineering, Ton Duc Thang University, Ho Chi Minh City, Vietnam
[2]Department of Civil Engineering, National Kaohsiung University of Applied Sciences, Taiwan

Corresponding author, E-mail address: phantoanhvu@tdt.edu.vn

Abstract. This study presents experimental tests to understand the undrained behavior of sand containing various fines contents. The specimens were prepared by the wet tamping method. The consolidated undrained triaxial shear tests were carried out by sands mixed with amounts of fines in ranging from 0 to 60%. The results showed that the deviator stress quickly reaches the peak value with an axial strain in a range of 0.5 to 2%, and then, the value drops significantly with further increases in the axial strain, the pore water pressure of all the sand-fines mixtures rapidly increases as the axial strain reaches a value in a range from 1 to 2% and then slowly increases and reaches a stable state when strain is greater than 8%. Peak deviator stress gradually decreases with an increasing fines content from 0 to 40%, thereafter, the peak deviator significantly increases with further increases in the fines content up to 60%, irrespective of confining pressure values using in these tests. Finally, the effective internal friction angles are remarkably greater than the total friction angles for various sand-fines mixtures.

1. Introduction

Natural sandy soil commonly contains sand particles and fines with different proportions. The engineering of these soils has been widely studied during the past few decades. Past studies have been demonstrated that the existence of fines significantly plays a major role on undrained behavior and on the engineering properties of sands [1, 2]. In fact, fine particles also cause to many complicated phenomena under undrained behavior, stress-strain relation, and liquefaction resistance [3, 4].

Some results showed that the liquefaction resistance either increased with increasing fines content in the mixture [5-7] or decreased with increasing fines content [8-10]. Other studies have found that the resistance of sand to liquefaction initially decreased as the fines content increased until some minimum resistance was reached and then increased as the fines content continued to increase [11-14]. Generally speaking, the existence of fines significantly affected the engineering properties as well as the behavior of a sand. The reason can be summarized, such as grain size distribution, the arrangement of fines in a mixture, the binary packing model, the plasticity index of fines, relative density, and the shear stress condition. Theoretically, the variation of void ratio with fines content is plotted in Figure 1. For 0 % fines content, the large grains produce a package with a global void ratio e_1, as indicated on the left side of Figure 1. Fine particles are then dropped within the void ratio among the large particles, as shown on the second inserted sketch in Figure 2. The global void ratio decreases when increasing the fines content. The minimum global void ratio e_{min} is obtained when fine particles fill up the void ratio between large particles. As the fine particles increase to surpass the minimum global void ratio, the large particles in the primary fabric are pushed apart and no large particles exist till fine grains reach up 100% (point C in Figure 1). For the stage of BC, the global void ratio increases linearly with an increasing fines content; furthermore, large particles become to separate and do not exist at the end with 100% fines corresponding with the void ratio e_2. Generally, e_2 is greater than e_1.

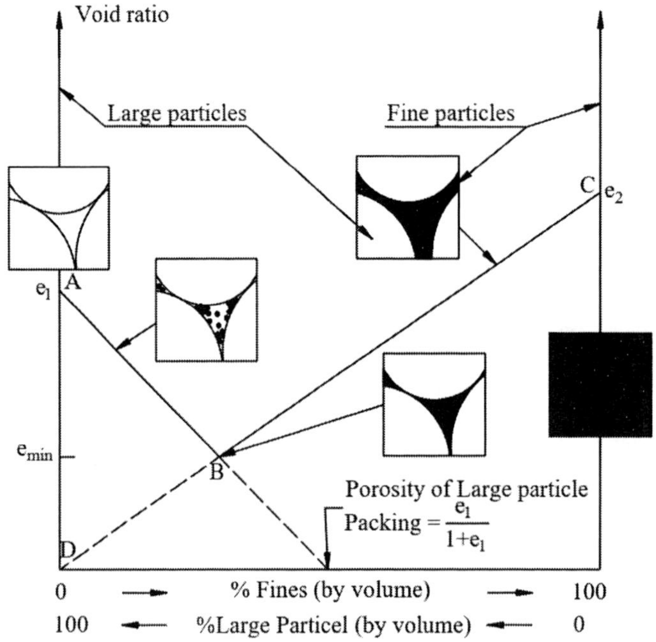

Figure 1. The global void ratio in relation to the fines content [15]

To answer the question whether the mixed soils behave sand-like or fines-like soil properties, a transitional fines content (TFC) is proposed and determined by certain studies [16, 17]. For instance, the TFC can be determined by Chang and Hong (2008) [17]:

$$TFC = \frac{e_1}{1 + e_1 + e_2} \tag{1}$$

where e_1 and e_2 are global void ratio of sand and fines, respectively.

Considering the portion of the fine particles, the active intergrain contacts are considered, TFC can be determined as follows [17]:

$$TFC = \frac{1}{(1-b)} \frac{e_1}{(1 + e_1 + e_2)} \tag{2}$$

The microstructure of a soil mixture, which can be composed of in various manners with various types of intrinsic contacts causes various undrained shear behaviors. For instance, considering a granular mixture consisting of two grain sizes, including fine particle (d) and large particle (D), which have different proportions. For this variation, Thevanayagam et al. (2002) [16] have already postulated three types of limiting categories of microstructure, including (a) primarily the coarse particles are in contact, (b) primarily the fine particles are in contact with each other. For type (a), three subsets are including: the fines are confined within the void spaces between the coarse grains with little contribution to supporting the coarse grain skeleton and it is called as case (i); the fines are partially supporting the coarse grain skeleton, case (ii); or they partially separate the coarse grains, case (iii). In type (b), for case (iv) the coarse grains are found completely to be separated the fine grain matrix. As indicated in Figure 2, some definitions are listed as follows:

$$e_s = \frac{1+e}{1-f_c} \tag{3}$$

$$e_f = \frac{e}{f_c} \tag{4}$$

Where b is a portion of the fine grains that contributes to the active intergrain contacts; e is global void ratio; FC is fine grains content; FC_{th} is threshold fine grains content; FC_L is limit fines content; m is reinforcement factor; R_d=D/d is a particle size disparity ratio

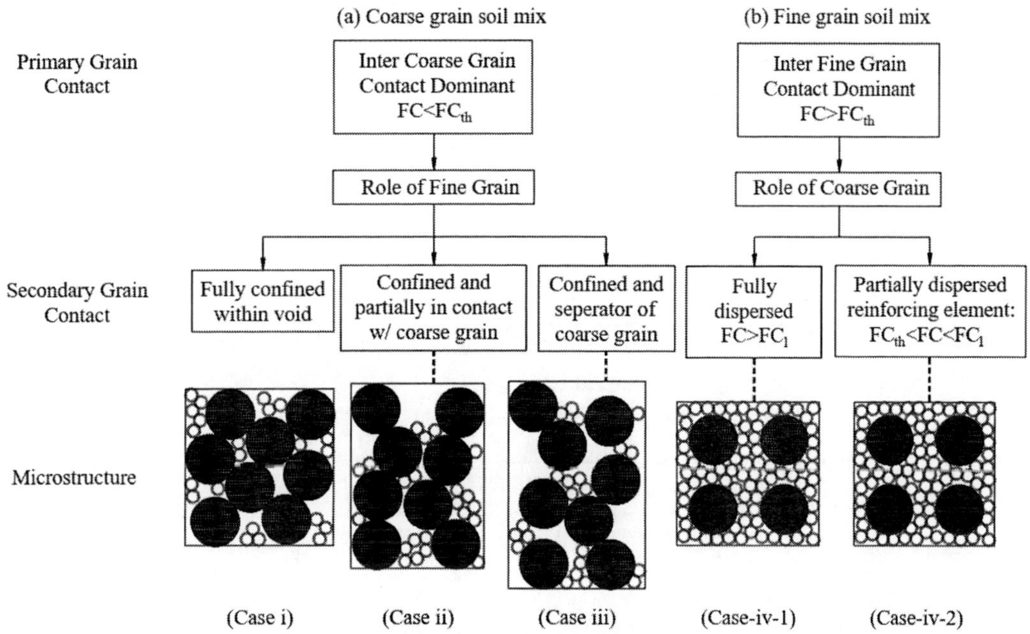

Figure 2. Framework concept of intergranular soil mix classification [16]

In this study, the test program employed triaxial shear tests on various sand-fines mixtures. The specimens were prepared at the same peak deviator stress of 290 kPa with a confining pressure of 100 kPa to understand the undrained behavior of sand containing various fines contents in ranging from 0% to 60%. The behavior of stress-strain relation, pore water pressure-strain relation, shear strength parameter (c, φ) were obtained and discussed. In addition, the relationship between the intergranular void ratio and the peak deviator stress, fines content, and confining pressure is also discussed. The experimental data is expected to have an insight about the behavior of sand containing various fines contents.

2. Materials tested sample preparation

Soil specimens were taken from Liouguei District, located in Kaohsiung city, Taiwan. A quantity of natural sandy soil was carefully sieved to separately obtain clean sand and pure fines. The fines particles are defined as the grain size of soil that is able to pass through a No. 200 (0.075 mm) sieve. The sand-fines mixtures were prepared from these two materials for the following combinations.

All tests were conducted with six sand-fines mixtures defined by dry weight: 100% sand plus 0% fines (0% FC), 85% sand plus 15% fines (15% FC), 70% sand plus 30% fines (30% FC), 60% sand plus 40% fines (40% FC), 50% sand plus 50% fines (50% FC), 40% sand plus 60% fines (60% FC). Based on the ASTM D422 [18] method, the specimen was carried out on specimens that had the same peak deviator stress of 290 kPa when tested in the triaxial test. To obtain the same peak deviator stress of 290 kPa, a series of trial mixes from 1 to 6 were gradually constructed with various void ratios and dry unit weights, and thereafter, each sample was tested with a consolidated drained triaxial shear test with a confining pressure of 100 kPa to obtain the peak deviator stress. Selected specimens had a peak deviator stress of 290 kPa, as explained in Hisao and Phan [19].

3. Results and discussion

A series of consolidated undrained triaxial tests were performed at an effective confining pressure of 50, 100, and 200 kPa using a strain-controlled system. The specimens were sheared in undrained compression at a constant strain rate of 0.5 mm/min. Past studies by Sladen et al. (1985) [20] concluded that there is no difference the results of either load-controlled or strain-controlled tests.

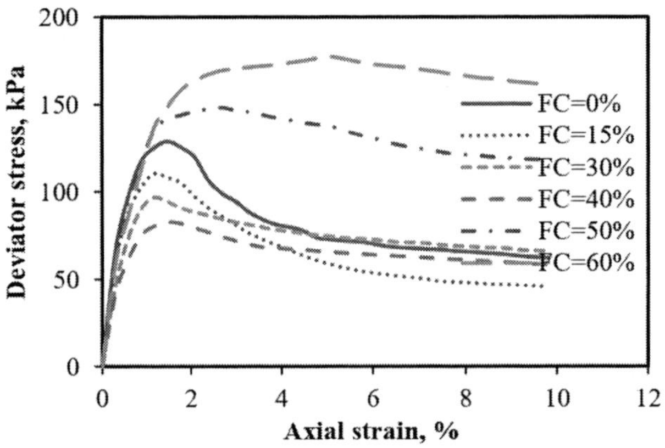

(a) Stress-strain relationship

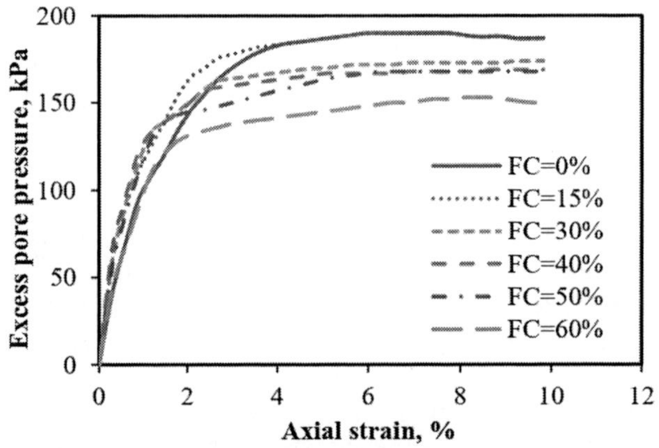

b) Excess PWP-strain relationship

Figure 3. Undrained behaviors of sand-fines mixture ($\sigma_3 = 200$ kPa)

Figure 3 typically presents the behaviors of stress and strain, the pore water pressure (PWP) and the strain of different fines contents at confining pressure of 200 kPa. The test results from a confining pressure of 50, 100 kPa produced the same tendency as the result from confining pressures of 200 kPa and thus are not shown. In general, the deviator stress quickly reaches the peak value, with an axial strain in the range of 0.5 to 2%, and then, the value drops significantly with further increases in the axial strain.

Figure 3(a) indicates that peak deviator stresses only decrease when the fines content increases from 0 to 40%, and then, the peak deviator stresses significantly increase with further increases in fines content. The peak deviator stress can reach a maximum deviator tress at a fines content of 60%. The obtained results are somewhat different from the observations of Pitman et al. (1994) [5] who studied the ravious types of of fines (kaolinite and crushed silica fines) and different fines contents and indicated a clear decreasing trend of a strength-sortening response with increasing fines content of up to 30% and noted that at 40% fines, there no defined peak and deviator stress continues to climb, flattening only at large strains.

Additionally, as shown in Figure 3(b), the PWP of all the sand-fines mixtures rapidly increases as the axial strain reaches a value in the range of 1 to 2% and then slowly increases and reaches a stable state with a strain greater than 8%. However, it is very difficult to distinguish the increasing tendency of the PWP among specimens contained various fines contents.

To understand the relationship between the peak deviator stress (PDS) and the fines contents, Figure 4 plots the variations of the peak deviator stress with various fines contents. In general, the PDS decreases with an increasing fines content of up to 40%. Figure 4 shows that the PDS gradually decreases with an increasing fines content from 0 to 40%, thereafter, the PDS significantly increases with further increases in the fines content of up to 60%, irrespective of confining pressure values using in this test.

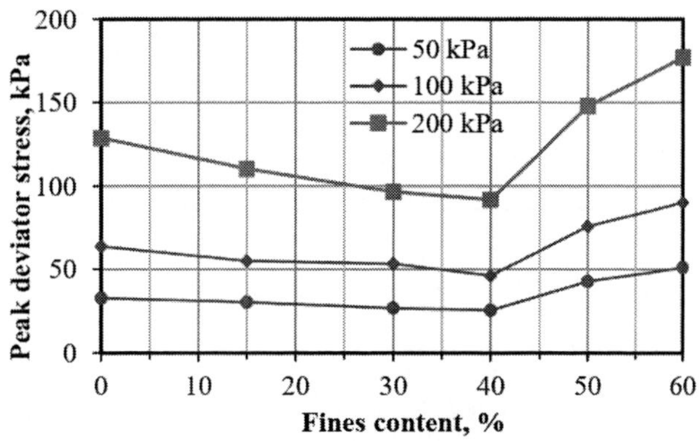

Figure 4. Peak deviator stress - fines content relation in the CU test

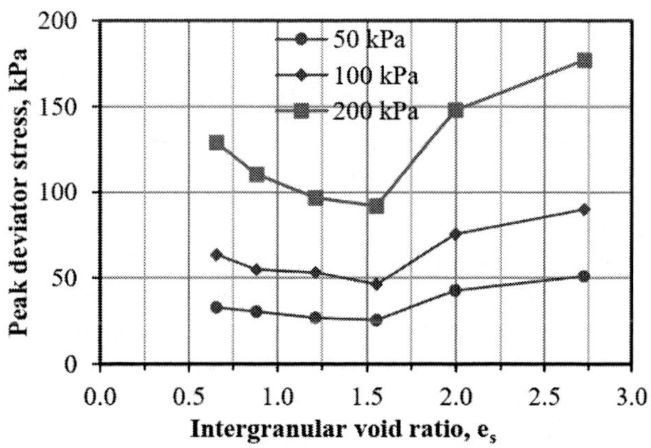

Figure 5. Peak deviator stress-intergranular void ratio relation in the CU test

In addition, the relationships between PDS and intergranular void ratio e_s of sand-fines mixtures are plotted in Figure 5. As a similar tendency presented in Figure 4, the intergranular void ratio seems to yield the same tendency with the fines contents, regardless of applied confining pressure. Figure 5 shows that the peak deviator stress is obtained with the intergranular void ratio of 1.55.

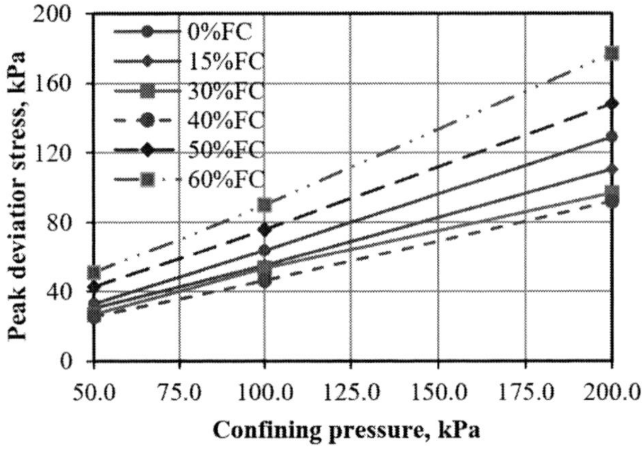

Figure 6. Various peak deviator stress and confining pressure in the CU test

Figure 6 plots the various peak deviator stresses with the confining pressures for various fines contents. In general, a higher confining pressure has a greater deviator stress, irrespective of the initial condition of the specimen. An increasing PDS with confining pressure is to be found. Moreover, the PDS line first moves downward until the fines content of 50% is reached and then PDS lines significantly moves above that of clean sand when added more fines.

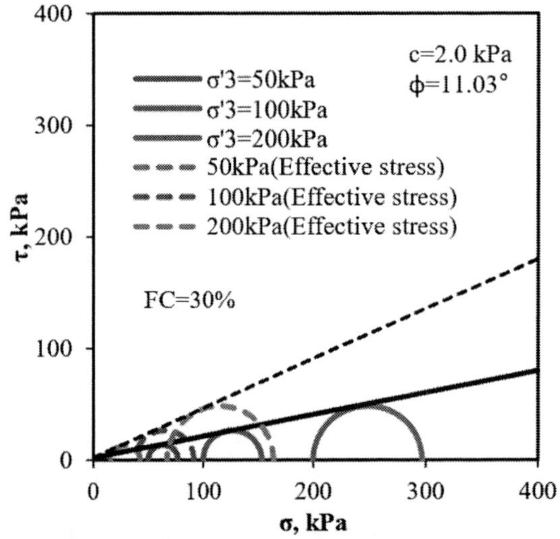

Figure 7. Mohr's circle of total and effective stress

Figure 7 presents the typical Mohr's circles of sand containing 30 % fines content. The Mohr's circles of total stress are presented by continuous lines, and the Mohr's circles of effective stress are shown by the dash lines. It is easy to recognize that the effective stress circle has the same diameter as the total stress circle and separates from it by the pore water pressure. In addition, the failure line of effective stress circle becomes steeper and causes the greater internal friction angle as comparing with its total stress circle. Finally, the results of internal friction angle and cohesion are also presented in Table 1. By comparing the results presented in Table 1, it is very obvious that the effective internal friction angles are remarkably greater than the total friction angles with different fines content. The reason can be explained by the role of excess PWP, the existence of PWP significantly causes a decreasing tendency in friction force between grain particles. This finding is well-fitted with the principle of geotechnical engineering.

Table 1 Test results in the consolidated undrained triaxial shear test

(a) Total shear stress parameters

0% FC		15% FC		30% FC		40% FC		50% FC		60% FC	
c_u (kPa)	ϕ_u(°)	c_u (kPa)	ϕ_u(°)	c_u (kPa)	ϕ_u(°)	c_u (kPa)	ϕ_u(°)	c_u (kPa)	ϕ_u(°)	c_u (kPa)	ϕ_u(°)
0	14.04	1.2	12.24	2.0	11.03	2.4	10.15	2.7	15.29	3.0	17.22

c_u: total cohesion; ϕ_u: total friction angle.

(b) Effective shear stress parameters

0% FC		15% FC		30% FC		40% FC		50% FC		60% FC	
c' (kPa)	ϕ'(°)	c' (kPa)	ϕ'(°)	c' (kPa)	ϕ'(°)	c' (kPa)	ϕ'(°)	c' (kPa)	ϕ'(°)	c' (kPa)	ϕ'(°)
0	31.5	1.7	30.4	2.6	31.2	2.9	30.5	3.7	29.3	3.0	28.9

c': effective cohesion; ϕ': effective friction angle.

4. Conclusions

Based on the obtained data from consolidated undrained trial shear tests of six sand-fines mixtures. Conclusions can be drawn as follows:

- The deviator stress quickly reaches the peak value, with an axial strain in the range of 0.5 to 2%, and then, the value drops significantly with further increases in the axial strain. The PWP of all the sand-fines mixtures rapidly increases as the axial strain reaches a value in the range of 1 to 2% and then slowly increases and reaches a stable state with a strain greater than 8%.
- Peak deviator stress gradually decreases with an increasing fines content from 0 to 40%, thereafter, the PDS significantly increases with further increases in the fines content of up to 60%, irrespective of confining pressure values using in this test.
- The intergranular void ratio seems to yield the same tendency with the fines contents, regardless of applied confining pressure.
- The effective internal friction angles are remarkably greater than the total friction angles with different fines content. This finding is well-fitted with the principle of geotechnical engineering.

References

[1] Yamamuro J and Covert K 2001 Monotonic and cyclic liquefaction of very loose sands with high silt content *Journal of Geotechnical and Geoenvironmental Engineering, ASCE* **127** 314-324

[2] Belkhatir M, Schanz T and Arab A 2013 Effect of fines content and void ratio on the saturated hydraulic conductivity and undrained shear strength of sand–silt mixtures *Environmental Earth Sciences* **70** 2469-2479

[3] Ni Q, Dasari G R and Tan T S 2006 Equivalent granular void ratio for characterization of Singapore's Old Alluvium *Canadian Geotechnical Journal* **43** 563-573

[4] Thevanayagam S and Mohan S 2000 Intergranular state variables and stress–strain behaviour of silty sands *Geotechnique* **50** 1-23

[5] Pitman T D, Robertson P K and Sego D C 1994 Influence of fines on the collapse of loose sands *Canadian Geotechnical Journal* **31** 728-739

[6] Salgado R, Bandini P and Karim A 2000 Shear strength and stiffness of silty sand *Journal of Geotechnical and Geoenvironmental Engineering, ASCE* **126** 451-462
[7] Amini F and Qi G 2000 Liquefaction testing of stratified silty sands *Journal of Geotechnical and Geoenvironmental Engineering, ASCE* **126** 208-217
[8] Lade P V and Yamamuro J A 1997 Effects of nonplastic fines on static liquefaction of sands *Canadian Geotechnical Journal* **34** 918-928
[9] Yamamuro J A and Lade P V 1997 Static liquefaction of very loose sands *Canadian Geotechnical Journal* **34** 905-917
[10] Chien L-K, Oh Y-N and Chang C-H 2002 Effects of fines content on liquefaction strength and dynamic settlement of reclaimed soil *Canadian Geotechnical Journal* **39** 254-65
[11] Koester J P 1994 The influence of fine type and content on cyclic resistance. In: *Ground failures under seismic conditions, Geotech. Spec. Publ. No. 44, ASCE, New York,*, pp 17-33
[12] Kuerbis R, Negussey D and Vaid Y 1988 Effect of gradation and fines content on the undrained response of sand *In Hydraulic fill structures. Geotechnical Special Publication No.21. Edited by VanZyl, D. J. A. and Vick, S. G., ASCE, New York* 330-345
[13] Polito C and Martin II J 2001 Effects of nonplastic fines on the liquefaction resistance of sands *Journal of Geotechnical and Geoenvironmental Engineering, ASCE* **127** 408-415
[14] Hsieh Y T 2014 Fines contents on the permeability and shear strength when static and dynamic loadings for silty sands. In: *Department of Civil Engineering*: National Kaohsiung University of Applied Sciences) p 132
[15] Lade P V, Liggro C D and Yamamuro J A 1998 Effects of non-plastic fines on minimum and maximum void ratios of sand *Geotechnical Testing Journal* **21** 336-347
[16] Thevanayagam S and Martin G R 2002 Liquefaction in silty soils—screening and remediation issues *Soil Dynamics and Earthquake Engineering* **22** 1035-1042
[17] Chang W J and Hong M L 2008 Effects of clay content on liquefaction characteristics of gap-graded clayey sands *Soils and Foundations* **48** 101-114
[18] ASTM D422-2007. In: *Standard Test Method for Particle-Size Analysis of Soils,* (100 Barr Harbor Drive, PO Box C700, West Conshohocken, PA 19428-2959, United States
[19] Hsiao D-H and Phan V T-A 2016 Evaluation of static and dynamic properties of sand–fines mixtures through the state and equivalent state parameters *Soil Dynamics and Earthquake Engineering* **84** 134-144
[20] Sladen J A, D'Hollander R D and Krahn J 1985 The liquefaction of sands, a collapse surface approach *Canadian Geotechnical Journal* **22** 564-578

IOP Conf. Series: Earth and Environmental Science **143** (2018) 012021 doi:10.1088/1755-1315/143/1/012021

Uncertainty qualification for the free vibration of a functionally graded material plate with uncertain mass density

Ta Duy Hien[1]*, Bui Tien Thanh[1] and Nguyen Thi Quynh Giang[2]

[1] Department of Civil Engineering, University of Transport and Communications, Ha Noi, Vietnam
[2] University of Education, Vietnam National University Hanoi, Vietnam

* Corresponding author, Email: tdhien@utc.edu.vn

Abstract. This work studies the response variability in the eigenvalue of free vibration of functionally graded material (FGM) plates with uncertain mass density. The mass density of plates is assumed to be a Gaussian random field on the plane of the plate. The formulation of free vibration of FGM plates is derived using isogeometric analysis based on higher-order shear deformation theory. The stochastic isogeometric analysis is established based on standard isogeometric analysis and a perturbation approach. Also, the formulation for variation of the eigenvalue of free vibration of the plate is issued. The present approach is validated with Monte Carlo simulation (MCS) based on a spectral representation method. The numerical examples investigated show good agreement between the present approach and Monte Carlo simulation. Based on the numerical results, the effect of a parameter of uncertain mass density on variability in the natural frequencies is drawn with respect to FGM plates.

1. Introduction

The stochastic problem based on an analytical and numerical approach with uncertain system parameters (e.g. material property, model geometry and loads) has been studied for many years. For dealing with uncertain parameters, most stochastic problems fall into two main categories; non-statistical approach and Monte Carlo simulation. Normally, Monte Carlo simulation repeats the deterministic problem by generating random parameters and it can be used for complex problems such as those which are nonlinear; however, it is time consuming. The non-statistical approach can produce statistics of responses, such as the mean and standard deviation of output parameters. The serval numerical approach for solving stochastic problems has been used for many years; for example, the stochastic finite element method [1-4], the spectral stochastic finite element method [5, 6], and the spectral stochastic element-free Galerkin method [7]. A number of researchers have worked on the stochastic dynamic problem. Zhu el al. [8] used the local averages of a random vector field for developing the stochastic finite element method for the eigenvalues problem. Achchhe, Lal and Singh [9] investigated the free vibration analysis of laminated composite plates resting on elastic foundations with the random variable of system method. Thuan and Noh [10] used Monte Carlo simulation to investigate the variability of the natural frequency of functionally graded beams with random fields of material properties. Kang *et al.* [11] studied the random vibration of laminated FRP plates with material nonlinearity. Chang [12] analysed the stochastic dynamics of bridge–vehicle systems subjected to random material properties and loadings using the finite element method.

This paper presents a stochastic isogeometric analysis (SIGA) for free vibration of functionally graded plates that involves uncertain mass density. The random field was discretized into a set of random variables with statistical properties obtained from the statistical properties of the random field. In conjunction with the standard isogeometric analysis, the first-order perturbation approach was performed to predict the variability eigenvalue of free vibration of FGM plates. Numerical examples are presented to validate the accuracy of the proposed method with those using Monte Carlo simulation.

2. Formulation of free vibration of FGM plate using non-uniform rational basis spline (NURBS)

2.1. FGM plate model

We consider an FGM plate, for which the geometry of FGM plate and the coordinate system is shown in Fig.1:

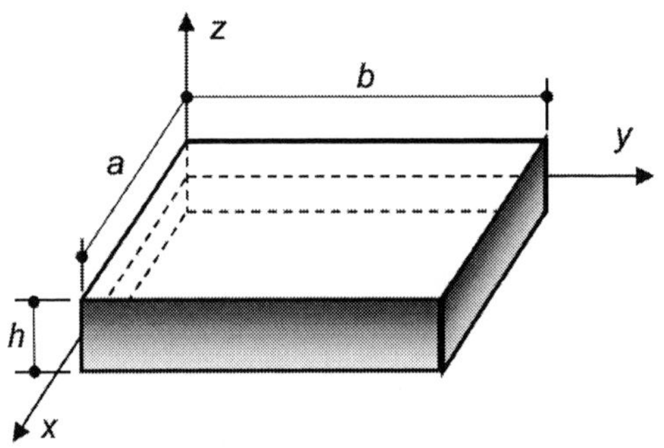

Figure 1. The Geometry of an FGM plate.

Using the higher-order shear deformation plate theory proposed by Hien *et al.* Noh [13], the displacement components at an arbitrary point *(x,y,z)* in the plate are rearranged as follows:

$$\begin{aligned} U(x,y,z,t) &= u(x,y,z,t) - z\frac{\partial w_b}{\partial x} + z\left[-\frac{1}{5}+\frac{8}{5\pi}\cos\left(\frac{\pi z}{h}\right)\right]\frac{\partial w_s}{\partial x} \\ V(x,y,z,t) &= v(x,y,z,t) - z\frac{\partial w_b}{\partial y} + z\left[-\frac{1}{5}+\frac{8}{5\pi}\cos\left(\frac{\pi z}{h}\right)\right]\frac{\partial w_s}{\partial y} \\ W(x,y,z,t) &= w_b(x,y,t) + w_s(x,y,t) \end{aligned} \tag{1}$$

where (u, v, w_b, w_s) are the displacement components on the mid-plane, respectively.

The linear strains can be obtained by differentiating Eq. (1) as:

$$\boldsymbol{\varepsilon} = \boldsymbol{\varepsilon}_0 + z\boldsymbol{\kappa}_b + \hat{q}\boldsymbol{\kappa}_s,\ \gamma = \hat{p}\boldsymbol{\varepsilon}_s \tag{2}$$

where:

$$\boldsymbol{\varepsilon} = \begin{Bmatrix} \varepsilon_x \\ \varepsilon_y \\ \gamma_{xy} \end{Bmatrix},\ \boldsymbol{\varepsilon}_0 = \begin{Bmatrix} \varepsilon_x^0 \\ \varepsilon_y^0 \\ \gamma_{xy}^0 \end{Bmatrix} = \begin{Bmatrix} \dfrac{\partial u}{\partial x} \\ \dfrac{\partial v}{\partial y} \\ \dfrac{\partial u}{\partial y}+\dfrac{\partial v}{\partial x} \end{Bmatrix},\ \boldsymbol{\kappa}_b = \begin{Bmatrix} \kappa_x^b \\ \kappa_y^b \\ \kappa_{xy}^b \end{Bmatrix} = \begin{Bmatrix} -\dfrac{\partial^2 w_b}{\partial x^2} \\ -\dfrac{\partial^2 w_b}{\partial y^2} \\ -2\dfrac{\partial^2 w_b}{\partial x \partial y} \end{Bmatrix},$$

$$\boldsymbol{\kappa}_s = \begin{Bmatrix} \kappa_x^s \\ \kappa_y^s \\ \kappa_{xy}^s \end{Bmatrix} = \begin{Bmatrix} -\dfrac{\partial^2 w_s}{\partial x^2} \\ -\dfrac{\partial^2 w_s}{\partial y^2} \\ -2\dfrac{\partial^2 w_s}{\partial x \partial y} \end{Bmatrix}, \ \gamma = \begin{Bmatrix} \gamma_{yz} \\ \gamma_{xz} \end{Bmatrix}, \ \boldsymbol{\varepsilon}_s = \begin{Bmatrix} \dfrac{\partial w_s}{\partial y} \\ \dfrac{\partial w_s}{\partial x} \end{Bmatrix}$$

The mechanical properties of FGM such as Young's modulus E and mass density ρ are assumed as

$$\begin{aligned} E(z) &= (E_c - E_m)\left(\frac{z}{h} + \frac{1}{2}\right)^n + E_m \\ \rho(z) &= (\rho_c - \rho_m)\left(\frac{z}{h} + \frac{1}{2}\right)^n + \rho_m \end{aligned} \tag{3}$$

where the subscripts m and c represent the metallic and ceramic constituents, respectively; and n is the power index of the volume fraction.

3. Stochastic isogeometric analysis for free vibration of FGM plates

3.1. A brief of isogeometric analysis

The isogeometric analysis was introduced in 2005 by Thomas Huge *el al.* [14, 15]. In this analysis, the non-uniform rational B-spline (NURBS) functions are used to construct a finite approximation.
The B-spline basis functions are defined recursively as follows [14-17]:

$$N_{i,0}(\xi) = \begin{cases} 1 & \xi_i \leqslant \xi < \xi_{i+1}, \\ 0 & \text{otherwise.} \end{cases} \tag{4}$$

and

$$N_{i,p}(\xi) = \frac{\xi - \xi_i}{\xi_{i+p} - \xi_i} N_{i,p-1}(\xi) + \frac{\xi_{i+p+1} - \xi}{\xi_{i+p+1} - \xi_{i+1}} N_{i+1,p-1}(\xi) \tag{5}$$

The Non-Uniform Rational B-Splines (NURBS) basis functions in two dimension, and can be constructed by taking the tensor product of two one-dimensional B-spline basis functions represented as [14, 16]:

$$R_{i,j}^{p,q}(\xi,\eta) = \frac{N_{i,p}(\xi) N_{j,q}(\eta) w_{i,j}}{\sum_{i=1}^{m} \sum_{j=1}^{n} \left\{ N_{i,p}(\xi) N_{j,q}(\eta) w_{i,j} \right\}} \tag{6}$$

where $N_{i,p}(\xi), N_{j,q}(\eta)$ are the B-spline basis functions of order p in the ξ, and order q in the η direction, respectively.

3.2. Stochastic isogeometric analysis for free vibration of FGM plates

Using the NURBS basis functions above, the displacement field of the plate is approximated as:

$$\begin{Bmatrix} \hat{u}_{0A} \\ \hat{v}_{0A} \\ \hat{w}_{0A}^b \\ \hat{w}_{0A}^s \end{Bmatrix} = \sum_{A=1}^{m \times n} R_A(\xi,\eta) \begin{Bmatrix} u_{0A} \\ v_{0A} \\ w_{0A}^b \\ w_{0A}^s \end{Bmatrix} \tag{7}$$

Substituting Eq. (7) into Eq.(2) the in-plane and shear strains can be rewritten as:

$$\begin{bmatrix} \boldsymbol{\varepsilon}_0^T & \boldsymbol{\kappa}_b^T & \boldsymbol{\kappa}_s^T & \boldsymbol{\varepsilon}_s^T \end{bmatrix}^T = \sum_{A=1}^{m \times n} \begin{bmatrix} (\boldsymbol{B}_A^0)^T & (\boldsymbol{B}_A^{b1})^T & (\boldsymbol{B}_A^{b2})^T & (\boldsymbol{B}_A^s)^T \end{bmatrix}^T \boldsymbol{d}_A \tag{8}$$

where $\boldsymbol{d}_A$ is displacement parameter at the control point A:

$$\boldsymbol{d}_A = \begin{Bmatrix} u_{0A} \\ v_{0A} \\ w_{0A}^b \\ w_{0A}^s \end{Bmatrix}$$

Using the virtual work principle, the weak form of free vibration of FGM plate can be obtained as follows:

$$\int_\Omega \delta\varepsilon_b^T \hat{\boldsymbol{D}}\varepsilon_b d\Omega + \int_\Omega \delta\gamma^T \boldsymbol{A}^s\gamma d\Omega = \int_\Omega \delta\bar{\boldsymbol{u}}^T \boldsymbol{m}\ddot{\bar{\boldsymbol{u}}} d\Omega \tag{9}$$

The formulation of free vibration of the FGM plate is expressed as follows:

$$\boldsymbol{Kd} + \boldsymbol{M}\ddot{\boldsymbol{d}} = \boldsymbol{0} \tag{10}$$

where ***M, K*** represent mass matrix and stiffness matrix, respectively.

The random field in mass density is assumed as homogeneous Gaussian in the plane of the plate:

$$\rho(x,y,z) = \left\{ (\rho_c - \rho_m)\left(\frac{z}{h} + \frac{1}{2}\right)^p + \rho_m \right\}(1 + r(x,y)) \tag{11}$$

The random fields are discretized into a set of Nr random variables with pre-assigned statistical properties as follows:

$$\boldsymbol{r} = \{r_1, r_2, \ldots r_{Nr}\} \tag{12}$$

Substituting random variables in Eq.(12) into the governing equation of free vibration, we obtain the discretization of the stochastic isogeometric analysis:

$$\left[\boldsymbol{K} - \omega^2 \boldsymbol{M}(\boldsymbol{r})\right]\boldsymbol{X}(\boldsymbol{r}) = \{\boldsymbol{0}\} \tag{13}$$

where ω and $\boldsymbol{X}$ are natural frequencies and displacement vector (eigenvectors).

The Taylor series expansion at $\boldsymbol{r}$=**0** gives:

$$\begin{aligned} \boldsymbol{M}(\boldsymbol{r}) &= \boldsymbol{M}_0 + \sum_{i=1}^{Nr} \boldsymbol{M}_{,i}^r r_i + \frac{1}{2}\sum_{i,j=1}^{Nr} \boldsymbol{M}_{,ij}^r r_i r_j + \ldots \\ \boldsymbol{X}(\boldsymbol{r}) &= \boldsymbol{X}_0 + \sum_{i=1}^{Nr} \boldsymbol{X}_{,i}^r r_i + \frac{1}{2}\sum_{i,j=1}^{Nr} \boldsymbol{X}_{,ij}^r r_i r_j + \ldots \\ \lambda(\boldsymbol{r}) &= \lambda_0 + \sum_{i=1}^{Nr} \lambda_{,i}^r r_i + \frac{1}{2}\sum_{i,j=1}^{Nr} \lambda_{,ij}^r r_i r_j + \ldots \end{aligned} \tag{14}$$

where $\lambda = \omega^2$ is an eigenvalue.

Substituting series in Eq.(14) into Eq.(13), solving Eq.(13) using the perturbation technique, we obtain the eigenvalues and their derivatives. The mean and variance of the eigenvalues are given as:

$$\begin{aligned} \mu_\lambda &= \lambda_0^k \\ Var_\lambda &= \sum_{i=1}^{NR}\sum_{j=1}^{NR} \lambda_{,i}^{r-k} \lambda_{,j}^{r-k} \boldsymbol{R}_{ij}^r \end{aligned} \tag{15}$$

The response variability COV of eigenvalue λ is computed as follows:

$$COV = \frac{\sqrt{Var_\lambda}}{|\mu_\lambda|} \tag{16}$$

For validating the present approach, the Monte Carlo simulation is performed using the spectral representation method as proposed by Shinozuka *et al.* [18, 19].

4. Examples

A simply supported FGM rectangular plate with side-to-thickness ratio a/h=40, dimension of rectangular a=0.4m, b=0.3m, and a power index p=2 is considered. The elastic moduli and mean of mass density are chosen to be the same as [20] $E_m = 70\ GPa$, ρ_m=$2707 kg/m^3$, $E_c = 380GPa$, ρ_c=$3800 kg/m^3$, with a Poisson's ratio of 0.3. The auto-correlation function for the stochastic fields function $r(x,y)$ of mass density is assumed as form:

$$R(\xi)=\sigma^2\exp\left[-\left(\frac{\xi_x}{d_1}\right)^2-\left(\frac{\xi_y}{d_2}\right)^2\right] \quad (17)$$

where ξ_x,ξ_y are components of the separate vector ξ in the plane Cartesian coordinate and d_1,d_2 are correlation distances along the corresponding directions.

The coefficient of variance of random field is assumed $\sigma=0.15$. We consider the correlation distance $d=d_1=d_2$.

Fig.2. shows the first mode shape and second mode shape of a simply supported rectangular FGM based on deterministic analysis.

Fig.3 illustrates the response COV of eigenvalue at the first mode and the second mode using both approaches: the stochastic isogeometric analysis method and Monte Carlo simulation. It is clear that the response COV increases when correlation distanced d increases. When correlation distanced d is quite large, the response COV of an eigenvalue is nearly equal to the coefficient of variance of the random field σ. Also, Fig.3 illustrates that there is good agreement between the results of the present approach and the Monte Carlo simulation.

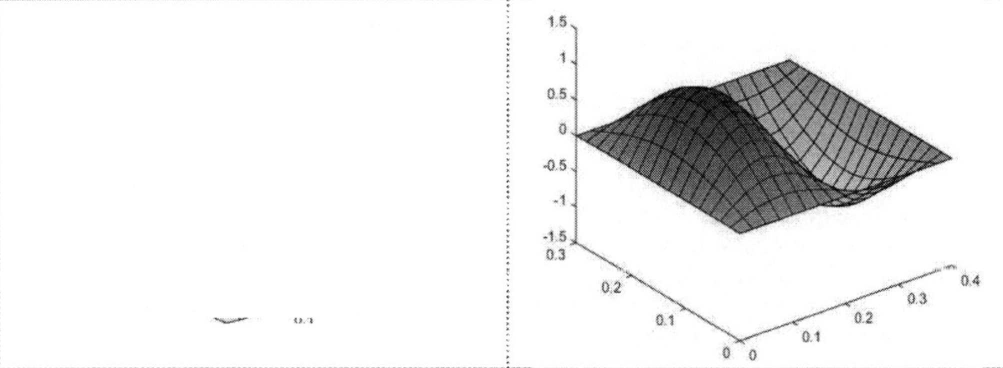

a) The first mode b) The second shape

Figure 2. The mode shape of a simply supported FGM plate.

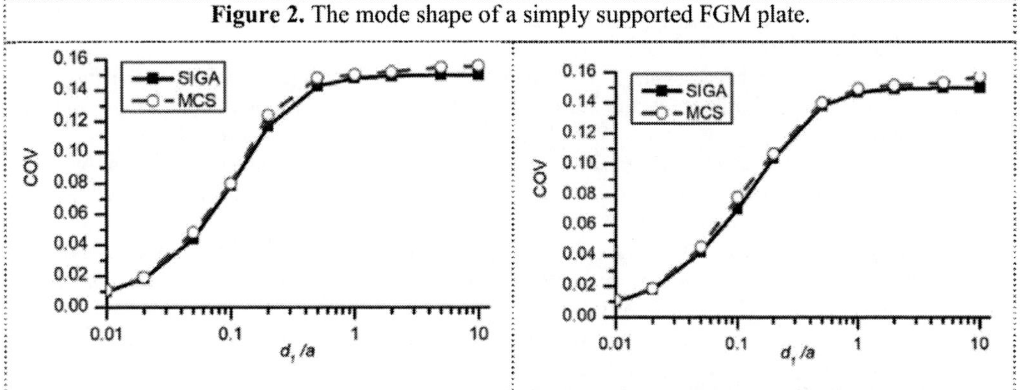

a) The first mode b) The second shape

Figure 3. Response COV of the eigenvalue λ & correlation distance d.

5. Conclusions

The statistics of the eigenvalue of free vibration of FGM plates such as the mean, standard deviation, and response variability of eigenvalue have been obtained by using the perturbation technique in conjugation with standard isogeometric analysis. The response COV of the eigenvalue obtained by the present approach are in good agreement with the Monte Carlo simulation approach. Form the results of numerical examples, we observed that the response COV is in the range of 0.9%–15% relative to the coefficient of variation of the random field, which is assumed to be 0.15. Also, the numerical examples clearly show that the effect of uncertain mass density on standard deviation and response COV are significant.

References

[1] Liu WK, Belytschko T, Mani A 1986 Random field finite elements *International Journal for Numerical Methods in Engineering* 23(10):1831-1845.
[2] Takada T 1990 Weighted integral method in stochastic finite element analysis *Probabilistic Engineering Mechanics* 5(3):146-156.
[3] Michael Kleiber, Hien TD 1992 *The stochastic finite element method: basic perturbation technique and computer implementation* Wiley.
[4] Noh H-C. 2011 Stochastic finite element analysis of composite plates considering spatial randomness of material properties and their correlations *Steel and Composite Structures* 11(2):115-130.
[5] Ghanem RG, Spanos PD 1991 Spectral stochastic finite-element formulation for reliability-analysis *Journal of Engineering Mechanics* 117(10):2351-2372.
[6] Chen NZ, Guedes Soares C 2008 Spectral stochastic finite element analysis for laminated composite plates *Computer Methods in Applied Mechanics and Engineering* 197(51–52) : 4830-4839.
[7] Kim HM, Inoue J 2004 A spectral stochastic element free Galerkin method for the problems with random material parameter *International Journal for Numerical Methods in Engineering* 61(11) : 1957-1975.
[8] Zhu WQ, Wu WQ 1991 A stochastic finite element method for real eigenvalue problems *Probabilistic Engineering Mechanics* 6(3–4) : 228-232.
[9] Lal A, Singh BN 2009 Stochastic nonlinear free vibration of laminated composite plates resting on elastic foundation in thermal environments *Computational Mechanics* 44(1):15-29.
[10] Van TN, Noh HC 2017 Investigation into the effect of random material properties on the variability of natural frequency of functionally graded beam *KSCE Journal of Civil Engineering* 21(4):1264-1272.
[11] Kang J, Harichandran R 1999 Random Vibration of Laminated FRP Plates with Material Nonlinearity Using High-Order Shear Theory *Journal of Engineering Mechanics* 125(9):1081-1088.
[12] Chang TP 2014 Stochastic dynamic finite element analysis of bridge–vehicle system subjected to random material properties and loadings *Applied Mathematics and Computation* 242(0):20-35.
[13] Hien TD, Noh HC 2015 Analytical solution for the dynamic response of functionally graded rectangular plates resting on elastic foundation using a refined plate theory *Applied Mathematical Modelling* 39(20):6243-6257.
[14] Hughes TJR, Cottrell JA, Bazilevs Y 2005 Isogeometric analysis: CAD, finite elements, NURBS, exact geometry and mesh refinement *Computer Methods in Applied Mechanics and Engineering* 194(39–41):4135-4195.
[15] Austin CJ, Hughes TJR, Bazilevs Y *2009 Isogeometric Analysis: Toward Integration of CAD and FEA* Wiley.
[16] Nguyen VP, Bordas S, Rabczuk T 2012 *Isogeometric analysis: an overview and computer implementation aspects.* arXiv:12052129 [csNA].
[17] Piegl L, Tiller W 1995 *B-Spline Basis Functions* The NURBS Book. Berlin, Heidelberg: Springer

Berlin Heidelberg p. 47-79.
[18] Shinozuka M, Deodatis G 1991 Simulation of Stochastic Processes by Spectral Representation *Applied Mechanics Reviews* 44(4):191-204.
[19] Shinozuka M, Deodatis G 1996 Simulation of Multi-Dimensional Gaussian Stochastic Fields by Spectral Representation *Applied Mechanics Reviews* 49(1):29-53.
[20] Hasani Baferani A, Saidi AR, Ehteshami H 2011 Accurate solution for free vibration analysis of functionally graded thick rectangular plates resting on elastic foundation *Composite Structures* 93(7):1842-1853.

Acknowledgments
The authors would like to acknowledge the support from the University of Transport and Communications under the basic research fund and the Vietnam National Foundation for Science and Technology Development (NAFOSTED) under grant number 107.01-2017.314.

IOP Conf. Series: Earth and Environmental Science **143** (2018) 012022 doi:10.1088/1755-1315/143/1/012022

Investigation of using steel slag in hot mix asphalt for the surface course of flexible pavements

Hien Q Nguyen*, **Dai X Lu and Son D Le**

Ho Chi Minh City University of Transport, No.2, D3 Street, Binh Thanh, Ho Chi Minh City, 700000, Vietnam

*Corresponding author, Email: hien.nguyen@ut.edu.vn , Tel : +84.28.3512.7781.

Abstract: The rapid development of heavy industry in Vietnam leads to the establishments of steel industry. Steel slag, a by-product of steelwork industry, under Vietnamese's law, was considered as a deleterious solid waste which needed to be processed and landfilled. However, this has changed recently, and steel slag is now seen as a normal or non-deleterious solid waste, and has been studied for reuse in the construction industry. In this study, steel slag was used, as a replacement for mineral aggregate, in hot mix asphalt. Two hot mix asphalt mixtures with an equivalent nominal aggregate size of 12.5 (C12.5) and 19 mm (C19) were produced using steel slag. In addition, one conventional hot mix asphalt mixture of C19 was produced using mineral aggregate for comparison purpose. Investigation in laboratory condition and trial sections was carried out on Marshall tests, surface roughness, skid resistance, and modulus of the pavement before and after applying a new surface course of hot mix asphalt. The study showed that all steel slag asphalt mixtures passed the Marshall stability and flow test requirements. The skid resistance of steel slag hot mix asphalt mixtures for the surface course satisfied the Vietnamese specification for asphalt. Moreover, the pavement sections with the surface course of steel slag hot mix asphalt showed a considerable higher modulus than that of the conventional one. Only the roughness of the surface course paved with C19 did not pass the requirement of the specification.

1. Introduction

Sustainable development is an vital goal of many countries in the world. Rapid development demands a great need of heavy industries which produce the required material. Vietnam is a developing country with a rapid development of heavy industries. In steel industry, there is a huge demand for steel production for the domestic use. This requirement results in a numerous new steel mills. As a result, more steel slag, a by-product from steel industry, is being produced. In Vietnam, a few years ago, steel slag was considered as a deleterious solid waste which needed to be processed and landfilled. However, nowadays steel slag is seen as a normal, non-deleterious solid waste, and has been studied for reuse in construction industry such as making concrete or brick block. So far, the recycling of steel slag in the road construction in Vietnam is still very limited.

Although steel slag is paid attention for reuse in the construction industry, the recycling of this material in road construction in Vietnam still lags behind compared to other countries. In USA, Japan, German and France, the reuse of steel slag is close to 100% [1]. The reuse of steel slag is highly concentrated to road construction in developed countries, with more than 50% of steel slag is used directly for road projects [2]. Until now, there are a number of official guidelines in English for the reuse of steel slag, which can be easily found by using online searching tools, such as [3] for using steel slag in roads published by Australian Slag Association, or in report [4] for the use of steel slag in pavements in Washington state, USA. Although there are many countries reusing steel slag as a pavement material, the application of the material in pavement is still a concern for developing countries like Vietnam. The reasons being are the suitability of the material to replace traditional mineral aggregate and especially whether the material is contaminating to the environment. The concern of contamination of the environment is understandable as steel slag, a by-product of steel production, is different based on different technologies applied in the production procedure. In a developing country like Vietnam the technology in steel industry may not be as advanced as in developed countries hence the by-product may contain chemical elements that are deleterious to the environment and human.

Study [5] investigated whether steel slag is suitable for road base shows that steel slag contained high percentage of CaO which is sensitive to water and may change the volume of the material. This study indicated that steel slag may lose stability if water is present due to volume expansion. After that, steel slag was conditioned in moisture for six months and grinded for experiments. At this stage, steel slag showed quite high modulus, even higher than mineral aggregate for road base. The study concluded that steel slag can be used for road base after moisture treatment to reduce the volume change possibility before application [5]. Another study [6] also investigated the posibility of using steel slag for road base and it concluded that the road base of 100% steel slag still worked well after a number of floods due to poor drainage system. The mentioned studies show that steel slag can be used for road base. However, in the manual on the use of steel slag in the United States, the alkalinity leach from steel slag is warn may exceeds 11 pH. Excessive pH levels of leachate can corrode sewers and affect aquatic ecological environment surrounding the road construction site [4]. Therefore, it is very important and necessary to determine the pH of steel slag prior to use for the road base.

Steel slag was also studied for use in asphalt concrete. Since the 1990s, steel slag has been used for asphalt concrete in Australia, USA [3] [4]. The advantages of steel slag compared to natural stone materials are the high roughness, high porosity and good strength. In 1991, Ali and his colleagues conducted a study on the stability of asphalt concrete in Canada using steel slag [7]. The research showed that higher steel lag content in asphalt concrete increased its elastic modulus, decreased deformation, increased compressive strength, and less susceptible to water [7]. However, using 100% steel slag would require a higher bitumen content of 25% than the mineral aggregate [7]. The research [7] did not mention the possibility of expanding the volume of steel slag affecting the actual work of asphalt concrete in fields. Therefore, a number of studies have been conducted to evaluate the effects of the volume expansion of steel lag on the performance of asphalt concrete [8] [9]. The results showed that high steel slag content in asphalt concrete caused cracking of asphalt concrete due to the effect of water [8] that causes expansion of steel slag [9]. The study [8] also recommended that a suitable test to check the expansion of the slag before use is needed.

2. Research aim and scope

This research aims to study the ability of recycling steel stag in hot mix asphalt for the surface course of road pavement in Vietnam. Firstly, the chemical and physical properties of steel lag were studied to evaluate the applicability of the material for production of hot mix asphalt. Asphalt mixtures were later prepared with steel slag and mineral aggregate. Two hot mix asphalt mixtures with nominal aggregate sizes of 12.5 and 19 mm were produced using steel slag, and one hot mix asphalt mixture with the nominal aggregate size of 12.5 mm was produced using mineral aggregate for comparison purpose. Investigation in laboratory condition and trial sections was carried out on Marshall tests, surface roughness, skid resistance, and modulus of the pavement, before and after applying a new surface course of hot mix asphalt.

3. Material and mixture preparation

3.1 Material

In the begin of the project, samples of steel stag were collected and evaluated on basic physical and chemical properties to check the applicability of using steel slag as a replacement material for mineral aggregate in production of hot mix asphalt. It should be noticed that the properties of steel slag presented here are for raw material which had not yet grinded and classified for gradation. The properties of steel slag were determined in accordance to Vietnamese standards as shown in Table 1.

Table 1: Physical and chemical properties of steel stag.

Order	Experiment details	Standard	Result	Unit
	Physical property			
1	Apparent Specific Gravity	TCVN 7572-4:2006	3.5	g/cm3
2	Bulk Specific Gravity	TCVN 7572-4:2006	3.24	g/cm3
3	Bulk Saturated Surface Dry (SSD) Specific Gravity	TCVN 7572-4:2006	3.32	g/cm3
4	Water absorption	TCVN 7572-4:2006	2.07	%
5	Bulk density	TCVN 7572-6:2006	1826	Kg/m3
6	Voids between aggregate particles	TCVN 7572-6:2006	46.86	%
7	Content of dust, mud and clay in aggregates	TCVN 7572-8:2006	0.95	%
8	Los Angeles abrasion	TCVN 7572-12:2006	20.8	%
9	Elongated particles	TCVN 7572-13:2006	0.74	%
10	Volume change due to water	22TCVN 333:06	0	%
11	CBR	22TCVN 333:06	90.68	%
12	Modulus of aggregate	22TCVN 211:06	230.7	MPa
	Chemical property (by weight of sample)			
1	SiO_2	TCVN 7131-2002	15.358	%
2	Al_2O_3	TCVN 7131-2002	7.1	%
3	Fe_2O_3	TCVN 7131-2002	34.7	%
4	CaO	TCVN 7131-2002	24	%
5	MgO	TCVN 7131-2002	5.09	%
6	Sunphat, sunphit	TCVN 7131-2002	0.088	%

Base on the results from Table 1, it can be seen that the specific gravity of steel slag (3.5 g/cm^3) is larger than typical mineral aggregates which are normally around 2.7 g/cm^3. Steel slag has higher water absorption but very stable with water as the volume change due to water is zero. The content of dust, mud and clay in steel slag is rather low, 0.95%. This may reduce the binder adhesion. However, as the test was carried out on the material before grinding, the amount of dust, must and clay can be adjusted depending on the purpose of the mix design. There is also an interesting notice that the voids in steel slag is high, and the surface of steel slag is rough which would increase the binder adhesion. Steel slag has a very small amount of elongated particles, 0.74%, meaning that the material satisfying the requirement of flat and elongate particles of aggregate for asphalt mixtures. Los Angeles abrasion resulted in 20.8% of aggregate mass abraded. The value satisfies the toughness requirement of aggregate for Hot Mix Asphalt (HMA). Basically the results shown here for physical properties of steel slag satisfied the requirement for aggregate for production of HMA. In terms of chemical properties, Fe_2O_3, SiO_2 and CaO dominate the mass of the material, accounting for roughly 74% of steel slag mass. Al_2O_3 and MgO account for around 12.2 % of the material mass. The component contents should be different for different technologies and source of mine.

3.2 Mixture design

Two hot mix asphalt mixtures with steel slag were designed then produced at an asphalt plant for placement and compaction at a trial section. The maximum nominal aggregate sizes of the two mixtures are 12.5 and 19 mm, abbreviated as C12.5 and C19 respectively. There was a control HMA mixture produced with mineral aggregate with the maximum nominal aggregate size of 12.5 mm. The gradations of the two asphalt mixtures aggregate are shown in Figure 1 and Figure 2.

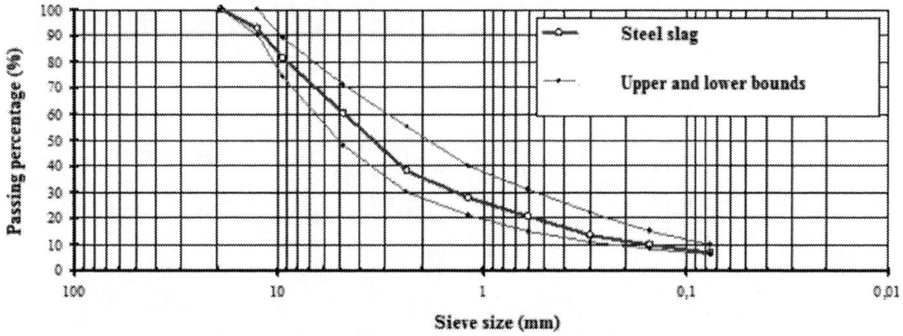

Figure 1: Aggregate gradation of HMA steel slag with nominal aggregate size of 12.5 mm

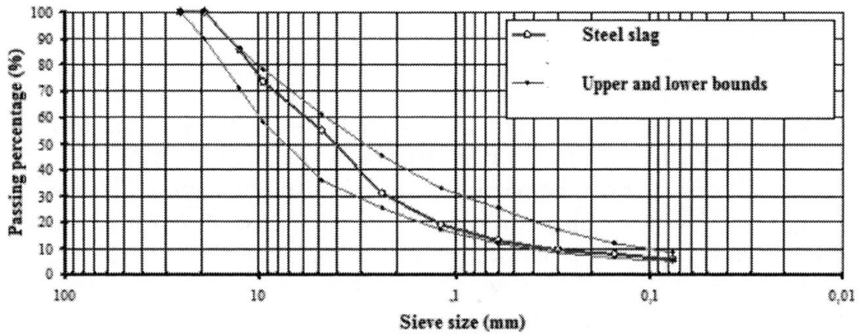

Figure 2: Aggregate gradation of HMA steel slag with nominal aggregate size of 19 mm

The optimum binder contents were designed according to Vietnamese specification. To determine the optimum binder content, five different binder contents were chosen to produce samples, and three replicas were prepared at each binder content. It should be noticed that the apparatus in this study followed Marshall mix design procedure, and samples with the diameter of 100 mm and height of 57 mm were prepared. The optimum binder contents for HMA using steel slag were 5.13 % and 4.92 % of the mixture weight for C12.5 and C19 respectively.

3.3 Mixture production

After designing the mixtures at laboratory condition, HMA mixtures with steel slag were produced at a local asphalt plant in Vung Tau, Vietnam. The mixing plant has an automatic batched mixing mechanism, originally from South Korea with the capacity of producing up to of 80 tons per hour. The mixing procedure of HMA using steel slag is similar to the mixing procedure of HMA using mineral aggregate. The heating temperature of steel slag was 170°C and bitumen was heated to 140°C for mixing. The mixing time was 40 seconds. After mixing, HMA steel slag was loaded on a specialised truck and transported to the trial section site for placement and compaction. Figure 3 illustrates main works at the mixing plant.

Figure 3: Main works of production of steel slag HMA at the asphalt plant

4. Trial section preparation and testing

4.1 Trial section preparation

The trial section road is the end section of a road in Quang Phu – Phuoc An, Chau Duc district, Ba Ria-Vung tau province, Vietnam. This road section had two lanes, and each lane is 3.5 m wide. The existing pavement was chip seal one which was planned to be upgraded to HMA surface course. The C12.5 steel slag HMA mixture was paved roughly 20 m long and C19 steel slag HMA mixture was paved approximately 15 m long (Figure 4). Before the construction, the modulus at 15 different positions in five cross section of the existing pavement were checked (Figure 4). The purpose of this is to have a comparison with the modulus of the pavement after surfacing which will be discussed in Section.

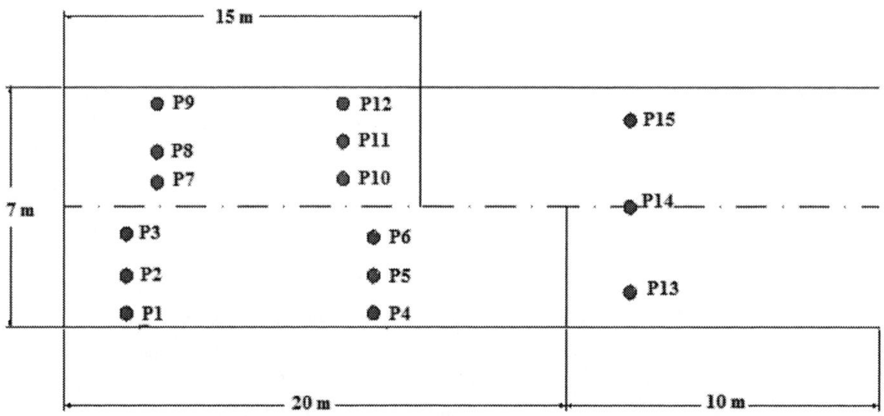

Figure 4: Trial section paving scheme and modulus checking positions

4.2 Testing

4.2.1 Surface roughness test.

The surface roughness test was carried out in this study by using a 3.0 m straight edge. The procedure was followed as described in Vietnamese specification TCVN 8864:2011. The straight edge was placed at three different positions at a chosen cross section as illustrated in Figure 5. The heights of the gaps between the straight edge and the road surface were determined by using a wedge designed as a ruler.

The gap heights are presetned in Tables 2 and 3. According to Vietnamese specification, there are three levels of surface roughness: very good, good, and medium. For surface course of hot mix asphalt, the level "very good" requires 70% of gap heights are not greater than 3 mm, and the rest 30% gap heights are not greater than 5 mm. For the level "good", 50% of gap heights should not be larger than 3 mm, and the rest 50% should not be larger than 5 mm. For the level "medium", 100% gap heights should not exceed 5mm.

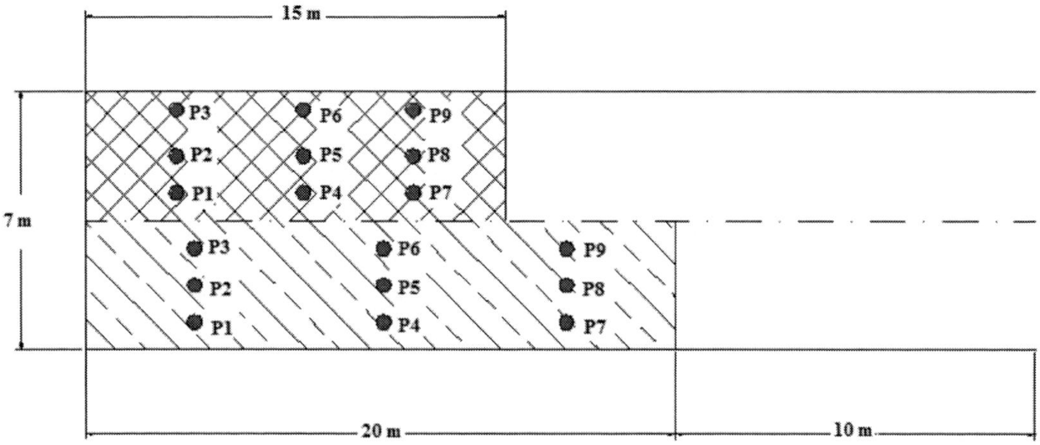

Figure 5: Surface roughness checking scheme

Table 2: Surface roughness test results of steel slag hot mix asphalt C12.5 surface course

Cross section	Checked Position	Number of gaps between the 3 m straight edge and road surface , x (mm)						
		x<3	3<x≤5	5<x≤7	7<x≤10	10<x≤15	15<x≤20	x>20
1	P1	6	1	0	0	0	0	0
	P2	4	3	0	0	0	0	0
	P3	5	2	0	0	0	0	0
2	P4	6	1	0	0	0	0	0
	P5	6	1	0	0	0	0	0
	P6	7	0	0	0	0	0	0
3	P7	3	4	0	0	0	0	0
	P8	6	1	0	0	0	0	0
	P9	6	1	0	0	0	0	0
Sum of checked position	63	49	14	0	0	0	0	0
% of total position		78%	22%	0%				

As can be seen from Tables 2 and 3, the percentage of gap heights for C12.5 satisfied Vetnamese specification for the level "very good". However, the results from C19 surface course pointed out that the surface roughness of the road did not satisfy the requirement of Vietnamese standard as there was 3% of gap heights were larger than 5 mm and smaller than 7 mm. Although the results may be affected by the construction of the layer, or by the suitability of the thickness chosen for C19 mixture, but it should be noted that the surface roughness of C19 needs to be checked again and have to satisfy the standard before applying the mixture for the surface course of flexible pavement. For C12.5, the mixture complies with the standard and well graded in terms of surface roughness.

Table 3: Surface roughness test results of steel slag hot mix asphalt C19 surface course

Cross section	Checked Position	Number of gaps between the 3 m straight edge and road surface , x (mm)						
		x<3	3<x≤5	5<x≤7	7<x≤10	10<x≤15	15<x≤20	x>20
1	P1	6	1	0	0	0	0	0
	P2	6	1	1	0	0	0	0
	P3	5	2	0	0	0	0	0
2	P4	4	3	0	0	0	0	0
	P5	5	2	0	0	0	0	0
	P6	7	0	0	0	0	0	0
3	P7	3	4	0	0	0	0	0
	P8	6	0	1	0	0	0	0
	P9	5	2	0	0	0	0	0
Sum of checked position	64	47	15	2	0	0	0	0
% of total position		73%	23%	3%				

4.2.2 Skid resistance test.
The skid resistance tests of the road surface used in this study was sand patch test method. The procedure was described in Vietnamese specification TCVN 8866:2011. This procedure in Vietnamese specification is similar to that in ASTM E 965. To carry out the test, dry road surface must be prepared. A known quantity of sand was poured onto the surface and spread in a circle. When the sand cannot be spread any more, the sand circle was measured and recorded in mm. As the sand quantity is known, the average texture depth is known by utilizing the measured diameter. Similar to the surface roughness test, three cross sections were chosen to test for each mixture. However, there were two positions checked per cross section rather than three as in the surface roughness test. According to the requirement of Vietnamese standard for skid resistance, the texture depth value of a newly built surface course constructed by hot mix asphalt should be equal to or greater than 0.45 mm. As can be seen in Table 4, the average values of texture depth are equally 0.47 for both C12.5 and C19, meaning that the skid resistance of the surface courses constructed by steel slag hot mix asphalt satisfied the requirement of skid resistance for hot mix asphalt.

Table 4: Sand patch test results for steel slag hot mix asphalt surface courses for C12.5 and C19

Mixture	Value determined in mm	Cross section 1		Cross section 2		Cross section 3		Average	CoV (%)
		P1	P2	P1	P2	P1	P2		
C12.5	Diameter of sand circle	262.8	259.5	260.9	258.6	262.1	261.3		
	Texture depth	0.46	0.47	0.47	0.48	0.46	0.47	0.47	1.61
C19	Diameter of sand circle	257.9	260.7	260.3	259.3	260.1	260.9		
	Texture depth	0.48	0.47	0.47	0.47	0.47	0.47	0.47	0.87

4.2.3 Marshall stability and flow test.
Cores from paved surface course of steel slag HMA mixtures were subjected to Marshall stability and flow test to check the compliance of the mixtures to the requirement of Vietnamese specification. Three samples were cored for each test purpose. Totally 6 samples were cored and tested and their locations

are shown in Figure 6 . The results of tests are shown in Table 5. The Marshall stability requirement is addressed in TCVN 8819-2011, in which the value should be equal to or greater than 6 kN. It can be seen that the Marshall stability of the two mixtures satisfied the requirement from of the Vietnames specification. Similar to the Marshall stability, the flow test of the samples from the two mixtures also passed the requirement of Vietnamese specification, TCVN 8819-2011, in which the results all fall into the requirement range of 2 – 4 mm.

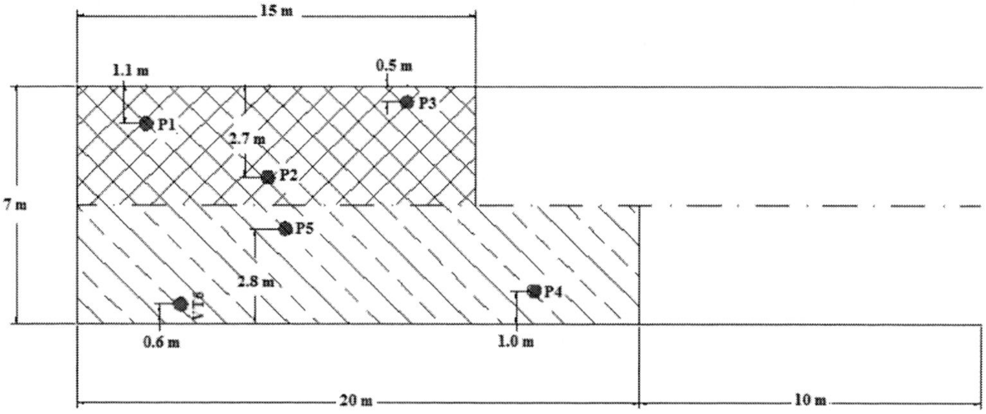

Figure 6: Coring scheme to prepare samples for Marshall stability and flow test

Table 5: Marshall stability and flow test results

Mixture	Sample	Marshall stability (kN)	Specification's Requirement (kN)	Satisfaction	Flow test (mm)	Specification's Requirement (mm)	Satisfaction
C12.5	1	11.08	≥ 6.0	Yes	3.81	2 - 4	Yes
	2	11.72	≥ 6.0	Yes	3.86	2 - 4	Yes
	3	11.62	≥ 6.0	Yes	3.78	2 - 4	Yes
	Average	11.47	≥ 6.0	Yes	3.82	2 - 4	Yes
C19	1	12.44	≥ 6.0	Yes	3.56	2 - 4	Yes
	2	12.36	≥ 6.0	Yes	3.51	2 - 4	Yes
	3	12.49	≥ 6.0	Yes	3.71	2 - 4	Yes
	Average	12.43	≥ 6.0	Yes	3.59	2 - 4	Yes

4.2.4 Modulus.

The Modulus of the pavement is used to evaluate the loading resistance capacity. To compare the load bearing capacity of steel slag HMA, the modulus of the pavement after surfacing steel slag was compared with the modulus of the pavement after surfacing the control HMA C12.5. It should be noted that the layer thickness of both steel slag and mineral HMA mixtures were both equaly 7 cm. Before surfacing, the modulus of pavement was checked as mentioned previously in this paper. The checking positions before surfacing can be seen as in Figure 4 and the positions after surfacing can be seen in Figure 7. The modulus of the pavement was checked by using Benkeman beam method following Vietnamese specification TCVN 8867 : 2011. The moduli of pavement before and after surfacing are shown in Table 6. Before surfacing, the average modulus of pavement of the road section with C12.5 surface course with steel slag HMA and mineral HMA are similar, while the average modulus at C19 mixture section is slightly higher than the others. After surfacing, as can be seen from Table 6 that the moduli of the pavement with steel slag HMA mixtures were higher than that of mineral aggregate HMA. It means that the steel slag HMA had higher modulus than that of mineral HMA. This makes steel slag HMA possible to replace for traditional HMA. Although the study did not investigate why the steel slag

HMA performed better than control HMA in terms of modulus, it can be explained that the steel slag has strong resistance to abrasion and the surface texture of steel slag allows better interlocking between particles and adhesion with bitumen. This resulted in higher modulus than that of mineral aggregate HMA mixture. In terms of steel slag HMA mixtures with different nominal sizes, the pavement with larger nominal size showed slightly higher modulus than that of smaller nominal aggregate size. However, as before surfacing, the average modulus of the pavement section with C12.5 was smaller than that with C19. Thus, it is thought that the nominal aggregate size did not significantly differentiate the moduli of C12.5 and C19 mixtures in this study.

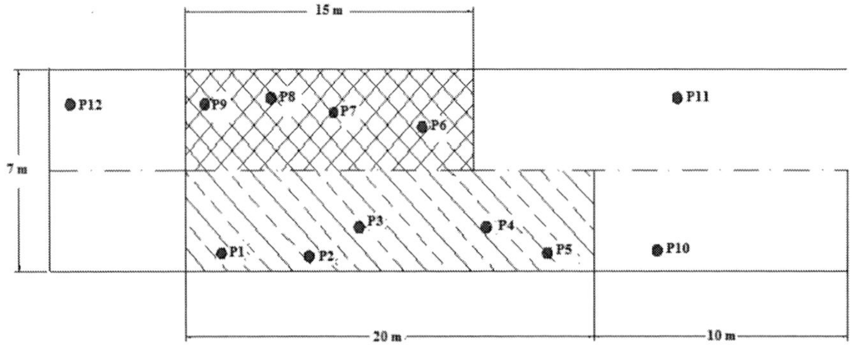

Figure 7: Checking scheme for modulus of pavement

Table 6. Pavement modulus results of two steel slag HMA and a mineral aggregate HMA

Mixture	C12.5 - Steel slag HMA				C19 - Steel slag HMA				C12.5 - Mineral aggregate HMA			
	Before surfacing		After surfacing		Before surfacing		After surfacing		Before surfacing		After surfacing	
	Position	Modulus (Mpa)	Position	Modulus (Mpa)	Position	Modulus (Mpa)	Position	Modulus (Mpa)	Position	Modulus (Mpa)	Position	Modulus (Mpa)
	P1	141.2	P1	209	P7	151.2	P6	213.1	P13	155.2	P10	191.2
	P2	152.5	P2	208.4	P8	173.4	P7	212.2	P14	165.3	P11	190.5
	P3	155.2	P3	209.6	P9	168.5	P8	209	P15	163.8	P12	190.2
	P4	173.4	P4	207.2	P10	173.4	P9	207.2				
	P5	177.1	P5	203.3	P11	162.1						
	P6	165.3			P12	177.6						
Average		160.8		207.5		167.7		210.4		161.4		190.6
CoV(%)		8.48		1.21		5.76		1.31		3.38		0.27

5. Conclusions

This paper presented the application of steel slag for hot mix asphalt to replace mineral aggregate in Vietnam. The results are very promising for future application and this work can be seen as a great effort in recycling waste materials in Vietnam. Some conclusions are drawn below:

- Steel slag has good and comparable physical properties to mineral aggregate to produce hot mix asphalt. Although the steel slag is heavier than general mineral aggregate, it is quite good in terms of water stability
- Steel slag used to be considered as a deleterious waste which needed to be processed and landfilled. However, the results show that the material is not deleterious and can be recycled as a mineral material
- Steel slag HMA mixtures complied with Vietnamese specifications in terms of Marshall Stability and flow test. The two steel slag HMA mixtures can be applied as surface courses in terms of skid resistance as they all passed the tests by using sand patch method

- Pavements with steel slag HMA surface course showed higher moduli than that with control mineral aggregate HMA indicating that the two steel slag HMA mixtures had greater moduli than HMA with mineral aggregate.
- The C12.5 passed the Surface's roughness test with level "very good" whereas the C19 failed in accordance to Vietnamese standard. It should be noted that except this test, all the other tests described in this paper passed the specifications' requirements. Thus, for future application of C19 as a surface course, further surface roughness needs to be checked on trial section before any practical application.
- Further research on steel slag is planned to be investigated in Vietnam at a more advanced level covering fatigue and rutting resistance of steel slag HMA mixture. From the preliminary results in this study, it can be concluded that recycling steel slag for hot mix asphalt is applicable and beneficial for sustainable development of Vietnam.

References

[1] H Yi, G Xu, H Cheng, J Wang, Y Wan, H Chen 2012 An overview of utilization of steel slag *Procedia Environmental Sciences* 16 p 791-801.

[2] Jt GAO, Sq LI, Yt ZHANG, Yl ZHANG, Py CHEN, S Ping 2011 Process of re-resourcing of converter slag *Journal of Iron and Steel Research International* 18 (12) p 32-39.

[3] Australasian Slag Association *A guide to the use of iron and steel slag in roads* Available at http://www.asa-inc.org.au/knowledge/technical-literature/technical-guides.

[4] Washington State Department of Transportation 2015 *Use of Steel Slag Aggregate in Pavements* A Report to the State Legislature In Response to 2ESHB 1299.

[5] L Rohde, W P Nunez and J A P Ceratti 2003 Electric Arc Furnace Steel Slag *Transportation Research Record* 1819 Transportation Research Board Washington DC pp 201-207.

[6] S A Aiban 2005 Utilization of steel slag aggregate for road bases *Journal of Testing and Evaluation* 34 (1) p 1-11.

[7] N Ali, J Chan, E Theriault, A Papagiannakis, A Bergan 1991 Sysco Electric Arc Furnace Slag As An Asphalt Concrete Aggregate Proceedings of *The Thirty-Sixth Annual Conference of Canadian Technical Asphalt Association.*

[8] B Farrand, J Emery 1995 Recent improvements in quality of steel slag aggregate *Transportation research record* 1468 p 137-141.

[9] A Coomarasamy, T Walzak 1995 Effects of moisture on surface chemistry of steel slags and steel slag-asphalt paving mixes *Transportation research record* 1492 p 85-95.

Effectiveness of multi tuned liquid dampers with slat screens for reducing dynamic responses of structures

T P Nguyen[1*], D T Pham[2] and K T Ngo[1]

[1] Ho Chi Minh City Open University, 97 Vo Van Tan St., Ho Chi Minh City, Vietnam
[2] Mientrung University of Civil Engineering, 24 Nguyen Du St., Tuy Hoa City, Vietnam

[*] Corresponding author, Email: phuoc.nguyen@ou.edu.vn

Abstract. Reducing vibration in structures under lateral load always attracts many researchers in during pastime, hence the mainly purpose of paper analyzes effectiveness of multiple-tuned liquid dampers for reducing dynamic responses of structures under ground acceleration of earthquakes. In this study, the multi-tuned liquid damper with slat screens (M-TLDWSS) is considered in detail for analyzing dynamic response of multi-degrees of freedom structure due to earthquake, which is more different previous studies. Then, the general equation of motion of the structure and M-TLDWSS under ground acceleration of earthquake is established based on dynamic balance of principle and solved by numerical method in the time domain. The effects of characteristic parameters of M-TLDWSS on dynamic response of the structure are investigated. The results obtained in this study demonstrate that the M-TLDWSS has significantly effectiveness for reducing dynamic response of the structure.

1. Introduction

Tuned liquid damper (TLD) has been used as passive control devices for controlling vibrations of structures under different dynamic loading conditions, such as wind, earthquake or other dynamics loads. A TLD is rigid tank, rigidly connected to the structure, with partially filled fluid that can be water or another fluid. The vibration motion of the structure due to the external excitation causes the sloshing motion of fluid inside the tank. By means of tuning the fundamental sloshing frequency of the TLD to the natural frequency of the structure, the inertia forces created would be approximately anti-phase to the external dynamic force, thereby reducing the dynamic response of coupling a structure with a TLD, which is similar to in a way increasing its effective damping [1-3].

The TLD has been attracted significant attention for mitigating vibration in many applications due to its many advantages over other conventional damping devices such as: require low maintenance and operating cost, easy to install in existing building structures, applicable for temporary use, non-restriction uni-directional excitations, and effectiveness even for small amplitude vibrations [4-8]. However, when studying TLD-structure system, there are a number of important design parameters that have to be considered such as: the mass ratio defined as the ratio of the mass of sloshing fluid to the generalized mass of the structure corresponding to the mode to be damped, the natural frequency of the liquid sloshing motion and the damping ratio defined as the ratio of the inherent damping of the TLD to the critical damping.

Expressions to determine the optimum inherent damping value depended on the value of the mass ratio as the optimum damping of a linear tuned mass damper (TMD), it can be adopted by replacing the solid mass of the TMD with the mass of sloshing fluid of the TLD. With a TLD without additional energy dissipating devices, the main source of inherent damping arises from viscous dissipation in the boundary layers at the walls and bottom of the tank and from free surface contamination. Hence, it is usually significantly less than the optimum inherent damping value, resulting in a less effective vibration absorber. To increase the energy dissipated by the sloshing fluid, several approaches have been implemented such as: roughness elements, surface contaminants, wave breaking in shallow water TLD, and nets or screens [9-15].

Recently, the addition of slat screens in TLD has been applied to increase the energy dissipated by the sloshing fluid. The damping devices consists of horizontal slats uniformly spaced apart in order to form screens. The use of this particular screen arrangement allows the space between the slats to be varied until the targeted amount of damping is achieved. Hence, the multi-tuned liquid damper with slat screens (M-TLDWSS) is used for reducing dynamic response of multi-degree of freedom structure due to earthquake, which is more different previous works. In sections 2 and 3 is formulated problem of the TLD with screens and coupling a structure with the M-TLDWSS such as the general equation of motion of the structure and M-TLDWSS under earthquake is established based on dynamic balance of principle and solved by numerical method in the time domain, respectively. In section 4 is presented the numerical investigation based on the formulation in the above sections, the effects of characteristic parameters of M-TLDWS on dynamic response of the structure are investigated. Finally, the main conclusions are derived concerning the efficiency of M-TLDWSS for reducing dynamic response of the structure based on the results obtained in this study.

2. Formulation

2.1. The TLDWSS model

Consider the rectangular tuned liquid damper with slat screens, having a length L, a quiescent fluid depth h, and a tank width b, is shown in figure 1. It is assumed that the fluid response amplitude is small compared to the fluid depth and the slat screens located at or near the centre of the tank do not significantly alter the overall flow of the sloshing liquid. The assumptions of inviscid, incompressible, irrotational flow and negligible surface tension are also made in this model [7, 10].

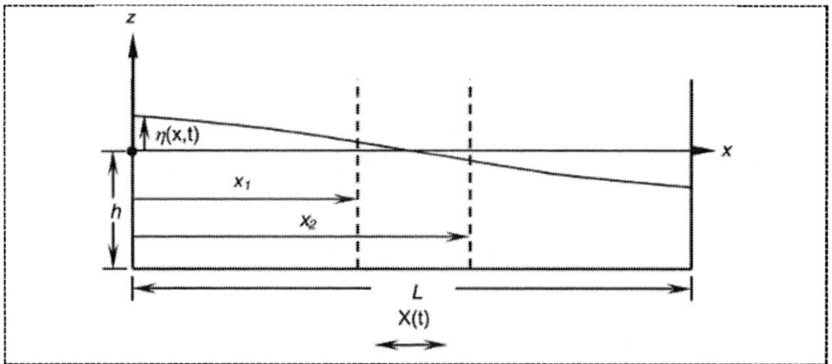

Figure 1. Definition sketch for liquid sloshing in TLD with slat screen.

The vibration motion of the structure due to the external excitation causes the sloshing motion of fluid inside the tank. Then, the energy of a system of standing waves of the simple harmonic type can be expressed in terms of the gravitational and kinematic potential as

$$V = \frac{1}{2}\rho bg\int_0^L \eta^2(x,t)dx; \quad T = \frac{1}{2}\rho b\int_{-h}^{0}\int_0^L\left[\left(\dot{X} + \frac{\partial \Phi}{\partial x}\right)^2 + \left(\frac{\partial \Phi}{\partial z}\right)^2\right]dxdz \tag{1}$$

where $\dot{X}$ is the horizontal velocity of the tank, $\Phi(x,z,t)$ is gradient of the velocity potential function.

Consider a number of screens is placed at discrete locations, x_j, shown in Figure 1. It is also assumed that constant C_m, C_d and C_l values denote inertia, drag, and loss coefficients, respectively. Then, it does contribute to the overall kinetic energy of the sloshing liquid, can be expressed as [7]

$$T = \frac{1}{2}\rho A_{cs} C_m \sum_{j=1}^{ns} \sin\left(\frac{n\pi x_j}{L}\right)^2 \int_{-h}^{0} \left[\frac{\cosh\left[\frac{n\pi(z+h)}{L}\right]}{\sinh\left(\frac{n\pi h}{L}\right)}\right]^2 dz\dot{q}_n^2 \tag{2}$$

Using the corresponding Lagrange's equations with Q_n the non-conservative forces, given by

$$\frac{d}{dt}\left(\frac{\partial T}{\partial \dot{q}_n}\right) - \frac{\partial T}{\partial q_n} + \frac{\partial V}{\partial q_n} = Q_n, \qquad n = 1, 2, 3, \ldots \tag{3}$$

The resulting equations of motion is expressed as

$$m_n^* \ddot{q}_n(t) + c_n^* \dot{q}_n(t) + m_n^* \omega_n^2 q_n(t) = \gamma_n^* \ddot{X}(t) \tag{4}$$

where m_n^* is the generalized mass and ω_n is the natural frequency for the nth sloshing mode as

$$m_n^* = \frac{1}{2}\frac{\rho b L^2}{n\pi \tanh\left(\frac{n\pi h}{L}\right)}; \quad \omega_n^2 = \frac{n\pi g}{L}\tanh\left(\frac{n\pi h}{L}\right) \tag{5}$$

and the generalized stiffness is also given as

$$k_n^* = m_n^* \omega_n^2 = \frac{\rho b L g}{2} \tag{6}$$

and the equation for the excitation factor can be expressed as

$$\gamma_n = \frac{\rho b L^2}{(n\pi)^2}\left[1 - \cos(n\pi)\right] \tag{7}$$

Based on the linearization of damping term for the fundamental sloshing mode ($n = 1$), the generalized damping coefficient in equation (4) is given by

$$c_n^* \equiv c_n^{eq*} = C_l \frac{4\rho b L}{3\pi^2}\Delta_n \Xi_n q_n \omega \tag{8}$$

and the corresponding generalized damping ratio can be expressed as

$$\zeta_n^{eq*} = C_l \frac{4}{3\pi}\tanh\left(\frac{\pi h}{L}\right)\Delta_n \Xi_n \frac{q_n}{L} \tag{9}$$

and the parameter Δ_n and Ξ_n are given by, respectively

$$\Delta_n = \frac{1}{3} + \frac{1}{\sinh^2\left(\frac{n\pi h}{L}\right)}; \qquad \Xi_n = \sum_{j=1}^{ns} \sin\left(\frac{n\pi x_j}{L}\right)^3 \tag{10}$$

where ns is total of slat screens and ρ is mass density of fluid.

2.2. Formulation of the structure-TLDWSS interaction

Consider the TLDWSS as the TMD model attached to the structure modeled as a generalized single degree of freedom system representing the mode of vibration being suppressed, is shown in figure 2.

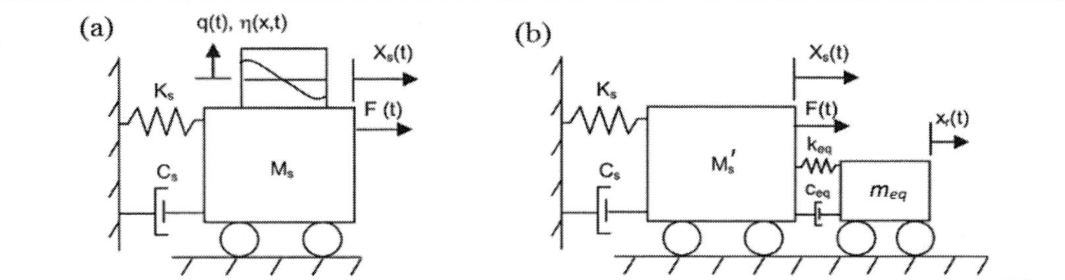

Figure 2. The model of the structure-TLDWSS interaction: (a) Structure-TLDWSS system, (b) Equivalent mechanical linearized structure-TLDWSS system.

As the TLDWSS tank moves with the structure due to the external excitation the contained fluid will exhibit a sloshing response motion. The equations of motion for a coupling the structure with TLDWSS as two degree of freedom system can be expressed as

$$\begin{bmatrix} M_s' & 0 \\ 0 & m_{eq} \end{bmatrix} \begin{Bmatrix} \ddot{X}_s \\ \ddot{x}_r \end{Bmatrix} + \begin{bmatrix} C_s + c_{eq} & -c_{eq} \\ -c_{eq} & c_{eq} \end{bmatrix} \begin{Bmatrix} \dot{X}_s \\ \dot{x}_r \end{Bmatrix} + \begin{bmatrix} K_s + k_{eq} & -k_{eq} \\ -k_{eq} & k_{eq} \end{bmatrix} \begin{Bmatrix} X_s \\ x_r \end{Bmatrix} = \begin{Bmatrix} F(t) \\ 0 \end{Bmatrix} \tag{11}$$

where $M_s^{'}$, C_s, and K_s are the total vibration mass to account for the non-participating component of the liquid corresponding with the fundamental sloshing mode, damping and stiffness of the structure, respectively, given by

$$M_s^{'} = M_s + m^{non} \tag{12}$$

where m^{non} is the mass of non-participating component of the liquid, can be expressed as

$$m^{non} = \rho bhL - m_{eq} \tag{13}$$

The equivalent mass, damping and stiffness of the TLDWSS, corresponding to the fundamental sloshing mode, which develops dynamic forces equal to the forces exerted by the sloshing fluid in the rectangular tank, are used to represent an equivalent mechanical system, given by [7]

$$\begin{aligned} m_{eq} &= \frac{8\rho bL^2}{\pi^3}\tanh\left(\frac{\pi h}{L}\right); \qquad & k_{eq} &= \frac{8\rho bLg}{\pi^2}\tanh^2\left(\frac{\pi h}{L}\right) \\ \omega_{eq} &= \left[\frac{\pi g}{L}\tanh\left(\frac{\pi h}{L}\right)\right]^{1/2}; \qquad & c_{eq} &= C_l\frac{256\rho bL}{3\pi^5}\tanh^3\left(\frac{\pi h}{L}\right)\Delta\Xi\omega x_r \end{aligned} \tag{14}$$

and the damping ratio can be also expressed as

$$\zeta_{eq} = C_l\frac{16}{3\pi^2}\tanh^2\left(\frac{\pi h}{L}\right)\Delta\Xi\frac{x_r}{L} \tag{15}$$

The aim of designing M-TLDWSS is to tune damper parameters to the modal parameters of selected modes of vibration. It means that the natural damper frequency (or a group of dampers) must be close to the natural frequency of a selected vibration mode of structure. Hence, the tuning ratio, Ω, is important parameter that strongly influence the performance of the M-TLDWSS which is defined as the ratio of the sloshing frequency to the structural frequency in the mode that the dynamic motions are to be suppressed

$$\Omega = \frac{\omega_{eq}}{\omega_s} \tag{16}$$

The optimal parameters of such a damper (or group of dampers) can be determined based on utilizing linear TMD theory, then the optimal frequency ratio and damping coefficient are given by

$$\Omega^{\text{opt}} = \frac{1}{1+\mu}(1+0.5\mu)^{1/2}; \qquad \zeta^{\text{opt}} = \left(\frac{\mu + 0.75\mu^2}{4+6\mu+2\mu^2}\right)^{1/2} \tag{17}$$

where the mass ratio μ is the ratio of the generalized participating fluid mass to the structural generalized mass expressed as

$$\mu = \frac{m_{eq}}{M_s^{'}} \tag{18}$$

Assuming that the mass ratio μ is known, the characteristic parameters of the M-TLDWSS can be obtained from the above formulae. And then, the equivalent system in the coupling structure with M-TLDWSS allows direct application of design tools. In order to achieve the optimal damping ratio for a given target response amplitude and corresponding relative response motion, the required screen properties and placement can be estimated.

3. Governing equation of motion

Consider a building structure with N stories as multi-degrees of freedom system and the M-TLDWSS installed at the top floor due to earthquake, shown in figure 3a and the equivalent mechanical model of coupling the structure with M-TLDWSS due to earthquake is described in figure 4. It is assumed that the building structure is treated as a discrete and linear elastic system, the mass is concentrated at the level of building floors and the beams of the frame are infinitely stiff, shown in figure 3b.

The system equation of motion of the coupling structure with M-TLDWSS due to ground acceleration of earthquakes $\ddot{u}_g(t)$ at time t can be expressed as

$$\mathbf{M}\ddot{u} + \mathbf{C}\dot{u} + \mathbf{K}u = -\mathbf{Mr}\ddot{u}_g(t) \tag{19}$$

where $\mathbf{M}$, $\mathbf{C}$ and $\mathbf{K}$ are the overall matrix of mass, damping and stiffness, respectively; u and $P(t)$ are the overall vector of displacement and force, respectively. The mass matrix $\mathbf{M}$ of the system is in the following form [16]

$$\mathbf{M} = \begin{bmatrix} \mathbf{M}_s & \mathbf{0} \\ \mathbf{0} & \mathbf{M}_{TLD} \end{bmatrix} \tag{20}$$

where $\mathbf{M}_s$ and $\mathbf{M}_{TLD}$ are the mass matrix of the structure and M-TLDWSS, respectively, given by

$$\begin{aligned} \mathbf{M}_s &= diag\left(m_1, m_2, ..., m_{N-1}, m_N^{'}\right) \\ \mathbf{M}_{TLD} &= diag\left(m_1^{eq}, m_2^{eq}, ..., m_{n-1}^{eq}, m_n^{eq}\right) \end{aligned} \tag{21}$$

where $m_N^{'}$ is total mass of the top floor and non-participating component of the liquid expressed as

$$m_N^{'} = m_N + \sum_{j=1}^{n} m_j^{non} \tag{22}$$

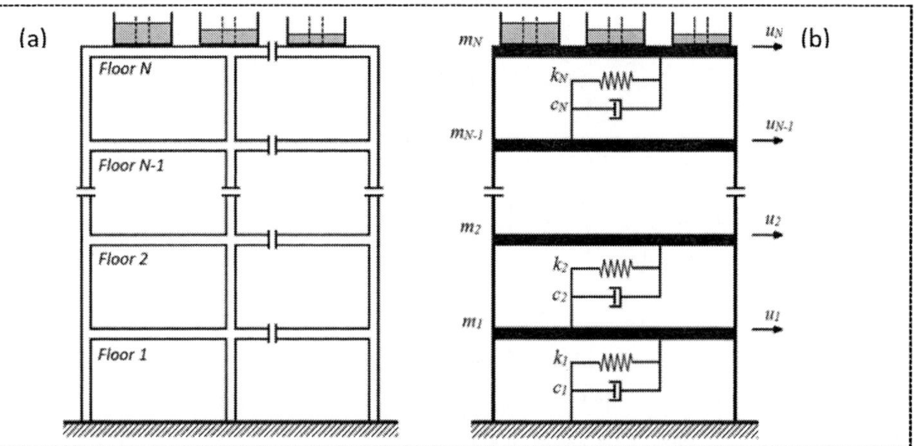

Figure 3. The coupling the structure with M-TLDWSS model under earthquake: (a) The real structure model, (b) The structure system model.

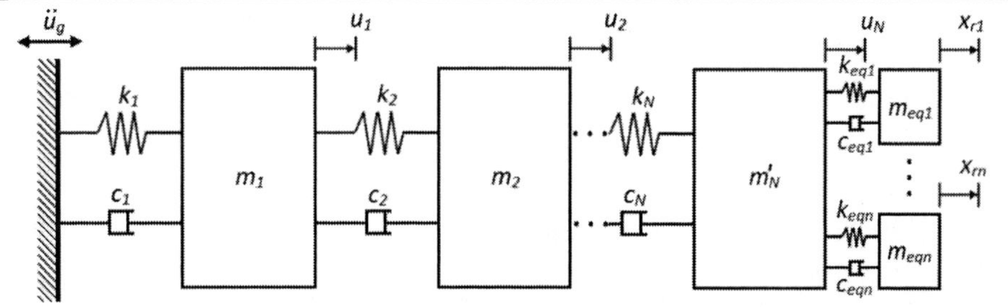

Figure 4. The equivalent mechanical model of coupling the structure with M-TLDWSS due to ground acceleration of earthquakes.

The stiffness matrix $\mathbf{K}$ of the considered system can also be shown [16]

$$\mathbf{K} = \begin{bmatrix} \mathbf{K}_s + \mathbf{k}_{TLD} & \mathbf{K}^* \\ \mathbf{K}^{*T} & \mathbf{K}_{TLD} \end{bmatrix} \tag{23}$$

where $\mathbf{K}_s$ is the stiffness matrix of the structure, given by

$$\mathbf{K}_s = \begin{bmatrix} k_1 + k_2 & -k_2 & 0 & 0 \\ -k_2 & k_2 + k_3 & -k_3 & 0 \\ 0 & -k_3 & \dots & -k_N \\ 0 & 0 & -k_N & k_N \end{bmatrix} \tag{24}$$

and the matrix $\mathbf{k}_{TLD}$ can be expressed as

$$\mathbf{k}_{TLD} = diag\left(0,\ 0,\ \dots,\ k_1^{eq} + k_2^{eq} + \dots + k_n^{eq}\right) \tag{25}$$

The matrices $\mathbf{K}_{TLD}$ and $\mathbf{K}^*$ are in the following form

$$\mathbf{K}_{TLD} = diag\left(k_1^{eq}, k_2^{eq}, ..., k_{n-1}^{eq}, k_n^{eq}\right)$$
$$\mathbf{K}^* = \begin{bmatrix} 0 & 0 & \dots & 0 \\ 0 & 0 & \dots & 0 \\ \dots & \dots & \dots & \dots \\ -k_1^{eq} & -k_2^{eq} & \dots & -k_n^{eq} \end{bmatrix} \tag{26}$$

The damping matrix of the system $\mathbf{C}$ is in a form similar to the stiffness matrix $\mathbf{K}$, defined as

$$\mathbf{C} = \begin{bmatrix} \mathbf{C}_s + \mathbf{c}_{TLD} & \mathbf{C}^* \\ \mathbf{C}^{*T} & \mathbf{C}_{TLD} \end{bmatrix}$$
$$\mathbf{c}_{TLD} = diag\left(0, 0, ..., c_1^{eq} + c_2^{eq} + ... + c_n^{eq}\right)$$
$$\mathbf{C}_{TLD} = diag\left(c_1^{eq}, c_2^{eq}, ..., c_{n-1}^{eq}, c_n^{eq}\right) \tag{27}$$
$$\mathbf{C}^* = \begin{bmatrix} 0 & 0 & \dots & 0 \\ 0 & 0 & \dots & 0 \\ \dots & \dots & \dots & \dots \\ -c_1^{eq} & -c_2^{eq} & \dots & -c_n^{eq} \end{bmatrix}$$

and the damping matrix of the structure $\mathbf{C}_s$ is determined based on Rayleigh damping as

$$\mathbf{C}_s = \alpha \mathbf{M}_s + \kappa \mathbf{K}_s \tag{28}$$

and $\boldsymbol{r} = (1, 1, ..., 1)^T$ is unit vector.

The equation (19) is used to analyze effects of the characteristic parameters of M-TLDWSS on dynamic response of the structure due to ground acceleration of earthquake. The flowchart of numerical procedure for analyzing this above problem is developed based on the Newmark's method.

4. Numerical investigation

In this section, considering real building parameters are given in Table 1 [16] and the damping ratio for first two modes of this building is taken 2%. The 1940 El-Centro earthquake is chosen to excite the building as shown in figure 5. The M-TLDWSS consists of 8 TLDWSS, each TLDWSS equipped with two damping screens located at $0.4L$ and $0.6L$, respectively, having a length $L = 6$ m, width $b = 6$ m, loss coefficient $C_l = 2.16$. To properly tune the M-TLDWSS to the structure, the tank considered as normal tank for living in the building was partially filled with water to a depth, h.

Table 1. The main parameters of the structure.

Story	Mass [kg]	Stiffness [N/m]
1	2.83×10^5	3.31×10^8
2-4	2.76×10^5	1.06×10^9
5-7	2.76×10^5	6.79×10^8
8-10	2.76×10^5	6.79×10^8
11-13	2.76×10^5	5.84×10^8
14-16	2.76×10^5	3.86×10^8
17-19	2.76×10^5	3.47×10^8
20	2.92×10^5	2.29×10^8

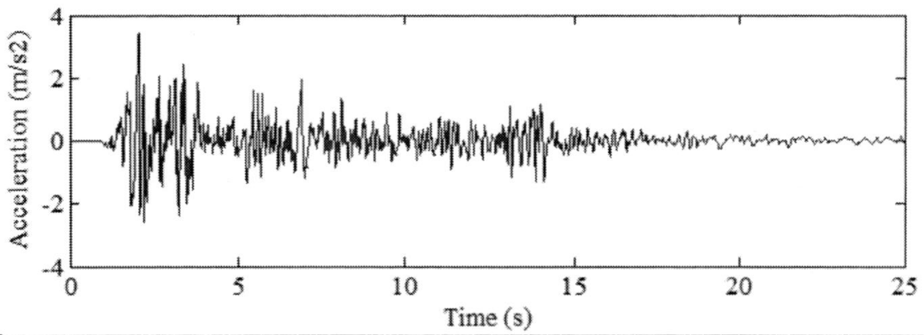

Figure 5. Ground acceleration for 1940 El-Centro earthquake.

It can be seen that the characteristic parameters of the M-TLDWSS depend on the tank length, L, width, b, water height, h, number, ns, and placement, x_j, of the slat screens. Those are important parameters for the tuning ratio and mass ratio that affect strongly on the performance of the M-TLDWSS. Hence, the effects of these parameters on reducing dynamic response of the structure due to earthquake will be investigated in detail in this section.

4.1. Effect of the ratio of depth per length $\frac{h}{L}$

From equations (14) to (15) show that this ratio effects directly on dynamic characters of the TLDWSS. Hence, this ratio is the most important parameter of the M-TLDWSS effecting strongly on reducing dynamic response of the structure. Figure 6a shows that with an increase of the ratio increase significantly the tuning ratio. It can be seen that an increase of tuning ratio is quite clear in the range of value of the ratio less than 1. With an increase of the ratio will tune the sloshing frequency to the structural frequency in the mode, then the effectiveness of M-TLDWSS for reducing dynamic response of the structure will be more significant than others.

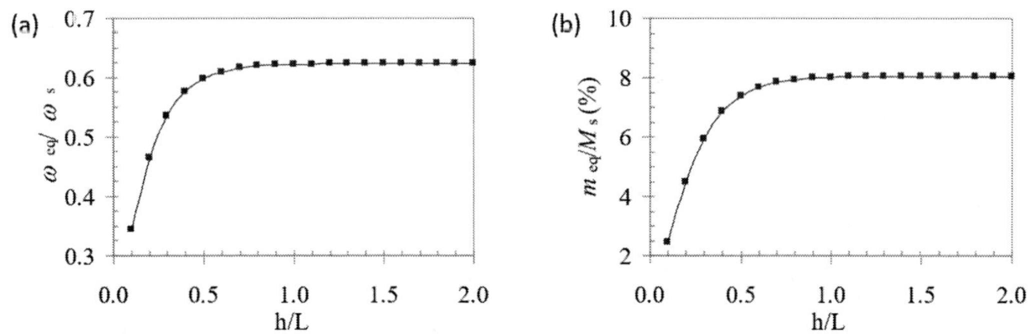

Figure 6. The effects of the ratios on: (a) the tuning ratio, (b) the mass ratio.

It can be also seen that with an increase of the ratio also increase the generalized participating water mass causing increase of the mass ratio, shown in figure 6 (b). With increase of the ratio in the range of value of the ratio less than 1, the mass ratio is significantly more increasing than others. When the mass ratio increases, the general vibration mass of the structure system is also increase causing increase building mass and various dynamic character of the structure system.

The influence of the ratio on damping coefficient and damping ratio are also studied, shown in figure 7. It can be seen that the ratio effects stronger on characteristic damping of the M-TLDWSS in the range

IOP Conf. Series: Earth and Environmental Science **143** (2018) 012023 doi:10.1088/1755-1315/143/1/012023

of value less than 1 than others. Additionally, with an increase of the ratio increases damping coefficient, shown in figure 7(a), but the damping ratio decreases with an increase of the ratio, shown in figure 7(b). It can be confirmed that this ratio effects strongly on dynamic characters of the M-TLDWSS such as tuning ratio, mass ratio, damping coefficient and damping ratio. Therefore, it also causes various dynamic character of the structure system. Hence, the effectiveness of the M-TLDWSS for reducing dynamic response of the structure due to earthquake is also different with various value of this ratio, shown in figure 8. In this real problem, the effectiveness of the M-TLDWSS is clear in the range of value of the ratio from 0.5 to 1 and 1.3 to 1.7. Reduction of displacement of the top floor and shear force of the first floor are more significant in this range than others. But, a comparison of increase of addition mass of the M-TLDWSS shows that the effectiveness of the M-TLDWSS is the strongest on both reducing dynamic response of the structure and addition mass in the range of value of the ratio from 0.6 to 0.8.

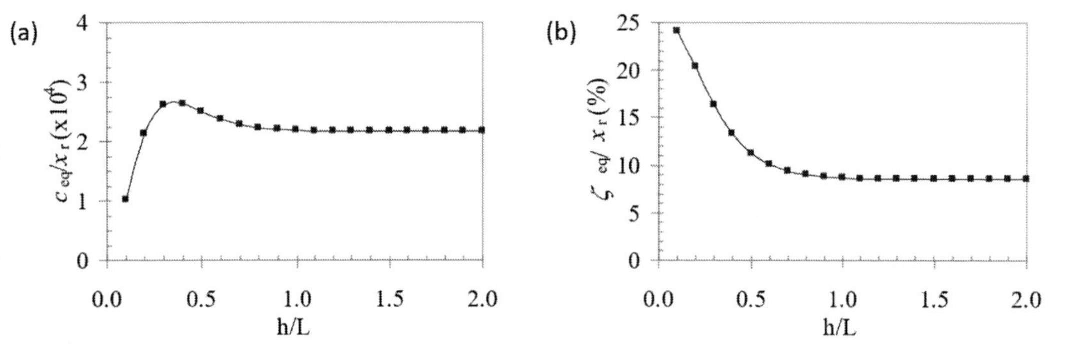

Figure 7. The effects of the $\frac{h}{L}$ ratios on: (a) the damping coefficient, (b) the damping ratio.

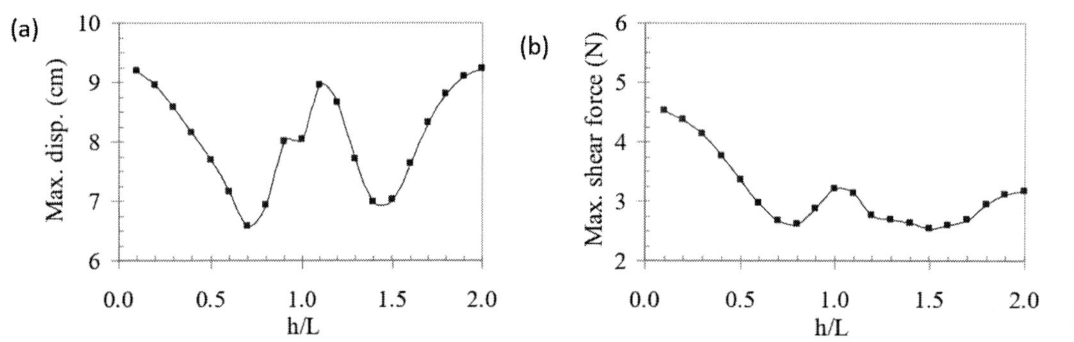

Figure 8. The effects of the $\frac{h}{L}$ ratios on: (a) the maximum displacement of the top floor, (b) the maximum shear force of the first floor ($x10^6$).

4.2. Effectiveness of the M-TLDWSS on reducing dynamic response

In this investigation, the value of the ratio of depth per length is chosen 0.6 and 0.7 to analyze effectiveness of the M-TLDWSS on the dynamic response, respectively. The time history of displacement at each floor and shear force of the first floor is presented in figures 9, 10, respectively. The maximum displacement and shear force of each story are shown in figure 11.

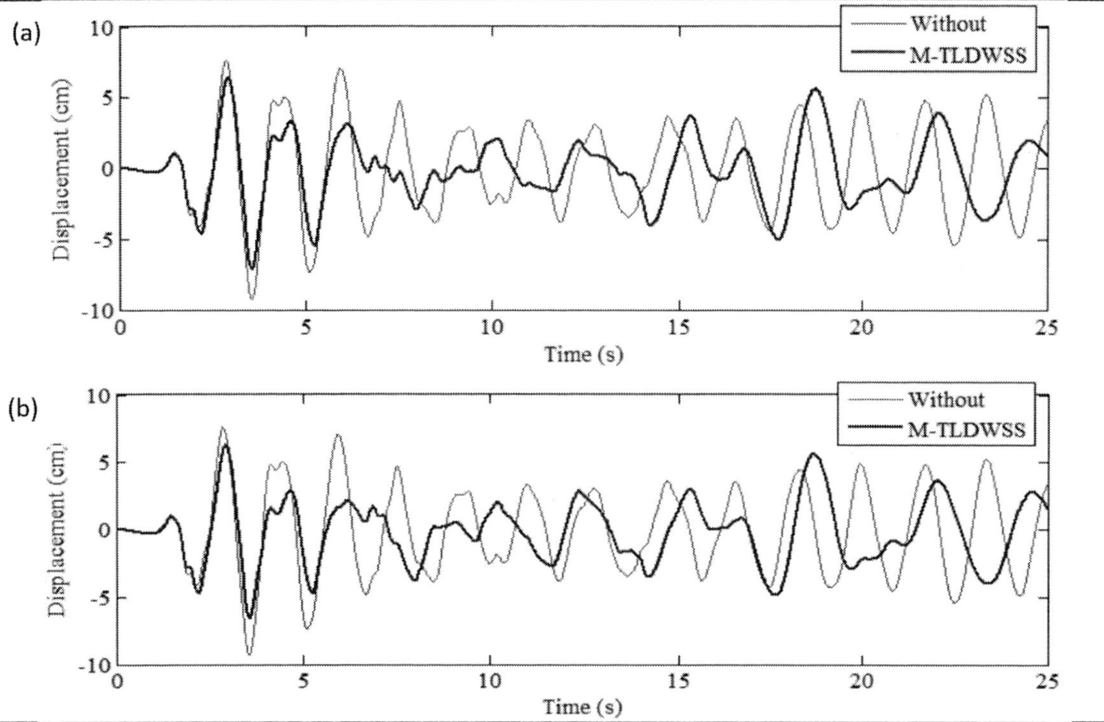

Figure 9. The time history of displacement of the top floor with: (a) $\frac{h}{L} = 0.6$, (b) $\frac{h}{L} = 0.7$

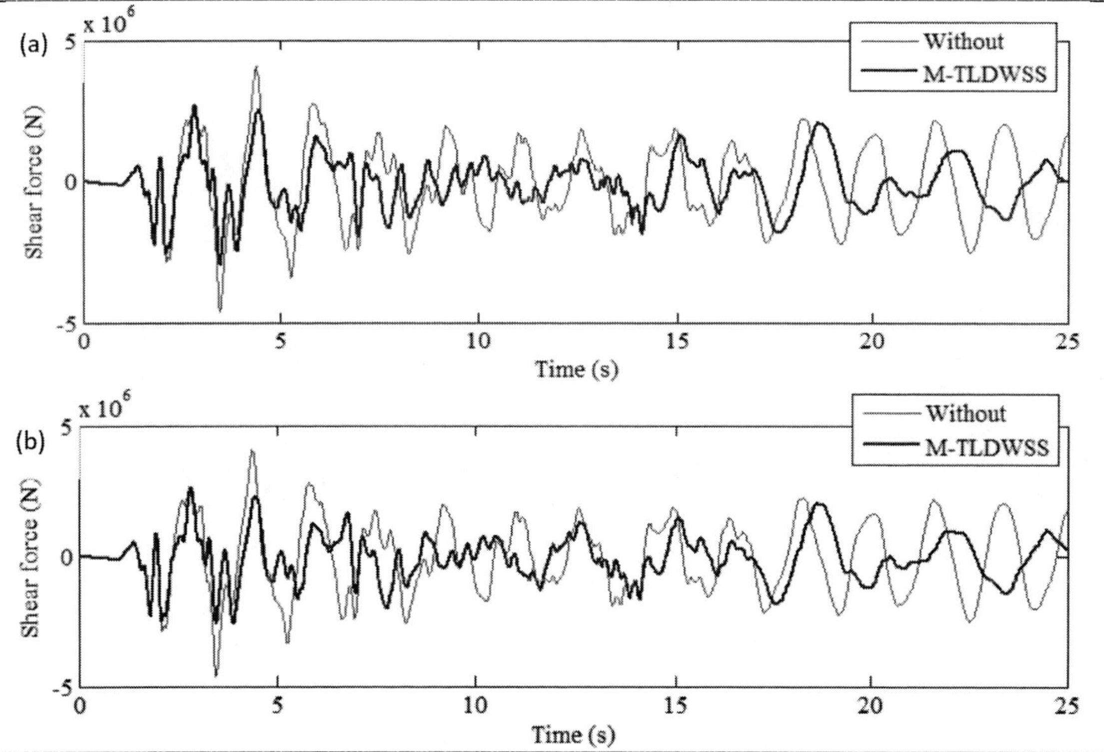

Figure 10. The time history of displacement of the top floor with: (a) $\frac{h}{L} = 0.6$, (b) $\frac{h}{L} = 0.7$

IOP Conf. Series: Earth and Environmental Science **143** (2018) 012023 doi:10.1088/1755-1315/143/1/012023

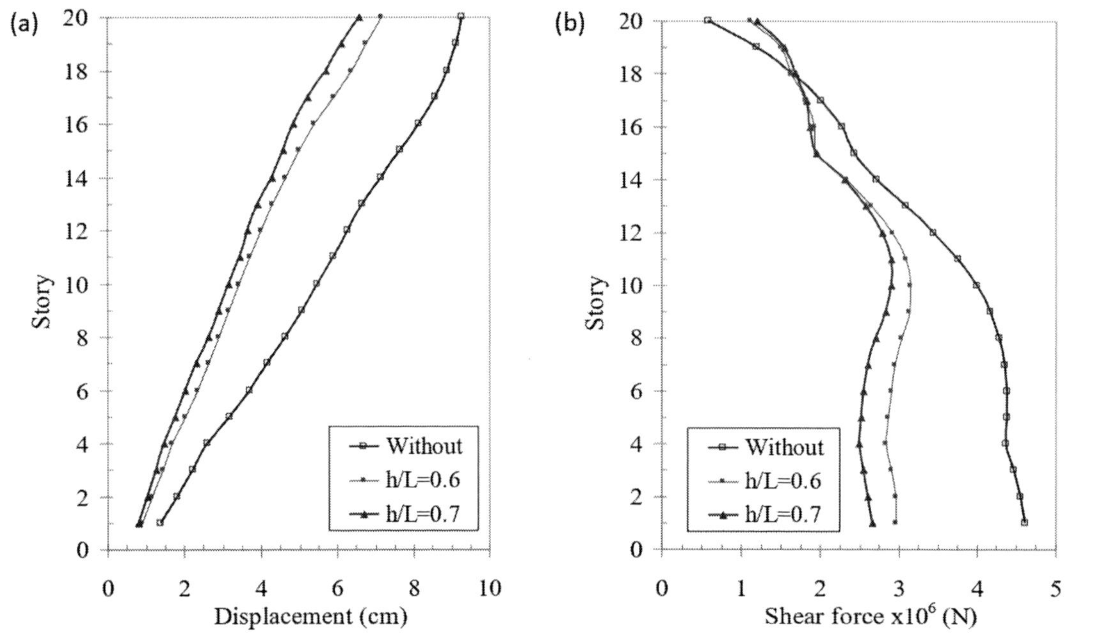

Figure 11. The effect of the $\frac{h}{L}$ ratio on the structure response: (a) the maximum displacement, (b) the maximum shear force.

It can be seen that the M-TLDWSS attached in the top floor had effectiveness for reducing dynamic response of the structure due to earthquake in this real problem. It is more decreasing significantly the time history of displacement at each story than without damper. These things are also similar to on the time history of shear force in the structure, shown in figure 9 and figure 10. When the tuning ratio increases nearly the optimal ratio, the M-TLDWSS will have strongly effects on the structure response. Hence, it is more decreasing significantly dynamic response of the structure than without damper, shown in figure 11.

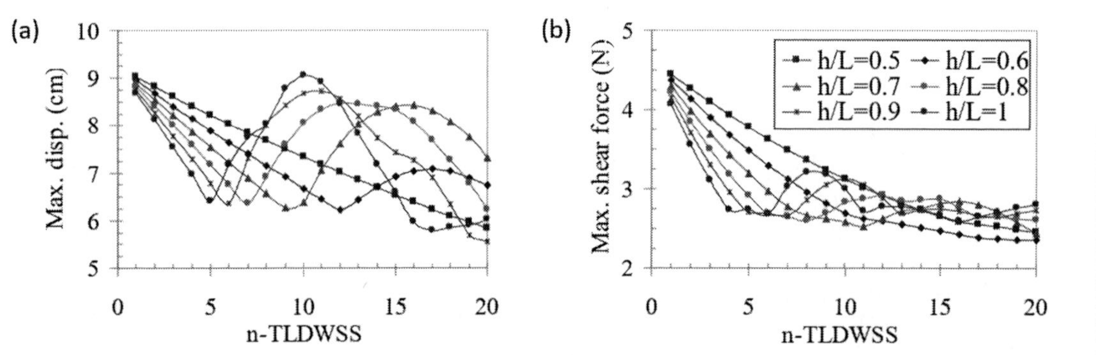

Figure 12. The effect of TLDWSS number: (a) the displacement, (b) the shear force (x10^{6}).

Continuously, the influence of number of TLDWSS on performance of the M-TLDWSS is investigated with various values of the ratio, presented in figure 12. This numerical result shows that the number of the TLDWSS has affect significantly on the structure response. The influence of the TLDWSS number is also different corresponding with each the ratio. But, a general estimate shows that the effectiveness of the M-TLDWSS does not have quite differences with various values of the ratio corresponding with various values of the TDWSS number. Therefore, it can be seen that the main performance of M-TLDWSS completely depends on both important parameters the ratio and number of the TLDWSS. Hence, in order to target for reducing dynamic response of the structure, the both above parameters must be considered to tune the tuning ratio to the optimal ratio, and then, the effectiveness of the M-TLDWSS for reducing dynamic response of the structure due to earthquake will be clear and more significant than others.

4.3. Suitability of M-TLDWSS with various earthquakes

The main purpose of this section verifies suitability of effectiveness of the M-TLDWSS for reducing dynamic response due to various random ground accelerations, shown in figure 13 and figure 14. The effectiveness of the M-TLDWSS for reducing dynamic response of the structure due to ground acceleration is studied with various values of the ratio, shown in figure 15 and figure 16, respectively. The numerical results show that the M-TLDWSS is more decreasing the maximum displacement and shear force at each story than without damper. Specially, in range of value of the ratio from 0.6 to 0.8, the performance of the M-TLDWSS for reducing dynamic response of the structure is clearer than others. Therefore, it can be confirmed that the M-TLDWSS has effectiveness for reducing dynamic response of the structure due to various ground accelerations in this real problem. It is more decreasing significantly structure response than without damper. Hence, the M-TLDWSS including many tanks as living water tanks in the building can be considered as a useful solution for designing earthquake resistance. And, one of the best important things for designing characteristic parameters of the M-TLDWSS is tune both the ratio and TLDWSS number to achieve target for reducing dynamic response.

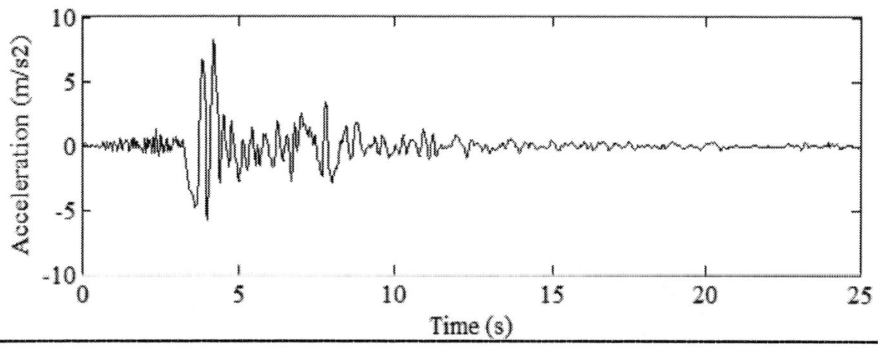

Figure 13. Ground acceleration for Northidge earthquake.

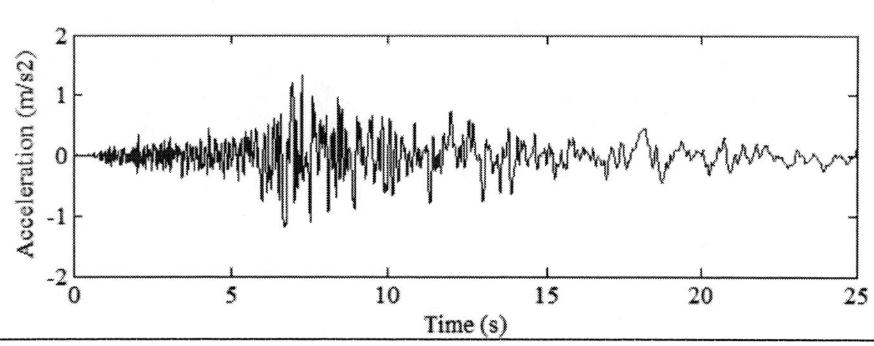

Figure 14. Ground acceleration for Superstition earthquake.

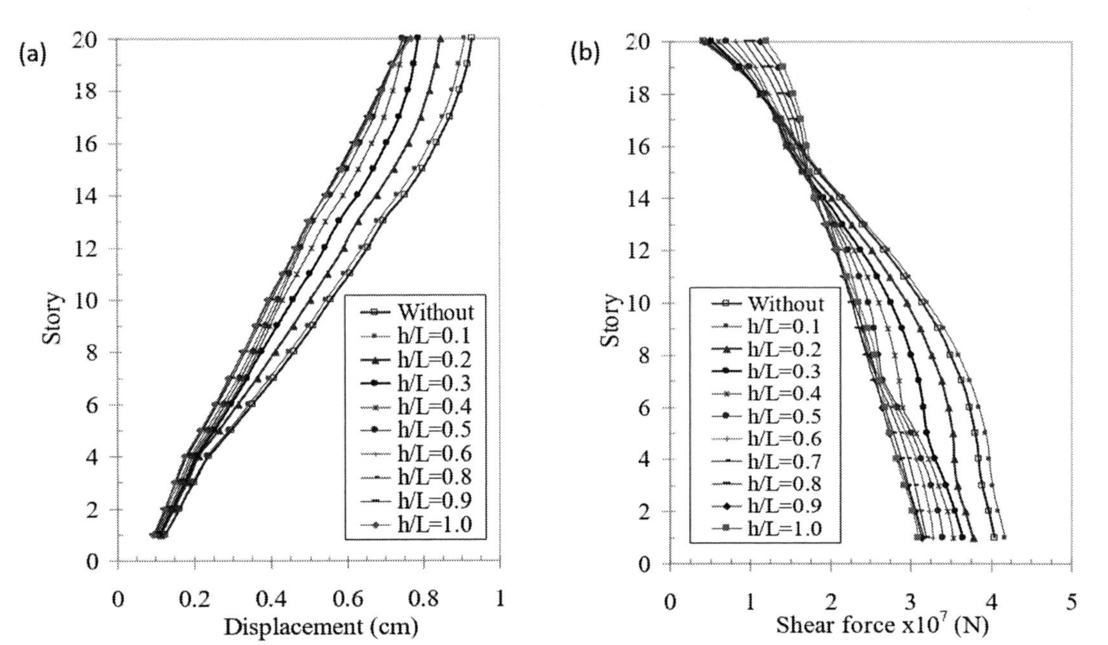

Figure 15. The effectiveness of the M-TLDWSS with Northidge earthquake: (a) the maximum displacement, (b) the maximum shear force.

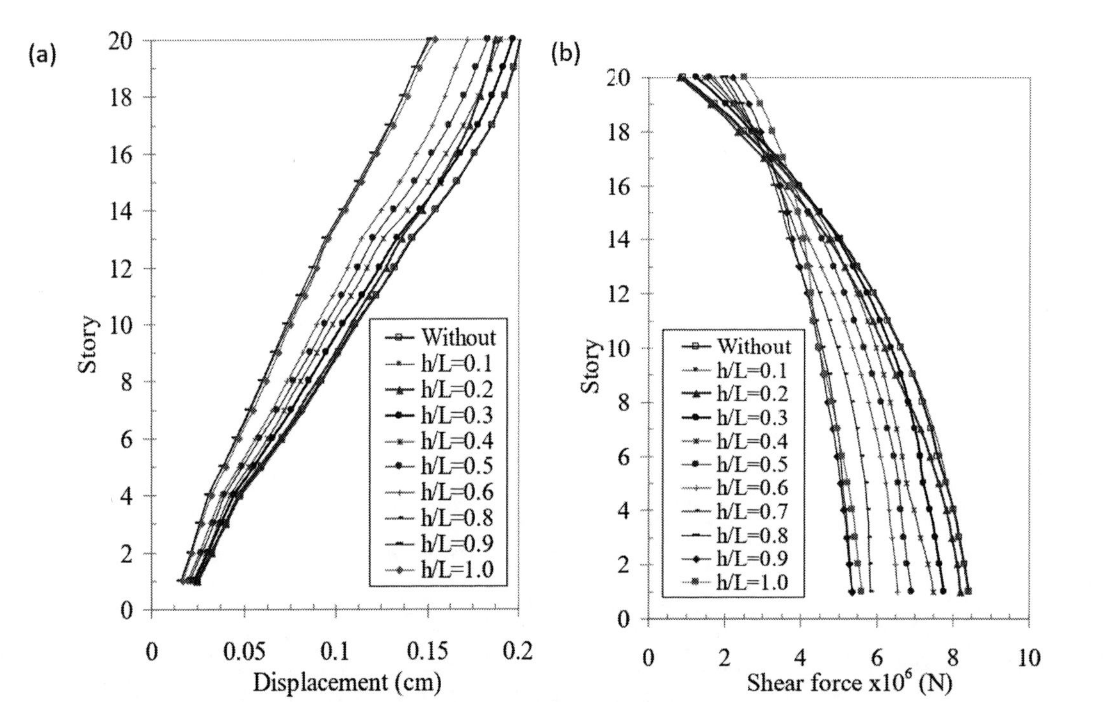

Figure 16. The effectiveness of the M-TLDWSS with Superstition earthquake: (a) the maximum displacement, (b) the maximum shear force.

5. Conclusion

The main concern of this paper is to analyse the effectiveness of the multi-tuned liquid damper with slat screens (M-TLDWSS) for reducing dynamic response of the structure. The M-TLDWSS includes many TLD which is added slat screens to increase the energy dissipated by the sloshing fluid. The character parameters of TLDWSS response depend on the response amplitude of the TLD, the water depth, tank length, damping screen location and damping screen loss coefficient. The response of the coupling structure with M-TLDWSS due to earthquake in the time domain has been analysed.

The following conclusions could be formulated from the numerical results as follows:

- The ratio of depth per length is affect strongly on dynamic character of the M-TLDWSS, it increases tuning ratio which is one of the most important parameters for performance of the M-TLDWSS on reducing dynamic response of the structure. In this real problem, the effectiveness of the M-TLDWSS for reducing structure response is more significant in the range of value of the ratio from 0.6 to 0.8 than others.
- The performance of the M-TLDWSS also depends on TLDWSS number. The influence of the TLDWSS number is also different corresponding with each the ratio. But, a general estimate shows that the effectiveness of the M-TLDWSS does not have quite differences with various values of the ratio corresponding with various values of the TDWSS number.
- The M-TLDWSS applied for reducing dynamic response of the structure can be suitable for various ground accelerations in this real problem. It is more decreasing significantly structure response than without damper.

The above parameters must be considered to tune the tuning ratio to the optimal ratio to achieve target for reducing dynamic response of the structure due to lateral loads.

References

[1] Sun L M, Fujino Y, Pacheco B M and Chaiseri P 1992 Modelling of Tuned Liquid Damper (TLD) *Journal of Wind Engineering and Industrial Aerodynamics*, 41-44 p 1883-1894.

[2] Wakahara T, Ohyama T and Fujii K 1992 Suppression of wind-induced vibration of a tall building using Tuned Liquid Damper *Journal of Wind Engineering and Industrial Aerodynamics* 43(1-3) p 1895-1906.

[3] Tamura Y, Fujii K, Ohtsuki T, Wakahara T and Kohsaka R 1995 Effectiveness of tuned liquid dampers under wind excitation *Engineering Structures* 17(9) p 609-621.

[4] Banciji P, Murudi, Shah A II and Popplewell N 2000 Tuned liquid dampers for controlling earthquake response of structures *Earthquake Engineering & Structural Dynamics* 29 p 587-602.

[5] Li H, Wang Y 2004 Theoretical and experimental studies on reduction foe multi-modal seismic response of high-rise structure by tuned liquid dampers *J. of Vibration and Control* 10 p 1041.

[6] Kim Y M, You K P, Ko N H and Yoon S W 2006 Use of TLD and MTLD for control of wind-induced vibration of tall building *J. of Mechanical Science and Technology* 20(9) p 1346-1354.

[7] Tait M 2008 Modelling and preliminary design of a structure-TLD system *Engineering Structures* 30(10) p 2644-2655.

[8] Deng X and Tait M 2008 Equivalent mechanical models of tuned liquid dampers with different tank geometries *Canadian Journal of Civil Engineering* 38 p 1088-1101.

[9] Tait M, Damatty A, Isyumov 2005 An investigation of tuned liquid dampers equipped with damping screens under 2D excitation *Earthquake Engineering and Structural Dynamics* 34(7) p 719-735.

[10] Morsy H, Marivani M, Hamed 2008 Effect of A Tuned Liquid Damper Screen configuration on structure response *in Nineth International Congress of Fluid Dynamics & Propulsion - ASME* Alexandria, Egypt.

[11] Love J S, Tait M J 2010 Nonlinear simulation of a tuned liquid damper with damping screens using a modal expansion technique *J. Fluid Struct* 26(7-8) p 1058-77.

[12] Cassolato M, Love J and Tait M 2010 Modelling of a Tuned Liquid Damper with inclined

damping screens *Structural Control and Health Monitoring* 18(6) p 674-681.
[13] Maravani M, Hamed M S 2011 Numerical modeling of sloshing motion in a tuned liquid damper outfitted with a submerged slat screen *Int. J. Numer. Methods Fluids* 65 p 834-855.
[14] Lovel J S and Tait M J 2015 Multiple tuned liquid dampers for efficient and robust structural control *J. Struct. Eng.* 141(12) p 04015045: 1-6.
[15] Love J S, Tait M J 2017 Estimating the added effective damping of SDOF systems incorporating multiple dynamic vibration absorbers with nonlinear damping *Engineering Structures* 130 p 154-161.
[16] Roman L., Justyna G. 2009 Dynamic analysis of structures with multiple tuned mass dampers *Journal of Civil Engineering and Management* 15(1) p 77-86.

Acknowledgements
This research is funded by Ho Chi Minh City Open University under the basic research fund 2018.

Vibration of functionally graded plate resting on viscoelastic elastic foundation subjected to moving loads

Ta Duy Hien [*], Nguyen Ngoc Lam

Department of Civil Engineering, University of Transport and Communications, Ha Noi, Vietnam

[*]Corresponding author, Email: tdhien@utc.edu.vn

Abstract. The dynamics of plates subjected to a moving load must be considered by engineering mechanics and design structures. This paper deals with the dynamic responses of functionally graded (FG) rectangular plates resting on a viscoelastic foundation under moving loads. It is assumed that material properties of the plate vary continuously in the thickness direction according to the power-law. The governing equations are derived by using Hamilton's principle, which considers the effect of the higher-order shear deformation in the plate. Transient responses of simply supported FG rectangular plates are employed by using state-space methods. Several examples are given for displacement and stresses in the plates with various structural parameters, and the effects of these parameters are discussed.

1. Introduction

The dynamics of beams and plates resting on foundations have been investigated during the past several decades. Beams and plate structures supported by soil and rock are widespread in civil engineering. To describe the interactions of beams, plates, and foundations, various models have been proposed. The simplest one is the Winkler model for elastic foundations, in which the foundation comprises separated springs with constant stiffness. Pasternak [1] proposed a new model by taking into account the shear interactions between adjacent Winkler spring elements. In addition, more efficient foundation models have been proposed; for instance, the modified Pasternak model [2] for reinforced soil, and the nonlinear foundation models [3, 4].

Numerous studies have solved the dynamics of plates resting on foundations using analytical, numerical or semi-analytical methods. Tang [5] investigated an infinite plate strip on a foundation of Kelvin material in terms of a moving load using the integral transform technique. Zaman *et al.* [6] proposed the finite element model for the dynamic response of thick plates resting on a viscoelastic foundation.

Huang *et al.* [7] used the finite strip method to develop a model for the dynamic response of plates resting on an elastic foundation in terms of moving loads. Trung *et al.* developed a cell-based smoothed three-node element for determining dynamics of composite [8] and Midlin plates [9] resting on a viscoelastic foundation. Barati [10] studied the analysis of functionally graded nanoplates with nanovoids on a viscoelastic substrate under hygro-thermo-mechanical loading.

In this work, analytical solutions for the dynamic response of FG rectangular plates resting on a viscoelastic foundation are employed. The FG rectangular plate is subjected to a concentrated moving load at the upper surface of the plate. The governing equations are solved by using state-space methods, and the effects of different parameters on the response of the plate are studied.

2. Governing equation of vibration of FG plate

2.1. FG plate model

We consider an FG plate resting on a viscoelastic foundation; the geometry of the plate and coordinate system is shown in Fig.1:

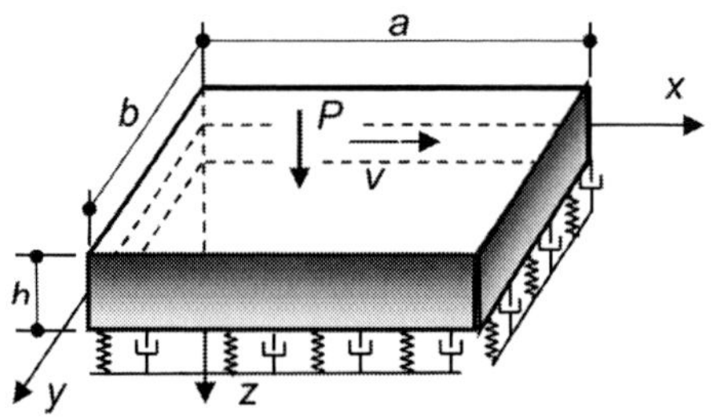

Figure 1. The Geometry of FG plate.

The viscoelastic foundation is modeled as a Winkler model with stiffness coefficient K_w and damping coefficient K_c.

Applying the higher-order shear deformation plate theory with another 'refined plate theory' proposed by Shimpi [11], the displacement components at an arbitrary point *(x,y,z)* in the plate are as follows:

$$\begin{aligned} U(x,y,z,t) &= u(x,y,z,t) - z\frac{\partial w_b}{\partial x} + \left[\frac{1}{4}z - \frac{5}{3}z\left(\frac{z}{h}\right)^2\right]\frac{\partial w_s}{\partial x} \\ V(x,y,z,t) &= v(x,y,z,t) - z\frac{\partial w_b}{\partial y} + \left[\frac{1}{4}z - \frac{5}{3}z\left(\frac{z}{h}\right)^2\right]\frac{\partial w_s}{\partial y} \\ W(x,y,z,t) &= w_b(x,y,t) + w_s(x,y,t) \end{aligned} \tag{1}$$

where (u, v, w_b, w_s) are the displacement components on the mid-plane.

The linear strains can be obtained by differentiating Eq. as:

$$\begin{Bmatrix} \varepsilon_{xx} \\ \varepsilon_{yy} \\ \gamma_{xy} \end{Bmatrix} = \begin{Bmatrix} \varepsilon_{xx}^0 \\ \varepsilon_{yy}^0 \\ \gamma_{xy}^0 \end{Bmatrix} + z\begin{Bmatrix} \kappa_x^b \\ \kappa_y^b \\ \kappa_{xy}^b \end{Bmatrix} + f\begin{Bmatrix} \kappa_x^s \\ \kappa_y^s \\ \kappa_{xy}^s \end{Bmatrix}, \quad \begin{Bmatrix} \gamma_{yz} \\ \gamma_{xz} \end{Bmatrix} = g\begin{Bmatrix} \gamma_{yz}^s \\ \gamma_{xz}^s \end{Bmatrix} \tag{2}$$

where:

$$\begin{Bmatrix} \varepsilon_{xx}^0 \\ \varepsilon_{yy}^0 \\ \gamma_{xy}^0 \end{Bmatrix} = \begin{Bmatrix} \dfrac{\partial u}{\partial x} \\ \dfrac{\partial v}{\partial y} \\ \dfrac{\partial u}{\partial y} + \dfrac{\partial v}{\partial x} \end{Bmatrix}, \begin{Bmatrix} \kappa_x^b \\ \kappa_y^b \\ \kappa_{xy}^b \end{Bmatrix} = \begin{Bmatrix} -\dfrac{\partial^2 w_b}{\partial x^2} \\ -\dfrac{\partial^2 w_b}{\partial y^2} \\ -2\dfrac{\partial^2 w_b}{\partial x \partial y} \end{Bmatrix},$$

$$\begin{Bmatrix} \kappa_x^s \\ \kappa_y^s \\ \kappa_{xy}^s \end{Bmatrix} = \begin{Bmatrix} -\dfrac{\partial^2 w_s}{\partial x^2} \\ -\dfrac{\partial^2 w_s}{\partial y^2} \\ -2\dfrac{\partial^2 w_s}{\partial x \partial y} \end{Bmatrix}, \begin{Bmatrix} \gamma_{yz} \\ \gamma_{xz} \end{Bmatrix} = \begin{Bmatrix} \dfrac{\partial w_s}{\partial y} \\ \dfrac{\partial w_s}{\partial x} \end{Bmatrix}$$

$$f(z)=-\frac{1}{4}z+\frac{5}{3}z\left(\frac{z}{h}\right)^2,\ g(z)=5\left[\frac{1}{4}-\left(\frac{z}{h}\right)^2\right]$$

The mechanical properties of FGM, such as Young's modulus E and mass density ρ, are assumed as

$$E(z)=(E_c-E_m)\left(\frac{z}{h}+\frac{1}{2}\right)^n+E_m$$
$$\rho(z)=(\rho_c-\rho_m)\left(\frac{z}{h}+\frac{1}{2}\right)^n+\rho_m \tag{3}$$

where the subscripts m and c represent the metallic and ceramic constituents, respectively, and n is the power index of the volume fraction.

Substituting Eq. (1) into Eq. (2) to obtain a strain expression. They can be used in constructing the strain energy and kinetic energy, and are dissipation expressions. Using Hamilton's principle to derive the equations of motion; we obtain the equations of motion of a plate as follows:

$$\delta u:\frac{\partial N_x}{\partial x}+\frac{\partial N_{xy}}{\partial y}=I_0\ddot{u}-I_1\frac{\partial \ddot{w}_b}{\partial x}-J_1\frac{\partial \ddot{w}_s}{\partial x}$$
$$\delta v:\frac{\partial N_{xy}}{\partial x}+\frac{\partial N_y}{\partial y}=I_0\ddot{v}-I_1\frac{\partial \ddot{w}_b}{\partial y}-J_1\frac{\partial \ddot{w}_s}{\partial y}$$
$$\delta w_b:\frac{\partial^2 M_x^b}{\partial x^2}+2\frac{\partial^2 M_{xy}^b}{\partial x\partial y}+\frac{\partial^2 M_y^b}{\partial y^2}-K_w(w_b+w_s)-K_c(\dot{w}_b+\dot{w}_s)+f$$
$$=I_0(\ddot{w}_b+\ddot{w}_s)+I_1\left(\frac{\partial \ddot{u}}{\partial x}+\frac{\partial \ddot{v}}{\partial y}\right)-I_2\nabla^2\ddot{w}_b-J_2\nabla^2\ddot{w}_s \tag{4}$$
$$\delta w_s:\frac{\partial^2 M_x^s}{\partial x^2}+2\frac{\partial^2 M_{xy}^s}{\partial x\partial y}+\frac{\partial^2 M_y^s}{\partial y^2}+\frac{\partial Q_{xz}}{\partial x}+\frac{\partial Q_{yz}}{\partial y}-K_w(w_b+w_s)-K_c(\dot{w}_b+\dot{w}_s)+f$$
$$=I_0(\ddot{w}_b+\ddot{w}_s)+J_1\left(\frac{\partial \ddot{u}}{\partial x}+\frac{\partial \ddot{v}}{\partial y}\right)-J_2\nabla^2\ddot{w}_b-K_2\nabla^2\ddot{w}_s$$

where

$$I_i=\int_{-h/2}^{h/2}\rho(z)z^i dz,\ J_i=-\frac{1}{4}I_i+\frac{5}{3h^2}I_{i+2}$$
$$K_2=\frac{1}{16}I_1-\frac{5}{6h^2}I_4+\frac{25}{9h^4}I_6$$
$$\nabla^2=\frac{\partial^2}{\partial x^2}+\frac{\partial^2}{\partial y^2}$$

The moving load is a concentrated moving load with constant speed and can be represented by the Delta Dirac function, as follows:

$$P=P(t)\delta(x-x_{mov}(t))\delta(y-y_{mov}(t)) \tag{5}$$

where $P(t)$ is the magnitude of the moving load, and $x_{mov}(t)$ and $y_{mov}(t)$ are the coordinates of the location of the load. In this work, the magnitude of the moving load is assumed to be constant $P=P_0$, $x_{mov}(t)=V_0t$.

Also, the moving load is assumed to move along a y-y direction at the central plate: $y_{mov}(t)=b/2$.

3. Analytical solution for vibration of FGM plate

In this work, the state-space representations of the dynamic systems will be used to analyze the transient response of a simply supported rectangular FG plate. The Navier approach is used to derive the closed-form solutions of equations of motion. The sinusoidal function is chosen to satisfy all boundary conditions, as follows:

$$\begin{aligned} u(x,y,t) &= \sum_{n=1}^{\infty}\sum_{m=1}^{\infty}\left\{U_{mn}(t)\cos\alpha x\sin\beta y\right\} \\ v(x,y,t) &= \sum_{n=1}^{\infty}\sum_{m=1}^{\infty}\left\{V_{mn}(t)\sin\alpha x\cos\beta y\right\} \\ w_b(x,y,t) &= \sum_{n=1}^{\infty}\sum_{m=1}^{\infty}\left\{W_{bmn}(t)\sin\alpha x\sin\beta y\right\} \\ w_s(x,y,t) &= \sum_{n=1}^{\infty}\sum_{m=1}^{\infty}\left\{W_{smn}(t)\sin\alpha x\sin\beta y\right\} \end{aligned} \tag{6}$$

where:

$$\alpha = \frac{m\pi}{a}, \beta = \frac{n\pi}{b} \tag{7}$$

Substitution Eq. (6) into Eq. (4) , the forced vibration of the functionally graded plate can be written as follows:

$$\begin{bmatrix} s_{11} & s_{12} & s_{13} & s_{14} \\ s_{12} & s_{22} & s_{23} & s_{24} \\ s_{13} & s_{23} & s_{33} & s_{34} \\ s_{14} & s_{24} & s_{34} & s_{44} \end{bmatrix}\begin{Bmatrix} U_{mn} \\ V_{mn} \\ W_{bmn} \\ W_{smn} \end{Bmatrix} + \begin{bmatrix} 0 & 0 & 0 & 0 \\ 0 & 0 & 0 & 0 \\ 0 & 0 & \hat{c}_{33} & 0 \\ 0 & 0 & 0 & \hat{c}_{44} \end{bmatrix}\begin{Bmatrix} \dot{U}_{mn} \\ \dot{V}_{mn} \\ \dot{W}_{bmn} \\ \dot{W}_{smn} \end{Bmatrix} + \begin{bmatrix} m_{11} & 0 & m_{13} & m_{14} \\ 0 & m_{22} & m_{23} & m_{24} \\ m_{13} & m_{23} & m_{33} & m_{34} \\ m_{14} & m_{24} & m_{43} & m_{44} \end{bmatrix}\begin{Bmatrix} \ddot{U}_{mn} \\ \ddot{V}_{mn} \\ \ddot{W}_{bmn} \\ \ddot{W}_{smn} \end{Bmatrix} = \begin{Bmatrix} 0 \\ 0 \\ F_{mn} \\ F_{mn} \end{Bmatrix} \tag{8}$$

where:

$$F_{mn} = \frac{4P_0}{ab}\sin(\alpha V_0 t)\sin\left(\frac{\beta b}{2}\right)$$

Eq. (8) must be rewritten in order to find a solution, as follows:

$$\dot{\boldsymbol{Z}} = \boldsymbol{AZ} + \boldsymbol{b} \tag{9}$$

where:

$$\begin{aligned} \boldsymbol{Z} &= \left\{U_{mn} \quad V_{mn} \quad W_{bmn} \quad W_{smn} \quad \dot{U}_{mn} \quad \dot{V}_{mn} \quad \dot{W}_{bmn} \quad \dot{W}_{smn}\right\}^T \\ \boldsymbol{b} &= \left\{0 \quad 0 \quad 0 \quad 0 \quad b_1 \quad b_2 \quad b_3 \quad b_4\right\}^T \end{aligned} \tag{10}$$

and $\boldsymbol{b}_i$ are the terms of the column matrix

$$\left\{b_1 \quad b_2 \quad b_3 \quad b_4\right\}^T = \boldsymbol{M}^{-1}\boldsymbol{F}^* \tag{11}$$

and block matrix A is

$$\boldsymbol{A} = \begin{bmatrix} 0 & I \\ -\boldsymbol{M}^{-1}\boldsymbol{S} & 0 \end{bmatrix} \tag{12}$$

The solution of is obtained as

$$\dot{\boldsymbol{Z}}(t) = e^{A(t-t_0)}\boldsymbol{Z}(t_0) + \int_{t_0}^{t} e^{A(t-\tau)}\boldsymbol{b}(\tau)d\tau \tag{13}$$

where t_0 is the initial time, $\boldsymbol{Z}(t_0)$ is the initial response, and the exponential matrix is $e^{A(t-t_0)}$.

This exponential matrix can be formulated in terms of the matrix of eigenvectors and eigenvalues associated with matrix A.

4. Examples

We consider a simply supported rectangular FG plate with side-to-thickness ratio a/h=40, rectangular dimensions of a=0.4m, b=0.3m, and power index p=2. The elastic moduli and mass density are chosen to be the same as [12]: $E_m = 70\,GPa$, ρ_m=$2707kg/m^3$, $E_c = 380GPa$, ρ_c=$3800kg/m^3$, with Poisson's ratio being 0.3. The moving load is written as $P_0 = 5000(N), v = 50\left(\frac{m}{s}\right)$.

For convenience, the normalized stiffness coefficient and the normalized damping coefficient of the foundation are defined as follows:

$$k_w = \frac{K_w a^4}{D_m}, k_c = \frac{K_c}{a^2 D_m} \tag{14}$$

where $D_m = \frac{E_m h^3}{12(1-v^2)}$ is the flexural rigidity of a full-metal plate.

In order to investigate the effect of the damping coefficient of the viscoelastic foundation, we investigate two cases of viscoelastic foundation:

$$\begin{bmatrix} k_w = 2 & k_c = 0.2 \\ k_w = 2 & k_c = 0.8 \end{bmatrix}$$

Fig.2 illustrates the transverse deflection as functions of time with a parameter $k_c = 0.2$, $k_c = 0.8$, respectively. The response of stress at the center of the FG plate is shown in Figs. 3–4. Two cases with different damping coefficients of viscoelastic foundation are employed (i.e., $k_c = 0.2$ and $k_c = 0.8$). In the forced vibration regime, it can be seen in Figs. 2–4 that the deflection and stresses predicted for FG plates with $k_c = 0.2$ are moderately larger than for the plate with $k_c = 0.8$. It is observed that when the damping coefficient increases, the deflection and stresses become smaller, as expected.

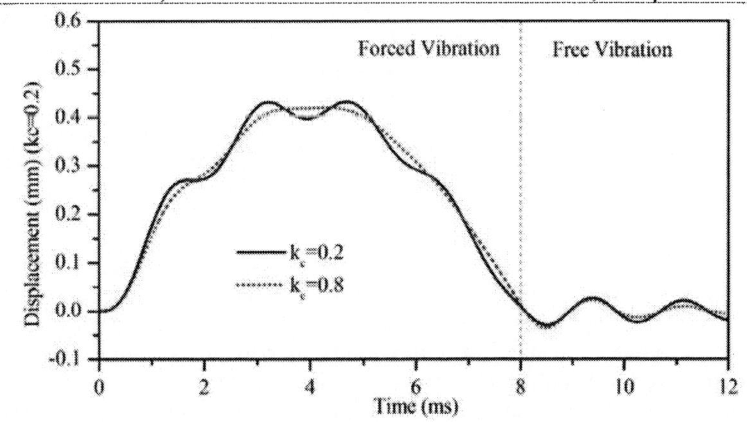

Figure 2. The deflection response at the central plate.

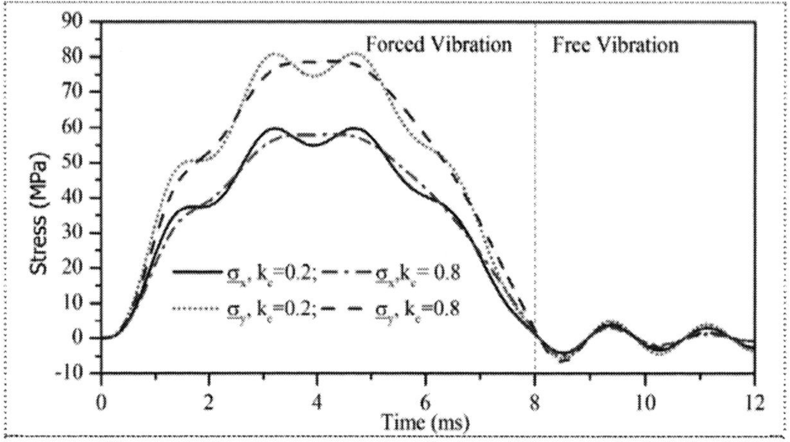

Figure 3. The stress response at the top of the central plate ($z=h/2$).

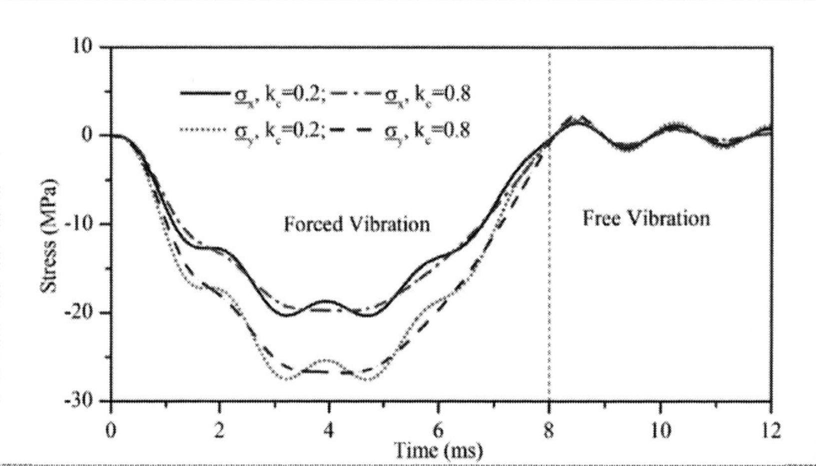

Figure 4. The stress response at the bottom of the central plate ($z=-h/2$).

5. Conclusions

This work applied a state-space method to determine forced vibration of FG plates resting on a viscoelastic foundation. The analytical solution for a dynamic FG plate was found to be a double sine series based on the Navier approach. The dynamic responses are considered for both forced and free vibrations. The results indicate that the damping coefficient of the foundation has significant effects on the dynamic response of the FG plate. The damping coefficient of the foundation causes the dissipation of energy, reducing ambient vibration at forced and free vibrations. This work contributes significantly to the field of engineering structures, and the results can be used in practical design structures.

References

[1] PL P 1954 On a new method of analysis of an elastic foundation by means of two foundation constants *Cosudarstrennoe Izdatelstvo Literaturi po Stroitelstvu i Arkhitekture.*

[2] Madhav MR, Poorooshasb HB 1989 Modified pasternak model for reinforced soil *Mathematical and Computer Modelling* 12(12):1505-1509.

[3] Beaufait FW, Hoadley PW 1980 Analysis of elastic beams on nonlinear foundations *Computers*

& Structures 12(5):669-676.

[4] Tsiatas GC 2009 Nonlinear analysis of non-uniform beams on nonlinear elastic foundation *Acta Mechanica* 209(1):141.

[5] Tang S-c 1967 Dynamic response of elastic plates on viscoelastic foundations to moving loads *Ingenieur-Archiv* 36(3):155-163.

[6] Zaman M, Taheri MR, Alvappillai A 1991 Dynamic response of a thick plate on viscoelastic foundation to moving loads *International Journal for Numerical and Analytical Methods in Geomechanics* 15(9):627-647.

[7] Huang M, Thambiratnam D 2002 Dynamic Response of Plates on Elastic Foundation to Moving Loads *Journal of Engineering Mechanics* 128(9):1016-1022.

[8] Phung-Van P, Nguyen-Thoi T, Luong-Van H, Thai-Hoang C, Nguyen-Xuan H 2014 A cell-based smoothed discrete shear gap method (CS-FEM-DSG3) using layerwise deformation theory for dynamic response of composite plates resting on viscoelastic foundation *Computer Methods in Applied Mechanics and Engineering* 272(Supplement C):138-159.

[9] Trung NT, Phuc PV, Anh TV, Chan NT 2014 Dynamic analysis of Mindlin plates on viscoelastic foundations under a moving vehicle by CS-MIN3 based on C-type higher-order shear deformation theory *Vietnam Journal of Mechanics* 36(1).

[10] Barati MR 2017 Vibration analysis of FG nanoplates with nanovoids on viscoelastic substrate under hygro-thermo-mechanical loading using nonlocal strain gradient theory *Structural Engineering and Mechanics* 64(6):683-693.

[11] Shimpi RP, Patel HG 2006 A two variable refined plate theory for orthotropic plate analysis *International Journal of Solids and Structures* 43(22–23):6783-6799.

[12] Hasani Baferani A, Saidi AR, Ehteshami H 2011 Accurate solution for free vibration analysis of functionally graded thick rectangular plates resting on elastic foundation *Composite Structures* 93(7):1842-1853.

Acknowledgments
The authors would like to acknowledge support from the University of Transport and Communications under the basic research fund.

Beam-column joint shear prediction using hybridized deep learning neural network with genetic algorithm

Zaher Mundher Yaseen[1*], Haitham Abdulmohsin Afan[1] and Minh-Tung Tran[2]

[1]Sustainable Developments in Civil Engineering Research Group, Faculty of Civil Engineering, Ton Duc Thang University, Ho Chi Minh City, Vietnam
[2]Faculty of Civil Engineering, Ton Duc Thang University, Ho Chi Minh City, Vietnam.

[*] Corresponding author: yaseen@tdt.edu.vn

Abstract. Scientifically evidenced that beam-column joints are a critical point in the reinforced concrete (RC) structure under the fluctuation loads effects. In this novel hybrid data-intelligence model developed to predict the joint shear behavior of exterior beam-column structure frame. The hybrid data-intelligence model is called genetic algorithm integrated with deep learning neural network model (GA-DLNN). The genetic algorithm is used as prior modelling phase for the input approximation whereas the DLNN predictive model is used for the prediction phase. To demonstrate this structural problem, experimental data is collected from the literature that defined the dimensional and specimens' properties. The attained findings evidenced the efficitveness of the hybrid GA-DLNN in modelling beam-column joint shear problem. In addition, the accurate prediction achived with less input variables owing to the feasibility of the evolutionary phase.

1. Introduction

One of the earliest studies defined the joint shear behavior in the beam-column established by [1]. Hanson and Connor described the joint shear of a free-body diagram at the mid-height of joint panel. Followed by [2], whom proposed a qualitative joint shear resistance mechanisms. Over the last two decades, numerous attempts have been conducted on the joint shear phenomena modeling using the probabilistic methodologies and prescribed codes expression [3–10]. All the forgoing researches demonstrated a good scientific progress on the shear joint problem. However, the motivation of studying this engineering problem more comprehensively is still ongoing topic for the civil and structural scholars. This is owing to the fact, the behavior of the joint shear beam-column is highly stochastic and non-linear and it is extremely difficult to be fully understandable. Especially, when there are various variables attributes influenced the joint shear value.

The capability of deep learning neural network (DLNN) learning techniques to overcome the challenges of the traditional algorithms that depend on hand-designed features has endeared to numerous researches of science and engineering, since their proposal by [11]. Based on the latest survey conducted by Liu and his collaborators [12], DLNN exhibited its potential in various disciplines including big data analysis, solving non-linear problems, pattern recognition, decision making system.

In short, the current research investigated hybrid evolutionary artificial intelligence model called GA-DLNN for predicting joint shear strength phenomena belong to RC beam-column connection. The main enthusiasm of developing the input selection algorithm is for the sake of the engineering optimization prospective. Where the major attributes are always motivated to be abstracted and focused to be studied. Whereas, the slight influenced variables can be eliminated from the mathematical procedure. On the other hand, establishing the data-intelligence model (i.e., DLNN) is magnificently for the understanding the non-linear mechanism between the predictors and predictand which is the joint shear behavior. Further, this is highly essential for the characterizing the joint shear behavior in general. Moreover, for the the structural engineering sustainability and stability.

2. Experimental data discerption

As mentioned earlier, in this research an experimental database with total number of 98 records belong to exterior beam-column connection was observed from acknowledged researches of the literature [13]. The modeled data set covered a wide range of parameters including: concrete strength, joint reinforcement ratio, reinforcement yielding stress, column axial average stress, beam height h_b, columns height (h_c), their ratio (h_b/h_c), beam width (b_b), columns width (b_c), (H), effective joint width (b_j), cylinder compressive strength of concrete (f_c'), (s), $(f_c^{0.5})$, (s/f_c), number of the influence parameters (n), beam area cross section (A_b), (f_{yb}), (f_{yh}), d_b, A_h, A_v, f_{yv}, and finally the horizontal joint shear V_{jh}. Note that the V_{jh} is the targeted output for all the input combinations used in the current research.

3. Methodology overview

There have been a massive of research interest in the field of deep learning techniques and literature evidenced that a great deal of success has been achieved [14]. The neural network models have provided a powerful framework for with deep architectures for supervised learning, supported by the rapid development of computation techniques. There is a hierarchical architecture in the deep learning algorithms with several layers. Each layer constitutes a non-linear unit for information processing [15]. This paper focused mainly on the deep neural network architectures. An increase in the number of layers and units in a single layer in deep neural networks can result in functions with higher complexity since they employ deep architectures.

3.1 Deep learning neural network

Due to the amalgamation of different functions in the feedforward neural networks, they are referred to networks. The model has a directed acyclic graph which describes how real architecture of the functions [16,17]. The training data provides the non-linearity information, and the approximate example of $f^*(x)$ evaluated at different training points. Each example x (input variables in the present case study) is accompanied by a label $y \approx f^*(x)$. The training examples directly points out what is expected from the output layer at each point x (must generate a value that is close to y). The training data does not specify the behavior of the other layers; the learning algorithm is responsible for deciding how those layers must be used to produce the desired output, though the responsibility of each layer is not specified by the training data. The learning algorithm rather must decide how these layers must be used for the best implementation of an approximation of f^*. Since the desired output of each of these layers is not specified by the training data, they are referred to the hidden layers (i.e., the deep learning processes). Figure 1 displayed the DLNN algorithm for establishing the predictive model.

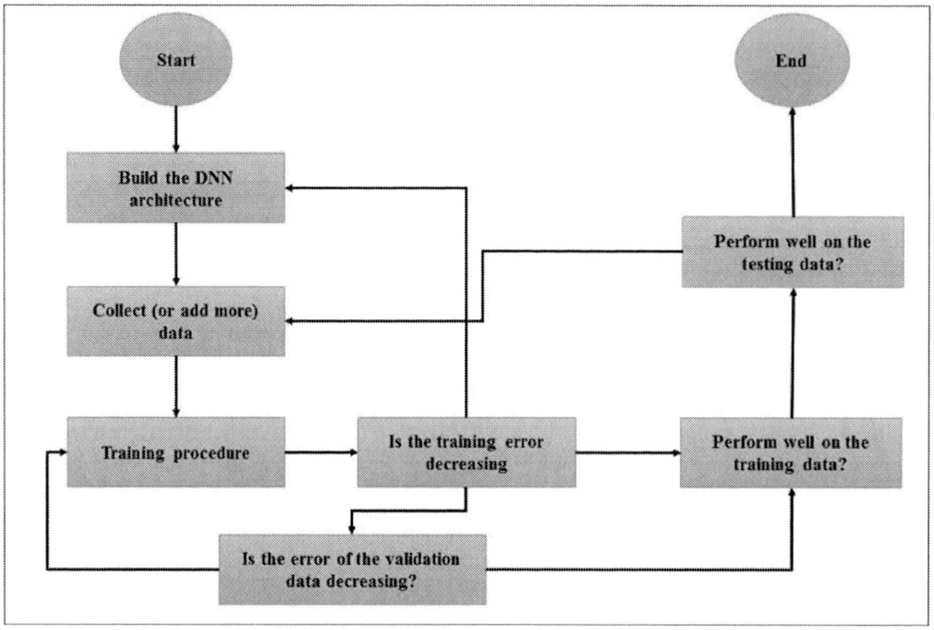

Figure 1. The deep neural network training processes

3.2 Genetic algorithm-input selection

GA is very well-known optimization technique that can classified as an evolutionary method based on biological process [18]. The effectiveness of this optimization approach discussed comprehensibly in term of solving the non-linearity and stochasticity by [19]. The main processes are involved the implementation of the heuristic GA are including reproduction of chromosomes, crossover, and mutation. Note these processes are applied to satisfied the probability of the discretization of the input variables that are coded into binary strings [20–22]. The GA processes are presented in Figure 2.

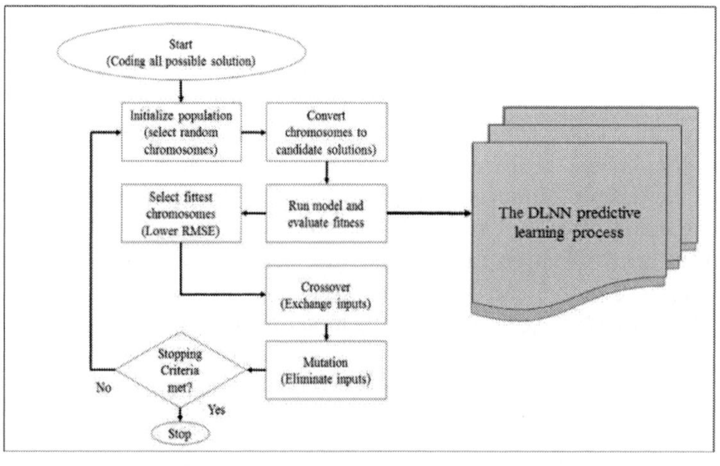

Figure 2. The evolutionary phase process that denote the genetic algorithm input approximation.

3.3 Modeling development

The modeling procedure of the current research conducted based on two phases: (i) the evolutionary phase, and (ii) the predictive model phase. The first phase was developed to identify the most correlated input attribute for the predictive model matrix in which presented as followed:

Model 1 $V_{jh} = f(b_c, A_v)$ (1-1)
Model 2 $V_{jh} = f(b_b, s, A_v)$ (1-2)
Model 3 $V_{jh} = f(b_c, s, h_c, A_v)$ (1-3)
Model 4 $V_{jh} = f(b_c, H, f_c', s/f_c, A_v)$ (1-4)
Model 5 $V_{jh} = f(b_c, s, A_b, A_h, A_v, f_{yv})$ (1-5)
Model 6 $V_{jh} = f(b_c, b_b, H, f_c', s/f_c, n, A_v)$ (1-6)
Model 7 $V_{jh} = f(b_c, s, h_b, h_b/h_c, b_b, A_b, d_b, f_{yv})$ (1-7)
Model 8 $V_{jh} = f(h_c, s, H, f_c', b_b, n, A_b, f_{yh}, A_v)$ (1-8)
Model 9 $V_{jh} = f(b_c, s, h_b, b_j, h_c, s/f_c, d_b, A_h, A_v, f_{yv})$ (1-9)
Model 10 $V_{jh} = f(b_c, h_c, b_b, h_b, h_b/h_c, f_c^{0.5}, s/f_c, n, A_b, A_h, A_v)$ (1-10)
Model 11 $V_{jh} = f(b_c, s, H, h_b/h_c, f_c^{0.5}, b_b, n, A_b, d_b, f_{yh}, A_v, f_{yv})$ (1-11)
Model 12 $V_{jh} = f(b_c, b_b, h_b, H, b_j, h_b/h_c, f_c^{0.5}, s/f_c, n, A_b, d_b, A_h, A_v)$ (1-12)
Model 13 $V_{jh} = f(b_c, h_c, h_b, H, b_j, f_c', s, n, A_b, f_{yb}, d_b, A_h, A_v, f_{yv})$ (1-13)
Model 14 $V_{jh} = f(h_c, b_b, h_b, H, b_j, f_c', f_c^{0.5}, s, s/f_c, n, A_b, A_h, f_{yh}, A_v, f_{yv})$ (1-14)
Model 15 $V_{jh} = f(h_c, b_b, h_b, H, b_j, h_b/h_c, f_c', f_c^{0.5}, s, s/f_c, A_b, f_{yb}, d_b, A_h, f_{yh}, A_v)$ (1-15)
Model 16 $V_{jh} = f(h_c, b_b, h_b, H, b_j, h_b/h_c, f_c', f_c^{0.5}, s, s/f_c, A_b, f_{yb}, d_b, A_h, f_{yh}, A_v, f_{yv})$ (1-16)
Model 17 $V_{jh} = f(b_c, h_c, b_b, h_b, H, b_j, h_b/h_c, f_c', f_c^{0.5}, s, s/f_c, n, A_b, f_{yb}, d_b, A_h, A_v, f_{yv})$ (1-17)
Model 18 $V_{jh} = f(b_c, h_c, b_b, h_b, H, b_j, h_b/h_c, f_c', f_c^{0.5}, s, s/f_c, n, A_b, f_{yb}, d_b, A_h, f_{yh}, A_v, f_{yv})$ (1-18)

Note that all these input combinations are function to the predicted value of the joint shear strength (V_{jh}). The prediction skills are evaluated in the current modeling using mean absolute error (*MAE*), root mean square error (*RMSE*), mean relative error (*RE%*), and correlation coefficient (R^2) [23,24]. The evaluation metric can be explained as follow:

$$MAE = \frac{1}{N}\sum_{n=1}^{n} |V_{jh(obs)} - V_{jh(pre)}| \quad (1\text{-}19)$$

$$RMSE = \sqrt{\frac{1}{N}\sum_{n=1}^{n} (V_{jh(obs)} - V_{jh(pre)})^2} \quad (1\text{-}20)$$

$$RE = [\frac{V_{jh(obs)} - V_{jh(pre)}}{V_{jh(obs)}}] * 100 \quad (1\text{-}22)$$

$$R^2 = \frac{\sum_{n=1}^{n}[(V_{jh(obs)} - V_{jh(Mobs)})(V_{jh(pre)} - V_{jh(Mpre)})]}{\sqrt{\sum_{n=1}^{n}(V_{jh(obs)} - V_{jh(Mobs)})^2 - \sum_{n=1}^{n}(V_{jh(pre)} - V_{jh(Mpre)})^2}} \quad (1\text{-}23)$$

where $V_{jh(obs)}$, $V_{jh(pre)}$, and $V_{jh(Mobs)}$, $V_{jh(Mpre)}$are the observed and predicted joint shear, (and the mean observed and predicted joint shear).

4. Joint shear prediction analysis

By recalling the main aim of the present study, this section is discussed the implementation of the employment of the hybrid GA-DLNN on the beam-column joint shear prediction. Indeed, the main inspiration of devoting this approach in the field of structural engineering in order to provide an easy and intelligent model that can be applied for real engineering practice. Based on the numerical statistical evaluation metrics tabulated in Table 1, it can be observed that the joint shear investigated in accordance of wide range of input combination. However, the attained results demonstrated variance outcomes. The best prediction results identified for Model 6 and Model 14 with correlation regression values 0.88 and 0.92, respectively. Model 14 could eliminated four variables that exhibited less influence to the targeted variable V_{jh}, in which provided a reasonable solution. The other excellent input combination,

Model 6 that indicated its prediction skills with including seven variables (i.e., $b_c, b_b, H, f_c', s/f_c, n, A_v$) only, that gave a perfect solution with engineering optimization prospective.

Table 1 Test results in the consolidated undrained triaxial shear test.

Models	*MAE*	*RMSE*	*RE%*	R^2
Model 1	209.1439	244.2575	11.8619	0.6031
Model 2	159.0401	210.7275	-10.4181	0.7459
Model 3	152.7922	202.9763	3.02179	0.7069
Model 4	166.6005	205.4624	-11.5576	0.7763
Model 5	161.4492	220.0787	-10.8532	0.7303
Model 6	**134.6139**	**158.0685**	**-11.4712**	**0.8823**
Model 7	206.8392	272.3324	-0.63218	0.4917
Model 8	181.7253	236.3841	-13.4925	0.7139
Model 9	163.7863	218.6789	-13.2947	0.7704
Model 10	147.7235	209.9443	-12.0954	0.7734
Model 11	193.5223	243.2407	-18.3117	0.7405
Model 12	159.4951	218.3748	-11.9957	0.7488
Model 13	205.3932	250.5302	-19.6313	0.7388
Model 14	**115.0746**	**141.4397**	**-12.2096**	**0.9238**
Model 15	217.8899	283.4964	-23.6467	0.7226
Model 16	201.3741	250.3991	-20.5874	0.7743
Model 17	173.5161	230.1902	-15.6410	0.7603
Model 18	209.3070	255.6957	-23.6298	0.8017

In more presentable vision, Figure 3 illustrated the scatter plots of the best prediction input combinations (Model 6 and Model 14). The figure visualized an acceptable agreement between the observed and predicted values around the 45° ideal line. Also, it can be observed, the prediction for the low, medium and high shear presented almost the same variance with slight diversion.

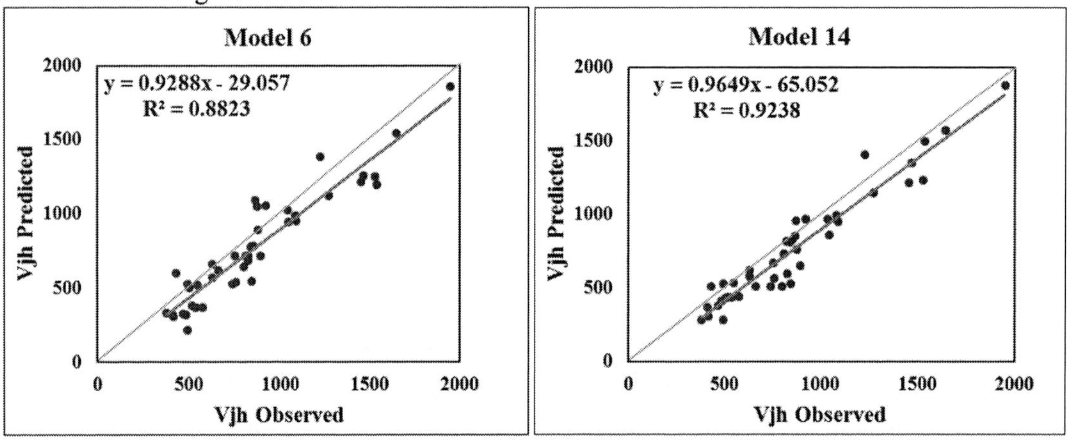

Figure 3. The scatter plots graphical presentation between the observed and predicted values of the joint shear over the testing phase.

Figure 4 displayed the relative error (*RE%*) distribution percentage graph over the testing phase period. The *RE* percentage pattern for Model 6 limited between -25% and +20% for the majority of the observation. Except, couple observation exceeded the -55% and +39%. Although Model 14 indicated higher R^2 than Model 6; yet, the *RE* pattern revealed higher under predicting percentages that exceeded the -20% for 50% of the testing phase. This is best can be explained due to the influence of the correlation coefficient to the high records. Whereas, the *RE* denoted error percentage for each record individually. In summary, the investigated model showed an optimistic finding that give a promising possibility for engineering implementation.

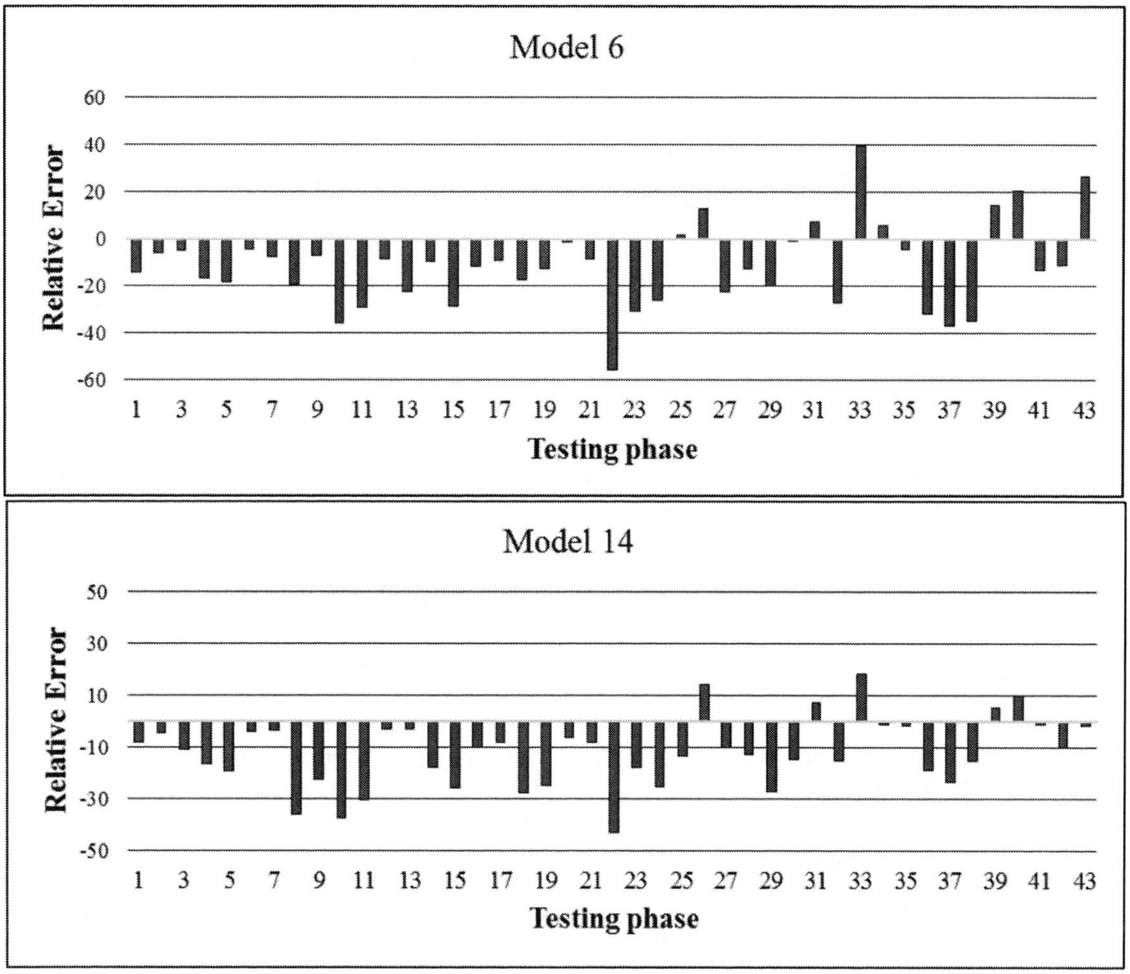

Figure 4. The relative error indicator percentage over the testing phase of the modeling.

5. Conclusions

In this research, new hybrid data-intelligence model called GA-DLNN is proposed to predict the joint shear strength of exterior beam-column structure frame. The modeling established using published experimental work by various scholar over the last two decades. A full details of these data summarized by [13]. The predictive DLNN model advanced with an input optimization method to select the most correlated input attributes to predict the joint shear strength. This is owing to conceptual fact, engineers always seek for optimal solution for any problem with minimum available information. The optimal input attribute combinations that attained the best prediction skill are Model 6 and Model 14 in which denoted seven and fifteen variables, respectively. Overall, the proposed integrated model showed an acceptable agreement between the observed experimental joint shear and the predicted value. In quantitative terms, the best models achieved with (*MAE-RMSE*) values (134.6-158.0) and (115.0-141.4) for Model 6 and Model 14, respectively. In general, the motivation of the applied methodology provided a robust and simple method for the structural engineering prospective.

References

[1] Hanson N W and Conner H W 1967 Seismic resistance of reinforced concrete beam-column joints *Journal of the Structural Division* **93** 533–60
[2] Paulay B T, Park R and Priestley M J N 1978 Reinforced Concrete Beam-Column Joints Under

Seismic Actions *ACI Journal* 585–93

[3] Kim J and LaFave J M 2008 Probabilistic joint shear strength models for design of RC beam-column connections *ACI Structural Journal* **105** 770–80

[4] Kim J and Lafave J M 2009 Joint Shear Behavior of Reinforced Concrete Beam-Column Connections subjected to Seismic Lateral Loading *Database* **61** 119–32

[5] LaFave J and Kim J 2011 Joint shear behavior prediction for RC beam-column connections *International Journal of Concrete Structures and Materials* **5** 57–64

[6] Kim J, LaFave J M and Song J 2007 A new statistical approach for joint shear strength determination of RC beam-column connections subjected to lateral earthquake loading *Structural Engineering and Mechanics* **27** 439–56

[7] Wang Y C and Hsu K 2009 Shear strength of RC jacketed interior beam-column joints without horizontal shear reinforcement *ACI Structural Journal* **106** 222–32

[8] Wang G L, Dai J G and Teng J G 2012 Shear strength model for RC beam-column joints under seismic loading *Engineering Structures* **40** 350–60

[9] Jeon J S, Desroches R, Brilakis I and Lowes L N 2013 Comparison of fragility curves for an older RC frame with column and beam-column joint shear models *11th International Conference on Structural Safety and Reliability, ICOSSAR 2013* pp 4423–30

[10] Jeon J, Shafieezadeh A and DesRoches R 2014 Statistical models for shear strength of RC beam-column joints using machine-learning techniques *Earthquake Engineering & Structural Dynamics* **43** 2075–95

[11] Hinton G E, Osindero S and Teh Y-W 2006 A Fast Learning Algorithm for Deep Belief Nets *Neural Computation* **18** 1527–54

[12] Liu W, Wang Z, Liu X, Zeng N, Liu Y and Alsaadi F E 2017 A survey of deep neural network architectures and their applications *Neurocomputing* **234** 11–26

[13] Tran T M, Hadi M N S and Pham T M 2014 A new empirical model for shear strength of reinforced concrete beam–column connections *Magazine of Concrete Research* **66** 514–30

[14] Liu W, Wang Z, Liu X, Zeng N, Liu Y and Alsaadi F E 2017 A survey of deep neural network architectures and their applications *Neurocomputing* **234** 11–26

[15] Bengio Y 2009 Learning Deep Architectures for AI *Foundations and Trends® in Machine Learning* **2** 1–127

[16] Huang W, Hong H, Song G and Xie K 2014 Deep process neural network for temporal deep learning *Proceedings of the International Joint Conference on Neural Networks* pp 465–72

[17] Schmidhuber J 2015 Deep Learning in neural networks: An overview *Neural Networks* **61** 85–117

[18] Holland J 1975 Adaptation in Natural and Artificial Systems *Ann Arbor, University of Michigan Press*

[19] Goldberg D E, Korb B and Deb K 1989 Messy Genetic Algorithms : Motivation , Analysis , and First Results *Engineering* **3** 493–530

[20] Golmohammadi D, Creese R C, Valian H and Kolassa J 2009 Supplier Selection Based on a Neural Network Model Using Genetic Algorithm *IEEE Transactions on Neural Networks* **20** 1504–19

[21] Yang J and Honavar V 1998 Feature subset selection using a genetic algorithm *IEEE Intelligent Systems and their Applications* **13** 44–9

[22] Bruderer E and Singh J V. 1996 Organizational evolution, learning, and selection: A genetic-algorithm-based model *Academy of Management Journal* **39** 1322–49

[23] Chai T and Draxler R R 2014 Root mean square error (RMSE) or mean absolute error (MAE)? - Arguments against avoiding RMSE in the literature *Geoscientific Model Development* **7** 1247–50

[24] Taylor R 1990 Interpretation of the Correlation Coefficient: A Basic Review *Journal of Diagnostic Medical Sonography* **6** 35–9

Influence of weld-induced residual stresses on the hysteretic behavior of a girth-welded circular stainless steel tube

Chin-Hyung Lee [1, 6], Vuong Nguyen Van Do [2], Kyong-Ho Chang [3], Jun-Tai Jeon [4] and Tae-Hwan Um [5]

[1] The Graduate School of Construction Engineering, Chung-Ang University, 84, Huksuk-ro, Dongjak-ku, Seoul 06974, Republic of Korea

[2] Sustainable Developments in Civil Engineering Research Group, Faculty of Civil Engineering, Ton Duc Thang University, Ho Chi Minh City, Vietnam

[3] Department of Civil and Environmental & Plant Engineering, Chung-Ang University, 84, Huksuk-ro, Dongjak-ku, Seoul 06974, Republic of Korea

[4] Department of Civil & Environmental Engineering, Inha Technical College, 100, Inharo, Nam-Gu, Incheon 22212, Republic of Korea

[5] Department of Civil Engineering, Chung-Ang University, 84, Huksuk-ro, Dongjak-ku, Seoul 06974, Republic of Korea

[6] Author to whom any correspondence should be addressed, ifinder@hanmail.net

Abstract. The present study attempts to characterize the relevance of welding residual stresses to the hysteretic behaviour of a girth-welded circular stainless steel tube under cyclic mechanical loadings. Finite element (FE) thermal simulation of the girth butt welding process is first performed to identify the weld-induced residual stresses by using the one-way coupled three-dimensional (3-D) thermo-mechanical FE analysis method. 3-D elastic-plastic FE analysis equipped with the cyclic plasticity constitutive model capable of describing the cyclic response is next carried out to scrutinize the effects that the residual stresses have on the hysteretic performance of the girth-welded steel tube exposed to cyclic axial loading, which takes the residual stresses and plastic strains calculated from the preceding thermo-mechanical analysis as the initial condition. The analytical results demonstrate that the residual stresses bring about premature yielding and deterioration of the load carrying capacity in the elastic and the transition load ranges, whilst the residual stress effect is wiped out quickly in the plastic load domain since the residual stresses are nearly wholly relaxed after application of the cyclic plastic loading.

1. Introduction

Stainless steels are becoming an attractive choice of material in construction which includes bridges and offshore structures attributed to the established benefits over carbon steels, i.e., they present favourable mechanical properties, high durability and excellent resistance to corrosion and fire. Stainless steel circular tubes are, in particular, gaining increasing usage as exposed structural applications owing to the combination of corrosion resistance and structural efficiency as well as refined appearance. During their period of use, these tubes experience diverse load types such as axial tension or compression, bending, torsion and cyclic loading. In most tubular structures, connection of the circular members is mainly implemented by girth welding. In a girth-welded circular tube, the existence of inevitable welding residual stresses is well known [1-6], which result from the highly localized, non-uniform, transient thermal expansion and shrinkage imposed during welding process. Weld-induced residual stresses increase the susceptibility to fatigue damage, stress corrosion cracking and brittle fracture [7-9]. It is well recognized that mechanical behavior of a girth-welded circular steel tube under monotonic loading is significantly affected by welding residual stresses [10, 11], i.e., the girth weld-induced residual stresses induce premature yielding and loss of stiffness and eventually lead to deterioration of the load carrying capacity.

A small number of research works have been devoted to examining the cyclic response of a girth-welded tubular member by experimental and/or numerical methods. Shao et al. [12, 13] investigated hysteretic performance of circular and square hollow section tubular T-joints under cyclic axial loading through the experiments and the corresponding finite element (FE) analyses. In the analyses, however, they did not consider the cyclic plasticity in the constitutive modeling and instead employed a bilinear stress-strain equation with isotropic hardening, even though the cyclic plasticity model is essential in the numerical reproduction of the cyclic behavior of the material, nor did they take account of the residual stresses generated by the girth welding due to the truly complex analytical procedure entailed in welding and subsequent cyclic mechanical loadings. Thus, quantitative assessment of the residual stress effect on the hysteretic behavior could not be attempted. Chang et al. [14] performed a numerical study on the hysteretic response of the welded joint in carbon steel pipe member under cyclic flexural loading taking welding residual stresses into account. Nevertheless, the accuracy of the result is doubtful since the performance of the cyclic constitutive model employed in the FE analysis, which is proposed based on the multi-surface hardening model, is not completely verified. Moreover, they failed to elaborate how the residual stresses affect the hysteretic behavior of the welded pipe joint. Hence, the structural response of a girth-welded tubular member subjected to cyclic loading has not been fully evaluated and is entitled to special attention.

This paper deals with hysteretic behavior of a girth-welded circular stainless steel tube under cyclic loading through the numerical investigation, focusing on the relevance of the weld-induced residual stresses to the cyclic performance. Thermal simulation of the girth butt welding process is first conducted to get the residual stresses by using the one-way coupled three-dimensional (3-D) thermo-mechanical FE analysis method. 3-D elastic-plastic FE analysis to scrutinize the residual stress effect on the hysteretic response of the girth-welded stainless steel tube subjected to cyclic axial loading is next carried out incorporating the residual stresses and plastic strains. A cyclic plasticity constitutive model for replicating the cyclic response is presented and its accuracy is confirmed by the experimental data, which is taken as the material constitutive equation in the FE analysis.

2. FE analysis of the residual stresses in the girth-welded circular stainless steel tube

FE thermal simulation of the girth butt welding process should first be carried out to obtain the weld-induced residual stresses and plastic strains, which are required input to the mechanical model for analysing the residual stress effect on the hysteretic behaviour of the girth-welded circular stainless steel tube under cyclic axial loading. In the present work, the process of welding was simulated by using the in-house FE code [15] developed by the authors based on the 3-D thermo-mechanical FE formulation, which has been extensively verified against experiments and numerical results found in the literature [5, 6]. The solution procedure for welding residual stresses and deformations is split into two parts:

1) A transient thermal analysis which solves for the temperature field and its history associated with the heating and cooling during welding.
2) A transient thermal-mechanical analysis that employs the temperature history solutions as the thermal loading for the evolution of residual stresses and deformations.

Since the thermal filed has a strong effect on the mechanical field with little inverse influence, the two analyses are one-way coupled and thus can be performed in sequence. Specific details on the thermo-mechanical FE analysis are well described in the forthcomings [5, 6].

2.1. Model configuration

The FE analysis model configuration and dimensions are schematically shown in figure 1(a). The welding start/end position ($\theta = 0°$) and the welding arc travel direction are also depicted in the illustration. Two circular steel tubes with V-groove between them were considered to be joined by a single-pass girth butt welding. The welding parameters employed for the present simulation were as follows: welding method, GTA

IOP Conf. Series: Earth and Environmental Science **143** (2018) 012026 doi:10.1088/1755-1315/143/1/012026

welding process; welding voltage, 22 V; welding current, 230 A and welding speed, 1.3 mm/s, respectively. The heat input and welding speed were derived from the experiment [16]. The 3-D FE mesh model is shown in figure 1(b). Note that only half of the tube geometry was modelled with eight-nodded isoparametric solid elements due to the structural symmetry along the weld centreline. The adequate mesh density is achieved when an increase in the number of elements has a negligible effect on the global results. This results in the smallest element size of 0.9 mm (axial) × 1.5 mm (thickness) × 12.1 mm (circumference) in the weld region. The mesh in the regions far away from the girth weld is relatively coarse and equally refined along the axis to save computing time. In order to facilitate nodal data mapping between thermal and mechanical models, the same FE mesh refinement is used with respective element types. The FE element type used in the heat transfer model is the one which has single degree of freedom, temperature, on its each node, whilst in the mechanical model the element type is the other with three translational degrees of freedom at each node. As the tube is not clamped during the welding process, no boundary conditions except those to inhibit rigid body motion of the work piece are applied. The detailed boundary conditions adopted in the mechanical model are illustrated in figure 1(b) by the arrows.

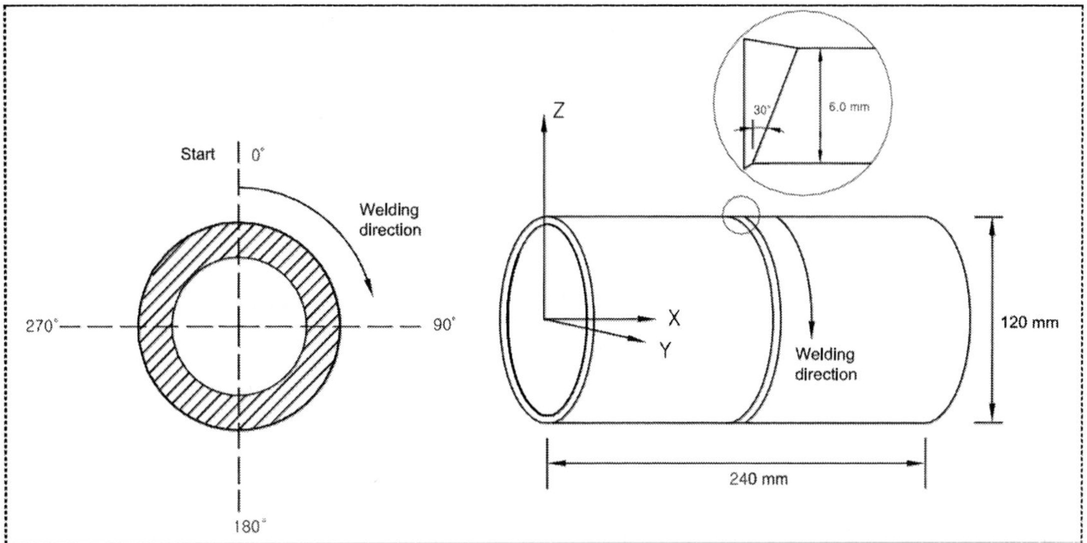

(a) Configuration of the analysis model and the welding arc direction

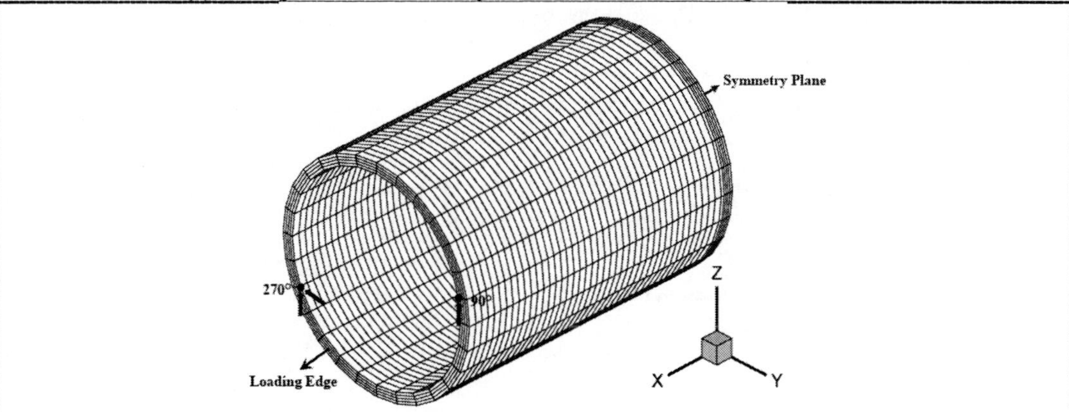

(b) 3-D FE model

Figure 1. Analysis model.

2.2. Material model

The base metal chosen for this work is SUS304 austenitic stainless steel. Its physical (e.g. thermal conductivity, heat transfer coefficient, specific heat and density) and mechanical properties (e.g. yield stress, Young's modulus, Poisson's ratio and thermal expansion coefficient) are dependent on temperature. The variation of the thermal and mechanical properties with temperature is considered and is given in [5, 6]. In this analysis, autogenous weldment was assumed [17], i.e. the same material properties were used for the weld metal. The strain hardening produced by the cyclic thermal loading during welding in the weld region and its neighbourhood was also considered with the temperature-dependent strain hardening rule [5, 6].

2.3. Results and discussion

Figures 2(a) and 2(b) show the axial residual stresses along the axial direction at different circumferential angles from the weld start/end position (see figure 1(a)) on the inside and outside surfaces, respectively. From the predicted results, it can be found that in and around the weld region, the axial residual stresses are tensile on the inside surface and compressive on the outside surface due to the bending moment attributed to the circumferential shrinkage after welding. The tensile axial residual stresses level out in compression and converge to zero for self-equilibration as the distance from the weld region increases, and vice versa on the outside surface. Production mechanism for the residual stresses is well defined in the references [5, 6]. Figures 2(c) and 2(d) compare the hoop residual stresses along the axial direction at the different locations on the inside and outside surfaces, respectively. With regard to the hoop residual stresses, their formation is affected by the axial residual stresses. This accounts for why on the outside surface, which undergoes axial compression, the hoop residual stresses are less tensile at the weld area and its periphery compared to those on the inside surface. Detailed explanations on the axial and hoop stress distributions are given elsewhere [5, 6]. It should be recognized that the axial and hoop residual stresses vary depending on the circumferential

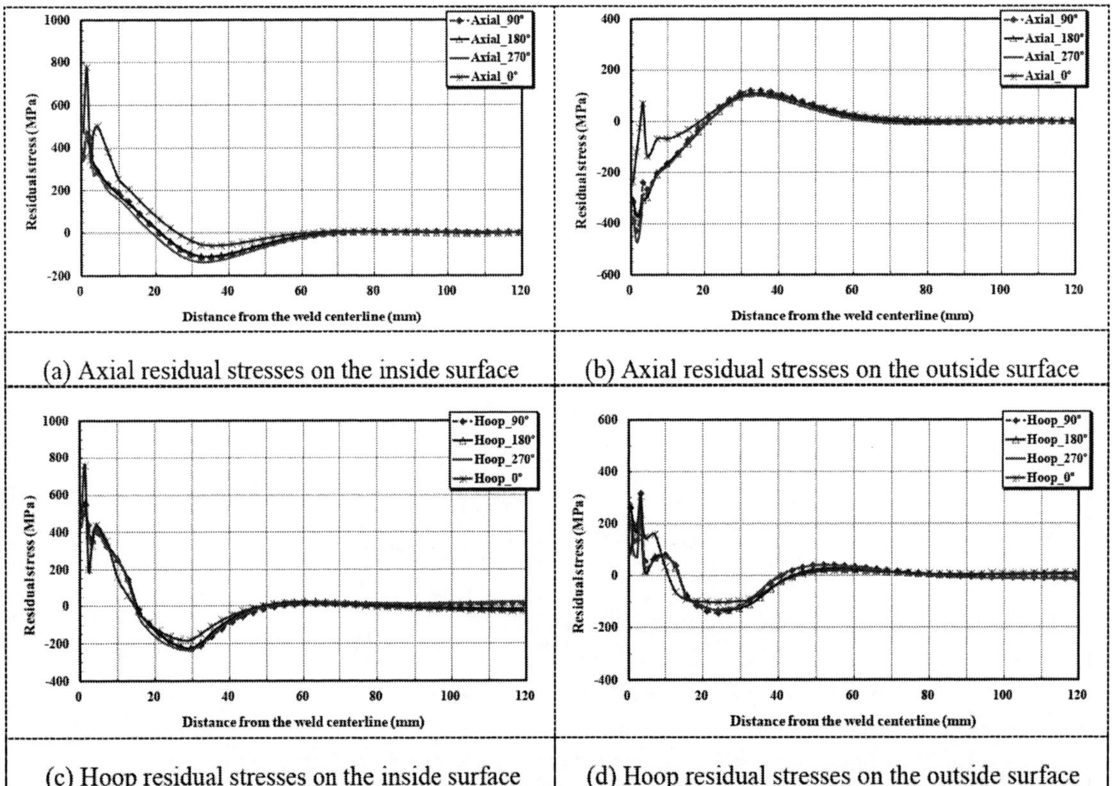

(a) Axial residual stresses on the inside surface

(b) Axial residual stresses on the outside surface

(c) Hoop residual stresses on the inside surface

(d) Hoop residual stresses on the outside surface

Figure 2. Residual stress distributions at locations with different circumferential angle θ.

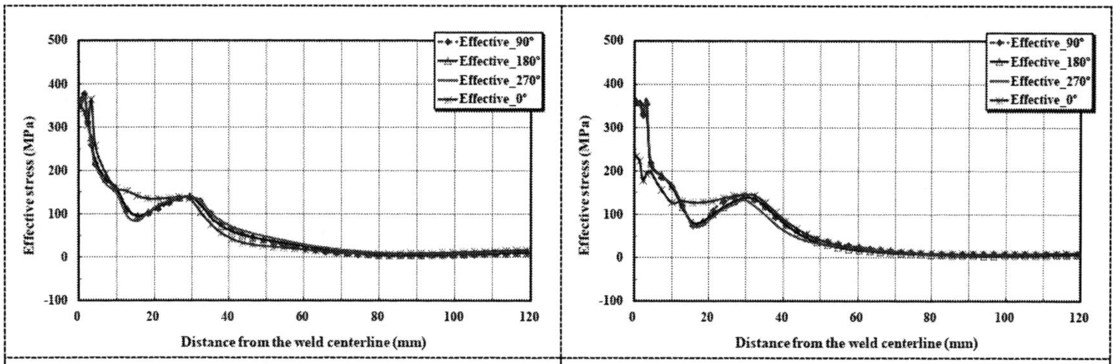

(a) Inside surface (b) Outside surface

Figure 3. Residual effective stresses at the four locations.

position. This is due to the spatial variation of the internal restraint during the girth welding process. A considerable change of the stress profile is also seen at the welding start/end position. The residual effective stress distributions are also depicted in figure 3. It is noteworthy that the maximum residual effective stress at the girth weld exceeds the room temperature yield stress, which is also reported in an experimental work such as [18]. The higher strain hardening rate could be regarded as the main factor to explain the maximum residual stress. Nevertheless, it is much below the room temperature ultimate strength of the stainless steel. The effective stress distributions also exhibit the 3-D effects in and around the welded zone.

3. Hysteretic behaviour of the girth-welded circular stainless steel tube

3.1. Cyclic plasticity constitutive model

Detailed descriptions on the cyclic plasticity constitute model utilized to quantify the inelastic cyclic characteristics of the base material which displays cyclic hardening are made in [19]. In this section, only the key constitutive relations are given for completeness. The rate-independent cyclic plasticity constitutive model which assumes isotropic hardening and J_2-type plasticity is considered and is based on the following three basic components:

(i) von Mises yield criterion: the yield criterion specifies the state of stress for which plasticity will occur. The yield surface is expressed by

$$f(\sigma-\alpha)=\left[\frac{3}{2}(\sigma'-\alpha')\cdot(\sigma'-\alpha')\right]^{1/2}-R-k^{*}=0 \qquad (1)$$

where f is the yield function, σ is the Cauchy stress tensor, α is the back stress tensor, σ' is the deviatoric stress tensor, α' is the deviatoric part of the back stress tensor, k^* is the initial size of the elastic domain and R is the isotropic hardening variable. The tensorial operator "$\cdot$" indicates the inner product between second-order tensors A and B as $A\cdot B=A_{ij}B_{ij}$.

(ii) Flow rule: the total strain increment is additively decomposed into the elastic and the plastic strain increment. The plastic strain increment is given by

$$\dot{\varepsilon}^{p}=\dot{\lambda}\frac{\partial f}{\partial\sigma}=\frac{1}{H}\langle\dot{\sigma}'\cdot n\rangle n,\ n=\sqrt{\frac{2}{3}}\frac{\partial f}{\partial\sigma} \qquad (2)$$

where $\dot{\lambda}$ is a positive scale factor of proportionality, H is the plastic modulus, $\langle\ \rangle$ represents the MacCauley bracket. n is the unit vector normal to the yield surface at current stress point.

(iii) Hardening rule: the hardening rule describes the yield surface change due to plastic deformation. In general, the yield surface can vary in size (the yield surface expansion by isotropic hardening), shape (the yield surface distortion) and/or centre location (the yield surface translation by kinematic hardening) in consequence of plastic deformation. In the present investigation, the yield surface variation with plastic loading is accommodated by considering the isotropic and the kinematic hardening rule. For the kinematic hardening rule which is the most favourite for simulations of a structural response induced by cyclic loading, the decomposed nonlinear kinematic hardening equation proposed by the authors [19] is adopted and incorporated into the cyclic plasticity constitutive model, which is expressed as

$$\dot{\alpha}'_{ij} = \sum_{k=1}^{l} (\dot{\alpha}'_{ij})_k = \sum_{k=1}^{l} \gamma_k \left\{ \frac{2}{3} c_k \dot{\varepsilon}^p_{ij} - \left(\frac{\bar{\alpha}'_k}{c_k} \right)^m \left[\varphi(\alpha'_{ij})_k + (1-\varphi)((\alpha'_{rs})_k \cdot n_{rs}) n_{ij} \right] \dot{p} \right\} \tag{3}$$

where m, φ, c_k and γ_k are material constants. $\bar{\alpha}'_k = \left({}^3\!/_2 \alpha'_k : \alpha'_k \right)^{1/2}$ is the asymptotic value of the kth back stress. The material parameters c_k and γ_k are determined using only the uniaxial stress-strain responses, while the additional constants m and φ are adjustable to fit the ratcheting experiment. In Eq. (3), $\dot{p}$ is the evolution of the accumulated plastic strain which can be expressed as

$$\dot{p} = \left| \dot{\varepsilon}^p_{ij} \right| = \left[\frac{2}{3} \dot{\varepsilon}^p_{ij} \dot{\varepsilon}^p_{ij} \right]^{1/2} = \dot{\lambda} \left[\frac{2}{3} \frac{\partial f}{\partial \sigma_{ij}} \frac{\partial f}{\partial \sigma_{ij}} \right]^{1/2} \tag{4}$$

With regard to the isotropic hardening rule, the following equation is adopted [19].

$$R = Q(1 - e^{-bp}) \tag{5}$$

where Q is the saturated constant value of the isotropic hardening variable under a specific cyclic loading and can be obtained from experimental data. b is the material parameter which describes the rapidity of the isotropic hardening.

Simulation of a cyclic response primarily depends on the accuracy of the plastic modulus calculation. The plastic modulus can be derived from the kinematic hardening rule through the consistency condition as follows [19].

$$\begin{aligned} H = {} & E_{ijkl} \frac{\partial f}{\partial \sigma_{ij}} \frac{\partial f}{\partial \sigma_{kl}} + \frac{2}{3} \sum_{k=1}^{n} \gamma_k c_k \frac{\partial f}{\partial \sigma_{ij}} \frac{\partial f}{\partial \sigma_{ij}} - \sum_{k=1}^{n} \gamma_k \frac{\partial f}{\partial \sigma_{ij}} \left(\frac{\bar{\alpha}'_k}{c_k} \right)^m \left[\varphi(\alpha'_{ij})_k + (1-\varphi)((\alpha'_{rs})_k \cdot n_{rs}) n_{ij} \right] \sqrt{\frac{2}{3} \frac{\partial f}{\partial \sigma_{kl}} \frac{\partial f}{\partial \sigma_{kl}}} \\ & + Qbe^{-bp} \sqrt{\frac{2}{3} \frac{\partial f}{\partial \sigma_{mn}} \frac{\partial f}{\partial \sigma_{mn}}} \end{aligned} \tag{6}$$

where E_{ijkl} is the fourth-order elastic modulus tensor. The incremental total stress can then be obtained as

$$\dot{\sigma}_{ij} = E_{ijkl} \left(\dot{\varepsilon}_{kl} - \dot{\lambda} \frac{\partial f}{\partial \sigma_{kl}} \right) = \left(E_{ijkl} - \frac{1}{H} E_{ijmn} \frac{\partial f}{\partial \sigma_{mn}} E_{opkl} \frac{\partial f}{\partial \sigma_{op}} \right) \dot{\varepsilon}_{kl} \tag{7}$$

Eq. (7) can be rewritten as

$$\dot{\sigma}_{ij} = D_{ijkl} \dot{\varepsilon}_{kl} \tag{8}$$

where the elasto-plastic modulus tensor is defined as

$$D_{ijkl} = E_{ijkl} - \frac{1}{H} E_{ijmn} \frac{\partial f}{\partial \sigma_{mn}} E_{opkl} \frac{\partial f}{\partial \sigma_{op}} \tag{9}$$

The path dependence on the plastic strain amplitude of the base material is not considered here owing to the lack of experimental data.

The cyclic plasticity constitutive model with three decomposed kinematic hardening rules ($l = 3$), which has three segments of a stable hysteresis loop, is used [20]. Therefore, the performance of the cyclic plasticity model requires a total number of eleven material constants, i.e. the material constant for the initial size of the yield surface, the material parameters of c_1, c_2, c_3, γ_1, γ_2, γ_3, m, φ for calculating the kinematic hardening rate and those of Q, b for computing the isotropic hardening rule are needed to simulate the hysteretic response. To determine the kinematic hardening parameters, uniaxial test data of the stable stress-strain hysteresis loop with pure kinematic hardening are required. Assuming that the cyclic hardening is only induced by isotropic hardening, the isotropic hardening evolution with respect to accumulated plastic strain through the strain-controlled uniaxial symmetric cyclic loading test can be gained. The effect of isotropic hardening on the stabilized stress-strain hysteresis loop in which both isotropic and kinematic hardening are mixed can be excluded by using the isotropic hardening evolution function. Then, the kinematic hardening parameters c_1, c_2, c_3 and γ_1, γ_2, γ_3 are acquired in accordance with the procedures specified in the reference [20]. The material constants Q and b can also be obtained from the isotropic hardening evolution curve [19]. The additional kinematic hardening parameters m and φ are identified from the ratcheting experiments by the trial and error method [19].

In order to confirm the validity of the cyclic constitutive model used in this investigation and for direct comparison between the analytical prediction and the test data, numerical simulation of the experimental work by Kang et al. [21] was conducted by using the FE analysis method. They measured the stress-strain response of SUS 304 stainless steel under the strain-controlled uniaxial symmetric cyclic loading. Specific details on the test are presented in the reference [21]. Figure 4 compares the simulated and the experimental hysteresis curves for the strain-controlled cyclic loads with the strain amplitude of 0.6%. All the material parameters for the cyclic constitutive model were obtained from the experiment and are presented in Table 1. The predicted stress-strain response curves show a high correlation with the measured hysteresis loops, and it can also be observed that the cyclic plasticity constitutive model accurately duplicates the cyclic hardening behavior. As a result, the cyclic plasticity constitutive model is believed to be appropriate for describing the inelastic cyclic behavior of the material.

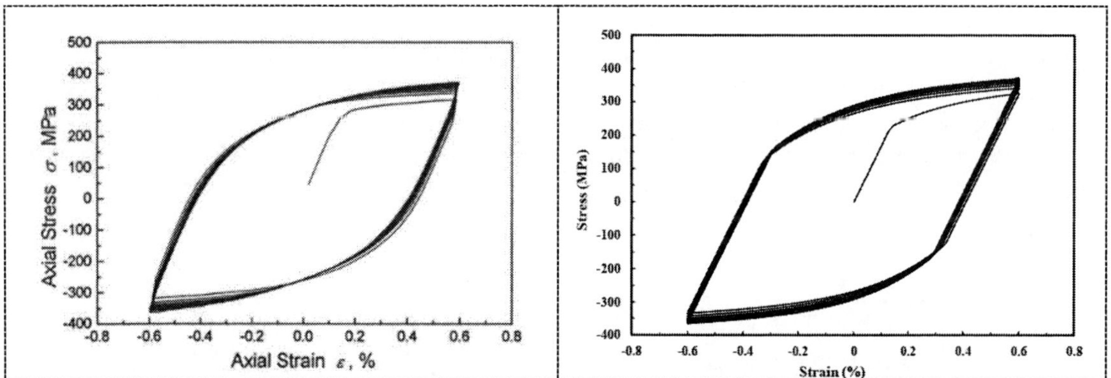

(a) Experiment (b) Simulation

Figure 4. Comparison of the predicted and experimental hysteresis loops.

Table 1. Material constants for the kinematic and the isotropic hardening rate.

m	φ	c_1	c_2	c_3	γ_1
0.5	1.0	30.11	82.71	70.8	3.466
γ_2	γ_3	Q	b	k^*	
350.0	53.19	50.0	12.5	225.0	

3.2. 3-D FE analysis model

3-D elastic-plastic FE analyses were performed to explore structural responses of the girth-welded circular steel tube exposed to cyclic axial loading through the in-house FE-code [15, 22] equipped with the cyclic plasticity constitutive model and the material parameters. The same FE mesh refinement scheme as the thermal-mechanical model, with dissimilar loading and boundary conditions, was adopted to expedite data mapping between the two structural FE models. The computational domain was discretized by 3-D eight-nodded isoparametric brick elements with three translational degrees of freedom at each node. The residual stresses and plastic strains obtained from the preceding FE thermal simulation were introduced as the pre-stress condition into the FE model. Then, symmetrical strain-controlled cycling was imposed on the edge of the girth-welded tube model (see figure 1(b)) parallel to the axial direction. Two loading patterns were considered, i.e. stepped uniaxial strain cycling in which the strain amplitude varies from 0.1 % to 0.6% and uniaxial strain cycling with constant strain amplitude of 0.6 %, which are shown in figure 5. The boundary conditions taken into consideration in the analysis assumed that the nodes at the loaded end were only free to move in the loading direction to allow for application of the strain cycling and were kinematically coupled to translate together in the same direction. As described earlier, autogenous weldment was employed during the girth welding. Therefore, the cyclic stress-strain relations for the base material can be used to capture the hysteretic response of the girth-welded circular stainless steel tube.

3.3. Results and discussion

Figure 6 compares the simulated cyclic strain-stress curves of the stainless steel tube with and without considering welding residual stresses for the stepped strain cycling with changing strain amplitude. It is immediately apparent that the cyclic behaviour is only affected by the residual stresses in the elastic (0.1%) and the transition load ranges (0.2 % and 0.3%), i.e., the residual stresses are not able to change the cyclic

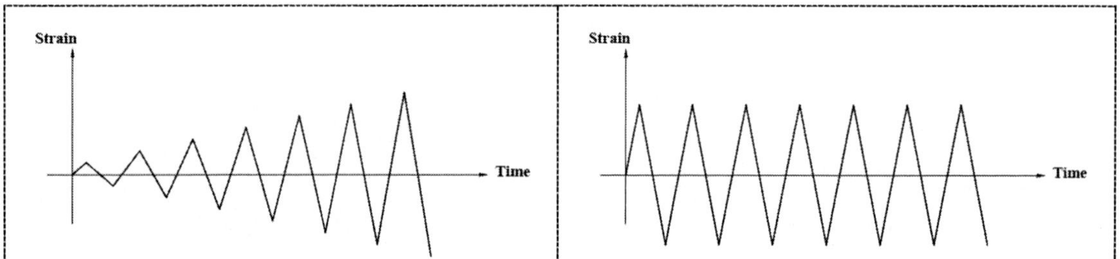

(a) Stepped uniaxial symmetric strain cycles (b) Constant uniaxial symmetric strain cycles

Figure 5. Loading histories.

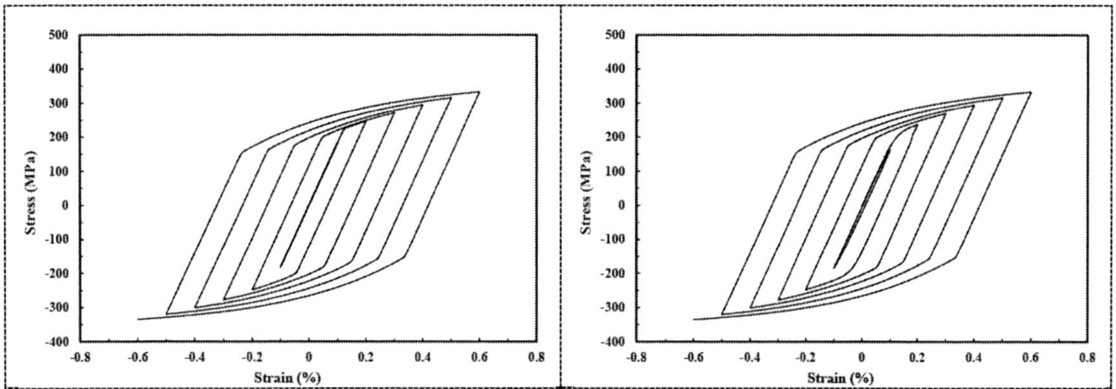

(a) Simulation without considering welding residual stresses

(b) Simulation with consideration of welding residual stresses

Figure 6. Cyclic strain-stress curves for the stepped strain cycling with changing strain amplitude.

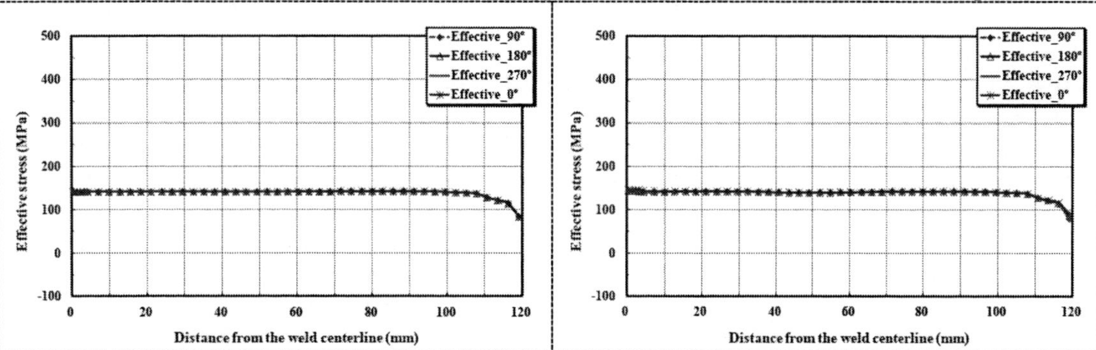

(a) Prediction without considering welding residual stresses

(b) Prediction with consideration of welding residual stresses

Figure 7. Comparison of the effective stresses on the inside surface after three cycles.

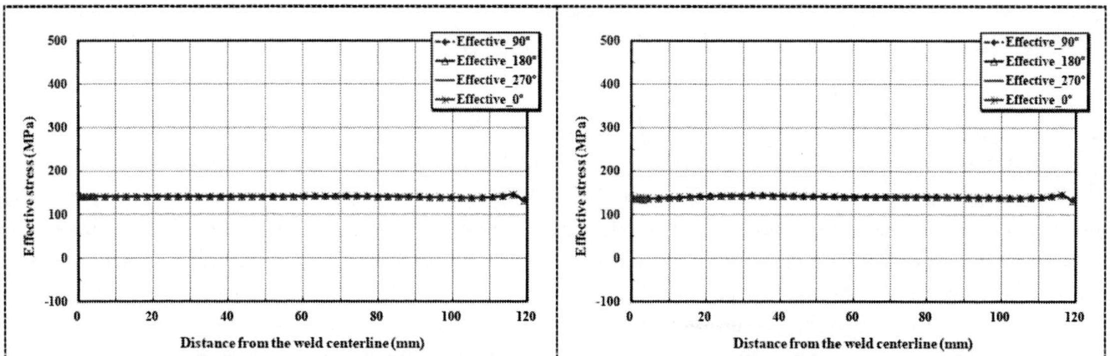

(a) Prediction without considering welding residual stresses

(b) Prediction with consideration of welding residual stresses

Figure 8. Comparison of the effective stresses on the outside surface after three cycles.

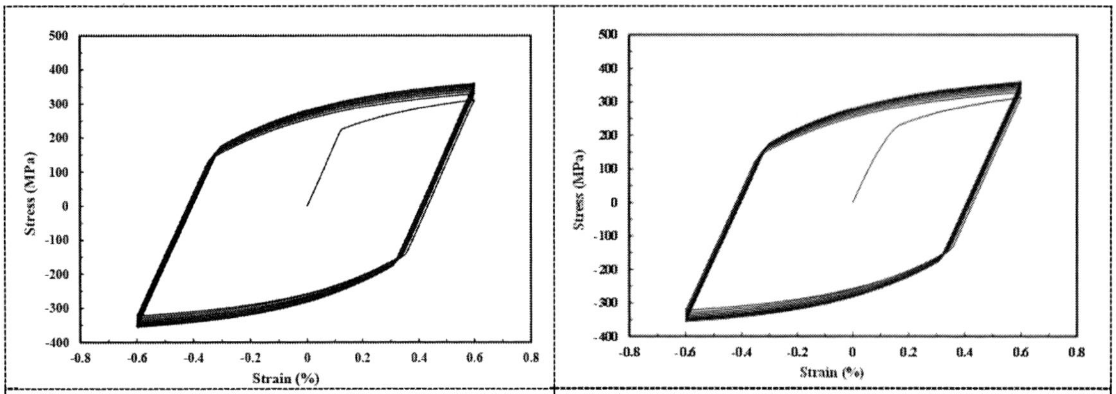

(a) Simulation without considering welding residual stresses

(b) Simulation with consideration of welding residual stresses

Figure 9. Hysteresis loops for the symmetric strain cycling with constant strain amplitude.

strain-stress histories in the fully plastic loading. In the elastic and the transition load domains, the residual stresses are not completely relaxed during the mechanical cyclic loading and the extent of the relaxation is mainly dependent on the stress amplitude applied [23]. The residual stresses cause premature yielding of the steel tube and result in deterioration of the load carrying capacity. In the plastic load range, however, the residual stresses are nearly wholly released by the cyclic plastic loading, thus do not influencing the hysteretic behaviour. Figures 7 and 8 represent the effective stress distributions after three cycles on the inside and outside surfaces, respectively with and without considering welding residual stresses. It can be recognized that both the effective stresses on the inside surface are nearly identical, and so do those on the outside surface. The simulated results corroborate the previous finding that the hysteretic behaviour of the steel tube is not affected by the residual stresses in the fully plastic load domain. Figure 9 depicts the predicted hysteresis loops of the tube model with and without considering weld-induce residual stresses for the symmetric strain cycling with the amplitude of 0.6 % which corresponds to the fully plastic load range. Careful observation of the analytical results reveals that effects of the residual stresses on the hysteretic behaviour of the girth-welded steel tube are confined to just the first cycle, i.e., the subsequent cyclic stress-strain curves are not altered in comparison with those of the tube model with no girth weld. The initial residual stresses bring about premature yielding and ensuing early onset of strain hardening, leading to a little higher load carrying capacity in the first loading phase. However, the residual stresses are almost completely relieved after the first cycle as a result of the cyclic loading, thus not capable of affecting the cyclic behaviour any more.

4. Conclusion

This study aimed to explore hysteretic behaviour of a girth-welded circular stainless steel tube subjected to cyclic mechanical loadings by numerical method and focused on characterizing the correlation of welding residual stresses to the cyclic behaviour. The one-way coupled 3-D thermo-mechanical FE analysis method was first utilized to simulate the girth butt welding process and hence to acquire the weld-induced residual stresses. Subsequently, the cyclic plasticity constitutive equation able to describe the cyclic response was presented and employed in the 3-D elastic-plastic FE analysis to scrutinize the effects of the residual stresses on the hysteretic performance of the girth-welded steel tube under cyclic axial loading. The weld-induced residual stresses and plastic strains gained from the thermal simulation were introduced as the initial condition for the FE analysis model. Two different loading types were considered to evaluate the influence that the residual stresses have on the hysteretic behaviour. Based on the results in this work, the following conclusions can be drawn.

- In the elastic and the transition load ranges, the residual stresses are not fully relaxed during the mechanical cyclic loading. The residual stresses induce premature yielding and deterioration of the load carrying capacity

- In the plastic load domain, the residual stresses are almost completely released by the cyclic plastic loading, thus do not affecting the hysteretic behaviour
- Accurate assessment of the residual stress effects on the hysteretic response under cyclic loading by the FE analysis method presented here can be helpful to understand the low-cycle fatigue behaviour of a girth-welded circular steel tube

Acknowledgments
This research was supported by Mid-career Researcher Program through the National Research Foundation of Korea (NRF) funded by the Ministry of Science, ICT & Future Planning (NRF-2015R1A2A2A01006390).

References
[1] Karlsson R I, Josefson B L 1990 Three-dimensional finite element analysis of temperatures and stresses in a single-pass butt-welded pipe *J. Press. Vessel. Technol.* **112** 76
[2] Brickstad B, Josefson B L 1998 A parametric study of residual stresses in multi-pass butt-welded stainless steel pipes *Int. J. Pressure Vessels Pip. 75* 11
[3] Yaghi A, Hyde T H, Becker A A, Sun W, Williams J A 2006 Residual stress simulation in thin and thick-walled stainless steel pipe welds including pipe diameter effects *Int. J. Pressure Vessels Pip.* **83** 864
[4] Deng D, Murakawa H 2006 Numerical simulation of temperature field and residual stress in multi-pass welds in stainless steel pipe and comparison with experimental measurements *Comput. Mater. Sci.* **37** 269
[5] Lee C H, Chang K H, Park J U 2013 Three-dimensional finite element analysis of residual stresses in dissimilar steel pipe welds *Nucl. Eng. Des.* **256** 160
[6] Lee C H, Chang K H 2014 Comparative study on girth weld-induced residual stresses between austenitic and duplex stainless steel pipe welds *Appl. Therm. Eng.* **63** 140
[7] Withers P J 2007 Residual stress and its role in failure *Rep. Prog. Phys.* **70** 2211
[8] Wang Z, Wang Q 2014 Fatigue assessment of welds joining corrugated steel webs to flange plates *Eng. Struct.* **73** 1
[9] Matos C G, Dodds Jr R H 2000 Modeling the effects of residual stresses on defects in welds of steel frame connections *Eng. Struct.* **22** 1103
[10] Lee C H, Chang K H, Park K T, Shin H S, Kim T 2013 Bending resistance of girth-welded stainless steel circular hollow sections *Thin-Walled Struct.* **73** 174
[11] Lee C H, Chang K H, Park K T, Shin H S, Lee M 2014 Compressive strength of girth-welded stainless steel circular hollow section members: Stub columns *J. Constr. Steel Res.* **92** 15
[12] Shao Y B, Li T, Lie S T, Chiew S P 2011 Hysteretic behaviour of square tubular T-joints with chord reinforcement under axial cyclic loading *J. Constr. Steel Res.* **67** 140
[13] Shao Y B, Lie S T, Chiew S P, Cai Y Q 2011 Hysteretic performance of circular hollow section tubular joints with collar-plate reinforcement *J. Constr. Steel Res.* **67** 1936
[14] Chang K H, Jang G C, Shin Y E, Han J G, Kim J M 2006 The behavior of welded joint in steel pipe members under monotonic and cyclic loading *Int. J. Pressure Vessels Pip.* **83** 846
[15] Lee C H 2005 A study on the mechanical characteristics of high strength steel for the application to the steel bridge *Ph.D. Thesis Chung-Ang University Korea*
[16] Sattari-Far I, Javadi Y 2008 Influence of welding sequence on welding distortions in pipes *Int. J. Pressure Vessels Pip.* **85** 262
[17] Teng T L, Chang P H, Tseng W C 2003 Effect of welding sequence on residual stresses *Comput. Struct.* **81** 273
[18] Paradowska A, Price J W H, Ibrahim R, Finlayson T 2005 A neutron diffraction study of residual stress due to welding *J. Mater. Process. Technol.* **164-165** 1099
[19] Lee C H, Do N V V, Chang K H 2014 Analysis of uniaxial ratcheting behavior and cyclic mean stress relaxation of a duplex stainless steel *Int. J. Plast.* **62** 17
[20] Bari S, Hassan T 2000 Anatomy of coupled constitutive models for ratcheting simulation *Int. J. Plast.* **16** 381

[21] Kang G Z, Gao Q, Yang X 2002 A visco-plastic constitutive model incorporated with cyclic hardening for uniaxial/multiaxial ratcheting of SS304 stainless steel at room temperature *Mech. Mater.* **34** 521
[22] Do N V V 2013 Finite element modeling of fatigue damage and its evolution in steel structures *Ph.D. Thesis Chung-Ang University Korea*
[23] Cho J, Lee C H 2016 FE analysis of residual stress relaxation in a girth-welded duplex stainless steel pipe under cyclic loading *Int. J. Fatigue* **82** 462

Behaviour of steel-concrete composite beams using bolts as shear connectors

Minh-Tung Tran[1]* Vuong Nguyen Van Do[1] Tuan-Anh Nguyen[1]

[1]Faculty of Civil Engineering, Ton Duc Thang University, Ho Chi Minh City, Vietnam;

* Corresponding author: tranminhtung@tdt.edu.vn

Abstract. The paper presents an experimental program on the application of bolts as shear connectors for steel-composite beams. Four steel- concrete composite beams and a reference steel beam were made and tested. The aim of the testing program is to examine which forms of the steel bolts can be used effectively for steel-composite beams. The four types of the bolts include: Type 1 the bolt with the nut at the end; Type 2 the bolt bending at 900 hook; Type 3 the bolt without the nut at the end and Type 4 the bolt with the nut at the end but connected with the steel beam by hand welding in other to be connected with the steel beam by bolt connection as in the first three types. The test results showed that beside the traditional shear connectors like shear studs, angle type, channel type, bolts can be used effectively as the shear connectors in steel-composite beams and the application of bolts in Types 1 and 2 in the composite beams gave the better performance for the tested beam.

1. Introduction

Because of the benefits of associating the advantages of the two materials, Steel- concrete composite (SCC) structures have been widely applied throughout the world for the construction of bridges and civil building [1, 2]. The majority application of this type of structure is in the kind of composite beam where the beam is formed by a steel beam, a concrete slab, and mechanical shear connectors. The shear connectors are the link between the concrete slab and the steel beam and are responsible for shear force transfer at the concrete-steel interface. Therefore, they are the key elements for ensuring the composite beam works effectively in sustaining bending moment. There are many kinds of shear connectors that have been applied and the welded headed shear stud is the most popular type of mechanical shear connector used in steel-concrete composite flexural members because of their relatively faster installation process [3]. However, headed shear studs are welded to the top flange of the steel beam using a special equipment that requires high voltage power for operation which may be dangerous in the installation process. Besides, in some cases, when large number of shear studs is needed to transfer safely the relatively large amount of lateral forces, the designers are challenged with the struggle of placing the sufficient number of shear studs on the steel beams [4]. Therefore, beside the welded headed shear stud, others shear connectors such as channel type, angle type, U type, perfobon and bolt have been developed.

The behaviour of SCC beam depended not only on the properties of the steel beam and the concrete slab but also on the degree of shear connection which was determined by the arrangement and the strength of the shear connectors. Based on the degree of shear connection, SCC beams can be divided as no shear connection beams, partial shear connection beams, and full shear connection beams [5]. To date, many experimental and analytical studies have been investigated to understand the behaviour of such beams, especially the partial shear connection beams. Sanghyo et al. [6] examined three SCC beams with different stud diameters. This study showed that the elastic stiffness, ultimate load, and yield load of the beams were affected by the degree of shear connection. Xing et al. [7] studied the behaviour of SCC beams with the variation of sectional dimensions, studs, and degree of shear connection and their study showed that a relatively larger degree of shear connection can lengthen the elastic stage and delay the development and spread of slip but may decrease the ductility of the SSC beam. Vasdravellis and Uy [8] tested 14 SCC beams and 1 steel beam. Their study showed that the main factors affecting the shear capacity of an SCC beam are the slab thickness and degree of shear connection.

Although a large number of experimental and analytical studies have been conducted, engineers are still facing with many challenges. Firstly, the installation of headed studs requires special skilled workers, which may take additional cost so that reducing the other advantages of composite beams. Secondly, the application of short channels or angles with their bottom flanges welded to the steel beam may help in reducing the cost associated with shear connector installation but controlling of the quality of the welded to ensure reliable shear connectors in this case may take the installing process more complicated. In addition, the use of channels or angles may cause congestion to the longitudinal steel bars in the concrete slab. Therefore, the application of bolts as shear connectors may be a merit solution for this problem and in this paper the results of an experimental program on SCC beams with using bolts as the shear connectors are introduced. The experimental program which was carried out at Ton Duc Thang structural lab included testing 4 SCC beams with using bolts as shear connectors and one steel beam refered as the control specimen. The results of the tests showed that bolts can work effectively as shear connectors in SCC beams and the application of this type of connectors is prospective.

2. Experimental program

2.1 Test specimens

Four SCC and 1 steel beams denoted as Specimen D_1 to D_4 and D_0, respectively were test under 4 points bending. The steel beam D_0 has a length of 4.5m with an I section of 150mm x 250 mm (Figure 1). It was composed of two flanges with the sectional area of 150 mm x 12 mm and a web with the sectional area of 250 mm x 8 mm. The SCC beams (D_1 to D_4) have the steel beams the same as Specimen D_0 but with a concrete slab having sectional area of 600mm x 120 mm connected with the steel beams at the top flange via four different types of bolts which were used as shear connectors.

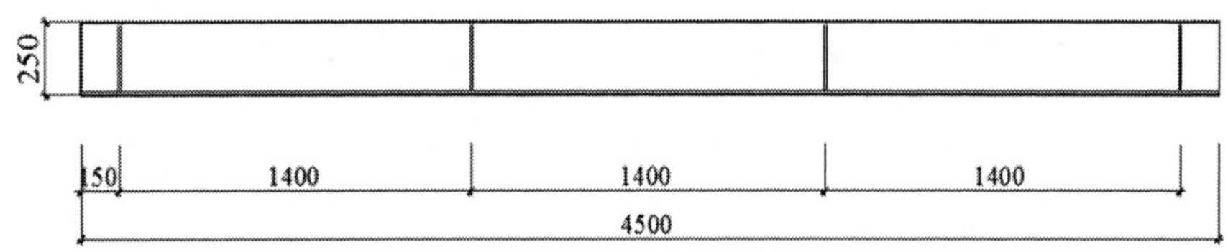

(a) Elevation

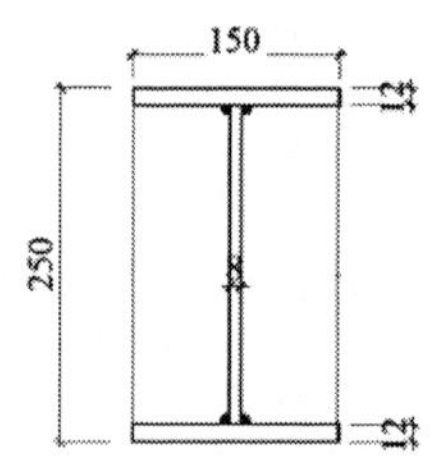

(b) Cross-section

Figure 1: The steel specimen

The dimensions of the SCC beams and the arrangement of the shear connectors are shown in Figure 2. A total of 42 bolts arranged at two lines with the space ranging from 160 mm to 240 mm were used as the shear connectors for each SCC beams. This amount of the shear connectors was calculated basing on the design code EC4 (1992) [9]. As shown in Figure 2, the top and bottom reinforcement of the concrete slab are similar and composed of seven deformed steel bars, for each layer, having a diameter of 12mm placed as the longitudinal reinforcement. The same steel bars were used as transverse reinforcement for the top and bottom layers but with a spacing of 100 mm.

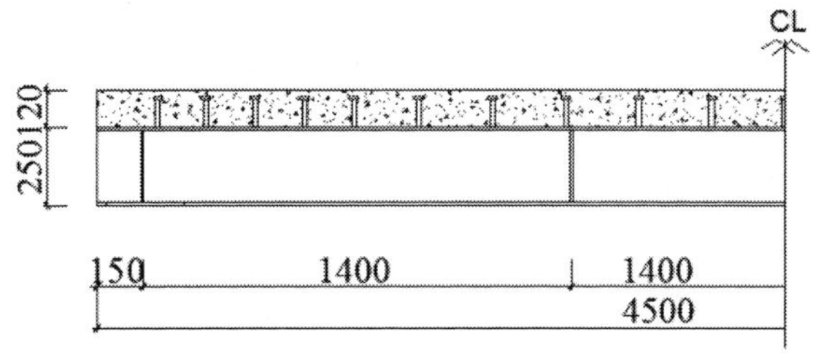

(a) Elevation

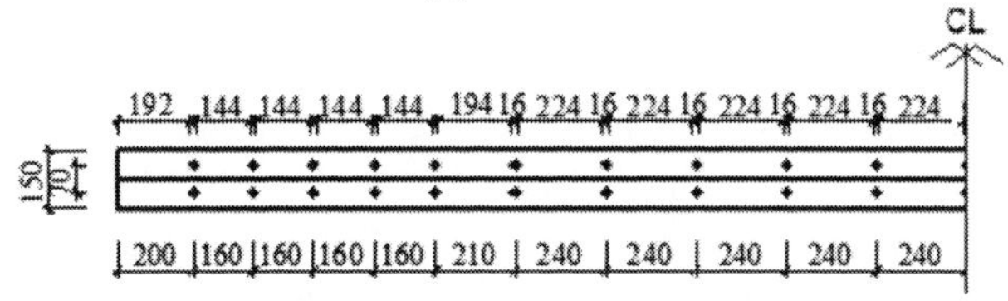

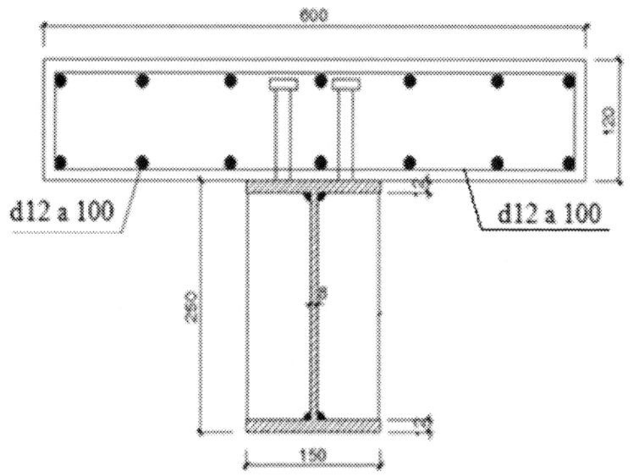

(b) Plane (c) Cross-section

Figure 2: Dimensions of the SCC beams and the arrangement of the shear connectors

(a) The shear connectors in Specimen D_1

(b) The shear connectors in Specimen D_2

(c) The shear connectors in Specimen D_3

(d) The shear connectors in Specimen D_4

Figure 3. The shear connectors in specimens from D_1 to D_4

The only difference between each specimen from specimens D_1 to D_4 is in the types of the bolts used as the shear connectors. Figure 3 illustrates the difference of these specimens. In Specimen D_1, M16x130 SD 8.8 bolts with diameter of 16mm and length of 130 mm were used, giving an embedded length of 100mm of the bolts in to the concrete slab (Figure 3a). In Specimen D_2, M16x250 SD 8.8 bolts with diameter of 16 mm and length of 250 mm were used, giving an embedded length of 220 mm of the bolts into the concrete slab. Difference from Specimen D1, in Specimen D2 the bolts did not have the nut but they were bended at 900 hooks along the beam as shown in Figure 3b. Specimen D_3 was also using M16 SD 8.8 bolts with a diameter of 16 mm as the shear connectors as in Specimen D_2 but the length of the embedded bolts in to the concrete slab is only 100 mm and the bolts without the nut at the end were also used. In all three specimens from D1 to D3, the bolts were connected with the top flange of the steel beam using bolted connections. Difference from the remaining specimens, Specimen D4 is the same as Specimen D_1 but the shear connectors were connected with the top flange of the steel beam via hand welded links (Figure 3d).

2.2 Material properties

The normal concrete mixing at the laboratory was used for casting all the SCC specimens. The contents of all the components for mixing a cubic meter of concrete included 420 kg of Holcim PC40 cement, 0.438 m3 of river sand, 0.845 m3 of stone aggregate, and 186 litres of water. All the SCC specimens were casted at same time and then curing at the same condition. The compressive strength of the concrete was determined by testing three 100mm x 200 mm cylinder samples which was casted using the same concrete with the one using for casting the specimens. At the age of 28 days the average compressive strength of the concrete was 25.4 MPa.

The tensile tests were taken for the plate steel which was used for composing the steel beams and for the reinforcement of the concrete slabs. There were two types of plate steel, including the 8 mm thickness and the 12 mm thickness. For each type of the steel, three similar samples having the width of 50 mm and the length of 500 mm were tested and the average yield strength of the steel plates was 343 MPa corresponding to an average yield strain of 1.9%. The average ultimate strength of the steel plates was 446 MPa corresponding to an average ultimate strain of 13.5%. Three samples of the steel reinforcement were tested and the average yield strength of the reinforcement was 335 MPa corresponding to an average yield strain of 1.6%. The average ultimate strength of the steel reinforcement was 476 MPa corresponding to an average ultimate strain of 21.5%.

3. Test setup and loading procedure

Four-point bending tests were implemented for all five Specimens. The test setup is illustrated in Figure 4. As shown in the figure, the beam was simply supported at the ends and two similar point loads which were created by two identical hydraulic actuators, were applied at its one-third points. The hydraulic actuators were set to work parallel at the same displacement speed and displacement control protocol during the test. The displacement speed of the hydraulic actuators was set to be 20 mm per minute until reaching the ultimate state. As the distance from the end of the beam to the supporter was 150 mm for both ends, the span of the tested beams were 4200 mm. For these setups, the beam segment between the point loads was mostly under pure bending while the remaining parts of the beam were mostly under constant shear and linearly increasing bending moment.

The loads were measured by using two load cells placed at the top of the hydraulic actuators and the displacements at the point loads were recorded by using two displacement transducers attached at the point load beneath the beam. The slip between the concrete slab and the upper flange of the steel beam at the end of the beam was measured by using one displacement transducers attached at the end of the beam. All the load cells and the displacement transducers were connected to a data logger which was set for collecting data every second.

Figure 4. Test setup

4. Test results and discussion

The relation between the load and deflection of the tested beams is shown in Figure 5. In the figure, the average values of the load and displacement at the two loading points are illustrated. For the steel beam D_0, the loading included two stages. In the first stage, the beam was loaded to the yielding load for checking the correction of the measured system and performance of the loading frame. The load was removed to zero and then the second loading stage was started. For the specimens from D_1 to D_4, only one loading stage was applied as the loading frame and the measured instruments were double checked and they worked reliably after testing the specimen D_0.

The the yield and ultimate loads about 76 and 121 kN, corresponding to the deflections of 17 and 40.5mm, respectively, were recorded for Specimen D_0. After reaching the maximum load, the beam was released from loading and an average of unrecovered deflection of 20.6 mm was recorded. The beam was failure due to the yielding of the steel at the bottom flange of the beam. The above testing results were very close to that of the modelling of the same beam using ABAQUS where the differences of the results ranking from 1 to 3 percent in the linear stage and about 5 to 10 percent in the non-linear stage.

In Specimen D_1, a maximum average load of 241 kN was recorded corresponding to an average deflection of 64.9 mm and the unrecovered deflection of this beam was 48.7 mm. Compared to Specimen D_0, the maximum and the yield loads of Specimen D_1 improved about 100% and 90%, respectively. The stiffness of Specimen D_1 also improved from 75% to 90% compared to Specimen D_0. During the test, when the load was reaching about 60% of the maximum load, cracks of the concrete slab at the mid-span of the beam were appeared leading to reduction in the stiffness of Specimen D_1. The slip between the steel and the concrete slab of Specimen D_0 at ultimate load was recorded about 2.75 mm. After testing, the concrete slab was carved out at some places to examine the working conditions of the shear connectors. These examinations showed that the bolts in Specimen D_1 were still working effectively without any failure to be recorded (Fig. 6a). These proofs indicates that the used bolts as the shear connector as tried in Specimen D_1 could be a reasonable scheme.

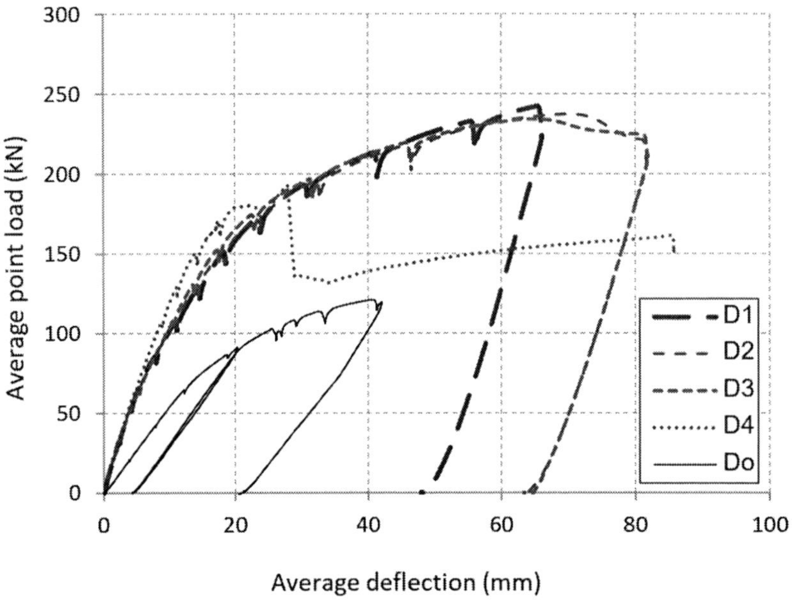

Figure 5: Load-deflection curves

The load-displacement curves of Specimens D_2 and D_3 during the loading were similar to that of Specimen D_1 but with the relatively lower ultimate loads of 237 and 234 kN, respectively, were recorded. The main difference in the behaviour of Specimens D_1 to D_3 were in the slips between the concrete slab and the top flange of the steel beam. These values for Specimens D_2 and D_3 were 2.12 and 3.22 mm, respectively compared to a value of 2.75 mm in Specimen D_1. The relatively larger slip of specimen D_3 could be caused by the relative slip of the bolts and the surrounding concrete as the bolts in specimen D_3 were not bended or had nuts as they were in Specimen D_1 and D_2. The failures of the concrete due to the slip of the bolts on the concrete slab were also recorded when carving out the concrete at the bolt position (Fig. 6c). These evidences prove for this point that there were relatively large slip betwwen the shear connectors and the concrete in Speciemn D_3 during the loading process. Thus, it could be concluded that the use of bolt as the shear connectors as in Specimen D_3 is not effective.

(a) Specimen D_1 (b) Specimen D_2 (c) Specimen D_3 (d) Specimen D_4

Figure 6: The bolts in the failure specimens

Different from Specimens D_1 to D_3, a relative poorer behaviour was recorded in which an ultimate load of 191.8 kN corresponding the displacement of 27mm were measured for Specimen D_4. This load is only 79% of that of Specimen D_1. The behaviour of this specimen at the first stage, when the load was below the ultimate value, was similar to specimen D_1 but at the later loading stage the load reduced rapidly together with large explosions coming from the concrete slab. The explosions occurred during testing this specimen can be caused by the breaking of the bolts connectors because when they appeared the load reduced rapidly and the slip between the concrete slab and the steel flange of the steel beam also increased hastily. The later examination when carving out the concrete (Fig. 6d) of this specimen also displayed the break of the bolts occurred near the welding position, whereas no failure was recorded for the bolts of the remaining specimens. This breaking showed that the welding process may cause a significant change in the properties of the bolts especially on their shear strength. A maximum slip between the concrete slab and the steel flange of 19 mm recognized for this specimen, more than three times for that of the specimens D_1 to D_3. This proved for the point that this type of the shear connectors was not effective.

5. Conclusions

The paper presented an experimental program in which five beams including four steel-concrete composite beams and a reference steel beam were made and tested. Four types of the bolts were implicated into the SCC specimens from D_1 to D_4. The behaviors of the tested beam were compared showing that the SCC beams had significantly better performances than the references steel beam. The results of the experimental program also indicated that bolts can be used successfully as the shear connectors for the SCC beam. The forms of the bolts as applied in specimens D_1 and D_2 gived the great performances of the beams so they can be considered for the implement in practice. The bolts without the nut at the end as represented in Specimen D_3 should not be used as the shear connector as the use of this kind of bolt can cause large slip between the concrete slab and the steel top flange of the beam. Using hand welding method for linking the bolts to the steel flange as applied in Specimen D_4 is also should be avoided be cause it can lead to the deterioration in the shear strength of the bolts.

References

[1] Liu J D, Fa-xing; Liu, Xue-mei; Yu, Zhi-wu 2016 Study on flexural capacity of simply supported steel-concrete composite beam *Steel and Composite Structures* **21** 829-47

[2] Tesser L and Scotta R 2013 Flexural and shear capacity of composite steel truss and concrete beams with inferior precast concrete base *Engineering Structures* **49** 135-45

[3] Baran E and Topkaya C 2014 Behavior of steel–concrete partially composite beams with channel type shear connectors *Journal of Constructional Steel Research* **97** 69-78

[4] Baran E and Topkaya C 2012 An experimental study on channel type shear connectors *Journal of Constructional Steel Research* **74** 108-17

[5] Wang B, Huang Q, Liu X and Li W 2018 Experimental investigation of steel-concrete composite beams with different degrees of shear connection under monotonic and fatigue loads *Advances in Structural Engineering* **21** 227-40

[6] Sang-Hyo K, Chi-Young J and Jin-Hee A 2011 Ultimate strength of composite structure with different degrees of shear connection *Steel & Composite Structures (1229-9367)* **11** 375

[7] Xing Y, Han Q, Xu J, Guo Q and Wang Y 2016 Experimental and numerical study on static behavior of elastic concrete-steel composite beams *Journal of Constructional Steel Research* **123** 79-92

[8] B V G a U 2014 Shear strength and momentshear interaction in steel-concrete composite beams *Journal of Structural Engineering* **140** 191-209

[9] Eurocode 4: design of composite steel and concrete structures : Part 1.1 : general rules and rules for buildings (together with United Kingdom Application Document) [London : British Standards Institute, 1992]

Discussion about effecting of stiffener in four bolts in a row end plate connection for long span and heavy load steel structures in Vietnam

Khang T Huong [1], **Cung H Nguyen** [2]

[1] PhD student, Faculty of Civil Engineering, Technology University of Ostrava, Ostrava 70833, Czech Republic
Email: thanh.khang.huong.st@vsb.cz, htkhang3012@gmail.com
[2] Department of Civil Engineering, Ho Chi Minh city of Industrial University, Ho Chi Minh 700000, Vietnam
Email: nguyenhuycung@iuh.edu.vn

Abstract. Nowadays, steel structure industry in Vietnam is in strong development. The construction of steel structure becomes larger span and heavier load. The issue spawned a number of issues arise from optimizing connections. Typical of steel connections in prefabricated steel structure that is an end plate (face plate) bolted connection. When the connection carried a heavy load, then the number of bolts is required much more. Increasing the number of rows bolts will less effective because can still not enough strength requirements, the bolts in row near rotational center will level arm reduction, then it cannot carry heavy loads. The current solution is doing multiple bolts in a row. Current standards such as EN [1], AISC [2] are no specific guidelines for calculating the connection four bolts in a row that primarily assumes the way works like a T-stub of the two bolts a row. Some articles studied T-stub four bolts in a row [3], [4], [5], [6] by component method but it has some components which weren't considered. In this paper, in order to provide a contribution to improve the T-stub four bolts in a row, the stiffener component in T-stub will be added and compared with T-stub without stiffener by the finite element model to demonstrate effectiveness in reducing stress and displacement of T-stub. It gives ideas for the economic design of four bolts in a row end plate connection in Vietnam for future.

1. Introduction

In steel structures, end plate bolted connections are used very often because of its usability. Currently, most of the connection design only considers the tension zone of bolt group for bolt design. The structural details of the connection bolts based on the conditions of use and requirements of the architecture are typically one bolt row will be located on the extended end plate and usually have one or more bolt located on the inside of a beam flange. Due to the requirements of the current works in Vietnam have crossed the rhythm and increasingly heavy loads, so use two bolts in a row in these situations is impossible because not enough meet requirement. Increasing the number of rows bolts will less effective because can still not enough strength requirements, the bolts in row near rotational center will level arm reduction, then it cannot carry high loading. The current solution is doing multiple bolts in a row. Based on the structural conditions of actual column flange, architectural requirements, and four bolts in a row connection are considered the optimal solution available today for the end plate bolted connection. Although many projects from around the world used this connection, but in Vietnam doesn't have many projects which used this connection. Solution of the steel company in Vietnam today is using four bolts in a row with two or multiple bolt rows connection, shear and compression zone of the connections are assumed in the standards EN 1993-1-8 [1] or AISC 2010 [2] like two bolts in a row. The ability of the tension zone depends on the ability of the components as the column flange, beam flange, end plate (there are no stiffener or have stiffener).

For these connections have two bolts a row, the capacity of connections will base on T-stub, however, if we have four bolts in a row and multiple rows, the traditional model of T-stub may not suitable.
As mentioned, currently had some preliminary research about the four bolts in a row connection, typically of Pisarek et al. [3], [4], Demonceau et al. [5], [6] using the component method and experiments to demonstrate the suitability of the model and experiments analysis. However, these studies only the thin plate with low intensity, the component method has yet to mention the other components such as the column web, the add stiffener in column flange and end plate. It means that the current design for this connection is not the most economic. Here is the case study that Vietnam' project applied, particularly the reinforcement in the extension of the plate due to Vietnam's manufacturing technology yet to control distortion when fabrication, stiffener at the extension of the plate will solve these problems when welding. In addition, the previous research before just the static analysis, dynamic or cycling load research required to apply for the construction of large wind or earthquakes. In the actual situation of current applied research T-stub, 4 bolts in a row with multiple rows is essential research. In the next step, in order to provide a contribution to improve the T-stub four bolts in a row model, the stiffener component in T-stub will be added and compared with T-stub without stiffener by the finite element model.

2. Typical steel structure projects in Vietnam.

As discussions, although many projects from around the world used this connection, but in Vietnam doesn't have many projects which used this connection. Solution of the steel company in Vietnam today is using the four bolts in a row with two or multiple bolt rows, shear and compression zone of the connections are assumed in the standards EN 1993-1-8 [1] or AISC 2010 [2] like two bolts in a row. We can refer the Jooco Vina project in Dong Nai province (figure 1), Vietnam [7]. This building is used for shoe factory. It is the large building with 96m width and 15m span spacing, while in normally of industrial building, its only 30m width and 6m to 8m span spacing. Because of large span, so use two bolts in a row in these situations is impossible because not enough meet requirement. Increasing the number of rows bolts will less effective because can still not enough strength requirements.

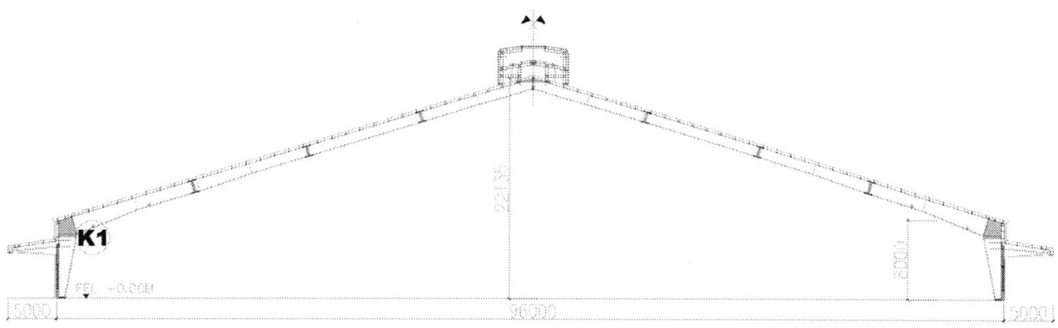

Figure 1. Cross section of main frame, Jooco Vina project [7]

IOP Conf. Series: Earth and Environmental Science **143** (2018) 012028 doi:10.1088/1755-1315/143/1/012028

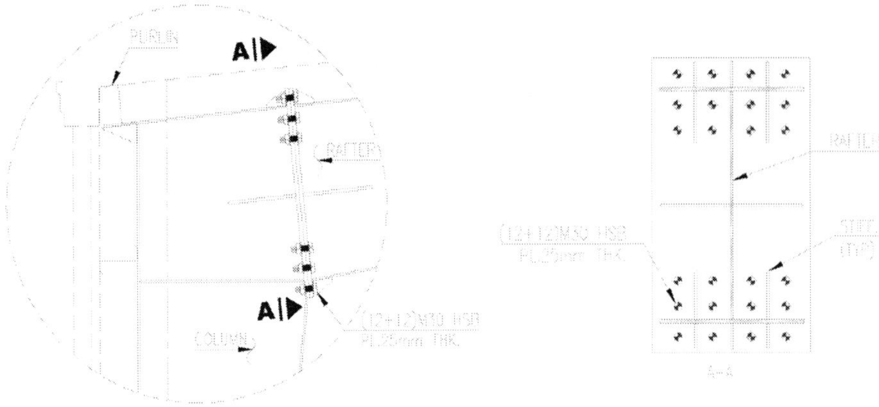

(a) **(b)**

Figure 2. Detail K1

a) The four bolts in a row connection b) Cross section A-A

The solution of designing is in detail K1 (figure 2), this connection is 24M30 bolt diameter with plate thickness 25mm. They use four bolts in a row end plate connection, which have enough space to locate many bolts with multiple rows to carry the large moment. They also provide stiffener to strengthen an end plate, this extended end plate with stiffener is so popular in Vietnam. The problem is this detail just for typical, they aren't allowed to consider stiffener to their calculation because lack of standard and researches about this component, for other purpose, Vietnam's manufacturing technology doesn't control distortion well, so stiffener at the extension of the plate will solve these problems when welding. But is has uncomfortable because of long time fabricate and harder to tightening the bolt when end plate has many stiffeners, another inconvenience, this is stiffener need be welded to end plate by hand, so it is hard to control the quality of welding. The next step, by the analyses FEM model of T-stub four bolts in a row, it has provided some evidence of stiffener effective in reducing stress and displacement in this connection.

3. Finite Element Analysis

3.1. Failure mode of T-stub

There were some studies of T-stubs with 4 bolts a row. In the EN1993 standard [1], the design of four bolts in a row would be assumed to be similar to two bolts a row, based on the predicted working on the T-stub two bolts in a row is the same as four bolts a row. Therefore, the results of the resistance of these T-stubs also depend on the type of failure mode such as T-stub 2 bolts. According to EN1993 [1], there are 3 main failure modes (figure 3). Where mode 1 and mode 2 are failure of the bolt with the complete or partial yielding plate (figure 3a and 3b), mode 3 is the failure mode of the bolts that the T-stub will not be effected (figure 3c). EN1993 [1] also proposed for further studies that the component method would be a widely used method in the development of T-stub research. Component method is the method in which the moment resistance of the connection is based on the capabilities of the connection elements such as end plate, column flange, beam flange, bolt and stiffener…

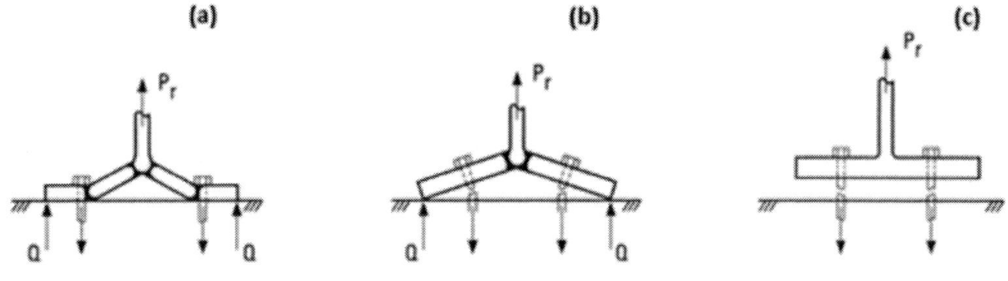

Figure 3. Three failure modes of T-stub [1]

a) Mode 1. Complete flange yielding b) Mode 2. Bolt failure with flange yielding c) Mode 3. Bolt failure

3.2. FEM Model and result for T-stub four bolts in a row with and without stiffener

In order to compare the effectiveness of the component stiffener to the T-stub four bolts in a row. The FEM model will with two samples consist of two T-elements bolted together by four M20 bolts. The mode of failure for such a connection depend on the tensile resistance of the bolts in relation to the resistance of T-element and it occurs one of three failure modes in figure 3. To reduce variable factor, the geometry and dimensions of two samples will be the same. The dimensions of the test samples are presented in figure 4. The T-elements were made of steel grade Q345, and the steel grade of bolt is grade 8.8 for both of two samples. The increasing tensile force 500N/mm2 on top of T-stub. The ABAQUS program software will be used for FE Model. The FE model has four parts: the rigid support at the bottom of T-stub web. Bolt and T-elements material is elastic-plastic isotropic model. The mesh size has been defined by cutting plane to divide mesh in circular and rectangle member to ensure it runs exactly. Interaction type of the two plates is surface to surface contact, normal direction is hard contact with friction coefficient is 0.2.

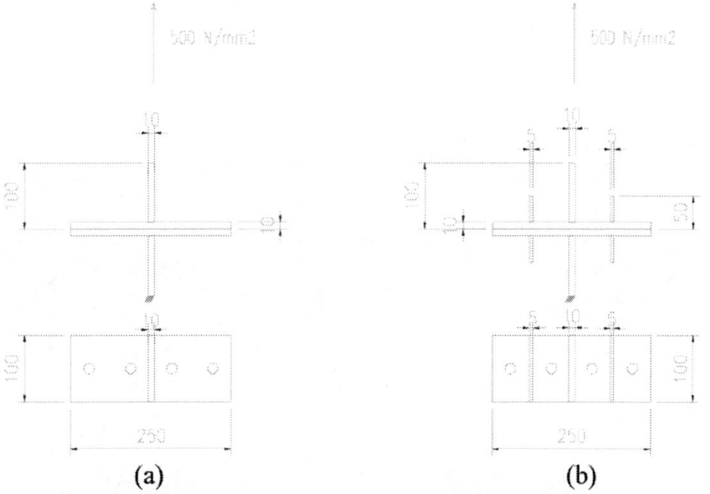

Figure 4. Samples of FE Model (dimension is mm)

(a) Sample 1: T-stub without stiffener, (b) Sample 2: T-stub with stiffener

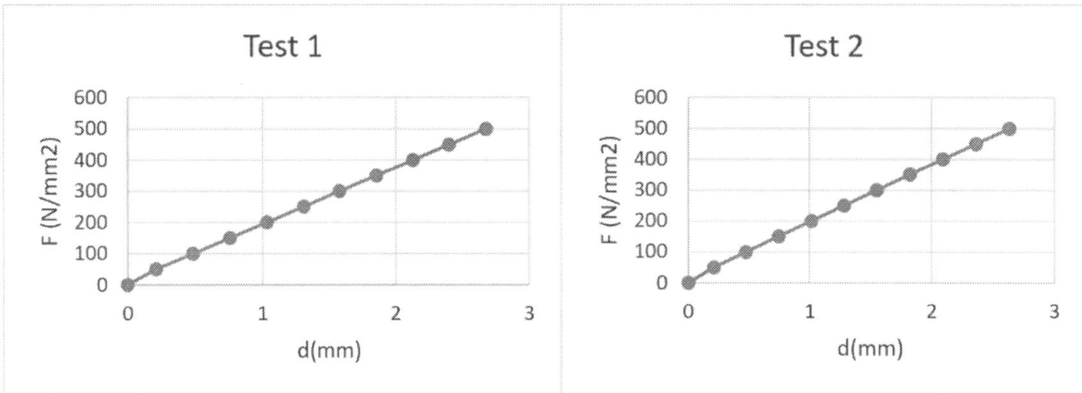

(a) **(b)**

Figure 5. Force-displacement graph for (a) sample 1, (b) sample 2

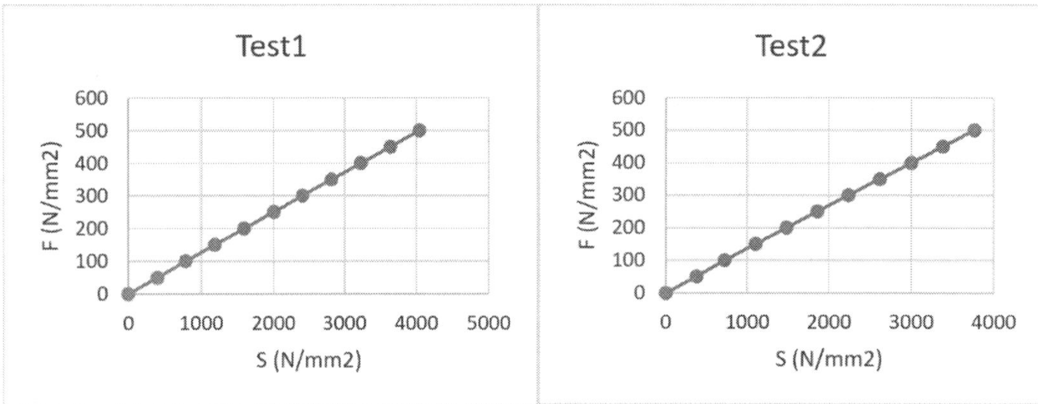

(a) **(b)**

Figure 6. Force-Von Mises Stress graph for (a) sample 1, (b) sample 2

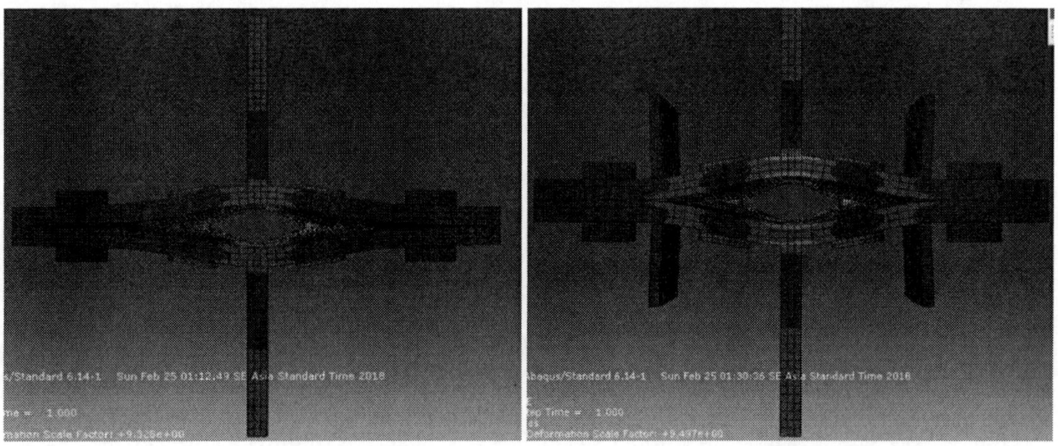

(a) **(b)**

Figure 7. Failure mode of (a) sample 1, (b) sample 2

The main aim of the test observed behavior of the T-stub under increasing loading to identify the mode of failure and compare the stress and displacement between two samples. In figure 5a, the highest displacement of sample 1 is 2.68mm and in figure 5b, the displacement of sample 2 is 2.63mm. The displacement of sample 2 is equal 98.1% displacement of sample 1. In figure 6a, the highest stress in sample 1 is 4044N/mm^2 while the highest stress in sample 1 is 3770N/mm^2 (figure 6b). The stress of sample 2 is equal 93.2% displacement of sample 1. The failure mode both of two samples are mode 1 (sample 2) and mode 2 (sample 1) (figure 7)

4. Conclusion

Nowadays, the connections of four bolts in a row in Vietnam usually have a stiffener in the end plate, however, it is only a typical type for easy fabrication. There have been no previous studies showing the effect of these stiffeners on the stress and displacement simulation of the four bolts in a row connections. With the increasing loading T-stub model in FEM shows evidence of significant effects on T-stub stress and displacement, thus opening up further studies of the effect of stiffener on the four bolts in a row connections, which will have more economic design.

References

[1] EN 1993-1-8 EN1993 3 2005 *Design of Steel Structures Part 1.8: Design of joints CEN*
[2] AISC (2010), *Code of Standard Practice for Steel Buildings and Bridges*, American Institute of Steel Construction, Inc, Chicago, IL
[3] Kozłowski A, Pisarek Z 2005 *Characteristics of bolted end-plate joints with four bolts in the row* Proc of 10th Scient Conf Rzeszow-Lviv-Kosice, State of Art, Trends of Development and Challenges in Civil Engineering, 11–13 September, 2005, Kosice, Slovakia
[4] Pisarek Z, Kozłowski *A End - plate steel joint with four bolts in the row Proc XIth Int Conf on Metal Structures* (ICMS-2006), Rzeszów, Poland, 2006, pp.257-266
[5] Demonceau J, Jaspart J, Muller C & Weinand, K, 2010. *Connections with Four Bolts per Horizontal Row*. Euro steel 2011, Budapest, Hungary
[6] Demonceau, J, Jaspart, J, Muller, C & Weinand, Klaus, Oerder and Muller *Application of Eurocode 3 to Steel Connections with Four Bolts per Horizontal Row* In SDSS', ed. *Colloquium on Stability and Ductility of Steel Structures* Rio de Janeiro, 2010
[7] https://pebsteelbuildings.wordpress.com/author/pebsteelbuildings/page/5/

Seismic fragility curves of bridge piers accounting for ground motions in Korea

Duy-Duan Nguyen[1,2*] and Tae-Hyung Lee[1]

[1] Department of Civil Engineering, Konkuk University, Seoul 05029, Korea
[2] Department of Civil Engineering, Vinh University, Vinh 461010, Vietnam
[*] Corresponding author, email: duyduan@konkuk.ac.kr

Abstract. Korea is located in a slight-to-moderate seismic zone. Nevertheless, several studies pointed that the peak earthquake magnitude in the region can be reached to approximately 6.5. Accordingly, a seismic vulnerability evaluation of the existing structures accounting for ground motions in Korea is momentous. The purpose of this paper is to develop seismic fragility curves for bridge piers of a steel box girder bridge equipped with and without base isolators based on a set of ground motions recorded in Korea. A finite element simulation platform, OpenSees, is utilized to perform nonlinear time history analyses of the bridges. A series of damage states is defined based on a damage index which is expressed in terms of the column displacement ductility ratio. The fragility curves based on Korean motions were thereafter compared with the fragility curves generated using worldwide earthquakes to assess the effect of the two ground motion groups on the seismic fragility curves of the bridge piers. The results reveal that both non- and base-isolated bridge piers are less vulnerable during the Korean ground motions than that under worldwide earthquakes.

1. Introduction

Fragility curve is a useful tool for seismic vulnerability assessment of infrastructures. In recent years, many researchers developed fragility curves for structures not only based on empirical methods but also analytical procedures. Developing fragility functions from empirical observations sometimes encounter challenges due to lack of sufficient damage data caused by past earthquakes. In an alternative way, fragility curves can be constructed by applying some analytical procedures such as elastic spectral analysis, nonlinear static analysis, and nonlinear time-history analysis methods [1-2]. Among these analytical methods, nonlinear time history analysis is the most widely used and the most reliable method for deriving fragility curves of structures [2-3]. Some authors developed fragility curves for bridge structures located in various regions applying the nonlinear time-history analysis such as Karim and Yamazaki [4] in Japan, Moschonas et al [5] in Greek, Choi et al [6] in the Central and South Eastern United States, Kibboua et al [7] in Algeria, Tavares et al [8] in Canada, Lee et al [9], Lee and Nguyen [10] in Korea, and Abbasi et al [11] in California. Moreover, the effect of base isolator devices on seismic fragility curves of bridge structures were also considered in elsewhere [12-14].

The previous studies developed fragility curves of different bridge types mostly using of ground motions from worldwide earthquakes. Recently, earthquakes have been occurred in South Korea with a higher level of intensity, such as the 2016 Gyeongju earthquake and the 2017 Pohang earthquake. This study focuses on deriving analytical fragility curves for RC piers of a typical continuous steel box girder bridge in South Korea considering ground motions in the region. The finite element framework, OpenSees [15], is utilized to perform nonlinear time-history analyses. A damage index is proposed in terms of the displacement ductility ratio [16]. Four damage states, namely slight, moderate, extensive, and collapse are defined based on the proposed damage indices. From the observation of damage levels, fragility curves for the bridge piers are obtained using the maximum likelihood estimation. The fragility curves are then compared with the curves obtained using a group of worldwide ground motions from history earthquake events.

2. Description of the studied bridge

The selected case study bridge is a seven-span continuous steel box girder bridge and supported by six circular RC single column piers. The length of the first and the last three spans is 50m, the length of the second, third and fourth spans is 52.8m, 62.0m, and 54.5m, respectively as shown in Figure 1. The columns height is varied from 10.4m for pier P1, 8.75m for pier P2, 6.8m for pier P6 and 6.5m for both pier P3, pier P4 and pier P5. The cross-sectional diameter of all pier-columns is 1.6m. The cross-sectional dimensions of the bridge girder and the column are depicted in Figure 2 and Figure 3, respectively.

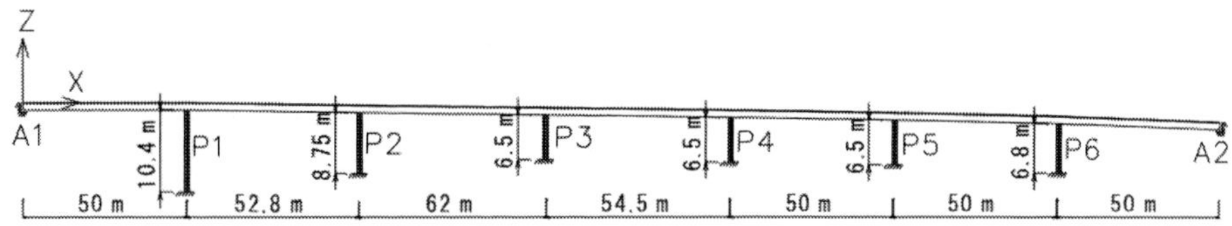

Figure 1. The view of the investigated bridge

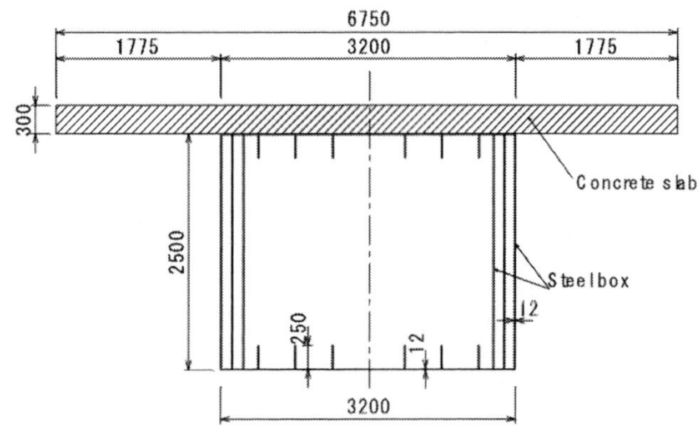

Figure 2. Cross section of the girder

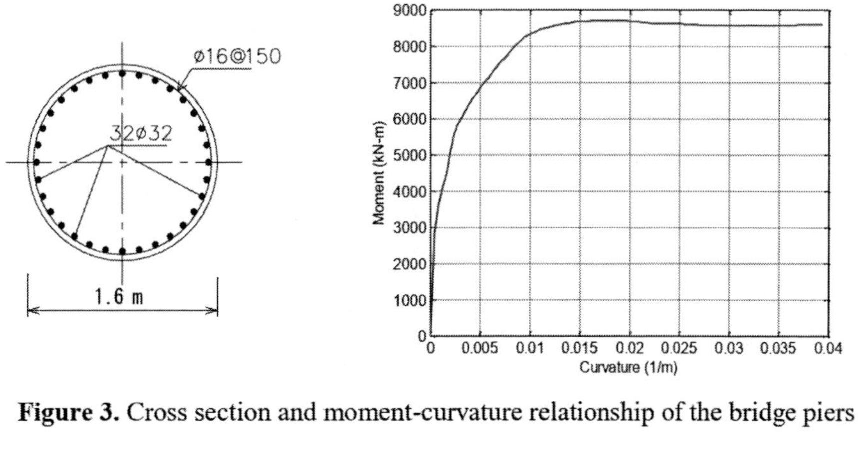

Figure 3. Cross section and moment-curvature relationship of the bridge piers

3. Seismic fragility analysis procedure

To develop the fragility curves for the bridges, nonlinear time history analyses are performed. The procedure for generating fragility curves of the structure can be implemented by basic steps as follows.

1) Set up the bridge models that take into account nonlinear materials as well as the inelastic features of the bridge columns.
2) Apply sets of ground motions with a wide range of peak ground acceleration (PGA) to the bridge models. In this study, two groups of ground motions are considered for nonlinear time history analyses. The first group consists of 20 worldwide ground accelerations from the historic earthquake events and the second one comprises 07 natural and man-made ground accelerations recorded in Korea. For each ground motion, PGA is scaled from 0.1g to 2.0g with intervals of 0.1g in order to consider a wide range of the seismic intensity measure. Consequently, the maximum displacement ductility of the pier columns is obtained.
3) Define the limit states of each component based on the damage index which expressed in terms of ductility ratio of columns.
4) Calculate the sample rate of reaching and exceeding the limit state at each PGA level.
5) Develop fragility curves of bridge piers based on fitting fragility functions to the observed data using the maximum likelihood estimate formulation [3].

A fragility function expresses the probability that the demand on the structure reaches or exceeds the structural capacity at a specific damage state. For this study, the fragility function is assumed as a log-normal cumulative distribution function expressed by

$$P[LS|IM = X] = \Phi\left(\frac{\ln X - \mu}{\beta}\right) \tag{1}$$

where $P[LS\,|\,IM]$ is the probability of a limit state (LS) at a given ground motion intensity measure (IM); X is the value of ground motion in terms of PGA; μ and β are the median and standard deviation of lnX, respectively; $\Phi(\text{-})$ is standard normal cumulative distribution function.

4. Numerical modelling

4.1. Material model

In OpenSees, there are some proposed models for concrete and steel materials. For this study, the *concrete02* model [17] is applied for confined and unconfined concrete, while the *steel02* model [18] is used for modeling of reinforcing bars in the bridge piers. These two models have taken into account the nonlinearity of materials. The moment-curvature relationship of all column sections is illustrated in Figure 3. Properties of the concrete and the steel are described in Table 1 and Table 2, respectively. In Table 1, f_{pc} is the concrete compressive strength at 28 days, ε_{co} is the concrete strain at maximum strength, f_{pcu} is the concrete crushing strength, ε_u is the concrete strain at crushing strength, f_t is the concrete tensile strength, and E_{ts} is the tension softening stiffness. In Table 2, F_y is the yield strength, E_s is the initial elastic stiffness, b is the strain-hardening ratio, R_0 is a constant between 10 and 20, C_{R1} and C_{R2} are coefficients. Figure 4 shows the stress-strain relationship of the concrete and steel material models.

Table 1. Concrete material properties

	f_{pc} (Mpa)	ε_{co}	f_{pcu} (MPa)	ε_u	f_t (MPa)	E_{ts} (GPa)
Confined concrete	40.8	0.004	34.2	0.02	4.0	0.55
Unconfined concrete	27.2	0.002	6.8	0.002	4.0	0.55

Table 2. Steel material properties

	F_y (Mpa)	E_s (GPa)	b	R_0	C_{R1}	C_{R2}
Steel	408	197	0.01	18	0.925	0.15

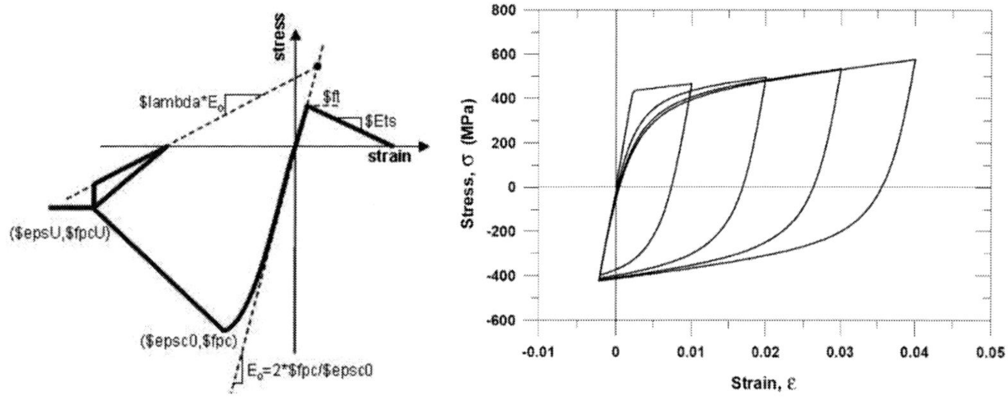

Figure 4. *Concrete02* (left) and *Steel02* (right) models

4.2. Elements modelling

During a strong earthquake, the bridge piers may suffer from experiencing inelastic behavior. For modeling the nonlinear behavior along the element length, the *nonlinearBeamColumn* element with the fiber section modeling scheme is used. The *concrete02* and *steel02* constitutive models are represented for concrete and steel fibers, respectively.

Additionally, the girder is assumed to remain elastic during seismic excitations. Hence, the *elasticBeamColumn* element is adopted to model the bridge girder components. Some characterized quantities of the girder section such as cross-sectional area, torsional moment of inertia of cross section, and second moments of area are calculated from the section in Figure 2. Moreover, the element mass per unit length of the superstructure is also imposed to implement the dynamic analysis procedure of the bridge.

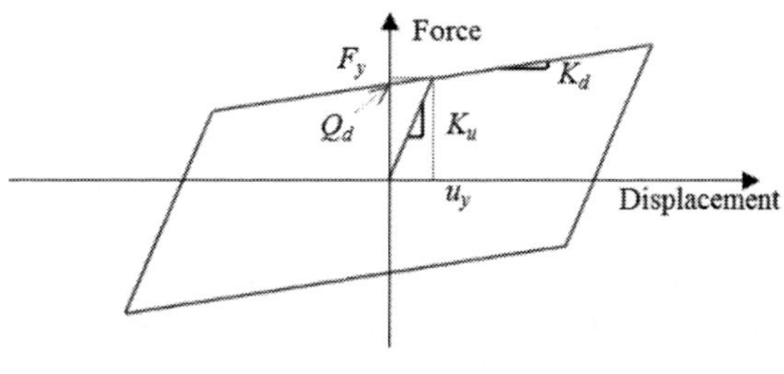

Figure 5. Force-deformation relationship of LRB

For the isolated bridge, one of the most important components of the structure is the connections between substructures and superstructure. On the pier-caps and abutments, the lead rubber bearings (LRBs) are installed to accommodate deformations of the superstructure during earthquake events. The behavior of an LRB under shear forces is assumed to be a bilinear model that is characterized by typical features: initial stiffness (K_u), post-yield stiffness (K_d) and yield strength (F_y) or characteristic strength (Q_d). Figure 5 shows the sketch of the *elastomericBearing* element, which was distinctly designed to apply for modeling a base isolator in OpenSees. Figure 6 shows the modeling of structural components of the base-isolated bridges.

For the non-isolated bridge, the continuous superstructure is rigidly connected to the columns. In OpenSees, this connection is modeled by rigid-link elements. It should be noted that the effect of abutments, as well as soil-structure interaction on performances of the bridge, is neglected in this research. Therefore, the piers are modeled to be fixed on the ground, while the abutment-superstructure connections are modeled as pinned restraints.

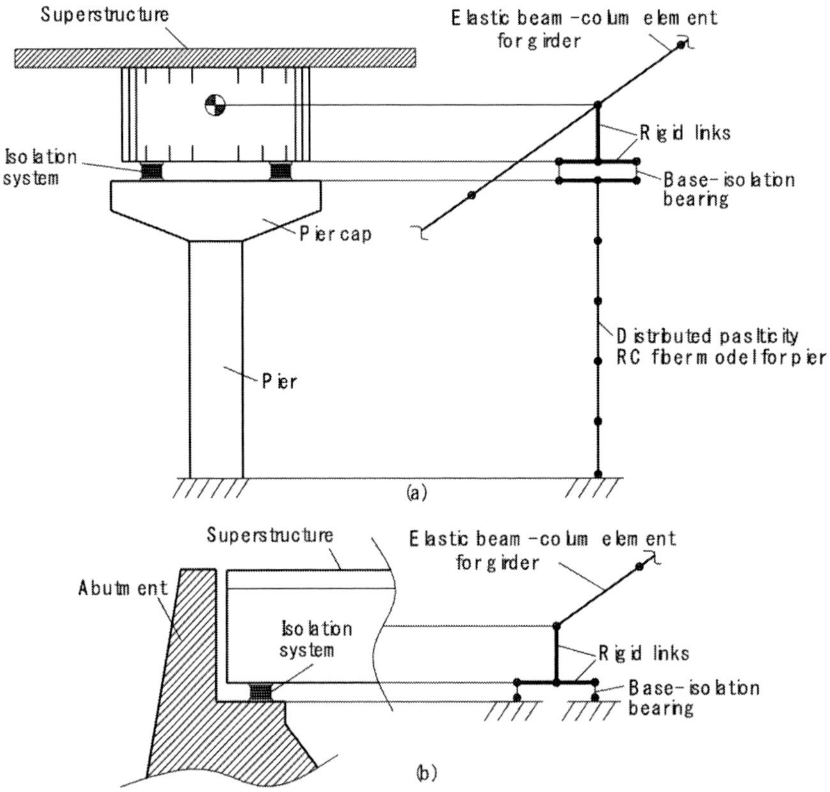

Figure 6. Modeling scheme of the isolated bridge in OpenSees (a) pier-girder and (b) abutment-girder connection

4.3. Vibration modes

Figure 7 shows the first three vibration mode shapes and corresponding natural periods of the studied bridges with and without based isolators. Due to large deformation capacity of LRBs, the inertial force in the girder is reduced and all vibration periods are approximately increased by 1.5 times compared with those of the non-isolated bridge. Table 3 shows the comparison of eigenvalue analysis between two FEM software, OpenSees and SAP2000 [19]. We can observe that the results are highly comparable.

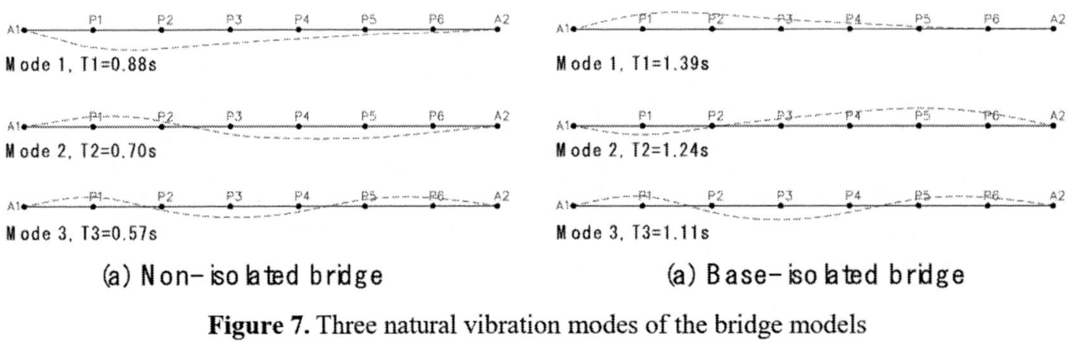

Figure 7. Three natural vibration modes of the bridge models

Table 3. Comparison of natural periods (sec) of the bridge between OpenSees and SAP2000

Mode	OpenSees		SAP2000	
	Non-isolated	Base-isolated	Non-isolated	Base-isolated
Mode 1	0.88	1.39	0.87	1.41
Mode 2	0.70	1.24	0.68	1.25
Mode 3	0.57	1.11	0.54	1.14

5. Ground motions

Twenty worldwide ground accelerations from historic earthquake events and seven natural and man-made ground accelerations in Korea are selected for analyses. The variation of the ground motion profiles has taken into account as a source of uncertainty for the seismic fragility analysis procedure. The magnitude and PGA of the suite of ground motions in Korea are described in Table 4. Figures 8 and 9 show the acceleration and displacement response spectra of the selected ground motions, in which the red bold curve indicates their average response spectrum. It can be seen that the acceleration response spectra in Korean ground motions are very small with periods T > 1.0 sec. In addition, the displacement response spectra in Korean motions are also much smaller than that of worldwide motions.

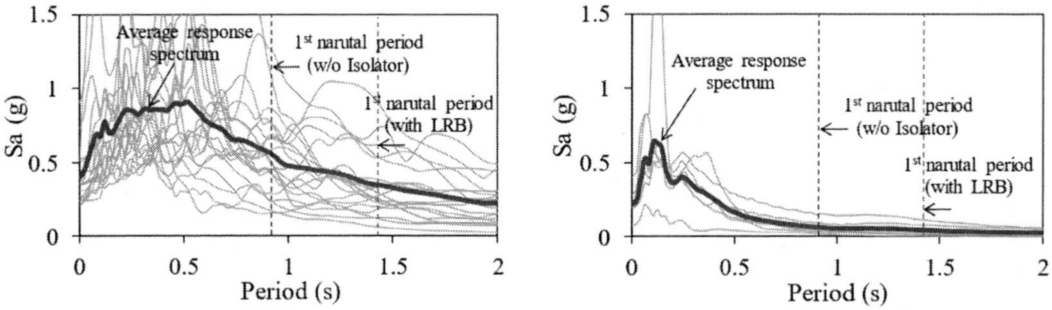

Figure 8. Acceleration response spectra of Worldwide (*left*) and Korean (*right*) ground motions

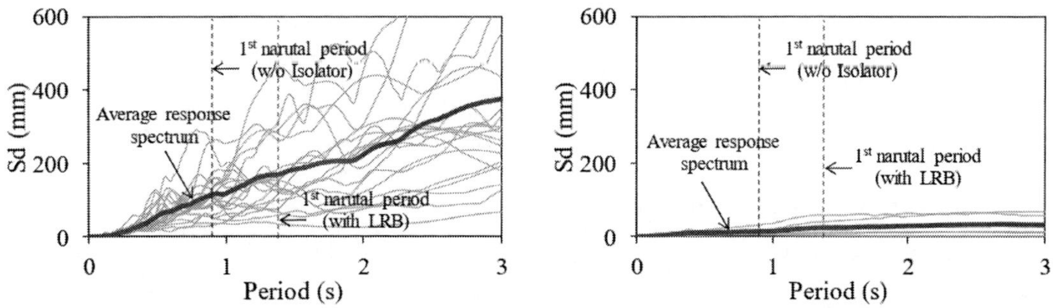

Figure 9. Displacement response spectra of Worldwide (*left*) and Korean (*right*) ground motions

Table 4. List of selected ground motion records in Korea

	Earthquake	Time	Station	Mag. (M)	PGA (g)
1	Gyeongju	2016/09/12	DKJ	5.8	0.078
2	Gyeongju	2016/09/12	MKL	5.8	0.284
3	Gyeongju	2016/09/12	USN	5.8	0.404
4	KESB	-	Artificial 1	-	0.200
5	KESC	-	Artificial 2	-	0.200
6	KESD	-	Artificial 3	-	0.200
7	KESE	-	Artificial 4	-	0.200

6. Seismic response of bridges

Nonlinear time history analyses are carried out for each ground motion in different scaled levels of PGA. The displacement at the top of bridge piers is monitored during seismic excitations. Figure 10 shows the lateral displacement time-history response of piers subjected to the 2016 Gyeongju earthquake for an illustration. Due to the presence of LRBs, the pier displacement at the top is significantly decreased compared with those of the non-isolated bridge piers. It should note that we do compare results of all cases of the bridge equipped with LRBs and non-isolated bridge models accounting for all selected ground motions.

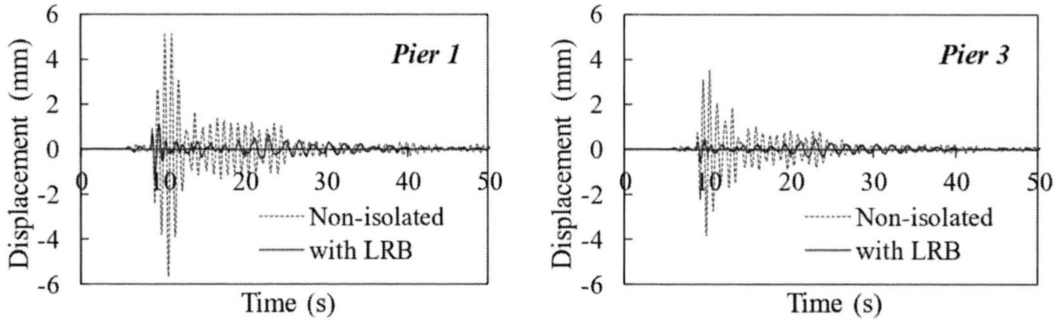

Figure 10. Displacement response of the bridge piers P1 & P3 with and without LRBs during the Gyeongju earthquake (DKJ station, PGA = 0.078g)

7. Damage states and fragility analysis results

Because of the extremely important role of piers, the vulnerability of a pier will affect the load bearing capacity of the entire structure. We only focused on the damage states definition and fragility analysis procedure for the bridge columns, which is the most crucial member of the bridge structure.

For this study, we have adopted the definition of damage states based on column curvature ductility ratio (μ_ϕ) and displacement ductility ratio (μ_Δ) proposed in FHWA [16] for constructing fragility curves of bridge columns. The curvature ductility ratio is expressed as following equation.

$$\mu_\phi = 1 + \frac{\mu_\Delta - 1}{3\frac{l_p}{L}\left(1 - 0.5\frac{l_p}{L}\right)} \quad (2)$$

Due to practice reasons, the damage states related to displacements are more facilitated than use of the curvature ductility. Equation (2) can be rewritten in terms of displacement ductility ratio as

$$\mu_\Delta = 1 + 3\frac{l_p}{L}\left(1 - 0.5\frac{l_p}{L}\right)(\mu_\phi - 1) \quad (3)$$

where L is the length of the column, l_p is the length of the plastic hinge given by the equation (4) hereafter, d_b is the diameter of the column longitudinal reinforcing bar. It should be noted that curvature ductility ratio (μ_ϕ) can be also defined as the ratio of the developing curvature in the piers due to bending moment to the curvature at the first yielding state of the reinforcement.

$$l_p = 0.08L + 9d_b \quad (4)$$

Accordingly, four damage states (DS), namely, slight, moderate, extensive, and collapse are defined in terms of the column displacement ductility ratio. Table 5 shows damage states of all bridge piers and the corresponding the damage indices. It should be noted that the piers P3, P4 and P5 are all the same height, thus, they are in the same value of the defined damage levels.

Figs 11 and 12 show fragility curves of the bridge piers P1 and P3 equipped with and without LRBs using Korean ground motions. It should be noted that piers P1 and P3 are the high and short piers in the bridge model, respectively. The result reveals that the bridge equipped with LRBs is less vulnerable than the case of without base isolators. Due to the energy dissipation capacity of the lead plug and the shear deformation capacity of the rubber layers in LRBs, lateral displacement at the top of the bridge piers is significantly reduced under seismic excitations. On the other hand, the short pier (i.e. P3) is more vulnerable than the high pier (i.e. P1) at a specific level of PGA.

Table 5. Proposed damage states of the studied bridges

Component	Demand Parameter	Threshold value			
		Slight (DS1)	Moderate (DS2)	Extensive (DS3)	Collapse (DS4)
Pier P1	Displacement ductility ratio	1.0	1.2	1.9	4.7
Pier P2		1.0	1.3	2.0	4.8
Pier P3, P4, P5		1.0	1.3	2.0	5.2
Pier P6		1.0	1.3	2.0	5.1

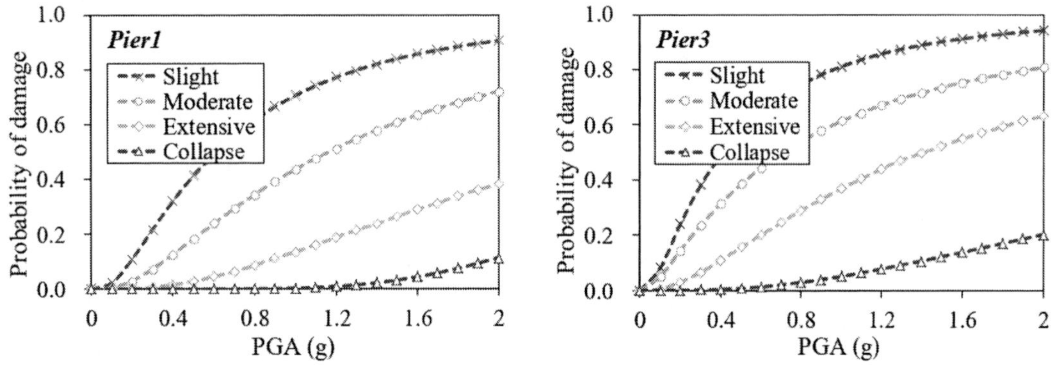

Figure 11. Fragility curves of the piers P1 & P3 for the base-isolated bridge

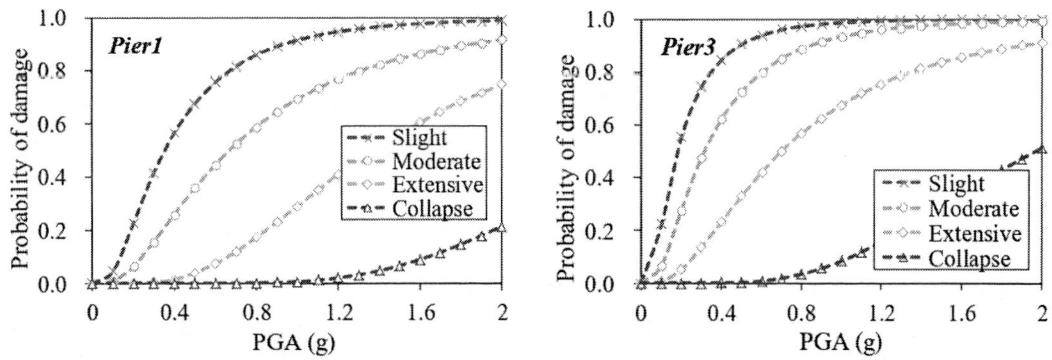

Figure 12. Fragility curves of the piers P1 & P3 for the non-isolated bridge

Figs 13 and 14 show a comparison of fragility curves of bridge piers applying worldwide earthquakes and ground motions in Korea for base-isolated and non-isolated bridges, respectively. Here, the "W" and "K" letters are abbreviated for Worldwide and Korean motions, respectively. It can be seen that at the same PGA level the bridges subjected to ground motion records in Korea are less vulnerable than imposed worldwide ground motions. This is attributed to the reason that the Korean response spectra at the fundamental period of structures are significantly smaller than those of the worldwide response spectra. In other words, the Korean motions are containing the high-frequency characteristics, while the fundamental frequency of structures is fallen in a small range. Consequently, the seismic responses of the piers during Korean motions are diminutive than those of under the global earthquakes. The high-frequency features of Korean ground motions can be referred in the literature [20].

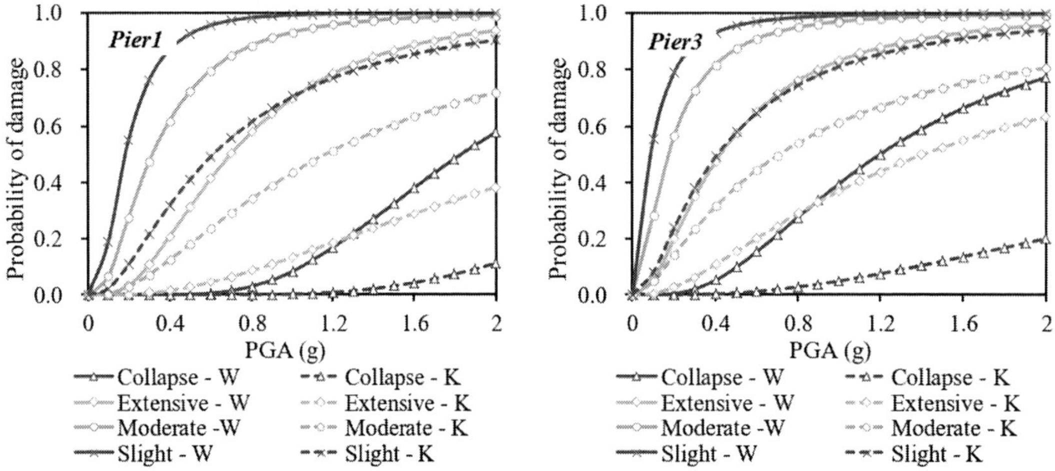

Figure 13. Comparison of fragility curves of the piers P1 & P3 for the base-isolated bridge

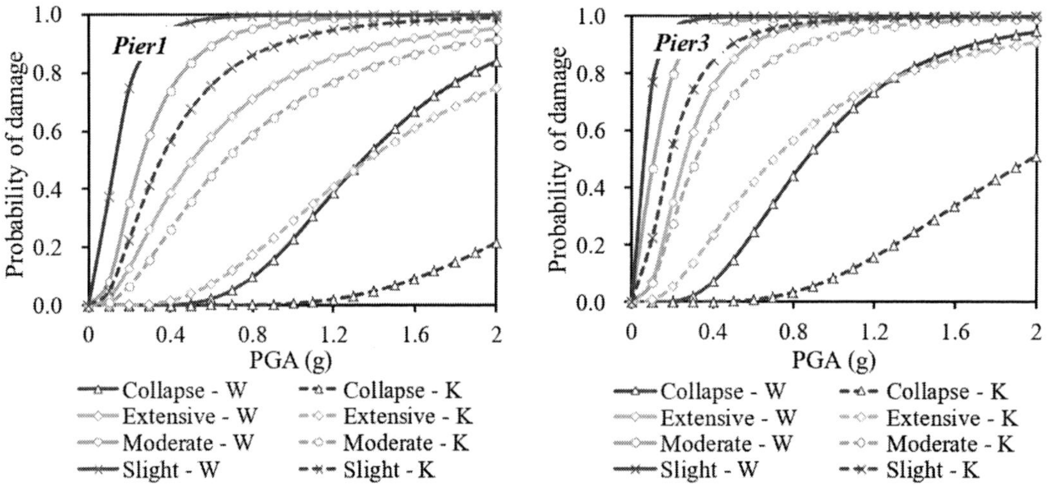

Figure 14. Comparison of fragility curves of the piers P1 & P3 for the non-isolated bridge

8. Conclusions

Seismic performance of a continuous steel box girder bridge with single-column RC piers was carried out using nonlinear time-history analyses. Two groups of ground motion records including worldwide earthquakes and Korean ground motions were considered in the fragility analyses. Based on the results, the main achievements are summarized as follows.

- The bridge piers are less vulnerable under Korean ground motions than that during worldwide earthquakes. The main cause is due to the high-frequency contents of the Korean ground motions.
- There is a significantly different seismic response between the piers at different positions of the bridge (e.g. at the ends and middle areas) in both cases with and without LRBs. Thus, the fragility analysis procedure should be conducted all the bridge piers for a sufficient assessment.
- The seismic damage probability of the non-isolated bridge is always higher than that of the base-isolated one.

A larger enough number of ground motions recorded in Korea is necessary in the fragility analysis to reduce the uncertainties of seismic loading. Additionally, the effect of soil-structure interaction phenomenon on the fragility assessment should be considered in the future study.

References

[1] Billah A H and Alam M S 2015 Seismic fragility assessment of highway bridges: a state-of-the-art review *Struct. and Infras. Eng.: Maint., Manag., Life-Cycle Des. and Perfor.* **11**(6):804-832
[2] Shinozuka M, Feng M Q, Kim H K and Kim S H 2000 Nonlinear static procedure for fragility curve development. *ASCE J. Eng. Mech* **126**:1287-1296
[3] Shinozuka M, Feng M Q, Lee J and Naganuma T 2000 Statistical analysis of fragility curves *ASCE J. Eng. Mech* **126**(12):1224-1231
[4] Karim K R and Yamazaki F 2003 A simplified method of constructing fragility curves for highway bridges *Earthq. Eng. Struct. Dyn* **32**:1603–1626
[5] Moschonas I F, Kapos A J, Panetsos P, Papadopoulos V, Makarios T and Thanopoulos P 2009 Seismic fragility curves for Greek bridges: Methodology and case studies *Bull. of Earthq. Eng.* **7**:439-468
[6] Choi E, DesRoches R and Nielson B G 2004 Seismic fragility of typical bridges in moderate seismic zones *Eng. Struct.* **26**:187–199
[7] Kibboua A, Naili M, Benouar D and Kehila F 2011 Analytical fragility curves for typical Algerian reinforced concrete bridge piers *Struct. Eng. and Mech.* **39**(3):411-415
[8] Tavares D H, Suescun J R, Paultre P and Padgett J E 2013 Seismic fragility of a highway bridge in Quebec *ASCE J. Bridg. Eng.* **18**(11): 1131-1139
[9] Lee S M, Kim T J and Kang S L 2007 Development of fragility curves for bridges in Korea *KSCE J. Civil Eng.* **11**(3):165-174
[10] Lee T H and Nguyen D D 2018 Seismic vulnerability assessment of a continuous steel box girder bridge considering influence of LRB properties *Sadhana* **43**, DOI: 10.1007/s12046-017-0774-x
[11] Abbasi M, Abedini M J, Zakeri B and Amiri G G 2016 Seismic vulnerability assessment of a Californian multi-frame curved concrete box girder viaduct using fragility curves *Struct. and Infras. Eng.: Maint., Manag., Life-Cycle Des. and Perfor.* **12**(12): 1585-1601
[12] Kim D K, Yi J H, Seo H Y and Chang C H 2008 Earthquake risk assessment of seismically isolated extradosed bridges with lead rubber bearings *Struct. Eng. and Mech.* **29**(6):689-707
[13] Ramanathan K, DesRoches R and Padgett J E 2010 Analytical fragility curves for multi span continuous steel girder bridges in moderate seismic zones *J. Transp. Research Board* **2202**:173-182
[14] Alam M S, Bhuiyan A R and Billah A M 2012 Seismic fragility assessment of SMA-bar restrained multi-span continuous highway bridge isolated with laminated rubber bearing in medium to strong seismic risk zones *Bull. of Earthq. Eng.* **10**:1885–1909.
[15] Mazzoni S, McKenna F, Scott M H and Fenves G L 2007 *OpenSees command language manual.* Pacific Earthquake Engineering Research Center (University of California, Berkeley, USA)
[16] Federal Highway Administration (FHWA) 2006 *Seismic Retrofitting Manual for Highway Structures: Part 1 – Bridges*, Publication No. FHWA-HRT-06-032 (Department of Transportation, McLean, Virginia, USA)
[17] Kent DC and Park R. Flexural members with confined concrete 1971 *ASCE J. Struct. Div.* **97**(7):1969-1990
[18] Menegotto M and Pinto PE 1973 Method of analysis for cyclically loaded reinforced concrete plane frames including changes in geometry and non-elastic behavior of elements under combined normal force and bending *Proc. IABSE Sym. of Resist. and Ult. Deform. of Struct. Acted on by Well-Defined Repeat. Loads* (Lisbon, Portugal) pp 15-22
[19] CSI 2011 *SAP2000 software*, ver15. Berkeley, California, USA
[20] Park H S, Nguyen D D, and Lee T H 2017 Effect of high-frequency ground motions on the response of NPP components: A state-of-the-art review *J. Korean Soc. Hazzard Mitig.* **17**(6) 285-294

Features of residual stresses in duplex stainless steel butt welds

Tae-Hwan Um [1], Chin-Hyung Lee [2, 5], Kyong-Ho Chang [3] and Vuong Nguyen Van Do [4]

[1] Department of Civil Engineering, Chung-Ang University, 84, Huksuk-ro, Dongjak-ku, Seoul 06974, Republic of Korea

[2] The Graduate School of Construction Engineering, Chung-Ang University, 84, Huksuk-ro, Dongjak-ku, Seoul 06974, Republic of Korea

[3] Department of Civil and Environmental & Plant Engineering, Chung-Ang University, 84, Huksuk-ro, Dongjak-ku, Seoul 06974, Republic of Korea

[4] Faculty of Civil Engineering, Ton Duc Thang University, Ho Chi Minh City, Vietnam

[5] Author to whom any correspondence should be addressed: ifinder@cau.ac.kr

Abstract. Duplex stainless steel finds increasing use as an alternative to austenitic stainless steel, particularly where chloride or sulphide stress corrosion cracking is of primary concern, due to the excellent combination of strength and corrosion resistance. During welding, duplex stainless steel does not create the same magnitude or distribution of weld-induced residual stresses as those in welded austenitic stainless steel due to the different physical and mechanical properties between them. In this work, an experimental study on the residual stresses in butt-welded duplex stainless steel is performed utilizing the layering technique to investigate the characteristics of residual stresses in the weldment. Three-dimensional thermos-mechanical-metallurgical finite element analysis is also performed to confirm the residual stress measurements.

1. Introduction

Duplex stainless steels are finding increased application in the chemical, oil and gas industries, petrochemical process plants, the pulp and paper industry, pollution control equipment, transportation and for general engineering thanks to their outstanding corrosion resistance and mechanical properties. Steel structures in engineering practice are generally fabricated by welding. During welding, residual stresses are unavoidably produced in the weld region and its vicinity. The existence of weld-induced residual stresses can be harmful to the structural integrity and the service behavior of the welded structure. Therefore, accurate assessment of the residual stresses would be very helpful to ensure the sound design and safety of the structure. The effects of solid-state phase transformation on welding residual stresses are investigated in this study. This study aims to present a FE modeling procedure for characterizing the residual stress of a duplex stainless steel, and focuses on the effects of solid-state phase transformation.

2. Residual stress measurement experiment of welded part

In this study, microstructural analysis and residual stress measurements of duplex stainless steel butt weld were carried out to investigate the microstructural changes in the weld region and HAZ, and to determine the full volumetric change strain associated with martensitic transformation, which is required input to the FE modelling of the phase transformation effects.

2.1. Specimen

The base material used in this study is duplex stainless steel plate with 10mm thickness. 'V' butt joint configuration was prepared as shown in figure 1. The joint was welded with five passes by gas tungsten arc (GTA) welding. Welding condition is presented in Table 1 and the mechanical properties of the base metal are given in Table 2. Table 3 shows the chemical composition of the base metal and weld metal.

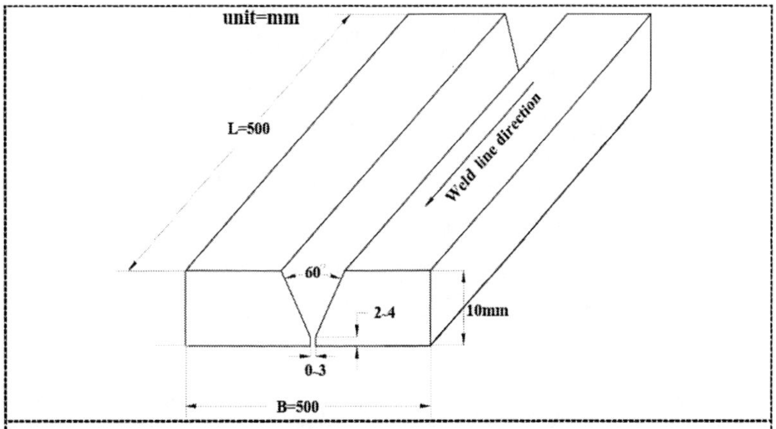

Figure 1. Specimen geometry and dimensions.

Table 1. Welding condition.

Pass	Welding time	Cooling time	Current (A)	Voltage (V)
1	9min 11sec	11min 59sec	160	12
2	4min 45sec	8min 3sec		
3	4min 50sec	10min 7sec		
4	6min 50sec	8min 58sec		
5	6min 15sec	18min		
6	8min 20sec	-		

Table 2. Material properties of the duplex stainless steel.

Duplex Stainless Steel	Yield Stress (MPa)	Tensile Stress (MPa)	Elongation (%)	Elastic Modulus (GPa)
Duplex 2205	731	824	15.8	227

Table 3. Chemical Components of Duplex and weld metal (mass, %).

Material	C	Mn	P	S	Si	Ni	Cr	Mo	N
Base metal	0.019	1.848	0.028	0.0004	0.468	5.065	22.255	2.535	0.1535
Weld metal	0.016	1.72	0.014	0.005	0.37	8.87	22.56	3.12	0.169

2.2. Microstructure and the results of measured hardness

After completion of welding, micro-structures were analysed using an OLYMPUS PME3 optical microscope. Samples used for micro-structural analysis were cut from the weld and the HAZ. Results are shown in figure 2. Figures 2(a) and 2(b), representing the weld region and the HAZ respectively, indicate the martensitic structures. Therefore, it is clear that the weld region and HAZ heated over the austenitic temperature during welding experience martensitic transformation. And Hardness is measured at each position. As shown in the table 4, the hardness of HAZ and welds is about 20 higher than the base metal due to the phase transformation.

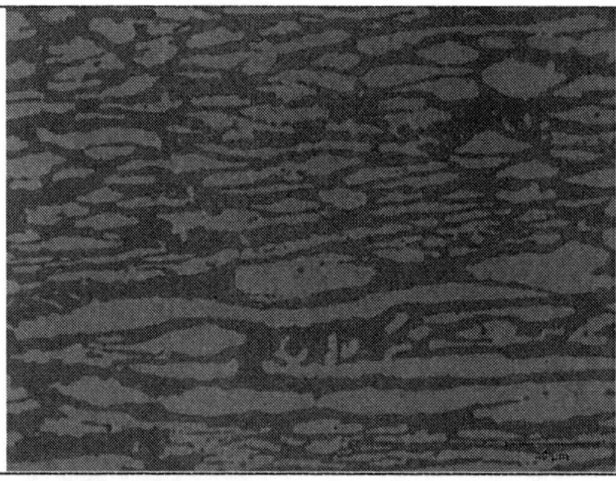

(a) Base metal (× 500)

(b) HAZ (× 500)

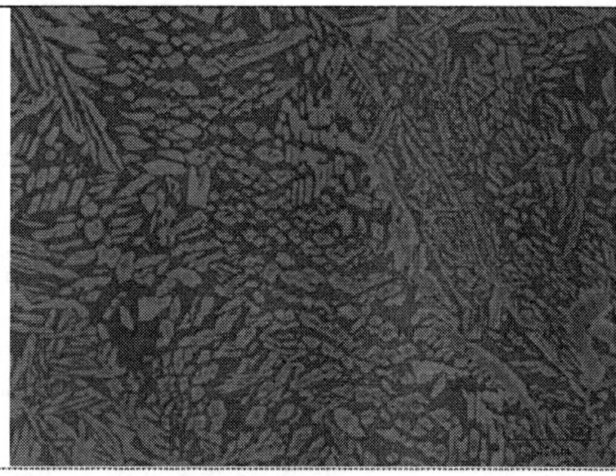

(c) Weld metal (× 500)

Figure 2. Microstructures of the base metal, the HAZ and the weld metal.

Table 4. Hardness at each position (unit: Hv).

Position		Base Metal			HAZ			Weld Metal			HAZ			Base Metal		
		1	2	3	1	2	3	1	2	3	1	2	3	1	2	3
Result	CAP	235	235	235	244	241	243	245	248	249	248	252	252	237	237	235
	MID	237	237	237	257	254	259	252	257	257	258	260	258	236	235	236
	ROOT	235	236	234	256	261	259	258	258	260	253	253	252	234	234	235

IOP Conf. Series: Earth and Environmental Science **143** (2018) 012030 doi:10.1088/1755-1315/143/1/012030

2.3. Test for residual stress measurement

Residual stress measurements were carried out on the two axis strain gauge with the layering technique. Measuring stresses by strain gauges using the layering technique can obtain the residual stresses on the surface of the structure to be evaluated. The detailed procedure for measuring the residual stresses on the surface of the specimen is as follows. First, strain gauges are attached on the bottom surface of the specimen. Residual stresses in the small pieces are released by cutting the specimen, and longitudinal released strain and transverse released strain are measured. Longitudinal residual stress and transverse residual stress can be obtained by Equations (1) and (2) using the measured strains.

$$\sigma_x = -\frac{E}{1-\nu^2}\left(\varepsilon_x + \nu\varepsilon_y\right) \tag{1}$$

$$\sigma_y = -\frac{E}{1-\nu^2}\left(\varepsilon_y + \nu\varepsilon_x\right) \tag{2}$$

where E is Young's modulus and ν is Poisson's ratio.

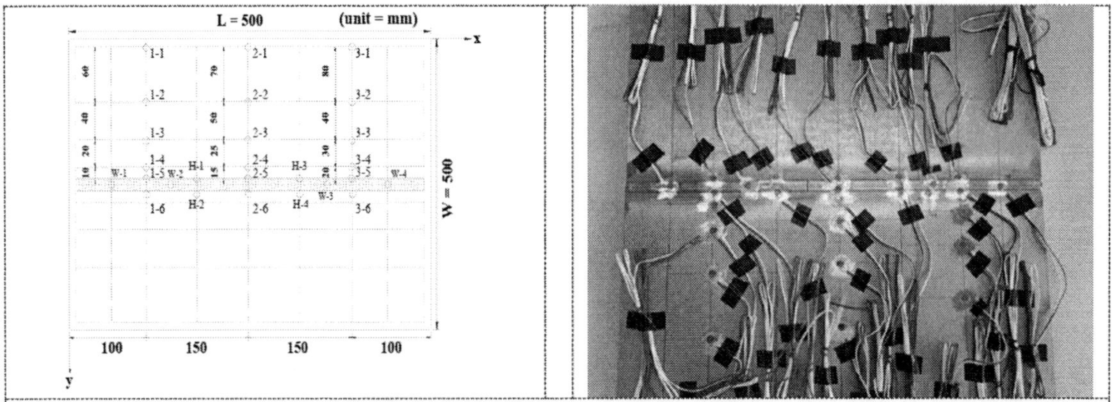

Figure 3. Locations of strain gages.

3. FE simulation of the welding process

Numerical simulation of the thermal and thermal-mechanical process associated with residual stress evolution during welding can be extremely complex since it needs to take account of the interactions between heat transfer, metallurgical transformation and mechanical fields. Complex numerical approaches are then needed to accurately model the welding process. However, as far as welding residual stress modelling is concerned, numerical procedures can be significantly simplified from the thermo-mechanical point of view, the heat input can be seen as a volumetric or surface energy distribution, and the fluid flow effect in the molten area can be simply taken into account by increasing the thermal conductivity over the fusion temperature.

3.1. Model geometry and material properties

FE simulation of the butt welding was performed on two duplex stainless steel plates with width of 250mm, length of 500mm, thickness of 10mm as shown in figure 4. The welding arc travel direction and welding start/stop position a real so shown in the figure. The type of welding process modelled was Gas Tungsten Arc Welding (GTAW) process, and six-passes welding was used to weld the plate. The 3-D FE model using eight-nodded isoparametric solid elements is shown in figure 5. Because of the symmetry with respect to the weld centreline, one half of the model was selected as the computational domain. Figure 6 shows the physical constants at high temperatures of duplex stainless steel and temperature-dependent thermo-mechanical properties of duplex stainless steel are presented in Figure 7.

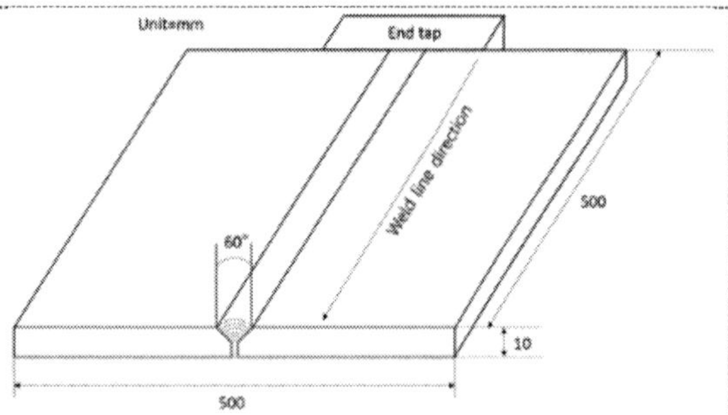

Figure 4. Specimen geometry and dimensions.

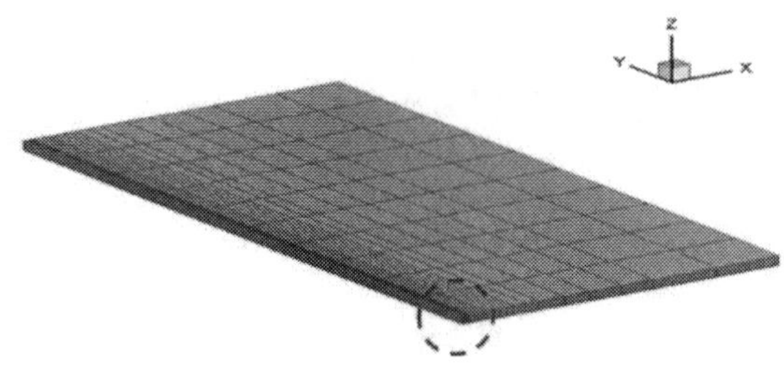

Figure 5. Three-dimensional FE model.

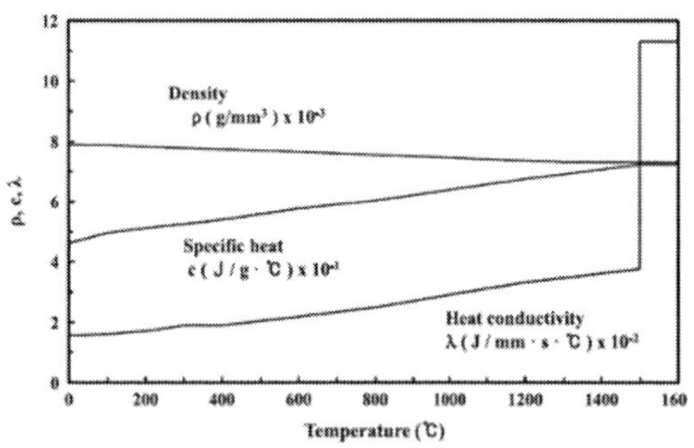

Figure 6. Temperature-dependent thermo-physical properties.

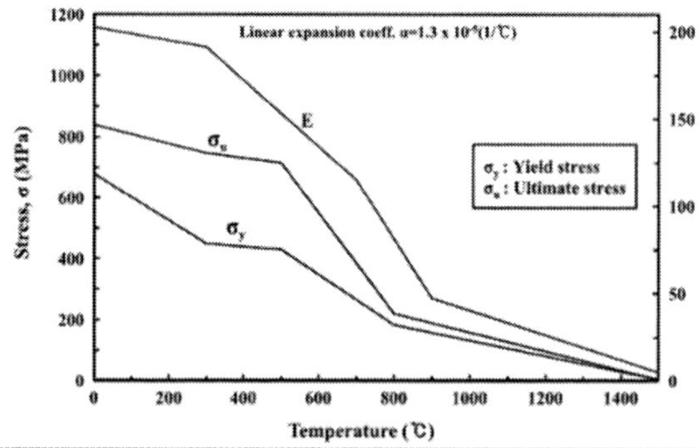

Figure 7. Temperature-dependent thermo-mechanical properties.

3.2. Thermal analysis

During welding, the governing partial differential equation for the 3-D transient heat conduction, with internal heat generation and considering ρ, and c as functions of temperature only, is given by the thermal equilibrium equation.

$$\frac{\partial}{\partial x}\left(K_x\frac{\partial T}{\partial x}\right)+\frac{\partial}{\partial y}\left(K_y\frac{\partial T}{\partial y}\right)+\frac{\partial}{\partial z}\left(K_z\frac{\partial T}{\partial z}\right)+Q=\rho c\frac{\partial T}{\partial t} \tag{3}$$

where T is the temperature, K is the thermal conductivity, c is the specific heat, ρ is the density and Q is the rate of moving heat generation per unit volume. According to the nature of arc welding, the heat input to the work piece can be divided into two portions. One is the heat of the welding arc, and the other is that of the melt droplets. The heat of the welding arc is modelled by a surface heat source with a Gaussian distribution, and that of the melt droplets is modelled by a volumetric heat source with uniform density. Heat flux distribution at the surface of the work piece within the arc beam radius r_0 is defined by the following equation:

$$q(r)=\frac{3Q_1}{\pi r_0^2}e^{-(r/r_0)^2} \tag{4}$$

where $q(r)$ is the local heat flux, r_0 is the arc beam radius and Q_1 is the heat input from the welding arc and given by

$$Q_1=\eta AV-Q_2 \tag{5}$$

where η represents the arc efficiency factor, which accounts for radiative and other losses from the arc to the ambient environment, I is the arc current, U the arc voltage and Q_2 is the energy induced by high temperature melt droplets. On the other hands, the heat from the melt droplets is applied as a volumetric heat source with a distributed heat flux (DFLUX) working on individual elements in the fusion zone.

$$DFLUX = \frac{Q_2}{V_p} \tag{6}$$

where V_p denotes the considered weld pool volume and can be obtained by calculating the volume fraction of the elements in currently being welded zone. The heat of the welding arc was assumed to be 40% of the total heat input, and the heat of the melt droplets 60% of the total heat input. The arc efficiency factor was assumed as 0.85 for the GTA welding process used in the present analysis. The heat flux was applied during the time variation that corresponded to the approach and passing of the welding torch. As for the boundary conditions during the thermal analysis, convection and radiation were both taken into consideration and their combined effect was represented in the following equation for the temperature-dependent heat transfer coefficient, h

$$h = \begin{cases} 0.0668\,T \ \ (W/m^2\,{}^\circ C) & 0\,{}^\circ C < T < 500\,{}^\circ C \\ 0.231\,T - 82.1 \ \ (W/m^2\,{}^\circ C) & T > 500\,{}^\circ C \end{cases} \tag{7}$$

3.3. Solid-state phase transformation

Solid state phase transformation is responsible for changes in both volume and yield stress in steel experiencing thermal cycles during welding. In duplex stainless steel, martensitic transformation which occurs during rapid cooling immediately after intense welding heat input, is the phase transformation associated with the significant volumetric change and the yield stress change. The metallurgical structure of the steel is assumed to be pearlite-ferrite before any thermal cycles take place. During welding, when the steel is heated over the temperature A_1 (cementite disappearance temperature), pearlite-ferrite partly transforms into austenite, and when the temperature is higher than A_3 (α -ferrite disappearance temperature), its microstructure completely changes into austenite. During austenite transformation, the steel undergoes a reduction in volume as can be schematically shown in figure 8. In this study, depending on the peak temperature that an integration point of an element reached during heating process, the decision was made whether the point underwent the martensitic transformation or not, i.e., all points whose peak temperature was higher than the temperature were considered to undergo martensitic transformation when cooling to the onset temperature M_s . When the temperature during cooling reached to the end temperature M_f , the martensitic transformation was considered to be full and complete.

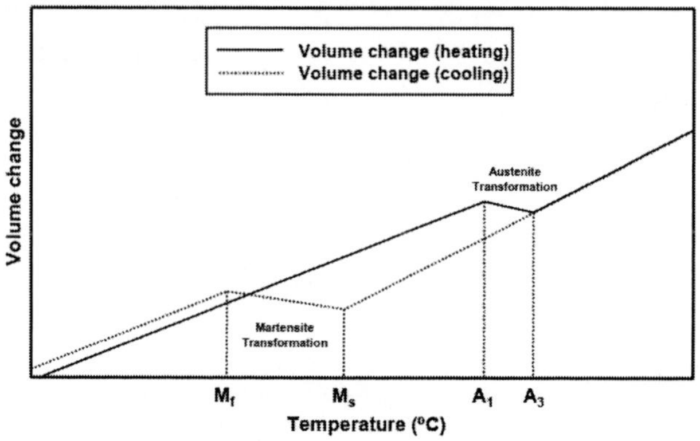

Figure 8. Schematic representation of the volume change due to the phase transformation.

For a steel undergoing solid-state phase transformation, total strain rate can be written as the sum of the individual components of the strain rate as follows.

$$\dot{\varepsilon} = \dot{\varepsilon}^{E} + \dot{\varepsilon}^{P} + \dot{\varepsilon}^{T} + \dot{\varepsilon}^{\Delta Vol} + \dot{\varepsilon}^{Trp} \tag{8}$$

where $\dot{\varepsilon}^{E}$, $\dot{\varepsilon}^{P}$, $\dot{\varepsilon}^{T}$, $\dot{\varepsilon}^{\Delta Vol}$, $\dot{\varepsilon}^{Trp}$ are the strain rate components due to elastic, plastic, thermal loading, volumetric change and transformation plasticity.

4. Results and discussion

In the thermal analysis, inter-pass temperature are taken into consideration. The inter-pass temperature was assumed to be 150°C. Figure 9 shows the thermal histories of a point in the weld area where the phase transformation occurs during the welding process. Two different cases, Case 1 and Case 2, are studied to evaluate the effect of the solid-state phase transformation on welding residual stresses. Case 1 does not consider the solid-state phase transformation. On the contrary, Case 2 takes the solid-state phase transformation into consideration. It is shown that the analytical results considering the phase transformation characteristics are similar to the experimental results. This result show that the austenite to martensite phase transformation is responsible for the significant reduction of the longitudinal tensile residual stresses in the weld region and HAZ.

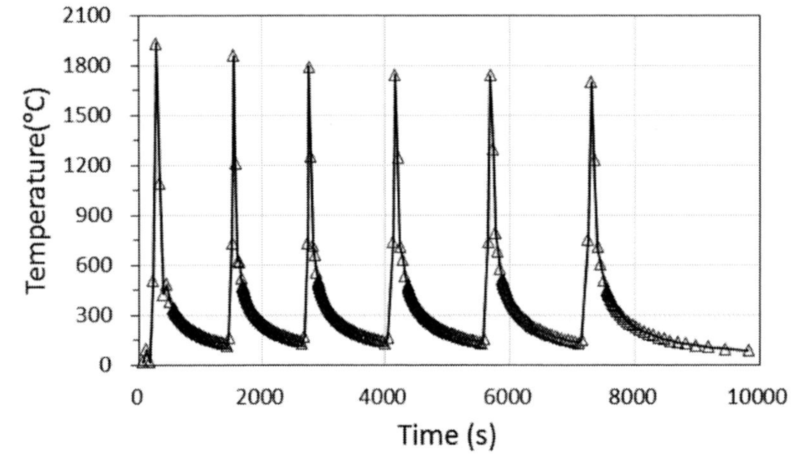

(a) Thermal histories

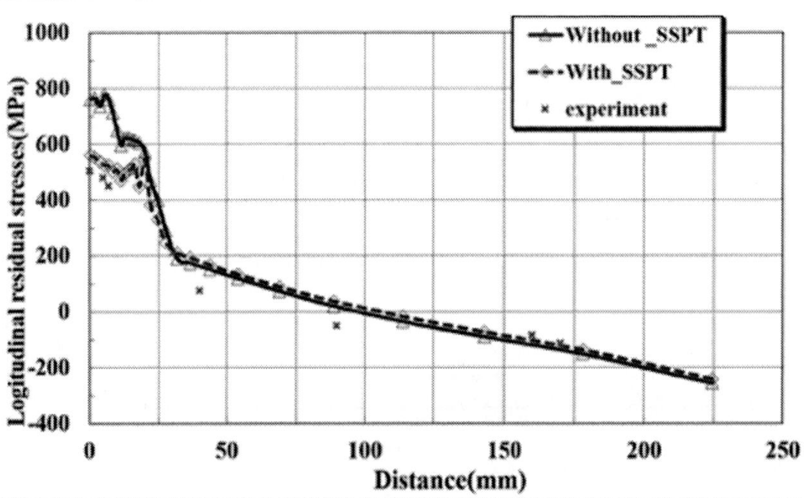

(b) Longitudinal residual stresses

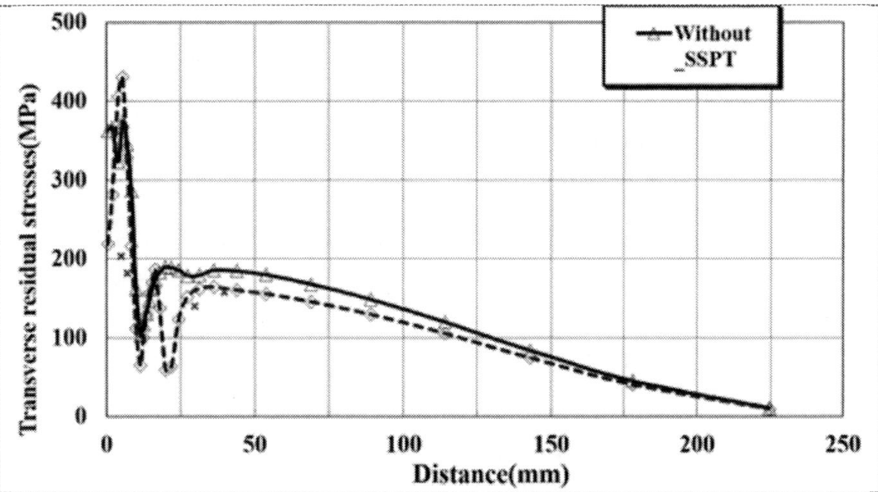

(c) Transverse residual stresses

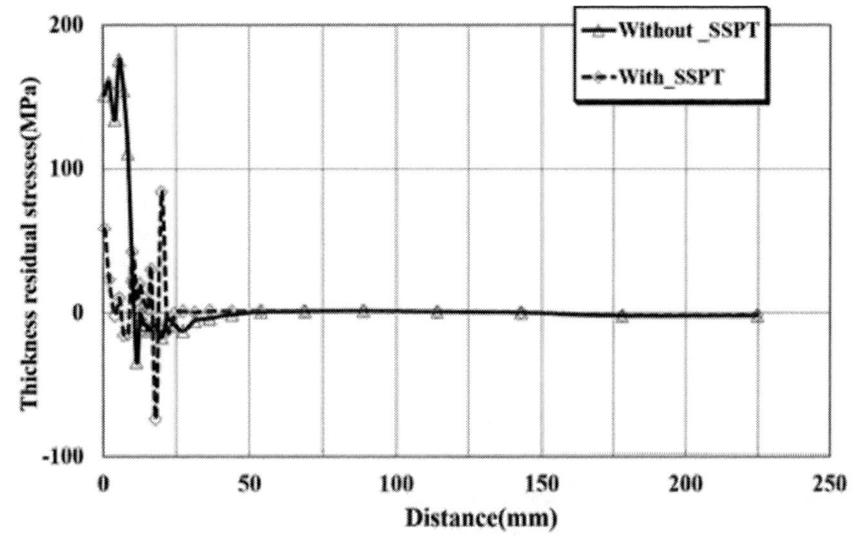

(d) Through-thickness residual stresses

Figure 9. FE analysis results.

5. Conclusion

In this study, FE thermal simulation of the butt welding process was performed using a sequentially coupled 3-D thermo-mechanical FE analysis method whose effectiveness was verified by the comparisons with the experimental measurements to identify and compare temperature fields and welding residual stress distributions in butt-welded duplex stainless steel plate. Based on the results, the following observations and conclusions can be written.

- FE model for the numerical simulation of welding residual stresses in duplex stainless steel butt weld fabricated by multi-pass welding is presented considering solid-state phase transformation.
- The effects of solid-state phase transformation on welding residual stresses associated with the changes in yield stress due to the austenitic to martensitic transformation are taken into account.
- The effects of solid-state phase transformation on welding residual stresses are clearly demonstrated in the large difference in peak value between the longitudinal residual stress

curves and the difference in shape between the curves of transverse and through-thickness residual stress components.

- The effects of solid-state phase transformation cannot be disregarded in the simulation of welding residual stresses in the duplex stainless steel butt weld

Acknowledgments

This research was supported by Mid-career Researcher Program through the National Research Foundation of Korea (NRF) funded by the Ministry of Science, ICT & Future Planning (NRF-2015R1A2A2A01006390).

References

[1] Anderson B A B 1978 *Thermal stresses in a submerged-arc welded joint considering phase transformation* Transaction of ASME **100** 110
[2] Díaz J J del Coz, Menendez Rodriguez P, Garcia Nieto P J, Castro-Fresno D 2010 *Comparative analysis of TIG welding distortions between austenitic and duplex stainless steels by FEM* Appl Therm. Eng. **30** 2448
[3] Iris A-A *2008 Duplex stainless steels: brief history and some recent alloys* Recent Pat. Mech. Eng. **1** 51
[4] Mourad A-H I, Khourshid A, Sharef T 2012 *Gas tungsten arc and laser beam welding processes effects on duplex stainless steel 2205 properties* Mat. Sci. Eng. A **549** 105
[5] Satoh K, Terasaki T 1976 *Effect of transformation expansion on welding residual stresses distributions and welding deformation* J. JWS **45** 560
[6] Satoh K, Terasaki T 1976 *Effect of welding condition on residual stresses distribution in welded structures materials* J. JWS **45** 150
[7] Withers P J 2007 *Residual stress and its role in failure* Rep. Prog. Phys. **70** 2211

Investigation on the Effect of Initial Welding Imperfection on Fatigue strength of Tubular Member by FEM

Kyong-Ho Chang[1*], WangSub Shin[1], Do Nguyen Van Vuong [2], ChinHyong Lee[3]

[1]*Department of Civil and Environmental & Plant Engineering, Chung-Ang University, 84, Huksuk-ro, Dongjak-gu, Seoul 06974, Republic of Korea*
[2] *Department of Civil Engineering, Ton Duc Thang University, 19 Nguyen Huu Tho, District 7,*
Ho Chi Minh City, VIETNAM
3 The Graduate School of Construction Engineering, Chung-Ang University, 84, Huksuk-ro, Dongjak-ku, Seoul 06974, Republic of Korea

* Corresponding author, E-mail: changkor@cau.ac.kr

Abstract. Steel tube structure is used for steel structure such as offshore platform, bridges and so on. Also, all circular members of tubular structures are mainly connected by welding. When the steel tubular structures are subjected to repeated loading, not only the load carrying capacity is reduced but also fatigue cracks may develop at the joint part of steel tubular members which are connected by welding.
Carrying out welding, welding initial imperfection such as welding deformation and residual stress are inevitably generated at weld part. It was well known that the effect of welding residual and welding deformation on fatigue strength. However, It's not clear which affects fatigue strength more. However, it's difficult to clear the effect on fatigue strength by experiment. To clarify these effect, fatigue analysis was carried out by FEM which is based on continuum damage mechanics. On the other hand, coupled three-dimensional non-steady heat conduction analysis, and the thermal elastic-plastic analysis was carried out to reproduce the initial weld state of tubular member. From the result, not only the fatigue strength of welded tubular member but also the fatigue life could be found by FEM fatigue analysis.

1. Introduction

Steel Pipe members such as K-type and T-type are widely used in civil and offshore construction and plant field. Also, pipe structure has longer length than its diameter and thickness. Thus, pipe structure needs weld joint. The initial imperfections such as welding residual stress and welding deformation of weld joint affect the fatigue performance of the structure. Also, it is well known as that mechanical behavior of a welded steel pipe under monotonic loading is significantly affected by welding residual stresses [1~7]. A few research works have been studied to examine the mechanical behavior of a welded T-joints pipe member under cyclic loading by experimental or numerical methods. In the analysis, even though a cyclic plasticity model is essential in the numerical study, however, they did

IOP Conf. Series: Earth and Environmental Science **143** (2018) 012031 doi:10.1088/1755-1315/143/1/012031

not employ the cyclic plasticity in the constitutive modeling. They used a bilinear stress-strain equation with isotropic hardening rule [8]. Also, nor did they consider the welding residual stresses. Thus, the mechanical behavior of a welded pipe subjected to cyclic loading has not been fully evaluated.

In this paper, it proposes a 3D fatigue high cycle damage analysis method. There are many parameters such as thickness, diameter, size, which affect fatigue performance of pipe member. This paper deals with fatigue behavior of a pipe subjected to cyclic loading through the numerical investigation, focusing on the effect of the welding imperfection such welding deformation and welding residual on the fatigue performance. The numerical investigation is carried out to clarify the effect of the initial imperfection on the fatigue life in the T-type steel pipe welding joint. Also, 3D analysis was carried out to obtain the welding residual stresses and welding deformation by using sequentially coupled three-dimensional non-steady heat conduction analysis and the thermal elastic-plastic analysis. Then, a FEM fatigue analysis was carried out to calculate the fatigue strength and fatigue life of the welded steel pipe. As a result, the fatigue behavior of the T-type steel pipe was clarified.

2. T-type Pipe Welding Analysis Model and Conditions

3D analysis is carried out to obtain the welding residual and welding deformation by using sequentially coupled three-dimensional heat conduction analysis and the thermal elastic-plastic analysis based on the in-house FE-code. The validity of 3D in-house FE-code has extensively been published by author [1~18].

2.1 Thermal conductivity, thermal elasto-plastic analysis

Chord diameter is D 500mm and its thickness t is 16mm. Brace diameter D is 400mm and it thickness t is 10mm. The FE analysis model geometry and dimensions are schematically shown in Fig 1. A V-groove butt welding of T-type circular which is made of carbon steel was considered. The 3-D FE mesh type with eight-noded isoparametric solid elements is used. The base material is SM490 steel used in civil engineering field. Temperature dependent thermo-physical and mechanical properties shown in Fig 2 were considered in the FE analysis. In welding analysis, a relatively fine mesh is used in near the weld centerline. The validity of these heat conduction analysis and thermal elastio-plastic analysis method are already verified by the published papers of author [9~18].

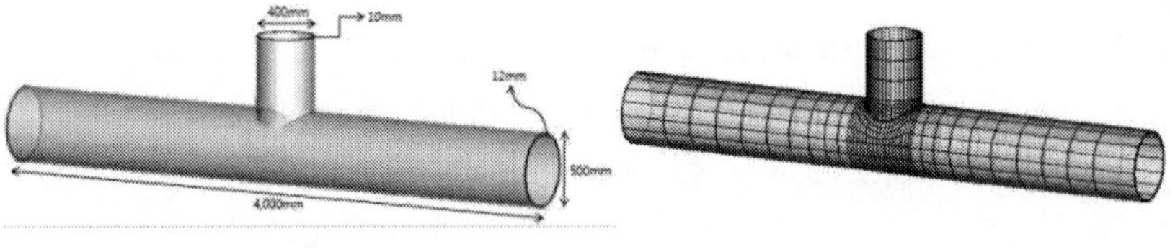

Fig 1. Configuration of T-type pipe welding

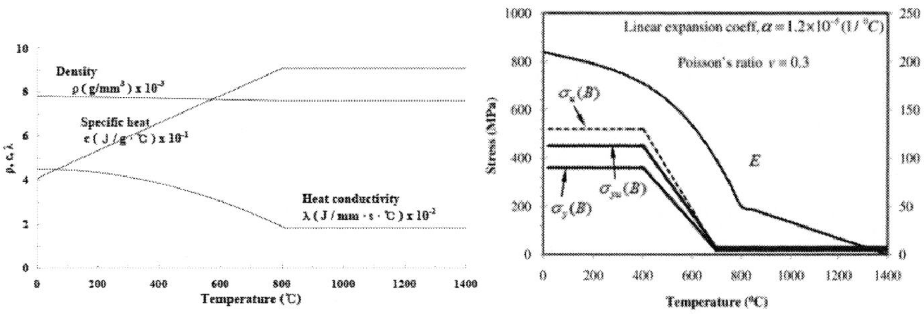

Fig 2. Temperature-dependent thermo- physical constants and mechanical properties.

Fig 3. Thermal heat conduction analysis

2.2 Residual Stress Distribution

Table 1 shows the mechanical properties of SM490 steel. These parameters are obtained by tensile and cyclic tests according to ASTM standard with smooth round bar [9].

Table 1. Mechanical properties of SM490 steel

Material	E (GPa)	ν	σ_y (MPa)	σ_u (MPa)	Elongation (%)	ε_{st}
SM490	204	0.3	365.	549.	30.6	0.0124

First, coupled three-dimensional heat conduction analysis was carried out to obtain temperature history in Fig 3. Then, the thermal elastic-plastic analysis was performed to get the residual stresses and welding deformation. The results are the residual stress of the inner and outer surface in Fig 4 and Fig 5. Fig 4 shows the residual stress distribution at inner surface of the upper thickness part in pipe. The residual stress distribution from the weld line is in three directions distribution of longitudinal, circumference and thickness. Circumferential stress generated a tensile stress in the welding joint.

Compressive stress was generated away from the welding joint. Longitudinal stress is tensile stress in the welding joint. In comparison with three direction stress, thickness direction stress generated tensile stress in the welding joint. However, its absolute value is very small. Fig 5 is a residual stress distribution at outside surface of the upper thickness. Circumferential stress showed a similar distribution to the inner side. But, tensile stress in the joint showed a relatively low value compared to the inner side. The axial direction residual stresses were generated small tensile stress on the outside surface compared to inside axial direction tensile stress. This is because the circumferential shrinkage during the welding process increases to a local inward deformation near the weld area, resulting in compressive stresses is acting on the outside surface. Considering this, it concluded that the axial direction residual stresses are sensitive to the circumference shrinkage. The overall thickness stress distribution is similar to the inner side.

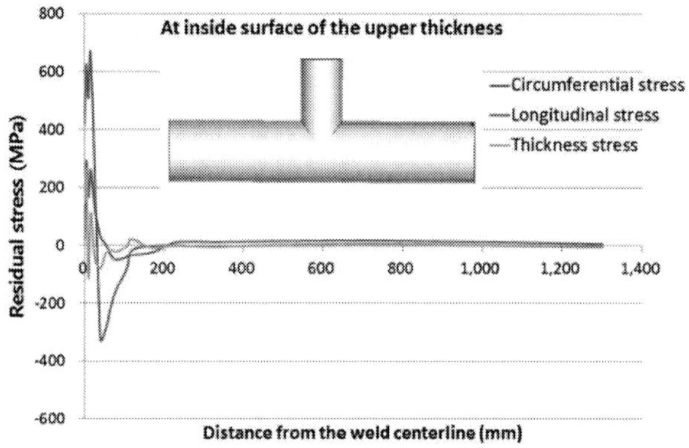

Fig 4. The residual stress distribution at inside surface of the upper thickness.

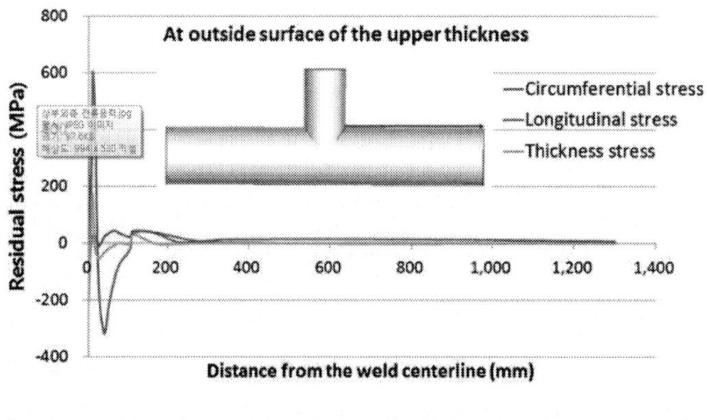

Fig 5. The residual stress distribution at outside surface of the upper thickness

2.3 Fatigue strength of Steel pipe in weld joint under cyclic loading.

To consider the effect of stress concentration and residual stress generated during welding process, the

IOP Conf. Series: Earth and Environmental Science **143** (2018) 012031 doi:10.1088/1755-1315/143/1/012031

initial stress and deformation should be incorporated with the FE model as an initial state. In the case of residual stresses are superposed into internal stress generated by external loading. In fatigue analysis, continuum damage mechanics was considered [19~32]. The damage variable is given : $D = 1 - \frac{A_D}{A_T}$, where, A_D is damaged area(micro-crack). The damage variable leads to the concept of effective stress, which can be expressed as following : $\tilde{\sigma} = \frac{\sigma}{1-D}$, where, coefficient (1-D) is a reduction factor associated with the amount of damage in the material. The value D = 0 indicates the undamaged state, D = Dc indicates the complete local rupture and D = [0 ~1] indicates the partial damage state. Under the number of cycles, the stress-plastic strain upgrade and damage variable was accumulated until reaching to critical value, at this time the crack initiation will occur in the structure. The damage at the cycle N : $D = 1 - \left[1 - \left(\frac{n}{N_f}\right)^{1/(1-\alpha)}\right]^{1/(1+\beta)}$. where, $N_f = \frac{1}{(\beta+1)(1-\alpha)}\left(\frac{A_{II}}{M}\right)^{-\beta}$. The damage is given an incremental form : $D_{N+\Delta N} = D_N + \Delta D$, where, D_N is the damage value at the cycle N, ΔD is the damage increment corresponding to the increment of the number of cycle ΔN . Let us define D = 0 corresponding to N = 0 and D = 1 corresponding to rupture of a material point.

The number of cycles to failure : $$N_f = \begin{cases} \frac{q\left(1/2(1-R)\right)^{2q}}{C(1+R^{2q})} \frac{1}{(2q+1)(1-\alpha)} \sigma_{af}^{-2q}, R < 0 \\ \frac{q\left(1/2(1-R)\right)^{2q}}{C(1-R^{2q})} \frac{1}{(2q+1)(1-\alpha)} \sigma_{af}^{-2q}, 0 \le R < 1 \end{cases}$$

For fully reverse loading case, R= -1 : $N_f = \frac{q}{2(2q+1)(1-\alpha)C_{-1}} \sigma_{-1}^{-2q}$, where, C_{-1}, q, σ_{-1} are the parameters determined from the experimental fatigue test [33~38]. The FE analysis was continuously repeated until the crack go through the entire length at cross-section of weld structure. All variables in FE code were simultaneously updated for every cycle. The fatigue life of welded structure under the cyclic load was determined by the number cycles of repetition until the calculation is finished. Related detailed analysis theory was published by author [33~38].

To analyze the deformation and stress behavior of the T type pipe welding joint under cyclic loading, 3D fatigue high cycle damage analysis was carried out. The welding residual stresses and welding deformation were obtained by the FE welding simulation as an initial condition. Figure 6 shows a load condition of the T type pipe. A load is given by the displacement control. A displacement magnitude is from 0.0(mm) to 10.0 (mm).

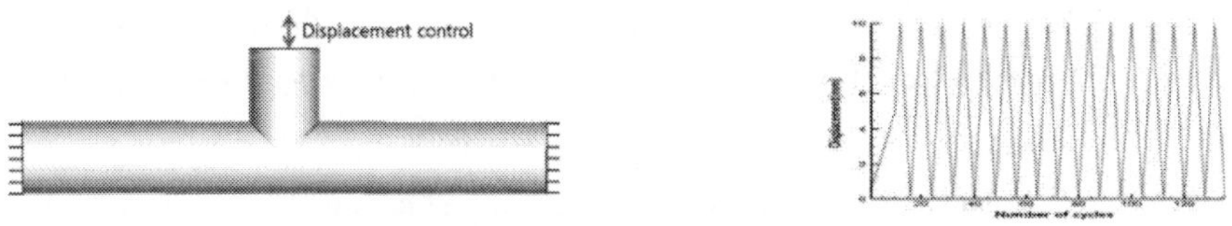

Fig 6. Load condition and Boundary conditions.

Fig 7 shows the fatigue behavior accompanied by crack. Initial crack is generated at the welding joint and crack is propagated. Because, it is the most important thing to find the position of crack initiation, crack propagation and fatigue life. Therefore, when crack is generating, maximum load displacement and number of cycles are compared as shown in Fig 8. Fig 8 shows Fatigue life comparing with or without residual stress and welding deformation. The effect of welding deformation on fatigue strength is greater than the effect of residual stress on fatigue strength.

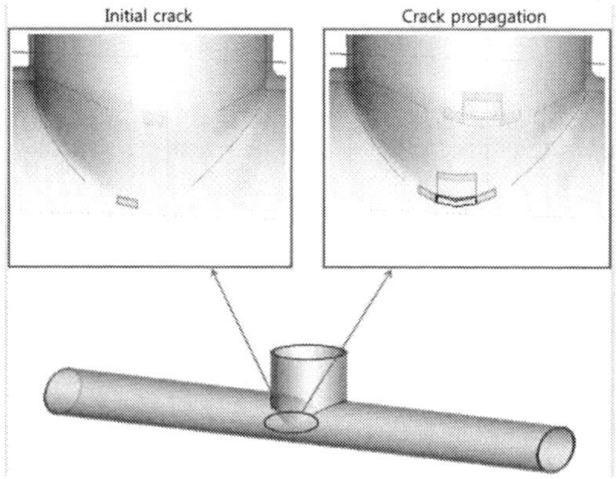

Fig 7. Fatigue behavior of the steel pipe accompanied by the crack.

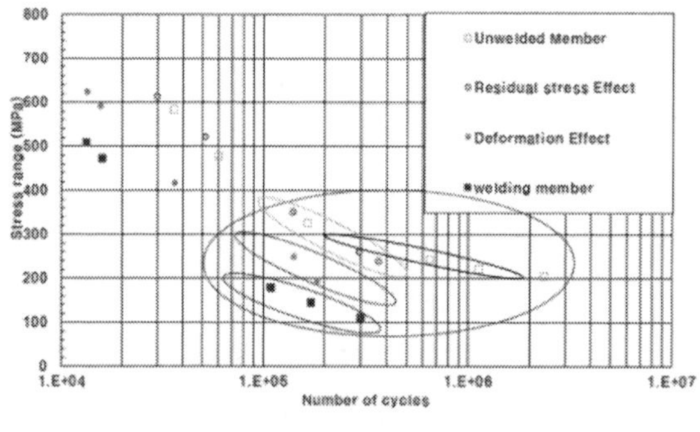

Fig 8. Fatigue strength comparing with or without residual stress and welding deformation

3 Conclusions

In this paper, a 3D fatigue analysis method was proposed from in-house codes based on continuum damage mechanics. To calculate the fatigue life of welded member, coupled three-dimensional heat conduction analysis, the thermal elastic-plastic analysis and 3D high cycle fatigue analysis was carried out. Also, there were many parameters such as stress concentration, residual stress, welding deformation, thickness, diameter, size, which affects fatigue performance of tube member. This paper dealt with fatigue behavior of a tube member subjected to cyclic loading through the numerical investigation, focusing on the welding deformation and residual stresses effect to the fatigue performance. Based on the results in this work, the following conclusions can be drawn.

1) It clarified the effect of the welding deformation and residual stress on the fatigue strength in tube weld member.
2) Also, it clarified a fatigue life of the T-type tube welding joint.
3) It clarified the initial crack position and crack progress direction in the steel member.
4) FE analysis method presented here is helpful to understand the fatigue strength of steel tube.

Acknowlegments

This Research was supported by the Chung-Ang University Research Grants in 2017. This research was supported by Basic Science Research Program through the National Research Foundation of Korea (NRF) funded by the Ministry of Education (NRF-2016R1D1A1B04930897).

References

[1] Chin-Hyung Lee, Kyong-Ho Chang *Failure pressure of a pressurized girth-welded super duplex stainless steel pipe in reverse osmosis desalination plants* Energy, 2013.11.01. Vol.61, pp.565-574
[2] Kyong-ho Chang, Gab-chul Jang *Rate dependent Cyclic Behavior Simulation of Cold Worked AISI 316L Stainless Steel Pipe Member* INTERNATIONAL JOURNAL OF FATIGUE Volume 39, June 2012, pp. 8-14,
[3] S P Chiew, S T Lie, C K Lee, and Z W Huang *Fatigue performance of cracked tubular T joints under combined loads. I: Experimental* Journal of Structural Engineering-Asce, vol. 130, pp. 562-571, Apr 2004.
[4] Kyong-Ho Chang, Chin-Hyung Lee *Behaviour of stresses in circumferential butt welds of steel pipe under superimposed axial tension loading*, Materials and Structures, Volume 42, Number 6, 2009.07, pp.791-801
[5] Chin-Hyung Lee, Kyong-Ho Chang *Three-dimensional finite element simulation of residual stresses in circumferential welds of steel pipe including pipe diameter effects*, MATERIALS SCIENCE AND ENGINEERING A-STRUCTURAL MATERIALS PROPERTIES MICROSTRUCTURE AND PROCESSING, Vol.487, Issues 1-2, 25 July 2008, pp.210-218
[6] K-H CHANG and G-C Jang *Comparative study on residual stress distribution of 600 MPa grade high tensile strength steel pipe weldment*, Materials Technology, Vol.27, Issue 1, 2012.02.05., pp.76-78
[7] P Gandhi, D S R Murthy, G Raghava, and A G M Rao *Fatigue crack growth in stiffened steel tubular joints in seawater environment*, Engineering Structures, vol. 22, pp. 1390-1401, Oct 2000.
[8] Kyong-Ho Chang, Chin-Hyung Lee *Finite element analysis of the residual stresses in T-joint fillet welds made of similar and dissimilar steels*, INTERNATIONAL JOURNAL OF ADVANCED MANUFACTURING TECHNOLOGY, Vol.41, No.3, 2009.03, pp.250-258
[9] Kyong-Ho Chang, Chin-Hyung Lee, Ki-Tae Park, Tae-Hwan Um *Experimental and numerical investigations on residual stresses in a multi-pass butt-welded high strength SM570-TMCP steel plate*, INTERNATIONAL JOURNAL OF STEEL STRUCTURES, Vol.11, No.3, 2011.09, pp.315-324
[10] C-H Lee, K-H Chang *Numerical investigation of the residual stress in strength-mismatched dissimilar steel butt welds*, Journal of Strain Analysis for Engineering Design, Vol.43 No.1, 2008, pp. 55-66
[11] Kyong-Ho Chang, Gab-Chul Jang, Chan-Min Park, Heung-Bae Gil *Strain rate dependence of mechanical behavior and hysteretic characteristics of TMCP steel (SM570-TMC) and its modeling*, Computational Materials Science, Volume 45, Issue 3, 2009.05, pp. 669-673

[12] Chin-Hyung Lee, Kyong-Ho Chang *A computational procedure for evaluating the J-integral for mode I crack in welds: A 3-D finite element analysis*, Computational Materials Science, Vol.46, Issue 1, 2009.07, pp. 186-192
[13] Kyong-Ho Chang, Chin-Hyung Lee *Characteristics of High Temperature Tensile Properties and Residual Stresses in Weldments of high Strength Steels*, Materials transactions, Vol. 47, No. 2, 2006.02, pp.348-354.
[14] Sang-Hyong Lee, Kyong-Ho Chang, Gab-Chul Jang *Features of thermal stress generated by welding and cutting of dissimilar steels*, Materials at high temperatures, Vol.23, No.3-4, 2006.08, pp.211-218
[15] G C Jang, K H Chang, and C H Lee *Characteristics of the residual stress distribution in welded tubular T-joints*, Journal of Mechanical Science and Technology Vol.21, No.10, 2007.10, pp.1714-1719
[16] Chin-Hyung Lee, Kyong-Ho Chang *Numerical Analysis of Residual Stresses In Welds of Similar or Dissimilar Steel Weldments Under Superimposed Tensile Loads*, Computational Materials Science, Vol.40 Issue 4, 2007.10, P.439-568
[17] K H Chang, G C Jang, C H Lee and S H Lee *Temperature and Thermal Stress Distribution for Metal Mold in Squeeze Casting Process*, Journal of Materials Science & Technology, Vol.24 No.3, 2008.06, pp.347-350
[18] Chin-Hyung Lee, Kyong-Ho Chang *Finite element computation of fatigue growth rates for mode I cracks subjected to welding residual stresses*, ENGINEERING FRACTURE MECHANICS, Volume 78, Issue 13, August 2011, pp.2505-2520,
[19] Chaboche JL, Rousselier G *On the plastic and viscoplastic constitutive equations-part I: rules developed with internal variable concept.* J Pressure Vessel Technol 1983;105:153–8.
[20] Lemaitre J A *continuous damage mechanics model for ductile fracture* J Eng Mater Technol 1985;107:83–9.
[21] Benallal A, Billardon R, Lemaitre J *Continuum damage mechanics and local approach to fracture: numerical procedures* Comp Meth Appl Mech Eng 1991;92:141–55.
[22] Bittencourt TN, Wawrzynek PA, Ingraffea AR, Sousa JL *Quasi-automatic simulation of crack propagation for 2D LEFM problems.* Eng Fract Mech 1996;55:321–34.
[23] Kintzel O, Khan S, Mosler J A *novel isotropic quasi-brittle damage model applied to LCF analyses of Al2024* Int J Fatigue 2010;32:1948–59.
[24] Nicholas T High cycle fatigue: *a mechanics of materials perspective* 1st ed Elsevier Science; 2006.
[25] Lemaitre J, Chaboche JL *Mechanics of solid materials* New ed Cambridge University Press; 1994.
[26] Sines G *Behavior of metals under complex static and alternating stresses* In: Sines G, Waisman JL, editors Metal fatigue New York: McGraw Hill; 1959. p.145–469.
[27] Simo JC, Ortiz M A *unified approach to finite deformation elastoplastic analysis based on the use of hyperelastic constitutive equations*. Comp Meth Appl Mech Eng 1985;49:221–45.
[28] Guo, Y B and Barkey, M E (2004a) *Modelling of Rolling Contact Fatigue for Hard Machined Components with Process-induced Residual Stress*, Int J Fatigue, 26: 605–613.
[29] Shao, YB, Li, T, Lie, S T, Chiew, S P, 2011 *Hysteretic behaviour of square tubular T-joints with chord reinforcement under axial cyclic loading* J Constr Steel Res 67, 140-149.
[30] Guo, Y,Wu, D, Ma, G, Guo, D, 2014 *Trailing heat sink effects on residual stress and distortion of pulsed laser welded Hastelloy C-276 thin sheets* J Mater Process Technol 214,2891–2899 Haz Metal, 2013 Stainless Steel sand their Properties.
[31] Finch D, Burdekin FM *Effect of welding residual stress on significance of defects in various types of welded joints* Eng Fract Mech 1992;41: 721–35.
[32] Stacey A, Barthelemy JY, Leggatt RH, Ainsworth RA *Incorporation of residual stresses into the SINTAP defect assessment procedure* Eng Fract Mech 2000;67:573–611.
[33] Chang, K H, Jang, G C, 2008 *A rate dependence of hysteresis model for structural steel(SM490) considering the coupled Effect of strain rate hardening and Temperature rise.*
[34] Kyong-ho Chang, Gab-chul Jang, S F Stiemer, N Loewen *Dynamic hysteretic characteristics of the high strength steel(posten 60, posten80) and applican of dynamic hysteretic model to FE analysis* Journal of Materials in Civil Engineering, ASCE, 2013. 10. vol.25, PP.1549 ~1557
[35] Vuong Nguyen Van Do, C H Lee, Kyong-Ho Chang *A constitutive model for uniaxial/multiaxial ratcheting behavior of a duplex stainless steel* Vol 65, January 2015, Pages 1161–1171.

[36] Vuong Nguyen Van Do, Chin-Hyung Lee, Kyong-Ho Chang *A nonlinear CDM model for ductile failure analysis of steel bridge columns under cyclic loading*, June 2014, Volume 53, Issue 6, pp 1209-1222 ,Computational Mechanics

[37] Chin-Hyung Lee , Kyong-Ho Chang , Vuong Nguyen Van Do *Finite element modeling of residual stress relaxation in steel butt welds under cyclic loading* Volume 103, 15 November 2015, Pages 63–71,Engineering Structures

[38] Vuong Nguyen Van Do, Chin-Hyung Lee, Kyong-Ho Chang *High cycle fatigue analysis in presence of residual stresses by using a continuum damage mechanics model* Volume 70, January 2015, Pages 51–62.International Journal of Fatigue

[39] Huh Y, Kim HI, Park JS, Seok CS *Fatigue characteristics of SM490A welded joints* Int J Mod Phys B 2006;20:4141–6.

[40] K Jae-Hoon and G Byung-Choon *A study on the fatigue life and the change of the strain during the fatigue fracture on the fillet welded specimens of SM490A* Railway Journal, vol 7, pp 131-135, 12 2004.

BIM authoring for an image-based bridge maintenance system of existing cable-supported bridges

N S Dang[1,]*, C S Shim[2]
[1] Ph.D. candidate, Department of Civil and Environmental Engineering, ChungAng University, Seoul, Korea;
[2] Professor, Department of Civil and Environmental Engineering, ChungAng University, Seoul, Korea

*Corresponding email: somdang.ksktxd@gmail.com

Abstract. Infrastructure nowadays is increasingly become the main backbone for the metropolitan development in general. Along with the rise of new facilities, the demand in term of maintenance for the existing bridges is indispensable. Recently, the terminology of "preventive maintenance" is not unfamiliar with the engineer, literally is the use of a bridge maintenance system (BMS) based on a BIM-oriented model. In this paper, the process of generating a BMS based on BIM model is introduced in detail. Data management for this BMS is separated into two modules: site inspection system and information management system. The noteworthy aspect of this model lays on the closed and automatic process of "capture image, generate the technical damage report, and upload/feedback to the BMS" in real-time. A pilot BMS system for a cable-supported bridge is presented which showed a good performance and potential to further development of preventive maintenance.

1. Introduction

Building information modeling (BIM)-based technology increasingly becomes the most innovative trend in the global construction industry. BIM presents in the process of generating and managing construction data during its life cycle [1]. The term BIM was popularized as a common name for these capabilities offered by several technology providers or vendors. On the other hand, BIM digitally represents the building process in terms of facilitating exchange and interoperability of the model information. It covers all geometric and non-geometric information, spatial or geographic constraint, quantity and physical characteristic of the structure [2] [3]. For the existing facilities, it has hardly been applied for the purpose of maintenance, refurbishment or deconstruction. This condition premises the huge benefits in terms of efficient resource management through development a solution to overcome uncertainties of building condition, as well as solving the problem of deficient documentation in existing facilities.

The conducted literature review of several (over 180) publications presented the state-of-the-art implementation and research of BIM in existing buildings, which focus on maintenance and deconstruction stages, highlighted the intensifying research approaches to harness BIM application in existing facilities as well as capturing and integrating data into the entire information system [4]. It also shows an increment of maintenance interfaces and new functionalities, however, it reveals that three major challenges which need to pay more attention from researchers: the automated data capturing and BIM creating, the timely update and completely maintenance of information, and the controlling of uncertainty data.

Bridge maintenance system (BMS) was paid special attention last decades. In the developed countries who are facing the aging of infrastructure, there are various BMSs adopted [5]. It engenders the demand in term of developing a national bridge inventory [6]. For existing bridges, in order to distinguish the difference between repairs and strengthening, estimating the progressive deterioration of performance for bridge members needs to be taken into account. Also, a balance between quality-maximization and maintenance-cost-minimization needs to be considered as the rehabilitation strategy for the bridge [7]. Automated system for bridge maintenance has always been a difficult problem for engineers. The first aspect lies in the way to generate 3D digital information model from modelers. In this regards, parametric modeling has been widely applied as an effective methodology to embed BIM expertise into the construction industry. Although the use of parametric modeling solution is not unfamiliar so far, it is still challenging and has some limitations in term of capturing and interpreting a tacit knowledge into computer-readable geometric and relationships. In an automated system, if the input/output data judgment and responsibilities are not strictly controlled, it increases the risk of propagating errors. System performance degradation also needs to be considered when large numbers of parameters and geometric constraints are included (Lee et al. 2006). The second aspect in order to automate the function of generating data for a BMS lies in the inspection process. Several researchers have been proposed the automatic workflows for the bridge inspection work which are based on the image processing technology [8] [9]. Adapting a good machine vision concept, nowadays engineer can detect the structural damage in real time and also has advanced utility functions for remote supervision. Among all inspection objectives, crack detection is the most important achievement and increasingly be paid significant attention [10] [11] [12]. Inspection data can be observed and directly uploaded onto the server by the aids of mobile device or UAV, or other wearable/AR devices, and timely analysis to find out the crack profile, width, and length, also propagation direction, significant support for decision-making process from a remote office.

2. Information requirement

2.1. Maintenance strategy and objectives

Recently, preventive maintenance has increasingly become the most important trend for operation and maintenance (O&M) phase of infrastructure system from worldwide developed countries. Comparing to the traditional O&M process of existing facilities, where the maintenance work presents as a "reactive activity" and be performed only when the structural deterioration and damage becomes dangerous to the safety in general. Consequently, the owner/government often faces a big hesitation in term of decision making for the repair/rehabilitation work. Nowadays, by a good cost-effective proactive strategy for maintenance purpose from the beginning of the project, it is possible to prevent unexpected risk during service-life of the structure, leading to minimizing the extra cost in O&M. Infrastructure is very diverse, depends on different types of bridge, the structural objective should be defined and highlighted as the main criteria when authoring the BMS. Maintenance work is the kind of continuous and repeated process, starts with the inspection and monitoring system, evaluates the structural behavior, proposes and performs the appropriate repair/rehabilitation work if needed, and finally, upgrade/feedback to the database. The following figure describes the configuration for a management system with consideration of BIM standard.

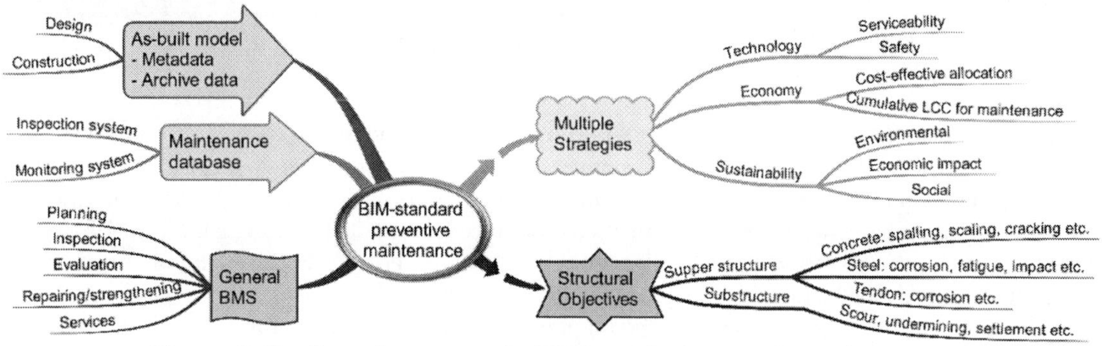

Figure 1. Configuration system for BIM-standard preventive maintenance.

2.2. Data schema for BMS

Basically, the bridge maintenance system using BIM model needs configuration of structured data schema, aiming to orient and form the model-based information system on later. BIM concept is the kind of object-oriented authoring, therefore, the 3D information model is separately defined element-by-element, and finally assembled together by specific ID and individual orientation (coordinates & constraints). In this regards, each structural element should be categorized into an inventory system according to either role or service in entire bridge system. For example, it can be inventoried as a superstructure or as a substructure. Noteworthy that among a distinguished category, structural elements should have the similar characteristics and properties, otherwise it is necessary to further classified into detailer hierarchy category, to form the feature of each category.

Since the inventory system and ID definition system is defined, the BIM-oriented information model can be generated by adding the corresponding information into specified object ID. Two main characteristics: "attribute" and "archive" are the backbone of the information system. The individual attribute feature consists of all physical information such as geometric and material property, orientation and role, quantity/cost estimating is moreover indispensable. The individual archive data, on the other hand, consists all related information following the EIR's aspects, for example, inspection plan/manual, damage/repair record etc. which is continuously obtained during the bridge lifecycle, from design and construction, to O&M stage, and specified in a list.

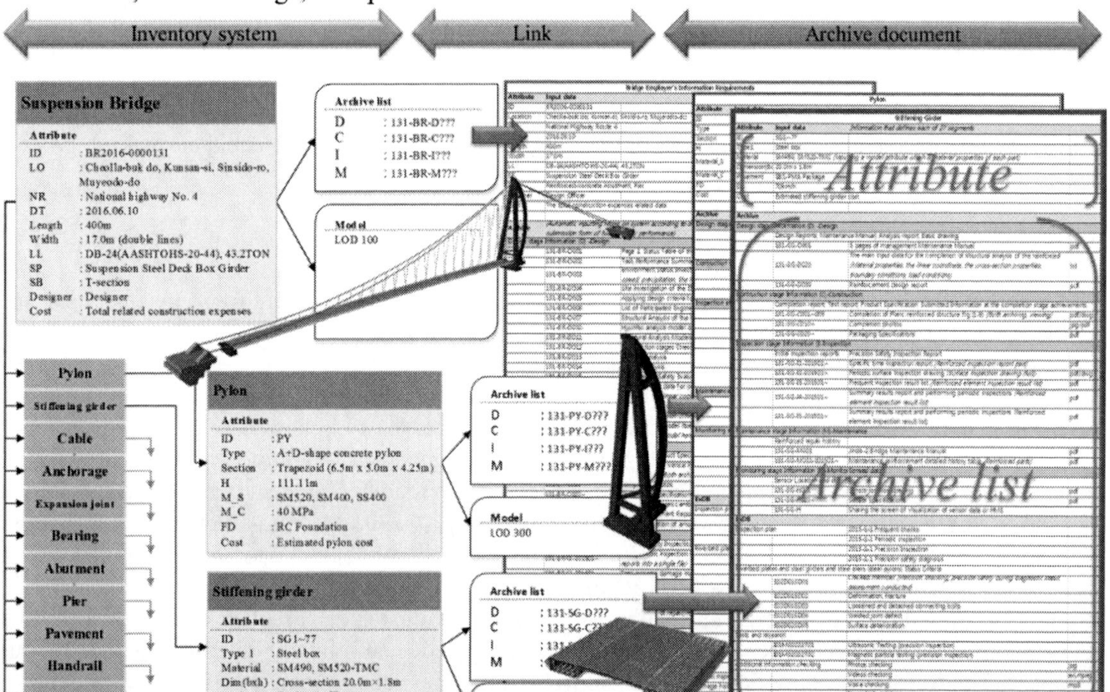

Figure 2. Shortened example for data-schema of a suspension bridge.

2.3. Information specification

Depending on the level of development (LOD) of information requirement, only needed information is adequately provided. For the physical 3D geometric model, each structural element is described by a minimized number of parameters with pre-setting by a relative value. Furthermore, this type of imported information for each element should be generated by a spreadsheet data type, so that information can be easily imported into the different application from each partner. For the archive data, in order to make the computer-readable feature, the standard report form should be considered. In this regards, the primitive file format is suggested such as .jpg for photo checking reports, .avi for video checking reports, or .pdf/.txt for document reports etc. With advantages of information technology nowadays, input information can be preserved, updated and stored throughout the project lifecycle, with the permitted access from rational stakeholders in each stage.

3. BIM authoring for the BMS

Figure 3. Modules of BMS using BIM model.

3.1. Module 1: Information management system using BIM-based model

3.1.1. 3D information model

In fact that almost existing older bridges which are in-use are based on 2D design, therefore generating the 3D digital model from 2D CAD-drawing is an important task in this regards. Moreover, for maintenance purpose, it is required a high interoperability such as linking the metadata, including all attribute and archive. Parametric modeling based on BIM technology was suggested and can be considered as the most innovative and modern technologies for designers. By the flexibility of design variables, all optimization and revision procedures can be automated and full-in-range controlled. During the design process, the geometric definitions or even conceptual definitions can be varied anytime. With the concept of parametric features, the design now can match more and more with the actual work as well as the manufacturing process. The following figure presents a flowchart for the concept of parametric modeling of the standard elements for a suspension bridge. In this regards, two types of the algorithm should be distinguished: one describes the shape and the other describes the position, orientation, or constraint of the geometric object. Corresponding to each parameter, a chain of the algorithm will be varied to decide the type of parameters. A geometric object is usually built from points, lines, and planes, and the parameters of only one type of these object features are used to specify the geometry. The algorithm to create the geometry of an object is based on linear equations of curves and surfaces in the Euclidean three-dimensional space, whereas most exterior orientation parameters are based on the matrix operations.

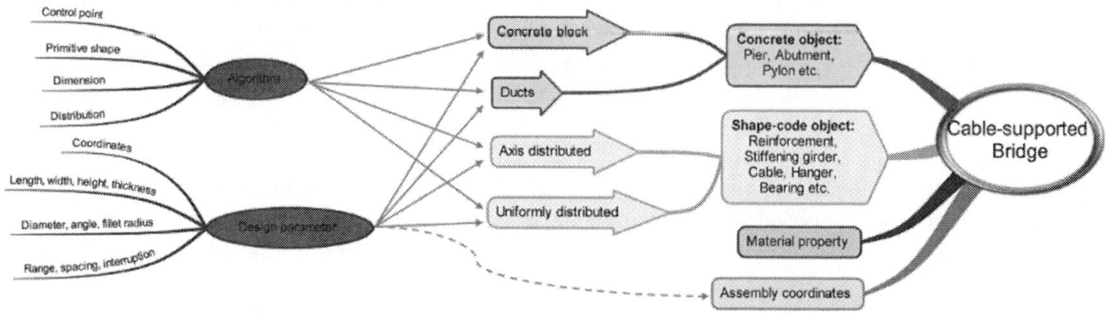

Figure 4. Parametric Modeling concept for structural element of cable-supported bridge.

3.1.2. API-based modeling

In order to enhance design performance, smart algorithm-based 3D information models should be possible to reuse for different construction practices from conceptual design to analysis, estimation, multi-dimensional simulation, and maintenance etc. This paper proposes an essential strategy for 3D simulation of the information system. An integrative work between a CAD program and a programing language was launched by the aided of application programming interface (API). The common API can ease the work of programming graphical user interface components by accessing databases system. It also can facilitate the integration of new features into existing applications especially if using an open-source programming language which strong and fundamental enough for the integrated purpose.

Moreover, an API is able to assist otherwise distinct applications with sharing data, which can help to integrate and enhance the functionalities of the applications.

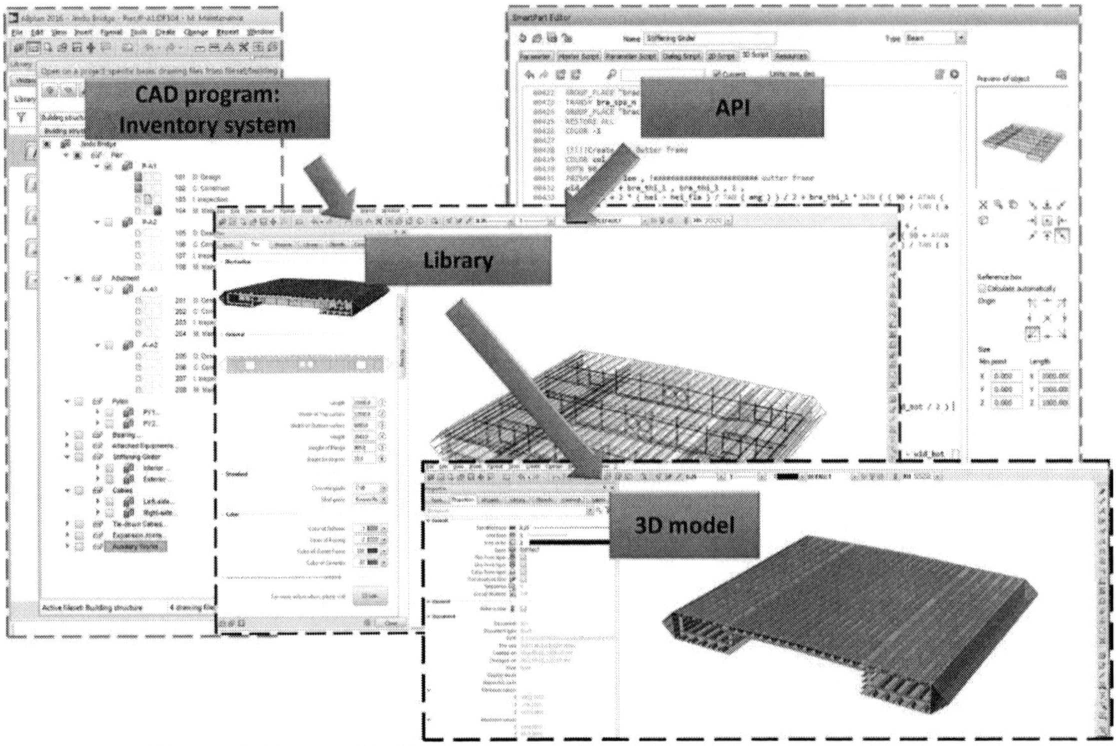

Figure 5. Concept for Digital mock-up by application programing interface.

3.1.3. Data management system

The common data environment (CDE) for a bridge stores all information related to the physical features of that bridge, which can be gathered from design and construction phase, and upgraded from inspection and repair records. As mentioned above, according to the information requirements in EIR, the database system is classified by structure element and their constraints, which is represented by element ID. The database system can be distinguished into three types: asset database, inspection database, and repair database. The BIM-based parametric model now can demonstrate its outstanding features in term of archiving, exchanging as well as delivering information, which absolutely preserves the accuracy of information. The inspection database can be collected primarily from regular or major inspection through an inspection contractor and synthesized into the bridge condition evaluation reports, which used as a basis for the proposed plan of maintenance work. After repair works, the repaired database including the date of repair works, repaired as-built model etc. will be upgraded to the initial asset information.

3.2. Module 2: Site inspection system

3.2.1. Damage classification

Depending on different types of a structural member of the bridge, the damage type of each member should be classified at first. For a concrete member, the regular damage type is surface crack, leakage etc. meanwhile the steel member commonly suffered rust, failure, loosening of bolt or damage to weld etc. Additionally the negative effect of abrasion or deformation, coating separation probably affect both.

3.2.2. Damage detection (image processing)

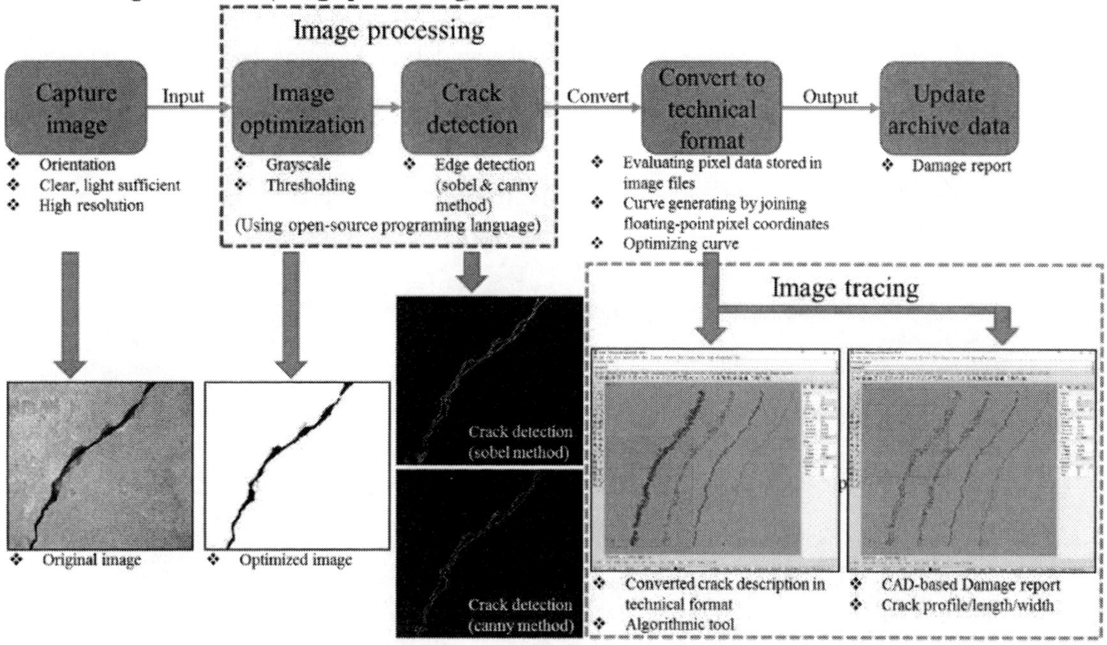

Figure 6. Concept of surface damage detection based on image processing.

Inspection works commonly based on the visual observation and the main activity is describing the damage of the structure by technical format. From the process of visual observation and hand-draw description or photograph, there may some inconsistent error occurred. In order to enhance the productivity of this work, the concept of image processing is proposed to make an automatic site inspection system. The above figure shows an example of surface-cracking inspection work, with the use of digital information in the whole process of inspection activities. The inspection engineers check the structure's conditions, whenever damage such as cracking is found, engineer captures the damage by camera or a mobile device. Damage data is sent to the site inspection system and the process of image processing is conducted. Starting with the original image, it can be optimized by transferring into grayscale and apply thresholding algorithm. This work aims to make a clear and edge-detectable input material. Hereafter, edge detection algorithms are used through an open-source programing language, which integrated into the site inspection system. The different approachable method is applied, the final result is chosen depending on the type of damage, e.g. canny method usually applied for crack detection whereas fuzzy-logic method used for area damage detection. Finally, by image tracing conversion (as so-called raster-to-vector conversion or vectorization), damage report is generated without any discontinuity or deficiency

3.2.3. Technical report (image tracing)

Since this kind of digital technical information in the vector graphic format is created, it is easy to collect the crack profile as well as calculating the length as well as the maximum width of cracks. The surface where crack occurred can be found from the initial 3D model by known specified object ID. Finally, the technical drawing of crack will be mapped into the target surface and update inspection information back to the initial 3D model in digital format. The site inspection system moreover can automatically generate inspection reports for bridge elements in 2D drawings. Noteworthy that, nowadays with the aid of mobile/wearable devices which integrated the data processing features as a mini-computer, e.g. an augmented reality device (or even a tablet with appropriate software), with one unique device but engineer now can perform all the inspection task easily. On the other hand, damage report in technical format can be quickly generated and upgraded back to the management system at the same time.

4. A pilot BMS application

This research is performed by BMS research team in Korea. The developed BMS is in the stage of installation to a cable-supported bridge in Korea. All the models and information were linked together and will be utilized for future maintenance work.

4.1. Module 1

From the proposed data schema and established data management system, the module of Information management system is created based on BIM authoring procedure.

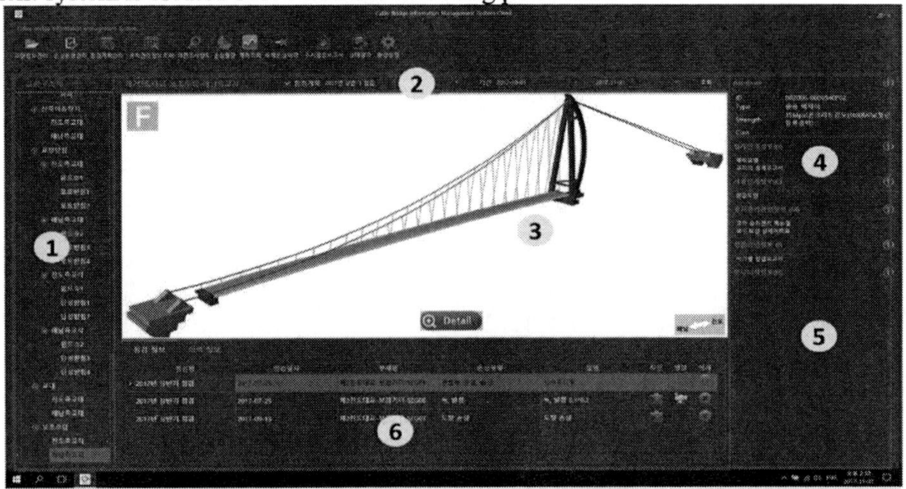

Section	Element	Description	Section	Element	Description
1	Inventory system	Structural element	4	Basic Information	Structure attribute
2	Inspection browsing	Selecting inspection case	5	Archive Information	Attached archive list
3	3D Viewer	3D model display	6	Inspection Detail	Inspection history with damages

Figure 7. BIM-based Information management system (Software interface).

4.2. Module 2

The second module focuses on the real-time data generation and utilization. Real-time collaboration between field engineers and office engineers is provided by this system. Robotic or drone-based inspection in the future was considered in the real-time sharing of the inspection image.

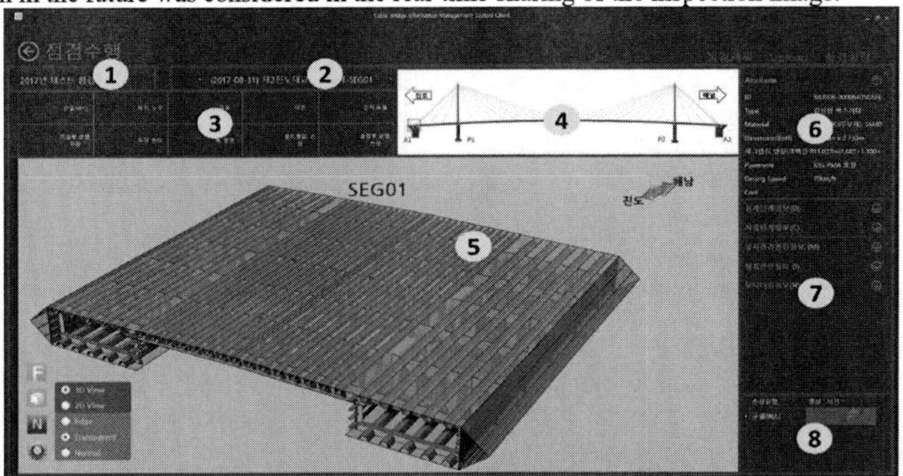

Section	Element	Description	Section	Element	Description
1	Inspection Browsing	Selecting inspection case	5	3D Viewer	3D model display
2	Segment Browsing	Selecting inspection parts	6	Basic Information	Structure attribute
3	Damage Remarks	Damage to apply	7	Archive Information	Attached archive lists
4	Key-Map	Approximate location of segment	8	Input Data Screen	Inputting data viewer

Figure 8. Automatic inspection system on site (Software interface).

5. Conclusion

In this research, a new framework for bridge maintenance system is developed using BIM concept with considering all aspects of maintenance task. One part of research work including a BMS application is applying into an existing cable-supported bridge in Korea and shows good potential for performance-enhancing of maintenance work. The bridge data management system was generated without any limitations by using the proposed BIM authoring and data schema which derived from EIR. Inspection work now can be easily performed and timely reported/upgraded the database in digital technical format, leading to significantly help engineer for saving time and cost. Moreover, the proposed framework can be adapted for various bridges or types of bridge. Based on this paper, the following conclusions were derived:

- In order to build up a bridge maintenance system, the data schema should be defined at first. By using the open-source BIM platform, all O&M manager or other stakeholders can easily access and handle any information within the BMS through object ID. Interoperability of 3D bridge information model should be considered. Information specification and data exchange need to be early specified. Parametric modeling has seen a key-refine for the object-based information system. By using only a few fundamental parameters and algorithms to build up some unique structural member as well as bridge alignment, the designer now can generate the entire bridge model with thousand members with the desired LOD.
- Inspection work needs to be a closed digital process. Surface damage inspection can be performed with just a few engineers through the aids of mobile/wearable devices. Damage report in technical format can be quickly generated and upgraded back to the management system in short time, which is more efficient than conventional one. Development of communications technology allows engineers to browse data wherever they have accessibility to BMS. BMS consists of two modules, one is main management module and the other is inspection module. Two modules can simultaneously share and exchange inspection or stored data. They are possible to handle their own work and to make decisions, even if they are not in the same place. Moreover, it helps to avoid work duplication and loss of data.

6. References

[1] Lee G, Sacks R, Eastman C M 2006 Specifying parametric building behavior (BOB) for a building information modeling system *Automation in Construction, 15(6), 758-776*

[2] Shim C S, Yun N R, Song H H 2011 Application of 3D Bridge Information Modeling to Design and Construction of Bridges *Procedia Engineering 14 (2011) 95–99*

[3] Lee K M, Lee Y B, Shim C S, Park K L 2012 Bridge information models for construction of a concrete box-girder bridge *Structure and Infrastructure Engineering, Vol. 8, No. 7, 687–703*

[4] Volk R., Stengel J, Schultmann F 2014 Building Information Modeling (BIM) for existing buildings - Literature review and future needs *Automation in Construction, 38 (2014) 109–127*

[5] Shim C S, Kang H R, Dang N S, Lee D K 2017 Development of BIM-based bridge maintenance system for cable-stayed bridges *Smart Structures Systems, 20(6): 697-708.*

[6] Hammad A, Yan J, Mostofi B 2007 Recent development of bridge management systems in Canada *2007 Annual Conference and Exhibition of the Transportation Association of Canada*

[7] Miyamoto A, Kawamura K, Nakamura H 2001 Development of a bridge management system for existing bridges *Advances in Engineering Software 32, 821-833*

[8] Oh J K, Lee A Y, Oh S M, Choi Y J, Yi B J, Yang H W 2008 Bridge inspection robot system with novel image processing *Bridge Maintenance, Safety, Management, Health Monitoring and Informatics – Koh & Frangopol (eds) © 2008 Taylor & Francis Group, London, ISBN 978-0-415-46844-2*

[9] Matsumoto M, Mitani K, Sugimoto M, Hashimoto K, Miller R 2012 Innovative Bridge Assessment Methods using Image Processing and Infrared Thermography Technology *Innovative Infrastructures - Toward Human Urbanism, 18th congress of IABSE, Seoul, 2012*

[10] Pereira F C, Pereira C R 2015 Embedded Image Processing Systems for Automatic Recognition of Cracks using UAVs *International Federation of Automatic Control, 48-10 (2015) 016–021*

[11] Bu G P, Chanda S, Guan H, Jo J, Blumenstein M, Loo Y C 2015 Crack Detection using a Texture Analysis-based Technique for Visual Bridge Inspection *Special Issue: Electronic Journal of Structural Engineering 14(1) 2015*

[12] Mohan A, Poobal S 2017 Crack detection using image processing: A critical review and analysis *Alexandria Engineering Journal, https://doi.org/10.1016/j.aej.2017.01.020*

Acknowledgments
This research was supported by a grant (17SCIP-B128570-01 2) from Smart Civil Infrastructure Research Program funded by Ministry of Land, Infrastructure and Transport of Korean government.

IOP Conf. Series: Earth and Environmental Science **143** (2018) 012033 doi:10.1088/1755-1315/143/1/012033

Development of weightage for criteria affecting in retrofitting of existing building in Higher Learning Institution with clean energy initiatives

Nur IzieAdiana Abidin[1], Eeydzah Aminuddin[2], Rozana Zakaria[3], Siti Mazzuana Shamsuddin[4], Shaza Rina Sahamir[5], Jam Shahzaib[6] and Darul Nafis Abas[7]

[1,6,7] *PhD Student, Faculty of Civil Engineering, Universiti Teknologi Malaysia*
[2] *Associate Professor, Faculty of Civil Engineering, Universiti Teknologi Malaysia*
[3] *Lecturer, Faculty of Civil Engineering, Universiti Teknologi Malaysia*
[4,5] *Lecturer, Faculty of Architecture, Planning and Surveying, Universiti Teknologi Mara, UiTM, 40450 Shah Alam, Selangor, Malaysia*
Corresponding email: rozana@utm.my

Abstract. Campus university building is the Higher Learning Institution (HLI) involves complex activities and operations, conserving the energy has become paramount important. There are several efforts taken by universities to improve its current energy use such as policy development, education, and adaption of energy conservation solution through retrofitting. This paper aims to highlight the importance of the criteria affecting in retrofitting of existing buildings with clean energy in order to achieve zero energy balance in buildings. The focus is given to the development of criteria for solar photovoltaic (solar PV), wind turbines and small-scale hydropower. A questionnaire survey was employed and distributed to the green building expert practitioner. Factor Analysis, Factor Score, and Weightage Factor were adapted as a method of analysis in order to produce the final result with weightage output for prioritization and ranking of the relevant criteria. The result performed assists to provide the stakeholders an overview of the important criteria that should be considered especially during the decision making to retrofit the existing buildings with clean energy resources. The criteria developed are also to establish a structured decision-making process and to ensure the selection of the decision or alternatives achieve the desired outcome.

1. Introduction

The built environment is known as one of the most dominant contributors to the greenhouse gas emissions. There has been an encouragement to have an attention towards the needs of retrofitting which is upgrading the existing building stocks because it has significant potential in reducing the emissions, energy consumption and also the risk related to climate change [1]. The issue related to sustainability has become the crucial subject to discuss in many campus universities across countries [2]. Besides, campus sustainability also has been discussed worldwide especially by the policy makers and planners of universities since they discovered that the campus operation, activities, size, populations, and its physical structures have an impact to the environment. Since that, many governments and international organization have started to have an effort to improve the existing building energy efficiency through retrofitting [3].

Retrofitting in university building is one of the efforts that have the potential to increase the existing building condition to become energy efficient. University buildings is also a strategic selection to start with the implementation of retrofitting construction program in comparing with other public buildings [4]. Towards the realization to minimize the reliance of fossil fuels, there has been an increase in use of smart technology in a campus building that able to fulfil the real-life condition such as the implementation of daylight, energy efficient lighting and HVAC [5]. In addition, since the energy depletion issues are gaining attention, it has brought a great motivation to investigate the sources from oil and coal, and to encourage building to utilize the renewable energy resources [6]. Despite implement the energy conservation, energy resources, and energy efficiency as strategies towards in achieving green building, it has been described that the reliance with the renewable energy is also considered as a big step to become a sustainable campus [7]. According to [8], eventually, people are encouraged to find the alternative to overcome the issue related to the fossil fuels supply which begins to dwindle. Thus, indirectly renewable energy has become the sources that should be considered or harness to transform it into energy. [9] Arguably that campus university building is described as a city but in a form of downscaled versions which has the biggest opportunity in exploring the currently available options to improve the inefficiency of energy use through the provision to use cleaner energy such as renewable energy.

2. The Implementation of Retrofitting in Higher Learning Institution

There are several studies conducted that covers the retrofitting of existing university buildings. The summary of the previous studies is presented in Table 1. Many studies have looked at different retrofitting strategies vary from lean energy (passive design), green technology and clean energy (renewable energy). Lean energy consists of initiatives such as improvement with the roofs, installation of skylights, replacement of existing windows, wall and roof insulation. For green technology, it varies from replacement of existing lighting, replacement of refrigerator, installation of VAV, VFD and occupancy sensors. Meanwhile, for clean energy it is mostly concern on the installation of solar photovoltaic. From the summary, it is also found that many retrofitting studies have much focused on the economic and environmental assessments. However, there is limited knowledge with the development of criteria for decision making in retrofitting. Therefore, this paper is to establish the important criteria for retrofitting of existing building that allows prioritization during the decision making the process by producing the Weightage Factor. The deliberation of this paper only focuses on the clean energy resources that consist of solar photovoltaic, wind turbines and small-scale hydropower.

Table1. Summary of Retrofitting of Existing Building in HLI and the Area Interest.

Retrofit Initiatives	Retrofit Initiatives	Area interest
University of Malaya [10]	Green	Energy consumption, operating cost, payback period, LCC
University of Mexico [11]	Green	Energy saving and CO2 emission
Melbourne University (Di Stefano, 2000) [12]	Green	Energy savings, payback time, NPV, cost of conserve energy
University of Seoul Korea [13]	Green +Lean	Reduction on energy consumption
Curtin University Sarawak [14]	Green	Annual energy and cost savings
Guangdong Colleges and Universities [15]	Green	Energy savings
Kingsville Texas University [16]	Green	Annual savings
University of Malaya [17]	Green	Energy savings and CO2emission
University of Mauritius [18]	Green	Energy saving
Stanford University [19]	Green	Energy saving and payback period
Highfield Campus of University of Southampton [8]	Clean	Energy consumption, capital cost, payback period, heat requirement
Hong Kong Universities [20]	Green +Lean	Energy saving, payback period, ROI and CO2 emission
University of Lagos [21]	Green	Energy savings
King AbdulAziz University [22]	Green	Power demand reduction

3. Clean Energy

Clean energy, commonly known as renewable energy is an initiative adapted to attain for a more energy reduction in buildings [23]. Clean energy consists of several sources such as solar photovoltaic and wind power that has the ability in meeting the building energy demand. This clean source has a large contribution to reduce the greenhouse gas emission since it assists in reducing the reliance towards fossil fuels to generate electricity [24]. The implementation of clean energy is one of the provisions and become a vital initiative in order to have clean operation or production of buildings which contribute to less environmental impact [9]. According to [25], the generation of electricity based on fossil fuels which are used by most countries in this global is known to have a direct impact to the greenhouse gas emission and indirectly make the atmosphere warm. Therefore, clean energy resources that can be utilized in a variety of alternatives such as sun, water, and wind is believed will help to make a large reduction in greenhouse emission.

4. Criteria Affecting in Retrofitting of Existing Building

4.1 Solar PV

Solar PV is the innovation by using cells to capture solar energy that able to convert it into electricity. It can be installed whether in rooftops, land or integrate with building designs [26]. It has been widely accepted due to its features which has silent operation, low costs of operation and maintenance and does not produce pollution to atmosphere. Solar PV is able to be installed with two purpose, whether for new building or retrofitting but requires major consideration with the design [27]. Besides, solar PV is known as clean energy which is beneficial to achieve large reduction of carbon emission, no fossil fuels requirement, has a long-term of energy resources [28]. According to [29], solar PV is the technology that promises to achieve a great sustainable development since it helps to reduce the burden related to fossil fuels, carbon emission and also manage to fulfil the electricity demand. Figure 1 below shows the five (5) main criteria that has been developed towards considering the retrofitting of existing buildings in Higher Learning Institution with solar PV. It consists of design, PV features, management, economic and environmental criteria.

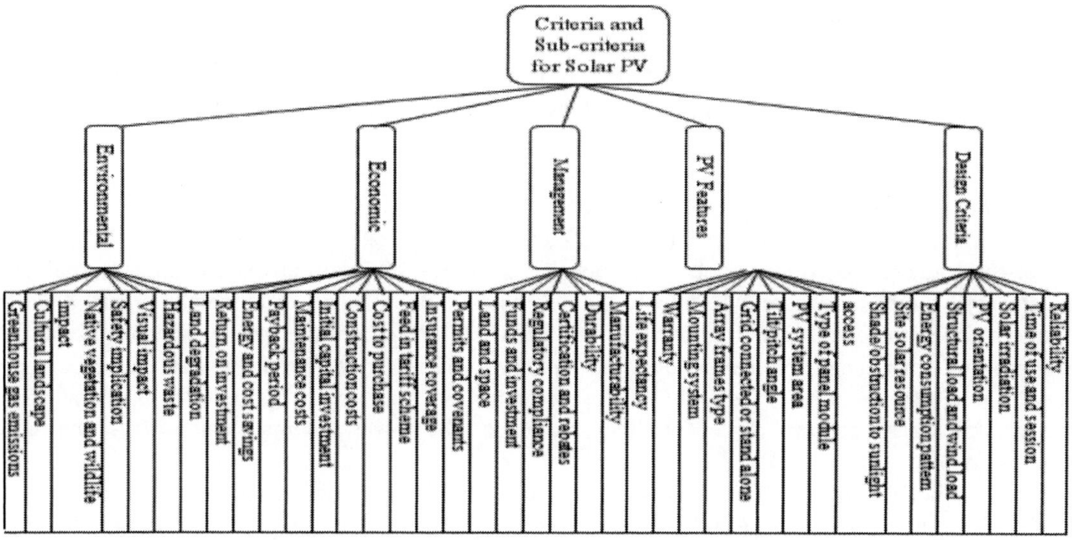

Figure 1. Criteria Affecting in Retrofitting of Existing Building with Solar PV.

4.2 Wind Turbine

Wind turbine is also known as a clean energy resource that currently has received attention by many due to realization to improve the environment [30]. Wind energy is also another alternative that able to reduce the reliance on fossil fuels and become a solution to the global ever-increasing energy needs [31]. According to [32], through the improvement of technology, wind turbine is currently reliable, efficient and cost-effective to be adapted in order to produce power. The ability of wind turbines to catch the air motion from natural wind to produce the rotational and electrical power does not create pollution or hazardous waste to the environment.

There are six (6) main criteria that have been established for the retrofitting of existing buildings with wind turbines. It covers on wind resource, management, economic, environmental, risk and design criteria as can be seen in figure 2.

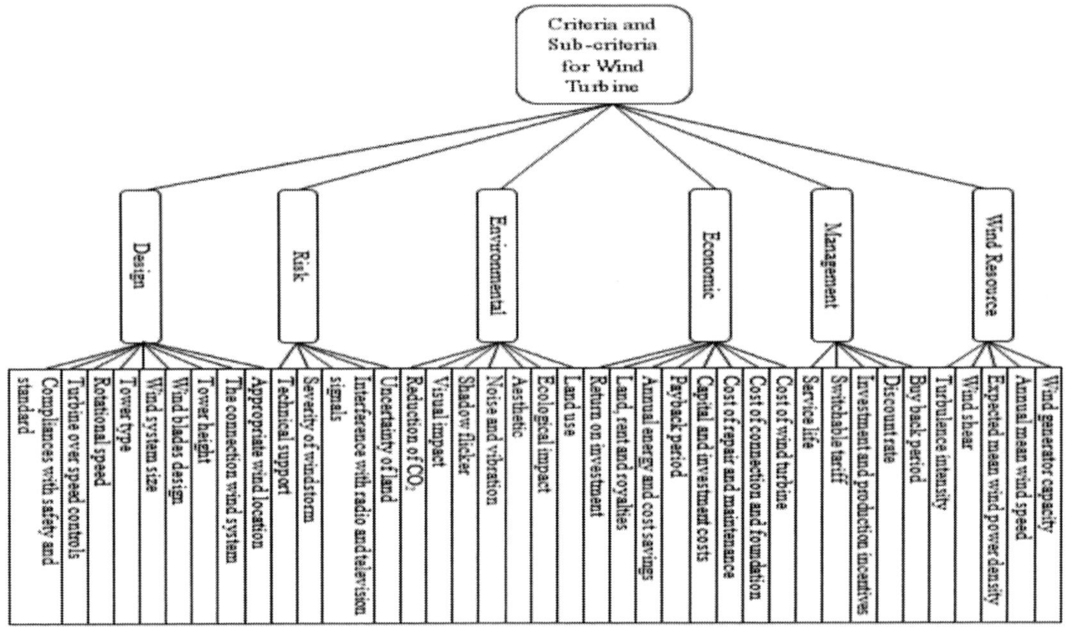

Figure 2. Criteria Affecting in Retrofitting of Existing Building with Wind Turbine.

4.3 Small-Scale Hydropower

Hydropower is another renewable energy and clean source of energy which is also an initiative to reduce the countries' dependency on fossil fuels [33]. Small-scale hydropower harness the small river and streams whereby the water will flow to the turbine component to generate the electricity. It has the ability to generate power from 5 Kilowatt to 100 Kilowatt, offers a high capacity which is about 50% compared to other renewable technology and has high efficiency at 70-90% which is considered as one of the best technologies (Nasir, 2014). According to [34], small-scale hydropower is operated with free from fuel, has a constant power for a long period and the power generated will increase during the rainy season.

Figure 3 below shows the establishment of main criteria and sub-criteria for the retrofitting of existing buildings with small-scale hydropower. It comprises of five (5) main criteria vary from site assessment, management, economic design, and environmental.

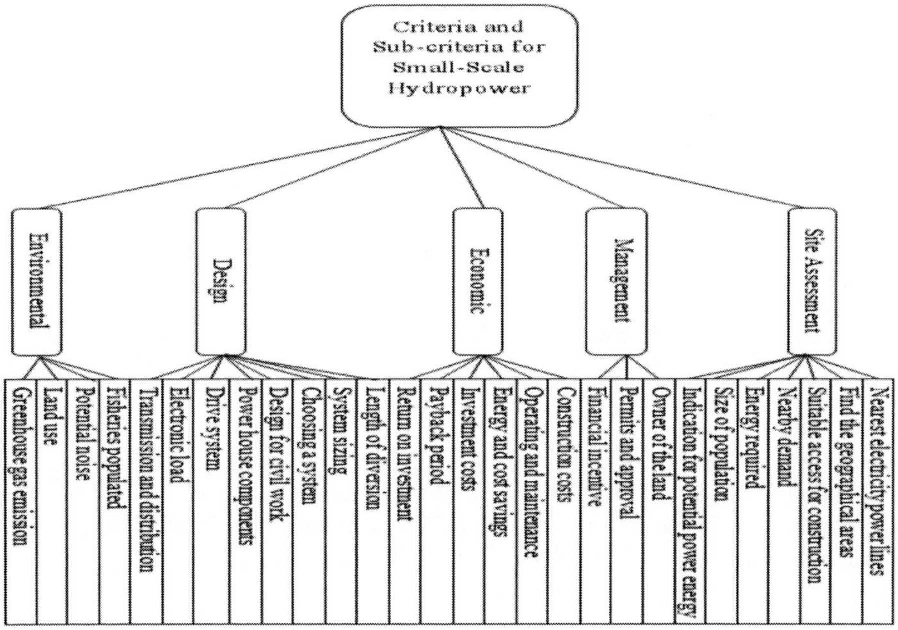

Figure 3. Criteria Affecting in Retrofitting of Existing Building with Small Scale Hydropower.

5. Methods

This paper employed a questionnaire survey to elicit the important criteria affecting in retrofitting of existing buildings in HLI with clean energy retrofit initiatives that encompasses of solar PV, wind turbines, and small-scale hydropower. Each of these initiatives has been established with the main criteria and sub-criteria to allow the respondents that consist of experts in green building design to provide the rate of importance by indicating the scale from 1-5. The establishment of criteria is based on the critical review collected through the secondary data. A total of 110 of the questionnaire were distributed, however, 54 of them has successfully respond which represents 21.26% of response rate. Factor Analysis, Factor Score, and Weightage Factor were adapted as a method of analysis. These three types of analysis are interconnected with each other because Factor Score can be performed once the data has been reduced into a small set of numbers through Factor Analysis and the Weightage Factor output is produced based on the continuation of the Factor Score analysis because they are mutually dependant. These three types of analysis are finally assisted to produce the weightage output that can prioritize and ranked in the form of arrangement according to its importance

5.1 Factor Analysis (FA)

FA is basically performed with IBM Statistical Package for Social Science (SPSS) 21. FA is a technique for data reduction by collapsing a large number of variables into several factors. It allows being generated in a manageable form prior to performing with other analysis [35]. FA involves several protocols, firstly, it is required to assess the Kaiser-Meyer-Olkin (KMO) to test the adequacy of sampling. KMO comes in a range from 0 to 1 and the acceptable minimum value is 0.50 [36]. Secondly, the statistical significance is tested through Barlett's Test of Sphericity with a significant value which is $P<.05$. Thirdly, the type of factor of extraction is chosen to allow it to produce the best and smallest number of factors. Then, the process continues with the factor rotation in order to interpret the number of variables to be a group of factor [35]. Lastly is the determination with the factor loading value whereby the minimum suggested value is 0.50 [37]. In this paper, the FA analysis was performed with Principal Component Analysis with Varimax Rotation as a method of extraction and the factor of rotation with the factor loading 0.50.

5.2 Factor Score (FS)

Once a large number of variables has been reduced through the adaption of Factor Analysis, the findings obtained were further analysed with FS and WF. FS was normally conducted after performing the FA with the intention to use it for further analysis such as to identify the detail rankings or to determine the score between one another or in a group of category (Di Stefano, 2009). Therefore, FS allows further justification with the variables developed based on the numerical value assign. The method to calculate the FS is by multiplying the Factor Loading (FL) from each of the variables with the same Mean value (average Mean value), Y. The multiplying process will produce Factor Score for sub-criteria, FSsc [38].

5.3 Weightage Factor (WF)

The WF analysis is carried out to further analyse the result obtained from FS to produce its own weightage. According to [39], WF allows to make a comparison or to see the influence of the variables in the group. The steps taken to measure WF is firstly by adding up all the value of FSsc obtained from the FS earlier to produce the ΣFSc. Then, each of the FSsc is divided with the ΣFSc value. After the division method is carried out, all values should be added and the summation of result should equal to 1 with a percentage 100 %. A basic formula for WF as shown in equation (1) below:

$$\pi \text{ subcriteria} = \frac{\text{\% Stratum in Variables (sub-criteria), FSsc}}{\text{\% Stratum in Criteria, FSc}} \qquad \text{Equation (1)}$$

Where;

FS_{sc} = Factor score for each item in the sub-criteria

ΣFS_c = Cumulative of factor score in the criteria

6. Findings

6.1 Solar Photovoltaic

Table 2 shows the results derived from the FA, FS and WF Analysis for Solar PV. For FA analysis, there are 39 variables loaded subjected to Principal Component Analysis with Varimax rotation. From the rotation, 3 of the variables have been eliminated due to the factor loading is less than the minimum value of 0.50. The FA result that reached 0.50 and above are only presented in the table. For WF analysis which is performed after the FS analysis, each of the sub-criteria that is gathered into its own main criteria category has achieve its own weightage value which is arrange according to its significance.

Firstly, under management criteria, certification and rebates have been at the first rank with a weightage of 16.643%, followed by funds and investment with a weightage of 15.873%, land and space with 15.367%, whilst lastly is feed in tariff scheme with a weightage of 11.889%. Secondly, for economic criteria, the third highest ranking consists of maintenance costs with the weightage of 16.512% and it is followed by capital investment with a weightage of 16.108% and payback period with 15.745%. Thirdly, under the environmental criteria, visual impact is at the highest weightage with a value of 18.953%, followed by cultural landscape with a weightage of 17.474%, native vegetation and wildlife impact with a weightage of 17.179%, safety implication to fire hazard with 16.951% value, hazardous waste with 16.587% value and lastly is greenhouse gas emission with 12.856% value. Meanwhile, under PV features criteria, life expectancy of PV module has been indicated as the most significant by respondents with the weightage of 11.035%. Then, manufacturability achieved at the second highest rank with a weightage of 10.875%, followed by mounting system with 10.539% value and PV system size with 10.525% value. Lastly, under the design criteria, the highest weightage is concerned with solar irradiation with the weightage value of 19.095%. Then, the second highest is achieved by site solar resource with the weightage of 18.679%, followed by orientation with 16.986% value, reliability with 16.264% value, time use and session with 14.488% value and lastly the ability to withstand structural and wind loading with 14.488% value.

Table 2. Results from Factor Analysis, Factor Score and Weightage Factor for Solar PV.

Main Criteria	Sub-Criteria	Factor Loading (FL)	Mean (X)	FSsc	Weightage Factor	%
Management	Certification, rebates	0.781	4.074	3.141	0.165	16.463
	Funds and Investment	0.753	4.093	3.028	0.159	15.873
	Land and space	0.729	4.019	2.932	0.154	15.367
	Regulatory compliance	0.677	4.093	2.722	0.143	14.271
	Permits and covenants	0.648	3.944	2.606	0.137	13.659
	Insurance Coverage	0.592	3.982	2.381	0.125	12.480
	Feed in-tariff scheme	0.564	3.944	2.268	0.119	11.889
	TOTAL		**4.021**	**19.077**	**1.000**	**100**
Economic	Maintenance cost	0.818	4.296	3.445	0.165	16.512
	Initial capital investment	0.798	4.111	3.361	0.161	16.108
	Payback period	0.780	4.204	3.285	0.157	15.745
	Construction cost	0.770	4.278	3.243	0.155	15.543
	Cost and energy saving	0.718	4.241	3.024	0.145	14.493
	Return on investment	0.535	4.111	2.253	0.108	10.799
	Purchase material	0.535	4.241	2.253	0.108	10.799
	TOTAL		**4.212**	**20.865**	**1.000**	**100**
Environmental	Visual impact	0.833	3.667	3.214	0.190	18.953
	Cultural Landscape	0.768	3.667	2.963	0.175	17.474
	Native vegetation and wildlife impact	0.755	3.833	2.913	0.172	17.179
	Safety implication to fire hazard	0.745	4.130	2.874	0.170	16.951
	Hazardous waste	0.729	3.944	2.812	0.166	16.587
	Greenhouse gas emission	0.565	3.907	2.180	0.129	12.856
	TOTAL		**3.858**	**16.956**	**1.000**	**100**
PV Features	Life expectancy PV module	0.757	4.074	3.027	0.110	11.035
	Manufacturability	0.746	4.019	2.983	0.109	10.875
	Mounting system	0.723	3.796	2.891	0.105	10.539
	PV system size	0.722	4.204	2.887	0.105	10.525
	Type of panel module	0.698	4.056	2.791	0.102	10.175
	Array frames type	0.687	3.704	2.747	0.100	10.015
	Warranty	0.678	4.000	2.711	0.099	9.883
	Durability and longevity	0.674	4.056	2.695	0.098	9.825
	Grid-connected and stand-alone	0.621	3.907	2.483	0.091	9.052
	Tilt/pith angle of PV array	0.554	4.167	2.215	0.081	8.076
	TOTAL		**3.9983**	**27.428**	**1.000**	**100**
Design	Solar Irradiation	0.688	3.926	1.889	0.191	19.095
	Site solar resource	0.673	0.714	1.848	0.187	18.679
	Orientation	0.612	3.889	1.680	0.170	16.986
	Reliability	0.586	3.944	1.609	0.163	16.264
	Ability of structure to withstand the structural wind loading	0.522	4.000	1.433	0.145	14.488
	Time use and session	0.522	4.074	1.433	0.145	14.488
	TOTAL		**2.7455**	**9.892**	**1.000**	**100**

6.2 Wind Turbine

Table 3 presents the results derived from the FA, FS and WF for the retrofitting of existing buildings with Wind Turbines. For FA analysis, there are 40 variables loaded subjected to Principal Component Analysis with Varimax rotation. From 40 variables loaded under 5 main criteria, 3 of the variables have been eliminated as the factor loading is less than 0.50. The summary result of strong factor loadings for wind turbine is as tabulated in the table below.

For WF which performed after the FS has been conducted, each of the sub-criteria has produced its own weightage. Firstly, under economic assessment criteria, the highest weightage achieved is concerned with a payback period with a value 14.713%, followed by return on investment with factor loading 13.764%, whilst the least is investment costs with a weightage of 9.379%. Secondly, for wind resource criteria, the wind speed changes are indicated by respondents as the most significant with a weightage of 22.119%. The second highest weightage is to determine the annual mean wind speed with factor loading with a weightage of 20.311%, and then followed by estimate the turbulence intensity, and lastly is wind generator capacity with a weightage of 18.475%. Meanwhile, under wind turbine design criteria, configuration achieved 13.478% of weightage which represents the highest ranking. For the environmental criteria, shadow flicker is indicated as the most significant by respondents with the weightage of 18.789%. For vibration, the weightage achieved is 17.374%, noise is 16.887%, visual impact is 16.585%, and lastly aesthetic is 14.243% of the weightage. Other than that, for risk assessment criteria, the highest weightage is the interference with radio and television signal with a weightage of 34.269%, whilst the least is the severity of windstorm with a weightage of 32.075%. Last but not least, under the management criteria, service life achieved 23.750% of the weightage, followed by the lowest buy back period with 21.128% value and discount tax rate with 19.665% value, switchable tariff with 18.079% value and investment and production incentives with 17.378% value.

Table 3. Results from Factor Analysis, Factor Score and Weightage Factor for Wind Turbine.

Main Criteria	Sub-Criteria	Factor Loading (FL)	Mean (X)	FSsc	Weightage Factor	%
Economic Assessment	Payback period	0.822	4.040	3.372	0.147	14.713
	Return on Investment	0.769	4.070	3.155	0.138	13.764
	Annual and cost energy	0.759	4.100	3.114	0.136	13.585
	Connection and foundations	0.721	4.130	2.958	0.129	12.905
	Land rents, royalties	0.687	3.950	2.818	0.123	12.296
	Cost of wind turbines	0.677	4.170	2.777	0.121	12.117
	Repair and maintenance	0.628	4.190	2.576	0.112	11.240
	Investment cost	0.524	4.170	2.150	0.094	9.379
	TOTAL		**4.103**	**22.921**	**1.000**	**100**
Wind Resource	Wind speed changes	0.783	3.910	3.118	0.221	22.119
	Mean wind power density	0.719	3.980	2.863	0.203	20.311
	Annual mean wind speed	0.715	4.130	2.847	0.202	20.198
	Turbulence intensity	0.669	3.850	2.664	0.189	18.898
	Wind generator capacity	0.654	4.040	2.604	0.185	18.475
	TOTAL		**3.982**	**14.096**	**1.000**	**100**

Design	Configuration	0.904	4.280	3.583	0.135	13.478
	Wind blades	0.894	3.800	3.543	0.133	13.329
	Tower type	0.869	3.780	3.444	0.130	12.957
	Tower heights	0.834	3.870	3.305	0.124	12.435
	Wind system size	0.768	3.800	3.044	0.115	11.451
	Turbine over speed controls	0.744	3.850	2.949	0.111	11.093
	Safety and performance standard	0.591	4.110	2.342	0.088	8.812
	Rotational speed	0.562	3.960	2.227	0.084	8.379
	Wind location	0.541	4.220	2.144	0.081	8.066
	TOTAL		**3.963**	**26.582**	**1.000**	**100**
Environmental	Shadow flicker	0.810	3.870	3.121	0.188	18.789
	Vibration	0.749	3.980	2.886	0.174	17.374
	Noise	0.728	3.850	2.805	0.169	16.887
	Visual Impact	0.715	3.780	2.755	0.166	16.585
	Ecological impact	0.695	3.940	2.678	0.161	16.122
	Aesthetics	0.614	3.700	2.366	0.142	14.243
	TOTAL		**3.853**	**16.612**	**1.000**	**100**
Risk	Interference with radio and television signal	0.781	3.770	3.002	0.343	34.269
	Technical support issues	0.767	3.910	2.948	0.337	33.655
	The severity of windstorm	0.731	3.850	2.809	0.321	32.075
	TOTAL		**3.843**	**8.759**	**1.000**	**100**
Management	Service Life	0.779	4.170	3.032	0.238	23.750
	Lowest buy-back period	0.693	3.930	2.697	0.211	21.128
	Discount of tax rate and duty rate	0.645	3.740	2.510	0.197	19.665
	Switchable tariff	0.593	3.860	2.308	0.181	18.079
	Investment and production incentives	0.570	3.760	2.218	0.174	17.378
	TOTAL		**3.892**	**12.766**	**1.000**	**100**

6.3 Small-Scale Hydropower

Table 4 presents the results derived from the FA, FS and WF for the retrofitting of existing buildings with small-scale hydropower. The FA has uploaded 28 variables subjected to the Principal Component Analysis. From the rotation, there is only 1 variable that was eliminated due to the factor loading is less than 0.50. The result of factor loading with a value 0.50 and above can be seen in Table 6.

Meanwhile, for WF which is performed based on the continuation of FS analysis, each of the sub-criteria has achieved its own weightage which also can be seen in Table 4. Firstly, the financial incentives, owners of the land, permit and approval process has achieved the weightage of 34.257%, 34.950% and 30.793% respectively under the management criteria. Besides, for the design criteria, properly design for civil works components is indicated as the most significant by respondents with a weightage of 14.943%, followed by transmission and distribution method with 15.100% value, power house component with 13.161% value, electronic load component with 12.887% value and drive system with 12.280% value. Under site assessment criteria, the highest factor loading ranked by respondents is to find the best geographical areas with a weightage of 19.230%, followed by indication for potential power and energy with a weightage of 18.179%, access for construction equipment with 17.695% value, a nearby demand with 17.129% value and lastly is the nearest electricity power lines with a weightage of 13.628%. Meanwhile, for cost and economic criteria, the respondents concerned towards the significant energy and cost saving which is at the highest rank with a weightage of 18.539%. Then, it is followed by payback period with a weightage of 17.482%, investment cost with 15.277% value, return on investment with 14.197% value, operating and maintenance cost with

IOP Conf. Series: Earth and Environmental Science **143** (2018) 012033 doi:10.1088/1755-1315/143/1/012033

18.585% value and construction cost with 15.920% value. Last but not least, under the environmental criteria, potential noise impact on nearby building is rated as the most significant with a weightage of 27.177%, followed by land use with a weightage of 26.323%, greenhouse gas emission with 25.367% value, and fisheries populated habitats with 21.133% value.

Table 4. Results from Factor Analysis, Factor Score and Weightage Factor for Small Scale Hydropower.

Main Criteria	Sub-Criteria	Factor Loading (FL)	Mean (X)	FSsc	Weightage	%
Management	Financial incentives	0.791	4.20	3.222	0.343	34.257
	Owners of the land	0.807	3.96	3.287	0.350	34.950
	Permits and approval process	0.711	4.06	2.896	0.308	30.793
			4.073	**9.405**	**1.000**	**100**
Design	Properly design for civil works components	0.763	4.13	3.132	0.149	14.943
	The transmission and distribution network	0.771	4.20	3.165	0.151	15.100
	Power house component	0.672	4.22	2.759	0.132	13.161
	Electronic Load Component	0.658	4.09	2.701	0.129	12.887
	Drive system	0.627	4.06	2.574	0.123	12.280
	Choosing a system	0.556	4.07	2.282	0.109	10.889
	Sizing of the system	0.554	4.09	2.274	0.108	10.850
	Length of diversion	0.505	3.98	2.073	0.099	9.890
			4.105	**20.960**	**1.000**	**100**
Site Assessment	Find the best geographical areas	0.714	4.30	2.926	0.192	19.230
	Indication for potential power and energy	0.675	3.98	2.766	0.182	18.179
	Access for construction equipment	0.657	4.13	2.693	0.177	17.695
	Power required	0.525	4.17	2.152	0.141	14.140
	A nearby demand for electricity	0.636	4.07	2.607	0.171	17.129
	Nearest electricity power lines	0.506	3.94	2.074	0.136	13.628
			4.098	**15.217**	**1.000**	**100**
Economic	Significant energy and cost saving	0.807	4.09	3.348	0.185	18.539
	Payback Period	0.761	4.06	3.157	0.175	17.482
	Investment cost	0.665	4.02	2.759	0.153	15.277
	Return on Investment	0.618	4.17	2.564	0.142	14.197
	Operating and maintenance	0.809	4.33	3.356	0.186	18.585
	Construction cost	0.693	4.22	2.875	0.159	15.920
			4.148	**18.058**	**1.000**	**100**
Environmental	Noise impact on nearby building	0.796	3.98	3.280	0.272	27.177
	Land use	0.771	4.02	3.177	0.263	26.323
	Greenhouse gas emission	0.743	4.37	3.061	0.254	25.367
	Fisheries populated habitats	0.619	4.11	2.550	0.211	21.133
			4.120	**12.067**	**1.000**	**100**

7. Conclusion

Clean energy is the main key driver that is beneficial not only for the nation, but also for the future in this globe. Since most of the power generation comes from the sources of fossil fuels, there has been an increase in concern with the global warming, and thus renewable energy has increasingly become an important source to generate energy. Considering the clean energy as part of the decision to retrofit existing buildings could become one of the great solutions to achieve for sustainable development and to become zero energy building. A building campus is one of the potential opportunity for retrofitting implementation and could serve as a pioneer in promoting as well as transforming the existing electricity generation through utilization of renewable energy. Basically, the implementation of retrofitting involves with the decision-making process in order to identify the alternatives or preferences. Therefore, the development of the clean energy criteria in this paper is hoping that it will be able to serve the stakeholders who is involved with the required information throughout the decision-making process. The findings from the weightage is also to provide a reference which criteria that is important to consider when implementing the clean energy initiatives.

References

[1] Bruce T, Zuo J, Rameezdeen R and Pullen S 2015. Factors Influencing the Retrofitting of Existing Office Buildings Using Adelaide, South Australia As A Case Study. *Structural Survey*. **33(2)**, 150–166. doi:10.1108/SS-05-2014-0019.

[2] Beringer, A, Wright, T and Malone, L 2008. Sustainability in Higher Education in Atlantic Canada. *International Journal of Sustainability in Higher Education*. ***9*(1)** 48–67. doi:10.1108/14676370810842184

[3] Sesana M M, Grecchi M, Salvalai G and Rasica, C 2016 Methodology of Energy Efficient Building Refurbishment: Application on Two University Campus-Building Case Studies in Italy With Engineering Students. *Journal of Building Engineering*. ***6*** 54–64. doi:10.1016/j.jobe.2016.02.006

[4] Carbonara E, and Tiberi M 2016 Assessing Energy Performance and Economic Costs of Retrofitting Interventions in a University building. *IEEE*,1-5.

[5] Ghadi Y Y, Rasul M G and Khan, M M K 2017 Energy Savings by Fuzzy Base Control of Occupancy Concentration in Institutional Buildings. *Energy Procedia* **105** 2850–2858. doi:10.1016/j.egypro.2017.03.628

[6] Berde P A, and Bhavani R G 2016 Investigation for Reducing Energy Consumption for A University Campus in Dubai Using Automation. *IEEE Recent Advances in Intelligent Computational Systems*. 358–363. doi:10.1109/RAICS.2015.7488442

[7] Alshuwaikhat H M and Abubakar I 2008 An Integrated Approach to Achieving Campus Sustainability: Assessment of the Current Campus Environmental Management Practices. *Journal of Cleaner Production*. **16** 1777-1785.

[8] Kalkan N, Bercin K, Cangul O, Morales M G, Saleem M M K M, Marji I, Metaxa A, and Tsigkogianni E 2011 A Renewable Energy Solution for Highfield Campus of University Of Southampton. *Renewable and Sustainable Energy Reviews*, **15(6)** 2940–2959.

[9] Kılkış Ş, Wang C, Björk F, and Martinac I 2017 Cleaner Energy Scenarios for Building Clusters in Campus Areas Based on the Rational Exergy Management Model. *Journal of Cleaner Production*. **155**, 72–82. doi:10.1016/j.jclepro.2016.10.126

[10] Mahlia T M I, Razak H A, and Nursahida M A 2011 Life cycle cost analysis and payback period of lighting retrofit at the University of Malaya. *Renewable and Sustainable Energy Reviews*. **15(2)** 1125–1132. doi:10.1016/j.rser.2010.10.014

[11] Escobedo A, Briceño S, Juárez H, Castillo D, Imaz M. and Sheinbaum C 2014 Energy Consumption and GHG Emission Scenarios of A University Campus in Mexico. *Energy for Sustainable Development*. **18(1)** 49–57. doi:10.1016/j.esd.2013.10.005

[12] Di Stefano J 2000. Energy Efficiency and the Environment: The Potential for Energy Efficient Lighting to Save Energy and Reduce Carbon Dioxide Emissions at Melbourne University, Australia. *Energy*. **25(9)** 823–839. doi:10.1016/S0360-5442(00)00015-3

[13] Chung M H and Rhee E K 2014. Potential Opportunities for Energy Conservation in Existing Buildings on University Campus: A Field Survey in Korea. *Energy and Buildings*. ***78*** 176–182. doi:10.1016/j.enbuild.2014.04.018

[14] Tang F E 2012. An Energy Consumption Study for a Malaysian. *International Journal of Environment, Earth Science and Engineering*. **(8)** 99–105.

[15] Zhou X, Yan J, Zhu J, and Cai P 2013 Survey of Energy Consumption and Energy Conservation Measures For Colleges And Universities In Guangdong Province. *Energy and Buildings*. **66** 112–118. doi:10.1016/j.enbuild.2013.07.055

[16] Ayala E, Kocurek J, Wright J, Sijansky L, Chang J, and Jr R C 2011 Energy Efficient Method Identification for an Education Building. *American Society of Continuing Education*. 235–243.

[17] Saidur R, Mekhilef S, Ali M B, Safari A, and Mohammed H A 2012 Applications of Variable Speed Drive (VSD) In Electrical Motors Energy Savings. *Renewable and Sustainable Energy Reviews*. **16(1)** 543–550. doi:10.1016/j.rser.2011.08.020

[18] Mohit O K, and Oree V 2013 *A*ssessing The Energy Savings Potential In Public Buildings Through Retrofit Measures In Tropical Climates – A Case Study In Mauritius. *IEEE*. 4673-5943.

[19] Flyn E 2011 *Stanford University the Building Energy Retrofit Programs*. Sustainable Endowments Institute.

[20] Wong W P, Fellows R F, and Liu A M M 2006 Use of electrical energy in university buildings: a Hong Kong case study. *Facilities*. **24(1/2)** 5–17. doi:10.1108/02632770610639161

[21] Abolarin S M, Gbadegesin A O, Shitta M B, Yussuff A, Eguma C A, Ehwerhemuepha L, and Adegbenro O 2013 A Collective Approach to Reducing Carbon Dioxide Emission: A Case Study of Four University Of Lagos Halls Of Residence. *Energy and Buildings*. **61** 318–322. doi:10.1016/j.enbuild.2013.02.041.

[22] Jomoah IM, Al-Abdulaziz AU, and Sreerama,K 2013 Energy Management in the Building of a University Campus in Saudi Arabia-A Case Study. *International Conf on Power Engineering, Energy and Electrical Drives*. 13-17 May. Istanbul, Turkey: IEEE, 659-663.

[23] Chee Yu J L 2014 *OTTV Calculation and Energy Simulation Technique for GreenRE Rating System*. Integrated Environmental Solutions.

[24] Danby D, Menter A, and Faludi J 2011 *Passive Design Strategies*. Retrieved From: http://sustainabilityworkshop.autodesk.com/buildings/passive-design-strategies

[25] Wu M H, Ng T S and Skitmore, M R 2016 Sustainable Building Envelope Design by Considering Energy Cost and Occupant Satisfaction. *Energy for Sustainable Development*. **31** 118–129.

[26] Stapleton G, Neill,S, Milne,G, Reardon C, and Riedy C 2013 Solar Photovoltaic. *Australian Government Department of Industry*. 375-380.

[27] Syarifah Fairuz SF, Byrd H 2012 *Energy and Building Control Systems in the Tropics*. 1st edition. Pulau Pinang: Universiti Sains Malaysia.

[28] Aman M M, Solangi K H, Hossain M S, Badarudin A, Jasmon G B, Mokhlis, H., and Kazi S. N 2015. A review of Safety, Health and Environmental (SHE) Issues of Solar Energy System. *Renewable and Sustainable Energy Reviews*. **41** 1190–1204. doi:10.1016/j.rser.2014.08.086

[29] Li D H W, Yang L and Lam J C 2013 Zero energy buildings and sustainable development implications - A review. *Energy*. **54** 1–10. doi:10.1016/j.energy.2013.01.070.

[30] Zhao Z and Huang W 2011 Multi-objective Decision-making on Wind Power Projects Based on AHP Method. *2011 International Conf on Computer Distributed Control and Intelligent Environmental Monitoring*. 242–245. doi:10.1109/CDCIEM.2011.430
[31] Rivkin DA, Anderson LD, and Silk L 2013. *Wind Turbine Control Systems*. United States. United States: Jones and Barlett Learning
[32] Wizelius T 2015 Developing Wind Power Projects: Theory and Practice. United States USA: Routledge.
[33] Supriyasilp T, Pongput K, and Boonyasirikul T 2009 Hydropower Development Priority Using MCDM Method. *Energy Policy*. **37(5)** 1866–1875. doi:10.1016/j.enpol.2009.01.023
[34] Uhunmwangho R, Ph D and Okedu E K 2009 Small Hydropower for Sustainable Development. *The Pacific Journal of Science and Technology*. **10(2)** 535–543.
[35] Pallant J 2014 *Spss Survival Manual. A step by Step Guide to Data Analysis Using SPSS for Windows (Version 12)*. (1st edi). New South Wales: Allen and Unwin.
[36] Williams B, Brown T Onsman A 2012. Exploratory Factor Analysis: A Five-Step Guide for Novices. *Australasian Journal of Paramedicine*, 8(3).
[37] Kaiser H 1974. An index of factorial simplicity. *Psychometrika*. **39** 31–36.
[38] Dien J, and Frishkoff G 2005 *Introduction to Principal Components Analysis of Event-Related Potentials*.1–42. Retrieved from: https://www.researchgate.net/publication/228449904_Introduction_to_Principal_Components_Analysis_of_Event-Related_Potentials
[39] Maletta H, and Aires B 2007 Weighting. *Spsstools*. 1–19. Retrieved from http://www.spsstools.net

Acknowledgments

This work was financially supported by Research University Grant 18H12, Research Management Centre, and research team members Green PROPMT. Thus, directly it is supported by Ministry of Education Malaysia.

The factors impacted to local contractor from Foreign Direct Investment in advancing economic hub development in Iskandar Malaysia

Muhamad Syafiq Salim[1], Rozana Zakaria[2], Eeydzah Aminuddin[3], Abdul Rahim Abdul Hamid[4], Redzuan Abdullah[5] and Jam Shahzaib Khan[6]

[1,6] *Postgraduate Student, Faculty of Civil Engineering, Universiti Teknologi Malaysia*
[2,5] *Associate Professor, Faculty of Civil Engineering, Universiti Teknologi Malaysia*
[3,4] *Senior Lecturer, Faculty of Civil Engineering, Universiti Teknologi Malaysia*
Corresponding email: rozana@utm.my

Abstract. Iskandar Malaysia is an advanced economic hub which is rapidly growing in the State of Johor. It has been an attractive place for Foreign Direct Investment (FDI) to invest. Many sectors are affected by the presence of FDI including the construction sector. This paper highlights the investigation on the effects of FDI to the local contractor in the Iskandar Malaysia Development. In this study, a questionnaire survey was carried out to gain the information on problems from internal factors and external factors that caused the limitation on involvement in FDI project by local contractors. 73 numbers of local contractor registered under CIDB in class G5, G6 and G7 are the respondents. Frequency analysis and Average Index Analysis are used for the results. This study provides the factors that impacted local construction players in Iskandar Malaysia Development. This study has portrayed that FDI plays a vital and significant role in spearheading the active involvement of local contractors in an urban sustainable development.

1. Introduction

Foreign Direct Investment (FDI) brought three major impacts on factors of productivity, employment creation and upgrading technological and managerial capabilities. Besides, FDI brings 'tsunami money' to the intention of economic development of a nation. FDI is essential at certain situations where the government sees it as a necessity for the development to be carried out. The arrival of FDI will encourage inflow of advanced production technology and input brought about by foreign-owned firms or through imports. These measures might be more effective to improve the efficiency and competitiveness at national economy [1]. One of the main impacts of FDI on local suppliers is the increase in productivity as a result of certain benefits of the relationship [2]. In addition to the presence of FDI, it will have more impact on local companies. This will increase the productivity, improve knowledge of management and local companies thus at the same time increase efficiency when dealing with foreign developers [1]. The deal later on resulted in many businesses and projects established in short and long run. For example, about 702,000 jobs have been created in various economic sectors, including the services sectors such as construction and property since the inception of Iskandar Malaysia [3]. [4] stated that foreign investment brings the technology that the local companies can learn and adapt. These measures might be more effective to improve the efficiency and competitiveness at national economy [1].

2. Iskandar Malaysia Economic Development Hub

On 30 July 2006, the 5th Prime Minister of Malaysia announced that Johor Bahru, Malaysia and its metropolitan region i.e The South-Johor Economic Region (SJER) was chosen to be the second economic corridor for the country after the Multimedia Super Corridor (MSC Malaysia) which was developed in Kuala Lumpur. A comprehensive development plan was put in place, together with a new-established regional development authority directly reporting to the cabinet of the Malaysian

Prime Minister. SJER was later renamed as "Iskandar Malaysia" honoring the name of the late Sultan of Johor [5]. Johor Chief Minister has committed to make Johor State to become more significant in delivering various socio-economic initiatives in Iskandar Malaysia. The Iskandar Malaysia development reflects Johor state's mission and commitment in providing inclusive and sustainable development for the citizen in various fields including social and economic aspects [6].

The investment in Iskandar Malaysia at first was estimated about RM 47 billion where their judgment is necessary to support the initial roll-out of the Iskandar master plan between 2006 and 2010, and a total of RM 4.3 billion was allocated by the Ninth Malaysia Plan in order to build the necessary infrastructures [6]. To complete the project by 2025, the plan requires another RM 335 billion from public and private investments. Private investments will be facilitated by UEM Land i.e the master developer of Nusajaya, other Government Linked Companies, and FDI. At end of 2016, a total cumulative committed investment of 40% (RM 86.71 billion) came from FDI. The top five countries with highest investments into Iskandar Malaysia from January to September 2016 are China, Singapore, USA, Japan and Spain [6]. The creation of IRDA plays a significant role in promoting and stimulating the Iskandar Malaysia Development as a trade, investment and logistics centre, duty free area and a tourist destination among FDI [7]. In opening year 2016, China remains the biggest investor in this rapid developing region economy. Iskandar Malaysia has investments amounted RM 22.2 billion so far by China. This investments from FDI provides employment opportunities to the local people and simultaneously produce high-skilled workers in various fields. Development that has been drafted indirectly will give opportunities to the birth of the country's human resources. The development will be completed in 2025 and that will put Iskandar Malaysia as one of the world-class sustainable city and make Johor as the country's economic hub.

3. The Research Problem Overview

Iskandar Region Development Authority (IRDA) is the regular authority mandated to plan, promote and facilitate development in Iskandar Malaysia. It is estimated that by 2025, the population in Iskandar Malaysia Development will increase to three million, and with this figure IRDA estimated at least 1.2 million housing units need to be built in this developing region [8]. By 2025, the population in Iskandar Malaysia will reach to three million, including 1.4 million in the workforce. Based on this assumption, the development of Iskandar Malaysia needs about 1.2 million housing units, including strata and landed properties. Today, the housing stock in Iskandar Malaysia is slightly less than 700,000, which is short of 500,000 units for the next 10 years [8]. This demand creates lots of opportunities to construction players for both local and foreign investors. The current foreign developers in Iskandar Malaysia are CapitaLand, China Country Garden and Walker Corp [9]. The Chairman of Tourism, Trade and Tourism Minister of Johor stated that about 11% of the RM 203 billion total investments in the region is contributed from the construction sector [10]. The development investment that has been announced given impacts to many sectors on Iskandar, especially when it comes to construction. With a large amount of investment, it provides job opportunities for local contractors. IRDA has awarded a total of RM 1.52 billion or 76% of the contract value of infrastructure projects in Iskandar Malaysia to Bumiputera contractors since 2006 [11].

In contrast, [12] claimed that multiple cases in developing countries shows that the admission of FDI does not give a lot of benefits to the local companies including construction players. The lack of spillovers to local firms was attributed to a number of factors, including limited hiring of local employees in higher level positions, very little labor mobility between domestic firms and foreign subsidiaries. Furthermore, *Yang Dipertua Persatuan Kontraktor Melayu Malaysia Negeri Johor* (PKMMNJ) stated that due to the development is now be developed by foreign developers, it caused problems to local contractors due to limited tender from government projects. This issue needs to be solved so that the local contractors can get construction projects. This problem should be considered and tackle by the local authorities [12].

The bigger demand on property development creates needs for FDI to involve in Iskandar Malaysia Development. Concurrently, FDI offers an important role to develop the creation of Iskandar Malaysia. Indeed, the arrival of foreign developers provide a wide impact to local construction industry players. Will the adaption of FDI give impacts on economic demographic in Iskandar Region, especially to local contractors who are the construction industry players? Since the data released by the IRDA is limited, whereby IRDA is still developing and collecting its historical data on the impacts of FDI to the development of Iskandar Malaysia economic hub. This study leads to the investigation of problems from internal and external factors faced by local contractors impacted from market demand of FDI in construction projects that spearheaded by foreign investors through FDI.

4. Literature Review

The presence of FDI to the Iskandar Malaysia will give impacts by increasing the existing investment and this situation will provide lots of opportunities for local players and one of the major impacts is construction. The presence of FDI in Iskandar Malaysia Development created competition between local contractors and foreign contractors. Many case studies show that in order to modernize, the local construction industries and clients in global construction markets, particularly in developing countries require project financing. Also a high level of technology and a foreign firm's advanced experience with knowledge and management skills are required [13]. Compared with foreign developers, local construction companies may have limited absorptive capacity to recognize and adapt the new technology or management skills from the multinationals [4]. This is one of the factors that reduce local contractors' competition to involve in FDI projects. In case of Iskandar Malaysia Development, many researchers found out that at developing countries, FDI ownership may influence the effects and tendency of local companies' sourcing [1]. That means FDI also can take action to prevent distribution of knowledge to local companies by ways of paying higher wages to foreign companies to prevent local companies from bidding.

In addition, local companies have limited capacity to adapt and recognize new technology and art of management skills from big multinationals foreign companies [4]. For internal factors, [14] stated that the major causes of construction company failures are: (1) Delay in collecting payments from clients (donors); (2) Closure; (3) Depending on banks and paying high profits; (4) Lack of capitals; (5) Cash flow management; (6) Lack of experience in the line of work; (7) Absence of construction regulations; (8) Low margins of profit due to competition; (9) Awarding contracts to the lowest price; and (10) Lack of experience in contracts. Throughout a project management life cycle, cost is the most considerable and prime factor of success. However, it is common to see often the project is finished with over budget [15]. This happens due to contractor's lack of experience; cost of material, fluctuation in the prices of materials, frequent design changes, economic instability, and high interest rates charged by banks on loan. This also triggered by mode of financing, bonds payments as well as fraudulent practices and kickbacks as a factor that causing cost overrun [16]. From the other view, [17] found out that the connection between poor site management system, financial problems, and design changes is the most significant reason that causes cost overrun. Things that have been discovered that because poor inspection program, lack of safety policy and education programmes, and unsafe working condition is caused by poor management system that impacts on cost overrun [17]. Similarly, [18] mentioned that one of problems faced by local contractors is unethical behaviour by the person involved in construction industry that gives the impact on quality of the project. The increase of unethical behaviour cause consequential decline in quality of project performance as shown by statistics from construction sites [19].

4.1. Strategy to Enhance Local Contractor Competitiveness

The construction companies nowadays are very dynamic, forcing the contractors to use right strategic decisions in doing the business [20]. Furthermore, their existence is depending on the contractor's capability to grab opportunities [21]. Local contractors' companies need to create mechanisms to increase their involvement in FDI projects. The mechanism is set in such a way that the local construction companies have to expose themselves to foreign developers' activities. At the same time, it can be observed that these firms' technologies imitate their way to operations and management practices that increase local construction productivity [22]. Contractors must create domestic linkages. This is because when companies create linkages, knowledge from these firms can be transmitted to the suppliers and distributors, and ultimately to domestic firms using the same suppliers and distributors [23]. In other perspective, the increase in competition that accompanies FDI entry can force domestic firms to increase their productivity by updating manufacturing technologies and adapting advanced management practices to meet these competitive challenges [24]. This is one of the strategies that can be adapted to increase the involvement of local companies in the development of the Iskandar Malaysia. The project participants realized that sharing of knowledge and information is one of the key elements of a successful contractual relationship. There appears to be no clear guide on the process of collaboration between the main contractor and subcontractor, therefore it makes it difficult to effectively interact and achieve a common project goals within the bounds of cost, quality and time [24]. [25] discussed collaborative relations from the angle of strategic alliances and noted that due to the ever-increasing pace of technological developments and access to new technologies, alliances have become a key success factor. Various things can be considered as a key factor for local contractor to involve in FDI projects.

5. Methodology

This research constitutes evaluation and analyses on the data collected though quantitative approaches by method of questionnaire survey delivered to the respondents which are constitutes of contractor Grade G5 to G7. The was questionnaires conducted to gain answer of the followings objectives (i) to identify the impacts of FDI to construction industry in Iskandar Malaysia and (ii) to investigate problem faced by local contractor from market demand of FDI. The results were analyzed using Frequency Analysis and Average Index.

5.1 Frequency Analysis

Frequency analysis is one of methods that is used to analyze data. This method of analysis is used for part A of questionnaire. Below is the formula for Frequency Analysis:

$$\textbf{Percentage (\%)} = \textbf{(n/N) x 100\%}$$

Where:

n = Number of respondent
N = Total numbers of respondents received

5.2 Average Index

Average Index (AI) is used to analyse the data collected from respondent. [26] stated that the formula for Average Index as follows:

$$\textit{Average Index (AI)} = \frac{\Sigma(\mu \times n)}{N}$$

μ = Weighting given to each factor by respondents (1 to 5);
n = Frequency of the respondents;
N = Total number of respondents.

Whereby the weighting would be:

$\mu 1$ = 1, frequency of "Definitely Disagree" response
$\mu 2$ = 2, frequency of "Mostly Disagree" response
$\mu 3$ = 3, frequency of "Moderate" response

$\mu4$ = 4, frequency of "Mostly Agree" response
$\mu5$ = 5, frequency of "Definitely Agree" response

[27] stated that the classification of the rating scale is like following:

1) Definitely Disagree	:	$1.00 \leq$ Average Index < 1.50
2) Mostly Disagree	:	$1.50 \leq$ Average Index < 2.50
3) Moderate	:	$2.50 \leq$ Average Index < 3.50
4) Mostly Agree	:	$3.50 \leq$ Average Index < 4.50
5) Definitely Agree	:	$4.50 \leq$ Average Index ≤ 5.00

There are 73 respondents representing local contractors that live and work within area of Iskandar Malaysia Development. Majority of them are in executive positions that have responsibility in managing their companies. About 60.3% of respondents are between age 20 to 29 and this represent the highest percentage of respondent and the lowest is 50-59 years old represent 1.4% of respondents. Based on the collected data, 69.9% of respondents represent the male and the rest are women. Majority of them have experience between 1 to 5 years that represent about 57.5% and the lowest percentage is respondents under categories 11 to 15 years and 20 and above which represent 4.1% each. Based on the demographic, study has found out that 64.4% of respondents are from grade G7 Construction Companies while for the G6 and G5 are 5.5% and 30.1% respectively. 70% respondents have experience working under foreign developers and others 30% has no work experience under foreign developers. Table 1 below shows the position of the respondents.

Table 1. Position of respondents

Position	Freq. (N)	(%)	Cum. (%)	Position	Freq. (N)	(%)	Cum. (%)
General Manager	4	5.5	5.5	**Quantity Surveyor**	3	4.1	89.0
Project Manager	8	11.0	16.4	**Director**	2	2.7	91.8
Site Manager	10	13.7	30.1	**Resident Engineer**	1	1.4	93.2
QA/QC Engineer	3	4.1	34.2	**Project Admin**	2	2.7	95.9
M&E Engineer	3	4.1	38.4	**Safety Officer**	1	1.4	97.3
Civil Engineer	11	15.1	53.4	**Development Officer**	1	1.4	98.6
Project Engineer	4	5.5	58.9	**Contract Admin**	1	1.4	100
Supervisor	13	17.8	76.7	**TOTAL**	**73**	**100**	
Site Engineer	6	8.2	84.9				

6. Results and Discussion

Table 2 shows the finding of respondents on the impacts of FDI to construction industry at Iskandar Malaysia and all the components resulted in more than 3.50 of the average index and all are categorized under the criteria of "mostly agree". The top impact on construction industry due to FDI is to create high competition among local contractors which is found as 4.07 followed by the increase in number of construction projects due to property demand invest by FDI that has a 3.93 average indexes. The lowest impact of FDI to construction industry at Iskandar Malaysia Development is to create opportunities to local contractors' involvement with an average index of 3.59 followed by an average index of 3.64 that is represented by more efficient working time management.

IOP Conf. Series: Earth and Environmental Science **143** (2018) 012034 doi:10.1088/1755-1315/143/1/012034

6.1 Impacts of Foreign Direct Investment

Table 2. Ranking for FDI Impacts

Ranking	The Impacts of Foreign Direct Investment	Description	Average Index
1	Create high competition among local contractors	Mostly Agree	4.07
2	Increase number of construction project due to property demand invest by FDI	Mostly Agree	3.93
3	Create job opportunities to local people	Mostly Agree	3.88
4	Transfer the knowledge of way of work	Mostly Agree	3.85
5	Create international network between local and international companies	Mostly Agree	3.85
6	Improve knowledge of management	Mostly Agree	3.84
7	Local professionals can involve in construction development	Mostly Agree	3.81
8	FDI bring technologies to local companies	Mostly Agree	3.75
9	Local construction suppliers can increases productivity	Mostly Agree	3.74
10	Systematic system flow of work	Mostly Agree	3.71
11	Developer companies bring skilled workers to local industries	Mostly Agree	3.68
12	Local companies can learn good work ethics from foreign company	Mostly Agree	3.68
13	More efficient working time management	Mostly Agree	3.64
14	Create opportunities to local contractors' involvement	Mostly Agree	3.59

In addition, Table 3 and Table 4 portrays the problems faced by Iskandar Malaysia local contractor by looking into the internal factors and external factors.

6.2 Problems Faced By Iskandar Malaysia Local Contractors

From Table 3, it can be seen that financial criteria scored the highest average index with a value of 3.54, followed by government criteria with an average index of 3.47, the human capital and policy criteria with average index value of 3.19 and 2.73 respectively. The fact that respondents are highly agreed that financial is the primary problem for internal factor faced by local contractors in Iskandar Malaysia Development.

For external factors shown in Table 4, government supervision is the criteria that has highest score with an average index of 4.07, followed by human capital with an average index of 3.90, the financial and culture criteria with an average index value 3.89 and 3.40 respectively. From the results, respondents often agree that subject related to government supervision criteria is the external problem faced by local contractors to participate in Iskandar Malaysia Development.

I. Internal Factors

Table 3. Problems from Internal Factors by Criteria

Criteria	Problems *(Internal Factors)*	AI	Description	Criteria AI
Financial	Financial difficulties of owner	3.64	Mostly Agree	3.54
	Local contractor lack of financial resources	3.49		
	Lack in capitals they need the support of foreign capitals	3.48		
Government Supervision	The government does not regulate the influx of foreign contractors in Iskandar Malaysia	3.47	Moderate	3.47
Human Capital	Local contractors lack of experience involve FDI projects	3.23	Moderate	3.19
	Do not have technical capacity	3.14		
	Poor safety policies and lack of safety education programmes	2.88		
Management Policy	Local contractors have bad past performance on quality workmanship	2.66	Moderate	2.73
	Local contractors lack in managing & coordinating the project site	2.64		

II. External Factors

Table 4. Problems from External factors

Criteria	Problems *(External Factors)*	AI	Description	Criteria AI
Government Supervision	Local contractors have to compete with foreign contractors	4.12	Mostly Agree	4.07
	Foreign companies come in with their own labour	4.05		
	Foreign companies prefer contractors originating from their own country	4.04		
Human Capital	Language problem as one of the largest issues dealt with foreign companies	4.22	Mostly Agree	3.9
	Foreign companies can afford to carry out such expense of high skills of worker	3.93		
	Foreign companies have technical capacity and needed no support team from local expertise	3.55		
Financial	Have strong financial resource by foreign companies cause foreign developers to have power of choosing their own sub-contractor	3.89	Mostly Agree	3.89

Culture	Foreign companies have perception that local companies are not suitable to be sub-contractors	3.40	Moderate	3.4

7. Conclusion

From the results and discussion, there are several conclusions that can be made. The higher impact of FDI to construction industry created high competition among local contractors to involve in development of Iskandar Malaysia. Other than that, the presence of FDI gives positive impacts by offering more job opportunities, bring new and latest technology (compared to current local application) to local industry, and good project management. The lowest impact is creating opportunities to local contractors' involvement. The highest criteria that become the primary problem to local contractors in terms of internal factors are related to financial reasons. For the external factor, the primary problem that was brought by FDI is related to lack of government supervision. The study provides factors impacted to construction industry in terms of social-economy among local construction players in Iskandar Malaysia Development. The result of this study helps as a reference to various parties to increase the participation of local contractors in the projects that are spearheaded by FDI. The study served a thought to create positive impacts to both local and foreign construction players in order to gain good outcome from the experienced development of Iskandar Malaysia as an economic hub.

8. References

[1] Liang, F, H, (2017). Does foreign direct investment improve the productivity of domestic firms? Technology spillovers, industry linkages, and firm capabilities. Research Policy, **Volume 46**, pp. 138-159
[2] Ghauri, P, N, & Firth, R (2015). The Impact of Foreign Direct Investment on Local Firms: Western Firms in Emerging Markets. The Future of Foreign Direct Investment and the Multinational Enterprise. pp 379-405
[3] Iskandar property outlook seen positive. (2017 Feb 14) The Star Online
[4] Blalock, G, Gertler, P, (2009). How firm capabilities affect who benefits from foreign technology. J. Dev. Econ. **90 (2)**, 192–199.
[5] Rizzo, A, & Glasson, J (2012). Iskandar Malaysia. *Cities, 29*, 417-427
[6] Iskandar Malaysia Secures Rm10.85 Billion In New Investments For Q3 2016. (2016, Nov 1) Iskandar Malaysia.
[7] Iskandar Regional Development Authority (IRDA), (2007). Information pack. JB. February.
[8] Growing population spur demand in Iskandar Malaysia (2017, April 20) New Straits Time
[9] IRDA, 2013. Official website of Iskandar Regional Development Authority (IRDA) retrieved from www.iskandarmalaysia.com.my. on 15 July 2013.
[10] PKMMJ minta kerajaan negeri pantau Iskandar (2014, July 23), Utusan Malaysia
[11] Iskandar Malaysia attracts total investment of RM32.15bil (2014, March 22) The Star Online.
[12] Aitken, J and Harrison, E (1999). Do Domestic Firms Benefit from Direct Foreign Investment? Evidence from Venezuela. The American Economic Review
[13] Mahalingam, A & Levitt, R E (2005). Understanding and mitigating challenges on global projects: the role of the freelance expatriate. In Proceedings of the ASCE Construction Research Congress 2005.
[14] Enshassi, A, Al-Hallaq, K and Mohamed, S (2006), "Causes of contractor's business failure in developing countries: the case of Palestine*", Journal of Construction in Developing Countries*, **Vol. 11 No. 2**, pp. 1-14
[15] Azhar, N, Farooqui, R U., & Ahmed, S M (2008). Cost overrun factors in construction industry of Pakistan. Proceedings from First International Conference on Construction in Developing Countries

[16] Ameh, J, Soyingbe, A, & Odusami, T (2010). Significant Factors Causing Cost Overruns in Telecommunication Project in Nigeria. *Journal of Construction in Developing Countries*, **Vol. 15(2),** 49–67, 2010
[17] Le-Hoai, L, Lee, Y D, & Lee, J Y (2008). Delay and Cost Overruns in Vietnam Large Construction Projects: A Comparison with Other Selected Countries. *KSCE Journal of Civil Engineering,* **12(6)**, 367-377.
[18] Adnan. H, Hashim N, Yusuwan M Y, & Ahmad N, (2011). Ethical Issues in Construction Industry: Contractor' Perspective. Asia Pacific International Conference on Environment-Behaviour Studies, Salamis Bay Conti Resort Hotel, Famagusta, North Cyprus, 7-9 December 2011
[19] Rahman, H A, Karim, S B A., Danuri, M S M Berawi, M A, Yap, X W (2007). Does professional ethic affects construction quality?. Quality Surveying International Conference. Kuala Lumpur, Malaysia.
[20] Izik Z Arditi, David. Irem Dikmen & M Talat Birgonul. (2010). Impact of Resources and Strategy on Construction Company Performance, *Journal of Management in Engineering*, **Vol. 206**, No.1, pp. 9-18
[21] Lansley, P R (2002). Coorporate Strategy and Survivel in The UK Construction Industry. *Construction Management Economic Journal*, **Vol. 5**, pp 141-155
[22] Blomstrom, M and Kokko A (1998) 'Multinational Corporations and Spillovers', *Journal of Economic Surveys*, **12(2)**: 1-31.
[23] Spencer, J W (2008). The impact of multinational enterprise strategy on indigenous enterprises: horizontal. Spillovers and crowding out in developing countries. Academy of Management Review, 33, 341–361
[24] Siti Hamidah A R, Intan R E, Nasruddin F, Soleyman P (2013). The Importance of Collaboration in Construction Industry from Contractors' Perspectives. International Conference on Innovation, Management and Technology Research, Malaysia, 22 – 23 September, 2013
[25] Douma, M U, Bilderbeek, J, Idenburg, P J and Looise, J.K. (2000), Strategic alliances: managing the dynamics of fit. *Long Range Planning*, **Vol. 33**, pp. 579-98
[26] Al-Hammad and Sadi Assaf (1996). Assessment of Work Performance of Maintenance Contractors in Saudi Arabia, *Journal of Management in Engineering*,**Vol. 12(2)**.
[27] Mc Caffer & Abd Majid (1997), Non-excusable delay in construction, cited from: https://core.ac.uk/download/pdf/2743478.pdf

Acknowledgments

This work was financially supported by Research University Grant 18H12, Research Management Centre, and research team members Green PROPMT. Thus, directly it is supported by Ministry of Education Malaysia.

IOP Conf. Series: Earth and Environmental Science **143** (2018) 012035 doi:10.1088/1755-1315/143/1/012035

Effect of in-situ disturbance within the soil mass on the stress-strain behaviour of silty soil

Sarah T Noor[1], SS Rabika Rahman[2] and Sabiqun Nahar[2]

[1] Department of Civil Engineering, University of Asia Pacific, Dhaka, Bangladesh
[2] Department of Civil Engineering, Military Institute of Science and Technology, Dhaka, Bangladesh

Email: sarah@uap-bd.edu

Abstract. To date, different techniques have been evolved to collect soil in undisturbed condition so that the in-situ soil behaviour can be determined by carrying out laboratory tests. For the same reason, the execution of undisturbed soil sampling in practice is given a lot of efforts. However, this study brings the fact into consideration that the in-situ soil condition may not remain constant, rather it might vary time to time, because of different internal or external reasons. For example, the internal stress state of soil layers, existing below or above the swelling soil layer, become modified during shrinking and swelling resulting from drying and wetting of swelling clay, respectively. Further, foundations of building may transfer cyclic loads (generated by vibration installed in the building) to the soil below the foundation. Therefore, this study investigates the effects of stress-strain behaviour due to the disturbances on the shear strength of the soil with respect to that of undisturbed specimens. The shear strength of disturbed soil shows deviation from that of undisturbed specimen depending on the different parameters defining the severity of disturbance.

1. Introduction

To date, problematic soils, such as liquefiable soil, sensitive clay, collapsible soil and expansive clay, have drawn the attention of researchers, as the performance of foundations on them is severely affected due to a change, such as occurrence of cyclic loading, variation in moisture content, etc. [1-4]. After 2018-Taiwan and 1995-Kobe earthquakes, structures were found to collapse due to bearing capacity failure of foundations. The disturbance that changes in-situ soil stress-state, might cause reduction in soil shear strength and accordingly such catastrophic failures. However, in the design of foundation, the stress-strain behaviour and shear strength parameters of soil in undisturbed condition (in-situ and before construction) are considered. Previous studies investigated the effect of disturbances during sampling [5], but the disturbances that might occur after construction has not been investigated in previous studies.

The effect of internal soil stress has significant influence on different aspects of geotechnical properties of soil [6-7]. The present study focuses the disturbances of soil condition (in-situ) that is not during sampling but while supporting the foundations of structures due to different internal or external reasons. The internal stress state of soil layers, existing below or above the expansive soil layer, becomes modified during shrinking and swelling of expansive soil resulting from drying and wetting, respectively. Further, foundations of building may transfer cyclic loads (generated by vibration installed in the building) to the soil below the foundation. Disturbances might be due to cyclic freezing and thawing during change in ambient temperature. Some research attempts regarding the investigation of cyclic freezing and thawing are found in the literature but mostly dealt with clayey soils [8-10].

The objective of the present study is to investigate the effect of disturbances on the shear strength, of the silt soil with respect to that determined in its undisturbed state. The severity and time dependency of disturbances in influencing strength reduction are considered in the experimental investigation carried out in this study.

2. Experimental Program

The present study conducts experimental investigation to study the soil shear strength in disturbed condition with respect to that of undisturbed soil.

2.1. Soil

Undisturbed grey soil samples were collected in Shelby tubes (open drive) from Ex-Marine Workshop ship building area in Barisal, the southern district of Bangladesh, while conducting standard penetration test. The depth of sampling was 4 m (approximately). Figure 1 shows the undisturbed soil in tube.

Figure 1. Undisturbed soil in the tube sampler

The soil was identified to be clayey silt with traces of sand. The soil properties, determined in the laboratory, are given in table 1. The soil is classified as low plastic silt (ML), according to Unified Soil Classification System (USCS).

Table 1. Soil properties

Variables	Value
Percent smaller than No. 200 sieve	> 90 %
Liquid limit (%)	32.5
Plastic limit (%)	28
Plasticity index (%)	4.5
Specific gravity	2.64 – 2.66
Natural moisture content	28 - 34 %
Bulk unit weight (kN/m^3)	17.3
Void ratio	0.51
Compression index	0.18

2.2. Methodology

In order to investigate the effect of disturbances on soil shear strength, the stress-strain behaviour and Mohr-Coulomb failure envelopes of disturbed specimens with respect to those of undisturbed specimens were examined. Therefore, the shear strength parameters of both undisturbed and disturbed

IOP Conf. Series: Earth and Environmental Science **143** (2018) 012035 doi:10.1088/1755-1315/143/1/012035

specimens were tested by conducting consolidated undrained (CU) Triaxial tests according to ASTM standard. The CU Triaxial tests were conducted on three undisturbed and three disturbed soil specimens. All the specimens were initially undisturbed and identical. Table 2 gives the details of the experimental program.

The disturbed specimens were prepared by cyclic freezing and thawing, and thus the degrees of disturbances were controlled. During freezing phase, the tube containing the soil that was collected in undisturbed condition, was kept in freeze (below 0° C) for 8 hrs. Further, the tube was left outside the freeze at room temperature of 22° C for about 5 hours. When some delay periods before testing or inducing further cycles of freezing and thawing were necessary according to the experimental plan, the tubes were kept in the refrigerator (temperature around 10°C).

Table 2. Experimental Program

Borehole No.	No of Test	Type of Specimen	Test Detail
01	2	Undisturbed	Failed by static loading by CU type test
02	1		
03	1	Disturbed	7 cycles of freezing & thawing, failed by static loading immediately after last thawing
	1		7 cycles of freezing & thawing, failed by static loading on 20th day after last thawing
	1		7 cycles of freezing & thawing, left for 20 days, again 5 cycles of freezing and thawing, then immediately failed by static loading

3. Results and Discussion

The deviator stress versus strain behaviour of both the undisturbed and disturbed specimens is compared at two different effective confining pressures (200 kPa and 300 kPa) in figures 2 and 3, respectively. The disturbance effects on the deviator stress at failure are found significant in both cases.

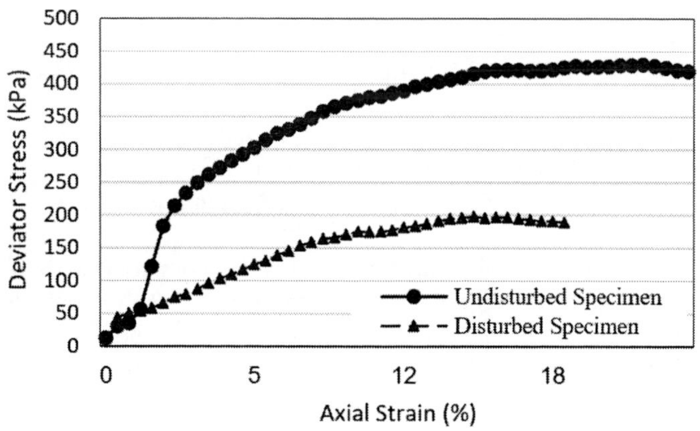

Figure 2. Effect of disturbances: Deviator stress versus strain at effective confining pressure of 200 kPa

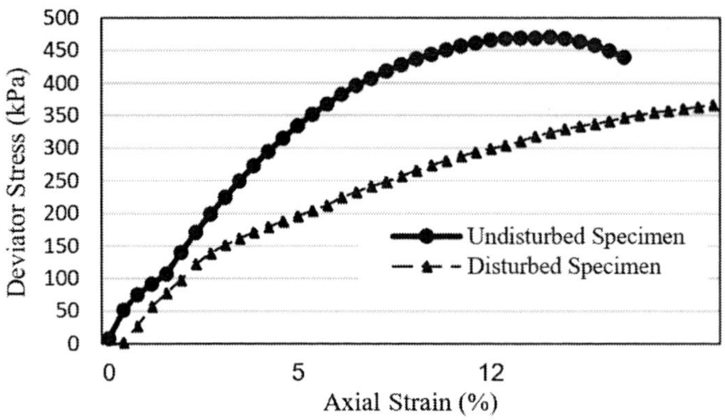

Figure 3. Effect of disturbances: Deviator stress versus strain at effective confining pressure of 300 kPa

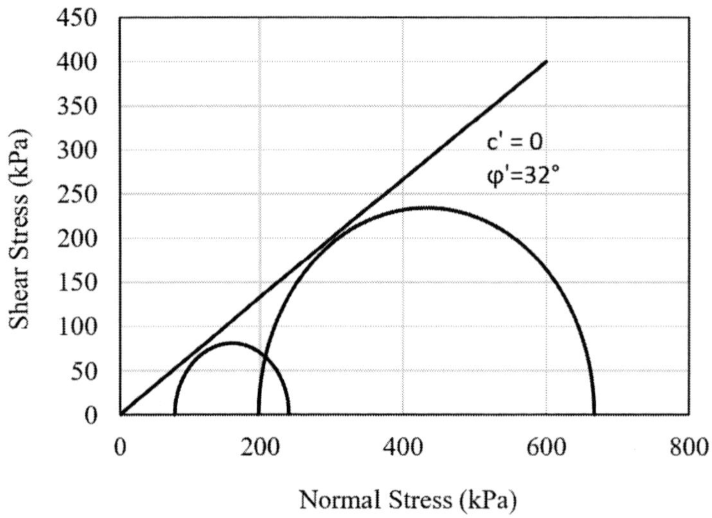

Figure 4. Mohr-Coulomb Failure Envelope in Undisturbed Condition

Further, the disturbance effects on the shear strength parameters are studied with respect to those of undisturbed soil. The Mohr-Coulomb failure envelope of the undisturbed soil used in this study is presented in figure 4. The effective angle of internal friction and the effective cohesion are obtained 32° and 0 kPa, respectively.

The Mohr circle of each disturbed specimen was separately compared with those, along with Mohr-Coulomb failure envelope, of undisturbed specimens in figures 5-7. It can be noted that reduction in shear strength parameters are greatly dependent on the degree of disturbances. Significant reduction in the effective angle of internal friction is observed especially when the specimen was tested after 20 days since last thawing (figure 7), while the effect of disturbance on shear strength parameter is found insignificant immediately after thawing (figure 5 and figure 6).

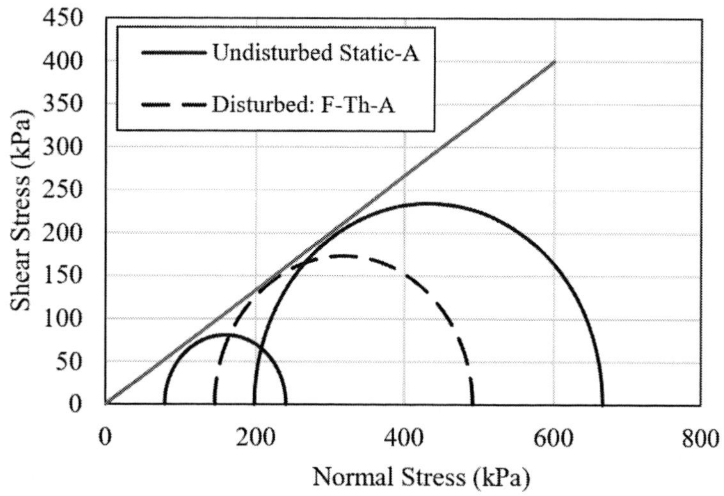

Figure 5. Effect of 7 cycles of freezing & thawing, Triaxial test conducted immediately after thawing

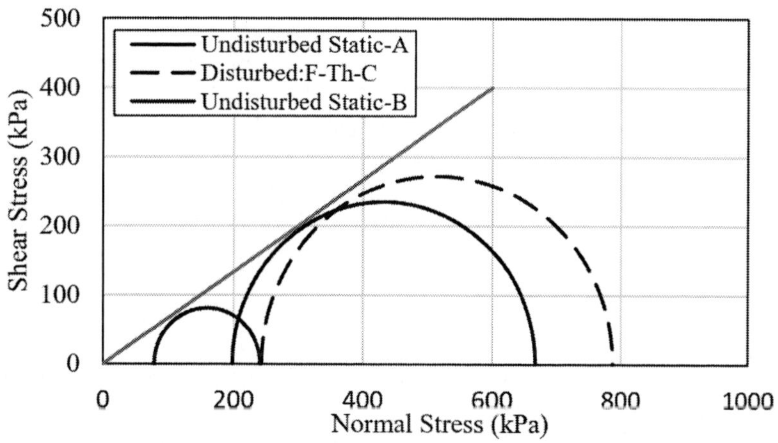

Figure 6. Effect of 7 cycles of freezing & thawing, then 20 days later again freeze and thawed 5 times, Triaxial test conducted immediately after thawing

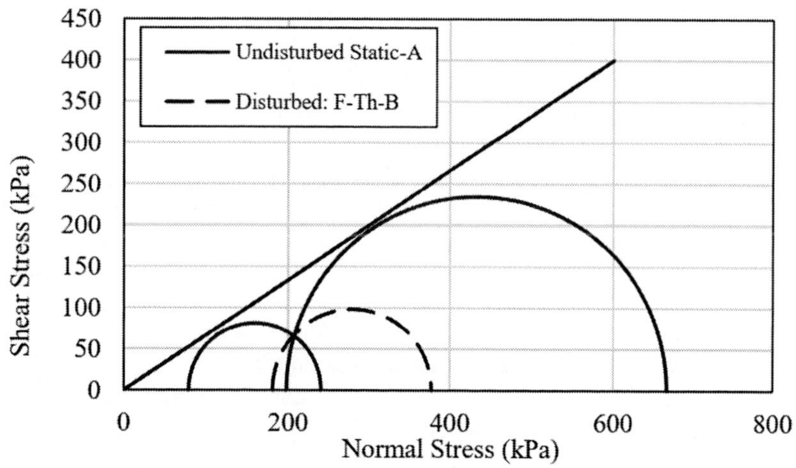

Figure 7. Effect of 7 cycles of freezing & thawing, Triaxial test conducted 20 days after thawing

Table 3: Effect of disturbance on effective angle of internal friction

Variable of disturbances		φ′ (°)
No of cycles	Time elapsed between last disturbance and test begins	
0	-	32
7	-	32
7-delay-5	-	27
7	20 days	18

The effect of different degrees of disturbances on the effective angle internal friction is summarized in Table 3. It is found that the greater the degree of disturbances, the greater the reduction of φ′ is. It can also be noted that the strength reduction continues with time to some extent. This implies that the structure will survive till the reduced soil strength could develop bearing capacity equalled the stress transferred to the soil from the foundation. The structure will collapse, when the strength reduction further continues beyond this point.

4. Conclusions

The paper explores the effects of in-situ natural disturbances on the shear strength of the soil supporting the foundation of structures. In this study, only low plastic silt soil was considered. Based on the experimental results and analysis, the conclusions of the study is summarized as below:

i. The greater the degree of disturbance, the greater the reduction of strength is.
ii. The time dependency of the strength reduction, caused by the disturbances within the soil mass, has been observed. Thus, the consequence of disturbances might not be hypothesized based on the stress-state observed immediately after the disturbances occurred.

Further investigation is necessary to explore other sources of natural disturbances with different variable parameters and time dependency of strength reduction of different types of soil.

Acknowledgments
The authors acknowledge Department of Civil Engineering, Military Institute of Science and Technology for the facility and support.

References

[1] Noor S T 2017 Discussion of "The Effects of Fines on the Liquefaction Behavior in Well Graded Material" Canadian Geotechnical Journal Vol. 54 1802-1803
[2] Noor S T 2017 Discussion of "Drag load on End-bearing Piles in Collapsible Soil due to Inundation" *Canadian Geotechnical Journal* Vol **54** 742-743
[3] Noor S T 2017 Numerical and Analytical Modelling for Predicting Drag Load Induced on Pile in Collapsible Soil because of Inundation *The Open Civil Engineering Journal* Vol **11**(1) 664-675
[4] Noor S T, Hanna A M and Mashour I 2013 Numerical Modeling of Pile in Collapsible Soil subjected to Inundation *International Journal of Geomechanics* Vol **13**(4) 514-526
[5] Horng V, Tanaka H, Hirabayashi H and Tomita R 2011 Sample disturbance effects on undrained shear strengths-study from TAKUHOKU SITE Sapporo *Soils and Foundations* **51**(2) 203-213
[6] Vanapalli S K, Fredlund D G and Pufahl D E 1999 The influence of soil structure and stress history on soil-water characteristics of a compacted till *Geotechnique* **49**(2) 143-159
[7] Boulanger R W and Idris I M 2007 Evaluation of cyclic softening in silts and clays. Journal of Geotechnical and Geoenvironmental Engineering **133** 641–652
[8] Ji-Lin, Q and Wei M A 2006 Influence of freezing-thawing on strength of over-consolidated soils. *Chinese Journal of Geotechnical Engineering* **28**(12) 2082-2086

[9] Qi J, Vermeer P A and Cheng G 2006 A review of the influence of freeze-thaw cycles on soil geotechnical properties *Permafrost and periglacial processes* **17**(3) 245-252
[10] Wang D Y, Ma W, Niu Y H, Chang X X and Wen Z 2007 Effects of cyclic freezing and thawing on mechanical properties of Qinghai–Tibet clay. *Cold regions science and technology* **48**(1) 34-43

IOP Conf. Series: Earth and Environmental Science **143** (2018) 012036 doi:10.1088/1755-1315/143/1/012036

Constitutive Models for Design of Sustainable Concrete Structures

J Brozovsky[1], R Cajka[1] and J Koktan[1]

[1] Faculty of Civil Engineering,
VSB-Technical University of Ostrava, Ludvika Podeste 1875/17,
708 33 Ostrava – Poruba, Czech Republic

E-mail: jiri.brozovsky@vsb.cz

Abstract. The paper deals with numerical models of reinforced concrete which are expected to be useful to enhance design of sustainable reinforced concrete structures. That is, the models which can deliver higher precision of results than the linear elastic models but which are still feasible for engineering practice. Such models can be based on an elastic-plastic material. The paper discusses properties of such models. A material model based of the Chen criteria and the Ohtani hardening model for concrete was selected for further development. There is also given a comparison of behaviour of such model with behaviour of a more complex smeared crack model which is based on principles of fracture mechanics.

1. Introduction

Large number of buildings in Europe (and also in other areas) is still based on reinforced concrete structures. There are many efforts to replace the traditional components of concrete by their renewable alternatives or by alternatives obtained by waste recycling processes. Even if we take into account these developments then the concrete still has to be considered as a material which in not easily renewable nor prepared from renewable sources. Thus there are obvious needs to limit its use as much as possible. So sustainable concrete should be used. By the term sustainable concrete, we mean concrete which not only uses secondary materials (slag in this case) instead of non-renewable ones as much as possible but which has long service life.

One of the possible approaches is use of advanced numerical models in the design phase. A high-quality numerical modelling can allow to design structures which should use as little concrete as possible. The most common structural design approach in the civil engineering is still use of linear elastic modelling for analysis of static and dynamic behaviour of bearing structure. Such approach is correct and it is recommended by current technical standards (the Eurocodes in Europe and other countries, for example). But these technical standards also allow to use more up-to-date approaches which should include, among others, assumptions of non-linear material behaviour. It is obvious that increased precision of material modelling should allow to predict more precisely the necessary amount of material and thus it should lead to more effective design.

The paper concentrates on models which should be compatible with author's effort on development of finite element code for design of reinforced concretes structures with enhanced modelling of soil-structure interaction. The developed algorithms are meant to improve numerical modelling on structural level and not only on level of individual structural members. Use of non-linear material models in combination of an additional non-linearity (the soil-structure interaction is a non-linear problem) requires to use algorithms which can be executed in parallel. Such approach can considerably speed-up the computations but it also defines further limitations for numerical models. [1]

2. Considerations

Design of structures like living houses and administrative objects should not allow to expose the used materials to load effects which may cause their progressive damage. For example, there has to be no plastic behaviour of steel structural members (except some special cases like behaviour of bolts and other connecting elements).

In the case of reinforced concrete there should be no progressive cracking nor crushing and material should not excess its peak bearing capacity. It is obvious that such behaviour should not occur in any stage of normal use of such objects.

Thus in macroscopic scale there should be no measurable softening. Obviously, on micro-level the cracking of concrete is inevitable. But presence of reinforcement should prevent the softening of the global reinforced concrete material. That it can be assumed that reinforced concrete structures of living and administrative objects can be designed with use of elastic-plastic material models which usually cannot model softening of material. There are of course exceptions, there are special cases when even more precise models are necessary: objects of high importance have to be designed to endure progressive damages caused by extraordinary loads (explosions, extensive fire and so) and in these cases more precise numerical models, which have to include softening effects [2] or visco-plastic behaviour, have to be used.

3. Material modelling

There are many computational models which utilise elastic-plastic material model. Some of them are implemented in well-known software packages (models by Willam and Warnke [3] or by Lee and Fenves [4], for example), some other models are often used in academic program codes (Chen [5], one of the modifications of the Chen model has been implemented by authors of this article [6]). All these models are represented by very similar plasticity conditions (see Figure 1 for an illustration).

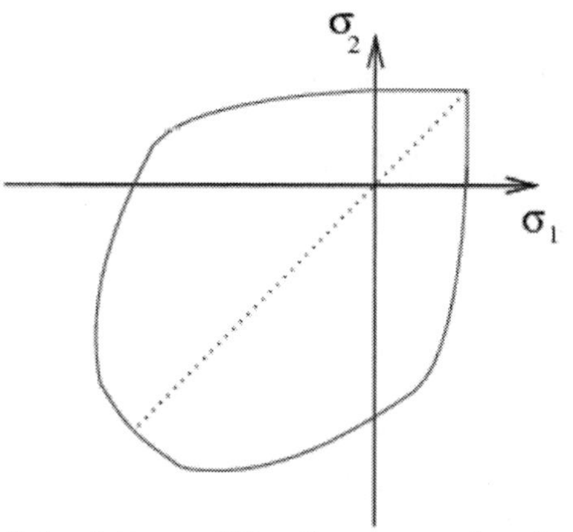

Figure 1. Typical shape of plasticity condition for concrete in principal stresses plane in 2D.

There are more advanced approaches which usually utilise approaches of fracture mechanics. [2] Those solutions often allow to simulate softening behaviour of concrete, sometimes at a cost of simplified modelling of undamaged material behaviour. The most basic approach is the smeared crack model [2]. It uses an assumptions that damaged concrete can be modelled as an orthotropic (or even anisotropic) material with different properties in different material axes. These axes are oriented in accordance with cracks direction. These models are known to be prone quality and size to finite element meshes and there are many approaches how to eliminate this issue [2]. One of the most common ways is to introduce additional material parameters, e.g. fracture energy [7].

In many cases material models use some combination of an elastic-plastic approach (which is used in initial stages of material behaviour) and the smeared crack approach (which is used for damaged material).

4. Elastic-plastic and smeared cracks material models

The authors of this paper have implemented a material model based on the Chen [5] plasticity and failure conditions. The model assumes an elastic-plastic material behaviour. Plastic behaviour is modelled as a hardening with different hardening functions for different stress states of material (an uni-axial compression, a bi-axial compression, an uni-axial tension). These functions are implemented by means of approximation by Ramberg-Osgood function [8].

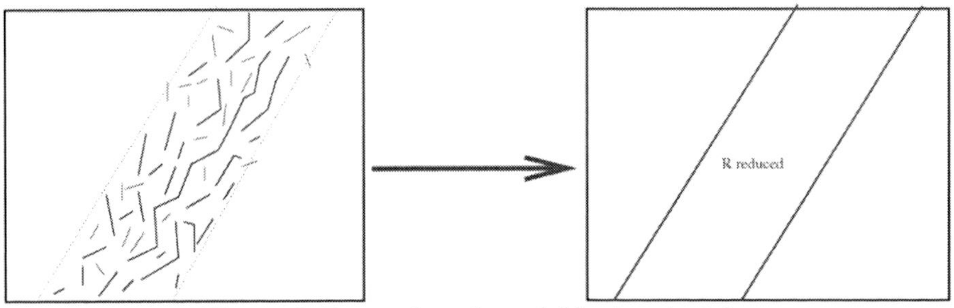

Figure 2. Smeared crack model basic principle.

The second model discussed here is a fracture mechanics based one. The assumptions are similar to [9]. The initial material properties are linear elastic. Material with cracks is modelled as orthotropic and its material axes depends on global cracks directions. But the area where the cracks are simulated remains to be continuous (Figure 2). The material axes directions (Figure 3) are obtained from orientation of principal stresses in the moment of crack initiation. This material model is a subject to finite element mesh-dependent behaviour, Thus the crack band model approach by Bazant [2] is used here. Due to nature of the model which does not simulate material softening it is possible to use the fairly simple Newton-Raphson non-linear computational procedure here.

This material model uses softening and it can result in softening on a structural level. Thus the computational strategy based on the Arc-Length method [10] is used in computations of these models.

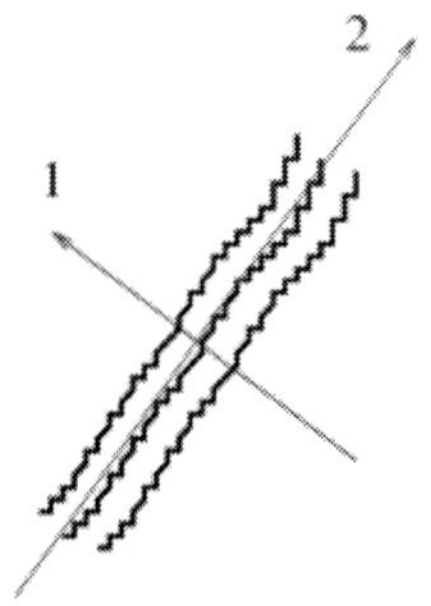

Figure 3. Material axes of smeared crack model.

5. Comparison of models behaviour

The first case that we are showing in this paper is a numerical simulation of four-point bending test of concrete beam. It was modelled with use of an elastic-plastic model described above (Figure 4) and the model based on fracture mechanics (Figure 5).

Both models can give comparable load-displacement curve below the peak of bearing capacity is reached. Even the area affected by non-linear material behaviour is very similar - the elastic-plastic model proposes larger affected area for the same level of deformations. Obviously, the elastic-plastic model cannot simulate post-beak behaviour (the process of material softening) which is thus only shown for the second model in Figure 5.

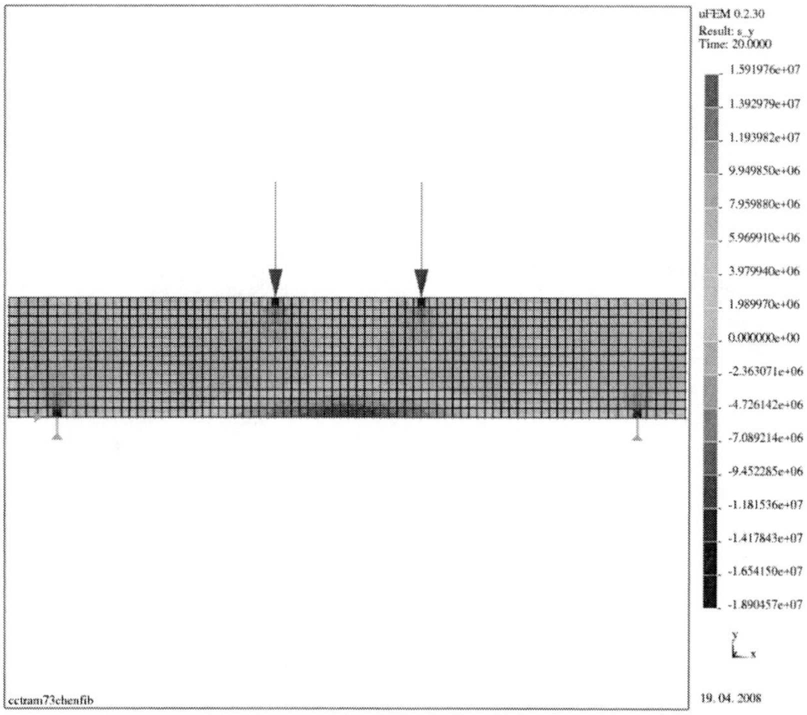

Figure 4. Plastic areas (red shades) of elastic-plastic model.

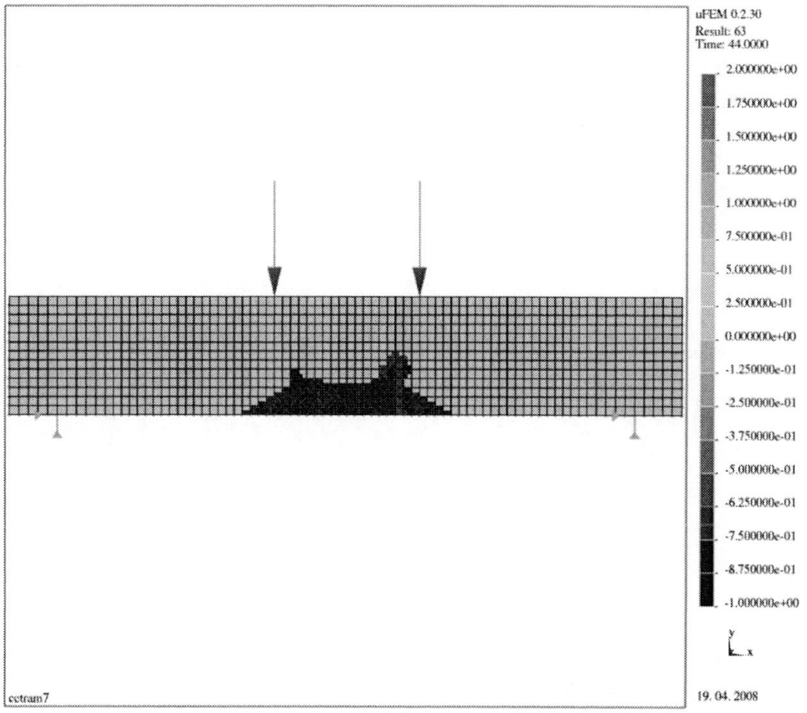

Figure 5. Development of area damaged by tension cracking.

IOP Conf. Series: Earth and Environmental Science **143** (2018) 012036 doi:10.1088/1755-1315/143/1/012036

The second model is a slender reinforced concrete arc of height 2 m. It was modelled with use of both abovementioned models. The Figure 6 shows areas where plasticity developed and the Figure 7 compares load-displacement curves obtained from these models, It is obvious that the elastic-plastic model dives more optimistic assumption of bearing capacity of the arc.

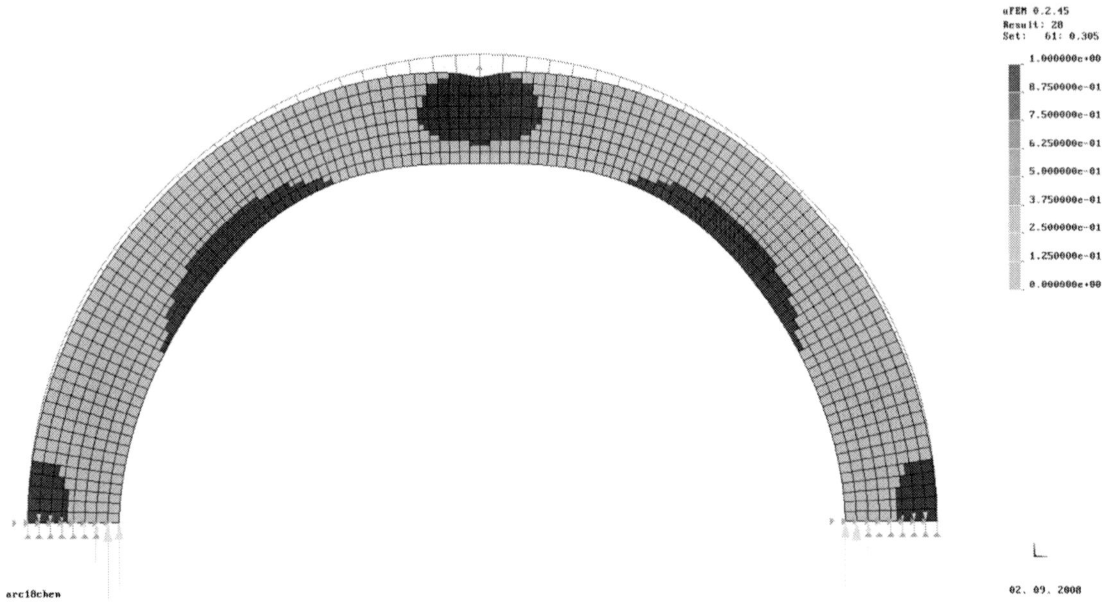

Figure 6. Plastic areas on arc model.

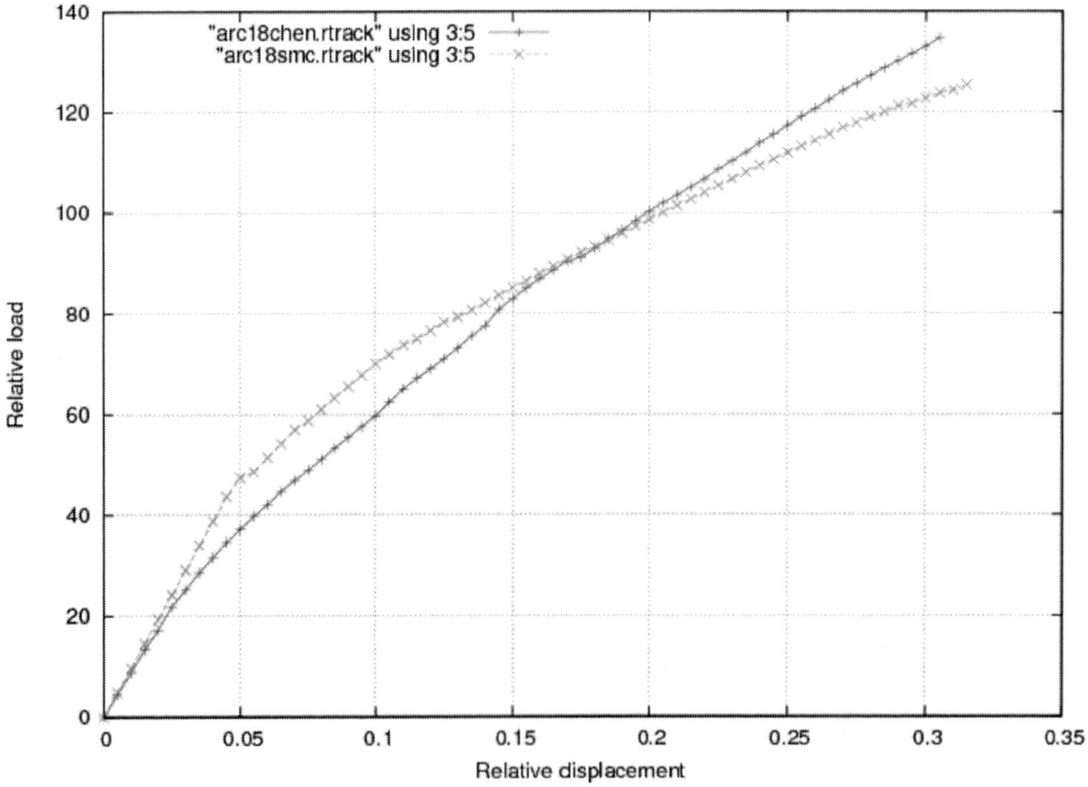

Figure 7. Comparison of elastic-plastic (red) and smeared crack (green) behaviour of the arc.

It is worth to mention that the smeared crack model required near 5-10x longer computational time than the elastic-plastic model with the same finite element mesh. Both discussed models used minimal reinforcement ration and thus thay are not the most typical examples of reinforced structure design but they are rather limit cases for such problems. For higher reinforcement ratios the result will be even better in favour of the elastic-plastic model.

6. Conclusions

The abovementioned examples illustrate that elastic-plastic models material can be used for more precise modelling of reinforced concrete structures. However, some points have to be taken into account: The assumption must be valid that the designed structure must not be exposed to progressive damage in any part of its lifetime.

The elastic-plastic models (even these which are included in commercial software packages) are dependent on size and shape of finite element meshes. Thus is is necessary to validate the input data on a numerical re-modelling of a basic material test of the actual reinforced concrete sample (a four-point bending test can be used here). The dependence of finite element mesh parameters can be minimised by using the same size (and shape) of finite elements both in model of such test and for the actual model of the designed structure.

For such reasons it can de stated that elastic-plastic models can be used as a basis for non-linear material modelling of reinforced concrete structures in the cases of sustainable buildings design.

7. References

[1] Fialko S 2015 *Comp. and Math. with Appl.* **70**, 2968–87
[2] Jirasek M and Bazant Z P 2002 *Inelastic Analysis of Structures* (Chichester: John Willey and sons)
[3] Willam K J and Warnke E P 1975 *Int. Assoc. Bridge Struct. Eng. Proc.* **35,** 1-30
[4] Lee J and Fenves G. L. 1998 *J. Eng. Mech.* **124,** 892-900
[5] Ohtani Y and Chen W. F. 1988 *Journal of the EDM ASCE.* **114** L11
[6] Brozovsky J and Krejsa M 2017 *Key Engineering Materials.* **738,** 319-28 |
[7] Karihaloo B L 1995 *`Fracture Mechanics and Structural Concrete* (Essex: Longman Group Limited)
[8] Sucharda O and Brozovsky J 2014 *Fratt. ed Integ. Strutt.* **30**, 375-82
[9] Cervenka V 1985 *Constitutive Model for Cracked Reinforced Concrete.* **82** L6
[10] Riks E 1979 *Int. J. Solids Struct.* **15** 524-51

Acknowledgments

This paper has been completed thanks to the financial support provided by the Czech Science Foundation under the project grant 16-08937S “State of stress and strain of fiber reinforced composites in interaction with the soil environment”.

Access to computing and storage facilities owned by parties and projects contributing to the National Grid Infrastructure MetaCentrum, provided under the programme "Projects of Large Research, Development, and Innovations Infrastructures" (CESNET LM2015042), is greatly appreciated.

IOP Conf. Series: Earth and Environmental Science **143** (2018) 012037 doi:10.1088/1755-1315/143/1/012037

The numerical high cycle fatigue damage model of fillet weld joint under weld-induced residual stresses

Vuong Nguyen Van Do[1*]

[1]Faculty of Civil Engineering, Ton Duc Thang University, 19 Nguyen Huu Tho St. Dist. 7, HCMC 700000, Viet Nam.

Corresponding author: donguyenvanvuong@tdt.edu.vn

Abstract. In this study, a development of nonlinear continuum damage mechanics (CDM) model for multiaxial high cycle fatigue is proposed in which the cyclic plasticity constitutive model has been incorporated in the finite element (FE) framework. T-joint FE simulation of fillet welding is implemented to characterize sequentially coupled three-dimensional (3-D) of thermo-mechanical FE formulation and simulate the welding residual stresses. The high cycle fatigue damage model is then taken account into the fillet weld joints under the various cyclic fatigue load types to calculate the fatigue life considering the residual stresses. The fatigue crack initiation and the propagation in the present model estimated for the total fatigue is compared with the experimental results. The FE results illustrated that the proposed high cycle fatigue damage model in this study could become a powerful tool to effectively predict the fatigue life of the welds. Parametric studies in this work are also demonstrated that the welding residual stresses cannot be ignored in the computation of the fatigue life of welded structures.

1. Introduction

Recently, welding steel structures are popularly used as productivity of bridge building, ships, aircraft, pipe line and thin walled structure, etc. The fatigue of these structures is complicated by many drawbacks to welded joint behaviours. The most important problems are the durability of welded structures under the active cyclic loading with responses of material changing at welding positions. The presence of cyclic loading and residual stresses makes the fatigue performance reduction of structures particularly in weld joints. For many defects influence to the fatigue corresponding to the structural integrity, residual stress is one of majority factors should be focused and studied. Residual stresses had significantly influence engineering properties of materials and structural component regarding to the fatigue stability, corrosion resistance [1]. Welding residual stress might lead to a dramatic reduction in fatigue strength of welded elements, therefore researching the distribution of residual stress in welded specimen and investigating the fatigue life corresponding to this effect are necessary in operating performance of materials parts and welded elements.

In practice, many studies for the residual stresses and welding deformation had been carried out both numerical and experimental manners. Deng et al [2] focused on considering the effects of thermal-mechanical welding to material properties at weld and base metals or HAZ region as well as the different response of similar or dissimilar weld zones. Kim et al. [3] had successfully developed the welding angular distortion in plate and welding local buckling. Lee and Chang [4] had carried out and investigated the residual stresses and the deformation with high accuracy computation in FEM as comparing to experimental data. This development in the computational method was significantly beneficial for taking account into calculating fatigue life and computational cost. In general, the results of residual stresses and deformation had played important role in the life cycle of welded structures due to mentioned conditions of welding defects. Moreover, residual stresses can be added to the stresses associated to applied loading, it induces the locally generated plastic strain as calculating the fatigue of welding structures. Many researchers had focused on the welding residual stresses to mechanical behaviours and fatigue life, however these have not been clearly explained until yet such as Hot spot stress method derived by Lotsberg and Sigurdsson [5], simple nominal stress method resulted by Fricke and Paetzold [7] and local strain estimation method developed by Kawin et al. [6]. They are among relatively successful methods in the recent decades and use with simple parameters obtained from the experiment results to acquire the fatigue life of materials. In addition, Teng et al [8] had developed the mathematical model to predict the effects of weld geometry parameters and residual stresses. However, the method is only calculated in 2D FE analysis to evaluate the generated residual stresses for the fatigue life of material derived through a strain life curve. Hence, the predicted results had insufficient accuracy to prove the precise residual stresses as well as the remaining fatigue of structures in only uniaxial loading cycle. Consequently, the mentioned methods were not adequately reflected the behaviours of residual stresses and internal variables such as stress, strain, elastic tensor by microstructure changes during plasticity and damage occur. Furthermore, through the experimental investigations in recent decades resulted in numerous hypotheses of high cycle fatigue, many proposals of such criteria could be found as Sine (1959), Crossland (1956) or Dang Van et al (1989) [9]. From the view point applied to numerical analysis, the convenience of first two was based on stress state invariants and the third one was based on the average value concept. These criteria were easily taken account into the FE analysis. The invariant formula always consists of quantities related to hydrostatic and octahedral stresses, then it can allow to determine the initiation point of fatigue crack when response of material reached to critical damage.

On the shortcomings of the present approaches to access the high cycle fatigue of weld structures, the investigation in the high cycle fatigue damage model is necessary. The fatigue damage model based on the continuum damage mechanics introduced the macroscopic damage phenomena and considered the localized at the microscopic scale to evaluate the fatigue life of steel specimen derived by Lemaitre et al [10]. This model presented quite well prediction physical phenomena related to high cycle fatigue in full reversed cycle loading, the effect of mean stress for high cycle fatigue was difficult to investigate. Shang and Yao had proposed the fatigue damage model for a nonlinear damage cumulative model [11], however these models only utilized for uniaxial fatigue, then they may not adequately determined much of behaviours of fatigue steel structures.

In this study, the three dimensional finite element simulation of computational fatigue damage model for welding structures is developed based on the continuum damage mechanics. The damage framework produced into elasto-plastic analysis is indispensable to illustrate the material degradation response as well as remaining fatigue life. On the basis of aforementioned criteria, the multi-axial high cycle fatigue formula are also derived. The residual stress and welding deformation are also determined by the studies in thermal-elasto-plastic analysis [4-12]. Three dimensional couple damage-elasto-plastic finite is implemented into a FEM code to derive the results. Update Lagrangian finite element formulation and a co-rotational system for the stress update algorithm were considered. Particularly, the proposed method has been the effective tool for analysis the fatigue and residual stresses of weld steel structures.

2. 3D couple thermal welding induced.

In order to consider the effect of welding-induced residual stresses and welding deformation which are determined by stress relaxation method, the computed residual stresses and deformation in welding process are used as initial data taken into fatigue damage FE model. The effective program should be developed to solve the couple elasto-plastic and damage mechanics for numerical evaluation of fatigue crack by FE simulation and damage evolution. Generally, there are two kinds of simulation methodologies: one is couple elasto-plastic damage analysis while another is uncouple between FEA and damage as considered a weak condition in fracture model. It is well recognized that the first with couple one is the high accuracy method, which means that full combination of damage process during the FEA [13]. The damage level and stress field of all representative volume element (RVE) are variable can be taken into account from the beginning to the end of the fatigue damage process. In detail, the initial Young's modulus should be updated and altered by effective Young's modulus at each computation step in FE analysis.

During the high cycle fatigue life computation of analysis FEM, the residual stress and welding deflection results will be considered into calculation due to the high residual stress concentration generating at the welding line position. This consideration is affected from the welding temperature and thermal-mechanical process. The material properties of specimen are also considered with different features of weld and base metals. From the view point of thermal mechanical simulation, the temperature is taken account as volumetric or surface energy distribution and thermal flow effect could be simplified by increasing thermal conductivity over the fusion temperature. Welding procedure is thermal mechanical process in which the temperature field has large effect and induces to residual stress field. On the other hand, these stresses have also weakly affected on thermal fields. Therefore, in order to simplify on this study, un-coupled analysis works of three-dimensional FE model are developed in consisting of two parts: a thermal model followed by a mechanical model.

2.1. *Thermal model*

The balance of energy equation for thermal simulation is given following as

$$\nabla(k\nabla T) + \dot{q} - \rho c\dot{T} = 0 \quad (1)$$

where T is the temperature, $\dot{T}$ is the rate change of temperature, k is the thermal conductivity, c is the specific heat, ρ is the density, $\dot{q}$ is the rate of moving heat generation and ∇ is the spatial gradient operator. As for the boundary conditions applied to the thermal model, convection and radiation are both taken into the consideration and their combined effects are presented via following equation for the total temperature-dependent heat transfer coefficient, h [14]

$$h = \frac{\sigma\varepsilon((T+273)^4 - (T_0+273)^4)}{T - T_0} + h_c \quad (2)$$

2.2. *Mechanical model*

The second step of the current analysis involves the use of temperature histories predicted by the previous thermal model as an input for mechanical model. During the welding process, the solid state phase transformation does not occur in the base and weld metals. Therefore, total strain can be expressed to the differential form of the total strain into three components as follows:

$$d\varepsilon_{ij} = d\varepsilon_{ij}^{e} + d\varepsilon_{ij}^{p} + d\varepsilon_{ij}^{th} \quad (3)$$

where $d\varepsilon_{ij}^{e}$ is the elastic strain increment and $d\varepsilon_{ij}^{p}$ is the plastic strain increment, $d\varepsilon_{ij}^{th}$ is the thermal strain increment. The elastic strain increment is calculated by using the isotropic hook's law with temperature-dependent Young's modulus and Poisson's ratio. The thermal strain increment is computed using the coefficient of thermal expansion. For the plastic strain increment, a rate independent elastic-plastic constitutive equation is considered with the Von-Mises criterion, temperature dependent mechanical properties and linear kinematic hardening rule. Kinematic hardening rule is considered as an important feature as material point typically undergo both loading and unloading in the course of welding process.

3. Proposed high cycle fatigue damage model

For high cycle fatigue damage is called quasi-brittle when the loading is a large number of repeated cycles and it is at low level below the conventional yield stress which subjected to quasi-brittle materials. The damage is localized in such the way that the damaged material occupies a small volume in comparison with the macro-scale of the structural component and even with the mesoscale of the RVE [15]. For the sake of the welded fatigue fracture, the fatigue fractures in welded structure had been known as the complicated phenomenon due to the accumulated damage evolution not be regarding to the plastic deformation at the macro-level while the damage growth and plasticity occurred at the micro-scale. The structural calculation is fully elastic even in the stress concentration zones. The brittle damage evolution obtained by micro-mechanics for particular mechanisms. In the brittle model of Kintzel et al [16] which novel damage indicator function was derived from Kuhn-Tucker complementation condition for low cycle fatigue calculation, the damage evolution as well as damage rate is evident as

$$\dot{D} = BY^{q-1}\dot{Y} \quad (4)$$

$$B = \frac{N^2}{S_2^2\left(B_\Gamma \frac{sign(Y^N - \Gamma)}{S_2}(\Gamma_\infty - \Gamma) + B_d(Q_{d\infty} - Q_d)\right)} \text{ and } q-1 = 2(N-1) \quad (5)$$

where Γ_∞ and $Q_{d\infty}$ are defined the implicit thresholds at which the damage rate accelerates, B_Γ and B_d are material parameters which have a similar meaning as the saturation constant materials. From above Eq.(5), the evolution of 1D fatigue damage model rewritten and introduced as following

$$\delta D = \left[1-(1-D)^{\beta+1}\right]^{\alpha(\sigma_M,\bar{\sigma})}\left[\frac{\sigma_{Max} - \sigma_m}{M(\sigma_m)(1-D)}\right]^\beta \delta N \quad (6)$$

with

$$M(\sigma_m) = M_0(1 - b_2\,\sigma_m/\sigma_u)$$

$$\alpha(\sigma_{Max}, \sigma_m) = 1 - a\left\langle\frac{\sigma_{Max} - \sigma_l(\sigma_m)}{\sigma_u - \sigma_{Max}}\right\rangle;\ \sigma_l(\sigma_m) = \sigma_{-1}(1 - b_1\,\sigma_m/\sigma_u) \quad (7)$$

Under a cyclic loading, the fatigue laws for a 1D problem depend on the mean stress value (σ_m) and the maximum stress (σ_{Max}). σ_{-1} is the fatigue limit which denoted as fully reversed loading and N is the number of cycles. For fatigue life prediction, the integration of Eq.(6) is performed with a number of cycles until crack initiation appears (D=1). The integration between $D=0$ and $D=1$ could achieve the number of cycle under the constant amplitude loading, the number of high cycle fatigue life is as following

$$N_f = \frac{1}{(1+\beta)aM_0^{-\beta}}\frac{\langle\sigma_u - \sigma_{max}\rangle}{\langle\sigma_a - \sigma_{af}(\sigma_m)\rangle}\left(\frac{\sigma_a}{1 - b_2\,\sigma_m/\sigma_u}\right)^{-\beta} \quad (8)$$

In this study, the process of welding was simulated using 3-D thermo-mechanical FE formulation and the fatigue damage model based on the in-house FE code, which has been extensively verified against numerical results found in the literature and experiments.

4. Numerical results and discussion

4.1. Verification for the numerical residual stress results

This section is concerned with parameters be determined by analyzing residual stresses and fatigue damage model for SM490 carbon steel. On the calibration, the basic parameters of SM490 steel are obtained by tensile and cyclic tests according to ASTM standard E8-96 and E606-92, respectively with smooth round bar specimen under displacement control. The material constants on the elastic-damage plastic model are investigated by utilizing the experimental tests as shown in Table 1 and Table 2

Table 1. Mechanical properties of SM490 steel

Material	E (GPa)	ν	σ_y (MPa)	σ_u (MPa)	Elongation (%)	ε_{st}

SM490	206	0.3	365	560.	30.6	0.0133

Table 2. Fatigue damage model parameters for the damage evolution

β	a	M_0	b_1	b_2
1.196	0.8362	2.082E+05	0.002085	0.000371

In order to confirm the accuracy of welding FE analysis for this calculation, the experiment work was carried out by Kim et al. [3] with the geometry and dimension of the fillet weld plate which is made of carbon steel is considered with the length, width and thickness of L=500mm,W=300mm, t=6mm, respectively. The temperature histories were recorded by thermo-couples and residual stress measurements were carried out on the two axis strain gauges with layering technique. Measuring stresses by strain gauges can obtain the residual stresses on the surface of the specimens to be evaluated. The development of 3-D thermo-mechanical analysis by using finite element method is implemented to verify the experimental test with the welding parameters as obtained in fabrication of specimen. It seen that the trend of FE simulation as shown in Figure 1 and experiment data agrees very well. Therefore, the welding formulation of Eq.(3) is considered as an appropriated method for analyzing and introducing the residual stresses into fatigue damage model.

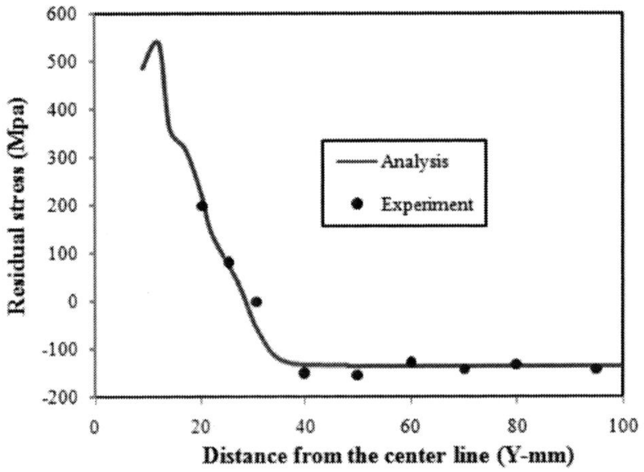

Figure 1. Verification of residual stresses for the numerical results with testing data

4.2. Fatigue damage study

Based on thermal residual stresses obtained from the welding process, the damage evolution was also determined through the damage growth rate Eq.(6). The fully coupled elastic-plastic damage is considered in analyzing structural damage interaction with effective stiffness degradation. Under the number of cycles the stress-plastic-strain upgrade, the damage variable was accumulated until reaching to critical value D_c at which the crack initiation will occur in the structure. The FE analysis was continuously repeated until the crack propagation processes and through the entire length at cross-section of the welded length. In addition, all variables in FE model were simultaneously updated for every cycle in calculation. The fatigue life of welded specimen subjected to the cyclic load is determined by the number cycles of repetition until the termination of the calculation.

A crack is described in the damage model by a region of totally damaged material. It is important to recognize that the local loss of stiffness in this region implies that the stresses drop to zero in deformation field. The material in the crack is given with a small residual stiffness, herein the damage variable is limited to a value which is slightly small than one, it makes to avoid the discrete equilibrium equations become singular. The increment led to the critical damage is then recomputed starting from the converged state in the previous increment, so that the damage accumulation is consistent with the current configuration under cyclic loading. The distribution of longitudinal residual stress profiles at the weld toe along the weld length indicated that the middle part high tensile residual stress appears meanwhile lower one exists in the edges. In spite of the relaxation occurring, the tensile stress in middle part is still high therefore the crack initiation will begin at mid position as displayed in Figure 2.

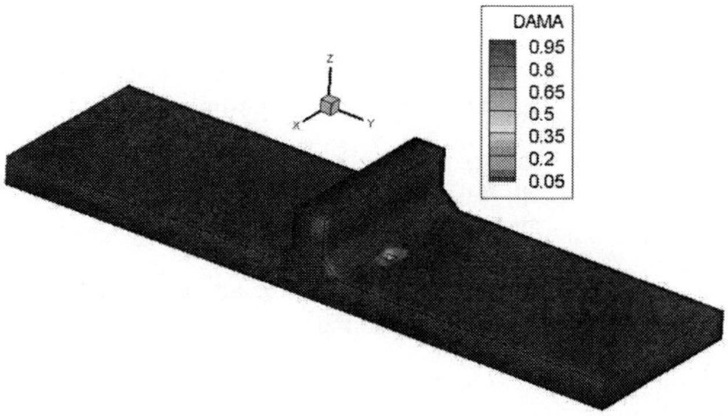

Figure 2. Failure analysis of fillet welding specimen as crack initiation and total crack processing

Initial imperfections such as initial defections and residual stress are explicitly considered since the reversed loading applied in the top of specimen. For most materials, compressive loadings are less critical than tensile ones, the damage evolution rate in compression case is lower than damage rate in tension as long as the structural instability is not taken placed, therefore crack takes place faster in tension case [15]. This is due to the phenomenon of micro-defects or micro-cracks closure when the material compressed. In fatigue, the existing of mean stress has plays role to reduce the fatigue life for a given stress amplitude. In this study, initial residual stress incorporating to high tensile part at the weld toe leads to increase mean stress effects as well as stress range. It is observed that the residual stress on FE model significantly speeds up both of crack initiation and propagation. As a result that the fatigue life of specimen is shown in Figure 3 has shorter life than base specimen and illustrates the high accuracy between analysis FEM and testing data with stress ratio R=0.1.

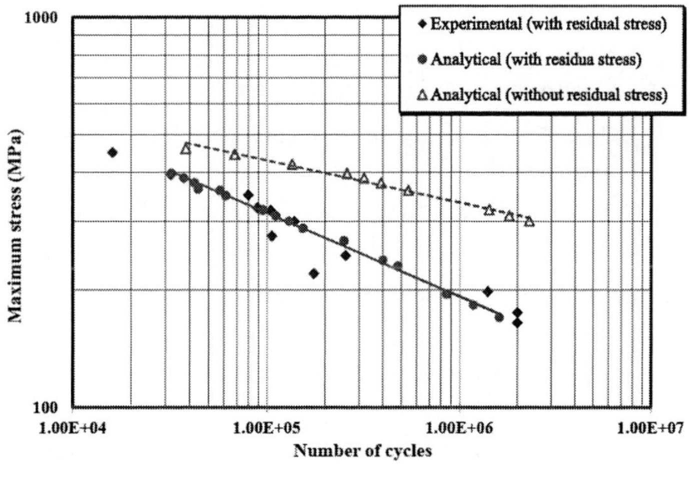

Figure 3. Fatigue damage analysis of fillet welding specimen as comparing to experimental data with/without residual stresses

5. Conclusion

This work aims to provide basic insight into the numerical simulation of multiaxial fatigue damage within considering the residual stresses on fillet welding structures. Material parameters included are used to describe high cycle damage evolution which are obtained from uniaxial experiments. The fatigue numerical simulation in which the residual stresses are incorporated into the fatigue damage model provides a simple step by step derivation of the consistent elasto-plastic modulus corresponding to FE algorithm and accesses the total fatigue life of undamaged welded structures. It can be drawn as following

a) The multiaxial high cycle fatigue damage model considered the effects of welding residual stresses be able to predict the total fatigue life of fillet welded joints under various cyclic load types with no initial cracks until to complete damage failure.

b) The temperature distribution and generating residual stresses in proposed damage model have been successfully developed in the welding process to indicate the highest intensity values at the welding joints.

c) The development of this fatigue damage model in accurately predicting the total high cycle fatigue life of welded structures can used as an effective method to reduce the cost and consuming time of fatigue test.

Acknowledgments

This research is funded by Vietnam National Foundation for Science and Technology Development (NAFOSTED) under grant number 107.02-2016.19.

Reference

[1] Jian L 2005 *Handbook on residual stress*. (Society for Experimental mechanics)

[2] Deng D 2009 FEM prediction of welding residual stress and distortion in carbon steel considering phase transformation effects *Materials and Design* 30 359–366

[3] Kim Y C, Lee J Y, Inose K 2010 Determination of dominant factors in high accuracy prediction of welding distortion *Welding in the world* 54 234-240.

[4] Lee C H, Kyong-Ho Chang Three-dimensional finite element simulation of residual stresses in circumferential welds of steel pipe including pipe diameter effects Materials Science and Engineering A 487 (2008) 210–218

[5] Lotsberg I, Sigurdsson G 2007 Hot Spot Stress S-N Curve for Fatigue Analysis of Plated Structures, *J. Offshore Mech. Arct. Eng.* 128 330-336

[6] Fricke W, Paetzold H 1995 Fatigue strength assessment of scallops- an example for the application of nominal and local stress approaches, Marine Structure. 8 423-447.

[7] Kawin S, Hanji T, Miki C 2012 Local strain estimation method for low and high cycle fatigue strength evaluation, *int J. Fatigue* 401-6

[8] Teng T L, Fung C P, Chang P H 2002 Effect of weld geometry and residual stress on fatigue in butt welded joints, int J.Pressure vessels and piping 79 467-482.

[9] Mrzyglod M, Zielinski A 2006 Numerical implementation of multiaxial high cycle fatigue criterion to structural optimization, *J. Theoretical and applied mechanics* 44 691-712.

[10] Lemaitre J, Sermage J P, Demorat R 1999 A two scale damage concept applied to fatigue. *J.Fracture* 97 67-81

[11] Shang D G, Yao W X 1999 A nonlinear damage cumulative model for uniaxial fatigue. *J. Fatigue* 21 187-194.

[12] Lee C H, Kyong-Ho Chang 2007 Numerical analysis of residual stresses in welds of similar or dissimilar steel weldments under superimposed tensile loads. *Computational Materials Science* 40 548–556

[13] Luccioni B, Oller S, Danesi R 1996 Coupled plastic-damaged model. *Comput. Meth. Appl. Mech. Engrg* 129 81-89

[14] Abid M, Siddque M 2005 Numerical simulation to study the effect to tack welds and root gap on welding deformations and residual stress of a pipe-flange joint. *Int.J.Press.Vessels Pip* 82 860-871.

[15] Desmorat R, Ragueneau F, Pham H 2006 *Continuum damage mechanics for hysteresis and fatigue of quasi-brittle materials and structures. Int.J.Numer.Anal.Meth.Geomech* 31 1-19.

[16] Kintzel O, Khan S, Mosler J 2010 *A novel isotropic quasi-brittle damage model applied to LCF analyses of Al2024. Int J.Fatigue* 32 1948-1959.

A Modified Kirchhoff plate theory for Free Vibration analysis of functionally graded material plates using meshfree method

Vuong Nguyen Van Do[1*]

[1]Faculty of Civil Engineering, Ton Duc Thang University, 19 Nguyen Huu Tho St. Dist. 7, HCMC 700000, Viet Nam.

Corresponding author (*): donguyenvanvuong@tdt.edu.vn

Abstract. In this paper, a modified Kirchhoff theory is presented for free vibration analyses of functionally graded material (FGM) plate based on modified radial point interpolation method (RPIM). The shear deformation effects are taken account into modified theory to ignore the locking phenomenon of thin plates. Due to the proposed refined plate theory, the number of independent unknowns reduces one variable and exists with four degrees of freedom per node. The simulated free vibration results employed by the modified RPIM are compared with the other analytical solutions to verify the effectiveness and the accuracy of the developed mesh-free method. Detail parametric studies of the proposed method are then conducted including the effectiveness of thickness ratio, boundary condition and material inhomogeneity on the sample problems of square plates. Results illustrated that the modified mesh-free RPIM can effectively predict the numerical calculation as compared to the exact solutions. The obtained numerical results are indicated that the proposed method are stable and well accurate prediction to evaluate with other published analyses.

Keywords: functionally graded material (FGM) plate, Mesh-free, modified Radial point interpolation, locking free, a normalization of cubic function.

1. Introduction

In the recent years, the functionally graded materials (FGM) were the first proposed by Bever and Duwez [1] and have been applied in a variety of engineering industries because of their distinctive material properties, which vary continuously through the thickness. These are the special types with their characteristics can be tailored for many application and various working environments. In the typically plates, the smoothness and continuously changing in microstructure from the top to the bottom layer can made the materials displayed in distinct phases regarding to ceramic and metal. The combination by two phases is demonstrated that the ceramic with low thermal conductivity can well resist the harms of thermal stress and surface corrosion due to the high temperature effects [2]. Meanwhile the metal-rich surface is well capable of the highly impacted loading on the structures. In addition, matrix materials can reduce and not be prone to the initial cracks growing into the material sections and debonding of fiber composites at extremely thermal forces. Due to the advantages of mechanical behaviours of functional graded materials (FGM), the significant number of researches had been conducted to examine the mechanical responses of FG shells and plates [3]. Therefore, the analyses of displacement, stress, dynamic problems of natural frequencies and buckling responses are really necessary for FG material behaviours.

Besides the advantages of material properties and behaviours of FGM plates, many plate theories have been researched. The typical method such as the classical plate theory (CPT) [4] with Kirchhoff-

Love assumptions for thin plate gave acceptable results. Nevertheless, the effects of shear deformations were not incorporated into the model, then it leads to totally unsuitable numerical calculations in thick plates. The large number of significant theories had been written in FEM [5] including the shear deformation to overcome the shortcomings of classical plate theory. For instances, the finite element formulations were derived by Reddy [6] for FG plates based on third-order shear deformation theory [TSDT]. The FSDT and kp-Ritz method were investigated by Zhao et al. [7] for bending and vibration analyses. The FSDT and four node finite element using the exact neutral surface position were derived by Singha et al. [8] to calculate the nonlinear bending responses of FG plate with different boundary conditions. However, these methods are over-stiffness when the plates become too thin. It is also known that the HSDTs with FEM are complicated and increased the higher variables in numerical analysis due to the use of linear shape function.

With these disadvantages of FE analysis, the mesh-free method developed by Belytschko [9] could be available to obtain the better accuracy results. Until today, the later various successful methods were established to capture the defects of conventional finite element computation during the meshing generation. In the present study, the cubic form taken account into the radial point shape function is the target to eliminate the dependence of coordinate meshes as well as disappearance of any coefficients in the chosen form. The high accuracy in analyzing for FGM plates of the dynamic vibrations is employed by a modified Kirchhoff theory. The verifications of numerical examples are shown a good validation between the proposed method and the other published from the various reported literatures. Results demonstrate that the mesh-free RPIM can effectively predict the dynamic vibration of the FGM isotropic plates.

2. The Kirchhoff theory for FGM plates

2.1. The Kirchhoff theory

Considering the Kirchhoff plate theory with the domain Ω presented in the R^2 given for the mid-plane of the plate, the displacement variables of u_0, v_0, and bending displacement w^b belong to three x; y; z directions. The conventional Kirchhoff model is not considered the effect of shear deformation, therefore it is only suitable for thin plate analysis. To improve this model for analysis both moderately thick and thin plates, a modified transverse displacement can be separated into bending and shear parts. A modified formulation of Kirchhoff theory is described by

$$\begin{aligned} u(x,y,z) &= u_0(x,y) - z\frac{\partial w^b}{\partial x} \\ v(x,y,z) &= v_0(x,y) - z\frac{\partial w^b}{\partial y} \quad , \quad \left(\frac{-h}{2} \le z \le \frac{h}{2}\right) \\ w(x,y,z) &= w^b(x,y) + w^s(x,y) \end{aligned} \tag{1}$$

The displacement field for proposed theory in Eq. (1) included four unknown variables of u_0, v_0, w^b, w^s. In comparison with the Reissner-Mindlin FSDT theory, the number of independent unknowns of the present theory are reduced one variable.

The relationship of strains and displacement equations is described by

$$\begin{aligned} \boldsymbol{\varepsilon}_p &= [\varepsilon_{xx}\ \varepsilon_{yy}\ \gamma_{xy}]^T = \boldsymbol{\varepsilon}_0 + z\boldsymbol{\varepsilon}_1 \\ \boldsymbol{\gamma} &= \left[\gamma_{xz}\ \gamma_{yz}\right]^T = \boldsymbol{\varepsilon}_s \end{aligned} \tag{2}$$

where,

$$\boldsymbol{\varepsilon}_0=\begin{bmatrix}\dfrac{\partial u_0}{\partial x}\\ \dfrac{\partial v_0}{\partial y}\\ \dfrac{\partial v_0}{\partial x}+\dfrac{\partial u_0}{\partial y}\end{bmatrix},\boldsymbol{\varepsilon}_1=\begin{bmatrix}-\dfrac{\partial^2 w^b}{\partial x^2}\\ -\dfrac{\partial^2 w^b}{\partial y^2}\\ -2\dfrac{\partial^2 w^b}{\partial x\partial y}\end{bmatrix}\text{ and }\boldsymbol{\varepsilon}_s=\begin{Bmatrix}w^s_{,x}\\ w^s_{,y}\end{Bmatrix} \tag{3}$$

The dynamic equation of modified Kirchhoff theory based on Ref.[11] can be obtained by the weak form as

$$\int_\Omega \delta\begin{Bmatrix}\boldsymbol{\varepsilon}_0\\ \boldsymbol{\varepsilon}_1\end{Bmatrix}^T\begin{bmatrix}\mathbf{A} & \mathbf{B}\\ \mathbf{B} & \mathbf{D}\end{bmatrix}\begin{Bmatrix}\boldsymbol{\varepsilon}_0\\ \boldsymbol{\varepsilon}_1\end{Bmatrix}\mathrm{d}\Omega+\int_\Omega \delta\boldsymbol{\varepsilon}_s^{\ T}\mathbf{D}^s\boldsymbol{\varepsilon}_s\mathrm{d}\Omega+\int_\Omega \delta\begin{Bmatrix}\mathbf{u}_0\\ \mathbf{u}_1\end{Bmatrix}^T\begin{bmatrix}I_1 & I_2\\ I_2 & I_3\end{bmatrix}\begin{Bmatrix}\ddot{\mathbf{u}}_0\\ \ddot{\mathbf{u}}_1\end{Bmatrix}\mathrm{d}\Omega=0 \tag{4}$$

where

$$\left(\mathbf{I}_1,\mathbf{I}_2,\mathbf{I}_3\right)=\int_{-h/2}^{h/2}\rho(z)\left(1,z,z^2\right)\mathrm{d}z,\ \mathbf{u}_0=[u\ v\ w_b+w_s]^T,\mathbf{u}_1=[\frac{\partial w_b}{\partial x}\ \frac{\partial w_b}{\partial y}]^T \tag{5}$$

and

$$\mathrm{A}_{ij},\mathrm{B}_{ij},\mathrm{D}_{ij}=\int_{-h/2}^{h/2}(1,z,z^2)\mathrm{Q}_{ij}\mathrm{d}z;\ \mathrm{D}^s_{ij}=k_s\int_{-h/2}^{h/2}\mathrm{G}_{ij}\mathrm{d}z \tag{6}$$

The material matrices are given as

$$\mathbf{Q}=\frac{E_e}{1-\nu_e^2}\begin{bmatrix}1 & \nu_e & 0\\ \nu_e & 1 & 0\\ 0 & 0 & (1-\nu_e)/2\end{bmatrix},\mathbf{G}=\frac{E_e}{2(1+\nu_e)}\begin{bmatrix}1 & 0\\ 0 & 1\end{bmatrix} \tag{7}$$

in which E_e, ν_e are effective Young's modulus and Poisson's ratio, respectively. Those parameters can be discussed in the next subsection. And k_s denotes the transverse shear correction coefficient because transverse shear strains in Eq. (2) are constant through the thickness of plate. For FGM, it can be defined by Ref.[16].

$$k_s=\frac{5}{6-\left(\nu_c V_c+\nu_m V_m\right)} \tag{8}$$

2.2. Problem formulation for isotropic FGM plates

The isotropic homogeneous plate is divided into two different material phases in which ceramic and metal are arranged at the top and bottom layers, respectively. In this paper, the homogenized formulation of the rule of mixture is utilized to homogenize the material parameters. The volume fraction of ceramic phase is described as [6].

$$V_c(z) = \left(0.5 + \frac{z}{h}\right)^n, \; z \in \left[-\frac{h}{2}, \frac{h}{2}\right] \; and \; V_m = 1 - V_c \tag{9}$$

where m and c are defined to the metal and ceramic. The scale parameter n is representative to the gradation of material properties varying through the thickness. The following estimation of the effective material characteristics is homogenized by

$$P_e = P_c V_c(z) + P_m V_m(z) \tag{10}$$

where P_c, P_m define the material constituents of the ceramic and the metal, respectively. The Young's modulus (E), Poisson's ratio (ν), and density (ρ) are material properties which are presented by Eq. (10).

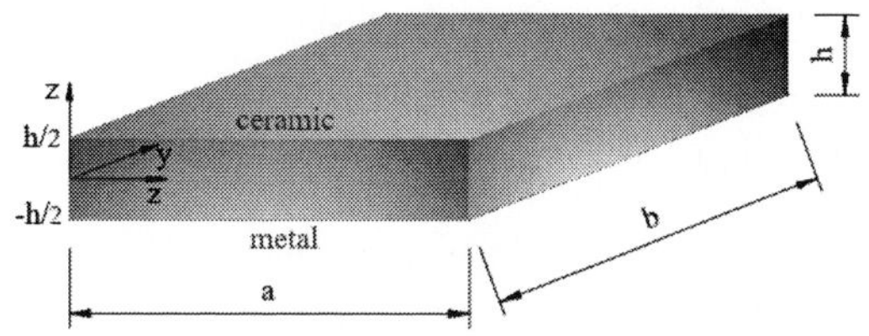

Figure 1. A typical FGM plate

3. Radial Point Interpolation Function

3.1. Introduction to RPIM shape functions

The brief description of radial point interpolation method was originally presented by Wang and Liu (2005) [10] is used to build the displacement field in mesh-free method. The shape function and their derivatives are integrated through the cloud of the nodal points based on the definition of sub-domain Ω_x in which $\Omega_x \in \Omega$. Considering a domain with a set of arbitrary scattered point $x_i (i = 1, 2, ..., n)$, where n is the number of the nodes in the local support domain Ω_x, the approximation can be built around the certain point x_k and the effect of nodes outside this domain is ignored. The radius of influence domain is determined by scaling factor and the maximum distance to cover the adequate neighboring nodes in this region. The radial point interpolation of $u^h(x)$ for every point x_i in the support domain Ω_x can be expressed as

$$\mathbf{u}^h(\mathbf{x}) = \sum_{i=1}^{n} R_i(\mathbf{x})\mathbf{a}_i + \sum_{j=1}^{m} p_j(\mathbf{x})\mathbf{b}_j \tag{11}$$

And the constraint condition

$$\sum_{i=1}^{n} p_j(\mathbf{x}_i)\mathrm{a}_i = 0, \quad j = 1, 2, ..., m \tag{12}$$

where $R_i(x)$ is the radial basis function (RBF) which describes the distance of interpolation point x and a node x_i, n is the number of nodes in the sub-domain around of x, $p_j(x)$ is the monomial in the space coordinates $\mathbf{x}^T = [x, y]$, m is the number of polynomial basis functions, coefficients of a_i and b_j are interpolation constants. These are determined by rewriting the Eqs. (11) and (12) in the matrix form.

$$Ga_0 = \begin{bmatrix} R & P \\ P^T & 0 \end{bmatrix} \begin{bmatrix} a \\ b \end{bmatrix} = \begin{bmatrix} u_s \\ 0 \end{bmatrix} \tag{13}$$

where $a_0^T = \begin{bmatrix} a^T & b^T \end{bmatrix}^T = (a_1, a_2, ..., a_n, b_1, b_2, b_3)$, $u_s^T = (u_1, u_2, ..., \mathrm{u}_n)$

To avoid calculating G^{-1} due to the symmetric matrices of R and G, the algorithm of Eq. (13) can be solved as following

$$a = R^{-1}u_s - R^{-1}Pb \tag{14}$$

Subtituting above equation to the Eq. (12), gives

$$b = S_b u_s \tag{15}$$

where $S_b = [P^T R^{-1} P]^{-1} P^T R^{-1}$

Subtituting Eq. (15) into Eq. (14), the expression can be written as $a = S_a u_s$, where $S_a = R^{-1}[I_{n\times n} - PS_b]$. Finally, the displacement field can be approximated by general basic form of radial point interpolation

$$u^h(x) = [R^T(x)S_a + p^T(x)S_b]u_s \tag{16}$$

In this study, the aim of the present work is to utilize a proposed interpolation function as polynomial form to build the shape function without any supporting fixing coefficients. The proposed normalized cubic function is formulated by

$$R(r_i) = \begin{cases} \frac{2}{3} - 4\left(\frac{\theta}{r_s}\right)^2 r_{ij}^2 + 4\left(\frac{\theta}{r_s}\right)^3 r_{ij}^3 & for\ 0 \le \frac{r_{ij}}{r_s} \le \frac{1}{2} \\ \frac{4}{3} - 4\left(\frac{\theta}{r_s}\right) r_{ij} + 4\left(\frac{\theta}{r_s}\right)^2 r_{ij}^2 - 4\left(\frac{\theta}{r_s}\right)^3 r_{ij}^3 & for\ \frac{1}{2} \le \frac{r_{ij}}{r_s} \le 1 \\ 0 & otherwise \end{cases} \tag{17}$$

where a denominator r_s is taken to be a maximum distance between a pair of nodes in the support domain $\Omega_{\mathbf{x}}$. In mesh-free method, a set of nodes are determined as a support domain. The circle support domain can be utilized with radius defined as in Eq.(18)

$$d_m = \alpha d_c \tag{18}$$

where d_c is a length parameter which relates to the distance of nodes, which is equal to nodal distance in regular mesh. α is the scale factor.

3.2. RPIM interpolation for discrete equations

Supposing that the number of total nodes converges in $\Omega_x \in \Omega$ is n and the amount of nodes n in the support domain corresponding to the interest node is at centered x with the radius d_m. The RPIM approximations for geometric shape and the displacement field u in the middle plane of the plate can be implemented following as

$$\mathbf{x}^h = \sum_I^n \phi_I(\mathbf{x})\mathbf{x}_I,\ \mathbf{u}^h(\mathbf{x}) = \sum_I^n \phi_I(\mathbf{x})\mathbf{q}_I \tag{19}$$

where $\mathbf{x}^T = (x \quad y)$ is the coordinate vector of the real geometry of plate. In Eq. (19), $\phi_I(\mathbf{x})$ is the normalized RPIM basis function, $\mathbf{q}_I = \begin{bmatrix} u_{0I} & v_{0I} & w_I^b & w_I^s \end{bmatrix}^T$ is nodal vector of the four unknown variables for an arbitrary node I and $\mathbf{x}_I$ is the nodal value.

The relationship between in-plane/shear strain and nodal displacements is obtained by substituting Eq. (19) into Eq. (3) by

$$\boldsymbol{\varepsilon}_0 = \sum_{I=1}^n \mathbf{B}_I^m \mathbf{q}_I;\ \boldsymbol{\varepsilon}_1 = \sum_{I=1}^n \mathbf{B}_I^b \mathbf{q}_I;\ \boldsymbol{\varepsilon}_s = \sum_{I=1}^n \mathbf{B}_I^s \mathbf{q}_I \tag{20}$$

in which

$$\mathbf{B}_I^m = \begin{bmatrix} \phi_{I,x} & 0 & 0 & 0 \\ 0 & \phi_{I,y} & 0 & 0 \\ \phi_{I,y} & \phi_{I,x} & 0 & 0 \end{bmatrix},\ \mathbf{B}_I^b = \begin{bmatrix} 0 & 0 & -\phi_{I,xx} & 0 \\ 0 & 0 & -\phi_{I,yy} & 0 \\ 0 & 0 & -2\phi_{I,xy} & 0 \end{bmatrix} \text{ and } \mathbf{B}_I^s = \begin{bmatrix} 0 & 0 & 0 & \phi_{I,x} \\ 0 & 0 & 0 & \phi_{I,y} \end{bmatrix} \tag{21}$$

Also, the displacement fields $\mathbf{u}_0$ and $\mathbf{u}_1$ by substituting Eq. (19) into Eq.(5), can be expressed as follows

$$\mathbf{u}_0 = \sum_{I=1}^n \mathbf{N}_I^0 \mathbf{q}_I\ ;\ \mathbf{u}_1 = \sum_{I=1}^n \mathbf{N}_I^1 \mathbf{q}_I \tag{22}$$

where

$$\mathbf{N}_I^0 = \begin{bmatrix} \phi_I & 0 & 0 & 0 \\ 0 & \phi_I & 0 & 0 \\ 0 & 0 & \phi_I & \phi_I \end{bmatrix} \mathbf{N}_I^1 = \begin{bmatrix} 0 & 0 & -\phi_{I,x} & 0 \\ 0 & 0 & -\phi_{I,y} & 0 \\ 0 & 0 & 0 & 0 \end{bmatrix} \tag{23}$$

The derivatives of the transverse displacements including components of w^b, w^s are also given by

$$\begin{Bmatrix} w^b_{,x} + w^s_{,x} \\ w^b_{,y} + w^s_{,y} \end{Bmatrix} = \sum_{I=1}^{n} \mathbf{B}_I^g \mathbf{q}_I \tag{24}$$

where

$$\mathbf{B}_I^g = \begin{bmatrix} 0 & 0 & \phi_{I,x} & \phi_{I,x} \\ 0 & 0 & \phi_{I,y} & \phi_{I,y} \end{bmatrix} \tag{25}$$

Finally, substituting Eqs. (20), (22) and (24) into Eqs.(4) for free vibration can be derived by following equations

$$\left(\mathbf{K} - \omega^2 \mathbf{M}\right)\mathbf{q} = \{\mathbf{0}\} \tag{26}$$

where $\omega \in R^+$ are the natural frequency and the critical buckling parameter, respectively.

where

$$\mathbf{K} = \int_\Omega \left[\begin{Bmatrix} \mathbf{B}^m \\ \mathbf{B}^b \end{Bmatrix}^T \begin{bmatrix} \mathbf{A} & \mathbf{B} \\ \mathbf{B} & \mathbf{D} \end{bmatrix} \begin{Bmatrix} \mathbf{B}^m \\ \mathbf{B}^b \end{Bmatrix} + \mathbf{B}^{sT} \mathbf{D}^s \mathbf{B}^s \right] \mathrm{d}\Omega$$

$$\mathbf{M} = \int_\Omega \begin{Bmatrix} \mathbf{N}_0 \\ \mathbf{N}_1 \end{Bmatrix}^T \begin{bmatrix} I_1 & I_2 \\ I_2 & I_3 \end{bmatrix} \begin{Bmatrix} \mathbf{N}_0 \\ \mathbf{N}_1 \end{Bmatrix} \mathrm{d}\Omega \tag{27}$$

4. Numerical results and discussion

This section reports several non-dimensional fundamental frequencies $\bar{\omega} = \omega h \sqrt{\rho_m / E_m}$ of simply supported (SSSS) plates to confirm the accuracy and the effectiveness of the proposed modified Kirchhoff theory using meshfree method. The material properties in which the first set is of Aluminum-Ziconia (Al/ZrO_2), while the second one is of aluminum and alumina (Al/Al_2O_3) can be found in **Table 1**. The effective characteristics of functional graded materials are assumed to homogenize by the rule of mixture. For the numerical calculations, FGM plate was discretized by 23×23 regular nodes with 23×23 quadrilateral background mesh for the integration and the scaling factor (α) of 2.4 was used to determine the size of the influence domain. The free vibration analyses for the isotropic Al/AL_2O_3 FG thin plate of length to thickness ratio, a/h= 100 tabulated in Table 2 have proved the comparison between the proposed theory and the conventional Kirchhoff plate theory.

As a result, the numerical solutions are acceptably given and effectively predicted by the present method, however, the higher discrepancy of conventional Kirchhoff plate theory is observed as validating to the analytical results [16]. Furthermore, considering the numerical results for the moderately thick FG Al/ZrO_2 plate of a/h=5 shown in Table 3, the computed frequencies from suggested theory are well matched to the exact 3D solutions [14] and slightly higher than others reported by SSDT[17] and Quasi-3D [15]. It is noting that the accuracy of the numerical results for the thin to moderately thick plates is dependent on the modified Kirchhoff theory irrespective of different FG material constituents, meanwhile, almost results from CPT theory are always over-estimated due to the shear part not be taken account into the calculation.

Table 1. Material properties

	Al	ZrO_2	Al_2O_3
E (GPa)	70	200	380
ν	0.3	0.3	0.3
ρ (kg/m^3)	2707	5700	3800

Table 2. The natural frequency $\bar{\omega}=\omega h\sqrt{\rho_m / E_m}$ of SSSS Al/Al_2O_3 square plate with a/h=100.

Model	a/b=1				a/b=2			
	n							
	0	0.5	1	2	0	0.5	1	2
Analytical [14]	115.8695	98.0136	88.3093	80.3517	72.3942	61.3313	55.1205	50.0743
Present	115.8383	98.0971	88.3998	80.3771	72.4684	61.3670	55.2979	50.2772
CPT	116.0355	98.2593	88.5425	80.5049	72.5024	61.3940	55.3212	50.2978

Table 3. The natural frequency $\bar{\omega}=\omega h\sqrt{\rho_m / E_m}$ of SSSS Al/ZrO_2 square plate with a/h=5.

Model	ε_z	n						
		0	0.5	1	2	3	5	10
Exact [14]		-	-	0.2192	0.2197	0.2211	0.2225	-
Quasi-3D [15]	≠0	-	-	0.2152	0.2153	0.2172	0.2194	-
SSDT [17]	0	-	-	0.2184	0.2189	0.2202	0.2215	-
SSDT [17]	≠0	-	-	0.2193	0.2198	0.2212	0.2225	-
Present	0	0.2462	0.2223	0.2190	0.2202	0.2220	0.2234	0.2223
CPT	0	0.2462	0.2542	0.2483	0.2477	0.2500	0.2529	0.2523

Table 4. Reveals the first ten normalized frequencies in the parametric study of length to thickness ratios such as a/h=5, 10, 20. The normalized frequencies $\bar{\omega}$ of the first mode are high accuracy and closed to exact results within thick to moderately thin plates, while higher modes are reasonable agreement could be observed. **Figure 2** illustrates the six dynamic vibration mode shapes of simply supported FG plate. Note that the deformations are magnified by 1000 times for clarity.

Table 4. The natural frequency $\bar{\omega}$ of SSSS Al/ZrO_2 square plate with various ratios a/h.

a/h	Model	Modes								
		1	(2,3)	4	5	6	7	8	9	10
5	Exact [14]	0.2192	-	-	-	-	-	-	-	-

	HOSNDPT [15]	0.2152	0.4114	0.4761	0.4761	0.582	0.6914	0.8192	0.8217	0.8242
	Present	0.2190	0.4108	0.4801	0.4801	0.5828	0.6978	0.8173	0.8174	0.8198
	CPT	0.2483	0.4331	0.4331	0.5939	0.5939	0.6144	0.8643	0.8645	0.9160
10	Exact [14]	0.0596	-	-	-	-	-	-	-	-
	HOSNDPT [15]	0.0584	0.1410	0.2058	0.2058	0.2164	0.2646	0.2677	0.2913	0.3264
	SSDT [17]	0.0596	0.1426	0.2058	0.2058	0.2193	0.2676	0.2676	0.291	0.3363
	Present	0.0596	0.1423	0.2054	0.2054	0.2194	0.2646	0.2647	0.2914	0.3352
	CPT	0.0635	0.1567	0.1567	0.2165	0.2165	0.2488	0.3050	0.3051	0.3072
20	Exact [14]	0.0153	-	-	-	-	-	-	-	-
	HOSNDPT [15]	0.0149	0.0377	0.0593	0.0747	0.0747	0.0769	0.0912	0.0913	0.1029
	SSDT [17]	0.0153	0.0377	0.0596	0.0739	0.0739	0.095	0.095	0.1029	0.1029
	Present	0.0153	0.0377	0.0598	0.0732	0.0732	0.0950	0.0950	0.1027	0.1027
	CPT	0.0160	0.0398	0.0398	0.0636	0.0784	0.0784	0.1026	0.1026	0.1083

Finally, parametric studies were performed to scrutinize the free vibration characteristics of the FGM plate. Table 5 and Table 6 respectively displays the effects, in that order, of the plate length-to-thickness ratio and boundary conditions of SSSC, SCSC. The letter 'S' is defined for simply supported and 'C' is denoted for the clamped edge. The material of FG Al/Al_2O_3 are also given in **Table 1**.

The first normalized natural frequencies $\bar{\omega}=\omega h\sqrt{\rho_c / E_c}$ are exhibited in Table 5 regarding to several gradient indices n(0; 0.5; 1; 4; 10) evidence accurate estimation of natural frequencies comparing to the Refs of FSDT-IGA, s-FSDT-IGA [12] and analytical results [11, 13], respectively. The non-dimensional frequencies are decreased when the index values n are increased. The similar trends could be seen regarding to different length to thickness (a/h) ratios. Furthermore, considering the effects of the different boundary conditions on the normalized frequencies, the three different length to thickness ratios $a/h = 5,10,20$ and inhomogeneous indices $n = 0,0.5,1,2,5,10$ of Al/Al_2O_3 FG plates are adopted. A good agreement can be found for the computed results of proposed method verified by the exact analytical solutions [13]. Remarkable influence of clamped increasing on the edges could make the higher dynamic vibration magnitudes as closed locking in Table 6 due to gradual increasing of the stiffness of the FG plates.

Table 5. The first normalized natural frequency $\bar{\omega}=\omega h\sqrt{\rho_c / E_c}$ of SSSS Al/Al_2O_3

a/h	Model	n				
		0	0.5	1	4	10
2	Analytical solution [11]	0.9400	0.8232	0.7476	0.5997	0.5460
	FSDT-IGA[12]	0.9265	0.8060	0.7330	0.6111	0.5640
	s-FSDT-IGA [12]	0.9265	0.8027	0.7267	0.6055	0.5620
	Present	0.9267	0.8062	0.7332	0.6113	0.5641
	CPT	0.9715	0.9467	0.9295	0.8839	0.8595
10	Analytical solution [11]	0.0578	0.0492	0.0443	0.0381	0.0364
	FSDT[13]	0.0577	0.049	0.0442	0.0382	0.0366
	FSDT-IGA[12]	0.0577	0.0490	0.0442	0.0382	0.0366
	s-FSDT-IGA [12]	0.0577	0.0490	0.0441	0.0382	0.0365

	Present	0.0577	0.0490	0.0442	0.0382	0.0365
	CPT	0.0593	0.0558	0.0545	0.0554	0.0556
20	Analytical solution [11]	0.0148	0.0125	0.0113	0.0098	0.0094
	FSDT[13]	0.0148	0.0125	0.0113	0.0098	0.0094
	FSDT-IGA[12]	0.0148	0.0125	0.0113	0.00981	0.0094
	s-FSDT-IGA [12]	0.0148	0.0125	0.0113	0.00981	0.0094
	Present	0.0148	0.0125	0.0113	0.0098	0.0094
	CPT	0.0149	0.0140	0.0137	0.0140	0.0140

Table 6. The natural frequency $\bar{\omega}$ of various boundary condition SSSS Al/Al_2O_3 square plate with various ratios a/h

Boundary condition	*a/h*	Model	Power law index (*n*)					
			0	0.5	1	2	5	10
SSSC	5	Exact [13]	5.9625	5.1188	4.6356	4.1996	3.8916	3.7146
		Present	6.0619	5.1957	4.7019	4.2605	3.9569	3.7818
		CPT	6.9319	6.4689	6.3196	6.3042	6.4355	6.4197
	10	Exact [13]	6.7751	5.7649	5.2039	4.7261	4.4462	4.2839
		Present	6.7853	5.7683	5.2051	4.7276	4.4525	4.2928
		CPT	7.0419	6.6263	6.4745	6.4660	6.6126	6.5978
	20	Exact [13]	7.0526	5.9810	5.3926	4.9019	4.6382	4.5443
		Present	7.0187	5.9501	5.3642	4.8764	4.6149	4.4630
		CPT	7.0876	6.6675	6.5150	6.5084	6.6591	6.6447
SCSC	5	Exact [13]	6.7663	5.8409	5.3039	4.8032	4.4127	4.1865
		Present	6.9847	6.0132	5.4538	4.9414	4.5585	4.3346
		CPT	8.3208	7.8387	7.6580	7.6392	7.7974	7.7779
	20	Exact [13]	8.5674	7.2715	6.5585	5.9612	5.6332	5.4423
		Present	8.4740	7.1879	6.4822	5.8931	5.5725	5.3853
		CPT	8.5964	8.0872	7.9026	7.8948	8.0773	8.0594

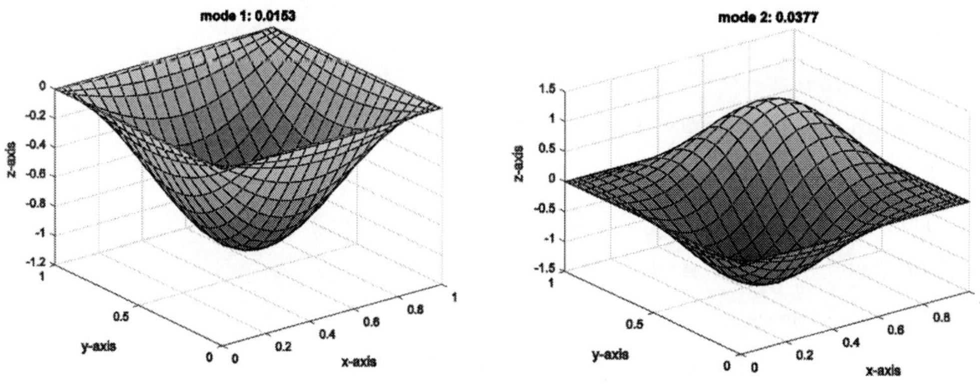

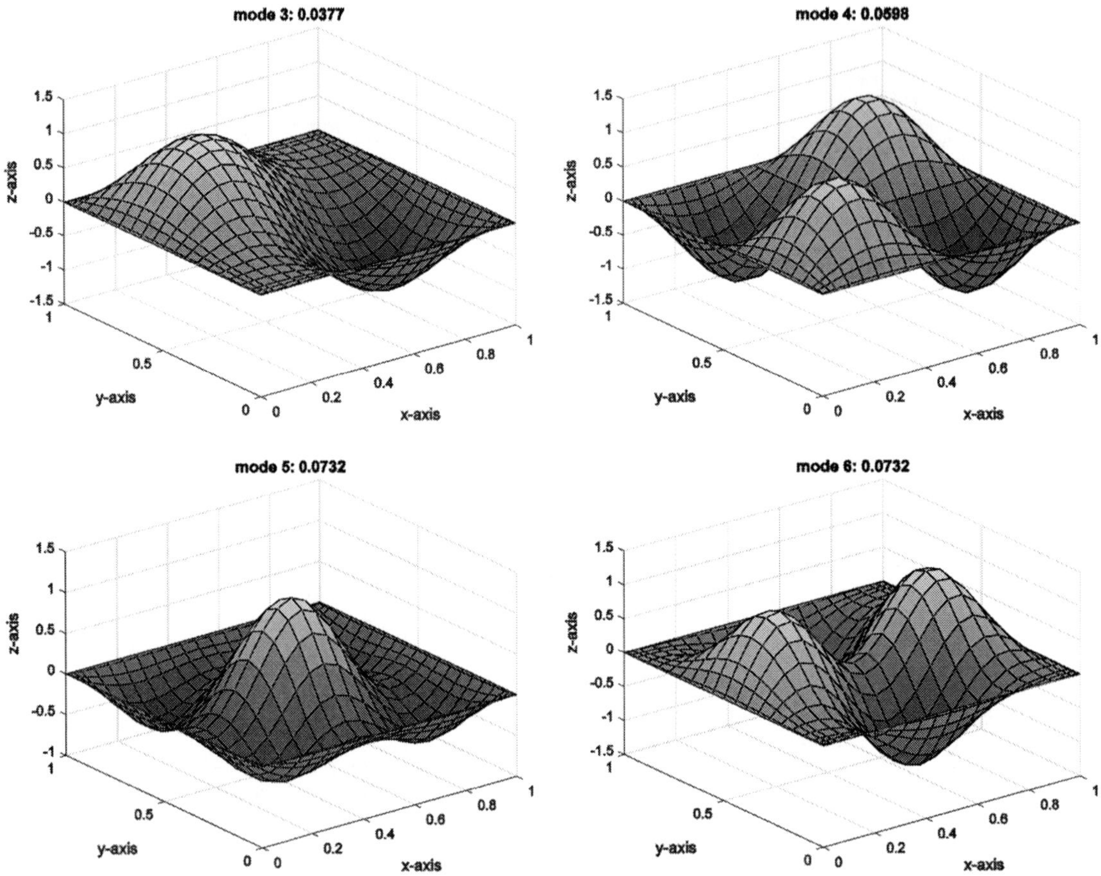

Figure 2. First six mode shapes of Al/ZrO_2 FG square plate (n=1, a/h=20)

5. Conclusion

In this paper, a simple improved Kirchhoff plate theory combines with mesh-free in which a cubic function using RPIM method for FGM isotropic plates has been proposed. Based on the results of this framework, the following conclusions can be drawn:

- The normalized cubic shape function in the radial point interpolation is chosen in the present method could give the stable solutions in the numerical analysis and better than the conventional Kirchhoff theory.
- The improved Kirchhoff theory of which the transverse deflection w divides into bending (w^b) and shear (w^s) components effectively predicts the FG plate from the moderately thick to thin plates.
- The numerical results in the parametric study approach to the exact solutions with closed form of the high order shear deformation theories such as HSDT, SSDT. It is worth concluding that the simple present theory is highly reliable to assess the results for many types of FGM plates.

Reference

[1] Bever M B, Duwez P E 1972 Gradients in composite materials *Mater Sci Eng* 10 1–8.

[2] Hasselman D P H, Youngblood G E 1978 Enhanced thermal stress resistance of structural ceramics with thermal conductivity gradient *Journal of the American Ceramic Society* 61 49-53.

[3] Shen H S 2002 Nonlinear bending response of functionally graded plates subjected to transverse loads and in thermal environments *Int J Mech Sci* 44 561–584.

[4] Dawe D J, LAM S S E, Azizian Z G 1992 Nonlinear finite strip analysis of rectangular laminates under end shortening, using classical plate theory *Int J Num Meth Eng* 35 1087-1110.

[5] Talha M, Singh B N 2010 Static response and free vibration analysis of FGM plates using higher order shear deformation theory *Applied Mathematical Modelling* 2010; 34 (12): 3991–4011.

[6] Reddy J N 1984 A simple higher-order theory for laminated composite plates *J Appl Mech* 51 745–752.

[7] Zhao X, Liew K M 2009 Geometrically nonlinear analysis of functionally graded plates using the element-free kp-Ritz method *Computer Methods in Applied Mechanics and Engineering* 198 2796–2811.

[8] Singha M K, Prakash T, Ganapathi M 2011 Finite element analysis of functionally graded plates under transverse load *Finite Elements in Analysis and Design* 47 453–460.

[9] Belytschko T, Black T 1999 Elastic crack growth in finite elements with minimal re-meshing *International Journal for Numerical Methods in Engineering* 45 601–620.

[10] Wang J G, Liu G R 2002 A point interpolation meshless method based on radial basis functions *Int J Numer Methods Eng* 54 1623–1648.

[11] Matsunaga H 2008 Free vibration and stability of functionally graded plates according to a 2D higher-order deformation theory *Compos Struct* 82 499–512.

[12] Yin S, Hale J S, Yu T, Bui T Q, *Bordas S P A 2014* Isogeometric locking-free plate element: A simple first order shear deformation theory for functionally graded plates *Composite Structures* 118 121–138.

[13] Hosseini-Hashemi S, Fadaee M, Atashipour S R 2011 A new exact analytical approach for free vibration of Reissner–Mindlin functionally graded rectangular plates *International Journal of Mechanical Sciences* 53 11–22.

[14] Vel S S, Batra R C 2002 Exact solution for thermoelastic deformations of functionally graded thick rectangular plates *AIAA Journal* 40 1021–1033.

[15] Qian L F, Batra R C and Chen L M 2004 Static and dynamic deformations of thick functionally graded elastic plate by using higher-order shear and normal deformable plate theory and meshless local Petrov-Galerkin method *Compos Part B: Eng* 35 685–697.

[16] Baferani A H, Saidi A R, Jomehzadeh E 2015 An exact solution for free vibration of thin functionally graded rectangular plates *Proc IMechE Part C: J Mechanical Engineering Science* 225 526-536.

[17] Neves A M A, Ferreira A J M, Carrera E, Roque C M C, Cinefra M and Jorge R M N 2012 A quasi-3D sinusoidal shear deformation theory for the static and free vibration analysis of functionally graded plates *Compos Part B: Eng* 43 711–725.

Acknowledgments

This research is funded by Vietnam National Foundation for Science and Technology Development (NAFOSTED) under grant number 107.02-2016.19.

IOP Conf. Series: Earth and Environmental Science **143** (2018) 012039 doi:10.1088/1755-1315/143/1/012039

Improvement in engineering properties of soft-soil using cement and lime additives: A case study of southern Vietnam

Vu To-Anh Phan[1*] Kha Ngoc-Anh Pham[2]

[1]Faculty of Civil Engineering, Ton Duc Thang University, Ho Chi Minh City, Vietnam;
[2]Vung Liem District Investment and Construction Project Management Unit

Corresponding author, E-mail address: phantoanhvu@tdt.edu.vn

Abstract. This paper presents the experimental results of using two additives to improve natural soft soil properties in southern Vietnam (i.g., cement and cement-lime mixture). The specimens were prepared by compacting method. Firstly, the natural soil was mixed with cement or cement-lime to determine the optimum water contents of various additive contents. Then, optimum water content was used to produce samples to test some engineering properties such as unconfined compressive strength, splitting tensile strength, and Young's modulus. The specimens were tested by various curing duration of 7, 14, and 28 days. Results indicated that using cement additive is suitable for improvement of soft soil in the local area and cement-soil stabilization can be replaced as the subbase layer of the flexible pavement according to current Vietnamese standard. In addition, a higher cement content has a greater compressive strength as well as tensile strength. Besides, the Young' modulus has significantly increased with a long-term curing age and more cement content. No evidences of increasing in strength and modulus are found with the cement-lime-soil stabilization. Finally, the best-fit power function is established by the relationships between unconfined compressive strength and splitting tensile strength as well unconfined compressive strength and Young's Modulus, with the coefficient of determination, R^2>0.999.

1. Introduction

The Mekong Delta is the region in Southern Vietnam; it has a total of approximately 39.734 km^2, of which 12 % is in Vietnam. The surface is low elevation, mainly from 0 to 4m above the mean sea level. Recently, certain studies have revealed that the Mekong delta was mainly covered by Holocene deposits due to the increase and decrease in sea level at the age from 20.000 to 4.000 years before present [1]. Therefore, soft soil is widely and deeply distributed in this region. The soft clay deposit is found from the surface to a depth varying from 10 m at Long An in the north to 50 m at Can Tho and approximately 20 m at Ca Mau in the South [2]. Actually, Mekong delta is also one of the biggest economical centrals in Vietnam. The infrastructure of this area remarkably is developed in recent years such as road, bridge, dam, industrial zone. Generally, any structure laid on soil is generally sustained to settlement, inclined settlement. But for soft soil, the settlement is more significant and greater. Excessive settlement is tipped to cause big problems as it often exceeds the permissible limits [3]. In fact, some constructions were unsafe as building on such soil condition such as settlement, inclined settlement, and disruption. Therefore, it is essential to find out a solution for treating the soft soil condition to ensure a good foundation design against failure as well as differential settlements.

Over the past five decades, many soil improvement techniques have been considered and employed in practice, including mechanical stabilization, stabilization using soft aggregates, bituminous stabilization, lime stabilization, cement stabilization, thermal stabilization, chemical stabilization, and electric stabilization. Various admixtures such as cement, fly ash, lime, blast-furnace slag cement, calcium chloride were used in different area on the world [4]. Particularly, in among them, cement and/or lime soil stabilizations were known as well-known method and most widely used. Compared to pure soils, soil stabilzation has more advantages such as low compressibility, high compressive strength, and its cost efficiency. Al-Rawas et al. (2005) [5] indicated that lime, cement and an artificial pozzolan can reduce significantly the swelling potential. Kavak and Akyarli (2007) [6] used lime to increased CBR from 16 and 21 times compared to initial soaked values for green and brown clays.
Due to these significant advantages. It is very essential to fully find the mechanical behavior of soil stabilization to satisfy various requirements in the field. So far, the current criterion to evaluate the mechanical behavior of soil stabilization mainly focuses upon such parameters as unconfined compressive strength, internal friction angle, and cohesion, and ignores the important role of curing stress. In general, based on the previous results, some conclusions were drawn such as when cement content increased, liquid limit decreased, shear strength parameters increased with cement content; unconfined compressive strength (UCS) and isotropic compressive strength remarkably increased with the increase in curing time, UCS significantly increases as vertical stress increases.

In Vietnam, although additives have popularly been used in practice, there were no obvious reports on the strength of using these additives in laboratory as well as in the field. The present study is addressed toward employing the different additives for improving the soft soil condition. Soil stabilization was conducted by different additives including Portland cement Blended and/or lime. The unconfined compressive strength (UCS), splitting tensile strength (STS), and Young's modulus of soil stabilization was tested. The results of study elucidated some engineering properties, some correlations from obtained data are established. The findings derived from this result are expected to contribute deeply the understanding the usage of additive as well as to meet the requirements of further design and construction purposes in the studied area and the Mekong Delta region.

2. Materials used

2.1. Natural soil

The excavated soil was collected from Vung Liem District, Vinh Long Province, located in the Mekong Delta, Southern Vietnam. Once the location has been located, the soil sample was collected from three trial pits that had been excavated to a depth of 0.5 to 1.0 m. Nearly 150 kg of natural soil was taken from these pits for this experimental program. The collected specimen was stored in a plastic bag to remain the natural moisture conditions. The basic properties of soft soil included such items as specific gravity, water content, unit weight, Atterberg limits, grain-size distribution were determined during the testing process and measured by following American Society of Testing Materials (ASTM) Standard D854, D2216, D4318, D421 and D422, respectively. Basic physical properties of samples are presented in Table 1. The grain-size distribution is plotted in Figure 1.

Table 1. Basic physical properties of excavated soil

Basic properties	Tested value
Specific gravity, G_s	2.60
Moisture content, %	35.15
Dry unit weight, kN/m^3	13.04
Liquid limit, LL, %	46.12
Plastic limit, PL, %	25.45

Plasticity index, PI, %	20.67

The consistency limits tests shows that the liquid limit (LL), plastic limit (PL), and plasticity index (PI) are 46.12 %, 25.45 % and 20.67 %, respectively. Based on ASTM D2487 classification system, the sample is classified as low plasticity organic clay (CL). According to the current design specification of Vietnam (22 TCN 211-06), the soil listed in Table 1 is soft soil and must be improved before building highway structures.

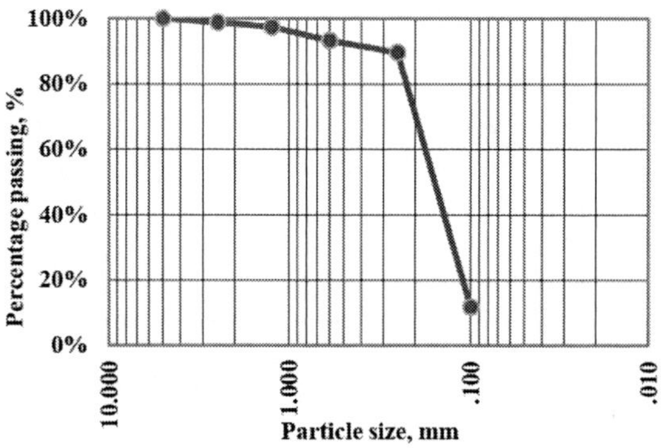

Figure 1. Grain size distribution of soft soil

2.2. Additives used

The additives used for the stabilization of soft soil are commercially available Portland cement blended and hydrated lime. Portland cement Blended is produced by mixing Ordinary Portland Cement (OPC) and mineral admixtures such as fly ash, slag, or silica fumes. PCB is now being used more popular than OPC in Vietnam due to some technical and environmental advantages. For example, the PCB significantly reduces water demand and water-cement ratio. The PCB is finer than OPC, and therefore the permeability of concrete is less and the result is to improve workability and durability. Furthermore, the manufacturing process of PCB save energy and reduce the precious minerals like limestone, clay, and silica due to using the waste products of thermal and steel plants. PCB is the most common stabilizer in treating the soft soil condition. The basic properties of PCB 40 are tabulated in Tables 2.

Table 2. Basic properties of PCB 40

Basic properties	Tested value
Compressive strength, MPa, 28 days	45.00
Setting time, min.	
Initial	120
Final	205
Fineness, cm^2/g	4050
Specific gravity	3.15

Quicklime is used in this study. The basic chemical properties can be found more than 85% of Calcium oxide, and the percentage passing of a 0.075m sieve is more than 90 %.

3. Specimen preparation

The disturbed soil specimens were dried for a duration of 24h and then pulverized using a wooden hammer. The pulverized soil was then completely dried by placing it in a drying oven at the temperature of 100 ± 5°C. The dry specimen was sieved through a 5.00 mm sieve to obtain a uniform soil mixture. The dry soil specimen was uniformly mixed with the predetermined percent of the additive. The first set was mixed with cement at 8%, 10 % and 12 % by dry weight and the other set was mixed with cement-lime at 8 % cement + 2 % lime, 8 % cement + 4 % lime, 8% cement + 6 % lime by dry weight of soil. The soil-cement and soil-cement-lime mixtures were mixed at optimum moisture content to obtain homogeneous mixtures. All specimens were cured at room temperature (27 ± 0°C) and relative humidity 90% for durations varying from 7, 17, 28 days. Thereafter, the specimens were placed in an airtight plastic bag to maintain the test environment. For each content of the additive, nine specimens were prepared for each mixture corresponding to the durations varying from 7, 14, 28 days. The obtained data are in mean value of three specimens. The optimum water content is determined by the Standard Proctor compaction test. Table 3 summarizes mix proportions for the experimental program. Figure 2 presents some typical equipments in this study, including compaction, UCS, STS, and Young's modulus tests.

Table 3. Mix proportions

Mix code	Mix proportions			
	Cement (%)	Lime (%)	Soil (%)	Water / (Additive +soil) (%)
1	0	0	100	11.40
2	8	0	92	11.10
3	10	0	90	10.90
4	12	0	88	10.80
5	8	2	90	10.90
6	8	4	88	11.00
7	8	6	96	11.20

(a) (b) (c)

Figure 2. (a) Unconfined compressive test; (b) Splitting tensile test; (c) Young's modulus test

4. Results and discussion

4.1. Compaction characteristics (UCS)

The variations in optimum moisture content (OMC) and maximum dry density (MDD) for various cement and lime-cement contents are plotted in Figure 3. It is clear that a decrease in MDD for cement-soil stabilization is obtained from 1.886 g/cm^3 to 1.839 g/cm^3; while for cement-lime-soil stabilization is to be decreased from 1.869 g/cm³ to 1.861g/cm³. However, the decrease in MDD of lime-cement treated soil is insignificant. The OMC of cement-soil stabilization increases from 10.7% to 11.1%, while in the case of cement-lime treated soil increases from 10.9% to 11.2%. The decrease in MDD can be attributed to the chemical reactions that play a role in reducing the density of the cement or lime mixtures. Additionally, the addition of stabilizers leads to an increase in the specific surface area of the mixture which thereby increases the OMC. The cation exchange reaction further intensifies the increase in OMC since the reaction utilizes higher amount of water to mobilize the Ca^{2+} ions that are introduced into the mixture by the stabilizers. The similar type of observations have been reported by other researchers [7].

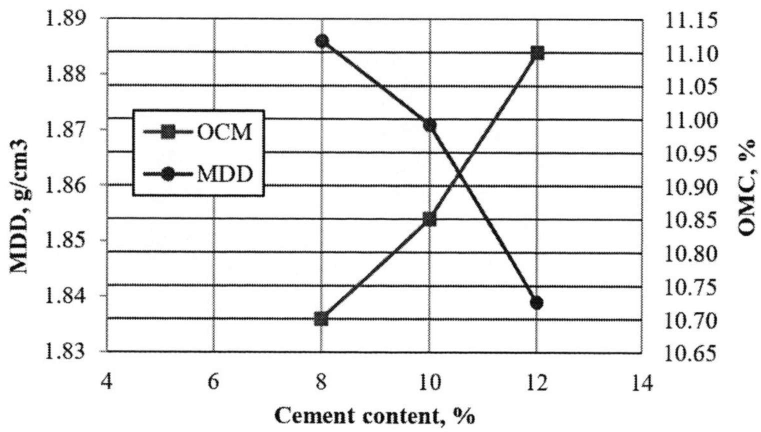

(a)

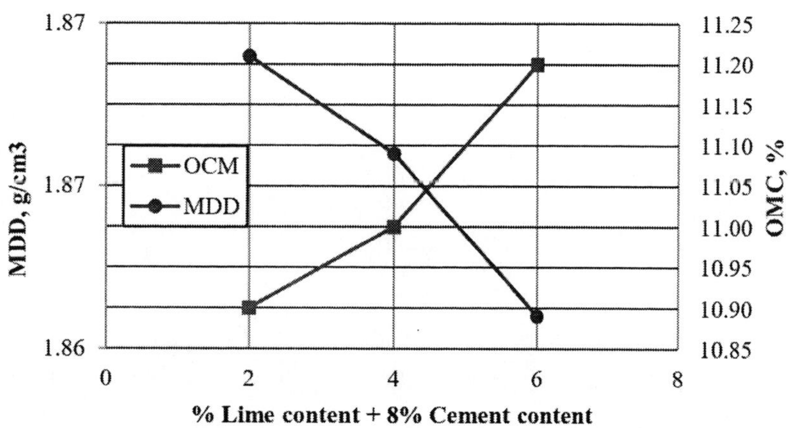

(b)

Figure 3. Compaction characteristics

4.2. Unconfined compressive strength (UCS)

The UCS obtained for all the mixing ratios can be plotted in Figure 4. This figure shows the effect of curing time on the unconfined compressive strength. In general, a cement-soil mixture increases the compressive strength when cement content increases. Adding 12% cement increases UCS from 3.94 MPa to 5.18 MPa, for the specimens have been cured for a duration of 28 days. The specimen improves the strength at long-term curing age. The strength development is thought as the calcium silicate hydrates (C-S-H) are established as the cement content is hydrated. As cement content increases, C-S-H gels form in greater amounts, increasing the compressive strength [8]. As expected, this behavior is also true for three cement contents. However, considering lime-cement treated soil, the compressive strength increases until 4% lime is added, then it slowly decreased with a further increasing in lime content. According to current Vietnamese Specification [9], 8% cement or more content can be substituted for crushed aggregate using in subbase layer of flexible pavement.

A comparative analysis between the effect of lime and cement suggests that the UCS of cement-soil stabilization is higher than the cement-lime stabilization. The reduction of strength is more prominent in the case of lime-blended soil as compared to soil-cement mixture. An increase in the addition of tabilizers does not favour the pozzolanic reaction. The phenomenon of decrease of UCS on the addition of large quantity of additives is more prominent in the case of lime [7].

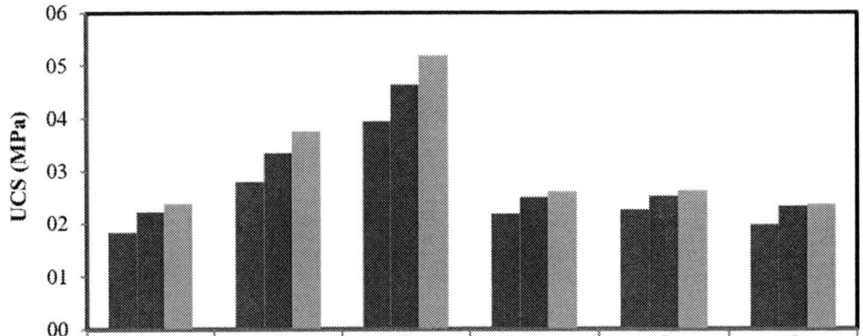

	1	2	3	4	5	6
■ 7-day	1.83	2.8	3.94	2.18	2.26	1.97
■ 14-day	2.22	3.34	4.63	2.5	2.52	2.33
■ 28-day	2.38	3.75	5.18	2.6	2.62	2.36

Figure 4. Unconfined compressive test: (1) 8%C, (2) 10%C; (3) 12%C; (4) 8%C+2%L; (5) 8%C+4%L; 8%C+12%L

4.3. Splitting tensile strength (STS)

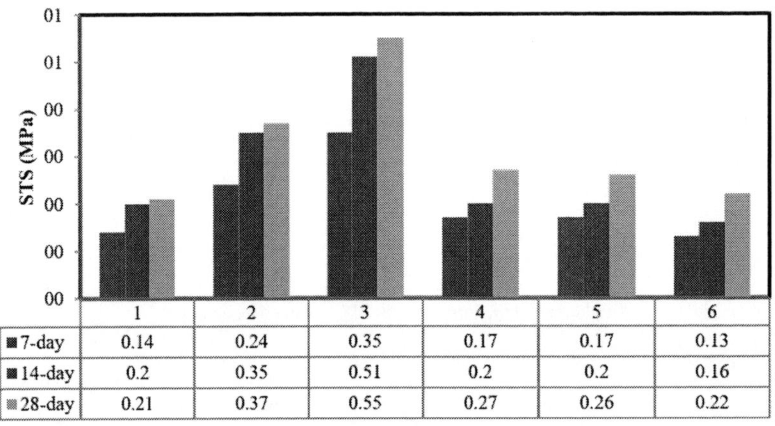

	1	2	3	4	5	6
■ 7-day	0.14	0.24	0.35	0.17	0.17	0.13
■ 14-day	0.2	0.35	0.51	0.2	0.2	0.16
■ 28-day	0.21	0.37	0.55	0.27	0.26	0.22

Figure 5. Splitting tensile strength: (1) 8%C, (2) 10%C; (3) 12%C; (4) 8%C+2%L; (5) 8%C+4%L; 8%C+12%L

The splitting tensile strength, mean value of three specimens, of different stabilized mixtures cured for 7-, 14-, 28-day are summarized in Figure 5. All cylindrical groups have the STS for curing age of 7-, 14-, 28-day in a range of 0.13-0.35 MPa, 0.16-0.51 MPa, and 0.21-0.55 MPa, respectively. In general, a longer age of curing is greater STS because of cemented-hydration. For instance, 28-day STS is 0.55, 0.37, and 0.21 MPa for 1-, 2-, 3-soil stabilization at curing age of 28-day, respectively. Accordingly, the closed relationship between the additive content and splitting strength of soil stabilization is realized such as the higher content of used additive, as shown in Figure 5, yields greater STS of soil stabilization. Similar to UCS, STS of cement-lime-soil stabilizations generally show the decreasing tendency when the lime content increases from 4% to 6%. This is likely because more lime is expected to reduce the reaction of C-H-S, and the result is to decrease the STS.

4.4. Young's modulus, E

Figure 6 presents the result of Young's modulus of soil stabilization with different additives at various curing duration of 7-, 14-, 28-day. The data presented in this figure are defined by the ratio of compressive stress and strain. As shown in this figure, the Young's modulus increases with the increasing in curing age and additive content. In general, when lime is added to the mixture, the Young's modulus slowly increases until the lime content reaches 4%, then it slowly decreases with the more lime content. In addition, cement-soil stabilization has larger Young's modulus than other cement-lime stabilizations.

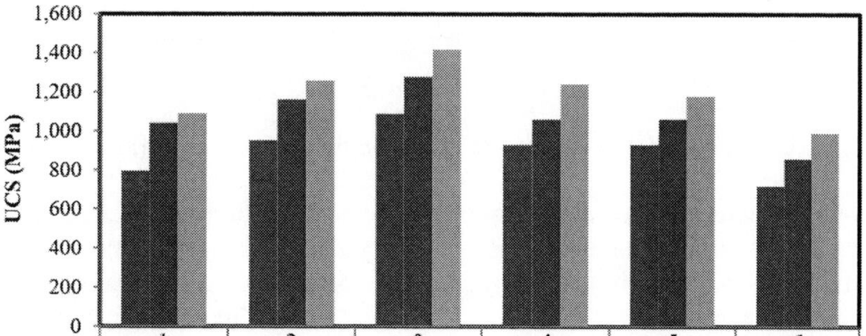

	1	2	3	4	5	6
■ 7-day	796	951	1088	930	931	721
■ 14-day	1041	1160	1275	1060	1061	858
■ 28-day	1089	1257	1416	1240	1179	990

Figure 6. Young's modulus: (1) 8%C, (2) 10%C; (3) 12%C; (4) 8%C+2%L; (5) 8%C+4%L; 8%C+12%L

4.5. Correlations from obtained data.

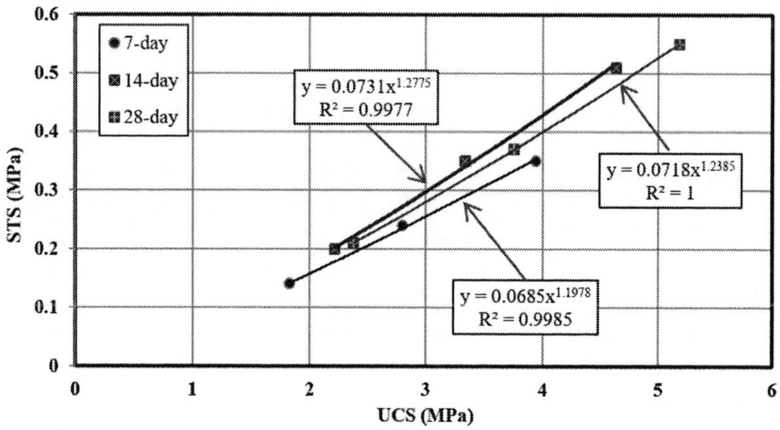

Figure 7. Relationships between Unconfined compressive Strength and Splitting Tensile Strength

The closed relationships between UCS and STS were realized, as plotted in Figure 7, higher UCS yields greater STS of soil stabilization. Also, there are best-fit power functions displayed the relationship between UCS and STS (R^2 ≥0.99). The results are suitable for three ages of 7-, 14-, 28-day.

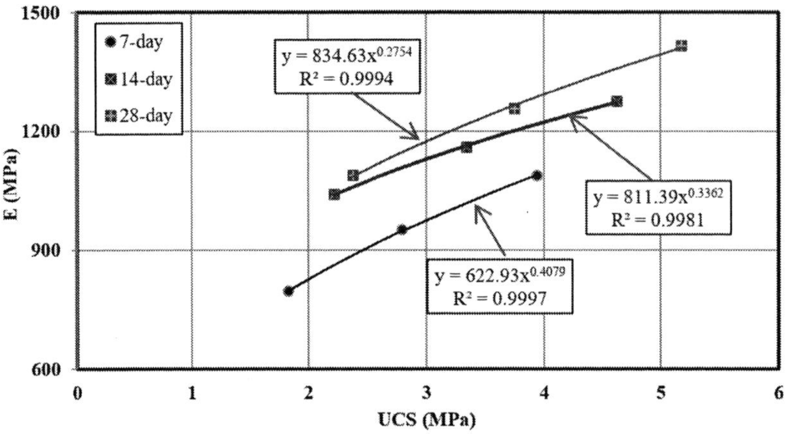

Figure 8. Relationships between Unconfined Compressive Strength and Young's Modulus

In practice, Young's modulus is usually employed to calculate the foundation engineering as well as bearing capacity of subgrade. There, it is essential to evaluate Young's modulus to give how much increase in Young's Modulus the unconfined compressive strength and answer is plotted in Figure 8. According to the obtained results, it concluded that the Young's modulus increases with the increase in unconfined compressive strength. Also, there are best-fit power functions displayed the relationship between Young's Modulus and unconfined compressive strength. Similar to Figure 7, Figure 8 presents relationships between Unconfined Compressive Strength and Young's Modulus for cement-soil stabilization. Also, there are best-fit power functions displayed the relationship between UCS and STS (coefficient of determination, R^2 ≥0.99). The results are suitable for three ages of 7-, 14-, 28-day. Based on the formula presented in Figures 7-8, it is easy to predict the STS and Young's modulus once UCS is obtained in the laboratory.

5. Conclusions

Based on the obtained data, some potentially important aspects of soil stabilization can be drawn as follows:

- The addition of stabilizers leads to an increase in the specific surface area and chemical reactions of the mixture which thus increase the optimum moisture content and reduce the density of the mixture.
- Portland cement blended can be considered as the suitable additive which can be applied to improve the soft soil condition in the study area. Adding lime content does not show the increasing in strength due to reducing the reaction of C-H-S in the mixture.
- The soil stabilization with the cement content of 8 % or more can be replaced for the subbase layer in flexible pavement design according to the current Vietnamese Specification.
- Some equations for predicting splitting tensile strength and Young's Modulus development based on unconfined compressive strength have been reasonably established. However, the suggested formula was obtained by limited results tested in the laboratory and they should be further considered and corrected in the practical uses.

References

[1] Ta T K O, Nguyen V L, Tateishi M, Kobayashi I, Tanabe S and Saito Y 2002 Holocene delta evolution and sediment discharge of the Mekong River, southern Vietnam *Quaternary Science Reviews* **21** 1807-1819
[2] Giao P H, Dung N T and Long P V 2008 An integrated geotechnical–geophysical investigation of soft clay at a coastal site in the Mekong Delta for oil and gas infrastructure development *Canadian Geotechnical Journal* **45** 1514-1524
[3] Craig R F 2004 *Craig's soil mechanic* (11 New Fetter Lane, London EC4P 4EE)
[4] Bardet J P 1997 *Experimental soil mechanics*: Prentice-Hall, Inc. United States)
[5] Al-Rawas A A, Hago A W and Al-Sarmi H 2005 Effect of lime, cement and Sarooj (artificial pozzolan) on the swelling potential of an expansive soil from Oman *Building and Environment* **40** 681-687
[6] Kavak A and Akyarlı A 2007 A field application for lime stabilization *Environ Geol* **51** 987-97
[7] Sharma L K, Sirdesai N N, Sharma K M and Singh T N 2018 Experimental study to examine the independent roles of lime and cement on the stabilization of a mountain soil: A comparative study *Applied Clay Science* **152** 183-195
[8] Barišić I, Dimter S and Rukavina T 2014 Strength properties of steel slag stabilized mixes *Composites Part B: Engineering* **58** 386-391
[9] TCVN 8858: 2011 Cement Treated Aggregate Bases for Road Pavement - Specification for Construction and Acceptance.

IOP Conf. Series: Earth and Environmental Science **143** (2018) 012040 doi:10.1088/1755-1315/143/1/012040

Analysis of Mechanical Behavior of the Callovo-Oxfordian Argillite

Huy T Tran

Faculty of Civil Engineering, Ton Duc Thang Univerity, 19 Nguyen Huu Tho, HCM city, Vietnam

Corresponding author. Email: trantanhuy@tdt.edu.vn

Abstract. This paper presents an overview of analysis the mechanical behavior of argillite; FE simulation uses an elastoplastic model that has been developed based on a competition between an anisotropic ductile behavior (direction of the subsidence's strain) and an anisotropic semi-fragile behavior (direction of the extension's strain), a nonlinear damage cumulative is considered to evaluate the creation of discontinuities due to material degradation. FE results matches well with the experimental results on the core samples. Consequently, the model shows that it can be applied to evaluate the retaining walls of future storage structures of nuclear waste.

1. Introduction

The ANDRA, French National Radioactive Waste Management Agency, has the role of designing, carrying out and exploiting centers of storage dedicating to the radioactive waste. This radioactive waste is very dangerous because of its high-level and long-lived property [1]. The research aims to prevent the dispersion of the radioelement in the environment and isolate them from any contact with the man for very long time (thousands of years). Therefore, the underground research laboratory (URL) Meuse and Haute Marne is built in Bure, France to evaluate the behavior of the argillite of Bure (Callovo-Oxfordian) that refers to as a geological barrier of the radionuclides containment [1, 2, 3]. The evaluation of the storage safety on the Meuse/haute Marne argillite requires mechanical behavior models of the rock, to calculate the distribution of the mechanical perturbations in terms of extension of damaged and fractured zones. This zone of damage rock, often known as the Excavation Damaged Zone (EDZ) is of importance for storage structure where the host rock plays a confining role.

In this context, the objective of this paper is limited to a model constitutive of the elastoplastic type associated an anisotropic damage to describe the mechanical behavior of argillite of Callovo-Oxfordian under mechanical loadings.

2. Method

2.1. Model formulation

According to previous results, conceptual model of Cox claystone is based on:

- Continuity of the stress-strain relation [1].
- Interaction between solid phase and interstitial fluid with an adaptation of Biot theory [2, 3].
- Internal reinforcement of the argillaceous phase [1].
- Elastoplastic behavior with an anisotropic damage [4, 5, 6].

In its initial state, the argillite is considered as undisturbed, with insignificant anisotropy and without damage. Under solicitation, the resulting behavior is a competition between the ductile behavior in the direction of the settlement deformations and the semi-fragile behavior in the directions where shear and extension take place [4]. The overall strain tensor is separated into the elastic strain and the plastic strain:

$$\underline{\underline{\varepsilon}} = \underline{\underline{\varepsilon^e}} + \underline{\underline{\varepsilon^p}} \tag{1}$$

Elastoplastic ductile behavior is characterized by a large elastic domain with a diagenesis yield surface F_D and a collective surface F_P. The argillite behaves elastically when P-Q relation is within the collective surface FP and elastoplastically when P-Q relation is beyond FP.

$$\tilde{F}_D = \tilde{G}_D = (p',q,\theta) = q^2 - 0.5L^2(p' + \tilde{p}'_{cohe})^2 a(1 - \frac{p'}{\tilde{p}'_{cf}}) = 0$$

$$\tilde{F}_p(p',q,\theta) = q^2 - L^2 \times (p' + \tilde{p}'_{cohe})^2 a \times (1 - \frac{p'}{\tilde{p}'_{cf}}) = 0 \tag{2}$$

$$\tilde{G}_P(p',q,\theta) = q^2 - (p')^2 \times \left(\left(\frac{p'_g}{p'} \right)^{1/2r} - 1 \right)^{2r} = 0$$

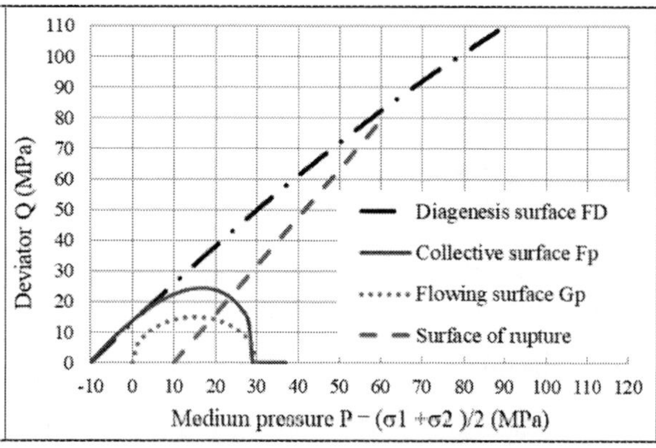

Figure 1. Surface of charges.

Semi fragile damageable elastic behavior is related to anisotropic damage. The damage is induced by solicitation concerning the carbonate phase, and by hypo-elasticity describing the behavior of the argillaceous phase [6]:

$$\varepsilon_{ij}^e = \left(S_{ijkl}\left(\underline{\underline{D}}\right) + S_{ijkl}\left(P'\right)\right) \times \sigma'_{kl} \tag{3}$$

The plasticity deformation derives from the surface of flowing GP with:

$$d\varepsilon_{ij}^p = dh \frac{\partial G}{\partial P} \tag{4}$$

2.2. Damage variables

The damage in the material is represented by a variable corresponding to an average material degradation affecting stiffness, strength, anisotropy. The anisotropic damage is described by a tensor $\underline{\underline{D}}$

$$\underline{\underline{D}} = \sum_{i=1}^{3} D_i n_i \otimes n_i \tag{5}$$

Where D_i and n_i are the principal value and the unit vector of principal direction of the tensor $\underline{\underline{D}}$, respectively.

2.3. New formulation of elastoplastic model with anisotropic damage

According to the microscopic observations on argillite samples before and after mechanical test [9], the argillite of Bure has an anisotropic damage induced by the existence of the strain in extension.
The damage is related to the existence of the strain in extension that is obtained from the spectral decomposition of the elastic strain tensor [7,8] such as:

$$\underline{\underline{\varepsilon}} = H(-n_i.\underline{\underline{\varepsilon}}.n_i)(n_i.\underline{\underline{\varepsilon}}.n_i)n_i \otimes n_i = \underline{\underline{P^+(\varepsilon)}} : \underline{\underline{\varepsilon}} \tag{6}$$

The quantity $\underline{\underline{P}}$ is a fourth–order tensor corresponding to a negative projection operator.

$$\underline{\underline{P}} = H(-n_i \underline{\underline{\varepsilon}} n_i)(n_i \otimes n_i \otimes n_i \otimes n_i) \tag{7}$$

Where n_i is the principal directions of the damage and H is the Heaviside function defined by

$$H(x) = \begin{cases} 1 & if \quad x < 0 \\ 0 & if : x > 0 \end{cases} \tag{8}$$

Elsewhere, one of the bases of the model is the coupling between the strain tensor $\underline{\underline{\varepsilon}}^+$ and the damage tensor $\underline{\underline{D}}$ by the energy of Helmholtz.

The free energy Ψ is the contribution of the decreased elastic energy of the dissipation energy due to the damage. Namely, we will postulate the following expression for the Helmholtz free energy [8,9]:

$$\rho\Psi = \frac{1}{2}\underline{\underline{\varepsilon}}^e : \underline{\underline{\tilde{C}(D)}} : \underline{\underline{\varepsilon}}^e + \Psi_p(\gamma_p, \underline{\underline{D}}) = \frac{1}{2}(\underline{\underline{\varepsilon}} - \underline{\underline{\varepsilon}}^p) : \underline{\underline{\tilde{C}(D)}} : (\underline{\underline{\varepsilon}} - \underline{\underline{\varepsilon}}^p) + \Psi_p(\gamma_p, \underline{\underline{D}}) \tag{9}$$

In the model, the Helmholtz specific free energy was used as potential, Ψ_p is a function translating the energy blocked by plastic hardening, $\underline{\underline{\tilde{C}(D)}}$ is the fourth order tensor or the tensor of damage material rigidity.

$$\underline{\underline{\sigma}} : \underline{\underline{\varepsilon}}^* - \rho\Psi^* \geq 0 \tag{10}$$

An elastic law of behavior is characterized by a relation between the stress vector and the strain vector. By definition, the stress is derived from potential

$$\underline{\underline{\sigma}} = \rho \frac{\partial \Psi}{\partial \underline{\underline{\varepsilon}}^e} = \underline{\underline{\tilde{C}(D)}} : \underline{\underline{\varepsilon}}^e = \underline{\underline{\tilde{C}}} : (\underline{\underline{\varepsilon}} - \underline{\underline{\varepsilon}}^p) \tag{11}$$

Thus, the damaged rigidity matrix is:

$$\underline{\underline{\tilde{C}(D)}} = \underline{\underline{M(D)}} : \underline{\underline{C}}^o : \underline{\underline{M(D)}} \tag{12}$$

With : $\underline{\underline{M(D)}} = \delta_{ik}.\delta_{jl}.\sqrt{(1-D_i).(1-.D_j)}n_i \otimes n_j \otimes n_k \otimes n_l$ and $\underline{\underline{C}}^0$ the initial tensor of rigidity, $\underline{\underline{C}}^0$ depends on the Young modulus E and the Poisson coefficient of the isotropic elastic linear material.

$$\Rightarrow \underline{\underline{\tilde{C}}} = \begin{bmatrix} C_1.(1-D_1)^2 & C_2(1-D_1)(1-D_2) & C_2(1-D_1)(1-D_3) & 0 & 0 & 0 \\ C_2(1-D_1)(1-D_2 & C_1.(1-D_2)^2 & C_2(1-D_2)(1-D_3) & 0 & 0 & 0 \\ C_2(1-D_1)(1-D_3) & C_2(1-D_2)(1-D_3) & C_1.(1-D_3)^2 & 0 & 0 & 0 \\ 0 & 0 & 0 & C_3(1-D_2)(1-D_3) & 0 & 0 \\ 0 & 0 & 0 & 0 & C_3(1-D_1)(1-D_3) & 0 \\ 0 & 0 & 0 & 0 & 0 & C_3(1-D_1)(1-D_2) \end{bmatrix}$$

$$C_1 = \frac{E.(1-\nu)}{(1+\nu).(1-2.\nu)}; C_2 = \frac{E.\nu}{(1+\nu).(1-2.\nu)}; C_3 = \frac{E.(1-2.\nu)}{2.(1+\nu).(1-2.\nu)}$$

The damage appears when the equivalent strain reached its limit, associated on the actual damage state.

$$\vec{f}(\hat{\varepsilon}, \underline{\underline{D}}) = \begin{cases} f^1(\hat{\varepsilon}, D_1) = \dfrac{D_1}{D_1 - 1} + \sqrt{\alpha(a.\hat{\varepsilon} + b.\hat{\varepsilon}^2)} = 0 \\ f^2(\hat{\varepsilon}, D_2) = \dfrac{\beta(a + 2b\hat{\varepsilon})}{\sqrt{a.\hat{\varepsilon} + b.\hat{\varepsilon}^2}} + \dfrac{1 - D_2}{1 - D_1} = 0 \\ f^3(\hat{\varepsilon}, D_3) = f^2(\hat{\varepsilon}, D_2) = 0 \end{cases} \tag{13}$$

Where α, β, a, b are four material parameters which depend on the medium pressure p and the elastic characteristics of intact argillite (E, ν).

2.4. Model parameters

Base on the experimental results of a cyclic oedometric path, we observe that the loading and unloading oedometric path does not modify the distribution of the pores and does not produce appearance of the cracks. The initial value of the Young modulus and the Poisson coefficient depend on the effective mean stress. We have:

$$E_i = 2400 + 5,3\frac{P_{conf}}{0,1} \qquad \nu = 0,25 - 0,00195\frac{P_{conf}}{0,1} \tag{14}$$

The damage coefficients D_i are:

$$D_1 = 1 - \frac{1}{1 + \sqrt{\alpha(a\varepsilon_2^+ + b(\varepsilon_2^+)^2)}} \qquad D_2 = 1 + (1 - D_1)\frac{\beta(a + 2b\varepsilon_2^+)}{\sqrt{a\varepsilon_2^+ + b(\varepsilon_2^+)^2}} \tag{15}$$

Determination of the material parameters damaged is simulated by the drained triaxial path.

$$\alpha = (0.017E)^2 \quad ; \quad a = 0.0047p + 0.0185 \qquad \beta = 0.5\nu \quad ; \quad b = -0.242p + 2.21 \tag{16}$$

3. Results and discussion

Figure 2 shows six curves, three of which describe the relation between the deviator stress and vertical deformation (ε_1 – Q) and the other three describe the relation between the deviator stress and horizontal deformation (ε_2 – Q). Three types of curves are presented: elastic model, elastoplastic model and the experimental data. The last is obtained from drained triaxial compression tests for two confining pressures and a value of null suction.

The results in simulation and experiment is agreement (34 – 37 MPa), but a little bit Deviator stress–Deformation curve in analysis is smaller than the experiment data because we do not mention about the change of parameters like Young modulus, compressibility factor, preconsolidation pressure … during the triaxial compression path.

When the elastic model is utilized, the Deviator stress– Deformation relation is linear. However, this relation in experimental context is consistent with the elastoplastic model. In addition, the deformation from the elastic model is less than the vertical deformation from the elastoplastic model and data because the plastic deformation and the anisotropic damage are not embedded.

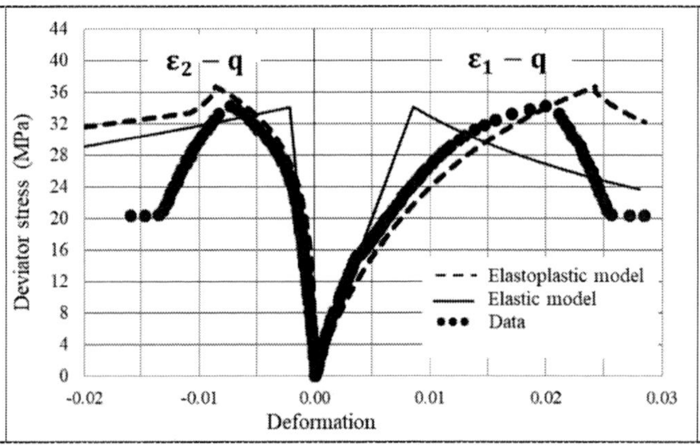

Figure 2. Evolution of deviator stress in the compression test.

To sum up, the suggested model is able to evaluate stress on the retaining walls of future storage structures of nuclear waste as shown in Figure 3.

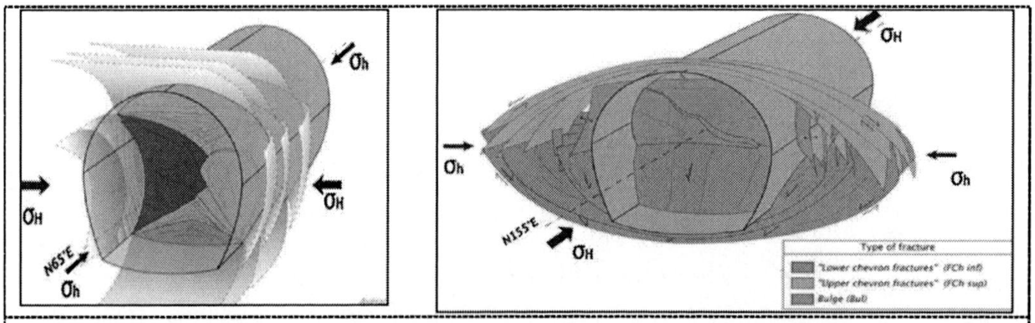

Figure 3. Discontinuity observed in research laboratory [4].

The retaining wall is created after excavation of circular tunnel of 2.6m radius. The process is at 517m of depth of Callovo-Oxfordian argillite [4].The research laboratory is then built up to evaluate the behavior of argillite, especially the long – term one like discontinuity and anisotropic convergence.

4. Conclusion

Argillite's behavior of Callovo-Oxfordian is very contrast: ductile in compression and semi-fragile in extension. A different model is produced to evaluate its mechanical behavior. The model presented is based on experimental results and microscopic observation, it reproduces the effects induced by degradations of the cementation by the extension strain. The model is elastoplastic characterized by two load surfaces, one of diagenesis and the other called collective that is non-associative plastic. The analysis data and experiment are quite similar. The difference between deviator stress of the model and that of the experimental context is only 6%. Therefore, the model is capable to evaluate stress and the discontinuity on the retaining walls of future storage structures of nuclear waste.

References

[1] Robinet J-Ch, Sardini P, Coelho D and al. 2012 Effects of mineral distribution at mesoscopic scale on solute diffusion in a clay-rich rock: example of the Callovo-Oxfordian mudstone *Water research resources* vol **48** W05554

[2] Gasc-Barbier M, Chanchole S, Bérest P 2004 Creep behavior of Bure clayey rock *Applied Clay Science* vol **26** pp 449–458

[3] Sémété P, Imbert C, Desgree P, Février B, Courtois A and Touzé G 2008 *Experimental study of the water permeability of a partially saturated argillite* Geomaterials and Applications pp 219–230

[4] Armand G, Noiret A, Zghondi J and Seyedi DM 2013 Short and long-term behaviors of drifts in the Callovo-Oxfordian claystone at the Meuse/Haute-Marne Underground Research Laboratory *Rock Mechanics and Geotechnical Engineering* vol **5** pp 221–230

[5] Zhang C and Rothfuchs T 2004 Experimental study of the hydro-mechanical behaviour of the Callovo-Oxfordian argillite *Applied Clay Science* vol **26** pp 325–336

[6] Supartono F and Sidoroff F 1985 Anisotropic damage modeling for brittle elastic materials *Archives of Mechanics* vol **37** pp 521–534

[7] Mazars J 1986 A description of micro and macroscale damage of concrete structures *Engineering Fracture Mechanics* vol **25** pp 729–737

[8] Lee DH, Hsein Juang C, Lin HM and Yeh SH 2002 Mechanical behaviour of Tien-Liao mudstone in hollow cylinder tests *Canadian Geotechnical* vol **39** pp 744–756

[9] Challamel N, Lanos C & Casandjian C 2005 Strain-based anisopic damage modelling and unilateral effects *Mechanical Sciences* vol **47** pp 459–473

Tentative to use wastes from thermal power plants for construction building materials

BUI Quoc-Bao*, PHAN To-Anh-Vu, TRAN Minh-Tung, LE Duc-Hien

Sustainable Developments in Civil Engineering Research Group, Faculty of Civil Engineering, Ton Duc Thang University, Ho Chi Minh City, Vietnam

* Corresponding author. Email: buiquocbao@tdt.edu.vn

Abstract. Thermal power plants (TPP) generates wastes (bottom and fly ashes) which become a serious environmental problem in Vietnam. Indeed, although in several countries fly ash can be used for cement industry, fly ash from actual TPP in Vietnam does not have enough good quality for cement production, because the fly ash treatment phase has not yet included in the generations of existing Vietnamese TPP. That is why bottom ash and fly ash purely become wastes and their evacuation is an urgent demand of the society. This paper presents an investigation using fly and bottom ashes in the manufacturing of construction materials. The main aims of this study is to reduce environmental impacts of fly and bottom ashes, and to test another non-conventional binder to replace cement in the manufacture of unburnt bricks. Several proportions of fly ash, bottom ash, cement, gravel, sand and water were tested to manufacture concretes. Then, geopolymer was prepared from the fly ash and an activator. Specimens were tested in uniaxial compressions. Results showed that the cement concrete tested had the compressive strengths which could be used for low rise constructions and the material using geopolymer could be used for non-load-bearing materials (unburnt bricks).

1. Introduction

Thermal power plants (TPP) generates wastes (bottom and fly ashes) which become a serious environmental problem in Vietnam. Indeed, although in several countries fly ash can be used for cement industry, fly ash from actual TPP in Vietnam does not have enough good quality for cement production, because the fly ash treatment phase has not yet included in the generations of existing Vietnamese TPP. That is why fly ash and bottom ash purely become wastes and their evacuation is an urgent demand of the society.

On the other hand, to reduce environmental impacts in the construction factor, Vietnamese government has demanded to replace clayed burnt bricks by other more "green" materials and unburnt bricks are the current strategy adopted by construction companies. However, unburnt bricks still have several difficulties to find their places on the construction site in Vietnam, due to several reason: economic, environment, tradition. One of the main reasons is the use of important cement amount in the unburnt brick manufacturing (about 25% by weight), this increased both the manufacturing cost and the carbon footprint.

This paper presents an investigation on the use of fly ash and bottom ash from a Vietnamese TPP (Vinh Tan) in the manufacturing of construction materials. The compositions of these fly and bottom ashes are analyzed following ASTM standard 0 and the results are presented in Tables 1 and 2 [2]. The aim of this study is to reduce environmental impacts of fly ash and bottom ash which are actually considered as waste in Vietnam. Several proportions of fly ash, bottom ash, cement, gravel, sand and water are investigated to manufacture low strength materials which can be used for non-load-bearing walls.

Table 1. Composition of fly ash studied, in %.

Carbon	Hydro	Nitrogen	Sulfur	Oxygen	SiO_2	Al_2O_3	Fe_2O_3	TiO_2	CaO	MgO	K_2O	Na_2O	P_2O_3	MnO	SO_3	Other
12.41	0.05	0.15	0.04	0.05	57.7	25.72	6.79	0.85	0.97	1.45	5.03	0.43	0.28	0.07	0.18	0.53

Table 2. Composition of bottom ash studied, in %.

Carbon	Hydro	Nitrogen	Sulfur	Oxygen	SiO_2	Al_2O_3	Fe_2O_3	TiO_2	CaO	MgO	K_2O	Na_2O	P_2O_5	MnO	SO_3	Other
10.44	0.09	0.14	0.05	0.07	57.54	23.47	8.76	0.70	1.57	1.27	5.12	0.21	0.20	0.11	0.12	0.93

2. Specimens manufacturing

2.1. Materials used

Gravels and sand are classical materials used for conventional concrete. The maximal diameters for sand and gravels are of 4mm and 20 mm, respectively. The optimum mix between Sand and Gravel was determined by using Dreux-Gorisse's method [3], which gave a ratio of Sand/Gravel = 2/3 (by weight and volume).

The cement used was a current Porland cement, named PC 30, where a specified compressive strength at 28 days of the cement mortar (1 cement : 3 sand by weight) was 30 MPa. Because the scope of this study was to manufacture low-strength materials with reasonable costs, other higher quality cements were not tested.

The specific densities of the fly and bottom ashes used were determined following Vietnamese standard. Results gave mean densities of 2.26 and 2.24 respectively for fly ash and bottom ash. The finesse modulus of bottom ash was also obtained from sieving analysis, giving a value of 2.7.

2.2. Concrete materials using fly ash and bottom ash

First, a classical concrete was formulated, as a reference for the comparison. The reference concrete corresponds to a 450kg of cement in 1 m^3 concrete. The composition of this referent concrete was calculated following the well-known Dreux-Gorisse's method. The water was chosen (W/C=0.55) to give a plastic concrete with a slump value of 7 ± 1cm, which is the current value for current concretes. The result is detailed in Table 3. This reference case is named "18%C" (18% cement).

Table 3. Composition of the referent concrete, by weight.

Gravel	Sand	Cement	Water
43%	29%	18%	10

From this reference case, other compositions were tested by replacing one part of cement by fly ash and bottom ash. The following compositions were tested: 7% cement + 14% Fly Ash (called "7%C + 14%FA"); 7% cement + 28% Fly Ash (called "7%C + 28%FA"); 7% cement + 28% Fly Ash + 14% Bottom Ash (called "7%C + 28%FA+ 14%BA").

Specimens were manufactured in steel cubic 15 cm x15 cm x15 cm molds (Figure 1). It is well-known that cylinder specimens gave more realist results [4] and the compressive strength obtained on cubic specimens should be multiplied by a corrector factor (about 0.9) to obtain the corresponding strength on cylinder specimens. However, cubic specimens were chosen because of the simplicity in their manufacturing and testing (no need of surfacing).

For each composition studied, nine specimens were manufactured: each three specimens were tested in uniaxial compression tests at 7, 14 and 28 days, respectively.

IOP Conf. Series: Earth and Environmental Science **143** (2018) 012041 doi:10.1088/1755-1315/143/1/012041

Figure 1. Cubic specimens: concrete specimens (on the right) have clearer colour than specimens with geopolymer (on the left).

2.3. Geopolymer materials using fly ash and bottom ash

Geopolymer is a product in which amorphous three-dimensional inorganic structures are formed through the reaction of alumino-silicate precursor, with an alkaline activator solution [5]. Geopolymer can be used as a binder in concrete to replace the cement. Since the fly ash possesses alumino-silicate in its composition (Tab. 1), it is interesting to test if the fly ash can be used for the creation of geopolymer and therefore the replacement of cement may be possible [6], [7]. The activators are typically sodium or potassium hydroxides or silicates.

In the present study, NaOH and Na_2SiO_3 were used as activator [8]. In the previous composition 7%C +14%FA, the cement was totally replaced by the activator (7% cement is replaced by 2%NaOH + 5% Na_2SiO_3). The values 2% and 5% were chosen to have the ratio Na_2SiO_3 / NaOH = 2.5 which is well known as an optimum ratio. So the composition tested was 14%FA + 2%NaOH + 5% Na_2SiO_3. It was observed in previous studies that for geopolymer, when the curing temperature increases, the compressive strength increases also. For this exploratory study, the specimens were cured in ambient conditions (in air), so the curing temperature was about 25°C. It is observed that the specimens manufactured with geopolymer had a colour darker than that of cement concrete (Figure 1).

2.4. Uniaxial compression tests

The specimens were tested at 7, 14 and 28 days after the casting, under uniaxial compression test (Figure 2). The loading rate was of 0.1mm/min. For each composition and at each moment of test, three specimens were tested.

Figure 2. Uniaxial compression test on cubic specimens.

3. Results

For each three specimens tested, the mean compressive strength were calculated and the obtained results are presented in Figure 3 and Figure 4, respectively for the case of concrete specimens (with fly and bottom ashes) and specimens using geopolymer.

From Figure 3, the mean compressive strength of the referent case (18% cement) after 28 days was about 30 MPa. It is worth noting that in this study, no plasticizer was used, if a plasticizer was used, the water/cement ratio could be reduced and therefore the compressive strength obtained could be better.

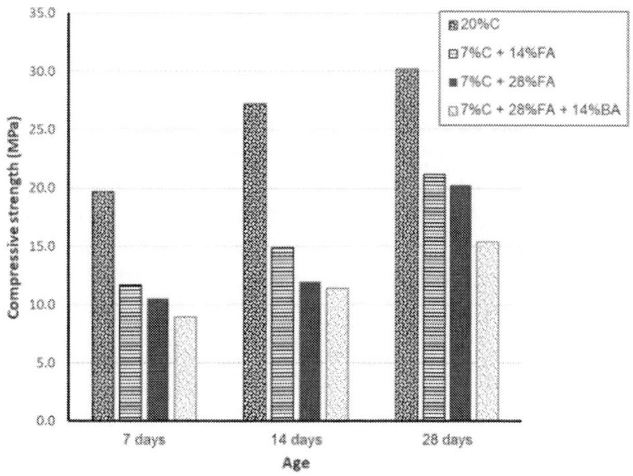

Figure 3. Compressive strengths of cement concrete using fly and bottom ashes.

From Figure 3, it is also observed that the compressive strength of all specimens increase following the time, from 7 days to 28 days. It is interesting to observe that the compositions 7%C + 14%FA and 7%C + 28%FA have similar compressive strengths after 28 days and these compressive strengths are about 21-22 MPa. These values can be used in Vietnam for concrete of individual houses where in many cases the specific compressive strength is only of 7.5 MPa or 11.5 MPa.

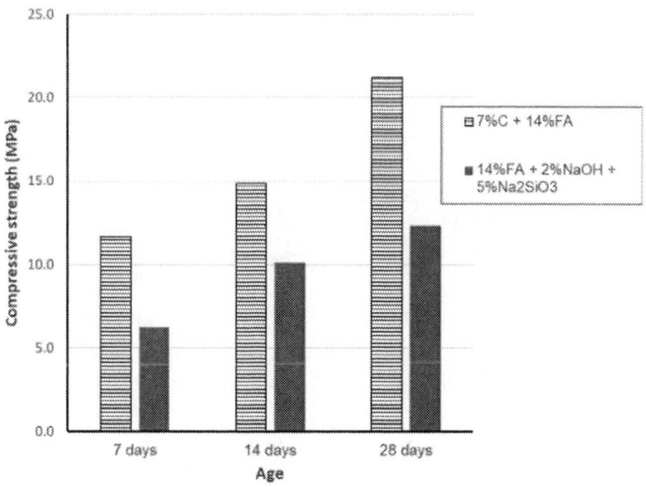

Figure 4. Compressive strengths of specimens using geopolymer.

Figure 4 presents the comparison between specimens using cement (7%C + 14%FA) and specimens using geopolymer. It is observed that geopolymer specimens has also the compressive strength increasing following the time, between 7 and 28 days. However the compressive strengths of the geopolymer specimens were relatively low (12 MPa at 28 days), compared to conventional concrete. So this material can be used for non-load-bearing material (bricks for example) but not for structural material (concrete). When this material is used for brick application, the temperature if the curing can be increased to improve its mechanical properties.

4. Conclusions and outlook

This paper presents an investigation using fly ash and bottom ash which are actually the wastes in Vietnam, in the manufacturing of construction materials. The first scope was to reduce the environmental impacts of these wastes and the second scope was to formulate non-conventional materials having low embodied energy by using less cement than the conventional concrete or current unburnt bricks. Several proportions of fly ash, bottom ash, cement, gravel, sand and water were investigated and finally two types of materials were selected to study in detail: the cement concrete with low quantity of cement, by using fly ash and bottom ash; the geopolymer concrete by replacing the cement by geopolymer, which is established from fly ash. The results showed that cement concrete using fly and bottom ashes had satisfying compressive strength for small construction projects (for example individual houses) while the mixes with geopolymer could only be used for non-load-bearing material (unburnt brick for example). The low compressive strength of the mixes with geopolymer could explained by the low quality of the ashes used, which is the general case for ashes from Vietnamese TPP. The effects of the curing temperature on the mixes with geopolymer will be investigated in the further studies. Further studies at the microscale of materials to investigate the potential reactions of fly and bottom ashes will also be interesting.

References

[1] ASTM D3176-15 2015 *Standard Practice for Ultimate Analysis of Coal and Coke* ASTM International West Conshohocken PA.

[2] Vinacontrol Ho Chi Minh City 2015 *Analysis results* 26/05/2015.

[3] Day KW 1999 *Concrete mix design, quality control and specification* 2nd edition E&FN Spon London.

[4] Bui QB, Morel JC, Hans S and Meunier N 2009 Compression behaviour of nonindustrial materials in civil engineering by three scale experiments: the case of rammed earth *Materials and Structures* **42** 1101-1116.

[5] Davidovits J 2008 *Chemistry and applications* USA: Institut Géopolymère.

[6] Cristelo N, Glendinning S, Miranda T, Oliveira D and Silva R 2012 Soil stabilisation using alkaline activation of fly ash for self compacting rammed earth construction *Construction and Building Material* **36** 727–735.

[7] Xu H and Deventer JSJV 2002 Geopolymerisation of multiple minerals *Minerals engineering* **15 (12)** 1131-1139.

[8] Prud'homme E, Ambroise J and Michel M 2015 Impact of alkaline solution and curing temperature on microstructure and mechanical properties of alkali-activated blast furnace slag *Ceramic Engineering and Science Proceedings* **36(8)** 61-75.

[9] Bui QB, Morel JC, Tran VH, Hans S and Oggero M 2016 How to use in-situ soils as building materials *Procedia Engineering* **145** 1119 – 1126.

Acknowledgement

This research is funded by Vietnam National Foundation for Science and Technology Development (NAFOSTED) under grant number 107.01-2017.22. Marie Bénard and Gabriel Guillet, students of

USMB in internship at TDTU, are warmly thanked for their helps during the specimens manufacturing and testing.

Optimum design of Geodesic dome's jointing system

Huy.T Tran

Faculty of Civil Engineering, Ton Duc Thang Univerity, 19 Nguyen Huu Tho, HCM city, Vietnam
*Corresponding author. Email: trantanhuy@tdt.edu.vn

Abstract. This study attempts to create a new design for joint connector of Geodesic dome. A new type of joint connector design is proposed for flexible rotating connection; comparing it to another, this design is cheaper and workable. After calculating the bearing capacity of the sample according to EC3 and Vietnam standard TCVN 5575-2012, FEM model of the design sample is carried out in many specific situation to consider the stress distribution, the deformation, the local destruction... in the connector. The analytical results and the FE data are consistent. The FE analysis also points out the behavior of some details that simple calculation cannot show. Hence, we can choose the optimum design of joint connector.

1. Introduction

Nowadays, the development of space frame is growing due to its great structural potential and visual, buildings that require large space such as sports arenas, exhibition pavilions, assembly halls, transportation terminals, airplane hangars ... frequently use Geodesic dome [1]. This dome has a huge bearing capacity, its "omnitriangulated" surface provides an extremely stable structure under the form of a large spherical volume with minimal surface area [1,2]. Though, Building up a Geodesic dome is very complicated due to its multiple types of members that are located in a three-dimensional space with different angle of rotation, and hence the force transfer mechanism is more complex. Jointing system is very important because more members are connected to a single joint; in the past decades, the effort expended on research and development of jointing systems has been enormous, and many different types of connectors have been proposed but they have not proved to be very successful mainly because of the complexity of the connecting method [3]. Our research aims to create a new optimized design of joint connector for the Geodesic dome to make it cheaper and easier to build up.

There are over 250 different types of jointing system suggested or used in practice like: Mero system (Disc node, Bowl node, Cylinder node, Block node, Space deck ....), Triodetic system, Unistrut system, Nodus system, Unibat system.... [2,3] These systems all have good bearing capacity but their drawback is they are only designed to connect fixed-numbers of bars with certain angle rotation; there are hundred to thousand bars in a Geodesic dome, so that the cost of the production of joints and the construction time are very large [2,3]. To resolve those problems, we propose a new optimized joint connector that ensures:

- Good bearing capacity
- Connected bars can rotate flexibly
- Easy and quick assembly
- Capability of mass production for all nodes at different positions

Base on different types of joint connector and its inconvenient, we propose a new joint connector that can solve these conditions. Numerical analysis methods have been used to analyse the behavior of the new connector. A 3D finite element model, which accounts for material behavior of the connector, is built in ABAQUS. The finite element model is verified by corresponding hand calculation results compare to Vietnam standard TCVN and EC3 specification.

2. Method

2.1. Proposed joint connector for Geodesic dome

New connector (as shown in *Figure 1*) consists:

- Ring : carries directly load
- Bolts

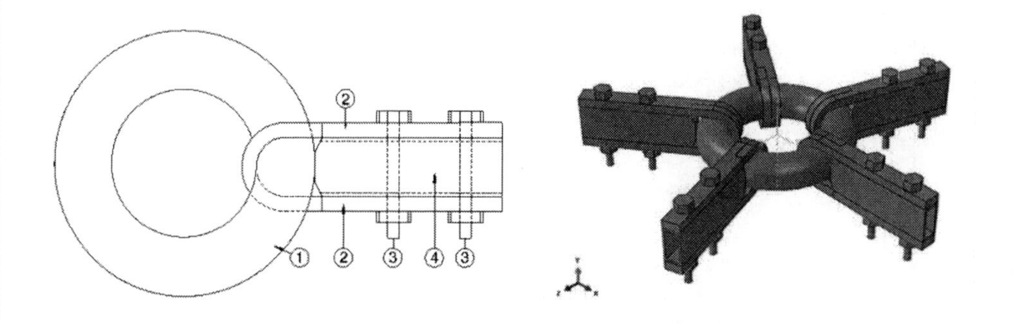

Figure 1. Design template of joint connector of Geodesic dome.

- Hook as shown in *Figure 2* : Links the ring and bars

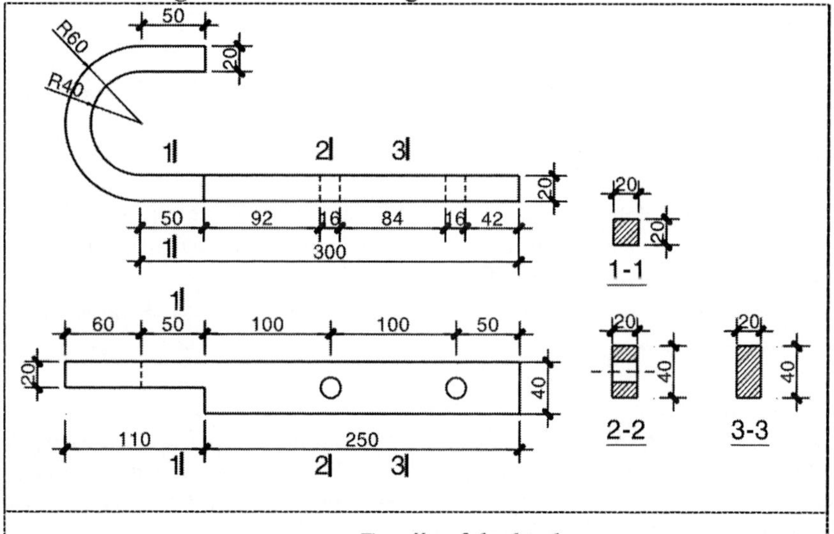

Figure 2. Details of the hook.

- Supporting box as shown in Figure *3* : receives and disperses stress on the top of the bar

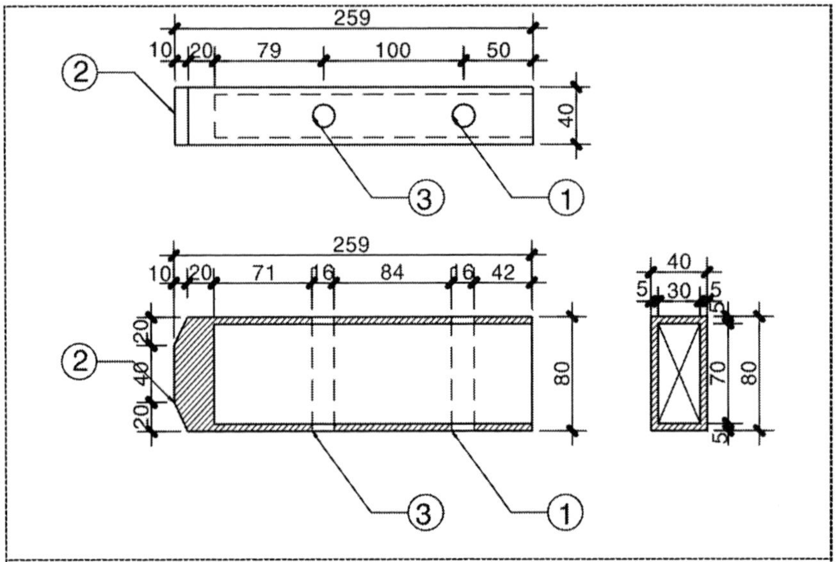

Figure 3. Details of the box.

2.2. Computing strength using design code

2.2.1 Vietnam standard (TCVN 5575 – 2012)
Assuming that the connection between elements of the template is perfect, the design connector is considered to be able to carry loads when the bearing capacity and the deformation of all elements and all contact points meet the demand of loading condition [4]. The followings need to be checked:

- The ring: shear and bearing strength.
- The hook: stress and deformation at its immediately reduced section; stress and deformation at contact point dangerous positions that contact the hook with the ring or bolts.
- The bolts: shear and bearing strength.
- The supporting box: stress and deformation at contact point; welded connection.

2.2.2 Component method of EC3 – 1993
In component method, the connections between elements are represented by deformation areas that are considered like springs. Those springs, based on the connection structure, are linked into a system, loads are transferred to the system through the springs and the system can show the capacity of the template. The component method assumes each element (component) in the joint is capable to carry the design load, thus we don't need to check element but the assembly of them when they work together [6].

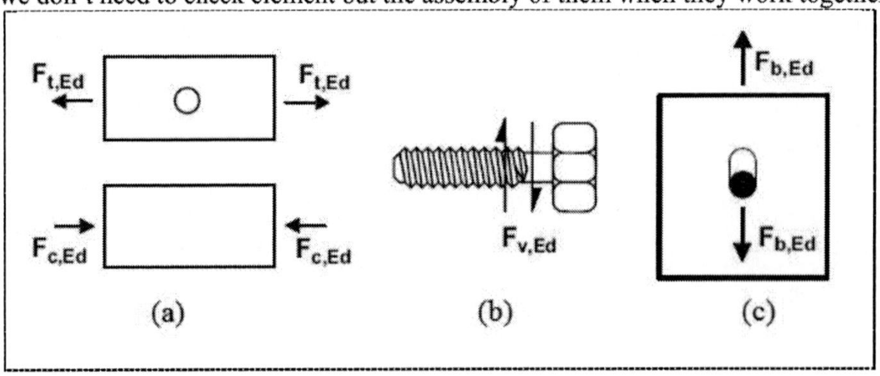

IOP Conf. Series: Earth and Environmental Science **143** (2018) 012042 doi:10.1088/1755-1315/143/1/012042

Figure 4. Components used in the sample: [6]

a-Plate in tension (compression) b- Bolt in shear c- Bolt in bearing.

There are 5 components in the design sample as shown in Figure *5*:
Component 1: ring
Component 2: hook
Component 3: bolt
Component 4: supporting box
Component 5: top of the box

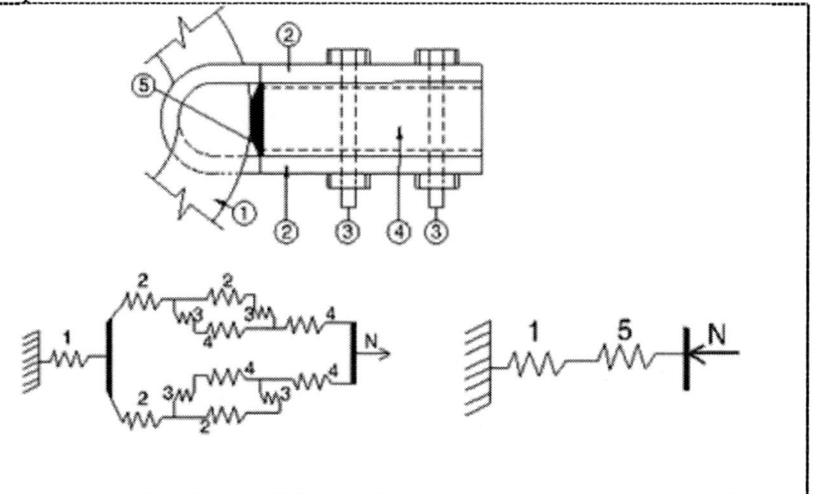

Figure 5. Assembled components checked of the sample.

In a serial system: $[F_{td}] = \min(F_i)$; $[S_{td}] = \left(\sum \frac{1}{si}\right)^{-1}$
In a parallel system: $[F_{td}] = \sum f_i$; $[S_{td}] = \sum S_i$
In a sample tension, the system has 2 parallel branches 2-3-4 in series with component 1.
The equivalent resistance of the sample: $[F_{Rd}] = \min(F_{Rd1}; 2F_{Rd2-3-4})$
In a sample compression, the system has component 1 in series with component 5.
The equivalent resistance of the sample: $[F_{Rd}] = \min(F_{Rd1}; F_{Rd5})$

2.3. FEM model

FEM model is built in ABAQUS to observe the behavior of the design joint connector in specific situation in reality; observing the stress distribution, the yield point, the local destruction … we verify the theoretical results and also study the behavior of details of elements that theoretical calculation can't response.

Table 1. Mechanical properties of the simulate sample.

	Steel	Bolt
Young modulus E (Pa)	2.1 x 10^{11}	2.1 x 10^{11}
Poisson ration v	0.3	0.3
Yield stress (Pa)	3.8 x 10^{8}	8.3 x 10^{8}
Ultimate strain	0.25	0.25

2.3.1 The hook in tension
Figure 6 shows the model of the hook in tension in ABAQUS.
Boundary condition: encastre at the section in touch with the ring.

Loads: Distributed load increases from 1 x 10^5 N/m^2 to 1 x 10^8 N/m^2

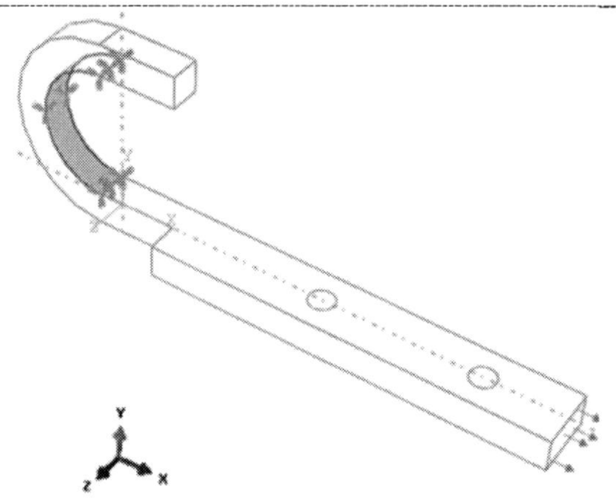

Figure 6. The hook in tension.

2.3.2 The box in compression

Boundary condition: encastre at the top.
Loads: Distributed load increases from 1 x 10^5 N/m^2 to 3.3 x 10^8 N/m^2

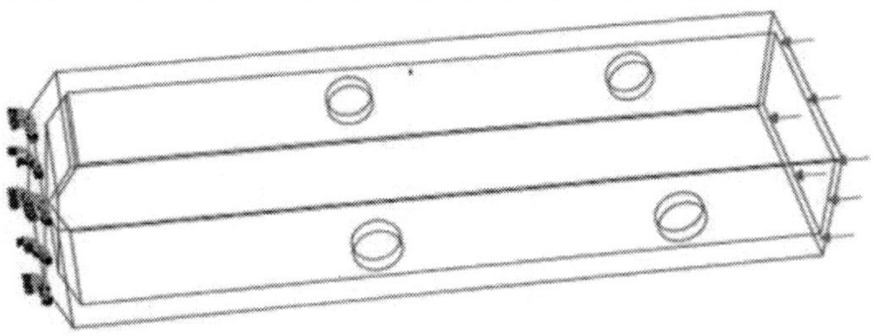

Figure 7. The box in compression.

2.3.3 The ring

The ring under pressure of 5 clusters of hook is modelled in with the boundary condition as shown in Figure *8*.
Loads: Distributed load increases from 1 x 10^5 N/m^2 to 1 x 10^6 N/m^2

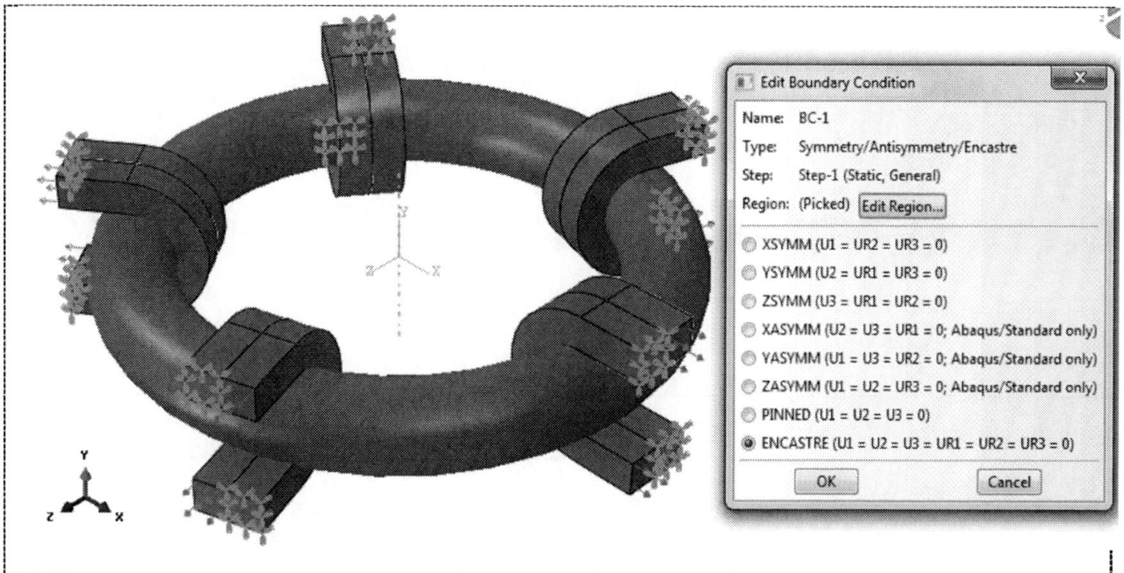

Figure 8. The ring under pressure of 5 clusters of hook.

2.3.4 The group of two hooks, two bolts and box in compression

We model the group with the boundary condition and loads as shown in Figure 9. Type of contact between elements is surface to surface.

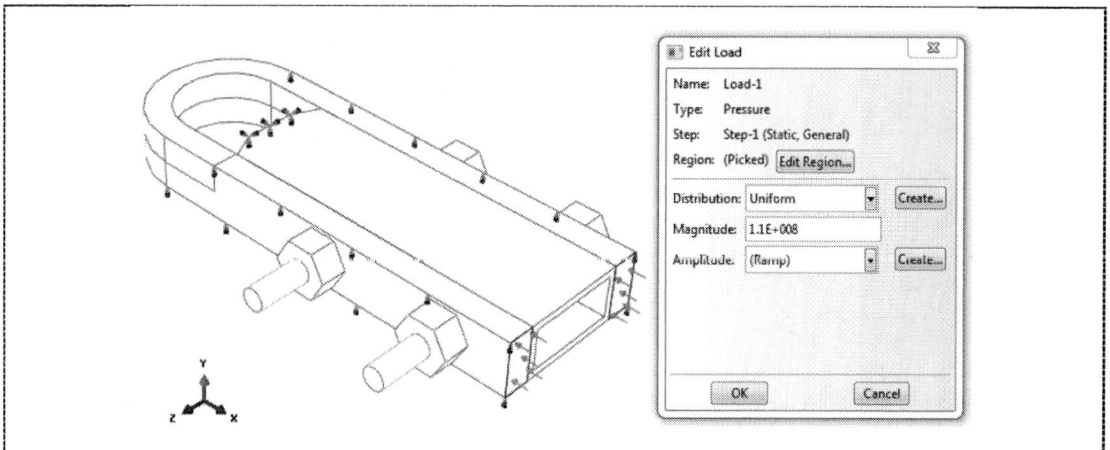

Figure 9. Group of two hooks, two bolts and box in compression.

3. Results and discussion

3.1. The hook in tension

By analytical method, tension resistance of the hook is calculated: [F_hook] = 92.4kN.

In the FEM model, the hook is damaged first at its points of reduced section and the damage force is about 97.1 kN, which is 5.08% different from theoretical calculation. The maximum deformation is also at this section as shown in Figure 10 .

It approves that the examined point in analytical method is compatible with the behavior of the hook in tension.

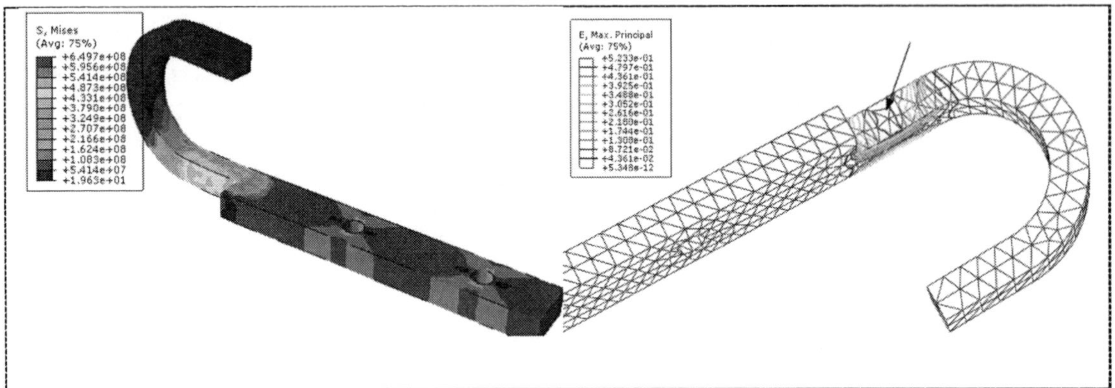

Figure 10. Contour of stress and deformation in the hook.

3.2. The box in compression

Observing FEM model, the box in compression is damaged when the force reaches 248 kN, which is 5.25% different from theoretical calculation (235.62 kN).

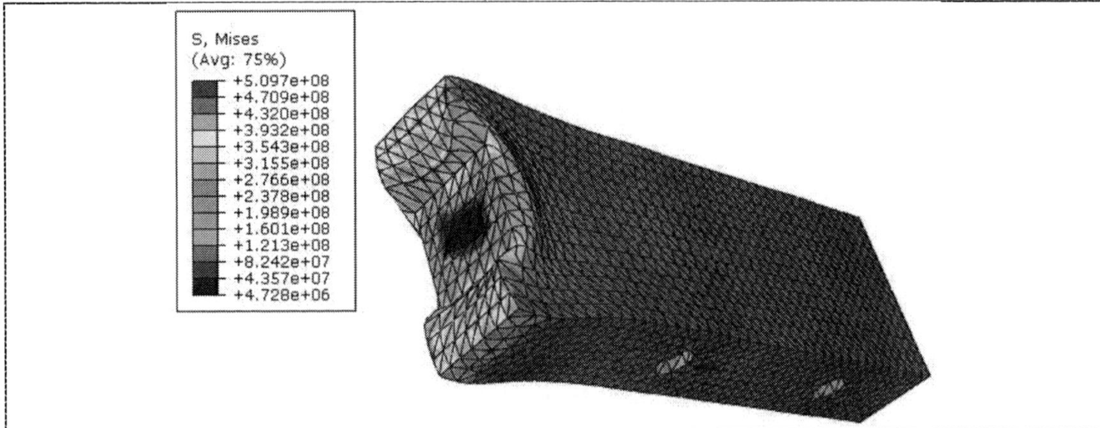

Figure 11. Contour of stress in the box.

3.3. The ring

Bearing resistance of the ring is calculated: [F_ring] = 651kN. Observing the ring under pressure of 5 clusters of hook in FEM model, the ring's bearing capacity is much higher than the hook's. Absolutely, the hooks are damaged first at the contact points with the ring.

IOP Conf. Series: Earth and Environmental Science **143** (2018) 012042 doi:10.1088/1755-1315/143/1/012042

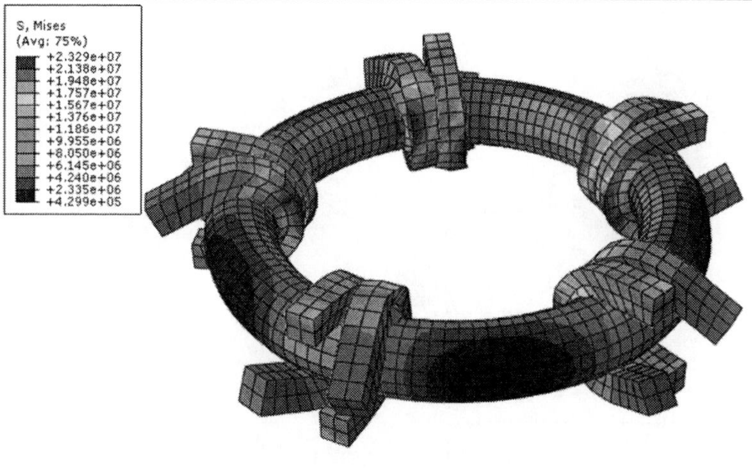

Figure 12. Contour of stress in the sample.

3.4. The group of two hooks, two bolts and box in compression

Observing the group two hooks – box – two bolts working together in compression, the box is damaged first at its contact point with the bolts. At that time, the hook still works in yield region and the bolts work elastically (Figure *14*). It is consistent to the partial damaged method used in theoretical calculation. We obtain the allowable compression force on the box is 48.9 kN. Comparing to the theoretical result (46.75 kN), 4.6% of deviation can be accepted.

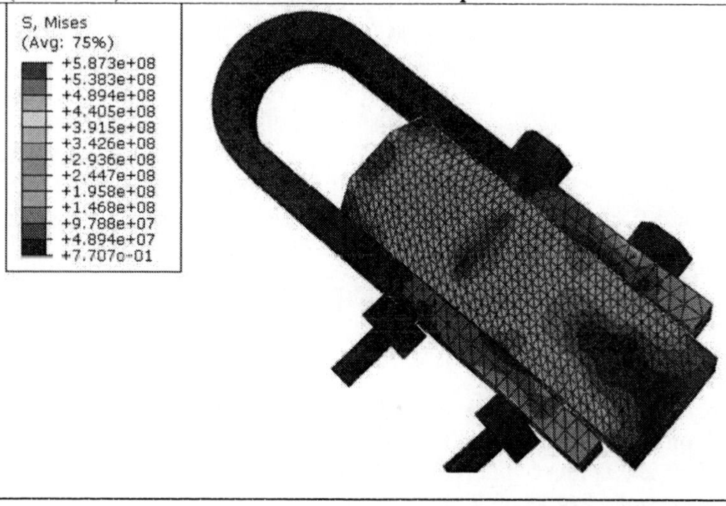

Figure 13. Contour of stress in the sample.

IOP Conf. Series: Earth and Environmental Science **143** (2018) 012042 doi:10.1088/1755-1315/143/1/012042

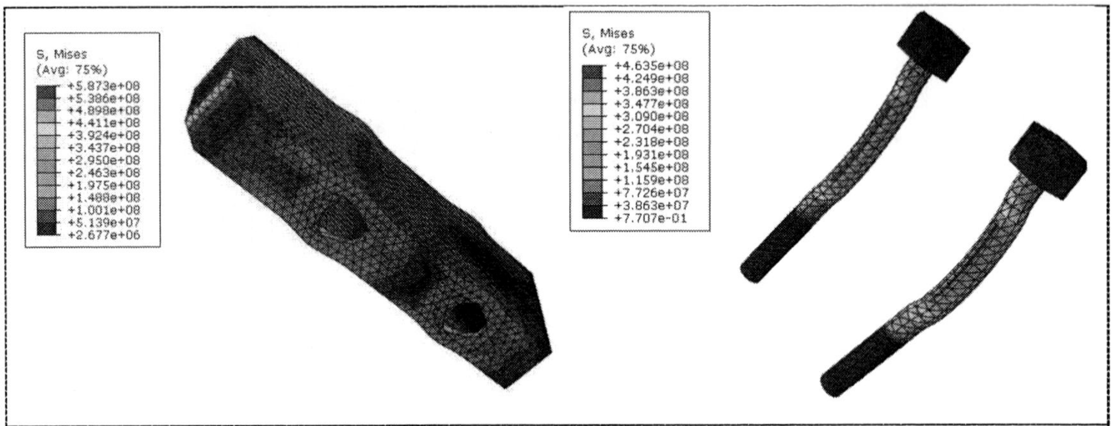

Figure 14. Contour of stress in the box and bolts.

4. Conclusion

A new joint connector for Geodesic dome is designed with 4 main elements (ring, hook, box, and bolt). The strength of each element is computed and examined by Vietnam Standard TCVN and EC3. The sample is modelled in ABAQUS with some specific situation. Theoretical calculations and FE analysis have an accepted deviation, the design sample may works efficaciously in reality.

Compare to existing connectors, this new connector is very flexible assembly, easy to produce and can be used for almost all types of nodes on the dome.

Expansive researches are proposed like: observe the behavior of the connector in more situations, verify the working of the connector in experiments, observe connector between dome and column ....

References

[1] Tien T Lan 2005 Space frame structure *Handbook of Structural Engineering* Section V **24** - 1

[2] Terri Meyer Boake 2013 Diagrid structures: systems, connections, details *Birkhäuser ed 2013*

[3] Hasançebi, O and Erbatur F 2002 Layout Optimization of Trusses Using Simulated Annealing *Advances in Engineering Software* vol **33** pp 681 – 696

[4] TCVN 2737-1995 *Tải trọng và tác động (Vietnam standard : Loads and Effects – Design standard)*

[5] TCVN 5575-2012 *Kết cấu thép – Tiêu chuẩn thiết kế (Vietnam standard : Design of steel structure)*.

[6] EN 1993-1-8 2005 *Eurocode 3: Design of steel structure - Part 1-8: Design of joints* CEN - European committee for standardization

[7] Mazars J 1986 A description of micro and macroscale damage of concrete structures *Engineering Fracture Mechanics* vol **25** pp 729–737

[8] Balling R J 1991 Optimal Steel Frame Design by Simulated Annealing *Structural Engineering* ASCE vol **117** pp 1780 – 1795.

[9] Choi S H and Kim S E 2002 Optimal Design of Steel Frame Using Practical Non-linear Inelastic Analysis *Engineering Structures* vol **24** pp 1189-1201.

IOP Conf. Series: Earth and Environmental Science **143** (2018) 012043 doi:10.1088/1755-1315/143/1/012043

A study on feasibility of super adobe technology –an energy efficient building system using natural resources in Bangladesh

Razia Kamal[1] **and Md. Saifur Rahman**[2]

[1]Department of Architecture, School of Science and Engineering, Southeast University, Dhaka-1208, 251/A and 252 Tejgaon Industrial Area, Bangladesh.
Email:adyarchi9@gmail.com
[2]Department of Information Technology, United Commercial Bank, Bulus Center, Plot - CWS- (A)-1, Road No – 34, Gulshan Avenue, Dhaka-1212, Bangladesh.
Email:titu.joaddar@gmail.com

Abstract. The inspiration and concept for the Superadobe system originates not from the modern architecture design experience, but from the influence of traditional rural buildings and landscape, together with a 13th century Persian poet named Jala Ad-Din Muhammad Balkhi, Rumi. The poetry sprit of Rumi, connects and enlightens the architectural theme of Nader Khalili with natural resources that anybody in the world should be able to build a home for his or her family with the simplest of elements: Earth, Water, Air and Fire. Therefore, to build a human shelter that will give maximum safety with low financial budget and minimum environmental impact with natural disaster resilient a Superadobe Technology has been adopted. The Superadobe, a form of earth bag construction using sandbag and barbed wire technology, is an economical, time efficient, energy efficient and ecologically friendly system developed by Iranian-born architect "Nader Khalili". The system connects the natural materials and rural traditions to create a new way to use natural materials such as mud, water, air and fire which can be finished in a short time without any large construction equipment. The goal of this study is to introduce the building system, analyse the ventilation, lighting and insulation of the prototype of Superadobe system replacing the contextual earth house in Bangladesh.

1. Introduction

The global need for housing includes millions of refugees and displaced persons - victims of natural disasters and wars. Iranian architect Nader Khalili believes that this need can be addressed only by using the potential of earth construction [1]. After extensive research into vernacular earth building methods in Iran, followed by detailed prototyping, he has developed the sandbag or 'superadobe' system. Superadobe is a super long sandbag, a super long adobe to construct a compression structure which uses layered long fabric tubes or bags filled with filled with clay (Katauskas, Ted.2007) [2]. It is a simple adobe with earth coil, an instant and flexible line generator. It uses the materials of war for peaceful ends, integrating traditional earth architecture with contemporary global safety requirements. Long or short sandbags are filled with on-site earth and arranged in layers or long coils (compression) with strands of barbed wire placed between them to act as both mortar and reinforcement (tension). It can be filled with almost any material available on earth from pure adobe (earth) to concrete incorporating stabilized earth and used materials. Stabilizers such as cement, lime, or asphalt emulsion may be added. The system connects traditional and natural resources such us mud, water, air and fire to create energy efficient houses in an effective way for an individual in a very short period of time .This concept was originally presented by architect Nader Khalili to NASA for building habitats on the moon and Mars, as "Velcro-adobe" however, the patented and trademarked (U.S. patent #5,934,027, #3,195,445) technology is offered free to the deprived of the world, and licensed for commercial use [3]. It comes from the concerned heart of someone who did not want to be bound to any one system of construction and looked for only one answer in human shelter, to simplify.

Superadobe is a kind of ecological and sustainable building, which can be coiled into vaults and domes, the way a potter coils a pot, with barbed wire reinforcement, to build structures that pass international earthquake codes [4, 5]. These structures can last for one season before returning to earth, or they can be stabilized, waterproofed, and finished as permanent houses. The system can be used for structural arches, domes and vaults, or conventional rectilinear shapes. The same method can build silos, clinics, schools, landscaping elements, or infrastructure like dams, cisterns, roads, bridges, and for stabilizing shorelines and watercourses. The whole family can build together, men and women, from grandma to the youngest child. As such, many years have been spent researching hands-on how to make the process simpler and easier. There is no heavy lifting or backaches, no expensive equipment, and is a flexible and fast construction. The bags are filled in place on the wall using small pots like coffee cans, or even kitchen utensils. It can be built alone or as a group.

In Superadobe, the ancient earth architecture of the Middle East using sun-dried mud bricks is fused with its portable nomadic culture of fabrics and tensile elements, not just through design and pattern, but also through the structure itself [6, 7]. Structural design uses modern engineering concepts like base-isolation and post-tensioning. The innovation of barbed wire adds the tensile element to the traditional earthen structures, creating earthquake resistance despite the earth's low shear strength. The aerodynamic forms resist hurricanes. The innovation of sandbags adds flood resistance, and easy construction, while the earth itself provides insulation and fire-proofing.

Sustainable Solution to Human Shelter, based on Timeless Materials (earth, water, air and fire) and Timeless Principles (arches, vaults and domes). The structural principles of the timeless forms of arches, domes, vaults are built with the materials of earth, sandbags and barbed wire using the engineering of single and double curvature compression shell structures, to reach the ultimate in strength and aesthetics. Because of self-supporting arched roof structure, it can be one single space, or form more spaces through merging multiple arch systems (Figure 1) [6-8]. This allows for flexibility and variability of the space. Because of its materials are native and recyclable as well as its structure needs no construction equipment, and its prototype has good ventilation, lighting and insulation.

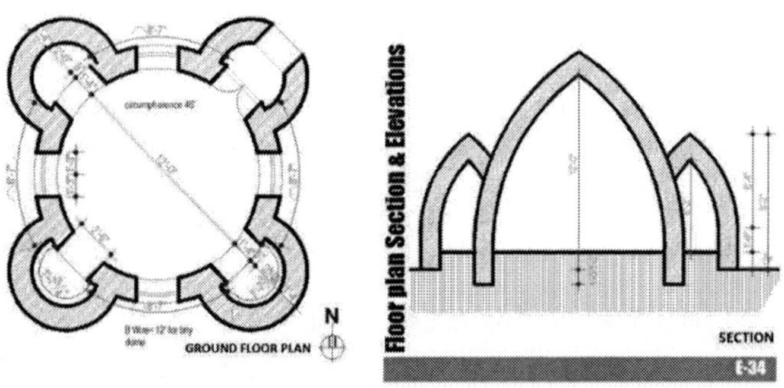

Figure 1. The plan of Superadobe.
Source: redrawn by the author according to HBRI

2. Literature Review

The technique's current pioneer is Nader Khalili who originally developed the superadobe system in 1984 in response to a NASA call for housing designs for future human settlements on the Moon and Mars. His proposal was to use moon dust to fill the plastic Superadobe tubes in layer. The Super Adobe Method is now use in Canada, Mexico, Brazil, Chile, Iran, India, Siberia, Thailand, as well as in the

U.S. Nader khalili earned his architectural design certification in California in 1970. Like many other architects, khalili had designed many skyscrapers and then realized that he should do something meaningful for those in need [3-7]. Khalili gave up his work as a modernism architect in the USA and travelled to the desert area in Iran for five years. When viewing all that with a new perspective, khalili saw wonderful, pure and valuable traditional rural architecture and affected and moved by tens of thousands of rural mud rooms. The adobe is stretched from history into the new century. It is like an umbilical cord connecting the traditional with the future adobe world.

Rumi conceived that elements such as water, air and fire in the land have a great power, which can cycle inside and outside of the life [6-8]. If these elements can be well balanced, the environment and all life in the world would have good status (Stephen. Holgate 2003). Rumi had said "The earth will become the gold of the wise". Affected by Rumi, Khalili realized the value of the earth and formed his design concept using clay to build (Figure 2) and fire to burn (Ceramic House) [7].

Figure 2. The outside view of Superadobe.
Source: www.calearth.org

Nader Khalili began to research how to use natural resources as much as possible and to study the possibility of no use of high technology to build house in modern world. The Superadobe system could satisfy above request and it has good seismic performance. It is a good option for the poor and is also suitable for the moon. In 1986, khalili founded a research organization called Cal-Earth which began practice in the California Hysperia area of the Mojave Desert. His main studies are about earth art and the technical problems of clay construction [7, 8]. The scope of his research includes building a house on the moon with technical innovations for NASA (National Aeronautics and Space Administration) as well as building housing for homeless people in the United States. The building philosophy seeks harmony between art and nature. The sandbag engineer prototype won the Aga khan Award in 2004 [8]. Cal-Earth has posted their concepts and methods on the Internet. The works and practices of Cal-earth have been studied by other research institutions as a model.

To build a sandbag shelter, the principle of construction with long or short sandbags is similar to the way a child puts his donut-like toy rings (Figure 3) on top of each other [8,9,10]. This geometry is one most appropriate for superadobe construction since the earth filled coils can lay flat on top of each other, and step them gradually inwards. This is called "corbelling". Equal steps will give us a conical shape, but to create a more spacious interior an egg-shaped volume has been used [8].

Figure 3. Donut-like toy rings.

The resulting beehive shaped structures employs corbelled arches, corbelled domes, and vaults to create single and double-curved shells that are strong and aesthetically pleasing [8,9]. It has received growing interest for the past few years in the natural building and sustainability movements. In the last century earth bag buildings have undergone extensive research and are slowly gaining worldwide recognition as an optimum solution to the global epidemic of housing shortages.

3. Methodology

Building with earthbags (sometimes called sandbags) is both old and new. Sandbags have long been used, particularly by the military, for creating strong, protective barriers, or for flood control [8-11]. The same reasons that make them useful for these applications carry over to creating housing. Since the walls are so substantial, they resist all kinds of severe weather (or even bullets) and also stand up to natural calamities such as earthquakes and floods. They can be erected simply and quickly with readily available components, for very little money.

Earthbag building fills a unique niche in the quest for sustainable architecture. The bags can be filled with local, natural materials, which lower the embodied energy commonly associated with the manufacture and transportation of building materials [8-12]. The fill material is generally of mineral composition and is not subject to decomposition (even when damp), attractive to vermin, or burnable...in other word it is extremely durable. The fill material is generally completely non-toxic and will not off gas noxious fumes into the building.

Many different materials can be used to construct Superadobe. Normally earth or sand, cement or lime, and Superadobe polypropylene tubing; bags can be polypropylene, or burlap. What is important is that they are UV resistant or else quickly covered in plaster [13]. Virtually any fill material will actually work including un-stabilized sand, earth, gravel, crushed volcanic rock, rice hulls, etc. If the fill material is weak the bags have to be really strong and UV resistant, or else plastered right away.

The material can be either wet or dry, but the structure is more stable when the tube's contents have been moistened. Other materials needed include water, shovels, tampers, wheel barrow, mechanical pumps, scissors, compass, large plugs or pipes (for windows), and small buckets or coffee cans for filling the sacks. If you decide to go the quicker way, then electric or pneumatic tampers can make the tamping easier, electric or gas powered bucket chain that can reach 7 meters or higher would eliminate the need of manual filling of sacks or tubing using coffee cans or small pails.

Mud, water, air and fire are the main materials of the Superadobe System. The building process is easy, time efficient and built a single dome following ten steps (Figure 4-13) [8, 9].

Firstly, collect some tools such as scissors, a rod, a shovel , a roll of woven sacks, short tubes, kegs or coffee tank, Materials-sand bags, Small buckets, gloves, tamper a roll of metal wire with barbed and a pair of pliers.

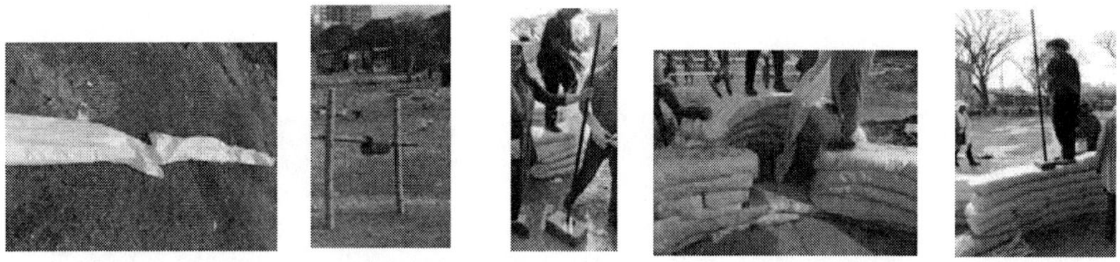

(a) (b) (c) (d) (e)

Figure 4. Step-1: Tools with materials (a) Sandbag. (b) Barbed wire. (c) Tamper. (d) Installing barbed wire inside bags. (e). Compacting earth bags using tamper.

Then dig mud from the base of the ground and mix some cement and water together, add cement or lime or an asphalt emulsion for reinforcement. Add enough water and squeeze it into a ball until it does not stick to your hand [8-13]. The location of the entrance should be chosen to avoid the rain and the winds.

(a) (b)

(c)

Figure 5. Step-2: (a) Dig -1. (b) Dig -2. (c) Soil test.

The dome layout should be based on landscape and climate [8, 9]

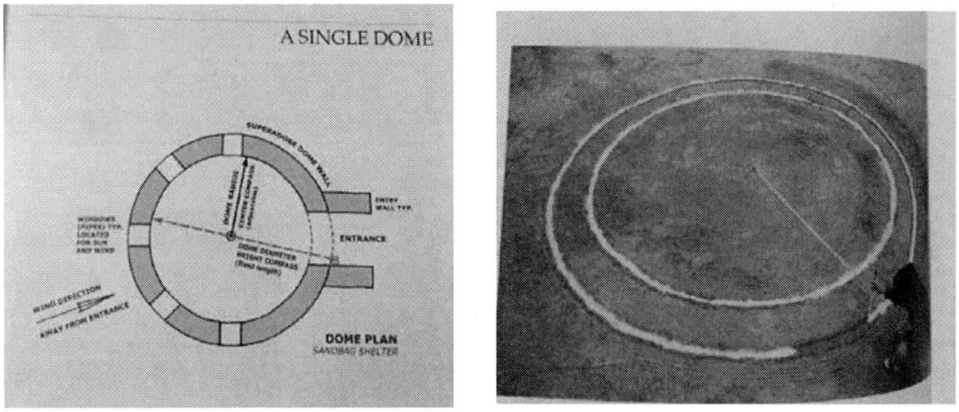

(a) (b)

Figure 6. Step-3: (a) Dome plan. (b) Dome Layout in construction phase.

The foundation for the structure is formed by digging a 30 cm deep circular trench with 2 to 4m diameter.

(a) (b) (c)

Figure 7. Step-4: (a) Theoretical dome layered foundation. (b) Centre scale on site. (c) Foundation on the constructed site.

Then, lay a bag along with the ditch. Fold the tail section to make it closed, and then it will be like a short rising column. Pour the soil into the bag and constantly shake it into the bottom. A good method inclining the bag to your feet in the helping of the gravity could be used to prevent the bags twisted, and withdraw your foot after the bag is well-filled [10-14]. Then correct its position by a compass. Make sure the tail section of the bag is well folded to close. A tamping is used to make the sandbag compacted. To make it smooth, fixed and uniform sealing good until they become strong enough. In order to make them stable, barbed wire is tied between the different layers of sandbags. First wire is 4 meters long, the next need more 65cm longer to make sure overlapping part when fracture occurred. Two or three layers of the filled polypropylene sand tubes (superadobe tubing) are set below the ground level in the foundation trench [15].

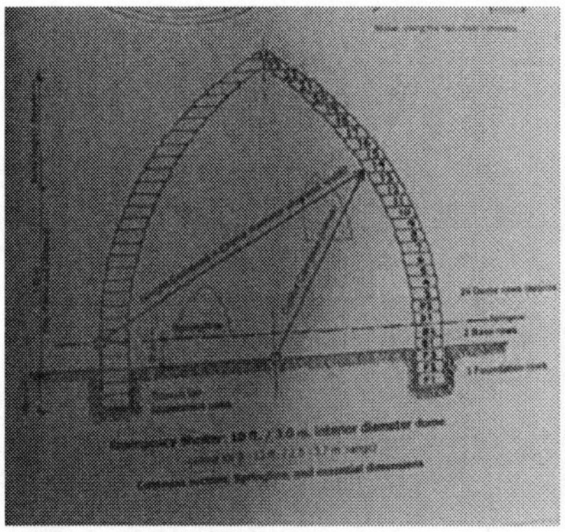

(a)

(b)

(c)

Figure 8. Step-5: (a) Theoretical Base Plan. (b) Base making phase on the site. (c) Base making progressive phase 2.

When build it in certain height using two compass to ensure the shape of the roof. The one is central compass formed with a wire rope or other chain unable to stretch and the ground center. The other is a height compass that can be increased according to the layer [16, 17]. And the sandbag should be rearranged if it's not matches compass trajectory.

A chain is anchored to the ground in the center of the circle and used like a compass to trace the shape of the base. Another chain is fastened just outside the dome wall: this is the fixed or height compass and gives you the interior measurement for every single layer of superadobe bags as they corbel ever higher. The height compass is exactly the diameter of the dome. The center chain/compass is used to ensure the accuracy of each new superadobe layer as it is laid and tamped [8-16]. The compasses must be made of non-stretchy material to ensure an accurate geometry.

On top of each layer of tamped, filled tubes, a tensile loop of barbed wire is placed to help stabilize the location of each consecutive layer: it plays a crucial role in the tensile strength of the dome - it is the 'mortar'. Window voids can be placed in several ways: either by rolling the filled tube back on itself around a circular plug (forming an arched header) or by waiting for the earth mixture to set and sawing out a pointed arch void. A round skylight can even be the top of the dome.

It is recommended not to exceed the 4m diameter design in size, but many larger structures have been created by grouping several "beehives" together to form a sort of connected village of domes. Naturally, this lends itself to residential applications, some rooms being for sleeping and some for living.

(a) (b) (c)

Figure 9. Step-6: (a) Theoretical dome scale determination. (b) Center scale and foundation phase. (c) Almost completed dome.

The next step is making the door and windows. Cutting parts of the sand bags to form openings. Excision and then tamping to make ensure that cutting soil not to stick in original sandbags. The door could be pushed when at least 5 layer sandbags is masonry [16, 17]. Insert the pipe into the sandbag as a window. Arched windows-make a form with loosely-filled bags or other materials, build the dome around the form and remove the form when the dome is completed. Make an 'eyebrow' over larger window openings. Pipe windows build the dome over the pipes for ventilation and light.
In order to prevent the rain inflow to inner, the pipe should be inclined to external. Keep the bag indentation in the top and form an adobe. Add the arched doorway to support and protect the entrance. The door way could be arched or sloped, higher or lower.

(a) (b)

Figure 10. Step7: (a) Doorway of dome on site. (b) Inside of dome on site.

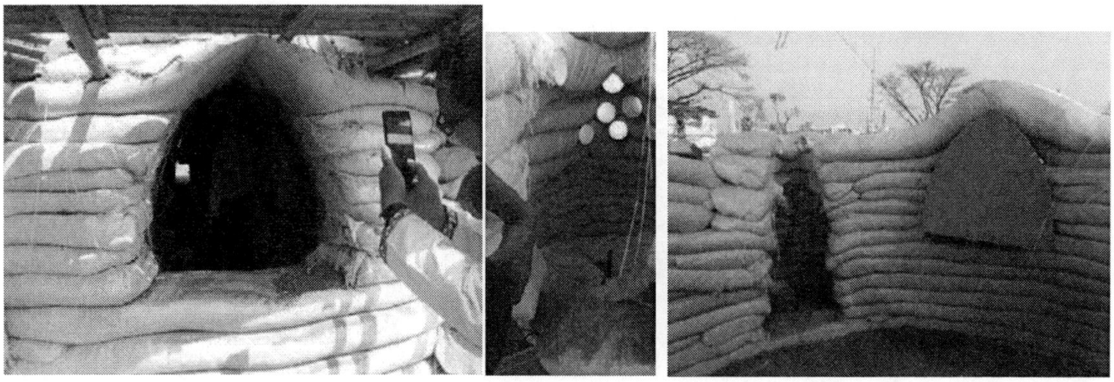

(a) (b) (c)

Figure 11. Step 8: (a) Windows making phase 1. (b) Windows making phase 2. (c) Windows making phase 3.

The buttress walls are built for the entry and connect to the dome at every row with barbed wire.

(a) (b)

Figure 12. Step 9: (a) Barbed wire in every rows of buttress walls used multipurpose (Kitchen, Toilet, Storage) (b) Buttress walls with ventilation and light with hollow plastics.

Eventual step is to use learning arches to complete a small entry vault to protect the doorway. Alternatively, use very short lintels for a 'mineshaft" entry after the upper dome is completed [8, 9, 10].

Figure 13. Step-10: Entry Vault.

Once the corbelled dome is complete, the work needs to protect. And cover waterproof material to let it to be moisture proof and anticorrosive. it can be covered in several different kinds of exterior treatments, usually plaster. Khalili developed a system that used 85% earth and 15% cement plaster and which is then covered by "reptile", a veneer of grapefruit sized balls of cement and earth [6, 7, 8]. Reptile is easy to install and because the balls create easy paths for stress, it doesn't crack with time. There are many different possibilities. Some Superadobe buildings have even been covered by living grass, a kind of green roof but covering the entire structure. Any exterior treatment and building details would need to be adapted to a region's specific climatic needs.

(a) (b)

Figure 14. (a) Exterior Outlook without ornaments. (b) Exterior Outlook with ornaments.

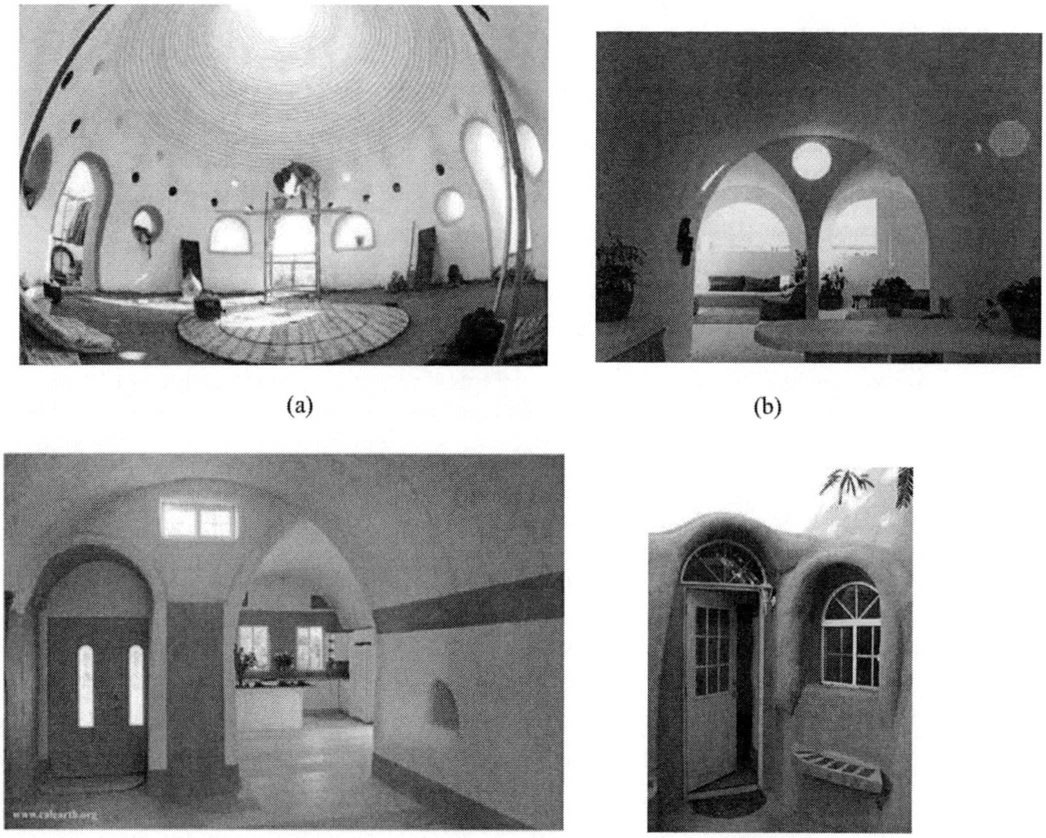

(a) (b) (c) (d)

Figure 15. (a) Dome with natural light. (b) Ventilation of Drawing room. (c) Kitchen interior with ventilation. (d) Interior of door and windows.

4. Some features of the Eco-Dome

- Built from local earth-filled Superadobe coils (earth stabilized with cement or lime) [7, 8].
- Tree free.
- Maximum use of space through alternative options. The main dome and four niches, depending on local code approval, can function as:
    - Main living room, entrance hall, kitchen, bathroom, bedroom (called "bed-womb" because of its small, organic form!)
    - Living room, entrance hall, and three bedrooms.
    - Living room, entrance hall, two bedrooms, and a bathroom.
- Self-contained single unit (potential for a guesthouse or studio apartment) or double unit (larger family residence).
- Can be repeated and joined together to form larger homes and courtyard houses.
- Can be built by a team of 3-5 persons.
- Designed with the sun, shade and wind for passive cooling and heating.
- Wind-scoop can be combined with a rated furnace unit, depending on local code approval. Solar energy and radiant heating may be incorporated.
- Interior furniture can be built-in with same material.

5. Advantages of Superadobe Domes

- Natural, reversible and recyclable building materials are used causing no harm to health and environment.
- Good thermal mass material: perfect heat storage capacity, regulation of temperature and humidity resulting comfortable interior microclimate [6-8].
- Good acoustic parameters.
- Statically strong, durable and resistant even to extreme weather conditions and natural catastrophes like flood, windstorm, fire, hurricane and earthquake.
- Wide range of use.
- Harmonic and diverse – traditional or modern forms and styles.
- Economic and environment friendly, easy and quick building with locally available materials, small waste production, minimal need for industrial background and use of machines, low shipping costs, low energy input and environmental pollution.
- Low housekeeping costs.
- Anyone can learn this building technology, the whole community from young to old.
- With building vaults and domes one can omit and minimize the use of wood, iron or reinforced concrete bond, beam system and roofs.
- Ideal for humanitarian purpose, strengthen communities and equal opportunities.

6. Disadvantages of Superadobe Domes

- It does take a lot of people to build a house by hands only.
- It gets difficult after several hours of lifting the heavy bags.
- It takes strength to lift and carry each bucket.
- No mention of them in building codes.

7. Analysis

In order to discuss the system's energy sufficient characters, it can be analysed from aspects of ventilation, lighting and insulation. The prototype be chosen is a single adobe. First of all, the dome roof has better ventilation effect than the flat one when they share the same inner wall length (Figure 17) [8-10]. when the model has been overlap together, the height from roof windows to the floor certifies that dome roof has accelerated the air fluent better because it forms a Stack effect.

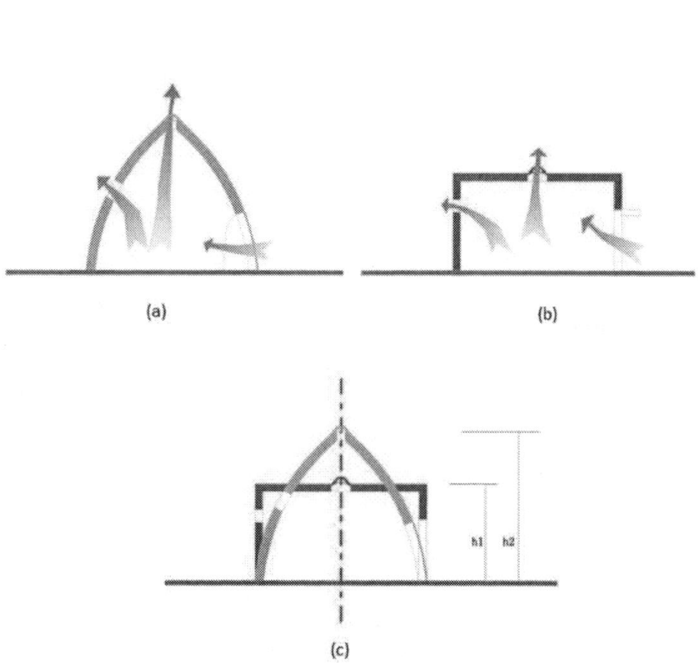

Figure 16. (a) Dome roof ventilation. (b) Flat roof ventilation (c) The different height from roof windows to floor with the same inner wall length

Designing comfortable homes without extensive energy use or associated cost for people in freezing temperatures is essential to the earthbag village.

Every material in a building has an insulation value that can be described as an R-value. Most builders think of R-value as a description of the ability of a structure or material to resist heat loss. This steady state value does not change regardless of the outside temperature variations that occur naturally on a daily and annual basis [8-10]. So why does an earthbag structure (or any massive earthen building for that matter) with an R-value less than 0.25 per inch (2.5 cm) feel cool in the summer and warm in the winter? Because this R-value can also be expressed as the coefficient of heat transfer, or conductivity, or U-value, which is inversely proportional, that is U=1/R. From this simple formula we can see that material with a high R-value will yield a low U-value. U-value (units of thermal radiation) measures a material's ability to store and transfer heat, rather than resist its loss. Earthen walls function as an absorbent mass that is able to store warmth and re-radiate it back into the living space as the mass cools. This temperature fluctuation is known as the "thermal flywheel effect." The effect of the flywheel is a 12-hour delay in energy transfer from exterior to interior. This means that at the hottest time of the day the inside of an earthbag structure is at its coolest, while at the coolest time of the day the interior is at its warmest. Of course, this thermal performance is regulated by many factors including the placement and condition of windows and doors, climatic zone, wall colour, wall orientation, and particularly wall thickness. This twelve-hour delay is only possible in walls greater than 12 inches (30 cm) thick.

Thermal lag is a time unit assigned to a material that indicates the amount of time it takes to heat up or cool down. Thermal mass is the material heated or cooled. Heating a material (and its effectiveness as thermal mass) depends on how much heat it can absorb (Specific Heat: Cp), the rate at which the heat can penetrate (thermal conductivity: k), the density of the material in question (density: ρ), and the thickness of the material [8-13]. The time that passes from heating one side to the other side heating up too is the "thermal lag" expressed by:

$$thermal\ lag = 1.38 * (thickness, m) * \sqrt{((Cp * \boldsymbol{\rho})/\boldsymbol{k})} \quad (1)$$

This means that when domes experience cyclical temperatures (day and night, seasonal) the temperature inside the domes lags behind.

The issue with earthbags is that they provide fantastic thermal mass (Figure 17) but truly are not a great insulator. The R-value of compacted earth is roughly R-1 per foot, so a standard earthbag wall might yield no better than R-2 (about the equivalent of a dual-pane glass window). Their insulating properties can be improved by adding materials that create air pockets inside them like volcanic rock, rice hulls, perlite, or vermiculite. This, however, has positive and negative elements.

Table 1. Comparison of Material R-Values.

Material	R-value/inch	R-value/15″
Rice hulls	R-3	R-45
Perlite	R-2.7	R-40
Vermiculite	R-2.13	R-32 to 36
Extruded polystyrene	R-3.6 to R-4.7	R-54 to R-70
Molded low-density polystyrene	R-3.85	R-58

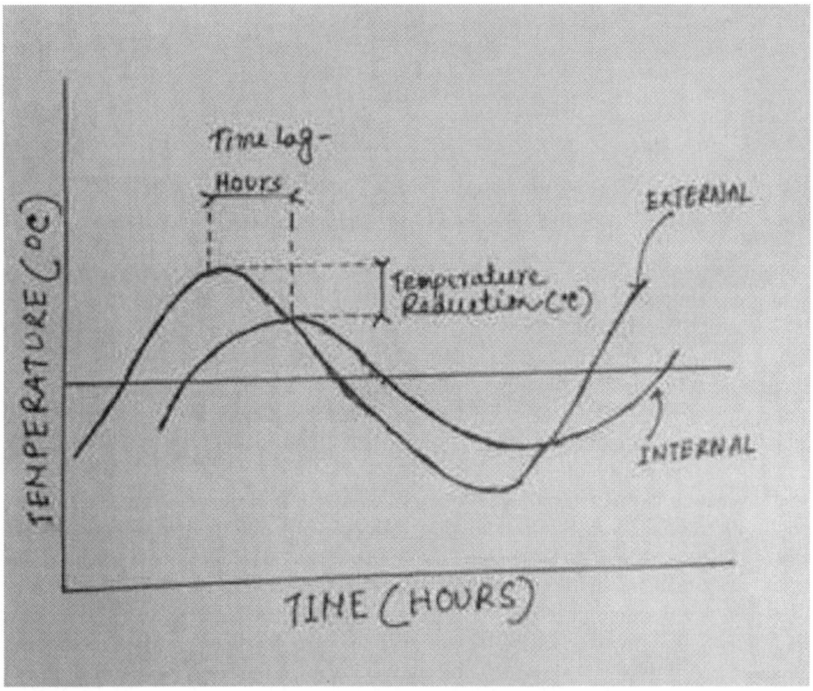

Figure 17. Thermal Lag Calculation

Simple thermal dynamics state that heat travels from warm spaces to cooler spaces and air pockets inhibit this heat transfer. If the goal is keeping heat in, inhibiting the transfer of heat through walls can be very good. The more you add air pockets or insulate the walls, the less heat will move through those walls [13, 18]. The downside of this is in the summer when heat being absorbed by the walls would have a cooling affect that would be lessened by this reduction of heat transfer. Here are the calculations used for a dome-shaped structure (Figure 18).

$$q = 2\pi k \frac{(t_i - t_o)}{ln(r_o / r_i)}$$

(a)

$$\dot{Q} = 4k\pi \frac{T_1 - T_2}{1/r_1 - 1/r_2} = 4k\pi \frac{(T_1 - T_2) r_1 r_2}{r_2 - r_1}$$

We take the formula for heat loss through a hollow sphere and divide this in half.

(b)

$$q = \pi r_i^2 (Uvalue) \Delta T$$

(c)

Figure 18. (a) Assuming the walls are essentially a hollow cylinder. (b)Assuming the roof is essentially a semi-sphere. (c) Assuming the floor is essentially an unbounded plane

Suggestions and options on what is the best approach vary and can be affected dramatically based on the building environment. For areas that are consistently cold, with a quality internal heating system, a person would do well with a solid earth-filled earthbag wall that has an external insulating layer [13, 15, 18]. This would provide thermal mass on the inside and a barrier to that thermal mass losing its heat to the outside. In areas that are consistently hot, however, this external insulation approach could create environments that warm up and are difficult to cool again, so building deeper in the ground and eliminating insulation is a better approach.

8. Conclusion

On the basis of the interpretation of the materials, the construction method and the achievements of the Cal-earth, the system can be analysed from four aspects.

First of all, it is material choice [8, 18]. The materials of Superadobe are almost all natural, such as mud, water, air and fire that can be easily gained. They are from nature and return to nature after construction when they must be abandoned. It can be circled in the ecosphere. That is sustainable. Besides, the thermal inertia of the earth makes the inner house to have good insulation as well as the material using make it have good air-tightness.

Then pay attention to its construction method. No large construction equipment be needed in building and it can be finished just by several men in a short time. No noise, non-pollution and little impact on the surrounding environment. On the other side, The Superadobe system provides a possibility for those who need temporary housing in case of emergency because it can be easily assembled and disassembled. Moreover, Go to its appearance, Superadobe follows the Islamic traditional building method, which was using dome. In addition, its construction method is also drawing lessons from the Persian Sassanid Dynasty (AD226-651) when the brick corbel method was popular [7, 8, 13]. The difference is he uses sandbags to corbel and each cross section is a circle. The advantages of the dome space they choose have been talked about in the front text from the angle of technology. Another explanation of the dome used by Khalili is that building just as smooth as human's body without edges and corner just like the nature itself. Its smooth surface gives a special view outside and a safe space to play for children inside. Finally, the scope of application of Superadobe system should be discussed. For signal building, its flexibility of space organization provides choice for different number of population. One family and more together (parent living with their children who are married) are both possible. It can expand space according to different situation. For building group, another question is if it suit to all cities or countries? This should be carefully discussed. First of all, the main materials of Superadobe system are clay. Not enough clay could be built in the density cities. For another the system itself can not satisfy the high-rise building requirement. However, it is suitable for the low-density rural area because of low cost, environment friendly and time efficiently [15, 18]. Meanwhile, Due to water permeability of the soil, Superadobe is more adapt to dry areas than wet regions, especially for desert districts.

In conclusion, Superadobe system is an economical, ecological and energy efficient system. The scope of its application is dry and low density rural areas. Superadobe connects the nature and tradition.

It is inexpensive technology. Natural, reversible, recyclable building materials are used which are not harmful to our health and the environment. It has low energy input and causes very less environmental pollution. Small waste production is there. It is environmentally friendly. It is statically strong, durable, and resistant even to extreme weather conditions and natural catastrophes like flood, windstorm, hurricane, fire, and earthquake. It reduces global warming and Speedy in construction. It can provide a very effective point of reference for our country construction in Bangladesh.

References

[1] L M Surhone, M T Timpledon and S F Marseken, 2010, Super Adobe, *Beta script Publishing.*
[2] T Katauskas 2007 Dirt-Cheap Houses from Elemental Materials *Architecture Week.*
[3] Stephen Holgate 2003 For Architect Khalili, Rumi Inspires 21st Century Housing Solutions “Super adobe” uses material at hand to build shelters on Earth and in space Special Correspondant *Washington File For USinfo.state.gov* (Farsi Website) http://www.calearth.org/whatnew/StateDept.htm 2007
[4] L Elizabeth and C Adams 2005 Alternative construction: contemporary natural building methods C Y Wu translate Perking *Machinery industry press*
[5] J Z Teslik and N M Vodicková 2013 Air-tightness and acoustic properties of SuperAdobe system International Conference on Buildings and Environment EnviBUILD 2 *Trans Tech Publications.*
[6] Ziling Zhaoa Qi Lub and Xinbo Jiangc 2015 *Procedia Engineering* **121** 1179 – 1185
[7] Piyush Sharma 2015 *International Journal of Research (IJR)* **2** 12
[8] Nader Khalili 2014 Emergency Sandbag Shelter and Eco-Village *Iliona Outram New_earth Uk London* Fine Arts Press Ltd.
[9] The Green Building Program 2006 *"Earth Construction"* Sustainable Building Sourcebook 14
[10] Hunter, Kaki and Donald Kiffmeyer 2004 Earthbag Building *Gabriola Island* BC: New Society Publishers.
[11] Kennedy and Joseph F 14 Feb. 2007 "Building With Earthbags" *Natural Building Colloquium* Networks Productions.
[12] One Community, Earthbag village Heating and Cooling http://www.onecommunityglobal.org/earthbag-village-heating-and-cooling/
[13] "Green Home Building and Sustainable Architecture: A Short History of Earthbag Building" Greenhomebuilding.com. 2007-08-18. Retrieved 2012-10-13.
[14] Freedom Communications, Inc. 22 Jan. 2007.
[15] *Alternative Construction: Contemporary Natural Building Methods Ed.* Lynne Elizabeth and Cassandra Adams New York: John Wiley & Sons, Inc. 2000.
[16] Husain and Yasha "Space-Friendly Architecture: Meet Nader Khalili." *Space.com* 17 Nov. 2000 19 Jan. 2007.
[17] Sinclair, Cameron, and Kate Stohr 2006 "Super Adobe" Design Like You Give a Damn *Ed. Diana Murphy, Adrian Crabbs, and Cory Reynolds.* Ney York: Distributed Art Publishers, Inc. 104-13.
[18] Kellogg, Stuart, and James Quigg 18 Dec. 2005 "Good Earth" *Daily Press.*

The effect of crack width on the service life of reinforced concrete structures

NGUYEN Van Hung[1], VU Viet Hung[2] and TRAN Bao Viet[3]

[1]University of Transport and Communications– campus in Ho Chi Minh City, VietNam, Master student
[2]University of Transport and Communications– campus in Ho Chi Minh City, VietNam, Lecturer; email: vvhung@utc2.edu.vn
[3]University of Transport and Communications– campus in Hanoi, VietNam, Lecturer; email: baoviet.tran@gmail.com or viettb@utc.edu.vn

Abstract. Reinforced concrete has become a widely used construction material around the world. Nowadays, the assessment of deterioration and life expectancy of reinforced concrete structure is very important and necessary as concrete is a complex material with brittle failure. Under the effect of load and over time, cracks occur in the structure, significantly reducing its performance and durability. Therefore, a number of models for predicting the penetration of chloride ions into the concrete were proposed to assess the durability of the structure. In the study performed by T B Viet (2016) [1], the author proposed a new theoretical model, especially considering the effects of macro and micro cracking on the diffusion coefficient of chloride ion in the cracked concrete. The following experimental results, in term of electrical indication of concrete's ability to resist chloride ion penetration, are used to calculate the lifespan of a reinforced concrete structure according to Dura Crete approach [8] with different crack widths to evaluate the accuracy and reliability of the above model in the range of concrete compressive strength of 30-70MPa.

1. Introduction

The durability of reinforced concrete (RC) structures in the marine environment can be understood as the ability to resist the intrusion of harmful substances, especially chloride ions from the external environment, through a network of pore and cracks. This results in the reinforcement corrosion, and eventually destroying the entire structure. There have been several scientific researches to determine the chloride ion diffusion coefficient in cracked concrete and have concluded that the pore structure of concrete affects decisively the penetration mechanisms and further the durability of the building [2-6]. Besides the pore system, during the fabrication and exploitation, concrete structures are also suffered from cracks (practically crack is considered as one of the intrinsic characteristics of concrete structure). Over time, under the effects of loads and environmental factors, cracks tend to expand and develop. It is believed that cracks reduce mechanical properties and significantly increase the diffusion of deleterious materials [1]. However, predicting the effect of cracking on diffusion of harmful substances in concrete, especially chloride ions is complicated and difficult [7].

Normally, when the structure is subjected to a certain state of load, macro cracks appear, extending from the exposed concrete surface to the reinforcement. In addition, microcracks also occur in the voids, internal defects and the interfacial transition zone (ITZ) between the aggregate and the cement paste in concrete. As the applied load is large enough, microcracks and macro cracks will connect together to form a crack network, affecting substantially the average diffusion coefficient of the structure [1].

The following study showed the obtained experimental results of electrical indication of concrete's ability to resist chloride ion penetration, and then to calculate the service life of reinforced concrete structures according to Dura Crete approach [8], with different crack widths to evaluate the accuracy and reliability of the model proposed by T B Viet [1] in the range of commonly used concrete compressive strength of 30-70MPa.

2. New approach for improving prediction of effective chloride diffusivity in cracked concrete structures [1]

With an isotropic material hypothesis, and on the basis of Fick's second law, the average chloride ion diffusion coefficient in the RC structure with parallel surface fracture is calculated as following:

2.1. Macroscopic structure model

The average diffusion coefficient of cracked concrete (D^e) is calculated based on the diffusion coefficient in the crack (D_1) and diffusion coefficient of the un-cracked concrete (D_0) as seen in Figure 1:

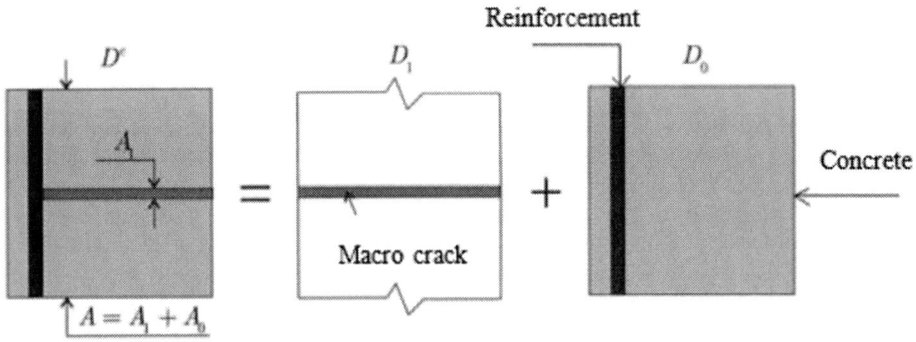

Figure 1. Illustration of the effect of macro cracking on the diffusion coefficient of concrete.

$$\frac{D^e}{D^b} = \frac{D_1}{D^b}\frac{1}{S} + K. \qquad (1)$$

In which: D^e: Average chloride ion diffusion coefficient of cracked concrete.

D_1: Chloride ion diffusion coefficient in crack.

$D_0 = D^b$: Chloride ion diffusion coefficient of concrete structure once only microcracks appear.

K: Coefficient depends on material characteristics.

$S = \frac{A_0}{A_1}$ in which A_0, A_1 are the cross-sectional areas of concrete structure and crack respectively.

2.2. Microscopic structure model

It was hypothesized that a microscopic crack is formed in elliptical shape with two basic geometric features: X=c/a (short axis and long axis ratio of the ellipse); The crack density parameter $e=Na^3$ with N is the number of cracks per unit volume.

Eshelby tensor is a relevant mathematical solution of the elliptic problem in an infinite domain bearing a constant load. [2].

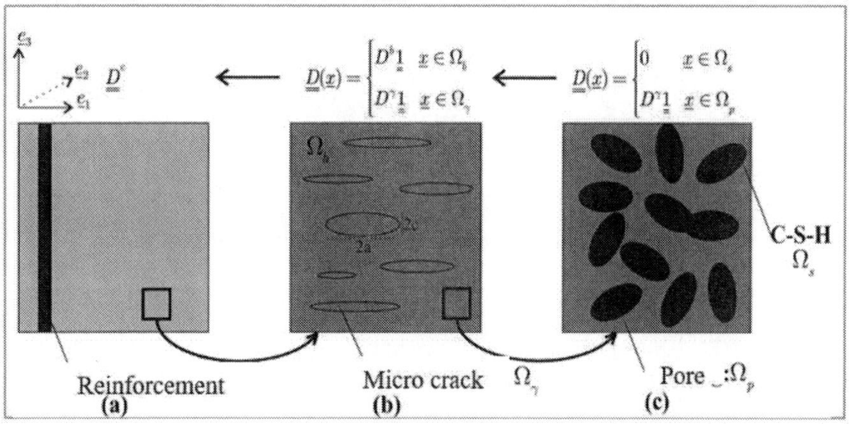

Figure 2. Illustration of the effect of micro crack on the diffusion coefficient of concrete structure.

$$K = \frac{D^e}{D^b} = 1 + f\left(\frac{D^g}{D^b}, e, X\right) \tag{2}$$

In which: D^g : Chloride ion diffusion coefficient of crack under saturated phase.

e : crack density parameter.

X: Short axis and long axis ratio of ellipse -shaped micro crack.

Thus, the chloride ion diffusion coefficient in the cracked concrete structure taking into account microcracks depends on not only the macro crack width (W), and depth (L), but also the pore structure characteristics, size and density of microcracks.

3. The service life of concrete structure according to Dura Crete [8]

According to Dura Crete approach [8], the initiation time of chloride induced corrosion of steel in concrete (t_1) is calculated as the following formula:

$$t_1 = \left[\left(\frac{2}{x^c - \Delta x} \mathrm{erf}^{-1}\left(1 - \frac{c^c_{cr}}{\gamma_{C_{cr}}} \frac{1}{A^c_{C_{s,cl}} \frac{\mathrm{w}}{b} \gamma_{C_{s,cl}}} \right) \right)^{-2} \frac{R^c_{0,cl}}{k^c_{e,cl} . k^c_{c,cl} . t_0^{n^c_{cl}} \gamma_{R_{cl}}} \right]^{\frac{1}{1-n^c_{cl}}} \tag{3}$$

In which:

erf: error function, stemming from solving Fick's second law of diffusion

c^c_{cr}: characteristic value of critical chloride concentration.

$c^d_{s,cl}$: design value of surface chloride concentration.

x_d: design value of concrete cover thickness.

R^d_{cl}: design value of chloride resistance.

Δx: margin for concrete cover thickness.

x^c: thickness of concrete cover.

γ_{Ccr}: partial coefficient for critical chloride concentration.

$A^c_{Cs,cl}$: regression parameter describing the relationship between chloride surface concentration and water & binder ratio (w/b).

$\gamma_{Cc,cl}$: partial coefficient for chloride surface concentration.

$R^c_{0,cl}$: resistance to chloride ingression determined on the basis of compliance trials.

$R^c_{cl,0} = \dfrac{1}{D_{0,cl}}$ in which $D_{0,cl}$: chloride ion diffusion coefficient.

$k^c_{c,cl}$: curing factor.

$k^c_{e,cl}$: environmental coefficient.

t_0: age of concrete when testing.

n^c_{cl}: ageing exponent.

γ_{Rcl}: partial coefficient of resistance for chloride intrusion.

4. Experiment to determine the effect of crack expansion on chloride ion diffusion coefficient

Due to time constraints and limited experimental conditions, as well as the difficulty in crack control, only one instance of artificial concrete cracks (W = 2 mm in width; L = 20mm in depth) were conducted and compared to un-cracked concrete sample. Thus, it can be found that these test results investigate qualitatively the effect of crack expansion on chloride ion diffusion coefficient.

4.1. Testing sample preparation

To examine the effect of microcracks (porosity, internal defects, ITZ between cement paste and aggregates) on the chloride ion diffusion coefficient, three types of concrete with different compressive strengths (f'c = 30, 50, 70 MPa) were used.

Moreover, in order to fully control the crack parameters and model evaluation, in this study an artificial cracking method was implemented by using a reciprocating saw machine to introduce a single parallel surface crack of 80 mm long, 2 mm wide and 20 mm deep.

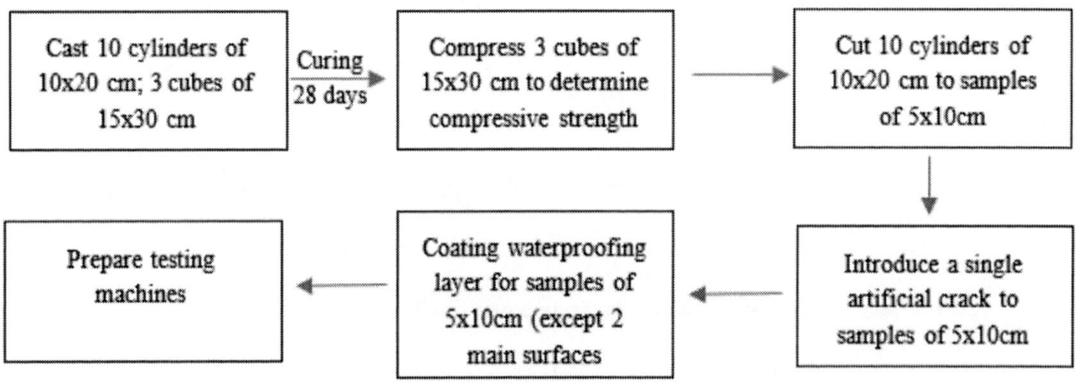

Figure 3. Test sample preparation procedure.

Table 1. Concrete mix proportion ($1m^3$).

Ingredient	$f'_c = 30$ *MPa*	$f'_c = 50$ *MPa*	$f'_c = 70$ *MPa*	*Unit*
D_{max}	19	19	19	mm
Cement	360	433	478	Kg
Fly ash	0	60	0	Kg
Silica fume	0	0	53	Kg
Water	183	162	161	Liters
Superplasticizer Sika Viscocrete 3000-20M	0	6.5	7.55	Liters
Sand	733	648	626	Kg
Gravel	1133	1108	1108	Kg

Figure 4. Slump test.

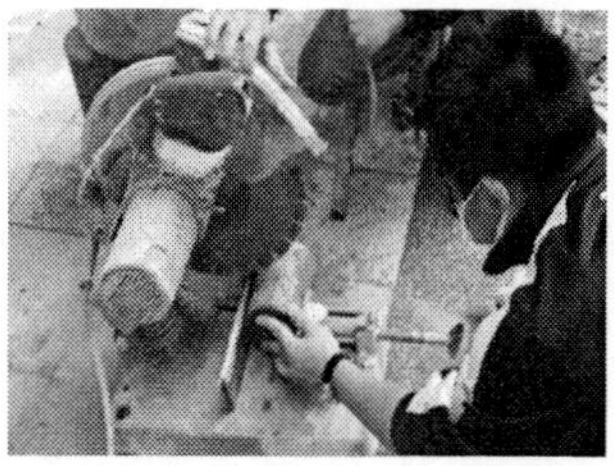

Figure 5. Cut cylindrical concrete sample of 5x10cm.

Figure 6. Samples of artificial crack.

Figure 7. Introduce waterproofing layer.

Figure 8. Vacuum machine.

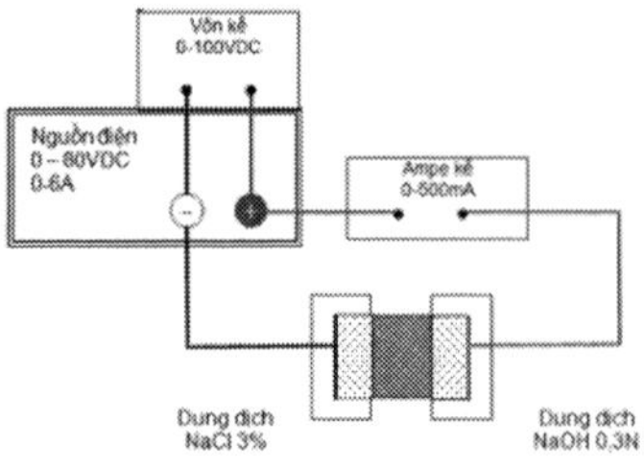

Figure 9. Electrical indication measurement.

4.2. Experimental approach or method (as TCVN 9337:2012- Heavy concrete -Method for electrical indication of concrete's ability to resist chloride ion penetration)

Measurement of electrical indication through a cylindrical concrete sample of 5x10cm was performed. The test was conducted in accordance with the principle of applying 60V DC to two sides of the test specimen, in which one side was exposed to a 3% sodium chloride solution connected to the cathode and the other was exposed to a sodium hydroxide solution connected to the positive electrode [9].

4.3. Formula to calculate the chloride ion diffusion coefficient

In this study, the chloride ion diffusion coefficient was determined by the electrical indication of concrete sample (Coulomb), based on the C-D relationship of some concrete references in the literature as following [10]:

- Normal concrete: D = 0.0056(RCP) – 8.45 (10^-8 cm2/s)
- Silica fume concrete: D = 0.0005(RCP) + 0.99 (10^-8 cm2/s)
- Fly ash concrete: D = 0.0019(RCP) + 0.86 (10^-8 cm2/s)

In which RCP: the amount of Coulomb transmitted through concrete sample (Coulomb).

4.4. Experimental results

After conducting a test of electrical indication transmitted through cracked and un-cracked samples, the obtained results of chloride ion diffusion coefficient across the samples are shown as following:

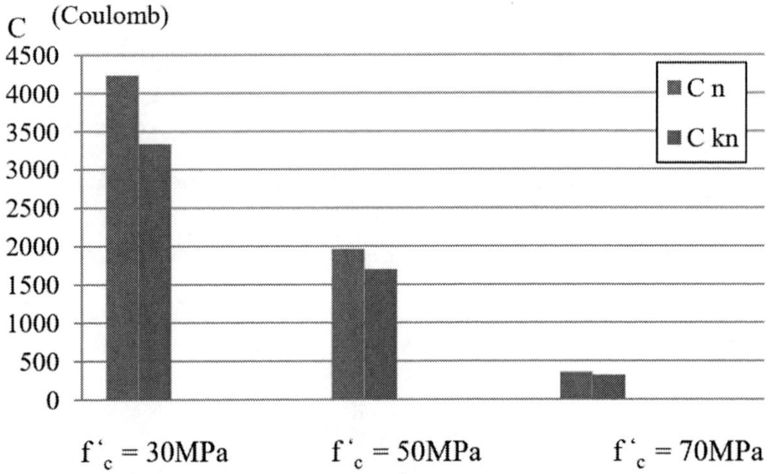

Figure 10. Electrical indication (Coulomb) passed through the concrete specimens (Cn: mean value of 18 samples, Ckn: mean value of 4 samples).

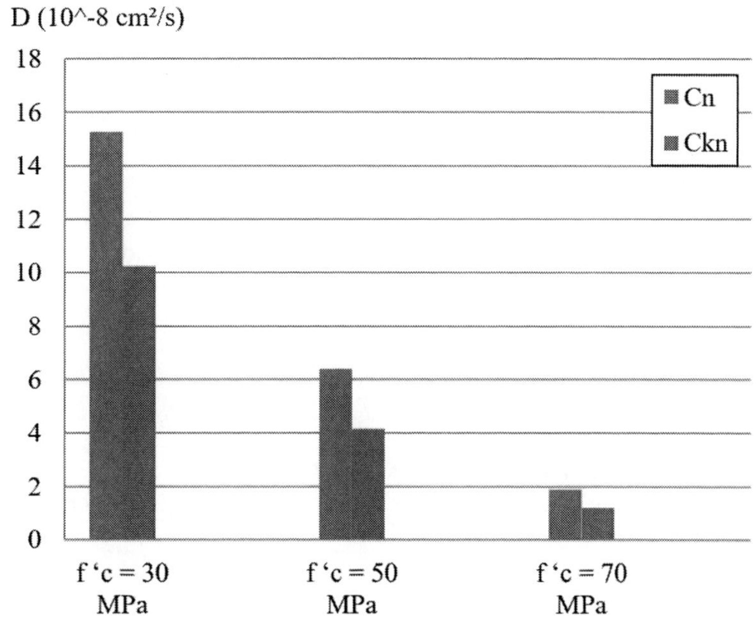

Figure 11. Diffusion coefficient of chloride ions (D) in different concretes.

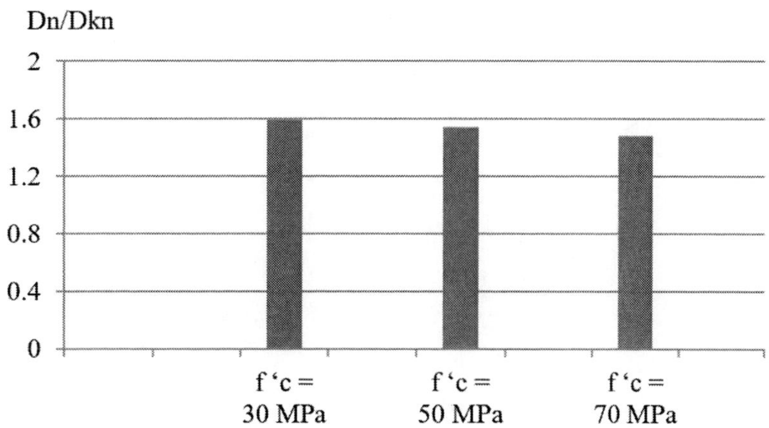

Figure 12. Dn/ Dkn ratio of different concretes.

Based on the above results, some preliminary comments can be drawn as following:

- In case of only considering micro cracking in un-cracked concrete with different strengths:

The higher compressive strength it is, the smaller chloride ion diffusion coefficient becomes. The reason is that in higher strength concrete with lower W/C ratio, the concrete structure becomes more solid and denser, thus the smaller porosity makes the penetration of chloride ions into the concrete more difficult (as seen in Figure 11).

- In case of considering macro cracking in cracked concrete and un-cracked concrete with the same compressive strengths:

The presence of macroscopic cracks results in a substantial increase of chloride ion permeability (Figure 11). Once cracks appear, introducing a free way for chloride ion from the external environment easily to penetrate into the concrete.

Dn / Dkn ratio decreases as the strength of concrete increases as shown in Figure 12. This obtained result is in contrast to the result of applying proposed T B Viet's model [1] for three types of concrete: normal concrete (OC), high performance concrete (HPC), and high performance concrete using silica fume (HPCSF) with the compressive strengths of 46, 75 and 85.5 MPa, respectively (experiment presenting the relationship between the macro crack width and the average diffusion coefficient was conducted by Djerbi and co-authors, published in Cement and Concrete Research in 2008 [11]). It can be explained that the crack width in this study is quite large (2mm in width). Therefore, the penetration of chloride ions from the external environment to the surface of steel reinforcement is easier due to the large crack width and high porosity in lower strength concrete. Hence, chloride ion diffusion coefficient of cracked low-strength concrete is much greater than that of non-cracking. As a result, the smaller strength of concrete it is, the larger Dn & Dkn ratio becomes.

5. An illustration of service life calculation for cracked and uncracked reinforced concrete (RC) beams

5.1. Parameters of reinforced concrete beams

The I-shaped cross section prestressed RC beams were used to calculate the life expectancy of cracked and un-cracked concrete with the following parameters:

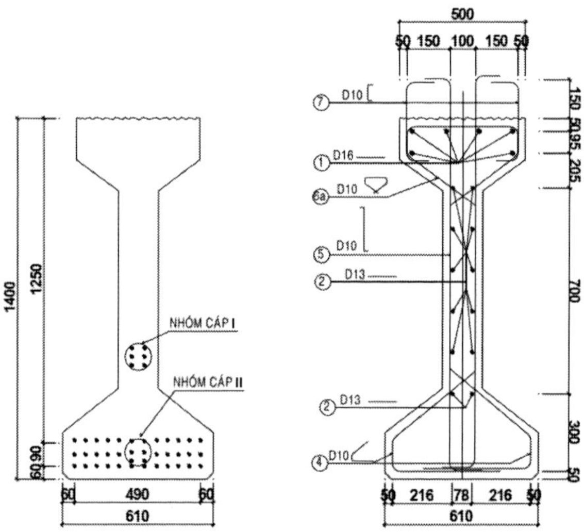

Figure 13. I-shaped cross section beam.

Table 2. I beam parameters.

Parameters	*Value (mm)*
Length x Height	33,000 x 1,400
Width of bottom flange	610
Width of top flange	500
Cover concrete	50
Compressive strength (MPa)	50 & 70

To evaluate the proposed model by T B Viet [1] as well as the above experimental results, the cracks appearing in these beams were assumed to be artificially cracked (crack width is not changed with crack depth).

For the purpose of investigating the effect of micro fracture or concrete structure characteristics, especially the shape and size of micro cracking on the serve life of cracked RC beams, two compressive strengths of concrete were applied (f'c = 50, 70MPa).

And, to examine the effect of macro-crack expansion on the life expectancy of cracked RC beams, different crack parameters, such as crack width (W) and depth (L), were introduced as seen in Figure 14.

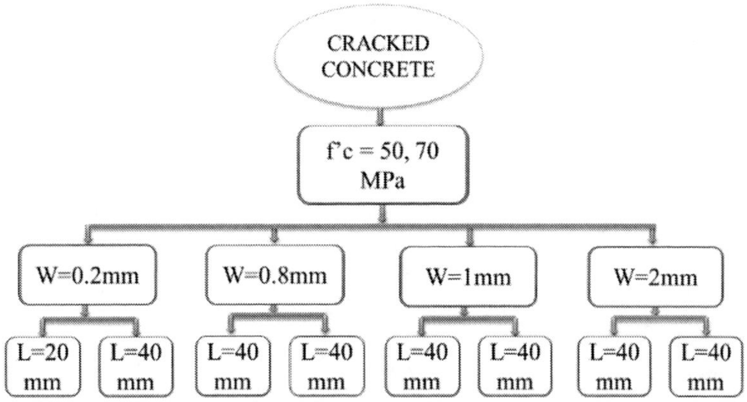

Figure 14. Calculated cases of cracked reinforced concrete beams with different strengths and crack characteristics.

Moreover, to validate the theoretical model with experimental results, the service life of RC beams should be calculated with diffusion coefficients according to the experimental and theoretical model results.

5.2. Results of calculation

After calculating the life expectancy of reinforced concrete beams, the results are shown as following:

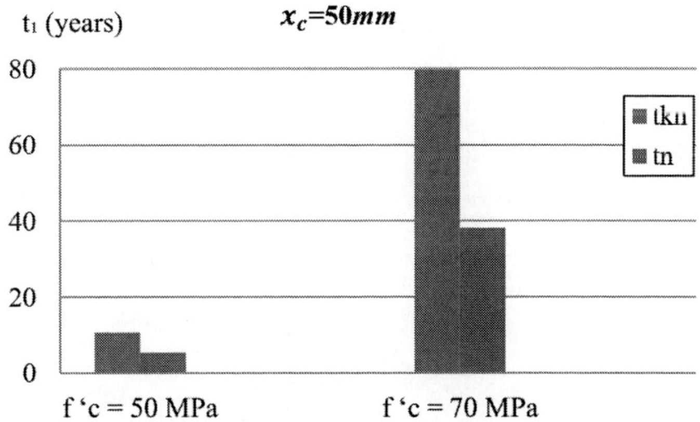

Figure 15. Service life of different types of RC cracked and un-cracked beams based on experimental results with crack characteristics of W = 2mm in width, L = 20mm in depth (tkn and tn: the un-cracked and cracked concrete service life) (x_c = 50mm: cover concrete depth).

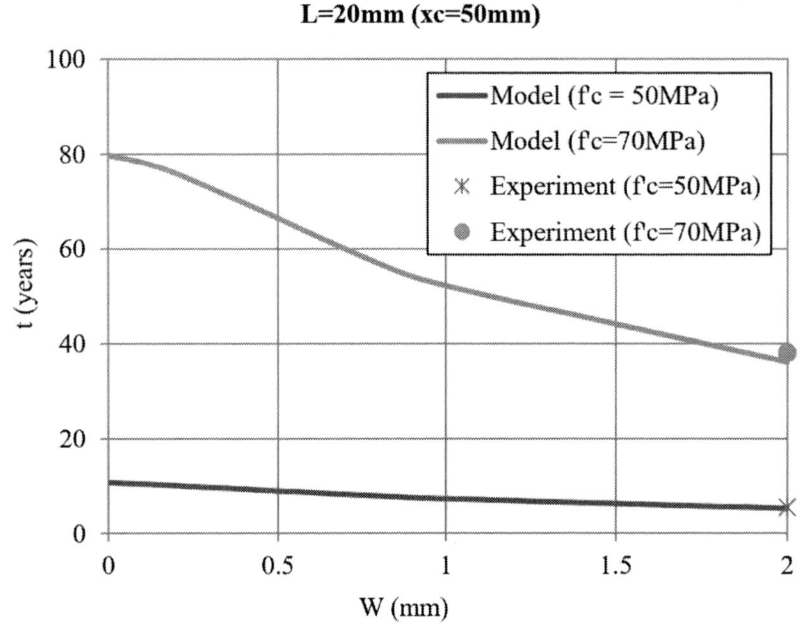

Figure 16. Comparison of the service life of cracked RC beams calculated according to the experimental results and model suggested by Viet (L = 20 mm, x_c = 50 mm).

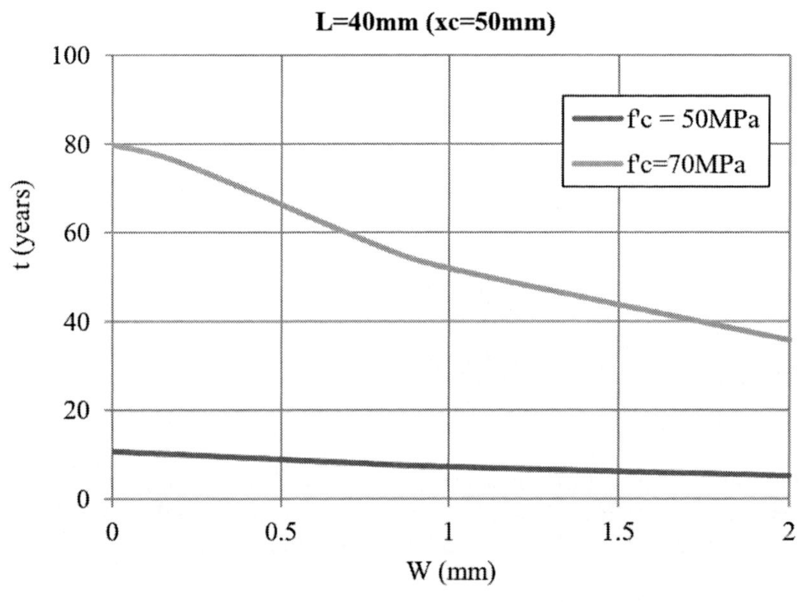

Figure 17. Relationship between the service life of cracked RC beams and crack width (L = 40 mm, x_c = 50 mm).

5.3. Discussions

The service life of RC beams depends on the strength of concrete. A higher strength concrete has a longer life expectancy as seen in Figure 15. This is due to the smaller chloride ion permeability in the higher strength concrete.

And, the service life of un-cracked RC beam is longer than that of cracked concrete (Figure 15). This is because the chloride ion permeability of cracked beams is much higher than that of un-cracked beams.

The difference in service life of RC beams, calculating based on the values of chloride ion diffusion coefficient obtained from the experiment and theoretical model is relatively small (less than 5%) (Figure 16). From this, it can be seen that the proposed model is relatively accurate by taking into account microcracks within the structure. The parameters of model can be estimated through the available information of microstructure or considered as free parameters experimentally determined.

The larger crack it is, the lower life expectancy becomes (Figure 16). The reason is that with the larger cracks, the higher possibility for chloride ion penetration into concrete occurs (larger diffusion coefficient), thus the shorter service life of the beam it is.

It is also found that with the same crack width, once the crack depth increases, the life expectancy of RC beams decreases. The reason is that the chloride ion diffusion coefficient depends not only on the crack width but also on the crack depth. When the crack depth is equal to the thickness of cover concrete, chloride ion penetrates freely or directly contact (through the crack) to the reinforcement surface. In the other hand, once a crack develops and not yet to reach the steel surface, the chloride ion penetration process requires much longer time to reach the steel due to remaining un-cracked concrete cover. However, it can be seen that the effect of crack depth is not significant compared to the crack width when calculating the service life of the beam as seen in Figures 16 & 17.

6. Conclusions and further research

The penetration of chloride ions into cracked reinforced concrete is directly affected by the characteristics of the macro crack (width and depth of crack). In addition, chloride ion intrusion depends on the characteristics of the microcracks (size and density) or structural characteristics of the concrete.

Based on the obtained results, it can be concluded that the model considering both the effect of macro and micro cracking to determine the chloride ion diffusion coefficient proposed by T B Viet (2016) shows a high accuracy and a useful approach for improving the prediction of effective chloride diffusivity in cracked concrete structures in case of parallel surface cracks.

In addition, further theoretical and empirical research should be continued to develop a computational model for practical cracks (taking into account the crack tortuosity and the reduction of the crack width in depth).

References

[1] T B Viet, M D Loc and L M Cuong 2016 Influence of cracks on chloride diffusion in concrete structures *Journal of Transportation Science* **54** (in Vietnamese)
[2] V Q Hoang, N T H Nhung and N N Thuy 2014 Prediction model of chloride ions dissipation into the cracks in reinforced concrete in the marine environment Ho Chi Minh City Open University *Journal of Science* **6 (39)** (in Vietnamese)
[3] T N Nguyen and V Q Hoang 2015 A Simple Approach to Modeling Chloride Diffusion into Cracked Reinforced Concrete Structures *Journal of Civil Engineering Research* **5**
[4] F A L Dullien 2012 Porous media: fluid transport and pore structure *Academic press Elsevier*
[5] L Dormieux, D Kondo and Franz-Josef Ulm 2006 Microporomechanics New York *John Wiley & Sons*
[6] K Maekawa, T Ishida and T Kishi 2008 Multi-scale modeling of structural concrete United States *CRC Press*
[7] M G Richardson 2003 Fundamentals of durable reinforced concrete United States *CRC Press*
[8] EU—B EuRam III 2000 DuraCrete - Final Technical Report Probabilistic performance based durability design of concrete structures The Netherlands *Document BE95-1347/R17(CUR)*
[9] *TCVN 9337:2012- Heavy concrete* - Method for electrical indication of concrete's ability to resist chloride ion penetration (in Vietnamese)

[10] H Kaesche 1959 Die Prüfung der Korrosionsgefährdung von Stahlarmierungen durch Betonzusatzmittel *Zement-Kalk-Gips* (in German)

[11] A Djerbi, S Bonnet, A Khelidj and V Baroghel-Bouny 2008 Influence of traversing crack on chloride diffusion into concrete *Cement and Concrete Research* **38(6)** pp 877-883

IOP Conf. Series: Earth and Environmental Science **143** (2018) 012045 doi:10.1088/1755-1315/143/1/012045

Effect of suction on the mechanical characteristics of uniformly compacted rammed earth

A El Hajjar, P Chauhan*, N Prime and O Plé

Univ. Grenoble Alpes, Université Savoie Mont Blanc, CNRS, LOCIE 73000, Chambéry, France

*Corresponding author. Email: parul.chauhan@univ-smb.fr

Abstract. Rammed earth, in the current environmental situation, is an alternative construction technique which can help in reducing energy and raw material consumption owing to its "sustainable" characteristics. To fully understand its behavior and properties, recent scientific investigations consider it as a compacted unsaturated material with suction as its one of the main sources of strength. Eathern constructions face, over their lifetime, variations in the suction state which have a significant impact on their mechanical characteristics. In the present contribution, unconfined compression tests are performed, with and without unload-reload cycles, on homogeneously compacted samples subjected to various suction conditions.This study shows that both the unconfined compressive strength and Young modulus reduce with the reduction of suction states. Suction also seems to influence the amount of plastic strains and damage phenomenon. Indeed, the soils analyzed are slightly active and shows both plasticity behavior and damage phenomenon.

1. Introduction

In the present scenario, it is crucial to reduce our energy and raw material consumption, notably concerning building construction which is responsible for large part of energy need and waste production. Traditional construction with raw earth responds to this concern. Rammed earth along with other raw earth construction techniques has more eco-friendly characteristics than steel or concrete. Indeed, it is recyclable and therefore inexhaustible, it needs low embodied and operational energy (thanks to thermo and hygro-regulator effects), and when manufactured properly can offer acceptable strength and numerous other advantages while remaining a low-cost material.

Despite these advantages, raw earth suffers from sensitivity to water with a consistency varying from brittle to plastic which has a direct influence on the mechanical capacity. Relative humidity variations thus lead to a variation of deformability and strength of the unstabilized rammed earth structure. Various researchers have studied this impact of relative humidity on the mechanical characteristics. Champiré et al. [1] performed unconfined compression strength tests, with and without unload-reload cycles on compacted earth samples conditioned at different relative humidities. They showed that compacted earth exhibits complex mechanical behavior, including plastic straining and mechanical damage both depending on the relative humidity at which the samples were stored and on the activity of the clayey portion. Bui et al. [1] also performed unconfined compression tests on samples from different soils (sandy, clayey, stabilized) and presented the variations with respect to water content and suction. This study showed that a slight increase of water content of rammed earth wall equilibrated in a classical ambient relative humidity is not followed by a sudden drop in the structural strength for a water content less than 4%.

IOP Conf. Series: Earth and Environmental Science **143** (2018) 012045 doi:10.1088/1755-1315/143/1/012045

Until recently, a new range of scientific investigation is being carried out considering rammed earth as highly unsaturated compacted material to fully understand its properties and behavior. In this frame, suction is considered to be one of the sources of strength. To establish a relation between strength and suction, unconfined compression tests at variable water contents achieved by air drying of the sample with suction measurement were done by Jaquin et al. [2]. Brittle behavior for low starting water content samples and ductile behavior for high water content samples was observed and a qualitative relation between stiffness of sample and suction was obtained.

The objective of this study is to extend the understanding concerning the influence of suction on mechanical behavior, and particularly on Young modulus, compression resistance and intensity of plastic strain, by mean of unconfined compression tests. We propose here to not work with classical rammed earth sample, organized in successive layers whose characteristics influence the mechanical response, but with homogeneously compacted samples which facilitate the result interpretation.

The organization of the paper includes the description of the earthen material tested, the sample preparation method, the hydric conditions imposition and finally the experimental procedures and mechanical response.

2. Material and Specimen Preparation

2.1 Material

The soil was collected from an existing construction site in the vicinity of Vienne in the Auvergne-Rhône-Alpes region in France. It was sieved at 5 mm. The particle size distribution by sieving and sedimentation analysis (figure 1) shows that the soil contains 40% sand, 53% silt, and 7% clay. Note that the particle size curve is not contained in the spindle proposed by British standard BS1377 [3] for earthen construction and is characterized by low clay content. Nevertheless, it is well known that the rammed earth material does not always follow this standard. The Atterberg limits are $w_l = 27.42\%$ (Liquid limit) and $w_p = 16.39\%$ (Plastic limit), leading to a plasticity index being $I_p = w_l - w_p = 27.42 - 16.39 = 11.03\%$. According to the French GTR classification ('Guide de Terrassements Routiers') for fine soils (more than 35% of grains passing 80 μm, no soil grain-size over 50mm), it is classified as a low plastic silt ($I_p < 12\%$).

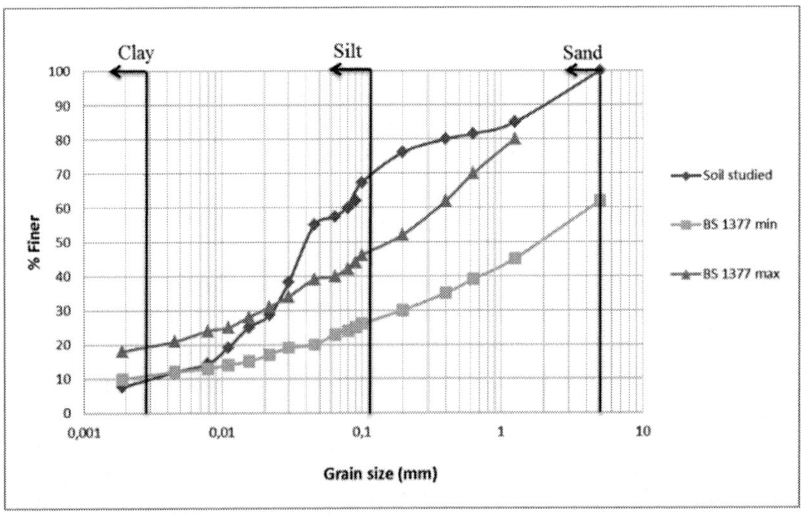

Figure 1. Particle size distribution and the British Standard BS1377 [3]

The activity of the soil (A_c), which depends on the mineralogical nature, is defined as the ratio of plasticity index and percentage passing 2μm sieve. It is here equal to 1.48 which can be considered in the active range ($1.25<A_c<2.0$). In order to more finely characterize the clays, the Cation Exchange

Capacity (CEC) and the specific surface area (S_{sp}) were determined. A CEC= 2.6 cmol/kg was obtained with a pH=7.63, which is considered very low value. So, this soil has a low organic matter content, low water retention capacity, and high sand content. The measurement of the specific surface area gave a low value: S_{sp} = 14.7 m^2/g. It is difficult to determine the nature of the clays contained in this soil with CEC and S_{sp} because of the heterogeneity of the soil, but it can be predicted that this soil either does not contain or contains a very low amount of smectites or vermiculites (swelling clays with very high surface area and CEC).

2.2 Specimen Preparation

The normal proctor test demonstrated that the optimum moisture content to allow the best compaction is between 11.8 and 13.4% (figure 2). Thus, the earth was mixed at 12.5% water content and placed in a sealed container for 24 hours to allow homogenization of water content.

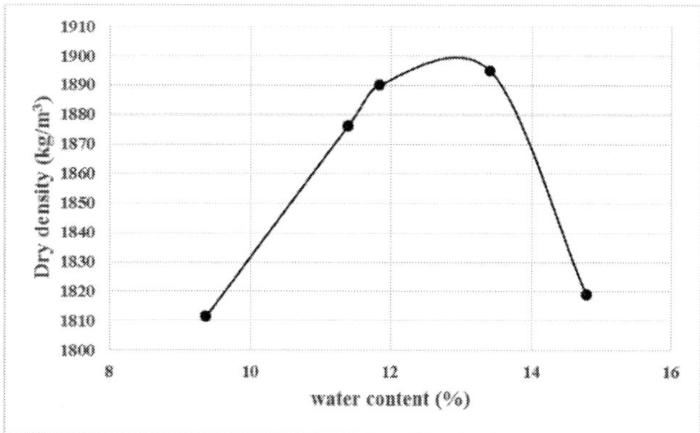

Figure 2. Normal Proctor compaction of soil studied

21 cylindrical samples were compacted in a stainless steel hollow cylindrical mold, designed to obtain specimens with a height of 10 cm and a diameter of 5 cm. The slenderness equal to 2 of the earth cylinders avoids the effect of shrinking (friction during the earth-press contact which hinders the deformation of the earth) and buckling. Indeed several studies have shown that for slenderness of about 2, the geometry does not influence the compressive strength obtained and depends only on the material and the method followed [4]. The samples were compacted by the means of two cylindrical pistons introduced in the upper and lower part of the mold. It was desired to achieve a compression pressure of 5 MPa, which is the pressure usually applied to press Compressed Earth Blocs (CEB). A displacement rate of 0.5 kN/s is chosen to allow air to escape and soil grains to be well managed. This pressure is applied in two stages, 80% of the stress by pressing on one side and 100% of this stress applied on both sides simultaneously, according to a methodology proposed by Bruno et al. [5]. This double compaction aims to increase the uniformity of the dry density in the material, in contrary to dynamic compaction which induces a density gradient due to lateral friction making the earth denser at the top surface of a layer and looser at the bottom. In addition, it has the advantage to perfectly control the pressure applied and thus the reproducibility of density is obtained (contrary to dynamical compaction used for rammed earth). Although this technique differs from real compacting processes (dynamic compaction for rammed earth, uniaxial static compaction for earth compressed brick), it leads to a realistic compacted state: the mean dry density obtained (1860 kg/m^3) is a classical value for rammed earth structures (values ranging between 1700 and 2200 kg/m^3 [7]. In this way, the material is like a model material which can be representative of CEB or Rammed earth structures.

3. Hydric Conditions

3.1 Control of Suction

The suction is controlled by liquid-vapor equilibrium method (also called relative humidity technique). In this method, the relative humidity of the atmosphere surrounding the sample is regulated by means of an aqueous solution of given chemical compound (different saturated saline solutions)[8]. Water is exchanged between the specimen and the surrounding in forms of water vapor according to the relative humidity of the air. When the equilibrium is reached, a particular value of suction is imposed on the sample. The relationship between the relative humidity of the air and the suction imposed at equilibrium is given by Kelvin's equation (1).

$$s = u_a - u_w = -\frac{R.T}{g.w_v}\ln(RH) \tag{1}$$

where s is the suction at a particular temperature T (K), u_a is pore air pressure, u_w is the pore water pressure, R is universal gas constant, g is acceleration due to gravity, w_v is the molecular mass of water vapour and RH is the relative humidity, which is the ratio of partial vapour pressure P in the considered atmosphere and the saturation vapour pressure P_o at a particular temperature.

The 21 cylindrical samples after compaction at optimum moisture content were distributed in batches of three samples within 7 different relative humidities 9%, 22.51%, 32.8%, 57.6%, 75.3%, 84.34% and 97.3% (see table 1) in order to achieve different moisture balance in each lot. Daily weighing was taken to follow the soil moisture content over time while saturation of the saline solution is checked every time, otherwise, the samples equilibrate at different relative humidity. Equilibrium is considered when the mass variation is less than 0.05% during 24 hours. It is noted that for low values of relative humidity (less than 60%), the samples reach the water balance in two weeks, whereas for high relative humidities they need more than a month to equilibrate, as shown in figure 3. From this graph, it appears that the batch equilibrated with $MgCl_2$ salt leads to a final water content which is greater than the batch equilibrated with NaBr salt, even if it applies a lower relative humidity. This observation being inconsistent, the saturation of the $MgCl_2$ saline solution has been re-checked a posteriori for all tests. It appears that it was not fully saturated, saturation being difficult to visualize with this salt. The suction applied being unknown in this batch, all the results obtained for this hydric condition have not been considered in the following analysis, and only 6 values of Relative Humidities are presented.

Table 1. Relative Humidity and Suction imposed by the saturated saline solutions

Salt	KOH	CH_3CO_2K	$MgCl_2$	NaBr	NaCl	KCl	K_2SO_4
RH (%)	9	22.51	32.8	57.6	75.3	84.34	97.3
Suction (kPa)	331.3	205.3	153.4	75.9	39	23.4	3.8

3.2 Soil Water Retention Curve (SWRC)

The phenomenon of condensation/evaporation and sorption/desorption favors the water exchange between the porous medium and the environment of the microstructure of earth [9]. The water retention curve is used to study the behavior of water in the porous medium ranging from a dry state (high suction) and saturated state (zero suction). Water retention is affected by the direct contribution of the soil components such as clays, organic matter, and oxides, the soil structure i.e. the nature of soil constituents and their organization [10]. To demonstrate the behavior of water in raw earth, three samples (dry mass between 5g and 8g), with the same compacted state than the 21 earth cylinders, were dried at ambient temperature and relative humidity (T=25°C and RH=62%). The 3 samples were then heated in the oven at 70°C for several days until reaching a constant mass of solid particles. Then, they were placed in box containing KOH saline solution (RH=9%). Daily weighing was taken and when the samples are equilibrated (mass variation of less than 0.05% for 24 hours), the samples were placed in the next relative humidity box. Once equilibrium is reached in the last box (RH=97,3%), the samples are transferred again toward the lower RH boxes.

To draw the retention curve, mean of water contents of 3 samples was done for each suction state since standard deviation (σ) was less than 6% (except for samples placed in RH=9% box). For both sorption and desorption path, Fredlund and Xing retention curve model [11] was used and fitting parameters were calculated using SWRC fit which is a nonlinear fitting program [12]. In the figure 4, θ_s and θ_r are the saturated and residual volumetric water content, a, m and n are fitting parameters. We note that the path traveled during desorption is above the sorption path, which highlights the phenomenon of hysteresis, which is due to the influence of electrostatic potential in the pores [13]. As the relative humidity increases from nearly zero, first there occur adsorption to a single layer and then to multiple layers of water molecules occurs within the pore structure of the material. The metastable groups of adsorbed water vapor molecule can spontaneously nucleate into a meniscus of liquid water that is in equilibrium with the relative humidity for a given pore radius.

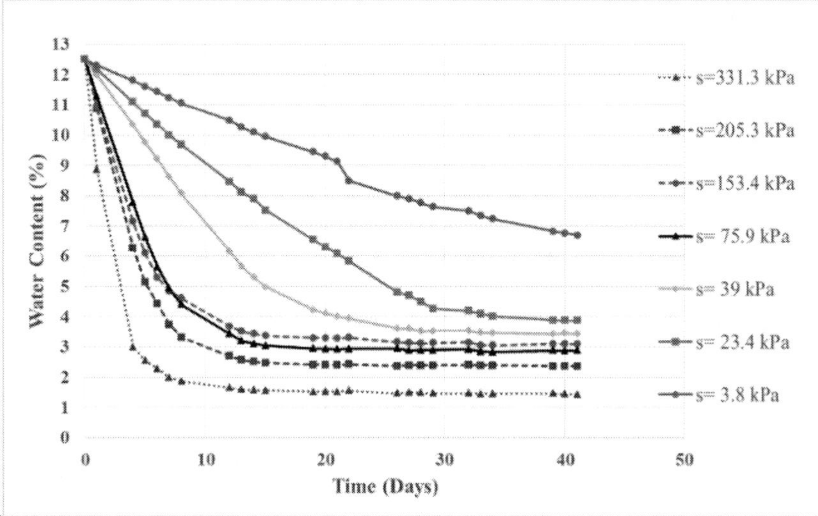

Figure 3. Evolution of water content with time during the equilibration period.

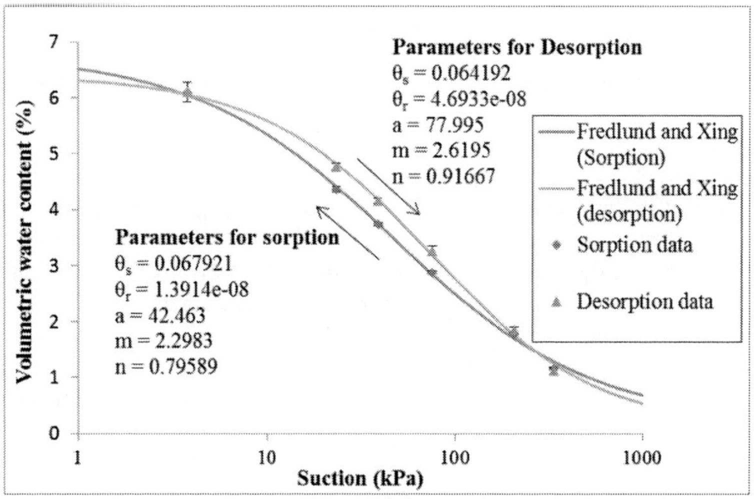

Figure 4. Soil water retention curve for sorption and desorption path along with Fredlund and Xing model

4. Mechanical Response

4.1 Unconfined Compression Strength

The 21 samples are submitted to unconfined compression tests. For each RH batch, 1 of the 3 samples is led to failure after performing 3 unload-reload cycles to calculate Young modulus. Unconfined Compressive strengths (UCS) are reported for the 21 samples in figure 5 with respect to the suction state.

First, it shows that UCS varies between 1.8 and 6.7 MPa. Then, it can be observed that a decrease of suction (an increase of relative humidity), induces a significant decrease of the resistance. In addition, it is noted that the samples which have undergone loading/unloading cycles, i.e. the samples represented in triangular marking in figure 5, have greater mechanical strength than those which have not undergone cycles. They appear in the graph as a curve outside the spindle formed by other samples. This must be due to the additional static compaction experienced by the samples during the three loading cycles, indicating that the mechanical strength of the raw earth is improved by the additional compaction. For classical relative humidity in our latitude (around 60%), the compressive strength is around 3MPa which is close to what is observed in the literature [14].

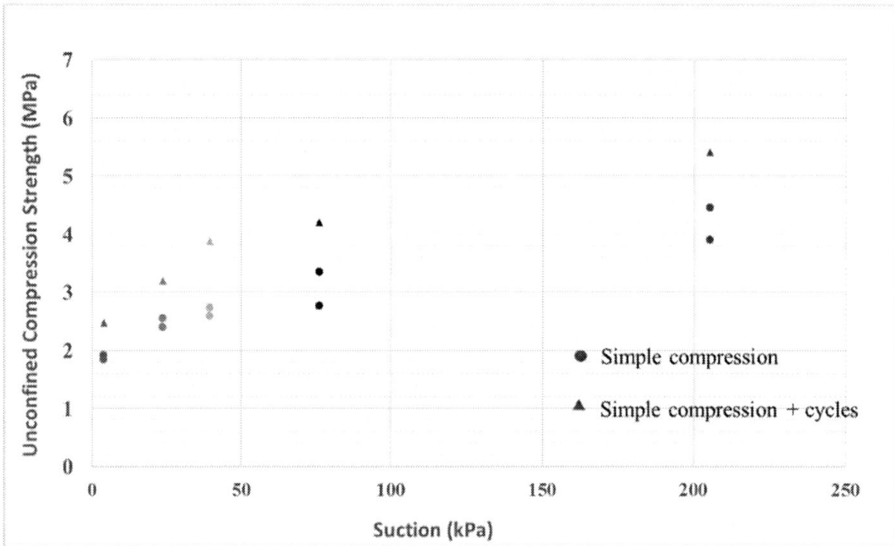

Figure 5. Evolution of UCS with suction for samples with and without loading-unloading cycles

4.2 Young Modulus of elasticity

Initial tangent modulus (E_{tan}) was determined from the unconfined compression test using the slope of initial linear part of the curve and secant Young modulus (E_{sec}) was determined from tests with unloading-reloading cycles. In figure 6, the variation of axial stress with deformation along with the determination of both moduli for one test is shown. The evolution of Initial tangent modulus (evaluated as the mean of the 3 tests made with the same suction state) with suction highlights that suction state has a large effect on E_{tan} (figure 7). A 3 fold decrease of initial tangent modulus is observed when the suction state at beginning of test changes from 331.3 kPa to 3.8 kPa (i.e. from a RH of 9% to 97.3). Thus, the initial Young modulus reduce as the suction state reduces which is consistent with literature [1],[1],[2][15],[16],[16].

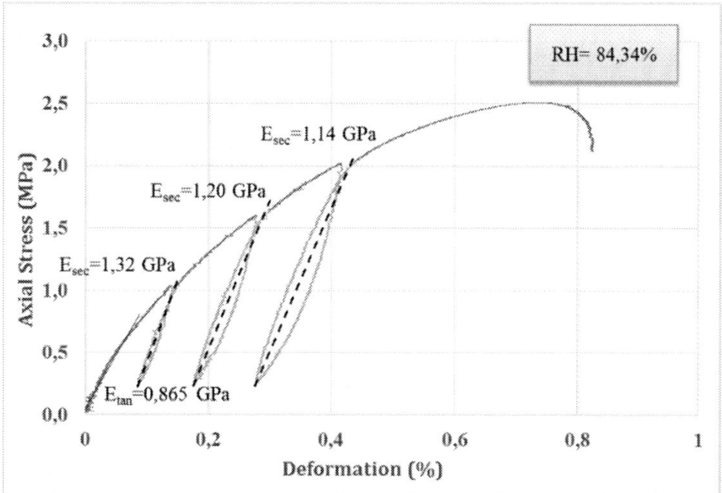

Figure 6. Determination of E_{tan} and E_{sec} with loading-unloading cycles for one test (s=23,4 kPa - RH=84,34%)

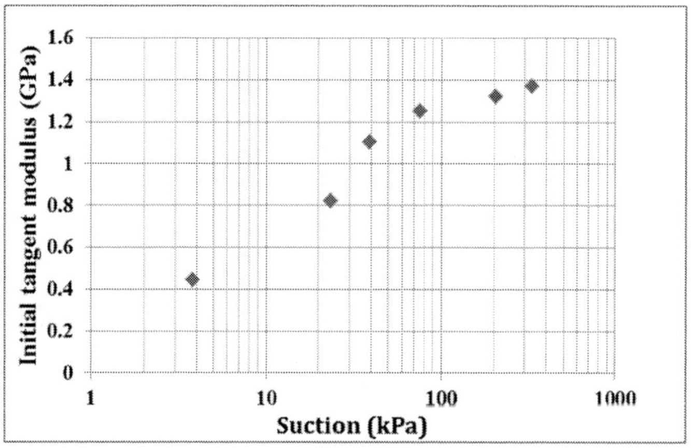

Figure 7. Variation of Initial tangent modulus with suction states (each point being the mean of the 3 values of the same batch)

The evolution of Secant modulus, once plastic behaviour is met, can be also characterized. In this objective, E_{sec} is plotted along the ratio of maximum axial stress previously experienced and UCS (figure 8a). The graph shows a global reduction of E_{sec}, i.e. damage phenomenon, with increasing axial stress (except for the sample conditioned at 331.3 kPa). The stiffness degradation seems to be reduced as the suction state of the sample decreases (from more than 20% for samples conditioned at 205.3 kPa to 13% for samples at 3.8 kPa suction). The dependency of damage phenomenon on suction can be clearly seen from the decrease of the slope in figure 8(a). It means that the dryer the earth sample, the more sensitive to damage it is.

4.3 Evolution of plastic strains

The evolution of the plastic strain (residual strain at the end of the unloading cycles) with the ratio axial stress/ UCSs (figure 8(b)) shows that residual strain increases with axial stress level. Besides, the lower the suction, more is the plastic behaviour (figure 8(b)). This effect becomes more and more important when the axial load approches the compression strength. It means that a humid pathology of an earthen structure leads to simultaneous degradation of the material stiffness correlated to an increase of the plastic strain.

The damage and plasticity behavior as discussed by Champire et al. [1] suggest that the nature of the clay, notably its activity, is a more determining parameter as compared to the content of clay. More active clays experience strong irreversible strains and less damage and vice versa. The soil investigated in this work is considered to be slightly active and thus shows both plasticity (characterized by residual strains) and damage (characterized by stiffness degradation.

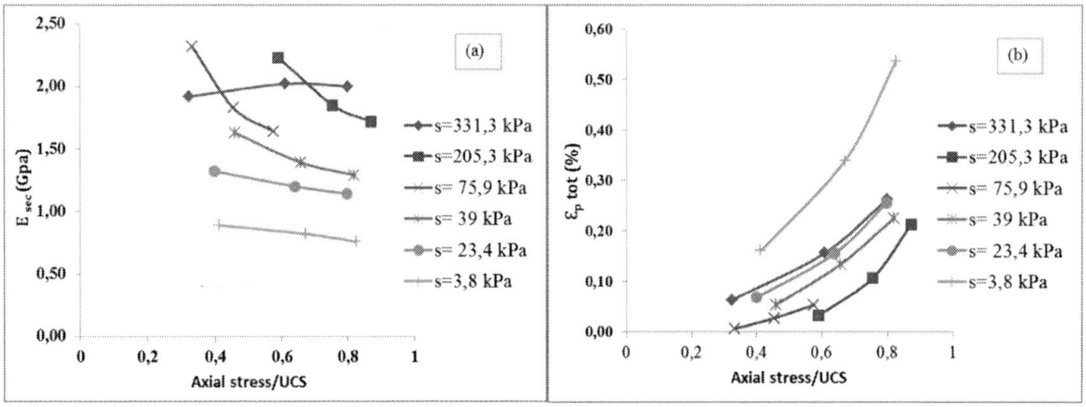

Figure 8. Secant Modulus (a) and residual strain after unloading (b) as a function of stress level

5. Conclusion

In this study, a consistent methodology to investigate hydric influence of the mechanical behaviour of compacted earth structures is proposed, based on a natural earth, compatible with construction project and being made with 7% of slightly active clay.

First, the double compaction process, with 5 MPa applied, allows to manufacture samples which are representative of raw earth structures, but with a reduced density variation inside the material. Then, matric suction is chosen as the state variable accounting for hydric state during the tests, since it represents the direct mechanical translation in the porous matrix of any kind of hydric solicitation applied. In this contribution, hydric state is chosen to be applied uniformly and permanently by mean of saline solution method, which makes it possible to apply 7 suction states, prior to the mechanical loading.

The unconfined compression tests, made with and without unload-reload cycles lead to the following conclusions with regards to mechanical strength, Young modulus, and plastic strain intensity. The samples which have undergone unload-reload cycle had greater mechanical strength indicating that the mechanical strength of the raw earth is improved by the additional compaction. As the suction state decreases, the resistance in simple compression, which is manifested by stress at break, decreases non-linearly, the slope being less steep for lower suction values. A global reduction of the initial Young modulus was observed as the suction state reduces. This highlights, that suction is one of the most important sources of strength and the influence of varying humidity conditions have an impact on the mechanical behavior. The evolution of the modulus during the test shows a global reduction. This stiffness degradation termed as damage phenomenon was significantly higher for samples conditioned at higher suction states.

The plastic straining in the material increased as the axial stress increased. This plastic behavior was more prominent in the samples at lower suction states. Generally, more active clays experience strong irreversible strains and less damage and vice versa. The soil analyzed in this work was slightly active and shows both plastic straining and mechanical damage.

Currently, the effect of hygrometery on the shear characteristics of the soil using Casagrande Direct shear test is being investigated and the effect of suction states on the parameters such as friction angle and cohesion will be analyzed.

References

[1] Champiré F, Fabbri A, Morel J C, Wong H and McGregor F 2016 May 1 Impact of relative humidity on the mechanical behavior of compacted earth as a building material *Construction and Building Materials* **110** 70-8

[2] Bui Q B, Morel J C, Hans S and Walker P 2014 Mar 15 Effect of moisture content on the mechanical characteristics of rammed earth *Construction and Building materials* **54** 163-9

[3] Jaquin PA, Augarde CE and Legrand L 2008 Jun 23 Unsaturated characteristics of rammed earth *First European Conference on Unsaturated Soils, Durham, England* pp. 417-422

[4] Standard *B.* British Standard 1377 1990 *Methods of Test for Soils for Civil Engineering Purposes* British Standard Institution London..

[5] Standard *A S* C1314 2010 *Standard test method for compressive strength of masonry prisms* West Conshohoken PA.

[6] Bruno A W, Gallipoli D, Perlot-Bascoules C, Mendes J and Salmon N 2015 May 26 Briques de terre crue: procédure de compactage haute pression et influence sur les propriétés mécaniques *Rencontres Universitaires de Génie Civi.*

[7] El Nabbouch R 2017 Mechanical behavior of rammed earth walls under Pushover tests *PhD thesis*, Université Savoie Mont Blanc.

[8] Delage P, Howat MD and Cui YJ 1998 Sep 30 The relationship between suction and swelling properties in a heavily compacted unsaturated clay. *Engineering geology* **50(1)** 31-48.

[9] Soudani L, Fabbri A, Morel J, Woloszyn M and Grillet A 2015 A coupled hygrothermal model for earthen materials. *Energy and Buildings*..

[10] Bigorre F. Influence de la pédogenèse et de l'usage des sols sur leurs propriétés physiques. Mécanismes d'évolution et éléments de prévision *PhD Thesis* Université Henri Poincaré Nancy.

[11] Fredlund DG and Xing A 1994 Aug 1 Equations for the soil-water characteristic curve *Canadian geotechnical journal* **31(4)**: 521-32.

[12] Seki K 2007 Feb 27 SWRC fit- a nonlinear fitting program with a water retention curve for soils having unimodal and bimodal pore structure. *Hydrology and Earth System Sciences Discussions* **4**(1): 407-37.

[13] Hall M and Allinson D 2009 Sep 30 Analysis of the hygrothermal functional properties of stabilised rammed earth materials. *Building and Environment* **44(9)**: 1935-42.

[14] Maniatidis V and Walker P 2008 Mar Structural capacity of rammed earth in compression. *Journal of Materials in Civil Engineering* **20(3)**: 230-8.

[15] Nowamooz H and Chazallon C 2011 Apr 30 Finite element modelling of a rammed earth wall. *Construction and Building Materials* **25(4)**: 2112-21.

[16] Olivier M 1994 Le matériau terre, compactage, comportement, application aux structures en bloc sur Terre *PhD thesis* INSA Lyon.

[17] Miccoli L, Müller U and Fontana P 2014 Jun 30 Mechanical behaviour of earthen materials: a comparison between earth block masonry, rammed earth and cob. *Construction and Building Materials* **61**: 327-39.

Acknowledgement

We would like to acknowledge ISTerre Grenoble and Chambery for the evaluation of Cation Exchange capacity (CEC) and Specific surface area of our material.

Influences of geological parameters to probabilistic assessment of slope stability of embankment

Qui T Nguyen[1,3*] Tuan D Le[2,3] and Petr Konečný[3]

[1] Faculty of Civil Engineering, Ton Duc Thang University, 19 Nguyen Huu Tho Str., Tan Phong Ward, Dist. 7, HCMC 700000, Vietnam
[2] Faculty of Civil Engineering, Saigon Technology University, 180 Cao Lo Str., Ward 4, Dist. 8, HCMC 700000, Vietnam
[3]VSB– Technical University of Ostrava, Faculty of Civil Engineering, L. Podeste 1875, 708 33 Ostrava–Poruba, Czech Republic
[*]Corresponding author. Email: nguyenthanhqui@tdt.edu.vn

Abstract. This article considers influences of geological parameters to slope stability of the embankment in probabilistic analysis using SLOPE/W computational system. Stability of a simple slope is evaluated with and without pore–water pressure on the basis of variation of soil properties. Normal distributions of unit weight, cohesion and internal friction angle are assumed. Monte Carlo simulation technique is employed to perform analysis of critical slip surface. Sensitivity analysis is performed to observe the variation of the geological parameters and their effects on safety factors of the slope stability.

1. Introduction

Evaluation of the slope stability of the embankment through the analysis of critical slip surface is a useful tool to determine safety factors of embankment structures such as roads, dams, dykes, etc. The evaluation depends on the geological conditions and failure profiles of embankment which may occur.

There are many random factors impacting on stability of a slope: scatter of geological profiles, variation of soil properties, natural conditions and stability assessment methods. These random factors then create the randomness of the slope stability evaluation problem. Because randomness is taken into account rarely, the result of analysis is single value or safe/unsafe verdict. Slope stability, therefore, should be analyzed under the scheme of probability in order to have a whole spectrum of possible results. This robust approach also allows to optimize design problems.

According to Vietnamese design standard 22TCN 262–2000 [1], stability analysis of slope can be carried out with two classic approaches on which input parameters are constant: fragmentation approach (Fellenius approach) and simplified approach (Bishop approach). The slope is considered stable when the calculated safety factor (K) is greater than the allowable safety factor (K_{min}). Values of K_{min} are different in the two mentioned methods: 1.2 in Fellenius method and 1.4 in Bishop one. In practice, mechanical properties of the soil layers such as unit weight (specific gravity) γ, cohesion c and internal friction angle φ always show high scatter. Thus analysis results from the two methods do not present exact performance of structures. As a result, many roads in operation are frequently encountered with slippage due to their instability. Figure 1 shows an incident of slope slip of a new–built road due to the instability of its slope [2].

Figure 1. A 140–meter road segment slipping due to instability in Hung Chau commune, Hung Nguyen district, Nghe An province, Vietnam (Photo taken on October 6th 2017, adopted from [2]).

To account for the randomness of the input parameters in problems of embankment's slope, probabilistic stability analysis method has widely been used. It was first introduced in the 1970s, for example, Alonso 1976 [3], Tang et al. 1976 [4] and Harr M E 1977 [5]. Since then, the concepts and principles of slope stability analysis using probabilistic theory have been continuously developed and improved. In 2002 and 2003, El–Ramly and his colleagues made probabilistic stability analyses for slope and a tailings dyke [6], [7].

Through using probabilistic method, risk level and hence reliability of safety factor of a particular structure can be quantified even in the inadequacy of representative data. For instance, an uncertainty analysis for a slope based on imprecise information was carried out in [8] and a reliability–based design procedure when the statistics of the random variables are incomplete was introduced in [9]. In addition, Finite element methods (FEM) in combination with probabilistic simulation techniques have become useful tools in solving probabilistic problems. FEM was used in probabilistic stability analysis of embankments in [10] and Monte Carlo simulation method was applied to reliability analysis of slope stability [11]. Another example is that Recursive finite element method (RFEM) was used in probabilistic analysis of reinforced slopes with respect to spatial variability of soil properties [12].

Study on critical slip surface for earth slopes was done in [13] based on First Order Reliability Method. Critical slip surface problems can be solved by the two mentioned classic approaches with certain limitations. Nowadays, a probabilistic analysis of stability of a slope can be conveniently done in SLOPE/W [14]. It is a powerful tool developed by Geo–Slope International Limited (Canada) on which Fellenius approach and simplified approach Bishop approach was integrated.

In this paper, influences of variation of geological parameters to slope stability of an embankment are considered through probabilistic analysis using SLOPE/W. Stability of a simple slope is evaluated in two scenarios: existence of pore–water pressure (PWP) and absence of pore–water pressure (non–PWP). Monte Carlo simulation technique was used to take into account of the variability of soil properties. Probability of failure and sensitivity of slope stability are evaluated with assumption of normal distribution of input data. Sensitivity of the analysis to the scatter of input parameters is also studied with three cases. Coefficient of variation of evaluated random input parameters is assumed to be 25%, 50% and 75% for cohesion; 17%, 33% and 50% for internal friction angle and 5%, 10% and 20% for unit weight in three cases, respectively.

2. Analysis of circular slip surface by classic approaches

Classic methods have been used for ages to analyze critical slip surface problems. In these methods, a slip surface is assumed as an arc and equilibrium of sliding mass is considered. Sketch of a typical problem are illustrated in figure 2.

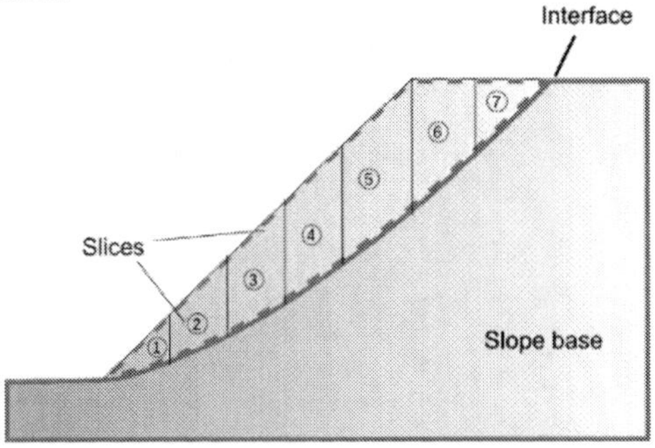

Figure 2. Sketch of a typical circular slip surface analyzed by classic approaches (adopted from [15]).

According to [16], safety factor of slope stability is defined as:

$$FS = \frac{T_f}{T} \quad (1)$$

where:

T_f: shear stress resistance (kPa);
T: calculated shear stress (kPa).

In practical calculations, sliding block is divided (by vertical surfaces, parallel to each other) into pieces and then its equilibrium is evaluated. The problem would have many parameters if interactive forces among pieces and deformation of the sliding block were taken into account. So, assumptions were added to simplify the problem [16], as follows:

- Fellenius method assumes that the sum of interactive forces equals to zero if projected to the axis perpendicular to the radius.
- Bishop method assumes that the sum of interactive forces equals to zero if projected to a horizontal axis.

Both these two approaches are simple quantitative methods on which the embankment's slope is evaluated as stable if the safety factor is greater than 1. The selected safety factor depends on the experience of the design engineers, design standards and other conditions. In addition, stochastic factors of input parameters are not taken into account in the problem. Consequently, the results do not allow accurate estimation of slope stability.

3. Stochastic factors in computation of slope stability of embankment

In evaluation of slope stability, stochastic factors include but not limited to the followings:

- Factors associated with the accident of nature;
- Randomness of input data, including errors in measurements, surveys, experiments, inconsistent data and errors in data processing;
- Randomness of calculation model and calculation method as well as empirical formulas describing the physical characteristics of the system;
- Randomness of the outputs due to imprecise input parameters.

In evaluation of sliding stability, there are three main random factors: mechanical properties of soil, randomness of calculation model as well as the variation of pore–water pressure.

3.1. Randomness of mechanical properties of soil

Random factors in the mechanical properties of soil may be arised by the spatial random distribution of different soil types or by the errors of the experimental process. Spatial variation is the change in the mechanical characteristics of the same soil layer between the boreholes. Random errors occur in the processes of measurement and testator's carelessness as well as laboratory equipment need to be eliminated before the calculation.

According to Lumb (1966) and Christian (1994), key random factor in circular sliding stability problem is the variation of mechanical properties of soil on which the following parameters are the main ones [17], [18]:

- Specific weight (unit weight), γ
- Cohesion, c
- Friction angle, φ

3.2. Randomness of calculation model
The difference between applied hypothesis to the calculation and reality generated random factors in the calculation model. These factors include the description of geological layers (different thickness of the layers, presence of weak geological sites) and water pressure (PWP and non–PWP).

3.3. Slope stability probabilistic analysis based on Monte Carlo simulation technique
Distribution functions describing the transformation of random parameters can be normal distribution functions, normal logarithm functions, uniform distribution functions, triangular distribution functions, and curved distribution functions. In this study, normal distribution functions are used to describe the transformation of three soil mechanical parameters (γ, φ and c). The normal distribution function is often referred to as the Gaussian distribution function, which describes the variation of inputs in probability analysis, which is expressed by the following expression:

$$f(x) = \frac{1}{\sigma\sqrt{2\pi}} e^{-\frac{(x-\mu)^2}{2\sigma^2}} \tag{2}$$

where:

f(x): the probability density functions of the normal distribution;
μ: the mean or expectation of the distribution;
σ: the standard deviation.

Variation of soil mechanical properties can be represented by coefficient of variation (COV). It represents the degree of randomness of the soil properties and is defined as:

$$COV = \frac{\sigma}{\mu}$$

(3)

Reliability function Z is described as:

$$Z = FS \tag{4}$$

Thus, probability of failure is defined as the probability at which FS is less than 1.0:

$$P_{Failure} = P(Z<1) \tag{5}$$

The reliability index (β) presents the slope stability through the number of times of standard deviations (σ) deviated from the mean safety factor (μ):

$$\beta = \frac{(\mu-1)}{\sigma} \tag{6}$$

Monte Carlo is a simulation technique based on a process of repeated random sampling to obtain numerical results. Its methodology includes: creating the sample, running the model and analyzing the data [19]. This method is adopted here to analyze slope stability of the embankment. The number of trials depends on the level of required reliability as well as the number of considered input parameters. To ensure accuracy, therefore, the more input parameters involve, the more Monte Carlo trials are needed. As stated in [20], number of trials is determined as:

$$n_{MC} = \left[\frac{\sigma^2}{4(1-\beta)^2}\right]^m \tag{7}$$

where:

n_{MC}: number of trials;
β: required reliability index (0% ÷ 100%);
m: number of input parameters.

4. Sample example of probabilistic evaluation of stability of circular sliding surface

Stability of an 7–meter high embankment's slope of 1H:1V as shown in figure 3 is analyzed in cases of PWP and non–PWP using SLOPE/W.

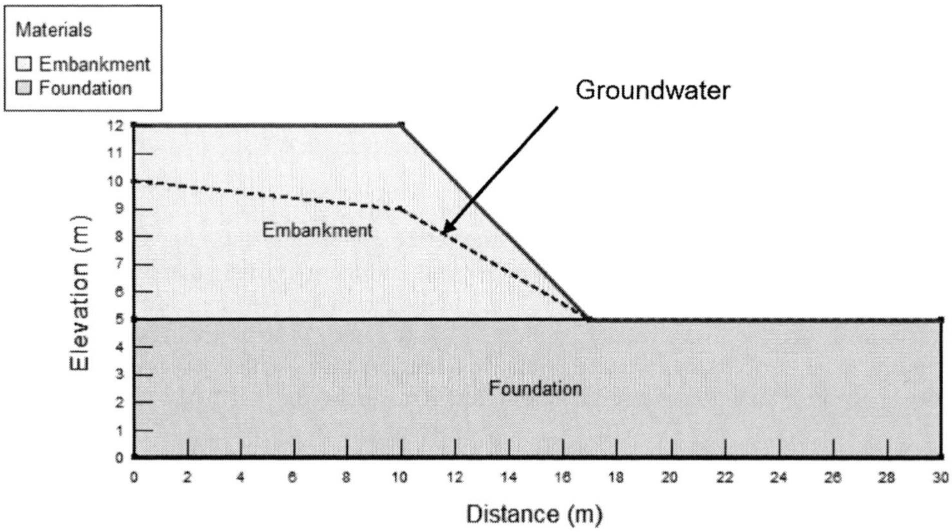

Figure 3. Detailed model of the analyzed slope (adopted and revised from [21]).

Monte Carlo simulation technique is used to cope with variability of random parameters. Critical slip surface is firstly determined by Bishop approach integrated in SLOPE/W software. Input parameters are mean values of soil mechanical properties. These parameters are then proceeded with variation of input parameters. The variability of the input parameters is assumed to follow the normal distribution function. Three cases of standard deviation of cohesion (c), internal friction angle (φ) and unit weight (γ) were considered as illustrated in table 1. Relationship between c and φ were evaluated without as well as with correlation factor of –0.5. Groundwater table (if any) fluctuates within 5 meters from the ground. From equation (7), necessary number of Monte Carlo trials is 15,625 in case of unit standard deviation and required reliability index of 90%.

Table 1. Input parameters used in the example.

Parameters	Mean value, μ	Standard deviation, σ		
		Case 1	*Case 2*	*Case 3*
Cohesion, c (kPa)	20	5 COV = 25%	10 COV = 50%	15 COV = 75%
Internal friction angle, φ (°)	30	5 COV = 17%	10 COV = 33%	15 COV = 50%
Unit weight, γ (kN/m³)	20	1 COV = 5%	2 COV = 10%	4 COV = 20%

5. Results

5.1. When c, φ and γ are constant

Table 2 presents safety factors computed by deterministic method i.e. c, φ and γ are constant parameters with PWP and non–PWP.

Table 2. FS calculated with Bishop approach and with constant parameters.

SF (non–PWP)	SF (PWP)
1.647	1.282

5.2. When c, φ and γ are inconstant and c–φ uncorrelation
It is noticeable that values of FS analyzed by probability method (table 3) are larger than those computed by deterministic method (table 2). In individual case, furthermore, FS in case of PWP is smaller than that of non–PWP. The probability of failure increases as the mechanical properties of the soil increase in values.

Table 3. Results from probabilistic method without c–φ correlation.

Parameters	Case 1		Case 2		Case 3	
	non–PWP	PWP	non–PWP	PWP	non–PWP	PWP
Mean of FS	1.654	1.285	1.707	1.312	1.875	1.492
Reliability index, β	3.088	2.152	1.581	1.152	0.933	1.336
Standard deviation of FS	0.212	0.132	0.447	0.271	0.937	0.368
Minimum of FS	0.785	0.733	0.359	0.426	0.297	0.776
Maximum of FS	2.963	1.998	8.757	4.829	63.807	4.941
Failure probability, P (%)	0.064	1.473	3.835	11.398	7.503	5.543
Number of trials	15,625	15,625	15,625	15,625	15,625	15,625

5.3. When c, φ and γ are inconstant and c–φ correlation
According to [22], c and φ are often negatively correlated with correlation coefficient ranges from –0.72 to 0.35. These resulted from laboratory tests on a wide variety of soils. In this example, a c–φ correlation coefficient of –0.5 is assumed. Results computed by probabilistic method are described in table 4.

Table 4. Results from probabilistic method with c–φ correlation.

Parameters	Case 1		Case 2		Case 3	
	non–PWP	PWP	non–PWP	PWP	non–PWP	PWP
Mean of FS	1.655	1.286	1.710	1.283	1.884	1.481
Reliability index, β	5.312	4.521	2.442	7.129	1.109	2.044
Standard deviation of FS	0.123	0.063	0.291	0.040	0.798	0.235
Minimum of FS	1.076	0.947	1.009	1.062	0.871	1.002
Maximum of FS	2.765	1.856	8.364	1.573	63.706	4.659
Failure probability, P (%)	0	0.038	0	0	0.125	0
Number of trials	15,625	15,625	15,625	15,625	15,625	15,625

5.4. Sensitivity of safety factor to c, φ and γ
The purpose of sensitivity analysis is to observe the variability of the input parameters and to evaluate their effects on the outputs. Results on this analysis with three parameters c, φ and γ of embankment soil layers are shown in figure 4 and figure 5.

As depicted in figures 4 and 5, the sensitivity of computed factor of safety versus the range of the parameters used in the computation was analyzed with 11 points in the range. It can be seen from the two figures that the rise in unit weight leads to a decrease in the safety factor (inverse relation). Meanwhile, when the cohesion or internal friction angle increases, the safety factor also increases (proportional relation).

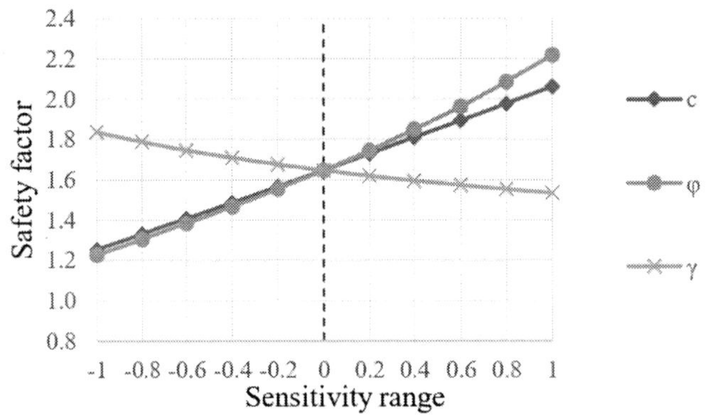

Figure 4. Results on sensitivity analysis in case of non–PWP.

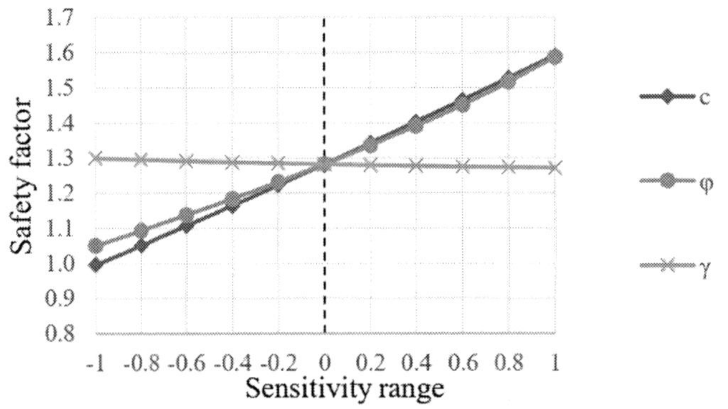

Figure 5. Results on sensitivity analysis in case of PWP.

For the absence of pore–water pressure (figure 4), the sensitivity of the internal friction angle to the safety factor is greatest, followed by the cohesion and finally the unit weight. With the presence of pore–water pressure (figure 5), however, the sensitivity of cohesion to the safety factor is greatest, followed by the internal friction angle and finally the specific weight.

6. Conclusions

Influences of geological parameters to slope stability of an embankment were studied through an example of probabilistic analysis using SLOPE/W. Results from the analysis revealed that the presence of pore–water pressure increased the sliding ability of the embankment. It is also worth noticing that probability of failure (slide) decreases if correlation between cohesion and internal friction angle are considered. In addition, the higher the standard deviation of the mechanical properties of the soil is, the greater the sliding probability of the embankment results.

The influence of the input parameters on the safety factor depends on the parameter properties itself. In this research, relationship between the input parameters and the safety factor may be an inversion relation (safety factor versus unit weight) or a proportional relation (safety factor versus cohesion/internal friction angle).

The sensitivity of the three considered parameters to the safety factor is at different levels in the two scenarios (non–PWP and PWP). Therefore, it is impossible to say that the nature of the parameter determines its overall sensitivity to the factor of safety.

It can be inferred from the study that embankment with high average safety factor does not mean that it is safe because there is possibility of existence of large failure probability. This means that there is no direct relationship between probability of failure and safety factor. This also reveals that solving the problem of circular sliding using probabilistic methods allows a better comprehensive view of the stability of the embankment.

Results from sensitivity analysis of the mechanical properties of the soil revealed that the cohesion had the largest and direct impact on the failure probability of the embankment. Nevertheless, this parameter did not affect mean value the safety factor.

This sample example demonstrates the effectiveness of the probabilistic method in solving the sliding stability problem in comparison with the classic deterministic approach. Embankment's slope stability analysis based on the Monte Carlo simulation technique allows the determination of the probability of failure, wide range of safety factor and its reliability index. This helps not only to minimize random errors in the calculation process but also to predict the fluctuation of safety factor in real conditions. As a result, safety factor and probability of failure could be selected based on nature of the work, specific design requirements to serve as the basis for design as well as evaluation of the embankment structures.

This study only investigated the most disadvantageous position of sliding arcs. In addition, external load, horizontal seismic force and variation of pore–water pressure have not been studied yet. In the future, therefore, this research should be developed with taking into account of external loads including horizontal seismic force and change of pore–water pressure.

References

[1] 22TCN 262–2000 Vietnamese standard: *Procedures for surveying and designing the road surface of embankment on weak soil* (Quy trình khảo sát thiết kế nền đường ô tô đắp trên đất yếu) Ministry of transport (in Vietnamese)
[2] https://vnexpress.net/photo/giao-thong/duong-gan-70-ty-chua-khanh-thanh-da-sut-lun-3651807.html
[3] Alonso E E 1976 Risk analysis of slope and its application to slopes in Canadian sensitive clays *Geotechnique* vol **26(3)** pp 453–472
[4] Tang, Yucemen and Ang 1976 Probability based short–term design of slopes *Canadian Geotechnical Journal* vol **13** pp 201–215
[5] Harr M E 1977 Mechanics of particulate media– a probabilistic approach (New York: McGraw–Hill)
[6] El–Ramly H, Morgenstern N R and Cruden D M 2002 Probabilistic slope stability analysis for practice *Canadian Geotechnical Journal* vol **39(3)** pp 665 – 683 https://doi.org/10.1139/t02–034
[7] El–Ramly H, Morgenstern N R and Cruden D M 2003 Probabilistic stability analysis of a tailings dyke on presheared clay–shale *Canadian Geotechnical Journal* vol **40(1)** pp 192–208 https://doi.org/10.1139/t02-095
[8] Rubio E, Hall J W and Anderson M G 2004 Uncertainty analysis in a slope hydrology and stability model using probabilistic and imprecise information *Computers and Geotechnics* vol **31(7)** pp 529 – 536
[9] Juang C H, Zhang J and Gong W 2015 Reliability–based assessment of stability of slopes IOP Conference Series: Earth and Environmental Science vol 26
[10] Xu B and Low B K 2006 Probabilistic stability analyses of embankments based on finite element method *J. Geotechnical and Geoenvironmental Engineering* vol **132(11)** pp 1444 – 1454
[11] Wang Y, Cao Z and Au S K 2011 Practical reliability analysis of slope stability by advanced Monte Carlo simulations in a spreadsheet *Canadian Geotechnical Journal* vol **48(1)** pp 162 – 172 https://doi.org/10.1139/T10–044
[12] Luo N and Bathurst R J 2017 Probabilistic analysis of reinforced slopes using RFEM and considering spatial variability of frictional soil properties due to compaction *Georisk: Assessment and Management of Risk for Engineered Systems and Geohazards* vol **5** pp 1 – 22
[13] Metya S and Bhattacharya G 2014 Probabilistic critical slip surface for earth slopes based on the first order reliability method *Indian Geotechnical Journal* vol **44(3)** pp 329 – 340

[14] GEO–SLOPE International Ltd Stability modeling with SLOPE/W – An engineering methodology June 2015 Edition Alberta Canada
[15] https://en.wikipedia.org/wiki/File:Slope_2d_mos_plain.svg
[16] https://www.geostru.eu/slope–stability–analysis/
[17] Lumb P 1966 The variability of natural soils *Canadian Geotechnical Journal* vol **3(2)** pp 74–97
[18] Christian J T, Ladd C C and Baecher G B 1994 Reliability and probability in stability analysis *J. Geotechnical Engineering ASCE* vol **120** pp 1071–1111
[19] Le D T and Konecny P 2017 *Chapter 8 – Simulation quality of the probability of the corrosion initiation reinforced concrete* in: Katzer J and Cichocki K *Research and modelling in civil engineering 2017* (Koszalin: Koszalin University of Technology Publishing House) Monograph 338 ISSN: 0239–7129 ISBN: 978–83–7365–474–7 pp 115–131
[20] Krahn J 2004 Stability modeling with SLOPE/W 1st edition (Alberta: GEOSLOPE/W International Ltd)
[21] https://www.geoslope.com/products/slope–w
[22] Lumb P 1970 Safety factors and the probability distribution of soil strength *Canadian Geotechnical Journal* vol **7(3)** pp 225–242

Acknowledgments
The article preparation was supported by sources for conceptual development of research, development and innovations at VŠB–Technical University of Ostrava for year 2018, which were granted by the Ministry of Education, Youth and Sports of the Czech Republic.

Semi top-down method combined with earth-bank, an effective method for basement construction.

Tuan B Q [1], Tam Ng M [2]

[1] Faculty of Civil Engineering, Ton Duc Thang University, Ho Chi Minh City, VietNam
[2] Faculty of Civil Engineering, Ho Chi Minh City University of Technology, Ho Chi Minh City, VietNam

buiquangtuan@tdt.edu.vn (Tuan B Q)

Abstract. Choosing an appropriate method of deep excavation not only plays a decisive role in technical success, but also in economics of the construction project. Presently, we mainly base on to key methods: "Bottom-up" and "Top-down" construction method. Right now, this paper presents an another method of construction that is "Semi Top-down method combining with earth-bank" in order to take the advantages and limit the weakness of the above methods. The Bottom-up method was improved by using the earth-bank to stabilize retaining walls instead of the bracing steel struts. The Top-down method was improved by using the open cut method for the half of the earthwork quantities.

1. Introduction:

The author's article limits the analysis from the Dung Tien contractor experience with deep excavation pits in HoChiMinh city and graduate thesis in HoChiMinh city University of Technology.

Nowadays, with high-speed development of major city, constructing high-rise buildings which has basements in inbuilt condition is exigent. Basement Construction Units meet two main difficulties in choosing construction method: *high cost for retaining the excavation and complicated excavation procedure in enclosed space.*

From manufactured reality, in order to improve these two difficulties, the authors propose "Semi Top-down combined with Earth-bank construction method": use open cut method for half of earthwork quantities and stabilize excavation's retaining walls with Earth-bank instead of internal bracing struts.

2. Summary about two popular basement construction methods:

Nowadays, there are two main methods for constructing basements: Bottom-up and Top-down. The following are explanations and some observations of the two methods:

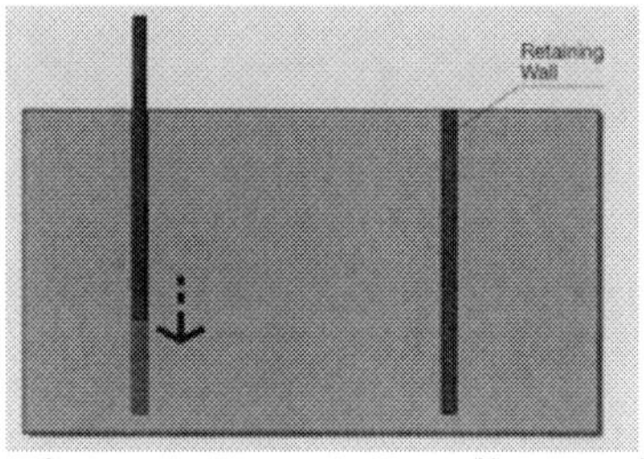

Figure 1. *Stage 1- Installation of Retaining walls*

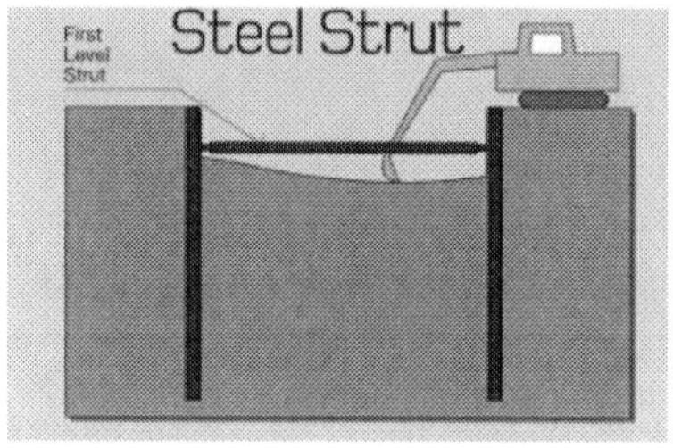

Figure 2. *Stage 2- Excavation and bracing steel strut installation*

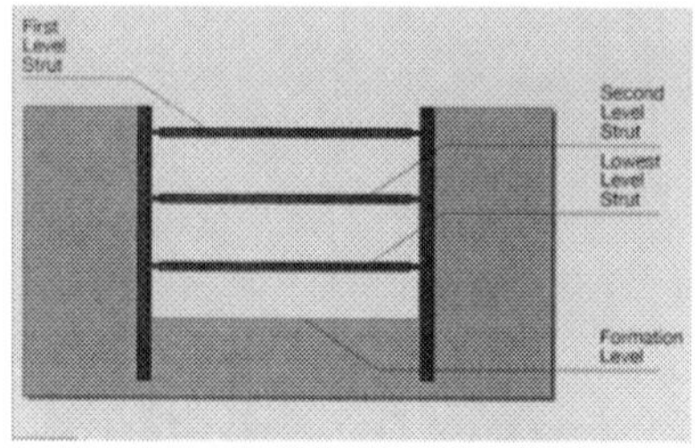

Figure 3. *Stage 3- Excavation and bracing steel strut installation*

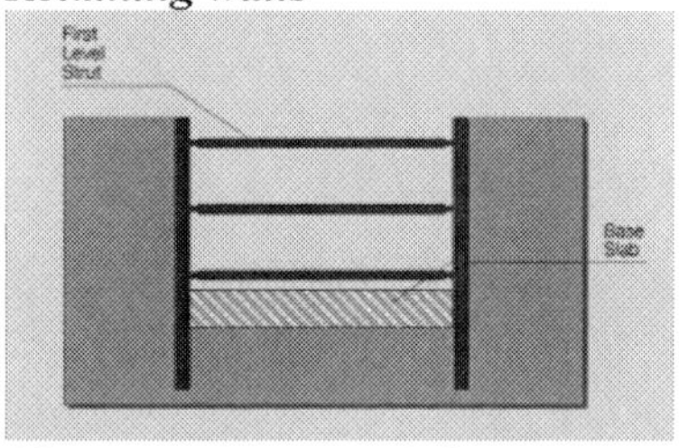

Figure 4. *Stage 4 - Substructure construction.*

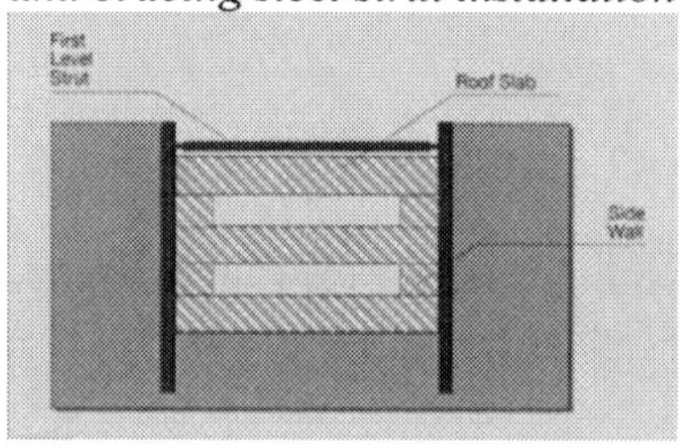

Figure 5. *Stage 5- Substructure construction*

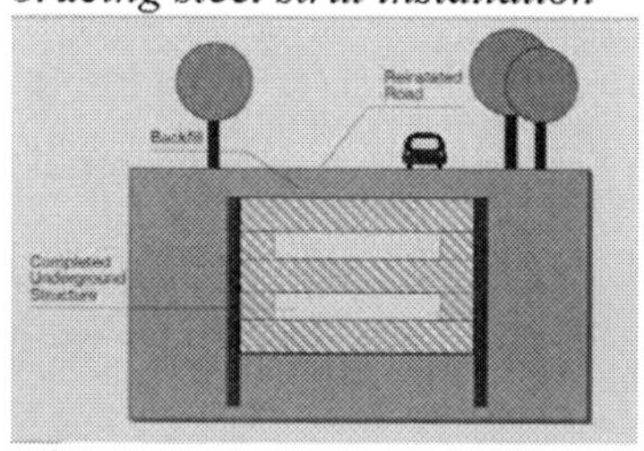

Figure 6. *Stage 6- Backfilling and superstructure construction*

- Stage 1- Installation of Retaining walls: Retaining walls are installed before excavation commences. Depending on site condition, soil type and excavation depth, retaining walls can be concrete walls, concrete bored piles or steel sheet pile walls.
- Stage 2- Excavation and bracing steel strut installation: The soil is excavated to the first strut level. The first level struts must be installed before excavation proceeds further
- Stage 3- Excavation and bracing steel strut installation: The soil is excavated to the next strut level and second level struts are installed. This process progresses until it reaches the final depth or formation level. The number of strut levels depends on the excavation depth
- Stage 4- Substructure construction: At formation level, base slab is constructed, then the lowest level struts are removed and side walls are constructed.
- Stage 5- Substructure construction: Next basement level slab is constructed and the struts at that level are removed. This process progresses upwards until roof slab of the basement is constructed.
- Stage 6- Backfilling and superstructure construction: After finishing roof slab construction, the soil is backfilled to the first strut level before removing first level struts or continuing superstructure construction.

2.1 Bottom-up construction method:

This method is popular and simple in design and construction. Excavation depth requirement depends on foundation base depth and lowest basement's floor level. After the soil was excavated, the building is constructed from the bottom or the lowest level of basement to the top (from foundation to roof).

While shallow cutting excavations can be stabilized by natural earth-bank, deeper excavations at narrow construction sites (inbuilt condition) require using retaining walls (concrete walls, concrete bored piles, steel sheet piles walls…) combined with internal horizontal section steel struts for ground stabilizing in construction process.

a. Construction procedure:

b. Advantages:

- Open cut method: simpler than other earthwork construction methods.
- Giving facilities for lowering groundwater level and constructing foundation.
- Basement waterproofing and technology infrastructure installation are more convenient.

c. Disadvantages:

- In inbuilt condition, it will be very hard to obtain required excavation depth and could cause adverse effects to adjacent buildings if the top soil layer is soft.
- The cost for resisting horizontal displacement along excavation boundary is high. This is also the major disadvantage.

Figure 7. *Excavation and bracing steel strut installation in Bottom-up method*

2.2 Top-down construction method:

This method assumes, that retaining walls (usually concrete walls), and bored piles are constructed firstly. Temporary support piles (section steel piles) and bored piles are simultaneously installed at roof slab level. After that basement floor beams are constructed on top of temporary raking struts. Natural soil surface is used as floor beam formwork. According to this, the soil is excavated to just below basement roof slab level with retaining walls and horizontal bored struts, which is necessary for high height basements. Stairwells are used for excavation and soil transportation from lower floors. After constructing formation level slab, side walls are constructed (or retaining walls are also used as side walls), and horizontal bored struts are removed. The access openings on the roof slab are sealed when superstructure is constructed. The basement will continue to be constructed along with superstructure construction.

a. Construction procedure:

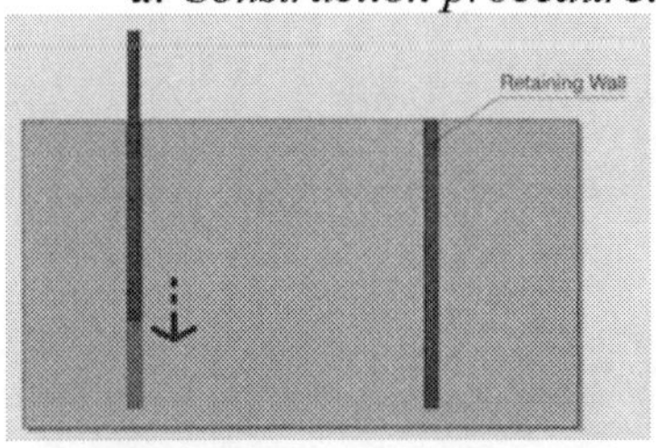

Figure 8. *Stage 1- Installation of retaining walls*

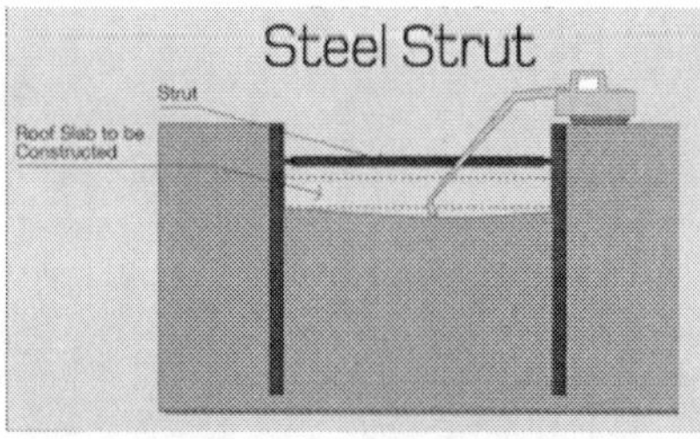

Figure 9. *Stage 2- Excavation and steel bored strut installation*

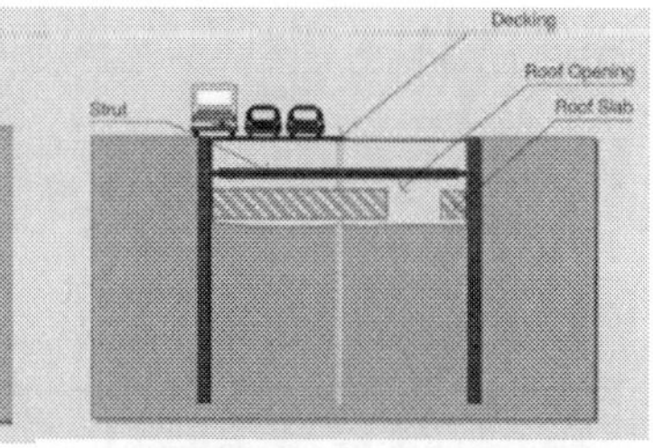

Figure 10.*Stage 3- Substructure construction*

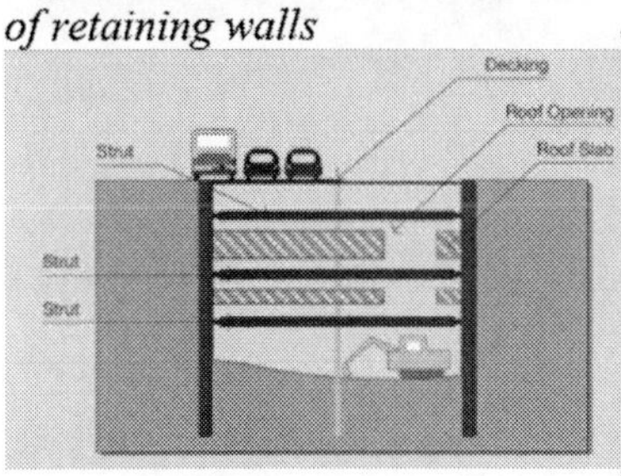

Figure 11. *Stage 4- Substructure construction.*

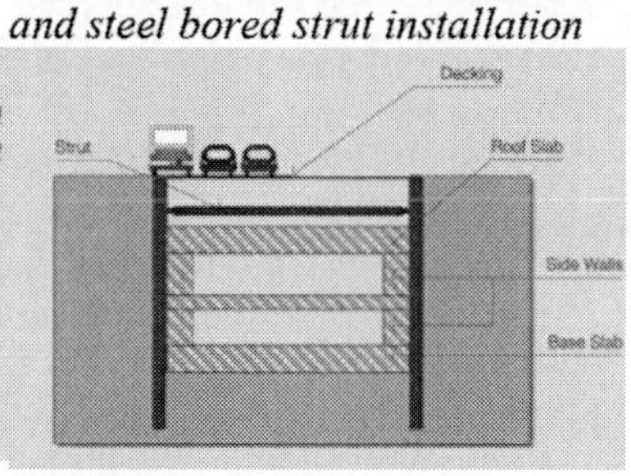

*.**Figure 12.** Stage 5- Substructure construction*

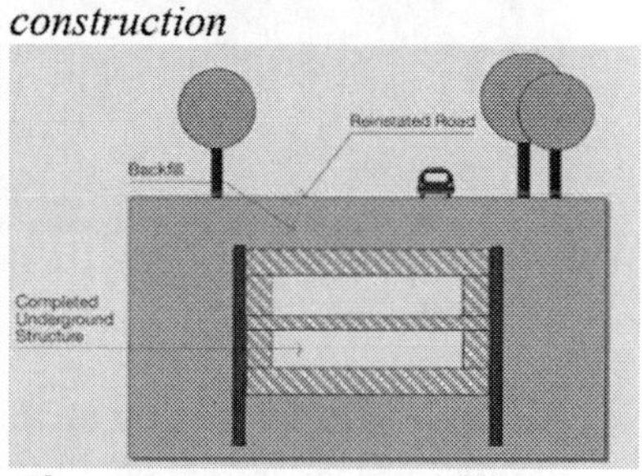

Figure 13. *Stage 6- Backfilling and superstructure construction*

• Stage 1- Installation of retaining walls: The retaining walls, which are usually concrete walls, is installed before excavation commences.

• Stage 2- Excavation and steel bored strut installation: The soil is excavated to just below roof slab level. Bracing struts are installed to retain the soil at the sides (when basement height is high).

• Stage 3- Substructure Construction: Roof slab and roof openings are constructed in order to proceed downwards. Roof slab not only resists horizontal displacement but also reduces the noise in construction progress

• Stage 4- Substructure Construction: This process progresses downwards until base slab is constructed. The operations on roof slab can keep working normally due to decking systems.

• Stage 5- Substructure Construction: The side walls are constructed upwards, after removal of struts between level slabs. Then the access openings are sealed.

• Stage 6- Backfilling and superstructure construction: After constructing basement, the soil is backfilled to the top strut level (for Metro station) before these struts are removed or superstructure construction is proceeded (for high-rise building).

b. Advantages:

• Economize scaffolds, struts and formwork frames by using natural soil surface.

• Use structural slabs for resisting horizontal displacement of retaining walls.

• Superstructure and basement construction can be simultaneously progressed due to pillar support.

c. Disadvantages:

• Support pillar designing process is more complicated: these pillars have to bear not only up loads but also different lateral loads in each stage of construction process, and also subordinate to computational scheme of the whole building.

• Detailing and constructing the connections for columns, beams and walls are more complicated.

• Sanitary facilities and labor safety conditions is hard to obtain (lighting, ventilation…).

• This semi-mechanical excavation mostly take place in enclosed spaces, therefore this method can only be used at soft soil layer sites.

Figure 14. *Kingpost (Decking) in Top-down method.*

Figure 15. *Access openings for excavating in Top-down method.*

3. Proposal of improving method:

In order to take advantage of the strengths and improve the disadvantages of these two methods, the authors propose to apply "Semi Top-down method combined with Earth-bank" in construction of a specific project: Sacombank building, 266 Nam Ky Khoi Nghia, District 3, Ho Chi Minh city.

- Improvement from Bottom-up method: Earth-bank is used to stabilize retaining walls instead of bracing section steel struts.
- Improvement from Top-down method: Using open cut method for half of the earthwork quantities.

3.1 Construction progress based on Semi Top-down method combined with Earth-bank:

Presentation of the construction process:

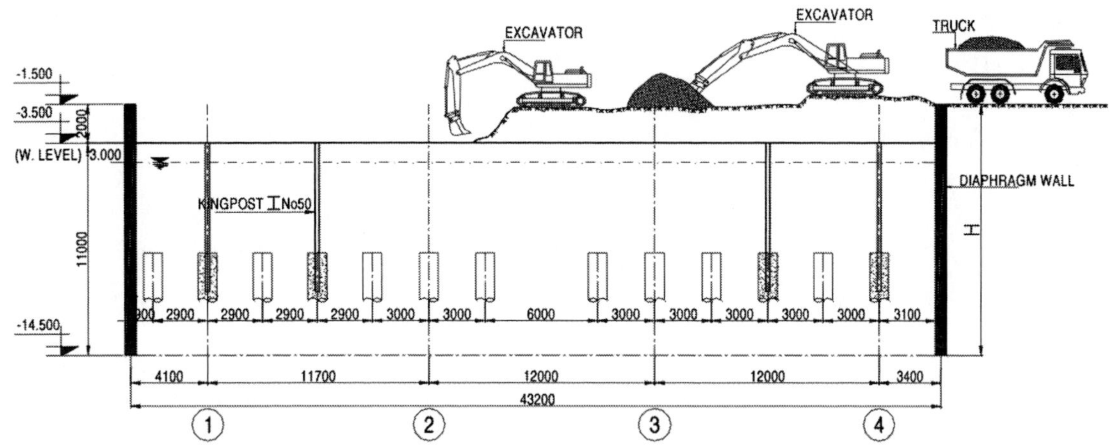

Figure 16. Stage 1: *The soil is excavated to first basement slab's bottom*

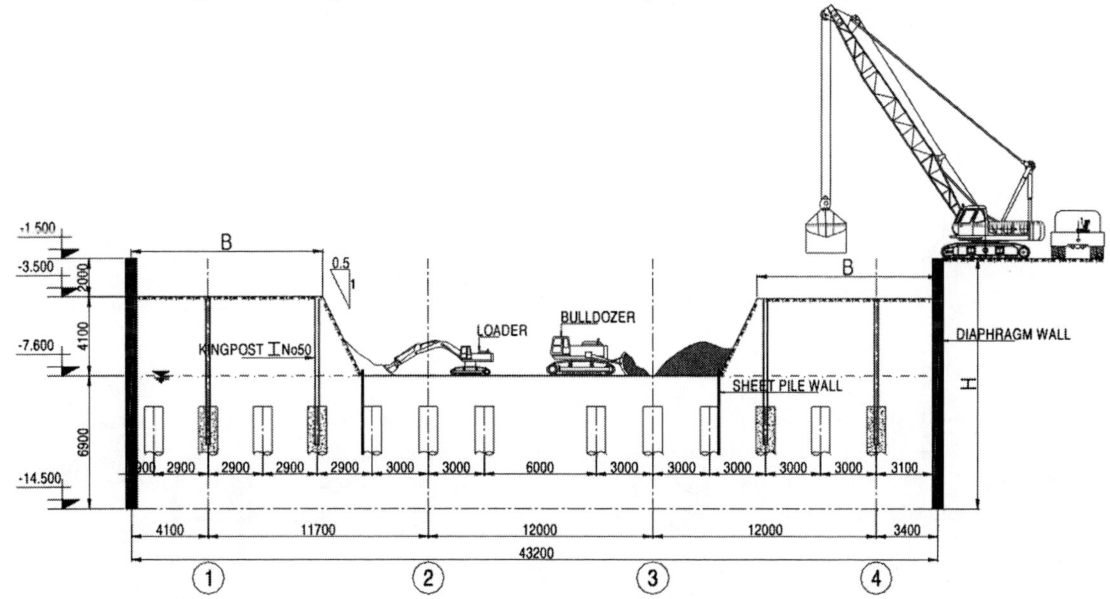

Figure 17. Stage 2: *Partially excavating and simultaneously lowering the water level to -7.600m. Stabilize the earth-bank and foundation pit using earth-bank and steel sheet wall (instead of bracing struts).*

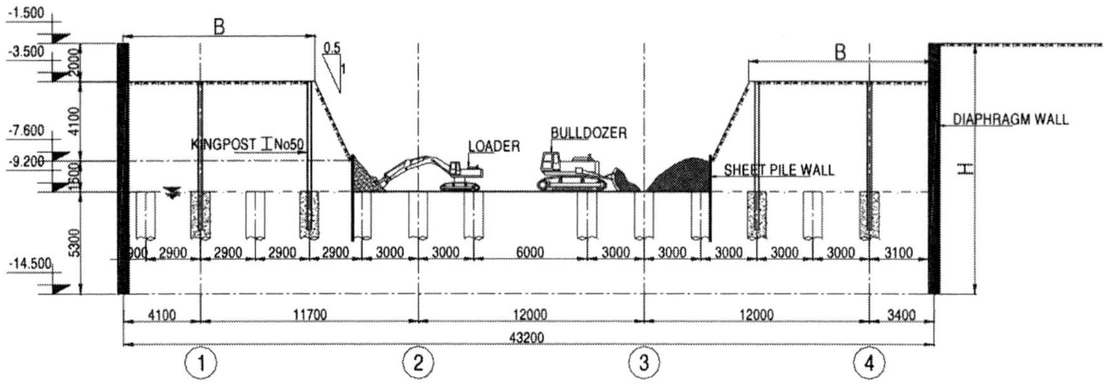

Figure 18. Stage 3: Excavated and groundwater level was lowered to formation level at -9.200m.

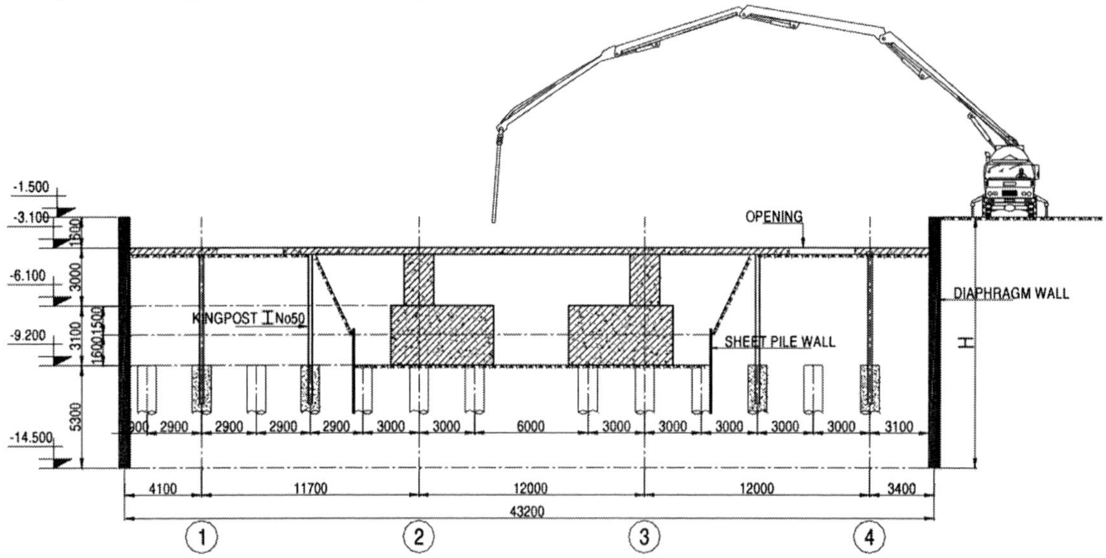

Figure 19. Stage 4: Constructing foundation and 1st basement floor slabs.

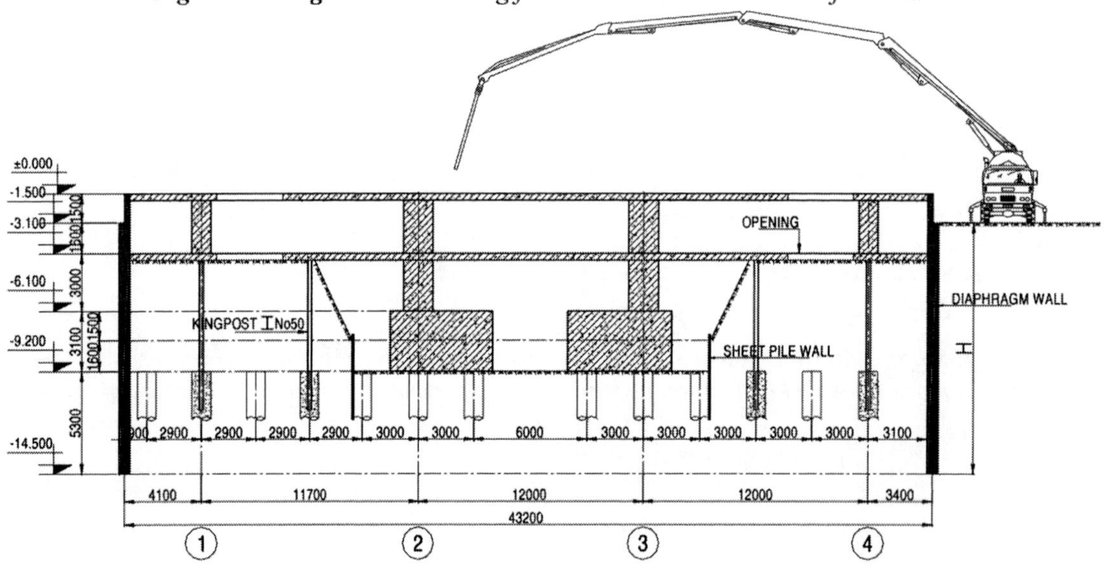

Figure 20. Stage 5: Constructing ground floor slabs.

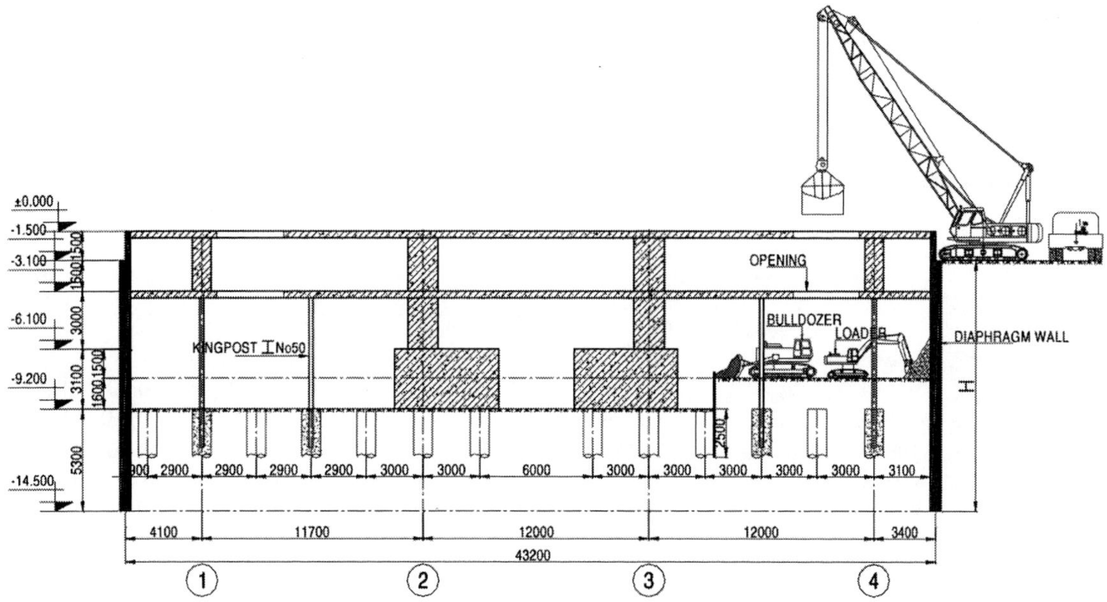

Figure 21. Stage 6: Excavating the rest of soil to formation level.

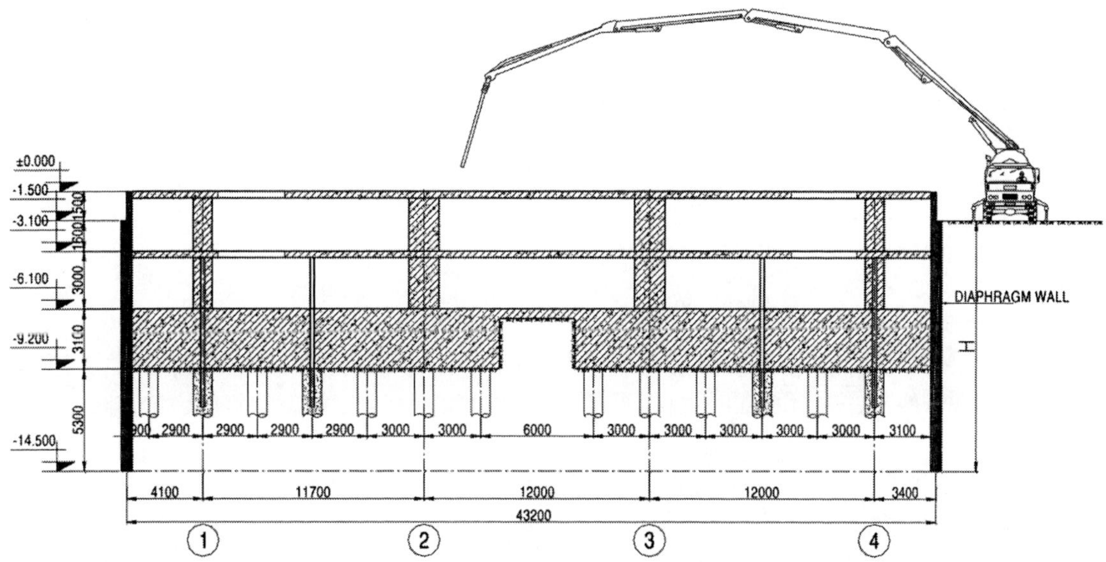

Figure 22. Stage *7: Constructing foundation beams, the rest of foundation and 2nd basement floor slabs*

3.2 Observations:

Through construction process, the authors realized that the stability of adjacent soil depends on the stability of side walls (or how deep that side walls were embedded). Along with that, earth-bank also helps to resist horizontal displacement of side walls and behaves as a gravity wall, so its stability depends on its own width (or B: earth-bank top width 0.5:1). Besides, steel sheet piles encompassed foundation pits will keep the earth-bank from bulging while the foundation is constructed.

These components formed a simultaneously operative system. It's very hard to explain exactly the behavior this structural system using explanation and theory for simplifying and fragmenting it. So in order to have the precise results in each construction period, the writer proposes using Finite Element Method (FEM) in Plaxis 7.2 to simulate the simultaneous operation of this structural system.

IOP Conf. Series: Earth and Environmental Science **143** (2018) 012047 doi:10.1088/1755-1315/143/1/012047

4. Calculation for construction method:

4.1 Material properties of soil model and components:

a. Properties of subgrade soil in Hardening-soil model:

Hardening-soil model is selected because it has the ability to distinguish between compression rigidity and dilatancy of soil. So it is compatible for simulating the process of excavation.

Table 1. *Properties of subgrade soil in Hardening-soil model.*

Parameter	Name	Layer 1 (SC)	Layer 2 (CH)	Layer 3a (SM/SW)	Layer 3b (SM/SW)	Layer 4 (CH)	Unit
Material model	*Model*	HS	HS	HS	HS	HS	-
Type of material behaviour	*Type*	*UnDr.*	*UnDr.*	*Drained*	*Drained*	*UnDr.*	-
Dry soil weight	γ_{dry}	22.80	22.70	20.90	20.50	20.70	kN/m³
Wet soil weight	γ_{wet}	22.85	22.75	20.98	21.34	20.76	kN/m³
E_{50}^{ref} (for p_{ref} = 100 kPa)	E_{50}^{ref}	4950	*7600*	5075	6125	10050	kN/m²
E_{ur}^{ref} (for p_{ref} = 100 kPa)	E_{ur}^{ref}	14850	22800	14625	18375	30150	kN/m²
E_{oed}^{ref} (for p_{ref} = 100 kPa)	E_{oed}^{ref}	*4950*	7600	6650	*6125*	10825	kN/m²
Cohesion	c'	28.3	*50*	1.4	0.5	80.5	kN/m²
Friction angle	φ'	18	*20*	34.53	33.58	18.75	°
Dilatancy angle	ψ	0	0	4.53	3.58	0	°
Lateral stress coefficient	K_0^{nc}	0.69	0.71	0.43	0.45	0.68	-
Poisson's ratio	ν_{ur}	0.2	0.2	0.2	0.2	0.2	-
Interface reduction factor	R_{inter}	0.7	0.7	0.7	0.7	0.7	-

Table 2. *Properties of Diaphragm wall.*

Structural element	Paramater	Name	Value	Unit
Diaphragm wall (plate)	Type of behaviour	*Material type*	Elastic	-
	Normal stiffness	*EA*	$1.86\ 10^7$	kN/m
	Flexural rigidity	*EI*	$5.58\ 10^5$	kNm²/m
	Equivalent thickness	*d*	0.6	m
	Weight	*w*	8	kN/m/m
	Poisson's ratio	ν	0.2	-

Table 3. *Properties of steel sheet pile walls around foundation pits.*

Structural element	Paramater	Name	Value	Unit
Sheet pile wall	Type of behaviour	*Material type*	Elastic	-
	Normal stiffness	*EA*	$1.97\ 10^6$	kN/m
	Flexural rigidity	*EI*	$1.15\ 10^5$	kNm²/m
	Poisson's ratio	ν	0.3	-

Table 4. *Properties of basement slabs.*

Structural element	Paramater	Name	Value	Unit
Basement slabs	Type of behaviour	*Material type*	Elastic	-
	Normal stiffness	*EA*	$1.09\ 10^7$	kN/m
	Flexural rigidity	*EI*	$1.11\ 10^5$	kNm²/m
	Equivalent thickness	*d*	0.35	m
	Weight	*w*	8.75	kN/m/m
	Poisson's ratio	ν	0.2	-

IOP Conf. Series: Earth and Environmental Science **143** (2018) 012047 doi:10.1088/1755-1315/143/1/012047

Table 5. *Properties of concrete columns.*

Structural element	Paramater	Name	Value	Unit
Concrete columns	Type of behaviour	*Material type*	Elastic	-
	Normal stiffness	*EA*	6.98 10^7	kN
	Spacing out of plane	L_s	8	m
	Maximum force	F_{max}	3.49 10^4	kN

Table 6. *Properties of King post.*

Structural element	Paramater	Name	Value	Unit
King post *(Decking)*	Type of behaviour	*Material type*	Elastic	-
	Normal stiffness	*EA*	2.1 10^8	kN
	Spacing out of plane	L_s	8	m
	Maximum force	F_{max}	2.3 10^5	kN

Table 7. *Properties of foundation footings.*

Structural element	Paramater	Name	Value	Unit
Foundation footing	Type of behaviour	*Material type*	Elastic	-
	Type of material behaviour	*Type*	Non-porous	-
	Unit weight	γ	25	kN/m^3
	Young's Modulus	E	3.1 10^7	kN/m^2
	Poisson's ratio	ν	0.2	-

Table 8. *Properties of bored pile.*

Structural element	Paramater	Name	Value	Unit
Bored pile	Type of behaviour	*Material type*	Elastic	-
	Type of material behaviour	*Type*	Drained	-
	Unit weight	γ	25	kN/m^3
	Permeability	$k_x=k_y$	0.05	m/day
	Young's Modulus	E	3.1 10^7	kN/m^2
	Poisson's ratio	ν	0.2	-

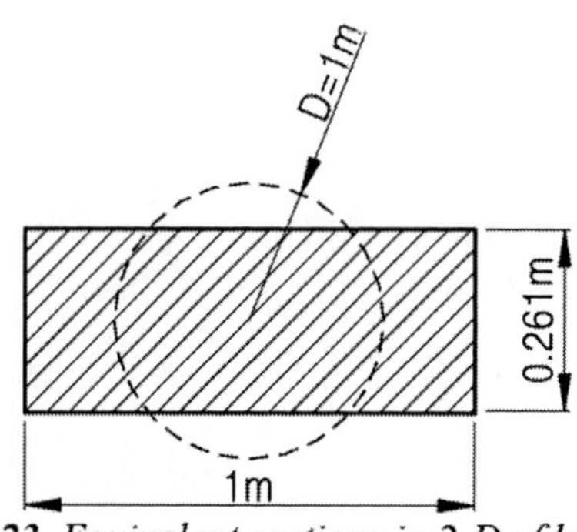

Figure 23. *Equivalent sections in 2-D of bored pile.*

Construction plants load is converted to equivalent load $q=10KN/m^2$.

IOP Conf. Series: Earth and Environmental Science **143** (2018) 012047 doi:10.1088/1755-1315/143/1/012047

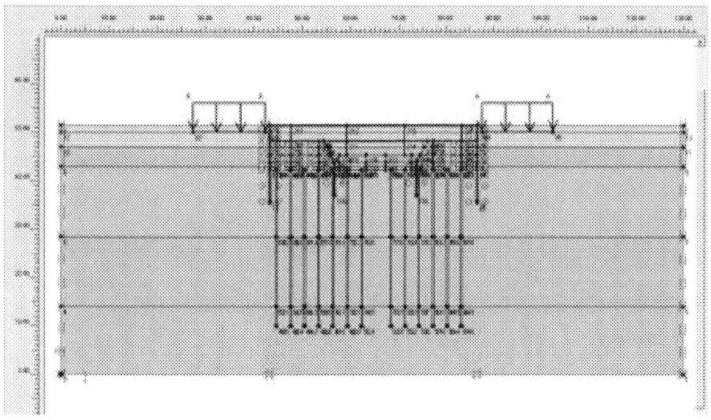

Figure 24. *Calculation model in Plaxis.*

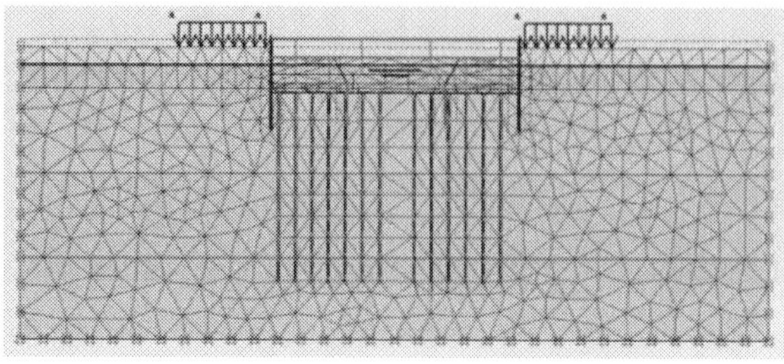

Figure 25. Stage 1: *The soil is excavated to first basement slab's bottom*

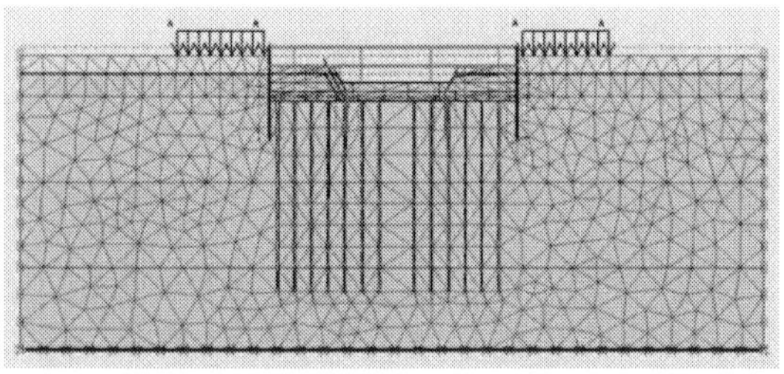

Figure 26. Stage 2: *Partially excavating and simultaneously lowering the water level to -7.600m. Stabilize the earth-bank and foundation pit using earth-bank and steel sheet wall*

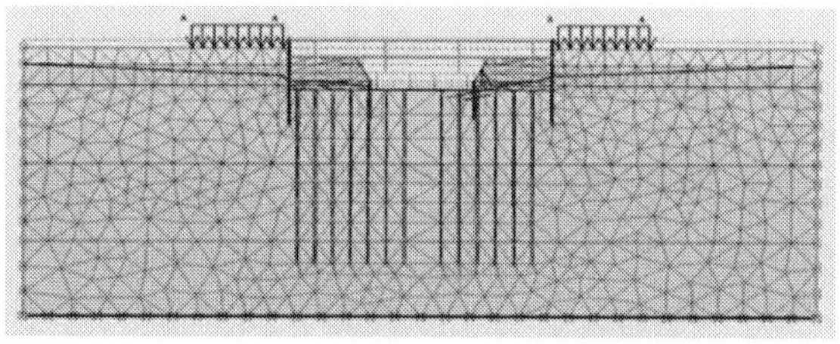

Figure 27. Stage 3: *The soil was excavated and groundwater level was lowered to formation level at -9.200m.*

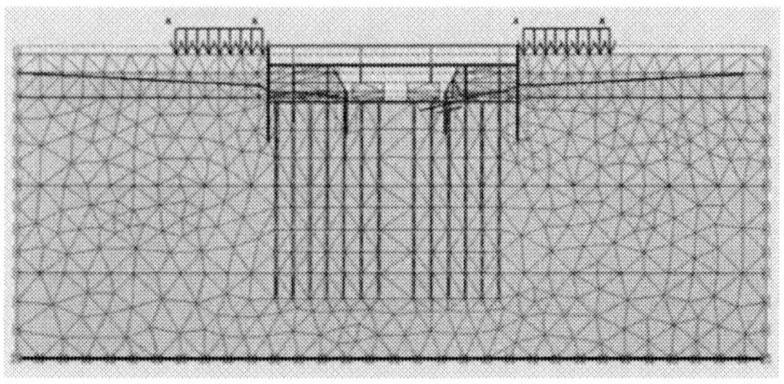

Figure 28. Stage 4: *Constructing foundation and 1st basement floor slabs*

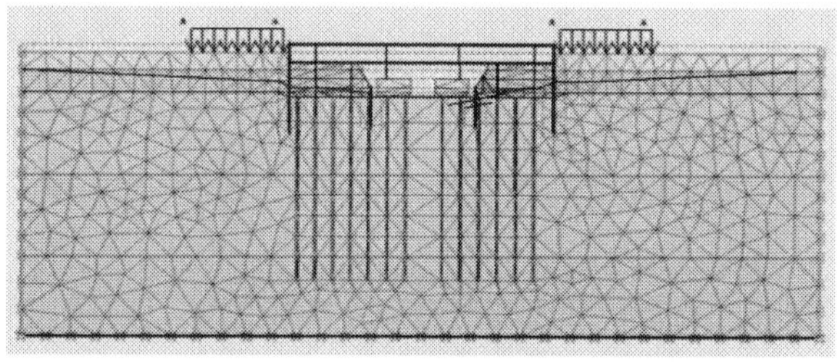

Figure 29. Stage 5: *Constructing ground floor slabs*

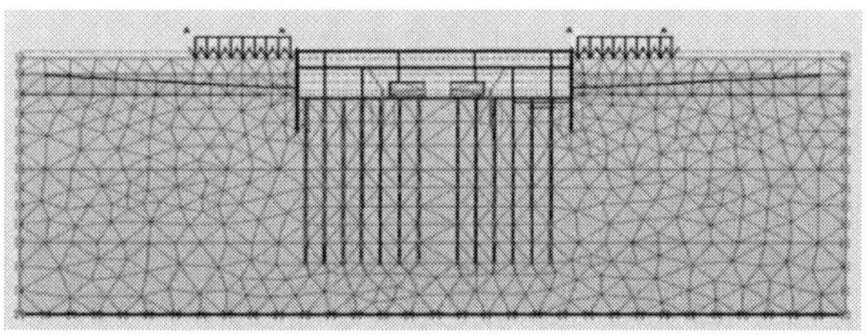

Figure 30. Stage 6: *Excavating the rest of soil to formation level*

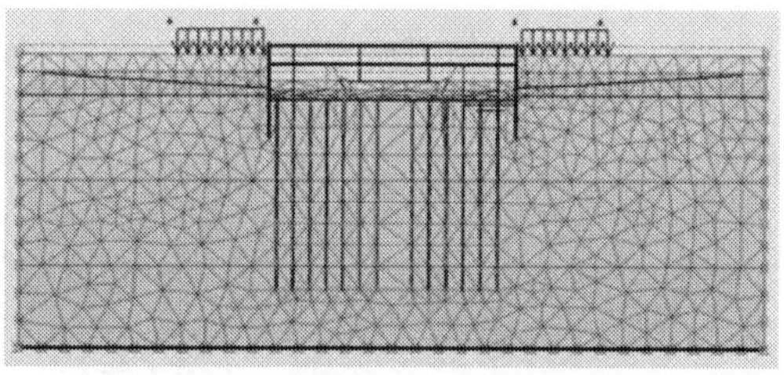

Figure 31. Stage 7: *Constructing foundation beams, the rest of foundation and 2nd basement floor slabs*

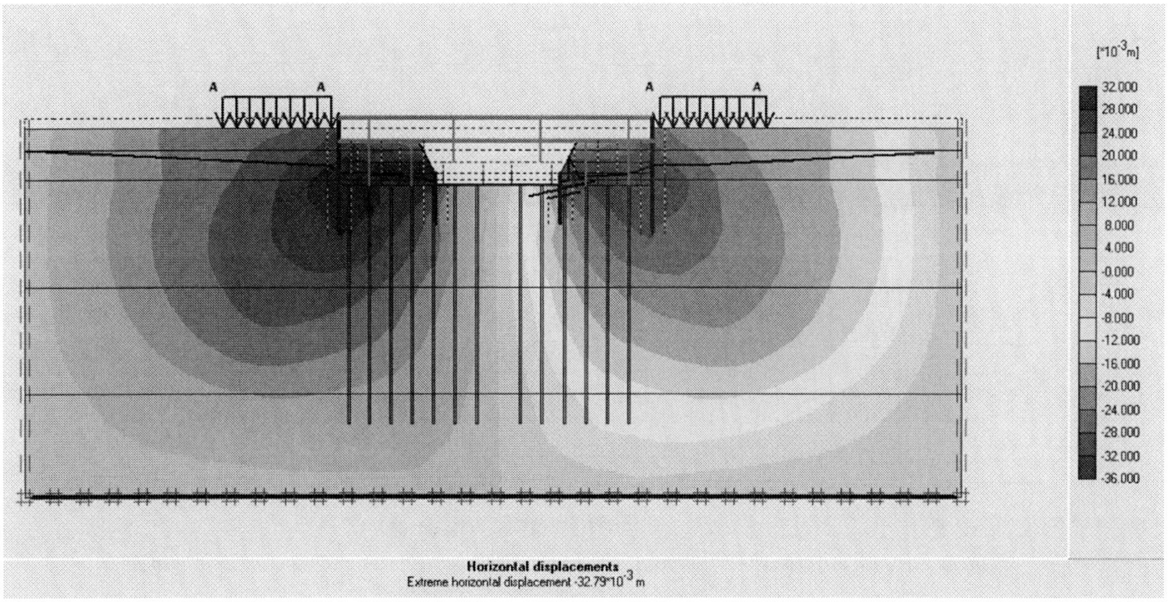

Figure 32. *Horizontal displacement.*

b. Observations:

Simulation of construction process shows that the horizontal displacement of side walls mostly occurred in stage 1,2,3 and 6. In which:

- Horizontal displacements in Stages 1, 2, 3 depend on H: wall depth and B: Earth-bank top width (0.5:1).
- In stage 6, horizontal displacement depends on H: wall depth and EI: flexural rigidity of the wall. Because wall thickness was restrained by construction plants, so horizontal displacement in this stage mostly depends on wall depth.

4.2 Analysing and selecting compatible side wall depth H(m) and optimal earth bank width B(m) for stabilizing excavation as well as restraining wall's movement.

a. Analysing and selecting optimal value for B and H in stages 1, 2, and 3:

- Horizontal displacement S_h(mm) of the left basement wall:

Table 9. *Horizontal displacement S_h(mm) of the left basement wall in relation to B and H.*

Wall depth H(m)	*Horizontal displacement S_h(mm) in relation to B*(m)						
	B=0	B=7	B=8	B=9	B=10	B=11	B=12
H=12	121.05	34.43	33.82	32.58	31.54	30.66	29.38
H=13	115.20	34.37	33.76	32.46	31.36	30.63	29.32
H=14	109.37	34.69	34.02	32.43	31.12	30.42	29.14
H=15	108.30	35.44	34.63	32.97	31.60	30.80	29.60
H=16	105.90	36.39	35.53	33.77	32.31	31.47	30.25
H=17	104.83	37.17	36.30	34.52	33.05	32.17	30.92
H=20	104.83	39.65	38.77	36.95	35.30	34.42	33.10

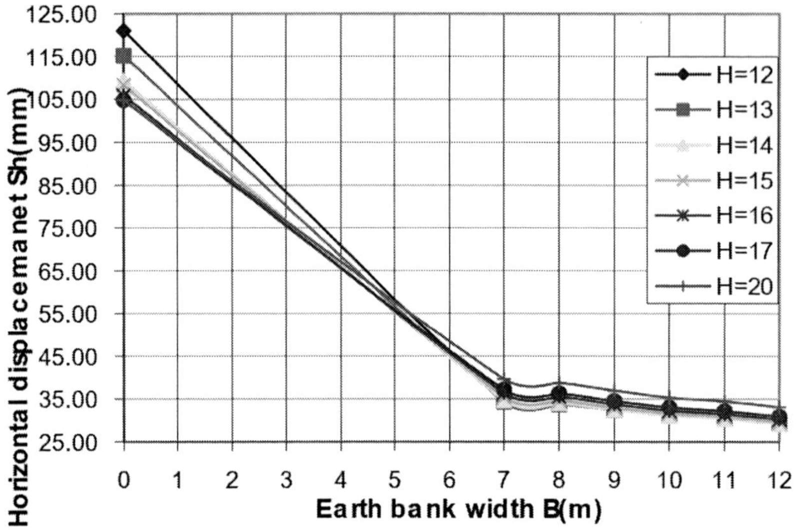

Figure 33. *Horizontal displacement S_h(mm) of the left basement wall in relation to B.*

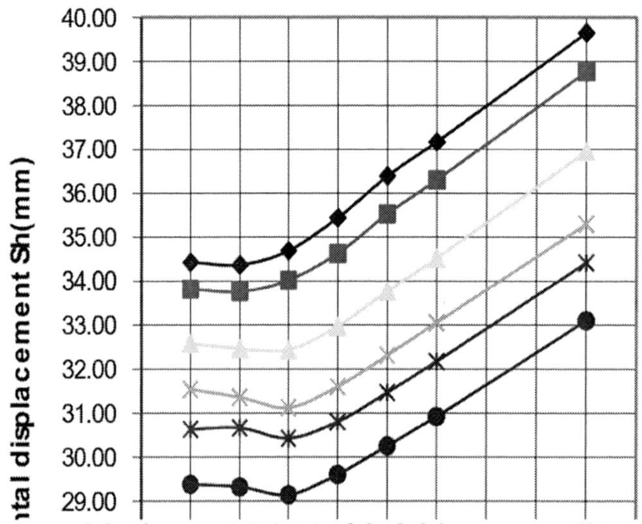

Figure 34. *Horizontal displacement S_h(mm) of the left basement wall in relation to H.*

- Horizontal displacement S_h(mm) of the right basement wall:

Table 10. *Horizontal displacement S_h(mm) of the right basement wall in relation to B and H.*

Wall depth H(m)	*Horizontal displacement S_h(mm) in relation to B*(m)						
	B=0	B=6.5	B=7.5	B=8.5	B=9.5	B=10.5	B=11.5
H=12	127.83	35.40	34.73	33.45	32.30	31.46	30.08
H=13	119.15	35.13	34.55	33.13	31.97	31.27	29.87
H=14	110.25	35.60	35.05	33.16	31.79	31.25	29.77
H=15	107.26	36.42	35.77	33.76	32.23	31.70	30.19
H=16	105.94	37.33	36.64	34.52	32.96	32.39	30.85
H=17	109.69	38.14	37.44	35.22	33.65	33.03	31.48
H=20	105.87	40.72	39.94	37.92	36.13	35.53	33.87

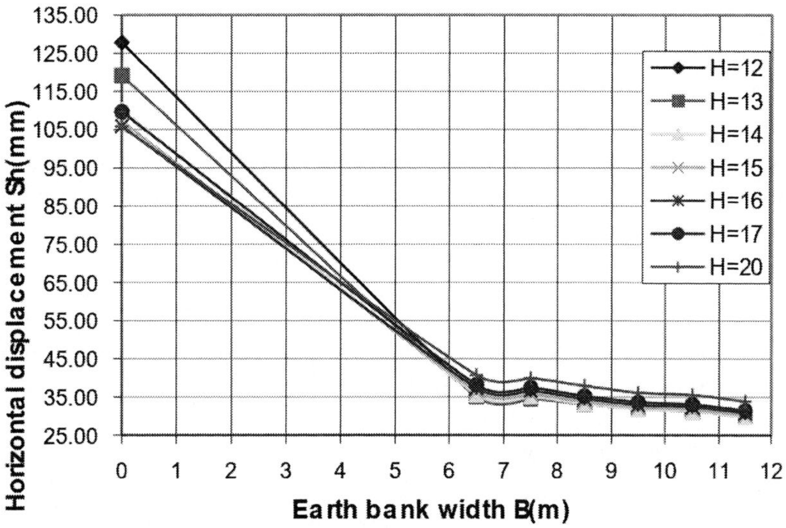

Figure 35. *Horizontal displacement S_h(mm) of the right basement wall in relation to B.*

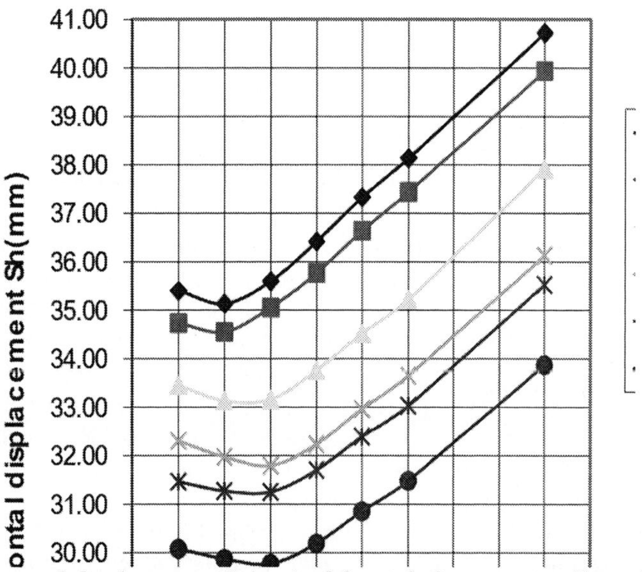

Figure 36. *Horizontal displacement S_h(mm) of the right basement wall in relation to H.*

Observations:

According to Charts 1 to 4, we can see the important role of earth-bank and basement walls in stabilizing as well as restraining horizontal displacement S_h of side walls.

Horizontal displacement of side walls increases very slowly when reducing B: the top width of earth-bank. But if the value of B decreases below 6 meters then horizontal displacement raises up very quickly. So earth-bank works as a bearer which also has displacement, and its displacement is totally subordinated to its own rigidity (or B: earth-bank top width).

- When remove the earth-bank (B=0), basement walls work as cantilever retaining walls and horizontal displacement depends on wall depth (H). Horizontal displacement decreases when increasing H and stops decreasing when H reaches a compatible value, then its value just depends only on flexural rigidity (EI) of the wall.

- In the opposite way, when earth-bank is included, basement walls work as buttress walls. Horizontal displacement will only decrease in a class interval of wall depth value (H). Outside that class interval, horizontal displacement (S_h) will increases when increasing or reducing H.

b. Analysing and selecting optimal value for H in stage 6:
According to constructive condition at site, and best horizontal displacement resistance conditions can be obtained, the writer proposes selecting excavation method in which top earth-bank width is 11 meters (0.5:1), and wall depth is 14 meters for stages 1, 2, 3.

After checking the options for stage 6, the writer realized that at stage 6, basement walls was supported by ground floor slab and 1st basement floor slab to resist horizontal displacement. According to table 11, horizontal displacement (S_h) just decreases in a class interval of wall depth value, outside of that interval, horizontal displacement will raise when increasing or reducing wall depth value (H). So it is logical to maintain wall depth value (H) by 14 meters for this construction stage.

Table 11. *Horizontal displacement S_h(mm) of retaining walls at stage 6 with B=11m and various values of H.*

Horizontal displacement S_h(mm)	*Various values of H(m)*						
	H=11	H=13	H=14	H=15	H=16	H=17	H=20
	34.79	33.90	33.72	34.59	35.53	36.26	38.85

c. Assessment of safety factor (FS) of stability during construction phases:

Table 12. *Safety factor (FS) stability during construction phases*

	Stage 1	Stage 2	Stage 3	Stage 4	Stage 5	Stage 6	Stage 7
FS	9.911	2.054	4.247	3.662	3.839	10.991	7.437

According to table 12, the safety factor of stability in the construction stages of excavation pit combined with earth-bank (stage 2) is smaller than the other construction stages. Hence, the overall stability of the pit depends mainly on the stability of the earth-bank. When applying this method, attention should be paid to measures to improve the stability of the earth-bank

5. Conclusion:

In using "Semi Top-down combined with Earth-bank" construction method to resist horizontal displacement of retaining walls, the stability of the earth-bank is the most important element that determine whether this method can be effective or not. Earth-bank works as bearer equivalent to a buttress in buttress wall. If this bearer is rigid enough (compatible earth-bank width), horizontal displacement value will change trivially when increasing or reducing earth bank width. But if earth-bank width is decreased too much (not rigid enough), then horizontal displacement will rise very quickly. Then, basement wall will work as cantilever wall, and horizontal displacement totally depends on flexural rigidity and the wall depth.

When there is requirement for horizontal displacement resistance, we shouldn't increase wall depth (H) of basement wall, just reinforce the lateral bracing at positions which has high horizontal displacement value before proceed further for the next stages.

This is an economically and technically effective construction method.

References:

[1] Dung Tien Construction Company Ltd 2005 *Analysing and selecting construction method for Sacombank Buiding's basement report.*
[2] Bui Q T 2007 *Graduate thesis "Analyze the behavior of diaphragm wall during excavation and construction of the basement floor"* HoChiMinh city University of Technology. p71-108
[3] Ke Ng B 2002 Design And Construction Of Deep Foundation Pit *Construction Publisher.*

[4] Lan Transport & Authority 2005 *Bottom-Up & Top-Down Construction Method.*
[5] http://dungtien.com/en/project/done-project
[6] Brinkgreve R B J, Vermeer P A 1998 *PLAXIS 2D Foundation Version 7 Manual* Balkema A A Rotterdam Brookfield Netherlands

IOP Conf. Series: Earth and Environmental Science **143** (2018) 012048 doi:10.1088/1755-1315/143/1/012048

Performance of Single Friction Pendulum bearing for isolated buildings subjected to seismic actions in Vietnam

N V Nguyen[1*], C H Nguyen[1], H P Hoang[2] and K T Huong[3]

[1]Department *of Civil Engineering, Industrial University of Ho Chi Minh City*, Vietnam
[2]The University of Danang, University of Science and Technology, Vietnam
[3]Ton Duc Thang University, Vietnam
[*]Corresponding author. Email: *nguyenvannam@iuh.edu.vn*

Abstract. Using structural control technology in earthquake resistant design of buildings in Vietnam is very limited. In this paper, a performance evaluation of using Single Friction Pendulum (SFP) bearing for seismically isolated buildings with earthquake conditions in Vietnam is presented. A two-dimensional (2-D) model of the 5-storey building subjected to earthquakes is analyzed in time domain. Accordingly, the model is analyzed for 2 cases: with and without SFP bearing. The ground acceleration data is selected and scaled to suit the design acceleration in Hanoi followed by the Standard TCVN 9386:2012. It is shown that the seismically isolated buildings gets the performance objectives while achieving an 91% reduction in the base shear, a significant decrease in the inter-story drift and absolute acceleration of each story.

1. Introduction

Earthquake is one of the major human dangers and the main cause of damage, collapse of constructions. Historically, there are many strong earthquakes occurring in many countries such as: Kanto (Japan, 1923), Sichuan (China, 2008)..., killing many people and causing heavy losses to their economies. Although Vietnam is not in the strong earthquake zone, strong earthquakes still occurred in history. Recently, earthquakes occur in Vietnam more frequently and stronger [1, 2]. The design of earthquake resistant buildings in Vietnam is limited with very simple theoretical models. The use of base isolation techonology in Vietnam is also rare. In this study, the use of a more modern isolation technology with SFP bearing for seismically isolated buildings in Vietnam is discussed.

Base isolation is a passive structural control technology, and is an effective approach to mitigate the damage caused by earthquakes to the structures. Recently, it has being studied and applied very commonly in developed countries such as USA, Japan.... The main idea of this technique is to isolate the superstructure from the ground by using flexible bearings, called seismic isolation bearings. It has a small horizontal stiffness and it is typically inserted between the foundation and superstructure to isolate the structure from the ground movement, detache the earthquake energy into the structure. With fixed-base structure, when earthquake happens, horizontal relative displacement between two stories is large. Hence, the risk of structural damage is high. Conversely, with base isolated structures, the superstructure is completely a rigid body and the inter-story drift of structure is significantly reduced. Base isolated structures will maintain the behavior in the elastic or near elastic region, and ensure that the structure is not damaged, as illustrated in Figure 1 [3].

Single Friction Pendulum (SFP) bearing is a popular device used in seismic base-isolation technology in recent years. It is first introduced in 1987 by Zayas [4]. It is made from stainless steel and it has a geometrical structure as shown in Figure 2. The bearing consists of an articulated slider sliding on a curved surface with a radius R. On the curved surface and slider are covered by a layer of Teflon material, which reduces the coefficient of friction between them (from 1% to 15%). The radius R in combination with the total weight of the superstructure W provides a horizontal stiffness of the bearing and restoring force to move the slider to the central position. At the same time, the friction coefficient of the bearing provides the initial horizontal stiffness to stabilize the bearing and dissipate a part of the earthquake energy when sliding occurs. The horizontal displacement capacity of the SFP bearing is d, as shown in Figure 2(b). The radius of the curved surface and coefficient of friction are important technical parameters of the bearing which decide the behavior of the bearing and the structure.

(a) (b)

Figure 1. Building subjected to earthquake excitation. (a) Non-isolated building, (b) Isolated building.

With this structure, the SFP bearing can be considered an improved type with many more advantages than previous types such as: rubber bearing, flat sliding bearing. The advantages include: durability and stability over time, resilience to the center at the end of earthquakes and easy adjustment to the technical parameters in the design.

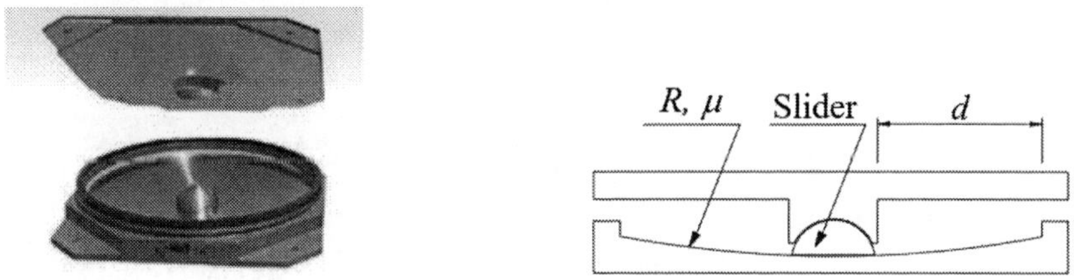

(a) (b)

Figure 2. Single friction pendulum (SFP) bearing. (a) Cutaway view, (b) Cross section.

2. Modeling of base-isolated buildings with SFP bearing

2.1. Force and horizontal displacement relationship of SFP bearing

The movement of bearing are described in detail in study of Zayas et al. (1987) [4]. When the ground motion is weak and the horizontal force in the bearing F is smaller than the friction force between the slider and the curved surface F_f, the sliding doesn't occurs. The base isolated structure, therefore, behaves like fixed base structure. Then the SFP bearing doesn't work, as shown Figure 2(b). When the ground motion is strong enough, the force F exceeds the friction force F_f, the sliding occurs, as shown Figure 3(a). During the motion, if the horizontal displacement u exceeds the displacement capacity of the bearing d, the circumferential hard boundary prevents the bearing displacement. The impact force F_r is provided to prevent sliding. The force components and horizontal displacements are shown in Figure 3(a). When the horizontal displacement of the slider is u, the relationship of force and horizontal displacement of the bearing is established on the basis of the force balance.

Considering equilibrium of the horizontal and vertical force at horizontal displacement u of the slider, the following relationships are obtained as follows, respectively:

$$F = F_n \sin\theta + F_f \cos\theta \tag{1a}$$

$$W = F_n \cos\theta - F_f \sin\theta \tag{1b}$$

where F_n is the reaction force from curved surface to slider, the friction force component is F_f and θ is the rotation angle of slider. They are determined by the following Equation (2) and Equation (3)

$$F_f = \mu F_n \tag{2}$$

$$\sin\theta = \frac{u}{R} \tag{3}$$

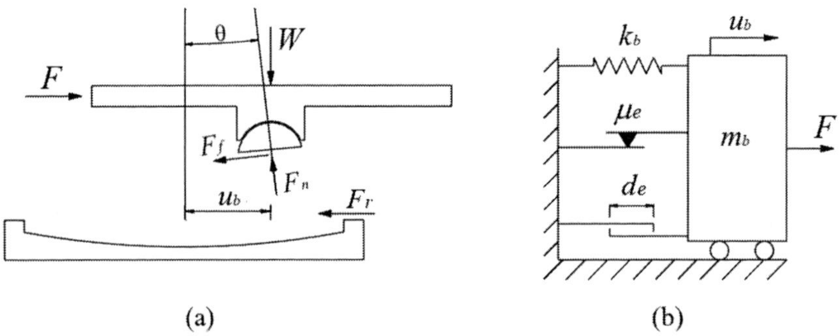

Figure 3. The model of SFP bearing. (a) The forces and displacement, (b) Numerical model

It is assumed that the displacement u is sufficiently small compared to the radius R so that the rotation angle θ is very small, approximately: $\mu\sin\theta = 0$ và $\cos\theta = 1$ and combining Equations (1a) and (1b) can be written as:

$$F = \frac{W}{R}u + \mu W \tag{4}$$

In Equation (4), the (Wu/R) component is the restoring force. Accordingly, the horizontal stiffness k_b of the SFP bearing is computed as follows:

$$k_b = \frac{W}{R} \tag{5}$$

And the natural period of the SFP bearing is determined by:

$$T_b = 2\pi\sqrt{\frac{m}{k_b}} = 2\pi\sqrt{\frac{W}{gk_b}} = 2\pi\sqrt{\frac{R}{g}} \tag{6}$$

Apply impact force F_r to Equation (4), the equation governing the force - horizontal displacement relationship of bearing as shown following Equation (7)

$$F = \frac{W}{R}u + F_f + F_r \tag{7}$$

The hysteresis loop is shown in following Figure 4.

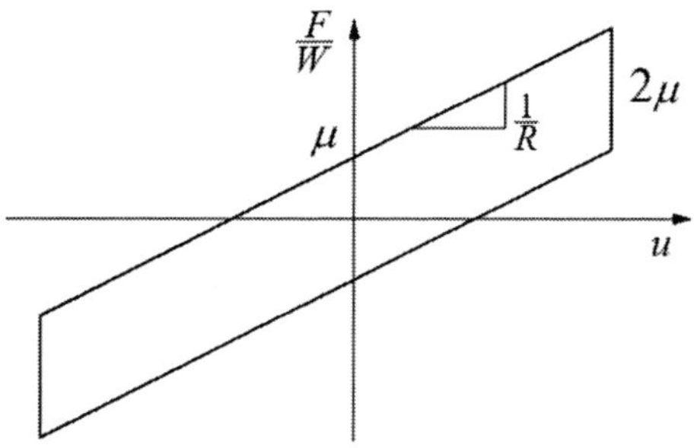

Figure 4. Hysteresis loop of SFP bearing.

With the inner force components of the bearing, the numerical model of bearing is an element with a combination of a spring element, friction element and gap element in parallel, as shown Figure 3(b). The linear spring stiffness k_b governed by Equation (5), friction element with friction coefficient μ_e and gap element governed by displacement limit $d_e = d$.

2.2. Modeling of seismic isolated structure with SFP bearing

The model of an n-story building is isolated by the SFP bearing. The slabs are assumed to be an absolutely rigid body. The structure is assumed to be symmetrical. Therefore, the rotational displacement effect is ignored. The 2-D model is applied to this structure. Each story have two degrees of freedom with the physical characteristics of: mass m_i, stiffness k_i, damping coefficient c_i and motion characteristics in two lateral directions of: displacement u_i, velocity $\dot{u}_i$ and acceleration $\ddot{u}_i$. The numerical model of base isolated structure has $(2n + 2)$ degrees of freedom and is shown in Figure 5. The other assumptions considered for this model are as follows: The superstructure is linearly elastic, the behavior of the SFP bearing is nonlinear.

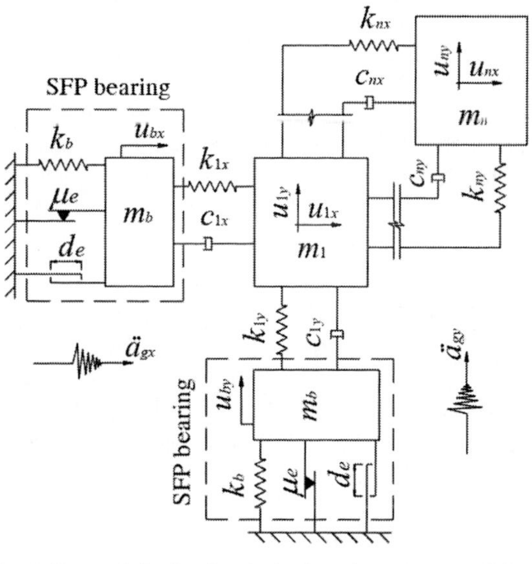

Figure 5. The 2-D model of seismic isolated structure with SFP bearing.

Mass m_b is the total mass of story above the bearing. It includes: base mass and equivalent to the mass of a floor.

Write the dynamic equations for each mass according to the D'Alembert principle, the equations of motion for the model in x-direction including $(n + 1)$ equations can be expressed by:

$$\begin{cases} m_b(\ddot{u}_{bx}+\ddot{u}_{gx})+k_{bx}u_{bx}+F_{fx}+F_{rx}+k_{1x}(u_{bx}-u_{1x})+c_{1x}(\dot{u}_{bx}-\dot{u}_{1x})=0 \\ m_1(\ddot{u}_{1x}+\ddot{u}_{gx})+k_{1x}(u_{1x}-u_{bx})+c_{1x}(\dot{u}_{1x}-\dot{u}_{bx})+k_{2x}(u_{1x}-u_{2x})+c_{2x}(\dot{u}_{1x}-\dot{u}_{2x})=0 \\ \dots \\ m_n(\ddot{u}_{nx}+\ddot{u}_{gx})+k_{nx}(u_{nx}-u_{n-1x})+c_{nx}(\dot{u}_{nx}-\dot{u}_{n-1x})=0 \end{cases} \tag{8}$$

The equations of motion for the model in *y*-direction is written similar to the *x*-direction.

The friction forces in the *x*-direction and *y*-direction are F_{fx} and F_{fy}, respectively. They can be written as follows [5].

$$\begin{cases} F_{fx}=\mu_e W Z_x \\ F_{fy}=\mu_e W Z_y \end{cases} \tag{9}$$

where the coefficient of sliding friction μ_e depends on sliding, which is modeled by the following equation [6]:

$$\mu_e=\mu_{\max}-(\mu_{\max}-\mu_{\min})e^{-\alpha|\dot{u}|} \tag{10}$$

where $\mu_{\max}$ and $\mu_{\min}$ are the sliding friction coefficients at maximum velocity and nearly zero sliding velocity, respectively, α (s/m) is a rate parameter for various conditions of Teflon-steel interface and pressure that controls the transition from $\mu_{\max}$ and $\mu_{\min}$ and $\dot{u}$ is the sliding velocity computed as [5]:

$$\dot{u}=\sqrt{\dot{u}_{bx}^2+\dot{u}_{by}^2} \tag{11}$$

Z_x and Z_y, which are bounded by the values ±1, are dimensionless hysteretic variables along the *x* and *y*-directions can be expressed as follows [5, 7]:

$$\begin{pmatrix} \dot{Z}_x Y \\ \dot{Z}_y Y \end{pmatrix}=\begin{pmatrix} A\dot{u}_{bx} \\ A\dot{u}_{by} \end{pmatrix}-\begin{bmatrix} Z_x^2(\gamma sign(\dot{u}_{bx}Z_x)+\beta) & Z_xZ_y(\gamma sign(\dot{u}_{by}Z_y)+\beta) \\ Z_xZ_y(\gamma sign(\dot{u}_{bx}Z_x)+\beta) & Z_y^2(\gamma sign(\dot{u}_{by}Z_y)+\beta) \end{bmatrix}\begin{pmatrix} \dot{u}_{bx} \\ \dot{u}_{by} \end{pmatrix} \tag{12}$$

where *Y* is the yield displacement, and *A*, β, γ and η are dimensionless constants and are determined from the experiment that control the shape of the hysteretic loop. These quantities can be specified as suggested by Constantinou et al [6], [8]: $Y = 0.25$ mm is the yield displacement, $A = 1$, $\beta = 0.1$, $\eta = 2$ and $\gamma = 0.9$.

And the impact forces F_{rx} and F_{ry} can be written as follows [9]:

$$\begin{cases} F_{rx}=k_r(|u_b|-d_e)sign(u_x)H(|u_b|-d_e) \\ F_{ry}=k_r(|u_b|-d_e)sign(u_y)H(|u_b|-d_e) \end{cases} \tag{13}$$

where *H* is the value of the Heaviside step function, k_r is the impact stiffness after contacting the maximum horizontal displacement which should be assigned a large value and u_b is the radial displacement of bearing ($u_b=\sqrt{u_{bx}^2+u_{by}^2}$).

The motion differential equations are solved by the fourth-order Runge-Kutta numerical method, using the ***ode15s*** function in Matlab [10] to find the results of the time-history analysis.

3. Numerical example

To illustrate the behavior of a structure seismically isolated by SFP bearing subjected to earthquake and evaluate seismic effectiveness of SFP bearing for construction in Vietnam, an analysis of a mid-rise steel building subjected to ground motion acceleration records with ground conditions in Hanoi is implemented.

3.1. Parameters of the structure and SFP bearing

The structure of the analysis is a 5-storey steel building with the structural parameters as shown in Table 1. The damping ratio $\xi = 2{,}5\%$ is taken from the steel material. This structure is taken from the full-scale experimental model of Ryan et al. [11]. It is compatible with seismic base isolation technology and can represent many similar constructions. This is the symmetrical structure. Therefore, the stiffness is the same in the x- and y-directions ($k_x = k_y$).

Table 1. Parameters of the structure.

Story	1	2	3	4	5
Mass (kN.s^2/mm)	0.0824	0.0814	0.0811	0.0801	0.1199
Stiffness (kN/mm)	131	105	93.3	76.2	61.1

The parameters of the bearing are chosen on the basis of assumptions with reference from previous studies. It is in accordance with structural and ground conditions at site. These parameters are chosen as follows: $R = 3000$ mm; $\mu = 0.02 - 0.04$; $d = 500$ mm.

3.2. Selecting and scaling ground motions

With the ground condition, according to the forecasts, earthquake in Hanoi is Level 8 followed by the Earthquake Magnitude Scale. According to Vietnam Standards TCVN 9386:2012, the design ground acceleration is taken $a_g = 0.24$g for ground types D (with the parameters: $v_{s,30} < 180$ m/s and NSPT < 15 blows/30cm) [12], and the values of the periods T_b, T_c and T_d and of the soil factor S describing the shape of the elastic response spectrum are determined: $S = 1.35$; $T_b = 0.2$; $T_c = 0.8$ và $T_d = 2$ [12]. Figure 6 shows the elastic response spectra for ground types D.

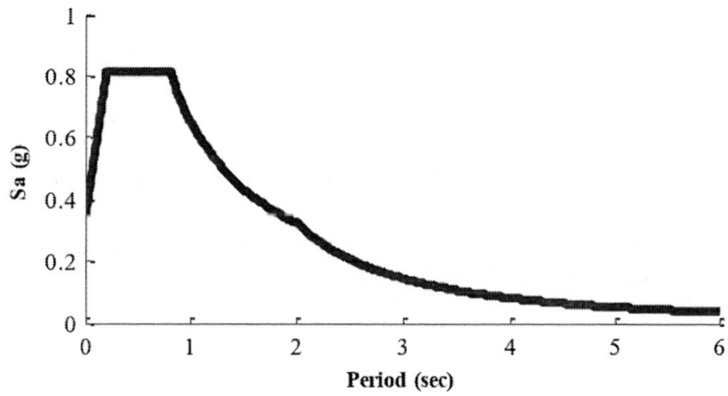

Figure 6. Elastic response spectra for ground types D.

The selection and scale of the magnitude of ground motion acceleration records in the time-history analysis is guided in ASCE 7-2010 (Section 17.3.2) [13] and is detailed by Naeim and Kelly [14]. Accordingly, the number of ground motion records is seven and the average value analysis will be selected as the design criteria. These values are combined by the square root of sum of squares (SRSS) method from x- and y-directions analysis results. In this study, the seven motion records selected from the The Pacific Earthquake Engineering Research Center (PEER) at the University of California, Berkeley [15], it is summarized in Table 2. These motions are recorded on ground types D.

The seven records will be multiplied by the *SF* factor to scale the average response spectrum matches the target spectrum according to Standards as shown in Figure 6. The *SF* factor is calculated as [16]:

$$SF = \frac{\int_{T_1}^{T_2} S_{aG} 1.3 S_{aC} dT}{\int_{T_1}^{T_2} S_{aG}^2 dT} \quad (14)$$

where S_{aG} and S_{aC} are response spectrum of motion and the target spectrum, respectively. T_1 to T_2 is effective periods range, is defined: with fixed-base structure $T_1 = 0.2T_f$, $T_2=1.5T_f$ where T_f is the natural period of the structure in the fundamental mode, with base isolated structure $T_1=0.5T_b$; $T_2=1.25T_b$ where T_b is the natural period of the base isolated structure.

Table 2. Ground motions for the analyses.

No.	Record Seq. #	Event (**ID**)	Year	Station	M_w	R_{rup}(km)	$V_{s,30}$ (m/s)	*PGA* (g)	
								X	*Y*
1	326	Coalinga-01 (**PAC**)	1983	Parkfield - Cholame 2WA	6.36	44.72	173.02	0.110	0.110
2	334	Coalinga-01 (**PAF**)	1983	Parkfield - Fault Zone 1	6.36	41.99	178.27	0.143	0.110
3	718	Superstition Hills-01 (**IVW**)	1987	Imperial Valley Wildlife Liquefaction	6.22	17.59	179	0.133	0.131
4	729	Superstition Hills-02 (**SUH**)	1987	Imperial Valley Wildlife Liquefaction	6.54	23.85	179	0.179	0.206
5	759	Loma Prieta (**FOC**)	1989	Foster City - APEEL 1	6.93	49.94	116.35	0.257	0.284
6	962	Northridge-01 (**WAT**)	1994	Carson - Water St	6.69	49.81	160.58	0.092	0.088
7	178	Imperial Valley-06 (**ELC**)	1979	El Centro Array #3	6.5	12.85	162.94	0.180	0.215

With the structure above, we calculate $T_f = 0.68$ s, T_b is assumed to be about 2.7 s. This value will be checked after having the analysis results. If it is not satisfied, must be assumed again. The results of response spectrum of seven records for 2 cases: fixed-base and base isolated structure are shown in Figure 7. The target spectrum value S_{aC} is multiplied by 1.3 for considering the ground motion in both the x and y directions [12].

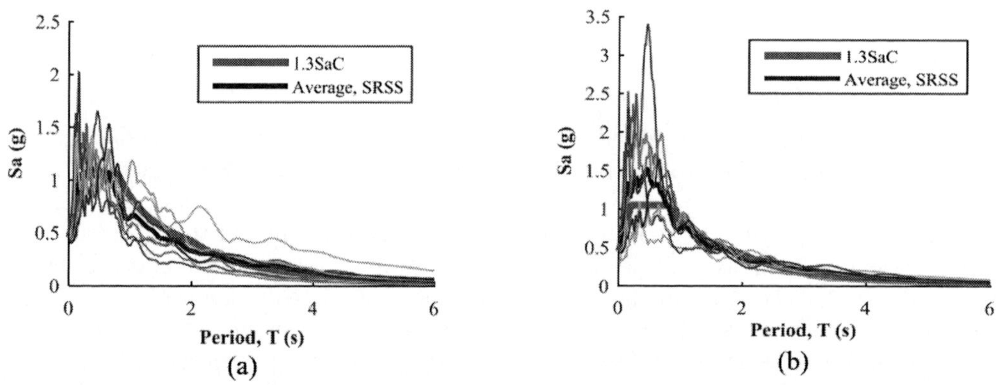

Figure 7. Target spectra compare to the median. (a) Fixed-base structure, (b) Base isolated structure.

3.3. Results and discussion

The Equations (8) present for the case of the base isolated structure. With the case of the fixed-base structure, the differential equations of motion are established similarly, it only consists of $2n$ equations (in this case, no mass m_b). The differential equations are resolved numerically using the Runge-Kutta method. However, for computational efficiency, the equations will be converted into the first order differential equations and solved by ODE functions (***ode15s*** solver) that are introduced in the research of Shampine et al. [10]. Accordingly, the results for the time history analysis of structure for two cases subjected to the seven ground accelerations are determined accurately. The average results are used to evaluate the behavior of the structure and the bearing.

Analytical values include: absolute acceleration of each floor, story shear force, total base shear and displacement of the bearing for two cases, and the effectiveness of the isolation system. These results are listed in Table 3.

Table 3. Isolation performance evaluation of SFP bearing.

Story		1		2		3		4		5	
		Val.	Red. (%)	Val.	Red. (%)	Val.	Red. (%)	Val.	Red. (%)	Val.	Red. (%)
Abs. acc. (g)	Fixed	0.67	64	1.14	82	1.64	88	2.14	91	2.58	90
	Isolated	0.24		0.20		0.19		0.18		0.25	
Inter. shear (.10^3 kN)	Fixed	7.27	94	6.84	94	5.98	93	4.70	92	3.03	90
	Isolated	0.42		0.41		0.40		0.36		0.29	
Base shear (.10^3 kN)	Fixed	7.27									
	Isolated	0.63 (Reduce: 91%)									
Base displ. (cm)	Fixed	0									
	Isolated	19.7									

The maximum absolute acceleration of each floor and the story shear forces of structure are shown in Figure 8. The data for this figure is the average value from 7 analysis.

The value of shear force in the 1st story in the time history analysis for two cases with ELC motion is shown in Figures 9 and 10. Similarly, the 5st story absolute acceleration is shown in Figure 11 and Figure 12.

The hysteresis loops represents the force - horizontal displacement relationship of the bearing is shown in Figure 13.

Figure 9 to 13 show the results of the analysis for ELC motion. With others, the results are similar, not shown here.

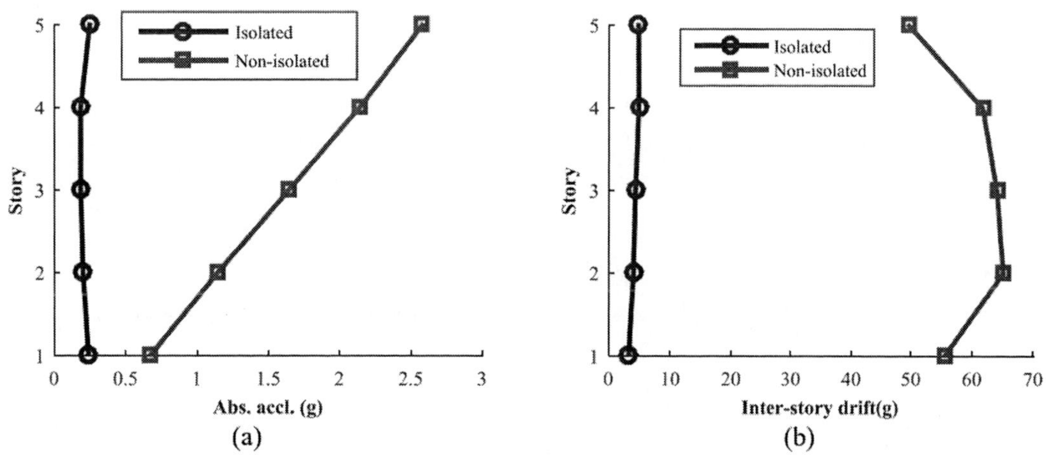

Figure 8. Effectiveness of SFP bearing. (a) Abs. acceleration of each story, (b) Inter-story drift

IOP Conf. Series: Earth and Environmental Science **143** (2018) 012048 doi:10.1088/1755-1315/143/1/012048

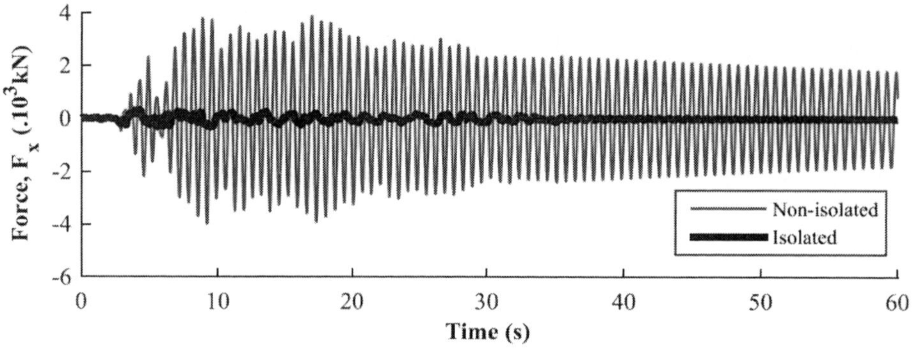

Figure 9. Shear force along x-direction F_x of 1st story with ELC motion

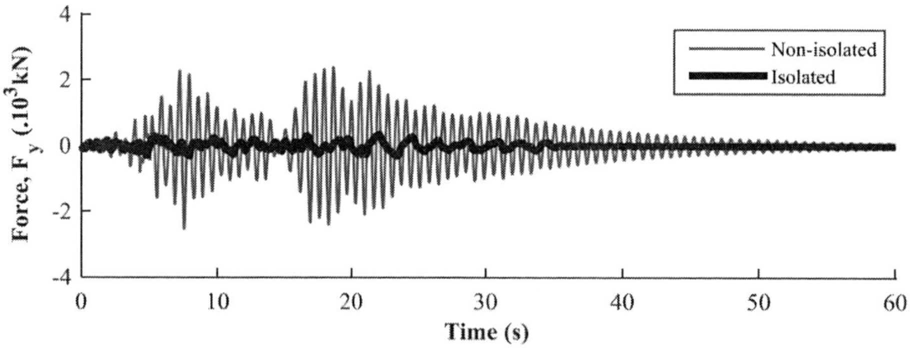

Figure 10. Shear force along y-direction F_y of 1st story with ELC motion

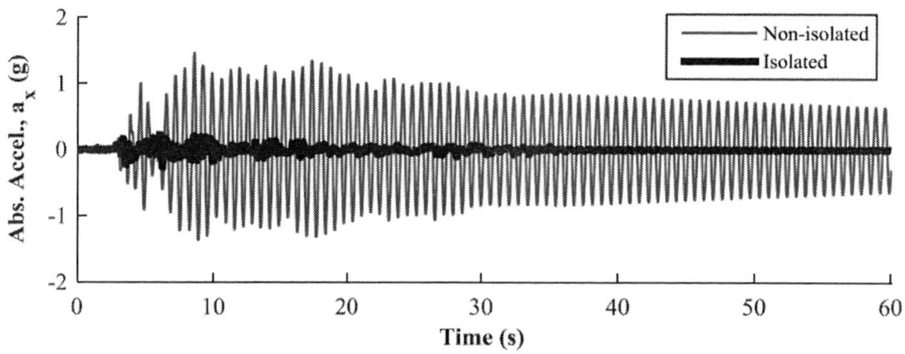

Figure 11. Absolute acceleration along x-direction a_x of 5st story with ELC motion

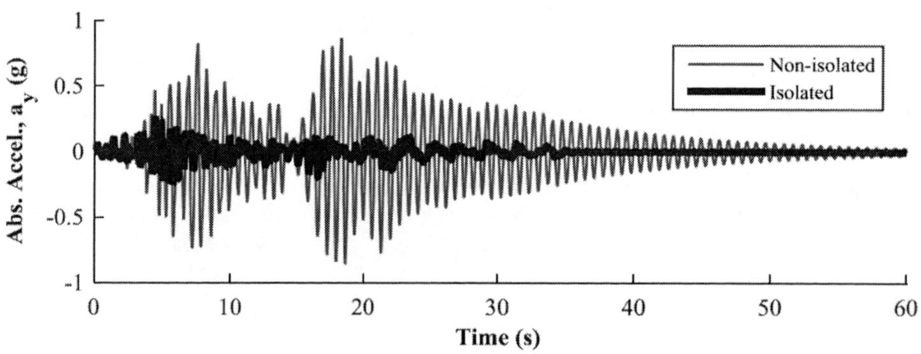

Figure 12. Absolute acceleration along y-direction a_y of 5st story with ELC motion

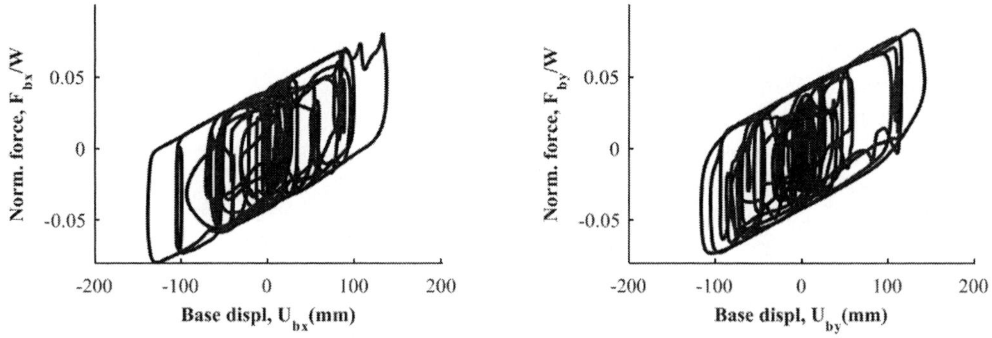

Figure 13. Hysteresis loops of bearing with ELC motion

According to the results shown in Table 3 and Figure 8, the seismic effectiveness of the SFP bearing is very high, approximately 85% for absolute acceleration in each floor and 90% for story shear forces. From the results, one can calculate the effective stiffness K_{eff} and effective period T_{eff} of the base isolated structure for each acceleration record according to the following formulas, respectively [17]:

$$K_{eff} = \frac{F_{\max}}{u_b} \tag{15}$$

$$T_{eff} = 2\pi\sqrt{\frac{W}{gK_{eff}}} \tag{16}$$

where F_{max} is the largest horizontal force in the bearing and u_b is the displacement at F_{max}.

The average value T_{eff} = 2.74 s will be compared with T_b = 2.7 s which is assumed initially. This result is slightly different, so the initial assumption value is reasonable, it doesn't need to repeat.

The displacement of bearing u_b = 19.7 cm is the basis for choosing the size of the bearing in the design. This displacement in base isolated buildings is usually large. This is a disadvantage of this technique. Therefore, the designer should note this problem.

With the time-history results in Figures 9 to 12, the seismic effectiveness is expressed throughout the time of the earthquake. At times with small ground acceleration, the movement of the bearing still occurs and seismic effectiveness is maintained. The shape of hysteresis loops in the analytical model shown in Figure 13 is similar with the results of research of Zayas [4], as shown in Figure 4.

4. Conclusions

The SFP bearing is very effective for earthquake-resistant design of buildings with the base isolation technology. This study propose a simple model to determine quickly the response of base isolated structures subjected to earthquakes using SFP bearing. Through this model, a time-history analysis is implemented for a 5-storey steel building subjected to ground motion acceleration records scaled to suit the ground conditions in Hanoi. The results point out that the effectiveness of the SFP bearing is very high, showing that the performance is over 85%. The dynamic responses of the structure show that absolute acceleration of each story, inter-story drift and base shear significantly reduce. Therefore, the use of this bearing in the seismic resistant design in Hanoi and in Vietnam general has promising perspective.

5. References

[1] Huong T 2014 *27 trận động đất ở Việt Nam trong nửa năm qua* http://vnexpress.net/tin-tuc/thoi-su/27-tran-dong-dat-oviet-nam-trong-nua-nam-qua-3035027.html
[2] Quang D 2011 *Hà Nội nằm trong vùng động đất cấp 7- 8* http://thanhnien.vn/thoi-su/ha-noi-nam-

trong-vung-dongdat-cap-78-432322.html.
[3] Wang P 2002 *Fundamentals of seismic base isolation* International training program for seismic design of building structures, sponsored by National Science Council, National Chiao-Tung Unversity, Hsinchu, Taiwan.
[4] Zayas, V, Low S and Mahin S 1987 *The fps earthquake resisting system: Experimental report* (*ucb/eerc* - 87/01) University of California, Berkeley.
[5] Nagarajaiah S, Reinhorn A M and Constantinou M C Nonlinear dynamic analysis of 3D-base-isolated structures *Journal of Structural Engineering* **117**(7) 2035 - 2054.
[6] Constantinou M, Mokha A and Reinhorn A 1990 Teflon bearings in base isolation II: Modeling, *ASCE Journal of Structural Engineering* **116**(2) 455 - 474.
[7] Park Y J, Wen Y K and Ang A 1986 Random vibration of hysteretic systems underbidirectional ground motions Earthquake *Engineering & Structural Dynamics* **14**(4) 543–557.
[8] Mokha A, Constantinou M C and Reinhorn A 1990 Teflon bearings in base isolation I: Testing, *ASCE Journal of Structural Engineering* **116**(2) 438 - 454.
[9] Fenz D and Constantinou M C 2008 Modeling triple friction pendulum bearings for response-history analysis *Earthquake Spectra* **24**(4) 1011-1028.
[10] Shampine F and Reichelt W 1997 The Matlab ode suite *SIAM Journal on Scientific Computing* **18**(1) 1 - 22.
[11] Ryan K and et al. 2013 *Full scale 5-story building with triple pendulum bearings at E-Defense* Network for Earthquake Engineering Simulation.
[12] Vietnam standards 2013 *TCVN 9386:2012 Design of structures for earthquake resistance* Publisher of Construction, Hanoi.
[13] American Society of Civil Engineers 2010 *Minimum Design Load for Buildings and Other Structures* ASCE 7-10.
[14] Naeim F and Kelly J 1999 *Design of seismic isolated structures: from theory to practice* John Wiley & Sons.
[15] Pacific Earthquake Engineering Research Center (PEER) 2017 *Ground motion database* http://ngawest2.berkeley.edu.
[16] Dao D 2012 *Seismic Response of a Full-scale 5-story Steel Frame Building Isolated by Triple Pendulum Bearings under Three-Dimensional Excitations* PhD Thesis University of Nevada.
[17] Fadi F and Constantinou M 2010 Evaluation of simplied methods of analysis forstructures with triple friction pendulum isolators *Earth. Engineering and Structural Dynamics* **39**(1) 5 - 22.

Acknowledgment
This research is funded by Industrial University of Ho Chi Minh City for research project under Grant No. 171.1051.

Disturbance of Traffic Noise: Evaluation on the Effects and Management on Road Corridors

Nur Hazliyana Abdul Mutalib[1], Nordiana Mashros[3], Eeydzah Aminudin[3], Rozana Zakaria[2], Zaiton Haron[2], Muhammad Hilmi Abd Talib[4], and Abdul Rahim Abdul Hamid[3]

[1,4] *Master's Student, Faculty of Civil Engineering, Universiti Teknologi Malaysia*
[2] *Associate Professor, Faculty of Civil Engineering, Universiti Teknologi Malaysia*
[3] *Lecturer, Faculty of Civil Engineering, Universiti Teknologi Malaysia*

Corresponding email: eeydzah@utm.my

Abstract. Several adverse of stimuli can cause annoyance due to the characterized by such effects of distraction on health and delay in activities. Noise annoyance mainly due to the increasing in traffic volume has recognized as important environmental stressor which associated to anxiety and depression. In this manner, it can impose serious damage to human wellbeing, human comfort ability and reduces labour productivity. Hence, this paper aims in evaluating the effects on traffic noise and managing the precaution on road corridors in order to reduce the traffic noise. This case study had been conducted at residential area which is located at Taman Mutiara Rini, residential area located in the southern region of West Peninsular Malaysia. The traffic noise index (TNI) and noise pollution level (NPL) were recorded for a whole day in order to evaluate noise performance with different time durations. From the study, it was shown that the noise level at the Mutiara Rini is above than 75 dBA at most of time which is exceed the permissible limit from the guidelines recommended by the World Health Organization (WHO) and Department of Environment (DOE). According to the guidelines, the maximum limitation for noise pollution during daytime at residential area is about 55 dBA. From the interviewed conducted, it shows that reduction on the traffic noise can be improved by proposing and providing the noise barrier which includes the restoration of trees and concrete wall which can reduce the effects on the traffic noise.

1. Introduction

Noise is one of the pollution that has always been an important environmental problem for human. It is characterized as 'undesirable sound', and it is seen as a natural stressor due to the annoyance from human daily activities [1]. The major form of noise is including transportation, industrial and neighbourhood sector. Transportation noise can come from a variety of sources including motorcycles, vehicles, aircraft and rail transport. Road traffic has become an important factor in societal development and economic progress due to increasing number of vehicles [2].Traffic noise has a tendency to be a dominant noise source in urban and rural environment, which has turned into a developing public concern. Based on the Report of World Health Organization, sound may cause hearing disability, sleep disturbance, performance loss, cardiovascular impacts, and interferences with social behaviour which are aggressiveness, protest and helpfulness. Traffic noise has also relates to the obstruction in speech communication and annoyance. Another perspective view on the economic consequences of these health impairments are property value. Loss in areas subjected to noise impact can bring down work performance of those influenced by noise [3] and medical expenses of improving the condition of health of those influenced by noise [4]. Its adverse effects on health and economy have forced communities to seek solutions to improve quality of life by reducing traffic noise. According to previous research [5] in figure 1, shows that 73% has pointed that traffic noise are the main source that contributes in urban area.

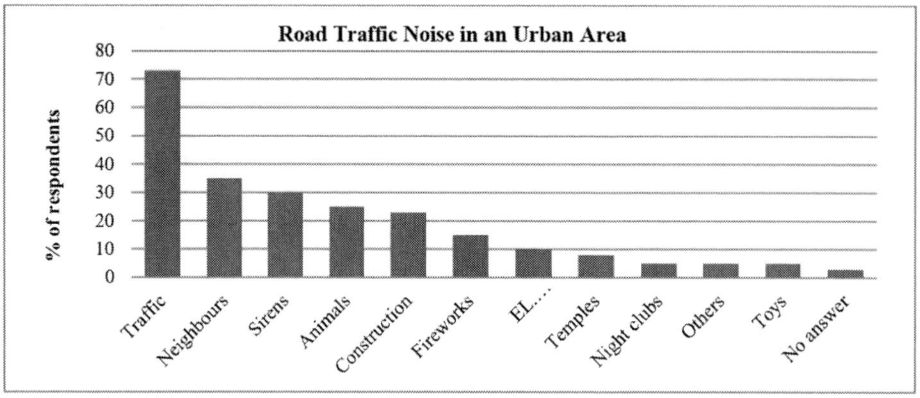

Figure 1. The histogram of road traffic noise in an urban setting [5].

2. Characteristic, Effects and Factors Affecting Traffic Noise on Human Behavior

2.1 Traffic Noise

Traffic noise typically results from the intersection of the sources of moving vehicle and the roadway. Traffic is the dominating source of noise [10] and is the major source of irritation and inconvenience. A considerable portion of traffic noise gets from the sound discharged by the ignition motors of these vehicles from the source to the receiver; noise varies both in level and frequency. The increasing number of living people and vehicles lead to the appearance of a noise pollution. Many noise reviews treating the issue of noise pollution in numerous urban communities through the world have been conducted. The noise impact was treated with as a stress inductor, and in consequence the part of sound as a risk factor for human health [11].

2.2 Traffic Noise Index

This building research survey is an analysis of the social nuisance caused by urban motor way sand their noise. The Traffic Noise Index is used to indicate traffic noises and their effects on architectural designs and planning, while suggesting the need for more and better window insulation and acoustical barriers [12]. Overall concern is for finding unacceptable noise levels, discovering how traffic noise is propagated, and determining effective traffic noise control for buildings. Graphs and charts are used to distinguish the relationship between noise levels and distances.

2.3 Noise Pollution Level

The environmental sound levels measured at a given location depend on a number of specific variables. In particular, many authors have found that the observed sound levels are mainly related to road traffic characteristics, and especially traffic volume, vehicle horns, rolling stock and tires, unruffled vehicles, etc. There is variation in the noise levels with the period of the day and the nature of the location. In general, there are high noise pollution levels (LNP) in the daytime compared with the night time except in the residential areas where the majority of the residents are not always at home during the working days of the week [13].

2.4 Effect on the traffic Noise on Human behaviour.

The Road traffic has turned an important factor in social improvement, especially in terms of living comfort and economic progress. With the expanding number in vehicles in Malaysia, traffic noise represents the most general source of community noise (Zannin et al., 2003). Besides, exposure to the noise can give some effect to human such as health problems, disturbance, and annoyance that probably can also affect work performance and quality of life. For instance, sleep disturbance is generally associated with low noise levels, and levels higher than 70 dB (A) can induce hearing impairment and ischemic heart disease (Nijland and Wee, 2005). According to (Marathe, 2012), effect of traffic noise can be classified under following categories. This includes subjective, behavioural, and physiological effect.

3. Case study: Taman Mutiara Rini

The main research focus is to identify TNI that is near to sensitive area in order to assess public annoyance. Once the highest level is being identify, a propose mitigation is being conducted in order to reduce the traffic noise. The data had been collected near the roadside residential areas at Taman Mutiara Rini, Johor. In the beginning, eight sites have been chosen in order to identify the most critical sites to be investigated. The critical three sites most had been selected which are Sekolah Agama Mutiara Rini (Site A), Presint Utama (Site B) and Taman Mutiara Emas (Site C). The data had been collected using sound level metre to measure noise level These site were chosen because of the data noise level is above than 55dBA which is exceeded the permissible limit as specified from the DOE and WHO. It can conclude that the average noise levels from three sites are above than 75 dBA. Besides that, these sites have four junctions which are possible the noise level will increase. Figure 2 site layout illustrates the location of the three surveys from A to C.

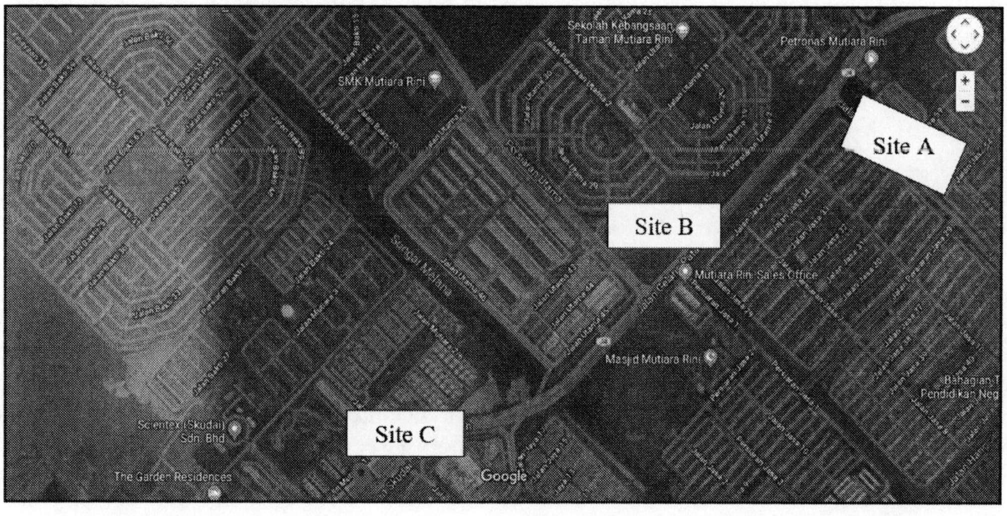

Figure 2. Site Layout illustrates the location of the three surveys from A to C

4. Methodology

4.1 Noise Data Collection

Sound level measurement will be performing by using the Sound Level Meter (SLM) starting from 7.00 am until 7.00 pm. The sound level meter is mounting on tripod closet to the noise source. SLM indicate as a sound pressure level (SPL) meter, decibel (dB) meter, noise meter or noise dosimeter. SLM use a microphone to capture sound. The sound is then assessed within the device and sound measurement values are shown. The most common unit of sound measurement is decibel (Db. It also can be permanently installed for constant monitoring of sound levels at a work or job site. The noise parameter are to measure in LAeq.

Figure 3. Installation of sound level metre (SLM)

4.2 Assessment of Traffic Noise

The equivalent noise (LAeq) values were converted in term of noise pollution level (LNP) and traffic noise index (TNI) in order to identify the noise pollution experienced by the residents in that areas. The LAeq and LNP were compared with Department of Environment (DOE) and World Health Organization (WHO) limits which is within 55 Dba. The TNI values were also compared with the limit of 74 dBA for residential area. The measurements were calculated using the equation as follows [22].

$$T_{NI} = 4(L_{10}\text{-}L_{90}) + L_{90} - 30 \text{ .... } (1)$$
$$L_{NP} = L_{Aeq} + (L_{10} - L_{90}) \text{ .... } (2)$$

4.3 Semi Structured Interview

Interview will be conducted with the panel expert in this case study in order to purpose the mitigation to reduce the traffic noise annoyance. This interview targeted to to Jabatan Kerja Raya (JKR), Majlis Perbandaran Johor Bharu Tengah (MPJBT) and Senior Lecturer from Faculty of Civil Engineering Universiti Teknologi Malaysia, (UTM). The focus interviewee is especially for exploring's people knowledge and experience on the research issue. Table 1 shown the interviewee profile background

Table 1. Interviewee Profile

Questions	Respondent 1	Respondent 2	Respondent 3	Respondent 4
Name of the Company	Majlis Perbandaran Johor Bahru Tengah (MPJBT)	Majlis Perbandaran Johor Bahru Tengah (MPJBT)	Jabatan Kerja Raya (Daerah) Johor Bahru, JKR(D)JB	Faculty Civil Engineer Universiti Teknologi Malaysia, UTM
Type of Organization	Pihak Berkuasa Tempatan (PBT)	Engineering	Senggara Jalan	Education
Position in Company	Assistant Engineer	Technician	Facility Civil Engineer	Senior Lecturer
Year's Working Experience	10 years	10 years	9 years	20 years
Gender	Female	Male	Female	Male

5. Result and Discussion

5.1.Noise Level Variation

Figure 4 is clearly indicated that the most of peak noise happened during morning peak hour afternoon peak hour and evening peak hour as stated by [25]. It can say that the average mean level of traffic noise at site A is 74.50 dBA. At site B, it indicated the mean level of traffic noise is 70.0 dBA. The average mean level of traffic noise at site C is 73.3 dBA. This particular scenario indicated that the increasing noise level can be respectively associated with some driver behaviour.

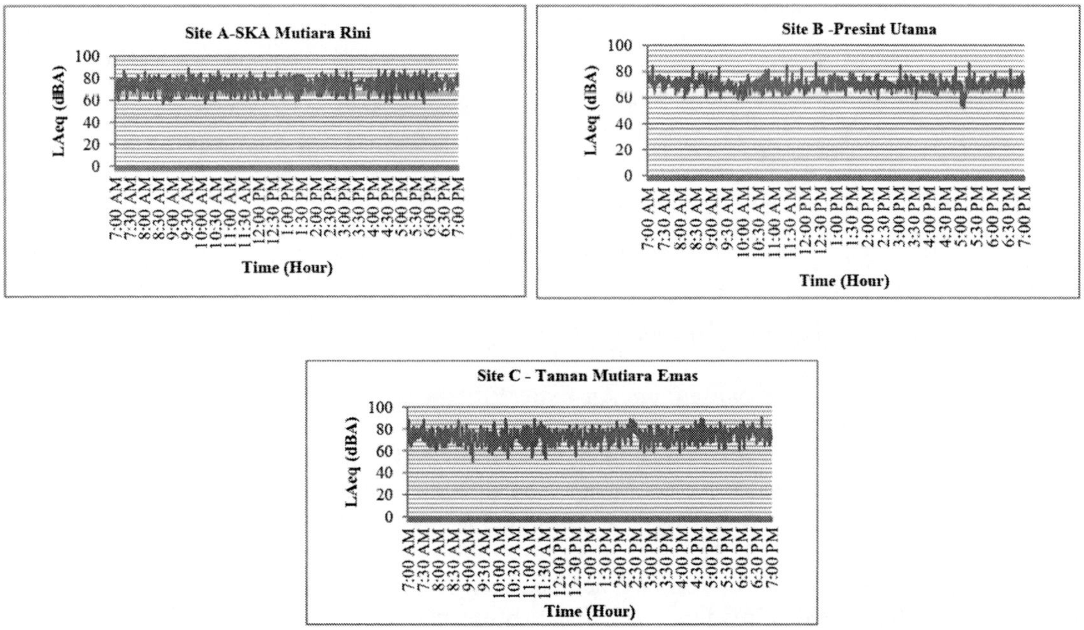

Figure 4. LAeq over 12 hour's period of sampling locations.

5.2 Percentage Noise Level in Study Area

Figure 5 shows the noise level at site A is high during morning off peak hour but it slightly increase 3% from the morning peak hour and 1% from afternoon peak hour which is 26%. At site B, the range from 65 dBA to 70 dBA the noise level is approximately same this is around 47 %. This can be concluded that Mutiara Rini area has fewer cars on the road commence from 8.00 a.m to 9.00 p.m. The noise level during morning peak hour is very high at range 70 dBA to 75 dBA which is 60% at site B. The pattern of the data at site C is likely same as at the site A. At the range 70 dBA to 75 dBA the noise level is almost same but it low during afternoon peak hour which is 13.33% [26] also proved that the noise level low during afternoon peak hour at residential area especially in low density residential areas because of majority of the residents are not always at home during the working days of the week.

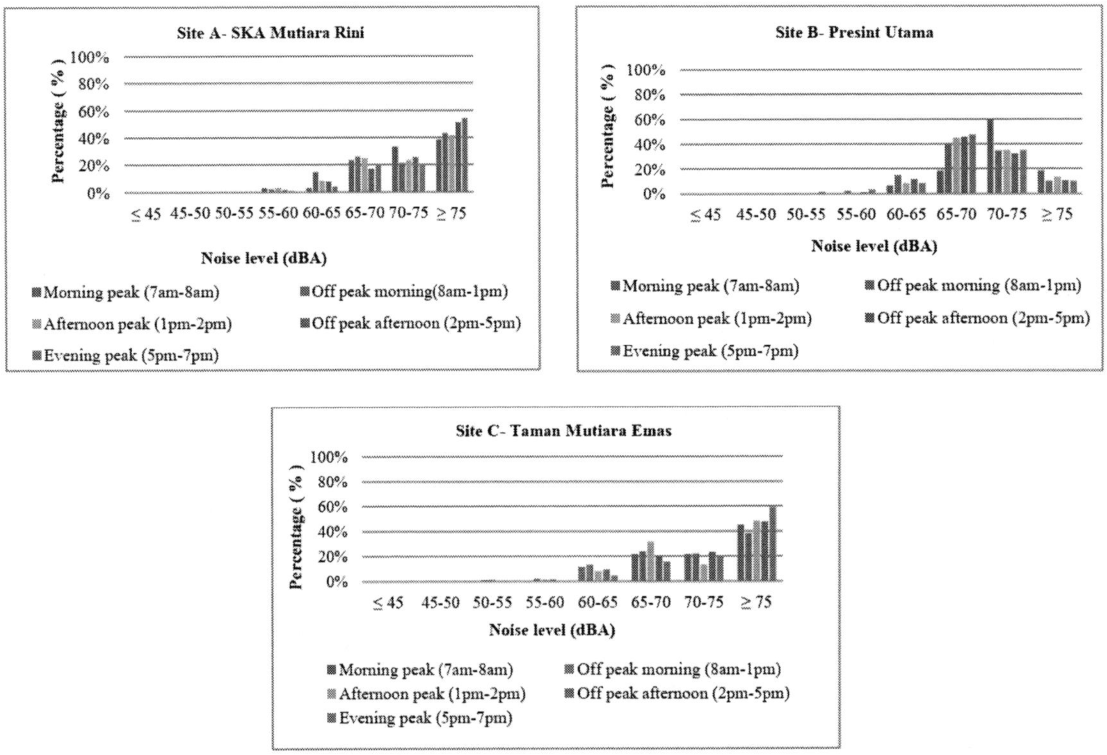

Figure 5. Histogram of percentage vs noise level

6. Mitigation to Reduce Traffic Noise from An Expert Interviewee

A proper mitigation to reduce traffic noise, interview survey had been done directly from an expert interviewee. The interviewee was selected among Jabatan Kerja Raya, JKR (D) Johor Bahru, Majlis Perbandaran Johor Bahru Tengah (MPJBT) and Senior Lecturer from Faculty of Civil Engineering Universiti Teknologi Malaysia, (UTM). Table 5 shows the different perspective from the expert in order to managing noise annoyance due to traffic noise. The results show that the installation concrete barrier, planting trees and providing the lush absorbent is the best method to be used.

Table 2. Analysis for the Best Mitigation That Can Reduce Traffic Noise

Question 1: What is the best mitigation that can reduce the traffic noise?								
Theme : Mitigation		Respondent				Frequency	Percentage	Remarks
	Code	1	2	3	4			
1.a	Noise barrier	1	1	1	1	4/4	100%	This question indicates the best mitigation that can reduce the traffic noise. The practitioners have concluded one of the mitigation that they can apply
1.b	Travel speed reduction				1	1/4	25%	
1.c	Signage	1				1/4	25%	
1.d	Quitter rumble strips					0/4	0%	
1.e	Travel Demand Reduction					0/4	0%	

7. Conclusion

Traffic noise impact (TNI) near to sensitive area has identified. There is a significance evidence of difference noise level produce in different peak and off peak hour. The noise level, noise pollution level (LNP) and TNI in residential area did not satisfy the road traffic noise limit recommended by WHO and DOE. This particular scenario indicated that the increasing noise level can be respectively associated with some driver behaviour such as honk from the vehicles, speed of the vehicles and also their vehicles itself has a turbo engine that can contribute to high noise level.

Second objectives stated that to assess public annoyance among residents. The analysis shows that, the respondents are agreeing with the level annoyance and sensitivity. Based on the traffic noise impact at all sites is moderately concern of worry even all the measured values are exceeding the permissible limits. This might be they are seldom in their house. Some of them are working from the morning until night. For them, there are not too seriously about the higher noise level. Furthermore, they just stay in their house and not effected too much. They are also feeling normal with this sound just sometime they feel annoyed when they are in the house. Moreover, this road is the main road to them drive to go everywhere.

From the third objectives, which is develop a proper mitigation that can reduce traffic noise annoyance from an expert interviewee has identified. They have same perspective that restoration of concrete barrier and trees are the best mitigation to reduce the traffic noise. However, they said that by installation this barrier it will give costs benefits that improve of the environmental performance. It means that, it can reduce the noise pollution. But, it has own barrier from this mitigation. They agreed that to install this barrier is very high in cost especially in installation of concrete barrier.

References

[1] Stansfeld S A & Matheson M P 2003 Noise pollution: non-auditory effects on health. *British medical bulletin*, *68*(1), 243-257 Chandratilake SR, Dias W P S 2013 "Sustainability rating systems for buildings: Comparisons and Correlations" Department of Civil Engineering, University of Moratuwa, Moratuwa, Sri Lanka

[2] Berglund B, Lindvall T & Schwela D H 2000 New WHO guidelines for community noise *Noise & Vibration Worldwide. 31*(4), 24-29.

[3] Su W 2009 September Urban Traffic Noise and Quieter Pavement Research In *Management and Service Science MASS'09 International Conference on* (pp. 1-6) IEEE.

[4] Öhrström E, Skånberg A, Svensson H & Gidlöf-Gunnarsson A 2006 Effects of road traffic noise and the benefit of access to quietness *Journal of Sound and Vibration. 295*(1), 40-59.

[5] Ouis D 2001 Annoyance from road traffic noise: a review *Journal of environmental psychology.* 21(1), 101-120.

[6] Aziz S Q, Lulusi F A, Ramli N A, Aziz H A, Mojiri A & Umar M 2012 Assessment of Traffic Noise Pollution in Bukit Mertajam, Malaysia and Erbil City, Iraq *Caspian Journal of Applied Sciences Research, 1*(1), 1-11.John Keung "Building planning & massing" 010 Building and Construction Authority,

[7] Department of Environment *The planning guidelines for environmental noise limits and control* (Ministry of Natural Resources and Environment, Malaysia, 2004)

[8] Davis M L & Cornwell D A 1998 *Introduction to environmental engineering* (Vol. 3) New York: McGraw-Hill.

[9] Li B, Tao S, Dawson R W, Cao J & Lam K 2002 A GIS based road traffic noise prediction model *Applied Acoustics.* 63(6), 679-691.

[10] Li B, Tao S & Dawson R W 2002 Evaluation and analysis of traffic noise from the main urban roads in Beijing *Applied Acoustics. 63*(10), 1137-1142.

[11] Langdon F J & Scholes W E 1968 The Traffic Noise Index: A Method of Controlling Noise Nuisance.

[12] Nijland H A & Van Wee G P 2005 Traffic noise in Europe: a comparison of calculation methods, noise indices and noise standards for road and railroad traffic in Europe *Transport Reviews. 25*(5), 591-612.

[13] Bruel K 1998 Technical Documentation—Integrating and Logging Sound level meterBK 2238 and BK 2260 *Denmark Naerum: Bruel and Kjaer.*

[14] Maschke C, Hecht K & Balzer H U 1999 Preventative medical limits for chronic traffic noise exposure *The Journal of the Acoustical Society of America. 105*(2) 1374-1374.

[15] Marathe P D 2012 Traffic noise pollution *IJED. 9*(1), 63-68.

[16] Alimohammadi I, Nassiri P, Azkhosh M & Hoseini M 2010 Factors affecting road traffic noise annoyance among white-collar employees working in Tehran *Iranian Journal of Environmental Health Science & Engineering. 7*(1), 25.

[17] Pal D & Bhattacharya D 2012 Effect of road traffic noise pollution on human work efficiency in Government Offices, Private Organizations, and Commercial Business Centres in agartala city using fuzzy expert system: a case study *Advances in Fuzzy Systems. 2012* 8.

[18] Saadatian O, Haw L C, Sopian K & Sulaiman M Y 2012 *Review of windcatcher technologies. Renewable and Sustainable Energy Reviews. 16*(3) 1477-1495.

[19] Laszlo H E, McRobie E S, Stansfeld S A & Hansell A L 2012 Annoyance and other reaction measures to changes in noise exposure—A review *Science of the total environment. 435* 551-562.

[20] Loukopoulos P 2005 *Future urban sustainable mobility: Implementing and understanding the impacts of policies designed to reduce private automobile usage.* Unpublished doctoral dissertation Göteborg University, Göteborg Sweden.

[21] Gärling T, Eek D, Loukopoulos P, Fujii S, Johansson-Stenman O, Kitamura R et al 2002 *A conceptual analysis of the impact of travel demand management on private car use* Transport Policy, 9, 59–70.

[22] Saadu A A, Onyeonwu R O, Ayorinde E O & Ogisi F O 1998 Road traffic noise survey and analysis of some major urban centers in Nigeria *Noise Control EngineeringJournal*, *46*(4),146-158.

[23] Cronbach L 1972 The dependability of behavioral measurements *Theory of generalizability of scores and profiles*.

[24] Majid M A & McCaffer R 1997 Assessment of work performance of maintenance contractors in Saudi Arabia *Journal of management in Engineering*. *13*(5) 91-9

[25] Garcia A & Garrigues J V 1998 24-hour continuous sound-level measurements conducted in Spanish urban areas *Noise Control Engineering Journal*. *46*(4) 159-166

[26] Mansouri N, Pourmahabadian M & Ghasemkhani M 2006 Road traffic noise in downtown area of Tehran *Journal of Environmental Health Science & Engineering*. *3*(4) 267-272.

[27] Dai L, Cao J, Fan L & Mobed N 2005 Traffic noise evaluation and analysis in residential areas of Regina *Journal of Environmental Informatics*. *5*(1) 17-2

Acknowledgments
The authors would like to express their deep gratitude to the Ministry of Higher Education Malaysia and Universiti Teknologi Malaysia for supporting financial grant (Ref:FRGS/1/2016/TK08/UTM/02/2), Cost Centre No: R.J130000.7822.4F867 & R.J130000.7322.4B291, research team members Green PROPMT and Department of Construction Management, UTM together with opportunity for the research.

A Review of Application Building Information Modeling (BIM) During Pre-Construction Stage: Retrospective and Future Directions

Santi Edra Nisa Lau[1], Rozana Zakaria[2], Eeydzah Aminudin[3], Chai Chang Saar[4] Aswadi Yusof[5], Che Muhammad Fatihi Hafifi Che Wahid[1]

[1] *Postgraduate Student, Faculty of Civil Engineering, Universiti Teknologi Malaysia*
[2] *Associate Professor, Faculty of Civil Engineering, Universiti Teknologi Malaysia*
[3] *Lecturer, Faculty of Civil Engineering, Universiti Teknologi Malaysia*
[4] *Lecturer, University of Reading Malaysia*
[5] *GIS Manager for Klang Valley Mass Rapid Transit Project, of Civil Engineering, Universiti Teknologi Malaysia*
Corresponding email: eeydzah@utm.my

Abstract. Pre-construction is one of the biggest areas of risk and uncertainty in construction project as it deals with subsurface ground conditions information. The amount of detail data needed in pre-construction especially for existing data modelling and site analysis should be sufficient enough to ensure that significant risks could not reasonably be anticipated. Current practicing method in interpreting data during this stage tasks reveal limitation. Construction industry faced many obstacles due to the depends on the traditional practice; paper-based document which missing and redundant data always happened. In recent years, there has been a shift in construction where people move to BIM application because of its potential to reduce the problem faced by infrastructure world. BIM has become a successful technology and widely popular in the construction world especially in developed countries because of its potential. Nowadays, people are moving one step ahead in BIM which is adoption of BIM during pre-construction stage. Thus, this paper review studies centered on BIM-integrated modelling during preconstruction stage. But there is lack of practical researches have been made during this stage. Although a large number of studies on BIM have been conducted in the past decade, a lack of consensus remains among researchers and practitioners regarding the applications of BIM during pre-construction stage, the availability of subsequent data integration tool for geotechnical activity. A comprehensive literature review was conducted for data collection and analysis. After in-depth review of journal articles widely cover the application of BIM, this study summarizes an overview and critical reflection of geotechnical data integration using BIM during pre-construction stage. The results are useful for the identification of research clusters and topics in the BIM community.

1. Introduction

A standard project typically has three phases; pre-construction phase, construction phase and post-construction phase [1]. Pre-construction is one the biggest areas of risk and uncertainty in construction project as it deals with subsurface ground conditions and geotechnical information. Ham et al. [2] summarized that this phase is where gathering necessary information is done in advance to set the direction of project as well as to start the construction. Tegtmeir [3] stated that any project that deals with geotechnical data consider as a complicated work as it begins with site investigation and data interpretation from public sources by sub-contractor before it can be integrated into 3D programme and simplifying it for design and construction purpose.

The building and construction industry today has been driven to use Building Information Modeling (BIM) in their project in conjunction with an idea to transform the construction industry to be highly productive, environmentally sustainable, with global competitive players while focused on safety and quality standards. Different people have different way in describing BIM [4] but they still share the same concept of and it is an important note that there is no single satisfactory meaning behind BIM. From Papadonikolaki et al. [5] perspective, BIM is an acronym to replace 'Building Product Modelling' (BPM) which widely used to describe the broad concept of IS in construction [6]. Surely, BIM is gradually changing the way of data being collected and analyse specifically during pre-construction phase; paper-based document to object-based modelling.

Subsurface aspect such as geology and geo-technical always being neglected and current practice do not provide sufficient data for realistic 3D modelling [4][7][8][9]. Currently less study conducted to evaluate and compare the data integration process during pre-construction phase for traditional system and BIM adoption. Such a study could help to understand the current trends of BIM adoption in geotechnical data integration and later on identify the research gaps. Therefore, this paper aims to compare the conventional and current practices of pre-construction development, to provide an analytical review of the current BIM adoption for geotechnical data integration during pre-construction stage, and identify the gaps in BIM development and adoption.

In order to address the above-mentioned research questions, this study conducts a critical review on geotechnical data integration process during pre-construction stage for construction using Building Information Modeling (BIM). The structure of this review paper is organized as follows. Chapter 2 specifically illustrates about the research methodology of this review. While, chapter 3 discussing the application of BIM during pre-construction stage.

2. Methodology

To review the geotechnical data integration during pre-construction stage using BIM, three steps were carried out. First step, a comprehensive literature review was conducted to increase the understanding of data integration process during pre-construction phase and the differences between traditional method and BIM-based method. Connecting BIM and pre-construction stage is the main concept of "geotechnical data integration" that which will be discussed in this review paper. This process involves an identification of a set of keywords before an intensive literature can be done later on. The main keywords that can represent this review are "BIM", "Building Information Modelling, pre-construction", "geotechnical data", "data integration", ''safety", ''quality",''time", ''cost", and ''budget". Once all the keywords being identified, the second step is to apply it in web search in online academic publication databases, i.e. ''Web of Science", ''Sciencedirect", ''Scopus", and ''Google Scholar". All of these web search engine is function for collecting academic and applied publications related to this topic. Only relevant reading materials such as conference papers, thesis, dissertations, journals and article can be accepted. Publication which include the keywords in the Title/Abstract/Keywords are selected to be reviewed. Most of the journal article found for this study were from leading journals of this area (e.g. Safety Science, Automation in Construction, International Journal of Project Management, Journal of Computing in Civil Engineering, Renewable & Sustainable Energy Reviews, Reliability Engineering & System Safety), publications from conference proceedings and other sources of professional associations, standard committees (e.g. HSE, ISO) and authorities. In the last step, all publications are analysed critically and being compare to identify current process, obstacles and future work to close these gaps. This strategy is also adopted by many previous studies, such as Zou et al. [10] and Li et al. [11].

3. Findings

3.1. Pre-construction Phase

The preconstruction phase of a project also known as planning and design phase [12]. Pre-construction phase is one of the important phases in construction that control the successful of project as stated in a study by Bakar et al. [13]. In response to that, many researchers have already acknowledged the significance of effective planning practices and integration of designer and contractor in the early stages of construction project life cycle [13]. Basically, the amount of detail data needed in pre-construction information especially for existing data modelling and site analysis should be sufficient enough to ensure that significant risks could not reasonably be anticipated. Gathering all information for the sake of designer and contractor really needed throughout the project before they can start work on any particular element such as design.

Kolltveita and Gronhau [14] stated that pre-construction or sometimes called as early phase is the process and activities that lead to, and immediately follow, the decision to undertake feasibility studies and to execute the main project. He continued to explain that the main tasks in this part of the early phase are feasibility studies, value analysis, formulation of project goals.

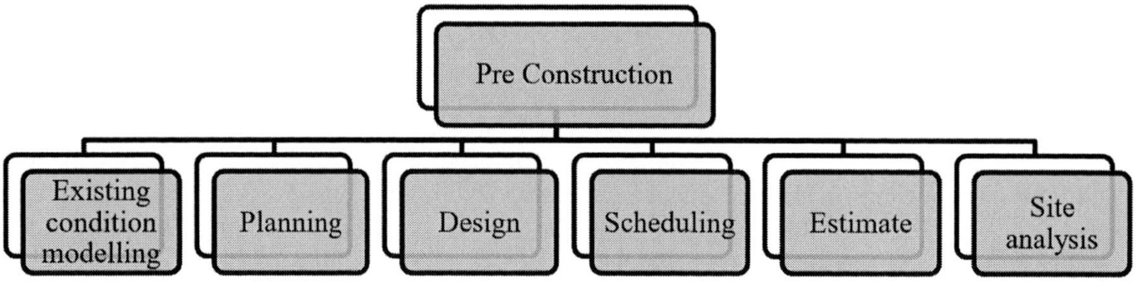

Figure 1: Activities in pre-construction phase by Latiffi et al. [1]

3.2. *Workflow of Traditional Practices*

Some countries still follow the traditional method of obtaining data for geotechnical information during pre-construction which is based on traditional surveying techniques such as total stations, physical measurement, and photography [15]. Then, all the data will be transferred to software before it can be used for design stage and the product of the integration is a paper-based document [14]. The method used to integrate data for geotechnical activity is via reports containing 2D site plans and log sheets and the engineer will think and visualize it in 3D [16]. Figure 1 shows the principle of geotechnical data journey laid out by Child et al [17] where the workflow is a linear waterfall process and will be ended up at national and local archive in form of paper-based document. The result data is passed to the following phase after the work in one phase is done [18].

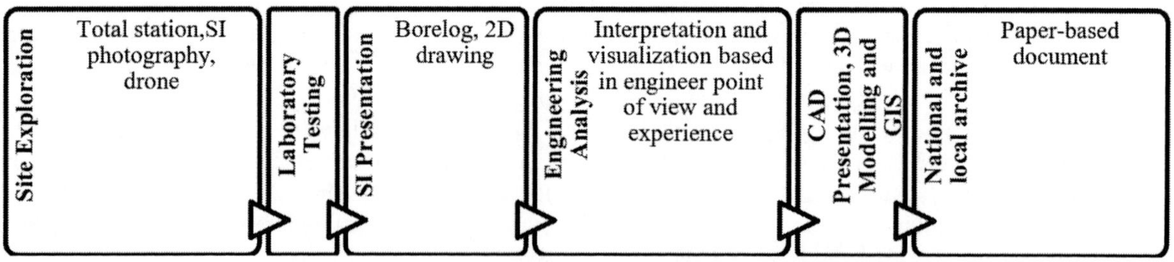

Figure 2: Traditional Approach by Child et al. [18]

3.3. *Workflow of BIM Practices*

3D scanning to modelling can be an impressive tool for facilities that are facing difficulties dealing with old and traditional systems like paper-based document in post-construction phase. Furthermore, particular technologies, such as the use of cloud computing [19], and 3D scanning [20] have facilitated adoption and added value to BIM. The integration of BIM through the model provided via 3D scanning ensure higher functionality [21], waste reduction, higher productivity and profit [22]. BIM provide benefit: overlap, reduce real project risks; irrelevant document eliminated; waste reduces; productivity increase; costs decrease; profit increase as improve product, improve services deliver, or expand market share [22].
Bentley [23] and Autodesk [16] are the example of a leading global provider of comprehensive software solutions that provide software to integrated data of geotechnical such as terrain, subsurface condition and etc and provide a modelling to represent surface condition and information. By that, a comprehensive project can be delivered. BIM application stops at the preconstruction phase with a limited amount of research regarding data collection of the construction process [24]. In recent years, most virtual 3D city models have been realised as purely graphical or geometrical models, neglecting semantic and topological aspects.
Previously, the data journey believed to be flat but nowadays it should be round (as in figure 3) and ingoing which available throughout the life of project and lead to the effective interpretation of ground conditions and characterization of the site [17]. Child et al. [17] continued to explained that visualizations being generated by tools and ended in the form of a digital archive.

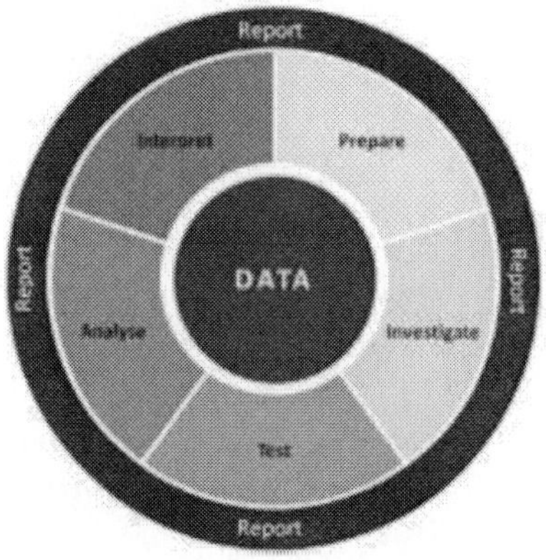

Figure 3: The data journey transformed by Child et al. [17]

3.4. Challenges with BIM Practices
The use of BIM for pre-construction phases is not as popular as other phases of the project despite these successful applications [25]. Even, the application of BIM in pre-construction phase is more evident [1][26] than during the construction and post-construction phases because there are many activities in pre-construction but the use of BIM for pre-construction phase still not gained a wide acceptance in industry [25]. BIM became compulsory for applying to the tenders in developed countries [26] although it took some time for the government to encourage the development of modelling [17] but rarely used in developing countries such as Malaysia [27] due to the comfortableness of using the old method and cost that need to be considered to change to new method. Current software available also has lacking as tabulated in table 1.

Table 1: Software available for data integration modelling

Name	Author/ Date	Using	Primitiv e	Entity object	Model base	Advantage	Disadvantag e
SSM	Zlatanova [28]	City visualiz ation	Node, Planar face	Point, Line, Surface, Body	Simplex concept	Abstract primitives easily, small storage	Update dynamicall y and modify are complex, constructi ve objects are multi-values
UDM	Coors [29]	City visualiz ation	Node, Triangl e	Point, Line, Surface, Body	Triangu lation	Abstract primitives easily, small storage, facial visualize	Update dynamicall y and modify are complex
OO3D	Shi Wenzhong, Yang bisheng [30]			Point, Line, Surface, Volume	OO modelin g	Complex object, LOD, Visualization	
GTP	Wu LiXin [31]	Geologic al engineer ing	Node, Line (TIN-L, side-L), Face (TIN-F, Side-F), GTP, diagona l	Point, Line, Surface, Body	Body divide	Topology completely, 3D expression of geology, based on drill hole data	Difficult to visualizat ion of complex geological object
TEN+Oct ree	Li Qingquan, Li Deren [32]	Geologic al engineer ing	Octree 、TEN		Octree express whole, TEN express part	Improve the precision of objects, reduce data storage	Difficult to build topology of spatial objects
CityGML	Tegtmeier, [3], Jayakody et al. [33]	Geotechn ical extensio n	3D-GEM, XML-based format		Geo-informa tion	Representativ e of 3D digital terrain models, buildings, vegetation, water bodies, transportation facilities or city furniture.	Less presentati ve to users
GIS	Hack [34]	BIM extensio			Point, edge,	3D visualization	No proper integratio

		n	polyline, polygon, 3d Solid	system of built environment and urban system	n of the boreholes and other geological and geotechnical data.
GeoBIM	Hack [34]	BIM extension		Integration of surface and subsurface data	Not a complete set of integrated formats to handle surface and subsurface data.
Holebase SI	Morin, 2014 [35]	Autocad Civil 3D Extension		HoleBASE SI and AutoCAD Civil 3D combine to generate and maintain comprehensive 3D models of the geology that can be integrated with any other project designed data.	

4. Discussion

An important aspect of this review is to find out research gaps in current BIM-based data integration through a systematic and critical review, which is discussed as follows:

According to the reviewed paper, the data needed and the software used are the main things that need to be considered. Everyone know that BIM is the latest innovation that considered as an intelligent 3D model because of it abilities to efficiently plan, design, construct, and manage buildings and infrastructure. Latest invention is reality capture where it will use drone to capture all the points and integrate it using BIM to obtain terrain of land make the process become easier. But, the integration data process is the most critical work. The review acknowledged the fact that the optimal software-BIM not yet be found [36]. Kessler et el. [9] also shared the same fact where geology part is still absent from most BIM models but a standard software that already available in market can be used to access and visual geotechnical data. but technology together with data transfer standards. From table 2 it can be seen that all the available software has their own benefits and disadvantages. All the software stated have something lack such as soil strength parameter, the bearing capacity characteristic and soil stratigraphy not being included in the model. It will be an interesting technology if a new BIM software can produce an intelligent model that consider these three main categories [37]. The development of standardized geotechnical data format and interpretation will give a detailed insight of the underground condition that can help the whole project life cycle.

5. Conclusion

As conclusion, a further research in the field needs to be focused towards data integration modelling, on finding the new technology or tools that can be work together with BIM to provide 3D topological modelling and semantic models. In addition, the research in the area also needs to study on the conditions of such an implementation, the prototype software components (developed, verified/validated during the research) demonstrated a technical proof-of-concept for the transfer of building information and the use of BIMs in facilitating these processes. The research results demonstrated that BIMs provide the required level and amount of (geometric and semantic) information (about the building) for the seamless automation of the data management in these process, and the research also showed that although there are still barriers in the transformation process it is possible to represent a high level of geometric and semantic information acquired from BIMs in the geospatial environment. Even though the research in the area is mainly focused on the transfer of geometric information into the geospatial environment, the success of such an effort will mainly depend on the efficiency in representing the semantic information in the geospatial environment.

6. References

[1] Latiffi A A, Mohd S, Kasim N, & Fathi M S 2013 Building information modeling (BIM) application in Malaysian construction industry *International Journal of Construction Engineering and Management*. *2*(A) 1-6.

[2] Ham N H, Min K M, Kim J H, Lee Y S & Kim J J 2008 – November A study on application of bim (building information modeling) to pre-design in construction project In *Convergence and Hybrid Information Technology, ICCIT'08. Third International Conference on* Vol. 1 pp. 42-49). IEEE.

[3] Tegtmeier W, Zlatanova S, Van Oosterom P J M, & Hack H R G K 2014, 3D-GEM: Geo-technical extension towards an integrated 3D information model for infrastructural development *Computers & geosciences*. *64* 126-135.

[4] Zhang J, Wu C, Wang Y, Ma Y, Wu Y & Mao X 2017 The BIM-enabled Geotechnical Information Management of a Construction Project *Computing*. (2) 1–17.

[5] Papadonikolaki E, Vrijhoef R, & Wamelink J W F 2015- September A BIM-based supply chain model for AEC In *1st Conference in Building Information Modelling (BIM) in Design, Construction and Operations* (pp. 181-193). WIT Transactions on The Built Environment.

[6] Cerovsek T. 2011 A review and outlook for a 'Building Information Model'(BIM): A multi standpoint framework for technological development *Advanced engineering informatics*. *25*(2), 224-244.

[7] Breunig M, & Zlatanova S. 2011, 3D geo-database Research: Retrospective and future Directions *Computers and Geosciences*. *37*(7), 791–803.

[8] Hack R 2010 - November Integration of Surface and Subsurface Data for Civil Engineering. *Information Technology in Geo-Engineering*. 37–49.

[9] Kessler H, Wood B, Morin G, Gakis A, Mcardle G, Dabson O, & Dearden R 2015 *Building Information Modelling (BIM) - A Route for Geological Models to Have Real World Impact* (Vol. 101).

[10] Zou Y, Kiviniemi A & Jones S W 2017 A review of risk management through BIM and BIM-related technologies. *Safety Science*. *97*, 88–98

[11] Li X, Wu P, Shen G Q, Wang X & Teng Y 2017, Mapping the knowledge domains of Building Information Modeling (BIM): A bibliometric approach. *Automation in Construction*, *84*(September), 195–206.

[12] Schierholz J 2012 Evaluating the preconstruction phase in a Construction Manager/General Contractor Project. *Journals in Business & Management*. *1*(2) 25–27.

[13] Bakar A Hamid A & Embi M R 2016, Review on Application of Building Information Modelling in Interior Design Industry *Journal of Business and Technology. 2 3*(2) 1–6.

[14] Kolltveit B J & Grønhaug K 2004 The importance of the early phase: the case of construction and building projects. *International Journal of Project Management. 22*(7), 545 551.

[15] Suchocki M 2017 THE BIM-FOR-RAIL OPPORTUNITY. *WIT Transactions on The Built Environment. 169* 37-44.

[16] AutoCad/Autodesk 2010 http://usa.autodesk.com/ .

[17] Child P, Grice C & Chandler R 2014 *The Geotechnical Data Journey.* 83–88.

[18] Zhang J, Wu C, Wang L, Mao X & Wu Y 2016. The Work Flow and Operational Model for Geotechnical Investigation Based on BIM *IEEE Access. 3536*(115) 1–1.

[19] Jiao Y, Wang Y, Zhang S, Li Y, Yang B & Yuan L 2013, A cloud approach to unified lifecycle data management in architecture, engineering, construction and facilities management: Integrating BIMs and SNS *Advanced Engineering Informatics. 27*(2)173-188.

[20] Arayici Y, Onyenobi T & Egbu C 2012 Building Information Modelling (BIM) for Facilities Management (FM) *International Journal of 3-D Information Modeling.* 1(1) 5573.

[21] Aziz N D, Nawawi A H & Ariff N R M 2016 Building Information Modelling (BIM) in Facilities Management: Opportunities to be Considered by Facility Managers *Procedia-Social and Behavioral Sciences. 234* 353–362. https://doi.org/10.1016/j.sbspro.2016.10.252

[22] Smith D K & Tardif M 2009 *Building information modeling: a strategic implementation guide for architects, engineers, constructors, and real estate asset managers*. John Wiley & Sons.

[23] Bentley. http://www.bentley.com/nl-NL/ .

[24] Goedert J D & Meadati P 2008 Integrating construction process documentation into building information modeling. *Journal of construction engineering and management. 134*(7) 509-516.

[25] Karan E P & Irizarry J 2015 Extending BIM interoperability to preconstruction operations using geospatial analyses and semantic web services *Automation in Construction. 53*, 1–12.

[26] Hadzaman H A N, Takim R, Nawawi A & Fadhil M 2016 An Exploratory Study: Building Information Modelling Execution Plan (BEP) Procedure in Mega Construction Projects An Exploratory Study: Building Information Modelling Execution Plan (BEP) Procedure *Malaysian Construction Research Journal. 1* (November) 29–40.

[27] Kassem M & Succar B 2017 Macro BIM Adoption: Comparative Market Analysis *Automation in Construction. 81*(May) 286–299.

[28] Zlatanova Siyka 2000 3D GIS for Urban Development *ITC Dissertation*

[29] Coors V 2007 3D GIS in Networking environments. CEUS Computers *Environment and Urban Systems*. 27: 345 –357.

[30] Shi W, Yang B & Li Q 2003 An object-oriented data model for complex objects in three dimensional geographical information systems. *International Journal of Geographical Information Science. 17*(5) 411-430.

[31] Wu L 2004 Topological relations embodied in a generalized tri-prism (GTP) model for a 3D geoscience modeling system *Computers & Geosciences. 30*(4) 405-418.

[32] Yanbing W, Lixin W, Wenzhong S & Xiaomeng L 2007 On 3D GIS spatial modeling In *ISPRS Workshop on Updating Geo-spatial Databases with Imagery & The 5th ISPRS Workshop on DMGISs. ISPRS.*

[33] Li X, Wu P, Shen G Q, Wang X & Teng Y 2017 Mapping the knowledge domains of Building Information Modeling (BIM): A bibliometric approach *Automation in Construction. 84*(September), 195–206.

[33] Jayakody A, Rupasinghe L, Perera K A M S, Herath H H P M, Thennakoon T M A & Premanath S U 2013 The Development of the CityGML GeoBIM Extension for Real Time Assessable Model (*Integration of BIM and GIS*).

[34] Hack R 2010 Integration of Surface and Subsurface Data for Civil Engineering *Information Technology in Geo-Engineering*. (November), 37–49.

[35] Morin G, Hassall S & Chandler R 2014 Case study–the real life benefits of geotechnical building information modelling *Toll, DG, Zhu, H. et al*, 95-102.

[36] Bryde D, Broquetas M & Volm J M 2013 The Project Benefits of Building Information Modelling (BIM) *International Journal of Project Management. 31*(7) 971–980.
[37] Tawelian L R & Mickovski S B 2016 The Integration of Geotechnical Data into the BIM Process *Procedia Engineering. 143* 734–741.

Acknowledgments
The authors would like to express their deep gratitude to the Ministry of Higher Education Malaysia and Universiti Teknologi Malaysia for supporting financial grant, Cost Centre No: Q.J130000.2522.19H53 & R.J130000.7322.4B291, research team members Green PROPMT and Department of Construction Management, UTM together with opportunity for the research

Towards a Competency-based Vision for Construction Safety Education

Akeem Pedro*, Pham Hai Chien, Chan Sik Park

Department of Architectural Engineering, Chung-Ang University, South Korea.

Corresponding Author, Email : lanrepedro3@gmail.com

Abstract. Accidents still prevail in the construction industry, resulting in injuries and fatalities all over the world. Educational programs in construction should deliver safety knowledge and skills to students who will become responsible for ensuring safe construction work environments in the future. However, there is a gap between the competencies current pedagogical approaches target, and those required for safety in practice. This study contributes to addressing this issue in three steps. Firstly, a vision for competency-based construction safety education is conceived. Building upon this, a research scheme to achieve the vision is developed, and the first step of the scheme is initiated in this study. The critical competencies required for safety education are investigated through analyses of literature, and confirmed through surveys with construction and safety management professionals. Results from the study would be useful in establishing and orienting education programs towards current industry safety needs and requirements

1. Introduction

Despite the consistent growth of global markets and emergence of cutting-edge technologies, construction sites remain among the most dangerous workplaces with extremely high accident and fatality rates [1]. Even though the industry only employs approximately 7% of the world's workforce, it reports incident rates which double industrial averages [2], and accounts for 30-40% of fatalities [3]. Addressing this problem requires continuous efforts not only industry practice, but in education and academia as well. Safety training and education play a crucial role in cultivating knowledgeable, skilled and competent personnel, ready to face the safety demands of construction work. Construction students who will take health and safety related positions after graduation will be required to interact with workers, equipment and working conditions, bearing a major responsibility in fostering positive attitudes and encouraging safety directives to ensure site safety [4]. A high percentage of construction accidents occur as a result of human errors which can be proactively prevented through safety education [5]. However, current training programs have been criticized for being deficient in developing and strengthening crucial safety competencies [6]. Moreover, safety education efforts in tertiary institutions have been inadequate and ineffective in preparing students to meet the safety demands of modern construction sites [7]. It is of crucial importance that students who will become project managers, operations managers, safety managers and site advisors have the knowledge, skills, and competencies necessary for fostering safe and healthy construction work environments, and contributing to accident reduction efforts in the industry. In this regard, there is a need for a reformative shift, towards competency-based construction safety education.

In today's increasingly complex and dynamic construction industry, graduates need to possess not only technical skills, but key competencies which contribute to organizational performance outcomes as well. Furthermore, with global competition and the multi-disciplinary nature of work in construction-related industries, it is important to educate new professionals to develop appropriate skills [8]. Considering the high risk nature of construction, and the poor safety record of the construction industry, educational programs should be designed to mirror the dynamic landscape of the construction industry, to ensure students are competent upon graduation [9]. Construction curricula typically include various outcomes derived to meet the perceived needs of industry [10]. However, since safety is usually a low priority among other learning outcomes [11], learners fail to develop the required safety competencies.

This can have grave consequences, as a lack of competency can lead to serious errors and negative results in terms of safety performance [12].

In light of these challenges, this paper proposes a vision for competency-based construction safety education, to bridge the gap between the competencies current pedagogical approaches develop, and those required for safety in practice. This study takes the first step to reorient construction pedagogy to build safety competencies and lead the industry towards a new construction safety paradigm. A research scheme is proposed, delineating how the competency vision can be achieved; and then critical competencies required for construction safety education are identified.

2. Towards Competency-based Construction Safety Education

This section discusses the vision of the study towards competency-based construction safety education. Firstly, a theoretical explanation of competency-based education is provided, then the vision for competency is described, illustrating the competency levels students are expected to reach, after graduation and industry experience. Lastly, a research scheme is provided, breaking down the steps and research issues which should be tackled in order to achieve the vision.

2.1 Theoretical Basis for Competency-based education

In line with an emphasis on "work-readiness" and learning outcomes, competence research has emerged from academic fields such as psychology and education [13]. The notion of "competency" refers to the knowledge, skills, attitudes and values that are necessary to successfully perform a particular activity or task [14]. Competency Based Education (CBE) is a learner centered, outcome based approach capable of reorienting educational processes towards demonstrated mastery and the application of knowledge and skills in the real world [15; 9]. CBE is performance oriented, placing emphasis on what students can do, as a result of learning [16]. Work in a range of disciplines has demonstrated that proficiency in particular competencies is associated with higher levels of performance [17]. Similarly, enhancing safety education to develop safety competencies would result in improved safety performance; and more desirable safety outcomes. In construction safety education, CBE can bridge the gap between industry safety requirements and graduates' knowledge and capabilities.

However, due to construction curricula lacking means of effectively disseminating safety knowledge, and assessing learner's safety proficiencies, it remains difficult to ascertain that construction graduates possess the competencies required for safe performance on the real jobsite. In response to these deficiencies, the CBE process is adopted, serving as a basis for establishing tertiary construction safety education. CBE is based on five core principles: 1) programs reflect valid competencies; 2) students can learn at variable pace; 3) effective learning resources are available; 4) competencies are explicitly mapped to learning outcomes and assessments; and 5) assessments are secure and reliable [15]. In order to reflect these principles, a competency-based program for construction safety education is: 1) based on specific industry-based competencies (which are identified in this study); 2) Web-based course delivery is proposed to afford flexible learning content accessibility; 3) effective learning resources are designed with a visualization-enhanced approach, to overcome the limitations of conventional educational approaches; 4) safety competencies are constructively aligned with learning activities and assessment measures; and lastly 5), web-based automatic and proctored assessment measures are proposed.

At the instructional level, CBE builds upon students' base knowledge and affords learning experiences which build content knowledge, cognitive skills, abilities and values. Next, CBE affords integrated learning experiences, whereby students are required to synthesize knowledge, skills, and abilities consistently [16], in a context resembling that required in industry. For safety education, CBE should facilitate not only the acquisition of explicit safety knowledge and skills, but the cultivation of the implicit abilities, attitudes and values pertinent to construction safety as well. In addition, assessment is a critical element, and learners should only advance after demonstrating the competency necessary to perform a task in the realistic context [17].

2.2 Vision for Competency in Construction Safety Education

As illustrated in figure 1, this study proposes a long-term vision for safety competency, focusing on improving safety education and safety culture in the construction industry. Currently, majority of tertiary construction education programs do not include safety courses; hence, graduates lack safety knowledge and practical safety experience after graduation. In essence, they enter the construction industry as incompetent safety practitioners. This study proposes a research agenda aiming to move learners from the entry level (incompetent) to level 2 (competent), prior to employment in the industry. For learners to achieve safety competency, first a basic level must be reached, whereby learners possess a core knowledge of safety rules, regulations and common hazards. Following the basic knowledge level, learners need to not only develop skills, abilities and competencies such as hazard identification and safety communication; but also they need to demonstrate them, at the desired level. The scope of the current study focuses on deploying a CBE strategy with the support of advanced visualization technologies, and web-based modalities to elevate learners to "become competent" and ready to contribute to improving safety in the construction industry.

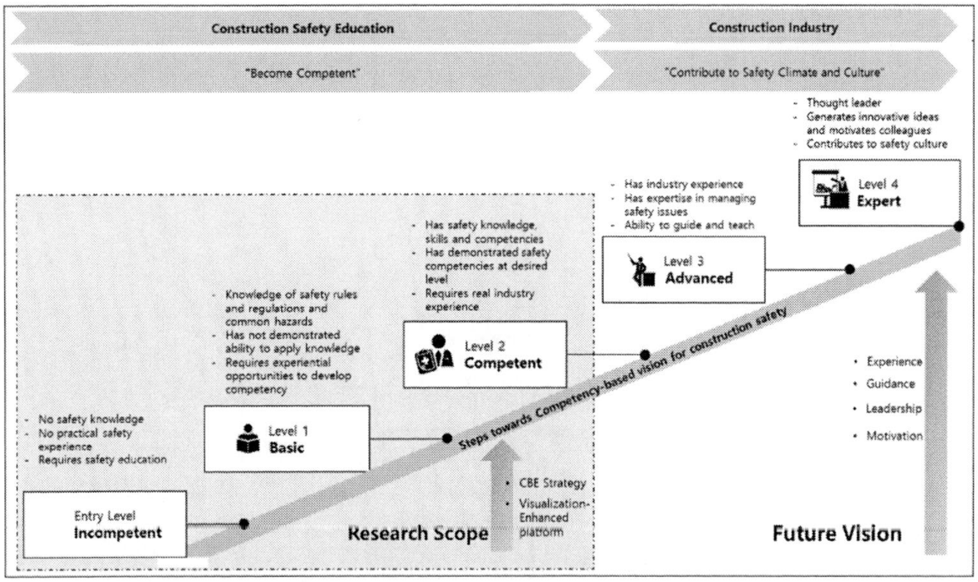

Figure 1. Vision for construction safety competency

Aside from tertiary education, continuous effort is also required in the actual industry, in order to reduce accidents and injuries; and improve the safety culture in the industry. As a future vision, construction graduates who have achieved competency (level 2), are expected to further influence organizational safety performance and safety outcomes, and become advanced safety practitioners (level 3), after they acquire further real industry experience, and develop the ability to guide and transfer knowledge in the workplace. Lastly, expert safety practitioners are expected to contribute even further to generate innovative ideas to tackle safety issues, motivate colleagues and workers to comply with safety requirements, and establish a safety culture at an organizational level.

2.3 Research Scheme to Achieve Vision

In order to achieve the vision for construction safety competency, this study proposes a research scheme to establish effective construction safety education. Based on the theory of constructive alignment [18], the scheme comprises of three key sections. Considering the lack of clarity and research on the safety requirements for construction graduates, the first section of the scheme focuses on identifying the safety learning outcomes (competencies) which are necessary to prepare construction management students for the safety demands of construction jobsites. This study focuses on executing this first step, and investigates the critical competencies for preparing learners for the real industry. Following this step, learning activities which would support students to acquire and develop the safety competencies need to be designed. Instructional methodologies need to be designed to ensure students develop the necessary safety knowledge, skills and capabilities. In order to develop an effective educational approach visualization technologies are considered, along with web-based knowledge dissemination modalities to afford experiential and interactive didactic activities. Lastly, the third section considers what type of safety assessment would be necessary for students to demonstrate that they possess the required safety competencies, before entering the real construction industry. A novel visualization-enhanced safety assessment method is proposed, in order to afford authentic, context-based assessment, whereby learners can effectively demonstrate their safety competencies in simulated environments.

This study takes the first step in achieving this vision, by investigating the specific competencies that should be incorporated and delivered in tertiary construction education in order to prepare students for the modern safety demands of construction jobsites. These competencies serve as the foundation upon which a competency-based learning and context-based assessment contents will be established in future studies.

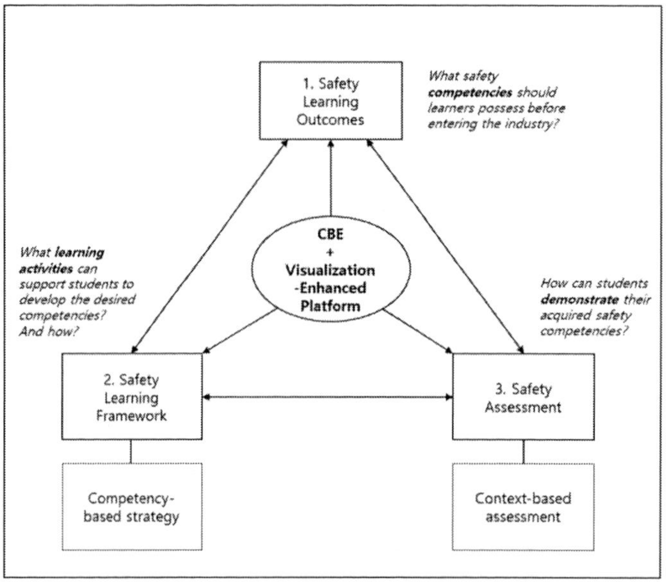

Figure 2. Research scheme for safety competency vision

3. Investigating Required Competencies for Construction Safety Education

In order to identify the required competencies for construction safety, literature, safety documents and accident records were analyzed; and then a survey was conducted to identify the competencies for improving safety outcomes. The investigation focused on the comprehensive range of competencies which are perceived to have an influence on construction safety performance. These included a broad spectrum of capabilities and expertise, such as technical (knowledge based) competencies, personal management, and interpersonal management competencies, and so on. The technical competencies such as Hazard Identification and Safety Inspection are based on Occupational Safety and Health Administration (OSHA) training requirements and standards [19], Korean Occupational Safety and Health Agency (KOSHA) rules and regulations, and the KOSHA Standard Risk Evaluation Model [20]. On the other hand, managerial and personal competencies were initially identified through related literature. Subsequently, surveys were conducted to ascertain their current significance.

As depicted in table 1, thirty experienced construction management and safety professionals participated in interview surveys to rate and identify the critical competencies, with a 5-point Likert scale (5 – Critical, 4 – Very important, 3 – Important, 2 – Unimportant, 1 – Completely Unimportant).

In order to improve the credibility and reliability of results, participants were allowed to maintain anonymity in their responses, and they were required to possess a minimum of 5 years practical experience in either construction or safety management.

Table 1. Specialization and experience of respondents.

Specialization		Average Experience
Construction Management	21 (70%)	8 Years
Safety Management	9 (30%)	10 Years

Table 2 presents the comprehensive list of twenty-nine competencies identified and the ratings for their perceived importance. Respondents identified hazard identification, safety inspection, compliance with safety regulations, site safety planning, accident avoidance strategies, hazard evaluation and control as the most important competencies for construction safety education, all with standard deviations less than 1. These are the competencies which have the highest impact on accident and injury prevention on the construction jobsite. The competencies perceived to have the lowest influence on safety included: conceptual safety skill, which is defined as the ability to see the organization as a whole, in the context of safety; relationship management; and analytical and conceptual thinking. Analytical and conceptual thinking was the only criterion predominantly considered unimportant (M: 2.97, SD: 0.77), while all other 28 competencies were deemed significant.

Table 2. Competencies for construction safety in order of importance.

Rank	Competency	Mean	S.D.	Mode
1	Hazard Identification	4.7	0.66	5
2	Safety Inspection	4.6	0.56	5
3	Compliance with Regulations	4.5	0.63	5
4	Hazard Control	4.5	0.68	5
5	Hazard Evaluation	4.47	0.86	5
6	Site Safety Planning	4.4	0.6	5
7	Self-Awareness	4.37	0.8	5
8	Accident Avoidance Strategies	4.33	0.75	5
9	Accident and Error Analysis	4.3	0.75	5
10	Safety Communication	4.3	0.79	5
11	Near-miss reporting	4.2	0.61	5
12	Emergency Management	4.13	0.73	5
13	Commitment to safety	4.06	0.69	5
14	Self-Control	3.97	0.85	5
15	Initiative	3.93	0.9	5
16	Interpretation Skills	3.9	0.85	5
17	Conflict Resolution	3.87	0.68	5
18	Knowledge of Work Tasks	3.7	0.74	5
19	Teamwork	3.67	0.66	5
20	Discipline	3.6	0.72	5
21	Self-Confidence	3.57	0.9	5

22	Teambuilding skills	3.57	0.77	5
23	Social Awareness	3.53	0.93	5
24	Leadership	3.47	0.81	5
25	Interpersonal Understanding	3.47	0.63	5
26	Corrective Work Allocation	3.47	0.77	5
27	Conceptual Safety Skill	3.27	1.11	5
28	Relationship Management	3.17	0.98	5
29	Analytical and Conceptual Thinking	2.97	0.77	4

There is an urgent need to incorporate the aforementioned safety competencies into construction education, in order to bridge the gap between real industry practice and academia, and equip students with the competencies necessary to ensure safety on the jobsite. In addition, noteworthy comments from respondents are as follows:

- Respondent 5: With effective instructional tools, strategies, and managerial support, the proposed vision for competency-based safety education can contribute to enhancing overall safety culture in the construction industry in the future.
- Respondent 7: "Current training programs in industry make use of pictures and videos. The proposed visualization-based approach could be useful in industry too, if it can cover the broad range of identified competencies
- Respondent 23: "It is necessary to incorporate the identified safety competencies into construction programs, especially competencies related to hazard identification, response and control. These have the most direct impact on safety performance."

4. Discussion and Conclusion

Construction projects are extremely complex, requiring highly skilled and competent professionals for their safe and timely execution. Hence, graduates who take up front-line positions such as project managers, safety managers and supervisors play a pivotal role in improving safety performance. However, safety knowledge, skills and proficiencies are currently severely underrepresented in tertiary construction curricula. As a result, graduates enter the construction industry without the competencies necessary to ensure jobsite safety. It is critical that construction education facilitates learning experiences which develop students' safety skills and competencies to the levels necessary to meet modern project safety requirements.

In response to these issues, this study proposes competency-based vision for construction safety education, focusing on cultivating the safety competency of future construction personnel based on industry specific standards and requirements. Competency-based education is introduced as a reformative theory to transform construction safety education in order to equip students with safety competencies necessary to improve construction safety performance. The competencies required for safety education are investigated and verified through interview surveys with construction and safety management professionals. These would serve as a basis for establishing competency-based instructional and assessment strategies, and educational content development in future studies. Future research will explore the critical affordances for effective competency based safety education and the instructional considerations for developing a competency-based safety course. In addition the development of "proof-of-concept" systems and frameworks will be necessary in order to evaluate the novel competency-based didactic approaches. Unless there are radical changes in construction education to enhance the safety competency of personnel, the construction industry will continuously be plagued by persistent accidents, injuries and fatalities. This paper provides an overarching theory as a new perspective for construction safety competency and also a framework to guide holistic reform of construction safety education. The frameworks and theories developed in this paper will aid educators and course developers in establishing safety education in tertiary construction curricula, and also help to align construction safety education with industry practice.

References

[1] Le Q T, Pedro A & Park C S 2014 A Social Virtual Reality Based Construction Safety Education System for Experiential Learning *Journal of Intelligent & Robotic Systems* 79 (3-4) 487-506.

[2] Pedro A, Lee D Y, Hussain R & Park C S 2017 Linked Data System for Sharing Construction Safety Information In ISARC Proceedings of the *International Symposium on Automation and Robotics in Construction* Vol 34 Vilnius Gediminas Technical University Department of Construction Economics & Property.

[3] Sunindijo R Y & Zou P X 2013 Conceptualizing safety management in construction projects *Journal of Construction Engineering and Management* 139(9) 1144-1153.

[4] Wybo J L & Van Wassenhove W 2015 Preparing graduate students to be HSE professionals *Safety Science* 81, 25-34.

[5] Pham H C, Pedro A, Le Q T, Lee D Y & Park C S 2017 *Interactive safety education using building anatomy modelling* Universal Access in the Information Society 1-17.

[6] Antonio R S, Isabel O M, Gabriel P S J & Angel U C 2013 A proposal for improving safety in construction projects by strengthening coordinators' competencies in health and safety issues *Safety science* 54, 92-103.

[7] Pedro A, Le Q T & Park C S 2015 Framework for Integrating Safety into Construction Methods Education through Interactive Virtual Reality *Journal of Professional Issues in Engineering Education and Practice*142(2) 04015011.

[8] Rahimian F P, Arciszewski T & Goulding J S 2014 Successful education for AEC professionals: case study of applying immersive game-like virtual reality interfaces *Visualization in Engineering* 2(1) 4.

[9] Benhart B L & Shaurette M 2014 Establishing New Graduate Competencies: Purdue University's Construction Management Curriculum Restructuring *International Journal of Construction Education and Research* 10(1) 19-38.

[10] Walther J & Radcliffe D F 2007 The competence dilemma in engineering education: Moving beyond simple graduate attribute mapping *Australasian Journal of Engineering Education* 13(1) 41-51.

[11] Jaeger M & Adair D 2012 "Construction Safety Simulations and Students' Perception of Stress." *40th SEFI Conference-Engineering Education 2020: Meet the future.*

[12] Chang S H, Chen D F & Wu T C 2012 Developing a competency model for safety professionals: Correlations between competency and safety functions *Journal of safety research* 43(5) 339-350.

[13] Azevedo A, Apfelthaler G & Hurst D 2012 Competency development in business graduates: An industry-driven approach for examining the alignment of undergraduate business education with industry requirements *The International Journal of Management Education* 10(1) 12-28.

[14] Dubois D D 1998 The competency casebook: twelve studies in competency-based performance improvement *Human Resource Development.*

[15] Johnstone S M & Soares L 2014 Principles for developing competency-based education programs *Change: The Magazine of Higher Learning* 46 (2) 12-19.

[16] Guthrie H 2009 *Competence and Competency-based Training: What the Literature Says* National Centre for Vocational Education Research Ltd. PO Box 8288 Stational Arcade Adelaide SA 5000 Australia.

[17] Hayton J C & Kelley D J 2006 A competency-based framework for promoting corporate entrepreneurship *Human Resource Management* 45(3) 407-427.

[18] Biggs J 1996 Enhancing teaching through constructive alignment *Higher education* 32(3) 347-364.

[19] Occupational Safety and Health Administration 2015 *Training requirements in OSHA standards and training guidelines* https://www.osha.gov/Publications/osha2254.pdf

[20] Korean Occupational Safety and Health Agency 2011 *Risk Evaluation Model* http://english.kosha.or.kr/english/main.do

Sustainability of existing areas of historic cemeteries in the city organism: A Czech case study

K Folíková Palánová[*], O Juračka

Department of architecture, Faculty of civil engeneering, VŠB - Technical University of Ostrava, Ludvíka Podéště 1875/17, 70833 Ostrava – Poruba, Czech Republic.

Corresponding author, Email: klara.frolikova@vsb.cz

Abstract. Old public cemeteries are integrated in urbanism as its obvious part what is given by its social function. The witness of these spaces express the relationship and respect to values created by generations ahead of us. There were a lot of forms of graves and cemeteries in a history. A big part of cemeteries preserved until today that we still have in our cities, were founded in 18th and 19th Century. Those burial sites that could be expanded fulfill its function till today. But some of them were closed in an urban organism and remained without use. The others were lost with an aggressive construction and were replaced by the settlements, factories or new parks in some cases, without etic rules. The story of many cemeteries is forgotten. There is a question: What is the fate of old cemeteries, which stopped fulfill their original function? What the investors, the monument care, the heads of cities would to do with it? The aim of the article is to compare the mutual influence of the urban structure of chosen cemeteries in Czech Republic and abroad. The results are obtained by analyzing and comparing the findings of the individual revitalized cemeteries that are still part of urban organism. It is necessary to specify the appropriate transformation of nonfunctional necropolis to become a full part of the city with respect to the dead, to the traditions, to the sacred space.

1. Introduction

Until 1918, the Czech Lands were part of the Austro-Hungarian Monarchy. The empire's laws influenced the burial and funeral tradition in the Czech Republic to such extent that the consequences can be seen in the local settlements to this day. The Josephine reforms in the late 18th century shifted necropolis sites to a distance of 15 minutes walk beyond settlements. Gradually, however, most of such graveyards turned to be part of the urban organism once again through becoming absorbed by the fast development of towns brought by the industrial revolution, albeit treating them as strictly separate "gardens of silence" is the general tendency. Due to the centuries-old, considerable impact of the Catholic Church, cemeteries form strictly circumscribed, sacred areas, separated from the rest of the public space of the city. The limited opening hours, which allows visiting the site by anyone, but within a defined time, exempts cemeteries from the common definition of *public space*. Formerly, the site sanctity, along with intercessory prayers right at the grave, was making cemeteries places of frequent visits, which today, in the secular and atheistic Czechia (Špatenková, Kovář & Peřinková, 2014) ceases to be current and is limited to less-frequent visits of loved ones or relaxing in greenery; sometimes, funeral tourism is also involved. The low frequency of visits - observed in other countries as well, including those with different historical backgrounds [1] - as seen particularly at sites found on the periphery of major cities of the Czech Republic, leads to vandalism and other unethical practices.

Cemeteries are often outside the attention of public life of the society and on the edge of interest of city's officials.

For graveyards found in villages and rather small towns, the situation is quite different. They are still managed and cared for by the local residents. They are mostly located near the church or in the countryside beyond the village. The reasons include the higher religiosity of the population as well as concerns as 'what people would say' if one does not take care of their family grave; there is also a better level of surveillance and checks and, last but not least, a higher degree of cultivation of traditions. Specific situations occur in the territory of the former Sudetenland, where there are family graves of residents dislocated after World War II as part of the German minority; this brings out specific questions and issues in terms of attitudes to caring for such sites.

On the contrary, larger cities have seen higher rates of secularisation, the settings appear to be more anonymous and indistinct, the walking distance logically increases, and it is often necessary to use personal means of transportation or public transport in order to reach the destination.

The high number of cremations, which in 2014 was 79.86% [2] of all burials in the Czech Republic and even 98% [3] [4] in the capital city, Prague, is also an indisputable aspect. Such a sharp increase occurred over the past hundred years - 2018 will mark 100 years from the first-ever cremation in this country. Many of the cemetery complexes are still not ready for new forms of burying that are envisaged through the phenomenon described above. The reason perhaps lies in the interest of clients in specific necropolises or the low attention of city/village officials toward cemeteries and transformation of the sites; the rather long-term process of the capacity to adapt the necropolises to the new demand may also be the factor. The existing cemeteries were mostly being set up in the 18th/19th century; in some, the process largely occurred in the first half of the 20th century. After WW II, rather urn groves were being established near new crematoria and funeral halls; this was generally underway in the 1970s and 1980s. The period following the 1989 "Velvet Revolution" has seen a minimum of new cemeteries being founded; rather, the smaller cemeteries tend to be enlarged, where the situation permits and backup areas are available - this mainly concerns rather small graveyards found in small settlements where they prefer burying into graves.

The aim of this paper is to (1) specify the problems of historical graveyards in towns and cities in comparison with burial grounds found in smaller settlements; and (2) indicate the potential of transformation according to current funeral habits and needs making use of case studies.

2. The transformation of perceiving final resting places by the society

In villages and small towns, cemeteries enjoy the favour and attention of the population. Located around churches in centres or on the outskirts, or, in some cases, outside the municipality in the open country, the bulk of the sites involves graves and tombs designed for skeleton burials; for interment of ashes, this is done using either the primary skeleton grave or the ash is stored on the surface, by a marker/monument of the skeleton grave the rental of which goes from generation to generation within the family. The fee for a burial place is usually paid for 10 years as an upfront payment; afterwards, the hire needs to be renewed or the grave needs to be passed to other interested parties. In small towns, urn groves are developed, mostly in connection with the construction of a mourning hall (predominantly in the second half of the 20th century) in close proximity to the existing cemetery.

For reasons of residents of small communities to take care of their cemeteries, there are certainly a few. Above all, they include (1) easier control and visibility of the complex, due to its size; (2) the concern of the local people in sense of their responsibility toward the tradition of their community; and (3) the amount of believer population in the municipality. The latest mentioned appears to be the most significant aspect, particularly in regions where they record rather large numbers of believers, i.e., the Vysočina, South Moravia and Zlín regions where believers from 20% to 30% of the population, closely followed by Olomouc and Moravia-Silesia regions with 15% to 19% believers. [5]. Here, Christian rituals support the practice of attending and looking after the cemetery. In rather large cities, in Central Bohemian Region and in the territory of the former Sudetenland, there we see lesser levels of religious populations. Besides other factors, this is related with the development of the particular region and the way of livelihood. In industrial areas of Bohemia, the departure from the traditional way of co-existence with laws of nature started the process of secularisation which in the agricultural areas - - those of South

Moravia and, economically underdeveloped, Vysočina Region - found its application at a later stage and to a much lesser extent. Secularisation was also boosted by the political regimes that were represented in Czechoslovakia in the course of the 20th century. First of all, the Communist regime had certainly its impact in providing for the "separation of church and state … which ... in democratic capitalism could lead to free and extremely intense development of the church, while communism sought to make it a means of achieving an irreversible 'death' of religion", p. 16 [6]. The consequences are something that the current society is facing in that it understands the relationship to the deceased and, particularly, to their place of final rest, with difficulties - particularly in the areas mentioned above, with the high degree of secularisation - or, sometimes, even controversially. Where socialism and communism offered "…well-being and comfort for everybody", which was "considered unlimited happiness for all. The 'trinity' of unleashed production, absolute freedom and unlimited happiness created a core for a new religion...", fulfilling its supporters with "energy, stamina, and hope." p. 14 [7]. In such a society, there is no room for contemplations about death and attending graves. Only agricultural municipalities, far from the centrally managed state, were making it possible to retain one's faith and traditions, albeit to a limited extent.

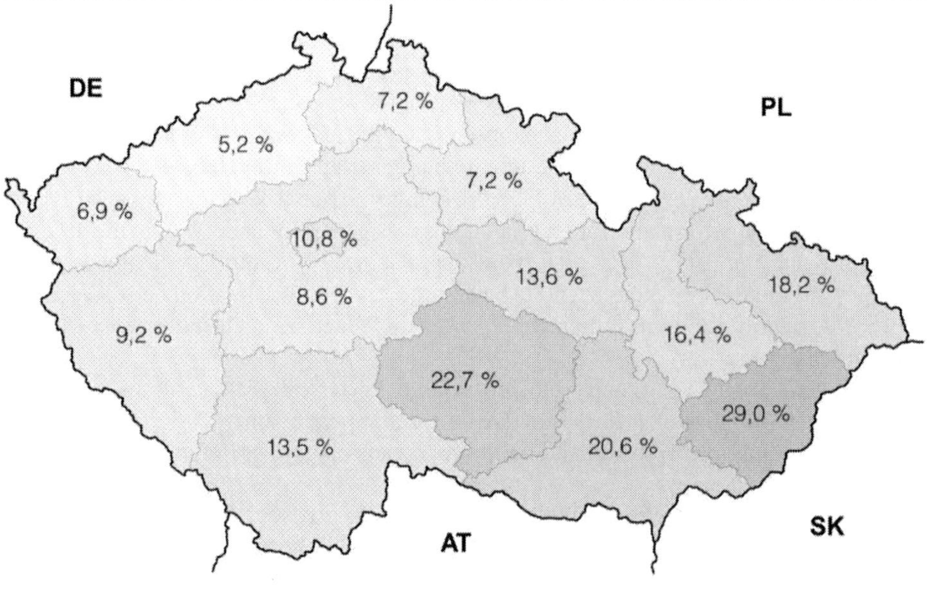

Figure 1. The proportion of believers declaring themselves to be part of church / religious society in the Czech Republic (Source: Czech Statistical Office).

If viewing the tabooisation of death, which is being pursued in the current society, from the perspective of Fromm, then the thought can be explained in two ways, depending on whether the modus of *TO BE* or *TO HAVE* is involved. From the perspective of the latter, even a modern human seeks to be immortal, wishing to leave something that commemorates them after their death. This is explained by Fromm as "biologically determined desire to live"; p. 101 [7]. In the current consumer society, however, it is just no longer a tombstone with a name; instead, the idea of immortality has modified into the form of glory or assets that can be bequeathed, i.e., something that is left after the individual who can be subsequently seen as immortal. On the other hand, the *TO BE* perspective involves a status when free ourselves from the crutch of property and exist independently of it. Without property, it would be possible to understand that very substance of life - thus death as well - in its naturalness; however, the idea of losing all the security which property provides is the same as "if someone is thrown into the sea and cannot swim"; p. 108 [7].

This approach is particularly manifest in the historical cemeteries set up in the 18th and 19th centuries where the cult of the death is shown in the achievements of high-.quality sepulchral arts and construction of burial vaults. The legacy of the 19th century is now something that is very hard to handle for the strict secular society of the 20th century; it has brought it forward into the current century when new solutions are necessary to find.

3. Historic cemeteries in the city organism

In the big cities of what is now the Czech Republic, particularly in Ostrava (289,158 residents on 1 October 2017) - chiefly in the larger districts of the city with rather high proportion of socially excluded sites, and in the peripheral areas of Prague (1,282,496 residents as per 31 March 2017) one can observe a low level of attention toward cemeteries, associated with low visitor numbers and reduced level of management or even vandalism in relation to „changing intervent practices such as 'no service' cremations, …, and informal burials" [8]. Factors for this status certainly include greater anonymity of the settings, increasing walking distance, larger rates of cremations and high levels of secularisation of the population, i.e. the low proportion of believers, as well as little interest on the part of city officials "despite the fact that it has a direct impact on land use planning and social integration." [8]. Cemeteries have thus become closed "gardens of silence", forming impermeable complexes in the middle of the city organism or isolated districts on the periphery. The above-mentioned secularisation - along with the high degree of cremation - significantly modifies funeral rituals. In the Czech Republic, as well as in the world, "there is a movement away from traditional burials toward cremations as the dominant interment choice", with the reasons being as follows: „lower costs to the consumer", as well as "the emergence of cremation factories; and the popularity of informal burials in location other than traditional cemeteries" - something that still has not been introduced in the CR [8]; an option of interment of ashes outside the sacred place of burial grounds, however, enjoys some degree of favour in this country. The level of so-called *social funerals* is also high - cases where the state secures the funeral instead of the bereaved and where burial is always performed by cremation and interment of ashes is done using a collective site of placement or a scattering lawn, where the ash is dispersed in an anonymous way. There is no need for the bereaved to visit or manage the burial site.

4. Sustainability of historical cemeteries in organisms of modern cities

The current status encourages reflection on the further direction of cemeteries should new features be added. Reasons such as reducing quantity of urban green areas, changed views of death and burials, and cultural factors supported by the upcoming multicultural population can potentially lead to changes in the use of public urban green areas, as well as cemeteries. When planning changes to existing historical cemeteries, one must first take a cemetery as a place of final rest and, further, consider other reasons for visiting the gardens of green in the organisms of cities. [1]. Grounds as well as frequency of visits differ when comparing weekdays against weekends or the cemetery locations. In addition to visits to graves, this may include walking, socialisation (meeting places), relaxation - plain sitting on a bench (Olšany Cemetery - the south-western portion is used for lunch break resting by employees of nearby offices), visiting cultural heritage (sepulchral arts, graves of significant personalities), etc. [1]. In In the historical setting of a strong influence of the Roman Catholic Church it is difficult to consider other reasons for visiting the sacred area of strictly defined cemeteries in Central Europe. In the Nordic countries of Europe - given the influence of the Lutheran church - cemeteries are more open-access, fenced using only a low wall or a hedge row, with gravestones of less formal make, set in a meadow-like arrangement. Paclová reports that in the west, north and south, one encounters greater austerity of the cemetery area than what we know in this country. It is mostly covered in a uniform surface of green, from which only markers or crosses protrude. P. 128-129 [9]. The list below shows more potential activities that in Central Europe are prohibited. They include cycling, walking dogs, skateboarding, etc., as well as activities that in the settings of cemeteries in the Czech Republic are unusual - such as jogging and picnicking [1]. Recently, a tendency has been evident in the management of cemeteries, funeral halls and crematoria to focus on the live who decide, as clients, whether or not to use the facilities and grounds [10].

From the perspective of city organisms, historical graveyards are strictly enclosed, fence- or wall-defined, grounds of isolated *gardens of silence*. They never or just minimally respond to adjacent public space. With their often limited, monitored access and controlled visitor hours, they become complexes with low or zero (evening hours) throughput. Given the low numbers of visitors, there is just a weak interaction with the nearby surroundings. The cemeteries alone do not exhibit sufficient amenities for visitors or participants in the funeral rites. Insufficient space for the live in Prague cemeteries is something recorded as early as in the writing by Jan Almer of 1928: For all of our cemeteries, … the lack of an adequate free space for coaches and audience to gather in rather large quantities is the significant flaw and conspicuous negative. Similarly, there is no catering for any form of shelter for the audience to hide from bad weather if any occurs suddenly, especially rain. pp. 5,6 [11]. The lack of amenities (as regards shelters for bad weather, as well as refreshment facilities) shows to be problematic particularly in cemeteries on the outskirts of major cities where one needs more time to arrive to the desired place and, given the vastness of certain grounds, even to walk across the cemetery. Cemeteries or their boundaries lack any defined space for the funeral ritual to fade out - e.g. condolences and mourning reception. Expressing condolences thus often takes place outdoors in unsheltered areas or even in the mourning hall. Clients also increasingly do not use mourning halls or churches as places of mourning rites, planning them outside the necropolis, or cancelling them completely. This not very good trend in terms of respect for the deceased and possible celebration of their life - as well as in terms of the possibility to believe the situation and go through the phases of grieving - is supported by the lack of preparedness of cemeteries [12].

5. Student proposals for amending facilities of burial sites and integrate the places into city organisms

Following the existing research of the Department of Architecture, VSB - Technical University Ostrava, students of Architecture and Civil Engineering devoted time, under the leadership of Klára Frolíková, Martina Peřinková and Petr Hrůša, to a study and their own proposals how to amend facilities of burial grounds as part of efforts to integrate them into the city organism.
Olšany Cemetery in Prague were selected for the case study along with cemeteries in Ostrava-Vítkovice, Ostrava-Mariánské Hory and Hulváky.

Olšany Cemetery in Prague was founded in the second half of the 17th century as a black-death graveyard to be subsequently declared a public cemetery after Josephine reforms of the late 18th century. In compliance with applicable laws [13], the grounds were then located beyond the town to become part of the city organism of Prague 3 - Žižkov later on as the city of Prague developed. At present, it is the largest Prague cemetery with a total area of 50 hectares; there are about 25,000 tombs; 200 burial vaults; 65,000 graves; 20,000 urn graves, 6 columbarium walls and two scattering fields [14].
The complex is firmly squeezed into the current city organism, in close proximity to a new shopping centre, and is lined by busy city roads from its three sides. Yet it is searched for its high-quality sepulchral art and visits to graves of major personalities and forms, with the mature trees which make the cemetery a rich urban green area, space favourable for relaxation. Currently, the grounds lack amenities for visitors, which involves, in addition to the bereaved, young people with pushchairs or working people who spend their lunch break in the area. The complex is split into two parts by a busy road. The bigger section is accessible via five entrances; the lesser part is a specific area with the military cemeteries, burial grounds for Eastern Orthodox church members, and scattering lawns; a strictly enclosed Jewish graveyard is also set in. This cemetery too just slowly responds to the new trend of burying, that is, cremation, which represents almost absolute majority of all burials in Prague [12].
Students defined key issues, responding to them by specifying functions to be applied to the current situation of the complex. It involved the creation of possibilities for the interment of ash, i.e. columbaria and new scattering lawns; lapidaria for storing unused artistically valuable gravestones; area for meditation and last farewell at sites of interment/scattering of ashes; scattering surfaces in front of the mourning hall, community meeting areas protected from bad weather; dining areas, sanitary facilities, flower souvenir shops.

The proposed new columbaria for storing urns with ash were located in the central part of the cemetery around the existing cross along the axis of the main entrance, i.e. at a very significant location in terms of composition; columbarium towers were also designed and situated in the oldest part of the cemetery in an area with many “defunct graves”, where they complemented the existing irregular structure of the remaining graves. [15] [16]. In addition, students paid attention to some of the entrances to ensure there is an appropriate switch-over from the daily rite of the public space to the silence of the cemetery garden of green. Entrances that involved only a small gate in a stone wall were replaced with an access platform to allow for the distance to be overcome when entering the cemetery and which, with its character, tunes up the visitor to the forthcoming experience. The access platform was completed with / defined by visitor facilities to be used as a visitor centre, gift-shop, café or lapidarium [16] [17] [18]. Areas to serve the updated forms of burial and meditation as well as sites for the last farewell when placing ash were mainly situated at the site of the oldest part of the cemetery, where there is a minimum of attended graves, which gradually transforms the place into a park. Subsequently, students proposed to remove any unused gravestones from the site and place some of them in the newly created lapidarium, which consisted of a covered arcade defining the area being designed; now the area so vacated was to be used for placing new scattering lawns. Using it as a sheltered area for ash interment rites as part of a meditation site was another option. [15][19][19]. The sheltered meditation area / arcade of the lapidarium responds to the need for a meeting place of the bereaved before/after the ceremony to provide a place for the live following the recommendation shared as early as 1928 in the publication by J Almer [11]. Following the crematorium in Heimolen, Belgium [21], as an example, one of the variants involved a restaurant for funeral guests designed by the cemetery wall near one of the entrances making sure that this item of the funeral rite agenda is not displaced from the place of burials.

The cemetery based in the city district of Ostrava-Vítkovice was founded in 1913 and consecrated a year later. The initial ambitious plans to build proper amenities and a surface area as needed had to give way to the events of World War I [22]. Today, therefore, there is only one wall out of those planned initially, along with the main entrance and already a sub-standard mourning hall. The cemetery alone is locked by busy roads as well as production and school complexes. Displaced from major pedestrian routes, the grounds no longer provide the necessary peace for the rites or for rest with the vicinity of the busy arteries. Burial places are gradually abandoned and the whole area decays. Students, therefore, seek to set up a new access platform for visitors in a way that it is not crossed by supplies to the mourning hall [23]. A new hall is being proposed at the site of the original building in the vicinity of the main entrance; the solution is not very good, but there was no other alternative with the absence of another free space. As a result, the hall is separated from the road by cemetery facilities behind the scenes and a meditational garden; everything is defined and shielded by walls. The mourning hall alone opens toward the cemetery within which there is a shelter for receiving condolences [24]. The draft by student Ondrej Vavro designs potential options for interment of ash which is currently limited within the cemetery. Reflecting the mourning hall mass along the axis of the main entrance, it offers three sheltered areas of columbaria. It also follows up the previous studies at the department of architecture (FAST, VSB TUO) [25] [26]; as a result, it proposes a basement area beneath the ceremonial hall, dedicated to the "path of the deceased" to the place of last farewell on the ground floor. This provides architectonisation not only in the area for the bereaved, but also in that for the deceased so that the conclusion of celebration of their life is not underway in a purely technological background area, but to ensure the ethics of treatment of the deceased including the possibility to be accompanied by the family and commemorated as part of last-goodbye ceremonies. This path is designed as a spacious hall, where in the ceiling there are masses of the ceremonial hall, the site of display and a small ceremonial hall for alternative ways of funerals. From here, the coffin is taken directly to the place of the funeral rite.

The Cemetery on the hill of Hulvácký kopec, Ostrava, was established 115 years ago and has seen no burial into the ground for almost 70 years. Almost half of the remaining graves is now "abandoned" with no one to take care of them. The paths and roads became overgrown and part of the cemetery was abolished earlier and used to construct a busy road and to enable development of Vítkovice Ironworks. Only a triangle was left from what was formerly a strictly composed rectangular shape with a central axis on which a cross was standing. The cemetery is fenced only by wire mesh; there is no structure; even no chapel and the original mortuary is also missing. The proximity of a socially excluded group of residents and the exclusion of the complex from the everyday life of the population, as well as the cut off of the part by the factory, which "turns its back" on the grounds, all of this triggered a wave of vandalism and indifference. In their designs, students define the cemetery using a solid wall into which names of the buried are inscribed so that they would never disappear from the thoughts of us, living. The students propose that the grounds are gradually transformed into a city park; the process should follow the ends of rental periods for the individual graves. There are already more such places in

Ostrava; however, there is a lack of any reference to the original purpose of the grounds, which, in this case, students replace by designs of meditation chapels or leaving the cross on the axis of the main entrance. Even this place calls for placing a lapidarium, e.g. in a form of a sheltered arcade placed by the newly built wall; the great artistic value of some monuments is the reason when nobody is interested in them. In this case, however, the complex cannot be assumed to welcome many visitors in the future; rather, it is to remain - with its location in an industrial area - a silent reminder of its time and a bounded area of urban greenery amidst a hostile setting of an industrial site.

6. Conclusion

The Department of Architecture of the Faculty of Civil Engineering (VSB - Technical University of Ostrava) has been engaged in studying cemeteries in the Czech Republic for 7 years. It seeks to apply the existing knowledge in the teaching activities so that future designers or city officials become interested in this issue which now stands on the brink of attention of the society, and have an idea of how to handle the legacy of the past centuries.

Primarily, historical cemeteries in large cities are the aim of the research; found at the edge of attention of the society as well as town officials, they remain isolated *gardens of silence* in the organisms of the cities.

With the centuries-old strong influence of the Catholic Church in Central Europe, necropolises are to continue to be enclosed complexes; however, new features were specified for them to be capable to become full-fledged parts of the cities, linked to the existing public areas, and, last but not least, enter awareness of ordinary people and political decision-makers.

This particularly includes ensuring the passability of the sites, adding services for visitors, improving their aesthetic aspects, treating their artistically valuable sepulchral heritage, providing a decent entrance area which connects the site with the remainder of the city organism, and enabling new forms of funerals following the current needs of the society.

Increased visitor numbers is something that cannot be provided at every location. If the grounds had been absorbed by an industrial site and eliminated from what is a residential zone, they will remain an enclosed site forever; nonetheless, they should be treated with dignity and the symbols which indicate the initial function of the place should not be omitted.

To conclude, focusing not only on the dead, but also on the living ones makes it possible to finally take historical graveyards into the current century and leave them to future generations, because cemeteries existed before us and they will continue even after us.

References:

[1] Evensen K H, Nordh H and Skaar M 2017 Everyday use of urban cemeteries: A Norwegian case study *Landscape and Urban Planning* 159 76-84.
[2] *International Cremation Statistics 2014*. In http://www.effs.eu/cms/fileadmin/members_only/documents/news_articles/Cremation_statistist_2014.pdf, published 15.12.2014, download 15. 11. 2017.
[3] Toman J 1983 Pokrokové ukládání zpopelněných ostatků [Advanced storage ashes] *Rozptyl a vsyp* Prague Společnost přátel žehu.
[4] Taufer F 1923 Krematoria [Crematoriums] *Věstník Spolku pro spalování mrtvých* Prague Vyšehrad.
[5] Podíl věřících hlásících se k církvi [The share of believers] 2014. In https://www.czso.cz/
[6] Fiala P and Hanuš J 2001 Katolická církev a totalitarismus v Českých zemích [Catholic Church and Totalitarianism in the Czech Lands] Prague Centrum pro stadium demokracie a kultury 216.
[7] Fromm E 2001 Mít nebo být? [To have or to be?] Prague Aurora 242.
[8] Davies P J and Benett G 2016 Planning, provision and perpetuity of deathscapes – Past and future trends and the impact for city planners *Land Use Policy* 55 98-107.
[9] Stejskal D, Šejvl D et al 2011 *Pohřbívání a hřbitovy* [Burials and cemeteries] Prague Wolters Kluwer ČR 464.

[10] Špatenková N et al. 2014 *O posledních věcech* [About the last things] Prague Galén 19-20
[11] Almer J 1928 *O potřebě nové úpravy pražských hřbitovů* [The need for a new adaptation of Prague cemeteries] Prague Důchody obce hlavního města Prahy 5-6.
[12] Nešporová O 2011 Last Rites: Secular, Religious or Rather No Funeral Ceremonies? *Sociální studia Prague Katedra sociologie FSS MU* 2 57-85.
[13] Winter E 1945 *Josefinismus a jeho dějiny: Příspěvek k duchovním dějinám Čech a Moravy 1740-1848* [Josefinism and its History: Contribution to the Spiritual History of Bohemia and Moravia 1740-1848] Praha Jelínek 383.
[14] Kovařík P 2001 *Klíč k pražským hřbitovům* [The key to Prague cemeteries] Prague Nakladatelství Lidové noviny 369.
[15] Nahodilová Š 2017 *Přestavba severní části Olšanských hřbitovů v Praze* [Reconstruction of the northern part of Olšanské cemeteries in Prague] Diploma thesis Ostrava VŠB-TUO.
[16] Dubovský L 2016 *Jihovýchodní brána Olšanských hřbitovů a kolumbárium* [Southeastern gate of Olšanské cemeteries and colubarium] Ostrava VŠB-TUO.
[17] Venusová D 2016 *Jihovýchodní brána Olšanských hřbitovů a kolumbárium* [Southeastern gate of Olšanské cemeteries and colubarium] Ostrava VŠB-TUO.
[18] Vavro O 2017 *Smutočná sieň Vítkovice* [Funeral hall Vítkovice] Ostrava VŠB-TUO.
[19] Bilová K 2017 *Smuteční síň, kolumbárium a meditační kaple – Olšanské hřbitovy v Praze* [Funeral hall, columbarium and meditation chapel] Ostrava VŠB-TUO.
[20] Eliášová J 2016 *Lapidárium a rozptylové loučky na Olšanských hřbitovech* [Lapidarium and scattering meadows at Olšanské cemetery] Ostrava VŠB-TUO.
[21] Claus en Kaan architecten 2010 Crematorium Heimolen. *A + U-Architecture and Urbanism* 118-123
[22] Kašing P 2009 *Ostravské hřbitovy* [Cemeteries in Ostrava] Ostrava: Repronis 59
[23] Celsing J 2016 The New Crematorium - The Woodland Cemetery *A + U-Architecture and Urbanism* 547 90-99.
[24] Bilová K 2017 *Smuteční síň Ostrava Vítkovice* [Funeral Hall Ostrava Vítkovice] Ostrava VŠB-TUO.
[25] Frolíková Palánová K et al 2017 Application of the current knowledge from research and development of the burial methods and their impact on designing or transforming contemporary cemeteries in the Czech Republic *Sborník vědeckých prací* 2 Ostrava: VŠB-TUO.
[26] Juračka O 2016 *Crematorium with innovated view at funeral culture, Nový Jičín.* Diploma thesis Ostrava: VŠB-TUO.

Acknowledgments:
This work was supported by means of the VŠB-TUO Student Grant Competition. Project registration number is SP2018/56. Thanks to Ing. arch. Jan Kovář for support and help.

IOP Conf. Series: Earth and Environmental Science **143** (2018) 012053 doi:10.1088/1755-1315/143/1/012053

The integration of traditional and advanced design in the formation of sustainable New Rural Housing solutions.

Van Khai Tran[1]

[1]Van Lang University, 54 Nguyen khac Nhu, District 1, Ho Chi Minh City, Viet Nam
Email: khaitv@gmail.com

Abstract: Socio-economic growth in Vietnam greatly depends on the new rural development process in this country; the work is considered a strategic position in national development. Thus, the study of the principles of the architectural design of new rural housing solutions in accordance with the New Rural Environment as required by the Vietnam Ministry of Construction has become urgent. Climate change has become a global concern, so the creation of significant impacts by architectural designs to respond to climate change and make the living environment of rural people better is a major demand. Experience has shown that the dogmatic application of current urban-type housing does not reach the requirements of New Rural Housing. This research intends to show that the solutions of the traditional Vietnamese rural house, which retains the advantages of traditional architecture with excellent characteristics that have been challenged over thousands of years, when combined with advanced design methodologies and technologies, will be the appropriate solution for 'New Rural Housing'.

Key words: Evolutionary, Adaptive, Responsive Architecture.

1. Introduction.

Recently, the Ministry of Construction promulgated the criteria for the design of the New Rural House as a more progressive model than the traditional one. But the dogmatic application of current urban-type housing does not reach the requirements of rural housing of the new era, as it cannot solve the dilemma of combining the traditional lifestyle with the natural surroundings while overcoming the impacts of climate change.

Literature reviews on Vietnamese traditional rural houses show that most of these houses have a good relationship with the surrounding landscape and that the composition of spaces inside and outside is arranged to be flexible, and therefore in harmony with the natural and social context. After thousands of years of evolution, traditional Vietnamese rural houses have learned how to respond to adverse climate conditions and gained valuable experience in climate adaptation. Furthermore, the copying of the urban house type, particularly tube-shaped houses, is causing the traditional rural aesthetic image to disappear.

Reviews on recent design methodologies conclude that recent design methodologies have the capability to respond to a number of requirements of contemporary housing. Firstly, the theories of Evolutionary Housing propose that the house is allowed to develop gradually in quality and quantity, and 'time' is regarded as a resource for housing development. Adaptive Architecture solutions show the ability of house structures and architectural components to transition from their initial function, in accordance with the need for new construction in the new developing environment. The application of 'Transformable Structures' can adapt their house shape according to changing circumstances, to solve new function demands or even move to new locations.

Vietnamese traditional rural houses have been formed to respond to adverse natural conditions and to collect valued experiences in functional, effective climate adaptation. In recent times of economic and social development, with the challenges of global climate change, the construction of rural houses on these valuable experiences, integrated with advanced design methodologies and technologies, is what it takes to look to New Green Rural housing models in the new era.

2. Definition of the concept of "New Rural Housing"

Based on the general policy of national development, the Vietnam Ministry of Construction adjusted some criteria of rural housing stipulated in Circular 41/2013/TT-BNNPTNT of the Ministry of Agriculture and Rural Development and supplemented additional instructions of implementation under this Circular. According to the demographic statistics of the state, in 1999 the total population of the whole country was 76,289,573 million people. Of these, 58,361,523 people lived in the countryside. In 2009, the total population was 85,846,977 people, but in spite of the rapid urbanisation, the rural population also rose to 67,200,000 people with smaller household size. This led to a great demand for rural housing.

The Ministry of Construction's stated main requirements on the solution of "New Rural Housing" was as follows: Ensure the rigidity of 3 parts: Fundament, Frame, Roof. The above-mentioned parts must be made of good quality material and not made of perishable, flammable materials. The tenure of use of residential buildings is from 20 years or more. The housing design must conform to local customs, habits and traditional lifestyle; ensure hygiene requirements; be convenient for daily life; inherit the traditional architecture style; connect in harmony with the natural surroundings; consume less energy; protect against the impact of sunshine, rain, wind and storms which are more likely following the impact of climate change; and resolve the close relationship between the production space, living space in each household.

3. Literature review

3.1. Review of the capability of Vietnamese traditional houses

Hoang Manh Nguyen [1] states that houses in the Northern rural area have evolved and learned how to respond to adverse climate conditions and put together valuable lessons in climate adaptation. Nguyen H M stated: "Surmount the challenges of recent global climate change, building on these valuable experiences and integrating with advanced technologies is what it takes to look to modern green rural housing models." He also claimed that traditional Vietnamese rural housing has a good relationship with the surrounding landscape, and that the coordination between the inside and the outside is arranged to create a close and flexible environment.

Nguyen Dinh Thi [2] stated that each housing model must be less than 150-200m2 with 2 or 3 stories surrounded by their own private traditional garden, conserving the traditional aesthetic appearance. Na L T H and Park J H [3] emphasize that, generally, a locality's traditional buildings offer the best examples of appropriate passive-mode design or bioclimatic design. Le Thi Hong Na and Jin-Ho Park cite that learning from Vietnam's traditional examples, it is important to adopt a design strategy that begins with a design of the built form by optimizing all the passive-mode strategies appropriate to the climate and ecology of that locality when the basic design of the traditional house and its construction methods give it great flexibility so that extensions to the house can be carried out whenever necessary.

3.2. Review of the concept of Evolutionary Housing
The concept of Evolutionary Housing was introduced in 1985 by Martinez E and Di Lullo R [4] through lectures on Low-cost Housing at the Institute for Housing Studies (IHS) in Rotterdam, The Netherlands as a basis for housing solutions to low-income people. Martinez E explained that when lacking initial investment resources and when initially unable to meet required standards, Evolutionary Housing allows the house to gradually develop in quality and quantity including in area, space and materials. In time, this housing will meet required standards. Both authors conclude that 'Time' is regarded as a resource in the projects of low-cost housing development for low-income people.

Arora S (USA) and Saxena S (India) [5] develop the concept of Evolutionary Housing towards applying Bio-climatic principles in order to create architecture and buildings in harmony with nature. Structures and components of the houses can change position, interact flexibly, alter or upgrade.

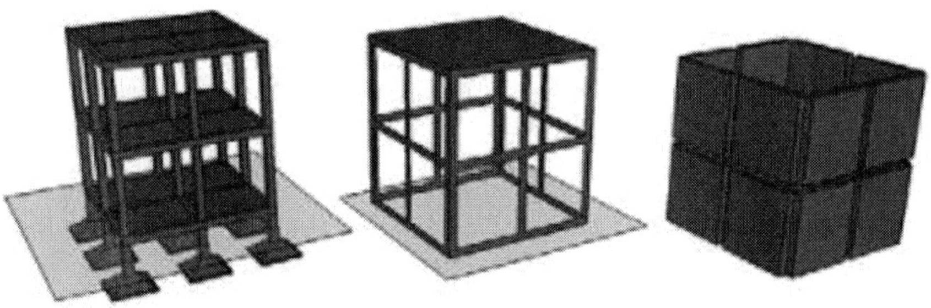

Figure 1. (a) Fixed stucture, (b) Exposed surface of fixed structures, and (c) Interactive structure.

3.3. Review of the concept of Adaptive Architecture
Schnädelbach H [6] cited that Adaptive Architecture thus has the capability to respond to a number of parameters with time. Time is an integral factor driving adaptation in architecture, thus Adaptive Architecture can be said to be Responsive Architecture evolving with time. Schnädelbach H further clarifies that Adaptive Architecture is concerned with buildings that are specifically designed to adapt (to their environment, to their inhabitants, to objects within them) whether this is automatically or through human intervention [6].

Verma S, Pradeep D [7] state that the environmental changes that occur in a given time, such as a day, can be a constant force of changes that need to occur in an architectural object. This leads to local adaptations to global climactic change occurring over the course of time and creates forces for architectural objects to change over the years in order to survive and sustain themselves. Furthermore, Sushant Verma and Pradeep Devadass explain that Adaptation in architecture is a long-term process that occurs with time and generations, where improvements in technology, economic support and human thought process contribute to the adaptive response.

Kim J J and Rigdon B [8] cite that: For the purpose of conceptual clarity, the life cycle of a building can be categorized into three phases: pre-building, building, and post-building, as shown in Figure 2 which illustrates how these phases are connected, and that the boundaries between them are not obvious. Analyzing the three phases of the building processes in this figure provides a better understanding of how after design, construction, and operation then disposal of a building could be inhibited to cause fewer negative effects to the ecosystem.

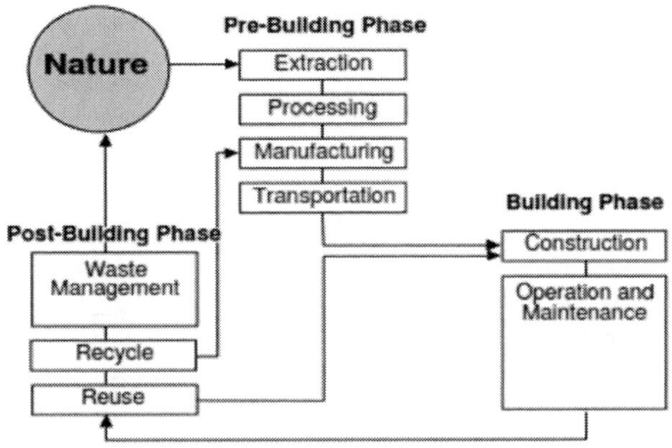

Figure 2. The sustainable building life cycle.

De Temmerman N, Mira L A, Vergauwen A, H. Hendrickx and De Wilde P [9], interpret the application of Transformable structures which can adapt their shape or function according to changing circumstances and to meet rapidly evolving needs, induced by a society which – progressively – embraces the concept of sustainable design and understands that structures are not designed in an end state, but in a transition state. The above authors further explained that the transformation of such structures, intended to meet changing requirements, is done by using mechanisms (deployable/foldable) or reconfigurable components (demountable kit-of-parts).

The above review shows that Adaptive and Transformable design can be summarised as a methodology which allows the building product to be formed in a main function but which can be adjusted to undertake other functions by a transformation mainly on the load-bearing structural system, external envelop and internal space. The Transformable factor creates sustainable development when possible to preserve the building and change it for the purpose of selling, exchanging and recycling the used or new components, instead of demolition, causing garbage and pollution.

4. Research Methodology

Integrative studies on the characteristics of the elements of traditional architecture carried out in this research were mainly based on the criteria of "New rural housing" launched by the Vietnam Ministry of Construction which is already the outcome of both qualitative and quantitative research done by state institutions. Quantitative research was further based on the data of demographic projection to provide a view of the large demand for rural housing and the appropriate scale of house models for rural households. Qualitative research was carried out by reviews of the contemporary trend of housing characteristics - sustainable housing and the concepts of Responsive Architecture, Adaptive Housing, and Evolutionary Housing which will provide a main part of the theory of housing design methodology. Qualitative research was also carried out on the traditional house architecture of Vietnam, on buildings that had met the challenges of the local social and natural conditions and survived through thousands of years of history. Therefore, their characteristics could be recognized amongst the most advantageous and suitable solution for the New Rural Houses. Issues of climatic and contextual restraints examined within this method will allow for the theoretical premise and unifying idea to direct the research. The studies will show that the column and truss structure of Northern traditional houses with excellent characteristics holds a rich history of thousands of years in responding to climate change and improving the living environment with better

IOP Conf. Series: Earth and Environmental Science **143** (2018) 012053 doi:10.1088/1755-1315/143/1/012053

qualities, including aesthetic appearance. Meanwhile, the structure of the Central and Southern traditional houses, with larger fixed elements, survive in a more favourable climatic conditions.

5. Materials on the current status of Vietnamese traditional rural houses

5.1. Categories of Vietnamese traditional rural houses

The architecture of Vietnamese traditional rural houses depends upon various local constraints including economic, social, religious, aesthetic, climatic, and regulatory. They have taken various spatial and structural forms at different areas over time, each of which leads to decision making at a particular stage in the design process. In an overall view, Vietnamese traditional houses consist of three main categories: Houses built on land, houses built on stilts and a minority of floating houses on the water surface.

5.2. Structural systems of Vietnamese traditional rural houses

Tran Thi Que Ha in the research "The origin and development process of Vietnam traditional houses" at Singapore National University, describes the forms of load-bearing structure systems of Northern, Central and Southern houses built on the land is illustrated in the charts of figure 3 and figure 4.

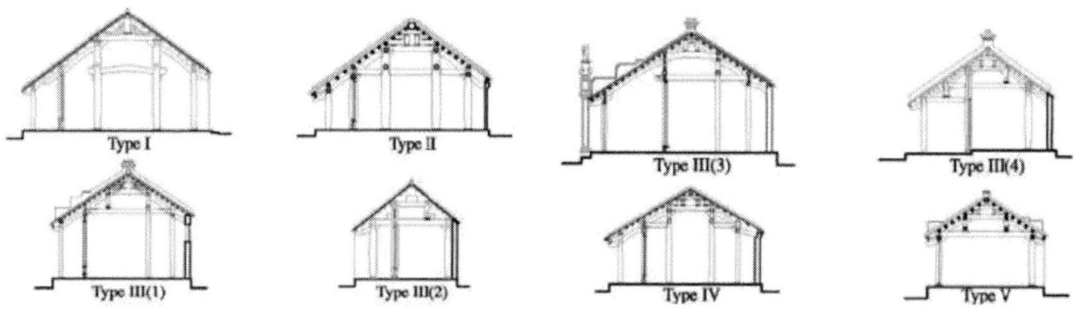

Figure 3. Column, truss system of Northern house

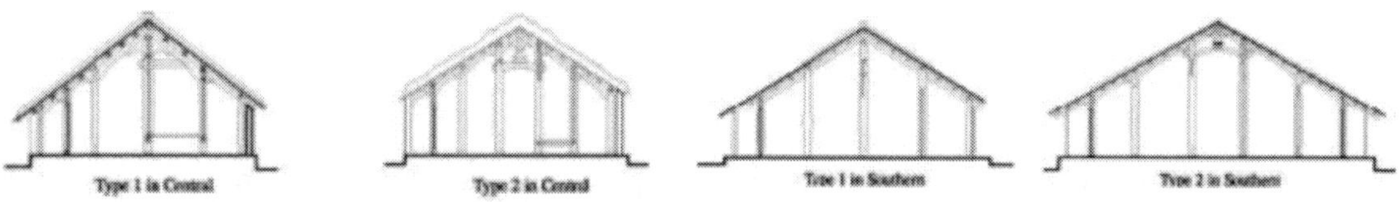

Figure 4. Column, truss system of Central Vietnam and Southern Vietnam

Materials, including sketches and illustrations of traditional architecture, are vital pieces of information to help understand practices of the vernacular and traditional. Written research results will also assist in analyzing the climatic and resource limitations, which may have influenced the development of the traditional and environmental architecture that may have affected the rural residential house solutions.

The above materials and reviews show that traditional Vietnamese rural houses have many characteristics suitable to the application of advanced design solutions, the maintenance of the traditional characteristics are not derived from the old-fashioned sense. Vernacular housing space retains the beauty of traditional architecture since these excellent characteristics have been challenged over thousands of years, and should not be constructed temporarily or poorly.

Figure 5. A traditional Vietnam house structural system

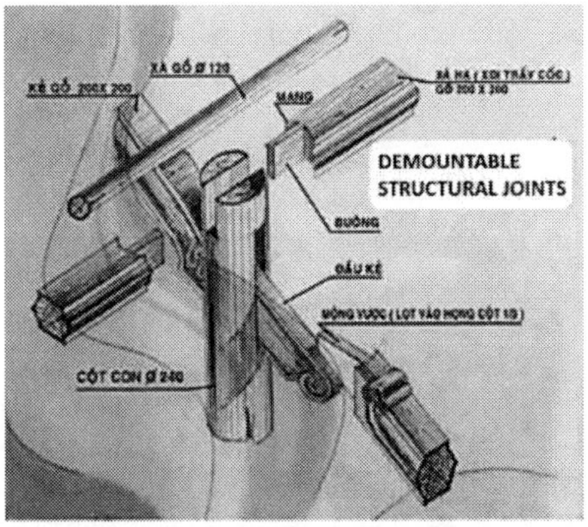

Figure 6. Demountable structural joints

Figure 7. House on stilts of Muong ethnic group

Figure 8. Northern traditional style house built on land.

5.3. The main difficulties and failures in the recent development solution of rural houses. The dogmatic application of current urban-type housing does not reach the effect of New Rural Housing. With the northern village-rural areas: people often think of the village gate, banyan trees, temple yard, bamboo ramparts, tree gardens and fish ponds. Now, those scenes are replaced mostly by urban tube house types, therefore requirements arise to resolve the acute problems in urban areas and to protect the natural environment. The recent construction trend of copying the urban housing tube model nationwide have many unsuitable characteristics for the rural environment and landscape, with a width of 4 - 5m, length of 20m, 1-3 floors, mostly in "district towns", poor natural lighting and ventilation. For failure of adequate attention being paid to the unfavorable environmental factors such as hot weather and floods, the trend of building village housing in an urban style does not save energy and contributes to a weak quality of life and architectural aesthetics, affects the rural landscape, and creates an urgent demand for tap water. The traditional source of materials for building traditional houses such as timbers, bamboo-knitted walls or burnt bricks to build walls are no longer abundant, which erodes the folk character. In contrast, the big problem herein is a demand for solutions to ensure a safe environment, especially a clean water source, address the demand for tap water and flood prevention from the climate change phenomena.

In comparison with rural houses which copied the recent urban house types, the traditional rural house of Vietnam is a remarkable option of adaptation to the vernacular climate conditions.

6. Results: Solutions for the New Rural House

6.1. Solution for Space Organisation
The house layout can still follow the traditional plan of 3 or 5 bays: 01 living room and worshiping area; 01 or 02 bedrooms or more, behind the advanced design perspective, it can be changed to be suitable for different types of families, or the existing or new multifunctional circumstances. Each house area should be 150-200m2 built with 2 or 3 stories surrounded by their own private traditional garden.

6.2. Common Technical Solutions
Due to the economic situation of the majority of family farmers, high technology may not yet been applied broadly. Therefore, appropriate and flexible technology, easy construction, using local manpower would be applied popularly. In detail:

- Load bearing structure is rigid frame therefore it should be the load bearing reinforced concrete frame.
- The roof is built in the form of "hard roof" roofing tiles or galvanized steel sheets with insulating layers of corrugated porous or insulating layers of natural leaves.
- The external walls should be made of unbaked brick, the partitions with light materials which can be changed flexibly in order to meet the transformation and renovation of utility requirements, combining the technology to solve the microclimate environment such as energy saving ventilation and natural lighting.
- The sanitation system includes good treatment of domestic wastewater, livestock wastewater and environmental protection.
- Pure rural housing: used for families mainly producing agriculture. This has existed for a long time so when designed it should retain the traditional architectural forms and space, combining the structural system of "stiff frame, hard roof", flexible wall and partition system.
- Type of household combined with family economic activity: with characteristics of a farm besides ensuring living conditions, associated with tree gardens, ponds, outbuilding and workshop.

6.3. Applying Adaptable architecture in the evolutionary process
The Vietnamese government considers that the socio-economic environment of the new rural areas should continue to improve living quality, while also enabling other productive requirements at high speed. So rural houses should have diverse forms that may transform continuingly. Thus, new rural housing must have the adaptability that has been described partly in the theory of "Evolutionary Housing" in order to meet the requirements of progress and diversity. The Adaptable Architecture includes some important features: Transformability; Mobility; Upgradability.

6.4. Applying the Solution of housing architecture allows flexible transformability
In order to reach the transformability, structure solutions of new rural housing may be:

The common reinforced concrete skeleton frame is still the suitable main structure.

- Some columns, beams, frames or joints can be demounted, connected, transformed and developed, so there are some connectable, removable steel compositions and joints on the construction elements.

- Some components of space may be subdivided or change function. The kitchen, bathroom and toilet sections can be removed partly or upgraded, so it should use the prefabricated assembly solutions.
- Not arrange the column network too thick to create space inside in order to change the space
- The structural and construction members are removable to create efficient ventilation and lighting.
- The dimension of the structural and architectural components should be modularized for changing those components easily, especially the inner partitions. From modularization, it is also easy to establish a market for trading and exchanging the components mentioned above, while at the same time being easy to replace and possibly refurbished, minimizing demolition-causing debris. The modularization of construction members and space is not new; it comes from the application of construction rules which existed long ago in traditional architecture such as the application of the "storey rod" which was introduced widely.

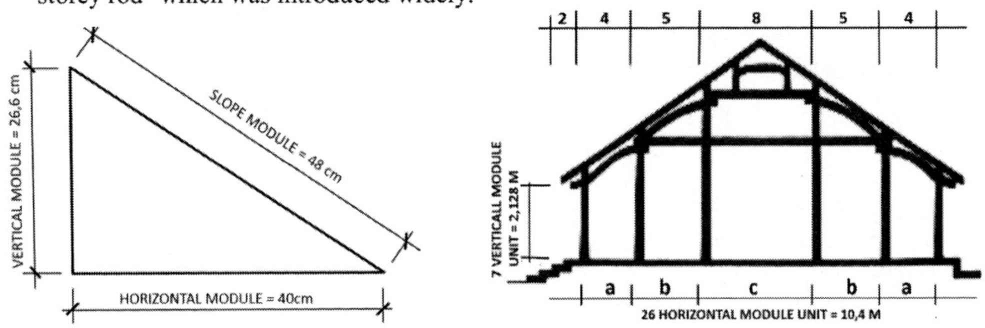

Figure 9: The modularization of construction members and space comes from the application of construction rules which exists long time ago in traditional architecture with demountable connections.

6.5. Advantages of the solution of mobile housing

The architecture allowing the change of construction location was applied long ago through the nomad tent. Its advantages are that it:

- Contributes to environmental protection when avoiding being demolished leaving garbage pollution.
- Easily meets the requirements of a keepers' hut at farm planning change, clearance or road extension...
- Responds to climate change circumstances, during floods it is possible to relocate to higher ground or float on the water's surface through some simple techniques mentioned in the following item.

The solution of a stiff frame structure, not using clamped joints with foundations buried deeply but using coupler joints on concrete or stone bases (typically, the form of traditional columns put on field stone bases). The above mentioned solution meets the requirement of movement when it is necessary to change construction location due to the movement of road boundary due to flooded ground or planning changes. In environment of new rural housing, it can be can applied even for multi-floored housing. If necessary, the houses can be relocated by the so called "folk lamp genies" who can move the buildings with common technologies without demolition

6.6. Architecture solutions for mobile housing

Solution of housing on stilts built on the ground or in a flooded area: This solution is the combination of the stilt house form of the Southeast Asian nations including Vietnam and five principles for designing

IOP Conf. Series: Earth and Environmental Science **143** (2018) 012053 doi:10.1088/1755-1315/143/1/012053

house on stilts of Le Corbusier. In the context of climate change occurrence, the ability to live with the water is great. The main structural system of Neo Rural house should be a rigid frame placed on a classical type base or one that can be removed from the foundation.

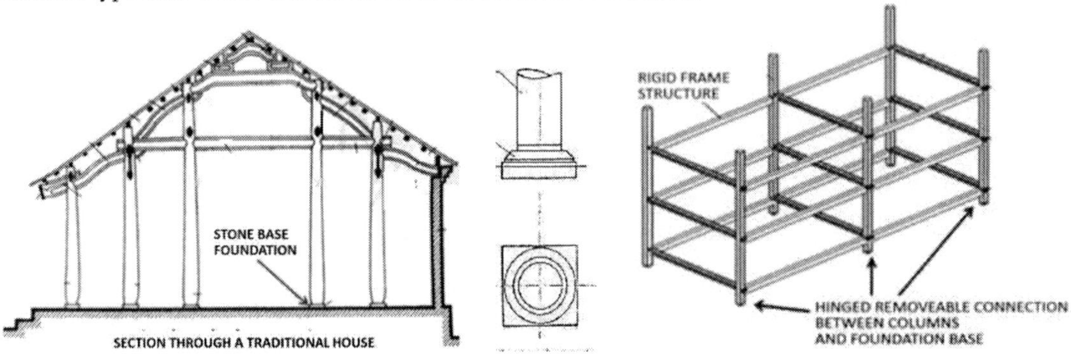

Figure 10. The main structural system of the Neo Rural House should follow the traditional house to be a rigid frame placed on a classical type base and connected by hinged structure.

6.7. Solutions creating effective energy saving

The application of advanced but simple technologies, matching with the economic ability of most farmer families to create effective energy saving mainly in two areas of lighting and ventilation, is feasible and so should be applied. The transmission of light deep into the house can be made with the use of reflective surfaces, such as mirrors put on the louvers to shield the direct sun against radiation, which then shines on the ceiling. The matte painted ceiling will disperse light in a scattering way deep into the house; this avoids the direct radiation causing glare and heat. The reflective surfaces can change the incline angle by the mechanisms of manual or automatic control in order to transmit light in the desired direction. It is possible to guide the sunlight into the pipe and transmit it by reflecting with the light guiding pipe wall. The use of Laser Cut Panel (LCP) is recommended to allow the light to be transmitted in a straight line, avoiding losses of light intensity due to many times of reflection.

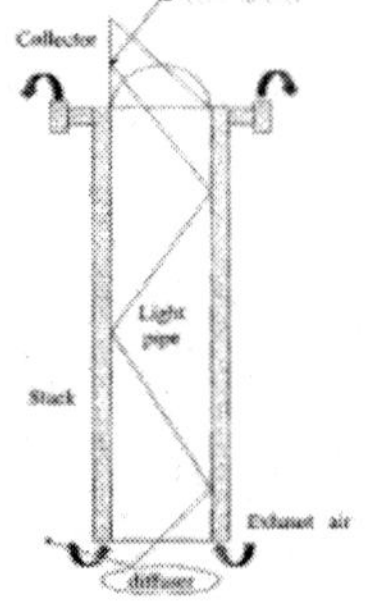

Figure. 11. Ventilation & light transmission.

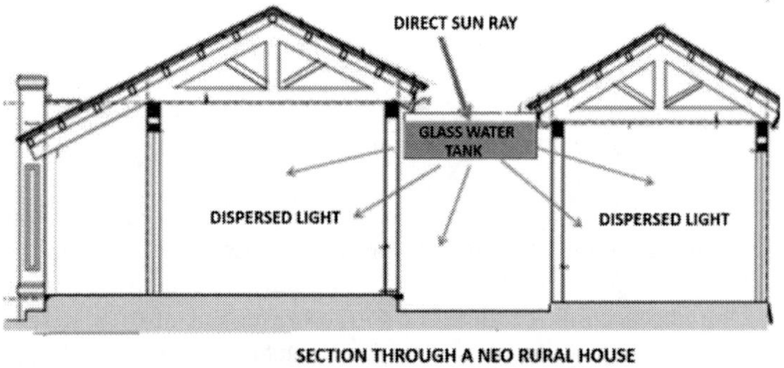

Figure 12 : Light transmission dispersed through water tank combined tube

The ventilation solutions are already presented by many authors, under two key principles: (1) Blowing straight, especially ventilating across the room, and (2) application of the stack effect to suck the airflow automatically to the top of the house or front door. The patio or staircases are effective in sucking the airflow up. The structural requirements of some architectural elements like the louvers or ceiling, partitions and mobile floor doors must be flexible for the guidance of airflow as this solution is again related to the theories of Transformable Architecture.

7. Discussion

To sum up, the design of a New Rural House for Vietnam is a remarkable solution of evolution and adaption of space and construction elements to specific natural conditions in the climate change era which potentially could be implemented into architecture of the new era. Choice of flexible transformable structural systems, construction details and application of materials impacts on sustainable development and economically beneficial selection.

8. Conclusion

The Vietnamese traditional folk house has many advantages in accordance with the view of Evolutionary Architecture and Adaptive Architecture; mainly, that it is possible to transform more flexibly than some stereotypical design styles of urban housing. The continuing promotion of Vietnamese traditional characteristics in order to qualify for the standard of a New Rural House is not only to preserve the old image, especially when combining with the new advanced construction technology system. With all the benefits that an Evolutionary and Adaptive rural house could bring, it should be considered as a solution to address related social issues such as the increasing demand for rural residential houses and natural challenges such as environmental problems and the negative impacts of climate change. It is recommended that the rural resident community and the local authorities need to give stronger attention to the integration of traditional and advanced design solutions for the Neo Rural house.

References

[1] Hoang Manh Nguyen. Dr.Arch, Institute of Tropical Architecture - Hanoi Architectural University, Hanoi, Vietnam. *Cultural Behavior: Climatic adaptive approaches of traditional housing in Vietnam Northern lowland area* © 2013 The Authors. Published by Elsevier Ltd.

[2] Nguyen Dinh Thi,. Assoc. Professor of Architecture, Ph.D, *Rural village green and sustainable planning* . Civil Engineering University, Hanoi. Mar 18, 2015.

[3] Le Thi Hong Na, Jin-Ho Park *Emphasis on Passive Design for Tropical High-rise Housing in Vietnam* Inha University, Department of Architecture Korea.

[4] Martinez E, Di Lullo R. *Evolutionary Housing.* The Institute for Housing Studies, Rotterdam.1985

[5] Arora S and Saxena S. *An Evolutionary Architecture: Adapted, interactive, and effectively integrated design* PLEA2009 - The 26th Conference on Passive and Low Energy Architecture, Quebec City, Canada, 22-24 June 2009.

[6] Schnädelbach H, MArch PhD. *Adaptive Architecture – A Conceptual Framework.* Mixed Reality Laboratory, Computer Science, University of Nottingham.

[7] Verma S, Pradeep D. *Adaptive skins: Responsive building skin systems. based on tensegrity principles.* Future Traditions 1ST eCAADe Regional International Workshop.AA School of Architecture - London, United Kingdom.

[8] Kim J J, Assistant Professor of Architecture and Rigdon B, Project Intern. *Sustainable Archtecture Module: Introduction to Sustainable Design* College of Architecture and Urban Planning,University of Michigan .

[9] De Temmerman N, Mira L A, Vergauwen A, De Wilde. *Transformable structures in architectural engineering.* Vrije Universiteit Department of Architectural Engineering https://www.researchgate.net/publication/271451822.

IOP Conf. Series: Earth and Environmental Science **143** (2018) 012054 doi:10.1088/1755-1315/143/1/012054

Recognition of a heritage in danger: rammed-earth architecture in Lyon city, France

Dorothée ALEX[1,2,3]
[1] Heritage Architect, Graduated from High Studies School of Chaillot, Paris, France
[2] Lecturer at Ton Duc Thang University, Faculty of Civil Engineering, HCMC
[3] Project manager at Academy of Architecture France Vietnam, HCMC, Vietnam
dorothee.alex@gmail.com

Abstract. Historically, man has built with what he had underfoot: if the earthen construction is a tradition of 11 000 years old, the technique of the rammed earth (earth rammed in a formwork) is relatively new since it appeared for the first time in Tunisia in 814 BC. Exported in France, rammed earth is established mainly in Rhône-Alpes Region which has an ideal soil, rich in gravel, sand, silt and clay. Traditionally associated with the rural world, rammed earth heritage is also present in urban centers. The city of Lyon is one of the rare European cities to concentrate a large proportion of rammed earth buildings: the inventories made show that they are present in almost all the districts as well as in the suburban areas. This high density could be explained by the phenomenon of rural exodus experienced by Lyon during the nineteenth century. The agricultural populations, attracted by the prosperity of the city, then settled at the doors of this city, bringing their know-how by building with the cheaper and easier material available: earth. Rammed earth buildings are therefore located on bounder areas between the countryside and the city. They are thus found on lands that were once outside the city walls. With the advent of the industrial era at the end of the 19th century and the appearance of concrete, rammed earth constructions gradually disappeared. The constructions that we see today are therefore prior to 1900. Varied, ranging from detached houses to the 6-storey buildings, they stand as a testimony to a know-how that finds a particular resonance today, while the environmental and economic concerns are at the forefront. Little known, they constitute a heritage in danger that should be rediscovered.

1. Introduction

Earth is the simplest building material, the most natural and thereby the oldest. Earthen constructions are indeed part of our built heritage, they are present worldwide, on all continents (see Figure 1 below), and in various forms. There are indeed different earth building techniques: rammed earth, mud, bung, adobe and CEB (Compressed Earth Block). Nowadays one estimates that 50% of the world's population lives in a mud construction, and it is even more interesting to note that 15% of the architectural works on the UNESCO World Heritage List are built in earth. Among these are, for example, the Great Wall of China or the Alhambra of Granada.

In France, the architecture of raw earth constitutes 15% of the architectural heritage, and all the techniques of construction can be found. Traditionally, earthen constructions are associated with the rural world. However, this is not the case in France, since it is found in bourgeois homes as well as large urban centers.

On the scale of the architectural heritage built in raw earth in France, the Rhône-Alpes region is an important and rich center of many rammed earth buildings. As the main area of study of the Grenoble laboratory CraTerre, this region has already been the subject of inventories, yet none has been made recently on the city of Lyon. In 1981-1983, Dominique Bertin and Anne-Sophie Clémençon, with the collaboration of Domar Idrissi, wrote a report for the Ministry of Urban Planning and Housing [1]. Thirty years later, part of the inventory has been destroyed, or deeply transformed. The work of the author was first to restrict the studied area to geographical areas where rammed buildings are numerous (see chapter 2); then, in this area, the buildings inventoried in 1983 were updated and then completed by the author's own inventory between 2011 and 2012.

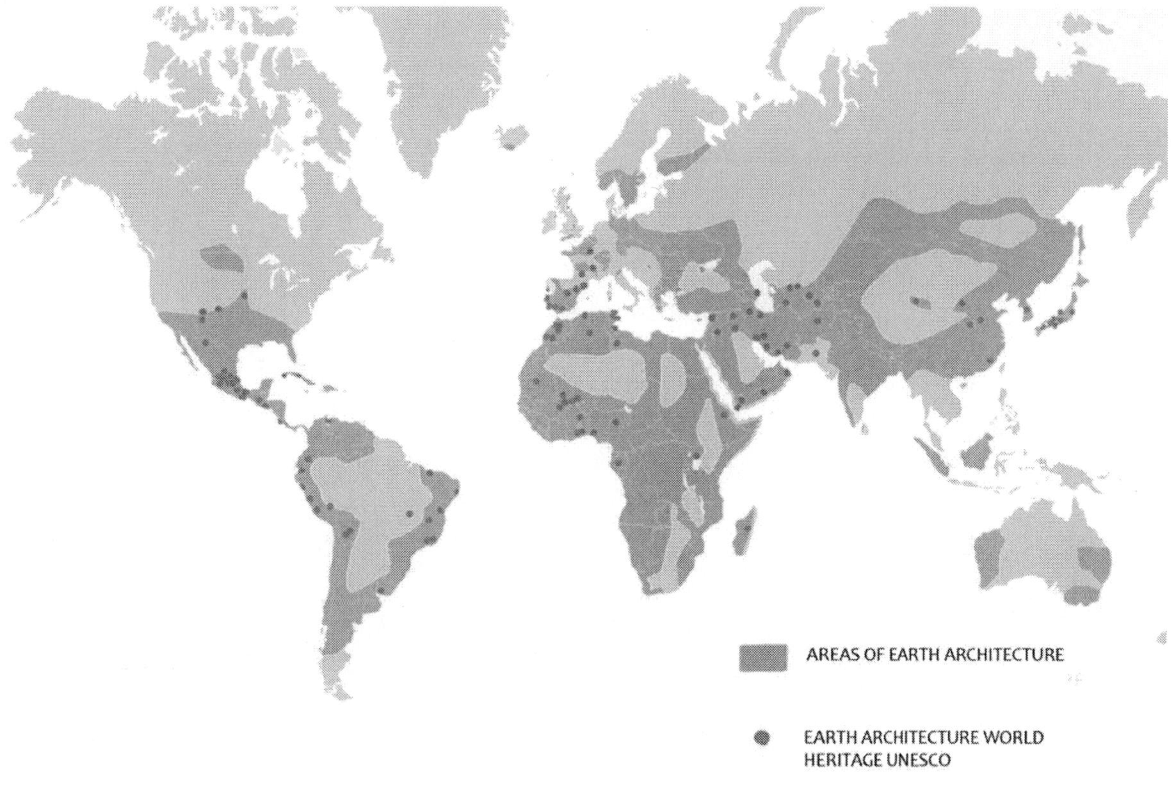

Figure 1. Half of Humanity lives in Earth construction [2]

2. Origins of earthen architectures in Lyon

2.1. Geological factors

The emergence of earth building is linked to a particular context based on the availability of a suitable material on the one hand, and converging human factors on the other. The special nature of the soil and subsoil of Lyon allows us to understand why rammed earth buildings were built, as well as their geographical distribution. Lyon stands at the crossroads of major geological regions, between the Massif Central in the west and the Alps in the east. In the Quaternary, the great glaciers of the Alps came to butter on the Cristalline Massif Central. This explains why it is on the heights of Lyon, towards the Fourvière hill, at the end of western Lyon and on the Croix-Rousse hill - promontory of the Dombes in the Northeast - that many houses realized in rammed earth coming from moraines were built. The silts lying on the alluvial basins of the left bank of the Rhone, in other words in the lowland areas east of Lyon, produce lower-quality buildings because these silts are composed of finer grains and are therefore poor in pebbles and coarse grains which usually ensure the cohesion of the masonry.

2.2. History of rammed earth constructions in Lyon

In the Gallo-Roman era, on the Fourvière plateau, the Romans already used the earth as building material. In one of the streets in the 5th District multiple testimonies of this type of construction [3] have been found. The excavations have made it possible to understand the different methods used. Molded raw earth brick (adobe) was commonly used for homes. Subsequently, at the end of the first and second century AD, adobe constructions were gradually replaced by half-timbered dwellings using mud as a filling. The specific technique of rammed earth is not attested at this time.

In the Middle Ages, buildings are in wood or stone. Earth constructions reappear in the mid-16th century with the increase of Lyon urbanization and they will really take off in the 19th century with the phenomenon of rural exodus. As it developed, Lyon attracted agricultural people who settled on arriving at the city's gates, near the main roads. They brought their know-how by building with the material available on site. As a result, rammed earth buildings are located mainly on frontier areas between the countryside and the city.

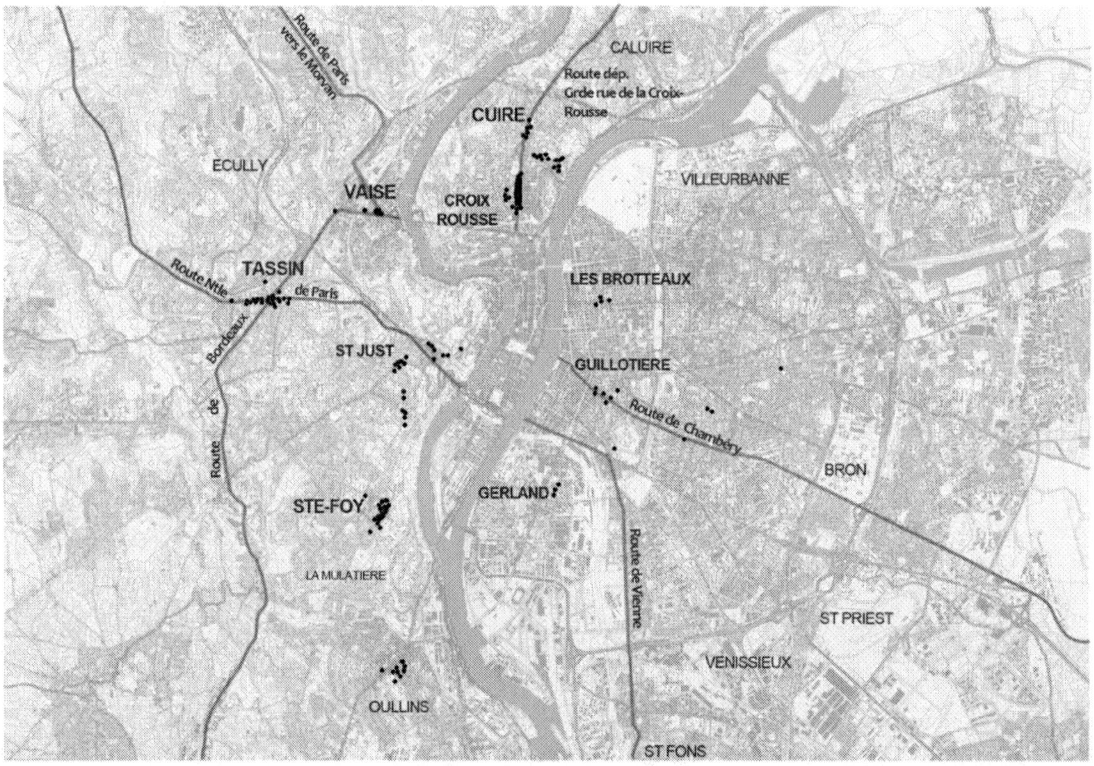

Figure 2. Distribution map of Rammed earth housings in Lyon, according to the inventory made to date [4] ©DA

We could find these constructions: in the North of Lyon, in the districts of Croix-Rousse, Caluire, on the axis of penetration of Dombes; in the West, at Vaise, at the crossroads of the Vienne and the Paris roads; in the fifth arrondissement, south-east of Lyon and Sainte-Foy which were formerly rural territories; finally, in the East, on the left bank of the Rhone, which, by its marshy nature, was only inhabited by a poor and rural population. Anne-Sophie Clemençon traced the history of rammed earth buildings in the context of urbanization in Lyon:

"From the end of the 16th century, began the urbanization of the Croix-Rousse, on the main street of this district, of which a large part is built of rammed earth and where complete sets remain today.[...] The urbanization began around 1550 but it is especially at the end of the 16th century and the beginning of the 17th century that the construction of small low houses of 1 to 2 floors intensified [...].Then, in the first part of the 19th century, the left bank of the Rhone, between the old town of Guillotière in the South and the new Morand district in the North, was urbanized" [5].

In the Brotteaux district, the land belongs to the *Hospices Civils de Lyon* (HCL) which, while waiting for them to increase in value, rent them at low prices, but with short-term leases (3, 6 or 9 years) to farmers who came to live in Lyon. As the land was to be released in the short term, buildings were often precarious. They were therefore erected quickly, for a short time, using earth as material, which requires no transport and is used directly, i.e. without transformation. Behind the facades of the middle-class housing consisting of high houses made of stone masonry, there are therefore warehouses, factories, workshops and small houses. Their main building material is rammed earth, to which wood, clinker, stone and sometimes brick are often added, as well as salvaged materials for reasons of economy. Rammed earth is considered as a material in itself, considered reliable especially for its fireproof qualities - unlike wood which was forbidden a long time before rammed earth because of the numerous fires which it favoured. Cheaper, economical and therefore "popular", rammed earth is used by the underprivileged classes, stone being reserved for the housing of the bourgeoisie and the nobility.

On several occasions, the administration tried to regulate buildings in earth. A decree issued in 1840 limits the height of masonry bases to 1.50 meters and buildings to five meters. However, it was only after the terrible flood of the Rhone of 1856 that submerged the left bank and destroyed in a day almost all rammed houses, that drastic measures were taken on June 19 of the same year by the Prefect Vaïsse. He proscribed «*the constructions of fat lime and clinker throughout the agglomeration of Lyon, including the town of Villeurbanne*» [6] and rammed earth walls with a masonry area of less than two meters. In its conclusion, the decree completely excludes rammed earth as masonry: « *Consequently, it will not be in the future, raised throughout the agglomeration of Lyon construction other than good masonry lime and sand* » [7]. The earth constructions, considered dangerous as a result of this flood, could not therefore be repaired and had to be demolished. Being forbidden to build with the only economic material at their disposal, the disadvantaged social classes were then forced to settle further east. They gave way to a middle-class that had become cramped in the peninsula. In 1872, a new decree lifted the ban on the construction of earth while regulating it. This relaxed was justified by the construction of dams and the Rhone development which removed the threat of floods. But the advent of the industrial era, towards the end of the 19th century and the beginning of the 20th century, brought other materials, such as reinforced concrete, which replaced the rammed earth and gave rise to its gradual disappearance.

Today, there are still buildings dating from before 1900 that remain in the urban landscape of Lyon. Although ignored by most writings on the architecture of Lyon, they are still numerous. They testify to a forgotten know-how that it is important to make known to save these historical remains.

3. Recognize rammed-earth

One of the first clues to locate rammed earth dwellings is their height. Built for popular and rural areas, these are often low-rise buildings not exceeding two or three storeys, with a total average height of eight meters. However, in the Croix-Rousse neighborhood, buildings can be up to seven storeys high. Generally a rammed earth building extends in length, forming along the streets bands wider than high and an average thickness of eight meters. The ratio of full and empty makes it possible to quickly locate a rammed earth construction. The openings are generally few and of low height.

In Lyon, rammed earth houses are usually covered by hollow terracotta tiles (canal). These tiles rest on batten nailed on rafters. The frame, traditionally, is made only of purlins, because there is a significant number of shear walls. These purlins, as well as the main parts of the floors, rest on the walls via wooden or mortar distribution blocks. They must not be removed during the work because they play an indispensable role in maintaining the floor.

Walls are built with earth drawn directly from the ground, without additional water supply. However, it is not rare to find in Lyon walls composed of a mix of various materials such as rubble, brick, pebbles, etc. The buried parts (foundations) and the basement, of a height above the ground between ninety-five centimetres and two meters, are stone masonry or pebbles. The upper parts are formed by successive layers of about eighty centimetres. A bed of mortar connects these layers together. The wall also has a fairly large fruit [8], which makes it more resistant to forces that could push it outward. Their thickness varies from forty to sixty centimetres, depending on the level. The angles are made of crossed forms reinforced with lime, spaced by twenty centimetres, or a masonry of brick and stones. The houses do not usually have chaining, this being ensured by the floors. This lack of chaining has frequently resulted in the use of tie rods because the angles were opening.

In urban areas, rammed earth buildings have often been coated to protect them from moisture, the roof overhang being generally less severe than in rural areas. These coatings play the role of a second skin, avoiding the penetration of water and therefore the dissolution of the cob wall. It is often easier to locate a rammed earth building when the coating covering it peels off in whole slabs. This coating detachment is the sign of its poor connection to the support and a gradual degradation due to bad weather. It is often caused by the composition used: a coating that is too hard, made of cement mortar with low water permeability [9], will cause swelling of the cob wall, which will crack it and take it off its support.

4. Architectural itinerary: discovering earth construction in Lyon metropolis

The rammed earth houses of Lyon are mainly located in four districts: Saint-Just, Vaise, Tassin and Croix-Rousse. Each of them has the particularity of being on a hill or at least on rocky massifs rich in moraines. So the land is good to build. Since the inventory of 1983, many have disappeared; but in some quarters, they still exist and are in good condition, proof that, despite the untimely use of cement plaster, these houses are resistant to the vagaries of time.

4.1. Saint-Just district

This district, built between the hills of Fourvière and Saint-Irénée, not far from Ste-Foy-lès-Lyon, was formerly mainly rural and largely located outside the ramparts of the 16th century. It constitutes an access to the west Lyon plateau. Its soil is rich in clay and pebbles, which makes it suitable for rammed earth buildings. Unlike the Croix-Rousse plateau, which quickly became part of the urban environment, the Saint-Just Saint-Irénée district has kept its typology of rural houses, with a heterogeneous alignment of low-rise buildings. Remains of Gallo-Roman buildings are still visible. Located high up, like that of Croix-Rousse, this district escaped the damage caused by the floods. François Cointeraux, father of the rammed earth construction in the 18th century, stayed with his family in a house located on Trion street, today disappeared.

Figure 3. < Hostel built by Cointeraux (Hotel des Macchabées), corner rue de Trion and Maccabees.
^ Building in rammed earth (today demolished) at the corner of Fossés de Trion str. and Appian str. ©DA.

4.2. Vaise district

This district, formerly attached to the fifth arrondissement of Lyon, is crossed by major roads that come from one side from the Morvan, and from Bourbonnais from the other. This "Crossroads district" was in the past badly connected to the city of Lyon. In 1840, a flood of the Saone covered all the low parts for several weeks. Place Valmy, the water reached 2.80 meters and stagnated for fourteen days. Nevertheless, rammed houses were not all damaged. This area is important because it has been a real experience ground for Cointeraux [10]. It is in Vaise that he lived and installed his second school of rural architecture, before leaving Lyon definitively to pursue his career in Paris. Today there is still a large group of rammed earth buildings.

Figure 4. House Cointereaux, build in 19th century for Lyon workers, today demolished ©DA.

4.3. Tassin district

Located in the western suburbs of Lyon, not far from the fifth district, Tassin remained until the period of the Revolution a small rural village. Meadows and vineyards stretched along the current Montée de Verdun. The area is crossed by the main roads of Bordeaux and Paris, which intersect at the current Place Pierre Vauboin. In the eighteenth century, it was only occupied in its northern part by a set of houses forming a half-moon, which gave - according to a legend - the name of the new district of "Tassin La Demi-Lune"(Half moon). On this square will be located post offices, hotels, shops and craftsmen's workshops, whose function and typology have little evolved since. All the dwellings presented here are characteristic of rammed earth houses: low height, few holes, corner chaining and lime plaster.

Figure 5. Place Vauboin and urban block with cob houses ©DA.

4.4. Croix-Rousse district

Located on the plateau of the hill of the same name, culminating at 250 meters, La Croix-Rousse is known for being the district of "Canuts", craftsmen who weaved silk. These "silkers" were numerous in the nineteenth century and they have deeply marked the urban planning of the district by their five or six storey buildings whose high ceilings allowed to accommodate the famous *Jacquard* loom.

In the eighteenth century, the hill was only occupied by religious congregations, as the Fourvière hill, as well as many traders whose low buildings are still visible today. To discover them, you have to take the Main street of the district, which runs through the fourth arrondissement. It connects the Place de la Croix-Rousse to the place Joannès Ambre and continue to Caluire by rue Coste. The work we did in this neighbourhood consisted of comparing the buildings identified in 1983 by Anne-Sophie Clémençon and what is now indicated by their location. Unlike other areas previously studied, it appears that the Croix-Rousse has undergone many changes. Indeed, of the fifteen buildings identified in 1983, nine are actually not in rammed earth, but stone. Four had been identified as buildings where the rammed earth was visible - often in the yards of the buildings. Today, these buildings have been renovated, re-coated, and the cob is no longer visible. However, as they present the characteristics of cob dwellings – low height, timber frame and floors, few openings and thick walls, it is unlikely that these constructions were modified in their structure. Finally, of the two remaining buildings, only one is in rammed earth, also visible on the facade, while the other could not be surveyed.

Figure 6. Rammed earth appears under coating in this construction main street of the district ©Craterre

5. Conclusion

This inventory work was not done in vain. It allowed to reveal architecture, and a mode of construction inherited from our forefathers, which we had forgotten. By going through the different Lyon districts where these constructions are present, we were able to give an account of their numerical importance: the first inventory of 1983 reported 88 rammed buildings. In updating this work and in pursuing it, we have raised more than a hundred in total, but it is certain that many constructions are still unknown to this day. On the one hand, this is because, in urban areas, and especially in Lyon, these houses are covered with a coating which makes them difficult to detect. On the other hand it is due to the fact that this material is little or not known by the inhabitants and by the public authorities. It has, however, obvious qualities, like Hubert Guillaud and Sébastien Moriset remind us in their work [11]; qualities that echo current environmental concerns. Unfortunately, the reality is quite different: because they are of traditional typology, generally of low height, they constitute a brake on urban development that encourages densification. And because they have undergone poor restoration (use of cement in coating in particular), these houses have lost their main qualities, and are now suffering the destruction of time. In fact, in the current state of things, they are doomed to disappear. For these reasons, they deserve all the more our attention.

The work started here in 2012 has been continued in 2015 by an architect and engineer, Emmanuel Mille, specialist of earth heritage: he synthesizes existing inventories and coordinates an inventory participatory [12]. Today sponsored by Aurhalpin Heritage Association, this project aims to change the public's view of the built environment of Lyon metropolis by recognizing the presence of rammed earth as part of its building heritage.

To know and to make know, this is how the value of a building is transmitted, and how the receiving of a heritage is to be born. According to Jean Davallon "*ultimately, heritage is what was saved from destruction, which should have disappeared and was saved from disappearance*" [13].

References

[1] Bertin D, Clemencon A.S, Idrisse D, “Earth Architecture, an urban way of construction, Lyon and its suburb case study”, in F. COINTEREAUX, 1740-1830, architecture de terre, Ministère de L’urbanisme et du Logement, Paris, March 1983 (last phase).
The author published « Guide of rammed earth architecture in Lyon », Craterre ed., October 2012.
See also Mille E participatory inventory in Lyon, in his memory at ENSA Grenoble, DSA «Earth Architecture ».
[2] Anger R, Fontaine L, 2009 *Bâtir en terre, Du grain de sable à l'architecture* Belin (Cité des sciences et de l'industrie)
[3] Desbat A, « Farges street : Gallo-roman earth construction in Lyonn », *Pignon sur rue*, monthly regional architecture-urbanism-construction information, nb30, May 1981 p 12
[4] Alex D, 2012 *Petit guide des architectures en pisé à Lyon* CRATerre éditions (Grenoble) 37 p
[5] Clemençon A.S Bertin D, 1981 Lyon : Pisé urbain, *Pignon sur rue* **30** p 18
[6] Extract from the decree of prefect Vaïsse, 19th June 1856
[7] *Ibid*
[8] The fruit designates in architecture a diminution of the thickness which one gives to a wall, on its external facing, as one raises it, thus advancing its base with respect to the plumb of the top.
[9] The artificial cement coating can be catastrophic for a rammed earth wall. This one, made up of water, needs to regulate its hygrometry rate with the outside. For this, it must be covered with a coating that allows it to breathe, a natural lime plaster for example.
[10] Cointeraux François, Xe conference, Paris, 1808-1809, p239 and next
[11] GUILLAUD, H., & MORISET, S. (2012, October). Lessons in sustainable architecture learnt from rammed earth heritage in France. In European Project VerSus, 1st International Scientific Workshop
[12] Website : http://patrimoine-terre-lyonnais.patrimoineaurhalpin.org/participer-au-projet/
[13] Original quotation: “ *à l'extrême, est patrimoine ce qui a été sauvé de la destruction, ce qui aurait dû disparaître et qui a échappé à la disparition, ce qui avait disparu et a "miraculeusement" réapparu*” in Le musée Canibale, 2002

Energy efficiency façade design in high-rise apartment buildings using the calculation of solar heat transfer through windows with shading devices

Ha P T H[1]
[1] Faculty of Architecture and Planning, National University of Civil Engineering (NUCE), 55 Giai Phong Road, Hanoi, Vietnam

Abstract. The architectural design orientation at the first design stage plays a key role and has a great impact on the energy consumption of a building throughout its life-cycle. To provide designers with a simple and useful tool in quantitatively determining and simply optimizing the energy efficiency of a building at the very first stage of conceptual design, a factor namely building envelope energy efficiency (K_{hqnl}) should be investigated and proposed. Heat transfer through windows and other glazed areas of mezzanine floors accounts for 86% of overall thermal transfer through building envelope, so the factor K_{hqnl} of high-rise buildings largely depends on shading solutions. The author has established tables and charts to make reference to the values of K_{hqnl} factor in certain high-rise apartment buildings in Hanoi calculated with a software program subject to various inputs including: types and sizes of shading devices, building orientations and at different points of time to be respectively analyzed. It is possible and easier for architects to refer to these tables and charts in façade design for a higher level of energy efficiency.

Keywords: façade design, high-rise apartment building, building envelope, energy efficiency, sun control, shading devices.

1. Introduction

An energy-efficient building design requires a holistic approach by integrating techniques, technologies and architectural innovation. The designers need to be well equipped with design strategies including active and passive design, as well as an appropriate combination of the two concepts. It is important at first to optimize all the opportunities of passive design based on the basis of special climate conditions of the site and indoor comfort requirements, then to enhance the combination of passive and active strategies, and finally to introduce active solutions of the M&E systems.

The passive strategies practiced by architects at the concept design stage are a key factor to the energy consumption of a building through its life-cycle. This is because the building design factors are almost unlikely to change, except in case of major renovation, whereas building M&E systems can be improved and replaced relatively often and better operated by occupants so that the building will become more energy-efficient [2].

An energy-efficient design comprises three stages of which stage no. 2 is supported by smart and complex energy simulation software programs such as Open Studio, Energy Plus, eQuest, Design Builder, Simergy, Trace, etc. (Figure 1).

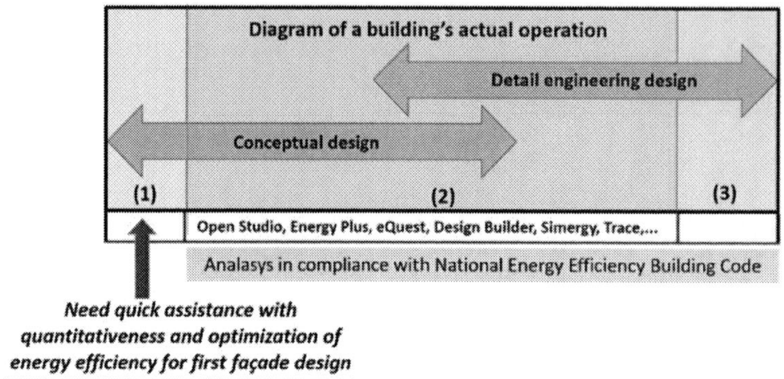

Figure 1. A design process for an energy-efficient building.

At present, when sketching the façades of a building at the early stage of conceptual design - stage no. 1, architects use the "shading mask" method to control direct radiation penetration into a building [14], which selects types and sizes of shading devices by hours that the windows need to be shaded completely. Therefore, this method cannot quantify the energy efficiency of the facade designed solutions. This article proposed a new building façade design method: selecting the type and the size of shading devices according to the requirements of controlling solar heat (both direct and diffuse radiation) penetrated through the shaded windows that could help architects quickly quantify and optimize the energy efficiency of building envelope.

2. Recommendation of building envelope energy efficiency factor – K_{hqnl}

According to data from a survey on civil buildings in Hanoi, Danang and Ho Chi Minh city conducted by the Ministry of Construction and International Finance Corporation (IFC) [9], energy consumption for cooling by air conditioners makes up a major part among the energy consumption within the building and is closely related to building envelope design solutions.

In order to support an architect in façade design at stage no.1, a factor namely building envelope energy efficiency (K_{hqnl}) for making a quantitative assessment of cooling energy efficiency in close relation to building envelope design should be proposed.

The factor is regarded as the ratio of the reduction of heat transfer through a building envelope, or in other words, is the rate of reducing cooling load on air-conditioner systems by means of architectural design to help prevent heat transmission from the outside to the inside.

The calculation of K_{hqnl} is proposed as follows:

$$K_{hqnl} = \frac{Q_0 - Q_{cn}}{Q_0} = 1 - \frac{Q_{cn}}{Q_0}; \quad (1)$$

$$Q_0 = Q_{cs.0} + Q_{m.0} + Q_{t.0}; \quad W \quad (1a)$$

$$Q_{cn} = Q_{cs.cn} + Q_{m.cn} + Q_{t.cn}; \quad W \quad (1b)$$

in which: Q_0 – Overall heat that may transfer through the building envelope (glazing, roof and external walls) into the building in case of no shading devices for glazed windows ($Q_{cs.0}$), no special isulation for the roof ($Q_{m.0}$), and no insulation layers inside the external wall ($Q_{t.0}$);

Q_{cn} – Overall heat that may transfer through the building envelope (glazing, roof and external walls) with shading devices for glazed windows ($Q_{cs.cn}$), special insulation for the roof ($Q_{m.cn}$), and ordinary or insulated external wall ($Q_{t.cn}$).

Calculation of the overall heat transfer values through certain types of building envelopes (roof, external wall and window) on a typical hot summer day in Hanoi for a high-rise apartment building - "17T10" - in Nguyen Thi Dinh street, Trung Hoa - Nhan Chinh new urban area supposing that the main facade of the 17T10 building orientate towards all eight directions (E, W, N, S, NW, NE, SW, SE) and the ratio of glazed window area to floor area (F_{cs}/ F_s) could vary in five cases 1/3.5; 1/4, 1/5, 1/6, 1/7 resespectively show that the percentage of heat transfer into the building (on average of three months in hot season of Hanoi) through glazed windows makes up from 86.96% to 91.94% of the overall heat transfer into the building through the building envelope [6]. Therefore, the building energy efficiency factor "K_{hqnl}" of a certain high-rise apartment building depends largely on how efficient its glazing could be in term of thermal insulation.

The most effective way of passive cooling is using shading devices for external walls, then choosing appropriate insulation glass patterns to reduce solar radiation. Thus, sun-shaded common-glass window is selected to analyze the establishment of "K_{hqnl}", and the formula (1) can be simplified for an approximation of "K_{hqnl}" in high-rise buildings as follows:

$$K_{hqnl} = 1 - \frac{Q_{cn}}{Q_0} = 1 - \frac{F_{cs}.SHGC.(K_{cn}.S_d + K_{bt}.D_d)}{F_{cs}.SHGC.(S_d + D_d)} = 1 - \frac{K_{cn}.S_d + K_{bt}.D_d}{S_d + D_d}; \quad (2)$$

in which: Q_0 - Overall heat transfer through glass windows without shading devices into the building.
Q_{cn} - Overall heat transfer through glass windows with shading devices into the building.
S_d - incident direct solar radiation on windows; D_d - sky diffuse radiation intensity on windows
K_{cn}; K_{bt} - lighting coefficient and sky diffuse radiation coefficient of shading devices.

3. Calculation of the factor "K_{hqnl}" for high-rise apartment buildings in Hanoi

It can be noted from formula (2) that the value of solar heat transferred (both direct and diffuse) through windows with shading devices must be calculated in order to identify the factor K_{hqnl}. The intensity of solar radiation (direct – S_d and diffuse – D_d) on-site is able to be determined from either on-site measurement data or theoretical calculation. Then, the biggest challenge to be dealt with in this case is how to determine the coefficients of both lighting (K_{cn}) and sky diffuse radiation (K_{bt}).

From the overview of various calculation methods for solar heat gain through glazed windows with shading devices currently applicable in Vietnam and the world, it is possible to come to the following conclusions:

- The calculation of solar radiation (direct and diffuse) through windows with either uninterrupted shading devices - horizontal, vertical as well as box-shaped - proposed by Dang P N [18] proves to be more comprehensive than other authors' methods. Despite the same principle, Dang P N's suggest (in 1978) is believed to come out the earliest in comparison to that of ASHARE [1], and by Chan T N [4], and it was then officially endorsed into СНиП [17]. The formula was first accepted in 1993 by ASHRAE and СНиП, then in 2008 by ISO, and most recently in 2011 by Chan T N, but the ISO's seems to be inaccurate. The most remarkable difference between Dang P N's formula and the others is K_{bt} - the hidden sky radiation coefficient from differential equation performing radiation heat exchange between the surface of shading devices and central point of a window.
- Very few studies on the calculation of solar radiation through windows with interrupted horizontal shading devices have ever been conducted. The first was carried out in 1983 by two American reseachers: Yandat R F and Jones R E [16], subsequently by two Iranian scientists: Raeissi S and Taheri M in 1998 [13] and most recent was Chan T N in 2011 [4].
- There has been still no outcomes to make an easy-to-use calculation of solar heat through interrupted horizontally shaded windows.

Therefore, K_{cn} and K_{bt} for uninterrupted (both horizontal and vertical) and box-shaped shading devices could be applicable with Dang P N's formula (see Table 1), while K_{cn} of interrupted horizontal shading devices is formulated and the K_{bt} is then developed from Dang P N's differential equation by the author.

3.1. Calculation for K_{cn} of windows with shading devices.

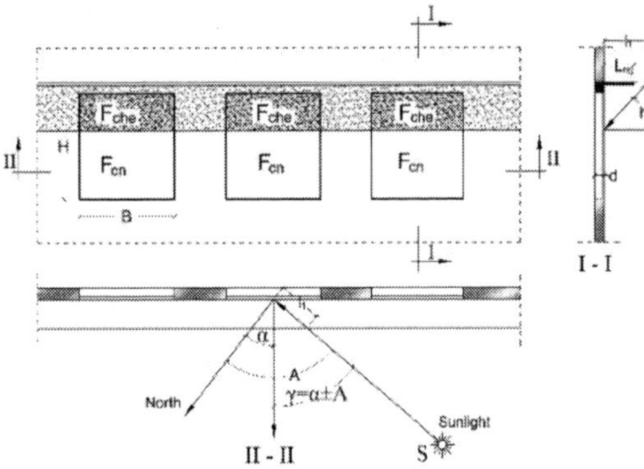

Figure 2. Illustration of indentifying K_{cn} of uninterrupted horizontal shading device

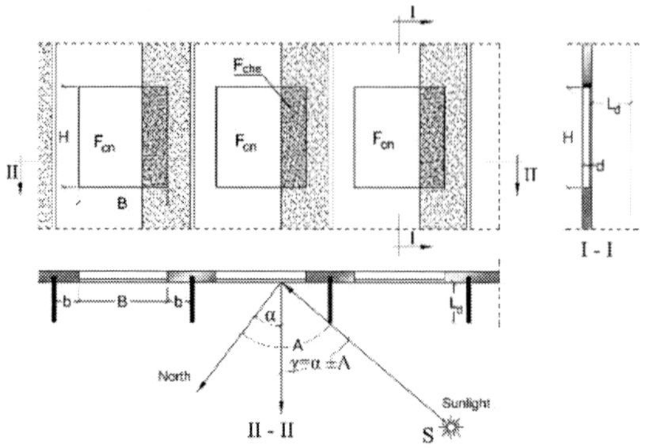

Figure 3. Illustration of indentifying K_{cn} of uninterrupted vertical shading device

Analyzing Figure 2 and Figure 3, it can be noted that:

$$K_{cn} = \frac{F_{cn}}{F_{cs}} = \frac{F_{cs} - F_{che}}{F_{cs}} = 1 - \frac{F_{che}}{F_{cs}} \quad (3)$$

in which: F_{cn} - area of exposure surface of window;
F_{che} – area of surface of window to be shaded;
F_{cs} – area of window, $F_{cs} = F_{cn} + F_{che}$.

Table 1. The calculation of K_{cn} for uninterrupted horizontal, uninterrupted vertical and box- shaped shading devices by Dang P N [5,18].

Uninterrupted horizontal shading device	Uninterrupted vertical shading device	Box-shaped shading device
$K_{cn.z} = 1 - (\frac{L_{ng}}{H} \cdot \frac{tgh_z}{\lvert\cos\gamma_z\rvert} - \frac{a}{H})$; (4)	$K_{cn.z} = 1 - (\frac{L_d}{B} \cdot \lvert tg\gamma_z \rvert - \frac{b}{B})$; (5)	$K_{cn.z} = K_{bt.ng.z} \cdot K_{bt.d.z}$; (6)

in which : L_{ng}, L_d, L_{ov} - horizontal and vertical depth of shading devices, respectively
H, B - height and width of window
a - distance from upper edge of a window to horizontal shading device
b - distance from side edge of a window to vertical shading device
γ - angle made by normal and solar radiation ray projection of a wall on horizontal ground
z - testing (measurement) point of time

For interrupted horizontal shading devices: Due to its complex form of shadow which does not just depend on the rate of outreach of shading elemens to the height of a window, but also widths of window and the length of shading devices, there can be nine types of shadow as shown in Figure 4.

The author has set up nine formulas for $K_{cn.ov.z}$ corresponding to nine different cases of shadow, and boundary conditions to determine the area of shadow for relevant use of particular $K_{cn.ov.z}$ formula, as shown in Table 2 below.

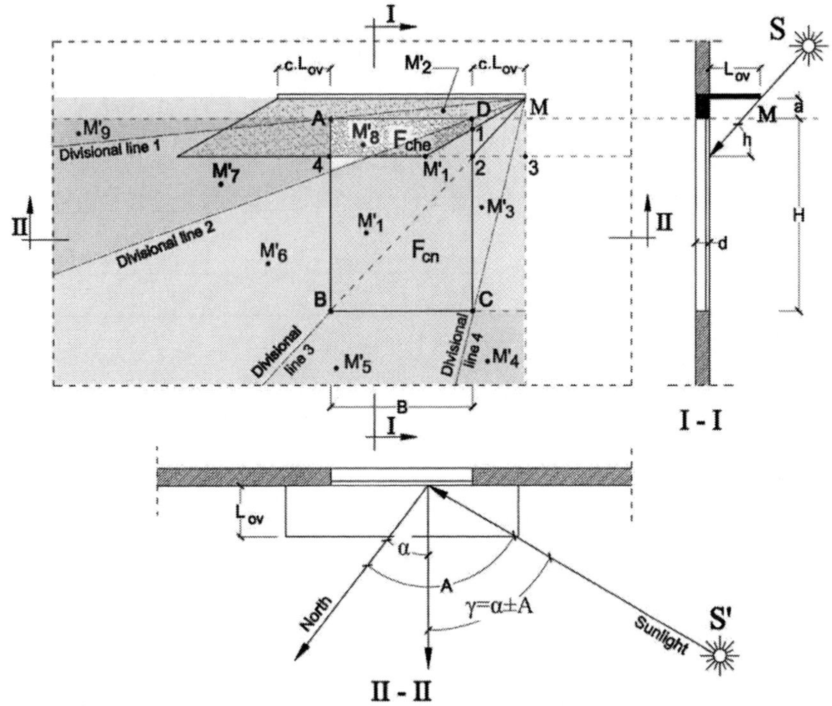

Figure 4. Nine types of shadow by interrupted horizontal shading devices

Table 2. Nine formulas of K_{cn} in corespondence to nine cases of shadow cast by interrupted horizontal shading devices

Case 1 (shadow of point M falls into the area of M'1)	$K_{cn.ov.z}=1-\left(\frac{L_{ov}}{H}.\frac{tgh_z}{\lvert\cos\gamma_z\rvert}-\frac{a}{H}\right)+\frac{1}{2}\left[\frac{L_{ov}}{H}.\left(\lvert tg\gamma_z\rvert-c\right)\right]^2.\frac{tgh_z}{k_b.\lvert tg\gamma_z\rvert.\lvert\cos\gamma_z\rvert};$ Boundary condition 1: $\frac{a/H}{c.L_{ov}/H}.\lvert tg\gamma_z\rvert\le\frac{L_{ov}}{H}.\frac{tgh_z}{\lvert\cos\gamma_z\rvert}\le(1+\frac{a}{H})$ Boundary condition 2: $c.\frac{L_{ov}}{H}\le\frac{L_{ov}}{H}.\lvert tg\gamma_z\rvert\le k_b+c.\frac{L_{ov}}{H}$	(7a)
Case 2 (shadow of point M falls into the area of M'2)	$K_{cn.ov.z}=1;$ Boundary condition: $\frac{L_{ov}}{H}.\frac{tgh_z}{\lvert\cos\gamma_z\rvert}\le\frac{a}{H}$	(7b)
Case 3 (shadow of point M falls into the area of M'3)	$K_{cn.ov.z}=1-\left(\frac{L_{ov}}{H}.\frac{tgh_z}{\lvert\cos\gamma_z\rvert}-\frac{a}{H}\right);$ Boundary condition 1: $\lvert tg\gamma_z\rvert\le c$ Boundary condition 2: $\frac{a}{H}\le\frac{L_{ov}}{H}.\frac{tgh_z}{\lvert\cos\gamma_z\rvert}\le1+\frac{a}{H}$	(7c)
Case 4 (shadow of point M falls into the area of M'4)	$K_{cn.ov.z}=0;$ Boundary condition 1: $\lvert tg\gamma_z\rvert\le\frac{tgh_z}{\lvert\cos\gamma_z\rvert}.\frac{c.L_{ov}/H}{1+a/H}$ Boundary condition 2: $\frac{L_{ov}}{H}.\frac{tgh_z}{\lvert\cos\gamma_z\rvert}\ge1+\frac{a}{H}$	(7d)

Case 5 (shadow of point M falls into the area of M'5)	$K_{cn.ov.z}=\frac{1}{2}\left[1+\frac{a}{H}-\frac{c.L_{ov}.tgh_z}{\lvert tg\gamma_z\rvert.\lvert\cos\gamma_z\rvert}\right].\left[\frac{\left(1+\frac{a}{H}\right).\lvert tg\gamma_z\rvert.\lvert\cos\gamma_z\rvert}{tgh_z}-\frac{c.L_{ov}}{H}\right];$ Boundary condition 1: $\frac{L_{ov}}{H}.\frac{tgh_z}{\lvert\cos\gamma_z\rvert}\geq 1+\frac{a}{H};$ Boundary condition 2: $\frac{tgh_z}{\lvert\cos\gamma_z\rvert}.\frac{c.L_{ov}/H}{1+a/H}\leq\lvert tg\gamma_z\rvert\leq\frac{tgh_z}{\lvert\cos\gamma_z\rvert}.\frac{k_b+c.L_{ov}/H}{1+a/H}$	(7e)
Case 6 (shadow of point M falls into the area of M'6)	$K_{cn.ov.z}=1+\frac{a}{H}-\left(\frac{k_b}{2}+\frac{c.L_{ov}}{H}\right).\frac{tgh_z}{\lvert tg\gamma_z\rvert.\lvert\cos\gamma_z\rvert};$ Boundary condition 1: $\frac{L_{ov}}{H}.\lvert tg\gamma_z\rvert\geq k_b+\frac{c.L_{ov}}{H}$ Boundary condition 2: $\frac{\lvert tg\gamma_z\rvert.\frac{a}{H}}{\frac{c.L_{ov}}{H}}\leq\frac{tgh_z}{\lvert\cos\gamma_z\rvert}\leq\frac{\left(1+\frac{a}{H}\right).\lvert tg\gamma_z\rvert}{k_b+\frac{c.L_{ov}}{H}}$	(7f)
Case 7 (shadow of point M falls into the area of M'7)	$K_{cn.ov.z}=1-\frac{1}{4}.\frac{k_b}{c}.\frac{a/H}{L_{ov}/H};$ Boundary condition 1: $\frac{\lvert tg\gamma_z\rvert.L_{ov}}{H}\geq k_b+\frac{c.L_{ov}}{H}$ Boundary condition 2: $\frac{a/H}{k_b+c.L_{ov}/H}.\lvert tg\gamma_z\rvert\leq\frac{tgh_z}{\lvert\cos\gamma_z\rvert}\leq\frac{a}{c.L_{ov}}.\lvert tg\gamma_z\rvert$	(7g)
Case 8 (shadow of point M falls into the area of M'8)	$K_{cn.ov.z}=1-\frac{1}{4}.k_b.\frac{a}{c.L_{ov}};$ Boundary condition 1: $c.\frac{L_{ov}}{H}\leq\frac{L_{ov}}{H}.\lvert tg\gamma_z\rvert\leq k_b+c.\frac{L_{ov}}{H}$ Boundary condition 2: $\frac{a}{H}<\frac{tgh_z}{\lvert\cos\gamma_z\rvert}.\frac{L_{ov}}{H}<\lvert tg\gamma_z\rvert.\frac{a/H}{c.L_{ov}/H}$	(7h)
Case 9 (shadow of point M falls into the area of M'9)	$K_{cn.ov.z}=1;$ Boundary condition 1: $\frac{\lvert tg\gamma_z\rvert.L_{ov}}{H}\geq k_b+\frac{c.L_{ov}}{H}$ Boundary condition 2: $\frac{a}{H}\leq\frac{tgh_z}{\lvert\cos\gamma_z\rvert}.\frac{L_{ov}}{H}\leq\lvert tg\gamma_z\rvert.\frac{a}{H}$	(7i)

Note: $k_b = B/H$, or $B = k_b.H$.
The extensions of shading device to the right and left of a window ($c.L_{ov}$) (c can be regarded as an arbitrary factor).

The calculation of K_{cn} has been computationally programmed based on inputs of four types of shading devices applied to eight orientations (horizontal, vertical, box-shaped and overhang) with 10 different dimensions, and it has been calculated in every five months of hot season in Ha Noi (including May, June, July, August and September). Some outcomes are going to be illustrated in Table 3 below.

Table 3. K_{cn} at particular point of time "z" of May in Hanoi for four kinds of shading devices with $L_{ng}/H = L_{ov}/H = L_d/B = 0.5$ facing Southwest

Type of shade	"z" – Point of time												
	6	7	8	9	10	11	12	13	14	15	16	17	18
Uninterrupted horizontal shading device	0	0	0	0	0	0	0	0	0	0.124	0.435	0.672	0.902
Uninterrupted vertical shading device	0	0	0	0	0	0	0.55	0.521	0.456	0.385	0.299	0.182	0.004
Box-shaped shading device	0	0	0	0	0	0	0	0	0	0.048	0.13	0.122	0.004
Limited horizontal shading device	0	0	0	0	0	0	0	0	0.039	0.224	0.521	0.741	1

Notes: K_{cn} - shown in table is determined supposing that a/H = 0.05 and b/B = 0.05; while those of limited horizontal shading devices are c = 0.5 and B/H = k_b = 1.2.

3.2. Calculation of K_{bt} for shaded windows

The coefficient of diffuse radiation over the shaded surface of a window is the ratio of diffuse radiation by unveiled sky dome to the surface of a window and that by overall visual sky dome through that window without any shading device.

$$K_{bt} = \frac{D_{d.cn}}{D_d}; \tag{8}$$

in which: $D_{d.cn}$: Diffusion of the radiation onto the surface of a window partially subtracted with an amount from the area protected from the sky dome by a shading device.

D_d: Diffuse radiation over a window without any shading device.

Since diffuse radiation rays are centripetal, it is necessary to set the origin on window surface for the calculation of K_{bt} (Figure 5).

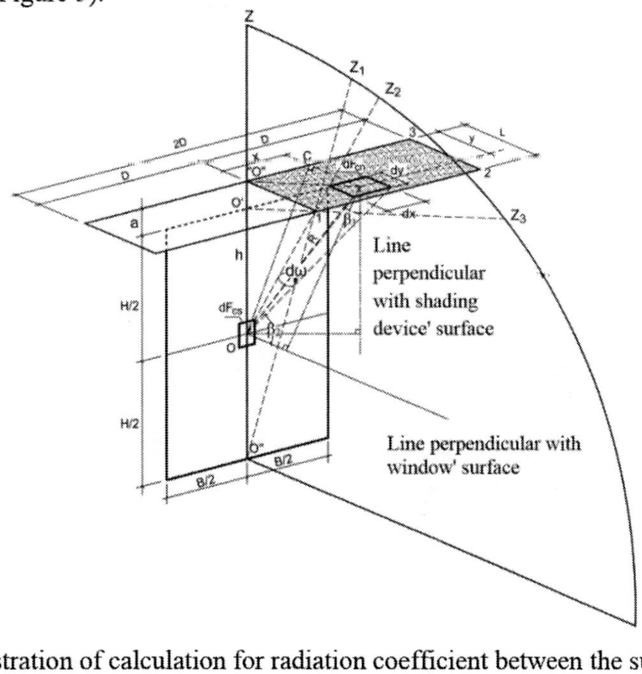

Figure 5. Illustration of calculation for radiation coefficient between the surface of the shading device and the center-point O of window [18].

Based on illustration on Figure 5, it is possible to notice that: taking mid-point of the upper edge of the window as the origin "O" causes an oversized sky dome Z-Z3, while the origin "O" as mid-point of the lower edge leads to an undersized sky dome; therefore, the center point of the window is set as point of calculation. According to [18] with the assumption that diffuse radiation is equally distributed, the calculation of K_{bt} will be:

$$K_{bt} = 1 - 2\psi_{cs-cn}; \quad (9)$$

in which: 1 - coefficient of radiation by the whole sky dome on the center point of the window;
ψ_{cs-cn} - coefficient of radiation by a shading device on the center point of the window.

To obtain radiation coefficient ψ_{cs-cn}, Dang P N made a ratio of differential of window's area at the center point (dF_{cs}) and differential of the area of the shading device (dF_{cn}) as shown in Figure 5, then drew a formula as follows:

$$\psi_{dF_{cs}-F_{cn}} = \frac{1}{2\pi} arctg\frac{D}{h} - \frac{h}{2\pi\sqrt{L^2+h^2}} arctg\frac{D}{\sqrt{L^2+h^2}}; \quad (10)$$

For a limitless horizontal shade incase D approaches Infinity, (6) will be: $\lim_{D\to\infty}(arctg\frac{D}{h}) = \frac{\pi}{2}$ and $\lim_{D\to\infty}(arctg\frac{D}{\sqrt{L^2+h^2}}) = \frac{\pi}{2}$; Therefore the calculation of K_{bt} for a uninterrupted horizontal shading device, uninterrupted vertical shading device and box-shaped shading device can be written in Table 4 below.

Table 4. The calculation of K_{bt} for uninterrupted horizontal, uninterrupted vertical and box-shaped shading devices by Dang P N [5,18].

Uninterrupted horizontal shading device	Uninterrupted vertical shading device	Box-shaped shading device
$K_{bt.ng} = \frac{1}{\sqrt{1+\frac{4L_{ng}^2}{(H+2a)^2}}}$; (11)	$K_{bt.d} = \frac{1}{\sqrt{1+\frac{4L_d^2}{(B+2b)^2}}}$; (12)	$K_{bt.hh} = K_{bt.ng} \cdot K_{bt.d}$; (13)

The formula for K_{bt} of interrupted horizontal shading devices is formulated with a differential equation by Dang P N as:

$$K_{bt.ov} = 1 - \frac{2}{\pi}\left(arctg\frac{\frac{k_b}{2}+\frac{c.L_{ov}}{H}}{\frac{1}{2}+\frac{a}{H}} - \frac{\frac{1}{2}+\frac{a}{H}}{\sqrt{\left(\frac{L_{ov}}{H}\right)^2+\left(\frac{1}{2}+\frac{a}{H}\right)^2}} .arctg\frac{\frac{k_b}{2}+\frac{c.L_{ov}}{H}}{\sqrt{\left(\frac{L_{ov}}{H}\right)^2+\left(\frac{1}{2}+\frac{a}{H}\right)^2}} \right); \quad (14)$$

Notes: arctg is presented in degree (°).

Values of K_{bt} for four types of shading devices with 10 different horizontal and vertical depths are shown in Table 5 below.

Table 5. Values of K_{bt} for 4 different types of shade with 10 various depth rates

Type of shading device	Outreach ratio of shading devices (L_{ng}/H, L_d/B và L_{ov}/H)									
	0.1	0.2	0.3	0.4	0.5	0.6	0.7	0.8	0.9	1.0
Uninterrupted horizontal shading device	0.987	0.949	0.896	0.834	0.771	0.71	0.654	0.603	0.558	0.518
Uninterrupted vertical shading device	0.987	0.949	0.896	0.834	0.771	0.71	0.654	0.603	0.558	0.518
Box-shaped shading device	0.974	0.901	0.803	0.696	0.594	0.504	0.428	0.364	0.311	0.268
Limited horizontal shading device	1	0.998	0.997	0.994	0.992	0.99	0.988	0.986	0.985	0.983

Notes: K_{bt} shown in the table is determined supposing that a/H = 0.05 and b/B = 0.05;

3.3. *Calculation of the factor "K_{hqnl}" for high-rise apartment buildings in Hanoi*

According to the research outcomes by Chan T N [3], Tuan N A [15] and Muon N V [12], the artificial cooling time in Hanoi ranges between 20% and 25% of the whole year. It is mainly in the daytime of the hot season (May, June, July, August and September). Therefore, the author has calculated the factor K_{hqnl} for high-rise apartment buildings in Hanoi according to formula (1), using solar radiation data recorded hourly for 10 years (1996 - 2005), it is based on the weather data of the World Meteorological Organization's Trace 700 software. This calculating has been carried out in the time that cooling system in Hanoi is needed (five months of hot season) supposing that the shading devices change in four different types (uninterrupted horizontal shadingdevice, uninterrupted vertical shadingdevice, box-shaped shadingdevice and interrupted shading device), 10 different outreach dimensions and main façade of the building could change in eight directions. The varied-by-hour K_{hqnl} can be drawn and performed in Hanoi's five summer months in tables (e.g Table 6) or charts (e.g Figures 6, 7, 8, 9).

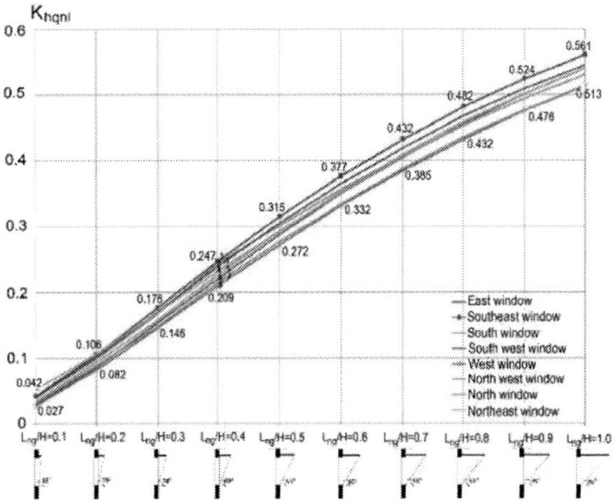

Figure 6. The variation of mean value of K_{hqnl} in 5-month summer subject to eight directions by uninterrupted horizontal shading devices

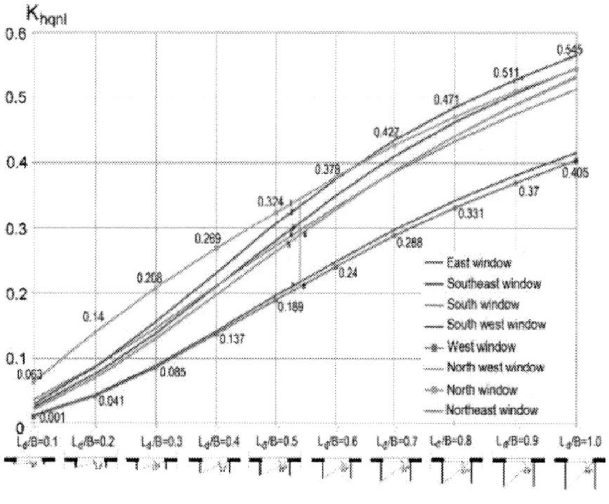

Figure 7. The variation of mean value of K_{hqnl} in 5-month summer subject to eight directions by uninterrupted vertical shading devices

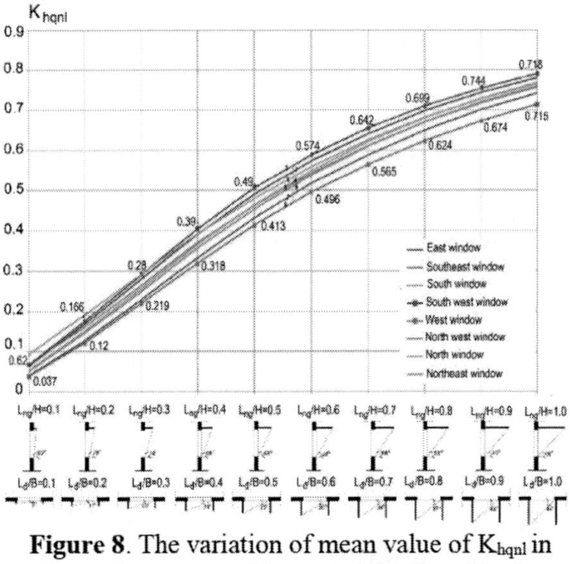

Figure 8. The variation of mean value of K_{hqnl} in 5-month summer subject to eight directions by box-shaped shading devices

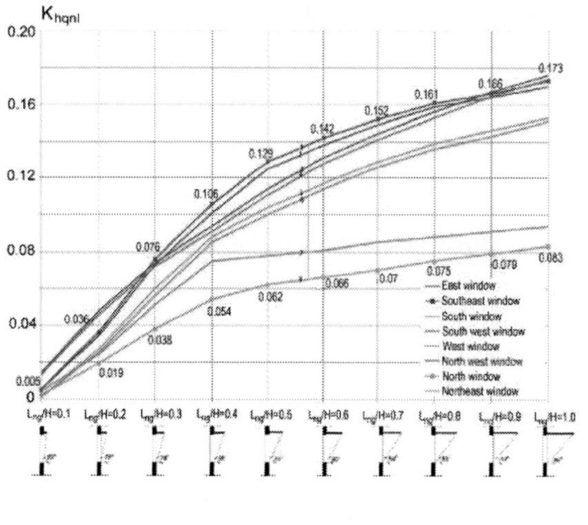

Figure 9. The variation of mean value of K_{hqnl} in 5-month summer subject to eight directions by interrupted horizontal shading devices

Table 6. Tables of K_{hqnl} for South elevation in summer months *(with a = 0.05H, b = 0.05B, while a interrupted horizontal shading device has c = 0.5 and B/H = k_b = 1.2)*

Ratio L_{ng}/H, L_{ov}/H, L_d/B	0.1	0.2	0.3	0.4	0.5	0.6	0.7	0.8	0.9	1
Uninterrupted horizontal shading device	0.029	0.086	0.153	0.220	0.287	0.350	0.405	0.455	0.501	0.540
Uninterrupted vertical shading device	0.011	0.043	0.088	0.141	0.196	0.248	0.297	0.342	0.382	0.417
Box-shaped shading device	0.039	0.125	0.229	0.332	0.431	0.518	0.589	0.650	0.702	0.743
Interrupted horizontal shading device	0.014	0.048	0.074	0.094	0.114	0.131	0.144	0.156	0.167	0.176

Based on the Figures 6, 7, 8, 9, it can be give some conclusions:

- The factor "K_{hqnl}" for high-rise apartment buildings in Hanoi varies considerably by time, by the direction of the building, by form and size of shading devices. Table 7 below present its average of the hot season.
- For four types of shading devices, when the dimension of shading device varies from 0.1 to 0.6, the slope of K_{hqnl} curve is larger when it varies from 0.6 to 1.0.
- With the same dimension, the K_{hqnl} of box-shaped shading device is the largest, assuming 100%, the K_{hqnl} of uninterrupted horizontal, uninterrupted vertical and interrupted horizontal shading devices are 67%, 60% and 18% corresponding. Obviously the box-shaped shading device is most effective in reducing solar radiation that transfers into the building during the hot season and has the highest building envelope energy efficiency factor.

Table 7. Variability of the K_{hqnl} on average of summer months depends on the type and size of shading devices and building orientation.

Type of shading device	Outreach ratio of shading devices (L_{ng}/H, L_d/B and L_{ov}/H)			
	0,3	0,5	0,8	1,0
Box-shaped shading device	0,292/ 0,219	0,507/ 0,413	0,71/ 0,624	0,791/ 0,715
Uninterrupted horizontal shading device	0,176/ 0,146	0,315/ 0,272	0,482/ 0,432	0,561/ 0,513
Uninterrupted vertical shading device	0,208/ 0,085	0,324/ 0,189	0,486/ 0,331	0,567/ 0,405
Interrupted horizontal shading device *(k_b = 1,2 and c = 0,5)*	0,076/ 0,038	0,129/ 0,062	0,161/ 0,075	0,174/ 0,083

Note: The above value corresponds to the direction of the building has got maximum value and the under value corresponds to the direction of the building has got minimum value.

4. Application of K_{hqnl} in energy-efficient façade design for high-rise apartment buildings in Hanoi

Charts and tables of K_{hqnl} can be used to determine façade features in etheir of the following two methods:

METHOD 1	METHOD 2
• **Step 1:** Determining the value of K_{hqnl} (for instance a façade with K_{hqnl} = 0.28 is able to reduce the total amount of heat transfer by 28%)	• **Step 1:** Sketching the form and façades of a building, then finding the appropriate forms of shading devices.
• **Step 2:** Sketching the form and façades of a building, then finding the appropriate forms of shading devices (such as uninterrupted horizontal shading devices for Southeast and Northeast façades as shown in Figure 10)	• **Step 2:** Referring to K_{hqnl} values of each façade with corresponding shading devices in Step 1 from charts or table of K_{hqnl}

• **Step 3:** Referring to a value of K_{hqnl} from tables or charts of K_{hqnl} based on the types, then determine dimensions of shading devices on each façade For example, determining the dimensions of shading for Southest and Northest elevations of an apartment block as shown in Figure 10. The angle of the shading device to be selected for Southest facade is tgα = L_{ng}/H = 0,44, while that of Northest is tgα = L_{ng}/H = 0,46 (Figure 11). The dimensions of the shadingdevice may be modifed as aesthetically required, or the initial shading device may be replaced and the K_{hqnl} can be looked up again in the chart for optimal dimensions.	• **Step 3:** Correct the shading devices if the value of K_{hqnl} is not adequate by: - Changing the dimensions of the shading device (K_{hqnl} will then change). - Modifying the initial concept sketch, until an optimal value of K_{hqnl} can be found. - Keeping the initial concept but later replacing with a new type of glass with low-SHGC factor (as a result the total cost will be higher).

Figure 10. Example of a concept sketch

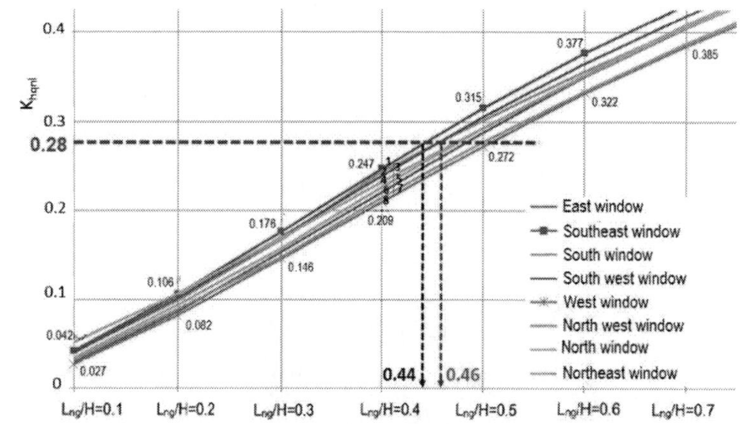

Figure 11. An illustration of how to refer to a value of K_{hqnl} for uninterrupted horizontal shadingdevice (0,28 is determined)

Here are several recommendations to optimize the shading devices when designing energy efficiency façade of high-rise apartment buildings in Hanoi based on K_{hqnl} results.

- *Box-shaped shading devices*

  With small size L_{ng}/H = 0.1 to 0.4 and $L_đ$ /B = 0.1 to 0.4, the K_{hqnl} is the largest in North direction. The shallow loggia or negative window solution is best in North elevation (Figure 12).

  With the size $L_{ng}/H \geq 0.4$ and $L_đ$ /B $\geq$ 0.4, the K_{hqnl} in Southwest direction is the highest. Shading device solution should be arranged in deep loggia or mixed style with large size in Southwest elevation.

  With the size Lng/H $\geq$ 0.3 and Lđ /B $\geq$ 0.3, the K_{hqnl} in Southeast direction is high. Therefore, flexible box-shaped shading devices such as negative window, mixed style shading device, any type and size of loggia is good in Southeast elevation (Figure 13).

- *Uninterrupted horizontal shading devices*

The K_{hqnl} of uninterrupted horizontal shading device is not much different and varies in all directions equally. In addition, the horizontal style of shading devices has a good effect of rain cover and low impact on natural ventilation, which is suitable for living building in Hanoi with the characteristics of heavy rain and long drizzly time. Therefore, for energy efficient high-rise apartment building, it is recommended that uninterrupted horizontal shading device design should be applied to windows in all directions (Figures 14).

- *Interrupted horizontal shading devices*

The K_{hqnl} of interrupted horizontal shading device in the normal case ($c = 0.5$) is very low. It is therefore recommended to use interrupted horizontal shading device with large extensions ($c >> 0.5$). This device in North and South direction have negligible energy efficiency and their main function is rain cover.

- *Uninterrupted vertical shading devices*

Uninterrupted vertical shading device is only effective in North direction with the ratio of L_d/B = 0.1 to 0.6, while in other directions, the K_{hqnl} is lower than uninterrupted horizontal shading device and much lower than the box-shaped shading device.

Figure 12. Example of small mixed-size shading device in North elevation

Figure 13. Example loggias with flexible in size and shape in Southeast elevation

Figure 14: Building facade with uninterrupted and interrupted horizontal shading devices

5. Conclusion

The diagrams and tables representing the factor of building envelope energy efficiency (K_{hqnl}) are expected to be useful for an energy-efficient façade design of high-rise apartment buildings in Hanoi at the stage of conceptual design, and it may offer designers a useful reference source when it comes to the designs of high-rise apartments in different parts of the Red River delta in Vietnam.

It would be also possible to introduce the factor of building envelope energy efficiency (K_{hqnl}) and the calculation can be applied to various types of high-rise buildings in general.

References

1. ASHRAE Handbook – Fundamentals 2009, Inch-Pound Edition, Chapter 15. Fenestration.
2. Baker N and Steemers K 2005, *Energy and Environment in Architecture. A Technical Design Guide*. Published in the Taylor & Francis, e-Library.
3. Chan T N 2004, Research Project "Meteorological data on temperature, humidity, solar radiation to be added to the Standard of construction climate data".
4. Chan T N 2015; Scientific base and inputs for Tables in National technical code QCVN 09:2013/Ministry of Construction – Energy effcient buildings, *Proceeding of "Urban infrastructure and sustainable urban development" conference* - Vietnam Association of Civil Engineering Environment; Construction Publisher (Hanoi), pp. 14-24.

5. Dang P N and Ha P T 2002, *Heat and Architectural Climatics*, Construction Publisher (Hanoi), pp. 101-123; 215-222.
6. Ha P T H 2016, A Concept for Energy-Efficient High-Rise Buildings in Hanoi and a Calculation Method for Building Energy Efficiency Factor, *Procedia Engineering of the International Conference on Sustainable Development of Civil, Urban and Transportation Engineering*; Ho Chi Minh city, Published by Elsevier, Volume 142, pp 153-159.
7. Ha P T H 2016, Method of calculation solar heat penetrated through shaded windows; *Proceeding of the International Conference "Indoor air and Environmental quality";* Published by Volgagrad University; pp 232-245.
8. Ha P T H 2016. Establishing calculation formula of direct solar radiation coefficient and diffuse radiation coefficient to calculate solar heat penetrated through windows with overhang device, *Proceedings of the 17th Science and technology Conference in National University of Civil Engineering*, Construction Publisher (Hanoi), Volume 2, pp 25 – 36.
9. International Finance Corporation (IFC-WB) 2014, *Guidance on the application of National technical code QCVN 09:2013/ Ministry of construction relating to energy efficient buildings*. Ministry of Construction.
10. ISO 13790:2008 (E). Energy performance of buildings - calculation of energy use for space heating and cooling, ISO copyright office. Published in Switzeland; Geneva.
11. Ministry of Construction, National Technical Norms QCVN 09:2013/BXD on energy efficient buildings, promulgated by the Ministry of Construction in accordance with Circular No. 15/2013/TT-BXD dated 26 September 2013.
12. Muon N V 2009, Assessment of indoor comfort with Fanger's PMV and neutral temperature in contexts of Vietnam, *Conference proceeding "Occupational health caring and safety in context of Global integration"* - Vietnam National Institue of Occupational safety and health.
13. Raeissi S, Taheri M 1998, Optimum Overhang Dimensions for Energy Saving. *Building and Environment,* Vol 33, N5, pp 293-302.
14. Steven V Szokolay 2008, *Introduction to Architectural Science: The basis of sustainable design*. Architectural Press, pp 35-37.
15. Tuan N A and Reiter S 2014, A climate analysis tool for passive heating and cooling strategies in hot humid climate based on typical meteorological year data sets, *Energy and Building,* Vol 68. pp 756-763.
16. Yandat R F and Jones R E 1983. Shading effects of finite width overhang on windows facing toward the equator. *Solar Energy,* Vol. 30, No.2, pp.171-180.
17. Industrial construction project, allowance 2.91 for SNIP 2.04.05-91 "Calculation of heat radiation tranfer into the room", Lvovsky I B and Barkalov BV, Moscow 1993.
18. Dang P N and Bogolovsky V N 1978 "The calculation total heat radiation tranfer into the room through the window", *Water supply and sanitary engineering*. USSR, No1.

Usage of Video Analysis of Traffic Conflicts for the Evaluation of Inappropriately Designed Building Elements on Intersections

Vladislav Krivda[1], **Jan Petru**[1]

[1] Department of Transport Constructions, Faculty of Civil Engineering, VSB – Technical University of Ostrava, Ludvika Podeste 1875, 708 33 Ostrava-Poruba, Czech Republic

Abstract. Human society is constantly evolving in all areas. The transport is no exception – especially development of road transport is considerable and visible on all levels. However, it is necessary to look for tools that can harmonize economic and social progress with the full preservation of the environment for future generations. The so-called sustainable development of road transport is also closely related to the issue of creating safe transport infrastructure. For example, roads must be in compliance with technical rules on one hand and on the other one they have to be clearly understandable for all road participants (drivers, pedestrians, etc.). Unfortunately, even if the designer of transport construction meets the technical requirements, the result is a dangerous place where very frequently the accidents or traffic conflicts take place. There exist statistics of accidents, but on the contrary, there are not any statistics of traffic conflicts. And simply monitoring the traffic conflicts can lead to early detection of problematic places in road infrastructure and thus increasing the safety in the road traffic. There are many methods used for monitoring traffic conflicts. The method used in the Czech Republic is described in the submitted paper. The original method, which monitored mainly and only the behaviour of drivers and their mistakes, was innovated, so that it could be used also for evaluation of inappropriately designed building elements of road constructions. Therefore, the paper deals with the description of so-called Innovated Video Analysis of Traffic Conflicts (IVATC) and with the usage of this method for one particular example of intersection, where larger vehicles (buses, trucks, etc.) had problems with passing through it. It points out to the fact that even a relatively small change can lead to the increase of the road safety.

1. Introduction

Each road transport construction, within its sustainability, must meet strict criteria for maintaining road safety. The reliability of such construction is influenced, inter alia, by the load which is caused by car weight (e.g. damage of roads, curbs, etc.) or by the sc. traffic load, i.e. the volume of traffic flows (high traffic volume can cause non-standard behaviour of road transport participants and hence increased number of dangerous situations or accidents).The most problematic places are intersections.

For the right design of intersections there are many sophisticated procedures and solutions based on legislation, standards, technical conditions, etc. Unfortunately, the issue of analysis of traffic conflicts is often omitted.

These analyses are generally referred to as Traffic Conflict Techniques (TCT). Amundsen and Hyden [1] defined traffic conflict as an observable situation in which two or more drivers approach each other in space and time to such an extent that there is a risk of collision if their movements remain unchanged. In the Czech Republic Folprecht dealt with this problematic and he called the traffic conflicts "conflict situations". According to Folprecht [2] the traffic conflict (conflict situation) is a moment or a situation in road traffic in which a higher than usual degree of risk arises or can arise for some traffic participants. A traffic conflict naturally precedes each accident. On the contrary, it can be stated that each accident is a result of such traffic conflict when the risk of collision was impossible to be averted. Thus, traffic conflicts are potential accident situations and their types then predetermine the types of accidents.

Drivers' behaviour is influenced by many factors which could have direct effect on the safety of road traffic. Generally, these three factors are the most significant: vehicles, human factors and the road and its surroundings. These factors also affect speeding which is one of the most common causes of an accident. For example, Atombo, Wu, Zhong and Zhang [3] describe other aspects that affect speeding (age, gender or nationality of drivers, differences between professional and non-professional drivers, etc.). This fact is also confirmed by other studies dealing with so-called working memory which changes with the driver's age and affects driver's behaviour (e.g. [4] or [5]).

However, it is necessary to realize that traffic conflicts can be caused also by the influence of some hidden factors, such as operating, construction, meteorological ones or others. These hidden factors can be revealed by the analysis of traffic conflicts. In other words, the designers of the transport construction can never consider their transport solution absolutely safe and satisfactory [6]. Using the analysis of traffic conflicts the doubts about their harmlessness can be avoided.

At present, a number of specialists deals with traffic conflicts in the Czech Republic (e.g. Krivda, Petru, [7]; Ambros, Kocourek [8], etc.). Based on their experience and knowledge the methodology for monitoring and evaluating traffic conflicts [8] was created. However, this methodology primarily deals with the analysis of traffic conflicts caused by drivers alone (as in the past Folprecht). This paper deals with the use of video analysis of traffic conflicts for the evaluation of inappropriately designed building elements on intersections, this is based on the original Folprecht's procedures.

2. Original Folprecht's methodology

Monitoring the traffic conflicts was introduced in the Czech Republic (or then in Czechoslovakia) by the associate professor Jan Folprecht, Ph.D. in the 1970s. In 1998 the cooperation with Mr. Folprecht started and the method of video analysis of traffic conflicts was further developed. At present, we keep using and developing this method within the professional activity of the Department of Transport Constructions (VSB –the Technical University of Ostrava, the Faculty of Civil Engineering). It is impossible to describe the whole methodology in this paper, so the text below is limited only to the essential facts (for details see [9]).

Folprecht based his methodology on the then studies (e.g. [10], [11], [12], [13] and [14]) and he began to use a video apparatus for recording the road traffic which was then analysed. This methodology is called Video Analysis of Traffic Conflicts (VATC). Folprecht marked each traffic conflict (TC) by a classification symbol exactly describing the given situation. The symbol has 3 parts [9]:

- number - description of participants of the traffic conflict, e.g. 1 = pedestrian, 2 = car, 4 = pedestrian and car, 6 = two or more cars, 9 = other (one cyclist, cyclist and car, etc.),
- one or more letters - description of the way of the traffic conflict, e.g. n = violation of the rule "Give way", f_p = influence of the pedestrian crossing, f_k = influence of the near intersection, jc = wrong ride of a cyclist, a = aggression, g = giving priority against rule, o = restriction or danger in driving by changing the lane or exiting the roundabout, D = danger of the collision from the rear, etc.
- number (1 - 4) - description of seriousness of the traffic conflict (1 - 4).

The seriousness of traffic conflict is described as follows [2]:

- 1st level - potential traffic conflict,
- 2nd level - traffic conflict when one or more participants are restricted in driving,
- 3rd level - traffic conflict when one or more participants are endangered,
- 4th level - accident.

For example the symbol 6a3 means aggression (a) between two drivers of cars (6) and 3rd level of seriousness (3).
The resulting numbers of traffic conflicts were recalculated to sc. relative conflict rate c_R [TC/100 veh]:

$$c_R = \frac{N_{TC}}{V}.100 \tag{1}$$

Where N_{TC} is the number of traffic conflicts (TC) per hour [TC/h] and V is an hour traffic volume [veh/h].

3. Innovated Video Analysis of Traffic Conflicts (IVATC)

Innovated Video Analysis of Conflict Situations allows the evaluation of inappropriately designed building elements. The main change in comparison with the original Folprecht's method is the addition of the classification symbol by other features that represent a new division of conflict situations (see 'Figure 1').

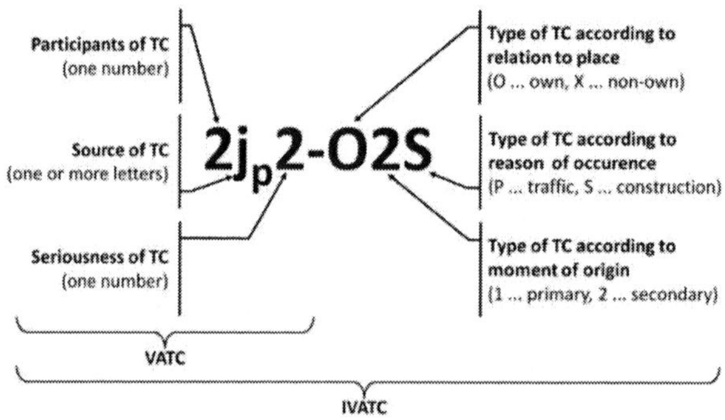

Figure 1. Innovated classification symbol of traffic conflicts (TC)

Some traffic conflicts were not caused by traffic on the intersection or by the intersection itself (i.e. its construction, traffic control, etc.). It can be e.g. a result of an action on a nearby intersection, which influences the traffic on the observed intersection (see video https://youtu.be/2CkoyNf9EFM - traffic jam on roundabout due to a queue of vehicles on nearby intersection with traffic lights /on the left out of the focus/), influence of a nearby parking area, etc. These are conflicts that are caused by external influences.
According to the origin of the given traffic conflicts the conflicts were newly divided into two groups (in the brackets there is a sign to be used in the classification symbol):

- own traffic conflicts (O) - traffic conflicts which are directly related to traffic on the observed intersection, to its construction, etc.,
- not-own traffic conflicts (X) - traffic conflicts which are not linked to the traffic on the observed place directly, its construction, etc. They can be originated outside the observed place.

Other traffic conflicts arise as a result of another traffic conflict. For example, the driver of the first vehicle must unexpectedly slow down (e.g. before a pedestrian crossing with careless pedestrian entering the road) and the driver of the other vehicle breaks violently in order to prevent a collision. The situation which affected the second vehicle driver, would probably not happen if it was not for the first conflict.

According to the moment of the origin of the traffic conflicts, we divide them into two groups:

- primary traffic conflicts (1) - traffic conflicts which are not caused by other traffic conflict,
- secondary (subsequent) traffic conflicts (2) - traffic conflicts which are caused by other traffic conflict.

Due to the fact that the reason of the occurrence of some traffic conflicts is not caused only by a wrong behaviour of the driver (or other road user), but e.g. building structure can be also blamed and responsible for these traffic conflicts. Therefore, it was necessary to divide the traffic conflicts according to the reason of its occurrence as follows.

- Operational traffic conflicts (P) - traffic conflicts caused only by the driver alone (or other road user),
- Construction traffic conflicts (S) - traffic conflicts caused not only by the drivers alone, but also (often primarily) by inappropriately designed building elements.

Into the group of operational traffic conflicts we can include, e.g. violation of rule "Give way", ride on red light, the risk of collision from the rear, etc. Construction traffic conflicts include, e.g. driving near curbs, moving into the opposite direction on intersection, etc., but only in case of conflict situations which are caused by inappropriately designed building elements (see video https://youtu.be/6clxD3-nmpo- small radius of intersection corner).

The resulting numbers of traffic conflicts are then recalculated by the new formula (weighted conflict rate c_{RW} [TC/100 veh]):

$$c_{RW} = \frac{N_{TC} . C_S}{V} .100 \tag{2}$$

Where N_{TC} is a number of traffic conflicts (TC) per hour [TC/h], C_S is a coefficient of seriousness of traffic conflict and V is hour traffic volume [veh/h].

The weighted conflict rates determined by this method were used e.g. in safety study about sc. U-turns (Meel, Brannolte, Satirasetthavee and Kanitpong [15] or Meel, Vesper, Borsos and Koren [16]).

4. Example of using IVATC

IVATC was used on an intersection with inappropriately designed building elements – i.e. e.g. small radii of corners and inappropriately situated refuge island (see 'Figure 2'). There were both operational TC and also construction TC. Attention was concentrated only on construction TC which were related to ride of buses or trucks. Measurements were made at many other intersections. The data given here is only a selection of many measurements and serves as an illustrative example of using the method described above.

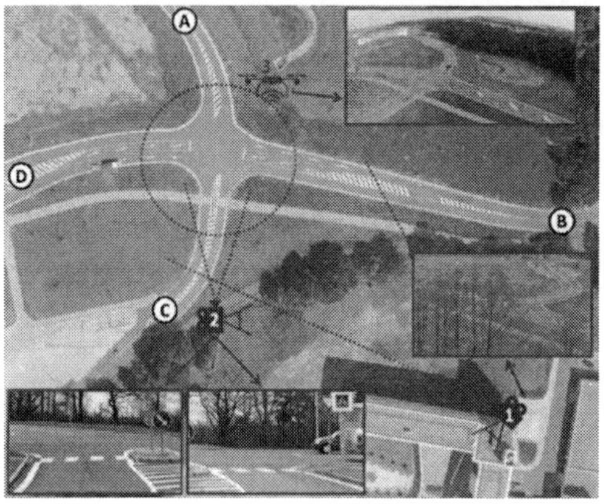

Figure 2. Monitored intersection and positions of cameras (1 and 2) and quadcopter (3) with demonstration of recorderviews.

The first camera was situated on the sixth floor of the university building (see 'Figure 2'– number 1) and the second camera was situated on the ground level of the intersection (see 'Figure 2'– number 2). Camera 1 monitored general traffic on the intersection (also for measuring traffic volume) and camera 2 was used for detailed view of monitored building elements (curbs or island). The third camera was installed on quadcopter (see 'Figure 2'– number 3). View from this camera on 'Figure 2' shows intersection after rebuilding (see below for more information).

Table. 1. Frequency of observed traffic conflicts (selection)

No.	Traffic conflict (TC) and direction (video on YouTube)	Time of measurement						Sum of TC	c_{RW}*
		13^{30}- -13^{45}	13^{45}- -14^{00}	14^{00}- -14^{15}	14^{15}- -14^{30}	14^{30}- -14^{45}	14^{45}- -15^{00}		
1	$2j_r1$-O1S (B→C)	4	2	3	4	4	1	18	58.6
	(from it $6j_r2$-O1S)	(1)	(1)			(1)		(3)	
	https://youtu.be/PfFuPsjrR5c								
2	$2j_r1$-O1S (D→C)	2	0	1	3	0	1	7	100
	(from it $6j_r2$-O1S)				(2)			(2)	
	https://youtu.be/b2XTrIEFE3A								
3	$2j_p1$-O1S (C→B)	6	4	5	4	6	4	29	100
	https://youtu.be/se3jYzzHUxo								
	or								
	https://youtu.be/6clxD3-nmpo								
4	$2j_p1$-O1S (C→D)	1	3	0	2	0	1	7	98.4
	https://youtu.be/AITK9S_JULY								
5	$2j_p1$-O1S (A→D)	0	0	1	0	2	0	3	88.7
	https://youtu.be/6clxD3-nmpo								
	and								
	https://youtu.be/wloPP2dhk8A								
6	$2j_p1$-O1P (A→B)	3	4	3	7	5	3	25	54.5
	https://youtu.be/8XVShHCYVok								
7	$2j_s1$-O1S (B→A)	3	1	3	3	5	7	22	94.3
	(from it $2j_s1$-O1S + $2j_p1$-O1P)	(1)				(1)		(2)	
	https://youtu.be/t-1zqY3YiUc								
	or								

	https://youtu.be/zPG0Enhf--4								
8	$2j_s1$-O1P(S) (D→A) (from it $2j_s1$-O1P(S) + $2j_p1$-O1P) https://youtu.be/KEbnTxYSf_k or https://youtu.be/A2i1xzp_7V0	7 (1)	3 (1)	3 (1)	6 (2)	8 (2)	1	28 (7)	89.6

j_r = moving of the vehicle to another lane
j_p = moving of the vehicle into the opposite direction
j_s = wrong way of driving (along visual island)
*) Coefficient c_{RW} [TC/100 veh] was calculated for all the buses of the monitored direction. The value "100" means that all the vehicles caused this traffic conflict.

One of the most dangerous traffic conflicts are situations marked as $2j_p1$-O1S. For example, the vehicle driving from C to B (see 'Table 1', No. 3 and 'Figure 3', blue vehicle) was using the opposite direction to prevent driving on the curb because this curb has too small radius (see 'Figure 4' or video https://youtu.be/6clxD3-nmpo).

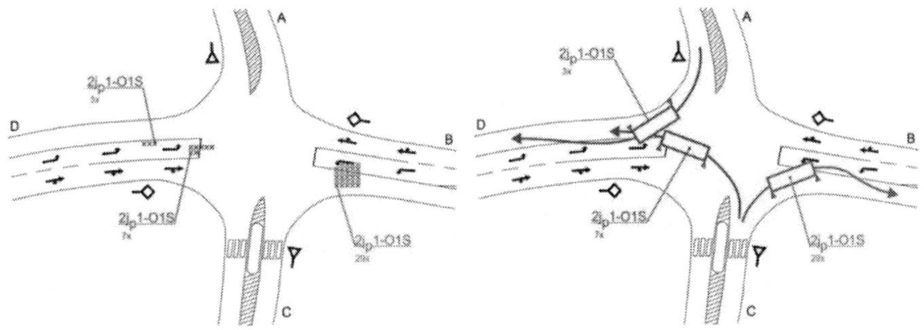

Figure 3. Scheme of location and course of traffic conflicts $2j_p1$-O1S

Figure 4. Photograph of traffic conflict $2j_p1$-O1S

The refuge island with pedestrian crossing (leg C) is another inappropriately designed building element of this intersection. Space between this island and curb of corner is relatively narrow for the buses exiting the intersection by this leg. For example, there was monitored traffic conflict marked as $2j_o1$-O1S (jo – driving near curb). The 'Figure 5' shows this conflict for bus which was driving from D to C (critical curbs are marked by red colour).

Figure 5. Scheme and photograph of traffic conflict $2j_o1$-O1S

These traffic conflicts were monitored by both the camera 1 (from the building) and by camera 2 (for detailed monitoring). 'Figure 6' describes important items for detailed video analysis of TC, i.e. the position of wheels and profile to guide strip ("a"+"b")and curb ("c"). Guide strip is divided into 2 parts – the line of horizontal marking ("a") and a part between the line and the curb ("b"). The recording was also divided into 2 parts (left and right) for more simplified and transparent analysis (see 'Table 2').

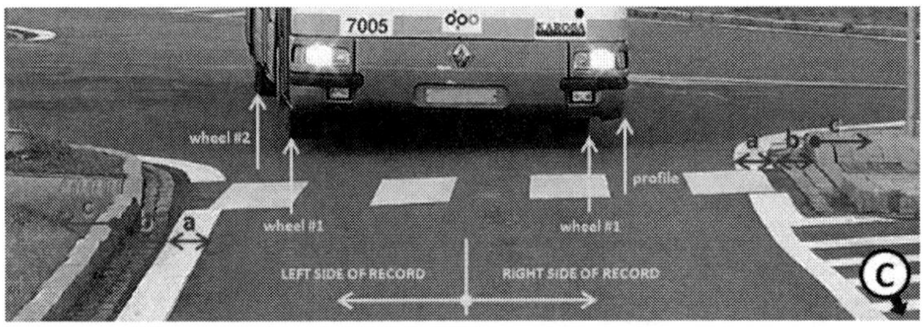

Figure 6. Description of monitored items for detailed analysis of traffic conflicts (camera 2)

'Table 2' shows selected vehicles and their movement through the intersection. The type of the vehicle (bus or truck, or car brand) is marked by colour. Red colour means traffic conflict j_r (moving of the vehicle to another lane) and blue colour means traffic conflict j_p (moving of the vehicle into the opposite direction). Some fields of the table are filled with red or orange colour. Red colour (in the "profile" column) means that the vehicle drove by profile (more than 50 %) above the guide strip (part "a" or "b") or above the curb (part "c" – see 'Figure 6'). Red colour (in the "wheel" column) means that the vehicle drove by the wheel (more than 50 %) the guide strip (part "a" or "b") or the curb (part "c"). Orange colour (in the "profile" column) means that the vehicle drove by profile (less than 50 %) above the guide strip (part "a" or "b") or above the curb (part "c"). Orange colour (in the "wheel" column) means that the vehicle drove by wheel (less than 50 %) over the guide strip (part "a" or "b") or the curb (part "c").

Table 2. An example of results of detailed video analysis of traffic conflicts

DIRECTION	RECORD	TRAFFIC CONFLICT	TIME	VEHICLE TYPE	STOPPED	LEFT SIDE OF RECORD												RIGHT SIDE OF RECORD											
						PROFILE			WHEEL 1			WHEEL 2			WHEEL 3			PROFILE			WHEEL 1			WHEEL 2			WHEEL 3		
						a	b	c	a	b	c	a	b	c	a	b	c	a	b	c	a	b	c	a	b	c	a	b	c
D→C	1	1	00:38	KAR											-	-	-										-	-	-
		27	13:36	JSP											*	*													
	2	51	00:21	REN											-	-	-										-	-	-
		89	22:03	REN											-	-	-										-	-	-
		100	27:19	SOR											-	-	-										-	-	-
		107	29:57	KAR											-	-	-										-	-	-
	3	164	29:21	KAR											-	-	-										-	-	-
C→B	1	4	01:51	KAR											-	-	-										-	-	-
		6	03:13	K-KAR																									
		7	03:33	KAR											-	-	-										-	-	-
		12	07:09	KAR											-	-	-										-	-	-
		24	11:39	KAR											-	-	-										-	-	-
		[illegible]	[illegible]	[illegible]																									

*) last wheel

It is evident the drivers were driving into the opposite direction (j_p) or in the wrong lane (j_r) to prevent driving on the curb (j_o). On the basis of this analysis the monitored intersection was rebuilt (see ‘Figure 7’). The pedestrian crossing and refuge island were demolished (see the red circle) and the turning lane for left turning (leg B) was rebuilt to connection lane for connection from the left (see the red arrow).

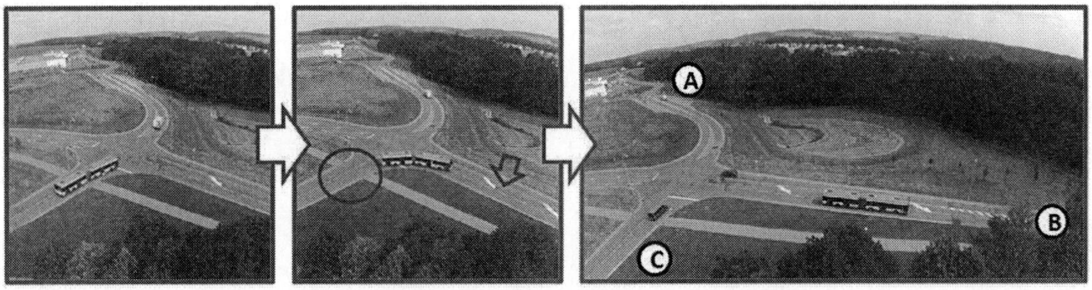

Figure 7. Monitored intersection after rebuilding

‘Figure 7’shows a situation when the bus is turning from C to B and it is driving into a new connection lane. This movement is correct and safe. During the analysis (IVATC) on this rebuilt intersection no buses (which were driving from or to leg C) caused the traffic conflict of type j_p or j_r or j_o.

5. Conclusion

The results of the video analysis of traffic conflicts point only to selected situations at monitored intersection and which are at least partially related to inappropriately designed building elements. These are the conflicts where the drivers drive into the opposite direction, use a wrong lane, drive near the curbs or drive on them, etc. Analyses have shown that the causes of these conflicts are small radii of corners and an inappropriately situated refuge island with pedestrian crossing.

The buses which used leg C, where the mentioned pedestrian crossing was situated, were monitored in the first place. The monitored traffic conflicts show relatively high values of weighted conflict ratec $_{RW}$ – it was often maximum, i.e. 100 TC/100 veh. As parameter V, the volume of all monitored vehicles (i.e. buses or trucks) was taken for each traffic conflict. This means that in the case of c_{RW} = 100 TC/100 veh, all vehicles created the particular traffic conflict. For all the traffic conflicts their seriousness was marked with a symbol 1 (potential traffic conflict) – this means that no other vehicle was affected at the time. The reason is that the bus drivers were aware that they will cause the traffic conflict (they will drive into the opposite direction, etc.). Therefore, they waited for an empty intersection without any other vehicles. This was also possible due to low traffic volume.

After the reconstruction of this intersection (demolition of refuge island and pedestrian crossing, and change of the turning lane to the connection lane), the previous traffic conflicts were not monitored anymore. To this is related the size of c_{RW} which equals zero. It can be said that the implemented changes contribute to the increase of the traffic safety at this intersection. However, it is always necessary to keep in mind that whichever change (even a positive one) can lead to another problem or several problems (which are often much more serious). For example, demolished pedestrian crossing can cause a traffic conflict with crossing pedestrian on leg C.

The conclusions of these analyses of traffic conflicts confirm the hypothesis that Innovated Video Analysis of Traffic Conflicts (IVATC) is the ideal tool for monitoring and evaluation of inappropriately designed building elements on intersections and this process has certainly its well-founded importance. It is always better to prevent problems (i.e. traffic accidents), than to handle their consequences later, when they can be (and usually are) of relatively serious character (death, injury, damage).

6. References

[1] Amundsen F and Hyden C 1977 *Proceedings of first workshop on traffic conflicts* (Oslo, Institute of Economics)

[2] Folprecht J 1995 Method for Monitoring and Analysis of Conflict Situations in Road Traffic and influence for road safety. In *International scientific conference on occsion of the 50th anniversary of VSB in Ostrava*. Ostrava: VSB – Technical University of Ostrava. pp 227-231. Czech

[3] Atombo C, WU C, ZHONG M and ZHANG H 2016 Investigating the Motivational Factors Influencing Drivers Intentions to Unsafe Driving Behaviours: Speeding and Overtaking Violations. In: *Transportation Research Part F: Traffic Psychology and Behaviour*. Elsevier Limited. Volume 43, 1 November 2016, Pages 104-121. ISSN 1369-8478. DOI: 10.1016/j.trf.2016.09.029

[4] Ross V, Jongen E M M, Brijs T, Ruiter R A C, Brijs K and Wets G 2015 The relation between cognitive control and risky driving in young novice drivers. In: *Applied Neuropsychology:Adult*. Volume 22, Issue 1, 2 January 2015, Pages 61-72. ISSN 23279095. DOI: 10.1080/23279095.2013.838958

[5] Cuenen A, Jongen E M M, Brijs T, Brijs K, Houben K and Wets G 2016 Effect of a working memory training on aspects of cognitive ability and driving ability of older drivers: Merits of an adaptive training over a non-adaptive training. In: *Transportation Research Part F: Traffic Psychology and Behaviour*, Volume 42, 1 October 2016, Pages 15-27. ISSN 13698478. DOI: 10.1016/j.trf.2016.06.012

[6] Folprecht J 2000 Current Developments and Perspectives of the Method of Monitoring of Conflict Situations on Road Traffic. Silnicni Obzor, Pages 39-44. ISSN 0322-7154. Czech

[7] Krivda V, Petru J, Mahdalova I and Zitnikova K 2016 *Evaluation Intersection Building Elements Using Video Analysis*. Ostrava: VSB - Technical University of Ostrava, Czech Republic, 184 p. Monograph. ISBN 978-80-248-3995-0. Czech

[8] Ambros J and Kocourek J 2013 *Methodology of Monitoring and analysis of Traffic Conflictes.* Brno: CDV Brno, ISBN 978-80-86502-62-5. Czech

[9] Krivda V 2013 Analysis of Conflict Situations in Road Traffic on Roundabouts. In: *PrometTraffic&Transportation: Scientific Journal on Traffic and Transportation Research.* Zagreb: University of Zagreb, Faculty of Transport and Traffic Sciences, Croatia, Volume 25, Issue 3, Pages 295-303. ISSN 0353-5320 (print), ISSN 1848-4069 (on-line). DOI: 10.7307/ptt.v25i3.296.

[10] Baker W T 1972 An Evaluation the Traffic Conflicts Technique. *Highway Research Record.* Washington, No 384

[11] Bennet C T 1971 Accident at Urban Junctions. In: *The Journal of the Institution of Highway Engineers*
[12] Hayward J C 1972 Near-miss Determination through Use of a Scale of Danger. *Highway Research Record*. Washington
[13] Perkins S R and Harris J I 1968 Traffic Conflicts Characteristics – Accident Potential at Intersections. *Highway Research Record*. Washington
[14] Rustam K and Sabey B E 1972 Accident and Traffic Conflicts at Junction. *TRRL Report*. Crowthorne, Bereshire
[15] Meel I P, Vesper A, Borsos A and Koren C 2017 Evaluation of the Effects of Auxiliary Lanes on Road Traffic Safety at Douwnstream of U-turns. In: *Transportation Research Procedia*. Elsevier B.V. Volume 25, 2017, Pages 1936-1950. ISSN 23521457. Conference Paper, In: *World Conference on Transport Research - WCTR 2016 Shanghai*, 10-15. DOI: 10.1016/j.trpro.2017.05.186
[16] Meel I P, Brannolte U, Satirasetthavee D and Kanitpong K 2017 Safety Impact of Application of Auxiliary Lanes at Downstream Locations of Thai U-Turns. In: *IATSS Research. International Association of Traffic and Safety Sciences*. Elsevier B.V. Volume 41, Issue 1, April 2017, Pages 1-11. ISSN 03861112. DOI: 10.1016/j.iatssr.2016.06.002

Acknowledgement
The work was supported from sources for conceptual development of research, development and innovations for 2017 at the VSB-Technical University of Ostrava which were granted by the Ministry of Education, Youths and Sports of the Czech Republic.

Psychological models for development of motorcycle helmet use among students in Vietnam

J Kumphong [1], T Satiennam [1,*], W Satiennam [1] and Tu Anh Trinh [2]

[1] Faculty of Engineering, Khon Kaen University, Khon Kaen 40000, Thailand.
[2] Faculty of Civil Engineering, Ton Duc Thang University, Ho Chi Minh 700000, Vietnam.
* Corresponding author. Tel.: +6681-974-4481.
E-mail address: sthaned@kku.ac.th

Abstract. A helmet can reduce head accident severity. The aim of this research study was to study the intention for helmet use of students who ride motorcycles in Vietnam, by Structural Equation Modeling (SEM). Questionnaires developed by several traffic psychology modules, including the Theory of Planned Behaviour (TPB), Traffic Locus of Control (T-LOC), and Health Belief Model (HBM), were distributed to students at Ton Thang University and University of Architecture, Ho Chi Minh City. SEM was used to explain helmet use behaviour. The results indicate that TPB, T-LOC and HBM could explain the variance in helmet use behaviour. However, TPB can explain behaviour (helmet use intention) better than T-LOC and HBM. The outcome of this study is useful for the agencies responsible to improve motorcycle safety.

Keyword: Helmet use, Motorcyclist, Traffic psychology, Structural Equation Modeling.

1. Introduction

Most prevention research has identified that helmets can reduce head injuries and reduce the chance of death [1], so wearing a helmet is a behaviour that should be observed when driving a motorcycle on the road. Most of the world has law enforcement about helmets for drivers. Vietnam is one country that has been successful in helmet law enforcement with a high helmet wearing rate. In addition to law enforcement, a campaign is an alternative that supports helmet use behaviour. According to the last studies on helmets in Ho Chi Minh City, Vietnam, traffic psychology factors (Subjective Norm and Perceived Behavioural Control etc.) have been associated with helmet use [2]. However, there is still a few of relating studies. Studies of traffic safety factors (TPB, HBM and T-LOC) tend to the creation of more effective campaign measures, and when done simultaneously with stronger traffic law enforcement, this will lead to a sustainable behaviour of helmet use. The aim of this research is to study the relationship between helmet use intention factor and other psychological factors by using the Traffic Locus of Control (T-LOC), Theory of Planned Behaviour model (TPB), Health Belief Model (HBM), on riders' behaviour toward helmet use.

2. Psychological Models and Hypotheses

Locus of Control (LC) was used to test our first model [3-4]. In this study, the LC model is composed of two elements which are internal and external LC belief modulation (see Fig.1) [4]. Internal LC indicates that events, such as accidents, are the result of someone's actions which can be controlled. On the other hand, external Traffic Locus of Control (T-LOC) orientation indicates that the events cannot be controlled by anyone. [5] Technically, it is easier to motivate people to use helmets associated with internal beliefs rather than external beliefs, because the belief of one's control has a direct correlation to the conscious decision to use a helmet to reduce risks of injuries from a crash.

As described in Fig.2, TPB states the reason for a person's intention to have certain behaviors. A person's attitude and subjective norms determine the behavioral intention. Attitudes are a person's behavioral evaluations such as thoughts about the benefit of using helmets [6-7]. On the other hand, subjective norms, composed of beliefs, also concern other people's beliefs [4, 8].

Threat perception and behavioral evaluation (see Fig. 3) are the two angles of health behavior that the HBM [4, 9] focuses on. There are two elements of threat perception which are delicateness of a cycling crash and anticipated severity of a crash's consequence, such as the possibility of getting a head injury. Based on the aforementioned literature review, the following are our study of the T-LOC model, TPB model and HBM model and our hypotheses are shown in Figures 1, 2 and 3. Therefore, we employ the T-LOC to explain the intention of helmet use. Accordingly, we propose the following hypotheses:

H_1: Internal variables are positively related to the intention of helmet use.

H_2: External variables are positively related to the intention of helmet use.

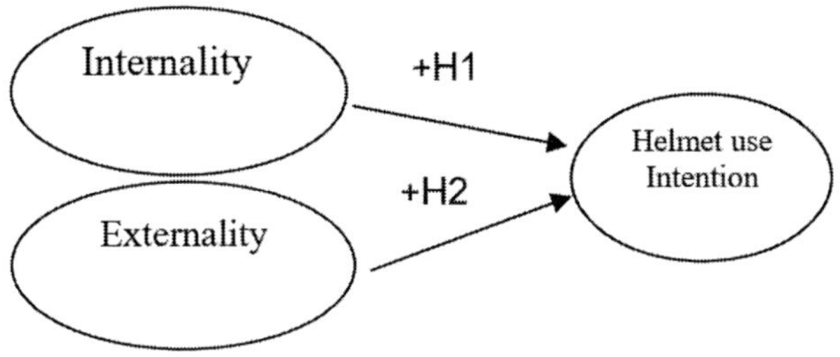

Figure 1. Traffic Locus of Control (T-LOC) framework.

We employ TPB to explain the intention of helmet use and helmet use behaviour. Accordingly, we propose the following hypotheses:

H_3: An Intention variable is positively related to the helmet use behavioural variable

H_4: An Attitude variable is positively related to the intention of helmet use.

H_5: A Subjective norm variable is positively related to the intention of helmet use.

H_6: A Perceived behavioural control variable is positively related to the intention of helmet use.

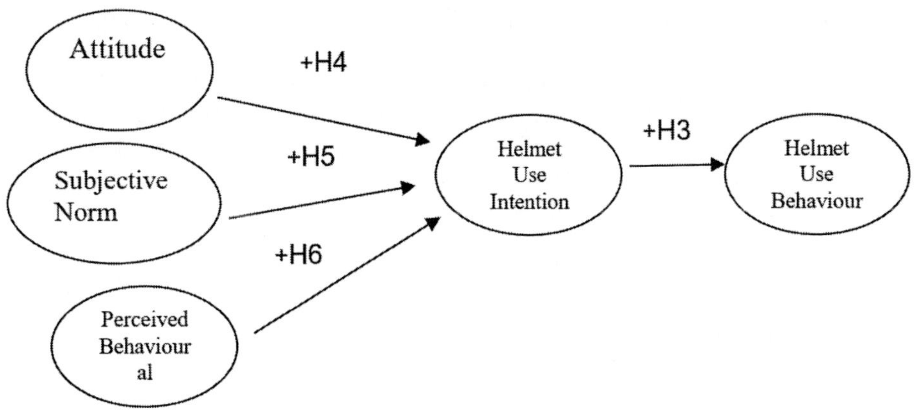

Figure 2. Theory of Planned Behavior model (TPB) framework.

Also, we employ the HBM to explain the intention of helmet use. Accordingly, we propose the following hypotheses:

H₇: A Perceived susceptibility variable is positively related to the intention of helmet use.

H₈: A Perceived severity variable is positively related to the intention of helmet use.

H₉: A Perceived benefits variable is positively related to the intention of helmet use.

H₁₀: A Perceived barriers variable is negatively related to the intention of helmet use.

H₁₁: A Motivation variable is positively related to the intention of helmet use.

H₁₂: A Cues to action variable is positively related to the intention of helmet use.

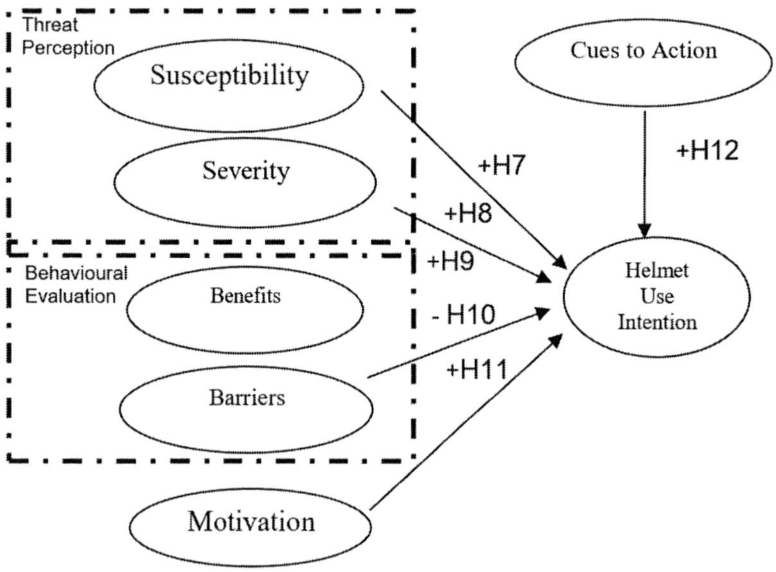

Figure 3. Health Belief Model (HBM) framework.

3. Methodology

3.1. Data collection and measures

Face-to-face interview where given to students for data collection. All of the total of 250 respondents were involved in this study survey. The time for an interview was about 10 min. The object of this study was to explore helmet use behaviour among students. The final sample size in this study was 210 (male = 104, female = 106). From the descriptive study, students were aged 17 to 28 years old (Average = 20 years old). Student riders had average experience of 3.8 (2.3) years. However, more than 70% of them had a motorcycle license and more than 90% of them had a motorcycle helmet. Table 1 shows the measurement items and scales. This study developed the psychological questionnaires examining behaviour using T-LOC, TPB and HBM as a frame of reference.

Table 1. Variables, Concepts of the Items and Scales

Variables	Items	Alpha	Example of item	Scoring
Traffic Locus of Control (T-LOC)				
Internality	2	0.26	I can reduce my accident risk.	1=Strongly Disagree /5 =Strongly Agree
Externality	2	0.31	Accidents depend mostly on shortcomings in other drivers' driving skills.	1=Strongly Disagree /5 =Strongly Agree
Theory of Planned Behavior model (TPB)				

Intention	2	0.61	For the next 3 months, I will wear a helmet when driving...	1=Strongly Disagree/ 5 =Strongly Agree
Behavioural	1	-	How often do you wear a helmet in the city?	1=Never/5= Always
Subjective Norm	2	0.79	People who are important to me think that I should wear a helmet.	1=Strongly Disagree/ 5 =Strongly Agree
Perceived Behavioral Control	2	0.53	I believe I have the ability to wear a helmet.	1=Strongly Disagree/ 5 =Strongly Agree
Attitude	3	0.73	Wearing a helmet is:	1=Bad/5= Good
Health Belief Model (HBM)				
Perceived benefits	2	0.50	Wearing a helmet protects me from getting a head injury in an accident.	1=Strongly Disagree/ 5 =Strongly Agree
Perceived severity	2	0.49	Being injured in an accident due to not wearing a helmet could lead to long term health problems, costs and income losses.	1=Strongly Disagree/ 5 =Strongly Agree
Perceived barriers	1	-	Wearing a helmet is uncomfortable when it is hot.	1=Strongly Disagree/ 5 =Strongly Agree
Health motivation	1	-	Nothing is as important as good health.	1=Strongly Disagree/ 5 =Strongly Agree
Perceived susceptibility	1	-	My probability of having an accident in the next 3 months is very small.	1=Strongly Disagree/ 5 =Strongly Agree
Cues to action	1	-	If more people would wear a helmet, then I would also wear a helmet more often.	1=Strongly Disagree/ 5 =Strongly Agree

3.2 Data analysis

The analysis of the results was divided into two parts (Adapted from the step study from Tankasem et al. (2016)) [10-11]. Overall, the model fit was evaluated against the number of recommended fit statistics and fit indices based on Hair et al. (2010) [12]. The first analysis was a factor analysis on latent variables given by questionnaire. Reliability of the latent variables was analyzed by three indices including Cronbach's α (Alpha). All variables were analyzed based on a hypothetical model, based on T-LOC, TPB and HBM, by confirmatory factor analysis (CFA). The final part used Structural Equation Modeling (SEM) to analyze all variables.

4. Results and Discussions

We present the indexes in the Structural Equation Model and factors influencing the indexes with standardized path coefficients. The most often indicated number of recommended statistics and indices in Table 2 are fitted for the SEM based on Hair et al. (2010) [12]. Therefore, the model fits between the theoretical constructs and observation constructs.

Figures 4, 5 and 6 show the results of structural models with standardized path coefficients for T-LOC model, TPB model and HBM model, respectively. Only the TPB model's fit passes a number of recommended fit indices.

Table 2. Explanatory power and fit index of models.

Model fit	Recommended value	Model $_{T\text{-}LOC}$	Model $_{TPB}$	Model $_{HBM}$
$\chi 2$		13.063	46.927	39.368
df		7	22	29
Chi-square/df	< 3.0	1.866	2.133	1.357
GFI	> 0.90	0.981	0.956	0.963
CFI	> 0.90	0.913	0.953	0.950
RMSEA	< 0.08	0.064	0.074	0.041

Note: The *p*-value is sensitive to sample size > 200

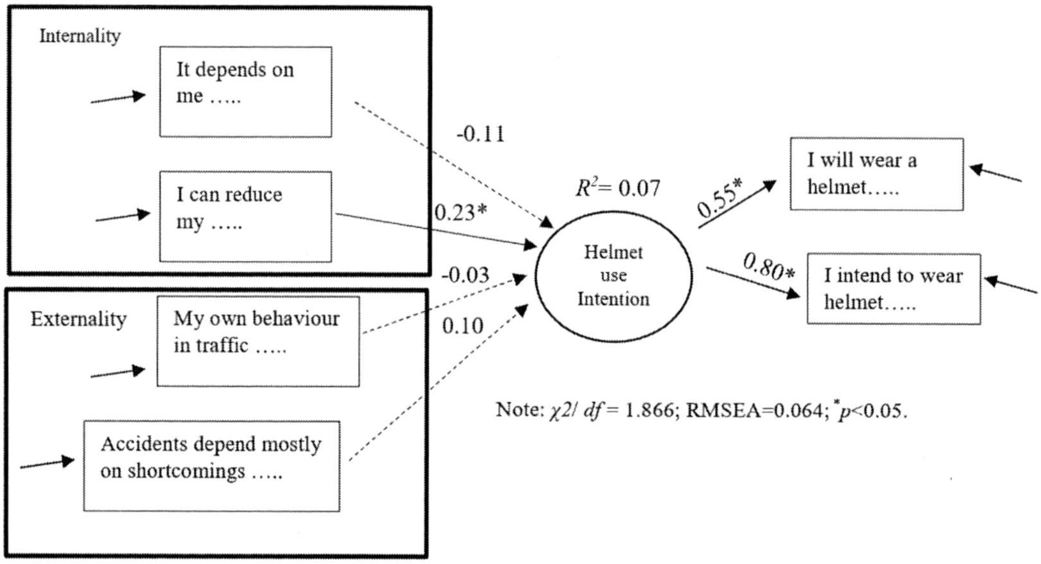

Figure 4. T-LOC model structure

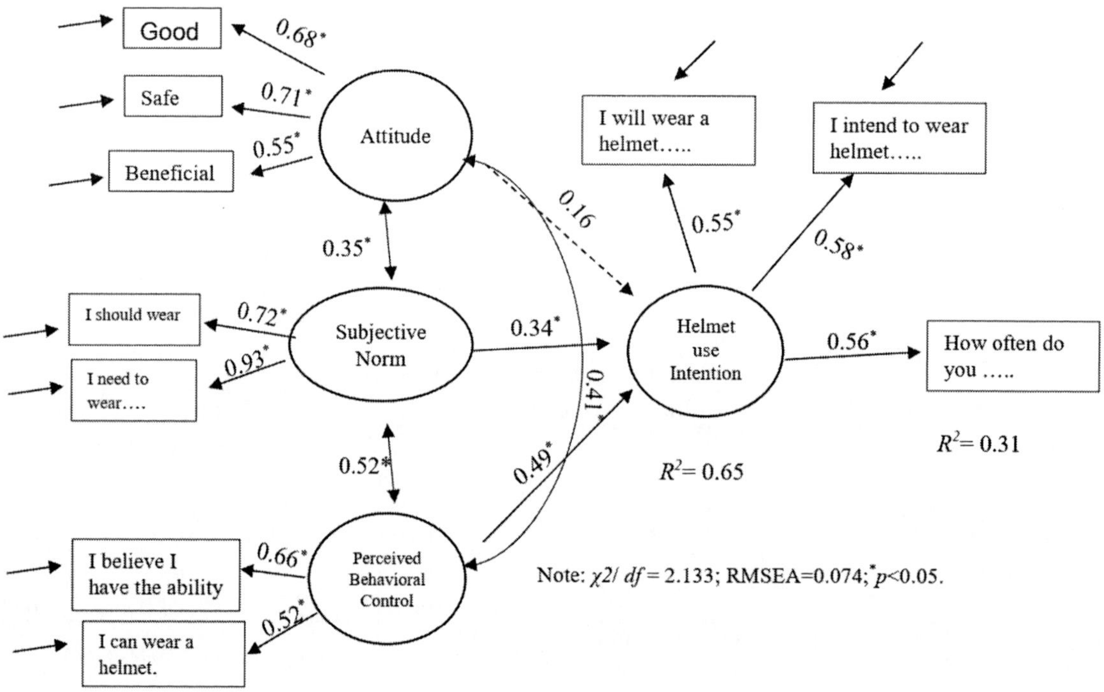

Figure 5. TPB model structure

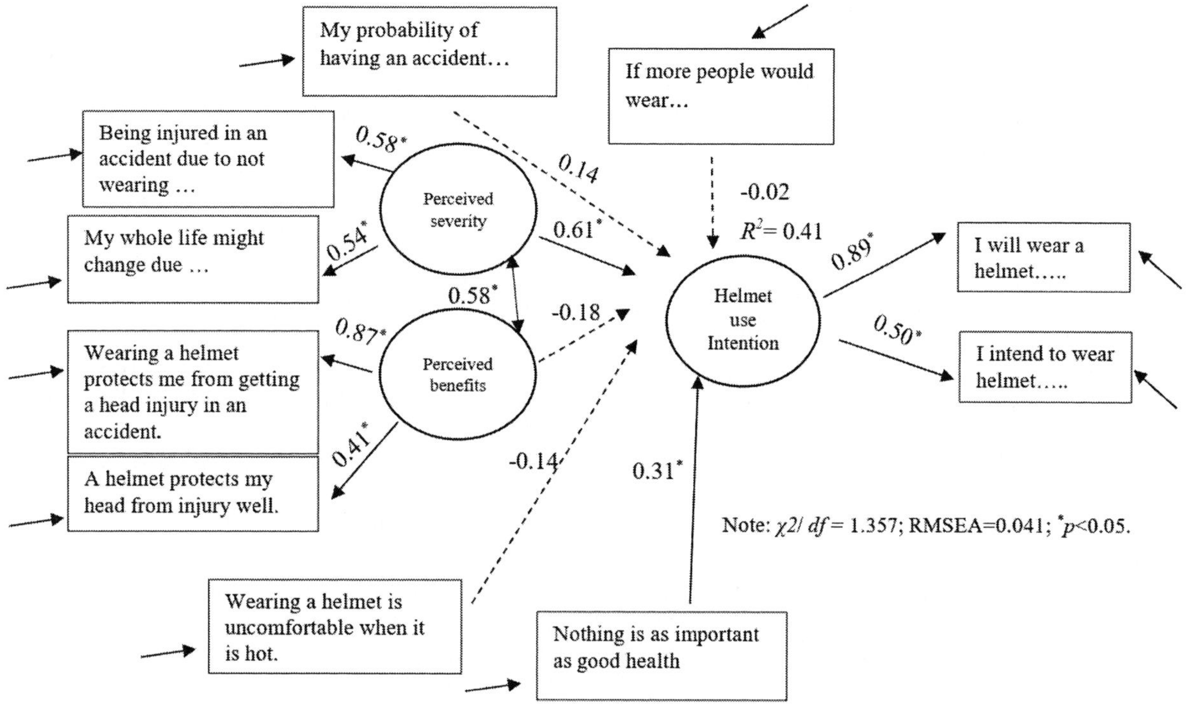

Figure 6. HBM model structure

These results confirm the hypothesis H_1 that for the psychological factors of T-LOC, the Internality factor is positively and significantly correlated with the helmet use Intention factor at the 95% confidence level. The measurement strategies for LC are general orientations in life or a particular domain limitation. Because general LC explains low variance in the amount of health behaviour, scales of particular domains have been improved [8, 4]. Multidimensional Health Focus of Control scale or MHLC, developed by Wallston et al. (1978), has been used the most as an instrument to measure the control of the health locus. [4] We discovered that MHCL is too broad, so we used the T-LOC measure for the motorcyclist's situation (Fig.4).

In this TPB study, these results confirm the hypotheses (H_4 and H_5) that the psychological factors of TPB, Subjective Norm factor and Perceived Behavioural Control factor are positively and significantly correlated with the helmet use Intention factor at the 95% confidence level (Fig. 5). According to TPB, the third Intention predictor is created by perceived behavioural control; this refers to one's insight about whether a behaviour is easy or difficult. [4, 6-8] A similar conclusion was reported by Brijs et al. (2014); Kumphong et al. (2017).

In the results of the Health Belief Model (HBM), these results confirm the hypotheses (H_8 and H_{11}) that the psychological factors of HBM (Fig.6), Perceived severity factor and Health motivation factor are positively and significantly correlated with the helmet use Intention factor at the 95% confidence level. Threat Perception is composed of two different belief sets related to perceived severity, such as "Being injured in an accident due to not wearing a helmet could lead to long term health problems, costs and income losses" to the use of helmet, as well as advantages like safety. Moreover, the HBM also includes cues to action and health motivation. "Health motivation" cites the stimulus of one to wear a helmet, while health motivation means someone is ready to worry about health [4, 16]. Health motivation is not only for cycling but determines the value that someone gives to safety and health. In this study the TPB could better explain behaviour than T-LOC and HBM.

5. Conclusions

Based on, the objectives of this study are to compare psychological factors influencing the helmet use intentions of motorcycle riders in university under the framework of T-LOC, TPB and HBM. These results confirm the hypothesis that the psychological factors (Internality, Subjective norms, Perceived severity and Perceived behavioral control) of all models have significant relations with the helmet use Intention factor. The TPB model can better explain behaviour than other models. These findings suggest that changing Social attitude, Subjective norms, Perceived severity and Perceived behavioral control for motorcycle riders about helmet use in another country can reduce the rate of fatalities among student motorcyclists.

6. References

[1] World Health Organization 2006 *Helmets :A road safety manual for decision-makers and practitioners* (Geneva: Switzerland)

[2] Trinh T A and Le T P L 2016 Motorcycle Helmet Usage among Children Passengers: Role of Parents as Promoter *Sustainable Development of Civil, Urban and Transportation Engineering Conference* (Procedia Engineering vol 142) pp10-17.

[3] Rotter J B 1966 *Generalized expectancies for internal versus external control of reinforcement* (Psychological Monographs)

[4] Lajunen T and Räsänen M 2004 Can social psychological models be used to promote bicycle helmet use among teenagers? A comparison of the Health Belief Model, Theory of Planned Behavior and the Locus of Control (*Journal of Safety Research* Vol 35) pp 115-123

[5] Warner H W Ozkan T and Lajunen T 2010 Can the traffic locus of control (T-LOC) scale be successfully used to predict Swedish drivers' speeding behaviour? (*Accident Analysis and Prevention* Vol 42) pp 1113–1117

[6] Ajzen I and Fishbein M 1980 *Understanding Attitudes and Predicting Social Behavior* (Prentice-Hall: Inc.: Englewood Cliffs: New Jersey)

[7] Ajzen I 2005 *Attitudes, Personality and behavior* (New York: USA)

[8] Conner M and Sparks P 1996 The Theory of Planned Behaviour and health behaviours *Predicting health behaviour* (Buckingham: Open University Press) pp. 121– 162

[9] Rosenstock I M 1966 *Why people use health services?* (Milbank Memorial Fund Quarterly Vol 44) pp 94– 124

[10] Tankasem P Satiennam T and Satiennam W 2016 Cross-cultural differences in speeding intention of drivers on urban road environments in Asian developing countries (*International Journal of Technology* Vol 7) pp 1187-1195.

[11] Tankasem P Satiennam T and Satiennam W 2016 Psychological factors influencing speeding intentions of car driver and motorcycle riders in urban road environments (*International Journal of Technology* Vol 7) pp 1179-1186.

[12] Hair J F Black W C Babin B J and Anderson R E 2010 *Multivariate Data Analysis a global perspective 7th ed.* (New Jersey: USA)

[13] Wallston, K. A., Wallston, B. S., & DeVellis, R. (1978). Development of Multidimensional Health Locus of Control (MHLC) scales (*Health Education Monographs* Vol 11) pp 160– 170.

[14] Brijs K Brijs T Sann S Trinh T A Wets G and Ruiter R A C 2014 Psychological determinants of motorcycle helmet use among young adults in Cambodia (*Transportation Research Part F Traffic Psychology and Behaviour* Vol 26) pp 273-290

[15] Kumphong J Satiennam T and Satiennam W 2017. A Study of Social Norms and Motorcycle Helmet Use Intentions among Student Riders in University: A comparison of the Theory of Reasoned Action and the Theory of Planned Behavior. *The 12th International Conference of Eastern Asia Society for Transportation Studies* (Sheraton Saigon Hotel & Towers: Ho Chi Minh City: Vietnam)

[16] Becker, M. H., Haefner, D. P., & Maiman, L. A. (1977). The Health Belief Model in the prediction of dietary compliance: a field experiment (*Journal of Health and Social Behaviour* Vol 18) pp 348–366.

Acknowledgments
This research received a Research Fund for Supporting Lecturer to Admit High Potential Student to Study and Research on His Expert Program Year 2016.

Establishing a green-and-blue network within an existing settlement in central Hanoi towards urban ecologicalisation

Quang Minh Nguyen[1]

[1] National University of Civil Engineering, Hanoi, Vietnam

* Corresponding e-mail address: ktsquangminh@yahoo.com

Abstract. As a mega city, Hanoi must solve huge problems in its urban development of which landscape is among the greatest and most urgent issues to deal with. Locally, within one settlement, landscape is often simply regarded by many city residents as a small park that helps improve environmental quality, especially in summer, and provides an open space for the local community to have a walk a few times a day. But actually, landscape plays a much more important role in one settlement as a complete unit and also as a miniature of a large city. Internationally, landscape is a complex concept as well as an integral part of urban ecology, and sustainability has been set at a much higher level and the final goal to reach. In central Hanoi, where the population and construction density is so high that the green area per capita is among the lowest in Asia, it is high time to rethink about the enhancement of living quality in a difficult situation. The academic article will focus on landscape ecology, rather than on landscape only, in central districts and propose an ecologicalisation for a typical settlement here towards a more liveable city concept.

1. Introduction

In 2015, Hanoi was selected along with 21 Asian cities for a city greening assessment survey. The green area per capita in Hanoi was just 11.2 m^2, significantly below the average figure (38.6 m^2) and incomparable to Singapore (66.2 m^2) and Guangzhou (166.3 m^2) [1]. The Government requires that by 2020 Hanoi and Ho Chi Minh City as two special metropolitan areas of the country will have secured 15 m^2 of green area per capita [2], including tree rows along streets while two tree rows are not conventionally counted into this indicator. Thus, it is still a long way for Hanoi to reach its modest goal towards 2020, before going further to match the continent's average level, seeing that there are only two years left.

In fact, the current green area is unevenly distributed across the city. As a result of the expansion in 2008, the whole province of Ha Tay and some part of Hoa Binh province were annexed to Hanoi. The total green area of Hanoi has increased manifold, because the expansion has taken all the forests, green belts, green corridors and farmland. In ten central districts, the greening indicator as of 2014 was extremely low, under 2 m^2 per capita [3].

Figure 2 shows an aerial view of a typical living quarter in a central district of Hanoi with a very high building density. All the roofs are made of either galvanised steel sheets or ceramic tiles while most of the ground area is covered with either concrete or asphalt as heat-absorbing materials. Green areas, if any, are scarce and scattered. Consequently, Hanoi is facing both urban heat island and flood as two great challenges, apart from traffic congestion, air pollution and water contamination. Over the past five years, the summer air temperature has increased from 40°C (in 2012) to 42.5°C (in 2017).

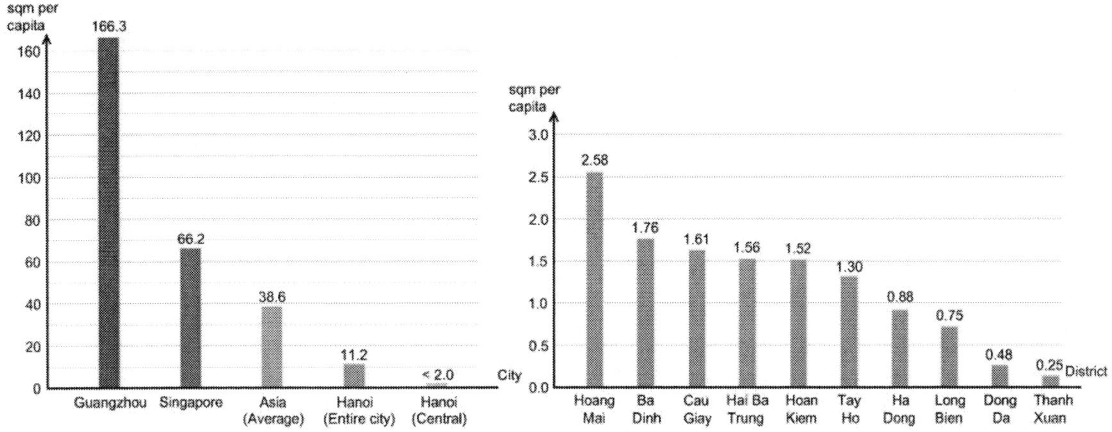

Figure 1. Green area per capita indicators in Asian cities [1] and in Hanoi's ten central districts [3]

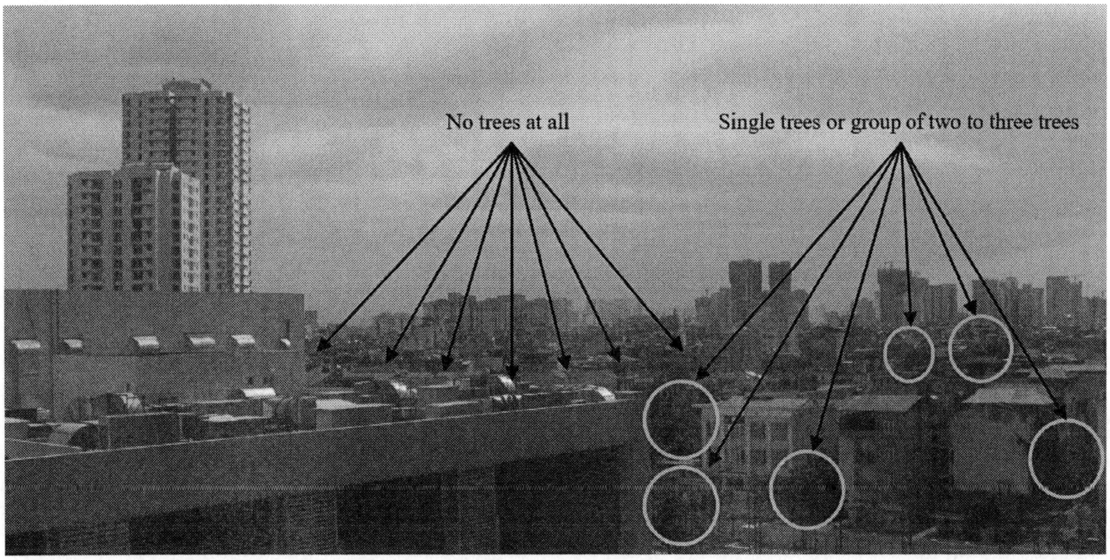

Figure 2. A typical residential quarter in Hai Ba Trung district (central Hanoi) [4]

Figure 3. Landscape of Nguyen Chi Thanh road in Hanoi in 2014 (left) and in 2016 [5]

Furthermore, in 2016 and 2017, the city authority decided to cut down approximately 8,000 tall shade trees along Nguyen Trai road, Pham Van Dong road and Nguyen Chi Thanh road (Figure 3) for sky train and urban traffic improvement projects, despite the petition of scientists and the protest of the public. It will take at least 20 years to have the same exuberant shade trees again. In the meantime, as climate change keeps on going and the summer air temperature continues to rise, Hanoi and its inhabitants will have to cope with an uncertain future.

In Vietnam, urban ecology has just been considered over the past 20 years in view of landscape as the first step of a multi-scalar approach. For example, in her master's thesis, Thu Trang Dam investigated into the history of landscape architecture in Vietnam and evaluated the quality of landscape (parks and street tree rows) in the historical centre of Hanoi and 17 socialist settlements between city ring road no. 1 and city ring road no. 2. She scrutinised the languages of trees and the arts of planting trees in living quarters and public building complexes and proposed a number of combining trees and plants of different species and sizes on three scales (single house, group of houses and entire living quarter) to improve the landscape and environment [6]. Choosing to study green areas only, she did not mention water surface as an important element in shaping landscape architecture and did not regard "bio-diversity" or "sustainability" as one of the key factors in securing a good living quality in cities in a broad sense either. Another expert, Tat Ngan Han, classified urban landscape in Vietnam into four main types (roads and streets, city squares, city parks and living quarters) and each main type consisted of several patterns. He emphasised the importance of water surface and recommended that it should be combined with green areas in several ways, depending on the typology and other natural conditions of the site, as well as on the socio-cultural context of the local community. Then he went into detail with the meaning and several design strategies of each pattern, but mostly in garden and park planning. He also proposed a list of 76 shade tree species and 53 ornamental plant species that can be grown in cities with characteristics of each species for reference, together with a step-by-step guidance for an urban landscape project [7]. Tat Ngan Han, just like Thu Trang Dam, did not take into account urban ecology as a complex system and as a biological resource that plays a vital role in both nature and the built environment. More recently, An Thinh Nguyen, published one handbook in landscape ecology in which he described a number of scientific terms related to landscape (including eco-zone, bio-diversity, multi-functional landscape, landscape matrix etc.) and addressed landscape ecology on territorial scale in a comprehensive manner by taking an interdisciplinary approach (combining biology, geography, life science and earth science) to ecology as the core of landscape in general [8]. However, he did not deal with landscape in view of urban planning or urban design as he specialises in environmental studies.

Today, in the context of environmental pollution, ecological degradation and rapid urbanisation with extremely high building density, the study on landscape ecology is a very important and urgent task, especially in theory, in order to pave the way for practical solutions that will help enhance the living quality for millions of inhabitants living in such a heavily polluted mega-city as Hanoi. The study becomes even more necessary, when most of research projects deal with landscape either on a large scale or in specific cases. Not only improving the landscape or protecting the environment, these studies should also accentuate the significance of ecology as a complex system and enrich bio-diversity as one of the most important natural resources to human beings.

The paper is primarily based on theoretical investigations into landscape ecology in combination with current tendencies in urban landscape development. Then, a preliminary study on integrating two landscape systems into one complex and complete network within a neighbourhood as a basic unit shall be undertaken, and further developed into a more detailed concept for a city-wide application.

2. Landscape in view of ecology

By nature, landscape has been closely connected with agriculture, forestry, geography, archaeology, ecology, climatology, environmental science, life science, earth science, etc. throughout its long history of development. One of the most outstanding and influential theories in urban landscape is Ebenezer Howard's book "Garden Cities for Tomorrow" published for the first time in 1898 in which he idealised a small city with 32,000 inhabitants planned on a concentric pattern with open spaces, public parks and six radial boulevards, 37 m in width each, extending from the centre. The garden city would be self-sufficient and when it reached its full population size, another garden city might be developed nearby. Howard envisaged a cluster of several garden cities as satellites of a central city of 58,000 people, linked by road and rail. With gardens, a city would remain a low building density while ensuring a high living quality [9].

Today, modern technology has broadened the horizons in sciences, including in landscape. Modern landscape has been integrated with and immensely supported by GIS, along with worldwide communication network. These effective tools facilitate landscape management and observation of landscape change over the years. The scope of work is not limited to a country or a continent any more, but developed on planet scale in order to find solutions to global issues, such as climate change and natural disaster mitigation.

One of the modern theories to be highlighted is multi-functional landscape. Multi-functional landscape is shaped to bring more benefits to local communities. Urban planners, landscape and building architects should be responsible for landscape design in order to meet the needs of residents and help the whole system of urban landscape work as smoothly and effectively as possible, from large-scale consideration (city) to small-scale investigation (building). A successful creation of high-performance and multi-functional landscape will accelerate the expansion of regional landscape with an integration of the metabolism within an eco-system and landscape design process. The aim of this task is to establish a strong eco-system needed for a liveable city and all of its settlements where wellness as the number one factor can be secured. A multi-functional landscape design solution must take into account various factors that may have long-lasting influences on that area and include the human intervention in natural cycles to a certain degree. Thereby it is possible to establish or re-establish both forms of environment: natural environment and built environment, so that social sustainability as the final aim can be achieved and reflected in the following aspects:

- Supporting physical and biological activities of an eco-system by protecting and enhancing the local biodiversity in aquatic environment and terrestrial environment, in both quality and quantity;
- Providing services, including energy supply and public facilities;
- Regulating technical plans, reducing waste and reusing waste materials from city parks;
- Organising socio-cultural activities [10].

The two terms - biotope and ecotope - were used and developed respectively Friedrich Dahl in 1908 and Carl Troll in 1945. Biotope is an area of uniform environmental conditions providing a living place for a specific assemblage of plants and animals while ecotope is defined the smallest ecologically distinct landscape features in a landscape mapping and classification system. As such, they represent relatively homogeneous, spatially explicit landscape functional units that are very useful for stratifying landscapes into ecologically distinct features for the measurement and mapping of landscape structure, function and change [11]. Both concepts show a respect for ecology and describe an eco-system as a natural context that is vital to all kinds of human activity. Therefore, biotope and ecotope work as important measurements of landscape and pave the way for landscape to become an integral part of modern urban planning in the era of environmental crises.

3. Landscape in view of sustainability

Sustainability is understood as the practice of maintaining processes of productivity indefinitely - natural or human made - by replacing resources used with resources of equal or greater value without degrading or endangering natural biotic systems, and has been officially defined in the Brundtland Report "Our Common Future" as "a development that meets the needs of the present without compromising the ability of future generations to meet their own needs. It contains two key concepts: the concept of 'needs', in particular, the essential needs of the world's poor, to which overriding priority should be given; and the idea of limitations imposed by the state of technology and social organisation on the environment's ability to meet present and future needs" [12].

Sustainability in urban planning and landscape design can be most clearly reflected in the efforts to maintain local biodiversity by focusing on vernacular flora and fauna as well as by closing the on-site water and material cycles. Moreover, renewable energy is encouraged to be generated and used on-site, most preferably and economically solar energy and wind energy. Similarly, rainwater as "gift from the sky" should be collected and used for various purposes instead of tap water. Saving for long-term use in the future is a guiding principle that should be applied to all other natural resources.

Socially considered, sustainability will not be achieved without the commitment of the public in any urban development project, including in landscape, because their behaviour has a strong influence on nature and landscape in general. It is the local people who decide how their living quarter may look in terms of urban design and landscape quality. Their awareness is a driving force for their participation and action programmes aimed at the protection of environment, improvement of landscape and development of ecology towards sustainability. Local people are expected to play a much more active role in the decision making, implementation and monitoring of implementation as well as the co-management of urban landscape along with the local authority within their neighbourhood as the first step in the three-level hierarchy of urban landscape [13].

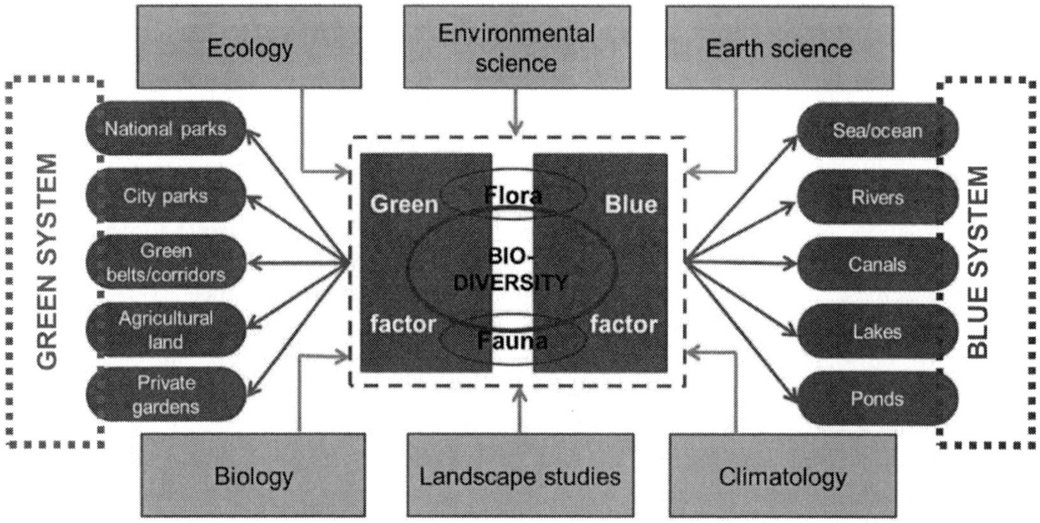

Figure 4. Bio-diversity as a central factor of urban landscape and ecology [14]

4. Green and blue network

Green-and-blue network is the interconnection of a green system and a blue system within one area in a city as shown in Figure 4. The advantage of developing such a broad network is apparent. As a single factor, parkland or water surface can reduce the summer air temperature by 3°C, but when they are combined, the effect will be greater, by 5°C [15].

The green factor in a green-and-blue network consists of all kinds of vegetation, from grass to tall shade trees. In reality, there are three patterns for a settlement to develop: patch (point or small area), corridor (line that links points or small areas) and matrix (lines that intersect one another and cover a large area) [16]. Forman's concept (patch-corridor-matrix) for the green factor will also apply to the blue element, with a fountain or pond (patch), canal (corridor) and lake or river (matrix).

As part of city greening, city farming is becoming popular again in Hanoi and other cities today, over three decades after the socialist plan economy and its self-sufficiency policy ended. The excessive use of pesticides, preservatives and chemicals in agriculture has risen up the public concern about food safety. That is the reason why more and more city residents have nowadays decided to grow vegetables and several kinds of fruits at homes. They have made full use of every square metre available in and around their houses/apartments for this purpose: pavements, corridors, loggias, balconies, courtyards, backyards, terraces and roofs are all used as mini or medium-sized gardens. So, the greening rate in concrete-and-glass cities has considerably increased. This is an unexpectedly positive aspect of food poisoning. In view of urban ecology and sustainability, city farming needs to be further developed and integrated into the spatial planning of a neighbourhood, especially those with a high building density and low greening rate in central districts.

Located in the tropical zone, Hanoi should take full advantage of its bio-diversity and high rainfall (1,700 mm a year) [17] to enrich urban landscape. This amount of rain water can be collected to water an abundant number of plants and trees. A small part will percolate into the ground and keep the underground water level stable. Rainwater is often seen as a major cause of flooding. However, this argument is not absolutely correct, because rainwater has not yet been properly used. Actually, to an urban eco-system, rainwater is much more important than most people often think. This is not just a landscape-shaping element or climate-regulating factor, but also provides an appropriate living environment for numerous species of plants and creatures. Grey water discharged from households and public buildings, after an on-site technical treatment process at a certain level of purification (about 70%), can be mixed with rainwater and stored in a retention pond or lake nearby where the natural cleaning process by means of phytoremediation continues to take place. Some of the remaining soluble substances, especially the ions of heavy metals will be absorbed by certain aquatic plants. These plants are quite diverse: some grow on water surface while others live entirely under water. It is recommended to raise different species of aquatic plants together in one place, since some can dissolve or decompose organic substances in water while others just absorb inorganic compounds. As a result, the purification will be more effective, and more importantly, biodiversity can be achieved. Once these plants absorb soluble substances in water to their maximum capacity, they can be removed and treated, and new plants should be put into that retention lake and another cleaning process will begin.

5. How to establish a green-and-blue network within a neighbourhood in central Hanoi?

Central Hanoi, as aforementioned, is characterised with high population density (14,700 inhabitants per km^2) [18] and also a high building density (up to 70% in some wards of ten central districts) [4]. That means there is not much space left for a green-and-blue network. In these wards, there are merely tree rows along the streets which should not be accounted in the green area per capita indicator. Only a few wards and neighbourhoods make good use of a city park or a city lake nearby. However, the local residents share it with people from other parts of the city. Therefore, such an open and public place is usually overcrowded. It is important to establish a green-and-blue network within each settlement for local residents themselves. In other words, they will have their own abundant eco-system within a short walk from their homes and enjoy what nature brings back to them.

This green-and-blue network is an open museum to show children how beautiful nature is and educate them how to protect nature as well as living environment. Apart from regulating the micro climate, storing rain water, improving the landscape and encouraging social communication, this network plays a much more essential role in restoring, maintaining or enriching bio-diversity and acts as a mini bio-sphere in city.

With this in mind, a settlement-wide network should be established in three levels: private (single house or building), semi-public (group of houses/buildings) and public (entire settlement). A broad range of plants can be applied and vernacular plants will be prioritised. Creating identity is a higher requirement in city greening. The plants which contain poisons or have unpleasant smells will not be selected for a safety reason. When the plants grow well enough, they may attract birds, insects and small animals. An open water network should be integrated and it is recommended to turn illegally occupied public places into water bodies which provide an ideal living environment for certain species of aquatic flora and fauna. In modern urban design, underground public buildings and parking areas are quite common in order to restore the ground to nature. Similarly, green roof and façade will also maximise the naturalisation and ecologicalisation in a densely constructed urban area.

Some residents may feel anxious about open water areas as larvae of pest insects that annoy them may proliferate in summer months. Therefore, no canals ponds or lakes may be designed within a living quarter. Nevertheless, the case of the Cheonggyecheon river in Seoul transformed into an underground sewage tube in the 1960's and restored as a beautiful man-made flow as well as a tourist attraction site 40 years later is still a good lesson to learn. In the food chain theory, larvae of pest insects are eaten by small freshwater fish. If fish are put into lakes and ponds, they can control the development of larvae.

The depth of a canal, a pond and/or a lake within a living quarter must be calculated to ensure that it will be safe for children. For a bio-retention pond or lake, the water depth should be 0.4 m within 5 m distance from the bank. A rainwater canal is normally 3 to 4 m in width and 1.5 to 2 m in depth in order to drain away rainwater in case of downpour. It is necessary to put up high inox railings all along the canal on both sides to prevent children from climbing and falling into water.

Another crucial point in landscape design for cities and towns to be noticed and practised is creating permeable ground surface, so that rainwater can percolate into the deep soil layers and add a certain amount to the level of underground water and help keep it stable. It is suggested to use grassland or sand surface on top of soft soil for children's playgrounds, grass-filled paving blocks for car-parking areas and pathways, and porous concrete or permeable asphalt for main roads.

Table 1. Recommendation of establishment and development of green-and-blue network within a neighbourhood in central Hanoi [19]

System / Category	Green and blue network													
	Green factor										Blue factor			
	Garden	Roof	Terrace	Frontyard	Courtyard	Backyard	Balcony	Loggia	Corridor	Façade	Aquarium	Canal	Pond	Lake
Level 1: Private														
Housing pattern 1: Villa or garden house	•	•	•				•	•		○	•		•	
Housing pattern 2: Shop house		•	•		•	•	•	•		○	•			
Housing pattern 3: Row house		•	•	•	•	•	•	•		○	•			
Housing pattern 4: Flat or apartment		•	•				•	•	○	•	•			
	Playground/mini-park					Tree row along pathway								
Level 2: Semi-public	•					•					•	•	•	
	Public park					Tree row along pathway								
Level 3: Public	•					•						•		•

Note: • Recommended ○ Optional

Table 2. Recommendation of species for green-and-blue network within a neighbourhood in central Hanoi [19]

Space level	Green area	Water surface
Private		
Flora	Vegetables, medicinal or spice herbs, flowers, creeping plants, ornamental trees, fruit trees, etc.	Water lily, algae
Fauna	Birds	Gold fish
Semi-public and public		
Flora	Grass, flowers, creeping plants, ornamental trees, fruit and shade trees, etc.	Algae, water lily, lotus, reed, coon tail, umbrella palm, water hyacinth, etc.
Fauna	Birds, butterflies, crickets, earth worms, ants, bees, beetles, lady birds, etc.	Fish, shrimps, crabs, snails, water spiders, frogs, etc.

Thus, the step-by-step implementation of ecologicalising a settlement with a high building density in central Hanoi will be recommended as follows:

- Step One: Conserving all the existing open public places, mostly green areas, for future use as mini-parks or flower gardens in semi-public areas and medium-size parks in public areas;
- Step Two: Acquiring the land area illegally occupied for various land-use purposes for future use as water surface, retention ponds in semi-public areas and lakes in public areas;

- Step Three: Planning two medium-size parks/ponds or four small parks/ponds may be considered, subject to the size of the land area, instead of one large park/lake;
- Step Four: In all cases, green and blue factors are coupled together and planned in a way that ensures an even distribution and easy access;
- Step Five: Parks and gardens are interconnected with rows of trees along roads and streets. Similarly, ponds and lakes form a water network with canals;
- Step Six: Selecting appropriate species of plants and animals to grow/raise in terrestrial and aquatic environments. Applying the concepts multifunctional landscape, biotope and ecotope. As time goes by, new plants will grow and more animals will come, hereby diversifying the local bio-sphere;
- Step Seven: Integrating city farming in each household, making full use of open areas available;
- Step Eight: Undertaking regular maintenance of parks and ponds/lakes.

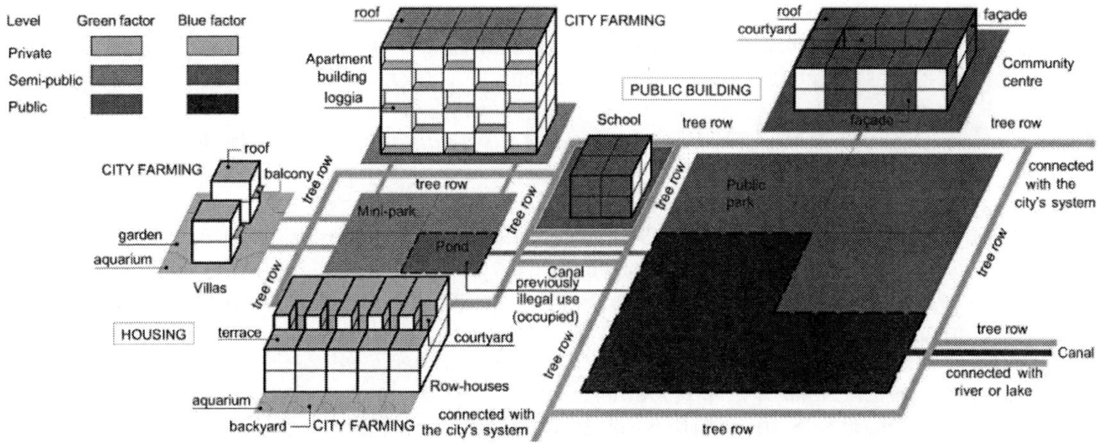

Figure 5. Interconnection of green and blue factors into an eco-network within a neighbourhood in central Hanoi [20]

6. Conclusions

Green and blue factors, as clarified, are indispensable elements in all cities and living quarters, because they will bring great benefits to the local communities, especially in central districts where the needs for more green areas and water bodies are stronger than in the suburbs. Today, a city and even a settlement must be a complete eco-system and a rich bio-sphere.

As two major components of this bio-sphere, green and blue factors can take various forms. They should be combined and flexibly applied to all open spaces and public places available, either on a small scale or a large scale. This green-and-blue network can be emphasised as an outstanding feature of a built environment that is worth living for people. Moreover, it will represent a new lifestyle, when everyone understands the importance of nature and ecology and will do his or her best to achieve the goal.

From a neighbourhood as the first step, this concept will be upgraded to a city-wide level as a more complex system which requires further investigations. In this long process, policy makers, landscape designers, life scientists and project managers need to work with local communities in realising such a holistic concept.

References

[1] United Nations ESCAP 2015 *Statistical Year Book for Asia and the Pacific - Facts and Trends as the Outset of the 2030 Development Agenda* p.119
[2] Prime Minister 2012 Decision No. 1659/QD-TTg, *Approval of the National Urban Development Programme in 2012 - 2020 period*

[3] Hanoi Single Member Park and Tree Co. Ltd. 2014 *Greenery Development Strategy for Hanoi until 2030*
[4] Nguyen Q M 2017 *City survey*
[5] Ministry of Information and Communication 2016 full weblink *http://vietnamnet.vn/vn/doi-song/duong-dep-nhat-viet-nam-truoc-va-sau-khi-bi-chat-cay-227236.html*
[6] Dam T T 1998 *Landscape design in living quarters in Hanoi* Master's Thesis National University of Civil Engineering Hanoi, pp 41-44, 96-119, 141-144, 177-179.
[7] Han T N 1999 *Landscape Architecture* Construction Publisher Hanoi pp 123-195, 196-205, 210-216
[8] Nguyen A T 2013 *Landscape Ecology* Science and Technique Publisher Hanoi pp 33-36, 57-58, 188-189
[9] Howard E 1902 *Garden Cities of Tomorrow* London S. Sonnenschein & Co. pp 2-7
[10] Yang B, Li M H and Li S 2013 *Design-with-Nature for Multi-functional Landscapes: Environmental Benefits and Social Barriers in Community Development* International Journal of Environmental Research and Public Health No. 10 pp 5433-5458
[11] Nguyen A T 2013 *Landscape Ecology* Science and Technique Publisher Hanoi pp 69-80
[12] United Nations Commission on Environment and Development 1987 *Our Common Future* Report
[13] Benson J F and Roe M H 2000 *Landscape and Sustainability* Routledge London and New York pp 74-80
[14] Nguyen Q M 2017 *Individual study*
[15] Pham D N, Nguyen T H and Tran Q B 2002 *Solutions in Bioclimatic Architecture for Vietnam* Science and Technique Publisher Hanoi p 182
[16] Forman R T T 2008 *Urban Region - Ecology and Planning Beyond the City* Cambridge University Press pp 17-18
[17] World Meteorological Organisation 2002 *Statistic Weather Data for Hanoi Weblink http://worldweather.wmo.int/en/city.html?cityId=308*
[18] State Department of Statistics 2016 Statistic Yearbook 2015 Hanoi p 37
[19] Nguyen Q M 2018 *Individual study*
[20] Nguyen Q M 2018 *Individual study*

Architectural design led approach to sustainable tourism for the waterfront development of Kunduchi in Tanzania

M Leus[1], P Winkels and E Hannes
Faculty of Architecture and Arts, University of Hasselt, Agoralaan Gebouw E, 3590 Diepenbeek, Belgium

[1] Corresponding author. E-mail address: maria.leus@uhasselt.be

Abstract. In Kunduchi, located in the Kinondoni district of the Dar es Salaam Region, it is of vital importance for the lives and livelihoods of the indigenous people to preserve the typical ecosystems and ensure the identity and economic resilience of these areas. The problem statement is as follows: In what way can sustainable tourism in Kunduchi serve as an engine for economic and social empowerment? How can Kunduchi be an inspiring example for the development of the coast of Dar Es Salaam in Tanzania that is threatened by large-scale tourist infrastructure? How to design sustainable solutions with respect for the local community and the local traditions? Firstly, a theoretical framework that connects sustainable tourism with the sustainable development of coastal areas is defined. Assumptions made on the basis of the literature review provide parameters that play an important role in the architectural concepts. Secondly, a research by design is presented in order to analyze and evaluate different scenarios to outline the opportunities of sustainable tourism on site of Kunduchi. Sustainable waterfront development is an obvious subtitle since the subtle spatial integration of these projects in the urban and water related context of Dares Salaam is of major importance.

1. Introduction

The World Ocean covers seven-tenths of the planet's surface and many societies depend on its health. The decline of fish populations and sea biodiversity occurs all around the world. In Tanzania approximately eight million people are living in coastal communities dependent on ecosystems typical for these coastal regions. Currently several coastal areas including Kunduchi, located in the Kinondoni district of the rapidly growing city Dar es Salaam Region, are under severe pressure because of frequent unsustainable use of natural resources. The inhabitants of Kunduchi depend on *fisheries* and coastal ecosystems. Various activities including overfishing, coral mining, mangrove-mining and waste water discharges permanently damage the ecosystems [1]. Kunduchi is an area with its own identity and with important natural resources. But the coastal zone of Kunduchi can also be considered as a deprived area. Deprivation is caused by a lack of social and economic development linked to three factors. First, there is a lack of economic activity; fishing is almost the only economic source. However, due to overfishing this is a vulnerable source. Secondly, in the future this area will be subject to an aggressive urban sprawl, so the identity of the site will be lost. Thirdly, there is a need for an additional economic development. Kunduchi has a unique mix of social engagement and aquatic knowledge to tackle these problems.

To develop this vulnerable coastal heritage site into valuable, livable spaces where experience and safeguarding of the local identity is essential, a more sustainable perspective to the waterfront development and tourism is inevitable. Sustainable tourism can be one of the possible strategies to overcome this deprivation, mainly economically and socially. Sustainable tourism in developing countries is even more crucial because of the lack of proactive legislations and policies [2]. Sustainability is based on the credo "people planet profit". Together these three dimensions represent the wealth and development potential of the society. They are interconnected with each other in such a way that any change in dimension will have an effect on the other ones.

Only by a strategic tourism action plan with an overall vision that adopts a holistic approach with a balance between the benefits for human, planet and economy, tourism can be sustainable.

2. Aim, a search for synergies

The problem statements are as follows: In what way can new developments associated with sustainable tourism in Kunduchi serve as an engine for economic and social empowerment? How to develop the coast of Kunduchi that is threatened by large-scale tourist infrastructure, in a sustainable way with respect for the local population and the local traditions? How can we empower the local communities that are part of the DNA of the Tanzanian coastline to counter the pressure of the fast growing city of Dar es Salaam?

In the design studio, the focus was not on tourism but undifferentiated phenomenon, but on types of tourism that are triggered by positive architectural interventions with respect for the cultural and natural identity and the host community in order to create a liveable site. The goal of the design studio was to create architectural interventions for a liveable community, building on its unique ecological and human capital and respecting the uniqueness, the identity of coastal areas. The design proposals must add value to the daily life of the local community. This objective can only be guaranteed when the local inhabitants will enclose the projects in their heart Therefore people have to understand the aim, by understanding they will appreciate and value it so they will take care of it, long after architects and contractors are out of sight.

The functions that should be developed concerning sustainable development must be based on the current and potential qualities of nature and the identity of the region. The objective is to protect the natural biodiversity along with the underlying ecological structure and the supporting processes in the environment as well as promoting education, innovative economic and sustainable concepts and recreation.

Sustainable development projects must unite the social community and the economic fabric of the area. Using the cultural biography as an instrument for revitalisation means "preservation through development". To do this, one needs to invest in the development and the enhancement of the identity of the site, in available knowledge, in the potential for economic activity in the region, and in its potentiality to attract tourism [3].

3. Methodology: Research by design as a tool for sustainable development

The material for this investigation is provided by a design course at the Faculty of Architecture & Arts of the University of Hasselt in Belgium and is part of a collaborative two-year research project funded by VLIR/UOS with the School of Architecture of Ardhi University of Dar Es Salaam, Tanzania. In the design studio, research by design is the guiding principle.

The theme of the design studio and the studio's official assignment is ecotourism in urban waterfront developments. According to the International Ecotourism Society ecotourism is defined as: "Responsible travel to natural areas that conserves the environment, sustains the well-being of the local people, and involves interpretation and education"[4]. Considering the urban context of Dar es Salaam the studio's assignment was redefined to sustainable tourism as explained by UNWTO "Sustainable tourism is tourism development with a balance between environmental, economic and socio-cultural aspects" [5].

Based on sustainable design and capacity building principles, this project addresses issues around sustainable tourism, sustainable economy, waterfront development and social responsibility. In 2014 Tanzania became a member of the World Green Building Council [6]. A green building is defined as a building that is energy efficient, resource efficient and environmentally responsible, incorporating design, construction and operational practices that significantly reduce or eliminate its negative impact on the environment and its occupants. Because the built environment plays such an important role in social welfare, climate responsive buildings can make a valuable contribution to the social aspect of a community.

4. Sustainable tourism, an opportunity for implementing sustainable development goals

Tourism can be considered as an important stimulator for the whole economic system. Tourism is one of the most important industries for the economy of a country and results in an increase of the gross national product and gross domestic product. It ensures an increase in added value creates jobs and plays an important role in improving the competition between local and international companies, which leads to investments of public services [7]. In many cases tourism is a source of income for the local community which can be an important lever for the general social situation. But tourist activities can also affect a community's social and natural sources. In order to avoid that the tourist capacity is exceeded, there is a need for an adapted and sustainable development of tourism in Kunduchi. The charter for sustainable tourism of Lanzarote states that "*Tourism* must *consider* its *effects* on the *cultural heritage and traditional elements, activities and dynamics* of *each local community"* [8]. The recognition of the identity and interests of the local communities must always play a central role in the design of tourism strategies. This charter also underlines that "To participate in sustainable development, tourism must be based on the diversity of opportunities offered by the local economy. It should be fully integrated into and contribute positively to local economic development". Tourism can support the specific character of this area, hence increasing the attraction of the site.

Notwithstanding all these benefits, tourism should not be taken for granted as 'a key to a golden future for the destination' [9]. Sustainable tourism should be 'culture and nature-friendly'. The protection of the natural-cultural heritage areas, biodiversity and eco-systems must be guaranteed. Secondly, sustainable-tourism must have a positive impact for the local community on both the socio-cultural and economic level. Sustainable tourism should respect the identity of the living environment, the socio-cultural authenticity of the host community, their heritage and traditional values and it should contribute to intercultural understanding and tolerance. Thirdly, the ecologically-sound way of opening up of the site by tourism must offer a high quality experience to the visitors.

Sustainable tourism is more than offering a banal market product. It is finding a delicate balance between daily life of an area and welcoming outsiders to visit and experience the uniqueness of the site. The identity of the coastal region is an important tourism product that should be safeguarded. The commitment to sustainability must be strengthened. A crucial element is the overall consciousness of the issue to ensure the coherence of the initiatives. Therefore education is desirable. Knowledge is a key in the whole process. Education for policy makers, the local community and tourist operators will encourage the understanding, appreciation and the care for this vulnerable site and the planet.

5. Case study: Kunduchi

Kunduchi is approximately 15 miles, to the northwest of the Dar es Salaam metropole and borders to the North on the river Tegeta and in the South on the Manyema Creek which is very dependent on tides [10]. Like every town, Kunduchi has to deal with general problems like urban sprawl of the city by pressures on greenfield development.

The research area in Kunduchi is located at the coastal line and includes different zones that are clearly separated: the Kunduchi Beach Hotel and resort, a fishing village, an abandoned university site and a green zone.

The Kunduchi Beach Hotel and Resort in Afro-Arabic style was built in 1971 and is recognized as one of the 100 best hotels in the world. During the 1980's erosion threatened the Kunduchi coastline as well as the foundations of the hotel. With the assistance of a number of international renowned engineers and a major financial investment, the land could be reclaimed and the hotel was restored to its original beauty and charm. Yet the hotel is fully walled and its beachline is privatized. Tanzania's largest waterpark, the Wet 'n Wild, is located in between the hotel's entrance pavilion and its lobby. A major parking lot separates the hotel from the surrounding and thus forms a psychological threshold for local inhabitants of Kunduchi.

The fishing village with houses constructed with waste materials is also not freely accessible. The fisherman communities along the Tanzanian coastline descend of the ancient Naga tribes [11]. With a rich past of trade, exchange of cultures and goods, an impressive knowledge of the sea and its fragile life, it is essential to transfer their (hi) story to future generations. When the fishermen obtain more economic stability they can preserve and strengthen the autarky system or self-sufficiency.

Figure 1. Different zones in Kunduchi.

The former university site north of the Hotel Resort is free to enter and is an opportunity for future development. At this moment the old abandoned and dilapidated university buildings create a desolate and hopeless atmosphere. Only a Muslim school is in use.

The green zone, west of the fishing village, the Hotel Resort and the university campus, is completely separated from the ocean. This zone, an open area without a defined function, is used for herding cattle and sports. In addition the area also acts as a meeting place for fishermen and villagers. The site is characterized by beautiful and diverse nature, densely forested areas with giant Baobab trees, open and dry grasslands and a vast mangrove forest that is flooded at times of high water. In the shadow of an impressive Baobab tree lie the historic ruins of an ancient muslim graveyard, as the witnesses of the former intercontinental trading routes [12]. The ruins are decorated with graphic patterns and fine china objects to visualize the status of its 'owner'. Situated at the highest point of this area, elevated above the water and somewhat hidden by the surrounding greenery, this site breathes respect and timelessness.

Figure 2. Beach hotel and resort

Figure 3. Fishermen village

Figure 4. University buildings

Figure 5. Green zone

Figure 6. Ruins

6. Research by design proposals.

The research by design starts from the future's perspective and is a provocation of the prevailing reality principles [13]. It provides opportunities that are currently hidden behind the reality with a view to a new reality. The three key drivers, a focus on the future, developing alternatives and stimulation of imagination, have a direct influence on the architectural thinking and ensure that problems can be redefined so that new insights are developed towards a foresighted alternative [14]

With an uncompromising naivety and open-minded view of the world, the studio projects of the faculty of architecture of the University of Hasselt aim to be a design inspiration for the Ardhi colleagues, and together with them, all the stakeholders and planners in Dar es Salaam.

Three case studies in three different areas are presented as inspirations for the sustainable development in relationship with eco-tourism of the coastal site of Kunduchi. The design proposals aim to find appropriate solutions for these areas that entail a balance between development and sustainability in terms of preservation of local culture, economy (sustainable tourism) and ecology. The first design proposal analyses local food production in order to dissolve social, spatial, and economic resilience. The second design proposal focusses on upgrading local life and nature through aquaculture. The third design examines the revival of the old aquatic sciences faculty site by creating a sea lab as an educational hub for scientists as well as for the host community.

6.1. Case 1. Food over Boundaries

The Food over Boundaries, in its focus on local food production, searches for a sustainable synergy between the hotel and the fishermen's community, between the tourists and the residents of Kunduchi. In this design, project local food production becomes the catalyst to dissolve social, spatial, and economic boundaries that are predominantly present at Kunduchi [15,16]

Developing countries face major challenges which are subject to several problems that continue to increase, for which urgent solutions have to be sought before they escalate. 'Urban farming' can offer a suitable solution for some of these problems such as food shortages, limited economies, vulnerability to crises, the disappearance of green areas. In addition to healthy and fresh food, urban farming gives the local community additional economic stability, making them more independent of the global food market. A very important development target in many African cities, including Dar es Salaam, is to provide a reliable source of food for everyone who lives in and around the city. If this is not the case, the society will be unstable and civil unrest will increase [17].

The goal of the project is twofold: firstly empowering local inhabitants by giving them more economic resilience by adding sustainable local food production to the site and secondly characterising the Kunduchi Beach hotel as a 'healthy living' hotel that offers its guests a high quality local food and natural swimming experience.

The design implements urban farming in the green area of Kunduchi with farmhouses at the centre, in between crops, the fishermen’s village and the hotel. The project focusses on three actions: working, learning and enjoying so that the urban farm becomes a meeting place, a place to work, a place to eat and relax, and a place to learn about sustainable farming techniques.

The aim of the project is that Kunduchi will be known as a respectful city quarter that functions as an example for the small plot farms spread over the entire city of Dar es Salaam. Thus food will be the engine to transcend the existing social, economic and spatial boundaries and it will create a synergy between the fishermen's community and the hotel, between local residents and tourists.

The project proposal is carried out in three phases. The first phase consists of the development of an agricultural project. The open green zone is reserved as an agricultural area that produces food for the hotel. The fishermen village will be a place for subsistence production by means of square-meter-gardens. In the second phase, the creation of a 'learning center’ or a consultation forum will bring fishermen and tourists together on basis of exchange of knowledge. This community project can be financed with the output of the agricultural projects. The third phase focuses on tourists who want to stay in eco-lodging units. The pavilions are characterised by constructive simplicity so that the buildings can be constructed by the local labour and local available materials. Also principles of climate response buildings are integrated to create low tech solutions. The future vision is that the Kunduchi beach hotel can become a ‘foody hotel’, some kind of health- and eco resort. When the existing Wet ‘n Wild waterpark will be replaced by an ecological swimming zone in harmony with the environment, this will mark the natural qualities of this area. By creating a synergy with the hotel, the urban farming concept is brought to an innovative level.

The integration of urban farming in architecture on a small scale provides the best chances of success. Vanderstadt concluded that urban farming on a small scale by inhabitants and users of the city is at least as important as urban farming on a large scale professional level [18].

Figure 7. Food over Boundaries [15, 16]

By investing in sustainable agricultural methods cities of developing countries are given the opportunity for the development of a sustainable living environment with a positive impact on climate change [19]..

6.2. Case 2. Fisherman and Sons

The second design project is based on research of the fishermen's economic situation and tries to upgrade local life and nature through aquaculture [20].The title 'Fisherman and sons ' refers to the English companies that often had the name of their founders or of their sons, that joined in the next generation. 'Fisherman and Sons' is about recreating a nature reserve in synergy with aquaculture activities, at the reef island, bordering the fishermen's village. The nature reserve and aquaculture are mutually beneficial. The project examines sustainable fishing, the preservation of corals reefs and the importance of ecological balance. At present, the financial gap between tourists and residents of the fishing village is wide. When this gap narrows and tourists support the local economy this will encourage mutual respect between tourists and local residents.

The design proposes to transform the island into an eco-economic hub with extensive fish farming with installations of breeding zones and labs in combination with seaweed farming. In order to minimalize the environmental impact of the intensification of the fish production, the Recirculating Aquaculture System (RAS) can be used [21].The existing wave breakers will be used as ponds. The shallow zones are perfect for installing seaweed fields while the deep areas can serve as fish growing ponds. The architectural concept is expressed by a series of pavilions, designed with a specific architectural appearance that facilitates the new activities: a hatchery building, a workmen's space, a workshop space, an information point, and a restaurant. Although the pavilions have the same materials and dimensions, each of these pavilions has a unique relation with its surroundings. The pavilions blend into the landscape and accentuate surprising views on the nature reserve and its surroundings.

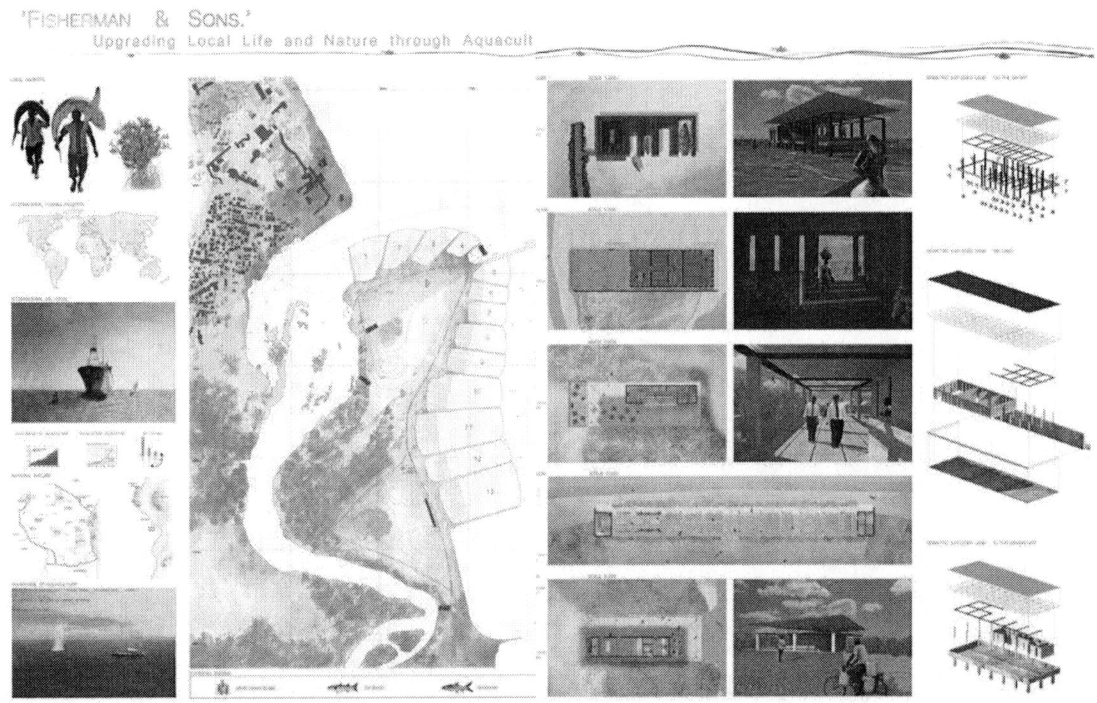

Figure 8. Fisherman and Sons [20].

To increase the coastal resilience mangroves are restored and replanted to prevent further coastal erosion. This action will also attract new wildlife. The created nature reserve invites visitors to experience local nature and wildlife. This gives them the opportunity to visit the Saadani National Park, to learn about the daily life of the Kunduchi fishermen and about native and contemporary fishing techniques.

6.3. Case 3. Sealab

The third design concept is highlighted by a sea lab as an educational hub, a kind of awareness center for the fragility of the oceans [22]. The goal is to revive the old aquatic sciences faculty site in Kunduchi into a place where the people of Kunduchi and scientists can combine their knowledge and work together to search for new sustainable solutions. This research centre, where academic and daily life knowledge of the fishermen are combined in a visitors centre, serves as a seamark. The architecture of the building represents a poetical ode to the sea. This seamark that brings the old site back into the spotlight will attract visitors to explore the unique places of Kunduchi and the Tanzanian Coast with locally trained guides.

The beauty of this project lies in the coherence of the message: creating awareness for the fragility of sealife and the necessity of finding a good balance between man and nature.The design evokes this awareness in its program as well as in the spatial experience, the architectural construction and the technical detailling.The concept of a circle reflects the settlements of remote communities. The pure form triggers curiosity. The outside and inside of the building create intense experiences. The circle shaped building allows a 360 degrees overview over the fantastic environment. On the inside of the circle an almost spiritual place arises where the sea and the sky are the main characters. The morphology of the lab was inspired by biomimicry and craftsmanship of the fishermen in relation with sustainable techniques. The nature's toolbox and processes are adapted in the design. The wooden construction is based on the fisherman's craftsmenship to build and restore boats. The roof reflects the elegance and structure of palm leafs and creates natural ventilation. The interior walls are based on the construction of a beehive, the techniques that bees use to cool their hive and protect the honey.
This awareness centre provides the inhabitants of Dar es Salaam with important and interesting information on their natural resources and will surely become a touristic point of interest in the netwok of unique visitable sites along the coastline.

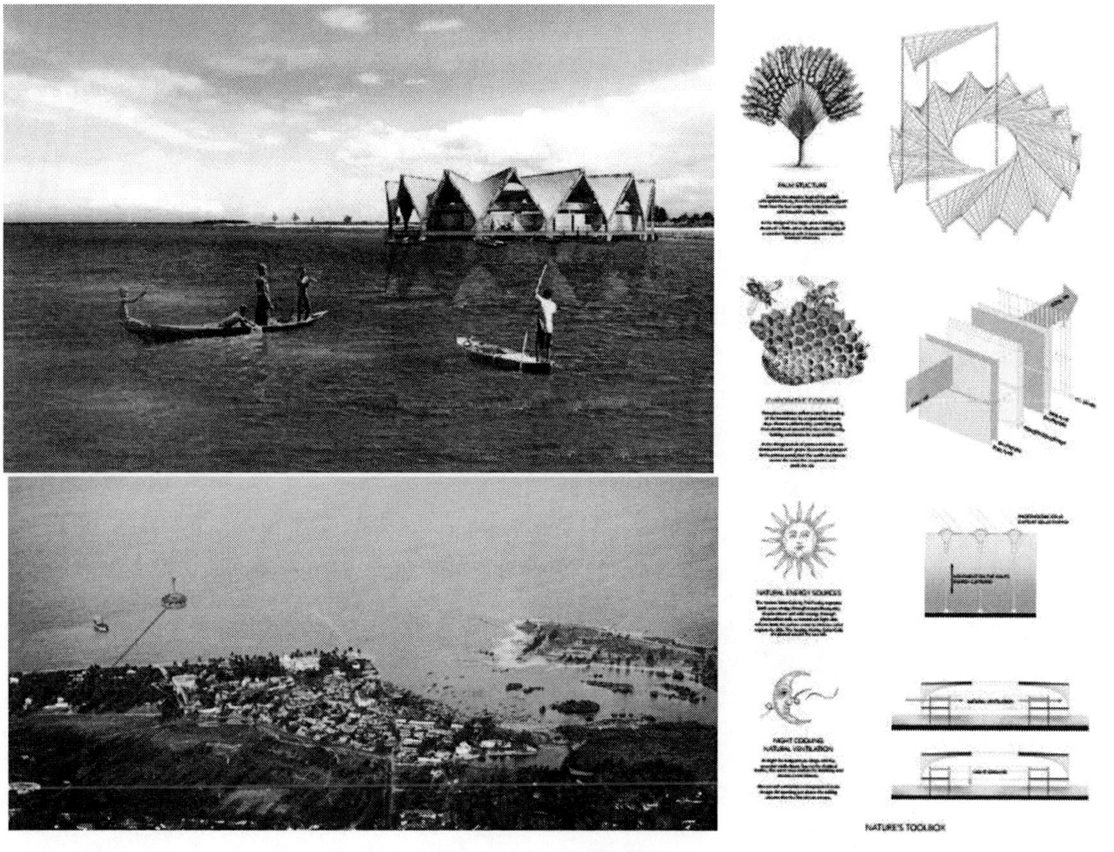

Figure 9. Sealab [22].

With this project, Frioni won the third prize of the Euregional Architecture Competition in 2015 The Euregional Prize for Architecture is a cross-border prize awarded annually to the best graduate project in one of the five participating schools of architecture in the so-called Euregion Meuse-Rhine, the European region spanning part of Germany, the Netherlands and Belgium. The jury report praised this project for its technical precision and consistency. "This very sensitive project touches on global issues in a very humble way. Local knowledge, vernacular and craftsmanship shape the training and teaching centre for local fishermen. The project addresses the topic of overfishing and polluting our oceans and the aquatic building itself has been worked out into the smallest detail" [23].

7. Evaluation of the design proposals and effects of the proposed scenarios

The criteria applied for the design evaluation of de design proposals were coherent, appropriated and captivating. The evaluation matrix in table1 gives information about the assessment of each case.

Table 1. Evaluation matrix

Criteria	Case 1. Food over Boundaries	Case 2. Fisherman and Sons	Case 3. Sealab
Coherent	The design decisions of this concept focus on creating a meeting point for all, for people of all walks of life in this city characterized by multi-culturalism and contrasts.	The design is the result of respectful social engagement and commitment. The project also pays attention to economic return for the local community and has a positive effect on the environment and the villages flood resilience.	This a well thought project, on all scales: from research to concept, from the urban level to the architectural project and from volume to detail. Sealab creates inspiring synergies, between all the stakeholders.
Appropriate	The creation of a series of 'places' by repeating and varying with one simple construction element and a sloped roof, makes this project affordable; easy to build and charming.	The project fits subtly in its surroundings and is respectful to the environment. The buildings are climate responsive in their concept and low-tech in their design.	Sealab accentuates the beauty and fragility of sea life by its architectural appropriateness. On the one hand it is a humble volume, inspired by biomimicry, on the other hand it draws attention by its unique and captivating formal appearance.
Captivating	In its well-thought location and simple but delicate structure, this project succeeds in respectfully linking all surrounding architectural elements. Being condensed to the elementary components, the design inspires by its formal simplicity and materiality.	The design proposal will be able to convince all viewers because of its humble and well-dimensioned pavilions that fit in naturally. The digital design skills of the designer make him visualize his narrative convincingly, poetically, even to the untrained (non-professional) viewer.	With its simple, archaic circular floorplan and palm leave-like roof structure, surrounded by the water, this pavilion immediately captivates the viewer. The drawing technique, inspired by Darwin's analytic sketches of plants and insects, strengthens the narrative of this Sealab.

The values of the design proposals were also evaluated based on effects on sustainability and the development of an adequate concept with a long-term vision.The three cases have an appropriate function and deploy an architecture highlighting to increase the sustainability performances and the revitalization of the specific sites. The effects of these sustainable design strategies are a direct reference to the problems of sustainability based on local, natural and economic aspects: connecting sites, landscapes and communities. As Deniz pointed out: "Environmental awareness and sustainability are not only environmental, but also ethical and social issues that new generations of designers should consider to minimize negative impacts of their works" [24]. Embracing environmental issues in design also will have a positive effect on students personal development in sustainable creative thinking because it strengthens potentials for comparative analysis and creative development of deep innovation.

In addition starting from the local socio-economic and environmental situation, these projects are firmly rooted in everyday reality of the site and the contemporary needs of the urban dwellers. Through low-tech interventions and by rephrasing the existing components, students turn this challenging situation into a positive opportunity for the city and its inhabitants.

8. Conclusion

Managing these fragile natural heritage sites according to the principles of sustainable tourism is a challenging task that starts from a holistic and forward-looking approach. The quality of a tourist destination is highly dependent on the natural and cultural heritage [7]. Being aware of the role of the cultural and natural heritage is an important element in strengthening the regional and global cooperation and a major step in the implementation of sustainable development [7].

Sustainable tourism has a particular focus on the preservation and protection of the environment with respect for the local culture. Chapter 28 of Agenda 21 for the Travel and Tourist industry emphasises that the local population should play a prominent role in projects for a sustainable tourism [25]. and cooperation between all actors' [26]. Hence sustainable tourism should aim to maximize the benefits of tourism to the host community. The host community can be described as a complete social structure, including the local community, the private and public sector and the educational institutes. An integrated and integral approach is necessary.

The proposed solutions for the problems of Kunduchi are positive steps to attract both the interest of the host community and the policy makers. Education and communication belong to the core of the tourist concept and are essential for a sustainable future [27]. The commitment to sustainability must be reinforced. An important part of strengthening the commitment is the awareness of the issues. Knowledge is a key in the whole process.

9. Evaluation of the education project

This cross continental cooperative service learning project, undertaken by the two faculties of architecture, works out to everyone's benefit and delivered a variety of experience. Dehabituation is one of the key-words in this academic experience for the Belgian students as well as for their Tanzanian counterparts. Everything the Belgian students think to know about urbanism, building and everyday life seems irrelevant in the Dar context. They have to start thinking 'out-of-the-box' which urges them to come up with surprising paths and innovative concepts. By empathically implementing new components, new synergies between present identities on site (people, buildings, natural resources and activities) take form to act as drivers for change. Because of this the projects cast new light on the existing qualities of the area and attract the attention of new public.

Resulting from a parallel design studio assignment, the positive effect of these projects lies not only in the content but also in the 'new eyes' that this intercontinental research by design collaboration brings. Multiple moments of reflection and interaction between the students during the academic year, stressed the added value of intercultural cooperation, improved and broadened the master projects and helped students to become better architects, prepared to cope with a changing society. By this project the sensitivity of the students for the social context is triggered. The architectural projects focus more on the ethical concerns averse of visual spectacle architecture but express a design with a sense of craftsmanship. This is consistent with the mission of the Faculty of architecture of the University of Hasselt which is strongly directed to sustainable and socially committed architecture. Because these projects had to deal with 'sustainable architecture' the exchange program can be seen as an adequate contribution to the intercultural dialogue about sustainability within the multicultural society of today. In addition, this is also a benchmarking project to evaluate its own functioning and operation.

10. References

[1] Mahika C, Makota V and Sallema R 2004 Monitoring shoreline change using Remote Sensing and GIS A case study of Kunduchi Area, Tanzania *Western Indian Ocean J. Mar. Sci.* **3** 1 pp 1-10
[2] Timothy D J and Nyaupane G P 2009 *Cultural Heritage and Tourism in the Developing World: A Regional Perspective* eds D J Timothy and G P Nyaupane (London: Routledge)
[3] Leus M 2008 The cultural biography of Pavia as a strategy for city branding. The Cathedral as a mayor component in the process of city branding *Il Duomo di Pavia, tra conoscenza, conservazione e valorizzazione* ed G Calvi (Firenze: Alinea) pp 227-240
[4] The International Ecotourism Society 2015 *Definition revised.* Retrieved from http://www.ecotourism.org/what-is-ecotourism
[5] United Nations World Tourism Organization (UNWTO) 2015a *Sustainable Development of Tourism*, Retrieved from http://sdt.unwto.org/content/about-us-5.
[6] World Green Building Council Report (WGBC) 2014 New GBC Launches in Tanzania. Retrieved from http://www.worldgbc.org/activities/news/africa-news/new-gbc-launches-tanzania/
[7] Ursache M 2015 Tourism – Significant Driver Shaping a Destinations Heritage *Procedia - Social and Behavioral Sciences* **188** pp 130-137
[8] CFST Charter for Sustainable Tourism 1995 by the World Conference on Sustainable Tourism, (*Lanzarote,* Spain) April pp 27–28
[9] Goodall B and *Ashworth* G J 1988 Marketing in the Tourism Industry eds B Goodall and G J Ashworth (Buckingham: Crook Helm)
[10] Almström B and Larsson L 2008 Shoreline change analysis for the Kunduchi beach area, Tanzania *Masterthesis. Lund Institute of Technology* (Lund: Lund University)
[11] Sumet J 1990 *Naga: Cultural Origins in Siam and the West Pacific* (Oxford: Oxford University
[12] Masele F 2012 Private business investments in heritage sites in Tanzania: Recent developments and challenges for heritage management *African Archaeological Review* **29** pp 51-65
[13] De Jong T 1992 *Kleine methodologie voor ontwerpend onderzoek* (Boom: Meppel)
[14] Geldof C and Janssens N 2007 *Van ontwerpmatig denken naar onderzoek. Achtergrond: Architect/Ontwerper / Onderzoeker?* (Antwerpen: Vlaams Architectuurinstituut)
[15] Valy S 2015 Food over bounderies *Masterproject. Faculty of Architecture & Arts* (Hasselt: University of Hasselt)
[16] Valy S 2015 Maakt 'urban farming' het verschil tussen honger en een goede maaltijd? *Masterthesis. Faculty of Architecture & Arts* (Hasselt=: University of Hasselt)
[17] Lynch K 1994 Urban fruit and vegetable supply in Dar es Salaam *The Geographical journal* **160** 3 pp 307-318
[18] Vanderstadt H 2012 *Urban Farming* Nieuwbrief 3 (Londerzeel: Het Autonome Huis)
[19] Pretty J *and* Bharucha Z P 2014 *Sustainable intensification* in agricultural systems *Annals of Botany* (Oxford: Oxford University Press) pp 1–26
[20] Gilissen L 2015 Fisherman and sons *Masterproject. Faculty of Architecture & Arts* (Hasselt: University of Hasselt)
[21] Dalsgaard J, Lund I, Thorarinsdottir R, Drengstug A and Arvonen P B P 2013 Farming different species in RAS in Nordic countries: Current status and future perspectives *Aquacultural Engineering*. 53 pp 2-13
[22] Frioni M 2015 Sealab *Masterproject. Faculty of Architecture & Arts* (Hasselt: University of Hasselt)
[23] Euregional Prize for Architecture (EAP) 2015 Retrieved from http://www.eap-pea.org

[24] Deniz D. 2016 Sustainable Thinking and Environmental Awareness through Design Education. Procedia Environmental Sciences, 34, 70-79.
[25] UN, United Nations 1992 *Agenda 21.* New York, UN. Retrieved from http://www.un.org/esa/dsd/agenda21/res_agenda21_28.shtml
[26] World Tourism Organisation (WTO) 1996 *Agenda 21 for the Travel and Tourism Industry*
[27] Munsters W 2007 *Cultuurtoerisme* (Antwerpen/Apeldoorn: Garant)

Acknowledgments
The authors would like to thank the Flemish VLIR-UOS Programme for Institutional Universal Cooperation IUC for supporting this investigation; also would like to thank M. Frioni, S. Valy and L. Gilissen for their excellent work on the design of this project.

Social housing for workers – A new housing model for Ho Chi Minh City, Vietnam

L M. NGO[1*]

[1] Faculty of Civil Engineering, Ton Duc Thang University, Ho Chi Minh City, Vietnam
* Corresponding author: ngoleminh@tdt.edu.vn

Abstract. Urbanization in Ho Chi Minh City (HCMC), Vietnam is rapidly increasing. Therefore, social housing for workers who work at industrial park and processing areas, is indispensable. There are difficulties and conflicts which still remain in developing the social housing for those people in HCMC. For example, the demand of social housing is high, however employers and/or business owners did not fully pay their attentions on social houses to support the workers. On another hand, even if they built the houses, these one seem not to be sufficient for the demands and/or unable to be competed to the rental housing market from private landlords. Building a social housing model for those workers is a vital importance, this aims to improve the quality of life for the workers; for examples, healthcare, personal safety, social relationships, emotional well-being, quality of living environment, etc. In this research, we study the investment, management, and operation of the social housing for workers in HCMC. This also seeks a new housing model which will adapt the criteria towards the sustainable economic development of HCMC.

Keywords: Social housing, Workers, New housing model, Ho Chi Minh City.

1. Introduction and overview on social housing

The experience of social housing initiated from Europe in the 19th century, and it then developed in some Asia countries such as Korea, Japan, and Singapore. The European view on social housing shows that this way of housing is connected with some groups of people who have some special needs, i.e. unemployed person, disabled person, seniors, alone parents with children etc. We expect that the majority of people (the workers) working in developing industrial areas are increased who are able to find and pay a housing by themselves without the support of government.

In Oxford Dictionary (2017), social housing is defined as: "*Housing provided for people on low incomes or with particular needs by government agencies or non-profit organizations*". So that, social housing is a house that can meet basic demands which are supplied by the government or non-profit organizations to low incomes or unremarkable incomes sector. Similarly, Cambridge Dictionary (2017) defined: "*Social housing is homes provided by the government for people with low incomes to rent cheaply*". Thus, Cambridge Dictionary stated that social housing should be supplied by government to low incomes. Comparing to the definition in Oxford Dictionary, social housing definition in Cambridge Dictionary has more specific limitation on investing party (Government) and the form of using social housing (for rent only) [1].

From these above two definitions, we can see that social housing is the type of house provided by the government or non-profit organizations to low incomes or unremarkable incomes sector for lease or free providing to the people.

In Vietnam, social housing has been defined in Housing Law in 2005. However, only from 2013 until now, with Decree numbered 188/2013/NĐ-CP (dated on November 20th, 2013) of the Government on developing and managing social housing, the definition of social housing was understood in the true social meaning, which is: "*Social housing is the type of house owned by state authorities (central or local) or the types of house owned and managed by the government and non-profit organizations built for the purpose of providing low price houses for some priority subjects in the society such as state officers with no stable houses, low incomes... and the houses used for lease at lower price than market price*" [2, 15]

In the world, besides the definition of the *social housing*, there are concepts such as *public housing* and *affordable housing*. *Public housing* is the house that states authorities, government, or locals are investors, sponsored by the government, this type of house is usually used for lease under the model of cooperatives. *Social housing* concept seems to be wider meaning, the investors are more diversify, but they are usually non-profitable, sponsored by different resources, the diversified ownership schemes could be in a periods and the purpose is to provide the houses for whom that cannot afford to buy commercial houses. *Affordable housing* is even a wider concept, which is applying to all type of the investors, usually has financial support. Affordable housing has diversified ownership schemes, and also targeting to broader subjects.

In fact, there is not a clear boundary for the distinctiveness among these concepts, especially, when comparing between the countries [3]. For instance, in Singapore, the concept of public housing includes the houses that being sold to all paid workers that contributed to the social funds. It seems similar to the concept "affordable housing" in America and many other countries [4]. In Vietnam "social housing" is supported by the government (with land and financial support) that offer lower price products for the workers, the tenants and the buyers; and it is different to "commercial housing" (completely working under market mechanism). Therefore, it could be claimed that Social housing concept in Vietnam also includes with "affordable housing" [5].

Generally, three kinds of social housing are the same on low price facet, but different on beneficiary and ownership facet, and usage form. The differences among these terms are expressed in the following Table 1:

Table 1. Comparing Social housing, Public housing, and Affordable housing

Indicator	Social housing	Public housing	Affordable housing
Price	cheaper than commercial housing, with price support from the government or non-profit organizations in order to lower the cost	mainly used for lease or free of charge staying	cheap, affordable for most people with housing demand
Owning object	Government, non-profit organizations or combinations of Government & non-profit organizations)	- Government	- Government - Non-profit organizations (NGO), - Combination of Government and non-profit organizations (Government & NGO) - Private
Use form	- Lease - Free of charge staying - Selling (few cases)	- Lease - Providing free of charge	- Selling - Lease (few)

▪ Characteristics	- Selling to the ones without purchase ability. - Balancing real estate market supply-demand - Adjusting, solving social problems in the urban area	- Housing for homeless persons - Providing free of charge for policy beneficiaries - Adjusting, solving social problems in the urban area	- Constructed with the target for majority of people with demand and suitable for the people's income tower for selling [5] - Collecting profit

From 2014, Vietnam Ministry of Construction has brought the model of safe house and house for workers into the social housing classification in order to get benefit from budget sponsorship for planning, and construction to encourage the enterprises which are participating in developing houses market for workers. This is the initial condition to solve the difficulties in building social housing for workers in HoChiMinh City.

Ho Chi Minh City is a big centre for economic, culture, social, communication, and international integration. Along with developing progress, many Processing zones and Industrial parks are formed and developed, which contributed remarkably in the City general economic growth, attracted hundreds of thousands of workers from other provinces. In the city, there are now about 2.59 million workers working in 17 processing zones - industrial parks (PZ-IP), High-tech zones, and 21 industrial zones.

The sustainable formation and development of PZ-IP in HCMC attracted a big volume of workers, especially young workers in neighbour provinces such as from the East and South West areas. The average demand to attract the workers for PZ-IP is about 100-150 workers/ha of industrial construction land, this is not included the supportive workers. Therefore, for each 100-200ha Processing zone or Industrial Park, there would be a new 10,000-30,000 workers urban area formed. If the number of supportive workers is added, there would be about 20,000 - 40,000 workers [2]. That is the reason leading to the rapidly increasing demand for housing among workers in Processing zone and Industrial Park areas in HCMC.

By the projection to 2020, the number of workers in PZ-IP, High-tech zones and Industrial zones would be reached about 377,389 workers [6]. In which, about 70% workers coming from other places and about 50% of them have demand for renting houses, equivalent to 132,000 workers in need for a place to stay. At the same time, number of workers who working in other places out of these zones is about 2.2 million workers, expectedly to 2020, it would be reached about 2.6 million workers, and 50% of which has housing demand, equivalent to about 1.3 million workers in need for a place to stay [6]. The city has built about 1 million plots, and there will still be short of 300,000 accommodations until 2020, this requires the City must try harder to develop residential places for new workers in order to satisfy the basic living conditions for them.

Therefore, building social housing model for workers to improve living quality, guarantee basic cultural, entertaining activities and sustainable development is necessary. It's an important mission for the City socio-economic developing progress. This research will propose a new model for investment, management, and operation of Social housing for workers in the regional PZ-IP of HCMC. This also aims to propose a new housing model, to adapt the criteria for the sustainable economic development of HCMC.

2. The status of social housing for workers in HoChiMinh City

HCMC has come up to the socio-economic developing achievements by the major contribution from over 2.59 million workers in PZ-IP, High-tech zones and Industrial zones. However, a massive amount of workers coming to work in PZ-IP has led to many urgent problems for the city and the districts which the Industrial parks located, especially the problem on social housing for average low incomes. Immigrated workers used to rent houses in or around the area that factories developed. These houses are mainly invested by households living there, usually with limited quality, no guarantee on sanitary and security with many potential risks for immigrated workers [7].

Since Housing law 2006 promulgated until now, in the HCMC, enterprises, households and individuals have invested and completed 4,027,726 m^2 of housing area, offering 1,009,794 accommodations for workers. The city has been constructing new social housing projects for workers such as worker home areas in Tan Thoi Hiep and Linh Trung I and Linh Trung II Processing zones, worker home area in Tan Tao Industrial Park, worker home area in Tan Thuan Processing zone (District 7, HCMC), and new industrial parks [8]. The situation of worker housing is expressed in Table 2 as follows:

Table 2. Worker housing construction situation in HCMC since 2006

Stage/Time	2006-2010	2011-2015	2016	Total
House construction in the HCMC (m^2 of floor)	1,359,200	2,198,972	469,554	4,027,726
House for workers (number of place)	435,000	469,994	104,800	1,009,794
Investment Enterprises (m^2 of floor)	137,300	108,840	51,386	297,526
	10%	5%	11%	**7.4%**
Households and individuals' investment (m^2 of floor)	1,221,900	2,090,132	418,168	3,919,574
	90%	95%	89%	**92.6%**

The Table 2 proves that the social housing rate belongs mainly to households and individuals (92%). In other words, the enterprises are losing big market share in social housing for workers (less than 8%).

The reasons for such fact are: 1) Big investment, long-lasted capital returning time (20 to 30 years), high grounding expenses and problems in operating process; 2) Land fund suggested by Districts for Social housing has low feasibility, and it is hard to operate the project; 3) Shortage of Land fund directly managed by the city (clean land fund) to deliver to enterprises investing in home area for workers; 4) Most PZ-IP only focus in developing the production to generate profit, and they are less focus or avoid the problem on housing for workers – the people directly making their products; 5) Enterprises, if investing in social housing construction, suffer from high interest, long capital returning cycle so there is a certain loss, let alone the fact that investors are not legally protected by the Government; 6) Social housing for workers environment is not suitable with the demand and living habit and conditions of workers, so it doesn't attract workers and they don't choose the model constructed by enterprises. Meanwhile, these problems are the advantages for households and individuals, since other than very competitive lease price, the leasing conditions, living conditions, and management in households' house for lease are all quite convenient for workers.

Besides formal housing production, up to 80% of the total private construction took place without any regulation or construction permit, due to the weak political control of free-market forces and the deficient planning and licensing system in Vietnam. In HCMC, it is estimated that more than 50% of residential buildings were built without official construction permit [9]. Therefore, households and individuals' house for lease always has potential socio-economic problems such as living conditions might impact on workers' health, the houses' living environment in long-term will cause pollution to the whole area around PZ-IP. Living in low income and minimum management rental houses, workers are always lack of entertaining, sport, relaxing, studying, and communicating conditions. Figures 1 and Figure 2 show the pictures are taken in two different residential areas, in District 7 of HCMC, where workers housing and private construction are located.

In fact, workers are the main beneficiaries for social housing but are not yet paid enough attention for understanding their cultural characteristics (actual demand, habit, regional characteristics), as well as their desires and wishes.

Figure 1. Worker housing in Tan Thuan Processing zone, Distict 7, HCMC

Figure 2. Worker rental home with steel sheet roof in Ly Phuc Man Street, District 7, HCMC

3. International experiences in developing social housing models

Overviewing some international experiences in developing the social housing in some countries, beside the refurbishment of the old houses, new social houses are also being invested in variety forms, these seek to suit with recent social policies, political economic mechanism, as well as innovation ideas in each country or local authority level. There might be three new social housing developing models: New concentrated construction model; Concentrated financial support model; and Diversified model toward individual and community role [10]. Some countries might focus in one developing model, while some others might develop many models at the same time with different priority levels.

- New concentrated construction model: using big state owned construction corporations, the Government builds the complete social housing areas with population scale of tens thousands people, they also supply basic amenity services such as education, medical, health, public transport, shopping… The Government plays a decisive role in distributing lands, introducing financial priority policies as well as setting up standards for social housing beneficiaries. The houses facility standards are improved constantly to meet modern social demand. This model has been applying in some countries such as Luxembourg, Singapore, China, Sri Lanka…
- Concentrated financial support model: A directly supporting policy for people who have housing demand instead of focusing the support in construction and social housing supplying units, the advantage is that it is a low cost solution comparing to supporting construction expense for social housing areas, lowering the building coverage ratio of low incomes in housing areas, then people can freely choose the location for their houses, and this lead to avoiding many corruption problems in construction investment. This model has been applying in many countries such as America, Brazil, Chile, and South Africa.
- Diversified model, pointing toward the role of the individuals and community: Social housing projects tend to be localized and privatized with more diversifying scale than before. The Government stimulates the local governments and Associations to decide on social housing projects in their area. Local associations and social housing construction companies operate on non-profit or low profit base. This model has been applying in many countries such as Australia, France, Germany, Holland, and Sweden [3, 11].

Figure 3. Social housing for workers in Luxembourg [12]

Figure 4. Social housing for workers in West Australia [13]

Figure 5. Social housing project for workers in Thailand [14]

Recently in HCMC , the housing for workers in PZ-IP is in three major models: houses built by the government, houses built by the enterprises, and houses built by the households and individuals [6]. In which, the fact showed that housing projects built by the government and enterprises cannot compete with rental houses supplied by households and individuals. In term of project investment, the households and individuals can utilize all existed advantages, capital returning cycle is two to three years, and collecting profit in the following years, the management problem is also very flexible. Therefore, if the rental price from the investors is not too much difference, it cannot compete with the households' one.

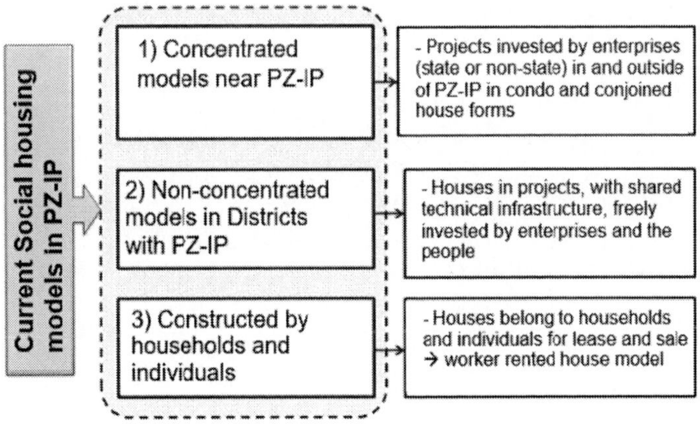

Figure 6. Current social housing models in Vietnam

Thus, what do the city government and investment enterprises find the solution to improve construction investment environment as well as raising the housing fund for workers in PZ-IP in

HCMC? Which developing model should be applied for social housing for workers that can guarantee the harmony of the investors and enterprises' benefit and suitable with actual conditions in HCMC?

4. Suggesting a model for investment and managing the social housing in HCMC

To reach the target for developing the social housing to adapt the housing demand for workers now an in the future, besides choosing and organizing reasonable planning space, the city must have a policy on general and synchronized investment, management, and operation in which the governments, investors, enterprises, people, and workers can really care and participate. We created three new models for investment, management, and operation the social housing for workers in HCMC, which are:

- Model A – concentrated remodelling social housing;
- Model B – new constructed social housing; and
- Model C – social housing for lease.

The purpose of these new social housing developing models is to improve the feasibility of social housing projects for workers, this also guarantees the conditions for basic cultural, entertaining activities and sustainable development of PZ-IP in the city.

1) Model A – concentrated remodelling social housing

General characteristics of the model:

- Concentrated remodelling social housing model is the model that being invested and constructed by the enterprises, the products are social housing apartment buildings;
- The model involves three stakeholders: the Government, the Enterprise, and the Households and individuals;
- The model applying scale should be near PZ-IP so it is easily approaching to the technical and social infrastructures;
- The Government directly steers the banks which managed by the Government (Social policy Banks and Commercial banks) to redirect the capital support for the workers (who have house purchasing and renting demand) to PZ-IP enterprises. The enterprises are considered as the intermediate party. Workers who want to approach the house purchasing support pack must come to the enterprises and follow enterprises' instruction to approach social houses that built by the enterprises in a certain time, social houses invested by enterprises must be guaranteed for standard quality, and the rental price should be equal or lower than rental price from the households and individuals. The households and individuals with land fund near PZ-IP will be rented or purchased by enterprises.

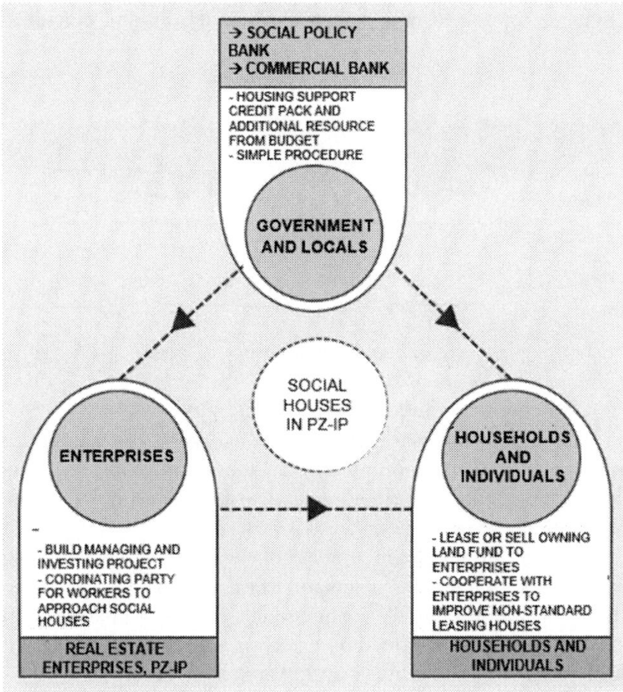

Figure 7. Model A – concentrated re-modelling social housing

In the above model, the third participant of the model is "Households or individuals", who have land funds near the PZ- IZ, they have different capacity, the land area of the individual is not the same. Therefore, the author proposed to divide into three solutions corresponding to the actual conditions of the households: a) Household has a large land area; b) Household has not enough land to build a condominium; c) Household has a small land area, but able to buy or rent conjoined lands.

a) In case the land which being purchased or rented is in large area or combining three-four conjoined land plots

- Based on the general operating method of the main model, the state instructs banks to redirect the credit support pack for workers with house demand to PZ-IP. The enterprises are considered as the intermediate party. The workers who want to approach the house purchasing support pack must come to the enterprises and follow enterprises' instruction to approach social houses that built by enterprises in a certain time. The households or individuals with land fund will be rented or purchased by the enterprises. In case the land plot is big enough to build apartment buildings or if it is small, it will be rented in combination of three-four conjoined plots to build apartment buildings. See Figure 8;
- If the model operated it would naturally bring workers who are living in the households and individuals' rented houses to the apartments, lowering the risk of creating a slum area from temporary accommodation. The households and individuals with rented houses which does not meeting the standards will have to choose between completely changing the rented house function, or cooperating with the enterprises to remodel the houses to meet the standards. When meeting the standards, PZ-IP enterprises will organize workers moving into these households and individuals' rented houses;
- The model also helps reasonably distributing single workers and workers who want to stay in family model (single workers can stay in the industrial parks' home). The enterprises in PZ-IP will be the coordinator to distribute workers in these two forms;
- So, this A1 secondary model will supplement many places to stay for workers, lowering the amount of workers living in low quality rented houses, lowering the number of potential slum

areas, remodelling low quality houses, helping to re-distribute the workers in and outside of PZ-IP in a more reasonable way.

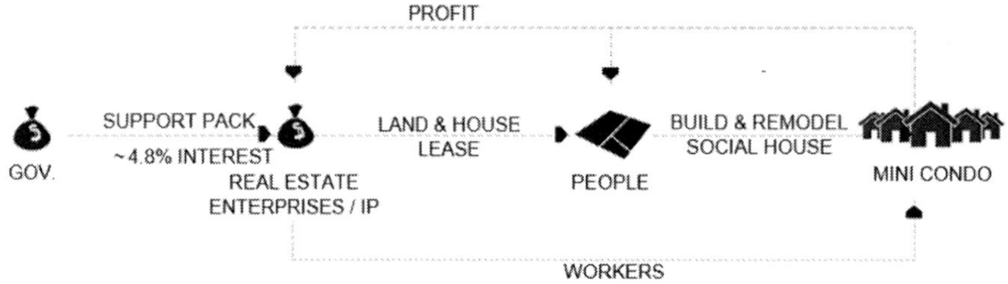

Figure 8. A1 model – social housing in rented or purchased large area of land

b) In case the land area is small and not big enough to build apartment buildings

- This model will utilize everything to create the places to stay for workers, the model still guarantees the flexibility in construction and transportation when the period ends; lowering potential slum areas, remodelling low quality rented houses, distributing workers in and outside of PZ-IP in the most reasonable way. See Figure 9;
- Based on the general operating method of the main model, the state instructs banks to converse the credit support pack for workers with house demand to PZ-IP. The enterprises are considered as the intermediate party. The households and individuals that own land fund but not big enough to build apartment building will be rented by enterprises in long-term and will be applied with flexible assembled houses to build rented housing area for workers.

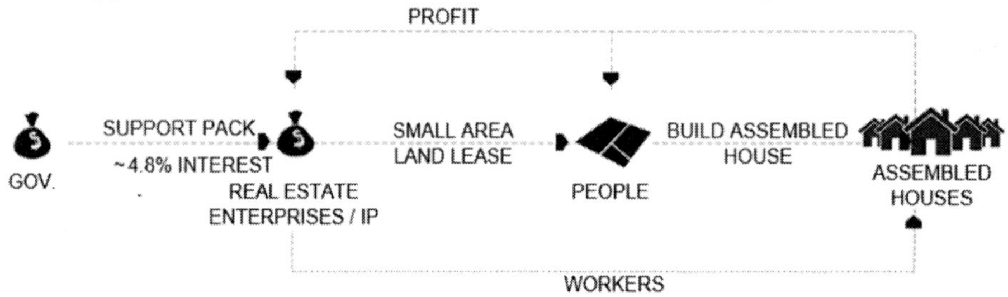

Figure 9. A2 model – social houses on small land plot that is not enough to build apartment buildings

c) In case the land area is small but able to buy or rent conjoined lands

- This model will utilize everything to create places to stay for workers, improving the situation of low quality slum areas around it.
- Based on the A1 model, the state steers the banks to redirect the credit support pack for workers with house demand to PZ-IP. The households or individuals with land that not enough to build mini-condo will be rented or purchased – both land and houses – by enterprises. It's applied with annual profit distribution by percentage of instalment payment for enterprises. Figure 10;

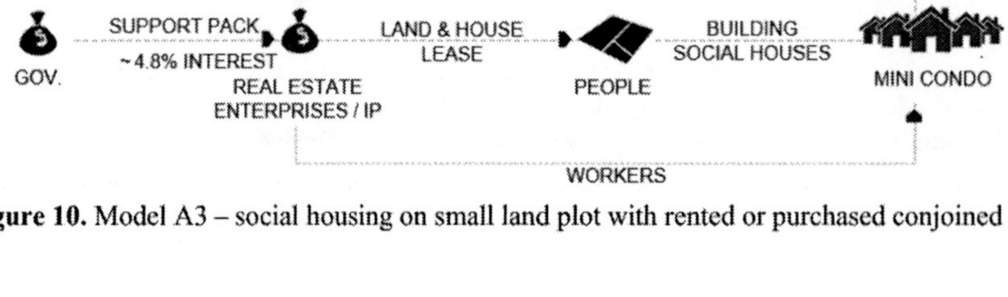

Figure 10. Model A3 – social housing on small land plot with rented or purchased conjoined land

2) *Model B – new constructed social housing*

The model's characteristics:

- The social housing model that being invested and constructed by enterprises with the approval on land used and land leased decision, final product is the social housing apartments for low income workers;
- The model combines two stakeholders: the government and the enterprises. See Figure 11;
- Applying area is near PZ-IP so it is easily approaching the technical and social infrastructures
- The state steers the banks to redirect the credit support pack for workers with house demand to PZ-IP. The social housing that being invested by enterprises must be guaranteed the standard quality, with equal or lower price than leasing price from the households and individuals. The enterprises can use or rent a part of the land fund planned for construction social houses for workers. See Figure 12;
- The model will utilize the usage of social facilities such as technical and social infrastructure of the area within the plan;
- So, this model will help supplement more places to stay for the low income workers, and the workers can also inherit the cultural infrastructure in the area, helping distributing workers in and outside of the PZ-IP in the most reasonable way.

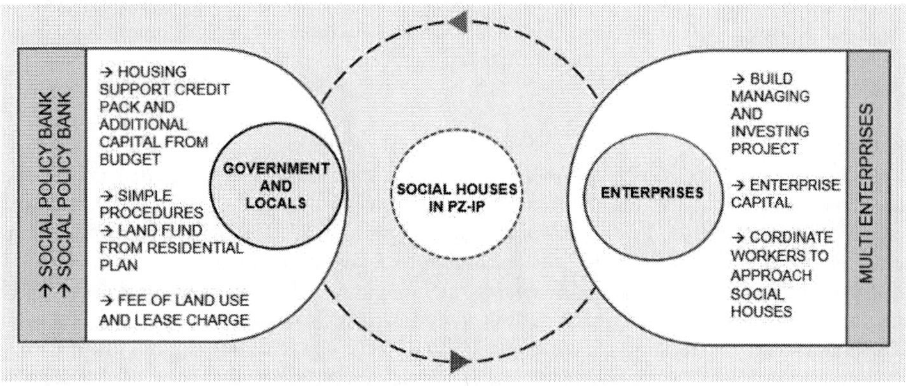

Figure 11. Model B – New constructed social housing

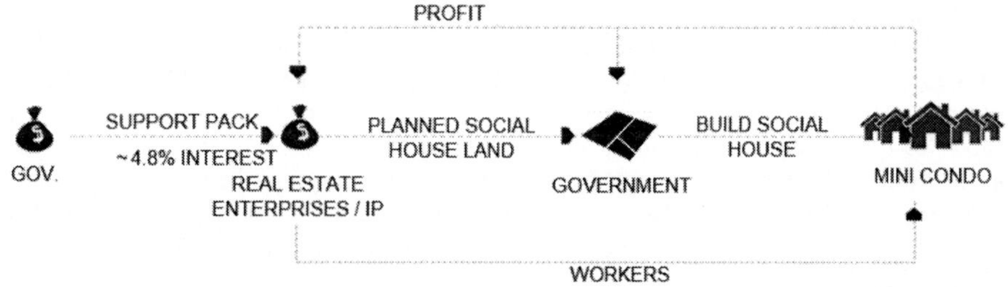

Figure 12. Model B – social housing on small land area with rented or purchased conjoined houses

So, the difference of model A (concentrated remodelling social housing) from model B (new constructed social housing) is that there is another stakeholder, the households and individuals. The households and individuals with land fund can be rented or purchased by enterprises, if the land is big enough for building apartment buildings or if the land is small, it can be rented or purchased in three-four conjoined land plots so it is enough to build apartment buildings. This model A will utilize the land fund, adding more places for workers to live, lowering the amount of workers living in low quality rented houses, lowering number of low quality slum areas around PZ-IP. Meanwhile, model B helps supplement more places for low income workers, and workers can also inherit the infrastructure and cultural basis in the area, helping to distribute workers in and outside of PZ-IP in a more reasonable way.

3) Model C –Social housing for lease

Leasing the social housing is the only way out for low incomes of less than 8 million VND a month in HCMC, especially before the immigration influx appears since HCMC is developed in a special mechanism and policy from January 15th, 2018. In the PZ-IP of the city, leased house demand will increase rapidly to millions of units. Meanwhile, the fact is that most of house for lease comes from the households and individuals, the model of conjoined houses divided into 15-20m^2 rooms in the centre, and conjoined houses at 20-30m^2 area in the suburb. Nevertheless, the investing enterprises still don't want to participate in this market because of the problems from policies, procedures, land fund, and financial resources.

Therefore, our research suggests a model for leased social housing as follow: a) for administration: City – Districts – Wards/communes; b) for participants: Government – Households and individuals. In which, the Government invests to build social houses for lease, combining with households and individuals with extra land fund.

In order to carry out this new model, the Government and Ministry of Construction must issue the policies, legal frames for leasing houses: regulations on designing standards (18-30m^2); raising planning indicators such as land used ratio, number or floors, number of citizens; having clean land fund used for leasing houses; investing in technical infrastructure; supporting enterprises financially like low interest rate, even having a special financial support pack for leasing houses in some big cities in Vietnam.

5. Conclusion

Based on the given results, which allow us to conclude that, the research created new models for investing, managing, and operating the social housing for workers in PZ-IP in HCMC, this work has never done before, and it's the first time having a model for HCMC. This stated the freshness in terms of science and the actual contribution of the research for social community.

Social housing model for investing, managing, and operating social housing for workers in HCMC includes: model for concentrated remodelling social housing, model for new constructed social housing, and model for leasing social housing. In which, the research highlights the model A (concentrated remodelling social housing), this model emphasizes the role of households and individuals. If it is carried out, the new models will contribute in improving the feasibility of social housing projects for workers in HCMC.

The above suggested models will support the investors and authorities to define their roles and positions, and also utilize the role of households and individuals in mobilizing and exploiting land fund around PZ-IP in HCMC for developing the social housing models for workers. This is also suitable for the inevitable trend of leasing house model for workers invested by the households and individuals around these PZ-IP in HCMC, creating the conditions for the people within the area to live and develop sustainably. The models suggested in this research might be reference for better manage worker leasing houses built by the households and individuals, and develop social housing fund for workers in HCMC.

For the community's participant in the social housing project, this research defines the roles of the Government, Banks (banks, investment funds, credit organizations) and Investors as the Supply stimulator, Demand stimulator, and Activator. Each party has its own role and function to stimulate the development, speed up the investment, construction, and operation of social housing projects. Only when all these three parties fulfil their roles in sync, the problem of social housing in HCMC will be resolved with right objects and effectiveness.

References

[1] Ngo L M 2017 Social housing for workers - Which development model is suitable for Ho Chi Minh City? *Journal of Architecture. Association of the Architects of Vietnam*, 09. Pg: 74 – 79

[2] Ho Chi Minh City Export processing and Industrial zones authority. http://www.hepza.hochiminhcity.gov.vn/web/guest/home
[3] Christine W, Scanlon K J 2007 *Social housing in Europe*. Denmark: Aalborg Universitet
[4] Housing and Development Board - Singapore 2000 *Facts on Public Housing in Singapore*. Singapore: HDB.
[5] Michael W, Ronald E, et al. 2007 *Housing for Low-income Groups in Ho Chi Minh City between Re-Integration and Fragmentation Approaches to Adequate Urban Typologies and Spatial Strategies*. Germany
[6] Pham D T 2016 *History of industrial architecture*. Hanoi: Construction Publishing House
[7] PADDI 2009 *Document Training course on social housing development*, HoChiMinh City.
[8] Department of Construction HCMC 2016 *Report on the implementation of housing development program for workers in the city*. HoChiMinh City
[9] Tran K P 2002 Overview of Housing for Low Income People in Ho Chi Minh City: Situation and Solutions. Paper presented on the ODAP Workshop No. 7 – *Housing for Low-Income Groups in Ho Chi Minh City: Standards and Solutions*, Ho Chi Minh City, Vienam.
[10] Le L H, Nguyen L P 2016 *The study of some social housing models in the world in stages of development and discussion about socail housing situation in Vietnam*. Proceedings of the workshop on "The Panorama of Social Housing - Challenges and Opportunities". Hanoi: University of Civil Engineering.
[11] Christiane D, Thomas K S 2014 *Social Housing in Germany. Social Housing in Europe, First Edition*. Edited by Kathleen Scanlon, Christine Whitehead and Melissa Fernandez Arrigoitia. John Wiley & Sons, Ltd.
[12] Government of the Grand Duchy of Luxembourg 2016 *Social housing for workers in Luxembourg Weblink: http://www.luxembourg.public.lu/en/vivre/logement/index.html*
[13] Government of Western Australia 2012 *Social housing for workers in West Australia Weblink: http://www.housing.wa.gov.au*
[14] The Jai Yen Country 2013 *Social housing project for workers in Thailand Weblink: https://thejaiyencountry.wordpress.com*
[15] Government Prime Minister 2013 *Decree No. 188/2013 / ND-CP*. Vietnam.

Acknowledgments

The author would like to acknowledge assistance from the Department of Science and Technology of Ho Chi Minh City, and Ton Duc Thang University. Especially, I would like to send my sincere thanks to the research team's members: M.Sc. LE Tan Hanh, MSc. HOANG Phuong Thao, MSc. NGUYEN Dinh Nam, and MSc. LAM Quy Thuong.

A Case Study on Sustainable Reuse of Abandoned Infrastructure at Seoul Station Overpass as Urban Park for the Design Strategies in Korea

Yeeun Boo[1], **Young-Sang Kwon**[1,*]

[1] Department of Civil and Environmental Engineering, Seoul National University, 1 Gwanak-ro, Gwanak-gu, Seoul, 08826, Republic of Korea.
* Corresponding author. Tel.: +82 2 880 8200
E-mail address: yskwon@snu.ac.kr

Abstract. As the 21st century, known for knowledge information era, many industrial infrastructures built as part of the 20th century urban development have been devastated functionally and new alternatives for them have been demanded nowadays. This study aims to discuss the strategies used in the design proposals of the International Competition for 'Seoullo 7017 Project', which was recently completed in May 2017, based on the sustainability of the deteriorate infrastructure as urban park. Through the competition brief, each proposal is analysed against the competition brief and the more generic approaches on the adaptive reuse of infrastructure are proposed. By examining the case in Korea, it is expected to explore the possibilities for the sustainability of abandoned infrastructure through adapting reuse as urban park in Korea, to propose design strategies that can be applied to the future adaptive use of deteriorated infrastructure in Korea, and to provide broader academic base to related works.

Keywords: Industrial Heritage; Adapting reuse; Seoul Station Overpass; International Design Competition

1. Introduction

As the age of knowledge information in the 21st century has come, many industrial heritages such as overpasses and railroads, which were once considered to be the symbols of urban development in the 20th century modernism era, are being devastated in their functional aspect and new alternatives are being demanded. Also, the number of underused infrastructures is growing globally, and managing them either by decommissioning or adapting is becoming important issue [1]. In this trend, there was an attempt to transform the abandoned highway into an urban park in Seoul Metropolitan, and the 'Seoullo 7107 Project', completed in May 2017, attracted the attention of people in that it was the first elevated walking path in Korea. More than that, this project is considered to be more noticeable in terms of sustainable use of urban park, instead of dismantling abandoned overpass.

At the heart of the project, there was an International Invited Design Competition to decide the future of Seoul Station overpass. This study, therefore, focusing on the external space and landscape, analyzed and discussed the design strategies reflected on the proposals. In this regard, this paper illustrated the variety of design strategies put forward by each competition team to meet the competition brief. Then these strategies were grouped into more generic approaches on the adaptive reuse of infrastructure. For the analysis, the report presented by Seoul Metropolitan Government and the websites to refer the drawings and the commentary were used.

By examining the case in Korea, it is expected to explore the possibilities for the sustainability of abandoned infrastructure through adapting reuse as urban park in Korea, to propose design strategies that can be applied to the future adaptive use of deteriorated infrastructure in Korea, and to provide broader academic base to related works.

2. Overview of international invited design competition of 'Seoullo 7017 Project'

2.1. History background of Seoullo 7017 project

The 'Seoullo 7017 project' in Seoul, which is an example of this study, is a project that has been completed recently in May 2017 and is significant in that it is the first high-level walking path in Korea. '7017' in the project name has several meanings: It means the 1970 and 2017, the years in which the structure was built and subsequently transformed and the height of 17m. Also, it has the meaning of being reborn with 17 different kinds of pedestrian road for the people, not for the car. This project was promoted as part of a city revitalization project to make high value-added areas, nearby Seoul Station including the overpass [2].

The overpass was built in 1970 to cope with rapid population growth and traffic congestion. However, concerns about its stability had been raised since the end of the 1990s, and in 2006, serious safety problems were raised, resulting in prohibition of vehicle traffic and decision on demolition. Although Seoul Metropolitan government has taken a uniform way to dismantle deteriorated structures so far, the city government chose to change its viewpoint and turned overpass into a walking path instead of removing it.

Figure 1. Seoul Station overpass in the past [2].

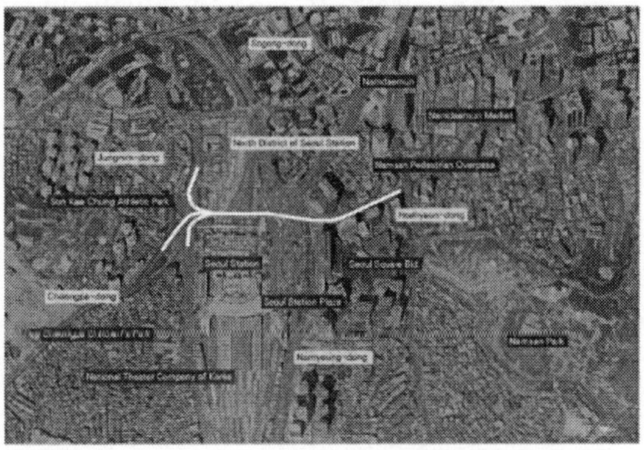

Figure 2. Current state of Seoul Station Overpass.

2.2. Introduction of the competition

The site overview

The site of the design competition is around Seoul Station including Seoul Station overpass. It has an area of 9,661m^2, a width of 10.3m, a length of 938m and a height of 17m. It is adjacent to major historical and cultural heritages in the downtown area of Seoul, and the abundant resources such as Namdaemun, Seoul fortress wall and Namsan Mountain are scattered around the area. (See Figure 2.)

The method of design competition

The design competition was conducted by invitation of designation, and seven architects and landscape architects from Korea and abroad were selected. Considering the structural problems of the overpass, it was promoted to form a consortium around the nominees in various fields of architecture, landscape, and bridge structure. The list of nominated participants is shown in Table 1. Because it is difficult to cover all the 7 entries within limited text, three entries, C, E, and G, that received a prize were selected as the subject of study.

Table 1. A list of entries of the competition.

Listing	Name	Company	Nation	Title	Notes
A	Chang Yung Ho	Atelier FCJZ	China	Seoul Mirage	
B	Juan Herreros	Estudio Herreros	Spain	Seoul Evergreen Terrace	
C	Winy Maas	MVRDV	Netherlands	The Seoul Arboretum	1st prize
D	Martin Rein-Cano	Topotek 1	German	Skyway	
E	Joh Sung-yong	Joh Sung-yong Urban Architecture	Korea	The Seoul-Yeok-Goga: Walkway for All	2nd prize
F	Jin Yanggyo	CA Landscape Design Firm	Korea	Slow, Soul, Seoul	
G	Cho Minsuk	Mass Studies	Korea	Continuous Landmark, Unified Hyper-Collage City	3rd prize

2.3. Guideline of the competition

According to the competition brief, the main purpose of the contest was to restore the urban memory and sovereignty of the citizens over place through conservation. In regard this, three main goals were proposed: conservation of memory, creation of new public space, and recovery of the walking network. In addition, it provided eight design requirements such as integrated design in the various fields, structure preservation, accessibility, ecological sustainability, and etc. Also design guidelines for each of three sections were provided to make use of local characteristics. (See Figure 3.)

Meanwhile, these guidelines are interrelated rather than independent, and there are overlapping parts in the guideline. In particular, the design requirement(b) contains elements that are not tied to the superordinate concept with weak specificity. In other words, there is inconsistency shown in hierarchical structure and bias is found in certain contents. It is expected that theses contents directly or indirectly affected the entries.

In Figure 3, although not all items are fully associative, the correlation between the parent items and child items is showed with lines. From the correlation in Figure 3, the items can be divided in to three parts according to the hierarchy from the most abstract part of concept and goal, to the spatial structure part and the detailed space design planning part like as shown in design guideline for sectors(c) for the analysis in Chapter 3. Also, in this study, only the physical elements that can be identified in the design will be evaluated. In other words, non-physical factors that are difficult to represent by design, for example maintenance management in design requirement, were excluded from the evaluation.

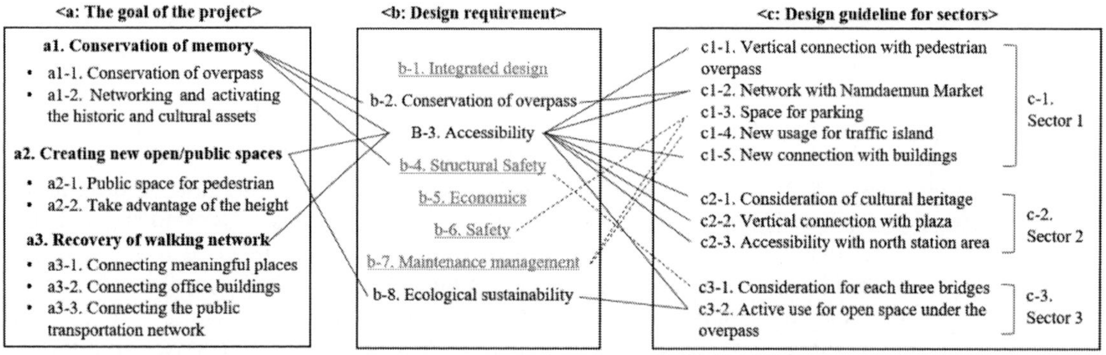

Figure 3. Correlation between guidelines of the competition brief

3. Analysis of relation between design strategy and guidelines

3.1. Analysis of the concept and main goal

In the case of the winning work C, which won the first prize, the main concept of the design is to create space for the pedestrians in the urban space through means of arboretum as revealed in the title. In other words, the goal is to make the overpass as a public garden and to transform the Seoul Station area into green space step by step. It can be seen that the design strategy is established by focusing on a-2 and a-3 among the three goals of the project.

In the case of work E, which won the second prize, it also values the place for the pedestrians as seen from the title, 'The road for everyone.' Its goal is to provides a public space that every citizen is entitled to enjoy by providing barrier-free design, design that touches different sensibilities and design flexible enough to suit changing needs. However, although C emphasis on the nature of the new public space used for the people, E emphasis on the history of the overpass(a1) and puts importance on the relationship between history and the place. It can be seen through diagrams indicated in the panel, that track down the changes in terrain and urban tissues over times.

Finally, in the case of the winning work G, it symbolizes the overpass as a new landmark, while at the same time reveals close connection with the urban tissues. The main premise is that of a new pedestrian-friendly experience and that of urban continuity. The dual characteristics of the existing structure, that of figure as ground and ground as figure has a potential of intervening and uniting the conditions of many neighboring areas. It devises design strategies with emphasis on the connection with the historical and cultural assets of the periphery like the goal in a1-2 and a3-1.

In the following proposals, it can be notified that all proposals take consideration of the main three goal, but they take different strategies in designing with different emphasis.

Figure 4. Perspective view of three entries with prize from first to third place in order [2]

3.2. Analysis of the basic spatial structure strategies

The formation of new public space as an arboretum in proposal C is structured as follows. It gives small deformation to the entrance and the side of the bridge while preserving the prototype of the bridge as much as possible. These deformations are mainly shaped like the branches of trees, which strengthens the possibilities of reaching around important historic and cultural assets such as Namdaemun Market, Junglim-dong and so on. Especially, various types of structures are used to make connections between up and down the bridge. It is significant not to end the design of the overpass itself, but to consider the connection with the many urban elements in the surrounding area. In addition, this connectivity is a design that can be changed and transformed continuously, so that flexible design can be made through consensus with the public and citizens in the future. Moreover, it takes a strategy of one integrated space with some of each parts used separately or in conjunction, not splitting or diving the whole space.

In the proposal E, the highest importance was given to the history value of the overpass and it appears as a design with little change of the overpass. Although it transforms the bottom part of the bridge, making new walking path, however, it has been criticized as a passive plan that rarely utilized the upper part of the bridge. It creates various public spaces by combining different elements with the part of the overpass. Although, it emphasis the historical value of the overpass itself, it seems that the connection with the surrounding historical and cultural assets is considered weakly compared to other proposals.

Lastly, the proposal G proposes the most sophisticated use of spatial composition and contains eight different strategic parts to connect with the areas adjacent to each part or segment, which have unique urban characteristics as well as specific requirements in their urban functions. However, it is criticized by over-transforming some parts of the bridge in designing it and it loses its historic value as Seoul Station overpass. It shows the role of successive landmarks in showing the connection to the widest area among the three proposals.

3.3. Analysis of detailed spatial planning and other planning

Connectivity

Along with the connectivity, the proposal C (Figure 5) is taking a design strategy that best shapes the relationship with the surrounding high-rise office buildings. This not only makes effective use of the height of the overpass, but also suggests the possibility that the floor of the connected building can be utilized as a public space. In the proposal E, vertical connection for all levels of movement is suggested in relation with Seoul station. Also, various structure such as sunken plaza is used to deal with the difference in level. In proposal G, various structures considering the level difference play a role of linking the public space with the bridge, but no specific connection to the building is seen.

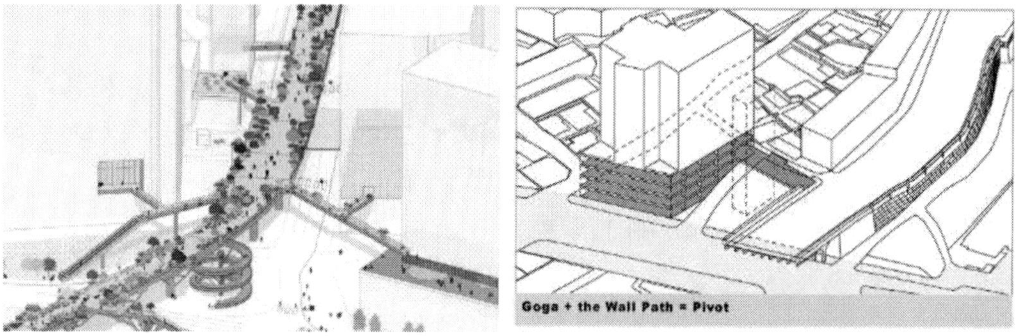

Figure 5. Some example of connectivity in proposal C and E [2].

Ecology

As the name of arboretum suggested, the proposal C proposes variety of ecosystem by ordering all possible plants in the order of the Korean alphabet from east to west, they cause different neighborhoods. These plants also create links to areas where local residents can be proud because of the plants they often see in their area. With the planting planning, C suggests plan for rainwater system. Although G specifies the possible plantations for eight specific sites, there is no mention for specific management that are appropriate for the climate of Korea. There is no planting plan for proposal E.

Management

Both C and E are open to the possibility of being finalized in the future, not limited to the present and focus on processes that are created by citizens. In C, it provides a design that can be flexibly transformed when management is needed. For example, it is possible to add and change the pots, new soil for pavement over time, when needed for maintenance. For proposal E, design strategy is established through participation of local residents, and private sector where public intervention is possible is presented. On the other hand, there was no specific description of management in the proposal G.

Program

Although there are no explicit programs in all three proposals, the activities that can occur in each place are presented in abstract words. In the proposal C, the space such as a rose plaza, a café and a flower shop are presented were activities can be occurred. In the proposal E, a space created by a combination of bridge is defined as a space character by attaching the word such as "Invigorate." This open program strategy allows for the flexibility of the needs of rapidly changing citizens.

3.4. Result of the comprehensive analysis of entries

Based on an analysis of the three submitted proposals described in section 3, the comprehensive analysis is shown in Table 2.

Table 2. Comprehensive analysis.

Contents		**C: first prize**	**E: second prize**	**G: third prize**	**Relation between design strategy and the guideline on the brief**
Concept and goal	The value of the overpass	As a new open space for the pedestrian	The historical value of the overpass	Connection with surrounding area	• Connectivity (a3-1) (b-3) • Priority of people over the car (a2-1)
	Design concept, and goal	Arboretum	Road for everyone	Continuity, landmark	
Basic spatial structure	Structure conservation	Well preservation, with little change	Conservation of the most of the overpass	Some bold change with the bridge	◎ Consider the historic meaning of the overpass (a1) (b-2)
	Accessibility	Various structures located up and down of overpass	Limited to nearby place	8 specific strategic parts as public spaces with nearby historic and cultural assets	• Using vertical structure (a2-2) (c3-2)
	Connectivity	Branch-like structure to connect with urban elements	Structure of Combination of public space and the bridge		• With different urban elements (a1-2) (a3-1) (b3)
	New open space	Viaduct as a new open space			◎ Creating open space using existing public space (a2-1) (b-8)
Detailed and other planning	Connectivity (Vertical)	Most active with office buildings with use of height	Seoul Station, nearby plaza, Path under the bridge	Broad range with surrounding area	◎ Use of existing buildings to be a new public space (a2-2) (a3-2) (c1-5)
	Ecology	Various ecosystem with rainwater system	-	Specific planting planning in each sector	◎ Less specification in ecosystem (b-8)
	Management	Flexible and changeable design over time with participation of citizens		-	-
	Program	Open program strategy			-

•: All applied, ◎: Partly applied, ○: Independently applied

4. Discussion design strategies of the competition and its implication

The three design proposals presented above are designed to integrate the disconnection of Seoul Station through the transition to the walkway of the overpass, and accept the will to contribute to regional revitalization. As looking through the competition brief, there was inconsistency shown in hierarchical structure and bias is found in certain contents. This factor affected the design proposed in the entries to emphasis on limited elements such as accessibility or connection. Also, although not outlined in the guidelines, the proposals include flexible programs as well as changeable design over the consensus with the citizens.

The design strategies that are common among the proposals are as follows:

- Horizontal continuity with the urban elements: It sees the spatial connection with the adjacent urban structures as practical design strategies.
- Reconstruction of the site context: It is the practice design strategies to emphasize the context of surrounding spaces, especially focusing on remembering memory by preserving the prototype of the overpass in various way.
- Interactive and flexible process within design: The non-deterministic spaces and programs that are capable of accommodating changes and the process design over time with the participation of various subjects are regarded as practical design strategies.

These strategies can be applied to other similar infrastructure, for example, New York Highline Park. Highline. While preserving the old railway in many aspects, it preserves some parts rather than reveals the traces of the past obviously, allowing people to notify that it was once a railroad. In the case of planting, it was not artificially planted, but it left wildflowers naturally grown while the railroad was left untouched. It also runs a variety of programs that allow members of the community to participate and that can be changeable through many suggestions from the members. In connection with the surrounding area, the possibility of being connected with the Highline structure is facilitated through appropriate regulation such as the floor are ratio bonus.

Meanwhile, the strong focus on the active interaction with urban elements and accessibility with surrounding elements, reflected in both design guideline and the proposals, is considered to be due to the feature of linear park itself as in this case study. According to Kullmann (2012), he paid attention to important features of linear parks such as connectivity with urban space, continuity of spatial experience, lateral permeability, linear inner circulation, and boundary design [3]. Because of its narrow and long characteristics, linear parks can be directly connected with large part of the urban areas than urban parks with the same area, and thus have excellent accessibility. In addition, since it has a close relationship with the periphery, it has a large influence on urban space.

In the case of large scale lands such as brownfield or landfill, they are often located far from the city. However, the linear infrastructures, such as railway or highway are often located within the city, so adaptive reuse of them will become a compelling issue in the future. Also, linear park has had increasing potential, as large scale network-type urban infrastructure such as rail ways, highways and overpasses are considered to be a good opportunity to provide parks for improving quality of life in existing urban areas [4]. In Korea, while making large scale parks was a mainstream in the past, nowadays, like the example of this study, many projects regarding reuse of linear infrastructures have been operated by the government. Therefore, it is believed that presenting the design strategy for the linear parks is significant. It is expected that better design will be achieved by using active joints with urban spaces, putting variable programs and operation plans, and providing varied sequent along the long space in dealing with linear infrastructure systems.

5. Conclusion

With the completion of the 'Seoullo 7017 project' in Seoul Metropolitan in May, 2017, the abandoned overpass that was used as a road for the cars was transformed into a park, being handed back to the citizens. The Seoullo 7017 project, which has attracted much interest before and after the project, is a case of the first high-elevated walkway in Korea with many significances. This study has been applied specifically to the international invited design contest of Seoullo 7017 project to examine the design strategies of the proposals against the competition brief. With the analysis, this study explored the possibilities of design strategies that can be applied to the future adaptive use of deteriorated infrastructure in Korea. However, there is a limit that this study only researches one specific example. Therefore, more profound examination for further study would be needed to develop the research.

References

[1] Doyle MW and Havlick, DG 2009 Infra and the environment *Annual Review of Environment and Resources* **34** 349-372
[2] Seoul Metropolitan Government 2015 *Seoul Station 7017 Project* Seoul Metropolitan Government(Seoul)
[3] Karl Kullmann 2012 Thin Parks/Thick Edges: Towards a Linear Park Typology for Infrastructural Sites *J. Landscape Architecture* **6** 70-81
[4] Karl Kullmann 2013 Green-Networks: Integrating Alternative Circulation Systems into Post-industrial Cities *J. Urban Design* **18** 36-58

Acknowledgments

This research was supported by the Creative-Pioneering Researchers Program through Seoul National University (SNU).

Web-based automation of green building rating index and life cycle cost analysis

Jam Shahzaib Khan[1], Rozana Zakaria[2], Eeydzah Aminuddin[3], Nur IzieAdiana Abidin[4], Shaza Rina Sahamir[5], Rosli Ahmad[6] and Darul Nafis Abas[7]

[1,4,6,7] PhD Student, Faculty of Civil Engineering, Universiti Teknologi Malaysia
[2]Associate Professor, Faculty of Civil Engineering, Universiti Teknologi Malaysia
[3]Lecturer, Faculty of Civil Engineering, Universiti Teknologi Malaysia
[5]Lecturer, Faculty of Architecture, Planning and Surveying, Universiti Teknologi Mara, UiTM, 40450 Shah Alam, Selangor, Malaysia
Corresponding email: rozana@utm.my

Abstract. Sudden decline in financial markets and economic meltdown has slow down adaptation and lowered interest of investors towards green certified buildings due to their higher initial costs. Similarly, it is essential to fetch investor's attention towards more development of green buildings through automated tools for the construction projects. Though, historical dearth is found on the automation of green building rating tools that brings up an essential gap to develop an automated analog computerized programming tool. This paper present a proposed research aim to develop an integrated web-based automated analog computerized programming that applies green building rating assessment tool, green technology and life cycle cost analysis. It also emphasizes to identify variables of MyCrest and LCC to be integrated and developed in a framework then transformed into automated analog computerized programming. A mix methodology of qualitative and quantitative survey and its development portray the planned to carry MyCrest-LCC integration to an automated level. In this study, the preliminary literature review enriches better understanding of Green Building Rating Tools (GBRT) integration to LCC. The outcome of this research is a pave way for future researchers to integrate other efficient tool and parameters that contributes towards green buildings and future agendas.

1. Introduction

The "green building" concept first emerged in the 1970's when, in an attempt to incorporate sustainable development principles, the construction industry began to design and construct more sustainable buildings [1]. "*Green building*" also known as the ecological construction that save energy and protects environment. Whereas, ecological construction or eco-construction refers to the clean construction that reduces the environmental impacts of building throughout life cycle[2]. Many researchers have provided the definition of green building and exclusively Kibert elaborated green building as a "... healthy facilities designed and built in a resource-efficient manner, using ecologically based principles" (p.9) [3]. Accordingly, it can be further taken as a building that has better quality of life with higher efficiency and satisfaction level. The green building trend has increased rapidly worldwide in recent decades [4]. It is considered as a means of addressing growing concerns over climate change and global warming to reduce the impact of the building industry on the environment [5].

It has been promoted that the environmental sustainability is the major ingredient of sustainable design and construction and it should be kept on priority as main element in green buildings, which will be responsive to sustainable development [6]. For instance, if a construction has satisfied environmental laws and regulation then it will be effective to the human that will directly impact to the well being socially, economically and environmentally. Conversely, [7] believes that all the aspects sustainable development principles are crucial in the nature and none of them can be left. Even though, if one aspect is completely satisfied that cannot bring the other to the upper level as desired by the concept of Triple bottom Line (TBL). Therefore, integration is necessary for green building movement to move towards the path of a successful paradigm shift. Green buildings not only integrates TBL elements but also incorporates too many other essential elements for better quality standards of life without compromising to the needs of future generations [8].

There have been extensive researches on the green building and green building rating tools, which have been developed in various countries from developed to developing world. However, lack of integration of green building rating tool and life cycle costing is found in existing body of knowledge. Such type integration plays essential part in discovering the streamline research. The aim of this research is to develop web-based automation of green building rating index and life cycle costing to attract more investors and developers with set objectives. This automated tool will eventually provide with the costing of green building, score and level of certification with return on investment to all the stakeholders. The objectives set for this study are to utilize MyCrest green building rating tool integrated with life cycle cost analysis and develop the automated web based program that results the above said output.

1.1 Background of Research Development

Malaysia has endured fast industrialization over the last 50 years, making between 5% to 9% annual economic growth [9,10] . This growth has urged the government, institutes and other stakeholders to bring up the epitome of development that is sustainable and takes holistic approach of green building development and its implementation [9]. The Malaysian government has a mission to promote green building within construction industry for sustainable built environment. Since the government of Malaysia is encouraging by developing policies providing platforms on green building, green technology and reducing carbon emissions. It has also been noticed that the green buildings costs higher compared to conventional buildings [11]. Since, Green buildings received a speculated claims that it is higher initial cost, therefore, it is difficult to attract building investors and developers to build green buildings [12].

In spite of Malaysian government strategic plans, building investors are not happily accepting this idea of green building [13]. Still there is a gap to attract and seek more interest of investors and developers in this challenging situation of fast moving era in decision making with advanced information technology. Some of the developed countries have paved their path for many decades toward this globalization of green building but developing countries are still moving rigorously to the same attention [8]. Countries individually are making efforts to meet the global challenges across various disciplines. World Green Building Council mentioned that there are more than 40 green building rating tools administered by their Green Building Councils that have 49% of green building certification systems in their countries' [14]. Malaysia has embarked on green building rating tool since 2009 and recently in 2016 Construction Industry Development Board of Malaysia has established Malaysia Carbon Reduction and Sustainability Tool (MyCrest) for green building assessment. This MyCrest tool provides critical evaluation of green building from design, construction to operation and maintenance stages.

The green building investors and facilitators need a centralized green building rating automation solution. Which is easy, efficient and workable on the web based system that justifies the score, level of certification, and Return on Investment (ROI), to handles their work more efficiently. To meet this challenge green building rating assessment tool and Life Cycle Cost Analysis are necessitated. This automated tool will provide a better solution towards a better green building decision making with economic solution for its development. However, the proposed tool with a centralized web based automation solution offering the integrated systematic solution between the green building assessment tool and life cycle cost analysis. This innovative tool moderates and indicates the potential level on the score that reflects the level of certification, and ROI. If an automation tool is available to uplift mind set for long term viability of green buildings then construction industry can be fully directed towards the well being of environment, economy, and society. This will also lead to transparent, efficient and in a quick time self-assessment. This new innovative tool will increase the investors and building developers interested to proceed with green building development in Malaysia and very potentially to be extended globally.

2. Literature Review

2.1 Integration of Green Buildings and Building Life Cycle

In global scenario, natural resources are depleting due to over usage by humans and their activities [15]. This dramatic change of human activities and style of living has devastatingly increased the green house gases and diminished many other natural resources such as water and soil. At the larger extent, the construction industry is considered as a major causative for this all environmental degradation [16]. Thus, it caused from design, construction and significantly to operation and maintenance stages, since O&M stage is considered as very crucial. The scientific research has admitted that the construction activities will affect the environment by direct or indirect implications [12]. These stages of activities involve a large number of trading from conception to demolition called building life cycle [17] furthermore, various processes involved in the building life cycle and every action is followed by some process. There is a total of six major phases starting from the conception when it's brought to the table as an idea of developing some infrastructure that's called inception, design, construction, commissioning, operation and maintenance, and demolition. Some of the researchers are saying that inception is a process of a brain storming, while some call as nurturing of idea besides that main owner and designer starts thinking about the product to be selected and used [18]. On the second and third phase, design and construction process start where most of the stakeholders are called in the eco-charrette meeting to navigate the potential pitfalls and a better understanding of conflicting priorities [17].

Due to increasing level of human luxury demands natural resources are depleting day by day and being the versatile industry, the construction industry is considered as one of the major contributors to the consumption of natural resources [20]. Green building is a holistic concept that starts with the understanding that the built environment can have profound effects, both positive and negative, on the natural environment, as well as on people who inhabit buildings every day. Green building is an effort to amplify the positive and mitigate the negative of these effects throughout the entire life cycle of a building [19]. The green building is taking a serious course of actions with in construction industry to emphasize the better, effective and efficient living. There has been lots of research carried on green buildings since the start of the conceptualization. The traditional way of building a building has started reaching towards its end of an era.

2.2 Triggers of the Green Buildings

In 2016, the smart market report of world green building trend was published with the partnership of Dodge Data and Analytics [21]. This report depicted the global picture of green buildings with the firm analysis of over 1000 respondent surveys from 69 countries of developed and developing world with statically significant results of 13 countries. The world green building report evaluated top triggers for green buildings and green building activities from the year 2008, 2012 and 2015 shown in Figure1. Four categories were kept, on the first category client demand was mentioned and it has increased by 1- 4% with the passage of time. Secondly environmental regulation that was found 2% decreased from 2008 to 2012 and 12% increased by 2012 to 2015; which is a positive approach and shows strict formation and follow of environmental protection of law.

However, the third category was a market demand that decreased by overall 2 to 3% from 2008 to 2012. The last and final category was the right thing to do where results show that it has declined from 2008 to 2015 cumulative 17%. The decrease in last category elaborates lack of knowledge and understanding in accepting concept of green buildings. In line with the said research, [22] defines that this pitfall is a lack of strategic green building development, and this can be reduced but providing knowledge and filling research between traditional and green buildings.

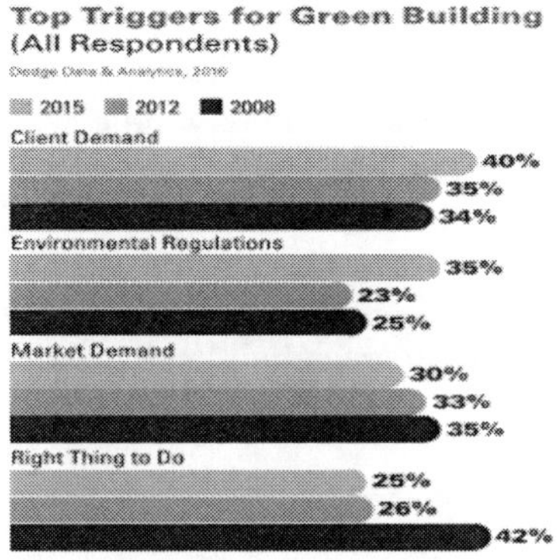

Figure 1. Shows Top Triggers of Green Building in 69 Countries (Source: [21])

After understanding the concept and triggers of the green building, scientists and researchers have started to evaluate the question *"what is the need of green buildings"*. From the emergence of the 1970's concept many ideas floated around and certainly some of the them were leading cause such as climate change, global warming, environmental (air, water and soil) pollution, degradation of natural resources, sea level rises, melting of glaciers and many more facts brought up the idea of green buildings [23] [3]. This thought breaking idea of green buildings emerged as a source of inspiration towards the achievement of set goals. Though, [24] remarkably argued that scientists have drawn their attention to the most common problems associated with the environmental, social and economic (Triple bottom line) perception of the global cause, where people, planet, and profit need to be balanced.

In line with the above statement, this people, planet and profit formula has been generally accepted in all the corners due to their correlative bonding with each other. Researchers have always seen this kind of tranquillity and mentioned that it is contesting in the better interest of all three elements forming a better and balanced eco-system [25]. Thus, formation of better eco-systematic should be balanced otherwise it not be called as balanced eco-system. Due to the deterministic approach of the widespread argument of the triple bottom line all the elements need to be taken together and every activity with in every field need to be more or less contributing in the better interest of TBL argued by [26]. However, [27] claimed that every project needs to be the green project and that resembles the concept of the green building where the green approach is taken into account towards any building. This green approach will reduce the impacts of the building with the increase in efficiency, water; energy and material usage. Nevertheless, it should not affect on human health, ecological foot prints and to the environment throughout its life cycle. The awakening of construction industry stakeholders towards sustainable development triggers the green building development; which not only confers its entities towards better quality of life but also maps prosperous way for future generations. Future generations needs to have natural resources saved therefore this generation has to be sustainable enough to save natural resources directly or indirectly. However, construction industry being one the major contributor to environmental and natural resources degradation has to establish a new trend of development that support rigorous paradigm. The paradigm is considered as *"Traditional Building development to Green Building Development"*.

2.3 Emergence and Development of Green Building Rating Tools

Due to the idealized and defined approaches of the all stakeholders belonging to construction industry forced to develop assessment system. Then, to rate the project for getting that sustainably certified and name the project as "Green Certified Project". The authors in the research [28] pointed out that hundreds of the building evaluation tools are available worldwide focusing on various aspects of green buildings and sustainable development. Around 382 registered building software tools were available by March 2010, for evaluating buildings' sustainability. These tools have been taking faster pace by time and moving the construction industry towards better and innovative level. Similarly, [24] mentioned in their research that assessment tools measures the progress towards the goal and objectives of the projects. Although, these tools vary in various contexts but still direction is towards the sustainable development. On the other hand [12] developed an argument that these tools enhance the capability of organization and stakeholders to work sustainably throughout the life cycle of the project. However, [12,24,28,29] evaluated in their researches that sustainability rating assessment tools vary by country to country, region to region, state to state and project to project with more or less difference in elements. Researchers also added that green building rating tools have potential to assess every project falling under the construction industry and potentially coping green buildings sustainability as shown in Figure 2.

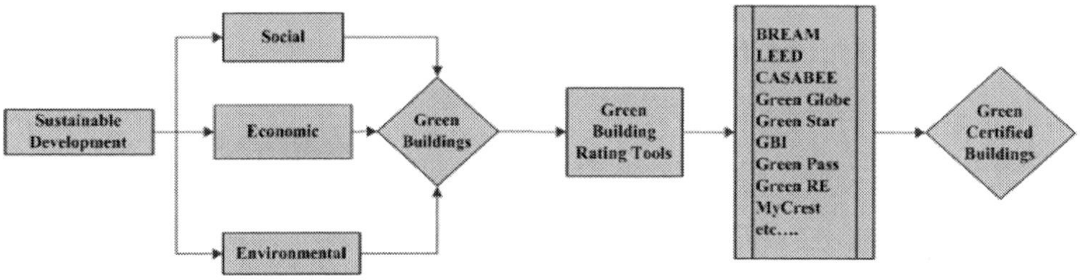

Figure 2. Emergence from Sustainable Development to Green Buildings and then towards Green Building Rating Tools

The green building rating tools were established to incorporate and help architects, designers, builders, owners, investors, developers and government bodies to understand the overall impact of each choice and solution [30]. The development of green building rating tools has evolved from time to time. For the past several decades there has been debates going on in construction industry for moving towards sustainable development through some strategies and action plans. These strategies and action plans has formed a way in 1990 in the essence of BREEAM green building assessment tool [31]. This BREEAM (Building Research Establishment's Environmental Assessment Method) was developed by the United Kingdom, followed by LEEDS (The Leadership in Energy and Environmental Design) developed by the United States [20]. Then Australia has initiated its green building movement of Green Star through GBCA (Green Building Council for Australia) [32]. Similarly, Green Mark was developed by Singapore for their green building assessment [33]; also Hong Kong has developed HK BEAM for the assessment of green buildings in Hong Kong [34]. Likewise, Malaysian construction has started developing their green building rating system with the help of industry professionals, academician, and researchers. In 2009, GBI (Green Building Index) was developed by Malaysia to assess the green building development [24,35]

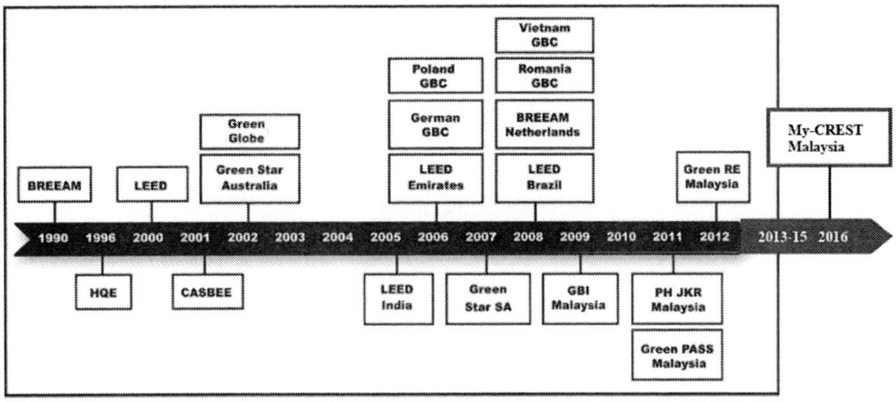

Figure 3. Evolution of GBRT in different countries (Source: [35,36])

The above Figure 3 shows that the in early 1990's there was not much on awareness of green building assessment tool or even building assessment tool. But as time has passed construction industry and researchers have drawn their attention to the development of building assessment tools to green building rating systems [37]. The above figure shows that from 1990 to 2000 there was not much of interest and development in GBRS (Green Building Rating Systems). Though, from 2005 to 2009 there is an exponential growth in the development of GBRS around the globe from developed to developing world this trend has remarkably achieved its peek [12,35,36,38]. The tool used in United Kingdom is called BREEAM; and it was the first green building rating assessment tool in the world developed in 1990. BREEAM was developed by BRE (Building Research Establishment) and this tool is widely accepted due to its versatilities [37,39,40]. All the other green building rating systems developed later such as LEED, GBCA, Green Mark, HK BEAM, GBI and MyCrest are deeply influenced by BREEAM [20].

2.4 The Life Cycle Costing (LCC) Analysis

Life cycle costing (LCC) is significant element with respect to cost related phenomena. LCC was initially introduced by the US Department of Defence for procurement purposes [41]; still it is widely used in most of the military and construction sectors as well [42]. This LCC tool provides prior information of the cost related items for every project and stage. Though the adaptation of the LCC has been found very minimum and low paced in many other industries due to the fact of low knowledge in the utilization of this tool [43]. In line with the above statement, low interest and lack of knowledge of this tool is one of the factors. Construction industry has been lacking behind in utilization this dynamic tool within construction projects. However, LCC has to be the firm choice in the built environment projects with the focus starting from the design phase to demolition phase and it must be utilized to optimize cost efficiency and assets value [18]. LCC not only incorporated to manage costs related items but also help to manage the project from start to end in a better way. Many researchers assert that the LCC is a focused approach of early stages that intends to examine total expenditure of a project by evaluating cost associated to all the phases throughout its entire life cycle [44]. Cost has been a premium factor within the construction industry and it plays superficial role in revitalizing the concept of green development by using more money to get better, effective and efficient building called as "Green Building". This adds up an outstanding and robust way of development that is sustainable for this and future generations as well. LCC takes into account three costs effectively design cost, construction cost and operation and maintenance cost. Thus, at very first research and development (R&D) stage to salvage stage it idealize the worth of the building [45,46] as mentioned in Figure 4. In accordance with the R&D stage LCC provide a venue to understand cost related to complete project and can compare green and traditional buildings. Chakravarty and Debnath [45] asserted that LCC performs its best in decision making processes for all the aspects in every project and industry.

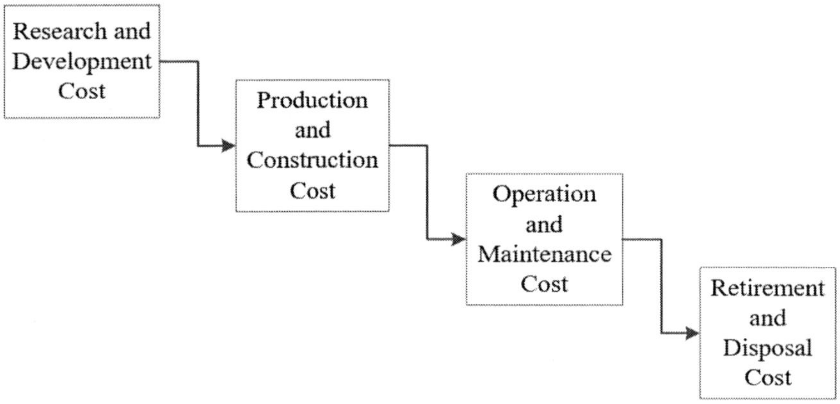

Figure 4. LCC Flow Chart (Source: [43])

The analysis of the cost in all the stages asserts the better understanding of the project value and further it can be used to compare cost variation between traditional building and green building [47]. However, this will lead to examine return on investment in green building, and thus it will also fetch interest of investors and developers towards more green building development. Green building rating assessment tools are available in wide variety but the LCC is not included and it has yet to be included [48].

3. Methodology and Research Development

The sustainable development has acquired a great applauses and success in past from developed to developing countries but soon after the economic turmoil throughout the world, investors are lacking interest in green building development. The in depth critical literature review of journal articles, thesis's, books, reports published, and conference proceedings lead to a problem formulation, *"how to attract more investors and developers towards green buildings"*. Thus, after problem formulation and identification it needs to be solved by a holistic approach. An approach that sustains the credibility of rules, agenda and strategies set by government agencies and provides a sustainable way on return on investment, level of certification and cost of a green building. To solve such a problem the initial steps were carried out at the beginning the research from identifying the problem, determining research area, aim, objectives and scope of the research study followed by significance and novelty of the research as mentioned in Figure 6.

In accordance with the aim to develop web based automation of green building rating index and life cycle costing. Some of the objectives were developed in which firstly MyCrest green building rating tool developed in 2016 has been selected for this research. Therefore, MyCrest elements and criteria are analysed that are to be integrated with life cycle costing elements (NPV, FV, and PV). Significantly, MyCrest tool has three certification stages, Design stage certification, Construction stage certification and Operation and Maintenance (O&M) stage certification. Solely, MyCrest tool is consist of 11 major elements all are emphasizing on the sustainable green building development. However, each element has other criteria to achieve points of the elements within three stages. Moreover, Design stage has 63 major criteria's followed by Construction stage certification that has 45 major criteria's and then O&M has 39 major criteria's. Though, some of the elements and their criteria in each section do not qualify as a costing factor therefore those will be segregated. The followings Figure 7 depicts the detailed flow chart of the research study methodology; this vital framework of the research is to achieve objectives set for study.

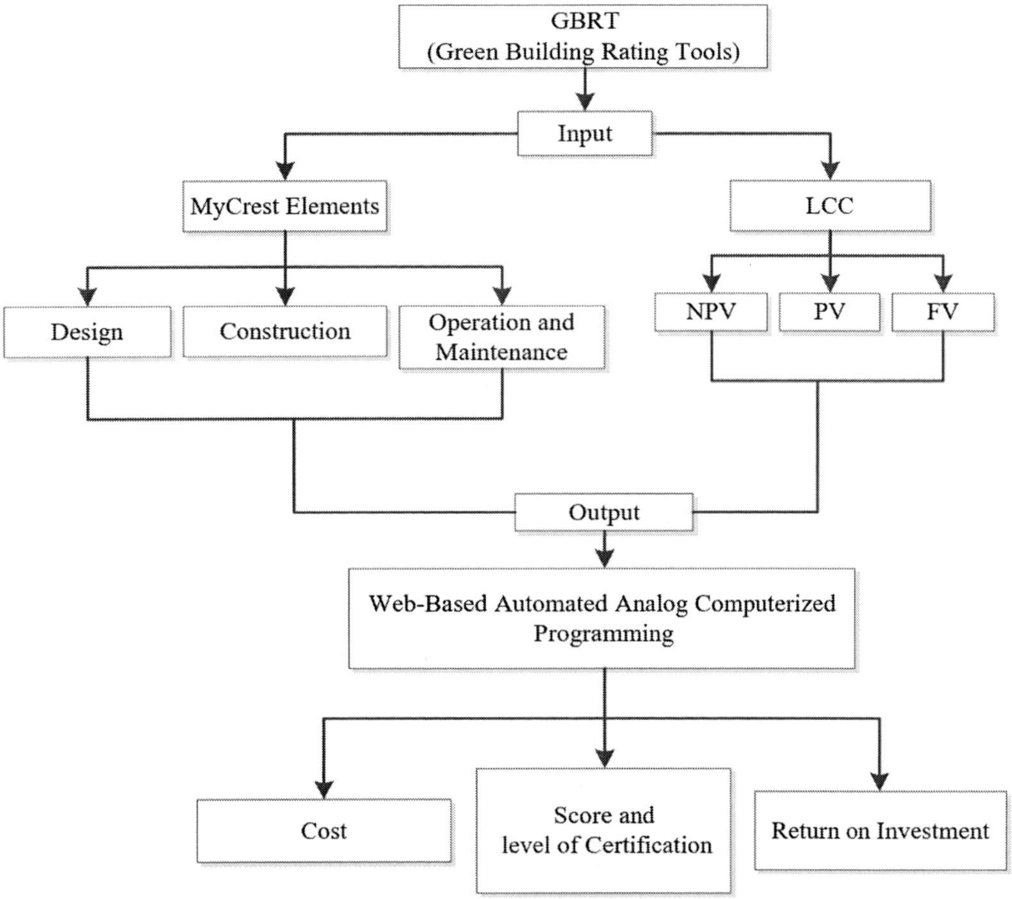

Figure 5. Interoperable Structural Functionality of the Research

Followed by the objectives, this research will tend to integrate LCC into the MyCrest tool; this will provide the novel of cost and score for the green building assessment. Once the green building score is achieved it will lead towards the level of certification for MyCrest (Green Building Assessment). This whole process will further append to accomplish return on investment element of the research which will help investors and developers to gain more attention towards green building development. In addition, proposed research will develop web-bases automated analog computerized program which help in decision making of a project from research and development (R&D) stage to final stage. The qualitative and quantitative methods have been considered as a research method for achieving the aim and objectives of the research. Under qualitative method, this research is seeking to identify and develop integrated web-based automation tool for the green building rating index MyCrest and life cycle costing which will be further validated by qualified professional. Though, for the quantitative method, content analysis method, critical literature review along with selective literature review will be used. However, the expert reviews by interview session will be done by the Qualified Professionals (QP) of the green building rating assessment tool recognized by the regulatory body to know applicability of the tool. Moreover, questionnaires will be distributed amongst industry practitioners for the design standard framework validation as the respondents that will elucidate the pertinence of the automated tool in real time projects. Once the tool is developed then it will be commercialized and if necessary a pilot project can be developed as a case study.

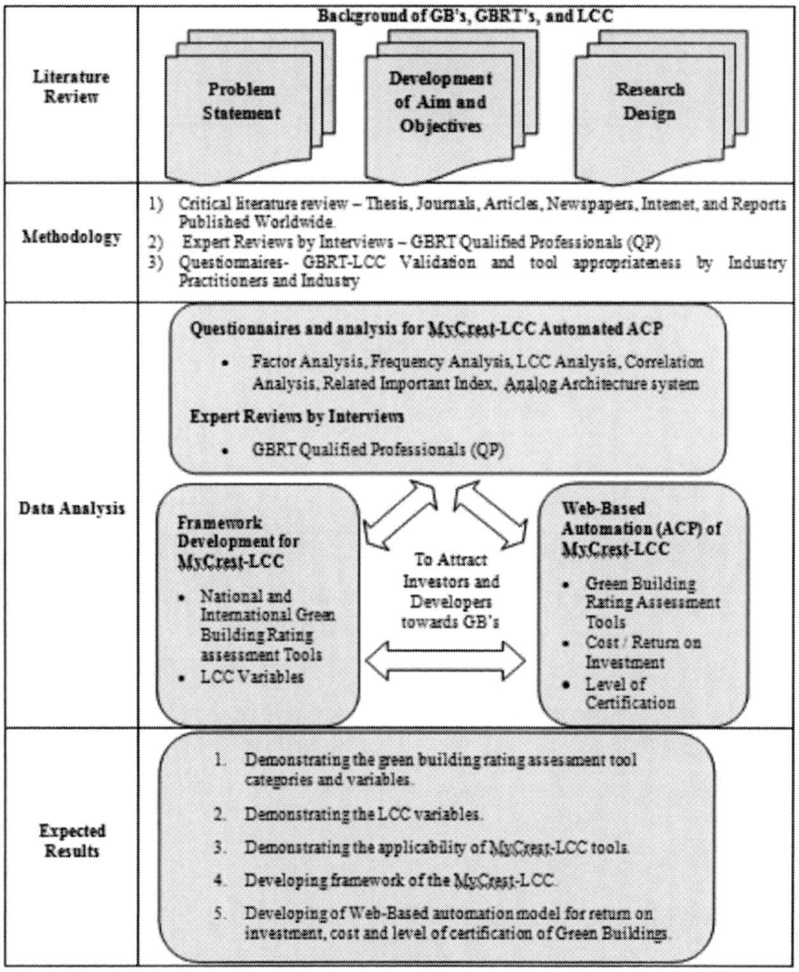

Figure 6. Detailed Flow Chart of the Research

4. Findings and Conclusion

There is a deficit of decisive actions to embark on green building at higher pace. Investors and developers are hesitated to move farther for green building development due to the lack of automated tools within construction industry. Significantly, there is lack of automated green building rating assessment tool integrated with life cycle costing in the available green building research. This research proposed to develop new framework tool for MyCrest-LCC transformed into web-based automated analog computerized programming which fills the gap of green building rating assessment tool and life cycle costing. This MyCrest-LCC tool will bring a new innovative approach towards the green building development and construction industry. This integrated tool will also give efficient, easy, quick and better platform to calculate cost, score, and level of certification more importantly return on investment. It will cut the project and green managers preliminary work, which effects on the reduction of building planning timeline and cost and will also provide better decision making and strategic planning. The research will develop new trend and technique of automation in green building technology in compliance with life cycle cost analysis. This automated MyCrest-LCC tool will also help investors in their decision making to calculate their initial and future costs in a systematic automation. Such type resilient green technology potentially can build human to move faster, robust, and transparent in strategizing projects decision making and can be adopted by governmental agencies locally and globally.

5. References

[1] Chew MYL, Conejos S, Asmone AS. Developing a research framework for the green maintainability of buildings. Facilities 2017;35:39–63.
[2] Li J, Liu Y. Research on the Construction of Multidimensional CIS Model in Green Building Based on Experiment Simulation. 2016 Int Conf Smart Grid Electr Autom 2016:234–6. doi:10.1109/ICSGEA.2016.29.
[3] Kibert CJ. Green Buildings: An overview of progress. J L Use 2004;19:491–501. doi:10.1097/ACM.0b013e3181e86dd4.
[4] Howe JC. Overview of Green Buildings. Natl Wetl Newsl 2011;33:10043–8.
[5] Xia B, Zuo J, Skitmore M, Pullen S, Chen Q. Green Star Points Obtained by Australian Building Projects. J Archit Eng 2013;19:302–8. doi:10.1061/(ASCE)AE.1943-5568.0000121.
[6] Sutton P. A Perspective on Environmental Sustainability. A Pap Vic Comm Environ Sustain 2004:27. doi:2.b 12.
[7] Leiper Q, Fagan N, Engström S, Fenn G. A strategy for sustainability. Proc ICE - Eng Sustain 2003;156:59–66. doi:10.1680/ensu.2003.156.1.59.
[8] Zuo J, Zhao ZY. Green building research-current status and future agenda: A review. Renew Sustain Energy Rev 2014;30:271–81. doi:10.1016/j.rser.2013.10.021.
[9] MGBC. Malaysia Green Building Confederation (MGBC) 2017. http://www.mgbc.org.my/ (accessed July 9, 2017).
[10] Economic Planning Unit. Malaysia Sustainable Development Goals Voluntary National Review 2017. 2017.
[11] Li Y, Chen X, Wang X, Xu Y, Chen PH. A review of studies on green building assessment methods by comparative analysis. Energy Build 2017;146:152–9. doi:10.1016/j.enbuild.2017.04.076.
[12] Deng Y, Wu J. Economic returns to residential green building investment: The developers' perspective. Reg Sci Urban Econ 2014;47:35–44. doi:10.1016/j.regsciurbeco.2013.09.015.
[13] Suliman LKM, Omran A. Sustainable development and construction industry in Malaysia. Econ Soc Polit Cult Probl Futur Soc 2009;11:76–85.
[14] WGBC. Rating tools | World Green Building Council 2017. http://www.worldgbc.org/rating-tools (accessed August 24, 2017).
[15] Olubunmi OA, Xia PB, Skitmore M. Green building incentives: A review. Renew Sustain Energy Rev 2016;59:1611–21. doi:10.1016/j.rser.2016.01.028.
[16] Hu H, Geertman S, Hooimeijer P. The willingness to pay for green apartments: The case of Nanjing, China. Urban Stud 2014;51:3459–78. doi:10.1177/0042098013516686.
[17] CIDB Malaysia. Construction Industry Transformation Programme 2016-2020 2016. http://www.citp.my/ (accessed May 11, 2017).
[18] Higham A, Fortune C, James H. Life cycle costing: evaluating its use in UK practice. Struct Surv 2015;33:73–87. doi:10.1108/SS-06-2014-0026.
[19] USGBC JK. What is green building? | U.S. Green Building Council. LEED 2014:/articles/what-green-building. https://www.usgbc.org/articles/what-green-building-0 (accessed August 2, 2017).
[20] Awadh O. Sustainability and green building rating systems: LEED, BREEAM, GSAS and Estidama critical analysis. J Build Eng 2017;11:25–9. doi:10.1016/j.jobe.2017.03.010.
[21] WGBT. World Green Building Trends 2016. 2016.
[22] Vadera S, Woolas P, Flint C, Pearson I, Hodge M, Jordan W, et al. Strategy for sustainable construction. Mater Sci 2008:64. doi:8731/2k/6/08/NP URN 08/973.
[23] Hopkins EA. Barriers to adoption of campus green building policies. Smart Sustain Built Environ 2016;5:340–51. doi:10.1108/SASBE-07-2016-0016.
[24] Poveda CA, Young R. Potential benefits of developing and implementing environmental and sustainability rating systems: Making the case for the need of diversification. Int J Sustain

Built Environ 2015;4:1–11. doi:10.1016/j.ijsbe.2014.12.003.
[25] Illankoon IMCS, Tam VWY, Khoa N. Le. Environmental , Economic , and Social Parameters in International Green Building Rating Tools. J Prof Issues Eng Educ Pract 2016;In Press:05016010–1. doi:10.1061/(ASCE)EI.1943-5541.0000313.
[26] IUCN/UNEP/WWF. Caring for the Earth. Environ Sci Policy Sustain Dev 1991;36:25–8. doi:10.1080/00139157.1994.9929182.
[27] Robichaud LB, Anantatmula VS. Greening project management practices for sustainable construction. J Manag Eng 2010;27:48–57. doi:10.1061/(ASCE)ME.1943-5479.0000030.
[28] Nguyen BK, Altan H. Comparative review of five sustainable rating systems. Procedia Eng., vol. 21, 2011, p. 376–86. doi:10.1016/j.proeng.2011.11.2029.
[29] Illankoon IMCS, Tam VWY, Le KN, Shen L. Key credit criteria among international green building rating tools. J Clean Prod 2017;164:209–20. doi:10.1016/j.jclepro.2017.06.206.
[30] Trusty WB, Horst S, Thena TA, Materials S, Street J, Box PO. Integrating LCA Tools in Green Building Rating Systems. Athena 2007:1–7.
[31] Ilhan B, Yaman H. Green building assessment tool (GBAT) for integrated BIM-based design decisions. Autom Constr 2016;70:26–37. doi:10.1016/j.autcon.2016.05.001.
[32] Australia G. Green Building Council of Australia 2017. http://new.gbca.org.au/ (accessed August 30, 2017).
[33] GREEN MARK Singapore. Building & Construction Authority Green Mark 2017. https://www.bca.gov.sg/GreenMark/green_mark_buildings.html (accessed August 30, 2017).
[34] HKGBC. The Hong Kong Green Building Council (HKGBC) 2017. https://www.hkgbc.org.hk/eng/ (accessed August 30, 2017).
[35] Hamid ZA, Zain MZM, Hung FC, Noor MSM, Roslan AF, Kilau NM, et al. Towards a national green building rating system for Malaysia. Malaysian Constr Res J 2014;14:1–16.
[36] Reed R, Bilos A, Wilkinson S. International Comparison of Sustainable Rating Tools. JORSE 2009;1:1–22.
[37] Sebake N (Csir). Overview of green building rating tools. Green Build Handb South Africa, Vol 1 (A Guid to Ecol Des 2008;1:27–34.
[38] Ahn YH, Jung CW, Suh M, Jeon MH. Integrated Construction Process for Green Building. Procedia Eng., vol. 145, 2016, p. 670–6. doi:10.1016/j.proeng.2016.04.065.
[39] Doan DT, Ghaffarianhoseini A, Naismith N, Zhang T, Ghaffarianhoseini A, Tookey J. A critical comparison of green building rating systems. Build Environ 2017;123:243–60. doi:10.1016/j.buildenv.2017.07.007.
[40] Sinha A, Gupta R, Kutnar A. Sustainable Development and Green Buildings. Drv Ind 2013;64:45–53. doi:10.5552/drind.2013.1205.
[41] White G, Ostwald P. Life Cycle Costing. Manag Account 1976;57:39-42.
[42] Woodward DG. Life cycle costing--theory, information acquisition and application. Internattonal J Proj Manag 1997;15:335–44.
[43] Korpi E, Ala Risku T. Life cycle costing: a review of published case studies. Manag Audit J 2008;23:240–61. doi:10.1108/02686900810857703.
[44] Sieglinde Fuller. Life Cycle Cost Analysis. Natl Inst Stand Technol 2010:1–5. http://www.nist.gov/index.html (accessed June 12, 2017).
[45] Chakravarty A, Debnath J. Life cycle costing as a decision making tool for technology acquisition in radio-diagnosis. Med J Armed Forces India 2015;71:38–42. doi:10.1016/J.MJAFI.2014.10.004.
[46] Lansink J. The benefits of applying the Life Cycle Costing method for the main-actors within the Dutch Commercial Real Estate sector. 2013.
[47] Gong Y, Tae S, Suk S, Chae C, Ford G, Smith ME, et al. Life Cycle Assessment Applied to Green Building Certification in South Korea. Procedia Eng., vol. 118, 2015, p. 1309–13. doi:10.1016/j.proeng.2015.08.493.
[48] Tam VWY, Senaratne S, Le KN, Shen LY, Perica J, Illankoon IMCS. Life-cycle cost analysis of green-building implementation using timber applications. J Clean Prod 2017;147:458–69. doi:10.1016/j.jclepro.2017.01.128.

Acknowledgments
This work was financially supported by Research University Grant 18H12, Research Management Centre, and research team members Green PROPMT. Thus, directly it is supported by Ministry of Education Malaysia.

Automatic helmet-wearing detection for law enforcement using CCTV cameras

P Wonghabut[1], J Kumphong[1], T Satiennam[1], R Ung-arunyawee[1,*] and W Leelapatra[1]

[1] Faculty of Engineering, Khon Kaen University, Khon Kaen 40000,Thailand.
* Corresponding author. Tel.: +6681-592-1947. E-mail address: rujchai@gmail.com

Abstract. The objective of this research is to develop an application for enforcing helmet wearing using CCTV cameras. The developed application aims to help law enforcement by police, and eventually resulting in changing risk behaviours and consequently reducing the number of accidents and its severity. Conceptually, the application software implemented using C++ language and OpenCV library uses two different angle of view CCTV cameras. Video frames recorded by the wide-angle CCTV camera are used to detect motorcyclists. If any motorcyclist without helmet is found, then the zoomed (narrow-angle) CCTV is activated to capture image of the violating motorcyclist and the motorcycle license plate in real time. Captured images are managed by database implemented using MySQL for ticket issuing. The results show that the developed program is able to detect 81% of motorcyclists on various motorcycle types during daytime and night-time. The validation results reveal that the program achieves 74% accuracy in detecting the motorcyclist without helmet.

Keyword: Road accident, Motorcyclist, Helmet, Law enforcement, Image processing

1.Introduction

Thailand has been known for its high rate of road accidents [1]. From all road accidents that cause death, 80% are from motorcycles as announced by World Health Organization (WHO) in 2015 [2]. As indicated by many researches, helmet wearing is an important factor that mitigates injury or mortality rates [3]. According to Thai lawsuit, motorcyclists are required to wear helmet, but the law has been poorly enforced. Policemen need to setup check points in order to charge motorcyclists without helmet [4], as shown in Fig. 1. Many times accidents occur at the check point because guilty motorcyclists attempt to escape. In this paper, we present a new approach that is more efficient for enforcing helmet wearing by using CCTV cameras and computer program to automatically detect motorcyclists without helmet. Our system is able to integrate with Thai Police Ticket Management System (PTM) for issuing ticket to motorcyclists who disobey the law.

Detection of helmet on motorcyclist has been shown to be possible with high accuracy as described in [5-7]. These works have been done but not reported to be used by any officials. In contrast, we have been working closely with local police department and government in order to utilize and integrate the developed program with existing police ticket system.

Figure 1 Helmet wearing enforcement at police check point [4]

2. Methodology

Our basic requirement is that the developed program must be able to capture images of motorcyclists without helmet and the motorcycles plate number. We utilized two types of CCTV camera installed at intersections: wide-angle and zoomed CCTV cameras. The wide-angle camera, which records overview image of intersection, is used for detecting motorcyclist and the zoomed camera is used for capturing plate number. We tested our program using pre-recorded video clips at different times of a day. The number of motorcyclists without helmet detected by our program is compared to the result that is manually counted by human. Fig. 2 illustrates methodology used in this research.

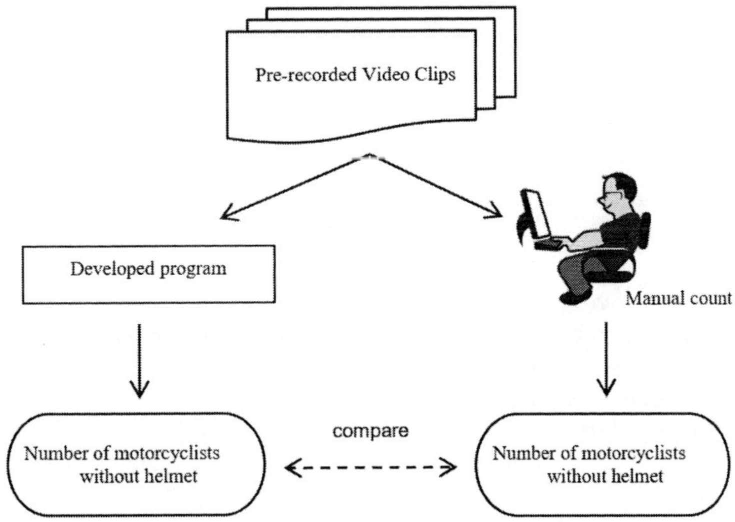

Figure 2 Methodology used in this research

2.1 Development of a program to automatically detect motorcyclist without helmet

The program is designed to monitor video frames recorded by wide-angle CCTV camera and examine areas of frame that motorcyclists are found. Then, the areas containing motorcyclists are analysed to determine whether helmet is found or not. If not, the program invokes video frames that are simultaneously recorded by zoomed CCTV camera to capture license plate of the motorcycle. The

images of motorcyclists without helmet and the image of plate number are saved in database as evidence and will be further used by Police Ticket Management System. We develop our program using C ++ language, OpenCV library and MySQL. Concept of the program is shown in Fig. 3.
The program is designed to operate in two steps. First, detect motorcyclist in video frames captured by wide-angle CCTV camera. Second, search all of the areas where motorcyclists are found to determine whether helmet is worn or not. Details of each step are as follows:

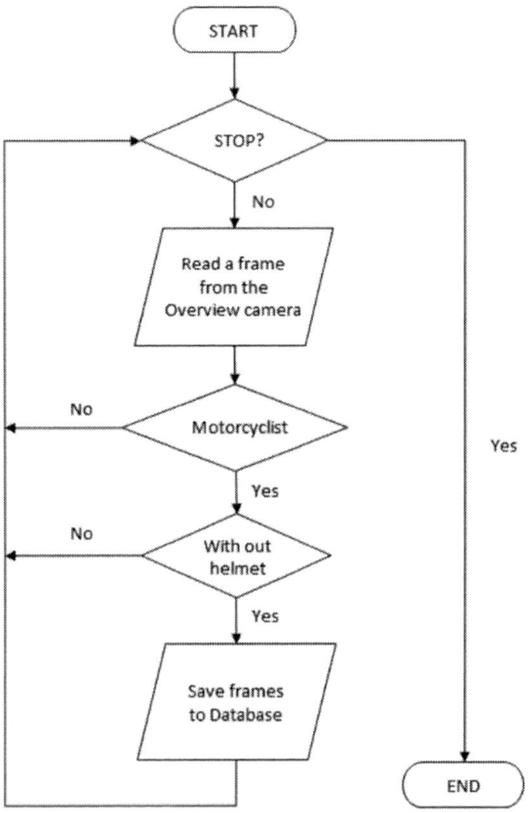

Figure 3 Program flow chart

2.1.1 Step 1: Detection of motorcyclist
In this paper, HAAR algorithm is used as descriptor for extracting object features. Output from this process are frames containing motorcyclists as shown in Fig.4 (b). These frames are *w* pixels width and *h* pixels height, where $w \leq 80$ and $h \leq 120$.
2.1.2 Step 2: Detection of helmet
This step determines whether the detected motorcyclists are wearing helmet or not. It is necessary to find head of a motorcyclist first. We consider 1/3 height and 1/2 width of the area where motorcyclists are found as shown in Fig. 4 (f).We considered two classes of motorcyclists: (i) motorcyclists without helmet (positive) and (ii) motorcyclists with helmets (negative). Also Fig. 5 illustrates system overall process from helmet detection to ticket issuing by police.

2.2 Detection results verification
To verify correctness of our program, we employed t-test to compare results produced by the developed program and the results manually counted with 0.05 significant.

IOP Conf. Series: Earth and Environmental Science **143** (2018) 012063 doi:10.1088/1755-1315/143/1/012063

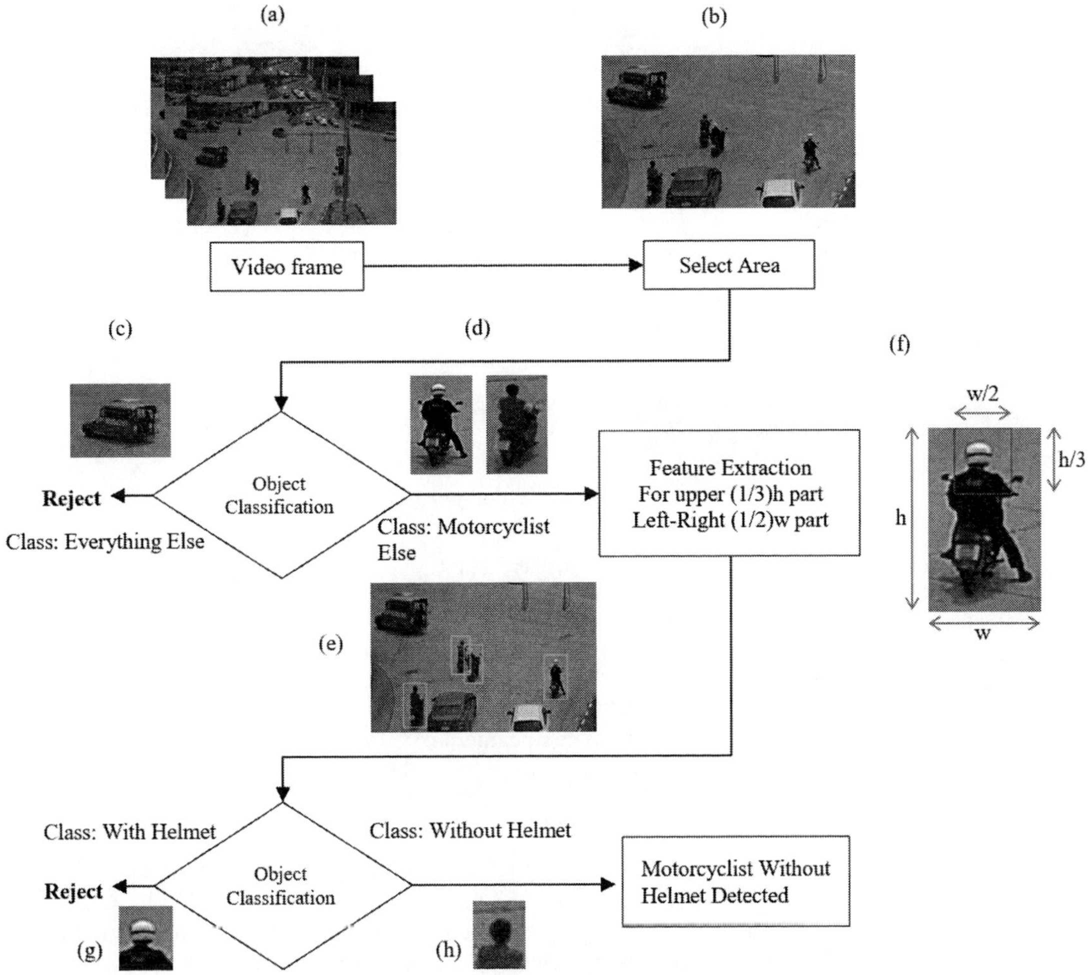

Figure 4. Diagram for detection of motorcyclist without helmet.
a) Input frame, b) Select Area A sample frame, c) Object classification as non motorcyclist, d) Object classification as Motorcyclist, e) Bounding box around Motorcyclist, f) Localized head of the Motorcyclist, g) Motorcyclist classified as 'with helmet' class, h) Motorcyclist classified as 'without helmet' class.

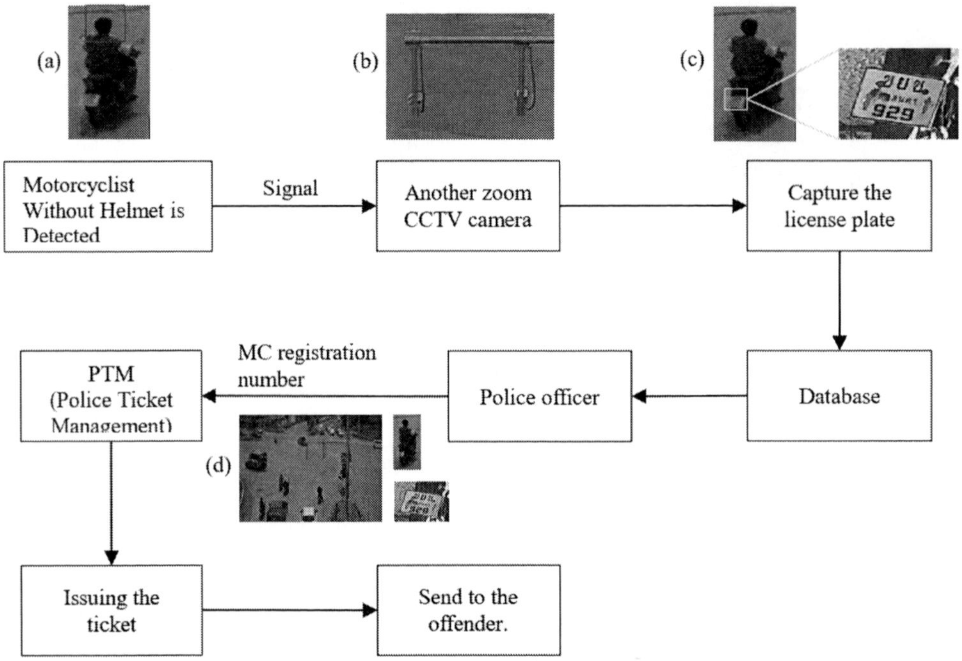

Figure 5. Diagram for issuing ticket. a) Detection of motorcyclist without helmet b) Invoke frames captured by zoomed camera c) Capture the license plate. d) Evidence submitted to police ticket system.

3. Results and Discussions

We tested our program using video clips recorded by CCTVs on August 19, 2017 during 06.00 am - 06.00 pm. Table 1 shows that there are total of 2,157 motorcycles and the program is able to detect 1,751 motorcycles. Our manually counting yields 1,526 motorcyclists who are wearing helmets and 631 are not. The program is able to detect 1,149 motorcyclists wearing helmets and 470 motorcyclists without helmets. The results show that our program can detect 81% of all motorcycles passing through the intersection and 74% of motorcyclists without helmet.

Table 2 shows detail of experiment results sorted by time of day. We found that detection performance of the program dropped during rush hours due to dense traffic, which causes overlapping of motorcycles image on video frames. This significantly reduces ability to identify motorcyclists of the program. Table 3 shows t-test significant level of differences between results obtained from manual count and results produced by program. As shown in the table, the significant levels are 0.00, which is less than 0.05, for all interested objects. Hence, the significant level of differences is said to be at 0.05. The t-test analysis indicated that the manual count data and program detection has 0.05 significant difference. However, we will continue to improve detection accuracy of the program.

Table 1 Overall accuracy of detection

Object	Manual count	Detected by program	Accuracy (%)
Motorcycles	2157	1751	81
Motorcyclists With Helmet	1526	1149	75
Motorcyclists Without Helmet	631	470	74

Table 2 Accuracy of detection sorted by time

Time	No. of Motorcycle			No. of Motorcyclist With Helmet			No. of Motorcyclist Without Helmet		
	Manual Count	Detected by program	Δ	Manual Count	Detected by program	Δ	Manual Count	Detected by program	Δ
6.00-7.00	74	57	17	49	35	14	25	15	10
7.00-8.00	197	161	36	148	119	29	49	34	15
8.00-9.00	317	248	69	252	231	21	65	47	18
9.00-10.00	138	117	21	88	62	26	50	38	12
10.00-11.00	124	84	40	79	47	32	45	36	9
11.00-12.00	145	121	24	102	75	27	43	28	15
12.00-13.00	198	159	39	125	89	36	73	58	15
13.00-14.00	147	126	21	113	88	25	34	22	12
14.00-15.00	120	95	25	78	49	29	42	34	8
15.00-16.00	184	157	27	131	92	39	53	37	16
16.00-17.00	216	188	28	164	128	36	52	37	15
17.00-18.00	297	238	59	197	134	63	100	84	16
Total	2157	1751	406	1526	1149	377	631	470	161

Table 3 Significant level of differences between results obtained from manual count and program detection.

Object	Result difference between manual count and program detection		t-test	Significant (p<0.05)
	Mean	S.D.		
Motorcycles	33.83	15.97	7.33	0.00
Motorcyclists With Helmet	31.41	12.10	8.99	0.00
Motorcyclists Without Helmet	13.41	3.14	14.77	0.00

4. Conclusions

We have developed a computer program for detecting motorcyclists who are not wearing helmets using CCTV cameras. Evaluation results show that the program can detect 81 % of all motorcycles passing through the intersection with accuracy of 74 % for detection of motorcyclists without helmet. The program has been integrated with official ticket system to produce essential evidences for police department.

5. Future improvement

The work presented in this paper is in developing process. We are attempting to improve accuracy of the program in order to increase efficiency of law enforcement in Thailand. Consequently, we will evaluate and report impacts after using this system in term of fatality of motorcyclists in the future.

6. References

[1] Burton J 2017 *Top 25 countries in car accidents,* URL:http:// www.worldatlas.com/articles/the-countries-with-the-most-car-accidents.html
[2] World Health Organization 2015 *Global Status Report on Road Safety* (Geneva: Switzerland)
[3] Alaa KA Ashraf FH and Fikri MAZ 2012 Does wearing helmets reduce motorcycle-related death? A global evaluation. (*Accident Analysis and Prevention* Vol 49) pp 249-252
[4] Images of police check points. URL: https://board.postjung.com/.918054html

[5] Silva R Aires K and Veras R 2014 Helmet detection on motorcyclists using image descriptors and classifiers *27th SIBGRAPI Conference on graphics, patterns and images* pp141-148.
[6] Dahiya K Singh D and Mohan C K 2016 Automatic detection of bike-riders without helmet using surveillance videos in real-time *2016 International joint conference on neural networks (IJCNN)* pp3046-3051.
[7] Vishnu C Singh D Mohan C K and Babu S 2017 Detection of motorcyclists without helmet in videos using convolutional neural network *2017 International joint conference on neural networks (IJCNN)* pp3036-3041.

Street Life as the negotiation process: case study of Sidewalk Informal Economy in Ho Chi Minh City

The Hien Dang[1]

[1]Urban and Regional Planning Department, Ton Duc Thang University, 19 Nguyen Huu Tho, Tan Phong Ward, District 7, Ho Chi Minh City, Vietnam
E-mail: dangthehien@tdt.edu.vn

Abstract. The study researches the issue of informal pavement economy in Ho Chi Minh City, a sector that has always been an important part of the city's economy but has not been recognized by the local state due to some certain facts. By using surveys and in-depth interviews in the cases of four streets and one alley in HCMC's center, combined with literature review, the paper does not only aim at sketching the portrait of HCMC's informal sector but also put it in the relationship with surrounding related sectors. The outcomes have some similarities to previous cases in the field of Hanoi and other parts of Global South, but at the same time reveal some unique characteristics of informal sector in HCMC. In which results, the key finding is that informal work is not a low and unproductive sector nor is it the obstacle to the deemed civilized and modern image of the city. In contrast, these activities are the suppliers of goods and jobs for the urban poor, creating diversity for the economy. Being regarded as an unrecognized sector by the Government, informal sector has no success in negotiating at high institute level, but succeeded at the lowest levels of the state's enforcement system relying on the social capital of long-time relationship and interaction. It is noticeable that informal sector is willing to contribute to the civilization and modernization. It also gains the acceptance and support of the society and media. All of these will open up opportunities for its future development.

1. Introduction

This paper aims to study the negotiation process between an informal sector and its related sectors. The main research objectives are: firstly to present, categorize, and analyze the informal activities on Ho Chi Minh City's pavements, in order to understand these activities' perspectives, especially the negotiation processes, which reveals their relationships to each other, to the government and to the nearby local citizens; secondly to specifically investigate further the point of view of Ho Chi Minh City citizens to informal economy issue, with studies coming from two directions of views of the informal sector: from inside-out and outside-in. The paper wants to sketch the portrait of informal in a truthful way, for further sustainable development in the future. Vietnam's informal economy has a thick history from the feudal times, undergoing many periods of the country's history, but until now the official definition of informality did not occur in any Vietnamese regulation text, there are also no current policies targeting the informal sector [1]. The campaigns of forbidding informal activities are under enforcement for more than ten years but those activities still continue overtly on everyday street life. All arguments about whether to accept or forbid informal economy are just on the media basing on opinions of some officers from the state and some informal interviews from unreliable or unverified sources. Unfortunately, these types of media are not accessible by informal sectors who are mostly poor and have low level of education. There is almost no official survey on what people in the informal business really do, how they feel or care about their work (in some cases, their only means of survival), their living condition, their future, or what the citizens of Ho Chi Minh City think about these economies on the pavements. Arguments are just arguments when everyone keeps their standpoints and this situation would not lead the issue to anywhere. On the contrary, surveys with real numbers will never tell lies and will create opportunities for further studies in the future. This research, working on both theoretical and empirical academic methods, with literature review of published papers, researches and small-scale surveys in the case study in Ho Chi Minh City, will therefore hope to achieve these two research objectives to give a first impression about pavement economy in Ho Chi Minh. Based on previous interesting results of Sarah Turner and Launa Schoenberger in Hanoi, combined with the local condition of Ho Chi Minh City, hypothesis has been made: "informal sector" is divided into two parts, one is itinerant street vendors, and the other is 'city residents operating small, 'fixed' pavement stalls. Both of which do not get along well with each other, when street vendors are considered as "outsiders" by the fixed stall sellers [2]. On the relationship with the government and local police, who directly have the duty to enforce regulations relating to informal sector, both itinerant vendors and fixed stall sellers have their own strategies of running out the area that the polices chasing them are in charge or hide their business tool [2]. They also use the social capital of long-time interactions or 'compassionate' (sự thông cảm) from local police and from there "opens up space for negotiation of laws and regulations between citizens and state" [3] .On the hypothesis for the citizens' view on informal sector, as mentioned above, there would also be two different types of opinions, one supports the state's idea of forbidding informal business, the other bolsters the existence of pavement economy. Empirical research by case study will examine which hypothesis is more acceptable, and from there suggests some adjustment or movements for informal economy in HCMC in the future.

The research is based mainly on the empirical results from observations, mapping, surveys as well as in-depth interviews with informal sectors, other citizens and the local police in the case study undertaken along four Ho Chi Minh streets: Pasteur (from Vo Van Kiet to Le Loi St), Nguyen Cong Tru (from Ho Tung Mau to Pho Duc Chinh), Nguyen Hue Street, Ton That Dam Street and "hẻm" (alley) 100-Nguyen Cong Tru Street. These streets are in the CBD area, in which there is the concentration of companies with numerous employees, forming a promising market for informal economy. On the other hand, the city's campaigns of street clearance are also strict here, as the CBD area is considered "the face of the city" or the "representative" of the modern and civilizing Ho Chi Minh City. With such a conflict, the area is a good sample to do research on and potentially make improvement suggestions. Moreover, the four chosen streets also have their unique characteristics: Nguyen Hue is the financial boulevard which represents the "civilizing" street of HCMC; Pasteur is one of the main streets with high volume of transport and is busy most time of a day. Ton That Dam is not a busy street with high transport volume and capacity, but these factors turn it into a street with rich street life during the whole day. Nguyen Cong Tru Street is famous for being the location of banking companies with most big banks in Ho Chi Minh City having a branch in the area; at lunch time, the street turns into a walking street with full of pedestrians and food stalls. The economy activities also take place in alleys with less transport and the campaigns have less effect. Activities on Hem 100 Nguyen Cong Tru with most stall owners being the locals living in the alleys are also counted on the result of an interesting type of negotiation. These streets belong to two different wards of Nguyen Thai Binh and Ben Nghe - District 1, with two different local management systems, so it would be a chance to compare how local governance influences the negotiation process on pavement economy. The observation takes place on the research area across 5 periods of times in a day: 6AM-10AM, 10AM-1PM, 1PM-7PM, 7PM-10PM, and 10PM-1AM to match with the main actives of a day: Breakfast time, Lunch time, After lunch time, Out for Dinner time, and Late night. There are three objects of observation: Food stall, Coffee stand (which not only sells coffee but also other drinks), and Other informal street economic activities. These three objects are chosen because it is generally observed that most pavement businesses are about food and drinks. The observation is aimed at getting the number of informal economic activities in the area, how they are distributed, in different periods of time in a day. The morphology of these informal businesses can itself tell the businesses' working times, targeted customers, and roles in the city's everyday life. After data are collected, the mapping method is applied to produce more visualized material for further analysis, with the maps of the streets, the number and location of these informal activities are illustrated, on different sets of time ranges.

Adding to the research material are the results of two types of survey questionnaires: one for business owners in the informal sector of the research area, the other is for customers living or working in the area, who represent the citizens and customers of Ho Chi Minh City and have their own thinking about these "informal activities". The questionnaires for informal sector are divided into five parts, the first part asks for basic information of the sellers to find out who they are, where they are from, information about their family and living condition, the second part enquires their work and economic issues that informal work brings to the owner, the third and fourth part are about the relationship of informal sector to political and social issues, and finally, part five is for those who do not usually have a chance to speak out the feeling about their job, their expectation and prediction about their own future. There is also uncertainty in the answers about income and the contributed percentage of this to their family's income. The income amount per day answered to the interviewers is usually changed after the conversation becomes friendlier. It is for some personal and social reasons that people tend to reduce their income from 100.000 VND to 150.000 VND in their answers. Furthermore, as most of the sellers in informal sector on HCMC's pavements are women, they tend to increase the share percentage of their husbands' income to the family's total income, as a way to show their respect to the men of the family, this social phenomenon should be considered when analyzing the result of survey.

In-depth interviews are also important to gain more in-depth information that questionnaires cannot obtain. These are the informal conversations with informal sectors, local citizens and officers, making them feel comfortable and leading the interviews to unexpected but valuable outcome. The main aim is to learn the relationship and interactions between these sectors and how they negotiate with each other. The local police refuse to answer questions related to the city's policies or campaigns. They also refuse to give opinion on public issues. The answer they usually give is "This issue is out of my competence, it depends on the guidance of the Party and the Government". Therefore, to understand what the local police and officers think, it can only be guessed from the way they react and manage different situations in the areas that they are in charge of.

The indispensable part of any research paper is the literature review which would be discussed more intensively in the later part of the paper. The aim is to review the main published work concerning this particular topic of informal sector. The review is not simply a description of what others have published in the form of a set of summaries, but also a critical discussion showing insight and awareness of different arguments, theories and approaches, from the very first definition of Hart in 1973 to the contemporary trends of studying about informal, especially the rare but interesting and valuable studies about informal economy in Vietnam. This knowledge is not just for placing the study in a historical perspective but also for creating the hypothesis for research questions, giving researchers a general direction for further research. With the two main research objectives of studying the informal activities in Ho Chi Minh City's pavements and the point of view of the city's citizens to informal economy issue, the main research question would be: "How is the informal pavement economy in HCMC?" and "How does the society consider informal pavement economy in the city's life and future?"

2. Literature review

First time the definition of 'informal' being realized and mentioned was in Hart studies about cases in Africa, in which he argued:" 'Formal' incomes came from regulated economic activities and 'informal' incomes, both legal and illegal, lay beyond the scope of regulation. I did not identify the informal economy with a place or a class or even whole persons." [4] .From that, there are numerous researches relating to the issue of informality, and the informality discourse becomes 'large, vibrant and expanding fast. But there is a certain conceptual incoherence to the literature. New definitions of informality compete with old definitions leading to a plethora of alternative conceptualizations' [5]. Despite that complexity, looking through the history of informal context, there are such a 'shifting conceptualizations of informality over time' [6] from more 'negative' perspective to 'neutral' in the last decades and becoming more in-depth studied in contemporary literature. Informality context as well is not only applied to the Third world, with the unregulated-sector as in its early days but also in more developed area, with more angles as an 'idiom of Urbanization' (Roy, 2009) an 'opportunity informality' [6] or even as 'a mode of flexibility serves as a strategy in locational policies to create a local competitive edge. Further, informality is used as tool for testing and learning'[7].

In the article "Why India cannot plant its cities: Informality, Insurgence and the Idiom of Urbanization", Anaya Roy(2009) analyzes the structure of urban informality in India through two sense of Indian Urbanization of Bangalore and Calcutta, to which cities are facing the urban crisis by the lack of infrastructure or strong social division within the city. Those rises are explained not by the failure of urban planning but 'an idiom of urbanization' in which the key feature is informality. 'While informality is often seen to be synonymous with poverty, this article makes the case that India's planning regime is itself an in formalized entity, one is that a state of deregulated, ambiguity, and exception. This idiom of urbanization makes possible of governance, justice, and development'[8]. Roy also stated that " Informality does not lie beyond planning; rather it is planning that inscribes the informal by designating some activities as authorized and others as unauthorized". Consequently, she came to the conclusion of informality as "a deregulated rather than unregulated system" and "informality lies within the scope of the state rather than outside it" [8]. If those 'informalized entity' of authority makes India 'cannot plan its city', in China, the informal "is increasingly used as a tool of flexibility, testing, and learning, whereas informality – used as a substitute to fill institutional gaps in the early phase of transition – has diminished over time in the course of the reform process" [7].

There are still few studies about Vietnam's informality despite its size and impacts on Vietnamese daily life, "this situation is due to a number of factors. Firstly, the concept of what constitutes informal is vague with a multitude of definitions having been put forward by different authors. Secondly, measuring the informal economy is problematic since it operates on the fringes of the economy. Thirdly, the informal economy suffers from a lack of interest on the part of the authorities as it does not pay (or pays little) taxes and is seen, especially in towns, more as a nuisance and a mark of underdevelopment, doomed to extinction by the country's economic growth. These elements explain why there has been no real significant effort to date to improve knowledge in this area"[1]. With little attention, the formal definition of Informality did not occur in any Vietnamese regulation text, there are also no currently no policies targeting the informal sector, however, the IOL (2002)'s recommendation is adopted as the definition:

"The informal sector is defined as all private unincorporated enterprises that produce at least some of their goods and services for sale or barter, are not registered (no business license) and are engaged in non-agricultural activities. Informal employment is defined as employment with no social security (social insurance). All employment in the informal sector is thus considered to be informal employment, as is part of the employment in the formal sector", "both the informal sector and informal employment are defined as belonging to the informal economy".

Lacking of data is also a problem for informal studies, recently, there is just a joint research project with the French Institute of Research for Development (IRD-DIAL) and The Labor force survey (LFS) which conducted the data national wide, following with Household business & informal sector survey (HB&IS) to calculate the account of informal with the economics of Hanoi and Ho Chi Minh City (HCMC) in also the two years of 2007 and 2009. Albeit the first and only official source of data, the number demonstrates the considerable size of informality in Vietnam, which is:

"In 2007, the informal sector accounted for almost 11 million jobs out of a total of 46 million. This represents nearly a quarter of all main occupations (24 per cent), with nearly half of nonfarm work found in the informal sector.

If we aggregate main and second jobs, a total of 12.4 million are held in the informal sector. On the whole, there are 8.4 million informal household businesses in Vietnam. At the national level, manufacturing & construction is the largest informal industry, (43 per cent of total employment in the informal sector), followed by trade, (31 per cent) and services, (26 per cent). Conversely, an impressive half of all industrial jobs in Vietnam are held in the informal sector. We estimate that the informal sector contributes to 20 per cent of GDP, without knowing what share is already included in the national accounts.

As expected, most employment (82 per cent) in Vietnam can be defined as informal employment. Informal employment is widespread in the economy, not just found in agriculture and the informal sector. The majority of jobs in domestic enterprises are also informal." [1]

Smaller scale with variety focuses also detect numerous valuable knowledge. Such as Jensen and M.Peppard (2003) work on Hanoi's street vendors has present who are the vendors, where they are from, how is their income and the close relationship between rural and urban area, where the streets vendors both live, and "they sell in the urban what the countryside produces"[9]. In another perspective, by observation and deep interviews with 40 local self-employed street traders, Sarah Turner and Launa Schoenberger (2012) reveals the motivation and interactions among Hanoi's Street vendors, the everyday politics and resistance to those policies. As the result, economy is the main motivation for people to choose street vending, there are also two main types of street vendors in Hanoi, one is itinerant street vendors, and the other is 'city residents operating small, 'fixed' pavement stalls'. Those fixed stall 'monopolized the best trading opportunities" and consider itinerant vendors as 'outsiders'[2], which also are the cause of overpopulation and transportation congestion (same point of view of the People's Committee of Hanoi, 2008). Taken a review on Hanoi's history of street trade, from the feudal times to Indochina war, the command economy, the early phase of transitional period or even the new goal of becoming an industrialized country, Sandra Kurfürst stated that informality street trade in Hanoi is from time to time change as being abandoned or tolerated by the state depends on the social- economic condition, and "the boundaries between formality and informality are not fixed, but constantly shifting"[3]. According to Kurfürst also as Roy, Kanbur, informality in Vietnam have close relation to the state.

Taken review on Vietnam Authority perspective, it is shown in most scholars that pavement economy is considered as an obstacle to modernization, which is the aim of the socialist state that Vietnamese cities would become 'civilized and modern city' (đô thị văn minh, hiện đại) or 'urban order' (trật tự đô thị) would be set. It is easily observed from the streets in Vietnamese Cities the propaganda panels of those civilizing campaigns from the government. As a result, street pending is declared illegal because it represents the relic from the past or the countryside, does not fit in the modern image of the city also as being the cause of traffic congestion. Although being prohibited in Hanoi since 1995, the presence of pavement economy is still strong. It can be explained by the "fuzzy concepts of formality and informality" and the "Mediation spaces and regulation enforcement" in Vietnam [3]. According to Kurfust, the regulations from the state is not homogeneous from time to time when prohibit informal sector in 1995, but allow the utilization of pavement for private business with fee in 2006, this situation is also described as "institutional gaps" by Waibel (2009) for the transition processes [10]. Moreover, with the transfer responsibility of public space management to the urban district level in 2006 in order to increase regulation application efficiency, the situation becomes more complex because of different level of regulation application, depending on the local situation. From the social capital of living in the same area for long time, or having long time of interaction, the fixed stall and street hawkers take advantage of the local official's 'compassionate' (sự thông cảm), from there "opens up space for negotiation of laws and regulations between citizens and state"[3]. It is also important to notice the social issue when doing research about Vietnam's public space, as it differs from Western experience in both direction of inside-out to inside-out, where from inside-out, people tend to use public space for their private purposes [11], and have the strong social feeling that the public space in front of their houses belong to themselves (Pham, 2010). From outside-in, "urban authorities with their socialist thinking that public works can touch everyone in everywhere at every moment, utilize naturally private space for public works or intervene directly into private life of urban people without any reflection" (Pham, 2010). Further, elderly people, war veteran, or close relationship are all social capital that can be used, and both itinerant vendors and fixed stall clearly understand the boundaries and have their own strategies to survive and build up resistance [2]. Still, there is no overt protests, because of "the state's strength in its ability to impose" [11], when it is "ordered and also supervised by higher institution of the municipality, informal agreements with local officials will immediately become obsolete" [3]. This makes an uncertain condition for the informal pavement economic, or in other words, their security "is nothing but borrowed" [12]. For the conclusion of her work, Kurfusrt argued that, informality, more than Roy and Kanbur discussion about unregulated, deregulated or the issue of enforcement, but "the state as active producer of informality", for "their objectives of becoming an industrialized country until the year 2020".

That can also explains the reasons for why "the campaign of 'green, clean, beautiful' (xanh, sạch, đẹp) city, or the campaign of so-called normal utilization of sidewalk, the results of those campaigns are always in the "đầu voi đuôi chuột" status(literally translation: head of elephant, tail of mouse-a Vietnamese proverb that describes works starting seriously and impressively, but ending so humbly with the unexpected modest results)" (Pham, 2010), the 'Ho con rua' case of 'civilization ' campaign of the state has thoroughly successes by replacing "rich street life where patrons would sit in outdoor cafés and turn their gaze towards the public activities of the street to indoors hi-end cafes" and "people do not feel loss" [13]. Those cases show the complexity of what the government would do or not, their ability as the "active producer of informality", as Vietnam "reflect China reform experiences more closely than the former Soviet Union and eastern Europe" [14], the situation is thus easily compared to the China case where "the state and its local representatives take a tactical approach in using their position as rule-setters. For example, they may accept informal bottom-up initiatives and subsequently formalize them, e.g., as an adequate response to newly emerging problems or niches, or they can forbid them and, as a consequence, ban informal institutions. Another more simple reason for informality in the PRD can be the lack of state power to efficiently filter through all spheres of institutions, markets, and society from the very beginning of the transitional process" [15]. No such paper can provide any solution nor long-sighted of Vietnamese informality on the street, especially when street vending is considered as a residual and non-productive activity and a cause of traffic congestion (People Committee of Hanoi, 2008), it is even harder to predict how the future of Vietnam's informal activities on the street would be.

3. Informal sector in HCMC: who are the stakeholders?

In the research area of four streets: Pasteur (from Vo Van Kiet to Le Loi St), Nguyen Cong Tru (from Ho Tung Mau to Pho Duc Chinh), Nguyen Hue Street, Ton That Dam Street and "Hẻm" 100-Nguyen Cong Tru Street, there were 60 interviewees in the survey, being chosen randomly while they were working, at any time period of the day, 40 people accepted to get interviewed by means of a standardized questionnaire, 20 others refused to answer. The main reason for refusal was that they said they were busy with their job while some others just did not want to tell about their life. Most of the actors in the informal sector in the survey area were women (37/40 interviewee), that counts for about 90 percent. It is not because of the chosen interviewees mostly are women, but that is the real percentage of men/women working on informal sector on Ho Chi Minh pavement according to the observation of the author. When it was asked if the husband of those informal-business women help them when they have free time, most of them would refuse even if they have time, because "it is embarrassing for men to go outdoor and selling things like that, men should do big things, or at least the manly work, selling things on the street is just for women" (said Mrs. IS-01, about her husband reason for not helping here). About the age of informal sector, the youngest age was 26, and the oldest was 70, there were even about 20 percent of the interviewees older than 56 years (the current age of retirement for women in Vietnam). Most of the informal sectors were in the age from 46-55, which counts for 50 percent, the age of 36-45 takes 20 percent. Eighty percent of interviewees have not reached the level of education of high school. More than half (55%) had only primary school level and about 45% had just junior high school level. The author interviewed one lady who had not been able to go to school, at all. As a result, she cannot read or write. There was one case of a person who had a university degree and who is now selling newspaper. All this makes clear that the informal street seller have a level of education which is significantly below the average of the urban population of HCMC.

Quite surprisingly, the main part of informal sector said that they were living within the administrative area of Ho Chi Minh City (95%). Almost 20% were even living in close proximity to the study area, and nearly 40% were living in the city center. Regarding the latter, it is within a reach of 15 minutes of travelling to the area their business takes place. When being asked about their home town, 60 percent said that they were local residents of HCMC, 15% moved from the North (15%), 12% from Central Vietnam (12%) and about 10% from Mekong Delta. As most people were born in HCMC, the proportion of staying time in HCMC more than 20 years is high (75%). Regarding the interviewees who migrated to the city, they at least have been staying in the city for 6 years, and most of them from 11-20 years (18%).

The first observation and interviews of four streets and one alley show that the informal sector activities in HCMC are vibrant and run openly as a part of the city's daily life. They are the source of jobs of many local people living there and they also contribute to the supply of services for thousands of local people. In general, food, drinks are the most common goods offered. The shops are small and portable in Ben Nghe Ward area but can also be bigger and overt in Nguyen Thai Binh Ward area. It is depends on the target customers that the informal sector operate their working hour. The food stalls which sell breakfast runs from 6AM to 10 AM, for lunch runs from 10AM to 1PM. The coffee stalls for local and passengers runs whole daytime, even on weekend, while other food stands serve mainly the employees in office buildings nearby will close on weekend. The shops are dynamic to utilize one space for other types of products in different time of a day. The sellers deal with each other to share the location, there are also the situation that customers can sit in a one shop can order food from other shops, in which the shop cooperate well for the best service. Negotiation process not only takes place within the shop owners but also between the shops to local police and residents, such as the Hem 100 case. All 8 food stalls are people living in the alley and it is not possible to run business on the surface of traffic, while the alley is narrow and that is public property. There is one important issue have to take into account is the presence of a small police spot where people can register their short stay in the area with the local police, which means those informal activities area directly in the eyes of local government. In that situation, the alley fee of 100.000 VND/ month and the guarantee from the shop owners to keep the alley clean and hygiene are the results of the negotiation process, which shows the support from others local citizen to the informal sector, and the contribution of those informal shops back to the local community. It is also worth mentioning, that people not living in the alley, cannot get permission to open shops on the alley surface. This is because longtime relationships and social capital are required for successful bargaining processes.

Those first observation results are adding to the outcomes of the questionnaire results such as about the opening hours. When interviewed about working hours per day, no shops work less than 3h/day, workings hours vary from shops to shops and the number of shops work for 3-5 h/day, 5-8h/day and 8-12h/day are nearly the same, while less people takes longer working time. Most people work the whole week (45%), 20% will close business on Sunday, and 30% will work from 04-05 days a week.

Most of people have worked in informal sector more than 7 years (80%), in which number of people work from 7-10 years count for 24%, from 10-20 years 18% and even 37% have been working on the street for more than 20 years. This number is opposite to what has been expected at the beginning of the research. Some literature suggests the idea that informal jobs are only of temporary nature and people tend to change to formal sector when they have the chance to do this. The empirical analysis shows however, that working in the informal sector is a stable and permanent: It is highly remarkable that a significant number of people have been working for more than 20 years working in the informal sector, already.

In regard of incomes, it is also surprising that people can make quite a lot of money from their economic activities in the informal sector: For the net income a day, 40% of informal shops earns from 150.000-200.000 VND a day, about 25% earn 100.000-150.000 and 20% can make from 200.000-300.000 VND from a working day. For average, it is about 140.000 to 200.000 VND can be add to the income from those pavement trading, if they do the business as normal employees work, with 4 weekends/ month, the monthly income would be from 3.080.000 to 6.000.000 VND, three to four time the basic salary from the state of 1.150.000 VND (basic salary in the year of 2013-following 66/2013/NX-CP).

It is also important to notice that people are likely to reduce their income in the survey from 50.000 to 100.000 VND, after give result to the questionnaire and have deeper conversation, when people are more open, most of answers about income said in the survey are adjusted. With the increased amount of 50.000, the monthly income of informal sector can be even four to five times more than basic salary.

Income from informal pavement business play most important role in their family economy. More than 80 percent of the interviewed persons contribute more than 50% to the whole family's income, in which there are up to 50% is the main income of their family (contribute from 70 to 100% the household income). For the spending, money from informal activities more less the main source of spending of those families, too, when there is up to 44% is the only spending possibility. The reality can be even more than that when the shop owners- who are mostly women- tend to reduce their contribution amount in order to increase the husband's part. Even the husband is not at the survey location, the women want to "save face to the husband, reduce my contribution a bit is good, man will sad if they can be the breadwinner" (said Mrs. IS-02, the 'com tam' seller). Mrs. IS-03, who had to sleep on the pavement when she first migrated from the North to HCMC, and has being working on the pavement for nearly 30 years also said: "The pavement feed my family, from working on the sidewalk, now I have my own house and my kids are all at their schools. My husband, he had formal job in the factory, but his salary is not enough for his own spending, sometime I have to give money to him, but you should write down I contribute just from 50-70% of the family income, poor him!"

There are variable reasons for people to choose working on the sidewalk; the main reason is that they do not have their own space for business (25%), lacking of education or simply like doing business on the pavement can also be strong motivation for choosing informal work (17%). To maintain family's business is also an inducement (15%), and for some others, informal work is just a means for living (10%). High agreements are also made (nearly 60%) that the business location was chosen because of short distance to the owner accommodation. A second criterion to choose business location is the number of target customer, which counts for nearly 20% of the interviewee's opinion.

The same criteria of location, quantity, quality, service and price are offered to both informal sector and their customers in the questionnaire to rate those criteria's importance following their opinion. The aim of this question is to compare the difference in evaluation of the mentioned criteria. There are five level of ranking from not at all important, slightly important, important, fairly important and very important, which is given the mark from 1 to 5. All sum the marks of each criteria will be divided by 40 for informal sector, by 60 for the customers and be put in the same column chart for easily comparison as below:

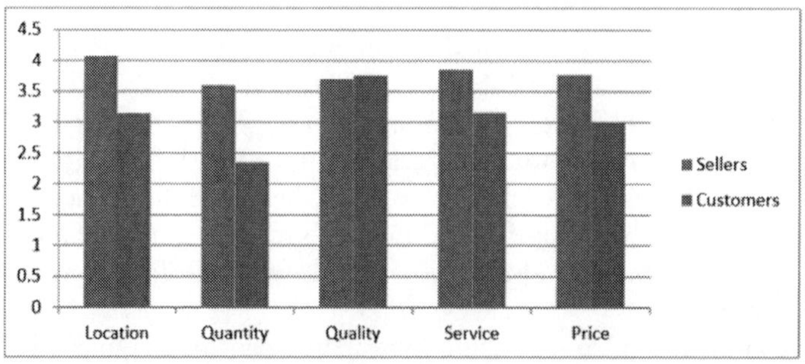

Figure 1. Consideration of Informal sellers and Customers to some service aspects.

4. Informal sector in HCMC: How is their relationship?

This part of the paper is to learn the relationship of informal business on the pavement to local authority, local citizens and to each other. Evaluation questions are putted in the survey to analysis these relationship. For awareness to the government, the general city's policy and local police is consider 'important' in the way of the duty of normal citizen the respect the national laws and the officers who enforce those laws. What influences and directly affect the business on the pavement is the city's campaign on clearance the sidewalk and the City's Order Team. As for most informal sector, if the campaign takes two weeks in a month, the damage it cost is much more than half of the income. The campaign usually cause the losing of customers because customers will look for or other supply for their need during the campaign time. Sometimes, the tools for doing the business such as table and chairs are confiscated, and the owner have to buy all the tools again or pay an amount of fee to get them back. The fine is also different from stalls to stalls. As it is not an official question in the questionnaire, the fine amount of each stalls are variable, from 70.000 to 400.000, the most common fines is 100.000 and 150.000 VND and there is no official range of fines, the shop owners just pay what they are told by the City's Order Team. As for the observation of the writer, there is a connection between the fines and the business's situation; the businesses which are working well with more customers are fined more than the smaller or the ones with fewer customers.

For the local police (Công an khu vực), who have the daily responsibility of managing the neighborhood's order and safety, not as other cases in Hanoi reviewed in chapter 2, they do not requires money from those informal sector runs business in their area. Nearly 100% of the interviewed people said during interviews that they do not have to pay any fee to the local police. Mr. LS-01 and Mr. LS-02, who are in charge in Nguyen Thai Binh and Ben Nghe Ward of the research area, gain the informal sector's respect. People showed their appreciation when they mentioned about them, as Ms. IS-04, who sell sweet soup (chè) said:" You should write good things about the local police here, he well knows that we are poor, so he never asks for money from us". Another opinion from Ms. IS-05, the mixed rice-paper seller: "In new year occasion, I brought him a box of beer as present, he asked me to bring it back home and tell me to save my money to buy milk for my little son". She continued: "But it is not easy to go on business in the area, as I was born here and Mr. LS-01 (the local police) has known me since I was a kid, so I am allowed to do sell things here, but I have to report to him when there are new comers or criminal activities on the pavement. If the knows that you have information about criminal and don't report, you are not allowed to run business in the area anymore, but that's all for the area's security". For a joke in another in-depth interview that the writer will open a drink stall next to a sugarcane drink stall, Mrs. IS-06 the owner said: "I would bet you whatever it takes that you cannot open a stall here, if it is that easy, the pavement will be full of shops but this situation!" To compare between local police with the City's Order Team, Ms. IS-07, the orange vendor stated: "It is easy with the local police because he sympathies, if the owner of the house let me to sit in front of the house to sell orange, he will allow me to sit there. For Order Team, they don't have compassionate, they chase us, take our tools or fine us." From the deep interview results, we can easily go to a conclusion that local polices in the research area do not exploit informality of the authority also as "institutional gap" (mentioned in chapter 2) as manner to gain personal profit, at least with the informal sector on the pavements. These polices know that informal trading is nearly the poor's only livelihood and therefore accept those activities as a part of daily life. Of course, this acceptance is accompanied with the social capital as most of informal sector in the research area are local people or have longtime interactions, this situation is similar to Hanoi case, which "opens up space for negotiation of laws and regulations between citizens and state" [3], the only difference is that the negotiation process is not for the police's personal benefit, but informal sector here is considered as an informal tool to enrich the local security, to prevent criminal and illegal activities. The necessary condition for these negotiation processes is the agreement of local citizens. As for the orange stall of Ms. IS-07, the police only allow her to sit on the pavement when "the owner of the house allow", or the case of eight lunch food stall in Hem 100-NCT, the police only let these shops run after their negotiation with the local within the alley comes out with a fee of 100.000 VND. From that, the key lesson is negotiation between informal sectors with authorities can only be success with the support of the citizen, or in other word, the 'society'. Moreover, it is not only the negotiation process but ethic and self-perception of enforcement sectors playing very important role in street life. With the same management system of transferring responsibility of public space management to the urban district level, other officers use it as a chance to get money from informal sector, the polices in the research area do not. For the City's Order Team, who "do not have compassionate" as the informal business owners, which means there is no negotiation process between these two sector, but the fines differ from one to others depends on the economic situation of the shops. The nearly 6 times difference 70.000 to 400.000 fines shows that although it is not negotiable, the Order Team adjust themselves with the situation, this is a kind of informality in enforcement process, which makes the difference between HCMC and Hanoi case, in which the scenario is brighter to the informal sector in HCMC.

In relationship to local citizens, informal sector in the research area do not have to pay to the owner of the house for running business and they have long time relationship, which is considered from 'normal' (70%) to 'positive' (nearly 20%) and 'very positive' (nearly 10%). Most of informal sector do not feel they have competitors (70%) as "everybody do their job, the customers prefer a shop than other because of their own characteristic, there is nothing to compete here!" Meanwhile, there are still some shop-owners consider other informal shops selling similar kind of goods (20%) and other informal shops selling different kind of good (10%) as competitors. No informal shops see formal or the international convenience stores as threat to their businesses. For the people who "have competitors", they label relationships to their rivals "normal" and those interactions usually do not affects their business situation (almost 90%).

5. Social perception on the Presence of Informal Economy

Generally, people accept the presence of informal activities on the sidewalk (60%), 15% do not agree and 25% are neutral. For those who agree with informal economy in HCM, the most popular reason is that trading on the pavement belongs to the city's culture (nearly 50%), other think that the trader of the informal sector and their customers are mostly poor people and this kind of services provide not only cheap products to customers but also jobs for sellers (25%). To the oppositions, activities on the pavement are not in order, hider pedestrians, makes the city lose its 'civilization' and the hygiene issue is not guaranteed. Discuss more about perspective that informal business on city's pavement ruins the image of civilized city, the proportion of 'totally agree' and 'disagree' are nearly the same (20%), nearly 1/3 'partly agree' ,14% 'quite agree' and there is up to ¼ of the survey have "no idea". Generally, we can see that HCMC' citizen do not have clear point of view about the relation between informal sector, almost 25% interviewees cannot give their opinion to the issue, most of the answer are also 'mediate answer'. There is also an undeniable trend that people tend to believe the connection between 'informal' and 'uncivilized', what the authority's propaganda. Further questions, 70% people concur that pavement economy is one of the key cultural characteristic of HCMC, 80% believe that those activities enrich the area's livability.

In regard to the future, there is a reduction on the number of people who see informal business on the sidewalk as one of the key cultural characteristic of HCMC (from 70% in the present down to 50%), 25% disagree and 25% do not have idea about this subject. Most of people also believe that pavement economy will reduce in the future (40%), another large amount see no changes in the informal business in the next years (30%), and just 10% predict that it would develop. When being asked about why people consider trading activities on the sidewalk one part of HCMC's culture do not think the same in the future, and predict that pavement economy will decrease, many agree to Mr. LS-03's opinion: "I see it interesting now to have the small stalls and street vendors, they are also be good suppliers of goods and services with our contemporary economic condition, but in the future, there should be some improvements. There is no guarantee about hygiene and products' origins, people will more and more care about their health and there would be less people use those services, just poor student use them because they have no money". Ms. LS-04, who is working as a cashier also said: "I bought food and drinks on the pavement quite often when I was a student, after some years of graduation, I just eat at the stalls that I can ensure that they make everything clean and hygiene as the food and drinks on the pavement are so good that I cannot repudiate them, but I have to reduce and chose carefully before using any pavement service". As for the younger generation, they think: "street food and other services on the pavement are good and especially cheap, to the students having very little money like us, they are our only chance to hang out with friends. Sometime we also aware of hygiene but we have no other choices, and using informal service on the street give us time to watch the city more carefully and love the city more!"

Other perspective from Mr. LS-05, leader of local citizen group (tổ trưởng dân phố): "I greatly sympathize with the sellers on the pavement, they are poor and work hard for their family, those are legitimate businesses and if some poor household in the group want to open an informal shop, no one will reposed. But for me, honestly, I would not allow my children use those services, hygiene is partly of the reason, and the other part is it does not look good to eat when we are standing on the sidewalk, with other pedestrians and activities. There are also street foods in other countries but it does not look untidy as we have here, I think we should learn from them to make it more clean and orderly".

Looking at the survey results, you can observe that most of HCMC's citizens are now supporting the presence of informal economy as it provides cheap products, convenience and partly give livelihood to the urban poor. Probably, the level of acceptance will decrease if there is no improvement on the weaknesses of HCMC's pavement economy, which is the hygiene, disorder and those activities are taking urban space away from pedestrians. For higher living condition in the future, the same appearance of pavement economy as present will not be considered the same by the citizens. For people now see it as necessitate to the poor, cultural characteristic of HCMC, they will not recognize the situation the same way, pavement economy will have to be adjusted, or it will decrease in the next few years.

6. Improvement Proposals

6.1 Existing suggestions for informal economy in HCMC

For more than ten years, the most common reaction of authorities to informal sector has been to forbid, to fine or to confiscate their business tools and vehicles. These activities will be strictly enforced by the city's campaigns, which also show the authority's point of view that informal is against the civilized and modern city. For many years, the result of these campaigns are still in a situation that Pham Thai Son called the "đầu voi đuôi chuột", in which informal sector is still in the street and the activities of the urban order team is not for clearance purpose, but as a daily barrier that informal sector has to get used to. This failure can be explained by certain facts of Ho Chi Minh City, as well as other Vietnamese cities:

- There is still the need for using informal sector as a cheap and convenient source of products and services for the poor citizens. Moreover, with most transportation means being individual vehicles, it is easy to stop by the street and buy some products from the informal sector. Informal services as the result are not only the needs of urban citizens but also the local habit of goods consuming, with strict connection to individual transportation. - The above mentioned informal sector, being dynamic and highly flexible, will use these characteristics as an advantage to cope with the government enforcement. Following the research of Ho Chi Minh city Institute for Development Studies, 65% of informal business activities will get back to their work immediately or one day after the street clearance campaigns and daily track quickly continues. - The "Institutional gap" leads to a situation in which enforcement process is not homogenous between areas of the city. Under command of The People's Committee, every District's Authority will establish Urban Order teams at Ward level. These Order Teams will cooperate with Local Polices to organize the urban pavements. While some wards do not allow any type of informal business on the pavement, some give permissions to draw a 1-meter line from the front of the houses within which range businesses can be conducted, and another 3-meter line for bike parking. Meanwhile, there is no cooperation between authority' offices. Sometimes informal activities allowed by some Ward government are prohibited by traffic inspectors, who work across multiple Wards having responsibilities of taking care of vehicle parking along the streets.
- Urban Order Team does not have enough staffs to manage activities on the street. As informal businesses are located nearly everywhere, a limited number of Urban Order staffs working mostly in office hours cannot control all these activities. Confiscating business tools from the informal sector cannot prevent them from buying new tools to continue their jobs, as these

working tools are usually cheap. Collecting them can even cause the authorities the problem of lacking storage space and the purpose to admonition fails. - The compassion and social capital of long-term interaction still strongly influences the enforcement process of the local police in Vietnam.

Another proposal to improve HCMC's informal economy was to concentrate all pavement activities to planned areas. This solution is usually both recommended by urban planners and most normal citizens, when asked, also suggest that there should be some specific areas for informal trading. With these areas, it would be easier for the management process both in orderliness and hygiene issues: shop owners and employees can be trained about service skills and hygiene to provide better services to customers. Because it is planned, the area will be in order and more "modern" and "civilized". In reality, there is a small number of Chợ Đêm (night markets) operated as described above, but these markets are small, limited and created for tourists rather than for daily needs of the local citizens. In the urban scale, there is virtually no such an area planned for informal business. This is due to some reasons:

- Informal sector has not yet been officially recognized by HCMC authorities but oppositely, has been considered the obstacle for the goal of becoming an industrialized city. For the authorities, informal trading is unproductive and represents something from the past. Therefore, planning for informal sector seems not to bring benefits to the city, there is little or no support from the government for this kind of projects, and without this support, it is impossible for planners to make plans for the informal sector.
- With high density, most land areas of Ho Chi Minh City are occupied, causing difficulty finding location for the planning. Wealthy investors will have more advantages for the rare land resources in the city as they provide more "modern projects" and give the city direct profit. - For the informal sector, they would prefer to keep the situation the same as how they daily do it, which means the planned areas are not interesting for them. According to a survey in 2004, fixed stall owners do not want to move to concentrated area as it will increase travel time from their house to work, and they will lose the current number of customers. For vendors, what scares them most is the fee they have to pay for renting location and they will lose the "flexibility" nature. If there are still large amount of unsold goods for a day, the vendors can currently travel further than usual, and from that they can reach new customers. Being gathered in an area will increase cost, reduce the trading opportunity, and possibly cause higher competition among sellers[16].

6.2 SWOT Analysis for Informal Sector in HCMC

In order to provide appropriate suggestion to Informal economy in HCMC, it is necessary to comprehend the strengths, weaknesses of informal sector as well as opportunities and threats from the outside. From that the strong characteristics can be developed, weak points can be eliminated. Certainly everything will be considered in the relationship to opportunities and threats. In this analysis, internal and external factors will be both considered in order to give informal sector a direction to fit in new needs and perspective of Vietnamese society in the coming years.

Strengths Weaknesses:

- Informal business provides stable jobs with stable income for the poor, reducing urban poverty. - Informal business is a source of cheap and convenient service for urban citizens. - Informal economy enriches social security by the close cooperation between informal
- The hygiene and product's origin of informal business is not qualified. - Some informal activities on the pavements are not well organized, creating disorder and confusion scene in the street. - With narrow pavements, informal activities occupy space of the pedestrians sector and the local police. They will in time inform local authorities whenever there are criminal activities. - Informal economy is flexible and dynamic, "the screen to absorb the economic crisis in Vietnam" (Saigon Times). - Informal business on the pavement is a part of street culture in Vietnam. - Some unique products and services cannot be found anywhere else but

on the street. - Informal sector is willing to show their responsibility and to contribute to the social, the urban "civilization" and "modern city image".

Table 1. SWOT analysis for informal sector in HCMC.

Strengths	Weaknesses
- Informal business provides stable jobs with stable income for the poor, reducing urban poverty. - Informal business is a source of cheap and convenient service for urban citizens. - Informal economy enriches social security by the close cooperation between informal sector and the local police. They will in time inform local authorities whenever there are criminal activities. - Informal economy is flexible and dynamic, "the screen to absorb the economic crisis in Vietnam" (Saigon Times). - Informal business on the pavement is a part of street culture in Vietnam. - Some unique products and services cannot be found anywhere else but on the street. - Informal sector is willing to show their responsibility and to contribute to the social, the urban "civilization" and "modern city image". and	- The hygiene and product's origin of informal business is not qualified. - Some informal activities on the pavements are not well organized, creating disorder and confusion scene in the street. - With narrow pavements, informal activities occupy space of the pedestrians and may cause danger to the traffic.
Opportunities	**Threats**
- Most of HCMC's citizens agree with the presence of informal business activities on the pavements. - People consider pavement activities to be one of the cultural issues of the city. - The media support the existence of informal business in the street.	- The authorities still consider informal business on the pavement as the "residual and non-productive activity and a cause of traffic congestion" (People Committee of Hanoi, 2008), and informality is against the image of a civilized and modern city. - Support of the local police will end when there are orders or supervision by higher institution of the municipality (Kurfürst, 2011). - Lacking education and legal recognition, informal sector has no chance to negotiate with higher institution. - The support and acceptance of the society will reduce in the future as there is a linkage between informality and uncivilized in the citizens' mind.

6.3 Some proposals for further improvements

With all strengths and profits that informal sector provides for urban life, it is nonsensical and irrational to consider informal sector as the obstacle to the "civilized and modern" city. Conversely, informal sector and informal economy is one important part of the city's economy, contributing both jobs and services which are necessary and fit well in Vietnamese current condition. In the diverse economy of Vietnam, no economic sector is as flexible and dynamic as informal sector. A high percentage of labor working in informal sector makes Vietnam "one of the only South East Asian emerging economies not to have gone into recession in 2009 in the wake of the world crisis" [1]. In the report on "How deep was the impact of the economy crisis in Vietnam", Cling argued that informal economy "absorbs the labor surplus which agriculture and the formal sector are unable to employ." Therefore, prohibiting informal activities like what the government is doing now is definitely not the way to develop the civilized and modern city, but a way to make this goal harder to achieve. As De Soto (2000) has mentioned, in this situation "the poor are not the problem but the solution" [17].

Despite having good strengths and contributions to Vietnamese economy, informal sector still has undeniable weaknesses. These weaknesses will make informal sector lose its current advantage of having support from the citizens and social media, which consequently make it lose development opportunities in the future. When life standard is raised, people require higher quality of hygiene in a more orderly and civilized city. Hence, informal sector needs to have some improvements in the coming years. As "the relationship between the state and civil society defines the character of informality" [18], these improvements need the support of both sectors. To understand by what instruments and at what level this support should reach, there needs to be deep studies and researches on the topic. However, this paper has limited scope and will only offer some brief suggestions on the direction of development that informal sector will go, and how to achieve it for the city's future. These suggestions are based on the results found by the author in this paper, in combination with recommendations from experts within the field of urban planning.

Ngo Viet Nam Son has expressed in an interview that the attitude of authority to informal sector in Vietnam is "from an extreme to another extreme", which is there are times when the state leaves informal sector freely developing, but there are also times when informality is strictly banned and controlled. With the experience gained by scholars and practitioners in working with informal sector in the third world countries such as Vietnam where there is "limited governance capability", "the existence of informal somehow instance the gap between legislation and normal life, and the more formalization the authorities try to apply, the larger gap they gain" [6]. Therefore, leadership of the state should be at the level of guidance and give direction rather than interfering deeply in informal activities, further improvements should be developed by informal sector. Flexibility and competitiveness which are considered opportunity informality [6] will open chances to informal sector. These are also the main principles for the development proposals for informal sector in HCMC: decentralizing from the state and exploiting the strength of flexibility and creativity of informal sector.

6.4 Suggestions for management system

The first suggestion is that informal economy needs to be recognized by the government. Without recognition or official definition, informal sector is not even a sector to negotiate and contribute to the entire economy system. To give informal sector a name and definition in legal system is to clear the fuzziness whether a type of business is formal or informal. With this declaration, the "grey zone" arrangements and negotiation of informal sector, as well as the discretionary decisions and harassment by public officials will be reduced. The enforcement of legal regulations will also be on a clear-cut basis.

The second proposal is to define the level of acceptance of informal activities on the pavements. There is not only one "extreme" to either prohibit or to freely allow but also other levels in between. Depending on specific situations of the streets or characteristics of the urban areas where these streets are located, there can be three types of streets in terms of informal businesses: - Non-informal trading streets, where there is no informal activity allowed. The punishment or fines for operating informal businesses in these streets must be hard. - Streets that allow informal business activities with fee, in which these business activities is organized in order and ensure accessibility of pedestrian activities. The 1-meter zone for business and 3-meter zone for parking can be applied on pavements which have enough space for these activities. - Streets that allow informal trading without imposing any fee as a strategy to reduce urban poverty in the transition period of Vietnam. Similar to the streets operating with a fee, the traffic paths for pedestrians, hygiene and street's orderliness has to be ensured.

With more specific policies and classification of streets relating to informal economy, the third proposal is that the system for controlling and managing of informal activities on the pavements needs to be re-organized in which the ranges of fines are unambiguous and homogenous across the entire system. These fines should also be high to admonish people to follow strictly. The current fine level is too low that informal sector can easily pay and continue their work despite the regulations. Compassion from officers, in particular the staffs who directly enforce the regulations, is not allowed in any type. Punishment is also applied to the staffs when there is any compromise with the informal sector.

6.5 Suggestion for informal sector
The biggest negative effects of informal sector in HCMC are the unqualified and unhygienic products; the chaos among business tools causing the situation of losing street scene of orderliness, reducing urban beauty. In some cases, this puts pedestrians in danger of traffic because they have to walk on the road way due to occupied pavements. The problems for pedestrians can be solved by the classification of streets mentioned above. About hygiene and quality of products, informal sector can be organized in self-manage teams. These teams can play the role as the coordinator between local authority and informal sector, where information can flow in both directions of top-down and bottom-up. With this type of organization, the authorities can manage informal economy without losing its flexible and dynamic nature. Beside the function of managing and sharing information, these self-managing teams can give informal sector some advantages of formalization such as access to credit, lower corruption risks and better economic performances. Training to improve business skills and hygiene knowledge for informal shops, especially the ones working in food and drinks area will also be an important task for these management teams. With the support from social media, these improvements will be recognized and welcomed by the society, from that, pavement economy can really be one of the key cultural characteristics of Vietnamese cities in the future.

7. Conclusion
The study was set out to explore negotiation process between informal sector on the pavement and other related sectors in Ho Chi Minh City via two main targets of understanding the performance of informal sector and getting to know how the society considers this specific part of urban economy. Being one of the biggest cities in Vietnam with rich and dynamic street life, Ho Chi Minh City still has not had any official study about informal sector, which has always been existent during the city's history. When mentioning about pavement economy in Hanoi, people have the imagination of the fixed stalls which are operated by the local people, whereas the itinerant vendors have strong connection with the surrounding rural areas of the city. These street vendors are not accepted by fixed stall sellers and are considered as "outsiders" although they are both struggling for the living and both have to use tactics to survive the authority's clearance campaigns. For HCMC case, the "portrait" of informal sector is fuzzy and the fact is that the experience gained in Hanoi case cannot be exactly applied in HCMC ambiance. With the empirical study in a small area of HCMC' center, the paper was aimed to create the general image of pavement economy in HCMC, in the relationship to social perspective. This target has been achieved by answering two main questions of "How is the informal pavement economy in HCMC?" and "How does the society consider informal pavement economy in the city's life and future?" As a conclusion, this chapter will review answers to the two questions and will also show the research limitation and further research direction.

In the imagination of people, informal sector working on the pavement is an ineffective economic part in which people gain low income and have bad working condition. Many scholars argue that people working in informal sector will transfer to formal when they have opportunities. Yet, this paper has presented some differences to the existing research results. Even though people in the informal sector have uncomfortable working condition on the street and most people consider their job as "hard" job in the survey, they are satisfied with the work and all informal sector wishes is just to keep their business running continuously in long term in the future. The reasons that informal sector does not want to change to formal can be explained in three important sections of the survey: one to reveal who they are, another is about their work and a third is about relationships related to their work. Besides the similarities to other papers such as most people working on the pavements are women, with low level of education and having few resources to operate a formal business, this paper's sections about the income and relationship of informal sector give us different findings. Informal sector is not as "unproductive" as the point of view of the authorities. They have stable income which is four to five times greater than the basic salary income defined by Vietnam law. This income is the main income and major source of spending for a normal family of four to five members. This information might be surprising but trustworthy since it is the result from the questionnaires combined with intensive interviews when people are more open and willing to share more sensitive personal information. Through these conversations, informality turns out to be a "rational choice", as it is the result of evaluating the resources and future business opportunities. When having an opportunity to decide whether to stay in informal or to change into formal business, informal sector usually decides to stay the same. There is also a huge difference between the cases of Hanoi and HCMC when informal sector in Hanoi have strong connection to surrounding rural areas, informal sector in HCMC are almost local people who have been living in the city for long time. Their business in informal sector is not temporary when about 40% of interviewees have been working in the field for more than 20 years. The observation and mapping process across various time periods of a day also provides lively images on informal activities in HCMC, showing how dynamic and cooperative the informal sector is in HCMC.

In terms of relationship, informal sector cooperates well with each other, with local citizens and even with the local Police. They do not consider other informal businesses as competitors or "outsiders". They also negotiate with local citizens to be allowed to run business in front of private doorways or to utilize public space of an alley. Similarly, they cooperate with the local Police in detecting and preventing criminals in the local area. These successful negotiation processes are usually based on the long term relationships or interactions which are called "social capital" and on the good attitude of willing to contribute to the area's civilization and cleanliness, such as by contributing to the public fund and committing to cleaning up their work space. In the negotiations, a key lesson learned is that it can only be successful with the support of the citizens, in other words, the "agreement" of the society. Ethic and self-perception of the stakeholders in a negotiation also play a very important role in street life. Because of different perception, with the same management system of transferring responsibility of public space management to the urban district level, some officers use it as a chance to earn money from the informal sector, while the Police of the area under research do not. At higher institution level, there has been no successful negotiation being made. It is due to the fact that informal is not even a realized sector by the law. On this point, Kurfusrt argues that the state of Vietnam acts as the "active producer" of informality for their objectives of becoming an industrialized country as of the year 2020 [3], and in the social-economic transformation, informality depends a lot on the ruling groups' vision of modernity. However, the intimacy and successful negotiation at the local level of informality with the local community and officers will open the potential for redefinition and interpretation of informal sector in HCMC. From these results, there is a high agreement to the point of view of Portes and Haller (2005) that "the relationship between the state and civil society defines the character of informality"[18]. Under Vietnam circumstance, the "civil society" even includes the staffs of the enforcement system. These staffs despite being on duty have their own social perception, and this would open spaces for negotiation.

To propose for further development of informal sector, in the specific social-economy condition of Vietnam where informality "is increasingly used as a tool of flexibility, testing, and learning to fill institutional gaps in the early phase of transition" [7], it is impossible and irrational to provide complete change in the recognition and management of informal economy. Moreover, being a non-formal sector, informal cannot be strictly managed by the authority. In this way it will destroy the sector's flexibility and competitiveness nature. According to Chiller (2012), "informality of external relations at the micro level may provide organizations with the amount of flexibility that is necessary to gain stability and competitiveness at the macro level"[6]. Hence, the solution is that informal sector should be approached by decentralization from the state, where the regulations about informal sector should be at the level of giving guidance and direction and put the hand-off on "semiautonomous communities" [19] within their leadership. From there, inner relationship at the local level should be managed and adjusted by informal sector itself. There should also be classification on different levels of acceptance of informal trading on the streets, which are: Non-informal trading streets, Streets allowing informal business activities with fee, and Streets allowing informal trading without imposing any fee. All these three types of streets need to be ensured walking paths for pedestrians, hygiene and street's orderliness. Because of this, enforcement process must be stricter than what it currently is. In accordance with this system of classification, informal sector needs to be recognized and planned homogenously across the entire city, with no compassion or exception. For informal sector at the local level, there should be self-managed teams acting as the center and coordination to share information, experience, knowledge and credit to the members within the teams. These teams are also the "formal" representative for the informal sector, improving negotiation opportunity of informal sector within the society, reducing the needs for official employees from the government to manage the city's orderliness and informal sector.

Since the research area is just one part of Ho Chi Minh City's Center, the result cannot reflect the general condition of the whole city's pavement economy. Besides, the problem of encountering denials for interviews from the authorities makes the research outcome lack the perspective of the governing sector. Therefore, further researchers can have a wider area of studying with more access from the government point of view as the active producer of informality. This sector has great influence on the presence and development of the informal sector in HCMC. However, within the scope of a Master's thesis, this paper hopefully has sketched the very first but interesting portrait of the informal sector of Ho Chi Minh City, in the relationship with the authority and the society. Contemporarily, informality in the city has not yet been recognized and the future of informal sector is unpredictable because of the continued shifting and lack of transparency of legal framework. However, the picture of informal sector in Ho Chi Minh City is not entirely bleak when they have their undeniable strength and great social and media support. With their flexible and dynamic nature, informal sector will find a way to exist as long as no other economic sector can replace this important and familiar source of daily necessity for a high percentage of the city citizens and, for many urban poor traders themselves.

References

[1] Cling J-P, Chi N H, Razafindrakoto M and Roubaud F 2010 How Deep Was the Impact of the Economic Crisis in Vietnam?

[2] Turner S and Schoenberger L 2012 Street vendor livelihoods and everyday politics in Hanoi, Vietnam: The seeds of a diverse economy? *Urban Stud.* **49** 1027–1044

[3] Kurfürst S 2012 Informality as a strategy: Street traders in Hanoi facing constant insecurity *Urban Informalities Reflect. Form. Informal* 89–111

[4] Hart K 1987 Rural-urban migration in West Africa.

[5] Kanbur R 2009 Conceptualising informality: regulation and enforcement

[6] Schiller D 2012 Informal dynamics of economic processes *Z. Für Wirtsch.* **56** 1–8

[7] Schröder F and Waibel M 2012 Urban governance and informality in China's Pearl River Delta *Z. Für Wirtsch.* **56** 97–112

[8] Roy A 2009 Why India cannot plan its cities: Informality, insurgence and the idiom of urbanization *Plan. Theory* **8** 76–87

[9] Jensen R and Peppard Jr D M 2003 Hanoi's informal sector and the Vietnamese economy: A case study of roving street vendors *J. Asian Afr. Stud.* **38** 71–84

[10] Waibel M 2009 The Model of the "Transitional Urban Development Cycle": Explaining complex interrelationships between urban informality, governability, and restructuring policies *Workshop "Urban Development in a Globalising World–Overcoming the Formal-Informal Divide", organized by the Irmgard-Coninx-Foundation in cooperation with the Humboldt-University and the WZB* pp 12–13

[11] Drummond L B 2000 Street scenes: practices of public and private space in urban Vietnam *Urban Stud.* **37** 2377–2391

[12] Keck M 2012 Informality as borrowed security: contested food markets in Dhaka, Bangladesh *Urban Informalities Reflect. Form. Informal*

[13] Harms E 2009 Vietnam's civilizing process and the retreat from the street: A turtle's eye view from Ho Chi Minh City *City Soc.* **21** 182–206
[14] Luong H V 2003 *Postwar Vietnam: Dynamics of a transforming society* (Rowman & Littlefield)
[15] ALTROCK U and WAIBEL M 2010 The governance of spatial and economic restructuring. Informal dynamics of the maturing mega-city. DFG Priority Programme 1233:"Megacities-Megachallenge: Informal Dynamics of Global Change" *Unpubl. Res. Proj. Propos.*
[16] Moustier P, Figue M, Tam P T G and Binh V T 2007 The participation of the poor in supermarket-driven chains in Vietnam *106th Seminar, October 25-27, 2007, Montpellier, France* (European Association of Agricultural Economists)
[17] De Soto H 2000 *The mystery of capital: Why capitalism triumphs in the West and fails everywhere else* (Basic Civitas Books)
[18] Smelser N J and Swedberg R 2010 *The handbook of economic sociology* (Princeton university press)
[19] Hart K 2005 *Formal bureaucracy and the emergent forms of the informal economy* (Research Paper, UNU-WIDER, United Nations University (UNU))

Biofuels Potential for Transportation Fuels in Vietnam: A Status Quo and SWOT Analysis

Tu Anh Trinh[1], Thi Phuong Linh Le[2]

[1]Urban and Regional Planning Department, Faculty of Civil Engineering, Ton Duc Thang University, 19 Nguyen Huu Tho, Tan Phong Ward, District No 7, Hochiminh city, Vietnam. Email: trinhtuanh@tdt.edu.vn
[2]Faculty of Airport, Vietnam Aviation Academy, 104 Nguyen Van Troi, Phu Nhuan District, Hochiminh city, Vietnam

Abstract. Petroleum consumption for road transportation is well-known as the largest source of CO2 emissions. Worldwide, biofuel is becoming more attractive as substitute for crude oil owing to the increasing demand for environmentally friendly energy and its contribution towards petro dependency reduction and climate change mitigation. This paper reviews the facts and prospects of biofuel production in Vietnam. A SWOT model is adopted to study the strengths, weaknesses, opportunities and threats of biofuels production. The conclusion is drawn that with advantages of weather conditions, soil conditions, the availability of biomass and commitment from government, the country has potential to develop biobuels for domestic consumption. However, threats to production are posed by social acceptance, land use, and technology. Thus, biofuels production still need more supports from government through robust policies, regulations, and institutional framework.

1. Introduction

Air quality problems are mainly caused by vehicle emissions [1]. It has been estimated that the global vehicle fleet will increase from around a billion to 3 billion by 2050. Due to the rapid growth in vehicle ownership, exhaust emissions in developing countries have been growing strongly resulting in an almost three fold increase of CO2 emissions from the global vehicle fleet. This growth adversely affects many populations [2] [3]. Vietnam, located in South East Asia, is highly vulnerable to the effects of climate change. Furthermore, the country is now facing real and urgent challenges in traffic-related air pollution. Vietnam has seen strong population, urban and economic growth in recent decades. In 2015, 30% of the population lived in urban area. The total number of vehicles in the country was approximately 47 million (2015), up by 8.4 million over that of 2012. The rapid population growth, urbanization, and motorization growth are associated with the high acceleration in transport demand and subsequently put immense pressure on energy resources for transportation purposes and the environment as well.

Urban air pollution is largely the consequence of the combustion of petroleum-based fuels for transport. Air pollution from combustion sources has adverse impacts on the structure and function of ecosystems as well as human health [1] [4]. Air pollution is not a distant threat: it is already costing the people in Vietnam. It has been estimated that the annual global mortality attributable to urban air pollution range from approximately 0.8 million to over 4 million [5]. This burden occurs predominantly in developing countries [6]; 360,000 premature deaths per year in Asia alone [7]. A study on the relation between air pollution and health showed that more than 90% of children under 5 years old living in HCMC (Vietnam) suffer different illnesses.

Recently, in order to achieve the goal of air pollution mitigation and reducing the dependence on petroleum fuels, much global attention has been focused on the development of cleaner, greener and environmentally friendly energy source which have to be technically feasible, domestically available and environmentally acceptable [8] [9]. Energy Policy Act of 1992 defines alternative fuels as ethanol, natural gas, hydrogen, biodiesel, electricity, methanol and so on. Many look to alternative fuels, particularly biofuels as promising complements to petroleum in the near-to-medium term [10]. They are being used to reduce dependence on fossil fuels, petroleum consumption and greenhouse gas emissions (GHG), in the transportation sectors. With potential to alter the transport sectors of decarbonizing societies [11], biofuels are poised to be competitive and there is a role for biofuels in the developing world [12]. In general, biofuels can be in solid form (fuelwood, charcoal, wood pellets or briquettes) or liquid (bioethanol or biodiesel) [13]. The most feasible biofuels for vehicles being considered globally are biodiesel and bioethanol [1].

Many developing countries have successfully produced biofuels [1]. In Southeast Asia, Indonesia, Malaysia, Philippines, and Thailand are the world's largest biofuels producer. There are currently 8 biodiesel plants in Indonesia and the country accounts for 44.5% of the world's production and 45.3% of global export of palm oil. The Philippines is well known for production of coconut oil, especially of coconut methyl ester. With an aggregate production capacity of about 4.5 million tons per annum, the Philippines has become the world's leading producer of coconut oil. In Thailand, palm oil is a biomass resource which has high potential as a renewable energy source for biodiesel production [1]. In 2008, the oil palm harvesting area in Thailand was 0.46 million ha and the fresh fruit bunch output was 8.68 million ton. In order to meet the demand for biodiesel production, the Thai government is planning to expand the oil palm cultivation area to 0.8 million ha by 2009 and to 1.0 million ha by 2012 [1]. There are countries in Asia following the global trend including India and China. Being the largest edible vegetable oil importer, India issued its biofuel policy in which biodiesel would meet 20% of the diesel demand beginning with 2011-2012. In order to achieve this, 11.2 million ha was planted with Jatropha by 2012 to produce sufficient oil seeds for biodiesel requirements [1]. In China, Government proposed the establishment of a biofuel industry in the Tenth Five-Year Planning (2000-2005). As a result, in 2000 fuel ethanol industry was created, and in 2001, a 200,000 ton capacity pilot ethanol production plant was established. By 2008, there were five licensed fueleethanol plants in operation. In addition, the fueleethanol production capacity reached 1.94 million tons. Currently, E10 (gasoline mixed with 10% ethanol) is being used in the transport sector in the five provinces and 27 cities [1].

Located in Southeast Asia, Vietnam is an agricultural country with approximately 71% rural population and 48.9% of the labour force working in the agriculture and forestry sector. Vietnam has high potential for crop cultivation, and for biofuels production due to the availability of arable lands and weather conditions. In addition, since 2007, the Vietnamese government has passed legal corridors, institutions, policies, and investment plan for biofuel development [14]. With the aim of meeting the transport demand, minimizing vehicle emissions and increasing sustainability, biomass-based transport fuels are considered an important alternative in the country. The development of alternative fuels, particularly biomass-derived fuels from locally available biomass, needs investigation.

This paper presents a desktop review of the liquid biofuel potentials to produce transportation fuels in Vietnam. SWOT analysis is adopted to assess the strengths, weaknesses, opportunities and threats of the liquid biofuel potentials. In addition, the status quo of country statistics, information on GHG emission from the transport sector and the role of biofuels in policy for climate change mitigation and sustainable transport in Vietnam are also discussed.

2. Status quo

2.1. Country Statistics

Vietnam has a total population of 92.7 million people (2016), 71% of which live in rural area. The total area of Vietnam is 332,000 square km; mountains occupy 75 percent while only 25 percent of the area is arable. Agriculture-forestry-fisheries are an important industries for Vietnam and accounted for 20 percent of the country's GDP, 54 percent of its employment, and 30 percent of its exports [15]. Since 2000 the sector has grown steadily at 3.4 percent per year. Vietnam has 9.4 million hectares under cultivation; 70 percent of its agricultural products are crops, of which rice is the most important. The forestry sector accounts for 1.2 percent of the GDP. In 2014, forests cover 37% of the land area which has increased from a coverage of 28% since 1995 due to government reforestation programs.

The transport sector in Vietnam much depends greatly on petroleum and, if fossil fuels continues to be used as previously, the country, especially mainly in large urbanized areas, will suffer from severe air pollution. Biofuels are an efficient energy source with no pollution emissions. It is considered that if biofuels are substituted for traditional fuels to meet the needs of travel demand, they can help tackle air pollution and improve urban health. Therefore, since 2007, studies on the feasibility of biofuels production in the country have been carried out [13] [16] [17] [18]. Also, the government has take steps to fight air pollution through policies and long-term strategies which take advantage of abundant agricultural biomass for biofuels production.

2.2. Fuel Consumption and GHG From Transport Sector

In 2011, the energy sector emitted about 33 million tCO2 or 0.37 tonnes per capita [19]. According to the 2014 Biennial Update Report from Ministry of Natural Resources and Environment, the sector produced a total of 141 MtCO2-eq, whereas 31.8 MtCO2-eq (23%) came from the transport sector. The biggest emission source within transport was by far road transport, which accounts for 68.5% (2010), followed by maritime transport (18.9%), aviation (6.8%) and railway (3.7%) (Figure 1) [13]. In terms of emissions by fuel type, diesel oil was the dominant emissions source, which represented 42.7% in 2010, followed by gasoline [19]. From 2000 to 2010, there was an increase in CO2 emission with the rate of, on average, 12.5%. Emissions from two-wheelers/motorcycles constitute the largest share of road transport CO2 emissions (46%) [13] (Figure 2).

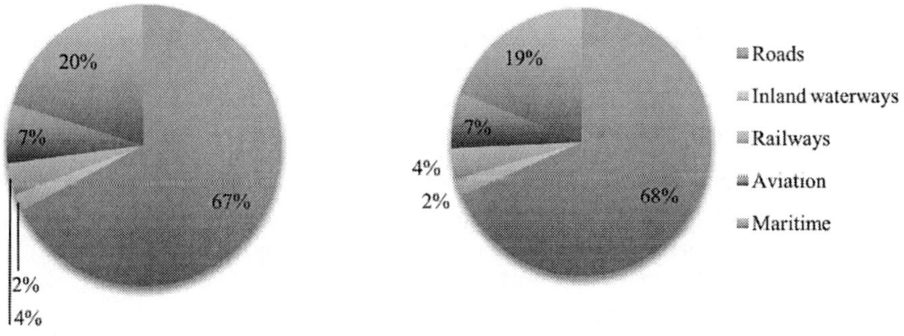

Figure 1. Emission source within transport - 2010

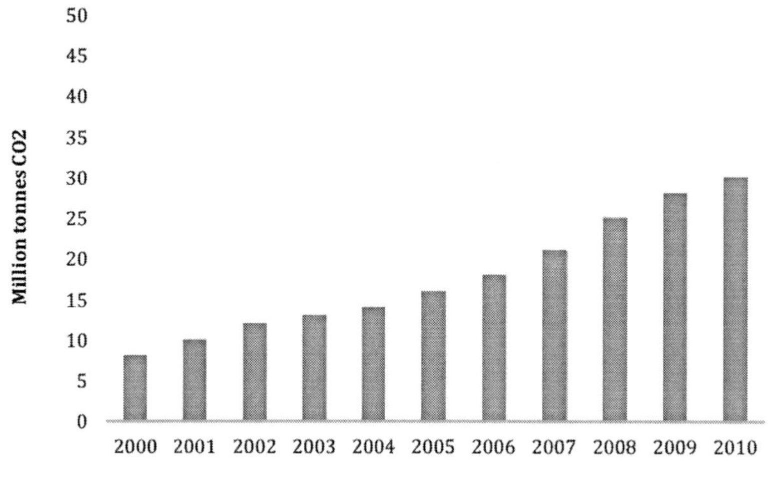

Figure 2. CO2 Emissions from road transport from 2000 – 2010

The transportation sector is the third largest consumer of petroleum-based fuels and accounted for about 22% of fuel consumption in 2010. Nearly all fossil fuel energy consumption in the transportation sector is from petroleum-based fuels (diesel fuel and gasoline fuel). Importantly, the sector consumed slightly more diesel fuel compared to gasoline fuel in the past two decades [19]. It is estimated that the proportion of diesel consumption will increase from 48% in 2010 to 71% in 2040 [14]. Despite having the petroleum refineries with capacity of 6.5 – 8 million tons of crude oil per year, the country still imports the majority of its refined petroleum. Expected primary energy imports will account for 36% of total primary energy consumption in 2020 and will increase to 57% in 2030 [20]. Gasoline and diesel oil are still the main driver of the country's petroleum import levels. As a consequence, CO_2 emissions are equally expected to rise as fuel use increases steadily each year. Therefore, biofuels that have been considered the major substitution of petroleum-based transportation fuels would bring about more sustainability in the transport sector.

2.3. Biofuels – A Vital Role in Regulatory Framework for Sustainable Transport

The world fleet of motorized vehicles emits millions of tons of pollutants, resulting in poor air quality conditions and causing, among other effects, human health deterioration and climate change [21] [22]. Therefore, emissions reduction targets, low-carbon and sustainable urban mobility objectives play an important role in many climate change programs.

Vietnam has a demonstrated commitment to fighting climate change and air pollution through a series of measures. Vietnam has ratified several international conventions, notably the United Nations Framework Convention on Climate Change (UNFCCC) and the Kyoto Protocol [19]. To achieve the objective of UNFCCC on the "stabilisation of greenhouse gas concentrations in the atmosphere at a level that would prevent dangerous anthropogenic interference with the climate system", a number of environmental protection laws and regulations have been released by National Assembly. In addition, some Action Plans formulated by the Ministries (e.g. National Target Program to Respond to Climate Change, National Green Growth Strategy, National Green Growth Strategy and National Green Growth Action Plan intend to mitigate emissions and to fight climate change. Apart from focusing on measurement identification, technological development, and infrastructure development, using "fuels of low greenhouse gas emission" or "renewable energy and less GHG emitted fuels" sets out the country's vision for combating climate change and partly guaranteeing the realization of sustainable transport:

"To introduce fuels of low greenhouse gas emission to means of transport, [...] so that 20% of these vehicles will use such energies by 2020 and 80% by 2050" (National Climate Change Strategy, 2011)

"Enhance the application of new technology, usage of renewable energy and less GHG emitted fuels in transportation. Implement a set of management in fuel, emission standards and vehicles maintenance"(In The National Green Growth Action Plan for the period 2014-2020, 2014)

In the transport sector, separate Avoid-Shift-Improve policies have been released ensuring sustainable transport for the future. "Avoid" policies aims to reduce travel demand, while "Shift" encourages people to switch to more efficient modes of transport and importantly, "Improve" promotes the adoption of green fuels and green vehicles (green technology) designed to utilize the fuels in both passenger transport and freight transport [19]. Table A1 (Appendix A) provides a summary of the biofuels policies in passenger and freight transport.

3. A SWOT Analysis for Potential Biofuels Production

SWOT analysis has gained widespread acceptance as a strategic way to analyze the internal and external competitiveness. This section analyzes the biofuel production in Vietnam (Figure 3).

3.1. Strength

3.1.1. Land use

There are 8 major land areas in Vietnam including Red River Delta, Northern midlands and mountain areas, North Central and central coastal areas, Central Highlands, South East and Mekong River Delta [23]. Each region has a characteristic zonal vegetation, soil and climate. Regarding agricultural land, it is divided into 4 types: Agricultural production land, Forestry land, Water surface land for fishing, Land for salt production (Table A2) (Appendix A). Of this, Agricultural production land covers 10231.7 thousand ha in 2014, whereas the figure for Forestry land, Water surface land for fishing and Land for salt production was 15845.2, 707.9, 17.9 thousand ha respectively. Table A2 indicates that from 2000 to 2015, the area of agricultural production land in the country has increased relatively, at an average rate of 114,000 hectares per year. This increase may come from the expansion of unused land, deforestation and forest land. It is expected that by 2020 the area of agricultural land will reach 27,038.09 hectares, an increase of 246.51 hectares compared with that of 2015. Producing biofuel in Vietnam comes from agricultural and forestry feedstock such as cassava, maize, sugarcane, jatropha; therefore, arable land is important in guaranteeing sources for biofuels production.

3.1.2. Biomass resources

Biomass including agricultural resources, agricultural residues, forest resources, waste (e.g. municipal solid waste, industrial waste, and algae) are often used for biofuels production. These materials, based on their sources of biomass, their limitations as a renewable source of energy, and their technological progress, are classified into four generations: first, second, third, and fourth [24]. First generation refers to the biofuels derived from agricultural products: sugar or starch-based crops and oilseeds [24], e.g corn, sugarcane for ethanol production, and soybean, jatropha, palm oil for production of biodiesel. Through well-understood technologies and processes, like fermentation, distillation and transesterification, first generation biomass feedstocks can be processed into bioethanol or biodiesel [24]. The present study focuses on the first generation. In addition, energy crops including: sugarcane, cassava, maize and jatropha are specifically discussed becaue they are currently the major feedstocks for emerging biofuel projects in Vietnam. Many schemes on development of biofuels in the country also identifies sugarcane, cassava and maize as the main raw materials for bioethanol production. Vietnam General Statistics Office (GSO) estimates in Table A3 (appendix A) show substantial cultivation of energy crops in the country over a ten-year period. There have been continuous changes in the areas of land harvested, yield and production of sugarcane, maize and cassava in Vietnam from 2005 to 2015.

	Strength	Weakness
	• S1: Land use • S2: Biomass resources • S3: Policy • S4: R&D and Production	• W1: Low social acceptance • W2: The shutdown of biofuels plants • W3: Technology barrier • W4: Human resource
Opportunity	**OS**	**OW**
• O1: Biofuel production and blended fuel combustion • O2: Biofuels Production from Residues	• O1-S4: R&D focusing on production and blended fuel combustion to gain effective social cost • O2-S2: Using agricultural residues for biofuels production	• O1-W1: Using campaigns to raise awareness of citizen about pros of biofuel • O2-W2: Using residues for biofuel plants
Threat	**TS**	**TW**
• T1: The shrinking of agricultural land • T2: Biomass Resource Reduction • T3: Food Security, Feedstock Priority and Policy	• T1,T2,T3-S3: Having policy capturing the issues of land use, biomass resources for production and food security • T3-S4: Utilizing wide range of feedstock and appropriate technology for biofuel production	• T2-W2: Explore policy support in terms of biomass development and production • T3-W2: Having feedstock priority program and releasing policy to avoid food versus fuel

Figure 3. SWOT Analysis

a) Sugarcane

Ethanol from sugar cane is well-known as the most attractive alternative to fossil fuel as it provides significant reductions in GHG emissions compared to gasoline and diesel fuel. Sugarcane is grown in the two main parts of Vietnam and its cultivation does not have any negative impacts on the environment. It is harvested once a year in the North and twice per year in the South. Sugarcane can be used by many manufactures [20].

Sugar, bagasse and molasses can be used for ethanol production. Currently, domestic bioethanol is currently produced from molasses (3-6%). In Vietnam, there is only one factory using molasses in combination with cassava for ethanol production. About 100 million tonnes of ethanol are produced annually (of which the molasses part is really small). It is expected that if the average yield is raised to 800 Quintal/ha by 2020, about 6.5 million tons of sugarcane would be available for ethanol production.

b) Cassava

Like sugarcane or maize, the humid tropical weather favours cassava's growth. Cassava is a tropical root crop, requiring at least 8 months of warm weather to produce a crop, and may be harvested between 10 to 14 months [25]. Thus, it shows high potential as industrial raw material for ethanol production in Vietnam. Cassava is considered as a first generation biofuel feedstock in direct competition with food. In addition, there is competition for the crop from other enterprises which process cassava for starch, produce animal feed, and export of cassava chips.

It is clear from the table 4 that cassava production has developed rapidly in Vietnam; from 6.7 million tons in 2005 to 10.6 million tons in 2015. It is the result of the expanding cultivation from 425.5 thousand hectares to 566.5 thousand hectares and the yield from 83.6 thousand tonne/ha in 2000 to 169.0 thousand tonne/ha in 2009. Furthermore, many kinds of cassava have been grown in different areas in the country (Table 4). They provide high yield, high quality, and high starch content (above 25%) [14]. Production growth at this rate is the result of the investment and application of technology in the selection and breeding of cassava [20].

Table 1. High-yield Cassava in Vietnam (tonne/ha)

Type	Cu M'gar (Dak Lak province)	Quang Hiep (Dak Lak province)	Tan Canh (Kon Tum Province)	Average
KM 140	38.7	35.5	32.0	35.0
KM 98-5	35.6	34.6	28.7	33.0
KM 98-1	24.3	25.5	26.7	25.5
KM94	31.5	28.9	26.0	31.0

c) Maize

Maize is the dominant feedstock for liquid biofuel production that accounts for about 8.4% of global ethanol production [26]. In Vietnam, maize is grown in the north-east (50%) and the south-east (10%) with the remainder scattered throughout other regions. Maize crops are planted and harvested at 2 major times: Dong Xuan (from December to April) and He Thu (from April to August). Over a 10 year span, agricultural areas harvesting maize increased from 1052.6 thousand ha in 2005 to 1179.3 thousand ha in 2015 with increases in production growing from 3787 thousand tonne in 2005 to 5281 thousand tonne in 2015.

d) Jatropha

Besides palm oil and soybean oil, Jatropha is widely considered as a potential cradle of biodiesel. Jatropha cultivation does not compete with food and other cash crops for arable land as it can be grown in all ecological zones and under harsh climatic conditions [24]. Jatropha's expected lifetime is about 50 years and the it starts yielding from the second year of plantation, but in limited quantity [26]. However, [27] reported that the plant can produce 4-5kg per tree from the fifth year onwards, if managed properly. In Vietnam, Jatropha has been grown since 2006. The planted area is approximately 3,358.8 in 2010. It was cultivated in many regions such as North West and North East (e.g Lai Chau, Cao Bang, Ha Giang, Yen Bai), North Central Coast (e.g Phu Tho), and South Central Coast (e.g Nghe An, Quang Binh, Quang Tri, Hue).

3.1.3. Policy

Appropriate policies play important roles in guiding and impelling change. Since 2003, the government of Vietnam has enacted a series of policies in energy efficiency and biofuel development.

a) Energy Efficiency and Conservation

As the energy consumption of Vietnam is projected to increase significantly within the next decade, Vietnam considers energy security top priority. The National Assembly and other administrative departments have issued policies and a legal framework for the guarantee of energy efficiency [19] [18]. The policies focus on targets including: (1) conserving resources, economical and efficient use of energy, (2) the application of science and alternate technology to the energy sector, e.g effective mining and utilization of existing energy resources, (3) the development and use of renewable energy sources (Table A5).

b) Biofuels Policy and Incentives

Several government-implemented policies encourage the production of biofuels. In 2007, Decision No. 177/2007/QD-TTg on the "Scheme on Development of Biofuels up to 2015 with the Vision to 2025" issued by Vietnamese government tends to "develop biofuels as a new and renewable energy to partially replace conventional fossil fuels, so that contributing to assuring energy security and environmental protection." To achieve this policy target, short-, medium-, and long-term strategies have been planned. In addition, the decision set four main missions, and they are (1) enhance scientific research and technology development, (2) firmly establish a thriving biofuel industry utilizing agricultural products as a means of improving the quality of automotive fossil-based fuels, (3) develop an effective framework and policy mechanism capable of attracting domestic and foreign investment in the biofuels industry, (4) promote international cooperation for the growth of the biofuel subsector. Each mission was accompanied with detailed tasks. The roles of various tiers of government to ensure an orderly development of the biofuels industry were also streamlined.

To guarantee the realization of program's target, different policies, regulations and incentives have been introduced (Table A6). The Decision No. 177/2007/QD-TTg opens new opportunities for local and international private sector. During 2007–2015, the government provide tax incentives on imported equipment and machinery and zero income tax. Additionally, companies also receive land-use concessions whose period is granted for 20 years. Different funding sources were also introduced to support basic and applied research and development [14].

3.1.4. Research and Development (R&D) and Production

a) R&D

The research on biofuels production in Vietnam started 10 years ago. The projects mainly focused on conventional biofuels, including ethanol from sugar or starch (Lam Son Sugar Co., Binh Tay Alcohol Co., Saigon Petro, Chi Hung Co.), and biodiesel from catfish oil (Agifish An Giang, Minh Tu Co., VAMCO Biochemical Co., Thien An Co.). In the recent two years, R&D activities have been expanded to the second-generation biofuels, with pilot projects on producing biodiesel from non-food crops (jatropha curcas L. tree) [13]. Under the drive of Biofuel Development Scheme, in 2009 a total state budget of 43,376 million VND was approved and granted to organizations and individuals responsible for implementing 22 R&D projects and 5 pilot biofuel production projects during period 2009-2011.

b) Production

Although biofuels production is still in the initial stage, under the drive of biofuels policies, some achievements have been attained. Various development projects are ongoing across the country [18].

From October 2007 to March 2009, with the total investment of over VND5.4 trillion (US$241 million), the Vietnam National Oil and Gas Group (PVN) has invested in three biofuel projects with a capacity of 100,000 m3 of ethanol per year (Table A7). The investment capital of the Dung Quat Bio Ethanol Plant amounted to nearly $92 million while the figure for Binh Phuoc Ethanol plant and Phu Tho plant was 114.2 million and ABC respectively. Construction began in 2009, Dung Quat Bio Ethanol Plant was the largest bio-fuel production plant in central Vietnam, primarily producing ethanol 99.8% from cassava with a capacity of 100 million liters per year. After 18 months of construction the plant opened its first production line of ethanol in February 2012. Binh Phuoc ethanol plant was built in 2010 and finished in December 2012. It produces approximately 300,000 liters of E5 per day. It was estimated that the plant would consume 240,000 tonnes of dried cassava annually. In April 2012, the plant came into effective operation with capacity of 1.4 million liters of bio-ethanol per month. Phu Tho Plant have not been finished due to the increase of capital investment. The Phu Tho bio-ethanol plant located in Phu Tho province was designed to produce 100 million litres of ethanol a year, using first-generation technology with cassava as feedstock.

Annually, Vietnam imported 15-16 million tons of fuels while consumption was growing at an annual rate of 10%. Therefore, biofuels will meet the domestic demand and simultaneously reduce the use of fossil fuels.

3.2. Weakness

3.2.1. Low social acceptance

Since 2009, biofuels (E5 and B5) have been sold in six major cities and provinces in Vietnam including Ho Chi Minh, Hanoi, Can Tho, Da Nang, Vung Tau and Nha Trang. Currently, there are only 13 out of 514 gasoline retailers selling E5 in Ho Chi Minh City, whereas the figure is 63 out of 170 in Can Tho. However, consumers in the cities keep using A92 petrol because the price of E5 petrol is not much cheaper than A92. It seems that end-users pay more attention to the economic benefit rather than the environmental benefit. End-users are reluctant to try a new energy source that has a higher cost. A preliminary survey revealed that not many people and gas station's owners have high perception of biofuels, e.g what biofuels is, how it can help to protect environment, whether it badly affects vehicle's engine and energy-savings. In addition, many biofuels' sellers frankly said that due to high evaporative losses, high cost of investment and tank cleaning, they were not willing to sell and introduce biofuels to customers.

3.2.2. The Shutdown of Biofuels Plants

The biofuels plants has already racked up millions of losses because of the low demand and the government has moved more slowly than expected to boost biofuel use. Since 2015, three biofuel plants (Table 6) have stopped production due to current difficulties in the market. Thousands of jobs and millions of dollars in investment will be at risk if the government does not have any mandates to speed up the use of green transport fuel.

3.2.3. Technology barrier

Biofuels production technology in Vietnam is far behind the foreign advanced level. Today, biofuels are not cost competitive with conventional fossil fuels in Vietnam because biofuels is still produced in old and obsolete plants resulting in high energy consumption, low productivity. Some biodiesel plants utilize obsolete filtration technology with low oil recovery rate, leading to higher price of immediate products and not economical for blending. Mastering blending technology is also a challenge to biofuels production in the country at present [13].

3.2.4. Human resource

The development of biofuels production requires experts and a skilled labor force. The lack of well-qualified workers restricts the research activities and the production of biofuels in the country. Many projects have been implemented; however, they have appeared on lost. The main reason is low ability (e.g techical skill and management skill) of biofuels producers. Human resource is one of the main reason that hinder the development of biofuels to a certain degree.

3.3. Opportunity

3.3.1. More attention should be paid to biofuel production and blended fuel combustion

In the research on cost-effectiveness of fuels, [16] stated that "the cost-effectiveness of using biofuel in comparison to fossil fuel depends on the efficiency of biofuel production and blended fuel combustion". The results of the research, for example, shows that "the biodiesel substitution would be cost-effective if the fuel consumption of B5 and B10, in terms of L km 1 compared to diesel, would decrease by more than 1.4% and 2.8% for B5 and B10 respectively at the discount rate of 4%" [16]. Therefore, the increased attention that has been paid by the government to production and blended fuel combustion provides a good opportunity to develop sustainable biofuels market.

3.3.2. Biofuels Production from Residues

Agricultural residues, organic materials produced as by-product in the course of harvesting and processing agricultural crops [24], can be used for biofuel production [26] [28] [29]. Such biofuels are generally considered sustainable as they use waste materials from food crop production, and do not compete with food crops for land [26] [28]. There are 2 types of agriculture residues: crop residues or field residues (materials left or burnt on farms after the harvest of desired crops, e.g. straw of grain crops) and agricultural industrial by-product or processing residues (materials produced after crop processing, e.g. husks, chaff, cobs or bagasse)

3.4. Threat

3.4.1. The shrinking of agricultural land

The development of biofuels, from 1st or 2nd generation, requires arable land for the production of biomass. However, available land for biomass production is limited in Vietnam due to existing policies on agriculture and forest sector. More importantly, agricultural land continues to decline due to industrialization and urbanization. According to the report from General Department of Land Management, average annual decrease of agricultural land is approximately 100 thousand hectares, especially in 2007, the figure was 120 thousand hectares, while every year about 400 thousand agriculture workers leave their occupation. Currently the unused land that can continue to exploit in the country is not significant. Meanwhile, climate change is likely to make the available arable land to be at risk of shrinking.

3.4.2. Biomass Resource Reduction

Although sugarcane is one of the main source for biofuels, the production of sugarcane in Vietnam has slightly decreased in the past 10 years. From 2005 to 2012, plantation areas increased from 266.3 to 301.9 thous.ha; however, the figure dropped to 284.5 thous.ha in 2015. As a result, both yield and production were also slightly declined to 644 Quintal/ha and 18320.8 thous.tons respectively in 2015. Despite the annual production of 100 million tonne of fuel grade ethanol, it was mainly for export as domestic demand is quite low. The cassava area has expanded in recent years because of high demand from starch factories and for the export of dry chips. Although land suitable for cassava cultivation is available, the government discourages cassava expansion because of the resultant soil degradation and water pollution from cassava processing. Cassava cultivation has been promoted strongly by the government; however, there is not an integrated master plan on land use planning and sustainable cultivation practice. Like sugarcance and cassava, maize has not been favoured as an ethanol feedstock because of concerns about competition with food. Moreover, maize uses large amounts of fertilizers and pesticides, and thus consumes fossil fuel energy.

Jatropha grows on almost any terrain; however, many experts stated that the economic feasibility of jatropha cultivation and biodiesel production from jatropha oil was not investigated enough. In Vietnam, production of biodiesel from jatropha oil might not beneficial because the oil pressed from 3kg of seeds is needed to make 1 litre of biodiesel with price of 10,000VND/litre, and the cost of 1kg of jatropha seeds in Vietnam is 3000VND. With that cost, the production biodiesel from jatropha oil might not be profitable. In addition, cultivation of jatropha alone may not be economically viable, well-management and inter-cropping with other plants & flowers could be a preferred option.

3.4.3. Food Security, Feedstock Priority and Policy

Although biofuels is an important player in the decarbonization of the transport sector and biofuel production helps create opportunities for agricultural development, increasing biofuel production is going to have an impact on world agricultural commodity prices and food security. For the poorest households, food accounts for a major part of their expenditures, and food prices directly affect their food security. Decision No. 177/2007/QD-TTg plays an important role in guiding the development of biofuels production in Vietnam; however, the issues of food security or the selection priority of agricultural feedstocks and residues for biofuels production has not been clarified. Today, as a consequence of the urbanization, population booming, climate change and now the loss of productive agricultural land for biomass, food security in Vietnam might be threatened in the long term. In addition, switching from producing low profitability food crops to producing high profitability biofuel crops may result in price increases for some essential food categories. Take maize and sugarcane for example. Maize is widely used as human food and feedstock for animals. Food insecurity is more urgent than the energy crisis; thus, food security takes priority over biofuel production with regard to maize. Sugarcane becomes a potential feedstock for biofuels as it does not require any specific type of soil as it can be successfully raised on diverse soil types. However, the policies for ethanol production from sugarcane or its by-products are incomplete at the present stage. Sugarcane based ethanol production is just in initial stage and other potentials have not been explored. Thus, it is critical to understand the trade-off of land use, price of feedstock supply, domestic food security when they promote biofuel programs and select feedstock to contribute to agricultural development [10].

4. Discussion and conclusion

4.1. Discussion

Biofuels offer many benefits and is defined as a substitute for conventional fuels. For energy-deficient countries, biofuels help to reduce the dependence on imported fuels and offer improved balance of trade and balance of payments. Fossil fuels produce large amount of greenhouse gases in the atmosphere which causes global warmth and climate change. If people start shifting towards biofuels, emissions could all be significantly reduced. The production of transportation fuels from agricultural resources in Vietnam requires sustinable feedstocks and new technology development. Although biofuels have latent contribution to sustainable development in Vietnam, they face challenges in connecting to food security, land requirements and availability, policies, knowledge, standards, awareness, participation and investment. Therefore, in the short term, petroleum will be still the main fuel used in Vietnam.

According to the SWOT analysis, although there are weaknesses and threats of biofuels in the present stage, the strengths and opportunities impel the development of biofuels. Therefore, the advent of biofuels in Vietnam needs the support of policy, technology, humanity, finance and many other internal and external factors. This paper gives some support measures listed below:

a) Land requirements and availability

Producing biofuels on a large scale could require huge tracts of land that causes potential land use conflict with food crops. Regarding the risk of diverting farmland or crops for biofuels production, much controversy surrounds the future of biofuels. Biomass utilise the same resources required for many forms of food production namely land, water and agrochemicals. Competition between food and fuel is perhaps the most prominent issue; however, competition may be avoided by careful planning for ecological conservation and sustainable production methods. A number of studies on land availability give wide-ranging results [30] [31] [32]; however, there is little research on the land use for biomass and food in Vietnam. Thus, study estimating land requirements to meet specified biofuel is needed.

b) Policy issues

Successful policy development and implementation requires a sound legal, regulatory and institutional framework. Vietnam has not had comprehensive biofuel policies, and the existing ones often driven largely by agricultural or energy considerations. Supportive government policies are essential to the development of biofuels. Policies are urgently required to capture a wide spectrum of activities involving food security transportation, environment, land-use change, forestry, agriculture, biomass priority and residues. In addition, worldwide interest in biofuels is certainly growing, which will hopefully spur investments. Therefore, it is essential to develop a robust investment policy to create a legal corridor to attract and support foreign interest and finance from private sector in the R&D and biofuels production.

At present, the "Scheme on Development of Biofuels up to 2015 with the Vision to 2025" issued by Prime Minister only provides a general orientation on the targets for each period and the prioritized list of activities. It is expected that a road map for biofuel development shall be completed soon in order to provide detailed instructions to achieve the set targets, and a workable implementing plan. Besides, it is also expected that the sources of funding for biofuel projects will be well defined.

d) Awareness and participation

Active government involvement plays a key role in successful biofuel programmes. Valuable lessons can be drawn from Germany and the United States. Germany and US are leaders in high-technology biofuel production, due to strong government commitment, viable policy and solid collaboration from the private sector. It is argued that one of the main reasons behind the shutdown of many biofuels plants in Vietnam is that the government has moved more slowly than expected to boost biofuel use. In fact, biofuel markets depend heavily on government incentives and mandates; therefore, the government should create a positive environment for biofuels production, and providing robust support (e.g. campaigns, PR) to raise biofuel awareness among citizens. Information including environmental impacts of petroleum, benefits of biofuels in terms of health, environment, vehicle engine,... may be used to communicate with community.

e) Investment

Lack of a skilled workforce, R&D and cutting-edge technology keeps Vietnamese biofuels production behind that of the West and other countries in the region such as Thailand, China, and India who have been innovating extensively to improve biofuels production. The government should invest in technology development and fund a broad portfolio of biofuels research programs because fundamental technology improvements and scientific breakthroughs are always necessary in both biomass optimization and the processing of biomass into fuels. The investment normally comes from national budget. Thus, the country should seek to encourage private involvement in biofuels production. In addition, the country could obtain funding from international financial institutions and regional and sub-regional development banks. Other development partners are now ready to participate in viable projects [20].

A qualified workforce is one of the key roles in the development of biofuels program. Currently, better-trained worker in Vietnam is inadequate Thus, the country should place greater emphasis on developing a skilled workforce. High eduction, professional training programs and other means shoud be designed to develop a skilled professionals capable of operating biofuels industry.

4.2 Conclusion

This study reviews the biofuels potential for production of biomass transport fuels in Vietnam. Transportation is the third largest single source of air pollution in Vietnam and with the rapid increase of vehicle fleet, air pollution is expected to worser in the near future. In addition, despite having high crude oil, Vietnam is still a net oil importer of transportation fuel and this makes the country vulnerable to volatility in global fuel prices and dependent on foreign exchange to meet its domestic energy needs. Vietnam is an agricultural country with tropical weather and has a wide variety of sources, such as energy crops, agricultural crop residues, forest resources which are suitable for biofuels development.

By using a SWOT analysis, the internal strengths and external opportunities in terms of land-use, biomass resources, and governmental is currently obvious and significant, but weaknesses and threats such as existing policy gaps, technology barrier, and contractor capability must be tackled in several ways in order to to realize biofuels as a major substitution of petroleum-based transportation fuels. Some solutions are suggested in the paper ranging from land requirements and availability, policies to investment. They are derived from the combination of strengths-weaknesses-opportunities-threats of avaibale resources in the country.

Acknowledgments

We wish to thank the reviewers' helpful suggestions in improving this paper. Thanks also to the officials who helped to provide information and data used in this paper.

IOP Conf. Series: Earth and Environmental Science **143** (2018) 012065 doi:10.1088/1755-1315/143/1/012065

Appendices

Table A1. A summary of the biofuels policies in passenger and freight transport

Policies	Date	Issuers	Main content
Law on Economical and Efficient Use of Energy	June 17, 2010	The National Assembly	• Provides economical and efficient use of energy; policies and measures to promote economical and efficient use of energy; and the rights, obligations and responsibilities of organizations, households and individuals in economical and efficient use of energy.
Decision No. 249/2005/QD-TTg	October 10, 2005	The Prime Minister	• Roadmap for application of emission standards for road vehicles
Decision No. 909/QD-TTg	June 17, 2010	The Prime Minister	• The exhausted gas emission control plan for in-used motocycles and mopeds in the provinces and cities
Decision No. 49/2011/QD-TTg	September 01, 2011	The Prime Minister	• The roadmap for application of exhaust emission standards to manufactured, assembled and imported brand-new cars and motorbikes
Decision No. 53/2012/QD-TTg	November 22, 2012	The Prime Minister	• Roadmap for application of ratios for blending biofuels with traditional fuels
QCVN 05:2009/BGTVT (for emission)	November 19, 2009	MOT	• National technical regulation on emission of gaseous pollutants from assembly-manufactured automobiles and new imported automobiles
QCVN 01:2009/BKHCN (for fuel)	September 30, 2009	MOST	• National technical regulation on gasoline, diesel fuel oils and biofuels
QCVN 86:2015/BGTVT (for emission)	July 24, 2015	MOT	• National technical on the fourth level of gaseous pollutants emission • for new assembled, manufactured and imported automobiles
QCVN 01:2015/BKHCN (for fuel)	November 11, 2015	MOST	• National technical regulation on gasoline, diesel fuel oils and biofuels
QCVN 04:2009/BGTVT	November 19, 2009	MOT	• National technical regulation on emission of gaseous pollutants from assembly – manufactured motorcycles, mopeds and new imported motorcycles, moped
QCVN 01:2009/BKHCN (for fuel)	September 30, 2009	MOST	• National technical regulation on gasoline, diesel fuel oils and biofuels
QCVN 77:2014/BGTVT (for emission)	October 3, 2014	MOT	• National technical regulation on the third level of gaseous pollutant emission for new assembled, manufactured and imported two-wheeled motorcycles
QCVN 01:2015/BKHCN (for fuel)	November 11, 2015	MOST	• National technical regulation on gasolines, diesel fuel oils and biofuels

Note: MOST (Ministry of Science and Technology), MOT (Ministry of Transport)

IOP Conf. Series: Earth and Environmental Science **143** (2018) 012065 doi:10.1088/1755-1315/143/1/012065

Table A2. Status quo of Agricultural land from 2000-2014

Type	Area (ha)				Land-use Change (ha)	
	2000	2005	2010	2014	2000-2005	2005-2010
Agricultural land	20,939,679	24,822,560	26,100,160	26,822,900	3,882,881	1,277,600
Agricultural production land	8,977,500	9,415,568	10,117,893	10,231,700	438,068	702,325
Forestry land	11,575,027	14,677,409	15,249,025	15845,2	3,102,382	571,616
Water surface land for fishing	367,846	700,061	690,218	707,9	332,215	-9,843
Land for salt production	18,904	14,075	17,562	17,9	-4,829	3,487
Others	402	15,447	25,462	20,2	15,045	10,015

Table A3. Crop cultivation in Vietnam from 2005 – 2015 (GSO Vietnam)

Crop	Cultivatable area	Element	2005	2006	2007	2008	2009	2010	2011	2012	2013	2014	2015
Sugarcane	Mekong River Delta, North Middle and Coastal Plain	Planted area	266.3	288.1	293.4	270.7	265.6	269.1	282.2	301.9	310.4	305	284.5
		Yield	561.3	580.3	592.9	596.4	587.7	600.6	621.5	629.9	648.5	649.9	644
		Production	14948.7	16719.5	17396.7	16145.5	15608.3	16161.7	17539.6	19015.4	20128.5	19821.6	18320.8
		Growth Index	102.6	103.4	102.2	100.6	98.5	102.2	103.5	101.3	103	100.2	99.1
Cassava	Central, North East, Mekong	Planted area	425.5	495.5	554	507.8	496.2	498	-	551.9	543.9	552.8	566.5
		Yield	179.4	162.5	160.7	-	169.0	-	-	-	-	-	-
		Production	6716.2	8192.8	9309.9	8530.5	8521.6	85956		9735.4	9757.3	10209.9	10673.7
		Growth Index	-	-	-	-	-	-	-	-	-	-	-
Maize	North East, South East	Planted area	1052.6	1033.1	1096.1	1140.2	1089.2	1125.7	1121.3	1156.6	1170.4	1179	1179.3
		Yield	36	37.3	39.3	40.1	40.1	41.1	43.1	43	44.4	44.1	44.8
		Production	3787.1	3854.6	4303.2	4573.1	4371.7	4625.7	4835.6	4973.6	5191.2	5202.3	5281
		Growth Index	103.9	103.7	105.2	102.2	100	102.5	104.9	99.7	103.1	99.3	101.6

Planted area (Thous.ha), Yield (Quintal/ha), Production (Thous.tons), Growth Index (%)

Table A4. Policy support for Energy Efficiency and Conservation

Policy	Date	Issuer	Main content
Decree No. 102/2003/ND-CP	September 03, 2003	Prime Minister	• Specifies roles and responsibilities for the various government agencies regarding energy efficiency; • Focuses on mitigating energy use by production establishments, buildings, and equipment through technological modificatio and choice of investment, energy-efficient design, and operations management; • Promote thrifty energy consumption of citizens' lifestyles.
Vietnam National Energy Efficiency Program	April, 2006	Prime Minister	• In line with EE&C decree, the Ministry developed specific strategies to promote energy efficiency • A targeted national program and the first-ever comprehensive plan to institute measures for improving energy efficiency and conservation across all sectors of the Vietnamese economy; • The overall aim is to make initial savings of 3–5% during the period 2006–2010 and a further 5–8% during the period 2011–2015 (Chau 2015)
Decision No. 1427/2012/QD-TTg	October 2, 2012	Prime Minister	• National Targeted Programme on Energy Efficiency and Conservation - Phase 2012-2015 • Specifies six objectives to promote and achieve energy efficiency
Decision No. 1885/QD-TTg	December 27, 2007	Prime Minister	• Strategy on National Energy Development up to 2020, with vision to 2050: • Develop refineries to meet the domestic demand for oil products, bringing the total refinery capacity to around 25-30 millior tons of crude oil by 2020; • Increase the proportion of renewable energy to about 3% out of total primary commercial energy by 2010; about 5% by 202(and about 11% by 2050
The Law on Energy Efficiency and Conservation	June 18, 2010	National Assembly	• Highlights energy labelling as an effective method to increase the use of vehicles and appliances with high efficiency, gradu: removing vehicles and appliances with outdated technology, reducing energy consumption in production. • Encourages citizen to be involved in the development of mass transit networks, to manufacture and use energy-saving vehicl exploit and expand the application of liquefied gas, natural gas, electricity, mixed fuels and biofuels in replacement of petrol

Table A5. Biofuels Policy and Incentives

Policy	Date	Issuer	Main content
Decision No. 177/2007/QD-TTg	November 20, 2007	Prime Minister	• Develop biofuels as a new and renewable energy to partially replace conventional fossil fuels, so that contributi: assuring energy security and environmental protection
Decision No. 3638/QĐ – BCT	July 17, 2009	MOIT	• Establish a task force to develop standards and technical regulations on the production, storage, distribution and of biofuels
Circular No. 20/2009/ TT-BKHCN	September 30, 2009	MOST	• The national technical regulation on gasoline, diesel fuel oils and biofuels following QCVN 1:2009/BKHCN
Decision No. 400/ QD-TDC	March 25, 2010	MOST	• The guidelines for standard-compliance certification of gasoline, diesel and biofuels following QCVN 1:2009/BKHCN
Decree 24/2007/ND-CP	February 14, 2007	Government	• The guidelines for the implementation of The Law On Enterprise Income Tax

Table A6. Three major Biofuels plants in Vietnam

Plant	Location	Invester	Investment	Capacity	Construction (Begin – Finish)	Remarks
The Dung Quat Bio Ethanol Plant	North Central and central coastal areas	PetroVietnam (PV Oil, BSR, PVFC, Petrosetco)	$92 million	100 million liters per year	2009-2012	Products: ethanol 99.7%, feedstock, purified CO2 for food industry, biogas Central market
Binh Phuoc Ethanol Plant	North Central and central coastal areas	Vietnam Oil Company (PV Oil), LICOGI 16 and ITOCHU Corporation (Japan)	-	- 106 million liters per year - Using 260,000 tons of cassava slices	2010-2012	Products: ethanol 99.7%, bio-fertilizer, purified CO2 for food industry, biogas Southern market
Phu Tho biofuel plants	Northern midlands and mountain areas	Vietnam Oil Company (PV Oil, a company under Vietnam National Oil and Gas Group – PetroVietnam) and SEA bank	$114.2 million	- 100 million liters per year - Using 260,000 tons of cassava slices	On-going (under construction)	Products: ethanol 99.7%, bio-fertilizer, purified CO2, for food industry, and biogas for Northern market.

References

[1] Liaquat, A. M., M. A. Kalam, H. H. Masjuki, and M. H. Jayed 2010 Potential Emissions Reduction in Road Transport Sector Using Biofuel in Developing Countries *Atmospheric Environment* 44 (32) Elsevier Ltd: 3869–77. doi:10.1016/j.atmosenv.2010.07.003
[2] Oanh, N.T.K., Thiansathit, W., Bond, T.C., Subramanian, R., Winijkul, E., PawArmart,I. 2010 Compositional characterization of PM2.5 emitted from in-use diesel vehicles *Atmospheric Environment* 44 (1), 15-22
[3] Oener, C., Altun, 2009 Biodiesel production from inedible animal tallow and an experimental investigation of its use as alternative fuel in a direct injection diesel engine *Applied Energy* 86 (10), 2114-2120.
[4] Faiz, A., Sinha, K., Walsh, M., Varma, A., 1990 *Automotive Air Pollution Issues and Options for Developing Countries* The World Bank
[5] Allen, Ryan W., Enkhjargal Gombojav, Baldorj Barkhasragchaa, Tsogtbaatar Byambaa, Oyuntogos Lkhasuren, Ofer Amram, Tim K. Takaro, and Craig R. Janes 2013 An Assessment of Air Pollution and Its Attributable Mortality in Ulaanbaatar, Mongolia *Air Quality, Atmosphere and Health* 6 (1): 137–50
[6] WHO, 2007 *Dioxins and Their Effects on Human Health* Fact Sheet No. 225
[7] Stone, E., Schauer, J., Quraishi, T.A., Mahmood, A., 2010 Chemical characterization and source apportionment of fine and coarse particulate matter in Lahore, Pakistan *Atmospheric Environment* 44 (8), 1062-1070.
[8] Bouaid, A., Martinez, M., Aracil, J., 2009 Production of biodiesel from bioethanol and Brassica carinata oil: oxidation stability study *Bioresource Technology* 100 (7), 2234-2239
[9] Han, M., Yi, W., Wu, Q., Liu, Y., Hong, Y., Wang, D., 2009 Preparation of biodiesel from waste oils catalyzed by a Brønsted acidic ionic liquid *Bioresource Technology* 100 (7), 2308-2310
[10] Koizumi, Tatsuji. 2015 Biofuels and Food Security *Renewable and Sustainable Energy Reviews* 52. Elsevier: 829–41
[11] Araújo, Kathleen, Devinder Mahajan, Ryan Kerr, and Marcelo da Silva 2017 Global Biofuels at the Crossroads: An Overview of Technical, Policy, and Investment Complexities in the Sustainability of Biofuel Development *Agriculture* 7 (4): 32
[12] Hubbard, P., 2010 Biofuels and the developing world. *Bologna Center Journal of International Affairs* 13/Spring 2010.
[13] World Energy Council 2016 *World Energy Resources Bioenergy*
[14] AFD 2009 *Biofuel Issues in Vietnam* Background Paper No. 1, no. 1: 1–20
[15] GSO 2013 *Statistical Year Book* Viet Nam
[16] Le, Loan T., Ekko C. van Ierland, Xueqin Zhu, Justus Wesseler, and Giang Ngo 2013 Comparing the Social Costs of Biofuels and Fossil Fuels: A Case Study of Vietnam *Biomass and Bioenergy* 54 Elsevier Ltd: 227–38
[17] Leinonen, Arvo, and Nguyen Duc Cuong 2013 *Development of Biomass Fuel Chains in Vietnam Introduction. Development of Biomass Fuel Chains in Vietnam* 134
[18] Nguyen, Trung H., Stephen Williams, and Keith Paustian 2017 Impact of Ecosystem Carbon Stock Change on Greenhouse Gas Emissions and Carbon Payback Periods of Cassava-Based Ethanol in Vietnam *Biomass and Bioenergy* 100. Elsevier Ltd: 126–37
[19] Contreras, Kathleen Dematera, Alvin Mejia, Phan Quynh Nhu, Mark Tacderas, Kaye Patdu, Loïc Daudey, Nguyen Anh Tuan, and Stefan Bakker 2015 Tracking Sustainable Transport in Vietnam : Data and Policy Review for Energy Efficiency and Climate Change 2015
[20] Zwebe, Dagmar 2012 Biomass Business Opportunities Viet Nam *SVN, Nertherland Development Organisation Vietnam* 1–85

[21] De Abrantes, R., Vicente De Assunção, J., Pesquero, C.R., Bruns, R.E., Nóbrega, R.P., 2009 Emission of polycyclic aromatic hydrocarbons from gasohol and ethanol vehicles *Atmospheric Environment* 43 (3), 648e654

[22] Lumbreras, J., Valdés, M., Borge, R., Rodríguez, M.E., 2008 Assessment of vehicle emissions projections in Madrid (Spain) from 2004 to 2012 considering several control strategies *Transportation Research Part A* 42 (4), 646e658

[23] GSO 2013 *Statistical Year Book* Viet Nam

[24] Ben-Iwo, Juliet, Vasilije Manovic, and Philip Longhurst 2016 Biomass Resources and Biofuels Potential for the Production of Transportation Fuels in Nigeria *Renewable and Sustainable Energy Reviews* 63. Elsevier: 172–92.

[25] Adeoti, O. 2010 Water Use Impact of Ethanol at a Gasoline Substitution Ratio of 5% from Cassava in Nigeria *Biomass and Bioenergy* 34 (7) Elsevier Ltd: 985–92

[26] Bindraban, Prem S., Erwin H. Bulte, and Sjaak G. Conijn 2009 Can Large-Scale Biofuels Production Be Sustainable by 2020? *Agricultural Systems* 101 (3). Elsevier Ltd: 197–99

[27] Vyas, D K, and R N Singh 2007 Feasibility Study of Jatropha Seed Husk as an Open Core Gasifier Feedstock 32: 512–17

[28] Duku, Moses Hensley, Sai Gu, and Essel Ben Hagan 2011 A Comprehensive Review of Biomass Resources and Biofuels Potential in Ghana *Renewable and Sustainable Energy Reviews* 15 (1). Elsevier Ltd: 404–15

[29] Dutta, Kasturi, Achlesh Daverey, and Jih Gaw Lin 2014 Evolution Retrospective for Alternative Fuels: First to Fourth Generation *Renewable Energy* 69. Elsevier Ltd: 114–22

[30] OFID. 2009

[31] Ogg, Clayton W. 2009 Avoiding More Biofuel Surprises: The Fuel, Food and Forest Trade-Offs *Journal of Development and Agricultural Economics* 1 (1): 6.

[32] Timilsina, Govinda R., John C. Beghin, Dominique van der Mensbrugghe, and Simon Mevel 2012 The Impacts of Biofuels Targets on Land-Use Change and Food Supply: A Global CGE Assessment *Agricultural Economics* 43 (3): 315–32.

The notions of resilience in spatial planning for drought - flood coexistence (DFC) at regional scale

Nguyen Trong Hoa[1], **Nguyen Quoc Vinh**[2]

[1] Hochiminh City Institute for Development Studies, 27 Le Quy Don St., Dist. 1, HCMC, Vietnam
[2] Hochiminh City University of Technology, B6 Building, 268 Ly Thuong Kiet St., Dist. 10, HCMC, Vietnam, Email: vinh.bmkt@hcmut.edu.vn

Abstract. The notions of urban resilience and resilient city has been developed in the 2000s [1], four decades since the first concept of ecological resilience was originally introduced in the 1970s by ecologist C.S. Holling [2]. However, they have attracted great attentions and interests, in both academia and urban governance, then in planning practice over recent years.
The first two sections of this paper examine the term resilience in ecological systems, urban systems, in spatial planning and in urban design. Specific attention of the paper, introduced in the third part, is to investigate resilience in the context of drought-flood coexistence (DFC), revolving two key objects and their interactions: DFC and urban at regional scale. Flood and drought events, in their turns intertwine in natural correlation, which is also reviewed. These relationships are literally investigated, to prove that they interplay mutually with each other, and that once a city develops in relation with water cycle at a regional context, in arid zone, not only hydrological drought could be regionally decreased, but human-induced floods could be ecologically regulated.
The study concludes in the fourth, together with lessons from relevant case studies in America, China, with some principles on spatial planning, resilient/adaptive to DFC, which could be ecologically managed in correlation with urban development on a sustainable pathway.

Key words: ecosystems, extreme weather events, drought - flood coexistence, sustainability, urban resilience, spatial planning, mitigate, adaptive cycles, adaptive capacity.

1. Introduction

Human beings have experienced a very challenging phase since the industrial revolution early 17th century. From then, we have witnessed amazing advances in social, demographic, and economic development, which have strongly caused urban pattern changes [3][4]. What challenges our artifactual environment more over recent decades is threats from climate change.

On the one hand, the climatic impacts in general on human systems are remarkable, such as: melting snow, altered hydrological systems, shifting species, migrating patterns, etc. On the other hand, global warming has caused specified consequences with more frequent occurrences of extreme climate disturbances, such as heat wave, drought, extreme precipitation, inland and coastal floods, landslides, cyclones, and other weather extremes [5]. Among these, flood and drought is one of the hottest issues, both damage and loss of life. Moreover, desertification is becoming more and more severely in riverine, coastal and urban areas across the world and in Asian countries, which has caused a severe problem of DFC [6].

In a review of drought under global warming [6], Dai indicated that human have experienced a tough time of drought since the 1970s. Climate models have increased aridity in the 21st century over most of Africa, Southern Europe and the Middle East, most of the Americas, Australia, and South-east Asia. Compared to some similarly affected areas in the central region of Vietnam, where most of them have been experiencing desertification ever since the 1970s [7], the similar findings show very high pressures of climate change impacts, such as: increased temperature, concentrated precipitation, flash flood in urban areas, coastal flooding, seal level rise, etc., particularly in Ninh Thuan province [8].

All impacts on human environment often come from natural systems, together with recent global changing climates. But more importantly, design is a very crucial task that would affect to the natural environment on either a good way or a bad way. Van Der Ryn and Cowan emphasized, in Ecological Design [9], that "*the current environmental crisis is caused by our design, which recently is just to serve human interest and ignore its relationships with other species*". It is true that there is significant lack of preparedness and readiness as well, for a city to deal with hazardous threats and to rebound back after perturbation, in our current planning and design practice. There have been different conceptions of mitigation, adaptation, and currently resilience to cope with our naturally shaking and deteriorating environment, unknown disasters, and any climate change. However, these terms, particularly resilience is still ambiguous in planning practice. The concept will be explored in four parts of the paper in a purpose of seeking an appropriate way for resilient spatial planning in the context of DFC at regional scale, including: (i) from ecological science to urban resilience; (ii) resilience in planning and design; (iii) DFC, urban and their relationships; and (iv) spatial planning resilient to DFC, together with lessons learnt from relevant case studies.

2. From ecological science to urban resilience

The prevalent notions of resilience, in theory and application over recent decades, is based on several concepts, including thresholds or tipping points, multiple stable states, regime shifts, complex adaptive systems, adaptive cycles, adaptive capacities, panarchy, etc., [10][11][12]. In this section, the definitions will be examined, firstly in ecological science, then interpreted in the context of understanding and managing social-ecological systems.

2.1. Ecological resilience

Originally, the concept of resilience was introduced by ecologist C. S. Holling in 1973 [2], where he defined as stability and the ability of ecological system to return to an equilibrium state after perturbations. Later on, it was defined as persistence or relationships within a system, and ability to absorb changes of state variables, driving variables, parameters, and still persist in its own basic structure and function, no need shifting into a different state. In some cases, a system can be very resilient, still fluctuate greatly and have low stability, or vice versa. Therefore, keeping balance between resilience and stability is a product of the evolutionary history of these systems in the face of the range of random fluctuations.

In an evolutionary literature relevant to the meaning from 1995 to 1997, many authors considered resilience as the time required for a system to return to a steady-state following a perturbation [13][14]. This conception implicitly means the system exists near a single or global equilibrium condition. Other authors considered the return time as a measure of stability. Among them, Holling described this return time as "engineering resilience" [2][15]. The figure 1 shows the engineering view of resilience, where a system deflected from an equilibrium point after a disturbance. After this emerging conception of resilience raised, most of the professional fields, including urban planning and design, adapt this concept through designing a system to be able to bounce back to stability.

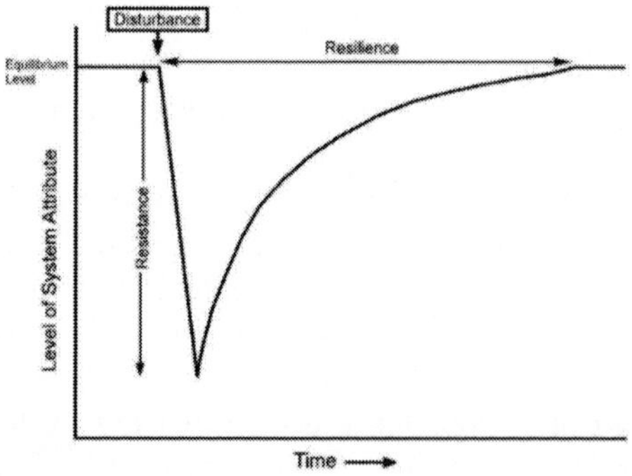

Figure 1. Diagrammatic illustration of "engineering resilience". A shock deflects a system from a steady state or an equilibrium point, to which a resilient system returns

Source: based on S. T. A. Pickett, B. McGrath, M. L. Cadenasso, and A. J. Felson, "Ecological resilience and resilient cities,", 2014.

The second movement, firstly termed ecosystem resilience [16], then broadly applied most interchangeable with the words ecological resilience [17][18][19] or resilience [2], has viewed ecosystems operating with multiple or far from equilibrium state, (which noticed their flexible changes of a system into another regime). In this case, resilience is measured by the magnitude of disturbance that can be absorbed before the system redefines its structure by changing the variables and processes that control their behavior. The term "ecological resilience" emphasizes on persistence, change, and unpredictability. It is contrast to "engineering resilience", focuses on efficiency, constancy, and predictability [17]. The non-equilibrium resilience seems to be paradigmatic in ecological science, and results in a rich body of work at the socio-ecological interface [11][20] afterward.

The later concept of resilience, which was resulted from more human-induced impacts to the state changes in ecosystems, is adaptive capacities [18], which is less ambiguous, and more easily applicable for the design practice. The concept "adaptive capacities" could trace through the adaptive cycles, (social groups are active, resources are released or captured) and all plays out in spatially heterogeneous urban areas [21]. In the biophysical realm, which is more relevant to the study concerning spatial planning, adaptive processes are addressed in a variety of sources [19].

In the more recent papers, many authors noticed resilience as "specific adaptations" (i.e., high adaptedness) of a system to known threats and more generic adaptability [1]. Academia tends to view adaptability as synonymous with adaptive capacity, or flexibility necessary for confronting unexpected hazards [22].

In short, the development of the term resilience is quite long process. It was originally defined as stability, then as engineering resilience, or as ecological resilience, which have been explored in academia and urban governance over recent decades, yet seem to agree on a clear definition but a descriptive concept. For that reason, the notions of resilience could not be easily adapted in many practical disciplines. The theory has, therefore, been continuously studied and updated on the direction of a more normative concept, such as "adaptive cycles" or "adaptive capacity". The latest concept appears as a step closer to what urban resilience would like to convey, which needs more applicable principles, with regard to planning practice, particularly with spatial planning.

2.2. Urban resilience

Since the first concept of resilience raised, it took four decades of development leading to the concept of urban resilience today.

To clearly understand urban resilience, it is necessary to first specify what is meant by 'urban', and how it is defined in ecological science. In a reviewing paper of the concept urban resilience, S. Meerow argued that most definitions of urban resilience are rather vague with respect to what constitutes an urban area of a city. Although there have been numerous definitions of urban, but she ended up with the holistic one: urban as a 'conglomeration of ecological, social, and technical components' [1][3]. In urban ecology, cities are often places where human and natural patterns and processes interact, evolving to form an "urban ecosystem" or an Social-Ecological-Systems SES [23] [24]. There have been two primary meanings of urban ecology [25]. One notices in designing human environment and prevalent in the urban planning field. While the other results from ecological science, which refers to studies of the distribution and abundance of organisms in and around cities, and on the biogeochemical inputs and outputs of urban areas [26].

Indeed, during the last decades, urban areas and their settlements have already been raised into another level in the discourse of ecological science [27], and city had been considered as ecosystem [28][29], at any scales, any levels. An ecosystem, based on ecological science, comprises a specified area or even size of the earth, in which the collections of organisms and the physical environment interact with each other [29][30]. In most ecosystems, some materials, nutrients or contaminants, will flow across boundaries. Therefore, it is important to know what flows will cross. According to ecosystem concept, the flow is metabolic transformations; while in landscape, spatial interaction in which organisms and fluxes engage is flowing. However, the contemporary ecology attempts to combine ecological approaches that had previously existed independently, to consider ecology of landscape and ecology of ecosystems reside on different special scales or organizational levels.

In the meantime, urban ecosystem also contains organisms, physical conditions and entities, and the interactions between them. Urban ecosystem, therefore, comprises not only human, institutional frameworks, their artifacts, but consisting of four dimensions: 1) biological, 2) social, 3) physical and 4) built components. They cannot be separated from one another, as shown on the figure 2 below [31]. "Urban systems" are conceptualized as complex, adaptive, emergent ecosystems composed of four subsystems - governance networks, networked material and energy flows, urban infrastructure and form, and socio-economic dynamic that themselves are multi-scalar, networked, and often strongly coupled [1][3]. Furthering this point, the four major subsystems of the urban system are "governance networks," "metabolic flows," "built environment," and "social dynamics."

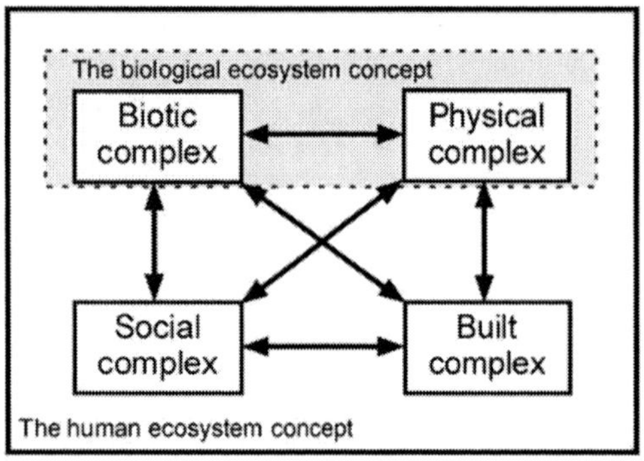

Figure 2. The human ecosystem concept was expanded from the bio-ecological concept of the ecosystem as originally proposed by Tansley (1935) in the dashed line. The expansion incorporates a social complex, which consists of the social and built environments.

Source: based on S. T. A. Pickett and J. M. Grove, "Urban ecosystems: What would Tansley do?," *Urban Ecosyst.*, vol. 12, no. 1, pp. 1–8, 2009

Urban regions of cities have not only been as an isolated component separately from other natural components. Scholars in urban ecological science have taken a big progress in this concern. They have no longer consider ecology IN the city such as green areas and suburbs, but combined with other social and other natural sciences. From which, a new conception of city-suburban-exurban (CSE) complexes as integrated systems generated and emerged from urban designs. It is, therefore, nature can be ignored as outsider in the CSE realm. CSE ecosystems must be seen as comprising organisms, including human social structures and processes of soils, waters, air, and energy, along with built environment and infrastructure. Urban ecology is now an ecology OF the city [32].

Together with the movement of considering urban as SES, as noticed earlier, SES scholars tend to view adaptability as synonymous with adaptive capacity, or flexibility necessary for confronting unforeseen threats [22][3]. While leading SES scholars do not contrast adaptability or adaptive capacity and adaptedness, they do emphasize the importance of maintaining "general" resilience to unknown hazards in addition to "specified" resilience to known risks [12]. Scholars focusing on resilience to climate change argued that urban resilience should focus on adaptive capacity rather than specific adaptations.

To sum up with S. Meerow's definition [1] *"Urban resilience refers to the ability of an urban system and all its constituent socio-ecological and socio-technical networks across temporal and spatial scales to maintain or rapidly return to desired functions in the face of a disturbance, to adapt to change, and to quickly transform systems that limit current or future adaptive capacity"*, where city operates in non-equilibrium, is viewed as a desirable state because it recognizes multiple change pathways (persistence, transition, and transformation) and emphasizes the importance of adaptive capacity and timescales.

In reviewing 29 papers and researches on the term resilience, which was derived from ecological science. Several multi-discipline studies, developed continuously from the year of 1973 to 2016, related to notions of resilience shows that the science of earth, ecology, biology, etc. has been developed and persisted for a very long time. Although there have been some contrasts among the consulted papers in this section, but within the content of the topic, which focuses only on the two key objects (natural events of DFC and urban) and on the term spatial planning, the literature is therefore considered as high confident documentary.

3. Resilience in planning and design

The earliest author concerned about environmental matters and natural conditions in urban planning is McHarg, who emphasized in his book "*Design with Nature*", published in 1969, that most cities in the world developed on their own processes and containing values, both natural and artificial [33]. However, the notions of resilience in relation to spatial planning just started in the late 1990s [34]. The main focus, nonetheless, was just on physical and infrastructure improvements to prevent disturbances.

Over recent years however, the challenges of climate change have become more prominent, the matter of mitigation, adaptation or resilience in planning, which quite coincides with a traditional approach to land use planning as a means of minimizing existing disturbances and reducing the risks and negative effects of possible disturbances, have been critically important. Resilience in urban planning realm is building a city or a community that could physically resists with external shocks or disturbances by properly controlling their interactions. The section will therefore, focus on natural intertwined interactions between key objectives: urban settlements, natural events of drought and flood, in seeking relevant ecological principles in planning and design.

3.1. Urban complex adaptive ecosystems, a changing process and form

A decade or so after the first concept of resilience being realized in urban planning, our world has now been threatened by the extreme changing climate, the literature on resilience in planning has widened to encompass mitigation strategies such as reducing greenhouse gas emissions and other unpredictable disturbances. In a smaller scale, the literature also focuses on adaptation. To move on the right track toward urban resilience, with the above-discussed concept which considers city as ecosystems, four mentioned environments (biological, physical, social and built environments) constituting a city should be carefully understood by architects or planners before starting the planning process. Among these four, biophysics is originally derived from its own geological and biological evolution and artifacts demonstrates the consequence of cultural and social adaptations of the people into the plan of the city.

It coincides with Lu, in planning practice, that the use of the term resilience is limited in scope and often considered as synonym of adaptation [3]. Lu expressed the notion of resilience into the concept of the adaptive cycle [35], which passes through four characteristic phases, as illustrated on the figure 3: (r) growth and exploitation; (K) conservation; (Ω) collapse or release; and (α) renewal and reorganization, in terms of a system's robustness and rapidity. He considered a system's robustness being determined according to its 'strength' to carry and absorb uncertain disturbances, whilst its 'rapidity' refers to the flexibility to rearrange itself into a new stable state after a collapse occurs [3]. He gave an example of this interpretation in dealing with a flood prone area, a city's resilience performance is represented in terms of preparations (robustness) before the occurrence of floods (e.g., building types in vulnerable areas, flood control systems, administration of flood risk management) and its reaction (rapidity) to and management of the disturbances (e.g., city drainage systems, rapid response, damage repairs, rescue services, financial support and future improvements). The new condition of a city may not necessarily be the same as the state before a disturbance occurred. However, his interpretation focused much on urban governance and policy making. In planning and design, the preparations should focus on considering natural conditions at regional scale, such as river basin, geography, topography, soil, slope, etc., which interfere and interact with each other closely and mutually. Based on this concern, planning and design could minimize or even avoid disturbances. In the context of DFC, the occurring time could be in different seasons, but the preparations and actions are quite the same in meanings, just some of them are different in characteristics and attributes.

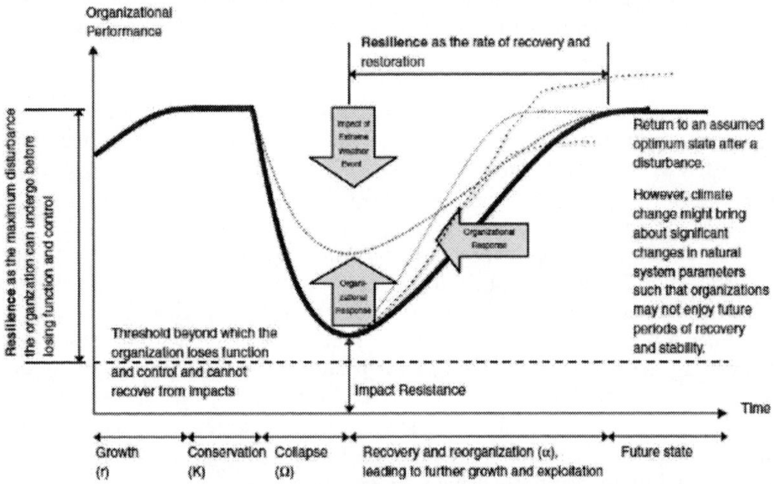

Figure 3. Resilience framework: a new stable state is not necessarily the same as the previous one. Source: based on Linnenluecke and Griffiths (2010)

Also following McHarg's suggestions, objectives and method should be well defined to pursue urban resilience. While objectives focus on identifying locally discrete elements, the method will develop principles and, finally, principles should be constructed into policies for realization. It is practically proven that most of the memorable and sustainable cities in the world are planned and designed based on their own incorporations with dramatic sites. According to Jabareen [36], studies of urban resilience often overlook the importance of multidisciplinary and complex nature of within that notion, and/or use the term without a deep understanding. Focusing on limited number of relevant factors can easily lead a city to bad performance toward resilient end.

3.2. Understanding river basin

The 19th century western explorer and geographer - John Wesley Powell, defined the term watershed as "*That area of land, a bounded hydrologic system, within which all living things are inextricably linked by their common water course and where, as human settled, simple logic demanded that they become part of a community*" [37], whilst McHarg emphasized river basin as a hydrological unit, an area of united water, and is permanent [38]. Both views are quite the same in defining boundary of water systems, which is an interacting process of diversified natural elements, as geography, topography, etc.

In the 1990s, the concept *Integrated Water Resources Management* (IWRM) with its emphasis on river basins as planning and management units was raised and encouraged, due to its failures in trying to intervene the unique natural system at large-scale river basin by building dams for the local context [39]. It reasoned out why McHarg emphasizing architects and planners should use river basin, for adequate understanding, to seek a more finite division of land, and designate appropriate land uses, and plan cities.

It is clear that, our world operates in its own rules, with its own interacting constituents, which reflects a half billion years of time in the interfaces of region, creating in its own local context. They are discrete, which is perfectly to investigate causes and effects between flood and drought events, and how to be affected by urban settlements growth.

3.3. Intrinsic suitability for urban use

As mentioned in the previous part about ecological knowledge on the nature of river basin system, McHarg stated that each ecological function within ecosystems will be compatible with an urban land use. For instance, subsurface geology, climate, soils, slope and thus drainage together with exposure, determine the appropriate types of agriculture that should be participated in the entire basin. There is the same logic for the functions of forest, grassland, wetland, and recreational areas. This concept coincides with Wenche E. Dramstad's [40] in proposing landscape ecology principles in landscape architecture and land use planning, of which the principles of landscape and regional ecology could be applied in any land mosaic, from suburban to agriculture and desert to human activity. Land mosaic will absolutely be a useful tool for planning and design a city toward resilience.

More importantly, defining sites for developing cities or urban settlements is much more complex. The number of assessment criteria should be set up. According to McHarg, the slopes must be less than five percent incline, not in the 50-year flood plain, and not in important aquifer recharge area. Adequate water supplies are logically a must. Similar to this concern, causes of flood in a city will be at first naturally controlled and then stored at conserved aquifer recharge areas. The context of DFC will be dealt with resiliently.

3.4. Degree of urban compatibility, a step forward to resilient cities

A defined system on intrinsic suitability for agriculture, forestry, recreation and urbanization indicates the values for each region and for the basin within each of the specified land uses. The system is not only defined for optimizing for a single urban land use, but for multiple compatible function [38].

Supporting the idea of considering river basin as a hydrological planning unit, subdividing into smaller units of watershed management [41][42], and studying aquifer recharge area and aquifers, flood plains [43] are also very fatal considerations for managing river basin ecosystem, a step forward to urban resilience. From theory to practice, there are many cases already applying principles of intrinsic suitability for urban use using GIS [44][45].

In 14 papers and researches collected and reviewed, that most of them are from urban planning field, some of them are from disciplines of environment, geology, ecology, and natural disasters. The oldest one published in 1992, and the newest in 2016, show that natural science is strongly developed.

4. DFC, urban settlements and their relationships

4.1. Flood, drought, DFC and interrelationships

A natural hazard is a threat of a naturally occurring event that will have a negative effect on people or the environment. Drought and flood is one of the killing hazardous events.

4.1.1. Flood, urban and relationships

Floods were the most devastating natural disaster of the twentieth century over the world, both damage and loss of life. Floods can occur at any regions, any seasons. Reasons of flood risk are caused by climate-related factors (water-containing capacity, characteristics of intense precipitation, etc.). Increased flooding can be attributed to two factors: 1) increased precipitation; and 2) changes in land use [46]. Inland floods are the result of storm water runoff that exceeds the capacity of their stream and river systems. The map below (figure 4) indicates the location of 3713 flood events during 1985–2010 recorded in the public global database maintained by G.R. Brakenridge at the Dartmouth Flood Observatory (http://floodobservatory.colorado.edu)

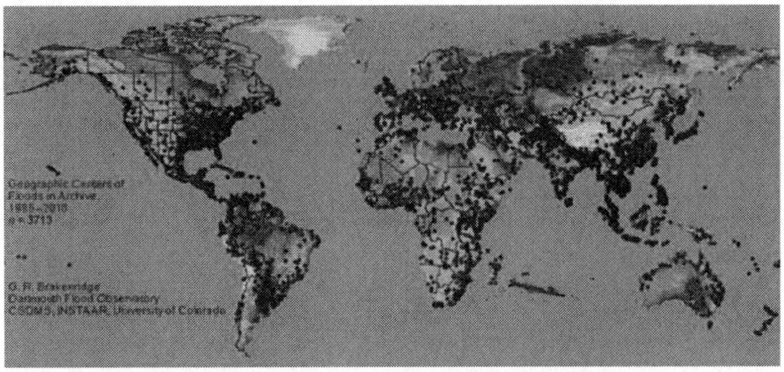

Figure 4. Geographic centers of large floods in the Dartmouth Flood Observatory archive. The map illustrates 3713 events over the period 1985–2010. Source: http://floodobservatory.colorado.edu

A number of current studies indicated that major floods around the world have raised the concern that river flooding is becoming more frequent and severe [49][50]. The frequency of major floods in many places around the world seems to be increasing; flash floods occurred throughout Europe in June 2016 [49]; the Elbe and Danube flooded in June 2013, just 10 years after the 2002 ''millennium'' flood; there was severe winter flooding in both 2013/2014 and 2015/2016 in England; and there are many more examples of fatal floods from all around the world [50]. River floods are affected by numerous processes and any changes in such processes may result in changes in the flood discharges. Merz et al. defined three groups of potential drivers related to the atmosphere, catchments, and the river systems [47].

Among flooding causes, natural flood plains are the most crucial driver. They are the most biologically productive and diverse ecosystems on earth. Globally, riverine flood plains cover more than $2x10^6$ km2, but they are the most threatened ecosystems. Its deterioration is closely link to the rapid decline in freshwater biodiversity; leads to habitat alteration, flow and flood control, species invasion and pollution. More than 90% of flood plains in Europe and North America are cultivated, functional extinction is unavoidable. In developing countries, the remaining disappearing at accelerating rate, due to changing hydrology. In 2015, increasing human population lead to further degradation of riparian areas, intensification of hydrological cycle, increase in the discharge of pollutants. In the future, South-East Asia will be the most threatened flood plains in the world [51].

Land use change is also among the major threats leading to floods to many countries worldwide [52], with more significant impacts on developing countries [53]. Land use change has, potentially, a very strong effect on floods as humans have heavily modified natural landscapes. Large areas have been deforested or drained, thus either increasing or decreasing antecedent soil moisture and triggering erosion. Hillslopes were modified for agricultural production, thus changing flow paths, flow velocities, and water storage, and consequently flow connectivity and concentration times. The intensification of agricultural practices has resulted in the formation of platy dense soil horizons with preferential lateral flow which may reduce and/or retard vertical infiltration in the soils, but cause an intensification of lateral mass flow besides the reduced filter and buffer processes in deeper soil horizons. It is likely that hydrologically significant changes will continue in the next decades due to loss of agricultural land and forests [54]. In all of these processes, however, the exact role of land use change in modifying river floods is still elusive [50].

In addition, processes of urbanization also lead to increased occupation of flood plains and, often, inadequate drainage planning, that is for sure that flooding events increased [55]. Donald Watson indicated that in order to regulate ecosystem's service of water, we need to plan the watershed, or to plan urban areas sitting on it, planners should follow the water, water cycle, and should completely understand water supply, annual rainfall, extreme weather events and water storage [46].

A coincidence to the literature, McHarg had theoretically and practically illustrated, in chapter 'The River Basin' [38], that there is clear evidence when we deeply understand how a river system works through history of geology and climate interacted upon the its basin, we will know how to find suitable areas for developing cities. By investigating climate and lithology, urban morphology will definitely explain the patterns of rivers and streams, the distributions of groundwater. This kind of information on the movements of sediments, some by fluvial processes and others from deposition, will reveal the pattern, distribution and properties of soils. More importantly, where rivers and their tributaries start and run from the mountainous areas to coastal plain could help professionals seeking the areas fitting for cities, for urban land uses, and for their urban intrinsic compatibilities, by incorporating all relevant information. As consequence, not only river flooding, but urban flooding is also well regulated and prepared for human beings activities.

4.1.2. Drought, urban and relationships

Among other natural events like flood or stormy surge, drought is one of the costly natural extremes to human system. Reasons for the occurrence of droughts are quite complex, since they depend not only in the atmosphere but also on the hydrologic processes, which feed moisture to the atmosphere. The lesser the relative humidity the less probable the rainfall becomes, as it will be harder to reach saturation conditions. Drought is a recurring extreme climate event over land characterized by below-normal precipitation over a period of months to years. Drought is a temporary dry period, in contrast to the permanent aridity in arid areas. Drought occurs over most parts of the world, even in wet and humid regions. This is because drought is defined as a dry spell relative to its local normal condition. On the other hand, arid areas are prone to drought because their rainfall amount critically depends on a few rainfall events [56][57]. That differentiated definitions between drought and aridity could explain why it is hard to visualize drought in Southeast Asia, Brazil, Western Europe, or the Eastern United States, the regions perceived by many to have a surplus of water.

Worldwide, economic damages attributed to natural disasters tripled from the 1960s (US$40 billion) to the 1980s (US$120 billion). The figures 5&6 below show economic, social, and environmental costs and losses associated with drought are also increasing dramatically, although it is difficult to quantify this trend precisely because of the lack of reliable historical estimates of losses [57]. The concerns about the trends in losses associated with natural disasters in developed countries are magnified when placed in the context of developing nations. Natural hazards result in significant loss of life and serious economic, environmental, and social impacts that greatly retard the development process.

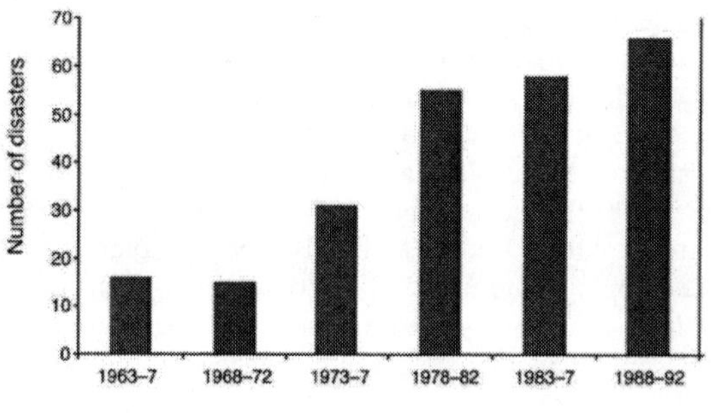

Figure 5. Number of disasters causing significant damage, 1963-92, and affecting one percent or more of the total annual gross national product.

Source: based on D. Wilhite, 'Drought as a natural hazard: Concepts and definitions', 2000

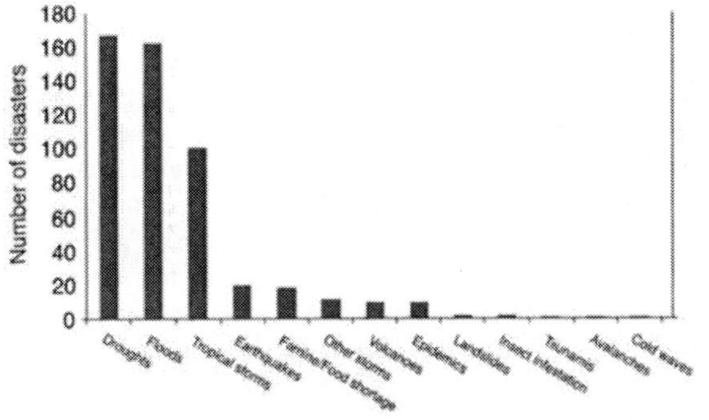

Figure 6. Disasters by type, affecting 1 percent or more of total population, 1963–92.

Source: based on D. Wilhite, 'Drought as a natural hazard: Concepts and definitions', 2000

Drought has a long-term history, that was examined by Dai in his review paper on drought under global warming challenge [6]. It reveals that large-scale droughts occurred many times during the past 1,000 years across the world: North America, Mexico, Asia, Africa, and Australia. In comparison to last multi-decadal droughts, the modern-day droughts in 1930s-1950s had similar intensity but shorter durations [58]. During this time particularly, the United States experienced one of the most devastating droughts in the last century, which covered almost two-thirds of the country, and parts of Mexico and Canada. S. Schubert et al presented that increased drought severity was caused by anomalous tropical sea surface temperatures and that interactions between the atmosphere and the land surface [58].

There have been several definitions of droughts. The more general one is as noted by the Intergovernmental Panel on Climate Change (IPCC), 'in general terms, drought is a 'prolonged absence or marked deficiency of precipitation', a 'deficiency of precipitation that results in water shortage for some activity or for some group' or a 'period of abnormally dry weather sufficiently prolonged for the lack of precipitation to cause a serious hydrological imbalance [59]. But the one more specific and more theoretical is a classification of droughts by type, as indicated on the figure 7: meteorological, hydrological, agricultural, and socioeconomic [57]. Meteorological (or climatological) drought is defined on the basis of the degree of dryness and the duration of the dry period. Agricultural drought links various characteristics of meteorological drought to agricultural impacts, focusing on precipitation shortages, differences between actual and potential evapotranspiration (ET), soil water deficits, and so forth. Hydrological droughts are correlated with the effects of periods of precipitation shortfall on surface or subsurface water supply (i.e., streamflow, reservoir and lake levels, groundwater) rather than with precipitation shortfalls [60]. Finally, socioeconomic drought associates the supply and demand of some economic good or service with elements of meteorological, hydrological, and agricultural drought. Some scientists suggest that the time and space processes of supply and demand are the two basic processes that should be included in an objective definition of drought [61].

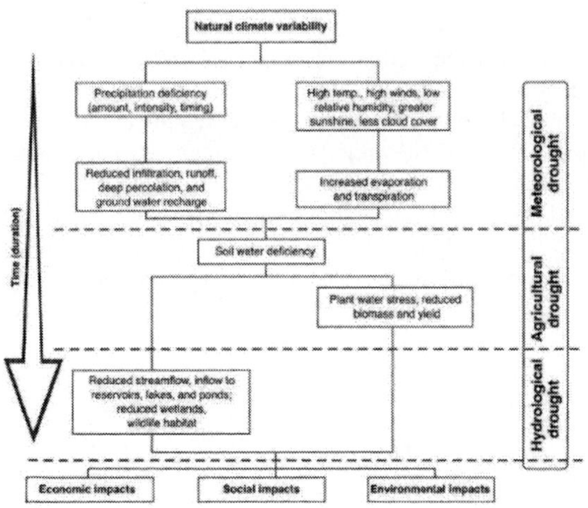

Figure 7. Relationship between various types of drought and duration of drought events.

Source: based on D. Wilhite, 'Drought as a natural hazard: Concepts and definitions', 2000

Natural drought events are increased under global warming pressure, that adds extra heat to the climate system and on land much of that heat goes into drying. It should therefore set in quicker, become more intense, and may last longer. Droughts may be more extensive as a result. Climate change may not produce droughts, but it could exacerbate them and it will likely expand their domain in the subtropical dry zone [62]. In another calculation, which was more realistic, based on the underlying physical principles8 that take into account changes in available energy, humidity and wind speed, author suggested that there has been little change in drought over the past 60 years [63]. This confirmation was based on his/her study on previously reported increase in global drought is overestimated because the Palmer Drought Severity Indices (PDSI) [64] uses a simplified model of potential evaporation that responds only to changes in temperature and thus responds incorrectly to global warming in recent decades. Many other scholars were with this view point [63][65][66].

Drought is an insidious natural hazard that is a normal part of the climate of virtually all regions. It should not be viewed as merely a natural or physical phenomenon. Rather, drought is the result of an interplay between a natural event and the demand placed on water supply by human-use systems. Drought should be considered relative to some long-term average condition of balance between precipitation and evapotranspiration.

Drought is not only a natural or physical event, but social as well. While we can do a little to alter drought occurrence, but social factors can do much to reduce drought-influenced vulnerability by several solutions, like prediction, monitoring, mitigation, preparedness, as shown on the figure 8 [57].

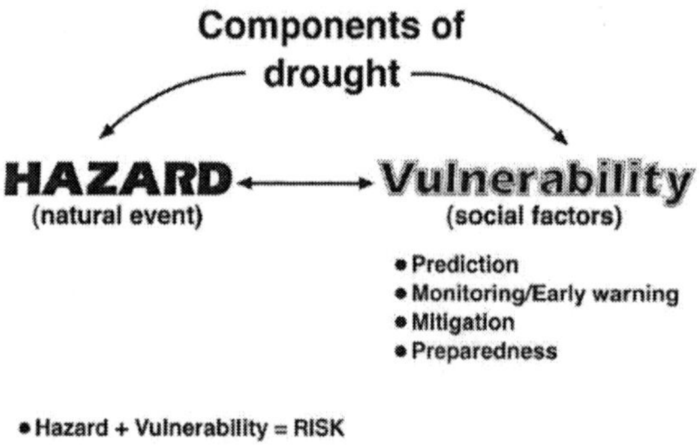

Figure 8. Components of drought.

Source: based on D. Wilhite, 'Drought as a natural hazard: Concepts and definitions', 2000.

Severe droughts in human-dominated environments, as experienced in recent decades in many parts of the world, cannot be seen as purely natural hazards. Anthropogenic changes to the land surface alter hydrological processes including evapotranspiration, infiltration, surface runoff, and storage of water and in this way affect the development of drought [67].

The sharp decreases in the PDSI and soil moisture from the late 1970s to the early 1990s mainly result from precipitation decreases in Africa and East Asia [6]. The recent drought in Africa is related to changes in the Atlantic and steady warming in the Indian Ocean. The warming in the Indian Ocean is likely related to recent global warming, which is largely attributed to human-induced GHG increases [59]. L. D. Rotstayn and U. Lohmann also suggested on their study that increased aerosol loading over the Northern Hemisphere may have played an important role in the recent drying over the Sahel and other tropical precipitation changes [68].

Traditionally, most definitions have viewed drought as a natural phenomenon and regarded human-made water shortage as a separate process. And the impacts of drought to socioeconomic and ecological systems are the result of this lack of water that is caused by the complex interaction of natural and anthropogenic processes [67]. In order to manage drought effectively in our profoundly human-influenced era, the Anthropocene, Jamie Hannaford suggested on her paper that we need to acknowledge that human influence is as integral to drought as natural climate variability [67].

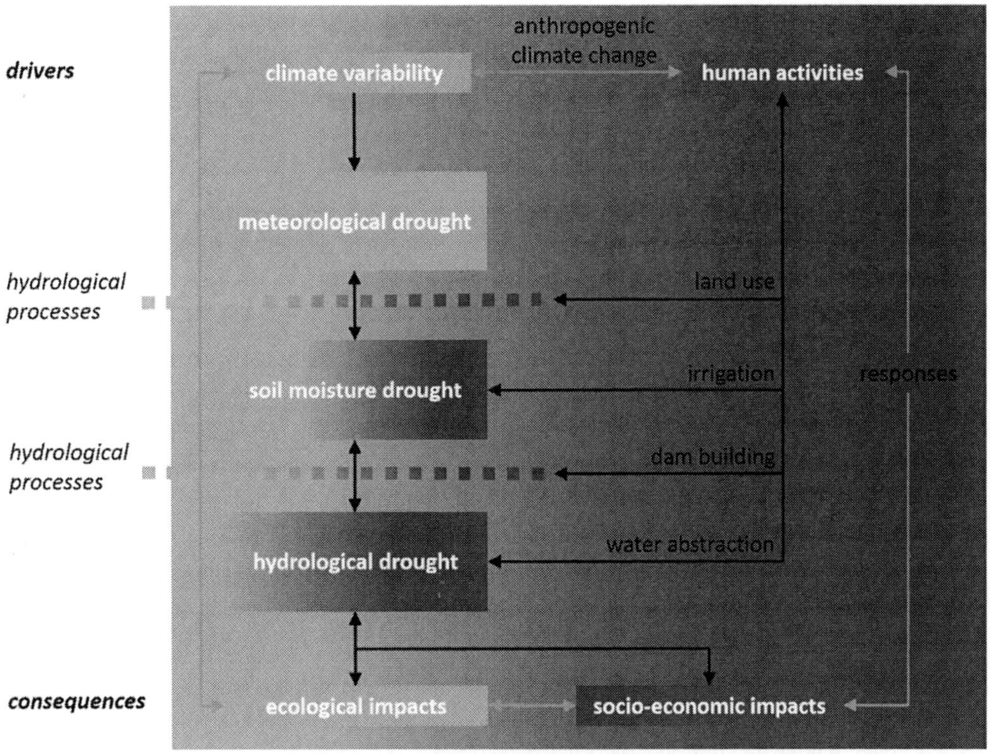

Figure 9. Drought propagation of Anthropocene.
Source: based on Jamie Hannaford, 'Drought in the Anthropocene', 2016

4.1.3. Drought - flood coexistence (DFC), urban and relationships

Many considers drought and flood are independent natural disasters, they still interrelate with each other, and exist in an interactive cycle. A cause in to drought as investigated by Ekwurzel, in the case occurred in Bonita Canyon in the Santa Catalina Mountains north of Tucson, Southwest of America, that during high temperature fires in the forest, organic debris in the top level soil vaporize, condense and block the soil pores to create a hydrophobic layer that reduces the permeability of the soil. This results in an increasing erosion and water runoff during a rainstorm [69]. Some studies before Ekwurzel's also suggested that accumulated fuel loads and drought created conditions favorable not just for the Aspen Fire, which happened in Bonita Canyon, but for post-fire flash flooding, and this situation is typical for most fires in the semi-arid Southwest. Flash flooding is always a danger in a desert. A drought, in turn, becomes a cause to flood.

In addition to Ekwurzel's analyses, Li also studied the occurrence of drought-flood coexistence in the middle and lower reaches of the Yangtze River [70]. The result shows that the frequency of DFC occurrence has increased remarkably in the past 50 years, especially since the 1980s. According to Li, there exists a significant correlation between DFC phenomenon and the intensity of East Asian Summer Monsoon, which could cause flood and drought in the same area.

The phenomenon exists in the central regions of Vietnam as well, where L. Sam defined four ecosystems spreads in a very thin strip along the complex topographic coastal zones, as indicated on the table 1 below [71]. The area suffers flash flood and storm annually, in addition to increasingly desertification.

Table 1 Four ecosystems in the central of Vietnam. Source: based on L. Sam,

Ecosystem of coastal sandy land	Ecosystem of central lowland	Ecosystem of central midland	Ecosystem of central highland
- Connect to the sea - No surface runoff - Sandy area	- Narrow area between sandy and midland - Flat, 5-100m high	- Large ecosystem, 35% of natural area. - Hilly, 100-500m high	- 40-45% of natural area - High mountains, belong to East Truong Son,

- Tough drought	- Affected by flash-flood down poured from highland - Tough drought, desertification due to storm water runs directly to the sea	- High precipitation, receives high runoff volume from highland - High water storage for the area	- High (> 500m high), slope, high precipitation - Water storage/reservoir for lower land

The discourse of typical DFC cases is unlimited across the world recently, whereas the world we are living now is not only in the era of extreme weather-induced hazards, but human-dominated era as well.

4.2. DFC, urban and relationship

Drought, flood and DFC have been identified not only natural events, but human-induced. Once water is hydrologically managed, drought could naturally be alleviated. However, the lack of freshwater in our current era is not only natural hazards, it is triggered by human interference as well, as it is resulted from extreme manipulation of urban aquatic systems and from changes of land-use patterns at large scale, particularly from land uses for urban settlements.

The study of Hualou Long about the correlation between drought-flood disasters and land use changes in West Jilin, China [72] has shown that the conversion rate of land use types (e.g. amongst forested land, grass land, wetland, and arable land) does not have much impact on the occurrence of drought-flood disasters, while the land use pattern changes do influence them. It is also proven in the research that the changes in landscape pattern affect floods more than droughts. Therefore, the land use pattern characteristics can be taken into account to mitigate the disaster or damages [72].

With holistic considerations of the two key objects (DFC, urban settlements), they interact with each other in environmental and ecological concerns. The analytical results will be a foundation for creating principles of resilient urban design.

In conclusion, by reviewing 29 studies and papers related to natural events of drought, flood, DFC, urban and their interrelationships, the multi-discipline literature ranged from earth science, climate, biology, hydrology, water resources, disasters, urban planning, etc., has a quite long history. The findings might not be collected adequately. However, the limited sources of literature indicate very remarkable results of natural extremes of drought, flood, DFC in their relations with human activities, particular in urban areas. From the oldest document prepared by W.C. Palmer in 1965, who suggested general methodology for evaluating the meteorological anomaly in terms of an index which permits time and scape comparisons of drought severity, there have not been many additional or supporting studies in the next two decades, until J. A. Dracup's 'On the definitions of droughts' in 1980. From 2000 to 2017, there have been numerous papers on the natural events. Since 2007, the problems of rapid urbanization and changing climates have been raised to an alarming level, that have taken great interests and attentions from scholars, that are shown on table 2 below.

Table 2. Synthesis of reviewed literature on Drought, Flood, DFC, Urban and their relationships
Source: author

No	Authors	Disciplines	Year	Findings
5	M. Roger et al.	Water Resources	2017	… fatal floods partly caused by land use change …
19	L. Alfieri et al.	Earth science	2017	… Natural drought events increased under global warming pressure … exacerbate them and it will likely expand their domain in the subtropical dry zone …
24	J. Hannaford, G. Di Baldassarre, A. J. Teuling, and L. M. Tallaksen	Hydrology Water Resources	2016	… Anthropogenic changes to the land surface alter hydrological processes: evapotranspiration, infiltration, surface runoff, and storage of water and affect the development of drought …
8	M. G. Miguez, A. P. Veról, M. M. De Sousa, and O. M. Rezende	Sustainability	2015	… significant impacts on developing countries …

4	J. Hall et al.	Hydrology	2014	… frequency of major floods in many places around the world increased …
7	W. Su, G. Ye, S. Yao, and G. Yang	Sustainability	2014	… Land use change is among the major threats leading to floods worldwide …
10	Z. W. Kundzewicz et al.	Hydrology	2014	… processes of urbanization lead to increased occupation of flood plains and, often, inadequate drainage planning, ... flooding events increased …
27	J. Li	Climate	2014	… exists a significant correlation between DFC phenomenon and the intensity of East Asian Summer Monsoon …
3	Z. W. Kundzewicz, I. Pińskwar, and G. R. Brakenridge	Hydrology	2013	
1	D. Watson and M. Adams	Urban Planning	2012	… two factors increase floods: 1) increased precipitation; and 2) changes in land use … Inland floods are the result of storm water runoff that exceeds the capacity of their stream and river systems …
2	B. Merz, S. Vorogushyn, S. Uhlemann, J. Delgado, and Y. Hundecha	Hydrology	2012	… river flooding is more frequent and severe … three groups of potential drivers: atmosphere, catchments, and the river systems
20	J. Sheffield, E. F. Wood, and M. L. Roderick,	Nature	2012	… little change in drought over the past 60 years …
29	L. Z. Wang, H. L. Long, H. Q. Liu, and G. H. Dong,	Disasters	2012	… the conversion rate of land use types … does not have much impact on the occurrence of drought-flood disasters ... … land use pattern changes do influence them
14	A. Dai	Climate	2011	
9	C. J. A. Bradshaw, N. S. Sodhi, K. S. H. Peh, and B. W. Brook	Biology	2007	… Land use change has, potentially, a very strong effect on floods as humans have heavily modified natural landscapes …
12	Y. Sun, S. Solomon, A. Dai, and R. W. Portman	Climate	2007	… arid areas are prone to drought because their rainfall amount critically depends on a few rainfall events …
16	IPCC		2007	Drought: 'prolonged absence or marked deficiency of precipitation', a 'deficiency of precipitation that results in water shortage for some activity or for some group' or a 'period of abnormally dry weather sufficiently prolonged for the lack of precipitation to cause a serious hydrological imbalance' ...
15	S. D. Schubert, M. J. Suarez, P. J. Pegion, R. D. Koster, and J. T. Bacmeister	Science	2004	… increased drought severity was caused by anomalous tropical sea surface temperatures and that interactions between the atmosphere and the land surface …
26	B. Ekwurzel,	Climate	2004	... drought causes fires in the forest … > flash flood ...
6	K. Tockner	Water Resources	2002	… natural flood plains are crucial drivers and the most threatened ecosystems …
25	L. D. Rotstayn and U. Lohmann	Climate	2002	… increased aerosol loading have played an important role in the recent drying in Northern Hemisphere …
13	D. A. Wilhite	Climate	2000	… Drought is a recurring extreme climate event over land characterized by below-normal precipitation over a period of months to years. Drought is a temporary dry period, in contrast to the permanent aridity in arid areas
11	I. McHarg	Environment	1992	… understand river system through history of geology and climate interact upon its basin,

				will know how to find suitable areas for developing cities …
17	J. A. Dracup, K. S. Lee, and E. G. Paulson	Water Resources	1980	… Hydrological droughts correlated with the effects of periods of precipitation shortfall on surface or subsurface water supply (i.e., streamflow, reservoir and lake levels, groundwater) rather than with precipitation shortfalls …
21	W. C. Palmer	Weather	1965	… Drought considered as a strictly meteorological phenomenon. … evaluated as a meteorological anomaly characterized by a prolonged and abnormal moisture deficiency …

5. Spatial planning resilient to DFC

As many other natural-caused disturbances like heat wave, drought, extreme precipitation, inland and coastal floods, landslides, cyclones, and other weather extremes, that have emerged over the last decades, DFC has been the most prevalent phenomenon. Although its impacts cannot be considered as disruptive as other disasters like storm surges, cyclones, tsunami, etc., but it has caused gradual impacts on both societies and ecosystems, that really need our proper attentions, including in urban governance realm, and planning discipline, on the way to urban resilience and sustainability.

In the notions of emerging ecological resilience, which have gained high attentions from most of professional disciplinary, spatial planning and urban design take a critically important role in designing and creating resilient cities [59]. Beside specified resilience (predictable and foreseen disturbances), defined for a concrete context, generic resilience (unknown and unpredictable threats) should be put at high consideration, in parallel with adaptive capacity [3].

Within the current climate awareness on spatial planning, it is a matter of course, raising it for DFC is also a critical task, and must be in accordance with the notions of resilience, includes: (i) considering urban regions as ecosystems, complex adaptive systems, that should be planned resiliently across scale, (ii) designing city or community adaptively, and (iii) accepting all unpredictable external and internal changes and disturbances.

5.1. Resilience in spatial planning

Urban area as a socio-ecological system, which is supported by ecosystem services, including: (i) provisioning services such as food and water; (ii) regulating services such as regulation of floods, drought, and disease; (iii) supporting services such as soil formation and nutrient cycling; and (iv) cultural services such as recreational, spiritual, religious and other nonmaterial benefits [73].

Regarding to indigenous approaches on city planning, South East Asia is quite an expertise in regulating water systems in urban areas. Chengde, China is a typical case [74], of which small-scale irrigation systems and terraced fields was originally created, with flood was fully controlled. However, in the current of coexistence of indigenous and modern techniques of water management, the distribution of water is highly imbalanced – both geographically and seasonally. The consequence, subsequently, is more frequent occurrences of drought and flooding events, and other social impacts. The reinvented waterscape urbanism in Chengde was developed to cope with new challenges of deteriorating ecosystems. The urban water landscape with a clear connection with the locally natural conditions, such as topography, hydrology and soils, fundamentally structures foundation of urban morphology. All urban water landscape becomes green infrastructures, which is less expensive and more ecological. More importantly, river basin boundary is considered for city physical planning and regulation.

In addition to the most prevalent concerns on natural events of flood and drought, which focus on how to mitigate situation at regional scale and to adapt at urban scale, a recent study from A. Buyantuyev has shown an interesting result. According to the study, urbanization of cities in arid environments can increase net primary production substantially [75], in the condition of introducing highly productive plant communities and weakening the coupling of plant growth to naturally occurring cycles of water and nutrients. The study stated that urbanization may not only increase regional net primary production and disrupt the coupling between vegetation and precipitation, but also increase spatial heterogeneity of NPP in this arid region.

In another research paper, Foster firstly reviewed and redefined resilience literature at regional scale as the ability of a region to anticipate, prepare for, respond to and recover from a disturbance, then proposed A Framework for Assessing Regional Resilience and a system of four stage criteria, including assessment, readiness, response, and recovery [76]. In addition to two key points of the study, he suggested another crucial element (i) consisting of two different connected types: preparation and performance, (ii) applies not only to a system, but to system elements (e.g. infrastructures, information, and physical environments), (iii) resilient on one element not necessarily resilient on another, and (iv) that resilience can be developed.

In regard to spatial planning, in particular with DFC, the service of regulating becomes the most critical one, whilst others become supporting factors. In urban ecosystems, the land used for natural conservation such as: forest, grassland, wetland, etc. take less attention than other economically-drawn beneficial land uses. To avoid deteriorating our ecosystems, such imbalanced and one-faceted thinking should definitely be put away from spatial planning. Additionally, to physically design a city resilient to DFC, defining river basin as a studying boundary is a must. Socio-economically, exploiting anthropogenic during urbanization, rather natural land covers, to increase net primary product.

5.2. *Adaptive capacity*

City as an ecosystem [29][77], and a complex adaptive system which consists of four basic environments. Adaptive solutions for city-scale resilience must follow natural flows from regional level and keep them interconnected (water, air, information, material, etc.), besides keeping ecology protected and intertwined in their mutual interactions.

The lesson from the Water Urbanism across scales in China, to solve the water crisis, is an integrative and adaptive solution. The key is to build an 'ecological infrastructure' based on water processes across scales, from the national and continental scale to the local and specific sites. Four ecosystem's services are accordingly provided in this water-based ecological infrastructure [42].

To achieve urban resilience, Siavash identified key urban planning and design principles and interventions that contribute to increasing as city's level of urban resilience [78]. A system of four key morphological elements consists of streets and spatial structures, green, blue and open space networks, building typologies and density, and patterns of uses. Whereas the first two elements (streets and spatial structures, green, blue networks and open spaces) she suggested on her article is kind of green infrastructures, which is quite similar to cross-scale ecological infrastructure project in China in solution. They, the street and overall spatial structures according to her consideration, are the most permanent components. Therefore, decisions relating to these structures should have long-term foresight, accounting for sufficient capacity and flexibility that can last for generations, and requiring adequate understanding of ecological knowledge.

In regarding to the concerns of DFC, based on a concrete context, adaptive solutions should be taken into account, aiming at achieving urban resilience.

5.3. *Prepare for uncertainties and unforeseen threats*

As defined by various scholars in ecological science, urban ecosystem going toward resilience could change its structure and functions, and lead to another stable state. Where fluctuations could be harsher or more disruptive, the system still exists and withstands with new equilibrium state. However, we really need ecological knowledge to accept all unknown threats and disturbances, and to adequately understand it as a process of four phases. The process is always moving as cycles, adaptive cycles. After the last cycle, another cycle will continue. After collapse, recovery is needed.

In short, 8 papers and researches have been collected and reviewed for the term spatial planning resilient to DFC as shown on the table 2. The figures show that there have not been many studies related to the spatial planning resilient to DFC. The findings are only limited from the year 2005 to 2014, which indicates that the topic is still open for coming studies.

Table 3. Synthesis of reviewed literature of spatial planning resilient to DFC. Sourse: author

No	Authors	Disciplines	Year	Key Findings/Concepts/Definitions
8	J. Ahern, S. Cilliers, and J. Niemelä,	Landscape/Urban Planning	2014	… A system of four key morphological elements consists of streets and spatial structures, green, blue and open space

				networks, building typologies and density, and patterns of uses …
2	P. Lu and D. Stead	Urban planning	2013	
7	S. T. Pickett,	Urban Ecology	2012	
5	A. Buyantuyev and J. Wu	Environment	2009	
4	Bruno De Mulder, Kelly Shannon	Urban planning	2008	… urban water landscape with a clear connection with the locally natural conditions, such as topography, hydrology and soils, fundamentally structures foundation of urban morphology …
1	IPCC		2007	
6	K. a. Foster,	Urban planning	2006	... ability of a region to anticipate, prepare for, respond to and recover from a disturbance ... ... four stage criteria, including assessment, readiness, response, and recovery ... (i) consisting of two different connected types: preparation and performance, (ii) applies not only to a system, but to system elements (e.g. infrastructures, information, and physical environments), (iii) resilient on one element not necessarily resilient on another, and (iv) that resilience can be developed.
3	A. K. Duraiappah and S. Naeem,	Biology	2005	

6. Conclusion

The world is a capsule [79], consisting of different ecosystems and being able to self-organize and actively adjust to cope with both internal and external changes and disturbances. City or community, according to ecological science, has been considered as an ecosystem as well [29], because of their dynamics and complexities in adaptive systems. Both are changing quickly in the pace of our rapid civilization over recent decades. Changes have currently not attributed to natural conditions, but have been caused by increasing human-induced activities. Global warming is one of the consequences of such activities. Some changes could be known and predictable, while others unforeseen and unpredictable. In addition to the changing problematic situation, increasing desertification across the world has become more and more severe [6]. All changes must be dealt with actively on their own adaptive cycles in order to create an environment designed and built to be resilient and sustainable.

As discussed, the two properties of resilience are in sharp contrast. The traditional equilibrium state presumes homogeneity, predictability, and inherent stability of ecosystems, focusing on keeping it at stasis. Meanwhile, the most contemporary resilience focuses on the multiple stable states with hierarchical patch dynamics paradigm, heterogeneity, nonlinearity, and concerning natural flows [80]. The ideas of the latter are not only essential in the theory of ecological resilience, but also crucial in applying them so we could understand and manage socio-ecological systems, including spatial planning and urban design.

Based on the utmost meaning of the term resilience, spatial planning and urban design professions must take crucial roles in the process to achieve success in building an ecologically resilient system. However, a deep consideration in ecological resilience is a must. Hence, our emphasis should be on creating and maintaining urban resilience – the ability of a city to persist without changing its structure and function when facing perturbations. Essentially, we could design urban regions or cities resilient to any changes, including coexistence of drought-flood, as mentioned earlier, only by considering them as complex socio-ecological systems with dynamics, cross-scale interactions, etc.

In the context of DFC within at regional scale, such as the central provinces of Vietnam, situating in specific characteristics of meteorology, geography, topography, etc., beside drought caused by local context, and worsen by desertification, it faces flash floods caused by deep slope in highland area, narrowness in lowland, in addition to high precipitation and n as well [71]. The areas suffer twice as hard. A possible solution is that the human systems' water supply could be regulated at the regional scale to avoid flash floods during rainy season and cope with drought during dry season [20]. The focal point of resilience perspective on spatial planning and urban design is proactively changing the situation of drought and flood, and possibly even dealing with the problem of rising sea level, instead of stasis – "to withstand change with adaptive change" [20]. Based on that theory, under carefully exploring local context at cross-scale interactions, all nature's services could be easily regulated and managed, resulting in urban resilience - a threshold toward to sustainable development.

References

[1] S. Meerow, J. P. Newell, and M. Stults, "Defining urban resilience: A review," *Landsc. Urban Plan.*, vol. 147, pp. 38–49, 2016.
[2] C. S. Holling, "Resilience and stability of ecological systems. Annual Review of Ecology and Systematics," 1973.
[3] P. Lu and D. Stead, "Understanding the notion of resilience in spatial planning: A case study of Rotterdam, The Netherlands," *Cities*, vol. 35, pp. 200–212, 2013.
[4] K. C. Seto and A. Reenberg, *Rethinking global land use in an urban era*, no. November. 2014.
[5] IPCC, "Climate Change, Adaptation, and Vulnerability," *Organ. Environ.*, vol. 24, no. March, pp. 1–44, 2014.
[6] A. Dai, "Drought under global warming: A review," *Wiley Interdiscip. Rev. Clim. Chang.*, vol. 2, no. 1, pp. 45–65, 2011.
[7] Belgian Development Cooperation, "Integrated Water Management and Urban Development in Relation to Climate Change in Ninh Thuan Province," *Tech. Financ. FILE.*
[8] L. Sam and Nguyen Dinh Vuong, "State of drought, desertification in Ninh Thuan Province - Causes and remediation solutions," *Vietnam Acad. Water Resour.*, pp. 45–52, 2008.
[9] S. Van Der Ryn and S. Cowan, *Ecological Design*. 2007.
[10] C. S. Holling, "Understanding the complexity of economic, ecological, and social systems," *Ecosystems*, vol. 4, no. 5. pp. 390–405, 2001.
[11] C. Folke, "Resilience: The emergence of a perspective for social-ecological systems analyses," *Glob. Environ. Chang.*, vol. 16, no. 3, pp. 253–267, 2006.
[12] B. Walker and D. Salt, *Resilience thinking: Sustaining Ecosystems and People in a Changing World*. 2006.
[13] M. G. Neubert and H. Caswell, "Alternatives to resilience for measuring the responses of ecological systems to perturbations," *Ecology*, vol. 78, no. 3, pp. 653–665, 1997.
[14] G. G. Mittelbach, A. M. Turner, D. J. Hall, J. E. Rettig, and C. W. Osenberg, "Perturbation and resilience: A long-term, whole-lake study of predator extinction and reintroduction," *Ecology*, vol. 76, no. 8, pp. 2347–2360, 1995.
[15] C. Holling and G. Meffe, "Command and control and the pathology of natural resource management," *Conserv. Biol.*, vol. 10, no. 2, pp. 328–337, 1996.
[16] C. S. Holling and L. H. Gunderson, "Resilience and adaptive cycles," *Panarchy Underst. Transform. Hum. Nat. Syst.*, pp. 25–62, 2002.
[17] C. S. Holling, "Engineering Resilience versus Ecological Resilience," *Eng. Within Ecol. Constraints*, no. 1996, pp. 31–44, 1996.
[18] L. H. Gunderson, "Ecological Resilience in Theory and Application," *Ecol. Syst.*, vol. 31, pp. 425–439, 2000.
[19] L. H. Gunderson, C. S. Holling, J. Pritchard Lowell, and G. D. Peterson, "Resilience of large-scale resource systems," *Resil. Behav. large-scale Syst.*, vol. 60, pp. 3–20, 2002.
[20] J. Wu and T. Wu, "Ecological Resilience as a Foudnation for Urban Design and Sustainability," in *Resilience in Ecology and Urban Design*, vol. 3, no. December 2013, 2013.
[21] J. Kauffman, "Advancing sustainability science: Report on the International Conference on Sustainability Science (ICSS) 2009," *Sustainability Science*, vol. 4, no. 2. pp. 233–242, 2009.
[22] C. Folke, S. Carpenter, T. Elmqvist, L. Gunderson, and B. Walker, "Resilience and Sustainable Development: Building Adaptive Capacity in a World of Transformations," vol. 31, no. 5, pp. 437–440, 2002.
[23] B. M. S.T.A Pickett, M.L. Cadenasso, "Ecology of the city as a Bridge to Urban Design," in *Resilience in Ecology and Urban Deisgn*, Springer, 2013, pp. 7–28.
[24] M. Alberti, J. M. Marzluff, E. Shulenberger, G. Bradley, C. Ryan, and C. Zumbrunnen, "Integrating humans into ecology: Opportunities and challenges for studying urban

ecosystems," in *Urban Ecology: An International Perspective on the Interaction Between Humans and Nature*, 2008, pp. 143–158.
[25] H. Sukopp, "Urban Ecology — Scientific and Practical Aspects," in *Urban Ecology*, 1998, pp. 3–16.
[26] S. T. A. Pickett *et al.*, "Urban ecological systems: Scientific foundations and a decade of progress," *J. Environ. Manage.*, vol. 92, no. 3, pp. 331–362, 2011.
[27] S. Pickett, M. Cadenasso, and B. McGrath, "Ecology of the City as a Bridge to Urban Design," in *Resilience in Ecology and Urban Design*, 2013, pp. 7–28.
[28] M. McDonnell, A. Hahs, and J. Breuste, *Ecology of cities and towns: a comparative approach*, vol. 129. 2009.
[29] S. T. Pickett, "Ecology of the City: a perspective from science," *Urban Des. Ecol.*, pp. 160–168, 2012.
[30] S. T. A. Pickett and M. L. Cadenasso, "The ecosystem as a multidimensional concept: Meaning, model, and metaphor," *Ecosystems*, vol. 5, no. 1, pp. 1–10, 2002.
[31] S. T. A. Pickett and J. M. Grove, "Urban ecosystems: What would Tansley do?," *Urban Ecosyst.*, vol. 12, no. 1, pp. 1–8, 2009.
[32] S. T. A. Pickett, M. L. Cadenasso, D. L. Childers, M. J. McDonnell, and W. Zhou, "Evolution and future of urban ecological science: ecology in, of, and for the city," *Ecosyst. Heal. Sustain.*, vol. 2, no. 7, p. e01229, 2016.
[33] I. McHarg, "The City: The Processes and Form," in *Design With Nature*, 1992, pp. 175–185.
[34] D. Mileti and a J. Henry, "Disasters by Design: A Reassessment of Natural Hazards in the United States," *Nat. Hazards*, pp. 1–372, 1999.
[35] G. Peterson, "The Adaptive Cycle and Resilience," 1999.
[36] Y. Jabareen, "Planning the resilient city: Concepts and strategies for coping with climate change and environmental risk," *Cities*, vol. 31, pp. 220–229, 2013.
[37] M. Wagner, J. Merson, and E. A. Wentz, "Design with Nature: Key lessons from McHarg's intrinsic suitability in the wake of Hurricane Sandy," *Landsc. Urban Plan.*, vol. 155, pp. 33–46, 2016.
[38] I. McHarg, "The River Basin," in *Design With Nature*, 1992, pp. 127–152.
[39] F. Molle, "Planning and managing water resource at the River-Basin Level: Emergence and evolution of a concept," 2006.
[40] W. Dramstad, J. D. Olson, and R. T. . Forman, "Landscape Ecology Principles in Landscape Architecture and Land-Use Planning." .
[41] R. Pitt, "Integrated Watershed Management," *World Environ. Water Resour. Congr. 2006*, pp. 1–10, 2006.
[42] Bruno De Meulder, Kelly Shannon, *Water Urbanims East*. Parks Books, 2013.
[43] I. McHarg, "Nature in the metropolis.," in *Design with Nature*, 1992, pp. 55–66.
[44] M. Mokarram and F. Aminzadeh, "GIS-based multicriteria land suitability evaluation using ordered weight averaging with fuzzy quantifier: a case study in Shavur Plain, Iran," *Jt. Int. Conf. Theory, Data Handl. Model. GeoSpatial Inf. Sci. 26-28 May 2010*, vol. 38, pp. 508–512, 2010.
[45] S. Jafari and N. Zaredar, "Land Suitability Analysis using Multi Attribute Decision Making Approach," *Int. J.*, vol. 1, no. 5, pp. 441–445, 2010.
[46] D. Watson and M. Adams, *Design for Flooding: Architecture, Landscape, and Urban Design for Resilience to Flooding and Climate Change*. John Wiley & Sons, Inc, 2012.
[47] B. Merz, S. Vorogushyn, S. Uhlemann, J. Delgado, and Y. Hundecha, "HESS Opinions: 'More efforts and scientific rigour are needed to attribute trends in flood time series,'" *Hydrol. Earth Syst. Sci.*, vol. 16, no. 5, pp. 1379–1387, 2012.
[48] Z. W. Kundzewicz, I. Pińskwar, and G. R. Brakenridge, "Large floods in Europe, 1985–2009," *Hydrol. Sci. J.*, vol. 58, no. 1, pp. 1–7, 2013.
[49] J. Hall *et al.*, "Understanding flood regime changes in Europe: A state-of-the-art assessment," *Hydrol. Earth Syst. Sci.*, vol. 18, no. 7, pp. 2735–2772, 2014.
[50] M. Rogger *et al.*, "Land use change impacts on floods at the catchment scale: Challenges and opportunities for future research," *Water Resouces Res.*, vol. 53, no. June 2013, pp. 5209–5219, 2017.
[51] K. Tockner, "Review of: Riverine Flood Plains: Present State and Future Trends," 2002.
[52] W. Su, G. Ye, S. Yao, and G. Yang, "Urban land pattern impacts on floods in a new district of

China," *Sustain.*, vol. 6, no. 10, pp. 6488–6508, 2014.
[53] M. G. Miguez, A. P. Veról, M. M. De Sousa, and O. M. Rezende, "Urban floods in lowlands-levee systems, unplanned urban growth and river restoration alternative: A case study in Brazil," *Sustain.*, vol. 7, no. 8, pp. 11068–11097, 2015.
[54] C. J. A. Bradshaw, N. S. Sodhi, K. S. H. Peh, and B. W. Brook, "Global evidence that deforestation amplifies flood risk and severity in the developing world," *Glob. Chang. Biol.*, vol. 13, no. 11, pp. 2379–2395, 2007.
[55] Z. W. Kundzewicz *et al.*, "Flood risk and climate change: global and regional perspectives," *Hydrol. Sci. J.*, vol. 59, no. 1, pp. 1–28, 2014.
[56] Y. Sun, S. Solomon, A. Dai, and R. W. Portmann, "How often will it rain?," *J. Clim.*, vol. 20, no. 19, pp. 4801–4818, 2007.
[57] D. A. Wilhite, "Drought as a natural hazard: Concepts and definitions," *Drought A Glob. Assess.*, pp. 3–18, 2000.
[58] S. D. Schubert, M. J. Suarez, P. J. Pegion, R. D. Koster, and J. T. Bacmeister, "On the Cause of the 1930s Dust Bowl," *Science (80-.).*, vol. 303, no. 5665, pp. 1855–1859, 2004.
[59] Ipcc, *Climate Change 2007: impacts, adaptation and vulnerability: contribution of Working Group II to the fourth assessment report of the Intergovernmental Panel*. 2007.
[60] J. A. Dracup, K. S. Lee, and E. G. Paulson, "On the definition of droughts," *Water Resour. Res.*, vol. 16, no. 2, pp. 297–302, 1980.
[61] V. Yevjevich, "An objective approach to definitions and investigations of continental hydrologic droughts," *Nature*, vol. 146, no. August, p. 1967, 1967.
[62] L. Alfieri *et al.*, "Global projections of river flood risk in a warmer world," *Earth's Futur.*, vol. 5, no. 2, pp. 171–182, 2017.
[63] J. Sheffield, E. F. Wood, and M. L. Roderick, "Little change in global drought over the past 60 years," *Nature*, vol. 491, no. 7424, pp. 435–438, 2012.
[64] W. C. Palmer, "Meteorological Drought," *U.S. Weather Bureau, Res. Pap. No. 45*. p. 58, 1965.
[65] G. Van Der Schrier, P. D. Jones, and K. R. Briffa, "The sensitivity of the PDSI to the Thornthwaite and Penman-Monteith parameterizations for potential evapotranspiration," *J. Geophys. Res. Atmos.*, vol. 116, no. 3, 2011.
[66] A. Dai, "Characteristics and trends in various forms of the Palmer Drought Severity Index during 1900-2008," *J. Geophys. Res. Atmos.*, vol. 116, no. 12, 2011.
[67] J. Hannaford, G. Di Baldassarre, A. J. Teuling, and L. M. Tallaksen, "Drought in the Anthropocene," pp. 1–5, 2016.
[68] L. D. Rotstayn and U. Lohmann, "Tropical rainfall trends and the indirect aerosol effect," *J. Clim.*, vol. 15, no. 15, pp. 2103–2116, 2002.
[69] B. Ekwurzel, "Flooding during a drought?," no. October, pp. 10–11, 2004.
[70] J. Li, "The summer drought-flood coexistence in the middle and lower reaches of the Yangtze River and analysis of its air-sea background ...," no. March, 2014.
[71] L. Sâm, N. V. Lân, and N. Đ. Vượng, "Phân vùng sinh thái, cơ sở khoa học để nghiên cứu xây dựng hệ thống hồ sinh thái ở miền trung," *Viện Khoa học Thủy Lợi Miền Nam*, no. 3, pp. 70–72, 2008.
[72] L. Z. Wang, H. L. Long, H. Q. Liu, and G. H. Dong, "Analysis of the Relationship between Drought-Flood Disasters and Land-Use Changes in West Jilin, China," *Disaster Adv.*, vol. 5, no. 4, pp. 652–658, 2012.
[73] A. K. Duraiappah and S. Naeem, "Millennium Ecosystem Assessment, 2005," *Ecosystems and Human Well-being: Biodiversity Synthesis*. p. 100, 2005.
[74] K. Shannon, *Water urbanisms*, vol. 1. 2008.
[75] A. Buyantuyev and J. Wu, "Urbanization alters spatiotemporal patterns of ecosystem primary production: A case study of the Phoenix metropolitan region, USA," *J. Arid Environ.*, vol. 73, no. 4–5, pp. 512–520, 2009.
[76] K. a. Foster, "A Case Study Approach to Understanding Regional Resilience," *Annu. Conf. Assoc. Coll. Sch. Plan.*, no. November 2006, pp. 1–45, 2006.
[77] J. Ahern, S. Cilliers, and J. Niemelä, "The concept of ecosystem services in adaptive urban planning and design: A framework for supporting innovation," *Landsc. Urban Plan.*, vol. 125, pp. 254–259, 2014.
[78] Y. S. Siavash, "Achieving Urban Resilience: Through Urban Design and Planning Principles," no. September 2016, 2017.

[79] I. McHarg, *Design with nature*. New York, USA: John Wiley & Sons, Inc, 1992.
[80] S. T. A. Pickett, V. T. Parker, and P. L. Fiedler, "The new paradigm in ecology: implications for conservation biology above the species level," in *Conservation Biology The Theory and Practice of Nature Conservation Preservation and Management*, 1992, pp. 66–88.

Acknowledgments
Authors wishing to acknowledge the assistance and encouragement from our colleagues in Hochiminh City Institute for Development Studies, in Hochiminh City University of Technology, especially from the Faculty of Civil Engineering and the Department of Architecture.

Investigating Proenvironmental Behavior: The Case of Commuting Mode Choice

Tu Anh Trinh[1], Thi Phuong Linh Le[2]

[1]Urban and Regional Planning Department, Faculty of Civil Engineering, Ton Duc Thang University, 19 Nguyen Huu Tho, Tan Phong Ward, District No 7, Hochiminh city, Vietnam. Email: trinhtuanh@tdt.edu.vn
[2]Faculty of Airport, Vietnam Aviation Academy, 104 Nguyen Van Troi, Phu Nhuan District, Hochiminh city, Vietnam

Abstract. The central aim of this article is to investigate mode choice behavior among commuters in Ho Chi Minh City using disaggregate mode choice model and norm activation theory. A better understanding of commuters' choice of transport mode provide an opportunity to obtain valuable information on their travel behaviors which help to build a basic for proffering solutions stimulating commuters to switch to public transport, which in turn contribute to deal with traffic problems and environmental issues. Binary logistic regression was employed under disaggregate choice method. Key findings indicated that Demographic factors including Age (-0.308), Married (-9.089), Weather (-8.272); Trip factors including Travel cost (0.437), Travel distance (0.252), and Norm activation theory (Awareness of consequences: AC2 (-1.699), AC4 (2.951), AC6 (- 3.523), AC7 (- 2.092), AC9 (- 3.045), AC11 (+ 2.939), and Personal norms: PN2 (-2.695)) had strong impact on the commuters' mode choice. Although motorcycle was the major transport mode among commuters, they presented their willingness to switch to bus transport if it had less negative impacts on the environment and their daily living environment.

1. Introduction

It is true that many environmental problems nowadays are the consequences of people behaviors. Transportation is environmentally significant behavior that should be changed in an environmental friendly direction. In Ho Chi Minh City (HCMC), the largest motorcycle dependent city in Southeast Asia with the population of more than 7,955 millions (2014), the overflowing of commuters traveling by motorcycle causes the high pressures on urban environment and urban transportation. 78% of total means of transport in HCMC is motorcycle, and the number of motorcycle increases 10% per year with around 675 motorcycles per 1.000 people. Although the public transport system has been improved, it has not kept pace with the growth of population and private vehicles there. Traveling by motorcycle is obviously flexible and convenience but has substantial negative consequences in terms of environment like such as air pollution, noise, high consumption of nonrenewable energy, and transportation like motorcycle accident, and traffic jam. Recent researches have found that total loss (e.g. fuel consumption, accidents, waste of time due to traffic jams) caused by motorcycle in HCMC is around 1.07 billion USD/year which is even higher than its GDP annual growth rate around 10%. Furthermore, 70% of urban air pollution is derived from motorcycle's exhaust fumes which are placed on alert.

Can travel mode choice become a proenvironmental behavior? [1] stated that travel mode choice is "among the most environmentally significant decisions faced by individuals". In daily life, commuters with different demographic and trip characteristics repeatedly face the selection of travel modes where their decisions lead to positive outcomes for themselves and negative consequences for the environment, or vice versa. To encourage proenvironmental behavior among commuters, a better understanding of psychological factor that influence their willingness to act in a proenvironmental manner is crucial.

Based on these facts expressed above, the main purpose of this article is to investigate model of the effects of demographic factor, trip factor and psychological factor on commuters' mode choice behavior in HCMC, Vietnam. In this study, the research model was derived from previous research, especially mode choice behavior was investigated by employing disaggregate model and psychological factor was originated in [2] norm activation theory (NAT).

2. Literature review

In reality, the exhaust from our automobiles is the number one cause of local air pollution and global climate change, so there could be environmental benefits if road users switch their transport modes, or simply traveled less [3]. In many cases, some types of journeys like leisure trip, educational travel might be reduced for some people, however commuting is one kind of travel that is more often than not unavoidable. Commuting mode choice have been examined in several studies [3] [4] [5] and in this work, it was also the main subject. In environmentally significant behavior research, [2] norm activation theory (NAT) was concerned as a type of classic psychological model [5] and it has been employed in several environemental behavior-related researches. Originally, NAT was developed to explain altruistic behavior, but it has been extended to proenvironmental behavior [6]. Both commuting mode choice behavior and NAT were the theoretical foundation of this work. This article attempted to investigate commuters' travel mode choice behavior by employing a disaggregate model containing variables from NAT. A brief description of choice model and NAT was presented below.

2.1. Disaggregate mode choice model

Disaggregate mode choice model is the subject in many transportation researches. In order to predict the preferences of travelers, a number of mode choice models have been developed; however, disaggregate mode choice model (or discrete choice model) based on utility is widely employed to predict an individual's choice [7]. The disaggregate mode choice model, whose benefits and shortcomings are studied and documented [8], are the second generation of modeling strategy [9] and it was capable of forecasting an individual's behavior in choosing a mode from various choice of available modes [10]. In fact, many researchers have applied disaggregate model to study the mode choice behavior [10] [11] [12]. Also, a wide range of variables impacting on mode choice behavior were considered from demographic factor (e.g. age, gender, education, occupation, family size, number of family workers, income level), trip characteristics (e.g. travel time, travel cost, travel distance) to personality traits (e.g. habit, attitudes, social norms, satisfaction, calm factor) [13] [14] [15] [16].

2.2. Schwartz's (1977) norm activation theory

Norm activation theory (NAT) developed by [2] explains altruistic and proenvironmental behavior. The theory holds that normative self-expectations experienced as feelings of obligation (personal norms, or PN) are the antecedent of altruistic acts and that personal norms are 'activated in individuals who believe that environmental conditions pose threats to other people, other species (awareness of consequences, or AC) and that actions they initiate could avert those consequences (ascription of responsibility to self, or AR)' [17]. AC and responsibility moderate PN's influence on behavior (Figure 1).

NAT has been applied to proenvironmental behavior with some successes [17] [18]. In the filed of transportation, NAT has also been employed and investigated thoroughly. In 1999, [19] added 'Personal norm' (PN) of NAT to the TPB model (Theory of planned behavior) to explain proenvironmental intentions for non-car travel modes. His regression analysis proved that R square increased after adding PN. In another research, NAT along with general environmental values were utilized to test a hierarchical model on general proenvironmental behavior. A Swedish sample of 1,400 individuals were used in a path analysis to test the model. The outcomes showed that 'the PN was derived from self-transcendent and ecocentric values; PN was activated by problem awareness and PN mediated the effects from general values, environmental values, and problem awareness on proenvironmental behavior'. For the purpose of examining the mode choice among students going to university, [5] also used German students' data to compare NAT, theory of planned behavior (TPB), and theory of interpersonal behavior (TIB).

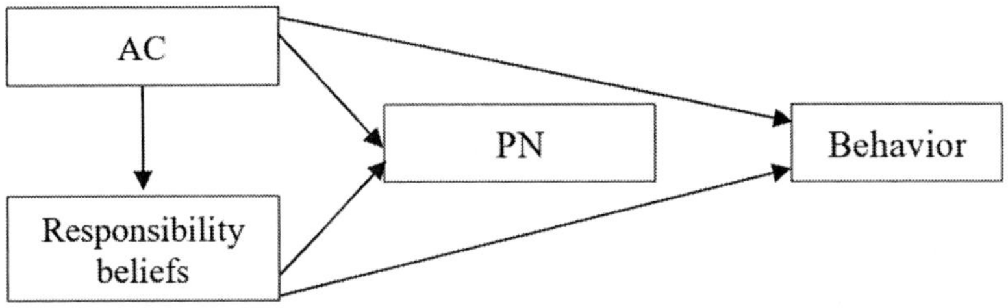

Note: AC = awareness of a behavior's consequences; PN = personal norm

Figure1. Norm activation theory [2]

3. Methods

[5] stated that the theories 'were developed in different research contexts and focus on different aspects of social behavior'. In this research, to examine commuters' travel mode choice behavior, research model included 'Demographic factor', 'Trip factor', and 'Psychological factor'. 'Demographic factor' and 'Trip factor' was derived from previous studies stated in literature review, and 'Psychological factor' was resulted from NAT. A mode choice model with 'Psychological factor' was tested based on NAT rather than using [2] exact formulation. NAT items which were adjusted to the current research context were relied on [2] definitions and previous environmentally significant behavior research [2] [5] (Appendix A).

We obtained the data for this study by Stated Preference (SP). Questionnaire was used for two modes of transports including motorcycle and public transport (bus transport). The survey was conducted in Ho Chi Minh City from Dec 2016 to Jan 2017. Both household survey and online survey were employed for data collection. Regarding household survey, we conducted the survey particularly in District 1, 3, 10, Phu Nhuan, Tan Binh and Binh Thanh which are the central urban areas of the city. The survey was piloted to make sure that the measuring instrument represented the goal of the study in terms of the information to be gathered [20]. Random samples of 20 observations attended the pilot test and the outcomes revealed that few questions needed to be modified. Hence, the questionnaire was amended based on the pilot test and was used for collecting the actual data for this study. A total of 407 respondents were interviewed in a period of three months. The number of valid samples for research was 345.

According to the central aim of this study, the binary logistic regression model under disaggregate choice methods was utilized for the purpose of determining the factors which would impact motorcycle riders to switch from traveling by motorcycle to selecting bus transportation. Generally, for the binary models, i and j are the two alternatives in the choice set of each individual:

$$U_{in} = V_{in} + \varepsilon_{in}$$
$$U_{jn} = V_{jn} + \varepsilon_{jn}$$

The probability that individual n chooses alternative i (P_{in}) is: $P_{in} = \dfrac{e^{V_{in}}}{e^{V_{in}} + e^{V_{jn}}}$, where V_{in} = the utility of alternative mode i to individual n.

In this model, the dependent variable was set to "1", if the commuters traveled by motorcycle and "0" for using bus transportation. The probability that an individual would choose the motorcycle could be written as:

$$P_{motorcycle} = \frac{e^{Vin}}{e^{Vin} + e^{Vjn}} = \frac{e^{\beta x_{motorcycle}}}{e^{\beta x_{motorcycle}} + e^{\beta x_{bustransportation}}}$$

where:

$$V_{motorcycle} = \beta_0 + \beta_1 X_{Gender} + \beta_2 X_{Age} + \beta_3 X_{Married} + \beta_3 X_{Children} + \beta_4 X_{Education}$$
$$+ \beta_5 X_{Motor\_owner} + \beta_6 X_{Income} + \beta_7 X_{Travel\_time} + \beta_8 X_{Travel\_distance} + \beta_9 X_{Travel\_cost} + \beta_{10} X_{AC} + \beta_{11} X_{PN}$$

The coefficients were approximated by fitting the data to the model (s). The maximum likelihood estimation technique was employed to optimize the probability produced by the observed individuals. Besides, Principal components analysis (PCA) was used to ensure that NAT's items were sufficiently correlated.

4. Results

4.1 Sample analysis

Motorcycle riders were significantly more likely to be young people, starting own family and having children. There was no significant motorcycle ownership difference between motorcycle rider (more than 92% of respondents having motorcycle), but there were differences in gender (62.5% female having a mean age of 29.3 years, and a mean family income of $315/month), and job type (58.7% of respondents were teacher/researcher, 34.9% were collar workers, and the remaining was blue worker). The number of commuter travelling by motorcycle accounted for 68.8%.

4.2 Reliability test

The assessment of scale reliability for NAT's variable was based on Cronbach's Alpha. Cronbach's alpha is a measure of internal consistency which indicates how closely related a set of items are as a group. Here, the result showed that the alpha coefficient for items (AC1, AC2, AC4, AC6, AC7, AC8, AC9, AC10, AC11, and PN1, PN2, PN3) were ≥0.6 and their correlated variables were >0.3, suggesting that the items had relatively high internal consistency. The item (AC3, AC5, and PN4) was rejected.

4.3 Principal component analysis

Next, to develop scales, NAT's items excluding AC3, AC5 and PN4 (Appendix A) were examined in a principal component analysis (PCA). The PCA is boot-strapped with 500 replications to construct bias-corrected confidence intervals for the eigenvalues [21] [22]. These value of confidence intervals will help to determine the number of factors. The principal factors analysis is utilized to identify items loading on a particular factor. If all items loading is above the value of 0.4 on a factor, they will be a part of the factor and will be included in scales constructed from that factor. In this article, we tested two variables from norm activation theory: 'Personal norms (PN)' and 'Awareness of consequences (AC)'. The result shows all item loadings is greater than 0.400.

4.4 Binary Logistic Regression Model

To investigate commuter mode choice, the binary logistic regression model was fitted to the data. The maximum likelihood estimates of binary logistic regression model are condensed in Table 2. Tested Model was estimated with 12 variables of NAT and 10 demographic variables. The coefficients for all the variables were estimated relative to the selected reference category (Table 1).

Table 1. Estimations from the binary logistic model

Model	B	S.E.	Sig.	Exp(B)	95% C.I.for EXP(B)	
					Lower	Upper
Gender (1)	-0.225	0.776	0.772	0.798	0.174	3.656
Age	-0.308	0.154	0.045	0.735	0.544	0.993
Job (1)	0.859	1.185	0.469	2.360	0.231	24.096

Job (2)	-0.293	1.185	0.804	0.746	0.073	7.601
Married (1)	-9.089	2.203	0.000	0.000	0.000	0.008
Income	-0.307	0.168	0.068	0.736	0.529	1.023
Weather (1)	-8.272	2.129	0.000	0.000	0.000	0.017
Travel time	0.015	0.018	0.389	1.015	0.981	1.052
Travel cost	0.437	0.102	0.000	1.548	1.268	1.890
Travel distance	0.252	0.089	0.005	1.287	1.081	1.532
Motorcycle ownership (1)	-1.497	0.863	0.083	0.224	0.041	1.214
AC1 (1)	-0.967	1.283	0.451	0.380	0.031	4.701
AC2 (1)	-1.699	0.799	0.034	0.183	0.038	0.876
AC4 (1)	2.951	1.111	0.008	19.122	2.165	168.874
AC6 (1)	-3.523	0.958	0.000	0.029	0.005	0.193
AC7 (1)	-2.092	0.843	0.013	0.123	0.024	0.644
AC8 (1)	0.131	1.561	0.933	1.140	0.053	24.298
AC9 (1)	-3.045	1.123	0.007	0.048	0.005	0.430
AC10 (1)	0.223	1.450	0.878	1.249	0.073	21.407
AC11 (1)	2.939	0.795	0.000	18.904	3.980	89.787
PN1 (1)	0.883	1.031	0.391	2.419	0.321	18.229
PN2 (1)	-2.695	0.868	0.002	0.068	0.012	0.370
PN3 (1)	1.914	1.401	0.172	6.783	0.436	105.566
Constant	8.776	4.239	0.038	6473.832		
-2 Log likelihood	73.094					
Model Chi-square	319.826, df 23, sig.=0.000					
Cox & Snell R^2	0.637					
Nagelkerke R^2	0.895					
N	317					

Reference category: bus transportation

$Ln = \frac{P}{1-P}$ = - 0.308 Age - 9.089 Married - 8.272 Weather + 0.437 Travel cost + 0.252 Travel distance -1.699 AC2 + 2.951 AC4 - 3.523 AC6 - 2.092 AC7 - 3.045 AC9 + 2.939 AC11 - 2.695 PN2

Chi-square omnibus tests of coefficients of the research model showing the value of 319.826, df 23, and sig.=0.000 confirmed that a test of the full model against a constant only model was statistically significant. The -2Log likelihood reflected the prediction deviation by the model. A smaller value indicated a better fit. With regard to model, the -2 Log Likelihood statistics was quite small (73.094). Cox and Snell's R^2 =0.637, and a Nagelkerke R^2 = 0.895, which explained 89.5% of the variation in the dependent variable. From the binary logistic regression model, the coefficient of variables had positive values meaning that motorcycle riders would not be likely to switch to bus transportation.

5. Discussion

5.1. Demographic variables significantly explained the mode choice behavior among commuters

The age had a significant contribution to explain commuter's mode choice behavior. The sign of the coefficient (-0.308) proved that old people tended to use their motorcycle than public transport. The Married coefficients for the motorcycle use was negative (-9.089), meaning that the couple was likely to use bus transport for their travelling rather than the single. Weather (-8272) affected the commuter's mode choice. If the weather condition was not good (e.g. flooding, heavy rain, storm), respondent was willing to switch to bus transportation.

With respect to cost, generally, cost is considered as one of the main factors affecting intercity mode choice (Manssour et al., 2014). Here in this article, travel cost (0.437) had strong impact on the travel decision of commuter. When travel cost increased and other variable was still unchanged, motorcycle riders did not choose bus service for their daily movement. Similarly, if travel distance (0.252) also increased, motorcycle riders would not willing to change to bus transport. The sign of both variables

(Travel cost, travel distance) was not as negative as expected. In Vietnam, despite the inexpensive of the travel cost of bus transportation (around 5000 - 10.000 VND/one ticket), the quality of bus service is quite low. Also, compared with motorcycle, bus transport does not have the advantages of the flexibility, the convenience, and it is mainly the means of transport for people who has low income and their travel distance less than 10 kilometers. It is these barriers that deter people who earn high income, considering time seriously from using bus service.

5.2. Norm activation theory provide useful insights into proenvironmental behaviors among commuters

5.2.1. Awareness of consequences

67% of respondents believed that 'flood water inner city is caused by motorcycle use' (AC2) and they expressed their willingness to change their transport mode (-1.699). In fact, flood is a major problems in HCMC in every rainny reasons and obviously, motorcycle is not a suitable means of transport for travelling when happening flood. Also, flood water inner city is the cause of many risks including motorcyle accident and traffic congestion. Hence, public transport seem to be much safer and convenient for commuting trips.

One of the solutions to tackle with traffic congestion and the growth of the number of motorcycles is building new roads or widening the old ones. In HCMC, the clearance of residential areas inner city faces various obstacles, therefore, many green space (e.g. green parks, ancient trees) has to yield to new big roads. Although, around 81% of respondents agreed that 'constructing new roads for the increasing number of motorcycle threatens the biosphere in my city' (AC4), however, respondents did not agreed to change their means of transport (2.951). In this research, it seemed to be dilemma for motorcycle riders whether to keep traveling by motorcycle or switch to bus service. Despite the degradation of biosphere due to the sacrifice of green space, widening roads would help to avoid many daily traffic problems, which was likely to encourage riders to use motorcycle.

Factors influencing motorcycle riders' quality of life directly affected their mode choice decisions. The findings showed that 'Motorcycle-related deteriorated biosphere reduces the quality of life in our cities (-3.523)', and 'Flood water inner city reduces the quality of life in our cities (-2.092)' were statistically significant with respect to commuter's choice of the public transport mode. Motorcycle is the main cause of polluted air in HCMC. The average emissions of motorcycle in rush hour or in traffic jam is too high compared with other vehicles making riders annoyed and also stimulating them to use bus transportation service.

In terms of the statement 'My transport choices can have an impact on the environment' (- 3.045), around 53% of respondents did not feel that their choices would be directly harmful to environment. They argued that their choices were just the intention and it only affected the environment when they decided to implement their decisions in reality. Consequence, they were not likely to switch to bus transportation. Then, over 80% of motorcycle riders confirmed that 'I contribute to pollution by commuting by motorcycle use' and expressing their positive attitudes towards switching to public transport use (2.939).

5.2.2. Personal norms

From their awareness about the impacts of motorcycle use on environment and the quality of their daily lives, commuters showed that they 'feel morally obliged to avoid using motorcycle to get to work place. The positive sign of the variable (- 2.695) implied that commuters were willing to use bus transportation as an expression of their moral responsibilities towards their own living environment.

6. Conclusion

The main purpose of this article is to investigate mode choice behavior among commuters in Ho Chi Minh City using disaggregate mode choice model and norm activation theory. The study showed that demographic factors (age, married), trip characteristics (weather, travel cost, travel distance) and NAT (awareness of consequences, personal norms) had strong impact on the commuters' mode choice.

Although bus service is crucial in urban transportation, since it affects all citizens and many social problems like reducing traffic jams, either improving energy efficiency or the environment, and mitigating private vehicles in urban areas, motorcycle is still the highest priority among commuters. This can be explained by the noticeably-deteriorated bus service quality (e.g. rude staff, low-quality buses, less safety, careless bus drivers and ticket controllers), which have hugely affected the quality perceived by the road users, created user resentment and encouraged them to travel by motorcycle. It is possible to encourage more commuters, particularly who are married and old people, to use bus transport when bus service quality is improved. Determinants like travel cost encourage/discourage people to choose transport modes, therefore, attractive travel cost (e.g. discounted ticket) need to be provided. Recently, the bus service is improved with professional training for bus drivers and ticket controllers, the availability of seats, air-conditioners and wastebaskets on board, communication like bus route information, and schedule timetable. However, it should be noted that just an improvement of the bus service will not strongly impact on the commuters' mode choice decision. Improving bus transportation needs to be incorporated with the development of other public services including road infrastructure, urban planning and the development of alternatives such as bus rapid transit system and Metro lines in order to reduce travel distance and achieve sustainable transportation development as well.

The findings also revealed that NAT better explain commuter's mode choice behavior. Here in this research, commuters revealed their awareness on environment problems and their perception on the consequences of motorcycle uses to environment, their quality of lives in terms of deteriorated biosphere, flood water. Although motorcycle is the major means of transport in Vietnam, motorcycle rider expressed their willingness to switch to bus service or other alternatives provided that it would help to protect the environment and increase their quality of life.

Appendix

Appendix A. Variable Definitions

	Variable	**Definition**
Demographic factor	Commute mode choice	Category variable, Motorcycle = 1, Public transport = 0
	Gender	Category variable, Female = 1, male = 0
	Age	Continuous variable, In years
	Married	Category variable, Married = 1, single = 0
	Children	Category variable, Having children = 1, Not having children = 0
	Job	Category variable, 1=Teacher/researcher, 2=Collar worker, 3=Blue worker
	Motorcycle ownership	Category variable, Having Motorcycle = 1, Not having motorcycle = 0
	Income	Continuous variable, millions VND/month
Trip factor	Travel time	Continuous variable, minutes
	Travel distance	Category variable, kilometers
	Travel cost	Category variable, thousands VND/one way
NAT	AC	AC1 - I believe that climate change is caused by motorcycle use. Agree/Disagree
		AC2 - I believe that flood water inner city is caused by motorcycle use. Agree/Disagree
		AC3 - I believe that dust pollution is caused by motorcycle use. Agree/Disagree
		AC4 - I believe that constructing new roads for the increasing number of motorcycle threatens the biosphere in my city. Agree/Disagree
		AC5 - Dust pollution reduces the quality of life in our cities. Agree/Disagree

	Variable	Definition
		AC6 - Motorcycle-related deteriorated biosphere reduces the quality of life in our cities. Agree/Disagree
		AC7 - Flood inner city reduces the quality of life in our cities. Agree/Disagree
		AC8 - Avoiding motorcycle use will help to solve wider environmental problems. Agree/Disagree
		AC9 - My transport choices can have an impact on the environment. Agree/Disagree
		AC10 - I can help to solve my town/city's transport problems by avoiding motorcycle use. Agree/Disagree
		AC11- I contribute to environmental pollution by commuting by motorcycle use. Agree/Disagree
	PN	PN1 - I feel personal responsibility for helping to solve my city's transport problems. Agree/Disagree
		PN2 - I feel morally obliged to avoid using motorcycle to get to work place. Agree/Disagree
		PN3 - I don't feel any personal responsibility for causing my city's traffic and environmental problems
		PN4 - People like me should feel morally obliged to avoid using motorcycle to get to work place

References

[1] Collins, C., & Chambers, S 2005 Psychological and situational influences on commuter-transport-mode choice *Environment and Behavior 37*(5) 640-661

[2] Schwartz, S 1977 Normative influences on altruism. In L. Berkowitz (Ed.), *Advances in experimental social psychology* 10 New York: Academic Press

[3] Department for Transport 2004 *The future of transport: A network for 2030* UK Government transport White Paper

[4] Steg, L. 2005 Car use: Lust and must. Instrumental, symbolic and affective motives for car use *Transportation Research Part A 39*(2-3) 147-162

[5] Bamberg, S., & Schmidt, P 2003 Incentives, morality, or habit? Predicting students' car use for university routes with the models of Ajzen, Schwartz and Triandis *Environment and Behavior 35*(2) 264-285

[6] Guagnano, Gregory A., Paul C. Stern and Thomas Dietz 1995 Influences on attitude-behavior relationships: A natural experiment with curb- side recycling *Environment Behavior* 27 699-718

[7] Ben-Akiva, M. and S.R. Lerman 1985 *Discrete Choice Analysis* MIT Cambridge Mass

[8] Watson, P.L 1972 Predictions of intercity model choice from disaggregate behavioral and stochastic models *Highway Res. Record* 446: 28-35

[9] Koppelman, F., G. Kuah and M. Hirsh 1984 *Review of Intercity Passenger Travel Demand Modeling: Mid-60's to Mid-80's* The Transportation Center and Development of Civil Engineering, Northwestern University, Evanston, IL, pp: 136

[10] Manssour A. Abdulsalam Bin Miskeen, Ahmed Mohamed Alhodairi 2014 Modeling of Intercity Travel Mode Choice Behavior for Non-Business Trips within Libya *Research Journal of Applied Sciences* Engineering and Technology 7(3): 442-453

[11] Ashiabor, S, H Baik and A Trani 2007a Logit Models to Forecast Nationwide Intercity Travel Demand in the United States *Transportation Research Record: Journal of the Transportation Research Board* No. 2007 Transportation Research Board of the National Academies, Washington, D.C., pp: 1-12

[12] Bhat, C.R., 1997 An Endogenous segmentation mode choice model with an application to intercity travel *Transport. Sci* 31(1): 34-47

[13] Tim Schwanen, Patricia L. Mokhtarian 2005 What affects commute mode choice: neighborhood physical structure or preferences toward neighborhoods? *Journal of Transport Geography* 13 83–99

[14] Klockner, C.A., Friedrichsmeier, T. 2011 A multi-level approach to travel mode choice – how person characteristics and situation specific aspects determine car use in a student sample. *Transportation Research Part F – Traffic Psychology and Behaviour* 14 261–277

[15] Zhou, J.P 2012 Sustainable commute in a car-dominant city: factors affecting alternative mode choices among university students *Transportation Research Part A – Policy and Practice* 46 1013–1029.

[16] Martin, S., S. Lee, and R. Lowry 2007 National Prevalence and Correlates of Walking and Bicycling to School *American Journal of Preventive Medicine* 33 (2) 98-105

[17] Stern, P. C., Dietz, T., & Guagnano, G. A 1995 The new ecological paradigm in social-psychological context *Environment & Behavior 27* 723-753

[18] Stern, P. C., Dietz, T., & Kalof, L 1993 Value orientations, gender, and environmental concern. *Environment & Behavior 25* 322-348

[19] Harland, P., Staats, H., & Wilke, H. 1999 Explaining proenvironmental intention and behavior by personal norms and the theory of planned behavior *Journal of Applied Social Psychology 29*(12), 2505-2528

[20] Aaker, D A, Kumar, V, & Day, G S 2001 *Marketing Research* (7th ed.) New York, John Wiley and Son Inc

[21] Hall, P 1988 Theoretical comparisons of bootstrap confidence intervals *The Annals of Statistics* 16 927-953

[22] Hamilton, L. C 1992 *Regression with Graphics: A Second Course in Applied Statistics* Belmont, California: Duxbury Press

Sustainable Strategies for the Dynamic Equilibrium of the Urban Stream, Cheonggyecheon

D Seo[1] and Y Kwon[1,*]

[1] Department of Civil and Environmental Engineering, Seoul National University, 1 Gwanak-ro, Gwanak-gu, Seoul 08826, South Korea
[*] Correspondence: yskwon@snu.ac.kr

Abstract. Cheonggyecheon, which had been transformed into a 14-lane urban highway and a large underground sewer system, was finally converted back to an urban stream again. Its transformation has been praised as a successful example of urban downtown regeneration and beautification. It is, however, obvious that there have not been prudent ecological considerations since the project's principal goals were to provide public recreational use and achieve maximum flood control capacity via the use of embankments. For a healthier and sustainable stream environment, Cheonggyecheon should be ecologically re-restored again, based on a dynamic equilibrium model. It must primarily establish a corridor of vegetation, an aquatic transitional zone, and install constructed wetlands nearby which support the water source. The upper streams of Cheonggyecheon should be further restored and supply natural waters. Furthermore, there ultimately needs to be de-channelization for hydrological sustainability. This would vary from merely increasing the sinuosity to thoroughly reconstruct a naturalized stream. Complete dynamic equilibrium of Cheonggyecheon can be accomplished through more fundamental sustainable strategies.

1. Introduction

The Cheonggyecheon restoration project of downtown Seoul, South Korea, was completed in 2005. "Cheonggye" indicates a geographic name of the stream and "cheon 천 (川)" means a stream in Korean. Korean normally calls the stream Cheonggyecheon [1]. We will call the Cheonggyecheon restoration project "CRP" briefly in this paper. CRP has been spotlighted as an example of successful downtown redevelopment and beautification through the stream restoration. The stream, which was covered by a multilayer infrastructure for vehicles during the 1960s, was reborn through demolishing a 6 km long, 10-lane concrete highway and transforming it into an attractive stream corridor (figure 1). While the CRP, however, has been regarded as a successful outcome for the regeneration of downtown Seoul [2-5], ecological and hydrological conditions in the restored stream have been criticized as an unsustainable product [6]. The enormous project was done in too short a time period, twenty-seven months, without far-sighted consideration of natural systems such as a self-sustaining dynamic over time and space, in what is defined as a 'dynamic equilibrium'. Although natural or human interventions to the stream could change environmental conditions, it allows the biological community to establish a dynamic equilibrium over time.

The dynamic equilibrium model was suggested by Huston. He argued that community diversity can be maintained by growth rates, a frequency of population reduction, or a low rate of competitive displacement. These variables are dependent and interactive among them. Through a trade-off in the dynamics, community diversity reaches a level of dynamic equilibrium, a point that the diversity may remain relatively constant over time. The dynamic balance is determined differently in different environments since the major parameters – frequency of reduction and rate of displacement – are showed with a wide range of variability there [7]. In this understanding, the dynamic equilibrium is mainly determined by the influence of species interaction to the environment rather than the relative competitive capability among the competing species [8]. In spite of frequent stream changes, a dynamic form of stability is persisted within a range of condition due to the dynamic equilibrium. For instance, although a flood event by heavy rainfall greatly influences the stream's environment and its ecosystems, a series of self-correcting mechanisms allows the ecosystems to adapt to sudden disturbances within a certain level. Human intervention, however, weakens or destroys self-sustaining tendencies of an ecosystem. Water contaminants, concrete channelization, and damming are formidable disturbances to keep from the natural resilience of the ecosystem. Therefore, stream restoration projects should remove the cause of the external stresses and disturbances and then ultimately achieve to regain self-sustaining ability and re-establish the desired state of the dynamic equilibrium. The dynamic equilibrium model has a significant implication for CRP which is struggling with ecological and environmental issues today. This paper aims to overview the CRP and its environmental and hydrological issues, and then proposes new sustainable ways to regain self-sustaining ability for the dynamic equilibrium in Cheonggyecheon.

Figure 1. CRP location in Seoul

2. Cheonggyecheon Restoration Project (CRP)

2.1. Project background and procedure

The Cheonggyecheon was historically and literally a 'valley of clean water'. As a tributary of the Han River, it has been a clean source in Seoul for several hundred years. It was a community place for housewives and children (figure 2). However, Seoul has remarkably developed with rapid urbanization and industrialization as a capital city of South Korea since the Korean War (1950-1953) and thus, the stream has become a large sewer and its banks turn to a place lined with the shanties of the poor (figure 3)[1].

IOP Conf. Series: Earth and Environmental Science **143** (2018) 012068 doi:10.1088/1755-1315/143/1/012068

Figure 2. Cheonggyecheon in the 1900s [1]

Figure 3. Cheonggyecheon in the 1950s [1]

In the end of the 1960s, Seoul proposed a plan to cover the stream and build a 10 lane-highway system on it. The construction of the new roads stimulated economic growth in the area and substantially contributed to the establishment of an industrial and commercial center there (figure 4). In the mid-1970s, elevated inner city highways of four lanes were additionally constructed on the roads to meet increasing traffic demands (figure 5) [9]. These infrastructures have played a significant role in the dramatic economic development of Seoul for several decades later. They have been considered a symbol of modern urban growth in Korea and other Korean cities have tried to apply this approach to the redevelopment of their contaminated urban streams.

Figure 4. Road construction in the 1960s [1]

Figure 5. Urban highway in the 1970s [1]

Despite the initial success of the CRP, the downtown started to decline due to the relocation of heavy industries to the periphery of Seoul, and air quality around the highway became highly polluted in the 1980s. Moreover, a critical security issue was raised with concerns about the structural instability of the aging highway and harmfulness of detrimental gases under the covered road [9]. A proposal to restore the stream, CRP, came into the public spotlight and was a major issue in the election of a mayor of Seoul in 2003. The city of Seoul launched the CRP soon after the election. The initial stage was to remove the elevated highway of 5.4 km (figure 6).

The city and the participating construction companies tried to carry out the project as quickly as possible, in an effort to minimize the potential of severe flood events, which frequently accompany heavy rainfall in the area during the summer season, to disturb the construction [6]. Nearby merchants conducted a series of demonstrations protesting economic losses to their businesses. Traffic control was also a critical issue since the existing fourteen-lane roads into just a four-lane road was too radical changes in the downtown accompanying a high traffic demand. Critics argued that CRP should be a long-term project in consideration of a financial cost, traffic congestion and restoration of historic relics under the covered road. Despite these issues, the CRP procedure was completed in 27 months by a strong leadership of the city mayor (figure 7). In the five years since the project was implemented, there have been remarkable environmental and ecological changes in and around Cheonggyecheon.

The air quality has been improved and urban heat island has been diminished[10]. In addition, the number of fish, bird, and insects has increased every year in the restored stream. However, there occurred hydrological issues with water quality problems.

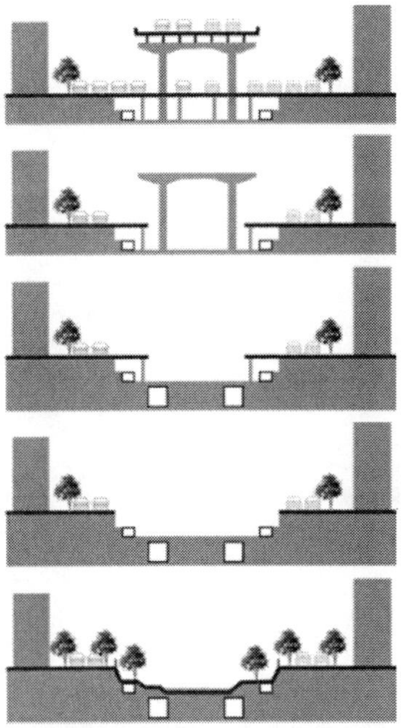

Figure 6. Road demolition and stream restoration procedure [11]

Figure 7. Transformation of the urban highway to the carnal[1]

2.2. Ecological and environmental issues

Historically, Cheonggyecheon was a seasonal stream with water flow in summer and mostly dry in the winter. However, the CRP planned year-round water flow to encourage human access to the stream and a variety of urban recreation activities. To keep consistent velocity with appropriate depth, a huge quantity of water was required. The primary water supply source of 98,000m3/day was conveyed through pipelines from the Han River, located 15 km away (figure 9). An additional 22,000 m3/day of groundwater, infiltrated into the existing subway tunnels, was collected and discharged into cheonggyecheon at several points through fountains and waterfalls. To minimize water losses from the restored channel by 3%, bentonite mats of fine clay layers were laid over the more permeable upstream riverbed and sheet pile walls were installed along each side (figure 8)[12]. In addition, combined sewer overflow (CSO) system was applied and it influenced the hydrological environment of the CRP. It results in high water levels and strong flows through the system after storms, creating flashy conditions that scour the bottom and affect habitat quality in the stream [13].

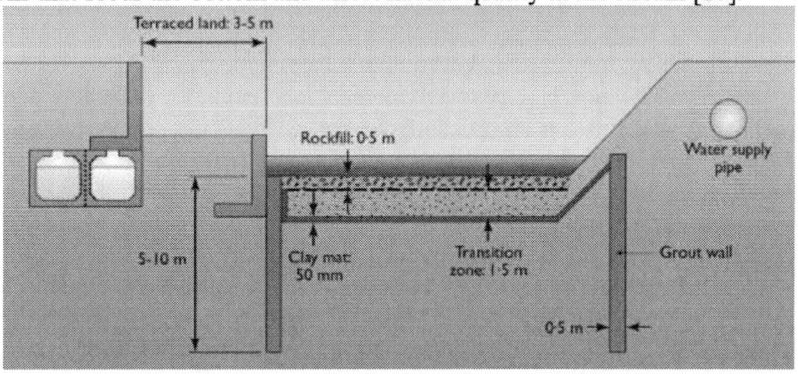

Figure 8. Section of the urban carnal in CPR [1]

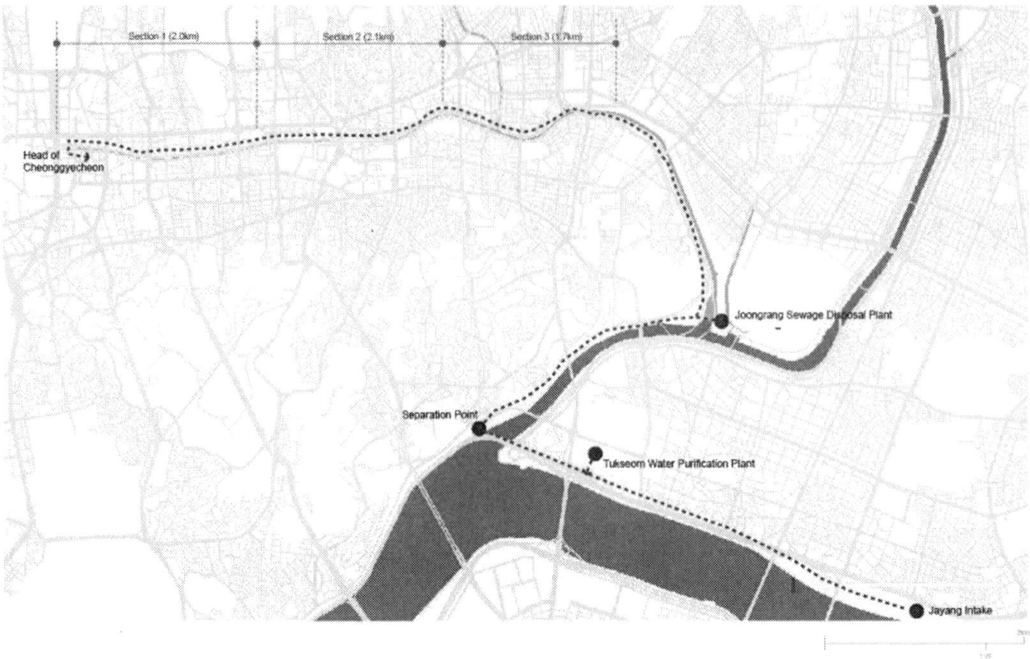

Figure 9. CRP water supply system [11]

Since CRP was completed in 2005, Cheonggyecheon has suffered from several problems with water quality[14]. Algae blooms have emerged over years (figure 10). The rapid accumulation of algae populations occurred in the whole water course in the spring and fall, March to May, and again October to November, and algae bloom were found in the upstream reaches in the summer and winter, from June to September and between December and February [15, 16]. This had negative impacts on the aquatic ecosystems. The algae blooms are closely associated with excess phosphorous and nitrogen levels in the water which are produced from sewage, wastewater treatment plants, and improperly functioning septic systems. Excess nutrients usually lead to eutrophication, resulting in ecological imbalances that cause the overgrowth of plants and animals and the discoloration of the water. It negatively impacts the aquatic environment, causing problems such as anoxia, water quality degradation, and reduction of fish and plant populations [17]. The main water source for Cheonggyecheon is not natural water from the mountains but rather chemically treated water that is piped from the Han River. And then the treated wastewater discharges into the upstream of the Cheonggyecheon intake point, and the water treatment process does not remove phosphorus and nitrogen. As a result, the algae blooms occur when they meet with sufficient sunlight into the stream. Furthermore, mass mortality of fish has occurred after heavy rainfall in the Cheonggyecheon area. According to the data from Seoul, it is estimated that the quantity of fish death reaches approximately 2,000 for last 10 years [18]. It has mostly resulted from the combined sewer overflows (CSOs) containing untreated wastewater discharges into the surface water. Combined discharges of untreated wastewater and storm water runoff overflowed into Cheonggyecheon whenever rainfall intensity exceeded approximately 0.02 mm/h. Dissolved oxygen is crucial for fish and other aquatic life. Dissolved oxygen concentrations are significantly impacted by CSOs in receiving water [19]. The mass death of Cheonggyecheon fish is closely associated with the reduction of dissolved oxygen level by CSOs [18].

The problems of fish kills and algal bloom create ecologically and aesthetically unpleasing conditions. There could be also public health concerns associated with continued fish kills and algae blooms. The problems with water quality and hydrology would prevent the stream community from continuing to recover to the extent that it could if those problems were not present. Improvements in both hydrological conditions and water quality could enhance the restored stream's value both as habitat and a recreational and aesthetic resource for the city's people.

Figure 10. Algae bloom in Cheonggyecheon [16]

3. Proposed strategies for sustainable Cheonggyecheon environment

Compared with the previous polluted and degraded environment of covered roads and elevated urban highways, the CRP has certainly had a significant positive effect, in that it has brought Cheonggyecheon back as a stream following the removal of highway infrastructure. In addition, it has been a successful example of urban regeneration with respect to air quality improvement and dramatic growth in real estate values around Cheonggyecheon. It is, thus, regarded as a successful urban regeneration by stream restoration. However, we argue that the CRP should be reconsidered with an understanding of ecological dynamics for sustainable nature. Confronting environmental and ecological issues such as algae blooms and mass mortality of fish, the CRP needs to reintegrate the lost natural dynamics of the stream and design with nature. It should ultimately pursue restoration of ecological function, which improves the Cheonggyecheon environment over time. We would propose two significant solutions to the current ecological problems of the CRP: the vegetation of the aquatic terrestrial transition zone (ATTZ) and the constructed wetlands for treatment of source water.

3.1. Vegetation of aquatic terrestrial transition zone (ATTZ)

The practice of vegetation in the ATTZ is one way to re-establish the dynamic equilibrium in the environment. Cheonggyecheon was channelized with a concrete material and enclosed with the purpose of intensive management for public recreation and flood control. Much of the ecological stream corridor was not considered and was simply confined to a narrow linear space [6]. Vegetation buffers are few between the upstream and midstream. The proposed vegetated corridor along the ATTZ can ecologically and environmentally benefit to communities. The vegetated ATTZ can serve as a meeting point for land and water, and terrestrial and freshwater organisms over time and space and it contributes to the richness of aquatic and terrestrial habitats [20]. Thus, the diversity and abundance of the species play a significant role to achieve an ecologically valuable self-recovering and self-sustaining system [21, 22]. While the population of wildlife species has been gradually augmented at Cheonggyecheon every year, its diversity is pretty limited and a majority has been discovered at downstream, located a short distance away from Han River [23]. Thus, networked vegetation of ATTZ should be created in the whole stream area in order to facilitate species movement and access leading to the dynamic equilibrium.

The vegetation corridor also functions as flooding control [22]. The CRP has been striving to control flooding events and their water flow in the summer season. The vegetated buffers in the floodplain have an effect to reduce the velocity of floodwaters. The roughness of their surfaces leads floodwaters to be temporarily detained and gradually released with infiltration to the bottom for groundwater recharge. As result, it diminishes the swiftness of the water flow and flood peaks. Water quality and temperature can be also protected by the vegetated ATTZ [24]. Current critical Cheonggyecheon issues, which are the algae blooms and the massive losses of fish by combined sewer overflows (CSOs) in the rainy season, have occurred due to the degraded water quality. As the vegetation buffer corridor can temporarily retain water and assimilate pollutants of surface water and CSOs, such as sediments, nutrients and pathogens, it helps to keep from contaminating the stream water. Moreover, shades of trees and shrubs lead consistent temperature of the water, which affects the algae bloom event and dissolved oxygen concentrations of Cheonggyecheon. When heated water meets excessive nutrients in a stream, it provides adequate conditions to multiply blue-green algae into a bloom, which usually displace aquatic food source. Meanwhile, the solubility of dissolved oxygen, which is necessary for fish, declines in a high temperature of the water since the oxygen level is inversely proportional to the temperature in the water. In this perspective, shading of the vegetation can play a significant role to deal with the critical issues of Cheonggyecheon by stabilizing the water temperature. The vegetated ATTZ will obviously strengthen the resilience of the ecosystem and make a healthy aquatic habitat in Cheonggyecheon.

There are many strategies and repair practices to ecologically enrich the ATTZ: vegetation establishment, root wad revetments, live crib walls, live stakes, and so forth [25]. Especially, Cheonggyecheon requires an effective vegetation type selection to consider the environmental solutions. The Table 1 indicates effectiveness of different vegetation types for specific buffer benefits. The trees have remarkable benefits to water quality, aquatic habitat, flood control, and visual effects. The groups of grasses particularly work effectively for filtering to improve water quality. The ATTZ of upstream and midstream in Cheonggyecheon, where is located nearby downtown and frequently produce algae blooms, should be vegetated with more trees and grass than shrubs in order to maximize filtering effects, aquatic habitat and visual interest.

Table 1. Effectiveness of different vegetation types for specific buffer benefits [26]

Benefits	Grass	Shrub	Tree
Stabilize streambank			
Filter sediments, nutrients, pesticides, and pathogens			
Filter nutrients, pesticides, and microbes from surface water			
Protect groundwater and drinking water supplies			
Improve aquatic habitat			
Improve wildlife habitat for field animals			
Improve wildlife habitat for forest animals			
Provide economically valued products			
Provide visual interest			
Protect against flooding			

3.2. Constructed wetland

As to the algae bloom issue, the critical problem is the water supply system which dominantly conveys Han River water to the head of Cheonggyecheon upstream and discharges into the stream after a chemical treatment in a sewage disposal plant. Within the current water circulation, there is no process to filter large amounts of phosphorous and nitrogen. High levels of phosphorus and nitrogen cause eutrophication and algae bloom in Cheonggyecheon [15, 16, 18, 27]. We propose installing constructed wetlands at discharging points of upstream water source in Cheonggyecheon, since the wetlands can play a remarkable role in removing BOD, TSS, organic nitrogen, and phosphorous from the source water, and can also provide a habitat and public recreational benefits. There is a substantial

IOP Conf. Series: Earth and Environmental Science **143** (2018) 012068 doi:10.1088/1755-1315/143/1/012068

body of literature about the construction and use of treatment wetlands for the improvement of water quality, and many well-implemented examples can be found [28-30].

Constructed treatment wetlands can be defined as engineered or constructed wetlands that enhance water quality through transforming or removing contaminants with physical, chemical, and biological mechanisms. It takes advantage of natural processes with a combination of wetland vegetation, soils, and their related microbial assemblages. A key role of the wetland is to improve water quality. The most significant concern for Cheonggyecheon's algae bloom is the treatment of nitrogen and phosphorous. While septic and primary tank effluents would include organic nitrogen and ammonia, the treatment pond is able to change reduced or oxidized forms of nitrogen, and phosphorous in a system of the constructed wetland. Nitrogen is removed by flocculation, sedimentation, filtration, and interception processes in the system (figure 11). The sustainable phosphate removal process is achieved by the accretion and burial of phosphate within the wetland sediments (figure 12) [28]. Treatment wetlands can be engineered to take care of a range of concentrations of contaminants, from removing relatively high levels of nutrients from the municipal wastewater to cleanse the storm water of sediments, metal, nutrients, and other contaminants.

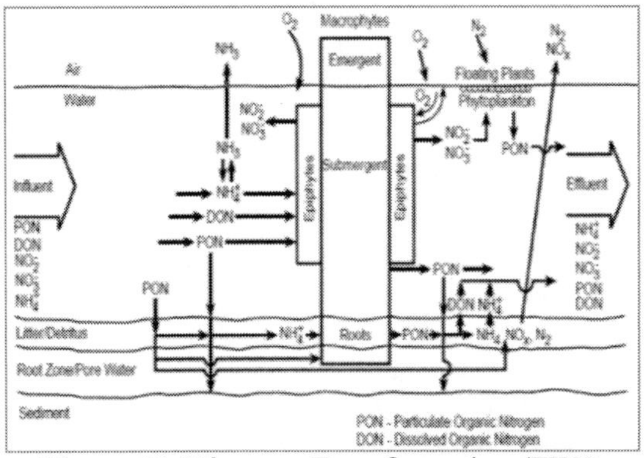

Figure 11. Nitrogen Transformation (FWS wetland)[28]

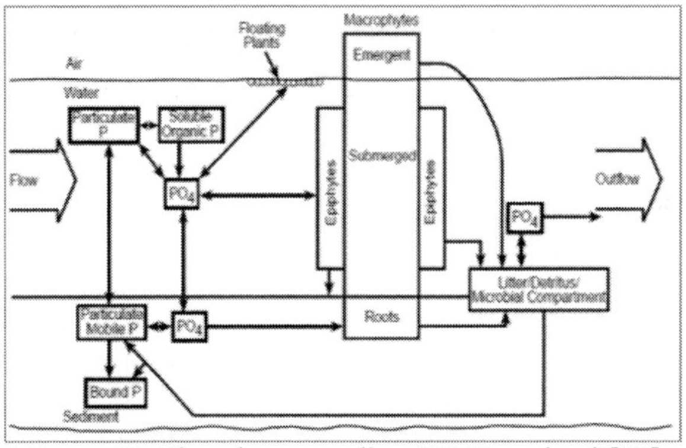

Figure 12. Phosphorus cycling (FWS wetland)[28]

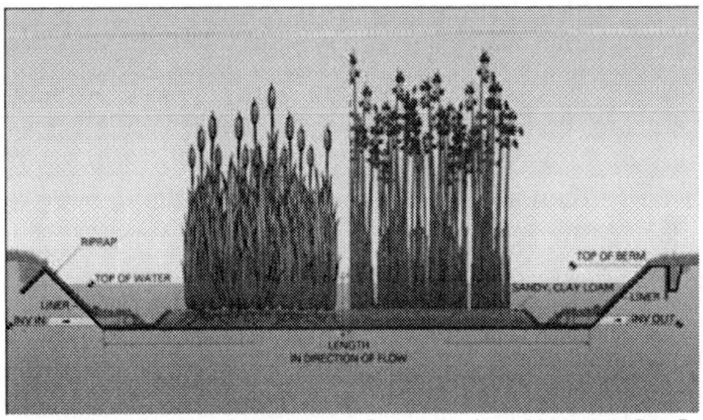
Figure 13. Free water surface wetland(FWS) [31]

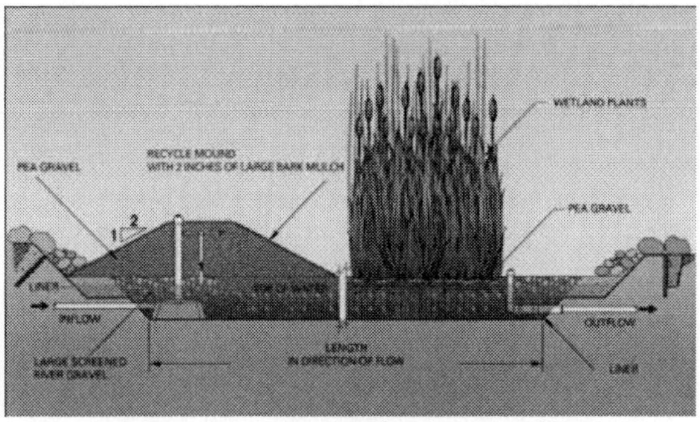
Figure 14. Vegetated submerged bed wetland(VSB) [31]

There are two sorts of constructed wetland: free water surface system (FWS) and vegetated submerged bed (VSB) system. A FWS system is typically similar to a natural wetland in terms of physical structure and performance. It consists of various combinations of open-water areas and fully vegetated surface areas, and outlet structures. To prevent water loss, subsurface is made by fine clay or impervious geotechnical materials and sufficient soils in order to support the growing vegetation. The FWS system is characterized with shallow water depth, low flow velocity, and linear channels for plug-flow conditions. The berms and inlet and outlet structures serve as a regulation and distribution of wastewater flow (figure 13). The VSB system, on the other hand, consists of gravel beds, which allow wetland vegetation to grow, with an impermeable clay layer of subsurface. Water flow is controlled by, not only the inlet and outlet structure like the FWS, but also gravel size and shape in a transverse channel (figure 14). The VSB has more diversity to create physical aesthetics of the wetland with the optional choice of gravel and vegetation. However, FWS has performed better in the water treatment process [28, 31].

Figure 15. Water source from upstream of Cheonggyecheon

Figure 16. Upstream of Cheonggyecheon

For the Cheonggyecheon issues, the constructed wetlands can be installed with a combination of FWS and VSB systems. The FWS system can be dominantly installed at a discharge point of the stream head since algae blooms have frequently occurred at upstream and midstream areas. More than half of the supplying water, which contains excess phosphorous and nitrogen, discharges into the stream head and requires high performance of a constructed wetland. The stream head area of Cheonggyecheon consists of an urban plaza with a hard floor and an underground reservoir for discharge (figure 15, 16). Although the plaza can offer a space for the cultural events of a downtown, more environmental benefits and visual interests can be provided by installing a constructed wetland in the plaza. In other words, it can accomplish as not only a water quality improvement through a natural filtering process, but plaza beautification with natural settings. Furthermore, the constructed wetlands can be established in connection with several CSOs discharge points along Cheonggyecheon. They can serve as detention ponds there and filter pollutants of CSOs. Seoul City has been struggling to physically remove increasing algae by cleaners and sprinkler trucks. It leads to a huge cost of the clean-up. Unless natural resilience can be restored, economic and ecological sustainability is hardly achieved in Cheonggyecheon. We assert that high performance of the constructed wetland helps to lead progressive advances of ecological function and dynamic equilibrium.

4. Conclusion

Cheonggyecheon, which had been transformed into a 14-lane urban highway, and then into a large sewer, was finally converted back to an urban stream again. Its transformation has been praised as a successful example of urban downtown regeneration and beautification. It is, however, obvious that there have not been prudent ecological considerations since the project's principal goals were to provide public recreational use and achieve maximum flood control capacity via the use of embankments. For a healthier and sustainable stream environment, Cheonggyecheon should be ecologically re-restored again, based on a dynamic equilibrium model. By confronting the current water quality issue, which causes algae bloom and does not allow a suitable environment for fish, Cheonggyecheon must primarily establish a corridor of vegetation, an aquatic transitional zone, and install constructed wetlands nearby which support the water source.

Furthermore, when pondering a long-term Cheonggyecheon restoration plan, more fundamental sustainable strategies must be applied. For instance, the upper streams of Cheonggyecheon should be further restored and supply natural waters. A combined sewer system should become a separate system equipped to handle both excess rainwater and sewage. In addition to these hydrological ideas, there needs to be de-channelization to create favourable conditions for riparian vegetation recruitment. This will help to encourage the biodiversity of riparian ecosystems, enhancing water quality, and sustaining the health of the stream environment. This would vary from merely increasing the sinuosity to reconstruct a naturalized stream thoroughly. The complete dynamic equilibrium of Cheonggyecheon can be accomplished through these strategies.

6. References

[1] Seoul_City 2006 The White Paper of Cheonggyecheon Restoration Project. (Seoul: Seoul)
[2] Busquets J 2011 *Deconstruction/construction: the Cheonggyecheon restoration Project in Seoul* (Cambridge: Harvard University Graduate School of Design)
[3] Song I-h and Chun W-y 2016 *Cheonggyecheon : flowing through Seoul and reflecting Seoul's history* (Seoul: Cheonggyecheon Museum)
[4] Lee J Y and Anderson C D 2013 The Restored Cheonggyecheon and the Quality of Life in Seoul *J Urban Technol* **20** 3-22
[5] Ryu C and Kwon Y 2016 How Do Mega Projects Alter the City to Be More Sustainable? Spatial Changes Following the Seoul Cheonggyecheon Restoration Project in South Korea *Sustainability* **8** 1178
[6] Cho M R 2010 The politics of urban nature restoration: the case of Cheonggyecheon restoration in Seoul, Korea *International Development Planning Review* **32** 145-65
[7] Huston M 1979 A general hypothesis of species diversity *The American Naturalist* **113** 81-101
[8] Resh V H, Brown A V, Covich A P, Gurtz M E, Li H W, Minshall G W, Reice S R, Sheldon A L, Wallace J B and Wissmar R C 1988 The role of disturbance in stream ecology *Journal of the North American benthological society* **7** 433-55
[9] Lee I K 2006 Cheong Gye Cheon Restoration Project. In: *Cheong Gye Cheon, Urban Revitalization and Future Vision International Symposium on the 1st Anniversary of Cheong Gye Cheon Restoration Seoul, Korea,*
[10] Jang Y K, Kim J, Kim H J and Kim W S 2010 Analysis of Air Quality Change of Cheonggyecheon Area by Restoration Project *Environmental Impact Assessment* **19** 99
[11] Rowe P, Kim S and Jung S 2011 A city and its stream: the Cheonggyecheon restoration project. (Seoul: Seoul Development Institute)
[12] Shin J-H and Lee I-K 2006 Cheong Gye Cheon restoration in Seoul, Korea. In: *Proceedings of the Institution of Civil Engineers-Civil Engineering,* (Seoul: Thomas Telford Ltd) pp 162-70
[13] Kim H, Noh S, Jang C, Kim D and Hong I 2005 Monitoring and analysis of hydrological cycle of the Cheonggyecheon watershed in Seoul, Korea. In: *Proc. of International Conference on Simulation and Modeling, Nakornpathom, Thailand,* pp 17-9
[14] Kim H J and Shin B K 2014 A Study on the Ecological Improvement Plan of the Cheonggye Stream, Seoul, Korea *Journal of Korean Society of Environment and Ecology* **24** 47-8
[15] Anh H 2009 Algae booms in Cheonggyecheon. In: *Kyunghyang Newspaper* (Seoul: Kyunghyang Newspaper)
[16] Yoo S 2009 Critical Issues of Algae booms in Cheonggyecheon. In: *Donga Newspaper* (Seoul: Donga Newspaper)
[17] Carpenter S R, Caraco N F, Correll D L, Howarth R W, Sharpley A N and Smith V H 1998 Nonpoint pollution of surface waters with phosphorus and nitrogen *Ecological applications* **8** 559-68
[18] Sonh H 2015 mass mortality of fish in Cheonggyecheon in last 10 years. In: *Hankook Newspaper* (Seoul: Hankook Newspaper)
[19] Field R and O'Connor T 1997 Control and Treatment of Combined Sewer Overflows *Control and Treatment of Combined Sewer Overflows* **134**
[20] Federal Interagency Stream Restoration Working G 1998 *Stream corridor restoration: principles, processes, and practices* (Washington, D.C.: Federal Interagency Stream Restoration Working Group)

[21] Hansen A J and DiCastri F 2012 *Landscape boundaries: consequences for biotic diversity and ecological flows* vol 92 (New York: Springer Science & Business Media)
[22] Décamps H and Naiman R J 1990 *The ecology and management of aquatic-terrestrial ecotones* vol 4 (New Jersey CRC Press)
[23] Choi J K, Byeon H K, Kwon Y S and Park Y S 2008 Spatial and Temporal Changes of Fish Community in the Cheonggye Stream after the Rehabilitation Project *Journal of Korean Society of Limnology* **41** 374
[24] Sand-Jensen K and Frost-Christensen H 1999 Plant growth and photosynthesis in the transition zone between land and stream *Aquatic Botany* **63** 23-35
[25] Schueler T R and Brown K 2004 Urban Stream Repair Practices. (Ellicott City: Center for Watershed Protection)
[26] Connecticut_River_Joint_Commissions 2000 Planting Riparian Buffers. (Charlestown: Connecticut River Joint Commissions)
[27] Shin M S, Kim B, Kim J K, Park M S, Jung S M, Jang C W, Shin Y K and Bae Y J 2008 The seasonal variation of water quality and periphyton in the Cheonggyecheon *Korean Journal of Limnology* **41** 1
[28] Brown D, Kreissi J, Gearhart R, Kruzic A, Boyle W and Otis R 1999 Manual of Constructed Wetlands Treatment of Municipal Wastewaters. (U.S. EPA: EPA-625/R-99/010)
[29] Hsu C-B, Hsieh H-L, Yang L, Wu S-H, Chang J-S, Hsiao S-C, Su H-C, Yeh C-H, Ho Y-S and Lin H-J 2011 Biodiversity of constructed wetlands for wastewater treatment *Ecological Engineering* **37** 1533-45
[30] Vymazal J 2011 Constructed Wetlands for Wastewater Treatment: Five Decades of Experience. (Easton: American Chemical Society) p 61
[31] Csaba P and Csaba J 2011 *Water resources management and water quality protection* (Szerzői jog: Debreceni Egyetem)

Acknowledgments
This research was supported by the Creative-Pioneering Researchers Program through Seoul National University (SNU).

The Association Between Risk-taking Behavior and Helmet Use Among Motorcyclist

Tu Anh Trinh[1], Thi Phuong Linh Le[2]

[1]Urban and Regional Planning Department, Faculty of Civil Engineering, Ton Duc Thang University, 19 Nguyen Huu Tho, Tan Phong Ward, District No 7, Hochiminh city, Vietnam. Email: trinhtuanh@tdt.edu.vn
[2]Faculty of Airport, Vietnam Aviation Academy, 104 Nguyen Van Troi, Phu Nhuan District, Hochiminh city, Vietnam

Abstract. The central aim of the research was to examine speeding behavior without helmet wearing among motorcycle riders in Ho Chi Minh City. The research model expanded the Theory of Planned Behavior by both psychological flow theory and sensation-seeking. 268 motorcyclists were involved in the research. A Confirmatory factor analysis and a Structural equation modeling were employed for model specification. The findings indicated a significant effect between predictors and the intention of speeding without helmet wearing. In addition, there was direct relation between intention and actual behavior. Also, sensation-seeking proved to be important in moderating predictors between the low-sensation-seeking motorcyclists and high-sensation-seeking ones. A deeper understanding of why motorcycle riders exceed the speed limit without wearing helmet played key role in changing their behavior.

1. Introduction

Due to the features of low price, high mobility, motorcycle riding has become popular and convenient means of transportation in many countries, including Vietnam. However, the increasing number of motorcycles are expected to lead to an increasing number of motorcycle accident. Various studies have demonstrated that motorcyclists were at higher risk of being killed or injured in a road traffic crash compared with others [1]. To prevent head injury from accidents, wearing a helmet while riding a motorcycle has been regarded as one of the best solutions. [2] confirmed that 'riders who have accidents while not wearing helmets were more likely to suffer critical head injury or death than were riders wearing helmets'.

Vietnam is a motorcycle-dependent country. Motorcycle-related accidents and deaths account for a substantial proportion of road accidents. In attempts to tackle with traffic accident, several policies and programs have been implemented such as the promulgation of the compulsory helmet legislation in December 2007 and annual national road safety campaigns; however, according to Vietnamese National Traffic Safety Committee, only in 2014, 25.322 traffic accidents related to motorcycles and more than 8.996 road users killed on the were recorded and most of these traffic accident deaths were attributed by risk-taking behaviors, particularly speeding without helmet wearing which accounted for 25%.

Although speeding without helmet wearing is the primary cause of traffic accidents, there have been no more research efforts to gain a better understanding of motorcycle helmet use and their correlation with speeding among citizen. This study aimed to fill this gap by specifically investigating speeding behavior without helmet wearing among motorcycle riders in Ho Chi Minh City (HCMC), Vietnam. To achieve the central aim of the research, the extended Theory of Planned Behavior proposed by Chen and Chen (2010) were employed as theoretical framework. By doing so, the research expected to better gain insights of motorcyclist's speeding behavior in conjunction with their neglects of helmet use that would help government implement solutions to change motorcyclist's risk behavior. The following section details the study's theoretical frameworks and then its hypothesized model.

1.1 Theory of planned behavior

The Theory of Planned Behavior (TPB) is a robust theory that has been used across multiple contexts to explain a wide range of behaviors [3] [4]. The TPB model assumes that behavior depends on intention, and that intention, in turn, is determined by a person's attitudes, subjective norms, and perceived behavioral control [5] (Figure 1).

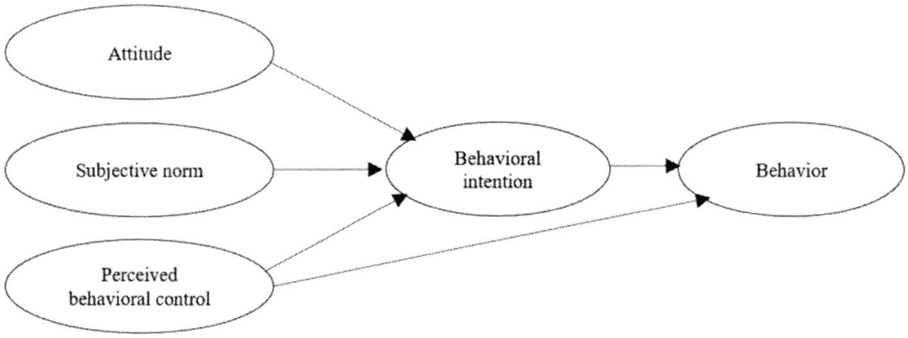

Figure 1. The theory of planned behavior

In this context, 'attitudes' refer to an individual's evaluation of a behavior, which are informed by behavioral beliefs concerning a behavior's outcome and behavioral beliefs pertaining to the experience of a behavior [4]. 'Subjective norms' refer to people's beliefs about how their significant others think they should engage in the behavior [4]. Last, 'Perceived behavioral control' refers to a person's perception of his or her ability and resources to perform a behavior and are reflected in an individual's control beliefs [5]. The Theory of Planned Behavior (TPB) has been successful applied to predict a variety of behaviors in the field of social science, health, education and transportation. In the field of traffic psychology many previous studies have used the TPB as a frame of reference [1] [6] [7].

1.2 The extended Theory of Planned Behavior

[8] was developed the extended Theory of Planned Behavior on the basic of Theory of Planned Behavior [5] to explore speeding behavior of heavy motorcycle riders by adding the variables of psychological flow theory to explore intrinsic motivations among motorcyclists, and (2) the sensation seeking and riding experiences to investigate differences of groups in the influence of affecting factors on speeding behavior (Figure 2).

Regarding Psychological flow theory, [8] argues that 'intrinsic motivations instead of extrinsic motivations might better explain and/or predict the related speeding behavior'. Thus, the author took into account 'perceived enjoyment' and 'concentration' to construct the flow of heavy motorcycle riders. According to [9] (as cited in [8]), 'Perceived enjoyment is defined as when, under conditions of playfulness, an individual will find an action intrinsically interesting, and thus will engage in it for pleasure and enjoyment rather than extrinsic rewards', and 'Concentration is defined as when an individual loses self-consciousness, becomes absorbed in the activity, and is more intensely aware of their mental processes when in a playful state'.

In terms of sensation-seeking and riding experience, both of them were regarded as an important factor influencing rider behavior. In his work, [8] stated that sensation-seeking individual tended to seek and enjoy high stimulus, exciting, novel and diverse experiences and stimuli. The greater sensation-seeking individual was, the higher possibility he was involved in risky driving behavior. Similarly, an experienced motorcyclist who felt confident in controlling their motorcycle due to past experience would be highly likely to exceed speed limit compared with those had less riding experiences. Hence, [8] did add two dimensions in his model to investigate their moderating role in heavy motorcycle riders' behavior. To examine the moderating effect of sensation-seeking, based on the mean level of sensation-seeking, the sample would be divided into two groups: high and low.

In the context of Vietnam - a motorcycle-dependent country, motorcycle is the major transport mode of citizen for various purposes; therefore, the current study only focused on high-sensation-seeking motorcycle riders and low-sensation-seeking ones. Respondents would be experienced motorcycle riders who have at least 5-year motorcycle driving experience. Novices (less than 5-year motorcycle driving experience) would be excluded in this study.

1.3 Hypothesis

Based on the literature review aforementioned above, the current study proposed a conceptual framework (Figure 2) with the following hypotheses. Specifically:
H1: Perceived enjoyment has a positive effect on attitude toward speeding without helmet wearing.
H2: Perceived enjoyment has a positive effect on intention of speeding without helmet wearing.
H3: Concentration has a positive effect on attitude toward speeding without helmet wearing.
H4: Concentration has a positive effect on intention of speeding without helmet wearing.

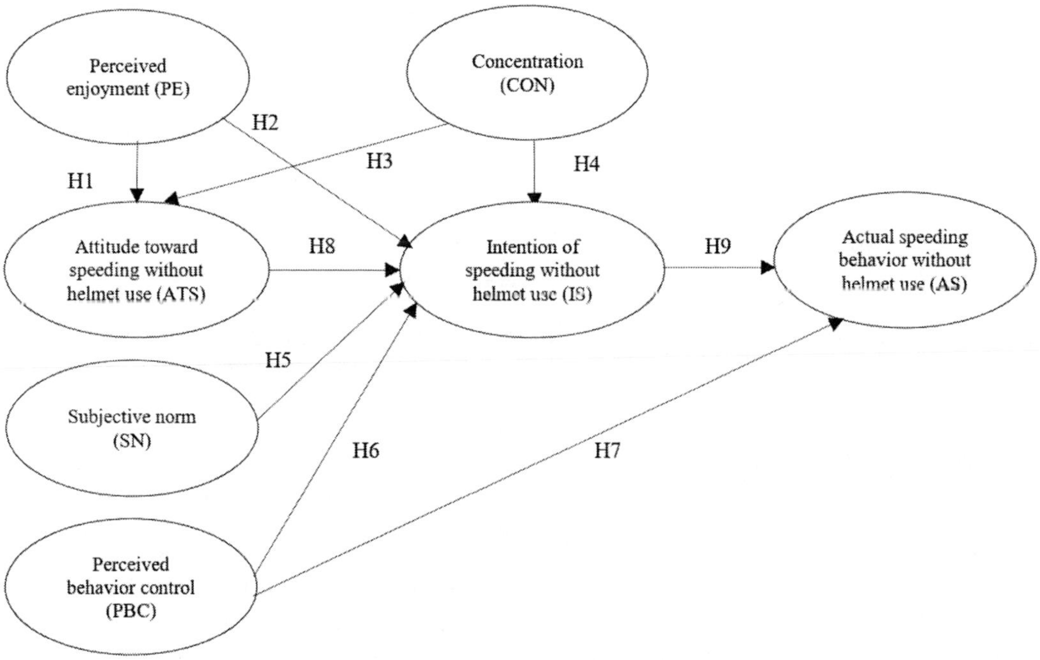

Figure 2. Conceptual framework

H5: Subjective norm has a negative effect on intention of speeding without helmet wearing.
H6: Perceived behavioral control has a positive effect on intention of speeding without helmet wearing.
H7: Perceived behavioral control has a positive effect on actual speeding behavior without helmet wearing.
H8: Attitude has a positive effect on intention of speeding without helmet wearing.
H9: Intention of speeding without helmet wearing has a positive effect on actual speeding behavior without helmet wearing.

IOP Conf. Series: Earth and Environmental Science **143** (2018) 012069 doi:10.1088/1755-1315/143/1/012069

2. Method

2.1 Data collection

The population of this study was selected at HCMC. The survey was conducted from June 2017-July 2017. Thirty minutes face-to-face interview were applied for data collection. The author visited the gas stations, shopping malls, super markets, bookshops and entertainment places to conduct interview. Pilot study were carried out to make sure that the measuring instrument represented the goal of the study. Total 300 respondents randomly selected were involved in this survey. The final sample size for the study was 268, whereas 32 observations (of the 300 collected) dropped due to incomplete informations.

There were 63.1% male and 36.9% female respondents. Riders aged 22–35 (66.3%) and 36–53 (33.7%) accounted for the majority of the sample. 14% of respondents had less than 5 year's experience, followed by 52.4% with 5-10 years' and 43.6% with more than 10 years' experience.

2.2 Measures

The dependent variable in the study was motorcycle rider's actual speeding behavior without helmet wearing. The predictors were perceived enjoyment, concentration, attitude toward speeding without helmet wearing, perceived behavioral control, subjective norms, and intention of speeding without helmet wearing. The dependent variable and predictors mentioned below were designed on the basic of literature review and items adapted from [8]. The variables measured with multiple items were examined with Cronbach's alpha reliability for internal consistency:

Actual speeding behavior was measured by asking rider about the frequency they speeded without wearing helmet during the six months before the interview. *Actual speeding behavior* was measured on a 5-rating scale ranging from 1 (very infrequent) to 5 (very frequent). The single item stated that "Within the past six months, how frequently did you exceed the speed limit when riding a heavy motorcycle?" (M=3.8, SD=1.3).

Attitude was assessed through items rated on a 4-point scale ranging from 1 (Totally disagree) to 4 (Totally agree) that asked the motorcyclist' position on speeding without helmet wearing (e.g. To exceed the speed limit when riding without helmet wearing (a) I would feel enjoyable, (b) I would feel comfortable, (c) I would feel acceptable) (Cronbach's $\alpha = .65$).

Subjective norm was measured with items rated on a 4-point scale ranging from 1 (Totally disagree) to 4 (Totally agree) (e.g. (a) Most people who are important to me think that I should not speed without helmet wearing, (b) Most people who are important to me do not support me speeding without helmet wearing, (c) Most people who are important to me will not speed without helmet wearing) (Cronbach's $\alpha = .64$).

Perceived behavioral control was measured with items rated on a 4-point scale ranging from 1 (Totally disagree) to 4 (Totally agree) (e.g. (a) When speeding without helmet wearing, I believe my driving skills can meet the challenge, (b) When speeding without helmet wearing, I feel in total control of what I am doing, (c) When speeding without helmet wearing, I am capable of manipulating the vehicle without thinking how to do it) (Cronbach's $\alpha = .72$)

Intention was measured by items rated on a 4-point scale ranging from 1 (Totally disagree) to 4 (Totally agree) (e.g. (a) When riding without helmet wearing, I am likely to speed, (b) When riding without helmet wearing, I intend to speed, if possible, (c) When riding without helmet wearing, I will try to speed) (Cronbach's $\alpha = .77$).

Concentration was measured by items rated on a 4-point scale ranging from 1 (Totally disagree) to 4 (Totally agree) (e.g. (a) During speeding without helmet wearing, I concentrate fully on the activity, (b) During speeding without helmet wearing, I am usually absorbed intensely in the activity, (c) During speeding, I am deeply engrossed in the activity) (Cronbach's $\alpha = .68$).

Perceived enjoyment was measured with items rated on a 4-point scale ranging from 1 (Totally disagree) to 4 (Totally agree) (e.g. (a) I find riding heavy motorcycles to be enjoyable, (b) I have fun riding a heavy motorcycle, (b) The process of riding a heavy motorcycle is pleasant) (Cronbach's $\alpha =$ .86).

Sensation seeking (SS) was assessed through items rated on a 4-point scale ranging from 1 (Totally disagree) to 4 (Totally agree) (e.g. (a) I would like to try bungee jumping, (b) I would love to have new and exciting experiences, even if they are illegal). (Cronbach's α = .65).

2.3 Data analysis
A Confirmatory factor analysis (CFA) and a Structural equation modeling (SEM) were employed in this study. CFA was used to confirm the relationship between the latent factors (e.g. psychological flow variables, TPB variables, intention) and their observed indicators (e.g. the measurement items). SEM was utilized to test the hypothesized relationships between the predictors and behavioral intention, actual behavior. Goodness-of-fit index (GFI), comparative fit index (CFI), and root mean square error of approximation (RMSEA) were used to examine SEM. For assessing goodness of fit, the author did follow the criteria proposed by [10] that values of GFI and CFI of 0.9 or above, RMSEA of 0.05 or less confirm a good fit between the model and the data.

3. Model Specification

3.1 Confirmatory factor analysis
CFA was employed in this stage to verify the factor structure. Table 1 present the results of CFA for the whole sample.

Table 1. Goodness of fit index

Model	χ^2	df	RMSEA	CFI	GFI	NFI
CFA	640.24	186 (p<.001)	.09	.959	.81	.96

3.2 Model Specification
After fitting the model, we utilized structural equation modeling to test the hypothesized model for the whole sample (n=268). The proposed model indicated a reasonably good fit (Table 2).

Table 2. Goodness of fit index

Model 1	χ^2	df	RMSEA	CFI	GFI	NFI
SEM	329.22	190 (P<.001)	.09	.960	.80	.95

As mentioned above, the current research focused on high-sensation-seeking groups and low-sensation-seeking groups. To examine the moderating effect of sensation-seeking, the sample was divided into two groups based on the mean level of sensation-seeking: high and low. Therefore, Model 2 and Model 3 were performed to examine the moderating effect of sensation-seeking (Table 3).

Table 3. Estimated results of the models

Path	Model 2 (n=116) (low SS)	Model 3 (n=152) (high SS)
PE → ATS	.27	.32
PE → IS	.33	.40
CON → ATS	.10	.13
CON → IS	.50	.37
SN→ IS	-.04	-.16
PBC → IS	-.10	.08
PBC → AS	-.05	.04
ATS → IS	.30	.13
SBI → AS	.59	.55

Path	Model 2 (n=116) (low SS)	Model 3 (n=152) (high SS)
Goodness of fit index	χ^2 = 386.42, df=192(p<.001) RMSEA=.08 CFI=.970 GFI=.70 NFI=.95	χ^2 = 395.64, df=192(p<.001) RMSEA=.09 CFI=.980 GFI=.81 NFI=.90

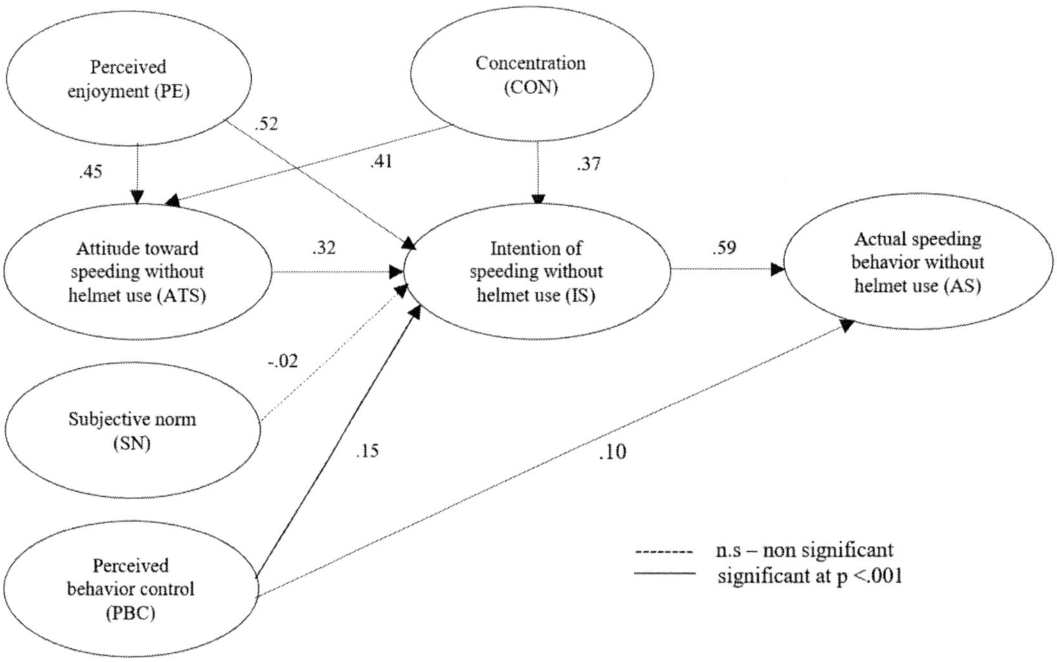

Figure 3. Structural equation model for whole sample

4. Discussion

Eight out of nine hypotheses in Model 1 were found to be supported. Regarding TPB variables, attitude had a positive effect on intention of speeding without helmet wearing. This finding suggested that to exceed the speed limit when riding without helmet wearing, motorcyclist rider would feel enjoyable, comfortable, and acceptable. In support of Hypothesis 5, the findings revealed a negative effect of subjective norm on intention of speeding without helmet wearing. The finding showed that the advice from people who are important to rider did not affect the intention of speeding among motorcycle riders. The effects perceived behavioral control on intention of speeding without helmet wearing was significant. This finding was proved to be consistent with those of previous studies that perceived behavioral control had generally been a significant correlate of behavioral intention. In this research, it was highly likely that when speeding without helmet wearing, respondent believed their driving skills that could meet the challenge, or when speeding without helmet wearing, they felt in total control of what they were doing. With respect to psychological flow variables, perceived enjoyment and concentration were significant predictors of intention of speeding without helmet wearing. During speeding without helmet wearing, respondent seemed to be absorbed intensely and deeply engrossed in the activity. Also, they found riding heavy motorcycles enjoyable and even more pleasant. The psychological flow variables did demonstrate their direct and indirect effects on intention of speeding without helmet wearing and motorcyclists' actual speeding behavior. The outcome supported previous researches that intrinsic motivations (perceived enjoyment and cognitive concentration) instead of extrinsic motivations confirmed better explanation of speeding behavior with/without helmet use among motorcycle riders. In terms of the correlate between intention and actual behavior, there was significantly direct relation between intention of speeding without helmet wearing and actual speeding behavior without helmet wearing. It revealed that when riding without helmet wearing, motorcyclist was likely to/intend to/and try to speed, if possible.

The findings of the moderating effect investigation of sensation-seeking between low-sensation-seeking and high-sensation-seeking group showed that there was a significant difference in the sensation-seeking level among the groups and the whole sample. The outcomes presented in Table 3 showed that the relationship between predictors was in line with Model 1, while it was inconsistent for low-sensation-seeking group. Regarding low-sensation-seeking motorcyclists, there was a positive effect of subjective norm on speeding without helmet wearing among them. In other words, the intention and actual behavior of low-sensation-seeking motorcyclists was highly likely to change owing to the advice or recommendation from people who were important to them (e.g. family, spouse, close friends). In addition, although perceived behavioral control (e.g. belief in skills that could meet the challenge, or in capability of manipulating the vehicle without thinking how to do it) seemed to motivate high-sensation-seeking group to be involved in speeding without helmet wearing, the same could not be said with low-sensation-seeking group.

5. Conclusion

In the context of speeding without helmet wearing among motorcycle rider, this study tested the mechanisms through which TPB variables, psychological flow variables and sensation seeking had a significant effect on intention of speeding without helmet wearing. The analysis revealed that (1) perceived enjoyment and concentration had a positive effect on attitude, (2) attitude, perceived behavioral control, perceived enjoyment and concentration were significant predictors of intention of speeding without helmet wearing, (3) subjective norm proved to have a negative effect on intention of speeding without helmet wearing, (4) there was significantly direct relation between intention of speeding without helmet wearing and actual speeding behavior without helmet wearing. Moreover, the findings indicated that sensation-seeking played important role in clarify the differences among the low-sensation-seeking and high-sensation-seeking group and the whole sample. The finding spurred implication for policy-makers to propose solutions on reduction of speeding without helmet wearing. Specifically, road safety interventions, driver education and training programs, or social campaigns should be designed aiming at (1) both motorcyclist and their close people, (2) perceived behavioral control characteristics to discourage them from the involvement of speeding without helmet, (3) highlighting the benefits of helmet wearing such as health protection, aesthetic values, fashion.

6. Limitation of the study

The current research possesses some limitations. Firstly, the survey data are limited, the data was collected in downtown and urban areas in HCMC while respondent living in suburban areas was not included in the sample. Secondly, the coefficients for the paths in question are very small leading to a less robust research finding. Additionally, the research did not consider the young generation aging from 18-21 although this group have tendency to involve in speeding without wearing helmet. Additionally, groups with different social and demographic characteristics like education background, occupation, income levels were not investigated in the study. In fact, group differences in speeding behavior without helmet wearing with respect to these characteristics are believed to be evident, but unfortunately are not investigated in this study due to data unavailability.

References

[1] Elliott MA, Baughan CJ, Sexton BF 2007 Errors and Violations in Relation to Motorcyclists' Crash Risk *Accid. Anal. Prev.* 39 491–499
[2] Rowland, J., F. Rivara, P. Salzberg, R. Soderberg, R. Maier, and T. Koepsell 1996 Motorcycle helmet use and injury outcome and hospitalization costs from crashes in Washington State *American Journal of Public Health* 86:41-45
[3] Armitage, C. J., & Conner, M. 2001 Efficacy of the theory of planned behaviour: A meta analytic review *British Journal of Social Psychology* **40** 471–499
[4] Fishbein, M., Ajzen, I. 2010 *Predicting and Changing Behavior: The Reasoned Action Approach* Taylor & Francis Group: New York, NY, USA
[5] Ajzen, I. 1991 The theory of planned behavior *Organizational Behavior and Human Decision Processes* **50** 179–211
[6] Wallen Warner, H., Aberg, L. 2006 Drivers' decision to speed: a study inspired by the theory of planned behavior *Transportation Research Part F: Traffic Psychology and Behaviour* 9 **6** 427–433
[7] Ambak, K., I. Rozmi, A.R. Atiq and B.M. Nazri 2011 Using structural equation modeling and the behavioral sciences theories in predicting helmet use. Proceeding of International Conference on Advanced Science *Engineering and Information Technology*, Bangi, Malaysia, pp: 639-645
[8] Chen, Ching-Fu, and Cheng-Wen Chen 2011 Speeding for fun? Exploring the speeding behavior of riders of heavy motorcycles using the theory of planned behavior and psychological flow theory *Accident Analysis & Prevention* **43.3**: 983-990
[9] Moon, J.-W., Kim, Y.-G 2001 Extending the TAM for a World-Wide-Web Context *Information & Management* 38 **4** 217–230
[10] Hair, J.F., Black, B., Babin, B., Anderson, R.E., Tatham, R.L. 2006 *Multivariate Data Analysis*, 6th ed. Pearson Education Inc., NJ.

Acknowledgments

We wish to thank the reviewers' helpful suggestions in improving this paper. Thanks also to the interviewees who were willings to complete the questionnaire used in the research.

IOP Conf. Series: Earth and Environmental Science **143** (2018) 012070 doi:10.1088/1755-1315/143/1/012070

Promoting sustainable mobility by modelling bike sharing usage in Lyon

T D Tran[1, 2, 3], N Ovtracht[1, 4]

[1] *Transport, Urban Planning and Economics Laboratory, Lyon, France*
[2] *Centre Institute for Economic Management, Hanoi, Vietnam*
[3] *E-mail: tiendung.tran@laet.ish-lyon.cnrs.fr and tiendungtran@mpi.gov.vn*
[4] *E-mail: nicolas.ovtracht@laet.ish-lyon.cnrs.fr*

Abstract. This paper aims to present a modelling of bike sharing demand at station level in the city of Lyon. Multiple linear regression models were used in order to predict the daily flows of each station. The data used in this project consists of over 6 million bike sharing trips recorded in 2011. The built environment variables used in the model are determined in a buffer zone of 300 meters around each bike sharing station. The results show that bike sharing is principally used for commuting purposes. An interesting finding is that the bike sharing network characteristics are important parameters to improve the prediction quality of the models. The present results could be useful for others cities which want to adopt a bike sharing system and also for a better planning and operation of existing systems. The approach in this paper can be useful for estimating car-sharing demand.

1. Introduction

During the last decade, bike sharing systems become more and more popular in the world. In the end of 2014, the bike sharing scheme is present in more than 50 countries with more than 700 systems and about 700 000 bikes (Larson et al. 2014). Bike sharing is considered as an alternative and complementary mode of transportation which has traffic and health benefits such as flexible mobility, physical activity benefits and supports for multimodal transport connections (Shaheen et al. 2010). Bike sharing can provide an alternative to traditional modes of transport or a complementary service for solving the "last mile problem" of getting from a public transportation stop to the final destination. In France, the bike sharing system is implemented in more than 35 cities in order to improve soft mode transportation and encourage the sustainable development. Vélo'v, the bike sharing system of the Metropole of Lyon, installed in May 2005 and operated by JC Decaux, is recognized as a success of bike sharing. In 2014, Vélo'v has 348 stations and more than 4,000 bicycles with about 58,000 long-term subscribers in 2014 for an average of 23,000 rentals per day (Lyon Capitale 2014). Although bike sharing is becoming popular in France and around the world, there are relatively few quantitative studies exploring the influence of built environment on the bike sharing usage. To ensure the success of bike sharing schemes, it is important to understand the influence of built environment on the bike sharing usage. Vélo'v in Lyon city is mature and successful bike sharing system that offers a unique opportunity for understanding the factors influencing on bike sharing usage. In this paper, we use bike sharing trips data of Vélo'v in 2011 and built-environment parameters to estimate the bike sharing's daily flow during a weekday at station level.

2. Literature review

Four generations of bike sharing have been introduced since the first bike sharing system in the 1960s in the Netherlands (De Maio et al. 2009; Shaheen et al. 2010). Bike sharing has become more popular since the introduction of the 3rd generation, with the automatic transaction kiosk at each station and identified bike sharing users. These bike sharing systems of the 3rd generation are relatively successful around the world. Currently, the majority of bike sharing systems are of the 3rd generation; there are some bike sharing systems of the fourth generation installed in Copenhagen and Madrid with improving docking stations, bike redistribution, integration with other transport modes (De Maio et al. 2009; Shaheen et al. 2010) and electrical bikes. In the Metropole of Lyon, the bike sharing system, called Vélo'v, belongs to the 3rd generation of bike sharing systems. Vélo'v bike sharing system is installed in the city of Lyon and Villeurbanne which cover an area of 60 kilometer square for more than 647 000 inhabitants (2013). The Vélo'v system aggregated more than 6.2 million trips in the 2011 and 8.3 million bike sharing trips recorded (Tran et al. 2014; Lyon Capitale 2014).

In recent years, many researches have used traditional surveys in order to determine the factors that may promote the adoptions of bike sharing by urban populations (LDA Consulting 2013; Melbourne BSS survey 2015). The automated data collected from docking stations constitutes a precious source of information to better understand the usage of bike sharing in the city. There have been several studies conducted using data from the Vélo'v system. These studies use bike sharing flow data obtained from stations to determine the typology of bike sharing users or to analyze the characteristics of bike sharing usage. They contribute to the literature by studying user behavior in response to bike sharing system and examining the characteristics of this system. Regarding the Vélo'v system, the average speed of bike sharing is 14 km per hour (Jensen et al. 2010) and the average duration of bike sharing trip is about 15 minutes. There are also several researches that determine a typology of bike sharing stations using the real bike sharing data trips (Froehlich et al. 2009; Lathia et al. 2012; Kaltenbrunner et al. 2010). They have determined that there are different groups of bike sharing stations according to their operation during a working day. The current paper contributes to literature by modelling the daily bike sharing demand at station level using the socio-economic parameters determined in a buffer zone around bike sharing stations using multiple linear regression models.

3. Mobility and bike sharing in the metropole of Lyon

The Metropole of Lyon is the second biggest agglomeration in France after Grand Paris with 1 324 637 inhabitants (2012) on a surface of 533.7 km square. The major public transport system (metro and tram line) of Lyon is mainly installed in the city of Lyon and Villeurbanne on a surface of 62.47 km square. There are 4 lines of metro, 2 funiculars and 5 tram lines. The surface public transportation is composed by 5 tram lines. The shift modal of bicycle in Lyon city is about 1.7% in 2006 according to the Household Travel Survey. The population of Lyon and Villeurbanne are relatively concentrated in the city center (see figure 1) while the employment density is important in the city center and some others zones far from the center (see figure 2).

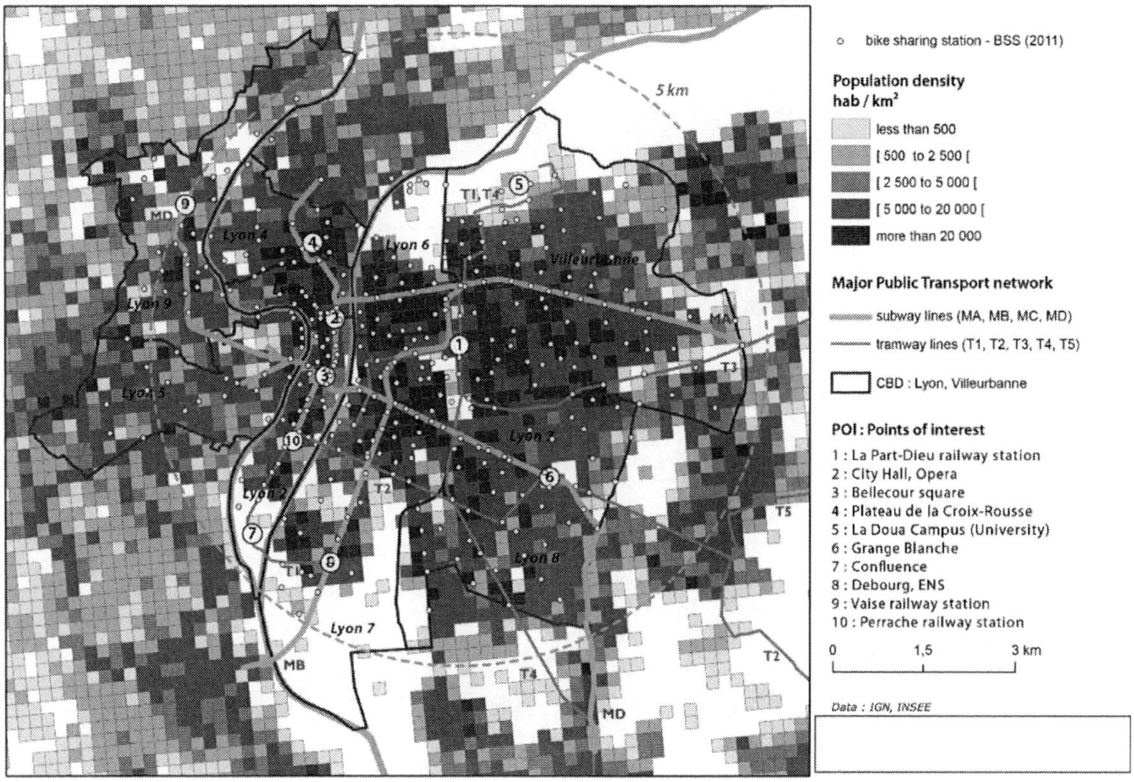

Figure 1. Bike sharing station and the population density by grill in the city of Lyon and Villeurbanne, source: JC Decaux, INSEE.

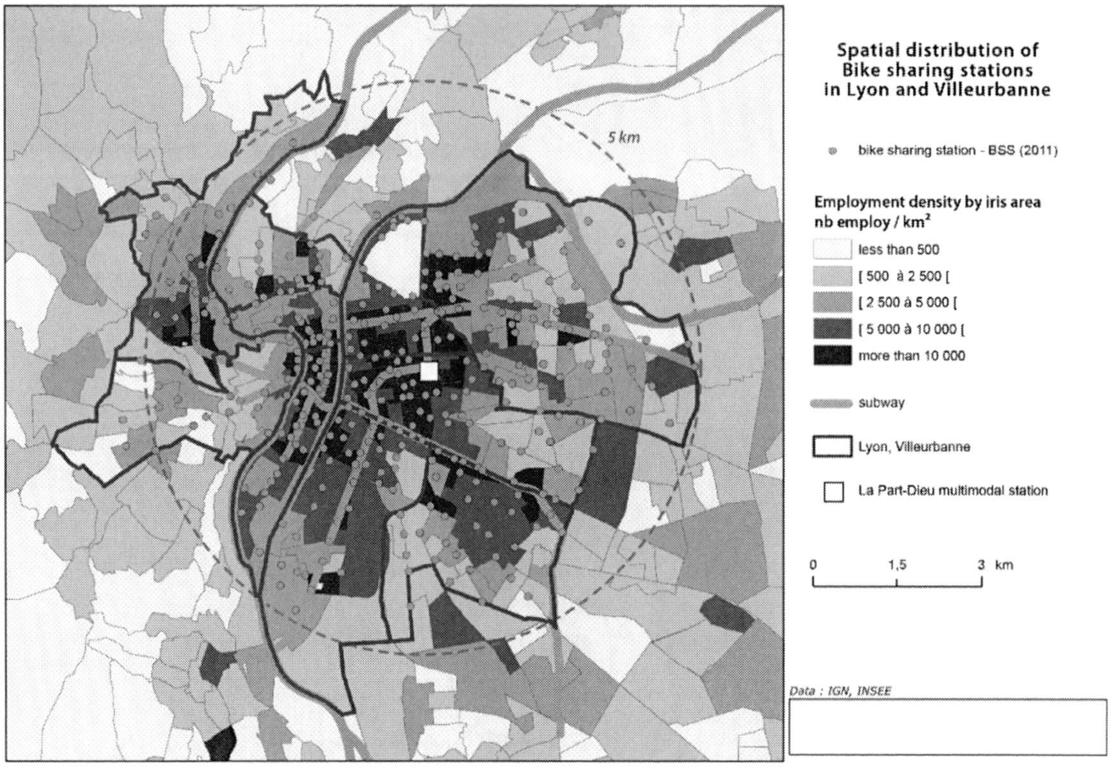

Figure 2. Bike sharing station and the employment density by IRIS zone in the city of Lyon and Villeurbanne, source: JC Decaux, INSEE.

The bike sharing system of the Metropole of Lyon, operated by JC Decaux, was installed in the city of Lyon and Villeurbanne, in May 2005 with 173 stations and 2 000 bikes. In 2014, this system has 348 stations and 4 000 bikes with more than 8 million trips. Bike sharing user can choose between three types of subscription: annual, weekly or daily with a minor cost equal to 25 euros, 5 euros or 1.5 euros respectively to be paid by a credit card. Bike can be rented from 0h to 24h and 7 days a week. The first 30 minutes is free for all subscribers; after the first free half hour users pay € 0.75 for every subsequent half hour, for a maximum of one hour. If the one hours limit is exceeded, the user is charged 1.75 euros per half hour. A 150 euros penalty is charged if the bike is not returned within 24 hours of its withdrawal from the station. The average number of daily trips is about 23 000 trips per day in 2014. By analyzing the data of bike sharing trips in 2011, we found that the level of bike sharing usage is not uniformly distributed during the days of week during the months of year: bike sharing is mostly used during working days and the usage level decreases considerably in vacation months (July, August) and in winter months. The difference of bike sharing usage by day and by month justifies our choice of bike sharing flows for modelling (see figure 3).

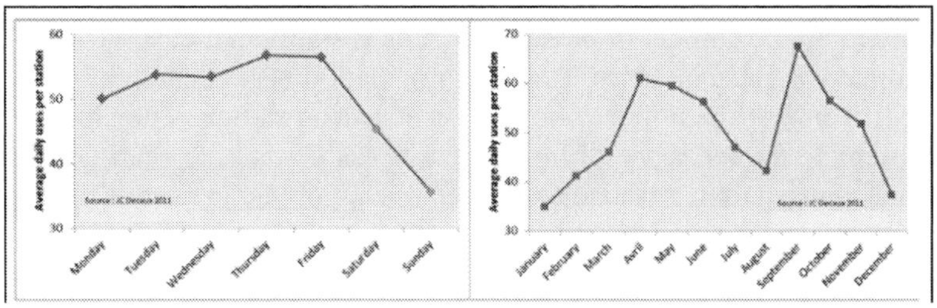

Figure 3. Average daily bike sharing trips per station by day (on the left) and by month (on the right) in 2011, source: using data from JC Decaux.

4. Data

The main objective of the study is to estimate the daily flows of bike sharing using the various socio-economic factors determined at station level. The data used in this study consists of all bike sharing trips in 2011: over 6 million trips and the built environment data geo-computed by the GIS-transport modelling platform MOSART developed by the researchers from Transport, Urban Planning and Economics Laboratory. Firstly, we are going to explain the processes of determination of bike sharing flows and socio-economic variables. Secondly, the multiple linear regression models are going to presented and used to estimate the daily flows of bike sharing.

4.1 Bike sharing trips data

In this study, we used the bike sharing trips data obtained from JC Decaux – operator of Lyon bike sharing system, for all stations during the year of 2011. Each trip gives us information about the departure and arrival stations, the date and hour of check in and check out and the type of subscribers. In order to calculate the flows of bike sharing, we aggregated the bike sharing trips per hour. All non-valid trips were eliminated. A non-valid trip is a trip less than 3 minutes or more than 2 hours. The data aggregated were calculated only for working days (from Monday to Friday and not during school holidays and public holidays). We eliminated also the bike sharing trips made during July and August because they are the months of vacations in France. Finally, 178 working days were counted for calculating bike sharing flows in 2011. We eliminated from our analysis 6 stations that were partially or totally inactive in 2011; 10 stations were combined to 5 couples of stations because they are situated very close to each other's. Finally, there are 331 bike sharing stations in our study.

Table 1. Descriptive on the daily usage of bike sharing per station in Lyon

	Sample	Minimum	Maximum	Average	Std. Deviation
Daily inbound flow	331	2	383	59	50
Daily outbound flow	331	1	345	57	46

4.2 Built environment variables

In order to analyze the influence of built environment on bike sharing usage, we determined built environment variables around each station. The explicative variables used in our analysis can be categorized in five groups: public transport variable, socio-economic variable, topographic variable, bike sharing network variable and leisure variable. Different sizes for a buffer zone were tested: 200m buffer zone, 300m buffer zone and 400m buffer zone. A 300 meter buffer zone was chosen because the explanatory variables determined within a 300m buffer zone are statistically most correlated to the bike sharing flows and it was to found to be an appropriate walking distance between Vélo'v stations according to the results of bike sharing survey of Lyon in 2007. In terms of public transit variables, the number daily passengers of metro and tramway; and railway stations near a Vélo'v station were generated to examine the influence of public transit on bike sharing flows. The variables of public transit were normalized by the number of passengers of each station per day.

The socio-economic variables included 3 factors: population between 18 and 64 years old, number of jobs and average income in the buffer zone of a bike sharing station. The elevation of each station was calculated to examine the influence of topographic variable on bike sharing usage. The length of bicycle facilities in the buffer zone was also calculated to capture the impact of placing Vélo'v stations near bicycle facilities on the usage of the bike sharing system. Leisure variables are also considered in our analysis: number of restaurants, number of cinema, and number of night clubs; the presence of a park and the presence of Embankment Road in the buffer zone of a bike sharing station. In terms of public transit system, there are 2 variables considered: the number of daily passengers of metro and tramway; and the number of daily passengers of railway station.

The number of bike sharing stations in a 3,500 meter buffer zone, the number of bike sharing stations in a 300-meter buffer zone around a Vélo'v station and the capacity of each Vélo'v station were computed to capture the effect of bike sharing network on the daily usage. The 3,500 meter buffer zone which covers about 95% bike sharing trips represents the opportunity of findings a bike sharing station of an user when he want to make a trip. The 3 variables of bike sharing network were added a posteriori of our modelling in order to improve the predictive quality. Finally, there are 15 socio-economic variables and 3 bike sharing network variables that we are using in our modelling of daily flows of bike sharing.

Table 2. Descriptive summary of explanatory variables

Continuous Variables	Min	Max	Mean	Std. Deviation
Population in 300m buffer	4	10977	4707	2478
Employment in 300m buffer	148	11828	2332	2111
Average income in 300m buffer per household	19963	36180	28559	2944
Elevation of bike sharing station (m)	164	289	181	28
Station capacity	10	40	19	6
Number of bike sharing stations in 300m buffer	0	9	1.8	1.8
Number of bike sharing stations in 3500m buffer	45	277	188	58
Number of restaurants in 300m buffer	0	28	3.1	5.3
Number of night clubs in 300m buffer	0	10	0.6	1.5
Number of cinemas in 300m buffer	0	4	0.25	0.68
Number of supermarkets in 300m buffer	0	6	0.7	1,1
Number of shoes and clothes shops in 300m buffer	0	223	13	33
Number of daily passengers of metro and tramway in 300m buffer	0	43562	5050	8659
Number of daily passengers of railway in 300m buffer (x 5000)	0	20	0.26	2.0
Length of bike facility in 300m buffer (m)	0	2835	1025	650
Binary variables	Min	Max	Percentage	
Park in 300m buffer	0	1	4%	
University in 300m buffer	0	1	8.5%	
Embankment road	0	1	8.5%	

5. Method

We use a multiple linear regression model to estimate dependent variables such as arrival and departure flows. Multiple linear regression attempts to model the relationship between two or more explanatory variables and a response variable by fitting a linear equation to observed data. In this study, the response variable is the daily flows of bike sharing and the explanatory variables are the socio-economic variables determined at station level.

The arrival and departure flows at a daily level for each station were used in the regression model. Let i = 1, 2..., 331 be an index to represent each station. The dependent variable (arrival or departure flow) is modeled using a multiple linear regression equation which has the following structure:

$$Y_i = \beta X_i + \varepsilon$$

Where Yi is the arrival or departure flow at the station i as dependent variable, Xi is a vector of explanatory variables determined around bike sharing station i. The model coefficients, β, are what we have to estimate. The random error term, ε, is assumed to have a normal distribution across the dataset. There are 4 principal assumptions that we have to verify in order to validate a multiple linear regression model: (1) normality of residuals, (2) autocorrelation of residuals, (3) homoscedasticity and (4) multicollinearity. If these assumptions are violated, a transformation of variables may be used. In order to ensure the validation of 4 assumptions and the robustness of the model, we have tested many different transformation methods. Finally, the power model seems to be the best choice to achieve linearity for our multiple linear regression analysis. The details of variables transformation are presented in the Table 3.

Table 3. Transformation of variables for multiple linear regressions

Variables	Transformation
Daily inbound flow of bike sharing	Log(Daily inbound flow of bike sharing)
Daily outbound flow of bike sharing	Log(Daily outbound flow of bike sharing)
Population in 300m buffer	Log(Population in 300m buffer + 1)
Employment in 300m buffer	Log(Employment in 300m buffer + 1)
Average income in 300m buffer per household	Log(Average income in 300m buffer per household)
Elevation of bike sharing station (m)	Log(Elevation of bike sharing station (m))
Station capacity	Station capacity
Number of bike sharing stations in 300m buffer	Log(Number of bike sharing stations in 300m buffer+1)
Number of bike sharing stations in 3500m buffer	Log(Number of bike sharing stations in 3500m buffer)
Number of restaurants in 300m buffer	Log(Number of restaurants in 300m buffer + 1)
Number of night clubs in 300m buffer	Log(Number of night clubs in 300m buffer + 1)
Number of cinemas in 300m buffer	Log(Number of cinemas in 300m buffer + 1)
Number of supermarkets in 300m buffer	Log(Number of supermarkets in 300m buffer + 1)
Number of shoes and clothes shops in 300m buffer	Log(Number of shoes and clothes shops in 300m buffer + 1)
Number of daily passengers of metro and tramway in 300m buffer	Log(Number of daily passengers of metro and tramway in 300m buffer + 1)
Number of daily passengers of railway in 300m buffer	Log(Number of daily passengers of railway in 300m buffer + 1)
Park in 300m buffer	Park in 300m buffer
University in 300m buffer	University in 300m buffer
Length of bike facility in 300m buffer (m)	Log(Length of bike facility in 300m buffer + 1)
Embankment road	Embankment road

6. Results and discussion

In this section, the results of multiple linear regression model estimation are discussed in order to understand the different effects of built environment on the bike sharing usage in the city of Lyon. We tested 4 models in order to estimate the daily inbound and outbound flow of bike sharing. For each type of flow, we present two models: a model using 15 socio-economics variables and the second model using 15 socio-economics variables and 3 bike-sharing network variables. The statistically significant results for these models are presented in Table 4 and Table 5.

Table 4. Model estimation results for daily inbound flows

Variable	Coefficient of model 2	Coefficient of model 1
Intercept	4.025	3.277
Park	0.006	0.160**
Embankment road	0.022	0.037
Night club	-0.014	0.113
Shoes and clothes shop	-0.012	-0.040
Cinema	0.095	0.123
Restaurant	0.045	-0.025
Population	0.065	0.212**
Employment	0.117**	0.317**
Income	0.945**	1.508**
Public Transport	0.022**	0.029**
Elevation	-4.656**	-4.661**
University	0.197**	0.251**
Railway station	0.011	0.035**
Bike facility	0.041**	0.065**
Supermarket	0.139**	0.166**
Number of bike sharing stations in 3500m buffer	1.237**	-
Station capacity	0.014**	-
Number of bike sharing stations in 300m buffer	-0.192**	-
Mean Squared Error (MSE)	*0.031*	*0.051*
Coefficient of determination R2-ajusted	*86.4%*	*78.0%*

***: significant variable at 99%; *: significant variable at 95%; -: variable non-utilized*

In terms of the Mean Squared Error (MSE) and the coefficient of determination of the two models: the model 2 has a better coefficient of determination and a less important mean squared error than the model 1. In this case, an F-test is conducted to see which model is statistically better. It gives a definitive answer and does not rely on arbitrary interpretation of an R2 value or mean squared error. The value of F-statistic is equal to 65.20 and the value of F-distribution is equal to 3.84; this means that the model 2 is statistically better than the model 1; and the 3 variables of bike sharing networks contribute to the improvement of the model. The same F-test was conducted to find the better model between model 3 and model 4; and we found that the model 4 with the presence of 3 bike sharing network variables is statistically better than the model 3. We are going to focus on the influence of explanatory variables on the daily inbound flows of bike sharing. Each group of variables is separately analyzed.

Table 5. Model estimation results for daily outbound flows

Variable	Coefficient of model 4	Coefficient of model 3
Intercept	-1,372	-2,734
Park	-0,020	0,172*
Embankment road	0,019	0,041
Night club	-0,068	0,089
Shoes and clothes shop	-0,025	-0,052
Cinema	0,114	0,132
Restaurant	0,069	0,002
Population	0,030	0,233**

Employment	0,105*	0,350**
Income	0,617*	1,393**
Public Transport	0,022**	0,030**
Elevation	-1,805**	-1,840**
University	0,187**	0,242**
Railway station	0,009	0,035**
Bike facility	0,052**	0,078**
Supermarket	0,107*	0,160**
Number of bike sharing stations in 3500m buffer	1,524**	-
Station capacity	0,013**	-
Number of bike sharing stations in 300m buffer	-0,169**	-
Mean Squared Error (MSE)	0.031	0.051
Coefficient of determination R2-ajusted	81.1%	65.8%

***: significant variable at 99%; *: significant variable at 95%; -: variable non-utilized*

6.1. Public transport and bicycle infrastructure variables

In terms of public transport variables, we observe that public transport is significant in the two models of regression. It means that public transport system (metro and tramway) has a positive impact on the daily inbound flows. It's about those who may combine bike sharing with public transport and/or those who may use bike sharing in order to avoid peak period usage in public transport. The variable Railway station is positively significant in the model 1 but is not significant in the model 2. An explanation for this fact is that the introduction of 3 bike sharing network variables hidden the influence of this variable because these 3 variables are important near the railway stations. The daily inbound flows of bike sharing station increase when there are more bicycle facilities nearby a station. This result is corresponding to the findings of Buck et al. 2012 and Faghih et al. 2014.

The 3 variables of bike sharing networks are significant in the model 2 and the model 4. It indicates the bike sharing network variables are important to bike sharing ridership. Bike sharing ridership increases when the capacity of station increases. The impact of the number of stations in a 3500 meter buffer zone and the number of stations in a 300 meter buffer zone need to be examined as a combination: when increasing the number of stations in a 300 meter buffer zone, the number of stations in a 3500 meter buffer zone also increases. So the effect of adding a station is the combination of these 2 variables. In fact, we can see that impact of number of bike sharing stations in 3500 meter buffer zone is 9 times bigger than impact of number of bike sharing stations in 300 meter buffer zone. It highlights that adding more stations is likely to increase bike sharing usage.

6.2. Land use and leisure variables

A bike sharing station located in an area with higher population and employment tends to have a higher daily inbound flows. This finding is compatible with the results of Rixey in 2013 and Wang in 2012. It is interesting to see the positive effect of income on the daily inbound flows of bike sharing. This result can be explained by the fact that bike sharing system is installed in the city and the need of credit card in order to be able to use this service. In terms of leisure and shopping variables, we observe that the variable Supermarket is positively significant in the two models. It indicates that bike sharing seems to be used for doing smalls shopping trips because it is difficult to transport a bulky stuff by bike. The other leisure variables like Cinema, Restaurant, and Night club are generally not significant in the four models, except the presence of park near a bike sharing station. This fact indicates that, bike sharing seems to be used mainly for utility trips on a weekday. Bike sharing stations near universities are likely to experience a higher volume of daily flows than bike sharing stations far from universities. The result is corresponding to findings in the Survey of bike sharing system of Lyon in 2007 that more than 30% of bike sharing users are student and in an agreement with the findings of Faghih et al. 2014 and Hampshire et al. 2012.

In terms of the Elevation variable, the number of bike sharing trips decreases with the increase of elevation. This variable is negatively significant in the model of inbound and outbound flow. This finding suggests that the inbound and outbound bike sharing ridership decrease in hilly zones. The negative effect of elevation on outbound bike sharing ridership can be explained by the fact it is more difficult to find a bike at stations situated in hilly zone due to unavailability of these stations. Bike sharing operators should take into account the effect of elevation for a better regulation of bike sharing.

7. Conclusions

Currently, bike sharing systems have become more and more present all around the world. This trend has led to build models in order to understand the influence of socio-economic characteristics on bike sharing usage and then to quantify the bike sharing flows by using these characteristics. The population, employment, university, park, shops, supermarket are statistically significant while the leisure variables such as cinema, restaurant, night club are not significant in the models suggest that bike sharing seems to be principally used for utility trips during a weekday.

This study also indicates that the bike sharing network variables such as: capacity, number of bike sharing stations in a 300m buffer zone and number of bike sharing stations in a 3500m buffer zone contribute to improve the model. This indicates that a dense and well amenities-connected bike sharing network with a good availability of stations contribute importantly to the increasing of bike sharing ridership.

The models estimated in this study can be directly used for determining the daily flows for an extension of current bike sharing system in the Metropole of Lyon. By eliminating some variables that appropriated to Lyon Metropole such as Embankment Road, these models can be used for positioning and dimensioning stations of a new bike sharing system with available socio-economic data.

Finally, the bike sharing data used in this study are based on mature-adopting bike sharing users; so, traffic planners should be aware that the results of modelling for a new bike sharing system could be different at the starting phase. The same process of analysis in this study can be used for an analysis on a car sharing system in order to understand the influence of different factors of built environment on car sharing usage and to quantify car sharing trips with some adjustments on the size of buffer zone and explanatory variables.

Although these models of bike sharing demand give a very good prediction quality, we can always bring several suggestions for improving the models. Firstly, the models are based on the bike sharing and socio-economic data in 2011. It is preferable to expand the study on a longer period in order to evaluate the temporal transferability of these models. It is interesting to be able to conduct a study on another bike sharing system in France with different sizes of system using the same method in this study. Secondly, a 300m buffer zone was adopted in this study with a hypothesis that the accessibility to a bike sharing station is spatially homogenous. In reality, a buffer zone that corresponds to a 5-minutes or 10-minutes of walking should improve the quality of explanatory variables. There may have other important explanatory variables for bike sharing trips that we cannot introduce in this study such as: weather variables, time of year, seasons, degree, precipitation, visibility of stations, etc. These variables may contribute not only to a better explanation of bike sharing ridership but also a more comprehensible influence of other parameters on bike sharing usage.

References

[1] Bargar, A., Gupta, A., Gupta, S., Ma, D., 2014. Interactive Visual Analytics for Multi-City Bike share Data Analysis. *The 3rd International Workshop on Urban Computing* (UrbComp 2014), New York, USA.

[2] Buck, D., Buehler, R., 2012. Bike Lanes and Other Determinants of Capital Bike share Trips. Paper #12-3539. *91st Annual Meeting of the Transportation Research Board,*

Washington, D.C., 2012.
[3] Crozet, Y., Mercier, A., Ovtracht, N., 2012, Accessibility: a key indicator to assess the past and future of urban mobility, in Geurs, K.T., Krizek, K., J., Reggiani, A., (Editors) *Accessibility Analysis and Transport Planning Challenges for Europe and North America*, Edward Elgar Publishing Limited. 263-279.
[4] DeMaio, P., 2009. Bike-sharing: history, impacts, models of provision, and future. *Journal of Public Transportation 12*, 41-56.
[5] Enquête Usagers Vélo'v 2007, (2008). Grand Lyon.
[6] Faghih-Imani, A., Elurua, N., El-Geneidy, A.M., Rabbat, M., Haq, U., 2014. How land-use and urban form impact bicycle flows: Evidence from the bicycle-sharing system (BIXI) in Montreal. *Journal of Transport Geography Volume 41*, December 2014, 306-314.
[7] Froehlich, J., Neumann, J., Oliver, N., 2009. Sensing and predicting the pulse of the city through shared bicycling. *In Proceedings of the 21st International Joint Conference on Artificial Intelligence*, IJCAI'09, San Francisco, CA, USA, 2009. Morgan Kaufmann Publishers Inc., 1420-1426.
[8] Gonzalez-Feliu, J., Morana, J. (2011), Collaborative transportation sharing: from theory to practice via a case study from France. In Yearwood, J.L. and Stranieri, A. (eds.), *Technologies for Supporting Reasoning Communities and Collaborative Decision Making: Cooperative Approaches, Information Science Reference*, Hershey, 252-271.
[9] Hampshire, R., Lavanya, M., Eluru, N., 2013. An Empirical Analysis of Bike Sharing Usage and Rebalancing: Explaining Trip Generation and Attraction from Revealed Preference Data. Technical Paper, Heinz College, Carnegie Mellon University.
[10] Jäppinen, S., Toivonen, T., Salonen, M., 2013. Modelling the potential effect of shared bicycles on public transport travel times in greater Helsinki: An open data approach. *Applied Geography*, 43:13-24, 2013.
[11] Jensen, P., Rouquier, J.B., Ovtracht, N., Robardet, C., 2010. Characterizing the speed and paths of shared bicycle use in Lyon. *Transportation Research Part D: Transport and Environment 15*, 522-524.
[12] Kaltenbrunner, A., Meza, R., Grivolla, J., Codina, J., Banchs, R., 2010. Urban cycles and mobility patterns: Exploring and predicting trends in a bicycle based public transport system. *Pervasive and Mobile Computing*, 6(4):455– 466, 2010.
[13] Larson J., 2014. Bike share fact sheet. *Earth Policy Institute*. Available from Internet: http://www.earth-policy.org/press_room/C68/bicycle_share_fact_sheet.
[14] Lathia, N., Saniul, A., Capra, L. 2012. Measuring the impact of opening the London shared bicycle scheme to casual users. *Transportation Research Part C: Emerging Technologies*, 22:88–102, June 2012.
[15] LDA Consulting, 2013. Capital Bikeshare 2013 Member Survey Report. Available from Internet:: http://www.capitalbikeshare.com/assets/pdf/CABI-2013SurveyReport.pdf
[16] Melbourne Bike Share survey results. Available from Internet: http://www.melbournebikeshare.com.au/survey-results.
[17] Politiques nationales en faveur du vélo – ISBN 92-821-2326-X - © CEMT, 2004.
[18] Record d'utilisation pour les Vélo'v en 2014. Available from Internet: http://www.lyoncapitale.fr/Journal/Lyon/Actualite/Actualites/Transports/Record-d-utilisation-pour-les-Velo-v-en-2014
[19] Rixey, R., 2013. Station-Level Forecasting of Bike Sharing Ridership: Station Network Effects in Three U.S. Systems. *Paper presented at the 92nd Transportation Research Board Annual Meeting 2013*, Washington, DC.
[20] Shaheen, S., Guzman, S., Zhang, H., 2010. Bikesharing in Europe, the Americas, and Asia: Past, Present, and Future. *Transportation Research Record 2143*, 159-167.
[21] Tran, T.D., Ovtracht, N., Faivre d'Arcier, B., 2014. Developing the bike sharing system to promote urban sustainable mobility: the case of Lyon city. *Polis Annual Conference*,

Madrid, Spain, 2014.
[22] van Lierop, D., Grimsrud, M., El-Geneidy, A., 2013. Breaking into bicycle theft: Insights from Montreal, Canada. *Forthcoming International Journal of Sustainable Transportation*.
[23] Wang, X., Lindsey, G., Schoner, J., Harrison, A., 2012. Modelling bike share station activity: the effects of nearby businesses and jobs on trips to and from stations. *Paper presented at the 92nd Transportation Research Board Annual Meeting 2012*, Washington, DC.
[24] Woodcock, J., Tainio, M., Cheshire, J., O'Brien, O., Goodman, A., 2014. Health effects of the London bicycle sharing system: health impact modelling study. *BMJ: British Medical Journal*, 348, 2014.

Acknowledgements
The authors thank JC Decaux, the ANR Vél'Innov project and the Metropole of Lyon for sharing data sets for use in this study.

Study on energy efficiency in Vietnamese row house - case study of Tang Nhon Phu A Ward, District 9, HCMC

H T P Thao[1, 2], N D Nam[3]

[1] Lecture of Regional and Urban Planning Department, Civil Engineering Faculty, Ton Duc Thang University, email: hoangthiphuongthao@tdt.edu.vn
[2] PhD. student, Urban Planning Department, National Cheng Kung University, Tainan, Taiwan, email: P28077016@email.ncku.edu.tw
[3] Lecture of Architecture Department, Civil Engineering Faculty, Ton Duc Thang University, email: nguyendinhnam@tdt.edu.vn

Abstract: District 9 is a new urban development area which is well-known as an industrial city in the rear of Ho Chi Minh City. This area has special characteristics which are different from central downtown. The housing of district 9 is a new urban area so that it is easier to orient particular architectural style accordance with climatic conditions and other conditions. However, at present, these new residential areas are rising with no clear management of architectural form. That is the reason lead to shortcomings such as: increasing the use of electricity, affecting to climate change and creating urban heat island, increasing costs of energy use. Those problems will be difficult to overcome in the future if we don't have the right attention on it.
By using a combination of multiple methods such as: data collection, case study analysis method, GIS, research by design, etc., this research topic will pay attention on analysis the row house in Tang Nhon Phu A Ward of district 9 and trying to propose some solution and management criteria to local government. The analysis results will ensure that the conclusions reflect the realities of the situation, those also become the basis for the proposed solutions to deal with the existence problems. The result of research may become an application research platform for the related research topics.

1. Introduction

Utilizing energy savings has always been the interesting topics for many scientists around the world especially in the context of climate change. [1]

In the welded and temperate countries, the model of Passive House was introduced in the 1990s under the management of Passive House Institute, Germany. Accordingly, a passive house must achieve a minimum heating energy consumption of 15kwh/m2. This theory becomes more and more popular. More than 25000 Passive House have been built around the world [2].

An example in European country is the Swedish Passive House. This building was built followed the Passive House Standard to reduce the using of energy (approximately 65-75%) [3].

In tropical countries, There are so many solutions to cost down the energy consumption such as creating a fresh micro environment by using the insulation material, using energy saving equipment, etc. [4]. The example of China, the solutions are replacing ordinary material with insulating material, using double-glazed windows, proposing the solutions for functional space suitable with the physical condition of the research site. This model is being widely used to promote energy saving in this country [5].

There are a lot of studies on energy saving on over the world. However, there are different climatic characteristics and citizen habits in different locations. So, further research is needed to meet the needs of differentiated areas.

In Vietnam, the traditional housing is considered to be in harmony with nature and people. Under the influence of global climate change and rapid urbanization, the urban housing have not yet ensured the comfort of the weather conditions.[6]. More than that, The ineffective use of household appliances plays a non-trivial role that affects installation, frequency of consumption, and energy savings [7]. Energy-saving housing models are almost unattended. A part of the architects in Vietnam began to notice the problem of the architecture but also limited in terms of research. Most of the solutions are only concerned with the development of external forms, compromise with microclimate by technology-based solutions, or have not been able to address practical problems. The principal findings of the study are in the concept of green architecture, but there are no specific solutions in the design but only in the general theory or in the exploration of energy efficient solutions (LED).[8]
For those reasons, the targets of research topics are to propose the changing of micro-climate by giving the suitable architectural design solutions. Moreover, the row house owner's behavior is also considered in this study.

This research topic will pay attention on analysis the Tang Nhon Phu A Ward of district 9 and trying to propose some solution and management criteria to local government. The result of research may become an application research platform for the related research topics. By applying GIS to the analysis of microclimatic data and visualization of results, it will become a positive support method for architects presenting their solutions and evaluating the results. This study is limited to one variable in temperature; other variables such as humidity, light intensity, etc. will be introduced in the next section.

2. Methodology

District 9 is a new urban development area which is well-known as an industrial city in the rear of Ho Chi Minh City. This area has special characteristics which are different from central areas of Ho Chi Minh City. The housing of district 9 is a new urban area so that it is easier to orient particular architectural style accordance with climatic conditions and other conditions. However, at present, these new residential areas are rising with no clear management of architectural form. That is the reason lead to shortcomings such as: increasing the use of electricity, affecting to climate change and creating urban heat island, increasing costs of energy use. Those problems will be difficult to overcome in the future if we do not have the right attention on it.

This study is planning to get the data from a set of research methods such as: data collection[9], case study analysis method [10,11], GIS[12,13], research by design, etc.

The data collection method will provide the primary data and secondary data from multiple sources. After that, case study analysis method will support the researcher a tool to make the analysis of the research area. Moreover, a set of ArcGIS's toolboxes also was applied to analyses the physical condition of row house in District 9, HCMC. The proposed solutions will apply on a real row house in the site to re-evaluate the effectiveness of the research results (research by design methods).

Through these methods, supportive research data will be collected, suggesting appropriate solutions through the research process below:

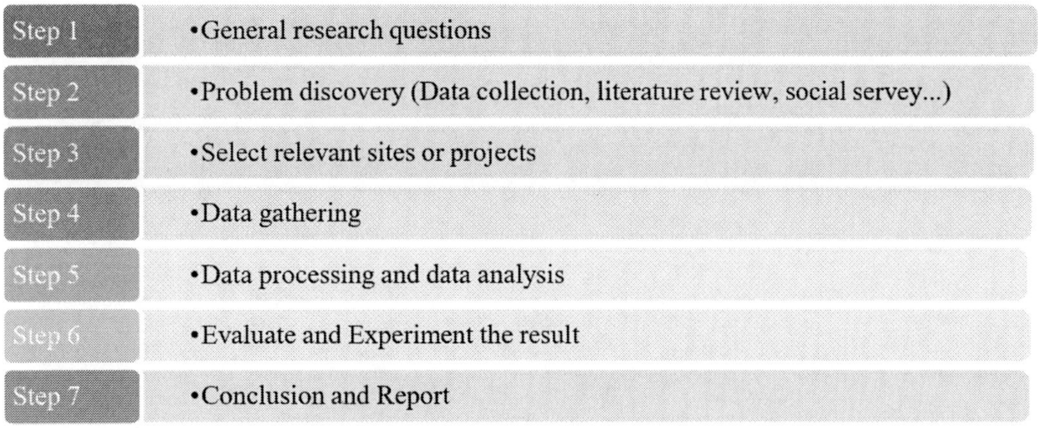

Figure 1. The research process

By using a combination of multiple methods, the analysis results will ensure that the conclusions reflect the realities of the situation that is the basis for the solutions proposed in the following sections.

3. Research results

3.1. Overview

As followed the information of HCMC Statistics Department in 2010, District 9 is about 7 km far from the city center along the Xa Lo Ha Noi Highway. There are 13 wards with an area of 114 km^2, population of 263,486 people, density of 2,311 people / km^2. Despite being urbanized since 1997, residents in District 9 are still relatively low in comparison with newly established districts such as Binh Tan District, Tan Phu District[14]. Hi-tech Park of Ho Chi Minh City located in district 9 has been built and put into operation. The new residential area of district 9 was formed under the positive influence of High Tech Park such as: Singa City, TDL Residence, HTReal 898, Tang Long River View, Valencia Riverside, Viet Nhan Villa Residence, Nam Khang Residence, Thai Duong Luxury, etc. Though, District 9 is the largest and least populated district in comparison with the rest of Ho Chi Minh City at present, this area is forecasted to grow very fast in the future with the development dynamics of local industrial zone and the synchronous technical infrastructure system.

Tang Nhon Phu A Ward (population: 23.158 people, area: 418.98 ha) is one of center ward of district 9, located near Hiep Phu Ward. Over the years with the impact of the market mechanism, the rapid increasing of industrial, service projects and the development of residential areas of Tang Nhon Phu A Ward has resulted in a lot of difficulties in the management of the land. Moreover, it also has a great impact on the social life of local residents.

IOP Conf. Series: Earth and Environmental Science **143** (2018) 012071 doi:10.1088/1755-1315/143/1/012071

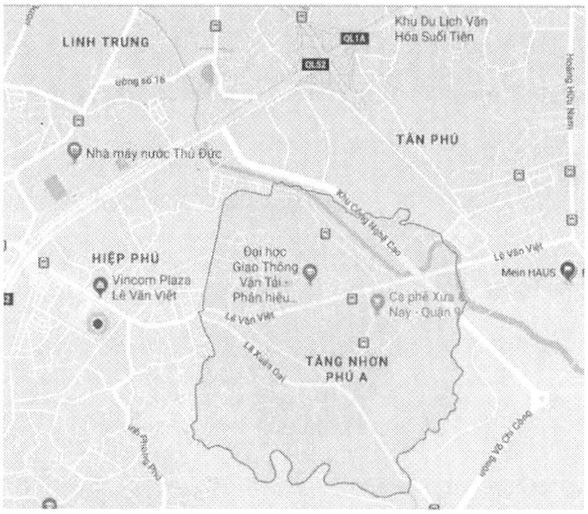

Figure 2. Location of Tang Nhon Phu A Ward in Google map

Figure 3. Location of Tang Nhon Phu A Ward in District 9's Master Plan from People's Committee of Tang Nhon Phu A Ward, District 9, 2017

Local climatic conditions have a great impact on each project as well as the design solution. According to general information about Ho Chi Minh City, the common characteristics of Ho Chi Minh City climate are high temperature (lowest average temperature is 25.7°C on the middle of December and January, the absolute high temperature is 38.6°C on April).

There are two seasons in Ho Chi Minh City. Rainy season begin from May to November, dry season begin from December to April next year. The rainfall is high, average 1.949 mm / year. The highest rainfall is 2,718 mm (1908) and the smallest rainfall was 1,392 mm (1958). The average number of rainy days is 159 days. Approximately 90% of annual rainfall is concentrated from May to November; June and September usually have the highest rainfall. Almost the north districts often have higher rainfall than the southern and southwestern districts.

The amount of radiation is plentiful, about 140 Kcal / cm2 / year. An average hour of sunshine / month is about 160-270 hours. The average air temperature is 27°C.

Relative humidity of air is 79.5% per year; average rainy season is 80% and absolute high values up to 100%; Average dry season is 74.5% and absolute low down to 20%.

Ho Chi Minh City is affected by two main wind directions and mainly monsoon winds in the West - Southwest and North - Northeast. The Western-Southwest winds from the Indian Ocean blew in the rainy season, from June to October, with an average speed of 3.6 m/s and the strongest winds in August, averaging 4.5 m/s. The North-East wind blows from the East Sea during the dry season, from November to February, with an average speed of 2.4 m/s. In addition, there is south - southeast wind, from March to May average speed of 3.7 m/s.

Due to extreme weather conditions, especially in terms of temperature and humidity in HCMC, housing projects need to be specially designed to reduce the use of artificial energy, creating comfortable for the people using the building.

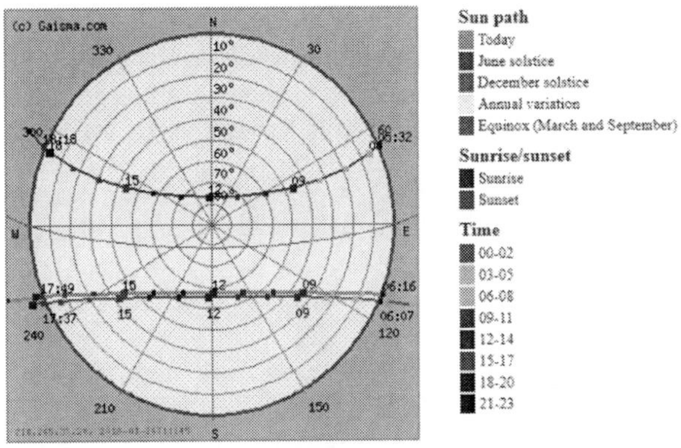

Figure 4. Ho Chi Minh City - Sun path diagram of Gaisma.com, 10/01/2018

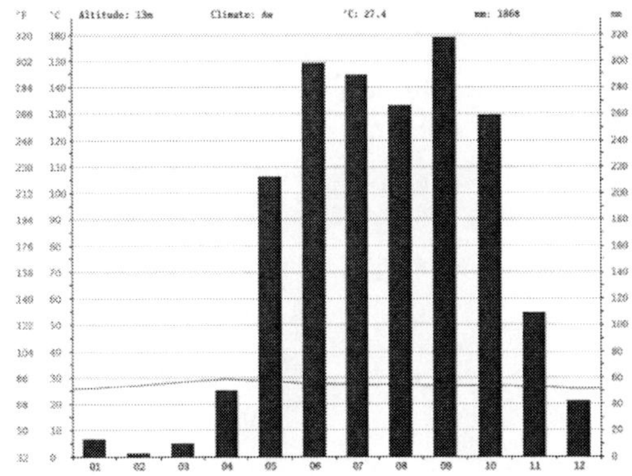

Figure 5. Climograph of Ho Chi Minh City from en.climate-data.org, 10/01/2018

Tang Nhon Phu A Ward has about 4,309 households, divided into 4 main types of home including: villas, apartment building, row houses, garden houses which are interspersed throughout the region. In the types of housing which was listed above, the type of row house occupied the majority (46.3%) of the dwellings in the area. In addition, the row house (also known as the townhouse, pipe house) has many limitations such as construction conditions, small area, etc. while the type of garden house and villas have the advantage of wide area of and easy to handle the architecture style or dealing with micro climate. This type of row house actually poses challenges for architects to create a design that assures efficiency in natural climate conditioning for energy savings.

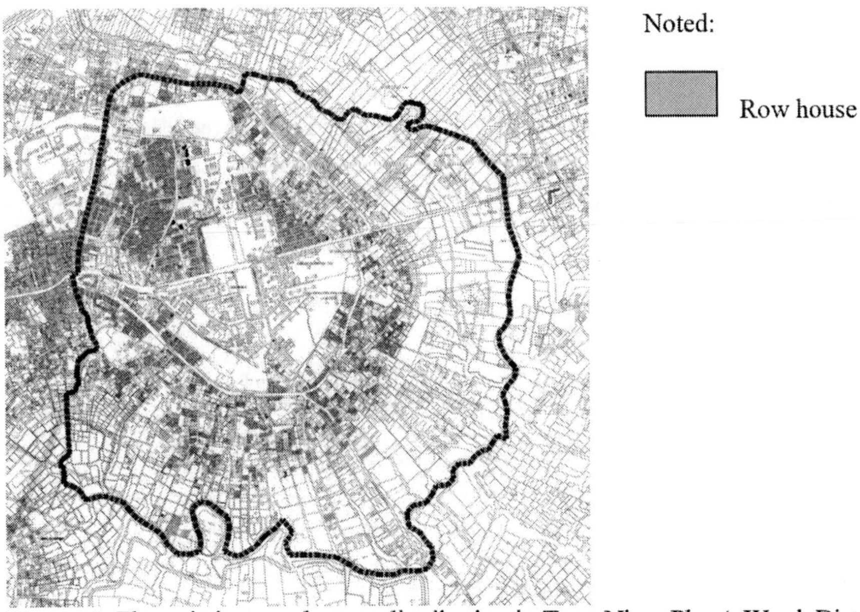

Figure 6. The existing row houses distribution in Tang Nhon Phu A Ward, District 9

Because of those reason, the research topic pay attention on the type of row house to give the suitable solution which may help improving the living quality and reduce the use of energy.

3.2. Case study analysis

There are two kinds of row house: one facade and two facade row house. The two-facade row house is usually stayed in the corner of each block. The amount of two- facade row house is less than the one-facade row house. The research team selected the sample for analysis. Sampling criteria should be the most commonly designed with local street-house designs. More than that, typical samples represent the type of one- facade row house and two-facade row house to carry out for analysis (see Figure below).

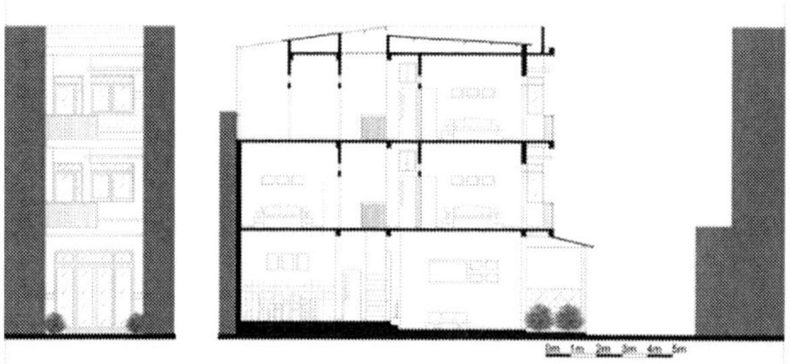

Figure 7. Sample of one-facade row house at No. 35, 449 Street, Tang Nhon Phu A, District 9, HCMC.

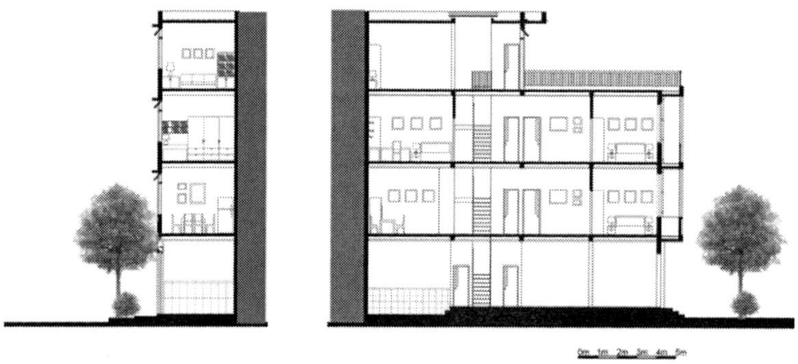

Figure 8. Sample of two-facade row house at No. 24, 447 Street, Tang Nhon Phu A, District 9, HCMC.

Each type of house has different advantages and needs to be considered separately, as shown in the table below:

Table 1. The type of row house comparison

Housing type	Strength	Weakness
One-facade row house	- On the side of the house is the neighbor's house, reducing the heat-resistant surface directly from the sun => lower indoor temperature	- No side windows => poor ventilation
		- Long house (~ 20-25m), only have 1 facade => the lack of natural light in the home if there are no suitable design solution

Two-facade row house	- Easy to obtain natural light	- Larger surface area to receive direct light => increase indoor temperature
	- Easy to get wind and light from the windows on both sides of the house	- Higher construction costs

The research team conducted a field survey and integrated the classification of row houses. After that, 2 typical row house samples are selected. Microclimate data and energy needs of households in Tang Nhon Phu A district were also collected and assessed by using appropriate tools.

Regarding the demand for energy, the research team used a random sampling method with 1% of total number of households in Tang Nhon Phu A Q9 ward (40 samples). The results of the survey show that 45.7% of the people are satisfied and above the level of satisfaction with their current home, 54.3% are not satisfied. The main cause of unsatisfied is high indoor temperature, small size of house and no ventilation.

The average household electricity price survey shows that monthly electricity bills range from 250,000VND to 750,000VND and up to more than 2 million VND. According to the landlord, electrical equipment caused most power consuming are: air conditioning, refrigerator, lighting equipment, washing machine, TV.

In addition, when asked about the criteria for the use of micro-climate control appliances including air conditioning and lighting equipment, the user said that they would start using the air conditioners when the room temperature approach 30°C. The temperature when using the air conditioner is at 19-26°C. The regular time frames at 10:30 - 14:00 or 19:00 - 22:00, full day use for mix-use row house. For lighting equipment, most people say they often go to work during the day so electrical equipment are used a lot at night. For household or home-based businesses, most homes lack natural light, and bulbs operate nearly 20/24 hours per day.

The survey results show that local people still have not met the needs in an optimal way. Most of them have to use temporary remedies, using artificial equipment such as air conditioning and lighting equipment to meet the daily needs of the family.

Data collection was conducted over a 1-year period. The data recorded is the data of temperature and illumination. These are two indicators directly affect the energy needs of the people. The article focuses on the analysis the temperature data (at the coolest and hottest month of year). Incorporation of data into analysis by using ArcGIS tool provides thermal interpolation diagram and the microclimate analysis diagram as follows:

3.2.1. One-facade row house

One-facade row house is often confined by neighbor buildings, so those building are unable to open the door for light and ventilation. There are many houses do not have enough length so they must to use the elevator space for creating air circulation. This also creates some of the effects on indoor microclimates

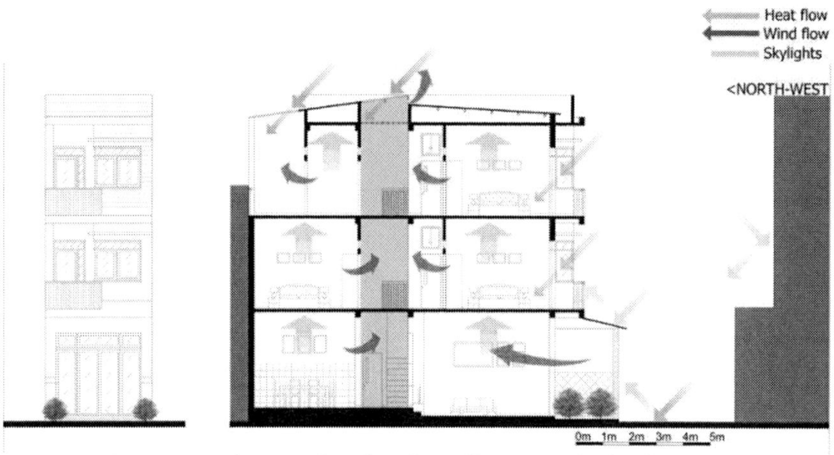

Figure 9. The one-facade microclimate analysis diagram

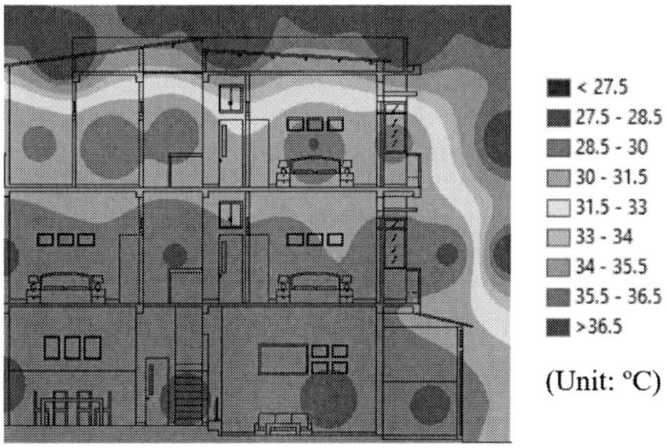

Figure 10. The one-facade thermal interpolation diagram on the coolest month of year

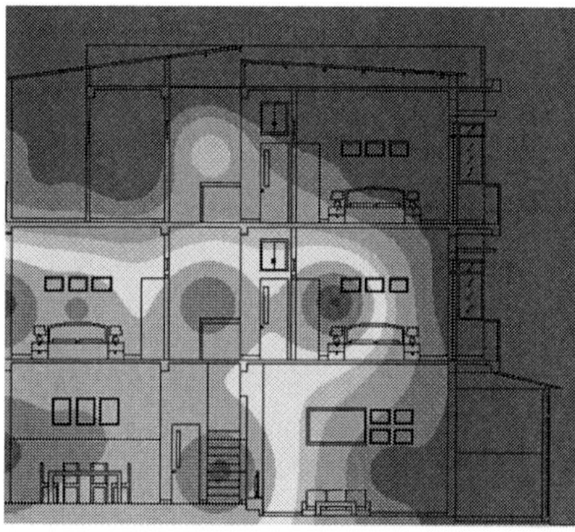

Figure 11. The one-facade thermal interpolation diagram on the hottest month of year

From the analysis results which were showed on the diagram above, we can easily find that:

- This project is short of length so it is not possible to build a vent through the floor, and poor circulation of air.
- No ventilation at the end of the house because of no window or back door
- The front bedroom is heavily impacted by sunlight while lacking in shielding.
- Indoor space is missing natural light.
- Looking at the heat spectrum on the thermal interpolation diagram, the temperature inside the room is almost under the influence of the sun. The fact that the building is surrounded by adjacent buildings, only the roof and facade is affected by the solar radiation.
- As the temperature of the outside air increases and the air circulation in the house decreases, the internal temperature also increases, the cool space in the house decreases dramatically.

In order to cool the interior space, the owners are usually equipped with air conditioning system, or fans ... These devices contribute to energy consumption and heat generation during the operation.

3.2.2. Two-facade row house

The two-storey row house type is usually built at the corner of the road, so there are many conditions to open windows for lightening and ventilation. Most of this type of house rarely was designed with skylight. The windows, balconies are usually arranged to get the natural light and release the air from inside to the outside.

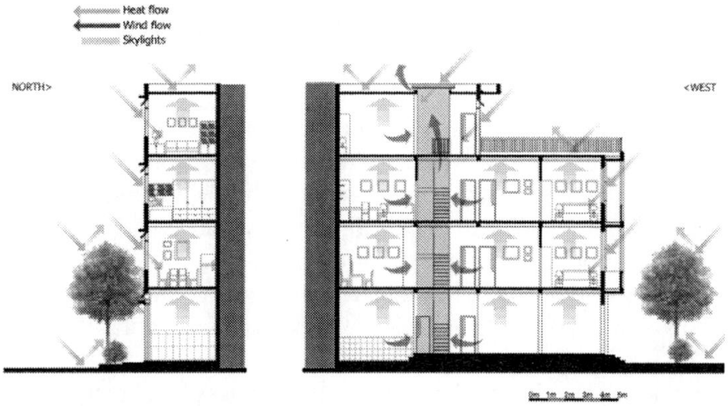

Figure 12. The two-facade microclimate analysis diagram

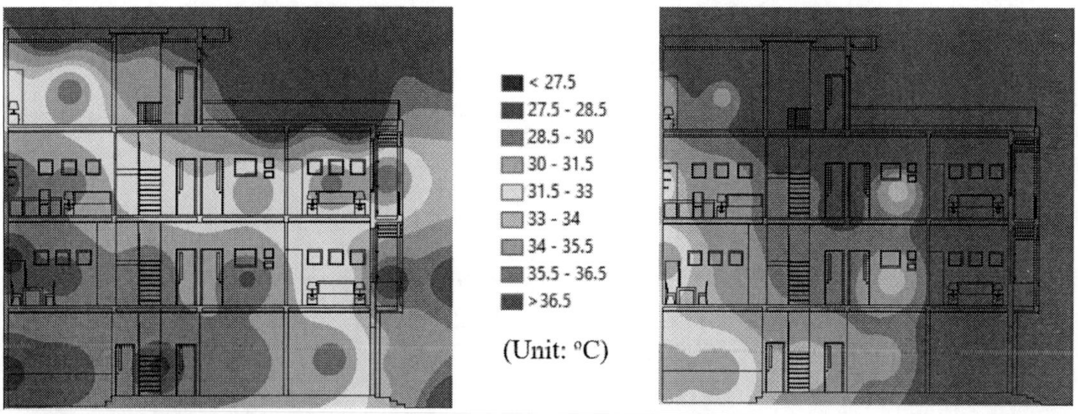

Figure 13. The two-facade thermal interpolation diagram on the coolest month of year

Figure 14. The two-facade thermal interpolation diagram on the hotest month of year

The analysis results of this case are:

- Because there are two facades exposed to the external environment, the building almost absorbed large amounts of radiation from the sun.
- The building is covered with a metal roof or a layer of reinforced concrete roof so that the sun affects directly to the roof, the terrace and indirectly affect the room inside.
- At sunrise, most of the interior space is heated. The inner temperature is always high (from 30-36oC)
- The windows are also affected by the sun, which increases the room temperature.

The two-facade buildings normally suffer more solar radiation than other types of facades, thus greatly increasing the room temperature. The traditional solution is to use a fan or air conditioning to reduce the heat.

3.3. The proposed solutions

In order to solve the thermal problems for the one and two-facades, the researcher proposes some solutions as follows:

- The one-facade row house:
- It is recommended to reserve a space behind the house for back yard or atriums (size may vary from 0.5 - 2m). Functional rooms inside the home should be designed in close contact with outside open space by windows or corridors. The roof is built in two layers (a layer of beneath concrete roof combined with one layer of tole roof above). The front surface should add a sunshade system to reduce exposure to the sun and to decorate.
- The one-facade row house:
- The building takes advantage of the two-facade building to get sunlight from both sides, but builds the side corridors, or balconies to minimize the heat. The roof is built in two layers (a layer of beneath concrete roof combined with one layer of tole roof above). Those layers have the appropriate distance to create an air cushion. The façade uses a system of sunshades, as well as a closed balcony structure. The staircase skylight is designed higher than usual with ventilation slots. The outer walls are constructed in three layers, consisting of two layers of bricks and one layers of air in the middle. The main doors and windows is designed high (2m high for window and 3m high for main door) with flaps for natural ventilation.

The above solutions are applicable to the same type of houses. The construction size is equivalent and the neighboring position. Those houses have the same direction and similar impact of the external environment.

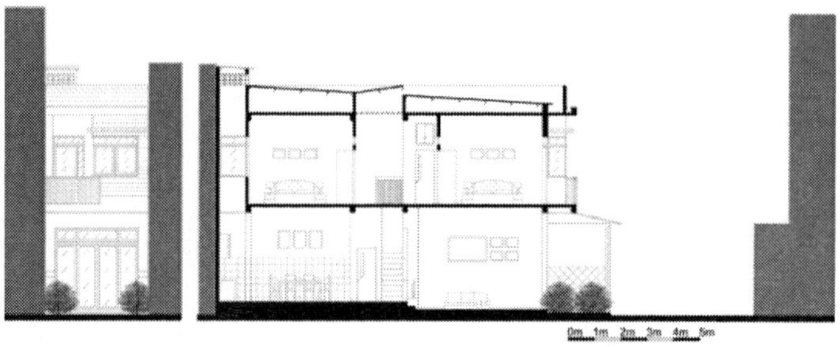

Figure 15. One - facade housing at No. C25/2, 449 street, Tang Nhon Phu A Ward, District 9, HCMC.

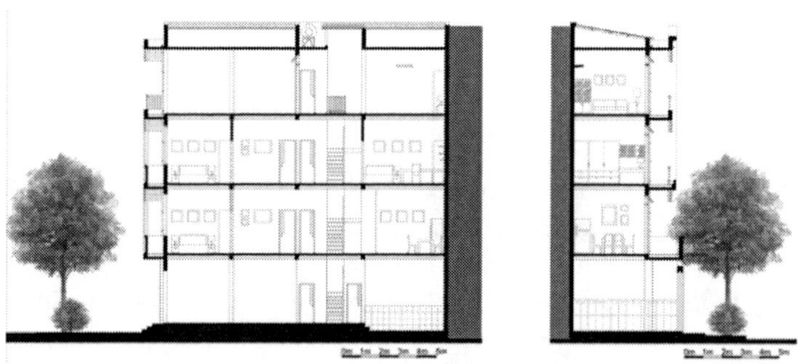

Figure 16. Two - facade housing at No. 74G, Lang Tang Phu street, Tang Nhon Phu A Ward, District 9, HCMC.

3.3.1. One-facade row house

This case study was built in internal road only 5m wide. The height of row houses is from 7m to 12m. The main road of the site is East - South. This house is designed with a small backyard and a skylight at the bottom of the staircase which is connect dining room with living room on the ground floor and 2 bedrooms on the 2nd floor.

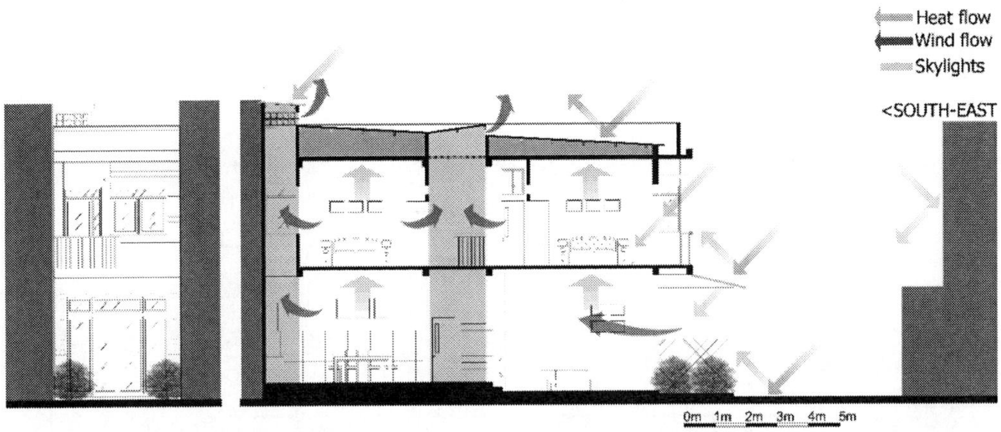

Figure 17. The two-facade microclimate analysis diagram
(Address: No. C25/2, 449 street, Tang Nhon Phu A Ward, District 9, HCMC.)

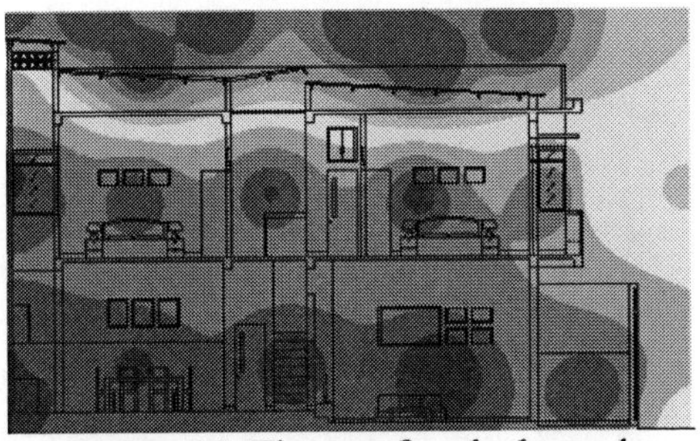

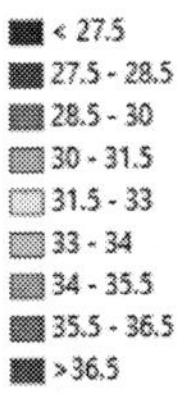

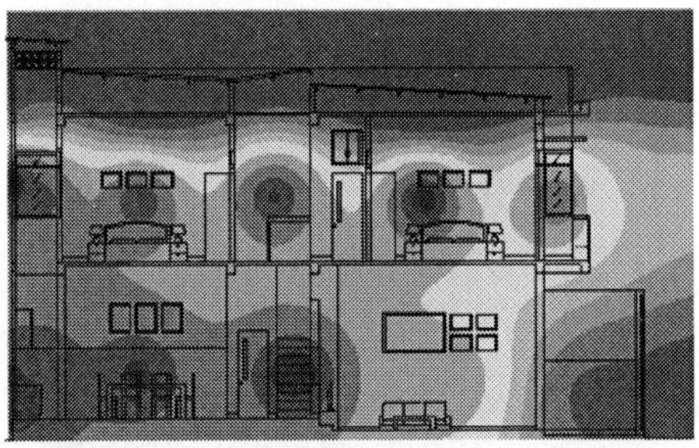

Figure 18. The one-facade thermal interpolation diagram on the coolest month of year

(Unit: °C)

Figure 19. The one-facade thermal interpolation diagram on the hottest month of year

- Due to the small backyard and a skylight at the bottom of the staircase create a continuously air circulates inside house. Therefore, room temperature is stable.
- The roof is designed in two layers so the room temperature is still less affected, and still gives the owner comfort in whole year especially in hottest season. At the coolest time, most of the house is heated only on the roof. At the hottest times the attic rooms are only slightly affected
- The bedroom at the front facade have the system of sunshades and flap which are limit the effect of heat radiation and increase the air circulation in the room.

This building creates pleasant feeling for the users, solves the problem of internal ventilation and has created effective insulation. No additional cooling devices such as fans and air conditioners are needed in this building. So, the power consumption is very low.

3.3.2. Two-facade row house

The research site is a 4-storey building located at the corner of the intersection. One facade looks towards the west, another looks towards the south. The two sides are adjacent to the existing neighbor's house, the two others side faced with an intersection of 10m and 20m road. The opposite buildings are only 2 - 3 floors.

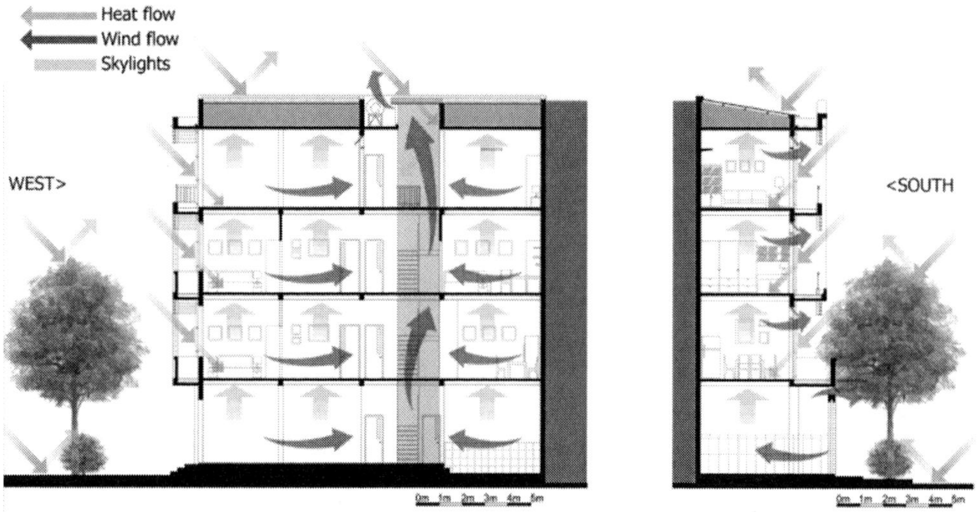

Figure 20. The one-facade microclimate analysis diagram
(Address: No. 74G, Lang Tang Phu street, Tang Nhon Phu A Ward, District 9, HCMC.)

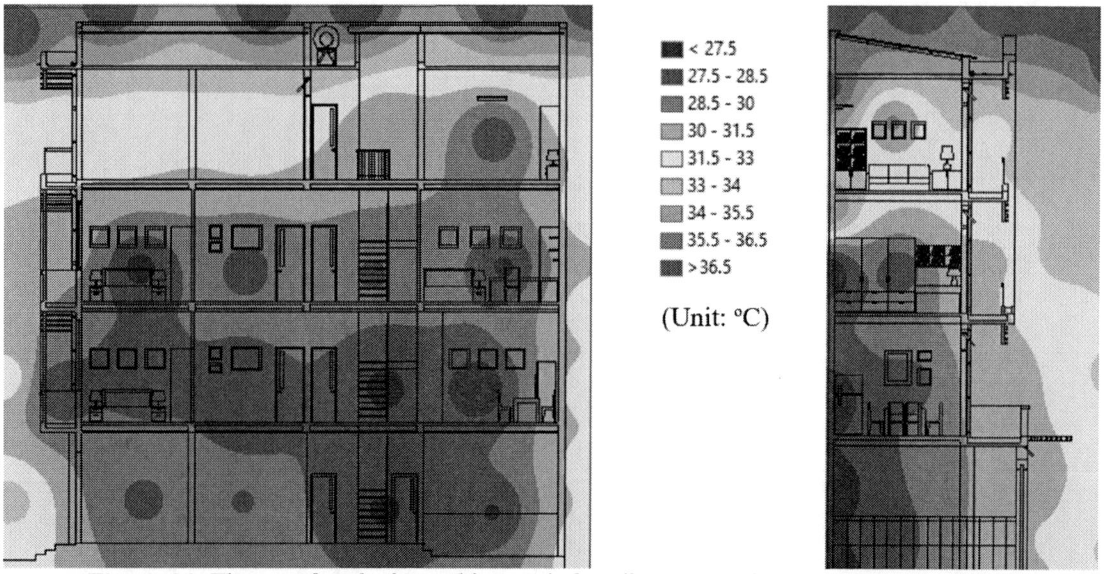

Figure 21. The two-facade thermal interpolation diagram on the coolest month of year

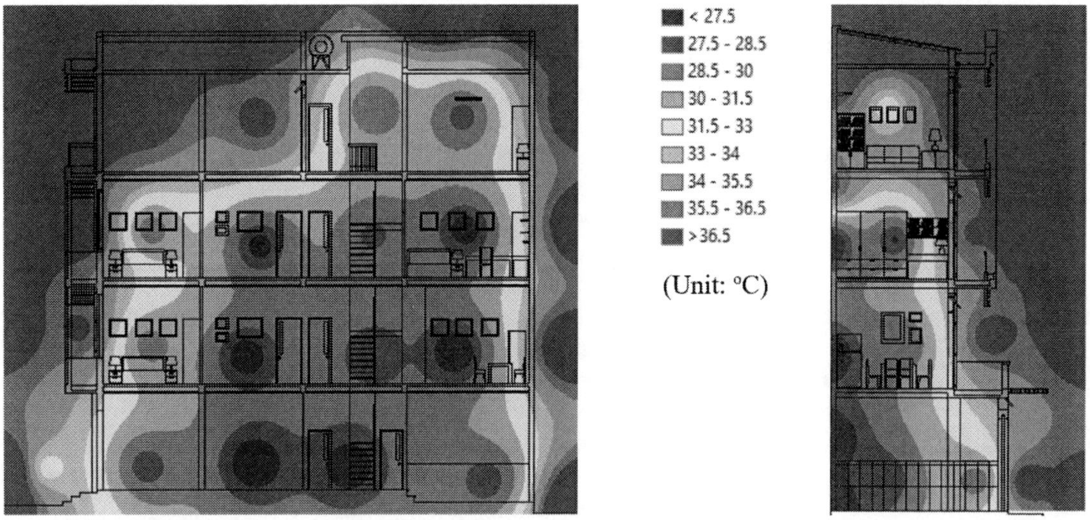

Figure 22. The two-facade thermal interpolation diagram on the hottest month of year

- Because the height of building is higher than the neighbouring buildings, the building easily catches the wind. The main wind direction is East - South and West - South.
- Because of the large, deep and closed balconies, sunlight at noon and in the afternoon only affects a small part of the wall surface. In addition, the walls are built in three layers for the higher thermal insulation.
- The windows and doors are designed to increase the air circulation in the room
- There is no wall on the rooftop. The function of this area is mainly for relaxation, planting trees. That is the reason why the thermal interpolation diagram shows the high temperature there. The indoor environment is slightly affected by heat; both are in the comfort zone.

Although the building is located in a disadvantageous micro climate position, the calculating disadvantage to give the suitable design is needed to overcome these weaknesses. Indoor lighting is reduced in capacity by utilizing natural light as well. Room temperature is always comfortable by the air circulation so that the owner does not have to use fans or air conditioning most of the time during the day. Therefore, the power consumption is always low in comparison with the same type of building.

4. Discussion

Through analyzing samples of row houses in Tang Nhon Phu A Ward, District 9, the authors summarize the issues affecting energy consumption and solutions as follows:

Table 2. The summarize of solutions and results of the row house in research area

Type of row house	Status quo	Solutions and results
One-facade row house	- Narrow width, short length. - No ventilation. - Lack of natural light. - One-layer rooftop.	- Add the skylight in the staircase space for natural lighting and ventilation - Add a small back yard - Installation of sunshade facade. - Design for a two-layer rooftop => Room temperature is stable, less affected from outdoor temperature, increased ventilation. No additional cooling devices such as fans and air conditioners are needed. The power consumption is very

		low.
One-facade row house	- Narrow width, short length. - Increasing the surface area of the building is directly influenced by solar radiation - One-layer rooftop	- Balcony surrounding the housing facade are needed, - Additional sunshade was proposed to add outside of building - Create a roof door at the staircase space to vent inside the house. - Use high window and door with the flap window above. - Use a thick 3-layer wall => High ability to receive wind. The large, deep and closed balconies have the effect of blocking sunlight at noon and in the afternoon, reduce heat absorption from the wall surface. => Room temperature is stable, less affected from outdoor temperature, increased ventilation. No additional cooling devices such as fans and air conditioners are needed. The power consumption is very low.

Because of many owners' financial conditions or habits, they are willing to eliminate the microclimate recommendations though they penetrate the impact of the environment on the space of the building. They accept the use of additional equipment to overcome the problem immediately. The result is the owners not only have to pay a big amount of money to operate but also for the maintenance cost, healthcare.

Actually, it is difficult to choose a place with favorable natural conditions to build a house. But when the designer and the owner share the same point of view, it is a good starting point to come up with the best solutions.

In order to offer the optimum solution consistent with the habits of using electrical appliances, it is necessary to evaluate the individual needs of the building, and then based on the actual conditions to provide the appropriate solution.

5. Conclusion

The results show that the effectiveness of the proposed solutions in comparison with the typical housing in research area. The interior space is always ensuring a proper indoor temperature. The gap between indoor and outdoor temperatures on the two samples of houses is very different. The rooms are all get the natural light.

The form and suitable material were selected carefully in accordance with the structural conditions and construction standards. These materials are mass-produced, and the installation method is not too complicated, so it is easy to apply commonly everywhere.

The weaknesses of this research are the data measurement by rudimentary tools, resulting in large errors in the measurement process; not expanded to other types of houses; is not going into the material properties for resistance to heat, or to analyze more deeply about the habit of daily use of household electrical equipment.

Nevertheless, the research topic also gives a studied direction in the global trend of energy saving. Besides that, this paper also suggests some research methods and some solutions for air circulation. This is also a good example of the application of GIS in analyzing the thermal interpolation diagram of a building.

References

[1] Simon Roberts 2008 Effects of climate change on the built environment *Energy Policy*

[2] Lang G 2010 International PassiveHouseDataBase. eu *Vienna: Pass-Net* 63

[3] Ekström T and Blomsterberg Å 2016 Renovation of Swedish Single-family Houses to Passive House Standard–Analyses of Energy Savings Potential *Energy Procedia* **96** 134–45

[4] Figueiredo A, Kämpf J and Vicente R 2016 Passive house optimization for Portugal: Overheating evaluation and energy performance *Energy and Buildings* **118** 181–96

[5] Liang X, Wang Y, Zhang Y, Jiang J, Chen H, Zhang X, Guo H and Roskilly T 2017 Analysis and Optimization on Energy Performance of a Rural House in Northern China Using Passive Retrofitting *Energy Procedia* **105** 3023–30

[6] Nguyen A-T, Tran Q-B, Tran D-Q and Reiter S 2011 An investigation on climate responsive design strategies of vernacular housing in Vietnam *Building and Environment* **46** 2088–106

[7] Hori S, Kondo K, Nogata D and Ben H 2013 The determinants of household energy-saving behavior: Survey and comparison in five major Asian cities *Energy Policy* **52** 354–62

[8] Jaffe A B and Stavins R N 1994 The energy-efficiency gap What does it mean? *Energy policy* **22** 804–10

[9] Sullivan-Bolyai S, Bova C and Singh M D 2014 Data-collection methods *Nursing Research in Canada-E-Book: Methods, Critical Appraisal, and Utilization* 287

[10] Yin R K 2006 Case Study Methods.

[11] Gomm R, Hammersley M and Foster P 2000 *Case study method: Key issues, key texts* (Sage)

[12] Malczewski J 1999 *GIS and multicriteria decision analysis* (John Wiley & Sons)

[13] Wolf P R and Dewitt B A 2000 *Elements of Photogrammetry: with applications in GIS* vol 3 (McGraw-Hill New York)

[14] Việt L V Sự phát triển đô thị và xu thế biến đổi khí hậu tại thành phố Hồ Chí Minh *Tạp chí KTTV* **558** 29–34

Van Quan new town in Hanoi and its socio-economic impacts on the four surrounding villages - towards a more sustainable urban development

Quang Minh Nguyen[1], The Trung Doan[1], Minh Tung Tran[1], Manh Tri Nguyen[1] and Quynh Hoa Ta[1]

[1] National University of Civil Engineering, 55 Giai Phong road, Hanoi, Vietnam

Abstract. Developed by Housing and Urban Development Corporation (HUD) over a decade ago as one of the pioneering and modelling urban housing projects in Hanoi, Van Quan new town can be regarded as a fairly successful scheme, because it provides people with a better living quality than most of the other new residential quarters planned and built at the same time, and even afterwards. However, the success would be greater, if the project developer paid due attention to some socio-economic aspects of mass housing construction - such as income, education, healthcare and social interaction as well as communication - in one of the city's fastest-growing and most typical areas characterised with the presence of traditional old villages where a rich rural culture can still be found and should be conserved in the rapid urbanisation and modernisation. Van Quan makes an even more remarkable case study, because there are four villages connected to the site with both similar and different features, rather than the other new towns with only one or two villages nearby. The research results drawn and the lessons learnt from Van Quan will be useful for other projects, as far as a harmony and a sustainable development between the old and the new factors are concerned.

1. Introduction

"New town" has been a different living concept in major cities in Vietnam from all the past models, even the closest one widely known as "socialist collective housing area" in the 1960's and 1970's (in the north) and in the 1980's (across the country) until the Renovation Policy came in late 1986. The most notable difference could be seen in the policy-making and the management system, as new town projects are run in the market economy where housing is recognised as an official market shared by both state and private providers, instead of only by state companies before. Another bold step is noted in the change of housing development from "plan-based" into "project-based". Developed by Housing and Urban Development Corporation (HUD), Linh Dam new town started in 1997 as the pilot project in Hanoi. Looking back at the 20 years of new town development, it is possible to confirm the advantages of this new living concept: shaping a brand-new image for a city, creating a better living environment, supplying a large number of housing units within a short time and offering good services to the residents. In general, new town is still a popular option for many people in cities today, despite of the fact that not every new town project is positively assessed. It seems that urban development managers in Vietnam have succeeded in trying to incorporate housing fabrication into complete and complex projects in view of function. This intention is even demonstrated with the effort of putting modernisation into universalisation by means of stakeholders (i. e. real estate corporations) in place of spontaneously building houses on a small scale (self-built houses) which can be difficult to control.

In order to support the development of a brand-new living concept, the National Assembly and state-owned agencies made a series of laws and policies in the 1990's and the 2000's, such as Housing Act passed in 1991, Land Law adopted in 1993, adjusted in 1998 and supplemented in 2001, Housing Law promulgated in 2005 and modified in 2009 and New Town Regulation enforced in 1996. In addition, the private housing ownership and then the transfer in land-use and real-estate rights have been accepted.

A broader context of new towns in Hanoi and the establishment of Van Quan as one of hundreds of new towns can be illustrated in Figure 1.

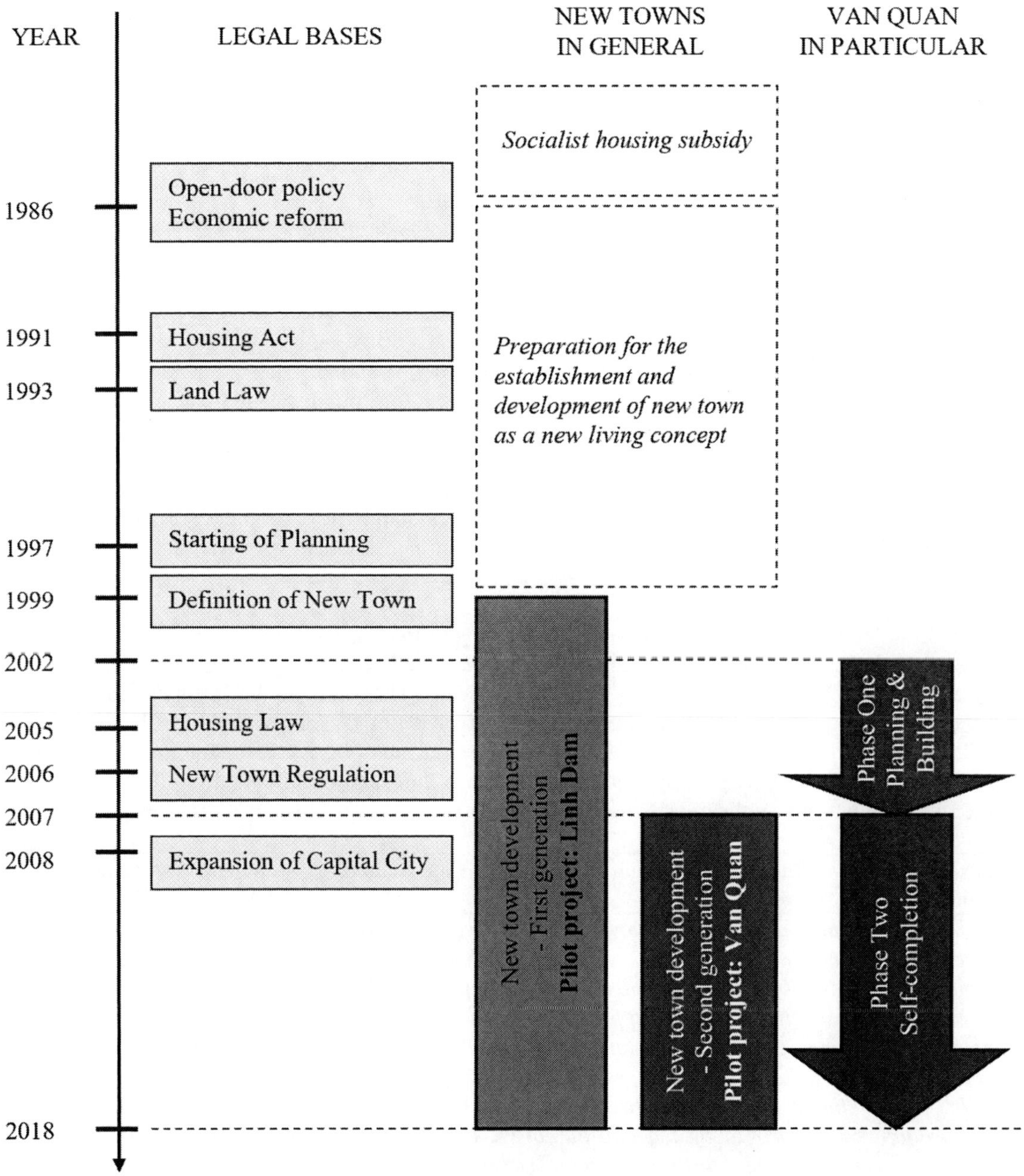

Figure 1. Diagram of new town development in Hanoi over the past 20 years. [1]

In terms of spatial planning, new towns have been constructed on the basis of synchronisation of the following three key factors: 1. Technical infrastructure, 2. Social infrastructure and 3. Housing units. One typical new town may be identified with four fundamental characteristics: 1. New elements in spatial urban design, 2. Relative independence of the city in terms of daily activities, 3. Systematic planning of infrastructure and 4. Specific structure of urban space [2]. Thus, theoretically, one new town can be regarded as a complete settlement that has its own lifestyle and is relatively independent of the city as well as its neighbouring residential areas. This independence enables that new town to self-provide everyday life services for local residents to a certain extent. Nevertheless, it may bring about an interruption in the relationship between that new town and some (or all) of the old villages nearby. Consequently, "break points" in both space and lifestyle are likely to happen. In order to attract a greater number of home buyers, that project developer has heavily invested in advertisement programmes. In the current project management system, the developer is merely responsible for internal affairs and not interested in anything else outside the boundary of such a project. As a result, just a simple mixture of living quarters or housing patterns has been created and that may not meet the needs of the inhabitants. A city developed this way lacks connection and communication. A symbiosis of old and new elements as a higher requirement seems to be out of reach.

Initially, in order to acquire land for new towns, the city authority decided to transform farmland in peri-urban areas into living quarters around existing urban cores. On the one hand, this initiative helps reduce the land price as an input of the subsequent urban development process. On the other hand, it may accelerate the urbanisation rate by enlarging the urban living space and hereby strengthening the attraction of new towns through physical aspects, such as size and population with some modern facilities. Most of the new towns have been located in peri-urban areas and surrounded by traditional villages where people still live on farming or handicrafts. A profound change in the fabric and land use within and around a project site can be clearly seen. Paddy fields and vegetable farms have been replaced by new urban housing blocks with a high building density and busy commercial streets. In other words, in the same territory, two seemingly opposite lifestyles coexist: new towns represent a modern and civilised lifestyle while traditional villages stand for a so-called "self-transformation" from old to new models under the strong impacts of urbanisation in general, and directly from the adjacent new towns in particular.

2. Looking back at the beginning of new towns for a better understanding

In 2016, the research team helped Assoc. Prof. Dr. Danielle Labbé from the University of Montréal as a member of a Canadian - Vietnamese new town research network conduct intensive interviews with six experts and professionals who were directly involved in the preparation for the planning and building of new towns in the early 1990's to understand much better the beginning of an important decision that would reshape the landscape of urban housing development in Hanoi. The main interview contents are summarised in Box 1 to Box 5.

Box 1. Interview with Mr. Ngoc Chinh Tran and Dr. Thu Thanh Ngo - Vietnam Association of Urban Planning and Development. [3]
In the pre-1986 period (commonly known as subsidised economy planned and centrally directed by the government), urban housing was entirely constructed, allocated and managed by the state on the basis of norm-based distribution. The shift in the policy towards an open market economy in which the private sector was officially accepted for the first time paved the way for a revolution in mind set and viewpoint regarding the investment and development in housing for city inhabitants whose housing demand had been held back for decades - before, during and after the war. *In 1996, Ciputra became the first joint-venture project in partnership with one Indonesian corporation ever approved by the city authority. From Ciputra project, there were some lessons learnt: providing decent housing for middle-class to high-income residents was considered an advantage while the shortcomings in operation, land rental mechanism and land-use right could not be ignored.*

Box 2. Interview with Mr. Ngoc Nghiem Dao - Chief Architect of Hanoi City. [3]
The First National Congress on Urban Planning and Development in 1990 proposed a number of new requirements for a radical change in urban planning orientation, city construction and housing development management for city inhabitants.   *In 1991, the state promulgated a Housing Act in which housing ownership would be recognised and legally protected. In addition, the participation of individuals and private organisations in housing development was encouraged. Thus, 1991 Housing Act made a significant re-direction in urban housing development policy, for instance the socialisation of housing development could be supported. Enterprises were allowed to invest in new living quarters designed in new town concepts for city inhabitants. This change in policy making came from reality and complied with the rules in social development.*   *The reform of Land Law in 1993 also resulted in a new housing policy. The 1993 Land Law began to separate urban land-use management rural land-use management which were both controlled by the Ministry of Agriculture. In this watershed law, land should be divided into two categories: farmland and urban land. One crucial adjustment in 1993 Land Law was that urban land would be discretely managed. One year later, the Prime Minister approved of the management in building investment whereby urban management, especially in urban planning, could be clearly determined.*   *The Second National Congress on Urban Planning and Development in 1995 adopted important changes in the orientation for urban planning and management. The national network of towns and cities expanded quickly in both land coverage and population size on account of an improved economy. The Prime Minister approved of the orientation for planning all the cities and towns nationwide and the adjustment in planning of Hanoi and Ho Chi Minh city. A variety of management and development policies for cities and towns has been quite comprehensively drafted and issued to attract more investment capital sources for building (including housing) projects. Another remarkable trend in urban planning was the changing role of stakeholders in the establishment, approval and implementation of planning projects in which the importance of private sectors would be emphasised. Also, the term "new urban area" or "new town" was used for the first time in 1995, when Hanoi began to realise the reverse side of the massive and spontaneous development of row-houses that spoilt the cityscape. The definition of "new urban area" or "new town" became officially endorsed with Decree No. 52/1999/ND-CP dated on July 8th 1999. According to this decree, "new urban area" or "new town" is a project-based new living quarter constructed as planned with a complete infrastructure system and connected with an existing city or town.*

Box 3. Interview with Prof. Dr. Dinh Viet Pham - National University of Civil Engineering. [3]
The Master Plan for Hanoi re-adjusted in 1998 took some revisions in both orientation and strategy: the city development should be based on both sides of the Red River with some industrial parks in the suburbs. In traffic planning, there would be three city ring roads while urban housing development was put to the forefront of the city development.   *With regard to living concepts, most of the new housing projects between 1992 and 1998 were small in size (less than two hectares, primarily constructed for state officers, using the capitals from both the state and state-owned agencies, with a partial contribution from the public) and those new housing projects failed to go in line with social/technical infrastructure systems. The following shortcomings of this kind of development might be noted: low-rise row houses made up an overwhelming part of all urban housing programmes up to that point of time, with inefficient land-use factor and consequently no balance between social and technical infrastructure could ever be achieved.*

Box 4. Interview with Mr. Trung Hieu Tran - Housing and Urban Development Corporation. [3]
Before 1998, the construction of new urban areas in Hanoi was rather limited to mini-projects as an exploration of the public demand. Typical new urban areas in that period, such as Linh Dam, Dinh

Cong, Trung Yen, Dai Kim, Nghia Do, etc. started to be planned and built on a small scale in terms of investment and with a mix of row-rise and high-rise housing types. Enterprises just focused on the investment in housing construction while the city authority was responsible for technical infrastructure development. Therefore, the city authority insisted that all housing developers should invest properly in technical infrastructure systems in accordance with housing quality to ensure a high living standard for all city inhabitants.

Box 5. Interview with Dr. Huu Phe Hoang - VINACONEX Corporation. [3]

From 1998 to 2001: this period started with Decision No. 108/1998/QD-TTg dated June 20th 1998 by the Prime Minister and ended up with Decision No. 123/2001/QD-UBND dated December 6th 2001 by Hanoi City People's Committee on the promulgation of the principles in investment and development of new urban areas, as well as housing renovation and improvement in Hanoi. These official documents determined the sharing of land for housing construction: approximately 60% for high-rise apartment buildings and 40% for villas and garden houses. The construction of new tube houses was not permitted.

The years 2001 to 2008: After Decision No. 123/2001/QD-TTg took effect, new urban areas were developed on the basis of the aforementioned land-use sharing. The planning of technical and social infrastructure systems as well as facilities for users in consideration of population size in each new urban area would be carried out, for instance in Trung Van (Tu Liem district in 2003), Sai Dong (Long Bien district in 2004), Nghia Do (Cau Giay district in 2004) and other areas from 2005 onwards. In typical projects, such as Linh Dam, South Thang Long, Trung Hoa - Nhan Chinh, Cau Giay, Me Tri, Viet Hung, etc., new building technologies were applied for high-rise apartment building construction. These new towns were regarded as modelling examples in terms of modern and comfortable living environment, in sharp contrast to poor housing quality in old collective residential quarters (in Kim Lien, Trung Tu and Giang Vo) or overcrowded and chaotic with low-rise self-built houses. Decree No.2/2006/ND-CP dated on January 5th 2006 was enacted and the regulations specifically for new towns were drawn up. A new town project was defined as "an investment project in the construction of a well-planned urban area with both technical and social infrastructure systems that will serve living quarters and public service buildings. It may be a continuity of an existing urban area or an independent part of a city with its own boundary and function determined in accordance with the city planning approved by a competent state agency".

From 2008 to date: Based on the Overall City Development Planning for Hanoi approved by the Prime Minister with Decision No. 1259/QD-TTg dated July 26th 2011, the zoning plan for urban districts (scale 1:2000) and for rural districts (scale 1:5000) previously approved in the 2000 - 2006 period referred to the size of a new urban area (from 20 to 200 ha) and the location (usually along main city roads). Other aspects included land-use planning, landscape architecture and technical/social infrastructure planning.

3. An in-depth view of Van Quan new town

Located in the southwest of Hanoi, 11 km from the city centre and just 2 km from Ha Dong district centre, Van Quan new town is accessible by car, by motorbike and by taxi from National Highway No. 6 (Nguyen Trai street) on the northwest side and from Provincial Road No. 70A (Phung Hung road) on the southwest side. Nguyen Trai street is noted for its heavy traffic flow and as one of the busiest bus routes in Hanoi with nine bus lines (No. 1, 2, 19, 21, 22, 27, 37, 39 and 57) and three bus stops near Van Quan new town. Along Nguyen Trai road, a sky train project is still under construction. So in the near future, it will be possible to go to Van Quan by tram.

Van Quan belongs to the second generation of new urban area development in Hanoi established a few years before 2006 - the year marked with the promulgation of development regulations applicable to new towns. That was a boiling period for all real estate enterprises and housing providers in the market as they had to cope with huge challenges caused by such a major change in policy making. Many companies embarked on developing new urban areas without any preparation or experience.

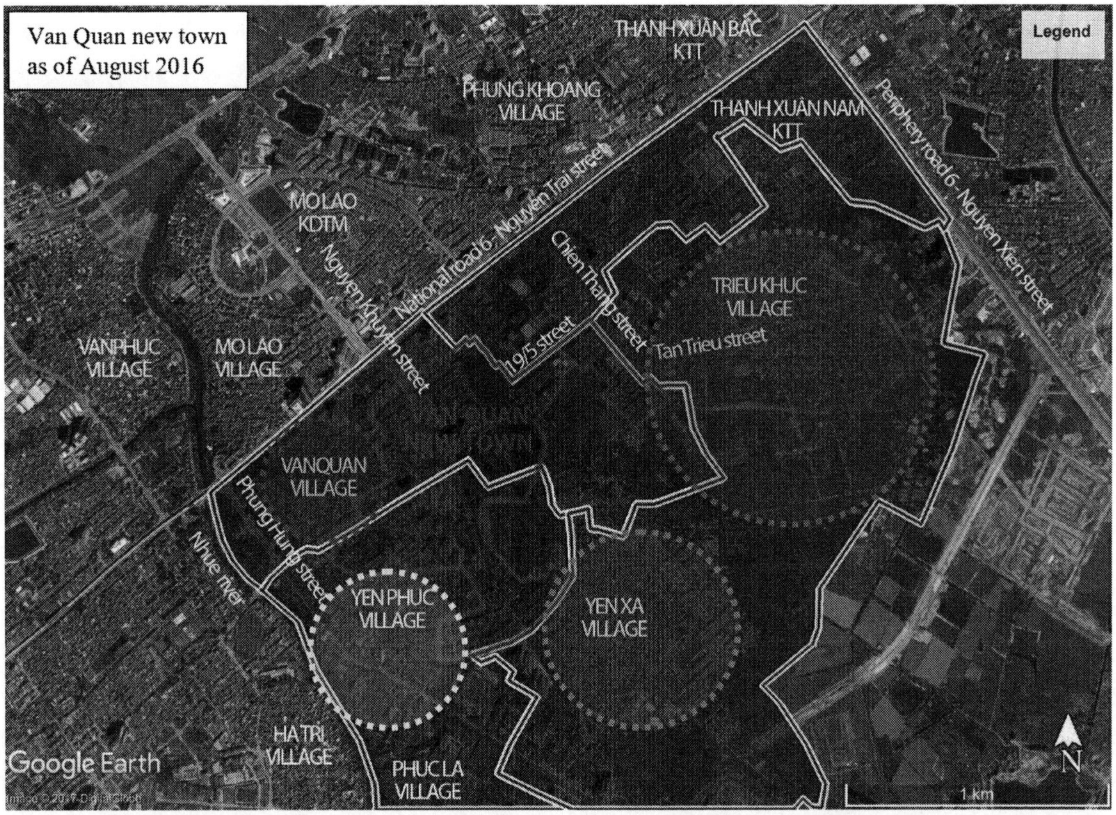

Figure 2. Van Quan new town and the four adjacent old villages. [4]

As a leading corporation, HUD eventually selected Van Quan - a new ward of Ha Dong town, which used to be an administrative centre of the former province of Ha Tay (currently becoming Ha Dong district of Hanoi city) - for a large modern housing project to begin a new phase of housing development. This location is actually quite strategic and realistic, because it is not so far from Hanoi city centre (for an effective management) and close enough (for easy access). In fact, Van Quan was the first site in Ha Dong district to be chosen and regarded as one of the kick-off projects for warming up the housing market in the southwest of Hanoi. Started in 2003, Van Quan today covers an area of 61 ha and provides home to approximately 14,000 inhabitants. Along with 1,412 row-houses and villas, there are three high-rise apartment buildings (over 25 storeys) and ten multi-storey residential blocks (11 to 21 storeys), providing 1,229 apartments. The total housing floor area exceeds 450,000 m^2 (32 m^2 per person) [5].

Over the last 15 years, Van Quan new town has been developed step-by-step and major changes can be found in some places: the boundary and total land area have been modified several times due to the difficulty in farmland acquisition for new town building purpose and for other reasons, part of the central public park has been used for restaurants and clubs, two six-storey apartment buildings have eventually become 27 and 30 storeys, some land plots initially planned for public buildings have been transformed into high-density housing areas, etc. As of 2017, it was still difficult to identify the final physical boundary of Van Quan new town, because the adjustments have not yet been updated on the map, and in reality, the boundary takes five different forms as follows:

- Road only: Normally, this is the edge of a village with a narrow road running parallel with that edge. When a new town is constructed, this road is turned into a clear boundary without buildings between an old village and a new town;
- Road with houses (usually row-houses): This is often an area planned for resettlement or relocation in a depopulation plan. This road will soon become a new street with shop-houses on both sides built with a certain number of housing patterns approved. As a result, an architecturally nice street can be designed in a good harmony with the surrounding areas;

- Houses only (row-houses in most cases): There is sometimes a conflict in functions, leading to the lack of activities along this edge;
- Public buildings (often the back façades/sides of the buildings): The boundary will then play the role of an enclosure wall, hereby creating no connection at all;
- Open places (possibly pagodas or public spaces with precincts in old villages, or grassland): This kind of boundary is quite flexible in enlarging or prolongating the space, and therefore highly applicable to fixing the boundary.

Surrounded by four villages (Trieu Khuc, Yen Xa, Yen Phuc and Van Quan), Van Quan new town makes a special case study and promises to give an interesting example of how a new town will interact with many adjacent villages at the same time. Both spatial and social transformations can be seen here and interpreted in an interdisciplinary approach. After almost 20 years, Van Quan has now become a populous and busy residential quarter, denoting that the project developer has succeeded in shaping a new concept for attractive living environment in a period when a majority of city inhabitants turned their back on high-rise housing because of errors in the management system. But the counter-effect cannot be ignored and thus it will require a thorough investigation. Land use, handicrafts, traditional rural lifestyles and customs, landscape, vernacular architecture and culture have either considerably or dramatically changed within a short time as a consequence of the rapid urbanisation reaching the area.

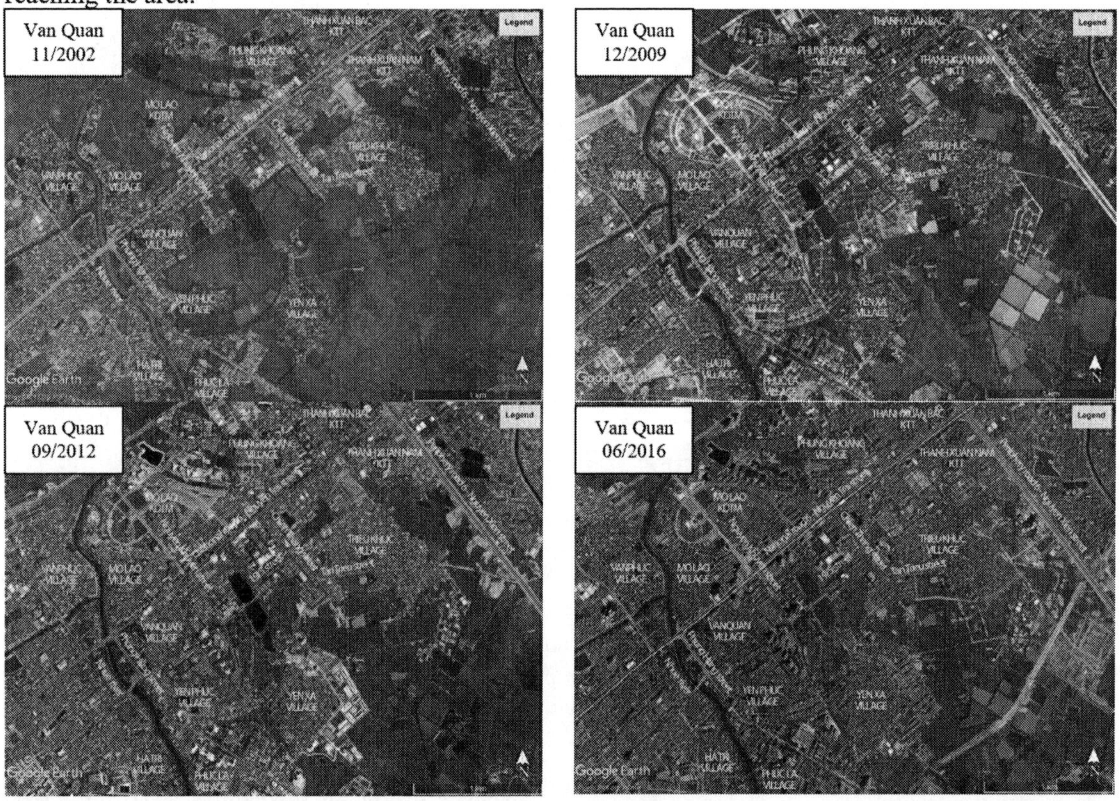

Figure 3. Transformation of Van Quan new town and the four adjacent villages over the years. [6]

4. Urban development in view of rural residents

The success of a new town project can be best measured in the level of satisfaction of its residents. However, the opinions of those living in the neighbouring area are also important and will be included in the evaluation of a new town development. In total, there are five categories related to the socio-economic development and facilities to investigate (Economy, education, healthcare, commerce and entertainment), and 308 inhabitants participated in the interviews (93 in Trieu Khuc village, 89 in Van Quan village, 75 in Yen Phuc village and 51 in Yen Xa village).

Table 1. Public opinions in four old villages under the impact of Van Quan new town. [7]

Category Village	Quan -tity	Economy			Education			Healthcare			Commerce			Entertainment		
		+	0	−	+	0	−	+	0	−	+	0	−	+	0	−
Trieu Khuc	**93**	50	31	12	78	10	5	58	22	13	67	17	9	56	14	23
	%	53.8	33.3	12.9	83.8	10.8	5.4	62.4	23.6	14.0	72.0	18.3	9.7	60.3	15.0	24.7
Van Quan	**89**	57	27	5	75	10	4	66	22	1	73	11	5	70	10	9
	%	64.0	30.4	5.6	84.3	11.2	4.5	74.2	24.7	1.1	82.0	12.4	5.6	78.7	11.2	10.1
Yen Phuc	**75**	52	22	1	66	8	1	63	12	0	66	6	3	56	7	12
	%	69.4	29.3	1.3	88.0	10.7	1.3	84.0	16.0	0.0	88.0	8.0	4.0	74.7	9.3	16.0
Yen Xa	**51**	38	12	1	41	9	1	39	10	2	33	13	5	32	17	2
	%	74.5	23.5	2.0	80.4	17.6	2.0	76.5	19.6	3.9	64.7	25.5	9.8	62.8	33.3	3.9
Total	**308**	**197**	**92**	**19**	**260**	**37**	**11**	**226**	**66**	**16**	**239**	**47**	**22**	**214**	**48**	**46**
	%	**64.0**	**29.9**	**6.1**	**84.4**	**12.0**	**3.6**	73.4	**21.3**	**31.3**	**77.6**	**15.3**	**7.1**	**69.5**	**15.6**	**14.9**

Note: (+) better than before (0) unchanged, the same as before (−) worse than before

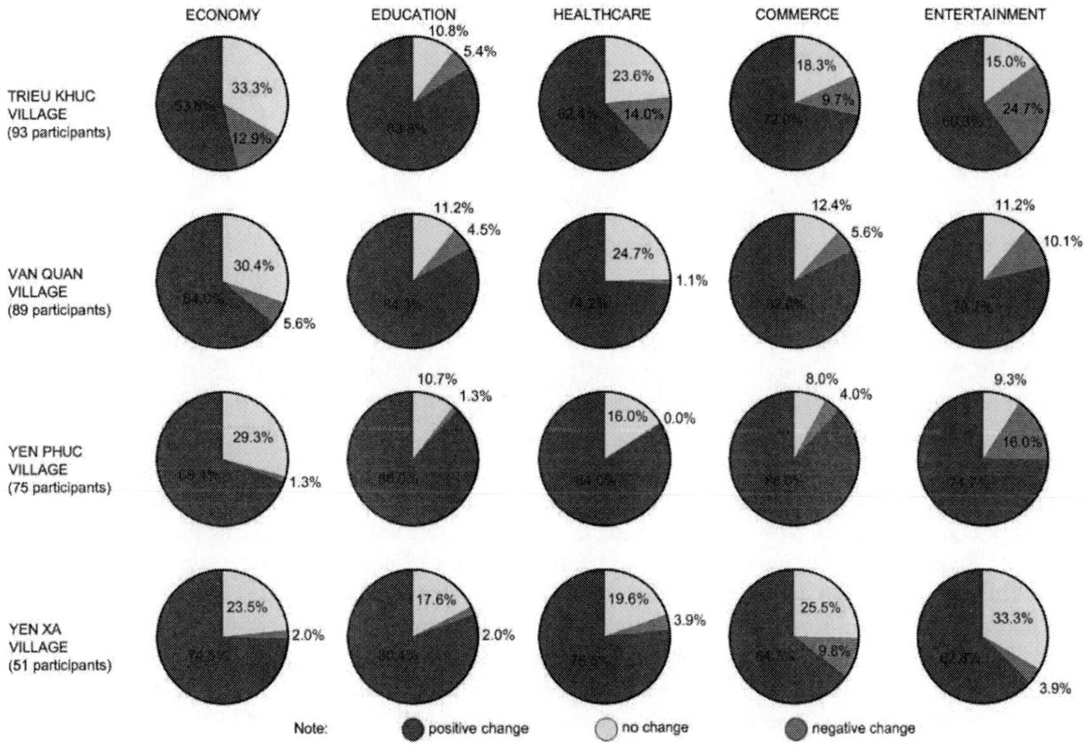

Figure 4. Graphic presentation of interview data. [8]

4.1. Economy

According to the statistics, 64% of the interviewed local residents in the four villages say that their income has been higher since the inauguration of Van Quan new town (Group 1) while 29.9% assume that their income stays the same (Group 2) and only 6.1% complain that they earn less money than they did before (Group 3).

Most of the people in Group 3 live in Trieu Khuc village (12 out of 19) and Van Quan village (5 out of 19). It is apparent that villagers in Yen Phuc and Yen Xa have gained more benefits from Van Quan new town project than those in the other two villages. Further studies reveal that most people in Group 1 are over 50 years old whose farmland has been entirely or partially confiscated for the construction of Van Quan new town. The remaining farmland area is too small, segmented, no longer fertile and even polluted by waste, therefore inappropriate for agriculture. Furthermore, watering is insufficient as the local irrigation systems do not work. Crops, if still cultivated, would be frequently destroyed by a larger number of rats in the field. In this circumstance, some farmers who still have farmland decided to stop cultivating and used the land for another purpose, such as car parking or temporary keeping building materials for the construction site. Those aged farmers are not as dynamic or versatile as the adolescents in finding new jobs when they no longer work in agriculture. Over 68% of the young people between 20 and 40 years old are satisfied with their higher income and the rest (approximately 32%) belong to Group 2, denoting that young people can adapt themselves quickly and more easily to major changes when their villages are urbanised. They are ready to attend professional training courses and meet the requirement in qualifications. In terms of occupation, workers and state officers are not so much affected by the establishment and development of Van Quan new town while farmers are among the underprivileged group in the course of urbanisation. After their farmland was confiscated, the farmers invested all or part of the compensation in certain kinds of small daily service, for example opening greengrocery shops and selling miscellaneous goods, which do not require a large initial capital investment. Some of their customers come from Van Quan new town and the others are their neighbours. Many villagers realised that building houses for rent would be highly profitable. Only by spending a modest amount of money from the compensation they have received before can they win a real fortune. The demand is always high, because the area is close to several universities and factories which attract many migrant students and workers. With a large garden area, about 300 m^2, a family in Yen Phuc village kept 100 m^2 in the middle for the family use and built ten rooms for rent (16 m^2 per room) in the front part of the garden. The landlord can monthly earn 15 million VND (660 USD) from rental fees and regain the initial investment in less than two years [9].

4.2. Education

Education is a great concern in Vietnam today, particularly in major cities. Parents always emphasise the importance of education and would like their children to graduate from university as the minimum requirement of qualification for a future well-paid job. Before Van Quan new town was built, there were some village schools in the area which did not offer children good facilities due to the limit of budget. Village teachers might not be as well qualified as those in central districts. Therefore, the quality of education was somehow influenced. However, a village school outperforms a city school in terms of having a larger playground with more trees. As a modelling project, Van Quan new town provides a better schooling even for village children. The survey shows that people in all four villages are content with the education opportunity that Van Quan new town has brought to them.

Out of 308 villagers, 260 (84.4%) agreed that the impact in education is positive, with better facilities, higher qualification of teachers and more beautiful school architecture. Their children or grandchildren have been accepted into new town schools. Those who continue to go to village schools are happy about the improvement in both quality and equipment. Most people have found that the roads to schools are upgraded for easy access. The ground floor has been elevated to avoid flooding, and new drainage systems have also been constructed in some streets and alleys. The levels of public satisfaction of schools in Trieu Khuc, Van Quan, Yen Phuc and Yen Xa are 83.8%, 84.3%, 88.0% and 80.4% respectively. The differences among the four villages in this case are minor [9].

There are 11 interviewees (3.6%) saying that the conditions for education become worse, most notably in Trieu Khuc village and Van Quan village (9 out of 11 cases). The main reasons for this negative conclusion are that they have to go a longer way to bring their children to school and pick them up again from school. The traffic congestions as they often encounter on the way make them feel inconvenient. School fees increase while they earn less money than they did before. These people come from a marginalised group (low-income farmers whose farmland has been confiscated). They received an amount of compensation from the project developer via the local authority without any assistance in shifting for their future jobs as well as for their earning. The neutral group consists of 37 residents (12%) and distribute quite evenly among the four villages: Trieu Khuc (10), Van Quan (10), Yen Phuc (8) and Yen Xa (8). In their opinion, the advantages and the disadvantages weigh equally, and basically there is no change [9].

4.3. Healthcare

Generally speaking, most of villagers in Trieu Khuc, Van Quan, Yen Phuc and Yen Xa are satisfied with the healthcare service today, as clearly reflected in the statistics. The highest level of satisfaction is noted in the case of Yen Phuc village (84.0%). Nobody feels that the local healthcare system and service quality at present are worse than in the past. In Yen Xa village, Van Quan village and Trieu Khuc village, the feedback is quite positive, 76.5%, 74.2% and 62.4% respectively. In healthcare sector, it seems that Van Quan new town has a good impact on the neighbouring villages, just like in terms of education [9].

New and modern medical centres recently built in and next to the villages together with the open treatment/health check-up policy explain why village people feel pleased with the service quality. Another convincing reason is the accessibility of the medical centres. However, when compared the locations of the medical centres within the area, it may suggest a new outlook. The investigation into the hospitals and clinics within 3 km from Van Quan lake proves that there are only two private clinics in Van Quan new town whereas ten are located in the four old villages. The others are built outside Van Quan new town. In other words, it looks as if the number of new medical centres and the access roads did not depend on the establishment and development of Van Quan new town. Moreover, taking into consideration of the dissatisfaction rate of service users, Yen Phuc village and Van Quan village are the lowest, 0.0% and 1.1% respectively. On the map, it is possible to see that these two villages are located along two main roads: Nguyen Trai road and Phung Hung road (70A road) which were constructed and/or upgraded in the early 2000's, thus not largely depending on the development of Van Quan new town. In the surrounding area, some of the most important hospitals can be found, for instance 103 Hospital (Army Hospital), National Institute of Burns and Hospital of Cancer No. 2. The dissatisfaction percentage in Trieu Khuc village is much higher - 14%, focusing on expensive hospital costs. This is only a question of income, and for some villagers the level of income has considerably decreased since 2007, when Van Quan new town welcomed its first inhabitants. This might be seen as an indirect consequence of Van Quan new town for the neighbouring villages [9].

4.4. Commerce

Residents may argue about the options or hesitate to give an answer to a question in other social aspects, but in commerce there is a remarkable convergence of opinions. The statistics show that an overwhelming majority of inhabitants appreciates commercial amenities and relies on the service quality, especially in Yen Phuc (88.0%) and Van Quan (82.0%). This can be demonstrated in the busy commercial activities, not only in each living quarter, but also in the bordering areas of Van Quan new town and adjacent old villages. In the other two villages - Trieu Khuc and Yen Xa - some people think that a supermarket stands for a modern and civilised city life and they like to go there, but they can hardly buy anything, because the prices are much higher than their spending power. Van Quan new town seems like a different world to them. Hence, it is understandable that only 64.7% of the surveyed people in Yen Xa village and 72.0% in Trieu Khuc village feel happy about the change. The rate of critical thinking is approximately 10% in each case [9].

The existing markets are often planned in the heart or along some main roads of a village, providing all kinds of goods for villagers. A wholesale market called "Van Quan night market for agricultural products" has taken place since 2008 in an open-air area (8,000 m^2) amid Trieu Khuc village, Yen Xa village and Van Quan new town. Over 300 households do their business there. Although it is named a "night market", it is also open at day as a residential market, serving local people and other residents from other settlements. With a strategic location and a large area, including parking lots and unloading areas, selling many kinds of goods as diverse as vegetables, meat, sea food, clothes and household wares, the market has always been busy and even become an "informal gate" for many residents [9].

This "informal gate" can be seen in the way people make a shortcut from their villages to the market. The previous paths between paddy fields made of rammed earth have been concreted, joining the villages to the market and connecting two villages together. Such a spontaneous development is nevertheless quite realistic, offering local people a much more convenient approach to the regional market where low-income people can buy popular food and household items at cheap prices.

A big Vinmart within the podium of a high-rise apartment building in Van Quan new town provides both the townsmen and the villagers with a perfect choice for high-quality commodities and modern products. That is why there is no need to build any other supermarket in the area. But mini-markets and convenience stores in street houses have been widely developed in the new town and in the villages as well which provide more shopping options. They often specialise in some certain kinds of goods, for example kid toys, candies and confections, fashion and garment, etc. In the villages, most of the convenience stores can be found along the village main roads.

It is apparent that Van Quan new town helps accelerate the delivery of various urban services, for example gastronomy, café, beauty salon, fitness training and English learning, etc. Sometimes, these services are concentrated in one street or one quarter, forming a complex of service buildings. Some other services are also available, for instance banking, office for rent and DNA testing centre. Normally, these services are "integrated" into the podium of a tower, the front room of a street house or even an entire ground floor of a villa in Van Quan new town.

In the bordering areas of Van Quan new town and the four surrounding villages, there are many agri-land plots transferred into urban housing land but still left unused. Urban farming concept shall not be applied. Instead, people are making full use of these land plots for certain purposes, for example car parking, car repair and washing, selling of building materials and growing flowers or ornamental trees. A few land plots are temporarily used as warehouses and handicraft workshops.

With convenience and quality, all the above-mentioned daily life services account for the satisfaction of villagers as beneficiaries. In addition to providing services and gaining benefits, the communication between sellers and buyers, particularly between townspeople and villagers, can be maximised. By means of social contact, they will be able to better understand each other's needs, preferences and lifestyles. This is important to achieve social sustainability.

4.5. Entertainment

A recent sociological survey reveals that villagers have had better access to public entertainment since Van Quan new town was built. In fact, 69.5% of the villagers agree with this. But the rate of satisfaction varies from village to village. Once again, Van Quan and Yen Phuc came first and second, with 78.7% and 74.7% respectively. The main reason is that some new clubs and entertainment centres have been in use over the past ten years. In Van Quan new town, one swimming pool, two tennis courts and one fitness centre have been built. Villagers can come, but not many of them are ready or happy to join, as they do not find the "sharing" in behaviour of townspeople or feel hesitant about the membership fees. This rate is lower, just 62.8% in Yen Xa and 60.3% in Trieu Khuc. If the current problems in traffic connections and social contact are solved, most notably in Trieu Khuc village, more local people, especially those living in the core, will take advantage of the development [9].

Most of the 14.9% of the villagers who suppose that they have fewer opportunities to gain access to entertainment services prefer that such services should be also available in their villages, not only in Van Quan new town. Public places in all four villages fail to meet the local people's needs, with only one simple and small cultural house in the middle of each village. If constructed, these new public places require a much better design, more modern facilities and an effective management system to ensure that the activities will be attractive and frequently organised [9].

Traditional villages symbolise the great cultural values of the Vietnamese coming down from generation to generation. In terms of culture, villages can be regarded as a miniature of the Vietnamese rural society and noted for having created a sustainable living environment in which spatial planning and social systems are closely connected. In one typical village, there are always public places for the local people to meet and to take part in many festivals throughout the year. A nice community house in the middle of a large yard with beautiful landscape is the most outstanding space in the village. Additionally, the villagers have a number of religious buildings, such as pagodas, shrines and temples for ceremonies and activities. The other special features include a village gate, a village market, a village water well, a village pond, a river boat harbour, etc. In all four villages - Trieu Khuc, Yen Xa, Yen Phuc and Van Quan - many of these symbolic buildings and sites still remain today. The communal buildings and activities have made a significant contribution to the conservation and promotion of the village culture, apart from meeting the needs of the local people for entertainment.

However, in the rapid urbanisation and the dynamic market economy, these heritage assets are threatened. Public places in all the four villages have been illegally occupied for various purposes. This negative trend, together with the non-original conservation or restoration of old public buildings, has led to the fading of cultural values and communal activities.

5. Conclusion: Looking forward to a more sustainable future development

The past ten years (2007 - 2017) should be considered Phase One of a multi-episode story about Van Quan new town as a special case study of new town development in Hanoi. The challenges between the old villages and the new town have been quite clearly reflected so far in a wide range of issues: land acquisition/confiscation/compensation, conflict of interests between social groups, farmers' adaptation capacity to major socio-economic changes, physical connection and social communication, loss of cultural identify, etc. Such problems need to be solved in order to secure a stronger social structure and to achieve a more sustainable urban development.

A new town does bring opportunities to adjacent old villages, as analysed in Section 3 from five perspectives: economy (with new jobs and income), education (with the offering of modelling schools and standard programmes), healthcare (with a better service quality), commerce (with more shopping options) and entertainment (with a broader range of activities). The crucial point is that some villagers have not yet been able to gain benefits from the urban development. Therefore, the first and foremost task for Phase Two - in the next ten years for Van Quan - is paying more attention to all the underprivileged/marginalised groups and then assuring equality. "Balance" and "harmony" will be two key words for a long-awaited social sustainability in the city where old and new factors are mixed and different minds/interests co-exist. The interrelationship between old villages and new town should be established and consolidated over the years, on the basis of a mutual understanding and a core value sharing among people from all walks of life. Some aspects as discussed and highlighted in the initial findings of this academic paper have already suggested interesting issues for further investigations.

Acknowledgements

The site survey with interviews was conducted by final-year students of seven classes at the Faculty of Architecture and Planning (the National University of Civil Engineering): 58KD1 and 58KD2 (Yen Phuc village), 58KD3 and 58KD4 (Trieu Khuc village), 58KDF (Yen Xa village), 58QH1 and 58QH2 (Van Quan village). The authors would like to thank all the students for their efforts and assistance.

Special thanks also go to the National University of Civil Engineering (NUCE), where the authors currently work, for the support and financial aid from the University's funding for prioritised research project (project code 193-2018/KHXD-TD).

References

[1] Nguyen Q M 2018 Research graphic work
[2] Tran M T 2016 *New urban areas in Hanoi, looking back at one living concept after two decades* (Hanoi: Construction Publisher) p 51
[3] Labbé D 2016 Intensive interviews with experts
[4] Tran M T 2017 Research graphic work with background map by Google Earth
[5] Housing and Urban Development Corporation 2017 Van Quan New Town Project Data Weblink http://www.hud.com.vn/content/du-khu-do-thi-moi-van-quan-yen-phuc-ha-dong-ha-noi
[6] Tran M T 2018 Research graphic work with background map by Google Earth
[7] Nguyen Q M, Tran M T, Doan T T, Nguyen M T and Ta Q H 2017 Data collection from surveys in the four villages of Trieu Khuc, Van Quan, Yen Phuc and Yen Xa (Hanoi, Vietnam)
[8] Nguyen Q M 2018 Research graphic work
[9] Nguyen Q M, Tran M T, Doan T T, Nguyen M T and Ta Q H 2017 Data analysis from surveys in the four villages of Trieu Khuc, Van Quan, Yen Phuc and Yen Xa (Hanoi, Vietnam)

An analysis of the development of port operation in Da Nang Port, Vietnam

T D H Nguyen[1,3,4] and M Cools[2]

[1]Research Center for Quantitative Methods and Operation Management, Management School, University of Liege, Rue Louvrex,14, B-4000 Liege, Belgium.
[2]Faculty of Applied Sciences, University of Liege, Chemin de Chevreuils 1, Bât B52/3, B-4000,Liege, Belgium.
[3]Faculty of Project Management, University of Science and Technology, The University of Da Nang, 54 Nguyen Luong Bang St, Danang, Vietnam.

Abstract. This paper presents the current operating status in Da Nang Port, Vietnam in the period 2012-2016. The port operation had positive changes that were reflected by a significant increase in total throughputs, especially containerized cargo volumes. Classical decomposition techniques are used to find trend-cycle and seasonal components of monthly throughput flows. Appropriate predictive models of different kinds of throughputs are proposed. Finally, a development strategy towards containerization and investment policies in facilities, equipment, and infrastructure are suggested based on the predictive results.

Keywords: Decomposition methods, monthly throughput forecasting, Da Nang port, seasonally adjusted model, container flows.

1. Introduction

Vietnam is a developing country and its economy is mainly based on marine related sectors. It is estimated that the economic contribution of sea and coastal areas occupied nearly half of the total gross domestic product (GDP) of Vietnam in the period 2010-2015 [1]. The target set by the 10th Communist Party's Central Committee adopted the Maritime Strategy in which the marine economy will contribute from 53% to 55% to Vietnam's GDP by 2020 [2]. Thus, the maritime sector plays a vital role in the development of Vietnam's economy. In other words, the economic performance of Vietnamese seaports can directly affect the country's prosperity.

The Vietnamese seaport system is divided into three main sub-systems including Northern, Southern and Central ports. In the north, Hai Phong Port plays a significant role as the main gateway to the sea of Northern Vietnam, accounting for 13% of total throughputs of Vietnamese ports in 2016. In the south, ports in Ho Chi Minh city are the main gateway to the sea of Southern Vietnam, accounting for 67% of total throughputs of Vietnamese ports at the same time. There are two major ports, Da Nang Port and Quy Nhon Port, and nine minor ports in the central region. In 2016, total throughputs in Central region occupied 11.65% of total throughputs. Da Nang Port is one of the main ports serving the import and export needs of Central Vietnam [3].

This paper focuses on Da Nang Port, which has a significant strategic geographic position since the hinterland of Da Nang Port is Da Nang city and the Central Vietnam Key Economic Region. Da Nang is the fourth largest city in Vietnam in terms of urbanization and economy and the city with the most socioeconomic and cultural conditions in the central region. The Central Vietnam Key Economic

Region covers Thua Thien-Hue, Quang Nam, Quang Ngai, and Binh Dinh provinces, and Da Nang city. The Key Economic Region has four nuclear economic zones including Chan May-Lang Co, Chu Lai, Dung Quat, and Nhon Hoi, and seven major urban areas - Hue, Da Nang, Hoi An, Tam Ky, Van Tuong, Quang Ngai, and Quy Nhon. Da Nang Port is located in Da Nang bay with an area of 12 km^2 and lies on a junction of four lanes to Da Nang international airport, national railway station, industrial zones in the area and national highway. Moreover, Da Nang Port is the final gateway of East-West Economic Corridor comprising four countries including Myanmar, Thailand, Laos and Vietnam [4].

The management board of Da Nang Port Company desires to develop the port as the leading trading gateway in the central region and the East-West Economic Corridor by focusing on the development of port operation and logistics services. Improvements have been made through investments by expanding its capacity, increasing the quality of services, warehouse systems with large yards, and good transportation system to serve the cargo transport smoothly. Consequently, the cargo volumes passed through Da Nang Port have increased sharply over the years, especially containerized cargo volumes [5]. Nevertheless, to the best of the authors' knowledge, there is a lack of academic research related to the development of port operation in Da Nang Port. There are just few technical reports on the general development of the port provided by Asian Development Bank (ADB) [6] and Japan International Cooperation Agency (JICA) [7]. These reports mostly focus on issues that are of interest to the donor organizations. They are generated from fact finding missions combined with available secondary data using different approaches depending on the intended goals of the donors.

The objective of this paper is to present a neutral approach that reflects the development of port operation in Da Nang Port from a multi-dimensional perspective. Firstly, this paper presents and analyses the current status of port operation in Da Nang Port in recent years to help the readers understand the overall picture of its business situation. Then, the classical decomposition method used in time series data analysis is employed to find the trend-cycle and seasonal factors of different kinds of monthly throughputs. This helps port managers understand the evolutions of the container, general cargo, and total throughput flows, which is important for port planning, development and operation. In addition, appropriate predictive models for these flows are provided with the aim to support decision makers to produce efficient long-term development plans. In particular, reasonable investment plans in facilities, types of equipment, and infrastructure for port operation are suggested based on the predictive results. Furthermore, the findings of this paper can be used as a starting point to initiate other studies on possible future development directions of Da Nang Port . While there are several academic studies on Hai Phong Port and those in Ho Chi Minh city such as [8], [9], [10], [11], and [12], no in-depth study on Da Nang Port can be found in literature so far. Lastly, it is also a useful document for foreign investors looking for investment opportunities in Da Nang Port.

The remainder of this paper is structured as follows. Section 2 presents the literature review and general overview of Da Nang Port. The current status of port operation in Da Nang Port is described in Section 3. Section 4 provides the details on data and methodology for analysing the development of port operation in Da Nang Port. Section 5 shows the results of this study. Finally, several discussions and conclusions are presented in Section 6.

2. Literature review and general overview of Da Nang Port

2.1. Literature review

In general, the tendency of the close relationship between ports and supply chain leads to increasing research area that is port hinterland intermodal development. The contributions of early studies focus on the optimization of container flow. Results show that future studies attend to the optimization of global intermodal container flow along with sustainable development [13]. The research towards sustainable port development is also discussed in [14]. The results of this study help identify significant factors of sustainable port development from the perspective of port authorities. The opportunities, challenges, and implications for management in Vietnamese port system are also referred to this study.

Furthermore, the study on the efficiency and competitiveness of Vietnamese port system is discussed in [15]. There are many negative factors in the Vietnamese port system such as outdated work practices, low efficiency, and poor competitiveness compared to other ports in the region. This research determines some of the issues in the Vietnamese port system and suggests some strategies to address them. Recently, a consideration for the developing a dry port system in Vietnam is considered as an emerging research. Dry ports are intermodal transport hubs in the inland, transhipment centres of sea cargo to inland destinations. [16] proposes conditions of the construction of a dry port system in the Indochina including Vietnam, Laos, and Cambodia. [8] presents the concept and classification of dry ports. It also examines and analyses the current status of domestic ports in the North and South of Vietnam. In addition, [10] introduces a multi-criteria approach to determine a dry port location in developing countries with application to Northern ports of Vietnam.

In addition, a strategic positioning analysis for Vietnamese ports is also a recent study direction. [9] uses dynamic portfolio analysis to analyse the competitive position of 20 seaports in five countries in Southeast Asia in six years from 2009 to 2014. This research reveals five ports having a dominant position that includes Tan Cang Sai Gon (Vietnam). [11] analyses the competitive position of seaports in North of Vietnam from 2005 to 2014 by applying the Boston Consulting Group matrix. Finally, [12] use SWOT and Porter's Five Forces methods to analyse the strengths, weakness, advantages, and disadvantages of domestic logistics service firms in Hai Phong Port. This study provides some solutions for these companies to develop better logistics operations and enhance global competitiveness.

As mentioned above, there are many studies on Northern and Southern ports of Vietnam, whereas it is hard to find academic literatures on Da Nang Port. There is only [17] that examines the quality of logistics services in Da Nang city. This study shows that Logistics Service Providers in Da Nang city need to improve Research and Development as well as develop Customer Care Programs. Most literatures related to Da Nang Port has been very much derived from technical assistance reports written by international organizations such as ADB, JICA and other donor organizations. These reports summarized the results of projects improving Da Nang Port such as [6], [7], [18]. Data on throughput flows in Da Nang Port are collected regularly (daily, monthly, quarterly, and annually). Managers in Da Nang Port often rely on statistical and descriptive reports on annual throughputs to give the investment plans in equipment, facilities, and infrastructure corresponding to their growth rate. There has been a gap between planning and actual cargo volumes since there is not any forecasting method used in planning in every year for port management in Da Nang Port [19]. One of basic approaches for forecasting is methods of time series analysis.

The time series analysis and forecasting models for sea port are suggested, for instance [20], [21], [22], and [23]. [20] forecasts the throughput flows in Hong Kong by estimating a co-integrated error correction model while [21] uses time series quarterly data between 1975 and 1998 to forecast the general cargo in Spanish ports. In addition, [22] uses the regression modelling to forecast the container traffic of Port of Rijeka whereas [23] analyses the evolutions of container flows through Vietnamese ports by applying Autoregressive Integrated Moving Average (ARIMA).

There are many methods of time series analysis for forecasting such as: spectral analysis, exponential smoothing, ARIMA modelling, Box-Tao modelling, and regression modelling. However, there is no unique or optimal method to forecast since each port and each type of cargo has its particularity. Furthermore, [24] emphasises that it is not necessary that the most sophisticated methods provide forecasting values more accurate than those of simple methods. Thus, we use regression modelling, a well-known forecasting method, to predict the trend-cycle components of throughput flows in Da Nang Port.

2.2. General overview of Da Nang Port

Da Nang Port is located at $16^0 17'33'$North latitude, $108^0 20'30'$ East latitude. Its pilot station is $16^0 10'$ North latitude and $108^0 11'$East latitude. It is located in semidiurnal tidal fluctuations which usually average 0.9 meters. Da Nang Port is the cargo gateway of Central Vietnam and plays an important role in the Vietnamese port system. Figure 1 describes the geographical location of Da Nang Port and its advantages on geography. Da Nang Port Company is officially operated as a joint stock company since July 25, 2014. Its business lines are loading and unloading of goods; storage service; service lease; tugging; supporting ships at the port; road and sea transportation; repairing of other means of transport; other maritime services. At present, Da Nang Port has two companies including Tien Sa Port Enterprise and Da Nang Port Logistics Joint Stock Company. Port operation and logistics services are taking place mainly in Tien Sa terminal, one of the three terminals, the others being Han river and Son Tra terminals [4].

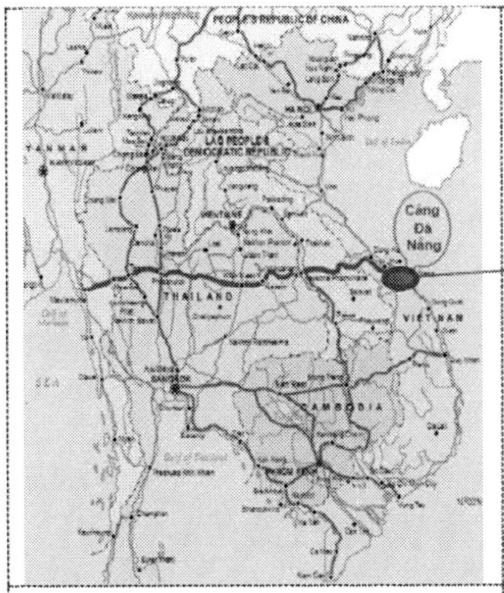

(a) Da Nang Port (blue circle) and the East-West Economic Corridor (red line).

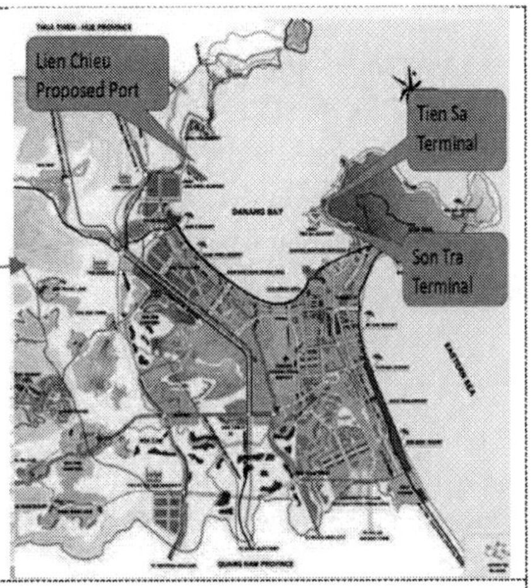

(b) Location of Da Nang Port within Da Nang bay.

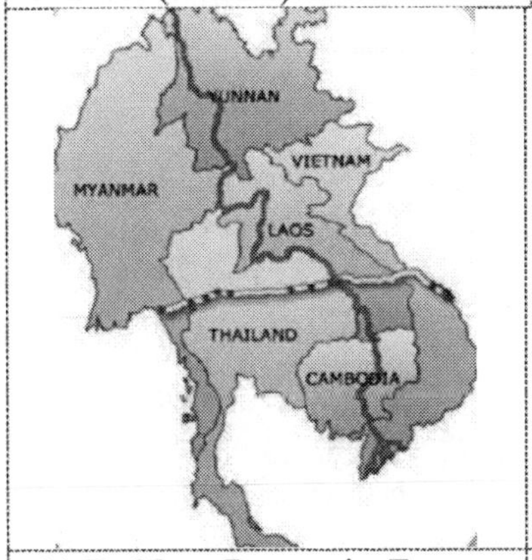

(c) Da Nang Port as the Eastern end of the East-West Economic Corridor (yellow line).

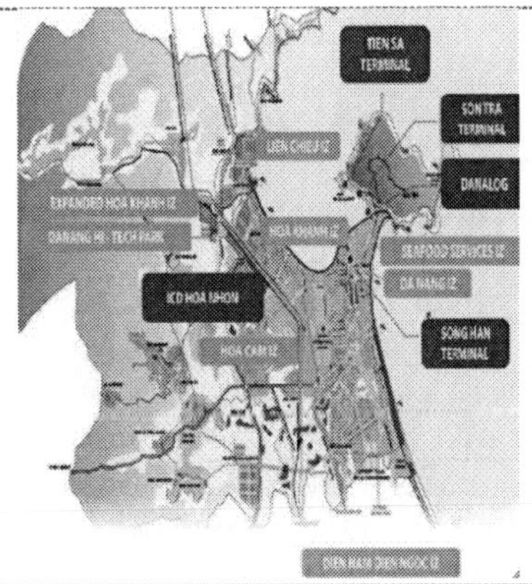

(d) Da Nang Port and Industry Zones (IZ) of Da Nang city.

Figure 1. Geographical location of Da Nang Port, Vietnam.
(Source: http://danangport.com/eng/)

Tien Sa terminal is a natural deep sea port located in Da Nang Bay. It accommodates big vessels from 35,000 to 50,000 Deadweight tons (DWT) and other specialized ships such as RORO ships, container ships, large and medium-sized passenger ships, super-length and super-weight cargo ships. Furthermore, Tien Sa terminal has total length of the berth of 1,193 metres and depth of the berth of 10-12 metres. The capacity of Tien Sa terminal is as follows: cargo ships from 35,000 to 50,000 DWT, throughput capacity of 8 million tonnes/year. It is expected to be 12 million tonnes/year in 2020. Tien Sa terminal has its space area of 229,414 square metres, its total warehouse area of 14,285 square meters, and its total yard area of 178,603 square metres. While Tien Sa Port Enterprise has big yards and warehouses, Da Nang Port Logistics Joint Stock Company has a modest yards and warehouses. It has space area of 51,037 square metres, total warehouses area of 12,225 square metres, and total yard area of 35,017 square metres . The equipment's capacity of Da Nang port is described in Table 1 [4].

Table 1. Equipment's capacity of Da Nang Port.

Equipment	Quantity(Units)	Capacity
Quayside gantry cranes exclusively used to load and unload containers at the Gantry	3	36 tonnes
Tyred gantry cranes (RTG) exclusively used to load and unload containers at the yard	4	36-40 tonnes
Fixed gantry cranes (Liebherr)	2	40 tonnes
Fixed gantry cranes (Liebherr)	2	25 tonnes
Forklift truck exclusively used to load and unload containers	5	42-45 tonnes
Wood scraper	2	5 tonnes
Automobile cranes	25	25-80 tonnes
Lifting trucks, excavators, bulldozers all kinds	24	1.5-7 tonnes
Tractor trucks, dump trucks, trailers	66	
Tugboat	7	500 -1,700 cv
Electronic scale	4	
Power supply system for refrigerated container all kinds	3	

(Source: http://danangport.com/eng/)

3. The current status of port operation in Da Nang port

The port operation is expressed by total cargo volumes passed through the port. Data on throughputs are provided by the sales and marketing department, Da Nang Port Joint Stock Company. At the present, the cargo volumes passed through Da Nang Port are divided into four kinds including domestic exports, foreign exports, domestic imports and foreign imports. Domestic exports are exported goods passed through Da Nang Port to other areas in the Vietnam's territory. Foreign exports are exported goods passed through Da Nang Port to other areas outside the Vietnam's territory. Domestic imports are imported goods passed through Da Nang Port from other areas in the Vietnam's territory. Foreign imports are imported goods passed through Da Nang Port from other areas outside the Vietnam's territory. These definitions are used in the statistical reports of Da Nang Port [19].

Considering the structure of cargo source, cargo volumes passed through Da Nang Port are divided into two main categories including container and general cargo. Based on the commodity classification of Vitranss2 [25] (Commodity classification is presented in appendix), general cargo volume passed through Da Nang Port includes commodity groups as follows: food, wood, steel products, building materials, cement, fertilizer, coal, and manufactured goods.

According to statistics from Da Nang Port, there was a significant increase in annual cargo volumes passed through Da Nang Port from 2012 to 2016. The throughput in 2016 was 7,255,000 tonnes, 64% higher than throughput in 2012 [19]. In the table 2, it can be seen that the average growth rate of total throughputs was 13.3% in the period (2012-2016) and the annual volumes of foreign import, domestic import, foreign export, and domestic export increased significantly from 2012 to 2016. The domestic export had the highest average growth rate in the period (2012-2016) of 48.4%, followed by foreign import, foreign export, and domestic import with their average growth rates of 26%, 8.7%, and 5.3%, respectively.

Table 2. Annual cargo volume passed through Da Nang Port from 2012 to 2016.

Kinds of throughputs (tonnes)	2012	2013	2014	2015	2016	Average period

						(2012-16)
I. Foreign import	908,767	1,345,060	1,576,963	1,901,601	2,249,948	26.0
II. Domestic import	1,287,096	1,005,176	1,414,038	1,142,963	1,388,065	5.3
III. Foreign export	1,990,236	2,361,018	2,285,033	2,421,946	2,749,704	8.7
IV. Domestic export	237,289	298,984	746,011	939,490	867,283	48.5
Total cargo volumes	4,423,388	5,010,238	6,022,045	6,404,000	7,255,000	13.3

(Source: Statistics of Da Nang Port)

Regarding the container and general cargo volumes, there was a steady increase in annual general cargo volume passed through Da Nang Port between 2,267,323 tonnes in 2011 and 2,793,844 tonnes in 2016. Meanwhile, the trend of container volume shot up from 1,601,222 tonnes in 2011 to 4,461,156 tonnes in 2016. Figure 2 depicted the tendencies of the container and general cargo volumes passed through Da Nang Port over the years. It is clearly shown that the share of container volumes was lower than the share of general cargo volumes from 2011 to 2013. However, it was higher than the share of general cargo volume since 2014.

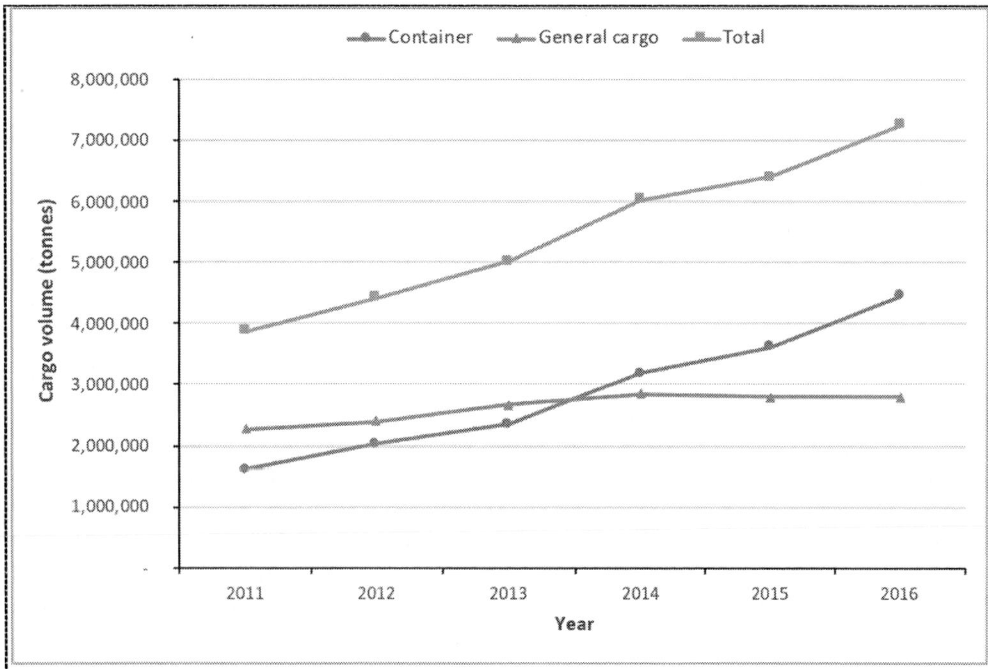

Figure 2. The tendencies of general cargo and container volumes passed through Da Nang Port from 2011 to 2016.

(Source: Statistics of Da Nang port)

Turning the annual cargo volumes by commodity group passed through Da Nang Port, container always had the highest value and a significant increase in total annual cargo volumes between 2012 and 2016. Table 3 illustrated the percentages of annual general cargo volumes by commodity group passed through Da Nang Port in the period (2012-2016); in addition the percentage of each commodity group at each year was illustrated. The wood group occupied from 24.58% to 30.91% between 2012 and 2016 that explained the highest percentages in total general cargo, whereas manufactured goods group explained the lowest proportions between 0.44% and 0.39% in the same period. The steel products group occupied the second important percentages ranging from 18.65% to 26.36% through years, followed by building materials group from 32.14% to 26.03%. The remaining groups such as cement, fertilizer, food and coal explained relatively small proportions, less than 10% of total general cargo volumes passed through Da Nang Port between 2012 and 2016.

In summary, Da Nang Port throughputs especially container had a significantly increased tendency during the period (2012-2016). To understand the development of port operation in the respective details, we will present data and methodology of research objectives in Section 4 and the results in Section 5 and some discussions and conclusions in Section 6.

Table 3. The percentages of annual general cargo volumes from 2012 to 2016.

	The percentages of general cargo volumes by commodity group (%)				
Commodity group	2012	2013	2014	2015	2016
Wood	24.58	37.46	31.48	29.56	30.91
Steel product	18.65	25.23	25.17	26.81	26.36
Building materials	32.14	20.24	25.88	25.47	26.03
Fertilizer	4.38	3.43	2.35	7.62	8.55
Food	5.52	4.35	4.04	2.04	2.10
Coal	6.53	4.91	9.04	6.13	4.17
Cement	7.76	3.70	1.78	2.05	1.50
Manufactured goods	0.44	0.25	0.25	0.32	0.39
Total general cargo	100	100	100	100	100

(Source: Statistics of Da Nang port)

4. Data and Methodology

To better understand the basic characteristics of throughput time series data, we use the classical time series decomposition method to find trend-cycle and seasonal components of monthly throughput time series data. Based on the results of the decomposition, appropriate forecasting models of the different kinds of throughputs are provided. In this section, we present data description in Subsection 4.1 and methodology in Subsection 4.2.

4.1. Data Description

The monthly throughput data from January 2012 to December 2016 are used to analyse time series decomposition. Like the classification of cargo volumes passed through Da Nang Port presented in Section 3, there are four main types of throughputs including Domestic Exports, Foreign Exports, Domestic Imports, Foreign Imports. Each type of throughputs is divided into two categories including container and general cargo. For example, Domestic Exports is divided into two kinds including ContainerDomesticExports and GeneralCargoDomesticExports. They are based on the throughput's classification of Sales & Marketing Department, Da Nang Port [19]. With such above throughput classification, we use 12-time series data given in Table 4 to analyse time series decomposition.

Table 4. Research Objectives and their definitions.

Ordered number	Objectives	Definition
1	ContainerForeignImport	The foreign imports passed through Da Nang Port are container
2	GeneralCargoForeignImport	The foreign imports passed through Da Nang Port are general cargo
3	TotalForeignImport	Total foreign imports passed through Da Nang Port
4	ContainerDomesticImport	The domestic imports passed through Da Nang Port are container
5	GeneralCargoDomesticImport	The domestic imports passed through Da Nang Port are general cargo
6	TotalDomesticImport	Total domestic imports passed through Da Nang Port
7	ContainerForeignExport	The foreign exports passed through Da Nang Port are container
8	GeneralCargoForeignExport	The foreign exports passed through Da Nang Port are general cargo
9	TotalForeignExport	Total foreign exports passed through Da Nang Port
10	ContainerDomesticExport	The domestic exports passed through Da Nang Port are container

11	GeneralCargoDomesticExport	The domestic exports passed through Da Nang Port are general cargo
12	TotalDomesticExport	Total domestic exports passed through Da Nang Port

(Source: Statistics of Da Nang Port)

4.2. Methodology
The technique for analysing monthly throughputs of the port is classical time series decomposition method. This method is developed in [26] and is described, for example, in [27]. A time series data can be broken down into separated components which are trend, cyclical, seasonal, and residual components. The trend component describes the long-term change in time series data. The cyclical component reflects the change in time series data that are a repeated change according to the business cycle. The seasonal component describes the change in time series data that depends on seasonal factors. The random component denotes the change in time series data that depends on random or irregular factors. It represents the residuals or the remainder of time series after other components have been removed from the time series data. The general form of decomposition technique is written as follows:

$$Y_i = f(T_{i,} S_i, e_i) \quad (1)$$

where Y_i is the time series (or observation) at time i; T_i is the trend-cycle component at time i; S_i is the seasonal component at time i; e_i is the error (or residual) at time i.

Equation (1) is represented in two different forms, that are multiplicative and additive decompositions. The multiplicative decomposition is described as follows:

$$Y_i = T_i * S_i * e_i \quad (2)$$

The additive decomposition is expressed as follows:

$$Y_i = T_i + S_i + e_i \quad (3)$$

The additive decomposition is used as seasonal variations do not change in level along the series. On the contrary, if the level of seasonal variations changes as the series goes, the multiplicative decomposition is preferred.

There are three steps to split an original time series data into separated components in both the multiplicative and additive decompositions as follows:

- Step 1- Calculating trend-cycle component: By applying a moving average method to the original data, the trend-cycle component in time series data is identified. If the number of observations is odd, m-MA smoother is used to obtain T_i. Otherwise, the trend-cycle component T_i is computed by using 2 x m-MA.
- Step 2- Computing seasonal component: Firstly, the detrended series is calculated. It equals $\frac{Y_i}{T_i}$ in the multiplicative decomposition and is equivalent to $(Y_i - T_i)$ in the additive decomposition. The seasonal indicator for a specific season is determined by taking the average of detrended values in the same time period. The seasonal time series can be obtained by repeating the seasonal index for the same periods in time series. This seasonal time series is denoted as S_i.
- Step 3- Calculating irregular component: After the trend-cycle and seasonal components are determined, the irregular component is calculated as $e_i = \frac{Y_i}{T_i * S_i}$ in the multiplicative decomposition and $e_i = Y_i - T_i - S_i$ in the additive decomposition.

5. Results
5.1. Results of decomposition method
Figures 3, 4, 5 illustrate the trends of container, general cargo, total throughput of draw data from January 2012 to December 2016. Magnitude of the seasonal patterns in these graphs are not so obvious that we tried to use both additive and multiplicative decompositions for separating time series data of throughput flows and choose the one with smaller forecast error such as using MAPE (Mean Absolute Percentage Error).

The tendencies of the trend-cycle component of the container, general cargo, and total throughput are shown in Figures 6, 7, and 8 by applying the moving average method to the original data. After the moving average, the data for the first 6 months of 2012 and the last 6 months of 2016 are the missing values. Thus, the trends of the trend-cycle component of throughput flows are illustrated from July 2012 to June 2016. Their seasonal indexes by multiplicative and additive decompositions are described in Tables 5 and 6. The seasonal indexes of the multiplicative decomposition revolves around the value 1 while those of additive decomposition revolves around the value 0.

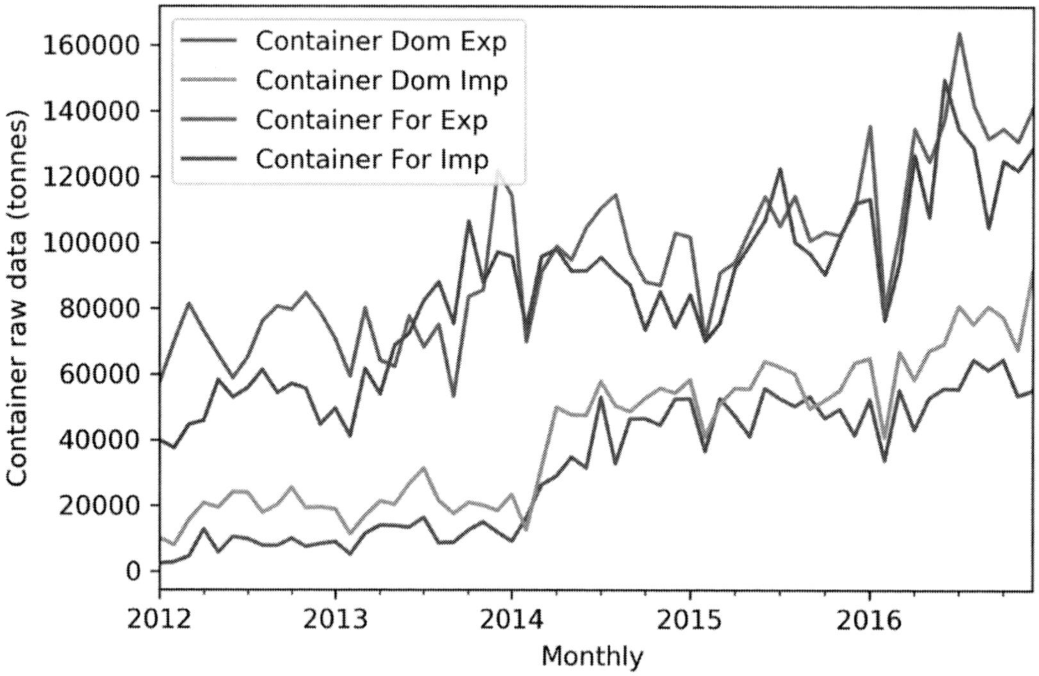

Figure 3. Raw data of monthly container throughputs from January 2012 to December 2016

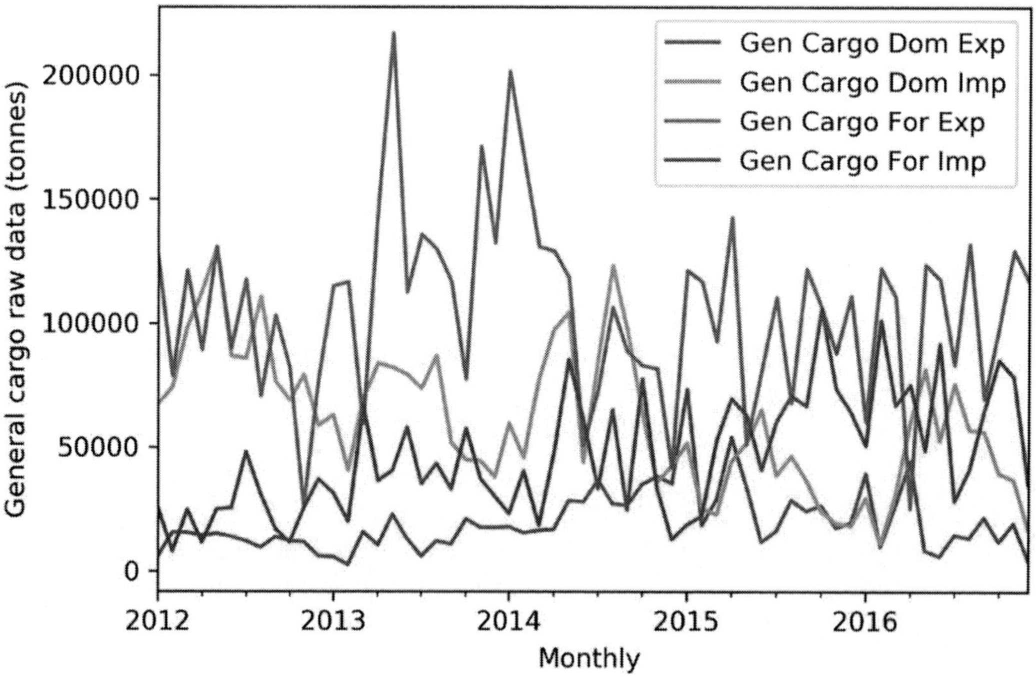

Figure 4. Raw data of monthly general cargo throughputs from January 2012 to December 2016

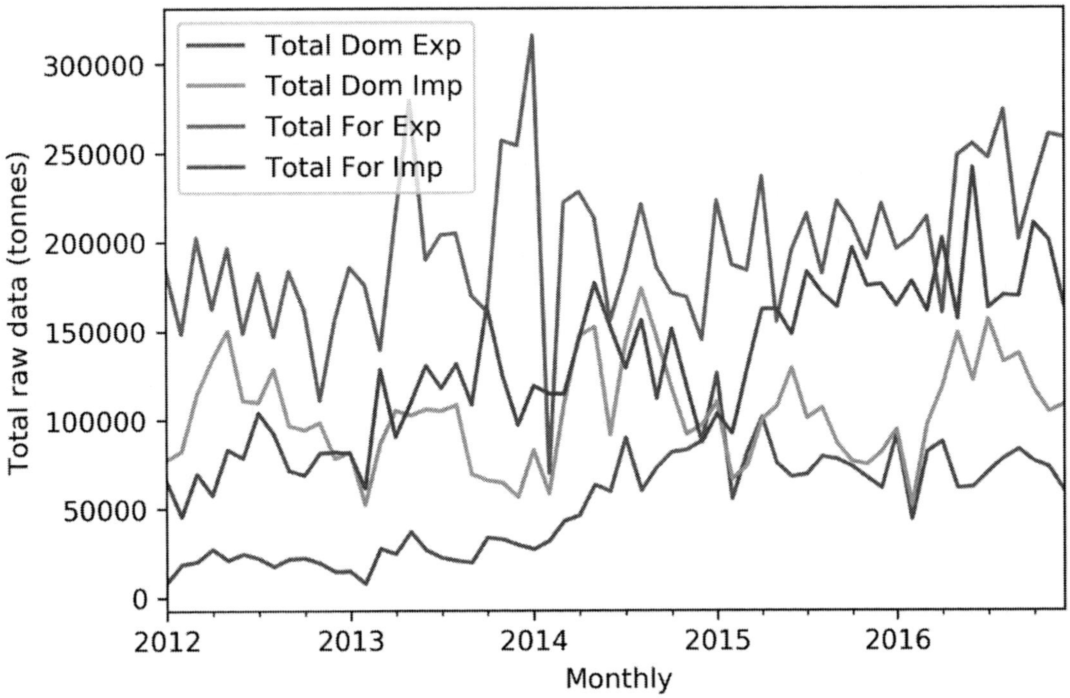

Figure 5. Raw data of monthly total throughputs from January 2012 to December 2016

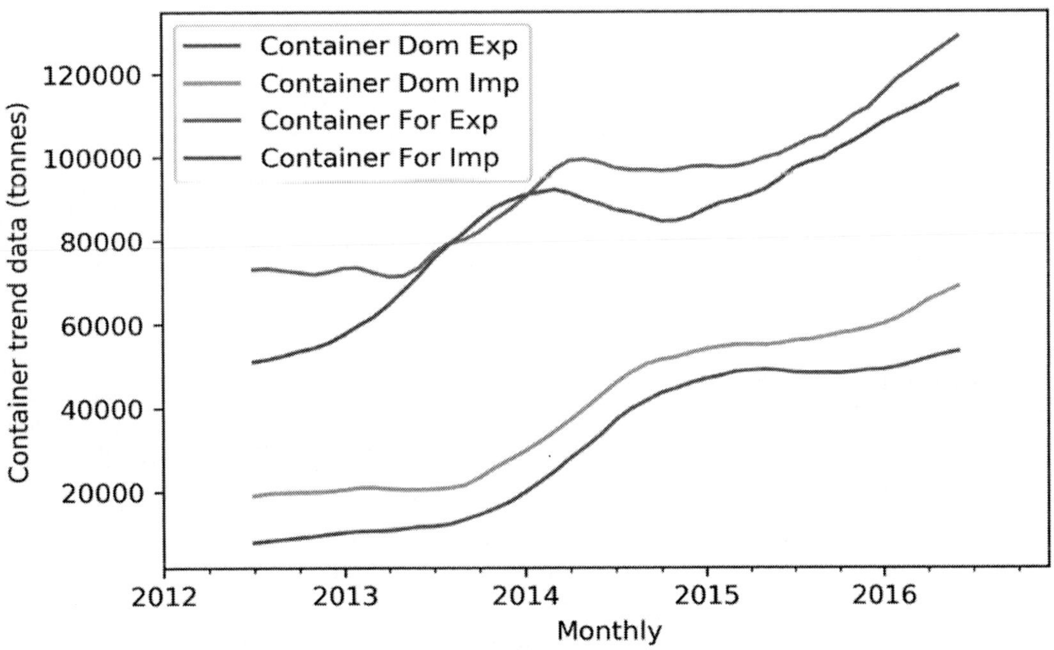

Figure 6. Trend data of monthly container throughputs from July 2012 to June 2016

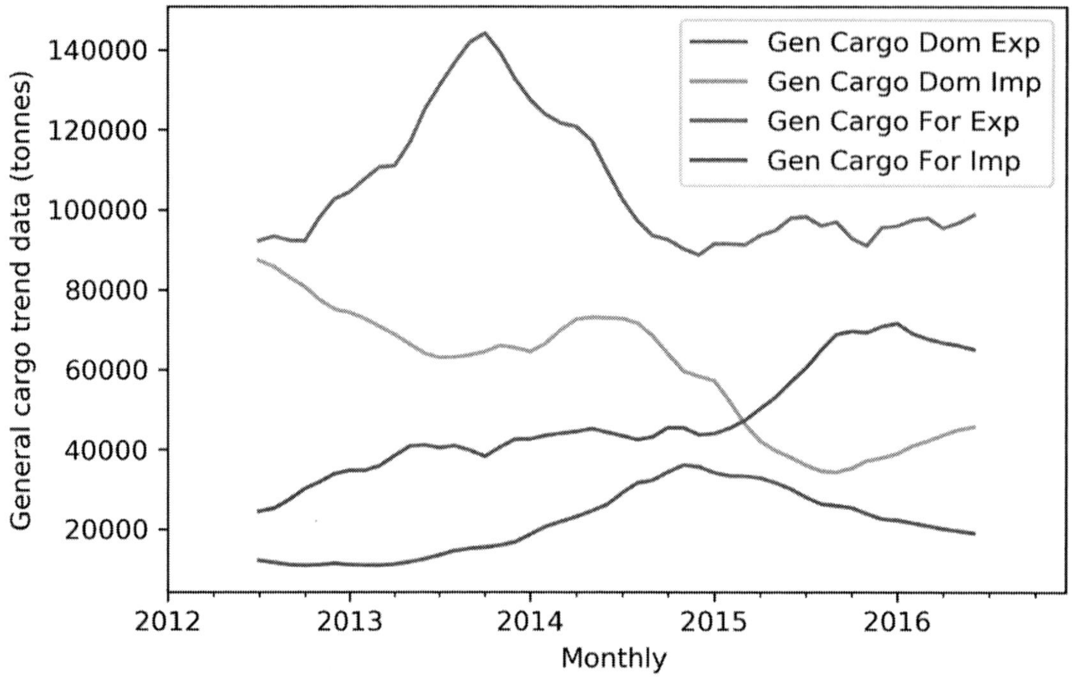

Figure 7. Trend data of monthly general cargo throughputs from July 2012 to June 2016

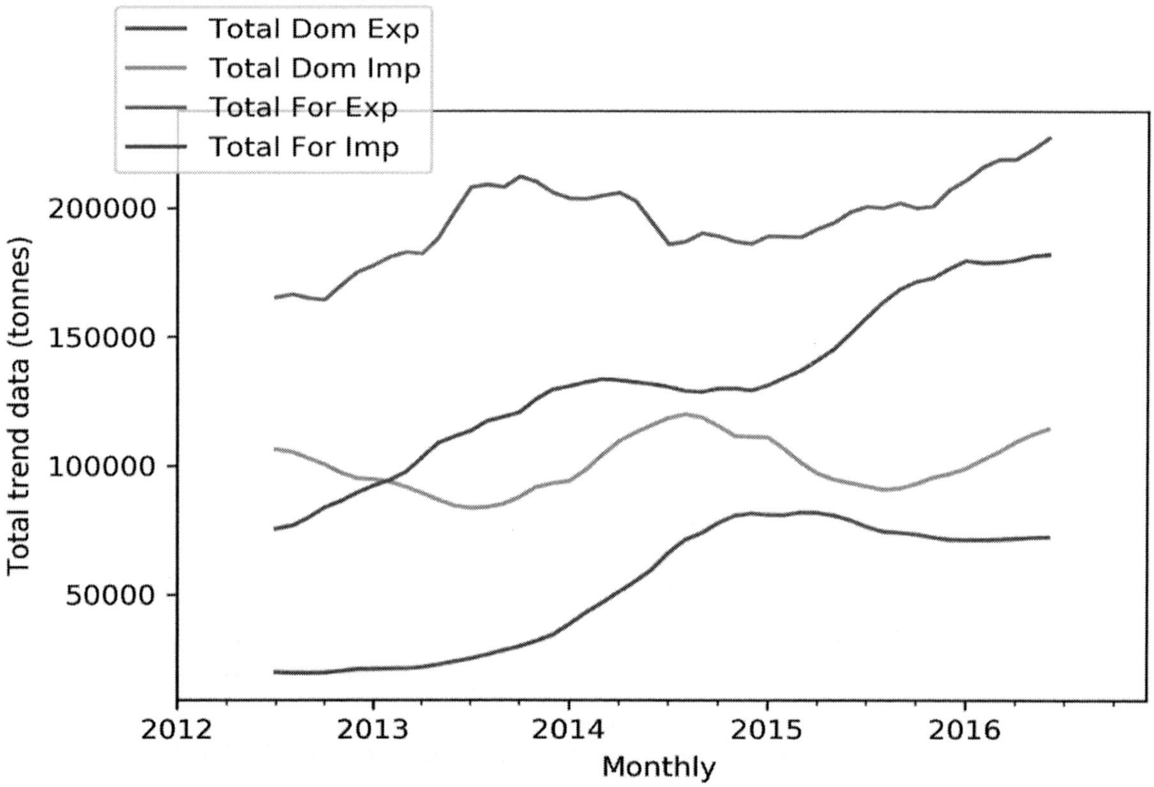

Figure 8. Trend data of monthly total throughputs from July 2012 to June 2016

Table 5. Seasonal indexes by multiplicative decomposition.

	Seasonal indexes (Multiplicative)											
Objectives	Jan	Feb	Mar	Apr	May	June	July	Aug	Sep	Oct	Nov	Dec
ContainerDomesticExport	0.90	0.68	1.10	1.06	1.08	1.09	1.32	0.90	0.97	1.03	0.96	0.90
ContainerDomesticImport	0.98	0.60	0.94	1.08	1.06	1.15	1.30	1.02	0.93	1.04	0.95	0.95
ContainerForeignExport	1.12	0.74	0.96	0.99	0.97	1.08	0.99	1.07	0.94	1.00	1.01	1.14
ContainerForeignImport	0.98	0.75	0.93	1.01	1.01	1.12	1.14	1.09	0.99	1.02	1.00	0.96
GeneralCargoDomesticExport	1.37	0.51	1.11	1.40	1.17	0.72	0.83	0.91	0.95	1.15	1.01	0.88
GeneralCargoDomesticImport	0.87	0.51	0.84	1.27	1.47	1.20	1.12	1.46	1.07	0.82	0.72	0.66
GeneralCargoForeignExport	1.17	1.25	0.95	1.05	1.18	0.85	1.04	0.88	1.04	0.88	0.85	0.85
GeneralCargoForeignImport	0.64	0.85	1.08	1.11	1.18	1.21	1.13	1.21	0.74	1.26	0.86	0.74
TotalDomesticExport	1.08	0.61	1.10	1.14	1.15	0.97	1.08	0.91	0.98	1.09	1.00	0.89
TotalDomesticImport	0.93	0.58	0.92	1.17	1.26	1.14	1.16	1.30	1.00	0.89	0.84	0.80
TotalForeignExport	1.18	0.81	0.95	1.06	1.12	0.97	1.04	0.99	1.01	0.93	0.94	1.00
TotalForeignImport	0.87	0.79	0.99	1.05	1.07	1.15	1.13	1.14	0.90	1.11	0.96	0.83

Table 6. Seasonal indexes by additive decomposition.

	Seasonal indexes (Additive)											
Objectives	Jan	Feb	Mar	Apr	May	June	July	Aug	Sep	Oct	Nov	Dec
ContainerDomesticExport	-418	-9305	2998	-1122	83	2537	7021	-1894	1447	463	-141	-1668
ContainerDomesticImport	740	-15341	-1465	2146	2434	5321	8853	1597	-2755	104	-1027	-605
ContainerForeignExport	11865	-25555	-5276	250	-2267	8493	224	7001	-5407	-519	-426	11617
ContainerForeignImport	-302	-22087	-7001	2748	473	12144	11250	6089	-1565	547	-145	-2153
GeneralCargoDomesticExport	12548	-9946	297	9603	1551	-7160	-2929	-1648	-2174	2182	-410	-1914
GeneralCargoDomesticImport	-7236	-26818	-6442	15125	24465	5689	6677	28570	3761	-9886	-14711	-19193
GeneralCargoForeignExport	20150	25720	-6562	6350	21525	-16992	3550	-11649	1803	-17663	-12418	-13815
GeneralCargoForeignImport	-18169	-3160	1458	6166	6990	10019	1054	8069	-10436	16178	-5447	-12723
TotalDomesticExport	12130	-19251	3295	8480	1633	-4622	4092	-3542	-727	2645	-551	-3582
TotalDomesticImport	-6497	-42159	-7908	17271	26899	11009	15530	30167	1006	-9782	-15739	-19798
TotalForeignExport	35477	-37913	-8377	10061	22720	-5037	7235	-1186	-142	-14720	-9381	1264
TotalForeignImport	-17842	-24617	-4914	9543	8092	22792	12934	14787	-11372	17354	-4963	-21794

5.2. Forecasting method and results

For predicting purpose, all data of the throughput flows including container, general cargo and total throughput are divided into initialization sets and test sets. The initialization sets include the sets of data between July 2012 and June 2015, which are used to find suitable models and parameters for the observations of study objectives. The evaluation of initialization sets is meaningful in finding the best models fitting observations. The selected models will be run on the test sets of the throughput flows for the last 12 months from July 2015 to June 2016. The evaluation of test sets denotes the accuracy and confidence of chosen models.

5.2.1. Trend- Cycle Forecasting

The general method of using the decomposition techniques in predicting the time series data of the throughput flows is to predict each factor (including the trend-cycle and the seasonal factors) into the future. Forecasting the seasonal component is not difficult since the seasonal indicators are constant over a certain time period. Thus, seasonal indexes in Tables 5 and 6 are repeated to produce forecasting values of the seasonal factor that corresponds to each decomposition method. However, the prediction of the trend-cycle factor is more complex. A common method for predicting trend-cycle factor is the parametric trend model, which is a method of finding the appropriate mathematical model to fit observations.

In this paper, we test the first, second and third degrees of polynomials on the initialization sets of trend-cycle data of throughput flows. We then use the selected models to run on the test sets of the trend-cycle data. There are two criteria for selecting the appropriate model, which are the a high coefficient of determination (R^2) on the initialization set and the small mean absolute percent error (MAPE) on the test set. Table 7 presents the results of the R-square of the first, second and third degrees of polynomials on the initialization sets and the MAPE of these model on the test sets of the trend-cycle data. The results shows that using the first-order models are the most appropriate for the trend-cycle data of container. In addition, the third degree of polynomials are the best models for the trend cycle data of general cargo and total throughput except GeneralCargoDomesticImport used by the first degree of polynomial. Table 8 illustrates the results of the selected mathematical models for the trend-cycle data of the studied objectives.

Table 7. The results of R^2 and MAPE of the first, second, and third degrees of polynomials.

	Order 1		Order 2		Order 3	
Objectives	R-square	MAPE (%)	R-square	MAPE (%)	R-square	MAPE (%)
ContainerDomesticExport	**0.92**	**19.50**	0.96	49.78	0.98	10.77
ContainerDomesticImport	**0.90**	**7.68**	0.94	30.55	0.98	20.68
ContainerForeignExport	**0.87**	**4.10**	0.87	6.70	0.95	36.70
ContainerForeignImport	**0.77**	**1.85**	0.92	24.76	0.92	24.41
GeneralCargoDomesticExport	0.88	83.88	0.89	107.97	**0.97**	**38.04**
GeneralCargoDomesticImport	**0.65**	**20.08**	0.67	21.61	0.94	173.34
GeneralCargoForeignExport	0.05	4.22	0.63	63.18	**0.75**	**11.56**
GeneralCargoForeignImport	0.84	17.74	0.87	26.19	**0.96**	**16.54**
TotalDomesticExport	0.91	38.82	0.94	67.11	**0.98**	**18.16**
TotalDomesticImport	0.16	10.95	0.17	16.13	**0.84**	**80.95**
TotalForeignExport	0.20	3.40	0.67	24.59	**0.76**	**2.91**
TotalForeignImport	0.87	6.90	0.95	23.55	**0.96**	**10.39**

(The bold numbers showed the better results of R-square and MAPE than those of other models)

Table 8. The results of the selected mathematical models using for trend-cycle forecasting of the study objective.

Objectives	Models	R^2	MAPE (%)
ContainerDomesticExport	$1439t$	0.92	19.50
ContainerDomesticImport	$1310t$	0.90	7.68
ContainerForeignExport	$1026t$	0.87	4.10
ContainerForeignImport	$1238t$	0.77	1.85
GeneralCargoDomesticExport	$-3t^3 + 177t^2 - 2047t + 16253$	0.97	38.04

GeneralCargoDomesticImport	$-893t$	0.65	20.08
GeneralCargoForeignExport	$7t^3 - 514t^2 + 10300t + 65374$	0.75	11.56
GeneralCargoForeignImport	$2.4t^3 - 144t^2 + 3034t + 20074$	0.96	16.54
TotalDomesticExport	$-5.8t^3 + 363t^2 - 4107t + 29842$	0.98	18.16
TotalDomesticImport	$-10t^3 + 570t^2 - 8397t + 122917$	0.84	80.95
TotalForeignExport	$5t^3 - 356t^2 + 8092t + 144026$	0.76	2.91
TotalForeignImport	$2t^3 - 191t^2 + 6060t + 62696$	0.96	10.39

5.2.2. Forecasting results

When the predictions of trend and cycle component are obtained, they are put back in the decomposition model in equation (2) (for the multiplicative model) and in equation (3) (for the additive model). They will be composed of the seasonal indicators presented in Table 5 (for the multiplicative model) and in Table 6 (for the additive model) to create forecasting values. The predicting values for monthly throughputs are computed as $\hat{Y}_t = \hat{T}_t * \hat{S}_t$ in the multiplicative decomposition and $\hat{Y}_t = \hat{T}_t + \hat{S}_t$ in the additive decomposition.

The forecasting models are run on test sets of study objectives from July 2015 to June 2016. The results of MAPE by multiplicative and additive decompositions are presented in Table 9. The MAPE values of forecasting results by additive method are lower than those of multiplicative method in most time series data. It is a clear evidence to determine which decomposition method for each time series data. Table 9 also presents the results of selecting the best decomposition models for time series data of throughput flows. The additive models are preferred for most time series data apart from ContainerDomesticExport, GeneralCargoDomesticExport, GeneralCargoForeignExport, and TotalDomesticImport used in multiplicative models.

Table 9. The results of selecting the best decomposition models for time series data of study objectives

Objectives	MAPE(%) in multiplicative (1)	MAPE (%) in additive (2)	Difference in MAPE (%) (3)=(1)-(2)	Chosen Decomposition Method
ContainerDomesticExport	*23.13*	*23.34*	*-0.21*	*Multiplicative*
ContainerDomesticImport	11.99	11.58	0.42	Additive
ContainerForeignExport	6.69	6.68	0.01	Additive
ContainerForeignImport	9.04	8.96	0.08	Additive
GeneralCargoDomesticExport	*49.85*	*59.95*	*-10.11*	*Multiplicative*
GeneralCargoDomesticImport	43.67	36.78	6.89	Additive
GeneralCargoForeignExport	*52.32*	*52.79*	*-0.48*	*Multiplicative*
GeneralCargoForeignImport	29.83	24.44	5.40	Additive
TotalDomesticExport	20.30	19.45	0.85	Additive
TotalDomesticImport	*79.31*	*83.74*	*-4.42*	*Multiplicative*
TotalForeignExport	12.80	12.43	0.37	Additive
TotalForeignImport	15.51	15.09	0.42	Additive

Based on the results of Table 9, the seasonally adjusted graphs of monthly throughputs of the container, general cargo and total throughput are plotted in Figures 9, 10, and 11. After removing the seasonal component from monthly throughput time series data, the container and total throughputs have significantly increase tendencies whereas those of general cargo volumes are not obvious.

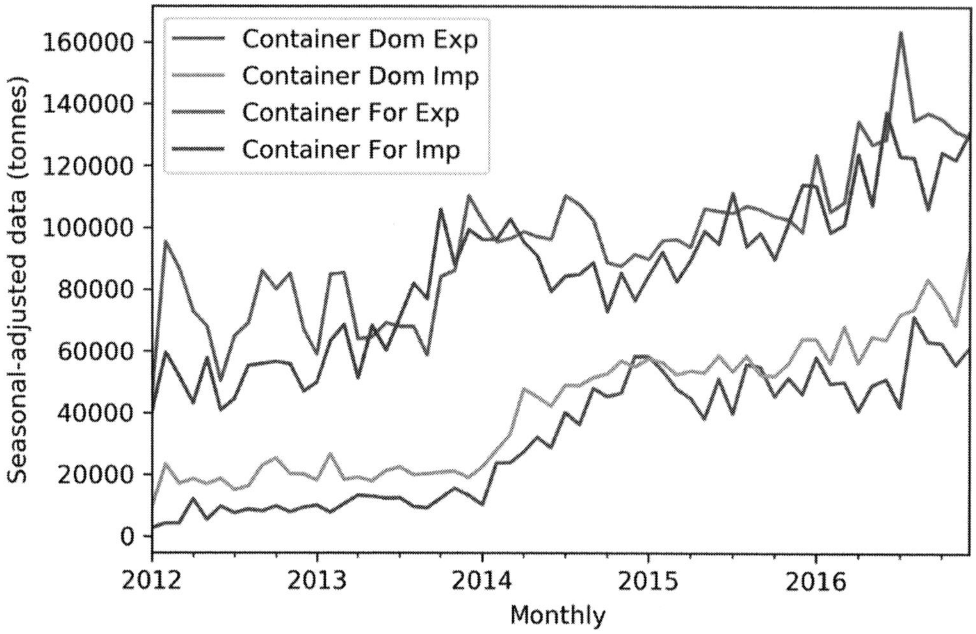

Figure 9. Seasonal-adjusted data of container from January 2012 to December 2016

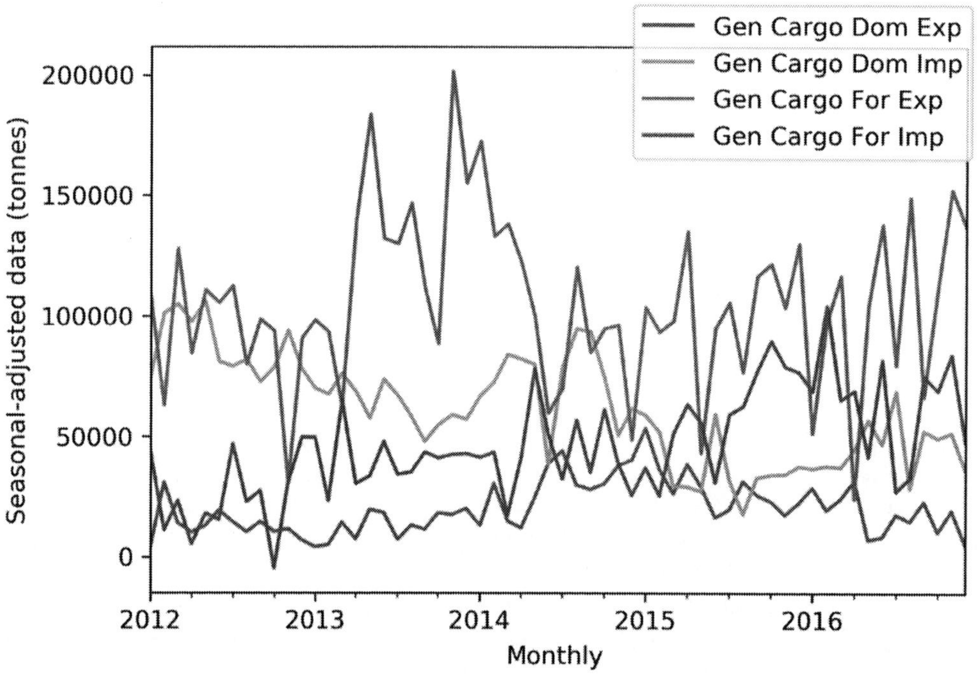

Figure 10. Seasonal-adjusted data of general cargo from January 2012 to December 2016

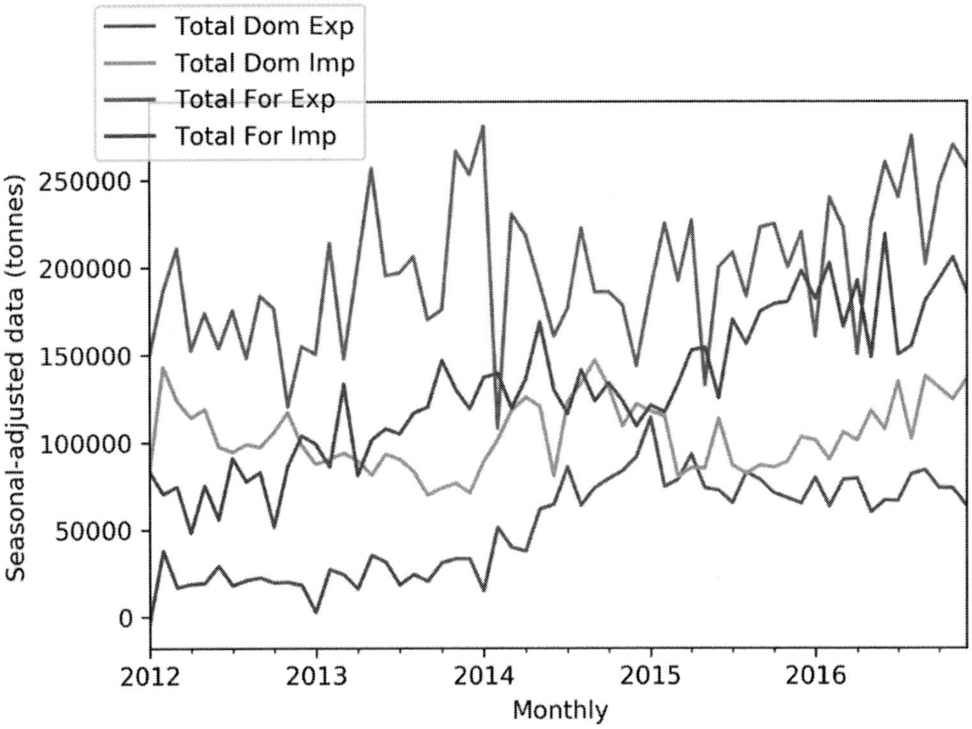

Figure 11. Seasonal-adjusted data of total throughputs from January 2012 to December 2016

6. Discussions and conclusions

6.1. Discussions

The results in Figure 6 and Table 8 show that container flows increase as first-order linear functions having positive and large slopes. Based on these predictive models, we can conclude that the volumes of containerized cargoes will rise significantly both in the growth rate and the magnitude of growth.

For the general cargo, Figure 7 and Table 8 describes the tendencies and models of its flows. The trend-cycle data of general cargo flows vary chaotically so that third-order polynomials are used to model their observations. In general, the growth trends of general cargo flows are unclear.

For total throughput flows, their trend-cycle components are the sum of those of the container and general cargo flows. As presented in Figure 8 and Table 8, the trend-cycle data of total throughput flows generally increase through time.

More interestingly, we have the tendencies of seasonal-adjusted time series data of throughput flows in Da Nang Port. Eliminating the seasonal component from the time series helps us judge the growth prospects of the container, general cargo, and total throughput flows in the long run. Figures 9, 10, and 11 show the seasonal-adjusted data of the container, general cargo, and total throughput flows, respectively. The seasonal-adjusted data of the container flows increase considerably over time, whereas those of the general cargo flows exhibit no clear trend. Finally, there is a steady increase in the seasonal-adjusted total throughput flows in Da Nang Port.

Based on the above discussions on the growth prospects of throughput flows, we provide some policy implications for the development of port operation in Da Nang Port. First, in terms of the development strategy, Da Nang Port should be towards containerization since the tendencies of container flows will increase rapidly in both the growth rate and the magnitude of growth. Second, managers in Da Nang Port should pay attention to the current status of facilities, equipment, and infrastructure. From this current status, there should be more investment in them to meet the growing demand of throughputs, especially for container. In addition, Da Nang Port should maintain general cargo flows although their growth tendencies are mixed up since these types of cargoes meet the consumer demand of Central Region of Vietnam.

6.2. Conclusions

This paper presents the current status of port operation in Da Nang Port. Through the analysis of monthly throughputs, the picture of the operating status in Da Nang Port between 2012 and 2016 becomes clear. The cargo volumes passed through Da Nang port exhibits a dramatically increased trend, particularly in terms of containerized cargoes. Furthermore, the decomposition of monthly throughput time series into trend–cycle and seasonal factors provides appropriate predictive models of container, general cargo, and total throughput flows. In terms of policy implications, this study suggests that the development of Da Nang port should be towards containerization and focus on improving its operation capacity by the investments in facilities, equipment, and infrastructure.

Appendix - Commodity Classification

Commodity code	Commodity group	Name of commodity
1	Food	Rice, cereals, wheat
2	Sugarcane	Sugarcane to be refined to produce candy, cake, and sugar
3	Wood	Wood to produce joinery and paper
4	Industrial plants	Industrial plants such as coffee, tea, rubber, pepper, bean..
5	Sea food	Various types of seafood being cached naturally and raised in the farm
6	Meat cattle	Pigs, cattle and poultry
7	Steel products	Type of steels including steel plate
8	Building materials	Stone, gravel, sand, brick, tile
9	Cement	cement
10	Fertilizer	Nitrogenous fertilizer and animal feeding
11	Coal	Coal as fuel for fertilizer industry, electricity and other industries
12	Petroleum products	Materials derived from crude oil (petroleum) as it is processed in oil refineries excluding petroleum
13	Manufactured goods	Equipment, machinery, chemicals, consumer goods…packed in containers or bulk goods

References

[1] General Statistics Office of Vietnam 2016 Reports of Maritime Economy.
[2] Resolution 09-NQ / TW 2007 Vietnamese Sea Strategies until 2020.
[3] Vietnam Seaports Association 2017 Reports of ports of Vietnam.
[4] Da Nang port's web (http://danangport.com/eng/).
[5] Nguyen Huu Sia 2016 Workshop Cooperation in developing logistics system and center in the central Vietnam Key Economic Region, Da Nang city.
[6] ADB 2008 Transport and Trade facilitation in the Greater Mekong sub region Time to shift Gears- Evaluation study (https://www.oecd.org/countries/mongolia/42228102.pdf).
[7] Japan International Cooperation Agency 2014 The Study for Tien Sa - Da Nang Port Improvement Project (http://open_jicareport.jica.go.jp/pdf/12250627.pdf).
[8] Nguyen L C , Notteboom 2016 *J of International Logistics and Trade* **14(1)** 65-88.
[9] Dang V L, Yeo G T 2003 *J of the Eastern Asia Society for Transportation Studies*, **5**.
[10] Nguyen L C, Notteboom 2016 *The Asian J of Shipping and Logistics* **32** 23-32.
[11] Pham T Y, Ma H M, Yeo G T 2016 *J of Korean Navigation and Port Research* **40(5)** 311-316.
[12] Deng Y L, Deng Y, Li X 2014 How can Domestic Logistics Enterprises Enhance Their Competitive Position in Haiphong Port City, Master's thesis, Yi Shou University.

[13] Jasmine S L L, Yimiao G 2013 *International J of Shipping and Transport Logistics* **5(3)** 257-281.
[14] Saeyeon R, Vinh V T, Yiik D W 2016 *The Asian J of Shipping and Logistics*, **32(2)**, 107-118.
[15] Vinh V T , Devinder G 2005 *Asia Pacific J of Marketing and Logistics*, **17 (1)** 3-31.
[16] Ngoc-Hien D, Ki-Chan N, Quynh-Lam N L 2011 *The flagship J of international shipping and port research* **38(1)** 1-9.
[17] Limbourg S, Ho T Q G, Cools M 2016 *Proc. Engineering Conf. on Sustainable Development in Civil, Urban, and Transportation Engineering* (Ho Chi Minh City) vol 1 p124-130.
[18] Japan International Cooperation Agency 2009 Ex-Post Evaluation of Japanese ODA Loan Project "Da Nang Port Improvement Project" (http://open_jicareport.jica.go.jp/pdf/12033940.pdf).
[19] Sales & Marketing Department of Da Nang Port Joint Stock Company 2017 Statistics of Da Nang port.
[20] Hui E C, Seabrooke W and Wong G K 2004 *J. Urban Planning and Development* **130(4)** 195–203.
[21] Coto-Millán P, Baños-Pino J, Sainz-González R, Pesquera-González et al 2011 *Maritime Economics & Logistics* **13(3)**237–249.
[22] Alen J, Svjetlana H, Tanja Poletan J 2011 *Promet - Traffic&Transportation* **23**.
[23] ThiAnhTam T, Mikio T 2015 *J of the Eastern Asia Society for Transportation Studies* **11**.
[24] De Borger B and Nonneman W 1981 *J of Transport Economics and Policy* **15** 155-165.
[25] Japan International Cooperation Agency 2000 The study on the national transport development strategy in the Socialist Republic of Vietnam (VITRANSS) final report : summary http://open_jicareport.jica.go.jp/700/700/700_123_11596749.html.
[26] Macaulay F 1938 *The Movements of Interest Rates, Bond Yields and Stock Prices in the United States* National Bureau of Economic Research, New York.
[27] Spyros M, Steven C, Wheelwright et al 1998 *Forecasting: Methods and Applications* John Wiley and Sons Third edition ISBN 0-471-53233-9.XIV+642pp.

Acknowledgment
The authors thank the officers in Sales & Marketing Department of Da Nang Port Joint Stock Company for providing the data used in this study.

IOP Conf. Series: Earth and Environmental Science **143** (2018) 012074 doi:10.1088/1755-1315/143/1/012074

Analysis of Child-related Road Traffic Accidents in Vietnam

Anh Tuan VU[1,3] and Dinh Vinh Man NGUYEN[2,4]

[1]Vietnamese-German Transport Research Centre
Telephone: +84 (0)274 2220990, Fax. +84 (0) 274 2220980
[2]Vietnamese-German Transport Research Centre
Telephone: +84 (0)274 2220990, Fax. +84 (0) 274 2220980
[3]E-mail: drtuan.va@vgtrc.vgu.edu.vn;
[4]E-mail: man.ndv@vgtrc.vgu.edu.vn

Abstract. In recent years, the number of road traffic accidents, fatalities and injuries have been decreasing, but the figures of children road traffic accidents have been increasing in Ho Chi Minh City of Vietnam. This fact strongly calls for implementing effective solutions to improve traffic safety for children by the local government. This paper presents the trends, patterns and causes of road traffic accidents involving children based on the analysis of road traffic accident data over the period 2010-2015 and the video-based observations of road traffic law violations at 15 typical school gates and 10 typical roads. The results could be useful for the city government to formulate solutions to effectively improve traffic safety for children in Ho Chi Minh City and other cities in Vietnam.

Keywords. Child traffic accident, accident pattern, accidents cause, traffic safety policy

1. Introduction

According to WHO, there were globally more than 186,000 children (aged under 18) died due to road traffic accidents every year (equivalent to more than 500 child deaths per day) [8]. In Vietnam, this figure was close to 2,000 children died because of traffic accidents per year, or equivalent to 5 children died per day [5]. In which, the proportion of child deaths who walked alone accounted for 36% and who rode a bike or a motorcycle accounted for about 20%. In the number of nearly 2,000 child deaths, nearly half of them had head injuries mainly due to the fact that they did not wear helmets. Vietnam's Ministry of Health reported that child traffic accident fatality rate of Vietnam was about 20 deaths per 100,000 children, while the figure was 7.4 deaths per 100,000 children in South East Asia and 4.2 deaths per 100,000 children in Europe [6]. It can be stated clearly that Vietnamese children are at the high risk of fatality when they participate in traffic.

There is a lack of in-depth and systematic studies on patterns and causes of traffic accidents involving Vietnamese children to support formulation of effective solutions to improving children road safety in urban areas. Up to now, there have been a few studies on traffic accidents involving Vietnamese children. The Ministry of Health published a traffic accident report by analyzing accident statistics from hospitals [6]. La Ngoc Quang investigated the status of helmet use by road users in six provinces of Vietnam, including Ho Chi Minh City [5]. Nguyen Thuy Quynh analyzed the relationship between helmet use and traumatic brain injury and the influences of personal and family factors on children's helmet use in Viet Nam [7].

This paper is to present key results of an in-depth analysis of children-related traffic accident data in urban areas. Accident patterns and causes are analyzed against a number of key factors, such as accident occurrence time, location, involved vehicle type, rider age, and rider gender. It is also aimed to capture the current situation of traffic law violations by pupils and their parents. Based on the results, the paper suggests some breakthrough solutions to improve traffic safety for children in Vietnamese cities.

2. Methodology and Data Collection

Figure 1 shows the analytical framework within this paper. To capture the trends and patterns of traffic accidents involving children, this study uses detailed traffic accident data collected from Ho Chi Minh City's Traffic Police Department (TPD) for a period of 6 years (2010-2015). To assess the situation and characteristics of traffic violations by children and their parents, video camera observations are conducted at 15 school gates and 10 typical roads in HCMC. To recommend children safety improvement solutions, the study reviews international and domestic experience in traffic safety policies for children.

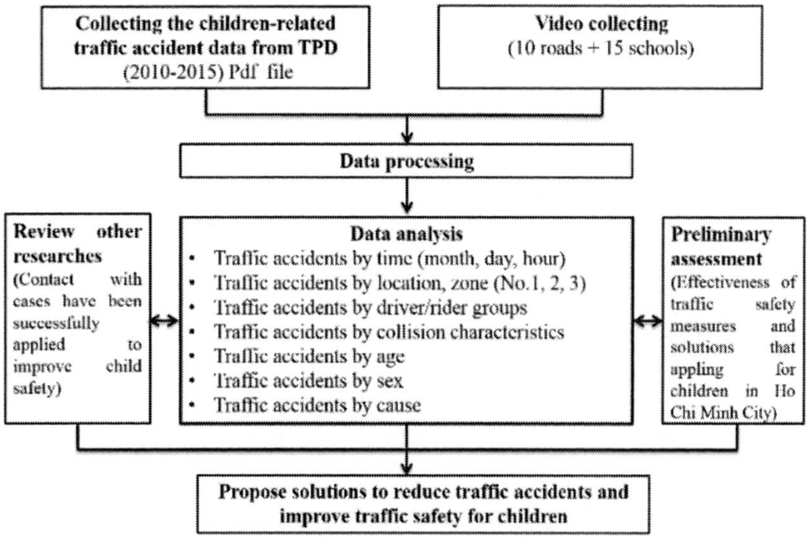

Figure 1. Analytical framework

3. Analysis of child-related traffic accident data

The data of 225 child-related traffic accidents occurred during the period 2013-2015 is analyzed and the key results are presented below.

3.1 General characteristics and contrasting trends

In the recent years, the data shows that despite of the fact that traffic accidents in HCMC had been improving in terms of the number of accidents, deaths and injuries, the accidents involving children had been worsening. For the period 2013-2015, the numbers of children killed and injured by traffic accidents increased by 217% and 260 % (Table 1). In term of the fatality rate, high school aged children are the most vulnerable group, with the fatality rate of 32.5 deaths per 100,000 children (Figure 2). It is about four times higher than the fatality rate of general road users in HCMC and 8 to 9 times higher than the children of the same age in developed countries, such as 3.05 in Italy, 3.55 in Belgium, and 4.55 in Greece [8].

Table 1. Traffic accidents HCMC, 2010-2015

Year	Total accidents	Total deaths	Total injuries	Child accidents	Child deaths	Child injuries
2010	1101	837	432	n.a	n.a	n.a
2011	1042	887	495	n.a	n.a	n.a
2012	958	824	388	n.a	n.a	n.a
2013	941	775	349	36	35	15
2014	850	701	322	85	61	55
2015*	771	692	268	104	111	54
Total 2013-2015	2562	2168	939	225	207	124
				8.8%	9.5%	13.2%

Source. *HCMC Department of traffic police (PC67), (2015)*

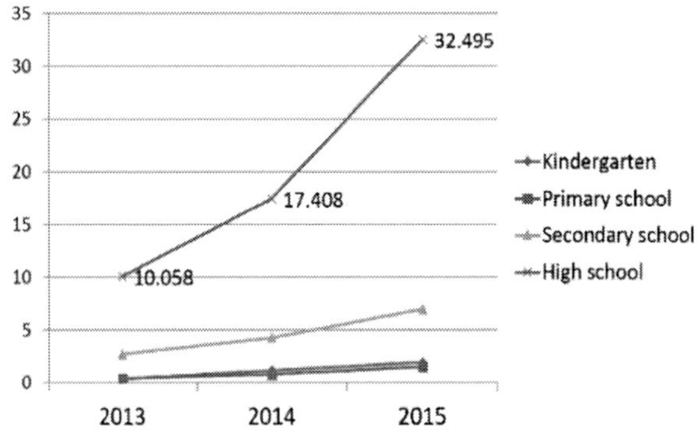

Source. *HCMC Department of traffic police (PC67), (2015)*

Figure 2. Fatality rate trends for children groups in HCMC

High school aged children involved in more than 70% of the total children traffic accidents in HCMC, followed by secondary school aged children (nearly 20%), and primary school aged ones (5%) and kindergarten children (5%) (Figure 3). Boys were more vulnerable than girls, being involved in more than 85% of the total children traffic incidents while girls involved in only 15% (Figure 4).

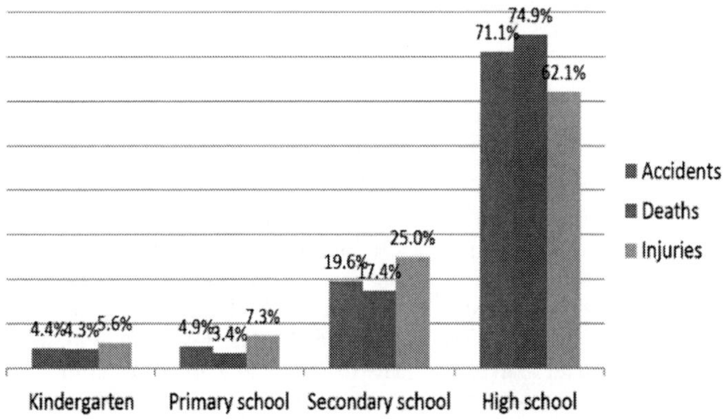

Figure 3. Children traffic accident, death, and injury distributions by age group

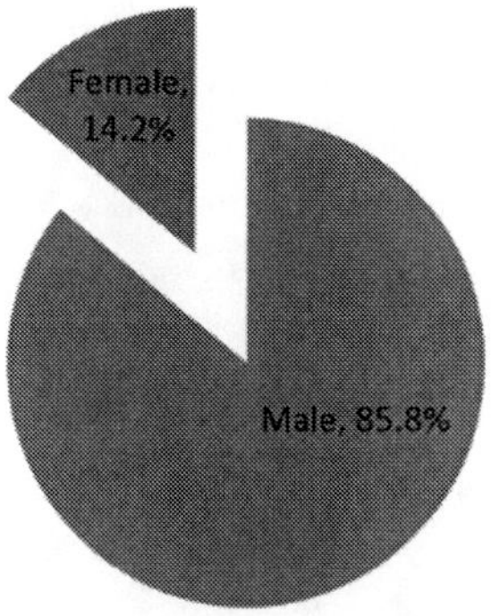

Figure 4. Children traffic accident distribution by gender

About 81.3% of the total children traffic accidents happened the children were driving, only 18.7% of the accidents occurred while the parents were driving and carrying their children. A look into accident distribution by child age group shows that all the kindergarten child related accidents happened while the parents or some adults were driving to carry them. Surprisingly, there was still 18.2% of the primary school aged child accidents occurring while the children were driving a bicycle or an e-bike. In contrast, about 80% of the accidents involving secondary and high school aged children happened while the children were driving a vehicle, either a bicycle, an e-bike or even a normal motorcycle despite of being illegal. To significantly reduce the number of traffic accidents involving children, the local government should focus their actions on the children at high and primary school age.

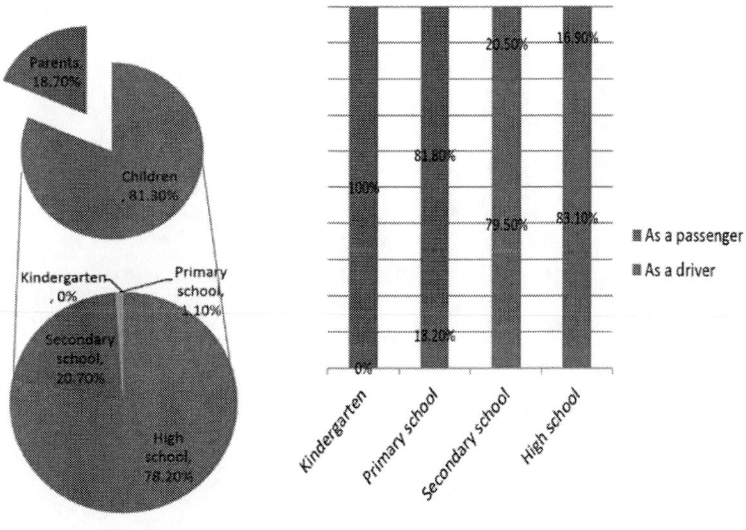

Figure 5. Percentages of children traffic accidents happening while they were a driver

3.2 Distribution characteristics of the child traffic accidents

Time distribution of the traffic accidents

During a day, more accidents occurred after the school hours. For examples, secondary school aged children tended to occur the period 11:00 am – 2:00 pm, the ones involving high school children were likely to occur after 6:00 pm. During a week, the accidents tended to increase at weekends. During a year, the accidents tended to happen during summer months (May – July).

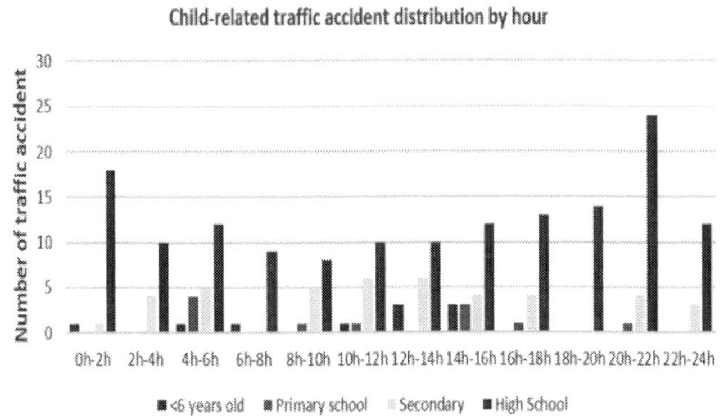

Figure 6. Child-related traffic accident distribution by hour of the day

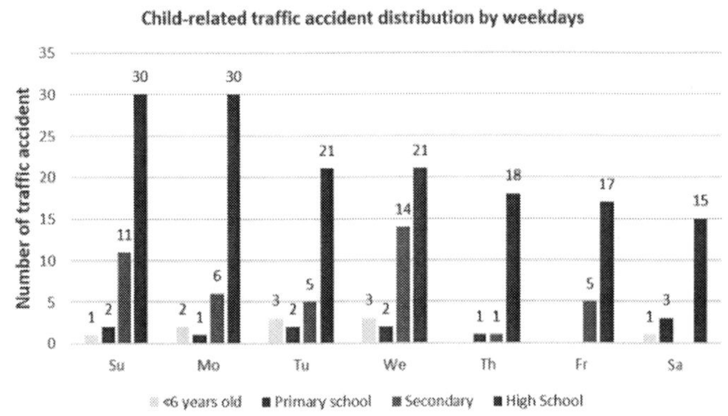

Figure 7. Child-related traffic accident distribution by day of the week

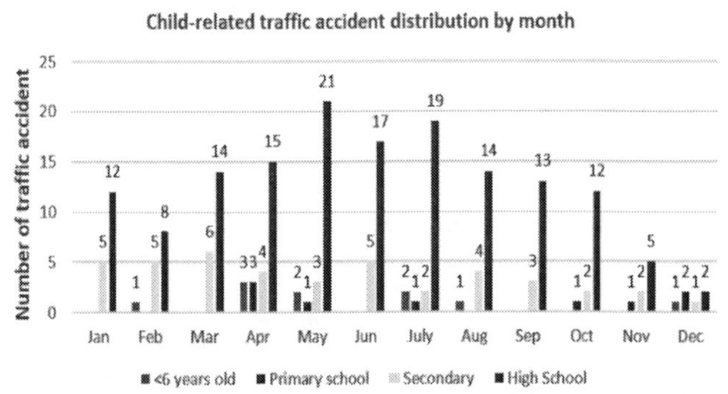

Figure 8. Child-related traffic accident distribution by month of the year

Location distribution of the accidents

The majority of the child-related traffic accidents occurred road sections (82%) and the rest occurred at intersections, including both signalized and non-signalized ones.

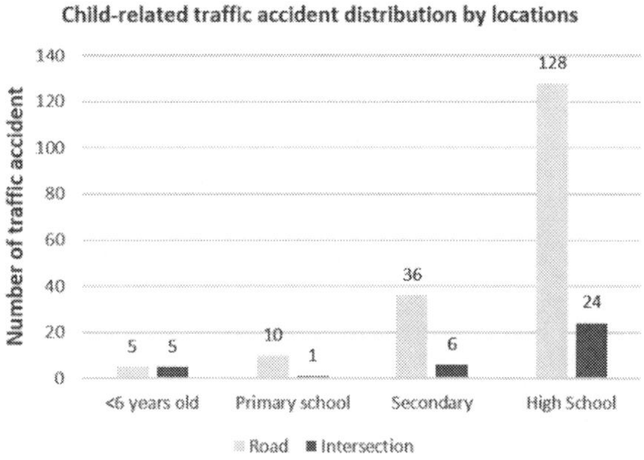

Figure 9. Child-related traffic accident distribution by location

Main cause distribution of the accidents

There were five main causes of the traffic accidents, including: (1) Wrong lane running (23.6%); (2) Dangerous change of direction (11.1%); (3) Speeding (8.9%); (4) Dangerous overtaking (6.7%); (5) Crossing road dangerously (6.2%). Interestingly, unknown cause accidents accounted for 23.1% and improper road infrastructure conditions and safety facilities may contribute largely to this.

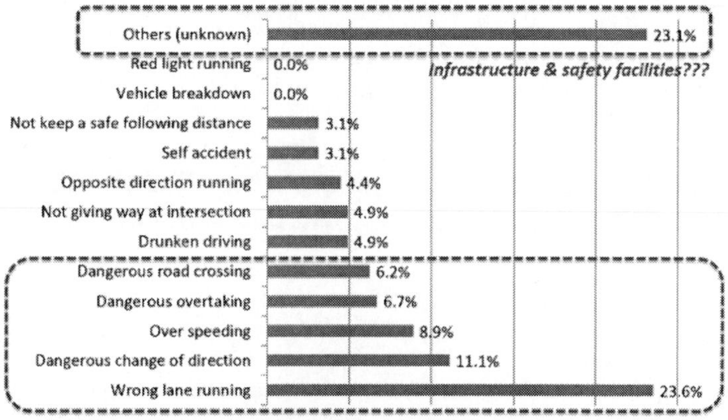

Figure 10. Child-related traffic accident distribution by main cause

4. Analysis of video based observation data

The study conducted a video based observation on traffic participation behaviors of the children and their parents at 15 school gates and on 10 typical roads in the city. Totally, 6,800 samples were collected for the analysis. Key results are shown below.

Modes used to school

For primary school children, walking and cycling to schools were very little (16.6% walking and 2.1% cycling) despite of the fact that primary schools are often located within walkable distances from home. For secondary and high school children, only about 5% of them went to school by bicycle or e-bike, which were mostly self-driven. More than half of them went to school by motorcycle, in which about one fifth illegally drove a motorcycle of more than 50cc to school (Figure 11).

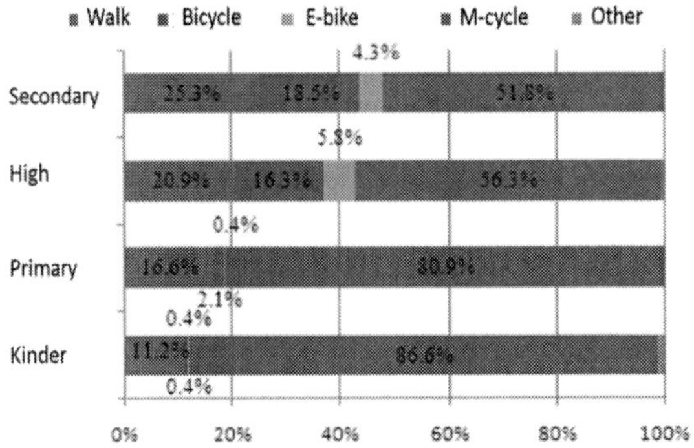

Figure 11. Distributions of transport modes used to go to school

Traffic law violations of the parents and children

The observations show the very high percentage of the children going to school without wearing a helmet. The figures were nearly 75% in the kindergarten children, more than 50% in primary school pupils, more than 35% in secondary school children, and over 15% in high school students (Figure 12).

For kindergarten children, the top violations are no helmet, no safety belt, seating dangerously, no signal while turning, and wrong lane running. For primary school children, the top violations are no helmet, carrying more than 2 children by motorcycle, dangerous pick-up/send-off, no signal while turning, wrong lane running. For secondary school children, the top violations are no helmet, dangerous road crossing, wrong lane running, carrying more than two children, dangerous pick-up/send-off. For high school children, the top violations are no signal while turning, wrong lane running, no helmet, dangerous pick-up/send-off, and dangerous road crossing.

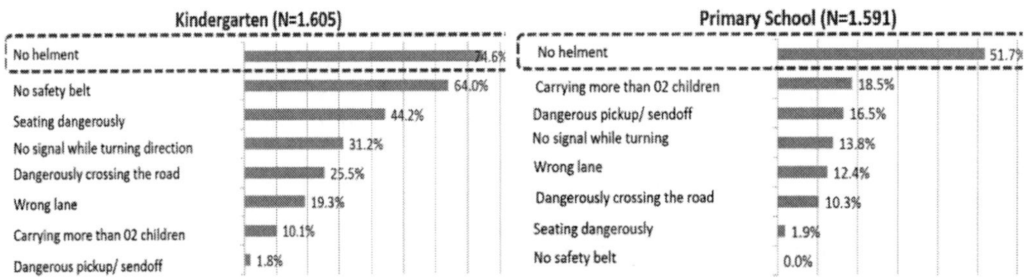

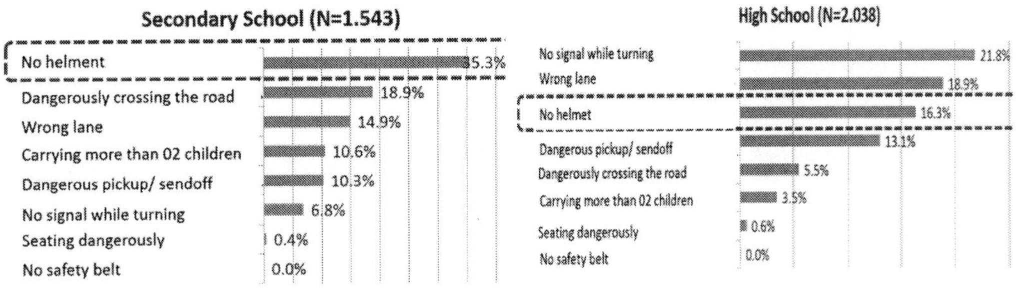

Figure 12. Percentage of students violating traffic safety regulations when going to school

5. Conclusions and recommendations

Specific key findings from the analysis and its policy implications are summarized as follows. First, while traffic accidents are generally decreasing, but child-related accidents are increasing, thus needs to be addressed carefully. Second, high school aged children are the most vulnerable group and thus need to be prioritized in traffic safety improvement actions plan. Third, as boys are much more vulnerable than girls and more than 80% of the accidents happened when the children were driving, thus more specific interventions should be advised for male children. Fourth, five leading causes of child-related traffic accidents are wrong lane running, illegal turning, speeding, illegal overtaking, and illegal road crossing. Educational, enforcement, and engineering measures should be formulated to specifically address such violations among the parents and children. Fifth, because the percentages of primary school children walking and bicycling to school are very low, actions must be taken to introduce safe routes to school in order to encourage walking and bicycling among the children. Sixth, more than 10% of high school students still drive a motorcycle over 50cc to school. This is illegal and thus needs to be enforced strictly by traffic police and school management. Seventh, the rate of no helmet wearing in children is very high, awareness and campaign programs should be continued and strengthened at schools and kindergartens. Last, as child-related accidents are likely to happen more after school hours, at weekends, and during summer months, programs and actions should be focused on such aspects. Traffic safety programs should be extended to households and other education centers other than schools and kindergartens.

Among the recommended actions based on the study results, the priority shall be given to three breakthrough solutions to improve road traffic safety for children in Vietnamese cities. First, it is important to amend the road traffic law and strengthen surveillance patrols to dramatically reduce traffic safety violations in children and parents. It should target at children driving a motorcycle, not wearing helmet, over speeding, and running in the wrong lane. The central government should consider amending the road traffic law so that helmets will be required for children of 3-5 years old. Second, it is necessary to promote culture of walking and cycling for children, especially primary and secondary school ones. To do that, it is critical to improve sidewalk pavement quality and set order on it, and build bicycle lanes so that children can ride bicycles safely and comfortably to schools. Last, it is right timing for implementing School Zone Safety Projects. Japan, Korea, Singapore and other European countries successfully implemented safe school projects, Vietnam can replicate such a program by setting up Traffic Safety Committee at school level headed by school principal. This committee is responsible for coordinating with traffic police and local authorities to re-organize school gate/entrance and nearby sidewalks, design waiting areas for pickup/sendoff, arrange safety equipment (such as signboards, paint lines, decks, etc.), arrange staff to regulate and supervise traffic order and safety in the area of school gates, and plan to provide school bus routes for students while the city is enhancing public transport services that connect to schools.

IOP Conf. Series: Earth and Environmental Science **143** (2018) 012074 doi:10.1088/1755-1315/143/1/012074

References

[1] Vu Anh Tuan (2015) Psychological survey of dangerous motorcycle driving behaviors in HCMC, *presented at the Vietnam Traffic Safety Conference* on July 8, 2015, Hanoi.

[2] Tran Quang Vuong and Vu Anh Tuan (2014) Traffic accident analysis at signalized intersections, *presented at International Conference ATRANS*, August 21, 2014, Bangkok.

[3] Vu Anh Tuan (2006) Understandings of the undesirable motorcycling behaviors in Vietnam, *Proceedings of the 12th International Student Seminar on Transport Research (ISSOT),* sponsored by International Association of Traffic Safety Sciences (IATSS), Bangkok-Thailand, October 2006.

[4] People's Police Newspaper (2015) Traffic accidents are the leading cause of death in children. *Web: http://cand.com.vn/Giao-thong/Tai-nan-giao-thong-la-nguyen-nhan-hang-dau-gay-tu-vong-o-tre-em-349433/*

[5] La Ngoc Quang (2013) The situation of helmets of motorbike traffic participants in 6 provinces, *Report presented at the Vietnam Traffic Safety Conference,* June 2013, Hanoi.

[6] The Ministry of Health (2009) Traffic Accident Report in 2009. *Web: http://moh.gov.vn/pctainan/pages/tintuc.aspx?CateID=4&ItemID=982*

[7] Nguyen Thuy Quynh (2010) *Analysis of traumatic brain injury in adolescents hospitalized for traffic accidents among children under 19 and factors related to child helmet use in traffic*, Research Report of Public Health University.

[8] World Health Organization (2015), *Global status report on road safety 2015*, 2015, Italy.